PHYSICS

For the Life Sciences

Second Edition

PHYSICS

For the Life Sciences

Second Edition

MARTIN ZINKE-ALLMANG
University of Western Ontario

KEN SILLS
McMaster University

REZA NEJAT
McMaster University

EDUARDO GALIANO-RIVEROS
Laurentian University

NELSON EDUCATION

NELSON EDUCATION

Physics for the Life Sciences, Second Edition

by Martin Zinke-Allmang, Ken Sills, Reza Nejat, and Eduardo Galiano-Riveros

Vice-President, Editorial Higher Education:
Anne Williams

Publisher:
Paul Fam

Executive Marketing Manager:
Sean Chamberland

Technical Reviewer:
Abdelhaq Hamza

Developmental Editor:
My Editor Inc.

Photo Researcher and Permissions Coordinator:
Mary Rose MacLachlan

Content Production Manager:
Jennifer Hare

Production Service:
Cenveo Publisher Services

Copyeditor:
Frances Robinson

Proofreader:
N. Manikandan

Indexer:
BIM

Production Coordinator:
Ferial Suleman

Design Director:
Ken Phipps

Managing Designer:
Franca Amore

Interior Design:
Peter Papayanakis

Cover Design:
Dianna Little

Cover Image:
Bongarts/Getty Images (speed skater); Eraxion/iStockphoto (brain); Exactostock/Superstock (arm); Medical RF/Superstock (leg); Eraxion/iStockphoto (active neuron)

Compositor:
Cenveo Publisher Services

Printer:
RR Donnelley

Printed and bound in the United States

3 4 5 6 16 15 14 13

For more information contact Nelson Education Ltd., 1120 Birchmount Road, Toronto, Ontario, M1K 5G4. Or you can visit our Internet site at http://www.nelson.com

Library and Archives Canada Cataloguing in Publication Data

Physics for the life sciences / Martin Zinke-Allmang ... [et al.]. -- 2nd ed.

Includes index.
1st ed. by Martin Zinke-Allmang.
ISBN 978-0-17-650268-3

1. Physics—Textbooks. 2. Life sciences—Textbooks. I. Zinke-Allmang, Martin, 1958-

QC23.2.Z55 2012 530.02'457
C2011-908433-3

ISBN-13: 978-0-17-650268-3
ISBN-10: 0-17-650268-8

Brief Table of Contents

Contents

Preface

"This is a home-grown Canadian product with a fresh perspective on first-year physics."

– Jason Harlow, University of Toronto

"This is the most comprehensive Canadian text devoted specifically to introductory physics for the life sciences that can be purchased on the market today."

– Lorne Nelson, Bishop's University

Physics for the Life Sciences is the result of a straightforward idea: to offer Life Sciences students a "Physics for the Life Sciences" course. Originally created at the University of Western Ontario in 1999, the course became very popular among students, and first-year Physics enrolments saw a corresponding—and unprecedented—dramatic increase. The course was also well-received by colleagues in Biology, Medicine, and other basic life sciences areas, who consequently added or strengthened the physics requirements in their own programs. The great level of interest expressed by colleagues in Canada and abroad motivated us to make this book available through a market-wide release.

An introductory-level university course must meet two key objectives: to provide a comprehensive synopsis of the subject matter relevant to the student's interests and career aspirations, and to present the material in a manner that encourages retaining acquired knowledge. In teaching physics to future life science professionals, meeting these objectives requires a major shift in content, order, and focus.

OUR APPROACH

We summarize this paradigm shift with our key principles that guided the development of the textbook:

- *Get the storyline straight and stick to it.* This is very important to not lose the student who concurrently takes courses in Biology and Chemistry, and will inadvertently focus on the material perceived as most relevant. Physics is a mature science with many branches and a one-size-fits-all approach does not even serve a physical science student well.
- *Keep it short and to the point.* This textbook has fewer than 900 pages. It still contains sufficient material for an instructor to pick and choose, but it does not contain endless pages that will never be covered in a two- or four-term approach to basic physics. Many textbook authors struggle with being selective because they try to write one-size-fits-all encyclopedic works; we have consciously chosen to focus on the relevant material because that is what the modern student requires and deserves.
- *Show the beauty of physics.* Physics represents an enormous body of knowledge and methodology, and almost all of it has a huge impact on understanding the Life Sciences. In a first-year course, particularly if it might be the only physics course some of the students will ever take, this impact of physical facts and thought has to be presented to the fullest extent possible; even students with other interests should appreciate at the end of the course that physics is relevant for their future careers and is an integral part of our shared human culture.
- *Build on what you have taught already.* The modern university student is faced with a tremendous body of knowledge and breadth of required skills that have to be acquired in a few short years in university. Success or failure in providing to these students the essential elements of physics has a particularly high impact in times when modern medicine relies increasingly on technology that provides advanced insights into the functioning of the human body by exploiting physics concepts. Yet physics has to compete with cell biology, biochemistry, and physiology for the student's attention. Here the old-fashioned physics course with its focus on pulleys, inclined planes, and simple electric circuitry with dimming light bulbs obviously fails miserably.
- *The sciences have become interdisciplinary and so must a modern physics course.* The times are long gone when it was justified to erect boundaries around subject areas such as physics, biology, and chemistry. Pretending that we still can study one of these sciences in isolation misrepresents what drives modern science and technology and deceives a student by putting a counterproductive condition on the selection of the material in a course.

ABOUT THIS EDITION

The second edition of *Physics for the Life Sciences* has preserved the strength of the original text, but has undergone revisions in the first seven chapters on the basic mechanical concepts. We also rewrote the latter part of the book, starting with a new chapter on magnetism. For this, an expanded author team of subject-matter experts was formed, including Ken Sills and Reza Nejat from McMaster University, and Eduardo Galiano-Riveros from Laurentian University.

So, what's new? Basically, the underlying philosophy of the text has been strengthened by adding a new objective. In its revised form, the textbook guides the Life Sciences student from the widespread applications of physics in his or her field of interest toward an appreciation of the rapidly developing field of Medical Imaging in Medicine. This is accomplished by extending the content in the sections on electricity and the atomic model, leading to new chapters on magnetism, radiation and its interaction with tissue, the therapeutic uses of ionizing radiation, and the key imaging modalities.

Revisions to the mechanics content allow the instructor to choose a more extensive focus on this area of physics when trying to follow a more traditional canon of topics. The early chapters have not expanded; rather, we adopted a dynamically structured approach to the entire text by presenting the material at three levels: mechanics (the first seven chapters) assumes that the student enters the academic realm with little physical sciences background from high school; the next group of chapters then presents material at a solid first-year university level; and the outlook toward medical physics is presented at a transition level toward upper-year academic studies. This allowed us to be very thorough in laying a foundation in chapters that are used at every university at the beginning of a first-year course and then provide a comprehensive study of the extensive knowledge in physics and its applications to the life sciences in the middle part of the book, while providing an exciting outlook in the direction of a major thrust in physics in the latter part, where, for example, we assume that the student has now picked up a sufficiently strong background in calculus to follow typical arguments using that mathematical method.

What then has not changed? The two key objectives were maintained: to provide a comprehensive synopsis of the subject matter relevant to the student's interests and career aspirations, and to present the material in a manner that encourages retaining acquired knowledge. Most instructors are well aware that meeting these objectives requires a major shift in content, order, and focus of the material presented compared to a standard physics course offered to physical sciences and engineering students.

FEATURE HIGHLIGHTS

Combining both an algebra-based and a calculus-as-needed approach, the second edition of *Physics for the Life Sciences* provides a concise approach to basic physics concepts by including a fresh layout, an exciting art program, and extensive use of conceptual examples, and by introducing the homework assessment program, Enhanced WebAssign®.

A Solid Base

There is increased treatment of basic physics content and more questions around this material.

Applicable Title

Since each topic is initially motivated by a pivotal application in physiology or the biological sciences, the title has been designed for both physics examinations and students preparing for the MCAT.

An Enjoyable Read

A consistent storyline was written within a life science context, containing relevant and applicable material with real values and quantitative figures.

Currency

New material is presented on medical diagnostic techniques such as PET and MRI. Students find these applications of physics to be particularly interesting, and that motivates them to better understand the first principles of physics.

Prepares Students

Physics presented as an integral part of the life sciences allows for a seamless transition into chemistry, biology, and physiology. As well, the end-of-text appendix, Math Review, ensures students are given a refresher on math concepts.

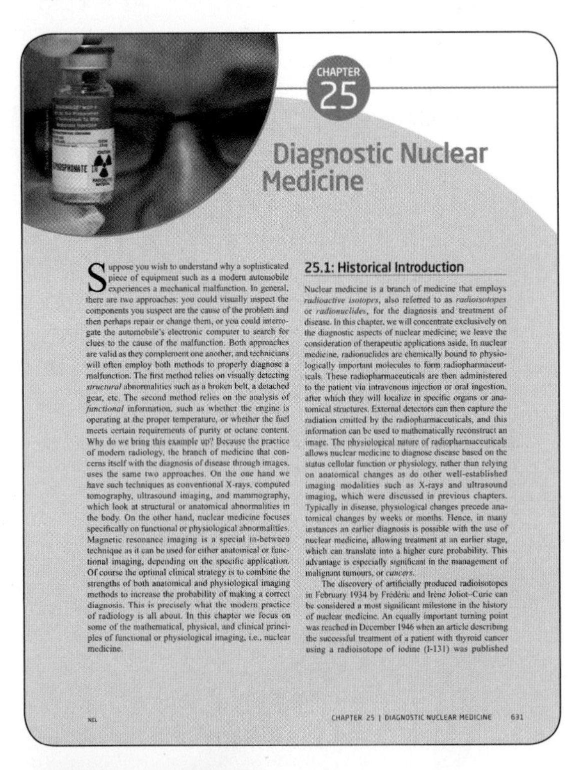

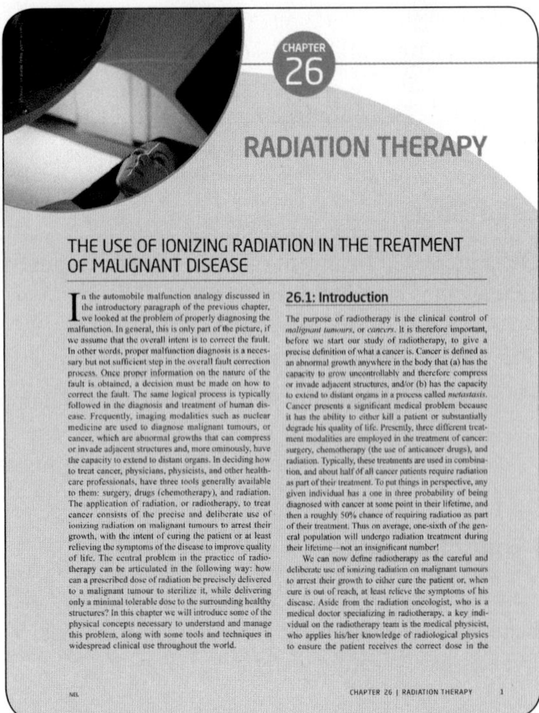

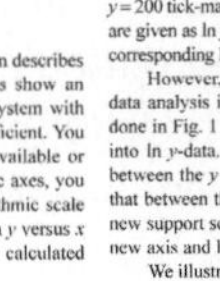

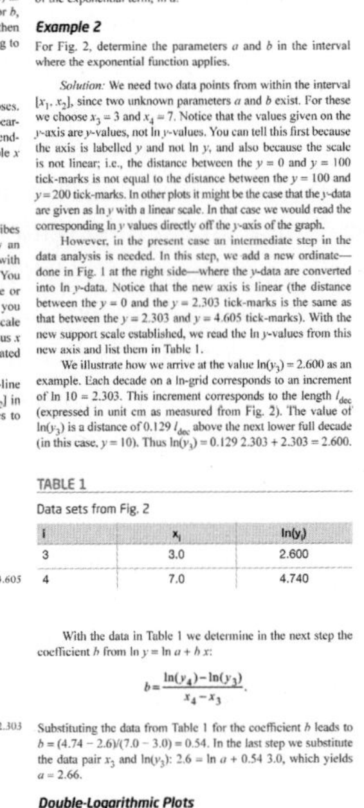

Updated Content

Increased coverage of biomechanical and electrical physics content allows a better understanding of locomotion and the physics involved in cardiac function.

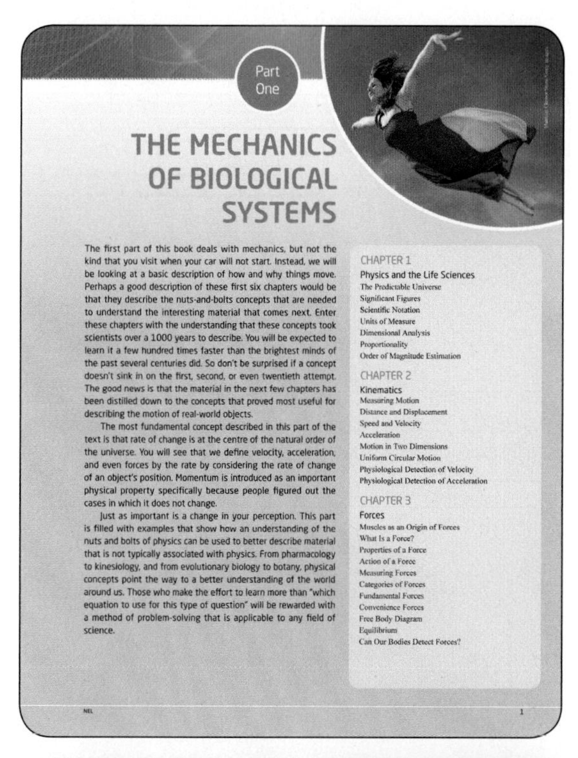

Part One

THE MECHANICS OF BIOLOGICAL SYSTEMS

The first part of this book deals with mechanics, but not the kind that you visit when your car will not start. Instead, we will be looking at a basic description of how and why things move. Perhaps a good description of these first six chapters would be that they describe the nuts-and-bolts concepts that are needed to understand the interesting material that comes next. Enter these chapters with the understanding that these concepts took scientists over a 1000 years to describe. You will be expected to learn it a few hundred times faster than the brightest minds of the past several centuries did. So don't be surprised if a concept doesn't sink in on the first, second, or even twentieth attempt. The good news is that the material in the next few chapters has been distilled down to the concepts that proved most useful for describing the motion of real-world objects.

The most fundamental concept described in this part of the text is that rate of change is at the centre of the natural order of the universe. You will see that we define velocity, acceleration, and even forces by the rate by considering the rate of change of an object's position. Momentum is introduced as an important physical property specifically because people figured out the cases in which it does not change.

Just as important is a change in your perception. This part is filled with examples that show how an understanding of the nuts and bolts of physics can be used to better describe material that is not typically associated with physics. From pharmacology to kinesiology, and from evolutionary biology to botany, physical concepts point the way to a better understanding of the world around us. Those who make the effort to learn more than "which equation to use for this type of question" will be rewarded with a method of problem-solving that is applicable to any field of science.

NEL 1

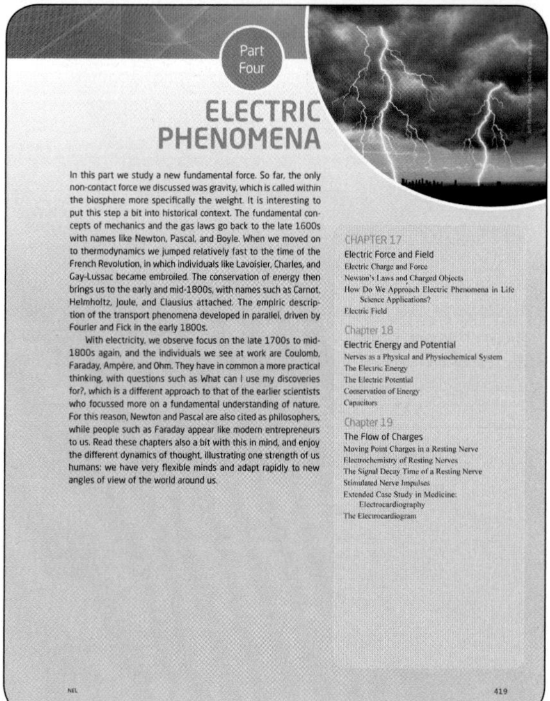

Part Four

ELECTRIC PHENOMENA

In this part we study a new fundamental force. So far, the only non-contact force we discussed was gravity, which is called within the biosphere more specifically the weight. It is interesting to put this step a bit into historical context. The fundamental concepts of mechanics and the gas laws go back to the late 1600s with names like Newton, Pascal, and Boyle. When we moved on to thermodynamics we jumped relatively fast to the time of the French Revolution, in which individuals like Lavoisier, Charles, and Gay-Lussac became embroiled. The conservation of energy then brings us to the early and mid-1800s, with names such as Carnot, Helmholtz, Joule, and Clausius attached. The empiric description of the transport phenomena developed in parallel, driven by Fourier and Fick in the early 1800s.

With electricity, we observe focus on the late 1700s to mid-1800s again, and the individuals we see at work are Coulomb, Faraday, Ampère, and Ohm. They have in common a more practical thinking, with questions such as What can I use my discoveries for?, which is a different approach to that of the earlier scientists who focussed more on a fundamental understanding of nature. For this reason, Newton and Pascal are also cited as philosophers, while people such as Faraday appear like modern entrepreneurs to us. Read these chapters also a bit with this in mind, and enjoy the different dynamics of thought, illustrating one strength of us humans: we have very flexible minds and adapt rapidly to new angles of view of the world around us.

NEL 419

Concept Questions

Concept questions appear throughout chapters to test students' knowledge of emerging topics.

CONCEPT QUESTION 2.1

Fig. 2.2 shows the *x*-position of an object as a function of time. If the object moves to the right, the *x*-position increases. If the object moves to the left, the *x*-position decreases. Which of the following best describes the motion of the object?

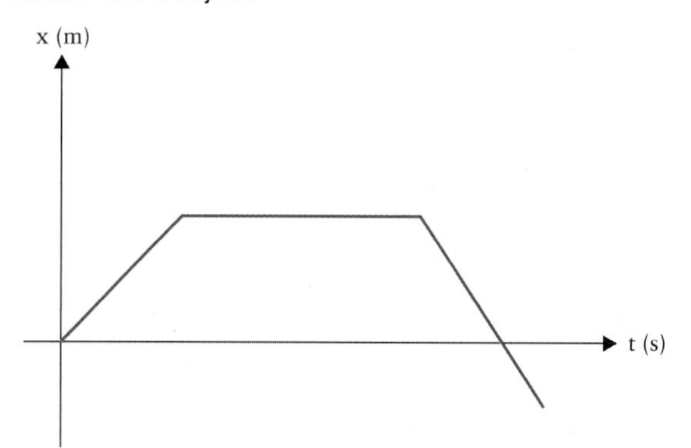

Figure 2.2 Graph of position versus time. The x-axis is positive to the right.

(A) The object moves to the left, stops, then moves to the right, ending up further to the right than where it started.

(B) The object moves up, stops, and then falls down lower than where it started.

(C) The object moves to the right, stops, and then moves to the left, ending up further to the left than where it started.

(D) The object moves up and to the right, then straight to the right, then down and to the right.

CONCEPT QUESTION 14.1

In Chapters 3 to 6 we discussed mechanical concepts using the assumption that forces are position independent. (a) Can we use this assumption for the force shown in Fig. 14.1? (b) Can we use it for the forces shown in Fig. 14.2?

Case Studies

Case studies are used for any concept question with a solution that requires material that has not been previously discussed in the text. This is a new element not used in the first edition.

CASE STUDY 10.2

(a) You put a room-temperature glass of water in a microwave oven and heat it to 50°C. Then you place it back on the table and let it cool down to room temperature. Neglecting a slightly increased rate of evaporation, did you perform a reversible process when heating the water in the microwave oven? (b) If your answer is no, can you name another process you can perform that is perfectly reversible? If you can't name such a process, why do we introduce the concept?

Answer to Part (a): *The process is irreversible. The fact that you recovered the system in its original state (a room-temperature glass of water on your table) does not prove reversibility of the heating process because the cooling process has not restored all the parameters of the environment of your system: electric energy available to do work before the experiment has been lost as thermal energy to the air and the table beneath the glass.*

Answer to Part (b): *There are no reversible processes you can actually perform. We have introduced idealized concepts before in this textbook with no match in the real world, such as point-like particles and the ideal gas. This is justified in each case because such idealizations allow us to model a physical situation or process under simplified conditions, which minimizes the mathematical complexity while preserving the key physical properties. In the current case, idealized reversible processes can be described with the equilibrium thermodynamics concepts we introduced earlier.*

CASE STUDY 21.6

The angle of refraction depends on the wavelength of the incident light. Shouldn't the angle of reflection also depend on the wavelength?

Answer: *No. Dispersion is the result of electric interaction between the light ray and the medium through which it travels. The reflected beam interacts to only a very limited extent with the material surface.*

Extensive Art Program

Over 600 full-colour illustrations, 200 photographs, and over 500 problems allow students to visualize the material and practice/apply what they have learned.

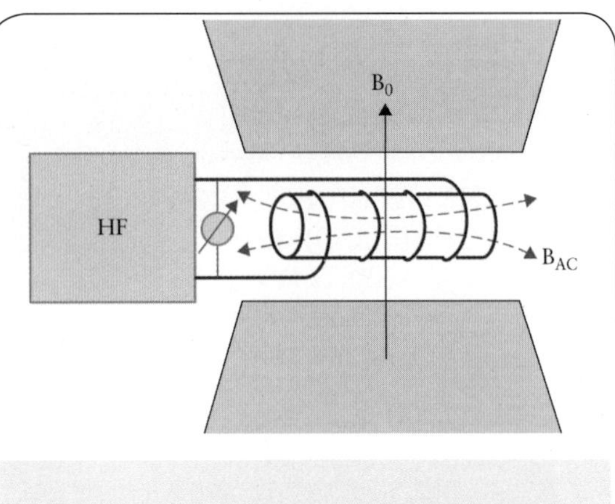

Figure 27.1 Experimental set-up of a nuclear magnetic resonance (NMR) experiment in chemistry.

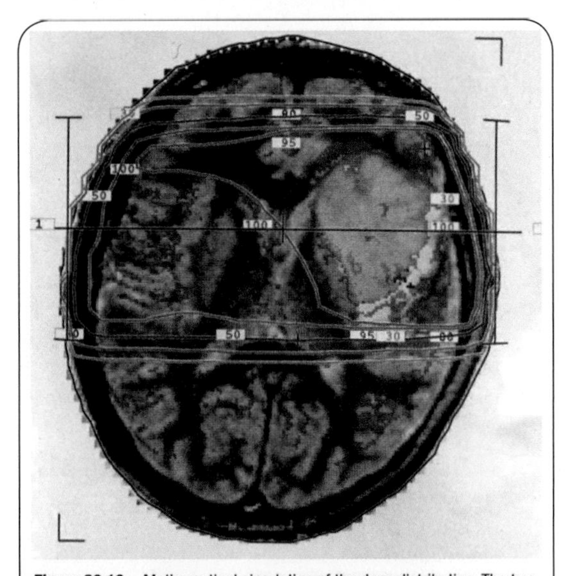

Figure 26.10 Mathematical simulation of the dose distribution. The two, parallel opposed, lateral, collinear beams are labelled as 1 and 2. Note the weighing of the dose distribution toward the left hemisphere, to better match the location of the tumour.

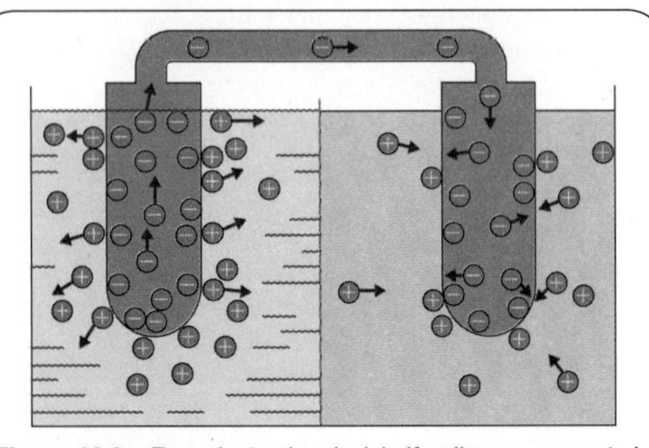

Figure 19.8 Two electrochemical half-cells are connected. Electrons are shown in the electrodes and their electric connection as small circles with – signs; metal ions are shown in the solutions surrounding the electrodes as circles with + signs.

Summary

At the end of each chapter is a summary of important topics covered, which will help students maintain information.

Definitions

Definitions appear at the end of each chapter to provide students with insight into newly discussed terms.

> **SUMMARY**
>
> **DEFINITIONS**
>
> - Density: $\rho = m/V$, with m for mass and V for volume
>
> **UNITS**
>
> Force and force components: N (newton), with $1\ \text{N} = 1\ \text{kg} \cdot \text{m/s}^2$
> Density: ρ: kg/m^3
>
> **LAWS**
>
> - Newton's first law (law of inertia) for an object in mechanical equilibrium:
>
> $$\vec{F}_{\text{net}} = \sum_{i=1}^{n} \vec{F}_i = 0$$

Multiple-Choice Questions

Multiple-choice questions allow students to practice for upcoming tests with similar material.

> **MULTIPLE-CHOICE QUESTIONS**
>
> **MC–3.1.** Consider a planet that has half the mass and twice the radius of Earth. How does the acceleration of gravity on the surface of this planet compare to that on Earth?
> (a) half that on Earth
> (b) four times that on Earth
> (c) twice that on Earth
> (d) one-forth that on Earth
> (e) one-eighth that on Earth
>
> **MC–3.2.** Consider two planets with radius r_1 and r_2 are made from the same material, so have the same density. What is the ratio of the acceleration of gravity (g_1/g_2) at the surface of these two planets?
> (a) r_2/r_1
> (b) r_1/r_2
> (c) $(r_2/r_1)^2$
> (d) $(r_1/r_2)^2$
> (e) $(r_2/r_1)^5$

Conceptual Questions

The intensive use of conceptual questions in addition to quantitative problems provides students with conceptual understanding and analytical skills that are developed with numerous in-text questions and examples. Many more quantitative problems have been included in the second edition.

> **CONCEPTUAL QUESTIONS**
>
> **Q–4.1.** An object hangs in the middle of a rope whose ends are fixed at the same horizontal level. Can the rope stay perfectly horizontal? Why? Explain your answer.
>
> **Q–4.2.** You are in a car, sitting next to the driver. If she suddenly pushes the gas pedal and accelerates rapidly, you feel a push backward against your seat. If she suddenly pushes the brake and comes to a stop, you feel a push forward toward the dashboard. Explain these situations according to the most suitable of Newton's laws.

Analytical Problems

Analytical problems encourage students to take their time and truly understand each question as they work through to find the answers.

> **ANALYTICAL PROBLEMS**
>
> **P–17.1.** We study three point charges at the corners of a triangle, as shown in Fig. 17.23. Their charges are $q_1 = +5.0 \times 10^{-9}$ C, $q_2 = -4.0 \times 10^{-9}$ C, and $q_3 = +2.5 \times 10^{-9}$ C. Two distances of separation are also given, $l_{12} = 4$ m and $l_{13} = 6$ m. Find the net electric force on q_3.

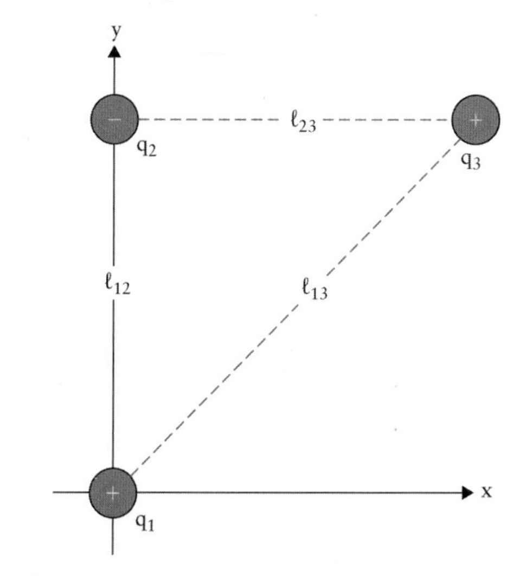

Figure 17.23

SUPPLEMENTS

About NETA

The **Nelson Education Teaching Advantage (NETA)** program delivers research-based instructor resources that promote student engagement and higher-order thinking to enable the success of Canadian students and educators.

Instructors today face many challenges. Resources are limited, time is scarce, and a new kind of student has emerged: one who is juggling school with work, has gaps in his or her basic knowledge, and is immersed in technology in a way that has led to a completely new style of learning. In response, Nelson Education has gathered a group of dedicated instructors to advise us on the creation of richer and more flexible ancillaries that respond to the needs of today's teaching environments.

The members of our Editorial Advisory Board have experience across a variety of disciplines and are recognized for their commitment to teaching. They include

Norman Althouse, Haskayne School of Business, University of Calgary

Brenda Chant-Smith, Department of Psychology, Trent University

Scott Follows, F.C. Manning School of Business, Acadia University

Jon Houseman, Department of Biology, University of Ottawa

Glen Loppnow, Department of Chemistry, University of Alberta

Tanya Noel, Biology Department, York University

Gary Poole, Director, Centre for Teaching and Academic Growth, and School of Population and Public Health, University of British Columbia

Dan Pratt, Department of Educational Studies, University of British Columbia

Mercedes Rowinsky-Geurts, Department of Languages and Literatures, Wilfrid Laurier University

David DiBattista, Department of Psychology, Brock University

Roger Fisher, Ph.D.

In consultation with the Editorial Advisory Board, Nelson Education has completely rethought the structure, approaches, and formats of our key textbook ancillaries. We've also increased our investment in editorial support for our ancillary authors. The result is the Nelson Education Teaching Advantage and its key components: *NETA Engagement*, *NETA Assessment*, and *NETA Presentation*. Each component includes one or more ancillaries prepared according to our best practices, and a document explaining the theory behind the practices.

NETA Engagement presents materials that help instructors deliver engaging content and activities to their classes. Instead of Instructor's Manuals that regurgitate chapter outlines and key terms from the text, NETA Enriched Instructor's Manuals (EIMs) provide genuine assistance to teachers. The EIMs answer questions such as *What should students learn? Why should students care?* and *What are some common student misconceptions and stumbling blocks?* EIMs not only identify the topics that cause students the most difficulty, but they also describe techniques and resources to help students master those concepts. Dr. Roger Fisher's *Instructor's Guide to Classroom Engagement (IGCE)* accompanies every EIM.

NETA Assessment relates to testing materials; not only Nelson's Test Banks and Computerized Test Banks, but also in-text self-tests, Study Guides and Web quizzes, and homework programs such as CNOW. Under *NETA Assessment*, Nelson's authors create multiple-choice questions that reflect research-based best practices for constructing effective questions and testing not just recall but also higher-order thinking. Our guidelines were developed by David DiBattista, a 3M National Teaching Fellow whose recent research as a professor of psychology at Brock University has focused on multiple-choice testing. All Test Bank authors receive training at workshops conducted by Prof. DiBattista, as do the copyeditors assigned to each Test Bank. A copy of *Multiple Choice Tests: Getting Beyond Remembering*, Prof. DiBattista's guide to writing effective tests, is included with every Nelson Test Bank/Computerized Test Bank package.

NETA Presentation was developed to help instructors make the best use of Microsoft® PowerPoint® in their classrooms. With its clean and uncluttered design developed by Maureen Stone of StoneSoup Consulting, *NETA Presentation* features slides with improved readability, more multimedia and graphic materials, activities to use in class, and tips for instructors on the Notes page. A copy of *NETA Guidelines for Classroom Presentations* by Maureen Stone is included with each set of PowerPoint slides.

Instructor Ancillaries

INSTRUCTOR'S RESOURCE CD (IRCD): Key instructor ancillaries are provided on the *Instructor's Resource CD* (ISBN 0176604456), giving instructors the ultimate tool for customizing lectures and presentations. The IRCD includes

- **NETA Engagement:** The Enriched Instructor's Manual was written by Eduardo Galiano-Riveros, Laurentian University. It is organized according to the textbook chapters and addresses six key educational concerns, such as the key points students need to learn, an outline of the chapter's learning objectives, a note why the chapter's topics are important in the real world, typical stumbling blocks students face and how to address them, and activities for instructors to enact student participation in class.
- **NETA Assessment:** The Test Bank was written by Vesna Milosevic-Zdjelar and Murray E. Alexander, University of Winnipeg. It includes over

530 multiple-choice questions written according to NETA guidelines for effective construction and development of higher-order questions. Also included are true/false and essay-type questions. Test Bank files are provided in Microsoft Word format for easy editing and in PDF format for convenient printing whatever your system.

The Computerized Test Bank by ExamView® includes all the questions from the Test Bank. The easy-to-use ExamView software is compatible with Microsoft Windows Operating System and Mac. Create tests by selecting questions from the question bank, modifying these questions as desired, and adding new questions you write yourself. You can administer quizzes online and export tests to WebCT™, Blackboard®, and other formats.

- **NETA Presentation:** Microsoft® PowerPoint® lecture slides for every chapter were created by Phil Backman, University of New Brunswick. There is an average of 30–40 slides per chapter, many featuring key figures, tables, and photographs from *Physics for the Life Sciences*, Second Edition. NETA principles of clear design and engaging content have been incorporated throughout.
- **Instructor's Solutions Manual:** This manual, prepared by Johann Bayer, University of Toronto, has been independently checked for accuracy by Abdelhaq Hamza, University of New Brunswick. It contains complete solutions to end-of-chapter multiple-choice questions, conceptual questions, and problems.
- **Image Library:** This resource consists of digital copies of figures, short tables, and photographs used in the book. Instructors may use these jpegs to create their own PowerPoint presentations.
- **DayOne:** Day One–Prof InClass is a PowerPoint presentation that you can customize to orient your students to the class and their text at the beginning of the course.

Enhanced WebAssign® is a powerful instructional tool that delivers automatic grading of solutions for physics courses, and reinforces student learning through practice and instant feedback. With the assistance of Abdelhaq Hamza, University of New Brunswick, for compiling appropriate questions, this proven and reliable homework system allows instructors to assign, collect, grade, and record homework assignments via the Web. Contact your Nelson sales representative for more information.

Student Ancillaries

The **Student Solutions Manual**, prepared by Johann Bayer, University of Toronto, and technically checked by Abdelhaq Hamza, University of New Brunswick (ISBN 0-17-660444-8), contains detailed solutions to all odd-numbered end-of-chapter questions and problems, exercises, and end-of-chapter supplemental problems.

ACKNOWLEDGMENTS

A large number of highly skilled people contributed to the success of this textbook as an integral part of a modern physics curriculum in the Life Sciences. We acknowledge the support of the following individuals:

At the University of Western Ontario we thank Victoria Boateng for extensive help in developing the manuscript; M. Rasche for her extensive original artwork; Mahi Singh (Physics and Astronomy), Jerry Battista (Medical Biophysics), and Dan Lajoie (Biology) for helpful suggestions; and Tom Haffie and Denis Maxwell (Biology) for discussions about their Nelson textbook in Biology.

Tremendous thanks is extended to Paul Fam, Publisher, for his continuing vision and support of the text and ancillary package. Thank you to Katherine Goodes, Senior Developmental Editor, and her team at My Editor Inc. for their guidance throughout the development process. We are grateful for the attention to detail from Frances Robinson, copy editor. Thank you to Jennifer Hare, Content Production Manager, and Sangeetha, Project Manager, for their assistance through the production process. We also extend our gratitude to our technical checker and our supplement authors.

We also wish to acknowledge the reviewers who offered guidance in the development process of the second edition text:

Johann Bayer, University of Toronto, Scarborough

David Fleming, Mount Allison University

Richard Goulding, Memorial University of Newfoundland

Jason Harlow, University of Toronto

Lorne Nelson, Bishop's University

Kenneth Ragan, McGill University

David Rourke, College of New Caledonia

We especially extend our gratitude to our families and friends for their support.

About the Authors

Martin Zinke-Allmang studied physics and chemistry at the University of Heidelberg in Germany. After completing his Ph.D. thesis at the Max–Planck–Institute for Nuclear Physics, he moved to New Jersey for a postdoctoral fellowship at AT&T Bell Laboratories and later settled at the University of Western Ontario, in Canada, where he currently teaches first-year Physics. He has published more than 80 scientific articles, including 2 major review articles in the journals *Surface Science Reports* and *Thin Solid Films*. He has supervised more than 25 M.Sc. and Ph.D. students and postdoctoral fellows, most recently in the Graduate Program in Medical Biophysics.

Ken Sills studied Astronomy at the University of Western Ontario and Saint Mary's University in Canada, and at Ohio State University in the U.S.A. For a change of scenery, he is studying Computer Engineering at McMaster University in Canada. He has also worked as a software engineer in England and as an industrial physicist in the U.S.A. Currently, he has a consulting company that designs scientific instruments that are used around the world. He has been published in the fields of astronomy, scientific instrumentation, and physics education. When he is not doing science or engineering he is either touring with his rock band or trying to eat the raspberries that grow in his backyard before his wife and kids get to them.

Reza Nejat was born in Iran and studied Physics at Tehran University (B.Sc., 1974). He then moved to the U.S to pursue graduate studies in the field of nuclear engineering, where he received his M.Sc. (1976) and his PhD (1980) from the University of Missouri-Rolla. Upon returning to Iran, he joined the department of Physics at Guilan University where he was a tenured faculty for 15 years. In 1996, he moved to Canada and continued his teaching career in the department of Physics and Astronomy at McMaster University in Ontario, where he has been teaching first-year Physics, Introduction to Modern Physics, classical mechanics, and nuclear physics. His current research interest is in developing the interactive method of learning and teaching Physics (algebra-based) to life sciences students.

Eduardo Galiano-Riveros was born and raised in Asunción, Paraguay, South America, where he attended elementary, middle, and high school. He went on to pursue university studies in the U.S.A., where he obtained B.Sc. (1983) and M.Sc. (1985) degrees in physics at Rensselaer Polytechnic Institute in Troy, NY. At around that time, he decided to pursue a career in medical physics and obtained an M.S. degree (1987) in Medical Physics from the University of Wisconsin-Madison under the supervision of the late Prof. Herbert Attix. He returned to Paraguay, where he worked as a clinical medical physicist at the National Cancer Hospital (1988–1992). He accepted a fellowship at the M.D. Anderson Cancer Center in Houston, TX, to pursue Ph.D. studies, which he completed in 1995. He returned to Paraguay to resume his clinical career at the National Cancer Hospital and, with a group of radiation oncologists, set up a private radiotherapy clinic in Asunción. He spent an academic year (1998–1999) at the University of Osaka, Japan, as a visiting scientist. In 2003, he accepted an offer to join the faculty at the Department of Physics at Laurentian University, in Sudbury, Ontario, where he is presently a tenured professor. Since joining Laurentian University, he has published approximately 25 peer-reviewed scientific papers and has supervised the M.Sc. theses of 5 students. His present research interests focus on nuclear imaging, as well as radiotherapy physics. Outside work, he enjoys flying both real and model airplanes, as well as mitigating the effects of aging by lifting weights and running.

Part One

THE MECHANICS OF BIOLOGICAL SYSTEMS

Matthias Clamer/Stone/Getty Images

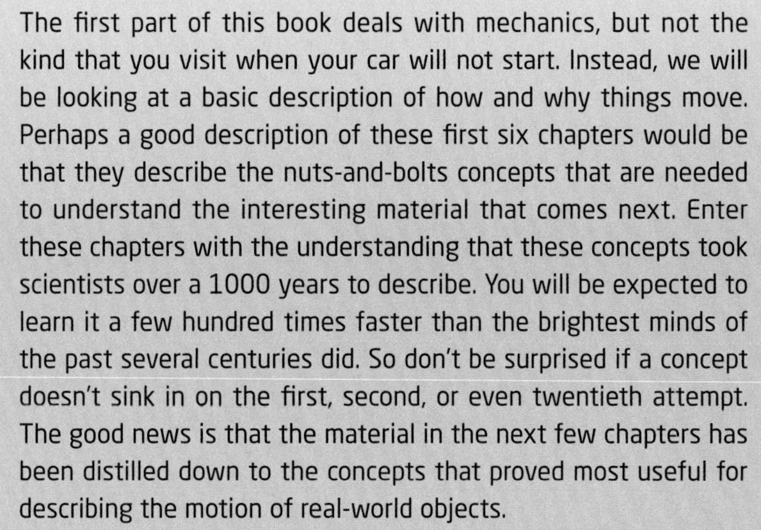

The first part of this book deals with mechanics, but not the kind that you visit when your car will not start. Instead, we will be looking at a basic description of how and why things move. Perhaps a good description of these first six chapters would be that they describe the nuts-and-bolts concepts that are needed to understand the interesting material that comes next. Enter these chapters with the understanding that these concepts took scientists over a 1000 years to describe. You will be expected to learn it a few hundred times faster than the brightest minds of the past several centuries did. So don't be surprised if a concept doesn't sink in on the first, second, or even twentieth attempt. The good news is that the material in the next few chapters has been distilled down to the concepts that proved most useful for describing the motion of real-world objects.

The most fundamental concept described in this part of the text is that rate of change is at the centre of the natural order of the universe. You will see that we define velocity, acceleration, and even forces by considering the rate of change of an object's position. Momentum is introduced as an important physical property specifically because people figured out the cases in which it does not change.

Just as important is a change in your perception. This part is filled with examples that show how an understanding of the nuts and bolts of physics can be used to better describe material that is not typically associated with physics. From pharmacology to kinesiology, and from evolutionary biology to botany, physical concepts point the way to a better understanding of the world around us. Those who make the effort to learn more than "which equation to use for this type of question" will be rewarded with a method of problem-solving that is applicable to any field of science.

CHAPTER 4

Newton's Laws

Newton's Laws of Motion
Free Body Diagram, Revisited
Newton's First Law
Newton's Second Law
Newton's Third Law
Application of Newton's Laws, Convenience Forces Revisited
Weight and Apparent Weight
Physiological Applications of Newton's Laws

CHAPTER 5

Centre of Mass and Linear Momentum

Centre of Mass Definition
Motion of the Centre of Mass
Newton's Third Law and Linear Momentum
Changes of Linear Momentum and Newton's Second Law

CHAPTER 6

Torque and Equilibrium

Force and Extended Object
Torque
Mechanical Equilibrium for a Rigid Object
Classes of Levers and Physiological Applications
Since When Did Hominids Walk on Two Legs?

CHAPTER 1

Physics and the Life Sciences

Physics and biology are two very different sciences. They differ not only in their respective objects of inquiry but also in their experimental and conceptual methods, in their history, and even in their contributions to culture and philosophy. Physicists explain the properties of the natural world on the basis of universal laws; biologists, instead, focus on diversity, singular events, the individual history of a species, and the evolution of specific traits. Why, then, should those interested in biology and the life sciences familiarize themselves with the concepts and methods used by physicists?

To provide a practical answer to this question, we first provide a description of how physicists model the behaviour of the world around them. Physicists research everything: from subatomic particles to the universe as a whole; from superconductivity to the permeability of cell membranes; from the motion of continents to the motion of professional athletes. These fields appear completely separate from each other, and it is true that the details of these fields are very different. It is the underlying similarity that defines what physics is.

1.1: The Predictable Universe

Physicists produce models that focus our attention on the most important properties of a system, while ignoring complexity that is unlikely to significantly change the outcome. These models must predict results that can be tested in the real world. Such a model is called a **physical model**. Fundamentally, physics is the science of identifying what can be ignored, while using the smallest possible set of rules to model that which cannot be ignored.

Imagine making a physical model to describe how the blood goes from your heart to your brain. One could imagine many possible ways that your heartbeat might cause blood flow. Perhaps every time your heart beats it sends out an ultrasonic vibration that is heard by an invisible, alien life form that inhabits Earth. Upon hearing your heart beat, the alien performs an operation on you that moves oxygenated blood from your heart to your brain. This process is then repeated approximately once every second.

Of course, this model requires a large set of rules: Aliens exist. They spend a large amount of their time on Earth. They are invisible. They enjoy surgery. Either they can operate very quickly, or they can suspend the flow of time. You might also wonder whether this process is the same for every human.

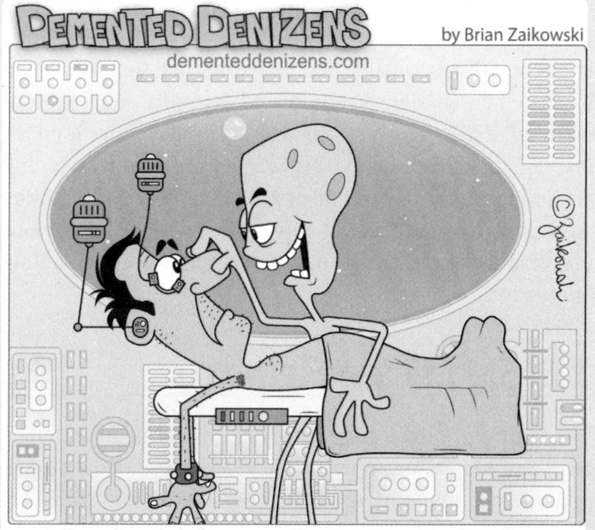

Figure 1.1 A physical model minimizes the number of assumptions that are required to explain an event.

Although it would be impossible to disprove this model, we have a sense that it is too complicated to be correct. Science is based on a belief that simple models are more likely to be correct models. This principle of parsimony is sometimes referred to as Occam's razor: "Entities must not be multiplied beyond necessity." In other words, if we can come up with another model for how blood goes from your heart to your brain, and that model involves fewer assumptions, it is a better model.

A physical model uses the smallest set of rules that is sufficient to describe the situation. These rules should be as fundamental as possible so that the model that describes how your blood goes from your heart to your brain also describes how every human's blood goes from his heart to his brain.

KEY POINT

A physical model is at its best when it invents no new rules at all.

If the model can explain the circulatory system with the same rules as it needed to explain how an airplane flies, then we believe that science has served us well.

A reasonable starting point for a physical model of blood flow in a human body might assume that blood is a fluid that behaves similarly to water. When you place your partially closed hand in water and then quickly squeeze your hand closed, you can observe water moving away from your hand. A heart muscle that squeezes and then relaxes seems like a good choice as the engine of our blood flow. By incorporating an observation of fluid flow that does not involve the heart, we can be assured that our physical model relies on rules that are more fundamental than the rules we invented that incorporate extraterrestrial surgeons.

The advantage of describing phenomena using a few fundamental rules is that it makes the universe predictable. If aliens are responsible for delivering oxygen to human brains, then humanity could end abruptly and without warning. There really is no way to predict how long an invisible alien will enjoy this activity. However, laws that are fundamental to the universe are unlikely to be fickle. And physical models that rely on a minimal number of universal laws are likely to be as useful tomorrow as they are today.

1.2: Significant Figures

The starting point for any physical model is an observation, and the ending point is a measurement that will test predications made by the new model. Any time measurements are used to test predictions, care must be taken to understand the numerical significance of both the measurements and the predictions. In our most basic physical model of blood flow, the heart squeezed blood and caused it to circulate. If this model is correct, we could use it to infer the strength of the heart muscle: it must squeeze with enough force to push blood all the way up to the brain. If a measurement states that this would require a force of 6.9 newtons, and our very basic physical model implies that the force should be 7 N, we need to be able to compare these numbers to see if they agree. If they do not agree, either the physical model is incorrect or the measurement is in error. In order to compare numbers, we need to understand the **accuracy** with which the numbers were stated. When concluding whether or not a measurement supports a particular physical model, it is important to understand the significance of the measurements and predictions. Because it is only quoted to one digit, 7 N could be as high as (just under) 7.5 N, or as low as 6.5 N. The measurement of 6.9 N is quoted to two digits, so it could be as high as (just under) 6.95 N or as low as 6.85 N. Although we are uncertain about these numbers, Fig. 1.2 shows that the measurement is within the uncertainty of the prediction of the physical model. That tells us that the measurement is consistent with the model.

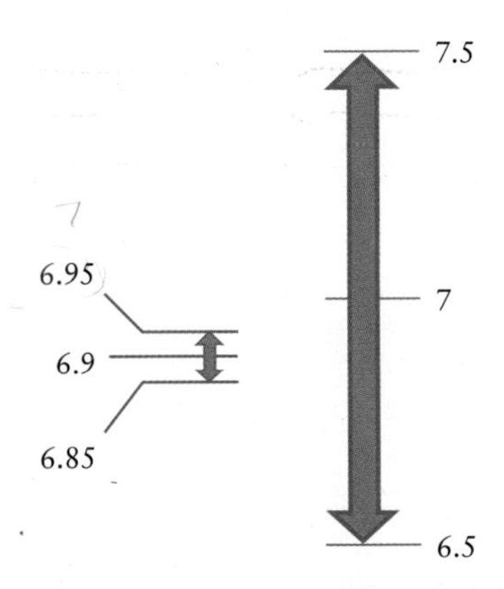

Figure 1.2 The amount of uncertainty in the values is conveyed by the number of digits used to express the value. The measurement is shown on the left and the value predicted by the model is shown on the right. This measurement falls within the uncertainty of the prediction.

For instance, if each of your textbooks is about 2 centimetres in thickness, how many can you fit on a bookshelf that is 103.3 centimetres in width? The answer is not 51.65 books because you only have a rough estimate of the thickness of each textbook. The answer should be quoted as 50 books but the actual number could reasonably be 54 or 47. We cannot be certain. The thickness of the textbook was estimated to be 2 centimetres, and that measurement has only one digit.

KEY POINT

The number of digits quoted in a result indicates how accurate the number is.

We call these digits **significant figures**. Notice that the number of significant figures in the measurement of the thickness is the same as the number of significant figures in the calculated number of books that would fit on the shelf. **When multiplying or dividing numbers, the result has the same number of significant figures as the least accurate number used in the calculation.**

EXAMPLE 1.1

How many significant figures are in the following numbers?

(a) 26.38
(b) 27
(c) 3.0880
(d) 0.00418
(e) 300
(f) 300.

Solution

(a) four significant figures
(b) two significant figures
(c) five significant figures: trailing zeros after the decimal point are significant if they are quoted
(d) three significant figures: leading zeros are not significant
(e) one significant figure: trailing zeros are not significant if there is no decimal point in the quoted number
(f) three significant figures: trailing zeros are significant if there is a decimal point in the quoted number

EXAMPLE 1.2

Express the results of the following calculations with the correct number of significant figures.

(a) $3.4 \times 0.982 \times 9000$
(b) $3.41 \div 0.99 \times 8950$
(c) $3.41 \times 0.981 \div 9000.$

Solution

(a) 30 000: The result has the same number of significant figures as the least accurate number used in the calculation. In this case, 9000 has only one significant figure, so the result must only have one significant figure as well.
(b) 31 000: A calculator would show something like 30.82777 as an answer. However, the least accurate number in the calculation (0.99) has only two significant figures. That means that the result of this calculation must be rounded to two significant figures.
(c) 0.000371: A calculator would show something like 0.000371311 as an answer. However, since the 9000 has a decimal point after the trailing zeros, the least accurate number in the calculation has three significant figures. That means the calculation must be rounded to three significant figures.

There is a different set of rules used when calculations involve addition or subtraction. For instance, suppose you carefully measure one of your textbooks with a very precise measuring device.

KEY POINT

The precision of a number reflects the size of the smallest digit.

If the precise measuring device measures with a precision of one-hundredth of a centimetre, you might find that your textbook is 5.53 cm thick. If you leave one page of study notes in the back of the text, how thick is the textbook with that page left inside it? If you measure the thickness of that page of study notes with the same precise measuring device, you might find that it is 0.01 cm thick. That number has only one significant figure. But, both measurements were made using the same precision measuring device. Both measurements are precise to 0.01 cm, so the sum of these measurements should be the same precision. So, 5.53 cm plus 0.01 cm gives a total thickness of 5.54 cm. The answer has more significant figures than the least significant number used in the calculation, but it has the same precision. **When adding or subtracting numbers, the result has the same precision as the least precise number used in the calculation.**

EXAMPLE 1.3

Express the results of the following calculations with the correct number of significant figures.

(a) $104.0 + 0.21342$
(b) $56.795 - 0.31$
(c) $100 + 25$

Solution

(a) 104.2: 104.0 is precise to the first decimal place, whereas 0.21342 is precise to the fifth decimal place. The result is as precise as the least precise number used in the calculation, so the result must be quoted to the first decimal place.
(b) 56.49: A calculator would show an answer of 56.485. The result must be quoted to the second decimal place and must be rounded up since the last digit is 5 or above.
(c) 100: The trailing zeros are not significant figures unless they are followed by a decimal point. The result must be quoted to the 100's digit and must be rounded down since the last digit is less than 5.

1.3: Scientific Notation

When studying the natural world, one of the things that you will quickly notice is that it is very, very large and it is made up of things that are very, very tiny. The observable universe is around a million billion billion billion billion times larger than the diameter of the nucleus of the hydrogen atom. Writing that number out in that way is cumbersome and would be even more so if I were to write out the number using zeros. One million billion billion billion billion is a 1 with 42 zeros after it. It would be time-consuming if you had to type that number into your calculator, especially given that you might accidentally miss 1 of the 42 zeros. Clearly, science requires a more efficient way to represent numbers that can vary by so many orders of magnitude.

The decimal system allows for a simple way to carry around all those extra zeros, which we will call **scientific notation**. Imagine multiplying 50 by 100. The answer is 5000. That could also be written as $5 \times 10 \times 10 \times 10$. Each of those multiplications by 10 would give us 1 of the zeros after the 5 in 5000. Of course, $10 \times 10 \times 10$ is 10 cubed, so $5 \times 10 \times 10 \times 10$ can be written as 5×10^3. The exponent on the 10 tells us how many zeros belong after the 5.

Writing the number in scientific notation also has another advantage. Suppose the number 5000 was the result of a measurement; perhaps it represents the number of metres between your house and your friend's house. You might tell someone that your friend lives five kilometres away from you. Five kilometres is the same thing as 5000 metres, but if you told someone that your friend lived 5000 metres away from you, it might be taken to imply that you measured the distance with a precision of just under one metre. However, it might also mean that you know the distance with a precision of just under 10 m, just under 100 m or just under 1000 m. How many of those trailing zeros are significant? In science, it is important to convey the precision of measurements, so it is best not to have the ambiguity caused by these trailing zeros.

If we write the distance in scientific notation, we can easily specify the precision. If the measurement is somewhere between 4500 m and 5500 m, but we are really not sure beyond that, we would write this as 5×10^3 m. If we know the measurement is somewhere between 4950 m and 5050 m, we would write it as 5.0×10^3 m. If we know the measurement is somewhere between 4995 m and 5005 m, we would write it as 5.00×10^3 m. Notice that the distance might be 5001.34 m in all the above cases; but unless you measured with something that will give you precision on the order of centimetres, you cannot be sure.

KEY POINT

The number of digits you use in your scientific notation communicates how reliable you think your measurement is.

These digits are referred to as significant figures.

Scientific notation can be used to describe numbers of any size, large or small. In general, any ordinary decimal number can be written in the form:

$$a \times 10^b$$

where the a is a real number between 1 and 10, and b is an integer that describes the order of magnitude of the quantity. To convert 5450 from ordinary decimal notation to scientific notation, count the number of digits the decimal place moves to make a a number between 1 and 10. In this case, the decimal place moved to the left by three digits. If the trailing zero is insignificant, this number has three significant figures. This means that $a = 5.45$ and $b = 3$; so in scientific notation we would write this as shown in Fig. 1.3.

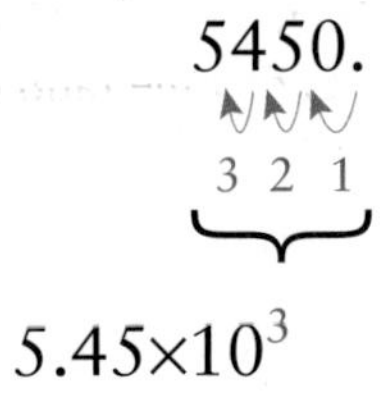

$$5450. \quad (3\ 2\ 1) \quad \rightarrow \quad 5.45 \times 10^3$$

Figure 1.3 Shifting the decimal to the left by three digits.

Scientific notation eliminates the trailing zeros and removes the ambiguity they can cause. We can now look at this number and immediately recognize two very important features: the number of significant figures and the order of magnitude.

For numbers where trailing zeros are significant, these numbers are included in the scientific notation for that number. For instance, the length of a 100-m football field could be written with one significant figure as 1×10^2 m, with two significant figures as 1.0×10^2 m, or with three significant figures as 1.00×10^3 m, depending on the precision of the measurement.

Ordinary decimal numbers that are between 0 and 1 can also be written in scientific notation following the same rules. To convert 0.024 to scientific notation, count the number of digits the decimal place moves to make a number between 1 and 10. In this case, the decimal place moved to the right by two digits. This number has two significant figures. This means $a = 2.4$ and $b = -2$, so in scientific notation we would write this as shown in Fig. 1.4.

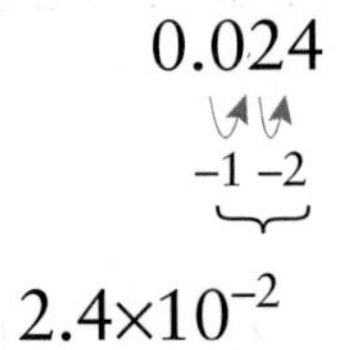

$$0.024 \quad (-1\ -2) \quad \rightarrow \quad 2.4 \times 10^{-2}$$

Figure 1.4 Shifting the decimal to the right by two digits.

Suppose the first number is the mass of a savannah male elephant, $m_{elephant} = 5.455 \times 10^3$ kg, and the second number is the mass of a long-tailed field mouse, $m_{mouse} = 2.4 \times 10^{-2}$ kg. We might want to know how much larger the elephant is than the field mouse. In order to get a ratio of their masses, we need to divide two numbers that are written in scientific notation. If we have two numbers written in scientific notation, $a_1 \times 10^{b_1}$ and $a_2 \times 10^{b_2}$, we can divide them using the following formula:

$$\frac{a_1 \times 10^{b_1}}{a_2 \times 10^{b_2}} = \frac{a_1}{a_2} \times 10^{b_1 - b_2}.$$

To divide powers of 10 is as simple as subtracting the exponents. This makes it very easy to quickly assess what power of 10 a calculation will result in. The power of 10 is called the **order of magnitude of a number**. So, the mass of the savannah male elephant is on the order of magnitude of 10^3 kg, and the mass of the long-tailed field mouse is on the order of magnitude of 10^{-2} kg. There are five orders of magnitude between those two masses, which immediately tells us that the ratio of those masses should be on the order of 10^5:

$$\begin{aligned}\frac{m_{\text{elephant}}}{m_{\text{mouse}}} &= \frac{5.45\times 10^3\ \text{kg}}{2.4\times 10^{-2}\ \text{kg}}\\ &= \frac{5.45}{2.4}\times 10^{3-(-2)}\\ &= 2.3\times 10^5.\end{aligned}$$

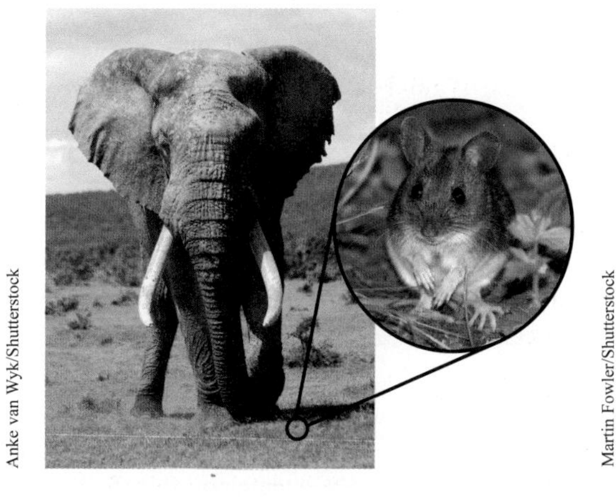

Figure 1.5 The mass of a male savannah elephant is five orders of magnitude larger than the mass of a long-tailed field mouse.

Note that the result of this division is written with only two significant figures. Any time you do a division or multiplication, your result will only have as many significant figures as the least accurate number used in the calculation. If we knew the mass of a long-tailed field mouse to only one significant figure, we cannot expect to know the relative mass to more than one significant figure.

A similar formula is used for the multiplication of two numbers. If we have two numbers written in scientific notation, $a_1\times 10^{b_1}$ and $a_2\times 10^{b_2}$, we can multiply them using the following formula:

$$(a_1\times 10^{b_1})(a_2\times 10^{b_2}) = (a_1\times a_2)\times 10^{b_1+b_2}.$$

Suppose we were going to ship exactly 100 long-tailed field mice to a nature preserve and we needed to know how much the package would weigh. If you calculator isn't handy, just rewrite 100 as 1.00×10^2 and you can quickly find that the total mass will be on the order of magnitude of 10^0 kg:

$$\begin{aligned}100\ m_{\text{mouse}} &= (1.00\times 10^2)(2.4\times 10^{-2}\ \text{kg})\\ &= (1.00\times 2.4)(10^2\times 10^{-2})\,\text{kg}\\ &= 2.4\times 10^0\ \text{kg}\\ &= 2.4\,\text{kg}.\end{aligned}$$

That's a lot of field mice but, all together, they still weigh less than a typical cat.

If we were to ship these 100 long-tailed field mice in a container with the savannah male elephant, would the mass of the 100 mice significantly change my estimate of the mass of a container that has the elephant and all the mice in it? The answer to that question depends on both the number of significant figures and the order of magnitude of both numbers. In this case, the mass of the savannah elephant is 5.45×10^3 kg. That number is significant to the 10^1 digit. In other words, the mass of the elephant is between 5.445×10^3 kg and 5.455×10^3 kg. The mass has a precision of 10^1 kg. The mass of 100 long-tailed field mice is 2.4×10^0 kg. Since the mass of 100 long-tailed field mice is smaller than the precision of the measurement of the mass of the savannah male elephant, the inclusion of the mice in the container does not significantly change the mass of the container. Later in this chapter you will see that order of magnitude approximations such as these are used to differentiate between interactions that can be ignored and interactions that significantly affect an outcome.

EXAMPLE 1.4

Box A has a length of 20.0 cm, a width of 105.00 cm, and a height of 10.450 cm. Box B has length of 2.15 cm, a width of 9.9 cm, and a height of 12 cm. The volume of both boxes together has how many significant figures?

Solution

First, let's find the volume of Box A by multiplying length × width × height:

$$V_A = 20.0\ \text{cm}\times 105.00\ \text{cm}\times 10.450\ \text{cm} = 21\ 945\ \text{cm}^3.$$

The length of the box has three significant figures, and the width and height both have five significant figures. When multiplying, our answer is only as significant as the least significant number in the product. That means that the volume of Box A should be written as $V_A = 21\ 900\ \text{cm}^3$.

Similarly, the volume for Box B is $V_B = 2.15\ \text{cm}\times 9.9\ \text{cm}\times 12\ \text{cm} = 255.42\ \text{cm}^3$. The length of Box B has three significant figures, but the width and height each have two significant figures, so the product should be written as $V_B = 260\ \text{cm}^3$. This number has two significant figures and was rounded up to 260 because 255.42 is closer to 260 than it is to 250.

Now that we have the volumes for both boxes, we need to add these volumes together:

$$V_{\text{Total}} = V_A + V_B = 21\ 900\ \text{cm}^3 + 260\ \text{cm}^3 = 22\ 160\ \text{cm}^3.$$

The sum cannot have more significant figures than the numbers in the sum. In this case, the least significant digit of V_A is the same order of magnitude as the most significant digit of V_B. This means we should write the sum of these volumes as $V_{\text{Total}} = 22\ 200\ \text{cm}^3$, which has three significant figures.

Note that if $V_B = 10\ 000\ \text{cm}^3$, then V_{Total} would only have 1 significant figure. That is because the most significant figure of V_B would be of the same order of magnitude as the most significant figure of V_A.

CONCEPT QUESTION 1.1

A typical student has a mass of 62 kg. Each student is carrying a book with a mass of 1.45 kg. What is the total mass of 152 of these book-carrying students?

(A) 9600 kg

(B) 9640 kg

(C) 9420 kg

(D) 9400 kg

1.4: Units of Measure

In order to discuss the world around us, it is useful to speak the same language. Imagine writing a book in a language that only you knew. Other people could read this book only if you also provided a means of translation. Even if you provided readers with a dictionary that translates from your language into a more common language, there would undoubtedly be errors of translation.

The world's most commonly used system of measurement is the International System of Units (abbreviated SI from the French le Système international d'unités).

KEY POINT

Physics, by and large, is communicated using SI units.

The reason for doing this is obvious: less will get lost in translation if we all agree on a common language. The consequences of working in non-standard units can be serious.

In 1999, the Mars Climate Orbiter was ripped apart by atmospheric stresses because its orbit was nearly 100 km lower in altitude than it was supposed to be. The NASA Mishap Investigation Board traced the problem back to the craft's thrusters. The force that they generated was 4.45 times greater than expected. The conversion between Imperial units of force and SI units of force is

$$1 \text{ pound of force} = 4.45 \text{ newtons}$$

Some of the spacecraft engineers used SI units; others used Imperial units. Over $100 million was lost in translation.

Such problems are not confined to space exploration though. For instance, if you were asked to administer 5 gr of the anticonvulsant phenobarbital, how many 0.1-gram doses would be required? The math may look simple until you realize that gr is the symbol for the grain, which is a

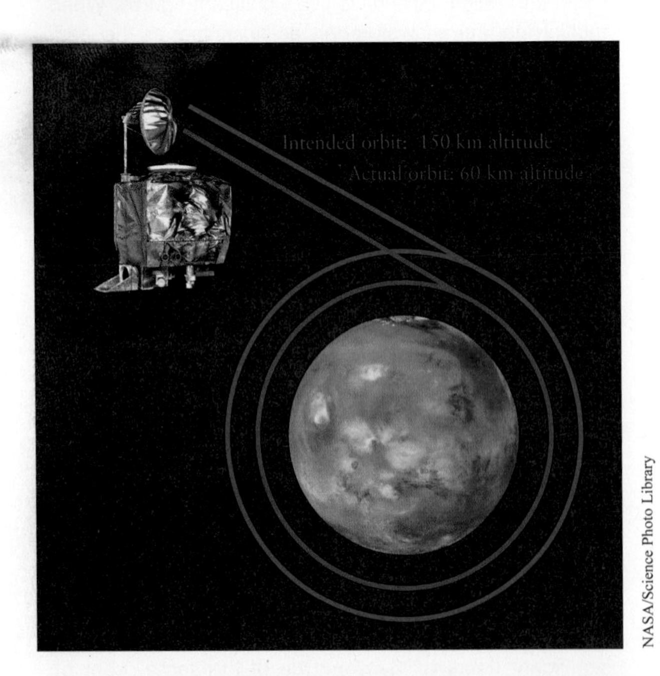

NASA/Science Photo Library

Figure 1.6 A unit conversion error is responsible for the loss of the Mars Climate Orbiter. *(Art not to scale)*

CASE STUDY 1.1

A child is prescribed 1 teaspoon of medication per day. How does that volume compare to 1 mL?

(A) 1 teaspoon is more than 1 mL.

(B) 1 teaspoon is less than 1 mL.

(C) 1 teaspoon is the same as 1 mL.

Answer: One teaspoon is approximately 5 mL in volume, so it is quite a bit more than 1 mL. The following article from the April 2006 newsletter of the Institute for Safe Medication Practice (ISMP) catalogues a few of the consequences of unit translation errors made while dispensing medication.

Time for a Change to Metric: Health professionals should be concerned about the number of mix-ups reported to ISMP and U.S. Pharmacopeia (USP) involving expressions of volume, specifically confusion between millilitre and teaspoonful. When these errors appear on pharmacy-generated labels, patients receive 5-fold overdoses or underdoses if undetected. In one report, a pharmacist labelled a prescription for Zithromax® (azithromycin) suspension with the directions to give "2 1/2 teaspoonfuls daily" (equivalent to 12.5 mL) instead of 2.5 mL daily. The entire content of the bottle was administered according to the labelled instructions, and the child developed diarrhea. In another case, an 8-month-old child was dispensed Zantac® (ranitidine) syrup to treat gastroesophageal reflux disease. The pharmacy label incorrectly instructed the parent to administer "0.5 teaspoonful three times daily" (equivalent to 2.5 mL) instead of 0.5 mL three times daily. The overdose was administered for two weeks and the child experienced tremors, excessive blinking, and inability to sleep. These reactions resolved after the medication was discontinued. Similar mix-ups between teaspoonful and mL have also involved drugs such as amoxicillin, amoxicillin/clavulanic acid, fluoxetine, citalopram, and fluconazole. To prevent the teaspoonful–millilitre confusion, volume expression on prescriptions and pharmacy labels must be standardized. Doses for oral liquids should be expressed only in metric weights and volumes (i.e., mg and mL).

historical system of mass units that is still occasionally used by physicians. The requested 5 gr is a little more than 300 mg. Administering 5 g of phenobarbital instead of 5 gr of phenobarbital would likely result in death. It is for just this reason that the Institute for Safe Medication Practices recommends health professionals express doses only in metric weights and volumes. SI units are internationally recognized as the standard units for metric measurements.

The base SI units are listed in Table 1.1. These seven base units can be combined together to describe any measurable quantity. Table 1.2 shows a list of some of the most commonly used combinations of the SI base units. These combinations of the base units are used frequently enough that constantly writing the combination of base units is inefficient, so they are named after someone who is known for historically significant work in the field that the unit is most closely related to. Most of the units used in this text will be on either Table 1.1 or Table 1.2.

When talking about the distance between two atoms, 1 m is many orders of magnitude larger than what you are measuring. Conversely, when talking about the distance between Brussels and Toronto, 1 m is many orders of magnitude smaller than what you are measuring. In both cases, the typical unit of measurement is a decimal multiple of the SI unit. These multiples are represented by SI prefixes and they allow the base unit to be scaled so that it more closely resembles the size of the measurement. Table 1.3 shows all the SI prefixes.

TABLE 1.1

SI base unit

Base quantity	Name	Symbol
Length	metre	m
Mass	kilogram	kg
Time	second	s
Electric current	ampere	A
Thermodynamic temperature	kelvin	K
Amount of substance	mole	mol
Luminous intensity	candela	cd

TABLE 1.2

SI named units

Base quantity	Name	Symbol	Base units
Celsius temperature	degree Celsius	°C	K
Plane angle	radian	rad	m m^{-1}
Frequency	hertz	Hz	s^{-1}
Force	newton	N	kg m s^{-2}
Pressure	pascal	Pa	kg m^{-1} s^{-2}
Energy	joule	J	kg m^{2} s^{-2}
Power	watt	W	kg m^{2} s^{-3}
Electric charge	coulomb	C	s A
Electric potential	volt	V	kg m^{2} s^{-3} A^{-1}
Capacitance	farad	F	kg^{-1} m^{-2} s^{4} A^{2}
Electric resistance	ohm	Ω	kg m^{2} s^{-3} A^{-2}
Magnetic flux	weber	Wb	kg m^{2} s^{-2} A^{-1}
Magnetic flux density	tesla	T	kg s^{-2} A^{-1}
Inductance	henry	H	kg m^{2} s^{-2} A^{-2}
Radioactivity	becquerel	Bq	s^{-1}
Dose equivalent	sievert	Sv	m^{2} s^{-2}
Absorbed dose	gray	Gy	m^{2} s^{-2}

TABLE 1.3

SI prefixes

Factor	Name	Symbol	Factor	Name	Symbol
10^{24}	yotta	Y	10^{-1}	deci	d
10^{21}	zetta	Z	10^{-2}	centi	c
10^{18}	exa	E	10^{-3}	milli	m
10^{15}	peta	P	10^{-6}	micro	μ
10^{12}	tera	T	10^{-9}	nano	n
10^{9}	giga	G	10^{-12}	pico	p
10^{6}	mega	M	10^{-15}	femto	f
10^{3}	kilo	K	10^{-18}	atto	a
10^{2}	hecto	H	10^{-21}	zepto	z
10^{1}	deca	da	10^{-24}	yocto	y

Any SI prefix may be used with any SI unit with one subtle exception: The SI base unit for mass is the kilogram. That unit already has an SI prefix, so when you are using SI prefixes to describe mass, you place the prefix in front of the gram. In other words

10^{-6} kg = 1 mg (one milligram),
not 10^{-6} kg = 1 μkg (one microkilogram).

EXAMPLE 1.5

Use the information in Table 1.3 to answer the following questions.

(a) How many milliseconds are in one kilosecond?

(b) How many teraseconds are in one picosecond?

continued

Solution

(a) One millisecond is 10^{-3} s and one kilosecond is 10^3 s. To find the number of milliseconds in one kilosecond:

$$\frac{10^3 \frac{\text{s}}{\text{ks}}}{10^{-3} \frac{\text{s}}{\text{ms}}} = \frac{10^3}{10^{-3}} \frac{\text{ms}}{\text{ks}} = 10^6 \frac{\text{ms}}{\text{ks}}.$$

(b) One terasecond is 10^{12} s and one picosecond is 10^{-12} s. To find the number of teraseconds in one picosecond:

$$\frac{10^{-12} \frac{\text{s}}{\text{ps}}}{10^{12} \frac{\text{s}}{\text{Ts}}} = \frac{10^{-12}}{10^{12}} \frac{\text{Ts}}{\text{ps}} = 10^{-24} \frac{\text{Ts}}{\text{ps}}.$$

CONCEPT QUESTION 1.2

How many kilograms are in one microgram?

(A) 10^9

(B) 10^6

(C) 10^{-6}

(D) 10^{-9}

The same logic can be used when converting from nonstandard units to standard SI units. For instance, our system for measuring time is based on the second, but the minute, hour, day, and year are not decimal multiples of the second. The sexagesimal system that is used to define the relationships between those quantities originated with the ancient Sumerians around 3000 B.C. and is based on multiples of 60. That same sexagesimal system was also used to describe angles, which explains why there are 360 degrees in a circle. Since people are slow to give up a system for communicating time that has functioned well for the past 5000 years, it is important to understand how to convert these historical units to base SI units.

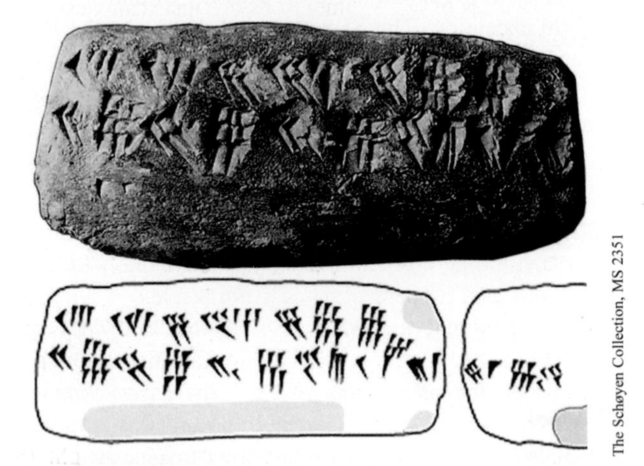

The Schøyen Collection, MS 2351

Figure 1.7 Extremely large 15-digit sexagesimal number recorded on a cuneiform tablet. Babylonia, 19th century B.C. Its decimal equivalent written in scientific notation is 1.048576×10^{23}.

EXAMPLE 1.6

Phytoextraction has attracted increasing interest within the scientific community following the discovery of hyperaccumulator plant species. Hyperaccumulators are plants that have an innate capacity to absorb metal at levels 50–500 times greater than average plants. They are often found in metal-rich regions, where this trait probably gives them a competitive advantage. Hyperaccumulators have a bioconcentration factor greater than one, sometimes reaching 50–100. Furthermore, they always have efficient root-to-shoot transport systems and have enhanced tolerance to metals, indicating increased capacity for detoxification. So far, more than 400 species of natural metal hyperaccumulators have been identified. The best-known hyperaccumulator is the pennycress *Thlaspi caerulescens*. This small plant can absorb zinc from the soil at a rate exceeding 40 kg per hectare per year.

If one hectare is 1×10^4 m^2, what is the rate of zinc accumulation in units of kg m^{-2} s^{-1}?

Solution

$$40\frac{\text{kg}}{\text{hectare} \cdot \text{year}} \cong 40\frac{\text{kg}}{(1\times10^4\ \text{m}^2)(365\ \text{days})}$$
$$= 40\frac{\text{kg}}{(1\times10^4\ \text{m}^2)(365(24\ \text{hours}))}$$
$$= 40\frac{\text{kg}}{(1\times10^4\ \text{m}^2)(365\times24(60\ \text{min}))}$$
$$= 40\frac{\text{kg}}{(1\times10^4\ \text{m}^2)(365\times24\times60(60\ \text{s}))}$$
$$= \frac{40}{1\times10^4\times365\times24\times60\times60}\frac{\text{kg}}{\text{m}^2\cdot\text{s}}$$
$$= 1.27\times10^{-10}\frac{\text{kg}}{\text{m}^2\cdot\text{s}}.$$

A similar situation exists with the litre: the litre is not an SI unit. The SI unit for volume is cubic metres. There are 10^3 litres in one cubic metre, so one litre is 10^{-3} cubic metres:

$$75\text{L} = 7.5\times10^1\left(10^{-3}\ \text{m}^3\right) = 7.5\times10^{-2}\ \text{m}^3.$$

When converting units that are raised to an exponent, extra care must be taken if the units have prefixes. For instance, 1000 L = 1 m^3, but 1000 mL $\neq$ 1 mm^3. That is intuitively clear if you remember that 1000 mL is one litre. One litre is a much larger volume than a cube that is one millimetre in length, width and height.

In order to correctly convert between units that are raised to an exponent, you need to consider how the prefix is affected by the exponent.

EXAMPLE 1.7

Convert the following squared and cubed units:

(a) How many cm^2 are in 1 m^2?

(b) Write 15 cm^3 in units of m^3?

Solution

(a) One metre is equal in length to 100 cm, but what we want here is the number of square centimetres per square metre:

$$\left(\frac{10^2\,\text{cm}}{\text{m}}\right)\left(\frac{10^2\,\text{cm}}{\text{m}}\right) = 10^4\,\frac{\text{cm}^2}{\text{m}^2}$$

(b) In order to represent this volume in terms of cubic metres, all we need to do it to rewrite the unit as a power of 10 multiplied by the base unit:

$$V = 15\text{ cm}^3 = 15\text{ cm}^3 \times \left(\frac{10^{-2}\,\text{m}}{\text{cm}}\right)^3 = 15\text{ cm}^3 \times \frac{10^{-6}\,\text{m}^3}{\text{cm}^3}$$

$$= 15 \times 10^{-6}\text{m}^3 = 1.5 \times 10^{-5}\text{m}^3.$$

When units are combined, the same process must be repeated for each unit that is to be converted.

EXAMPLE 1.8

The density of lead is 11.35 g/cm^3. If a bar of gold that is 250 mm by 70 mm by 35 mm weighs 11.8 kg, is gold more or less dense than lead?

Solution

There is more than one way to solve this problem. Let's start by finding the density of gold in the units given:

$$\rho_{\text{gold}} = \frac{m}{V} = \frac{11.8\text{ kg}}{250\text{ mm} \times 70\text{ mm} \times 35\text{ mm}} = \frac{11.8\text{ kg}}{610\,000\text{ mm}^3}$$

$$= 1.9 \times 10^{-5}\,\frac{\text{kg}}{\text{mm}^3}.$$

Since the density of lead is given in units of g/cm^3, to compare the density we found it is probably fastest if we convert the density of gold into units of g/cm^3:

$$\rho_{\text{gold}} = 1.9 \times 10^{-5}\,\frac{\text{kg}}{\text{mm}^3} = 1.9 \times 10^{-5}\,\frac{(10^3\text{g})}{(10^{-1}\text{cm})^3}$$

$$= 1.9 \times 10^{-5}\,\frac{10^3\text{g}}{10^{-3}\text{cm}^3} = 19\,\frac{\text{g}}{\text{cm}^3}.$$

So, gold is actually more dense than lead, not to mention much more expensive.

1.5: Dimensional Analysis

If it takes you half an hour to ride your bicycle a distance of 5 km, what is the mass of your bicycle? Can that be answered with only the information given? Regardless of whether or not the time is in seconds, or years, and regardless of whether the distance is in kilometres or Egyptian royal cubits, there appears to be no way to combine a time with a distance to get a mass. In order for the question to make sense, we would need to know how the bicycle mass is related to the time to ride that distance.

If you are coasting down a hill for 10 km, and air resistance is causing the bicycle travel at a speed that depends on its mass, for instance 2 km per hour per kilogram, then we can see how some combination of those three quantities can give us information about the mass of the bicycle:

$$\text{mass} = \frac{\dfrac{5\text{ km}}{0.5\text{ h}}}{2\dfrac{\text{km}}{\text{h}\cdot\text{kg}}} = 5\text{ kg}.$$

Using the information contained in units to gain insight into a relationship between quantities is called **dimensional analysis**, but the dimensions that we are analyzing are not length, width, and height. The reason that we use length, width, and height to measure the volume of an object is that those measurements are independent of each other, which means that I could increase the length without changing the width. If I draw a rectangle on a piece of paper, regardless of what length and width I choose, it will never have a height.

The dimensions that we used to rule out a relationship between how much time you spent bicycling, how far you bicycled, and the mass of the bicycle were time, length, and mass. Time, length, and mass are independent quantities. If you ride a bicycle for an hour, you cannot guess how far you went without information that includes a distance.

KEY POINT

Length, mass, and time are fundamentally different than each other, so we refer to them as dimensions.

The units of these dimensions are just scales used to measure the length along these independent axes. If our species were to communicate with a sentient species from another planet, it is unlikely that we would agree with them on the units of time, length, and mass. The second, the metre, and the kilogram are products of our cultural heritage and have no intrinsic value beyond being an adopted standard. However, the concepts of time, length, and mass are far more fundamental, and are likely to form a shared basis for perception of the universe.

Dimensional analysis is a blunt tool that does not attempt to verify details such as whether the scales are consistent. It is only concerned that the dimensions are consistent for any values that are to be added, subtracted, compared, or equated. In other words, it is alright to divide the dimension of length by the dimension of time to find a speed, but you cannot add five seconds to three metres, nor can you say that five seconds is greater than three metres. Dimensional analysis is used to check the plausibility of derived equations. If you check the dimensions at every step of a computation, you can be certain that you have made an error in any step that introduces inconsistent dimensions. Similarly, mathematical functions such as $\sin(x)$, $\cos(x)$, e^x, and $\ln(x)$ take a dimensionless argument and return a dimensionless number. So, the x in $\sin(x)$ must be dimensionless, and $\sin(x)$ will evaluate to a dimensionless number.

CONCEPT QUESTION 1.3

The position of mass attached to the end of a spring is given by the equation $y = 25\sin(\omega t)$, where y is measured in metres and t is measured in units of seconds. What must the unit of ω be?

(A) s

(B) m

(C) m s^{-1}

(D) s^{-1}

Table 1.4 lists the dimensions that physicists typically use when applying dimensional analysis to a problem. Quantities such as energy, force, or even magnetic field strength can be described using combinations of these five dimensions. When writing the dimension of a quantity, it is common to surround the dimension with square brackets. For instance, the dimension of speed would be represented as [speed] = $[\mathrm{L\ T^{-1}}]$. In other words, the dimension of speed is length divided by time.

TABLE 1.4

Base dimensions

Dimension	Symbol
Length	L
Mass	M
Time	T
Electric current	I
Thermodynamic temperature	K

EXAMPLE 1.9

Calculate the dimensions for the quantities in the following equations:

(a) area = length * width

(b) density = mass/volume

(c) acceleration = ½ distance/time^2

(d) force = mass * acceleration

continued

Solution

(a) Area is found by multiplying together two quantities that have dimension of length:

$$[\text{Area}] = [L] \times [L] = [L^2]$$

(b) Density is found by dividing a mass by a volume. The dimension of volume is found by multiplying three lengths together:

$$[\text{Density}] = \frac{[\text{Mass}]}{[\text{Volume}]} = \frac{[M]}{[L][L][L]} = \left[\frac{M}{L^3}\right].$$

(c) Acceleration is found by multiplying a dimensionless constant by a length and then dividing by time squared:

$$[\text{Acceleration}] = \frac{[L]}{[T][T]} = \left[\frac{L}{T^2}\right].$$

(d) Force is found by multiplying mass by acceleration. Since we just found the dimension of acceleration, this is straightforward:

$$[\text{Force}] = [\text{Mass}][\text{Acceleration}] = [M]\left[\frac{L}{T^2}\right] = \left[\frac{ML}{T^2}\right].$$

Force combines three of the base dimensions. Be sure not to confuse the dimension $[m]$ for the unit of length, the metre.

1.6: Proportionality

When making physical models of the natural world, it is rarely the case that you know all the necessary details. Often, you can make important inferences without having all of the information. For instance, if experience shows that your body can metabolize alcohol at the rate of one pint of beer per hour, a simple physical model in which the rate of metabolization is linearly dependent on body mass would lead you to believe that a friend with twice your body mass is safe to drive even when drinking at

Figure 1.8 Is the rate of metabolization linearly proportional to body mass?

the rate of two pints of beer per hour. However, if it turns out that the rate of metabolization is independent of body mass, your friend might be making a poor decision getting behind the wheel of a car.

Proportionalities of this type occur everywhere in nature. For instance, although it is conceivable that a mammal could have a tiny brain and a huge body, there is no evidence that such a creature exists. Fig. 1.9 shows a graph of brain mass versus body mass for a wide variety of mammals. There is a distinct correlation between brain mass and body mass. Since Fig. 1.9 is a double-log graph, the slope of the graph is the exponent in the equation that relates brain mass to body mass:

$$\text{brain mass} \propto (\text{body mass})^{0.68}$$

That relationship is representative for over six orders of magnitude of body mass, and knowing the exponent of the proportionality allows for reasonable approximations to be made about the brain size of any mammal. If a mammal has a body that is twice as massive as our own, it should have a brain that is approximately $2^{0.68}$, which is 1.6 times as large as ours. Deviation from this trend is used to infer evolutionary adaptation.

For instance, *Homo floresiensis* is the species name for the "Hobbit-like" hominids discovered in 2003 on the island of Flores in Indonesia. This species is believed to have become extinct around 20 000 years ago, which is very recent history with respect to hominid evolution. Compared to *Australopithecus afarensis*, the most famous fossil of which ("Lucy") dates from around 3.2 million years ago, *Homo floresiensis* could be considered modern. However, *Homo floresiensis* is known less for its modern age than it is for its exceptionally small size, in body and brain. But was *Homo floresiensis* smart enough to make stone tools? Remains of *Homo floresiensis* have been found with stone tools that appear to date from the same time period, but its brain mass is closer to that of a chimpanzee or *Australopithecus afarensis*, neither of which are known to fashion stone tools.

Knowing the proportionality between brain mass and body mass helps to predict the likelihood that *Homo floresiensis* was crafting stone tools. Simply having a small brain mass is not enough information to make inferences about intelligence. A commonly used tool to compare the intelligence of one species to another is a ratio between brain mass and body mass, called the Encephalization Quotient (EQ). Intelligent species are likely to have an above average amount of brain for a given body size, and thus a larger EQ than species that have below average ratios of brain mass to body size. We can see from Fig. 1.9 that

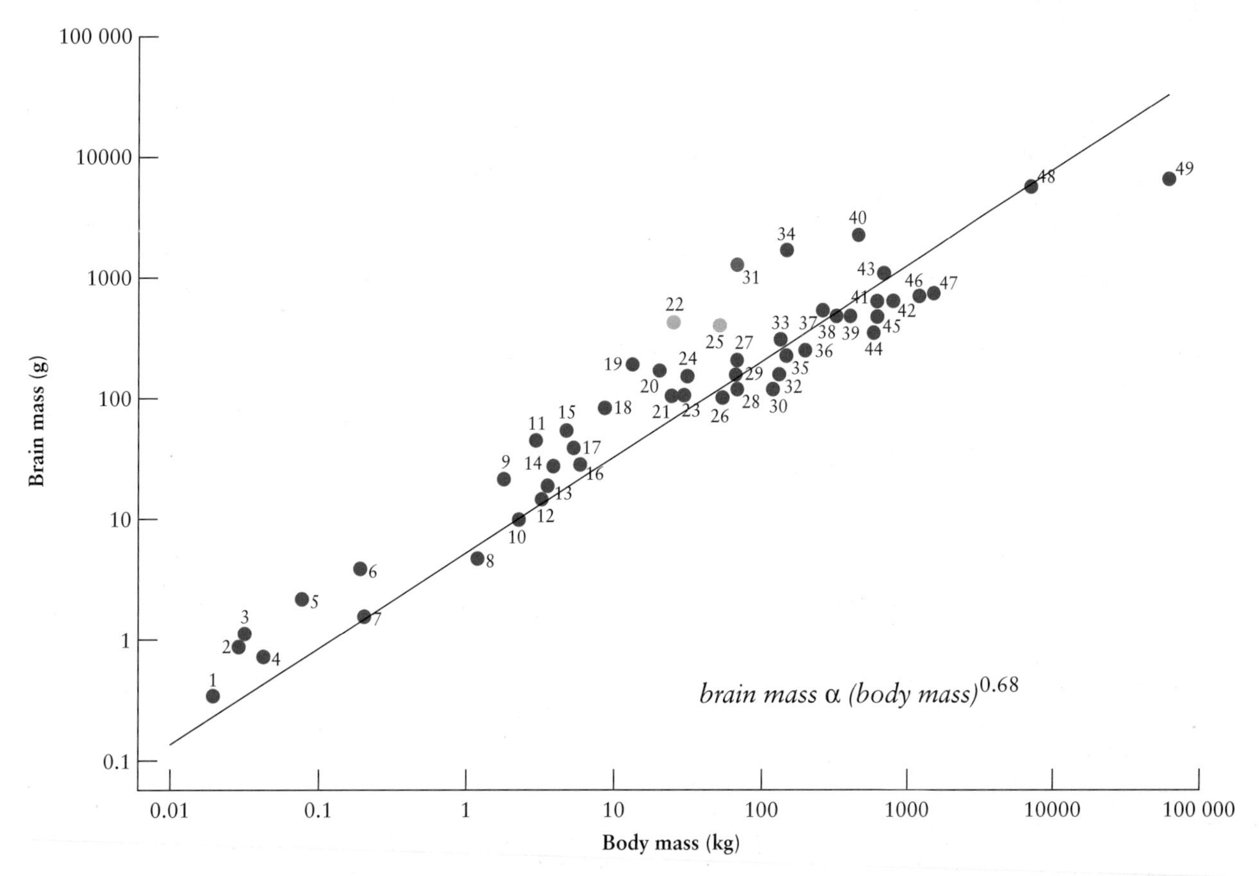

Figure 1.9 Graph showing brain mass as a function of body mass for the mammals listed in Table 1.5. Notice that this graph is a log–log graph, and that the slope of the straight-line fit represents the exponent of the power law proportionality.

TABLE 1.5

Brain mass and body mass for a wide variety of mammals

	Common name	Body mass [kg]	Brain mass [g]
1	Mole	0.0188	0.352
2	Vampire bat	0.028	0.936
3	Brown lemming	0.032	1.126
4	Meadow mouse	0.0413	0.7635
5	Chipmunk	0.075	2.22
6	Red squirrel	0.183	3.97
7	Norway rat	0.197	1.61
8	Opossum	1.147	4.8
9	Ring-tailed lemur	1.725	21.8
10	Skunk	2.26	10
11	Jackal	2.85	46
12	Three-toed sloth	3.121	15.1
13	Porcupine	3.41	19.15
14	Domestic cat	3.778	28.37
15	Red fox	4.625	53.3
16	Raccoon	5.175	40
17	Beaver	5.83	29.52
18	Coyote	8.51	84.24
19	Elk	13.61	194.2
20	Baboon	19.51	175
21	Greyhound dog	24.49	105.9
22	*Homo floresiensis*	25	433.2
23	Goat	27.66	115
24	Timber wolf	29.94	152
25	*Australopithecus afarensis*	50.6	415
26	Sheep	52.1	106.5
27	Chimpanzee	56.69	440
28	Spotted hyena	62.37	175
29	White-tailed deer	65.09	210
30	Warthog	65.32	125
31	Modern man	65.8	1310
32	Pig	113.2	123.9
33	Caribou	128.47	306
34	Porpoise	142.43	1735
35	Grizzly bear	142.88	233.9
36	Lion	190.8	258
37	Zebra	254.99	541
38	Polar bear	317	507
39	Hackney pony	362.87	504
40	White whale	441.31	2349
41	Holstein cow	574	415
42	Percheron stallion	635.04	662
43	Walrus	667	1126
44	Buffalo	759	653
45	Rhinoceros	763	655
46	Giraffe	1220	700
47	Hippopotamus	1351	720
48	Elephant	6654	5712
49	Blue whale	58059	6800

[Source: Crile and Quiring (1940), except for Australopithecus Afarensis: “An Introduction to Human Evolutionary Anatomy”, p.191, Aiello & Dean, 1990, and Homo Floresiensis: Nature 431, 1055–1061, 2004.]

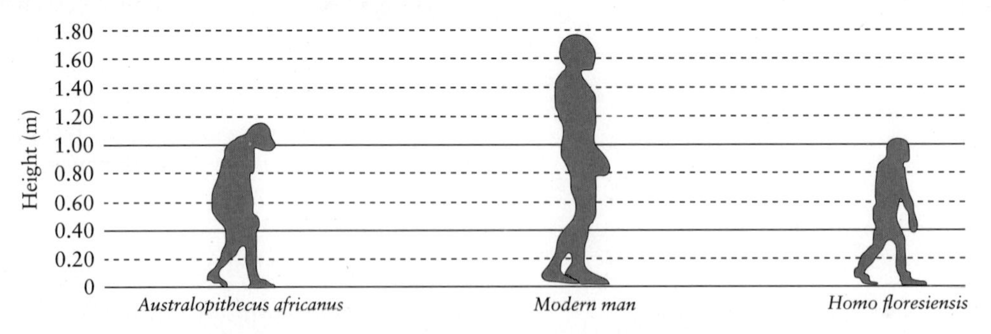

Figure 1.10 Height in metres for *Australopithecus africanus*, *Homo sapiens sapiens*, and *Homo floresiensis.*

Homo floresiensis is further away from the line of best fit than either the chimpanzee or *Australopithecus afarensis*. This means that its EQ is higher than either of those spe- ···es, and it is possible that it used that extra brain mass to ··· ··ut complex tasks such as fashioning stone tools.

KEY POINT

In physics, proportionalities can be used to make inferences about quantities without going through a complete analysis of all of potential variables.

13

CASE STUDY 1.2

Figure 1.11 The dugong is a marine mammal of the order Sirenia. How can we use proportionalities to estimate its brain mass?

A dugong is the only strictly herbivorous marine mammal and it is sometimes called a sea cow. Its body mass is typically around 400 kg. If it is about as intelligent as a Holstein cow, what would you expect the typical dugong brain mass to be?

(a) 500 g

(b) 400 g

(c) 300 g

(d) 200 g

Answer: Fig. 1.9 shows that the power law relationship between brain mass and body mass. We can write this proportionality as brain mass μ (body mass)$^{0.68}$. In this question, we are told to assume that the sea cow is about as intelligent as its terrestrial namesake, the Holstein cow. If intelligence can be estimated by the distance from the line of best fit on Fig. 1.9, the sea cow should be a little below the line. A line that passes through the data point for the Holstein cow, and also has same slope as the line of best fit shown on Fig. 1.9, must also pass through the data point for the sea cow. It is a good approximation that all mammals on that line have the same intelligence. For a sea cow,

$$\text{brain mass}_{\text{sea cow}} \propto (\text{body mass}_{\text{sea cow}})^{0.68}.$$

For a Holstein cow,

$$\text{brain mass}_{\text{Holstein cow}} \propto (\text{body mass}_{\text{Holstein cow}})^{0.68}.$$

If we divide those two proportionalities, we can get an expression for the brain mass of the sea cow in terms of masses that are either given in the question or are in Table 1.5:

$$\frac{\text{brain mass}_{\text{sea cow}}}{\text{brain mass}_{\text{Holstein cow}}} = \frac{\left(\text{body mass}_{\text{sea cow}}\right)^{0.68}}{\left(\text{body mass}_{\text{Holstein cow}}\right)^{0.68}}.$$

Which can be rewritten as:

$$\text{brain mass}_{\text{sea cow}} = \text{brain mass}_{\text{Holstein cow}}\left(\frac{\text{body mass}_{\text{sea cow}}}{\text{body mass}_{\text{Holstein cow}}}\right)^{0.68},$$

$$\text{brain mass}_{\text{sea cow}} = 415\,g\left(\frac{400\ kg}{574\ kg}\right)^{0.68} = 324\,g.$$

But the mass of the sea cow was just an approximate number that was quoted to only one significant figure, so the most precise answer that we can give for the brain mass of a sea cow is 300 grams.

For instance, the mass of an object is linearly proportional to its density and linearly proportional to its volume. That means that if the volume is doubled and the density is held constant, the mass of the object increases by a factor of two. This is true for any volume and any density, so proportionality is an efficient way to infer the change in density without having to use a calculator. For instance, if the volume of an object is proportional to the cube of its radius, the mass of the object is also proportional to the cube of its radius. If we consider a set of cylinders that have a radius that is made to increase as the square of the cylinder length, the mass of the object is proportional to length to the sixth power.

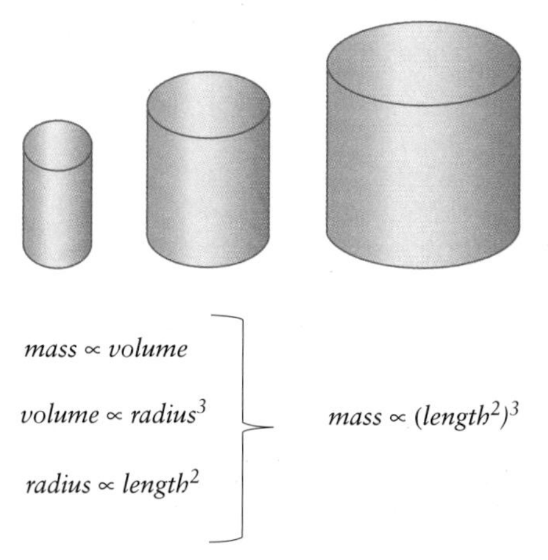

Figure 1.12 These three cylinders were drawn such that the radius is proportional to the square of the length. That means the cylinders will get fatter more quickly than they will get taller. Putting it all together, the mass of each cylinder is proportional its length to the sixth power.

EXAMPLE 1.10

3,4-Methylenedioxymethamphetamine (MDMA; commonly called ecstasy) is a synthetic compound that is increasingly popular as a recreational drug. One of the side effects of using this drug is an increased diastolic blood pressure (BP). The relationship between BP and MDMA dosage (D) is nonlinear. Assume the following formula describes BP (in the non-SI units mmHg) as a function of MDMA dosage (in mg).

continued

$$BP \propto D^{0.32}$$

If a 50 mg dose of MDMA causes a BP of 65 mmHg, what diastolic blood pressure would you expect with a dose of 150 mg?

Solution

The nice thing about proportionalities is that you do not even need to figure out what the constant of proportionality is. Instead, write out the equations for the 50-mg dose and the 150-mg dose:

$$BP_{50} \propto D_{50}^{0.32} \quad \text{and} \quad BP_{150} \propto D_{150}^{0.32}.$$

Divide the first equation by the second equation to get the ratio of the diastolic blood pressures in terms of the ratio of the dosage:

$$\frac{BP_{150}}{BP_{50}} \propto \frac{D_{150}^{0.32}}{D_{150}^{0.32}} \Rightarrow \frac{BP_{150}}{BP_{50}} \propto \left(\frac{D_{150}}{D_{50}}\right)^{0.32} \Rightarrow \frac{BP_{150}}{BP_{50}} \propto (3)^{0.32} = 1.4.$$

This means that if you double the dose, the diastolic blood pressure rises by almost a factor of one and a half. This proportionality predicts that the diastolic pressure you would expect with a 500 mg dose of MDMA is around $65.(10)^{0.32} \cong 136$ mmHg.

Data showing the nonlinear pharmacokinetics of MDMA in humans can be found in the *British Journal of Clinical Pharmacology* 2000 February; 49(2):104–109.

1.7: Order of Magnitude Estimation

The techniques developed so far are all tools that can help you make decisions about which parts of a complex system are primarily responsible for its behaviour and which parts can be ignored. That process is the largest barrier to making a good physical model. **Order of magnitude estimation** is the first tool that should come out of your toolbox when developing a physical model. It allows you to make educated decisions about what needs to be included in your physical model. At times, order of magnitude estimation is little more than justified guessing, and it should always be kept in mind that the answers derived from order of magnitude estimation are only as good as the estimates used to produce the answers. It is a little like asking for directions from random people on the street. If you asked how to get from where you are to a particular place in the next city, you might get someone who would tell you all the necessary street names, turns, and distances in order to get there. Order of magnitude estimation is more like having a person tell you that it is about 30 km south-west of where you are. You know the general area but might lack the details to get you exactly where you want to go.

Enrico Fermi popularized the use of this order of magnitude analysis in the middle of the 20th century. He was one of the scientists at Los Alamos in 1945, when the first [illegible]ic bomb was tested. On a windless, early morning in [illegible]n, he stood about 15 km away from the nuclear [illegible]y scraps of paper before, during, and after the shock wave from the blast past his position. He could have waited for more sensitive instruments to give him a more exact number for the energy release from the world's first nuclear weapon, but the distance that these papers drifted, combined with his distance from the blast, allowed him to immediately estimate that the explosion was the equivalent of 10 kilotons of TNT. His number was within a factor of two of the more precise measurement, but was well within the same order of magnitude. In other words, the real answer was much closer to 10 kilotons than it was to 1 kiloton, or 100 kilotons. In order to understand just how much more energy was released from that relatively primitive nuclear bomb, it is useful to compare it to the most destructive non-nuclear bomb currently in existence. The Aviation Thermobaric Bomb of Increased Power, a Russian-made bomb that is currently touted as the world's most powerful conventional (i.e. non-nuclear) bomb, releases energy that is the equivalent of 44 tons of TNT. Fermi's method of calculating the energy released by the world's first nuclear bomb showed that it released about 10^3 times as much energy as a conventional bomb. The factor of two is not relevant to the discussion. The nuclear bomb was orders of magnitude more destructive. This is the basic idea of order of magnitude estimation: find the power of 10 for the most important effect and ignore effects that are smaller by more than one order of magnitude.

It is important to note that absolutely anything can be approached with this technique. If you want to know how many steps your feet will take in a lifetime, you only need to estimate your lifespan, your typical walking speed, and the fraction of a day that you spend walking. If you want to know how many balloons you would need to fill your friend's bedroom, you only need to estimate the volume taken up by a typical balloon and the free volume of the bedroom. This may not be science, but the technique for coming to a solution is certainly fundamental to the way physicists think.

KEY POINT

Order of magnitude estimation is a technique that every scientist should use regularly as a coarse test of the validity of research results.

EXAMPLE 1.11

In SI units, the energy released from the bomb that Fermi observed was around 10^{14} joules. A 10 watt light bulb uses 10 joules of energy every second. The energy from that nuclear blast would provide the electricity for a city of one million people for how long?

Solution

Remember, we are interested only in an order of magnitude estimate, so we do not really need to know the exact power usage for every individual in the city. Chances are very good that the average per capita electricity usage in your neighbourhood is no more than twice your usage,

continued

and no less than half your usage. Your electrical consumption depends more on which country you live in. Per capita electricity consumption is around 3000 J per second for an average person living in Iceland, whereas it is only about 300 J per second for an average person living in China. In Somalia, the average person uses only 3 J per second. This means that the choice of country will be important for getting an order of magnitude estimate, and that the person-to-person differences within a country are something we can ignore.

If I look at my electricity consumption for my family home, my electricity bill says I average around 500 kwh per month. That unit is kilowatt–hours per month; therefore, the first step will be for me to convert that to an SI standard unit:

$$500\,\frac{kW\,hr}{month} \cong 500\,\frac{kW\,hr}{month} \times \frac{\left(10^3\,\frac{W}{kW}\right)hr}{\left(\frac{30\ days}{month}\right)\left(\frac{24\ hr}{day}\right)}$$

$$= \frac{500\times10^3}{(30)(24)}\frac{W\,hr}{hr} = \frac{5\times10^5}{(30)(24)}W.$$

continued

My average electricity usage $\cong (5\times10^5)/(30\times24)$ W $\cong$ 700 W.

One watt is the same as one joule per second, so I can say that my family uses around 700 J per second on average. To within a factor of 2, I can say that every person in my city uses around 1000 J per second. One million people using 1000 J per second gives a rate for the whole city of one million persons' usage $\cong 10^6$ people $\times$ 10^3J/s $\cong 10^9$J/s.

The bomb released 10^{14} J of energy, so this would power the homes in this city for time:

$$\cong \frac{10^{14}}{10^9\,\mathrm{J/s}} \cong 10^5\,\mathrm{s} \cong 1\,\mathrm{day}.$$

If I wanted a better estimate, I would have to take into account the fact that there are four people in my family, so my individual residential consumption is closer to 200 J per second. I would also have to estimate the fraction of electricity that is consumed outside residences; for instance, at your place of work. Chances are good that these two effects would partially offset each other and together would not change the answer by an order of magnitude.

CASE STUDY 1.3

Intergovernmental Panel on Climate Change (IPCC), Working Group II, Fourth Assessment Report: Climate Change 2007

The Himalayan Glaciers: Himalayan glaciers cover about 3 million hectares or 17% of the mountain area as compared to 2.2% in the Swiss Alps. They form the largest body of ice outside the polar caps and are the source of water for the innumerable rivers that flow across the Indo-Gangetic plains. Himalayan glacial snowfields store about 12 000 km^3 of freshwater. About 15 000 Himalayan glaciers form a unique reservoir which supports perennial rivers such as the Indus, Ganga and Brahmaputra which, in turn, are the lifeline of millions of people in South Asian countries (Pakistan, Nepal, Bhutan, India and Bangladesh). The Gangetic basin alone is home to 500 million people, about 10% of the total human population in the region.

Glaciers in the Himalaya are receding faster than in any other part of the world and, if the present rate continues, the likelihood of them disappearing by the year 2035 and perhaps sooner is very high if Earth keeps warming at the current rate. Its total area will likely shrink from the present 500 000 to 100 000 km^2 by the year 2035 (WWF, 2005).

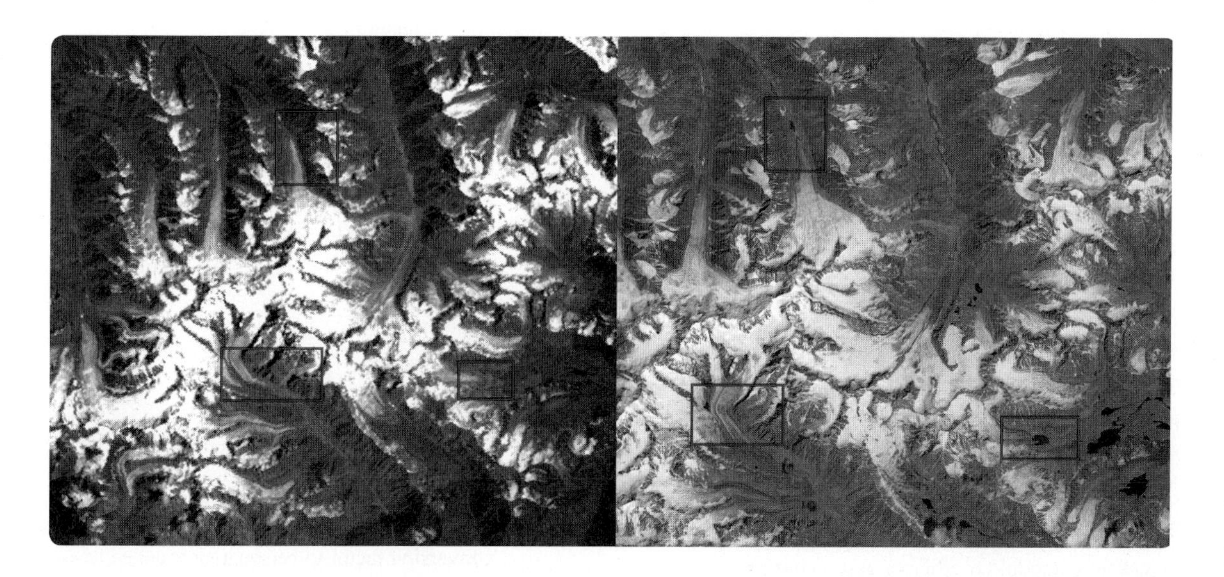

Umesh K. Haritashya · Michael P. Bishop · John F. Shroder · Andrew B. G. Bush · Henry N. N. Bulley: Space-based assessment of glacier fluctuations in the Wakhan Pamir, Afghanistan/Courtesy of Gregory Leonard, NASA/JPL and GLIMS/University of Arizona

Figure 1.13 Glacier retreat from 1976 to 2003 in the Wakhan Pamir, Afghanistan. The red boxes highlight areas of notable glacier terminus retreat or new lake formation.

continued

Answer: The IPCC set off a storm of criticism over the two paragraphs above. The two paragraphs were buried deep in one of its thousand page reports on climate change, but the relatively short time scale of the estimated glacial disappearance was picked up by the media and used to provide an incentive for quick action on climate change issues. However, those same paragraphs state that the glaciers in the Himalayas hold a frozen equivalent of 1.2×10^4 km^3 of water, and that the glaciers cover about 3 million hectares. A hectare is a metric measurement of area that may be used with SI but it is not an SI unit. To keep things straight, it is best to convert that into an SI unit of area, in this case, km^2:

$$1 \text{ hectare} = 10^4 \text{ m}^2 = 10^{-2} \text{ km}^2.$$

So, the glaciers cover around:

$$3 \times 10^6 \text{ hectares} = 3 \times 10^4 \text{ km}^2 = 30\ 000 \text{ km}^2.$$

But that immediately raises a red flag. The second paragraph states that the glaciers in the Himalayas "will likely shrink from the present 500 000 km^2 to 100 000 km^2 by the year 2035." Their own numbers show that the Himalayan glaciers are an order of magnitude smaller.

Further, we can estimate the average thickness of the Himalayan glaciers by dividing the volume by the surface area:

$$\text{average thickness} \cong \frac{1.2 \times 10^4 \text{ km}^3}{3 \times 10^4 \text{ km}^2} \cong 0.4 \text{ km}.$$

In the years between 1961 and 2003, the Institute of Arctic and Alpine Research measured the average specific mass balance of the Himalayan glaciers to be -0.41 ± 0.04 m per year. That means that the glaciers lost an average of around 40 cm of thickness per year over that 39-year period. So, using the average thickness and the average rate that the glaciers have been thinning recently, we can find the number of years that it would take for the glaciers to have an average thickness of zero:

$$t_{melt} \cong \frac{0.4 \text{ km}}{40 \text{ cm/yr}} \cong \frac{4 \times 10^2 \text{ m}}{0.4 \text{ m/yr}} \cong 10^3 \text{ yr}.$$

In order for the glaciers to melt completely between the date of the report and 2035, the rate would have to be approximately:

$$\text{rate} \cong \frac{0.4 \text{ km}}{2035 - 2007} \cong \frac{0.4 \text{ km}}{28 \text{ yr}} \cong \frac{4 \times 10^2 \text{ m}}{2.8 \times 10^1 \text{ yr}} \approx 10 \text{m/yr}$$

For the Himalayan glaciers to be losing 10 m/year over the next three decades, the rate of melting would have to have to be at least 25 times higher than what it has averaged over the past 40 years.

The IPCC shared a Nobel Prize in 2007 for its "efforts to build up and disseminate greater knowledge about man-made climate change, and lay the foundations for the measures that are needed to counteract such change." Unfortunately, these two careless paragraphs in the IPCC report have been used to discredit the work of hundreds of scientists around the world who have credibly shown that the glaciers are indeed thinning, and that human activity is causing climate change on Earth. The outcome has been to erode confidence in the predictive ability of science in general. The extra step of checking units and then using order of magnitude estimation to verify the validity of an answer can have a profound impact on every field of science.

SUMMARY

DEFINITIONS

- Physical model: a description that is focused on the most significant properties of a system. The description comes from observations of the world around us, has a minimum number of underlying assumptions, and predicts testable outcomes.
- Significant figures: the number of digits that a result is expressed with. Trailing zeros do not count unless the number has a decimal point. Leading zeros never count.
- Accuracy: the number of significant figures quoted in a result should represent the accuracy of the result.
- Precision: the smallest power of 10 quoted in a result should represent the precision of the result.
- Order of magnitude: the closest power of 10 for a number with one significant figure.
- Scientific notation: an efficient and unambiguous way to ...ess the number of significant figures and the order of ... of a numerical result.
- Order of magnitude estimation: a technique that provides solutions to complex problems that are accurate to about one order of magnitude.

CONCEPTS

- Significant figures:
  - Multiplication and division: the result has the same number of significant figures as the least accurate number used in the calculation.
  - Addition and subtraction: the result has the same precision as the least precise number used in the calculation.
- Units of measure:
  - Always use SI units.
  - Use metric prefixes to make numbers more manageable.
- Unit conversion:
  - Converting units that are squared or cubed requires the conversion factor to be squared or cubed as well.

- Dimensional analysis:
  - The dimensions of onc side of an equation must be the same as the dimensions of the other side of that equation.
  - Mathematical functions such as the trigonometric functions have dimensionless arguments.

MULTIPLE-CHOICE QUESTIONS

MC–1.1. If m_{brain} and M_{body} were directly proportional to each other, i.e., $m_{\text{brain}} \propto M_{\text{body}}$, the slope of the line in Fig. 1.1 would have to be
(a) zero
(b) –1
(c) +1
(d) –2
(e) +2

MC-1.2. If we re-plot Fig. 1.1 with the brain mass shown in unit gram (kg), the slope
(a) increases.
(b) decreases.
(c) remains unchanged.
(d) cannot be predicted before the re-plotting of Fig. 1.1 is completed.

MC-1.3. Which of the following numbers has the highest precision?
(a) 3×10^6
(b) 13.55
(c) 0.003124
(d) 5.1762394×10^{18}
(e) 1×10^{-18}

MC-1.4. Which of the following numbers has the highest accuracy?
(a) 3.6772×10^6
(b) 13.55
(c) 0.003124
(d) 5.1×10^{12}
(e) 1×10^{-18}

CONCEPTUAL QUESTIONS

Q–1.1. Can you ever be certain that a physical model is correct?

Q–1.2. We use the decimal system for representing numbers with scientific notation. This is referred to as a base-10 counting system. However, using ten symbols as a basis for counting has more to do with physiology (we have ten fingers) than math or physics. Assume for the moment that an evolution eventually causes humans to develop two extra thumbs (for faster text messaging). Now we are more likely to need twelve symbols as a basis for counting. Represent the number 10 by the symbol ♦, and for the number 11, use the symbol ♥.
(a) Express the base-10 number 11 as a base-12 number.
(b) Express the base-10 number 3 498 572 as a base-12 number.
(c) Express the base-12 number ♦6♥3 as a base-10 number.

Q–1.3. Estimate the volume of the room that you are in. Find that number in units of mm^3, cm^3, m^3, and km^3. Which one of these units is best suited to measuring volumes of that order of magnitude?

Q–1.4. In an 1897 article in *The American Naturalist* called "The Cricket as a Thermometer," physicist Amos Dolbear expressed an empirical relationship between air temperature and the rate of a Snowy Tree Cricket's chirp. To estimate temperature, T_F, in degrees Fahrenheit using the number of chirps per minute, N:

$$T_F = 50 + \left(\frac{N-40}{4}\right).$$

The formula is accurate to within around a degree Fahrenheit when applied to the Snowy Tree Cricket, and still somewhat accurate for many other common crickets. Rewrite this formula so that you find temperatures in degrees Celsius. This formula is dimensionally incorrect. Does that mean that it is cannot correctly predict the temperature? What does it mean?

Q–1.5. Fig. 1.9 shows a relationship between brain mass and body mass for a large variety of mammals. If we make the assumption that birds are less intelligent than mammals, would you expect the slope of a graph of brain mass versus body mass for a large variety of birds to be different from the slope that fits the mammalian data? If so, why? If not, what parameter would you expect to characterize the difference?

Q–1.6. Repeat the order of magnitude estimates made in Example 1.11 using an estimate for your own electricity usage. Can you think of a better way to estimate the energy usage of a city of 1 million people?

ANALYTICAL PROBLEMS

P–1.1. Express the following numbers in scientific notation:
(a) 123
(b) 1230
(c) 12 300.0
(d) 0.123
(e) 0.00123
(f) 0.00000123000

P–1.2. How many significant figures do the following numbers have?
(a) 103.07
(b) 124.5
(c) 0.09916
(d) 5.408×10^5

P–1.3. Express the following products in scientific notation:
(a) 123×0.00456
(b) 1230×0.456
(c) 0.0012300×4560.0
(d) 0.01230×456.00

P–1.4. Express the following quotients in scientific notation:
(a) $123 \div 0.00456$
(b) $1230 \div 0.456$
(c) $0.0012300 \div 4560.0$
(d) $0.01230 \div 456.00$

P–1.5. Express the following sums and differences in scientific notation:
(a) $123 + 456$
(b) $1230 + 0.456$
(c) $123.456 - 123.123$
(d) $123.45678 - 123.123$

P–1.6. Express the following in scientific notation:
(a) $123 + 456 \div 123$
(b) $123 + 456 \times 123$
(c) $123.456 - 456 \div 0.00123$
(d) $0.000123 \times 0.045678 + 0.0000012345$

P–1.7. Express your height in scientific notation in units of (a) nm; (b) mm; (c) cm; (d) m; (e) km. Which of these units is best suited to measuring lengths of that order of magnitude?

P–1.8. Your weight is measured in units of newtons and can be approximated by multiplying your mass in kilograms by 10 m/s^2. Express you weight in scientific notation in units of (a) μN; (b) mN; (c) N; (d) kN; (e) GN. Which of these units is best suited to measuring weights of that order of magnitude?

P–1.9. In your life, you will spend around 7 million seconds brushing your teeth. Express this number in (a) minutes; (b) hours; (c) days; (d) months; (e) years. Which of these units is best suited to measuring times of that order of magnitude?

P–1.10. Assume that you can run a 42.195 km marathon in 2 hours, 2 minutes, and 11 seconds (congratulations on setting a new world record). Use dimensional analysis to find an expression for your average speed. Express your average speed in scientific notation in units of (a) km/h; (b) m/s; (c) km/s; (d) m/h; (e) μm/ns.

P–1.11. Gravity accelerates objects toward the ground at around 10 m/s^2. Express this number in (a) mm/s^2; (b) m/ms^2; (c) km/h^2; (d) Mm/yr^2; (e) $\mu m/ms^2$.

P–1.12. Find the area of one side of a piece of paper that is 30 cm in height and 20 cm in width. Express this number in scientific notation in units of (a) cm^2; (b) mm^2; (c) μm^2; (d) m^2; (e) km^2.

P–1.13. The volume of a sphere is approximately $4R^3$, where R is the radius of the sphere. Find the volume of a sphere that is 6378 km in radius. Express this number in scientific notation in units of (a) cm^3; (b) m^3; (c) km^3; (d) dm^3; (e) litres.

P–1.14. Use dimensional analysis to find a formula for the density of the sphere described in problem P–1.13, given that the density depends only on the mass and the volume of the sphere. If the mass of the sphere is 5.9742×10^{24} kg, what is the density of the sphere? Express this number in scientific notation in units of (a) g/cm^3; (b) kg/m^3; (c) Mg/km^3; (d) $\mu g/pm^3$; (e) $ng/\mu m^3$.

P–1.15. Which of the following equations are dimensionally correct? (a) $A = 4\pi R$, where A is an area and R is a radius; (b) $x_2 = x_1 + v_1 t^2$, where x_2 and x_1 are lengths, v_1 is a speed, t is time; (c) $V = xyz$, where V is a volume and x, y, and z are lengths; (d) $f = (1/2)\ m + 7$, where f is the optimal age for a woman to get married and m is the optimal age for a man to get married. (Can an equation be both dimensionally correct and wrong?)

P–1.16. Assume that the speed of an object is inversely proportional to the square of the time it has travelled. After 4 seconds, the object is traveling 10 m/s. How fast is it traveling after 8 seconds?

P–1.17. The pygmy three-toed sloth *(Bradypus pygmaeus)* is a mammalian species that is endemic to Isla Escudo de Veraguas. The species was first identified as distinct in 2001. Through insular dwarfism, this species evolved to have a 40% smaller body mass than its mainland brethren *(Bradypus variegatus)*. The average body mass of the pygmy three-toed sloth is approximately 3 kg. Use the proportionality shown by Fig. 1.9 to estimate the brain mass of this mammal. What would you conclude if you were told that its brain mass was measured to be 200 grams?

Figure 1.14 The pygmy three-toed sloth is a newly discovered mammal.

P–1.18. We develop an empirical formula connecting the wingspan and the mass of some species able to fly. Then we evaluate a few interesting consequences. (The first to make these considerations was Leonardo da Vinci.) (a) Use the data in Table 1.6 to draw a double-logarithmic plot ln W versus ln M where W is the wingspan and M is the mass. Determine the constants a and b in a power law relation $W = a\ M^b$. (b) The largest animal believed ever to fly was a late Cretaceous pterosaur species found in Texas and named *Quetzalcoatlus northropi*. It had an 11-m wingspan. What is the maximum mass of this pterosaur? *Note:* The largest wingspan of a living species is 3.6 m, for the wandering albatross. (c) Assume that a human wishes to fly like a bird. What minimum wingspan would be needed for a person of 70-kg to take off?

TABLE 1.6

Wingspan data for birds spanning four orders of magnitude in body mass

Bird	Wingspan [cm]	Mass [g]
Hummingbird	7	10
Sparrow	15	50
Dove	50	400
Andean condor	320	11 500
California condor	290	12 000

P–1.19. Lashof and Ahuja show that methane has, per mole, a global warming potential 3.7 times that of carbon dioxide (*Nature* 1990; 344:529–531). In one day, a cow produces the same mass of methane as one typical car produces of carbon dioxide. Which of these has a more significant impact on global warming?

P–1.20. Estimate how much gasoline you use driving a car from one gas station to the next closest. What price difference, in cents per litre, would be required to justify "hunting for cheap gas"?

P–1.21. Some astrologers say that the position of the Moon and planets during your birth can affect the outcome of your life.

If the force of gravity is linearly proportional to mass, and inversely proportional to the square of the distance between you and that mass, rank in order from smallest to largest the relative magnitudes of the gravitational force exerted by the Moon, Mars, the Sun, Jupiter, and your mother.

P–1.22. You have been wrongly convicted of a crime and need to escape from prison in order to set the record straight. How long would it take to dig a 100-m long tunnel using a teaspoon ($V_{spoon} \cong 5\ cm^3$)?

P–1.23. Approximately how many times will your heart beat during your lifetime?

P–1.24. Approximately what is the average rate of growth of human fingernails? Express this in units of m/s, μm/day, and cm/year.

P–1.25. A white sand beach is 100 metres long and 10 metres wide. How many grains of sand are there if the beach is sandy to a depth of (a) 4 m and the average grain of sand has a volume of 1 mm^3? (b) 2 m and the average grain of sand has a volume of 1 mm^3? (c) 4 m and the average grain of sand has a volume of 2 mm^3? Why are the answers for (b) and (c) the same when in (b) a linear measurement is doubled and in (c) a cubed measurement is doubled? How is this different than converting cubed units?

P–1.26. How many people can fit on the dry surface of Earth without any person being able to touch the person next to them?

P–1.27. How many more years will it take before the dry surface of Earth is covered with graves? (Assume that people are buried lying horizontally in a coffin that is the same size as the person.)

P–1.28. If the density of all mammals is approximately the same as the density of water, what is the ratio of the volume of a hippopotamus to the volume of a sheep? What is the ratio of the volume of a lion to the volume of a chipmunk? Could 2 of every modern land mammal on Table 1.5 be housed in an ark that is 300 cubits long, 50 cubits wide, and 30 cubits in height? (Assume 1 cubit is approximately 45 cm.) Could 2 of every modern land mammal on Table 1.5 be housed in the volume equivalent to that of a blue whale?

ANSWERS TO CONCEPT QUESTIONS

Concept Question 1.1: (A). Always complete the calculation with as many digits as possible, and then round to the correct number of significant figures when quoting the result.

Concept Question 1.2: (D). A kilogram is much larger than a microgram, so the number certainly must be less than 1, eliminating two of the possible answers.

Concept Question 1.3: (D). Since the product ωt is being used inside a trigonometric function, it must be dimensionless. If ω has units of s^{-1}, the product ωt is dimensionless. All the other choices result in the product having a dimension.

CHAPTER 2

Kinematics

Motion is quantified through the physical parameters of position, time, velocity, and acceleration. Relationships exist between these four parameters. An object's velocity describes its change of position with time, and its acceleration describes the object's change of velocity with time. Position, velocity, and acceleration are *vectors,* because motion is described by both magnitude and direction. The term *speed* is used for the magnitude of the velocity, but travelling at high speed will not get you to your destination if you are heading in the wrong direction.

An intuitive feel for the relationship between position, velocity, and acceleration can be developed by graphing the motion of objects as a function of time. The slopes of these graphs can be used to relate position to velocity, and velocity to acceleration. These graphs can then be related to the basic equations that describe motion. These equations are called the kinematic equations.

The wide range of motion patterns seen in the natural world depends on an object's acceleration. We will see in later chapters that acceleration is caused by forces. That makes acceleration special, so this chapter is devoted to developing an intuitive feel for acceleration.

2.1: Measuring Motion

Eventually, we would like to be able to characterize any type of motion, whether it is a child skipping down the street, or a protein folding into its native state. However, in order to understand the parameters that characterize motion, it is best to start with the least complex type of motion. That can later be used as a framework to understand more complicated motion.

A beginning point for a physical model of motion should start by removing all but the essential ingredients of motion. If motion describes something moving from one place to another, then what is the minimum amount of information needed to characterize that motion? In the case of a child skipping down the street, if we only want to know how many times the child has jumped up and down, it is not immediately relevant that the child was moving down the street while jumping. Conversely, if we want to know how far down the street the child is after two minutes of skipping, we might not have to describe the up and down motion in order to answer that question.

The most basic motion that we could describe would be constant motion along a straight path. If someone came up to you after class and asked where to catch the bus, what is the minimum amount that you could say while still giving sufficient direction? To start with, you need a reference point. Your instructions have to begin somewhere. If the person is familiar with campus, you could give them instructions that start at the student centre. If the student is new to campus, it might be best to give directions that start at your current location. In either case, this reference point is called the *origin* of your directions: the place that your directions will be measured from.

If you told the student that the bus stop is 100 metres away from the student centre, the student would still have to hunt around all sides of the student centre to find the bus stop. You could save the student a lot of time by indicating that the bus stop is 100 metres to the south of the student centre. If the student doesn't know which way south is, you could specify the heading many other ways:

- Start at the student centre and walk for one minute toward the music building.
- Start at this exit and walk one kilometre along the road that runs up the hill.
- It's on the road just outside this building.

All three of these statements have everything that is required to establish what we will call a *coordinate system.*

KEY POINT

A coordinate system needs an origin, a scale, and a positive direction.

The positions are measured relative to the origin, the scale indicates what units the numbers are associated with, and the positive direction allows the student to move toward the bus stop instead of away from it. Although the scale can be given in any measurable unit, there is less confusion if SI units of length are used. For instance, walking for 1 minute produces different distances for walkers of different speed, and saying something is "just outside" is a subjective description. Does "just outside" mean 1 metre, or 50 metres?

KEY POINT

A line that passes through the origin and is parallel to the positive direction of motion defines the position axis.

Since we only have one axis for the moment, we will call this the x-axis. The x-axis starts at an x-position of negative infinity and continues to an x-position of positive infinity. The origin is the position at which $x = 0$. If you can move only along this line, we call your motion one-dimensional motion. This motion is restricted—for example, you can never travel along a curved path—but it is a very good starting point for a physical model of motion.

Even this one-dimensional motion requires one more parameter for quantitative evaluation: time. We have already said that the origin is the position where $x = 0$. Let's now further call it the instant in time that we define $t = 0$. Time is a second axis. It has an origin, a scale (for instance, seconds or years), and a direction that indicates increasing values of time. Each instant along this axis is assigned a numerical value. Motion is then defined by the change of your position as a function of time.

EXAMPLE 2.1

Let's go back to the idea of giving directions for finding a bus stop. Take the x-axis to be parallel to a line that goes directly from the student centre to the bus stop, and positive in the direction of the bus stop. Take the origin of the x-axis to be at the student centre. To complete the coordinate system, we need to decide on a scale for the x-axis. In this case, we'll measure the x-position in units of metres. Fig. 2.1 shows a graph of a student's x-position as a function of time. If the bus stop is 500 m from the student centre, does the student get to the bus stop at some point in the time interval shown in Fig. 2.1? Where does the student end up at time $t = 10$ minutes?

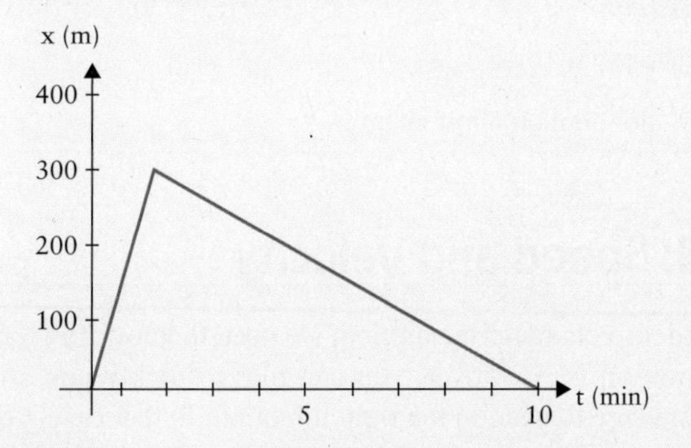

Figure 2.1 A graph showing the position of a student as a function of time.

continued

Solution

The student goes to a maximum x-position of 300 m before turning around. He never gets to the bus stop, and at $t = 10$ minutes, he is right back where he started.

CONCEPT QUESTION 2.1

Fig. 2.2 shows the x-position of an object as a function of time. If the object moves to the right, the x-position increases. If the object moves to the left, the x-position decreases. Which of the following best describes the motion of the object?

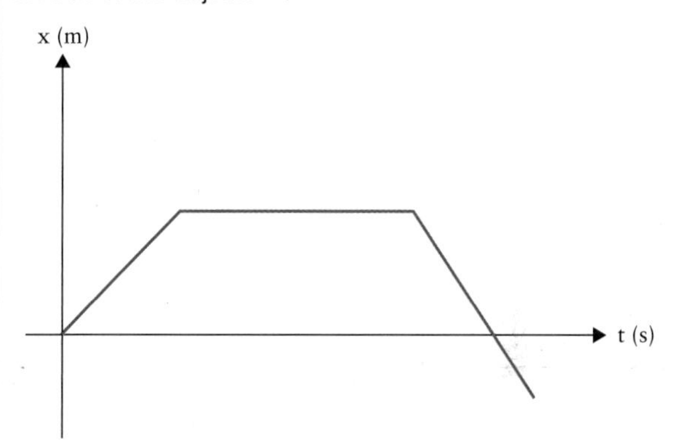

Figure 2.2 Graph of position versus time. The x-axis is positive to the right.

(A) The object moves to the left, stops, then moves to the right, ending up further to the right than where it started.

(B) The object moves up, stops, and then falls down lower than where it started.

(C) The object moves to the right, stops, and then moves to the left, ending up further to the left than where it started.

(D) The object moves up and to the right, then straight to the right, then down and to the right.

2.2: Distance and Displacement

The amount that an object's position changes over a given amount of time describes its motion during that time interval. To quantitatively describe motion, we have to be careful about how we define the position of an object. A coordinate system needs an origin, a scale, and a positive direction. So, a positive x-position tells you that the object is on the positive side of the origin, and a negative x-position tells you that the object is on the negative side of the origin. The numerical value of the x-position tells you how far away the object is from the origin. We call the numerical value of the x-position its *magnitude*. This means that a position on the x-axis has both a magnitude and a direction.

KEY POINT

Quantities that have directions associated with them have a special name in physics: vectors.

That means that positions measured along an x-axis are one-dimensional position vectors. The position vector $\vec{x}_1$ indicates where on the x-axis an object is at time t_1.The arrow indicates that the position contains information about both the direction (is it on the positive or negative side of the origin?) and the magnitude (how far is it from the origin?). When we write x_1 without an arrow, it refers to the magnitude of vector $\vec{x}_1$. A magnitude is always positive.

If you know the position vector $\vec{x}_1$ for an object at initial time t_1, as well as the position vector $\vec{x}_2$ at some later time t_2, you can find how far away the object is from where it started. We call that the displacement, which is a vector defined by the equation:

$$\Delta\vec{x} = \vec{x}_2 - \vec{x}_1. \qquad [2.1]$$

Notice that the displacement tells you only how far away the object is from where it started. It does not tell you how far the object travelled between time t_1 and time t_2. For example, Fig. 2.1 shows the position of a student as a function of time. At time t_1 = 1 minute, the student is at position vector $\vec{x}_1$ = 150 m. At time t_2 = 6 minutes, the student is at position vector $\vec{x}_2$ = 150 m. Remember that the sign of a one-dimensional vector gives its direction. In this case, both of these position vectors are positive numbers, which means that they are both on the positive side of the origin of the x-axis. The displacement is $\Delta\vec{x} = \vec{x}_2 - \vec{x}_1$ = 0 m, meaning that the object is zero metres from where it started, but the graph shows that the student is moving during that whole time interval. In order to calculate the speed of the student, we need to have a quantity that measures the distance travelled regardless of direction.

KEY POINT

Displacement describes the length between two locations measured along the straight-line path connecting them.

Distance describes the length between two locations measured along the actual path used to connect them. The distance travelled by the student between time t_1 = 1 minute and t_2 = 6 minutes is d = 300 m: 150 m in the positive direction, plus 150 m in the negative direction. Because distance does not depend on the direction, both these numbers are positive and do not add to zero.

EXAMPLE 2.2

Using Fig. 2.1, determine both the distance and the displacement of the student between time t_1 = 1 minute and t_2 = 10 minutes.

continued

Solution

At time t_1 = 1 minute, the student is at position vector $\vec{x}_1 = 150\text{m}$. At time t_2 = 10 minutes, the student is at position vector $\vec{x}_2$ = 0 m. The displacement is $\Delta\vec{x} = \vec{x}_2 - \vec{x}_1$ = 0 m − 150 m = −150 m. Once again, remember that the sign of a one-dimensional vector gives its direction. In this case, the displacement is a negative number, which means that the change in position is in the negative x-direction. Earlier, we said that the x-axis is measured positive in the direction of the bus stop. The negative displacement indicates that at t_2 the student is further away from the bus stop than at t_1. The magnitude of the displacement tells us the distance between the student's position at t_1 and the student's position at t_2.

The distance the student travelled can be found by breaking the path into two sections: the time interval between 0 minutes and 2 minutes, and the time interval between 2 minutes and 10 minutes. Since the direction of travel does not matter when determining the distance, we have to look at the distance travelled in the positive direction, and then add that to the distance travelled in the negative direction. If, as in this example, there is a change in direction of travel, the distance travelled will be more than the magnitude of the displacement. The distance travelled can never be less than the magnitude of the displacement. In this case, the student travelled a distance of 150 m over the first time interval, and then travelled a distance of 300 m over the second time interval. In total, the distance the student moved between time t_1 = 1 minute and t_2 = 10 minutes is 450 m.

CONCEPT QUESTION 2.2

Over a 5-minute time interval, an object moves from an initial position vector $\vec{x}_1$ = −150 m to a final position vector $\vec{x}_2$ = 0 m. What is the total distance travelled during that time interval?

(A) 150 m

(B) 0 m

(C) −150 m

(D) not enough information

2.3: Speed and Velocity

In order to characterize motion, we need to know how fast the motion is. However, you can move quickly and still not manage to head in the right direction. In that case, you would not end up any closer to your intended destination. This is another case where there is a distinction between the vector quantity and the more common concept that describes the magnitude of that vector.

KEY POINT

Velocity is a vector that describes how much motion an object has in a particular direction.

The more commonly used term, **speed,** is a positive scalar that describes how much motion an object has, without any associated sense of direction. Both of these quantities can be viewed as an average during a finite time interval, in which case they are called average velocity and average speed.

If you ride your bicycle from home to school and then back home again, your average velocity is zero because, by the end of the journey, you ended up at the same place on the x-axis where you started. However, you were moving, so you had an average speed. The average velocity depends on the displacement, whereas the average speed depends on the distance travelled. Notice that we are now talking about average velocity and average speed. These quantities are measured over a time interval in which velocity and speed might not be constant. For instance, if you walked 1000 m to get from your house to your friend's house, and that walk took you 500 seconds, you could find your average speed by dividing the distance you walked by the time taken:

$$\vec{v}_{\text{av}} = \frac{\Delta x}{\Delta t} = \frac{x_2 - x_1}{\Delta t} = \frac{1000\,\text{m}}{500\,\text{s}} = 2\ \text{m/s}.$$

At times during that walk, you likely walked somewhat faster than 2 m/s, and at other times you walked somewhat slower than 2 m/s. The average speed doesn't give you any information about your speed at particular instants in time; it simply tells you that you would have been able to walk that distance in that amount of time if you had walked at a constant rate of 2 m/s. The average speed also has no information about the directions you travelled during your walk. If you could walk in a straight line to get to your friend's house, it might be less than 1000 m away from your house; but you might have had to walk around the block to get there. Average speed describes how fast you were moving along your actual path.

If the straight-line distance between your houses is 500 m, you can find your average velocity by dividing your displacement by the time taken:

$$\vec{v}_{\text{av}} = \frac{\Delta \vec{x}}{\Delta t} = \frac{\vec{x}_2 - \vec{x}_1}{t_2 - t_1}. \qquad [2.2]$$

Using Eq. [2.1] with our values of displacement and time taken, we find:

$$\vec{v}_{\text{av}} = \frac{\Delta \vec{x}}{\Delta t} = \frac{500\ \text{m}}{500\ \text{s}} = 1\ \text{m/s},$$

where we are defining positive to be in the direction of your friend's house. Your average speed is larger than your average velocity, but your average velocity describes how quickly you are moving toward your goal of reaching your friend's house.

CONCEPT QUESTION 2.3

Can your average speed ever be lower than the magnitude of your average velocity?

(A) yes

(B) no

EXAMPLE 2.3

At 3:14 p.m., a train is 3 km east of the city centre. At 3:56 p.m., the train is 26 km west of the city centre. What is the average velocity of the train during this time interval?

Solution

Step 1: Draw an axis that shows the initial and final positions of the train.

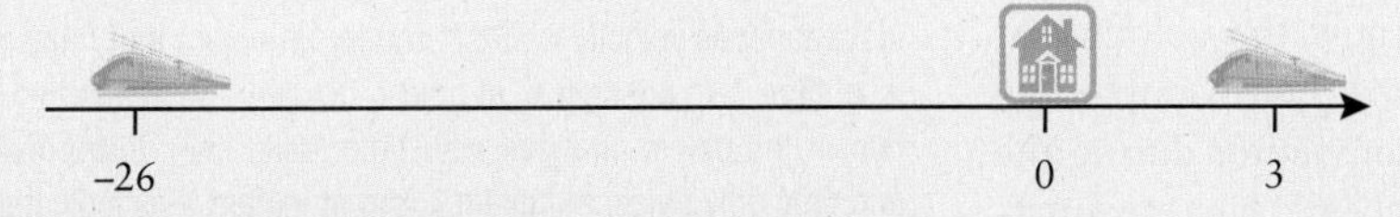

Figure 2.3 The x-axis showing the position of the train as measured from the city centre, which is taken to be the origin. Distances are in units of kilometres and I am choosing east to be the positive direction, which means that distances west of the city centre are negative.

Step 2: Find $\Delta \vec{x}$ and Δt:

$$\Delta \vec{x} = \vec{x}_2 - \vec{x}_1 = 3\ \text{km} - (-26\ \text{km}) = 29\ \text{km}$$

$$\Delta t = t_2 - t_1 = 3{:}56\ \text{p.m.} - 3{:}14\ \text{p.m.} = 42\ \text{minutes}.$$

continued

If we want to find the average velocity in units of kilometres per hour (km/h), which is a typical unit for describing the motion of such things, we need to convert 42 minutes to hours:

$$\Delta t = \frac{42\ \text{min}}{60\ \text{min/h}} = 0.70\ \text{h}\ .$$

Now, using Eq. [2.1], we can find the average velocity:

$$\vec{v}_{av} = \frac{\Delta \vec{x}}{\Delta t} = \frac{29\ \text{km}}{0.70\ \text{h}} = -41\ \text{km/h}\ .$$

The average velocity during that time interval is 41 km/h in the west direction.

We can develop the concept of motion along the x-axis further by looking at two examples from track and field events: 100-m sprint and long jump. The box in Fig. 2.4(a) represents the sprinter at the starting line (position x_1) and at the finish line (position x_2), or the long jump competitor at the start of the approach (position x_1) and at the point of takeoff, position x_2). Fig. 2.4(b) is a graph of x-position as a function of time. The solid curve in Fig. 2.4(b) is a possible description of the motion of each of the athletes as they get from position x_1 (starting position) to position x_2 (final position).

During the early stage of the sprint or approach, i.e., near time t_1, the athlete covers a comparably smaller distance per time unit than during the late stage, near time t_2. In the case of the sprinter, this is due to increasing the velocity toward a maximum value; in the case of the long jump competitor, this is done to achieve maximum takeoff velocity.

Equation [2.1] corresponds graphically to the dashed line in Fig. 2.4(b). We call the slope of the dashed line the average velocity of the object between time t_1 and time t_2. The dashed line in Fig. 2.4(b) doesn't follow the solid line, meaning that the actual velocity of the athlete varies with position and is, at most points, different from the average velocity.

To discuss the usefulness of Eq. [2.1], the sprinter and the long jump competitor are compared more carefully. The definition of velocity in Eq. [2.1] is useful for the sprinter's coaching team. For example, they could plot the average velocity of a 100-m sprint, a 200-m dash, and a 400-m race for the same sprinter to obtain information about the endurance performance of the leg muscles. However, we more often want to know the velocity of an object at a particular instant in time. For example, the average velocity of the long jump competitor during the approach run is useless for the coaches. They are interested in the instantaneous velocity at takeoff.

The velocity at a particular instant in time is called the **instantaneous velocity.** The magnitude of its instantaneous velocity is an object's **instantaneous speed.** This is why speed is a positive scalar. When we talk about velocity and speed at a given time, we typically leave off the word instantaneous. However, since instantaneous speeds and velocities can change over a period of time, it is important to recognize that average speed and velocity can differ from instantaneous speed and velocity.

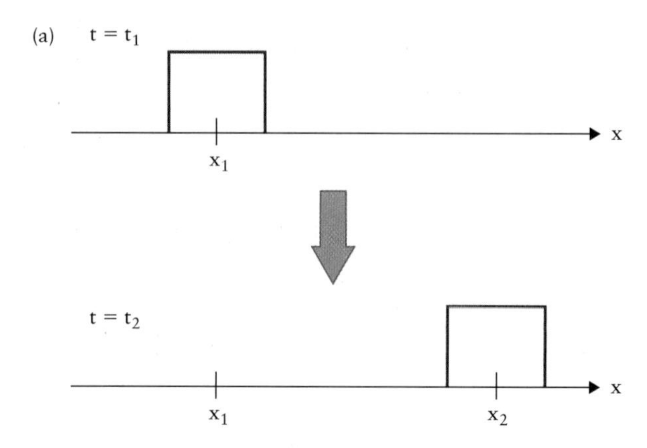

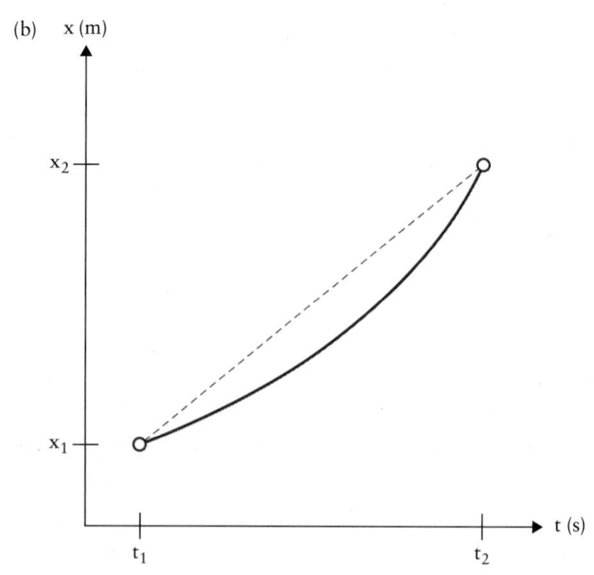

Figure 2.4 (a) The difference between the average velocity and the instantaneous velocity is illustrated for an object that moves from position x_1 at time t_1 to position x_2 at time t_2. (b) A graph illustrating how the position of the object changes with time (solid line). If we measure position and time only twice, at instant 1 and at instant 2 as indicated by two open circles, then we obtain the average velocity as the displacement, $\Delta x = x_2 - x_1$, divided by the elapsed time, $\Delta t = t_2 - t_1$ (dashed line).

The instantaneous velocity can be obtained from the velocity definition in Eq. [2.1]. We start with the end points shown as open circles in Fig. 2.4(b) and assume they provide a rough estimate of the instantaneous velocity at time t_2. Then we improve that estimate of the velocity by determining the average velocity for shorter and shorter time intervals, each ending at time t_2. This is illustrated in Fig. 2.5: the slopes of the dash-dotted, dashed, and solid blue lines are successively better estimates of the slope at time t_2.

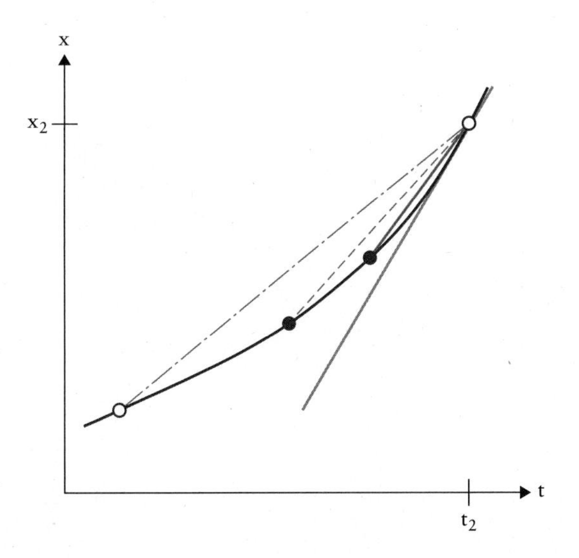

Figure 2.5 Repeat of Fig. 2.4 with the initial instants (open and solid dots) chosen successively closer to the final instant. As the time interval approaches zero, the line through the two respective points on the curve approaches the tangent at the final instant.

The instantaneous velocity at time t_2 is written as $\vec{v}$ and follows when we extrapolate to $\Delta t = 0$, as shown by the slope of the green line in Fig. 2.5. In the limit that $\Delta t = 0$, this line is the tangent to the solid curve at position x_2, and the value of the instantaneous velocity is the slope of the tangent:

$$\vec{v} = \lim_{\Delta t \to 0} \frac{\Delta \vec{x}}{\Delta t}. \quad [2.3]$$

We use the mathematical limit notation in Eq. [2.2]; i.e., $\vec{v}$ is the limit of the function $\Delta\vec{x}/\Delta t$ as Δt approaches zero. This is an approximation because we cannot simply set $t_2 = t_1$ in Eq. [2.2], as division by zero is undefined.

CONCEPT QUESTION 2.4

For the motion of a particular object we find that the maximum instantaneous velocity is equal to the average velocity. What can we conclude from this observation?

(A) The time interval must be very long.

(B) The object is moving at a constant velocity.

(C) The object is decelerating.

(D) The object is accelerating.

The tangent to the position-versus-time curve at time t_1 is the slope at that point. That means that we can use information about the slope of a position-versus-time graph to produce a corresponding velocity-versus-time graph.

KEY POINT

The slope of an object's position-versus-time graph gives the velocity on the object's velocity-versus-time graph.

Figure 2.6 shows a position-versus-time graph for an object moving along the x-axis over a time interval from 0 s to 10 s. The average velocity for the interval from 0 s to 4 s can be found by calculating the slope of the curve in that time interval using Eq. [2.1]:

$$\vec{v}_{av} = \frac{\Delta\vec{x}}{\Delta t} = \frac{\vec{x}_2 - \vec{x}_1}{t_2 - t_1} = \frac{600\text{ m} - 0\text{ m}}{4\text{ s} - 0\text{ s}} = 150\text{ m/s}.$$

The slope for that interval is a constant, which immediately tells us velocity must be constant during that time interval. That also means instantaneous velocity is equal to average velocity for that time interval. This is shown in the velocity-versus-time graph in Fig. 2.6. Velocity is 150 m/s for all instants between 0 s and 4 s. It is generally true that a region of constant slope on a position-versus-time graph will produce a flat region on the velocity-versus-time graph.

The velocity for the interval from 4 s to 10 s can also be found by calculating the slope of the curve in that time interval using Eq. [2.1]:

$$\vec{v}_{av} = \frac{\Delta\vec{x}}{\Delta t} = \frac{\vec{x}_2 - \vec{x}_1}{t_2 - t_1} = \frac{0\text{ m} - 600\text{ m}}{10\text{ s} - 4\text{ s}} = -100\text{ m/s}.$$

The slope for that interval is also constant, which means the velocity is –100 m/s for all instants between 4 s and 10 s.

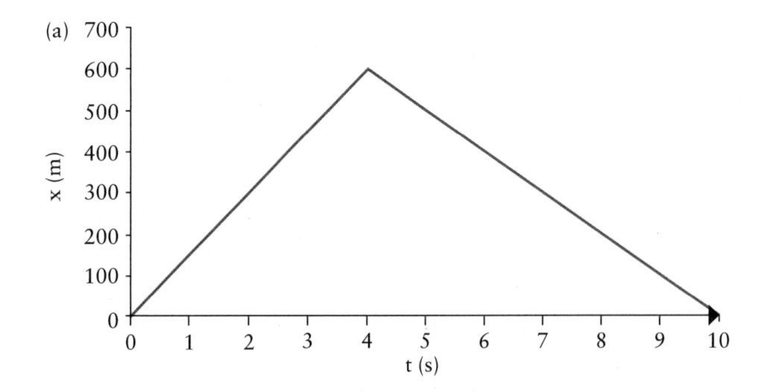

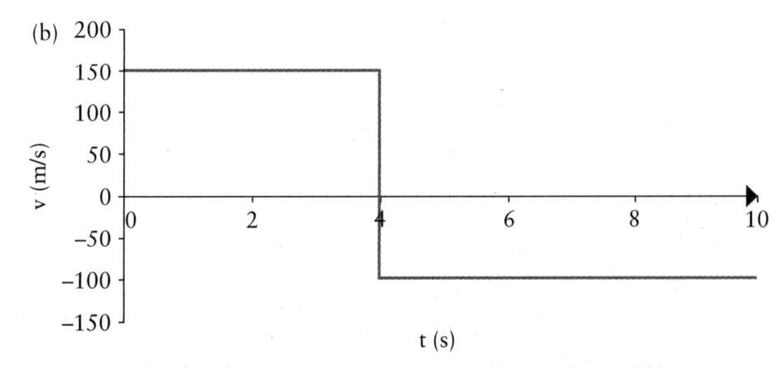

Figure 2.6 A position-versus-time graph can be used to infer velocity.

EXAMPLE 2.4

The position of a particle as a function of time is given by the curve shown in Fig. 2.7.

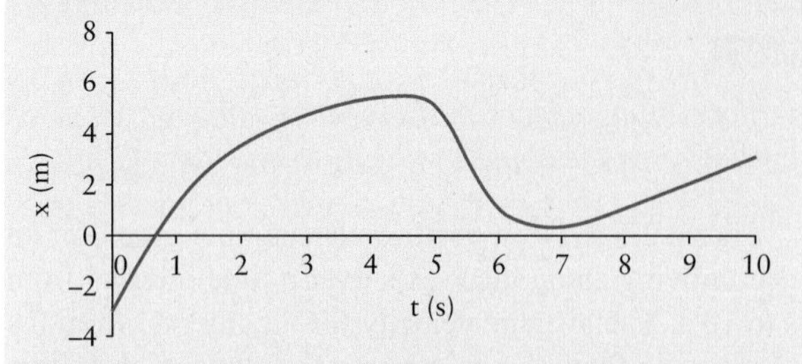

Figure 2.7 A position-versus-time graph for a non-constant velocity.

(a) When does the velocity have its most positive value?

(b) Is the velocity ever negative?

(c) When is the speed the greatest?

(d) When is the speed zero?

(e) What is the instantaneous velocity at time $t = 2$ s?

(f) What is the average velocity between $t = 3$ s and $t = 6$ s?

Solution

(a) The velocity has its most positive value when the tangent to the curve is most positive. This happens at around 0 s.

(b) The velocity is negative when the tangent to the curve is negative. This happens in the time interval between 5 s and 6 s.

(c) The speed is greatest when the slope has its largest magnitude. This happens some time between 5 s and 6 s. Even though the velocity is negative in that time interval, the speed is the magnitude of the velocity, and is always positive.

(d) The speed is zero when the slope of the curve is zero. That happens twice: at around 4.8 s and again at around 7 s.

(e) The instantaneous velocity at $t = 2$ s can be found by drawing a tangent to the curve at $t = 2$ s, as shown in Fig. 2.7. The slope of that tangent can be found by calculating the rise over the run, which in this case is:

$$\vec{v}(t = 2\text{ s}) = \frac{\Delta\vec{x}}{\Delta t} = \frac{\vec{x}_2 - \vec{x}_1}{t_2 - t_1} = \frac{7.0\text{ m} - 0.0\text{ m}}{4.0\text{ s} - 0.0\text{ s}} = 1.8\text{ m/s}.$$

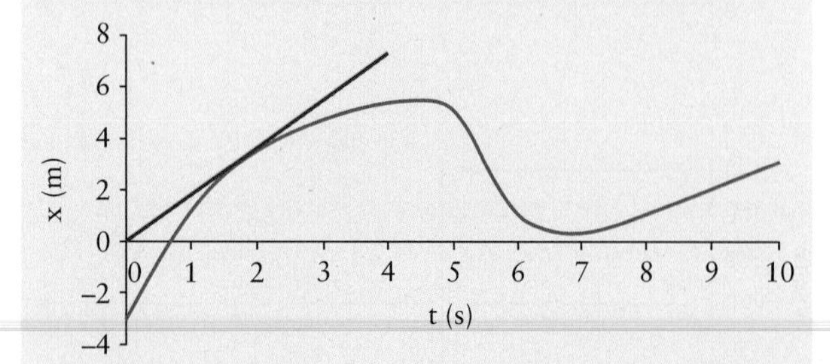

Figure 2.8 The slope of the tangent to the curve at $t = 2$ s represents the velocity at $t = 2$ s.

continued

(f) The average velocity during the interval between $t_1 = 3$ s and $t_2 = 6$ s can be found using Eq. [2.1]:

$$\vec{v}_{av} = \frac{\Delta\vec{x}}{\Delta t} = \frac{\vec{x}_2 - \vec{x}_1}{t_2 - t_1} = \frac{1.0\text{ m} - 5.0\text{ m}}{6.0\text{ s} - 3.0\text{ s}} = -1.3\text{ m/s}.$$

2.4: Acceleration

Figure 2.9 The cheetah is the world's fastest land animal; but its acceleration is what makes it such an accomplished predator.

The cheetah and the gazelle are two of the fastest land animals on Earth. The gazelle is also a staple in the cheetah's diet. The cheetah's maximum speed is around 110 km/h. The gazelle's maximum speed is around 75 km/h. It would seem as if the cheetah could have an easy meal any time it found a gazelle. However, running at such high speeds quickly exhausts the cheetah. If it does not catch the gazelle within the first 15 seconds, it has to give up or risk brain damage due to overheating. We can rewrite Eq. [2.1] in a form that we can use to find out how far the cheetah would travel in those 15 s:

$$\vec{v}_{av} = \frac{\Delta\vec{x}}{\Delta t} \Rightarrow \Delta\vec{x} = \vec{v}_{av}\Delta t$$

Recalling that $\Delta\vec{x} = \vec{x}_2 - \vec{x}_1$, we can rewrite this as:

$$\vec{x}_2 = \vec{x}_1 + \vec{v}_{av}\Delta t. \qquad [2.4]$$

This can be used to find the final position of an object that starts at initial position vector $\vec{x}_1$ and travels with a constant velocity $\vec{v}_{av}$ for some time Δt. We can use this to find how far the cheetah would travel before exertion would likely cause brain damage. First we have to convert speed into units of metres per second in that calculation so that we could multiply it by time in units of seconds:

$$\vec{v}_2 = \frac{(110\ \text{km/h})(1000\ \text{m/km})}{(60\ \text{min/h})(60\ \text{s/min})} \cong 30\ \text{m/s}.$$

Because we are only interested in the distance the cheetah travels from its initial position to its final position, we can set $\vec{x}_1 = 0$ and solve for $\vec{x}_2$:

$$\vec{x}_2 = \vec{x}_1 + \vec{v}_{av}\Delta t = 0\ \text{m} + (30\ \text{m/s})(15\ \text{s}) = 450\ \text{m}.$$

We find that the cheetah can run approximately 450 m before it has to give up the chase.

We made the assumption here that the cheetah travelled with a constant velocity over that period of time. In reality, the cheetah starts from rest, accelerates quickly, and then runs at its maximum speed. In order to examine what difference that makes to the cheetah's ability to catch a gazelle, we need to understand what acceleration is.

KEY POINT

Acceleration describes how much an object's velocity changes in a given amount of time.

Much like velocity described how much an object's position changes in a given amount of time, acceleration describes how much an object's velocity changes in a given amount of time. We can write this in a mathematical formula that is very similar to the one used to define average velocity:

$$\vec{a}_{av} = \frac{\Delta\vec{v}}{\Delta t} = \frac{\vec{v}_2 - \vec{v}_1}{t_2 - t_1}. \qquad [2.5]$$

So, if a cheetah can accelerate from 0 km/h to 110 km/h in 3 s, we can use this equation to find its average acceleration. Recall that 100 km/h is around 30 m/s:

$$\vec{a}_{av} = \frac{30\ \text{m/s} - 0\ \text{m/s}}{3\ \text{s} - 0\ \text{s}} = 10\ \text{m/s}^2.$$

The average acceleration of a cheetah is around 10 m/s^2. That means that, on average, it increases its speed by 10 m/s every second.

On the other side of this dining arrangement, the gazelle is able to accelerate at around 5 m/s^2. This pales in comparison to the cheetah, but it is still more acceleration than any human sprinter can produce. If the gazelle is starting from rest, we can rewrite Eq. [2.4] to find how much time it takes for the gazelle to reach its top speed of 75 km/h. Again, the first step is to convert velocity into units of metres per second:

$$\vec{v}_2 = \frac{(75\ \text{km/h})(1000\ \text{m/km})}{(60\ \text{min/h})(60\ \text{s/min})} \cong 20\ \text{m/s}.$$

Solving Eq. [2.4] for Δt yields:

$$\Delta t = \frac{\vec{v}_2 - \vec{v}_1}{\vec{a}_{av}} = \frac{20\ \text{m/s} - 0\ \text{m/s}}{5\ \text{m/s}^2} = 4\ \text{s}.$$

That means that the gazelle takes 4 s to go from rest to its maximum speed. Note that we can only divide vectors if they are one-dimensional. Dividing vectors of more than one dimension is not defined (what would North divided by East be?). But as we will see later, we can break up any problem into a set of one-dimensional problems, so we need not worry.

We must remember that biomechanical systems such as animals cannot continually accelerate. If we can rearrange Eq. [2.5] to solve for final speed, we can find a formula that easily relates final speed to initial speed, acceleration, and time interval:

$$\vec{a}_{av} = \frac{\vec{v}_2 - \vec{v}_1}{\Delta t} \Rightarrow \vec{v}_2 = \vec{v}_1 + \vec{a}_{av}\Delta t. \qquad [2.6]$$

We can use this to show what would happen if the gazelle were able to maintain that acceleration for even a short time. For instance, assume the gazelle starts at rest and then accelerates at 5 m/s^2 for 100 s. Using Eq. [2.6] we can find the resulting final velocity of the gazelle:

$$\vec{v}_2 = \vec{v}_1 + \vec{a}_{av}\Delta t = 0\ \text{m/s} + (5\ \text{m/s}^2)(100\ \text{s}) = 500\ \text{m/s}.$$

This is considerably higher than the speed of sound in air. If a gazelle were able to accelerate at 5 m/s^2 for anything more than approximately 1 minute, the cheetah would be greeted with a sonic boom every time it tried to get dinner.

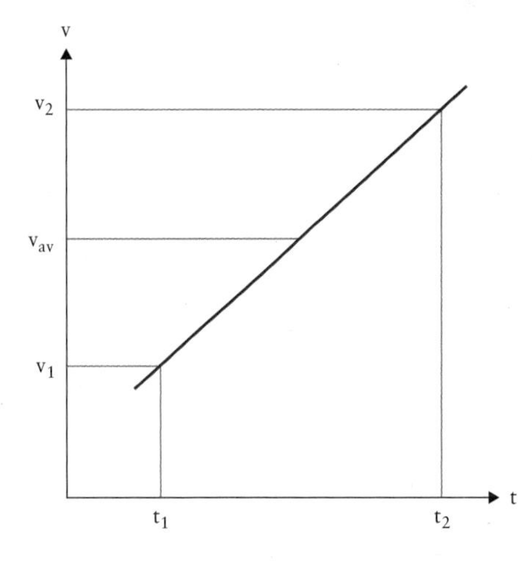

Figure 2.10 An object moves with constant acceleration. Acceleration is defined as rate of change of velocity with time. Constant acceleration results in a linear relationship between velocity and time.

We have now done as much as we can with Eq. [2.6]. In order to find out how far the cheetah travels during its 15 s of exertion, we need to find a way to relate acceleration to position. As always, we will start with a simple physical model: when the cheetah is accelerating, its acceleration is constant. This is obviously not true in that the cheetah's acceleration must gradually reduce to zero as it approaches its maximum speed. However, it is a reasonable first approximation. Fig. 2.10 shows a graph of velocity versus time for an object moving with constant acceleration. Notice that velocity is increasing linearly with time. This linearity allows for an easy definition of average velocity:

$$\vec{v}_{av} = \frac{\vec{v}_1 + \vec{v}_2}{2}. \qquad [2.7]$$

But we already have seen in Eq. [2.1] that:

$$\vec{v}_{av} = \frac{\vec{x}_2 - \vec{x}_1}{\Delta t}.$$

That means the right-hand side of Eq. [2.1] must be equal to the right-hand side of Eq. [2.7]:

$$\frac{\vec{x}_2 - \vec{x}_1}{\Delta t} = \frac{\vec{v}_1 + \vec{v}_2}{2} \qquad [2.8]$$

If we substitute Eq. [2.5] in for the $\vec{v}_2$ in Eq. [2.8], we have an equation that relates position to acceleration:

$$\frac{\vec{x}_2 - \vec{x}_1}{\Delta t} = \frac{\vec{v}_1 + (\vec{v}_1 + \vec{a}_{av}\Delta t)}{2}.$$

Reorganizing this equation and solving for $\vec{x}_2$, we find:

$$\vec{x}_2 = \vec{x}_1 + \vec{v}_1\Delta t + \frac{1}{2}\vec{a}\Delta t^2, \qquad [2.9]$$

where acceleration is written without the "av" subscript because acceleration has been assumed to be constant. That means average acceleration is equal to instantaneous acceleration at every instant in time, so we will call acceleration $\vec{a}$ when using that equation.

We can use that equation to find how far the cheetah travels if it starts from rest at position vector $\vec{x}_1 = 0\,\text{m}$ and accelerates at 10 m/s^2 for the 3 s it takes to accelerate to its maximum speed:

$$\vec{x}_2 = \vec{x}_1 + \vec{v}_1\Delta t + \frac{1}{2}\vec{a}\Delta t^2$$
$$= 0\text{ m} + (0\text{ m/s})(3\text{ s}) + \tfrac{1}{2}(10\text{ m/s}^2)(3\text{ s})^2 = 45\text{ m}.$$

In the first 3 s, the cheetah travels 45 m. In order to find how far the cheetah travels in the 15 s it has to catch the gazelle, we will have to break the problem into two parts: the accelerating phase and the constant velocity phase. During the acceleration phase, it travels 45 m. We can use the same equation to find how far the cheetah is from where it started after it travels the remaining 12 s of its chase. The acceleration during that 12-s time interval is zero, the initial position is 45 m (where the cheetah is at the end of the accelerating phase), and the initial velocity is 30 m/s (the velocity of the cheetah at the end of the accelerating phase). This gives us the following:

$$\vec{x}_2 = \vec{x}_1 + \vec{v}_1\Delta t + \frac{1}{2}\vec{a}\Delta t^2$$
$$= 45\text{ m} + (30\text{ m/s})(12\text{ s}) + \tfrac{1}{2}(0\text{ m/s}^2)(12\text{ s})^2 = 400\text{ m}.$$

This means that, at the end of the 15 s chase, the cheetah has run a total distance of around 400 m. That is 50 m less than the distance it would have covered if it ran the whole time at its top speed; so, taking the accelerating phase into account made more than a 10% difference to the result.

Eq. [2.5] and Eq. [2.9] can be combined to eliminate the variable time. The resulting equation can sometimes be used to quickly solve problems in which time is neither given nor of interest:

$$v_2^2 - v_1^2 = 2a(x_2 - x_1) \qquad [2.10]$$

Much like Eq. [2.5] and Eq. [2.9], this equation is useful only for situations in which acceleration is constant.

EXAMPLE 2.5

How much distance is required for the gazelle to reach its maximum speed?

Solution

To find where the gazelle is after the accelerating phase, rearrange Eq. [2.10] to solve for the final position x_2:

$$x_2 = x_1 + \frac{v_2^2 - v_1^2}{2a} = 0\text{ m} + \frac{(20\text{ m/s})^2 - (0\text{ m/s})^2}{(2)(5\text{ m/s}^2)} = 40\text{ m}.$$

At the end of the accelerating phase, the gazelle is 40 m from where it started. Notice that we don't know how long it takes a gazelle to get to its maximum speed. So long as we are not interested in that number, Eq. [2.10] is the fastest way to solve for distance.

Just like we did for velocity, the instantaneous acceleration at time t_2 is written as $\vec{a}$ and follows when we extrapolate to $\Delta t = 0$:

$$\vec{a} = \lim_{\Delta t \to 0} \frac{\Delta \vec{v}}{\Delta t} \qquad [2.11]$$

Instantaneous acceleration is represented by the slope of the tangent to the curve on the graph of velocity versus time. This means we can find the velocity-versus-time graph from the position-versus-time graph, and we can find the acceleration-versus-time graph from the velocity-versus-time graph.

KEY POINT

The slope of an object's velocity-versus-time graph gives the acceleration on the object's acceleration-versus-time graph.

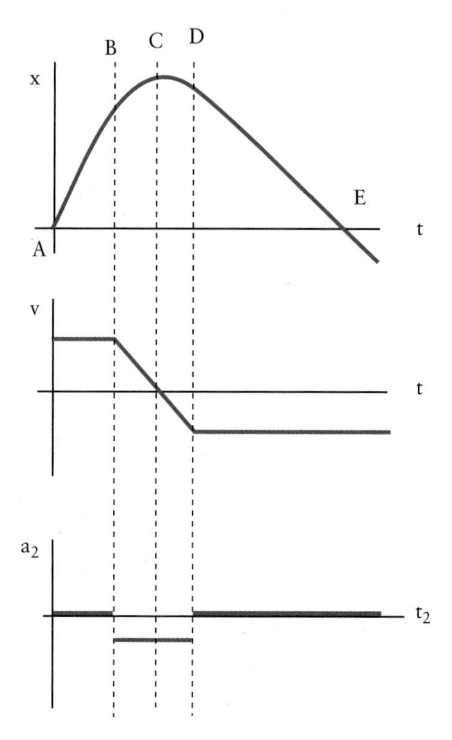

Figure 2.11 The velocity-versus-time graph can be inferred from the slope of the position-versus-time graph. The acceleration-versus-time graph can be inferred from the slope of the velocity-versus-time graph.

Fig. 2.11 shows the position of an object as a function of time. Between points A and B, the slope of the position-versus-time curve is positive and constant, which means the velocity is positive and constant and the acceleration is zero. Between points B and C, the slope of the position-versus-time curve is positive and decreasing, which means the velocity is positive and decreasing and the acceleration is negative. Between points C and D, the slope of the position-versus-time curve is negative and getting more negative, which means the velocity is negative and getting more negative and the acceleration is negative. Between points D and E, the slope of the position-versus-time curve is negative and constant, which means the velocity is negative and constant and the acceleration is zero. Without carefully measuring the slope in the region between points B and D, it is hard to know if the slope of the velocity-versus-time graph is constant, but in this chapter we will consider only constant acceleration, whether positive, negative, or zero. Constant acceleration produces straight lines on the velocity-versus-time graph.

Most biological systems do not produce constant acceleration, although it might be a reasonable approximation. However, there is one type of acceleration that we are all familiar with and that is well approximated as being constant. The acceleration due to gravity at Earth's surface is something you notice any time you drop a pencil or throw a ball.

Acceleration due to gravity points down toward Earth, and the magnitude of the acceleration is the same for all matter, regardless of size, shape, or mass. Because of this, we have a special symbol for acceleration due to gravity.

KEY POINT

Acceleration due to gravity is represented by the vector $\vec{g}$.

The magnitude of the acceleration due to gravity changes slightly depending on where you are, but on Earth's surface, it is approximately 9.8 m/s^2.

You have likely noticed that if you drop a sheet of paper and a heavy textbook from the same height, they do not typically reach the floor at the same time. This means

CONCEPT QUESTION 2.5

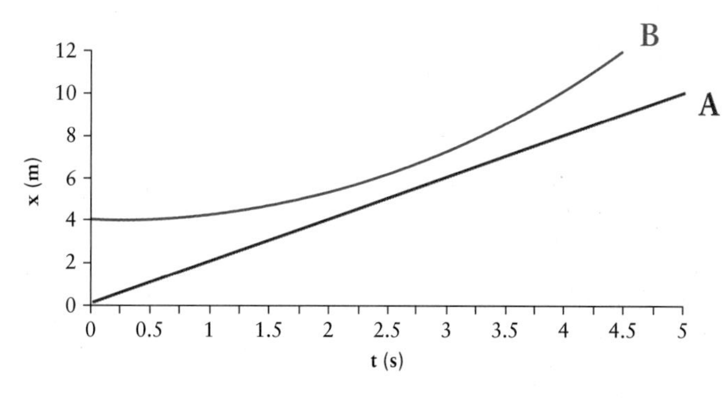

Figure 2.12 Position as a function of time for object A and object B.

At time $t = 1$ s, which of the following correctly describes a comparison of the velocity and acceleration of objects A and B?

(A) The velocity of A is larger than B. The acceleration of A is larger than B.

(B) The velocity of B is larger than A. The acceleration of A is larger than B.

(C) The velocity of A is larger than B. The acceleration of B is larger than A.

(D) The velocity of B is larger than A. The acceleration of B is larger than A.

that the accelerations of the sheet of paper and the textbook are different. However, if you place the sheet of paper directly on top of the textbook and then drop both, you will notice they fall at the same rate. As you might expect, the difference between the two experiments has to do with air resistance. When you place the textbook in front of the sheet of paper, the textbook pushes the air out of the path of the sheet of paper and allows it to fall with considerably less air resistance. Objects that have a smaller surface for the air to interact with experience less air resistance, and air resistance has a harder time resisting very massive objects. We will ignore the effects of air resistance for now and talk about motion that occurs when only gravity is allowed to accelerate an object at 9.8 m/s^2. When that occurs, the object is said to be in freefall. We can use the equations for motion in one dimension to predict the motion of objects experiencing freefall acceleration. We simply use acceleration due to gravity, $\vec{g}$, in place of acceleration, $\vec{a}$.

EXAMPLE 2.6

A freezing rainstorm on March 2, 2007, resulted in a thick layer of ice coating the top of the CN Tower (see Fig. 2.13). When this ice began to melt, it fell to the ground below. If ice fell from a height of 550 m, how fast would it be travelling when it reached the ground, if you ignore air resistance?

I. Kolesnik/Shutterstock.com

Figure 2.13 People often worry that a penny dropped from the top of Toronto's CN Tower would reach a speed high enough to kill someone on the ground. It is a good thing air resistance prevents this from being a threat.

continued

Solution

Assume initial velocity is zero and the x-axis is vertical and positive in the upward direction. Also assume the ice is experiencing freefall acceleration $\vec{g}$ = 9.8 m/s^2. Rewriting Eq. [2.10] to solve for final velocity $\vec{v}_2$, we have:

$$v_2^2 = v_1^2 + 2g(x_2 - x_1) = (0 \text{ m/s})^2 + (2)(-9.8)(0 \text{ m} - 550 \text{ m})$$

$$= 11\,000 \text{ m}^2/\text{s}^2.$$

Taking the square root, we find that the speed is a little over 100 m/s. This is more than 300 km/h! It is a good thing air resistance kept it to a small fraction of that speed, otherwise more damage would have occurred. As it was, police evacuated the area and only windows were damaged.

Notice that Eq. [2.10] takes the square of the velocity. That means the solution can be positive or negative. This equation can be used only to determine the magnitude of the velocity, in other words, the speed. Acceleration due to gravity points toward Earth though, so the velocity must also have been pointing in that direction.

2.5: Motion in Two Dimensions

In the real world, not everything travels along a one-dimensional axis. The good news is that the same equations we developed for use in describing motion in one dimension can also be used to describe motion in two dimensions. All we have to do is break the two-dimensional problem into two one-dimensional problems. Going back to the child skipping along the sidewalk, that two-dimensional motion is a combination of up-and-down motion and motion along the sidewalk. We can imagine an x-axis that could be used to describe motion in the horizontal direction, and a y-axis that could be used to describe motion in the vertical direction. Since we chose axes that are perpendicular to each other, motion along the x-axis is independent of motion along the y-axis. This allows us to treat each axis independently using the equations that were developed for one-dimensional motion.

For instance, imagine walking 10 km in a straight line in the direction of 30° north of east. If this walk took you 2 hours, then your velocity was 5 km/h in the northeast direction. However, if you only care about how far east you managed to get in that hour, you could use trigonometry to figure out that you are now 10 cos(30°) km east of where you started. If you wanted to know how much further north you were, you could use the same technique to find that you are 10 sin(60°) km north of where you started. Because north and east are perpendicular to each other, the displacement in the east direction doesn't change the displacement in the north direction. If you chose axes that were not perpendicular, for instance, east and northeast, then motion in the east direction would also cause some motion in the northeast direction. Keeping the axes perpendicular simplifies the problem, allowing for independent analysis of motion along the x-axis and the y-axis.

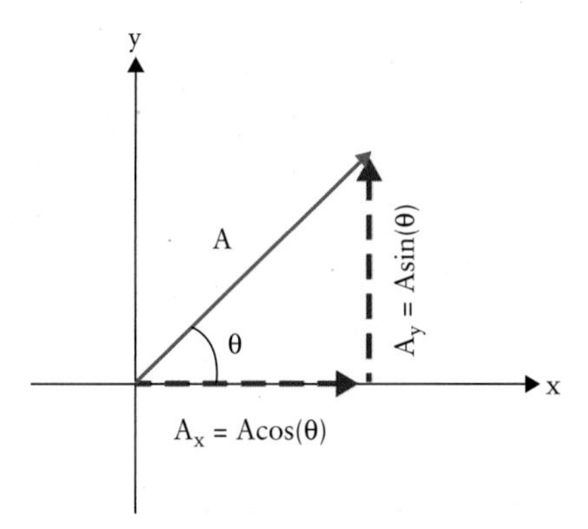

Figure 2.14 A vector can be projected onto a perpendicular axis to produce *x*- and *y*-components. The sum of these vector components is identical to the original vector.

Fig. 2.14 shows this more generally for a vector of length A that makes an angle θ to the x-axis. The projection of that vector onto the x-axis gives a vector component that has length $A\cos(\theta)$ and points in the positive x-direction. The projection of that vector onto the y-axis gives a vector component that has length $A\sin(\theta)$ and points in the positive y-direction. If the vector described a displacement from the origin to point (10, 10), the component of the vector in the x-direction would describe a displacement from the origin to point (10, 0), and the component of the vector in the y-direction would describe a displacement from point (10, 0) to point (10, 10). The sum of the two component vectors produces the same displacement as the original vector. The original vector is often called the resultant vector because it is the result of the sum of the component vectors.

KEY POINT

Breaking vectors into components allows a complex motion to be analyzed as if it were a combination of simple one-dimensional motions.

In general, breaking vectors into components does not change the result. It simply allows a more complex motion to be analyzed as if it were a combination of simple one-dimensional motions. This remains true even if the vector of length A in Fig. 2.14 represents velocity or acceleration. Breaking motion into x- and y-components is almost always the first step for solving motion in two dimensions.

EXAMPLE 2.7

A person jogs in a straight line across a park at a speed of 2.0 m/s in the direction of 60° south of east. After 5.0 minutes of jogging, how much further south is the person?

Solution

The first step of most problems in physics is to draw a diagram. Fig. 2.15 is a diagram that shows the velocity projected onto a two-dimensional coordinate system. The choice of axes is arbitrary; however, they must be perpendicular. Because the question uses cardinal directions (north, east, south, and west) that are perpendicular to each other, it makes sense to align our axes such that they are parallel to these directions. Fig. 2.15 is drawn such that the positive *y*-direction is north and the positive x-direction is east.

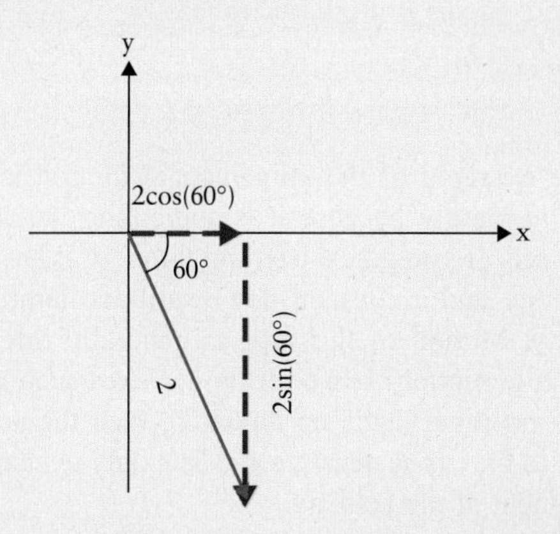

Figure 2.15 The velocity vector of the jogger is shown in blue. The components of the velocity in the *x*- and *y*-directions are shown in red.

We want to know how far the person jogs in the south direction in 5 minutes, so we can ignore the component of the velocity that points in the east direction. The velocity in the south direction is $\vec{v}_y = -2.0\ \sin(60)\text{m/s} = -1.7\ \text{m/s}$. Notice that the answer is negative because the *y*-axis is positive in the north direction, and the *y*-component of the jogger's velocity is in the south direction. To find how far the person travelled in the south direction over 5 minutes (300 seconds), use Eq. [2.9] and set the initial position and acceleration to zero:

$$\vec{y}_2 = \vec{y}_1 + \vec{v}_y \Delta t + \frac{1}{2}\vec{a}_y \Delta t^2$$

$$= 0\ \text{m} + (-1.7\ \text{m/s})\ (300\ \text{s}) + \frac{1}{2}(0\ \text{m/s}^2)(300\ \text{s})^2 = -510\ \text{m}.$$

This tells us the jogger has moved 510 m to the south during that time interval.

Using the same technique, we can also find how far to the east the jogger moved. In this case, the velocity is $\vec{v}_x = 2.0\ \cos(60)\text{m/s} = 1\ \text{m/s}$, or 1 m/s in the east direction. We can find the *x*-component of the position vector using Eq. [2.9] with the initial position and acceleration set to zero:

$$\vec{x}_2 = \vec{x}_1 + \vec{v}_x \Delta t + \frac{1}{2}\vec{a}_x \Delta t^2$$

$$= 0\ \text{m} + (1\ \text{m/s})(300\ \text{s}) + \frac{1}{2}(0\ \text{m/s}^2)(300\ \text{s})^2 = 300\ \text{m}.$$

Notice that the distance as the sum of magnitudes of the components of the position vectors does not add up to the distance that the jogger actually travelled.

continued

Because the components are perpendicular, they must be summed using the Pythagorean theorem:

$$d = \sqrt{(-510)^2 + (300)^2} \cong 600 \text{ m}.$$

This is the distance that a person would travel if their speed was 2 m/s and they jogged for 300 s. (The approximately equal to sign is used because we used only two significant figures in these calculations.)

One example of two-dimensional motion is worth looking at simply because it is something we all have experienced. If you take a ball and throw it, it has an initial velocity and a constant downward acceleration due to gravity. Motion of this type is generally referred to as projectile motion. If we set up our axes such that the y-axis is positive in the up direction, then the acceleration due to gravity is negative and acts only to change the y-component of the velocity.

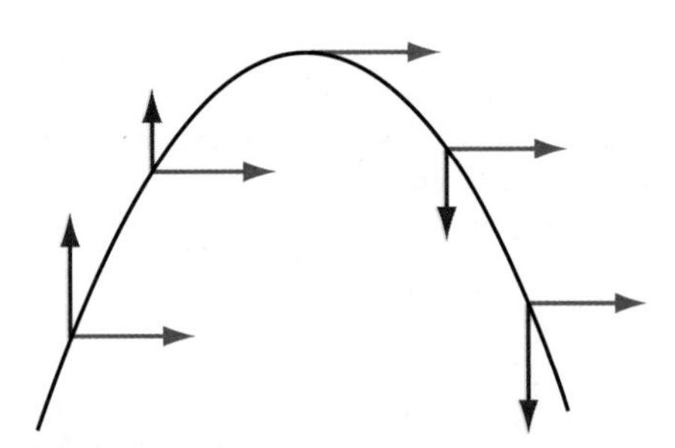

Figure 2.16 The horizontal and vertical components of velocity for a projectile. Notice that the horizontal component of the velocity is constant; whereas the vertical component is accelerated downward by gravity.

Figure 2.16 shows the x- and y-components of the velocity at different points along the path of a ball thrown at around a 45° angle to the horizontal. Notice how the ball starts out with a positive y-component of its velocity; at the peak of the path, the y-component of the velocity is zero and after that the y-component of the velocity is increasingly negative. In marked contrast, the x-component of the velocity is unchanged over the whole path. The equations used to describe projectile motion are the same as our standard kinematic equations (Eq. [2.5] and Eq. [2.9]), except we now have to include subscripts to identify which axis we are describing:

$$\begin{aligned} x\text{-direction: } \vec{x}_2 &= \vec{x}_1 + \vec{v}_{1x}\Delta t + \frac{1}{2}\vec{a}_x\Delta t^2 \\ \vec{v}_{2x} &= \vec{v}_{1x} + \vec{a}_x\Delta t \\ y\text{-direction: } \vec{y}_2 &= \vec{y}_1 + \vec{v}_{1y}\Delta t + \frac{1}{2}\vec{a}_y\Delta t^2 \\ \vec{v}_{2y} &= \vec{v}_{1y} + \vec{a}_y\Delta t. \end{aligned}$$

However, simplifications can be made immediately. For projectile motion, we know $\vec{a}_x = 0$ and $\vec{a}_y = -g$. That also means that $\vec{v}_x$ is constant, so there is no need to refer to $\vec{v}_{1x}$ or $\vec{v}_{2x}$. So we can rewrite the equations that describe projectile motion as:

$$\left.\begin{aligned} x\text{-direction: } x_2 &= x_1 + v_x\Delta t \\ v_x &= \text{constant}, \\ y\text{-direction: } y_2 &= y_1 + v_{1y}\Delta t - \frac{1}{2}g\Delta t^2 \\ v_{2y} &= v_{1y} - g\Delta t. \end{aligned}\right\} \quad [2.12]$$

EXAMPLE 2.8

A ball is thrown at an angle of 60° to the horizontal with an initial speed of 10.0 m/s. With its initial position taken to be the origin, find the position vector that describes the position of the ball 3.0 s later.

Solution

The first step in a problem such as this is to draw a diagram.

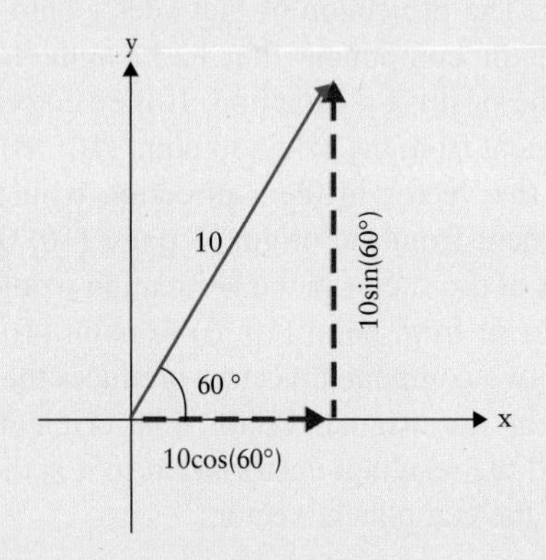

Figure 2.17 The initial velocity of the ball is broken into x- and y-components.

We now have all the information needed to use Eqs. [2.12] to solve for $\vec{x}_2$ and $\vec{y}_2$:

$$\vec{v}_x = 10.0 \cos(60°) = 5.00 \text{ m/s}$$

$$\vec{x}_2 = \vec{x}_1 + \vec{v}_x\Delta t = 0 \text{ m} + (5.00 \text{ m/s})(3.0 \text{ s}) = 15 \text{ m}$$

$$\vec{v}_{1y} = 10.0 \sin(60°) = 8.66 \text{ m/s}$$

$$\vec{v}_{2y} = \vec{v}_{1y} - g\Delta t = 8.66 \text{ m/s} - (9.8 \text{ m/s}^2)(3.0 \text{ s}) = -21 \text{ m/s}$$

$$\begin{aligned} \vec{y}_2 &= \vec{y}_1 + \vec{v}_{1y}\Delta t - \frac{1}{2}g\Delta t^2 \\ &= 0 \text{ m} + (8.66 \text{ m/s})(3.0 \text{ s}) - \frac{1}{2}(9.8 \text{ m/s}^2)(3.0 \text{ s})^2 \\ &= -18 \text{ m}. \end{aligned}$$

That means the final position vector of the ball is (15 m, −18 m). This means that, after 3 s, the ball is lower than where it started by 18 m. Perhaps it was thrown off of a 5th floor balcony?

CASE STUDY 2.1

The current theory of the evolution of flight requires intermediate species to benefit from gliding off tree branches. We can observe such behaviour in animals such as the Malayan Wallace's tree frog or the flying dragon, a Southeast Asian lizard shown in Fig. 2.18. These animals have membranous extensions in various places along their bodies, which they can unfold to sustain a gliding fall after jumping off a tree. As a reference, (a) what formula describes the falling of an animal that jumps off a tree but cannot glide, e.g., a human, and (b) how must this formula vary for a Wallace's tree frog or a flying dragon for us to accept that the animal displays a successful glide?

Answer to Part (a): *The first step we take in solving this problem is a simplification that is almost always justified in mechanical problems: we identify a two-dimensional plane in space in which the relevant physical parameters vary and the resulting motion occurs. This plane is defined as the xy-plane, allowing us to use Eqs. [2.12].*

Figure 2.18 The flying dragon has a membrane that is stretched between elongated ribs and acts as a strut.

We can draw an initial sketch of what happens based on our everyday experience. This sketch is shown in Fig. 2.19(a). The object starts at an initial position with a given initial velocity vector. If distance matters to the jumping animal it will choose to make this velocity vector point above the horizontal. We calculate the actual path without gliding in this part. If a person jumps off a tree and you observe the event looking at the plane of motion from the side, you observe the path sketched in Fig. 2.19(a).

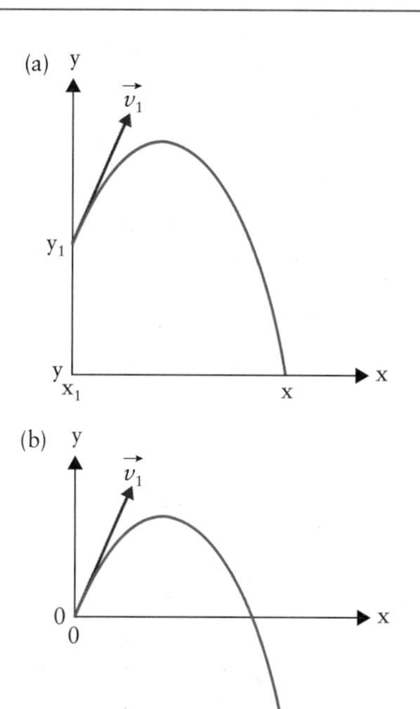

Figure 2.19 (a) Sketch of the path of an object released at position (x_1, y_1) with an initial velocity vector $\vec{v}_1$. The object does not accelerate horizontally and moves with a constant gravitational acceleration downward. The mathematical treatment of the motion yields a parabolic path called a projectile trajectory. (b) The calculations are greatly simplified by choosing the position (x_1, y_1) as the origin (0, 0) in this case.

We define the horizontal direction as the x-axis and the vertical direction as the y-axis. Once the animal has jumped from the tree, the object is no longer accelerated horizontally ($a_x = 0$) and accelerates vertically with $a_y = -g$, as discussed. The y-component of the acceleration is negative because the gravitational acceleration is pointing downward, i.e., in the negative y-direction. For convenience, the initial position is chosen as the origin. This means that we modify the sketch we made in the schematic approach as shown in Fig. 2.19(b) such that $(\vec{x}_1, \vec{y}_1) = (0, 0)$. We can now substitute the two acceleration components in Eqs. [2.12]:

$$\text{x-direction:} \quad x = v_{\text{initial},x} t$$

$$\text{y-direction:} \quad y = -\frac{g}{2} t^2 + v_{\text{initial},y} t.$$

These two formulas in are now combined to eliminate the time variable. This produces a relationship that expresses the y-position as a function of the x-position:

$$y = -\frac{1}{2} g \left(\frac{x}{v_{\text{initial},x}} \right)^2 + v_{\text{initial},y} \frac{x}{v_{\text{initial},x}}$$

$$= \frac{v_{\text{initial},y}}{v_{\text{initial},x}} x - \frac{g}{2 v_{\text{initial},x}^2} \cdot x^2$$

continued

which is mathematically the form of a parabola ($y \propto x^2$). Thus, an animal or a person that cannot glide will travel along a parabola from the branch to the ground. This path is shown in Fig. 2.19 and is an example of projectile motion, because any object under the exclusive control of gravity follows it.

Answer to Part (b): Any path with $y(x) > y_{projectile}(x)$ means the object is at a greater height than a freely falling object. Thus, that object is gliding.

2.6: Uniform Circular Motion

Uniform circular motion is illustrated in Fig. 2.20. The line labelled "path" is the line connecting all successive positions of the object, which is shown as a blue dot. At the instant shown, the object moves with a given velocity $\vec{v}$, as illustrated in Fig. 2.20(a), and a given acceleration $\vec{a}$, as illustrated in Fig. 2.20(b).

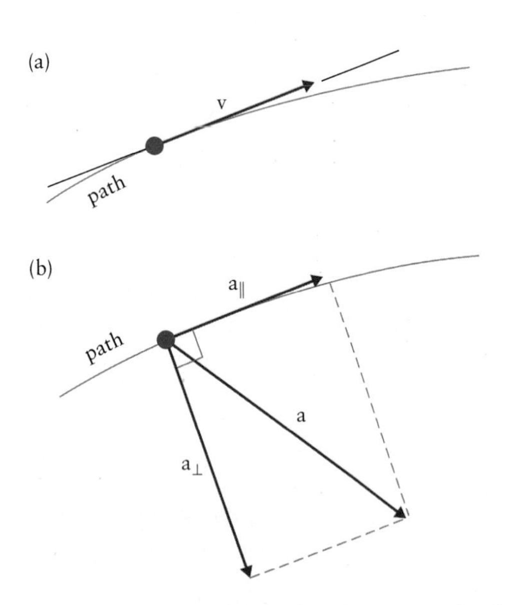

Figure 2.20 An object (blue dot) moving along a curved path. The indicated velocity components (a) and acceleration components (b) are derived for an object-centred coordinate system. The subscript || is used to refer to the tangential components of the motion, and the subscript ⊥ is used to refer to the perpendicular components of the motion.

Velocity is defined as the change of position with time. Thus, the direction of the velocity is tangential to the path of the object; i.e., it has a component along the path but is not perpendicular to it. One way to write this is to use a coordinate system that introduces fixed x- and y-axes in space:

$$\lim_{\Delta t \to 0} \frac{\Delta x}{\Delta t} = v_x$$

$$\lim_{\Delta t \to 0} \frac{\Delta y}{\Delta t} = v_y$$

In that coordinate system, both x- and y-components of the velocity change with time: $v_x = v_x(t)$ and $v_y = v_y(t)$. Alternatively, we can substitute the Cartesian coordinates with a coordinate system that is attached to the object. Instead of using the fixed x- and y-axes, we can set an axis along the direction tangential to the path of the object, leading to a velocity component $v_{||}$ that is equal to the magnitude of the velocity and pointing in the direction of motion, and an axis perpendicular to the direction of motion, leading to a velocity component $v_{\perp}$ that is of zero magnitude. This coordinate system therefore allows us to reduce the number of time-dependent variables to one, $v_{||}$.

KEY POINT

Velocity can change even when the speed is constant.

Velocity is a vector, with both a magnitude and a direction, so velocity can change even when the speed is constant. To produce a curved path, the direction of the velocity vector must change, which means that acceleration is necessary for an object to follow any curved path. This acceleration is illustrated in Fig. 2.20(b). To quantify it, we start again with the general definition of acceleration based on our standard (x, y)-coordinate system,

$$\lim_{\Delta t \to 0} \frac{\Delta v_x}{\Delta t} = a_x$$

$$\lim_{\Delta t \to 0} \frac{\Delta v_y}{\Delta t} = a_y$$

in which both components are a function of time, $a_x = a_x(t)$ and $a_y = a_y(t)$, if an object is following a curved path. We can again use the object-centred coordinate system, which leads to the two components shown in Fig. 2.20(b): $a_{||}$ and $a_{\perp}$. Choosing this coordinate system represents a great simplification for uniform circular motion, as we will see next.

Objects that travel on curved paths are best described by an object-centred coordinate system with one axis parallel and one axis perpendicular to the path of the object. In this case, velocity and acceleration are written as $\vec{v} = (v_{||}, 0)$ and $\vec{a} = (a_{||}, a_{\perp})$.

For now, we will limit our discussion to cases in which the tangential acceleration is zero; i.e., the object moves with a constant tangential velocity component. We will call circular motion that has zero tangential acceleration uniform circular motion. The term uniform refers to the fact that the magnitude of the velocity of the object does not change. The term circular indicates the path becomes a circular path. You can convince yourself that a constant $a_{\perp}$ results in circular motion using an object that is attached to a string. When you swing the object horizontally, the fixed length of the string forces the object onto a circular path around your hand. Throughout the swing you need the same magnitude of force to keep the object on its path; the constant magnitude of force is associated with a constant magnitude of acceleration. Note, however, that the acceleration is not constant, as the vector is continuously changing directions.

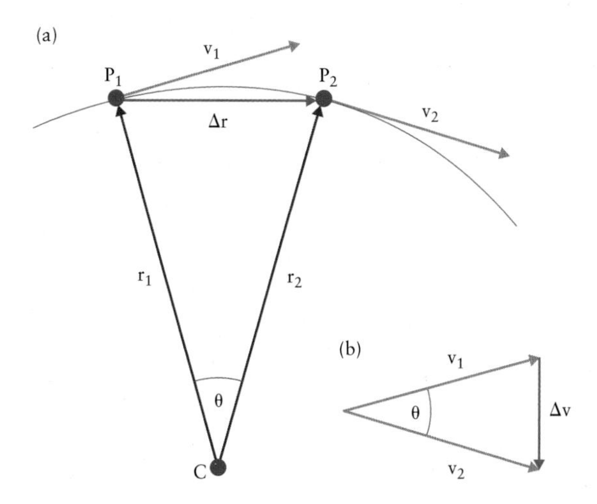

Figure 2.21 Position and velocity components for an object moving along a circular path, shown at two different times that we label with indices 1 and 2.

We want to further quantify the relations between the various parameters describing a uniform circular motion. To do so, we specify several parameters in Fig. 2.21. Part (a) shows two points, P_1 and P_2, along the circular path with centre C. The two points are characterized by position vectors $\vec{r}_1$ and $\vec{r}_2$, which subtend an angle θ. The two points on the path are separated by a displacement $\Delta\vec{r}$. An object at point P_1 has velocity $\vec{v}_1$, and at point P_2 it has velocity $\vec{v}_2$. Sketch 2.21(b) illustrates the corresponding relation between the two velocity vectors: they also describe an angle θ due to the geometrical fact that $\vec{r}_1 \perp \vec{v}_1$and $\vec{r}_2 \perp \vec{v}_2$. The difference between $\vec{v}_1$ and $\vec{v}_2$ is defined as $\Delta\vec{v}$.

We relate the perpendicular acceleration component $a_\perp$ to the velocity and the radius of the path using the two sketches in Fig. 2.21. For this we assume only a very short time Δt has elapsed as the object moved from point P_1 to point P_2. That means both $\Delta\vec{r}$and $\Delta\vec{v}$ are very short vectors and θ is a small angle. Using trigonometry we find from the two parts of Fig. 2.21, respectively:

$$\sin\theta = \frac{|\Delta\vec{r}|}{r}$$
$$\sin\theta = \frac{|\Delta\vec{v}|}{v} \qquad [2.13]$$

where $|\Delta\vec{r}|$ represents the magnitude of $\Delta\vec{r}$, and $|\Delta\vec{v}|$ represents the magnitude of $\Delta\vec{v}$.

In Eq. [2.13], we simplified the notation to $|\vec{r}_1| = |\vec{r}_2| \equiv r$ and $|\vec{v}_1| = |\vec{v}_2| \equiv v$ since both position vectors and both velocity vectors have the same magnitude for an object moving with uniform circular motion. Next we combine both formulas in Eq. [2.13]:

$$|\Delta v| = \frac{v}{r}|\Delta r|. \qquad [2.14]$$

We divide Eq. [2.14] on both sides by the elapsed time interval Δt:

$$\frac{\Delta v}{\Delta t} = \frac{v}{r}\frac{\Delta r}{\Delta t}$$

Taking the limit $\Delta t \rightarrow 0$ on both sides, we get:

$$a_\perp = \frac{v^2}{r}, \qquad [2.15]$$

in which the limit of $\Delta v/\Delta t$ is $a_\perp$, and the limit of $\Delta r/\Delta t$ is v, because $a_\perp$ and v are instantaneous values. This acceleration is called centripetal acceleration because it is an acceleration that points toward the centre of the circular path. Centripetal is directly translated from Latin as centre seeking.

KEY POINT

Any acceleration that points toward the centre of curvature is a centripetal acceleration.

It is important to note again that the acceleration in Eq. [2.15] is not an x- or a y-component in the xy-plane of the circular motion; with respect to fixed x- and y-axes, the components of the centripetal acceleration continuously change.

The period T is the time required to complete a full cycle. Since the circumference of a circle has length $2 \cdot \pi \cdot r$, this definition allows us to rewrite velocity as $2 \cdot \pi \cdot r/T$. Substituting this in Eq. [2.15] leads to a second formula for centripetal acceleration:

$$a_\perp = \frac{4\pi^2 r}{T^2}. \qquad [2.16]$$

Centripetal acceleration is larger for an object moving around the centre with a shorter period, and it is larger for an object at a greater distance from the centre of the path.

CONCEPT QUESTION 2.6

The rotation axis of planet Earth is a line that joins the North and South Poles. Your distance to that axis determines the acceleration that you must experience in order to continue rotating with the planet. Where on planet Earth is $a_\perp$ equal to zero?

(A) the equator

(B) the South Pole

(C) the North Pole

(D) both the North and the South Pole

EXAMPLE 2.9

Calculate the centripetal acceleration due to the rotation of Earth of a person in New York. New York is located at 40.8° northern geographical latitude. Note that this acceleration is independent from the gravitational acceleration caused by the attraction between the two objects Earth and person. Use R_{Earth} = 6370 km for the radius of Earth.

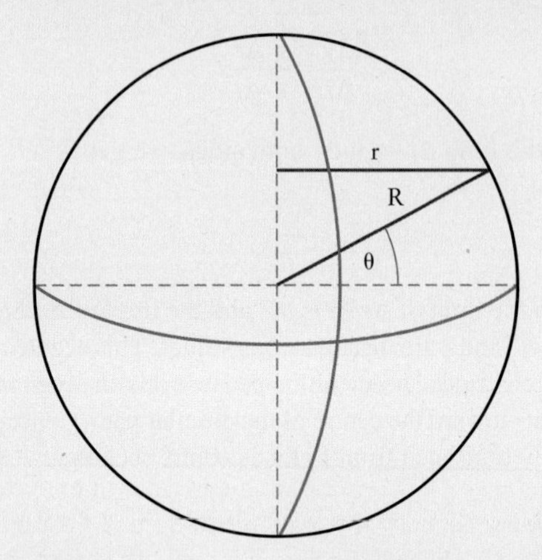

Figure 2.22 Illustration of the position of New York, at latitude θ = 40.8°. Note that the distance to the axis, *r*, is different from Earth's radius, *R*.

Solution

The period of Earth is 1 day, which corresponds to 86 400 s. As noted in Concept Question 2.6, a point at θ = 40.8° above the equator moves on a circle with a radius $r = R_{Earth} \cdot \cos\theta$, as illustrated in Fig. 2.22. This allows us to calculate the centripetal acceleration required to keep a person in New York on the ground:

$$a = \frac{4\pi^2 R_{Earth} \cos\theta}{T_{Earth}^2},$$

which, with the specific values of the example text, leads to:

$$a = \frac{4\pi^2 (6.37 \times 10^6 \mathrm{m}) \cos(40.8°)}{(8.64 \times 10^4 \mathrm{~s})^2}$$

$$= 0.026 \frac{\mathrm{m}}{\mathrm{s}^2}$$

Note that this result is much smaller than gravitational acceleration. Thus, gravity provides a more than sufficient downward force to keep us from floating off the ground.

EXAMPLE 2.10

How fast would Earth have to spin to have a person in New York float apparently weightless across the room?

continued

Solution

This would happen if centripetal acceleration were equal to gravitational acceleration. In this case the entire gravitational pull would be needed to keep us on a circular path as Earth spins, and no fraction of gravity would be left to pull us down onto the ground. We rewrite Eq. [2.16] to determine the period of Earth required for this case:

$$T = 2\pi \sqrt{\frac{r}{a_\perp}} = 2\pi \sqrt{\frac{R_{Earth} \cos\theta}{g}},$$

in which we can substitute the specific values given in the example text:

$$T = 2\pi \sqrt{\frac{(6.37 \times 10^6 \mathrm{~m}) \cos 40.8°}{9.8 \frac{\mathrm{m}}{\mathrm{s}^2}}}$$

$$= 4410 \mathrm{~s}$$

This result means Earth would have to spin around once every 1 hour and 13.5 minutes, which would be the length of a day in this case. Such fast-spinning planets do not exist in our planetary system. The fastest is Jupiter, with 0.41 Earth days for one revolution.

2.7: Physiological Detection of Velocity

Given the importance of motion to our everyday life, it is interesting to look at one of the ways our body detects velocity. Fig. 2.23 shows a light micrograph of a human fingertip, including Meissner's corpuscles just below the epidermis. That they are very near the surface of our palm is an important feature for the functional role of Meissner's corpuscles as velocity detectors. We further

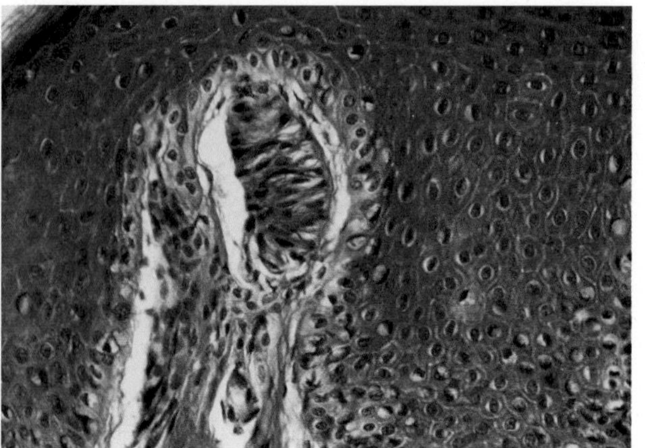

Figure 2.23 Light micrograph of a Meissner's corpuscle in the skin of a human fingertip. Shown is an oval-shaped corpuscle of approximately 100 mm length, with cells stacked in an alternating arrangement of the cell nuclei. The dendrite is intertwined between the cells. The corpuscle is surrounded by numerous densely stained epidermal cells.

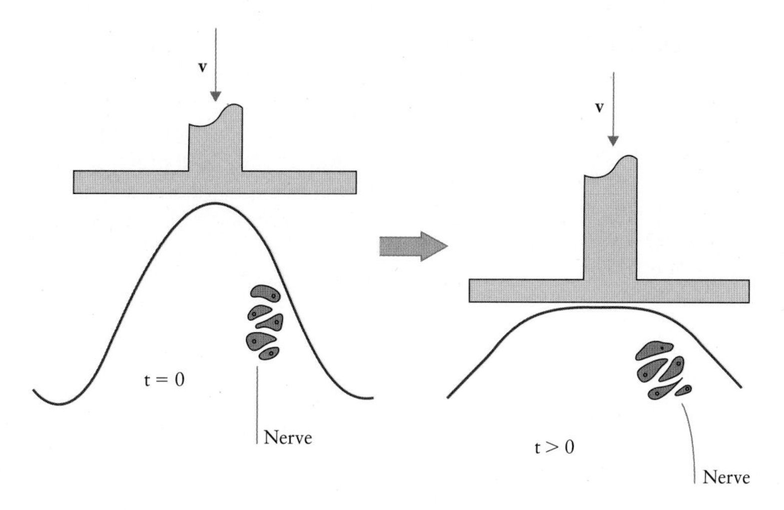

Figure 2.24 Mechanism of Meissner's corpuscles. At left, a piston is shown at time zero approaching the skin surface with velocity $\vec{v}$. At right, the same area is shown after the piston has pushed the skin a certain distance down. The piston is assumed to move with constant speed. Meissner's corpuscles get deformed in the process.

note that they are located near the steepest slope of the ridge–valley pattern in the fingertips. Meissner's corpuscles measure the speed of objects that come into contact with the skin. Each corpuscle consists of a stack of cells in the shape of an ellipsoid, with the more rigid cell nuclei located on alternating sides from cell to cell. Dendrites are intertwined between these cells.

The mechanism by which a Meissner's corpuscle detects velocity is illustrated in Fig. 2.24. Shown is a piston just before and while pressing the skin of the finger. The corpuscle gets squeezed, which leads to a change in the stacking structure of the cells. As a consequence, the intertwined dendrite is bent, causing nerve impulses to the brain. The impulse rate, which is the number of impulses sent per second, is higher when the dendrite is bent faster.

CASE STUDY 2.2

Fig. 2.25 shows both a linear plot and a double-logarithmic representation of the relationship between velocity and nerve impulse rate for Meissner's corpuscles. What can you conclude from (a) the linear plot and (b) the double-logarithmic plot in Fig. 2.25?

Answer to Part (a): *The linear plot shows variations in the steepness of the curve. The regions of greatest steepness represent the regions of greatest velocity sensitivity. Thus, the linear graph shows the greatest sensitivity of Meissner's corpuscles occurs at very small speeds,*

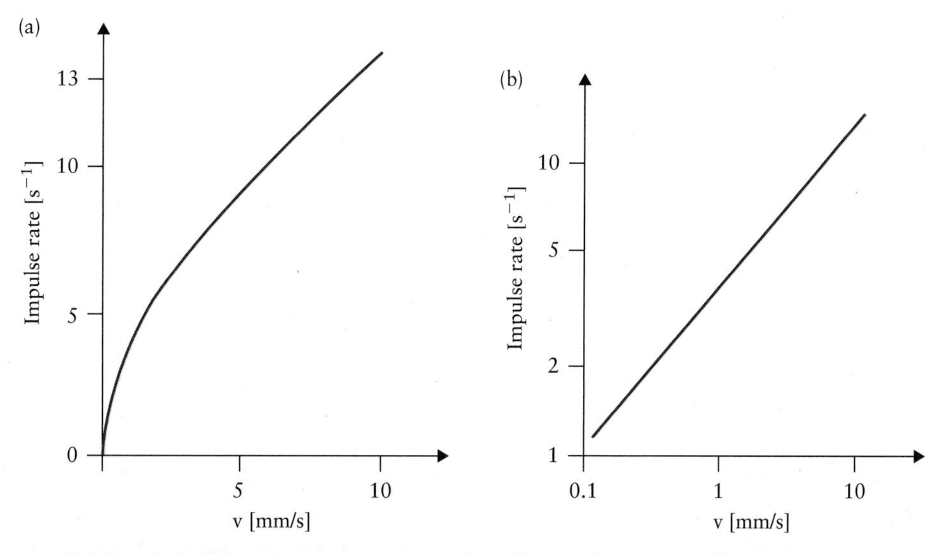

Figure 2.25 The nerve impulse rate for Meissner's corpuscles as a function of the speed of an approaching object. The impulse rate is given in unit impulses per second and the speed is given in the units millimetres per second. (a) A linear plot of the data showing the greatest sensitivity to speeds less than 1 cm/s. (b) In the double-logarithmic plot, the same data fall on a straight line.

continued

less than v = 1 cm/s. They are suitable for distinguishing forms of gentle touching, not for measuring the speed of an incoming baseball!

Answer to Part (b): *The double-logarithmic plot allows us to distinguish whether one or more mechanisms are involved in a process because it shows straight-line segments for each power law. In the specific case of Fig. 2.25(b), the power law for the nerve impulse rate, P, is written in the form:*

$$P = av^b,$$

in which v is the speed of the object making contact with the skin. If the pre-factor a or the exponent b in this power law is not constant, deviations from a straight line in the double-logarithmic plot occur. Any deviation from a straight line is, therefore, an indication that additional physical explanations are needed. In the present case, however, no such deviations occur, suggesting the single mechanism illustrated in Fig. 2.24 is sufficient to explain Meissner's corpuscle.

2.8: Physiological Detection of Acceleration

Since acceleration plays a key role in kinematics, particularly in predicting the motion of an object, it is crucial for organisms to detect and measure accelerations. Interestingly, vision is not particularly effective for this purpose. You know this from your everyday experience: if somewhere in your field of vision a small insect moves, you immediately notice it. Compare this to your visual sensitivity for accelerations. When you sit in an airplane you feel the thrust of the aircraft immediately when the pilot receives clearance for takeoff (as the acceleration pushes you into your seat). But, if you watch the initial acceleration from an observation deck, you have the impression that the airplane takes forever to get off the end of the tarmac. Thus, our vision is not sensitive to acceleration, only to velocities.

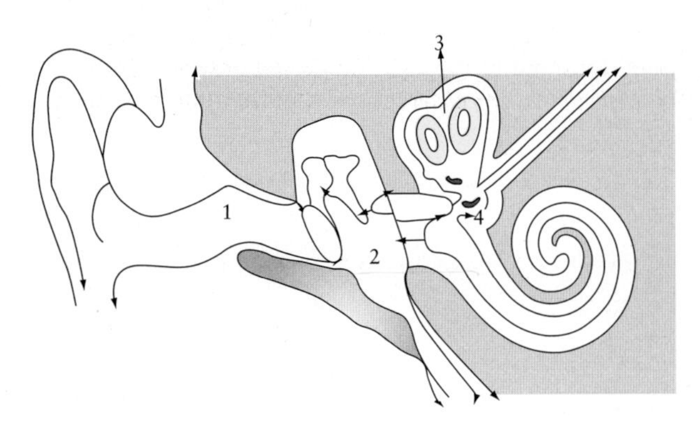

Figure 2.26 Overview of the human ear. We can distinguish three main sections of the ear: the outer ear, with the auditory canal (1) ending at the eardrum; the middle ear, with the three ossicles, the hammer, anvil, and stirrup (from left; 2), and the inner ear, with the vestibular organ. The vestibular organ includes the semicircular canals (3), which we discuss in the context of acceleration detection in the head, and the maculae (4), which we discuss in the context of gravity detection.

The human body is capable of detecting accelerations in two ways: accelerations of our own body, particularly the head, and accelerations of objects in contact with our skin. Detection of the latter is achieved by Pacinian corpuscles, which are acceleration sensors located in the skin. The sense that tells us if our body is being accelerated is located in our head as part of the ear. Fig. 2.26 presents an overview of the ear. The vestibular organ consists of two components:

- the orthogonal semicircular canals, highlighted in yellow, which allow the measurement of accelerations, and
- the maculae (plural of *macula*), which are 0.3-mm-wide spots located just below the semicircular canals. The maculae are indicated in Fig. 2.26 as elongated red bars. The upper one is called the utricular macula and the lower one the saccular macula, because they are located in small chambers called the *utricle* and the *saccule* respectively. Both maculae measure the orientation of the head relative to the direction of gravity.

Accelerations of our head are measured by the semicircular canals of the vestibular organ. The semicircular canals consist of three orthogonal, crescent-shaped tubes filled with a fluid called the endolymph. The orthogonal orientation of the three tubes provides the brain with acceleration detection along the three Cartesian coordinates: the acceleration components right and left, back and forth, and up and down.

The mechanism of the semicircular canals is illustrated in Fig. 2.27. Resting on the crista is the swivel-mounted cupula. Dendrites reach from the neuron embedded in the crista into the cupula. While the head is motionless, the endolymph surrounding the cupula is at rest. When the head accelerates parallel to the orientation of a semicircular canal, the inertia of the endolymph results in flow of the endolymph in the direction opposite to the direction of the acceleration. This phenomenon can be simulated by holding a half-full glass of water in your hand and suddenly accelerating it. Its inertia causes the water to stay behind. The flow of the endolymph pushes the cupula to the side so that the tilting is sensed by the nerve endings in the cupula and is communicated to the brain. Note that the semicircular canals cannot sense speed because motion with constant speed does not cause an acceleration of the endolymph against the cupula. If you move your head with constant speed, the endolymph is at rest and the cupula retains an upright position, in the same fashion as if the head were motionless.

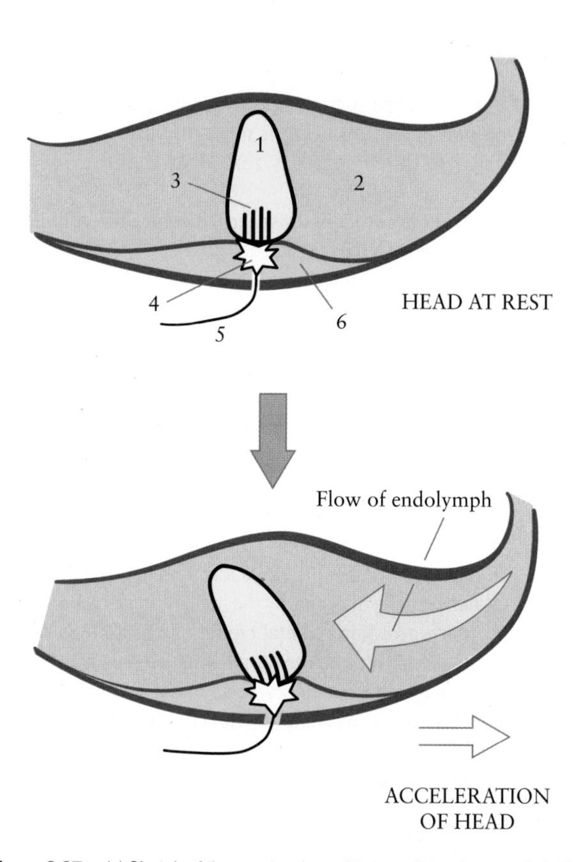

Figure 2.27 (a) Sketch of the mechanism of the semicircular canals in the vestibular organ of the inner ear. The top part shows the head motionless. When the head accelerates toward the right, as shown at the bottom, the endolymph (2) flows because of its inertia toward the left. This pushes the cupula (1) resting on the crista (6). The cupula tilts, bending the dendrites (3) that belong to a neuron (4) embedded in the crista, triggering a signal in a nerve (5) to the brain.

Acceleration detectors developed early in animals and are widespread among vertebrates. The lateral line system in fish is illustrated in Fig. 2.28. You see this canal system as lines running the full length of the fish, from the gills to the tail along both sides (trunk canals); the pattern at the head is highlighted in Fig. 2.28(a). The canals lie below the scales of the fish. When the fish accelerates, water pushes against neuromasts (see Fig. 2.28(b)), which are the equivalent component to the cupula of the semicircular canal system in the human ear. Water flowing past the bendable neuromast causes dendrites that reach into the neuromast to bend and trigger a nerve signal to the brain of the animal. The lateral line system allows fish to monitor not only their own accelerations, but also water pressure changes due to other moving objects (such as predators or prey) and low-frequency sounds carried through the water.

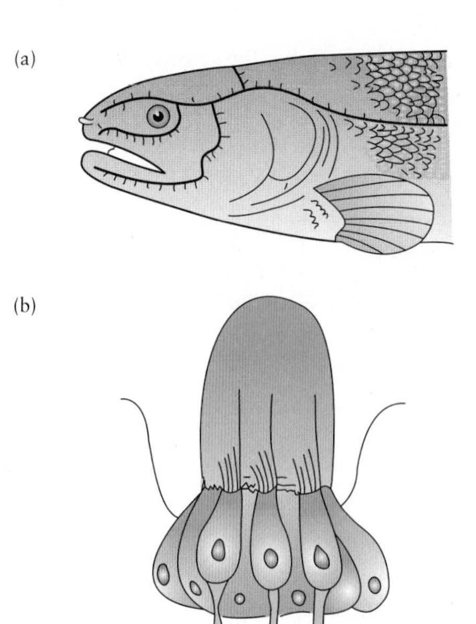

Figure 2.28 The mechanism of the lateral line system of fish. Fish can detect the acceleration of the water passing past their body. (a) The lateral line system consists of a system of canals with external openings that run below the fish's epidermis between the scales. The line system in the head region is highlighted by thick lines. Acceleration receptors are located at various points along these canals, indicated by small tick marks along the highlighted lines. (b) The receptor consists of a very similar arrangement to that of the cupula in the semicircular canals of the vestibular organ: the cupula rests on hair cells with hair-like sensory extensions reaching into the base of the cupula. When these sensory hairs are bent, a nerve signal is sent to the centre of the nervous system of the fish.

SUMMARY

DEFINITIONS

- One-dimensional motion: motion that takes place along a straight line.
- Vector: a measurement that has both a magnitude and a direction.
- Position vector: a measurement showing the position of an object relative to the origin.
- Displacement: the magnitude and direction of the straight-line path between two position vectors:

$$\Delta\vec{x} = \vec{x}_2 - \vec{x}_1$$

- Distance: the length between two locations measured along the actual path used to travel between two position vectors.
- Velocity: the vector that describes how fast an object is moving.
- Speed: the magnitude of the velocity vector.
- Average velocity: the constant velocity that would produce a given displacement:

$$\vec{v}_{av} = \frac{\Delta\vec{x}}{\Delta t} = \frac{\vec{x}_2 - \vec{x}_1}{t_2 - t_1}$$

- Average speed: the constant speed that would produce a given distance of travel.

- Instantaneous velocity: the velocity at a particular instant in time:

$$\vec{v} = \lim_{\Delta t \to 0} \frac{\Delta \vec{x}}{\Delta t}$$

- Tangent: the best straight-line approximation to a curve at a given point.
- Acceleration: the vector that describes the change in velocity over a time interval.
- Average acceleration: the constant acceleration that would produce a given change in velocity:

$$\vec{a}_{av} = \frac{\Delta \vec{v}}{\Delta t} = \frac{\vec{v}_2 - \vec{v}_1}{t_2 - t_1}$$

- Instantaneous acceleration: the acceleration at a particular instant in time:

$$\vec{a} = \lim_{\Delta t \to 0} \frac{\Delta \vec{v}}{\Delta t}$$

- Acceleration due to gravity: the acceleration caused by an object's mass. On Earth's surface, this acceleration can be approximated as constant.
- Freefall: motion where the only acceleration acting on an object is the acceleration due to gravity:

$$\vec{g} = 9.8 \text{ m/s}^2$$

- Resultant vector: The vector that results from the addition or subtraction of multiple other vectors.
- Projectile motion: freefall motion where a component of the initial velocity is perpendicular to the acceleration:

x-direction: $\vec{x}_2 = \vec{x}_1 + \vec{v}_x \Delta t$ and $\vec{v}_x$ = constant, and

y-direction: $\vec{y}_2 = \vec{y}_1 + \vec{v}_{1y} \Delta t - \frac{1}{2} \vec{g} \Delta t^2$ and $\vec{v}_{2y} = \vec{v}_{1y} - \vec{g} \Delta t$

- Uniform circular motion: motion with zero tangential acceleration along a circular path of constant radius.
- Centripetal acceleration: acceleration that points toward the centre of curvature:

$$a_{\perp} = \frac{v^2}{r}.$$

UNITS

- Displacement $\vec{x}$: m
- Velocity $\vec{v}$: m/s
- Acceleration $\vec{a}$: m/s^2

KINEMATIC FORMULAS

- For constant velocity: $\vec{x}_2 = \vec{x}_1 + \vec{v}_{av} \Delta t$
- For constant acceleration: $\vec{v}_2 = \vec{v}_1 + \vec{a}_{av} \Delta t$

$$\vec{v}_{av} = \frac{\vec{v}_1 + \vec{v}_2}{2}$$

$$\vec{x}_2 = \vec{x}_1 + \vec{v}_1 \Delta t + \frac{1}{2} \vec{a} \Delta t^2$$

$$v_2^2 - v_1^2 = 2a(x_2 - x_1)$$

MULTIPLE-CHOICE QUESTIONS

MC–2.1. A student walks from home to school and back again. Which of the following is most correct?
(a) The student has zero displacement and zero average speed.
(b) The student has zero distance travelled and positive average velocity.
(c) The student has zero average velocity and zero distance travelled.
(d) The student has zero displacement and positive average speed.

MC–2.2. An object has zero acceleration in the x-direction? In the y-direction, its acceleration must be which of the following?
(a) positive
(b) negative
(c) zero
(d) Any of the previous answers is possible.

MC–2.3. A father grabs his child by the hands and spins her around him in a nearly horizontal circle. The daughter's motion is clockwise around the father when viewed from above. The daughter's acceleration points in a direction that is pointed in approximately a direction that goes from the centre of her body to her
(a) right.
(b) left.
(c) father.
(d) feet.

CONCEPTUAL QUESTIONS

Q–2.1. If an object travels with positive velocity in the x-direction, can it have a negative acceleration along the x-axis at the same time?

Q–2.2. If the average velocity of an object is zero in a given time interval, what do we know about its displacement during the same time interval?

Q–2.3. An object is thrown vertically upward. (a) What are its velocity and acceleration when it reaches its highest altitude? (b) What is its acceleration on its way downward half a metre above the ground?

Q–2.4. A dog and its owner walk toward their home. When they are within 1 km of the door ($t = 0$), the dog starts to run twice as fast as the owner walks, back and forth between master and door until the master reaches home. How far did the dog run after $t = 0$?

Q–2.5. (a) Can an object accelerate if its speed is constant? (b) Can an object accelerate if its velocity is constant?

Q–2.6. Is there any point along the path of a projectile where its velocity and acceleration vectors are (a) perpendicular to each other or (b) parallel to each other?

Q–2.7. An object is thrown upward by a person on a train that moves with constant velocity. (a) Describe the path of the object as seen by the person throwing it. (b) Describe the path of the object as seen by a stationary observer outside the train.

Q–2.8. An object moves in uniform circular motion when a constant force acts perpendicular to its velocity vector. What happens if the force is not perpendicular?

Q–2.9. Describe the path of a moving object if its acceleration is constant in magnitude and (a) perpendicular to its velocity vector or (b) parallel to its velocity vector.

Q–2.10. An object moves with uniform circular motion. (a) Is its velocity constant? (b) Is its speed constant? (c) Is its acceleration constant?

Q–2.11. A bucket of water can be whirled in a vertical loop without any water being spilled. Why does the water not flow out when the bucket is at the top of the loop?

ANALYTICAL PROBLEMS

P–2.1. (a) What is the sum of the two vectors $\vec{a} = (5, 5)$ and $\vec{b} = (-14, 5)$? (b) What is the magnitude and direction of $\vec{a} + \vec{b}$?

P–2.2. If vector $\vec{a}$ is added to vector $\vec{b}$, the result is the vector $\vec{c} = (6, 2)$. If $\vec{b}$ is subtracted from $\vec{a}$, the result is the vector $\vec{d} = (-5, 8)$. (a) What is the magnitude of vector $\vec{a}$? (b) What is the magnitude of vector $\vec{b}$?

P–2.3. The origin of the x_2-coordinate axis is located at $x_1 = 2$ as measured on the x_1-coordinate axis. What is the position vector of the red dot in Fig. 2.29 when measured using the x_1-coordinate axis? What is the position vector of the red dot in Fig. 2.29 when measured using the x_2-coordinate axis?

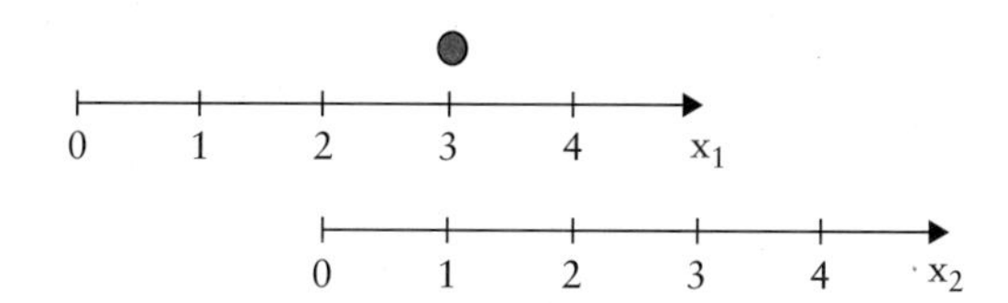

Figure 2.29 Two one-dimensional coordinate systems x_1 and x_2.

P–2.4. The origin of the (x_2, y_2)-coordinate system is located at (4, 1) as measured on the (x_1, y_1)-coordinate system. What is the position vector of the red dot in Fig. 2.30 when measured using the (x_1, y_1)-coordinate system?

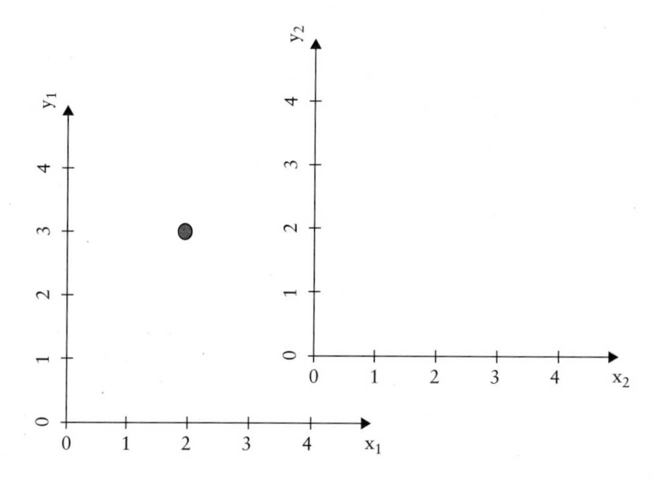

Figure 2.30 Two two-dimensional coordinate systems (x_1, y_1) and (x_2, y_2).

What is the position vector of the red dot in Fig. 2.30 when measured using the (x_2, y_2)-coordinate system?

P–2.5. A competitive sprinter needs 9.9 seconds to run 100 metres. What is the average velocity in units metres per second (m/s) and in units kilometres per hour (km/h)?

P–2.6. Fig. 2.31(a) shows a shear fracture of the neck of the femur. In a shear fracture, opposite fracture faces have slid past each other. Fig. 2.31(b) shows a sketch of a fracture with the net displacement *AB* along the fracture plane. (a) What is the net displacement *AB* for a horizontal slip of 4.0 mm and a vertical slip of 3.0 mm? (b) If the fracture plane is tilted by $\theta = 20°$ to the plane perpendicular to the bone, by how much have the two bones moved relative to each other along the bone's axis?

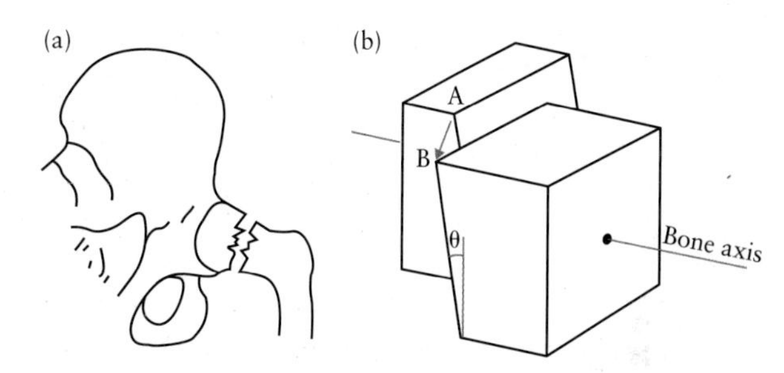

Figure 2.31 (a) A shear fracture of the neck of the femur and (b) its physical model.

P–2.7. Fig. 2.32 shows a back view of an adult male and an adult female human (accompanied by two children). (a) For a typical male, the vertical distance from the bottom of the feet to the neck is $d_1 = 150$ cm and the distance from the neck to the hand is $d_2 = 80$ cm. Find the vector describing the position of the hand relative to the bottom of the feet if the angle at which the arm is held is $\theta = 35°$ to the vertical. (b) Repeat the calculation for a typical female with $d_1 = 130$ cm, $d_2 = 65$ cm, and the same angle θ.

Figure 2.32 Back view of a family holding hands.

P–2.8. Fig. 2.33 shows (left) a front view and (right) a side view of a human skull. Two perpendicular projections such as these are often used to determine distances and

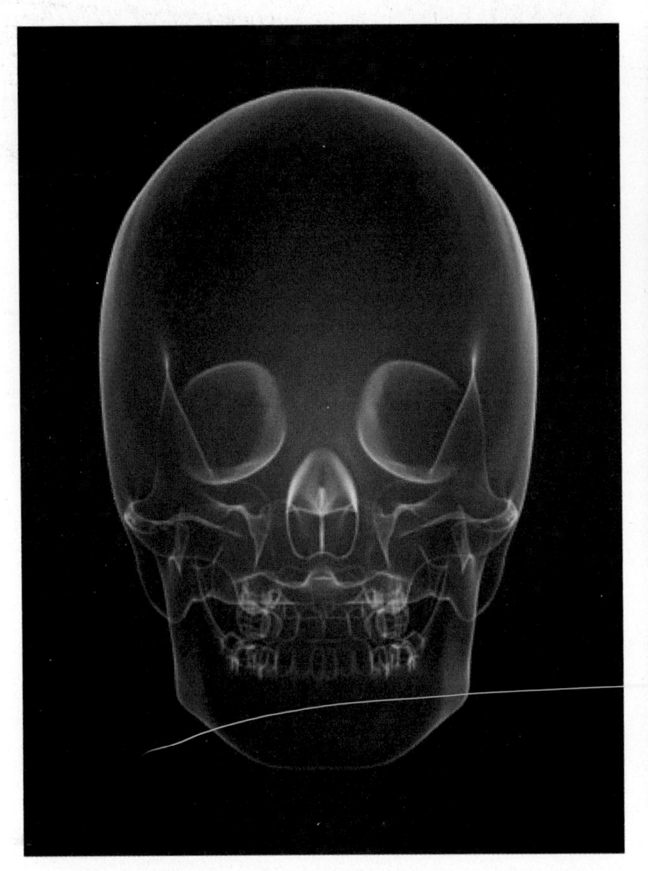

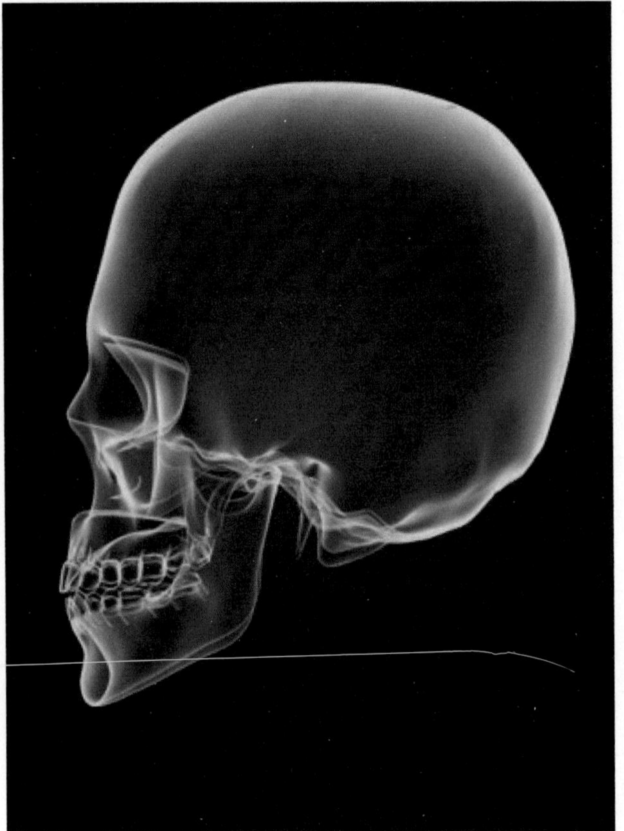

Sebastian Kaulitzki/Shutterstock

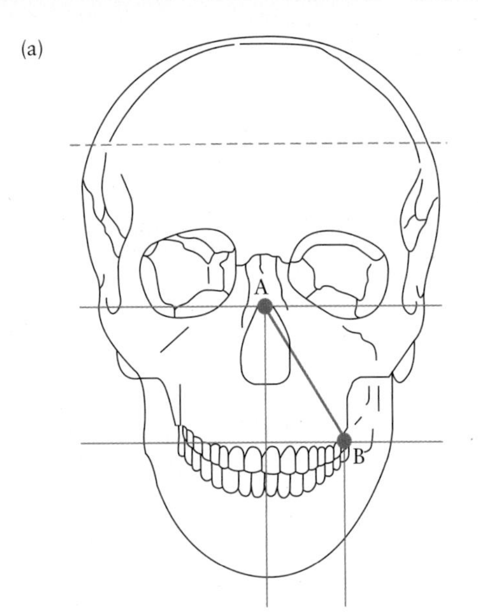

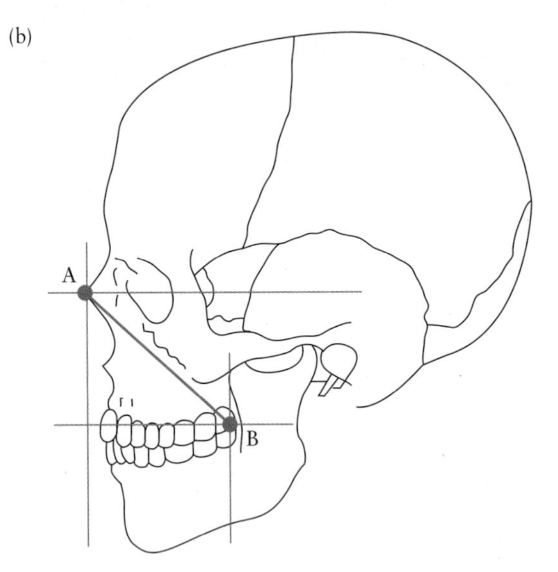

Figure 2.33 Two perpendicular views of a human skull.

angles in three-dimensional bodies, e.g., for focussed radiation therapy with high-energy X-ray beams. (a) Assuming that the diameter of the skull at the dashed line shown with the left-hand skull in Fig. 2.33 is 16 cm, determine the distance from the tip of the nasal one (point A) to the centre of the last molar in the upper jaw (point B). (b) Determine the angle between two lines connecting the point halfway between the two central maxilla incisor teeth and the last maxilla molars on either side. (c) Compare the result in (b) with the result obtained from Fig. 2.34, which shows a top view of the permanent teeth.

P–2.9. A bacterium moves with a speed of 3.5 μm/s across a petri dish with radius r = 8.4 cm. How long does it take the bacterium to traverse the petri dish along its diameter?

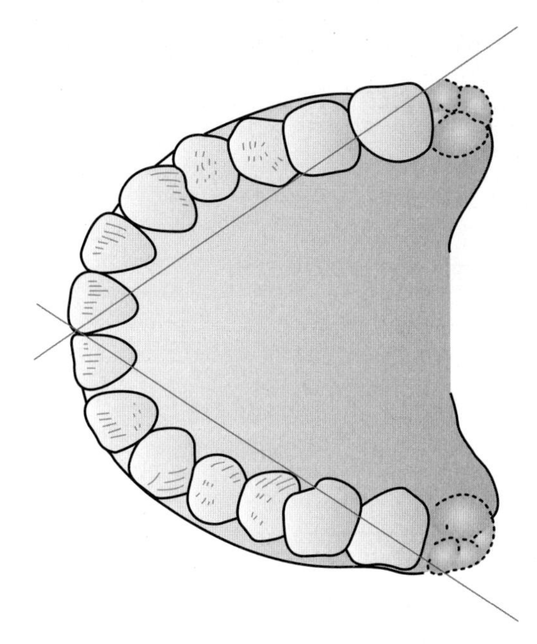

Figure 2.34 Top view of human permanent teeth.

P–2.10. Draw the one-dimensional position-versus-time graph for the time interval between $t = 0$ s and $t = 10$ s for an object that starts at initial position $x_1 = 5$ m with an initial velocity of –5 m/s and a constant acceleration of –10 m/s^2.

P–2.11. Draw position-versus-time graphs for both the x-position and the y-position for the time interval between $t = 0$ s and $t = 10$ s for an object that starts at initial position $(x_1, y_1) = (2$ m, 3 m$)$ with an initial velocity (–3 m/s, 2 m/s) and a constant acceleration (1 m/s^2, 2 m/s^2).

P–2.12. Fig. 2.35 shows the position-versus-time graph for a moving object. The positions above the time axis are to the right of the origin.
(a) At what point(s) does the object have the largest speed? (b) At what point(s) does the object have the smallest (non-zero) speed? (c) At what point(s) does the object have the smallest (non-zero) velocity? (d) At what point(s) does the object have zero velocity? (e) At what point(s) is the object moving to the left?

P–2.13. If an object has an initial position $ax_1 = 12$ m, and an initial velocity $v_1 = -3$ m/s, what is the final x-position of the object if its acceleration is described by the graph in Fig. 2.36?

P–2.14. If an object has an initial position $(x_1, y_1) = (4$ m, 5 m$)$, and an initial velocity $(v_{1x}, v_{1y}) = (5$ m/s, –3 m/s$)$, what is the final position vector (x_2, y_2) of the object if its two-dimensional acceleration is described by the graphs in Fig. 2.37?

P–2.15. Fig. 2.38 shows the velocity-versus-time graph for an object moving along the x-axis. Find the average speed and the average velocity over the 8-second time interval shown.

P–2.16. Sitting beside a friend on a park bench, you grab her hat and start running in a straight line away from her. Over the first 15.0 m, you accelerate at 1.0 m/s^2 up to your maximum running speed. You then continue at your maximum running speed for 15.0 s more before being your friend catches you. How far from the bench did you get before being caught? How long did it take for your friend to catch you?

P–2.17. In 1865, Jules Verne suggested sending people to the Moon by launching a space capsule with a 220-m-long cannon. The final speed of the capsule must reach 11 km/s. What acceleration would the passengers experience and would they survive the launch?

P–2.18. An object is released at time $t = 0$ upward with initial speed 5.0 m/s. Draw an $x(t)$ plot for the time period until it returns to its initial position.

P–2.19. An object is thrown vertically upward with an initial speed of 25 m/s. (a) How high does it rise? (b) How long does it take to reach this highest altitude? (c) How long does it take to hit the ground after it reaches the highest altitude? (d) What is its speed when it returns to the level from which it was initially released?

P–2.20. An object is dropped from rest from a height of 10 m. What is its average acceleration if it hits the ground with a speed of 1 m/s?

P–2.21. An object initially at rest falls vertically from a tree branch a height $h = 4.0$ m above the ground. Neglecting air resistance, how long does it take for the object to hit the ground? What is its speed the instant before it hits the ground? If the tree is twice as tall, by what factor does the time taken to fall increase? If the tree is four times as tall, by what factor does the speed at the instant before it hits the ground increase? If the object starts off with an initial downward velocity of 1 m/s, what is its speed when it hits the ground? If the object starts off with an initial downward velocity of 2 m/s, what is its speed when it hits the ground? Is the relationship between initial speed and final speed linear (i.e., did doubling the initial speed result in a doubling of the final speed)?

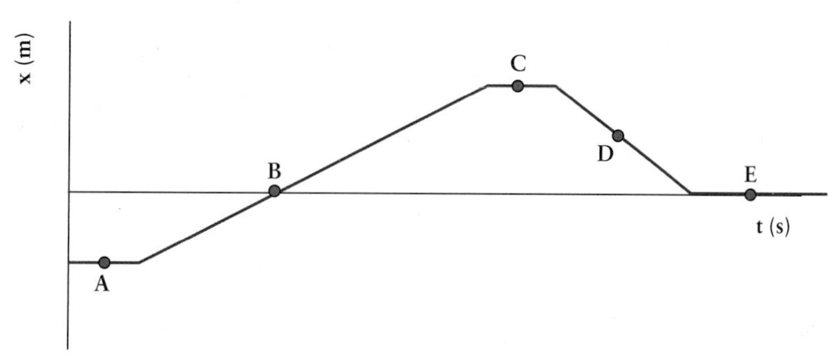

Figure 2.35 Position-versus-time graph of an object.

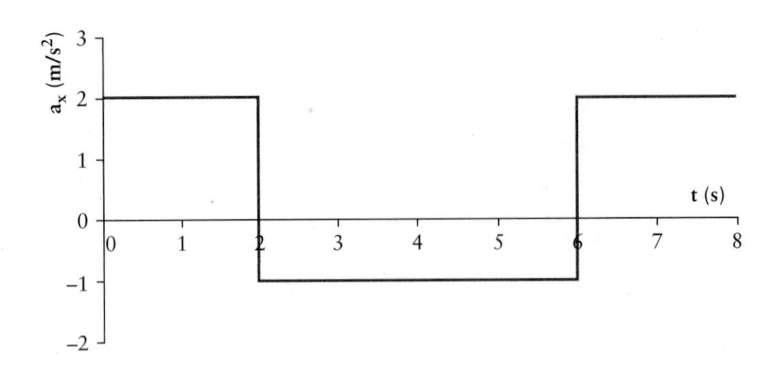

Figure 2.36 Acceleration-versus-time graph of an object.

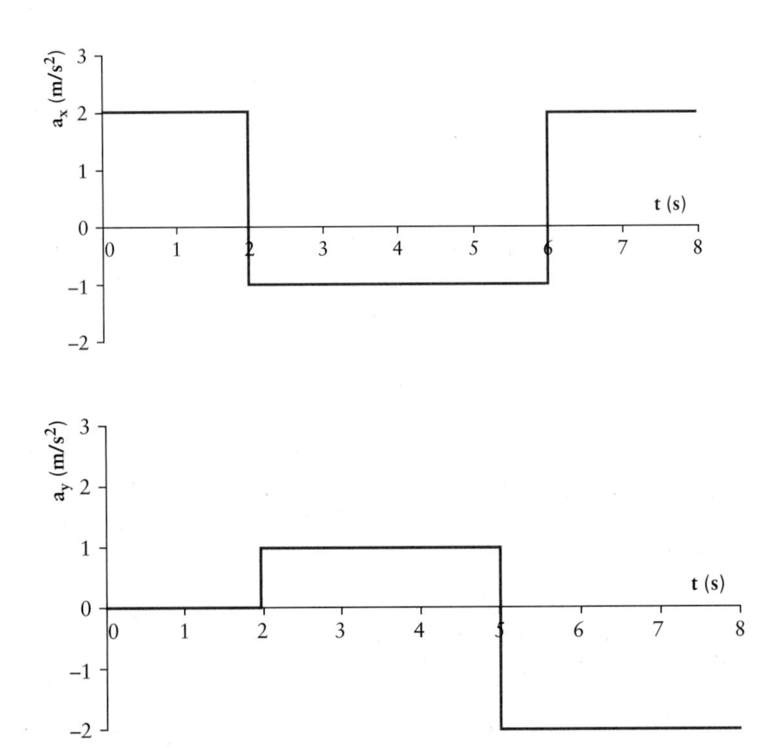

Figure 2.37 Acceleration-versus-time graphs of an object travelling in two dimensions.

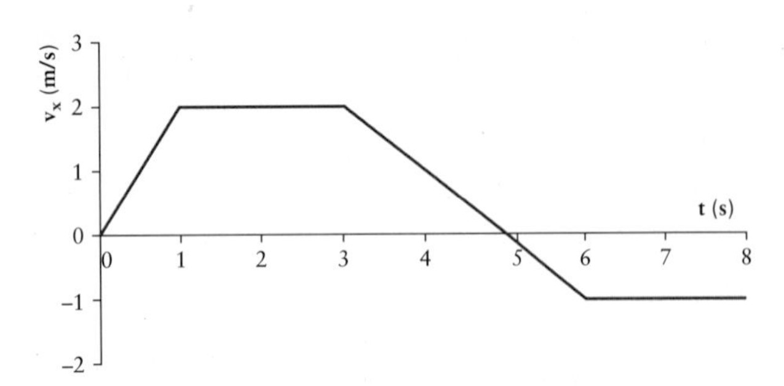

Figure 2.38 Velocity-versus-time graph for an object.

P–2.22. A rock is dropped (v_1 = 0 m/s) straight down from a bridge and steadily speeds up as it falls. It is moving at 30 m/s when it hits the ground under the bridge. If the motion can be modelled as freefall acceleration, draw the position-, velocity-, and acceleration-versus-time graphs for the motion. Take up to be the positive direction. There is enough information given to draw graphs that have quantitatively correct values for all slopes and intercepts.

P–2.23. A long-jumper leaves the track at an angle of 20° with the horizontal and at a speed of 11 m/s. (a) How far does the athlete jump, disregarding arm and leg motion and air resistance? (b) What is the maximum height reached during the jump?

P–2.24. The best major league baseball pitchers can throw a baseball with velocities exceeding 150 km/h. If a pitch is thrown horizontally with that velocity, how far does the ball fall vertically by the time it reaches the catcher's glove 20 m away?

P–2.25. In American football, place-kickers often decide a game with an attempt to kick the football from as far as 40 m through goalposts with a lower crossbar that is 3 m above the ground. If the ball is kicked with an initial velocity of 20 m/s at an angle of 53° to the horizontal, does the ball clear the crossbar?

P–2.26. Fish use various techniques to escape a predator. Forty species of flying fish exist—such as the California flying fish, which has a length of 50 cm. These animals escape by leaving the water through the surface, propelled by their tails to typical speeds of 30 km/h. If the flying fish could not glide, (a) how far would they fly through air if they left at 45°? (b) They can travel up to 180 m before re-entering the water. Do they use their wing-like pectoral fins to glide?

P–2.27. We become uncomfortable if an elevator accelerates downward at a rate such that it reaches or exceeds a velocity of 6 m/s while travelling 10 floors (30 m). What is the average acceleration of such an elevator?

P–2.28. An airplane flies in a horizontal circle with a speed of 100 m/s. The maximum acceleration of the pilot should not exceed $7g$, where g is acceleration due to gravity. What is the minimum radius of the airplane's circular path?

P–2.29. A child at an amusement park takes a ride on a horizontal wheel that has a radius of 9.0 m and spins 4 times per minute. What is the child's centripetal acceleration?

ANSWERS TO CONCEPT QUESTIONS

Concept Question 2.1: (C). Motion to the right is shown by increasing x-values. Motion in the left is shown by decreasing x-values. Stopping produces constant x-values.

Concept Question 2.2: (D). Start and end points can give information on displacement, but not on the distance travelled between those points.

Concept Question 2.3: (B). The magnitude of the average velocity is the average speed for the straight-line path between the initial and final points. Since no path from the initial to final points can be shorter than a straight line, any deviation from that path makes the average velocity smaller and the average speed larger.

Concept Question 2.4: (B). If the maximum is equal to the average, the minimum must also be equal to the average. For example, if your highest test mark in this class is 80%, and you have an average of 80% in this class, you must not have had any marks higher or lower than 80% in this class.

Concept Question 2.5: (C). The slope of A is larger than the slope of B, but the slope of B is increasingly positive, whereas the slope of A is constant.

Concept Question 2.6: (D). Follow Example 2.9 but use 90° north or south as the latitude.

CHAPTER 3

Forces

A force is defined by the interaction between separate objects. The term force is used to describe and measure the interaction. Force has several properties but is mainly characterized by its two quantitative properties: magnitude and direction. The most important objective of this chapter is to demonstrate the different ways an object interacts with its surroundings and the forces that result from the interaction. Forces are categorized as either contact forces or contact-free forces; for example friction and gravity respectively. There are four fundamental forces: (1) gravity, (2) electromagnetics, (3) strong nuclear force, and (4) weak nuclear force. These forces are contact-free forces that do not require any direct contact between objects; they are called **field** forces. These fundamental forces are essential for the way our world has been established: they describe how the universe works. Together they allow life to arise and form in its current state. All forces that we deal with daily, except gravity, are non-fundamental forces and are called convenience forces. These convenience forces are contact forces, such as tension, normal, or muscle forces. All of these forces have microscopic electromagnetic origins.

The laws of mechanics (kinematics and dynamics) allow us to understand, in the animal kingdom, the anatomical design and physiological function of muscles as a source of forces, and the skeleton as the frame on which these forces act. Not only can forces be exerted by living systems, they can also be detected by them. Organisms have a range of receptors that detect external forces directly or by measuring the resulting acceleration. These receptors are called mechano-receptors.

The widely accepted definition of life consists of three necessary conditions: (1) metabolism and growth; (2) recognition of external stimuli, combined with the ability to respond; and (3) reproduction. It is interesting to note that the second condition does not specify the response to stimuli as locomotion (which is motion from one place to another). From a physical point of view, locomotion is not a direct response of an organism to external stimuli, but is one possible consequence of the primary response, which is to **exert a force**. Thus, exerting a force is a direct response to external stimuli. The relationship between this primary response, **force**, and motion in general and locomotion specifically, will be discussed in next chapter. Indeed, organisms more often exert forces to *prevent* motion, e.g. when we hold objects or keep our body in a particular posture.

The predominance of forces over locomotion is seen in the hierarchy of specialized tissues that have developed in the animal kingdom: muscle tissue serves the specific purpose of exerting forces; locomotion in turn is achieved when several muscles and other tissues, such as bones, cooperate. We therefore need to start our discussion with the basic muscle tissue and the properties of the forces it generates.

We turn our attention in the current chapter also to the methods of **mechanical stimulus detection**, and the question of why we are sensitive to environmental forces such as gravity and the various contact forces.

To fully describe forces requires that we measure both their magnitude and their direction; i.e., we require vector algebra in this chapter as a mathematical tool. The mathematical definition of a vector and an overview of vector algebra operations are given in Chapter 2.

3.1: Muscles as an Origin of Forces

Muscle tissues are distinguished in anatomy on the basis of their structural differences, and in physiology on the basis of their functional purposes. Both approaches lead to the same three categories, reinforcing the close relation between design and function in living organisms: **skeletal muscles** that are attached to bones, **smooth muscles** that surround abdominal organs and blood vessels, and **cardiac muscles** that operate the heart. Defining muscle function based on the tissue onto which a muscle exerts a force is justified because of the underlying physics: physics is primarily concerned with the object on which a force acts. Even though all three muscle tissues share a common mechanism, we limit the current discussion to skeletal

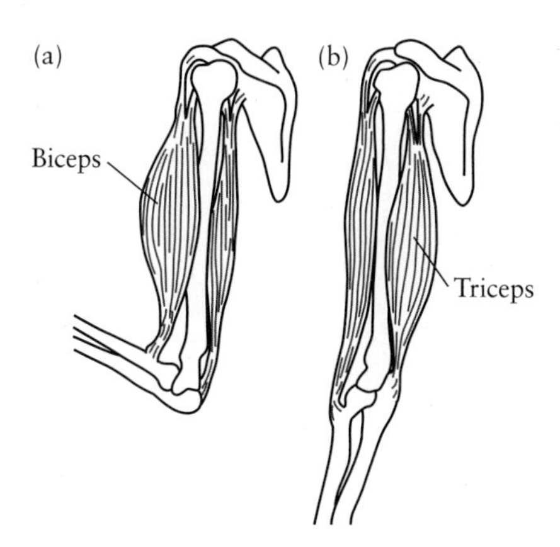

Figure 3.1 Biceps and triceps muscles. (a) The biceps is contracted. (b) The triceps is contracted.

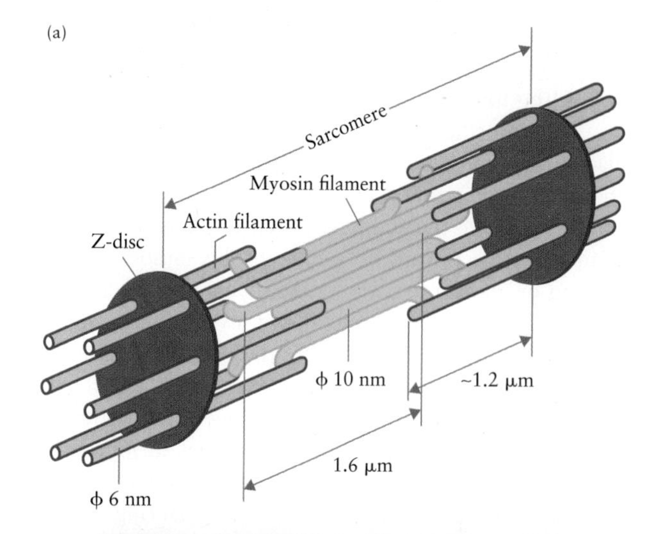

Figure 3.2 (a) The sarcomere is the contractile unit of the myofibrils in muscle cells. The length of the sarcomere is defined by the distance between two adjacent Z-discs. This length varies during muscle action when the myosin filaments crawl along the actin filaments, which are connected to the Z-discs. (b) Coloured transmission electron micrograph (TEM) of a section through a skeletal muscle. The muscle myofibrils are orange and run from upper left to bottom right. Z-discs are red and mitochondria are red ovals, e.g. at the lower right.

muscles because their actions relate most directly to the everyday experience with our bodies. The action of smooth muscles is briefly discussed in Chapter 14, and the action of cardiac muscles is discussed in more detail in Chapter 19.

Skeletal muscle is muscle attached to the skeleton and is responsible for skeletal movement. Most skeletal muscle is attached to bones by tendons. When these muscles contract, the bone does not always move. The contraction of skeletal muscle is under voluntary control. Fig. 3.1 illustrates two skeletal muscles, biceps and triceps, in the arm. When the biceps contracts it bends (flexes) the joint. The biceps is called a **flexor**. Simultaneously, the triceps is pulled, making it longer. As the triceps contracts, the arm is straightened (extends) at the elbow joint. The triceps is called an **extensor**. Simultaneously, the biceps is pulled, which makes it longer. Note that these two muscles work together as a pair. All muscles, regardless of their type, work in pairs because they can contract but cannot stretch themselves. Muscle can pull but cannot push.

The need for two complementary sets of muscles is rooted in the physical mechanism of muscle action: a muscle can only contract actively but then has to be stretched passively as another muscle contracts. In biology, this is called *antagonistic action*. While the biceps contracts, the triceps relaxes; while the triceps contracts, the biceps relaxes. They always act in opposite ways. Together, the biceps and triceps muscles compose an antagonistic pair. It is amazing that a muscle can produce force. Keep in mind that although the muscle action in Fig. 3.1 always exerts a force on a component of the skeleton, it does not necessarily cause locomotion. For example, while you hold a book in your hand, you exert force to hold it.

Skeletal muscles consist of bundles of *fibres* running the length of the muscle. Each fibre is a *cell* that is subdivided into smaller repetitive units called *myofibrils*. Muscle cells each contain about 100 myofibrils. The myofibril is divided in the elongated direction into *sarcomeres*, the basic contractile units of the muscle. The sarcomere is the smallest structural unit in a muscle that can contract.

A schematic of a sarcomeric unit is shown in Fig. 3.2. It is confined at both ends by stiff Z-discs. Sarcomeres are made up of 2 types of protein filaments, *actin* and *myosin*, which run alternatively parallel to each other but are not joined. *Actin filaments* are anchored in the Z-discs and extend on both sides. *Myosin filaments* bridge the gap between the actin filaments of two adjacent Z-discs. Myosin consists of a tail and a specialized binding head called *myosin head*; actin contains binding sites for myosin. Each myosin filament is surrounded by six actin filaments. Actin proteins allow the cell to bear tensile (pulling) forces. The myosin protein acts as a motor molecule by walking along the actin rods. The microscopic mechanism of muscle contraction is called the **sliding filament model**.

As a muscle contracts, the myosin filaments bind via the myosin heads onto the actin filaments, forming chemical bonds called cross-bridges. After the myosin heads hook onto the actins, the myosin filaments pull the actin filaments toward the centre of the sarcomere. The binding of myosin to actin by the myosin heads causes a release of energy that results in the myosin head swivelling, pulling the myosin filament over the actin filament. As the actin filaments slide over the myosin filaments, the length of the entire sarcomere is shortened. The collective shortening of the sarcomeres in a number of myofibrils causes muscle contraction. These sliding motions cause the muscle to contract, producing **force**. During muscle contraction, each sarcomere gets shorter, so the whole muscle gets shorter. However, the filaments do not contract and so do not get shorter; instead, they slide on each other. After the myosin heads relax, they release their hold on the actins, allowing them to slide back to their relaxed position. The processes of hooking the myosin head onto the actin and pulling it alongside, then releasing it back to relaxed position are caused by activation/deactivation stimuli instructed by the nervous system. Thus, contraction of muscle as a result of combining myosin and actin causes the muscle to produce **force**.

The limit of contraction of sarcomeres can be determined from Fig. 3.2. When the myosin filament hits the Z-discs on both sides, a further shortening of the muscle would require filament crumbling, which does not occur. Actually, each sarcomere, and therefore the entire muscle, can shorten by slightly more than 20%. Mechanisms discussed in Chapter 14 protect muscles against overstretching, which occurs when a sarcomere is elongated by more than 35%.

In vertebrates, muscles are not directly connected to bones but extend as connective tissues called tendons, which are attached to the bones. Thus, the force of the muscle is transferred to the bone via a tendon. Tendons act like extremely strong strings that are flexible but do not stretch. They are made of large strands of white, fibrous proteins called *collagen*. This is different from the muscle tissue but originates within the muscle to provide maximum strength. You may have noticed these anatomical features when eating chicken legs.

An example of the relative arrangement of muscle, tendon, and bone in humans is shown in Fig. 3.3 for the *Achilles tendon*, the thickest and strongest tendon in our body. It extends from the calf muscle to the heel bone.

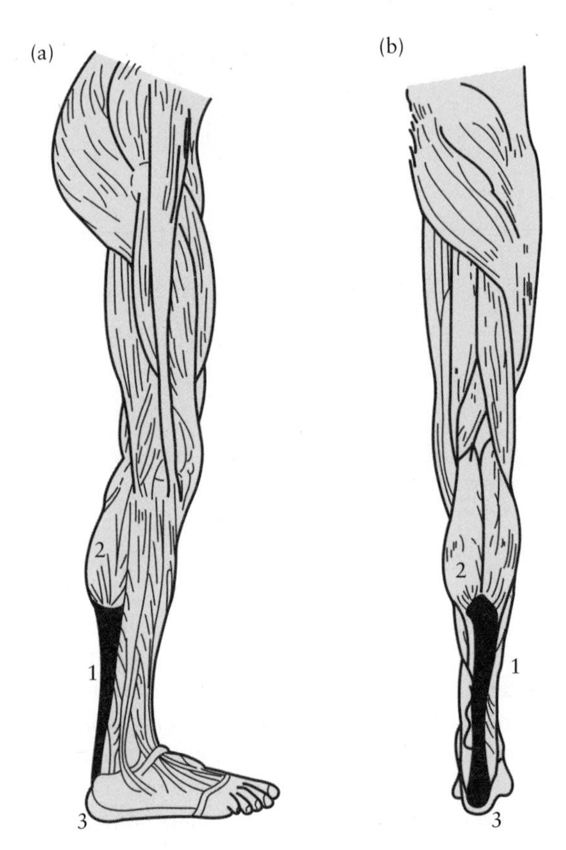

Figure 3.3 (a) Side view and (b) rear view of the lower leg of a human showing the Achilles tendon (1) connecting the calf muscle (2) to the heel bone (3). The Achilles tendon (highlighted) stretches as a narrow band along the lower one-third of the back of the lower leg. You can easily feel the Achilles tendon in your own leg because it runs shallowly below the skin.

3.2: What Is a Force?

The muscle, tendon, and bone in Fig. 3.3 are three separate tissues that interact as a particular functional group. We want to introduce the term force for this interaction. In the sciences, we first generalize such observations before formulating fundamental definitions and laws. The most inclusive definition follows when essential features have been separated from system-specific properties. The muscle and the bone from Fig. 3.3 are replaced by the more general term **object** because we want to apply the force concept to other systems as well, e.g., the action of myosin on actin filaments in Fig. 3.3, or the action of a person on some equipment in a gym. It is common, though, to always identify two distinguishable objects: we never think of an object acting on itself, and later will see why we have to exclude this specifically.

3.2.1: Force

Whether pushing a grocery cart or pulling a wagon, a force is being exerted. In simple terms, force is a push or a pull exerted on an object, resulting from the interaction between two objects, such as the person and the cart.

Pushing or pulling a box and throwing or catching a ball are examples of interactions between objects. Force is applied on objects whenever there is an interaction between them. Since the interaction is between two objects, there are two forces, one on each object. As Fig. 3.4 shows, a force is exerted on the box and simultaneously the box exerts a force back.

Figure 3.4 A man pushing a box and simultaneously the box pushing the man back.

CONCEPT QUESTION 3.1

(a) How many interactions can you identify when you throw a ball toward a wall and catch it after it hits the wall? (b) What are those interactions?

CONCEPT QUESTION 3.2

How many interactions can you identify when you throw the same ball upward and catch it when it returns to your hand? What are those interactions?

When throwing a ball toward a wall, you push it away from yourself while the ball is still in contact with your hand. After you release it, there is no interaction between you and the ball, so there is no force on the ball by you or on you by the ball. When the ball is in the air, neither you nor the wall have any effect on the ball. As the ball makes contact with the wall, there is an interaction between them. These kinds of interactions require contact. In other words, if there is no contact, there will be no interactions, and as a result there will be no forces exerted. These interactions are contact-based interactions, as illustrated in Fig 3.5.

In the second case, when throwing a ball upward or dropping it downward, it falls toward you or the ground. While in the air, there is an interaction (gravitational interaction) between the ball and Earth. Earth is pulling the ball toward itself. Earth exerts a force (gravitational force) on the ball. In this gravitational interaction between the ball and Earth, the ball simultaneously pulls Earth toward itself. Thus the ball simultaneously exerts a force (gravitational force) on Earth. The interaction between the ball and Earth is the same type of interaction that keeps Earth orbiting around the Sun. This kind of interaction does not require contact between two objects and is called a contact-free (or action –at a distance) interaction; they are known as field interactions. Similarly, when you hold a book in your hand, you exert an upward force and gravity exerts a downward force on the book. Thus, simultaneously, there is a contact force and a non-contact force on the book (Fig. 3.6).

The interaction between two objects is quantified by a force. A force can exist only if there is an interaction. In the broadest definition:

KEY POINT

A force represents the interaction of two distinguishable objects. Actually, force is only one of the models for describing the interaction between two objects. Momentum and energy are other models and they will be studied in later chapters.

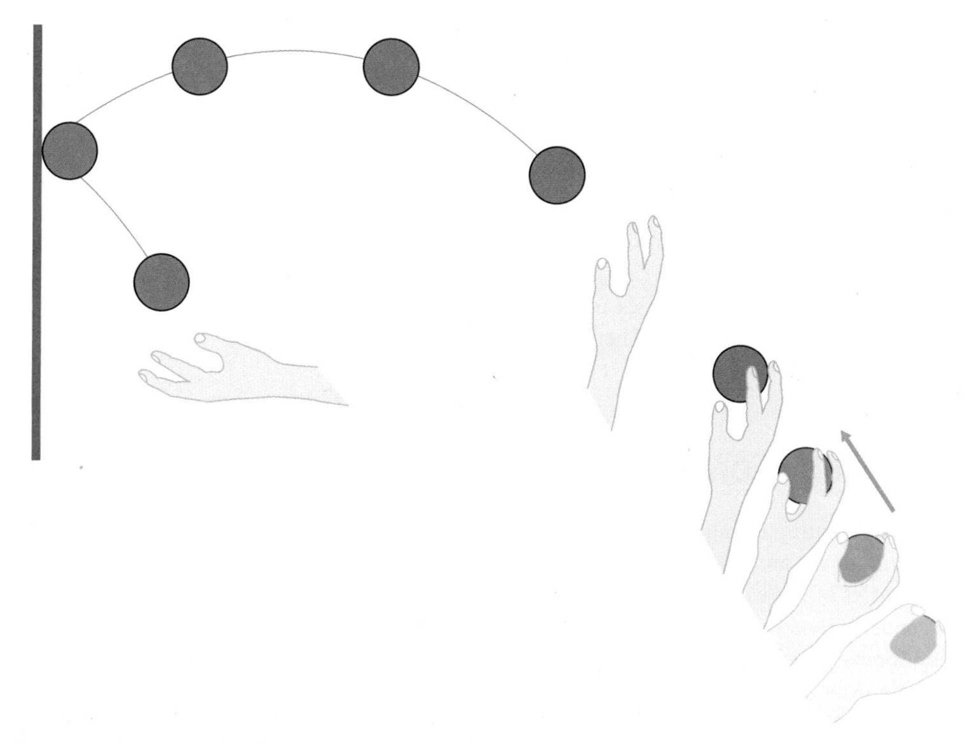

Figure 3.5 A ball is thrown toward a wall, flies on the air, hits the wall, and returns.

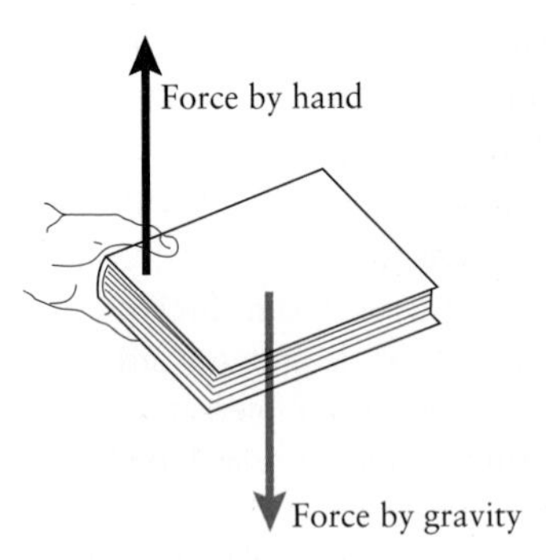

Figure 3.6 Holding a book on your hand. There are two forces on the book: the contact force you apply on the book while holding it, and contact-free gravity, the force that Earth exerts on the book, pulling it down.

All forces are placed into two broad categories: (1) contact forces, which require physical contact between two interacting objects; and (2)contact-free forces, which do not require contact between two interacting objects and act over a distance. These forces may be long-range forces. Contact-free forces are also called field force.

Thus, contact forces act only when physical contact between the objects is established; contact-free forces act over a distance. So, the calf muscle cannot exert a contact force on the heel bone in Fig. 3.3 because the muscle and the bone are not in contact with each other. The tendon creates this contact; thus, the calf muscle exerts a force on the Achilles tendon, and in turn the Achilles tendon exerts a force on the heel bone.

When two bodies interact, they exert force on each other. In other words, you cannot touch anything without being touched by it in return. In the same manner, you cannot exert a force on an object without a force being exerted on you by the object. There is no single isolated force. Therefore, forces always act in pairs. As Fig. 3.4 illustrates, two forces of interaction are applied on two different objects that are interacting. It is crucial to keep in mind that these **forces of the pair are acting on two different objects**. They are called ***interaction pair*** forces. Fig. 3.4 shows a pair of contact forces. The contact-free forces also act as a pair. The gravitational forces between Earth and the ball or Earth and the Moon are good examples of pairs of contact-free forces. Two charges also exert force (electromagnetic force) on each other. They too act as a pair of forces.

3.3: Properties of a Force

Throwing and catching a ball, pushing and pulling a box, throwing a ball up and down are examples that all illustrate one of the properties of forces. These examples illustrate that a force must be exerted on a material object. In addition, these forces are not applied spontaneously. Something is pushing or pulling these objects. Therefore, another property of a force is that it has to be applied by a material object. Note that, for any force, you should be able to identify both the object that exerts the force and the object upon which the force is exerted.

You can push or pull an object hard or gently and in different directions: up, down, left, right, or any other direction. This can be illustrated with a simple test. Hold a ping-pong ball in your hand and bring it up to your shoulder. To do that, a force should be exerted on the extremely light ping-pong ball, which you barely felt. Now exchange the ping-pong ball with this book and repeat the test. This time you can feel the force you exert on the book since it is larger than before. Again, repeat this test with the book, bringing it up parallel to your shoulder but taking it to the right side and then to the left side of your body. Note that you need the same amount of force holding the book in these tests, but the book ends up in different places: in front of you, on your right side, and on your left side. Hence, force acts in different directions. Force has not only magnitude, it also has direction. Quantities that have both magnitude and directions are vectors. Force is a **vector quantity**.

Macroscopically, as discussed earlier, some forces require contact and some forces do not require any kind of touch or contact. Contact forces act only when physical contact between objects is established, while contact-free forces act over a distance.

KEY POINT

The general properties of all forces may be outlined as follow:

- *Force is a pull or a push.*
- *Force acts on a material object.*
- *Force is applied by a material object.*
- *Force is either a contact force or a contact-free force.*
- *Force is a vector, having both magnitude and direction.*
- *Forces are paired. Two interacting objects exert a force on each other simultaneously.*
- *Force is additive. The effect of 2 simultaneous forces on the same object is the same as that of a single force equal to the addition of the forces. The addition is vector addition and the resultant force is called the net force.*

3.4: Action of a Force

What does a force really do? Consider you are playing volleyball. An opposing player hits the ball; it passes the net and reaches you. Before it hits the ground, you hit the ball back. Let us look at the brief contact between your hand and the ball. Your hand pushed the ball back during contact. Your hand exerted a force on the ball as it was coming toward you, changing its motion and sending it back. So force changes the state of the motion of an object. ***Force changes the velocity of the object upon which it is exerted*** as illustrated in Fig. 3.7.

Valeria73/Shutterstock

Figure 3.7 Interaction between a player's hands and the ball during a volleyball game changed the velocity of the ball.

CONCEPT QUESTION 3.3

Velocity is a vector. (a) Can a force change the magnitude of velocity? (b) Can it change the direction of velocity? (c) Can it change both the magnitude and the direction of a velocity?

CONCEPT QUESTION 3.4

(a) Give some examples that a force changes the direction of velocity of an object upon which it is exerted. (b) Give some examples that a force changes the magnitude of the velocity of an object upon which that force is exerted.

Consider another case. Assume you are watching your friend while he is kicking a football. While your friend's foot is touching the ball, the ball's shape is changing. The change will last as long as your friend's foot and the ball are in contact. Thus the change lasts as long as his foot applies force on the ball. Fig. 3.8 shows the change of the shape of ball and the foot while the ball is being kicked.

Figure 3.8 While kicking a ball, the ball's shape is changing, as is the shape of your foot, which is not as noticeable as the change in the ball.

This change of shape or deformation is allowed because of the molecular bonds between the object's atoms. Molecular bonds between atoms and molecules in a rigid object are not inflexible: they behave like a spring force. As an accepted physical model, consider there are tiny springs between atoms, connecting them and keeping them together in the object as illustrated in Fig. 3.9. If you push or pull on a spring, it pushes or pulls back on you. This spring force is called ***restoring*** force, which brings the object back to its unstretched/uncompressed position. Consider two atoms are held together by a tiny spring in their undisturbed position. If you push them closer together they repel each other; similarly, if you pull them apart they attract each other. If they get displaced from their regular position and then are released, they will get back to their regular position.

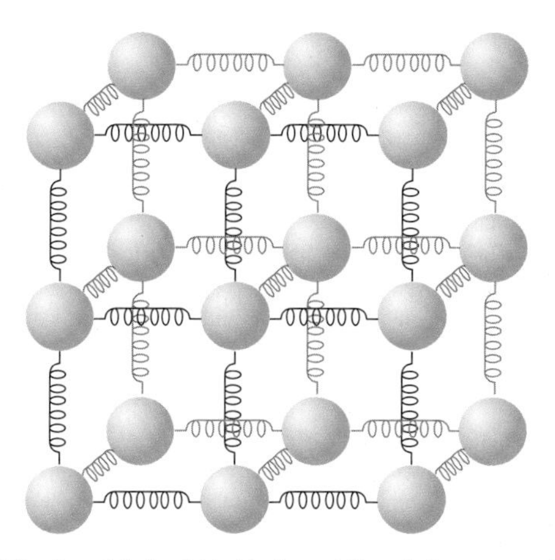

Figure 3.9 A model of a rigid object consisting of atoms connected with tiny springs.

The action of a force is outlined and illustrated in Fig. 3.10.

Forces operate in pairs. In an interaction between two objects, each object simultaneously exerts a force on the other. The change of velocity and/or shape also occurs simultaneously for both objects, although not necessarily in same magnitude.

CONCEPT QUESTION 3.5

When you kick a ball, the ball will fly away from the front of your foot. What happens to your foot? Does it fly back?

CONCEPT QUESTION 3.6

When you unsuccessfully push your desk to move it to another corner of your room, what changes are made, on both you and the desk?

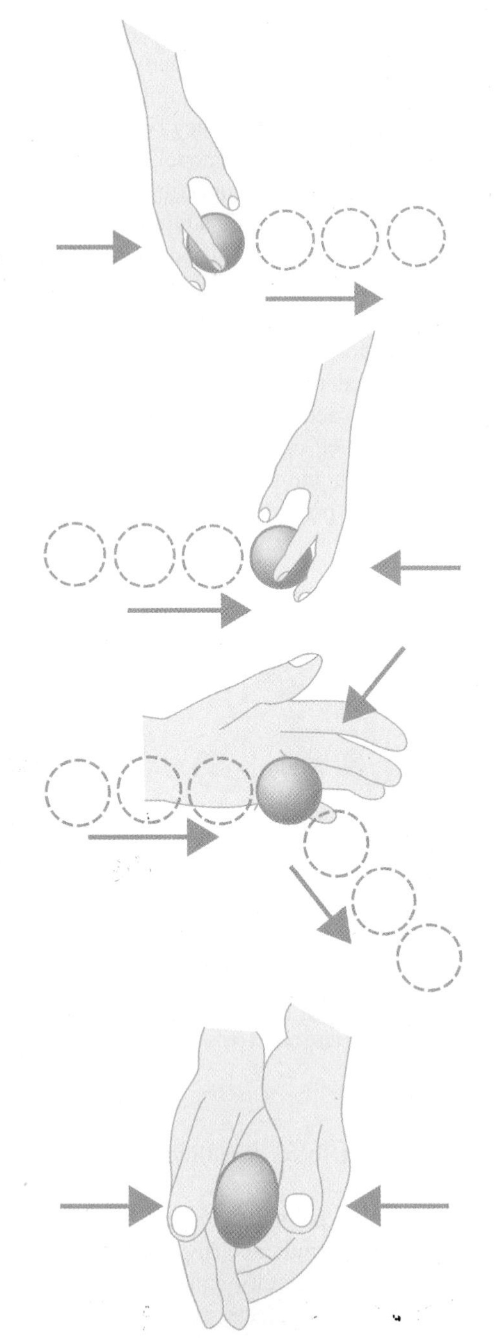

Figure 3.10 Action and effect of a force on a ball. (a) Pushing the ball gives it speed. (b) Pushing against the ball stops it. (c) Pushing the ball changes its direction of motion. (d) Squeezing the ball changes its shape.

3.5: Measuring Forces

Similar to any physical quantity, there should be a procedure for measuring a force. Generally, the description of a procedure to measure a physical quantity is known as the **operational definition** of the quantity. For example, force can be defined operationally by the extension or compression of a spring. If a spring is hung from the ceiling and the free end is pulled, the spring will stretch. Pulling the spring harder will cause it to stretch more. There is a linear relation between the applied force on the spring and the amount of its extension. This is known as Hooke's law, which will be discussed in more detail later in this chapter. An amount of force which produces a certain amount of extension in the spring is considered a unit for force. So two times this certain extension length in the spring stands for two units of force; similarly, half the extension presents half a unit. The unit of force is called the newton (N). The newton is a derived (SI) unit. In terms of fundamental units, it is $1 \text{ N} = 1 \text{ kg. m/s}^2$.

3.6: Categories of Forces

All forces can be classified into 2 categories: **fundamental forces** and **convenience forces**. Fundamental forces are contact-free forces that cannot be explained as any composite of other forces. Convenience forces can be explained in terms of fundamental forces.

3.7: Fundamental Forces

Four fundamental forces are identified in physics:

- gravitational force (gravity)
- electromagnetic force
- strong nuclear force
- weak nuclear force

Gravity and electromagnetic forces are two forces that have profound impacts in our everyday lives.

3.7.1: Gravity

If you toss a ball upward in the air (near Earth's surface), it will return and you can catch it. Any object dropped near the surface of Earth falls because Earth exerts a gravitational pull on the object. The force that Earth exerts on the object, pulling it down toward Earth, is called gravitational force, or gravity. The gravitational force of Earth on any object is commonly called the object's **weight**. A simple test may illustrate the effect of the gravitational force. What happens to a ball when you throw it horizontally? After leaving your hand, the ball will move away from you and it will drop down toward the ground, with some kind of tilting from horizontal. The ball is pulled to the ground by gravity. What happens if you throw it faster? It will go further before it is pulled to the ground. Either way, it will eventually get pulled down to the ground.

Any two objects, large or small, exert gravitational force on each other. Gravitational force is responsible for keeping Earth and the planets orbiting the Sun (Fig. 3.11). Similarly, Earth's gravitational force on the Moon keeps it in orbit around Earth. The gravitational force is one of the forces that attract one object to another. Interestingly, *the gravitational force between celestial objects and between terrestrial objects are the same*. Even though gravity is responsible for holding planets in their orbits, it is also the weakest of the fundamental forces.

Detlev van Ravensway/Science Photo Library

Figure 3.11 In our solar system, all planets orbit the Sun because of the gravitational force between the Sun and each planet.

CONCEPT QUESTION 3.7

(a) What effect does the Moon's gravitational force have on Earth's oceans? (b) Does the Sun's gravitational force have any effect on Earth's oceans?

Sir Isaac Newton showed that gravity also acts on bodies other than Earth. It is an attractive force that exists between any two masses. He stated that

KEY POINT

There exists an attractive force between two objects that is proportional to the product of their masses and inversely proportional to the square of the distance between them (Fig. 3.12). This is known as ***Newton's law of universal gravitation****.*

Mathematically, the force of gravity is given by

$$F = G\frac{m_1 m_2}{r^2}, \qquad [3.1]$$

where m_1 and m_2 are the masses of the objects, r is the distance between the two objects, and the constant G is the universal gravitational constant, and is equal to $G = 6.674 \times 10^{-11}\ \text{N m}^2/\text{kg}^2$.

The force of gravity increases as the masses increase and/or as the distance between them decreases. In the mathematical representation of gravity, interacting objects should be assumed as point objects, even if the distance between two objects is large with respect to their sizes. As Newton's universal gravitational law shows, gravity decreases gradually as the distance between the two objects increases. Thus the force of gravity is a long-range force. In the general form given in Eq. [3.1], the law of gravity is applied primarily in astronomy. We will use it in that form for some discussions of the effect of weightlessness on the human body in Chapter 20. When gravity is observed at or near Earth's surface, which is the origin of the force, it is called **weight**.

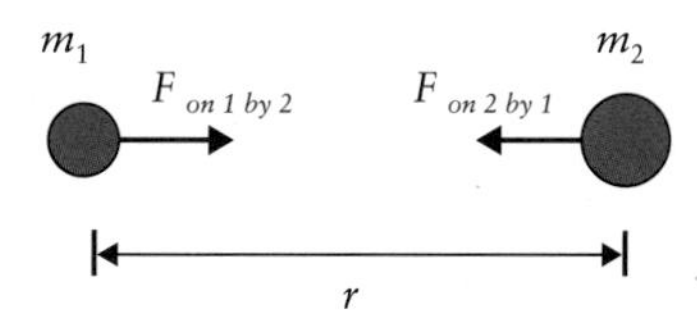

Figure 3.12 Gravitational force between two masses is proportional to their masses and inversely proportional to the square of the distance between them.

EXAMPLE 3.1

Two spheres, each having 20 kg mass, are placed at the centre-to-centre distance of 30 cm from each other. What is the force of gravity acting on each sphere?

Solution

Using Eq. [3.1], gives

$$F = G\frac{m_1 m_2}{r^2} = (6.67 \times 10^{-11}\,\text{N m}^2/\text{kg}^2)\,\frac{(20\,\text{kg})(20\,\text{kg})}{(0.3\ \text{m})^2}$$

$$= 2.9 \times 10^{-7}\ \text{N}.$$

EXAMPLE 3.2

Coccal bacteria are spherical in shape with a mass of approximately 9.5×10^{-13} g and a radius of approximately 0.5 μ. Estimate the force of gravity between two coccal bacteria that make diplococcal bacteria.

Solution

Using Eq [3.1] gives

$$F = G\frac{m_1 m_2}{r^2} = (6.67 \times 10^{-11}\,\text{N m}^2/\text{kg}^2)$$

$$\frac{(9.5\times10^{-16}\,\text{kg})(9.5\times10^{-16}\,\text{kg})}{(2\times0.5\times10^{-6}\ \text{m})^2} = 6.0\times10^{-29}\ \text{N}.$$

Consider an object with mass m at or near Earth's surface. Take the mass of Earth M_E and its radius R_E. The force of gravity on this object, which is the **weight** of the object and is labelled w, can be calculated using Newton's law of universal gravitation:

$$F_g = w = G\frac{Mm}{R^2}. \qquad [3.2]$$

By rearranging this equation, we determine the weight of the object:

$$F_g = w = m\left(G\frac{M}{R^2}\right) = mg \qquad [3.3]$$

$$w = mg, \qquad [3.4]$$

where g is the gravitational acceleration. Its magnitude at the or near Earth's surface is calculated using Eq. [3.3]:

substituting for R, Earth's radius, $R_{\text{Earth}} = 6.37\ 10^6$ m, and for M, its mass, $M_{\text{Earth}} = 5.98 \times 10^{24}$ kg:

$$g = \left(G\frac{M}{R^2}\right) = \frac{(6.67\times10^{-11}\,\text{N m}^2/\text{kg}^2)(5.98\times10^{24}\,\text{kg})}{(6.37\times10^6\,\text{m})^2}$$

$$= 9.8\ \text{m/s}^2.$$

Weight, w, is a gravitational force and a vector. Thus g ($g = w/m$) is also a vector and is in the same direction of gravitational force, always downward toward Earth. The vector $\vec{g}$ is called the gravitational field, which is defined as the gravitational force per unit mass:

$$\vec{g} = \frac{\vec{w}}{m} = \frac{\vec{F}_g}{m} \qquad [3.5]$$

Therefore, the gravitational field $\vec{g}$ exerts the force $\vec{F}_g$ on the mass m. The magnitude of g, and hence the force of gravity, varies as $1/r^2$.

EXAMPLE 3.3

Calculate the weight of the 20-kg sphere of Example 3.1.

Solution

Using Eq. [3.4]

$$w = gm = (9.8\ \text{m/s}^2)(20\ \text{kg}) = 196\ \text{N}.$$

EXAMPLE 3.4

What is the ratio of the gravitational force between two coccal bacteria to the weight of a coccal bacterium of Example 3.2?

Solution

Using Eq. [3.4],

$$w = gm = (9.8\ \text{m/s}^2)(9.5\times10^{-16}\ \text{kg}) = 9.3\times10^{-15}\ \text{N}$$

$$\frac{F}{w} = \frac{5.4\times10^{-29}}{9.3\times10^{-15}} = 5.8\times10^{-15}.$$

CONCEPT QUESTION 3.8

What does the small magnitude of the ratio of the gravitational force between a pair of bacteria to the weight of a bacterium show?

CONCEPT QUESTION 3.9

What is the ratio of the gravitational force between two ordinary persons to the weight of one of them? How do you interpret the result?

EXAMPLE 3.5

The weight of an astronaut on Earth is 833 N. What is his weight on Mars if Mars has a mass of 6.42×10^{23} kg and a radius of 3.40×10^6 m?

Solution

Using Eq. [3.3],

$$w_{\text{M}} = m\left(G\frac{M_{\text{M}}}{R_{\text{M}}^2}\right) = mg_{\text{M}}$$

$$w_{\text{M}} = mg_{\text{M}} = m(6.67\times10^{-11}\ \text{Nm}^2/\text{kg}^2)\frac{(6.42\times10^{23}\ \text{kg})}{(3.40\times10^6\ \text{m})^2}$$

$$= (3.7\ \text{m/s}^2)\,m$$

$$w_{\text{E}} = g_{\text{E}}\,m = (9.8\ \text{m/s}^2)\,m = 833\ \text{N}$$

$$m = \frac{833\ \text{N}}{9.8\ \text{m/s}^2} = 85\ \text{kg}$$

$$w_{\text{M}} = (3.7\ \text{m/s}^2)\,m = (3.7\ \text{m/s}^2)(85\ \text{kg}) = 314.5\ \text{N}$$

$$\frac{w_{\text{M}}}{w_{\text{E}}} = \frac{314.5}{833\ \text{N}} = 0.38.$$

The force of gravity is constantly pulling our bodies. There are no forces other than gravity that have a continuous effect on our body. Gravity affects all parts of our body, from our sense of balance to our sense of orientation to blood flow. Although the human body has adapted to gravity, it is still influenced by it in many ways. The obvious effect of gravity is a compression of the spine, causing moisture loss in the sponge-like disks of the spine. This causes loss of height over a lifetime. As people get older, gravity influences their internal organs by causing them to undergo prolapse, in which organs drop from their usual places. Suffering from varicose veins, swollen feet, or an aching back can all be the result of being constantly under the influence of gravity.

A simple test illustrates the influence of gravity on the blood's circulation: keep one arm lifted up for 3 minutes and then compare it with the other arm. The lifted arm will appear paler. This is due to imbalance in blood flow, which itself is caused by the influence of gravity.

CONCEPT QUESTION 3.10

Why do infants sleep bottoms up?

CONCEPT QUESTION 3.11

When you are tired, why does bending forward and placing your head between your knees (or leaning forward on your lap) help?

CONCEPT QUESTION 3.12

After sitting on a chair for a long time, some people prop their legs and feet on their desk. What does this action do? How does it help?

While in orbit, an astronaut's body experiences floating inside the spacecraft, which is called free fall (microgravity). Floating impacts the astronaut's body, since the human body is raised under the influence of gravity. The effects of floating and/or experiencing zero gravity, are summarized as a change in circulation of the blood, which results in the "puffy –face," "bird leg" look shown in Fig. 3.13, and in the long-term changes to bones and muscle structure, and disturbance of balance and sense of orientation. The effect of microgravity (weightlessness) on the human body will be discussed in a later chapter.

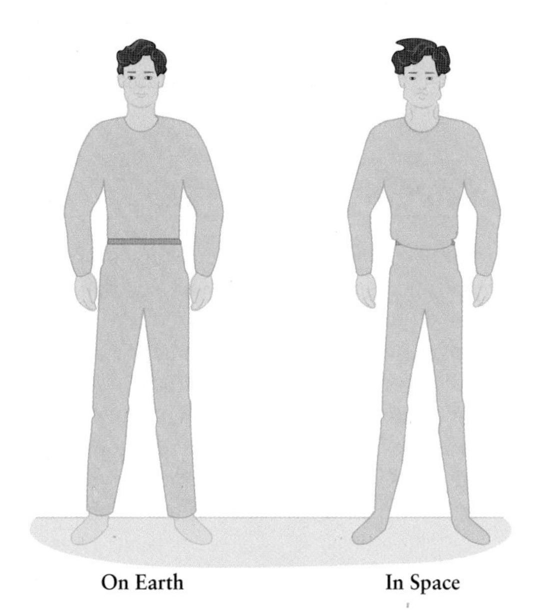

Figure 3.13 Puffy-face and bird-leg look.

3.7.2: Electromagnetic Force

Two charged objects (particles) interact and exert electric force on each other. Two magnets also interact and exert magnetic force on each other. Two moving, charged objects (particles) exert electromagnetic (electric and magnetic) force on each other. Depending on the type of charges on objects, or magnetic poles, the electromagnetic forces may be attractive or repulsive.

This can be illustrated with a simple test. Place a positively charged object close to another positively charged object, and then replace it with a negatively charged object. You will see two charged objects repel each other or attract each other. You may repeat this experiment with two magnets using different poles. Either way, you will get attractive interaction with two opposite charges (poles) and repulsive interaction with two like charges (poles) as shown in Fig. 3.14. By increasing the distance between two charged objects, or two magnets, the force between them decreases gradually. In any case, electromagnetic force is a long-range force that acts at a distance.

The electric force between two (point) charges was first introduced quantitatively by Charles-Augustin de Coulomb. He stated that

KEY POINT

There exists an electric force between two charged objects proportional to the product of the magnitude of their charges and inversely proportional to the square of the distance between them.

This is known as **Coulomb's law** (Fig. 3.15).

Mathematically the electric force is given by

$$F = k\frac{q_1 q_2}{r^2}, \qquad [3.6]$$

where q_1 and q_2 are the magnitudes of the charges of each object (in coulombs, "C"), r is the distance between two objects, and the constant k is the electric force (or Coulomb) constant and is equal to $k = 8.99 \times 10^9$ N.m^2/C^2.

CONCEPT QUESTION 3.13

(a) What kind of fundamental interaction is in charge of a muscle force on tendon? (b) On a sailing boat, what kind of fundamental interaction is in charge of a wind force on the sail?

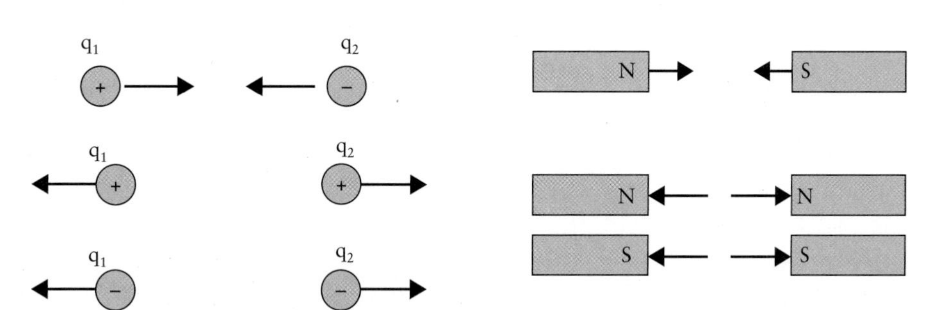

Figure 3.14 Electric force between two opposite or two like charges is attractive or repulsive respectively; similarly, magnetic forces between two opposite or two like magnetic poles are attractive or repulsive respectively.

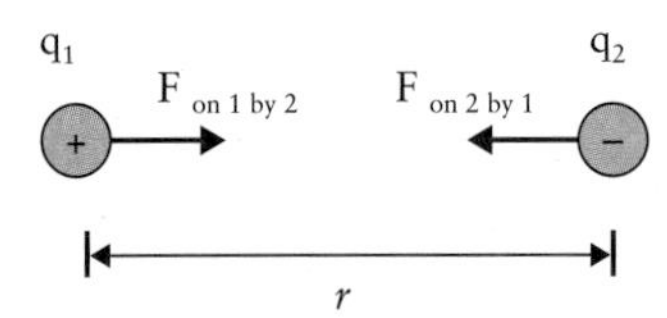

Figure 3.15 Electric force between two charges is proportional to their charges and inversely proportional to the square of the distance between them.

The force that holds electrons to nuclei forming atoms and binds atoms together in molecules is the electromagnetic force. Chemical bonds such as covalent, ionic, hydrogen, metallic, polar, and van der Waals are all attractive electromagnetic interactions.

Electromagnetic interactions play very important roles in our everyday lives. All contact forces are electromagnetic in nature. When two objects are in contact, the molecules (atoms) in their surfaces establish an electromagnetic force between each other. The electromagnetic interaction between atoms is the cause of all contact forces. Electromagnetic interactions and their influence on the human body are discussed in more detail in later chapters.

CONCEPT QUESTION 3.14

Water strider can stand on water (Fig. 3.16). What fundamental force holds the water surface intact, preventing the water strider from sinking?

Figure 3.16 Water strider standing on water.

CONCEPT QUESTION 3.15

Why do compass needles orient themselves toward Earth's North Pole?

3.7.3: Strong Nuclear Forces

An atomic nucleus consists of positively changed protons and neutral neutrons. Protons repel each other by electric force. The nuclear force holds protons and neutrons together in the nucleus. Since it overcomes the repulsive electric force, it should be strong. Actually, the nuclear force is the strongest of the fundamental forces, but it has no significant effect outside the atomic nucleus. So nuclear force is very strong and a short-range force (10^{-15} m).

CONCEPT QUESTION 3.16

Two protons, a positively charged particle of $q = 1.6 \times 10^{-19}$ C and $m = 1.67 \times 10^{-27}$ kg, each in a nucleus are separated by a distance of approximately 2.4×10^{-15} m. How can you show that the nuclear force is strong?

3.7.4: Weak Nuclear Forces

Weak nuclear force has a role in the disintegration of certain radioactive nuclei. The emission of a beta particle (electron in nature) involves the weak nuclear force. Not only it is much weaker than the strong force, its range (about 10^{-17} m) is even shorter than the range of the strong nuclear force.

3.8: Convenience Forces

All non-fundamental (convenience) forces are contact forces. At the microscopic level, contact forces in nature are electromagnetic forces. They are the result of the interaction between the charged particles contained in atoms and molecules, which make molecular bonds between them. As discussed in Section 3.4, these molecular bonds are flexible and act like spring forces (Fig. 3.18). In this atomic model of an object, the tiny springs between atoms are very stiff, but pushing or pulling on them compresses or stretches these tiny springs respectively, on molecular level, a very small extent. As a result, they push or pull back on the pushing or pulling objects. Convenient forces such as surface (normal and friction), tension, spring, drag, and viscous forces can be described by this atomic spring model. They will be discussed later.

3.8.1: Surface Force

Surface forces are applied by the surface of an object on the surface of another object. They are contact forces, which can be identified by a visible contact between two interacting objects. There are two types of surface forces: normal force and the force of friction.

3.8.1.1: Normal Forces

Normal force is a contact force that the surface of an object exerts perpendicularly on an object that is pressing against the surface. Thus it is a repulsive force. Normal forces always act perpendicular (normal) to the contact surfaces of two interacting objects. They are the result of compression of molecular bonds of the surface. A simple test can illustrate it. When you sit on the seat in your car, the springs

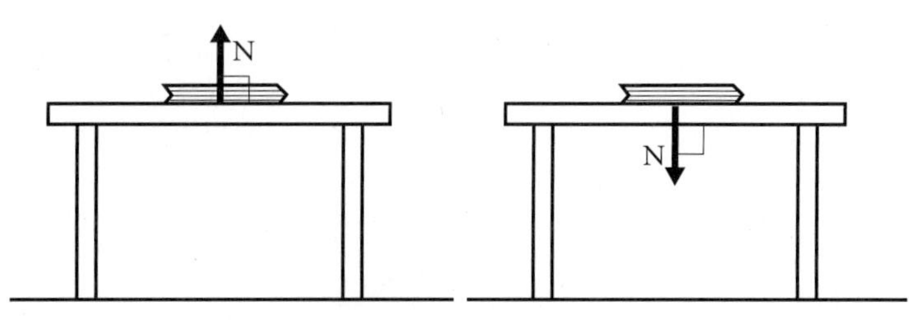

Figure 3.17 A book on a table: normal force on the book and normal force on the table.

in the seat are compressed by you. As a result, they exert an upward force back on you to hold you up. Otherwise, you would sink right through the seat. Based on the atomic model of an object having tiny stiff springs between its atoms (Fig. 3.9), these molecular tiny springs push back and apply forces when being compressed, almost in the same way that the spring in the car seat exerts a force back on you.

Consider a book placed on a table. The table exerts an upward force, $\vec{N}$, on the book at the contact surface, which prevents the book from falling to the ground. Similarly, the book exerts a normal force, $\vec{N}$, on the table. These forces are shown in Fig. 3.17.

CONCEPT QUESTION 3.17

What effect does the weight of the book have on the normal force? Can we say that the greater the weight of the book placed on the table, the greater the normal force is exerted by the table on the book?

The normal force originates in the electromagnetic interaction between atoms. This interaction is responsible for the maintenance of the shape of a solid object. While the book is on the table, it pushes the table down, compresses the tiny molecular springs between the atoms of the table, and causes an invisible small compression of the surface of the table. Consequently, these tiny compressed molecular springs push back in the reverse direction. Thus, in turn, the table pushes the book up, opposing the downward compression. This is a general model applicable to different cases, such as when you stand on the floor or lean against a wall. In all these cases, the molecular springs between the atoms are compressed and in return they push back, which is why you can stand on a floor or lean against a wall without going right through them. As an example, consider a person standing upright with both legs on the floor. When the person is standing, the floor pushes the body up through the contact point at each foot, exerting two normal forces: one on each foot. In the absence of these normal forces, the floor would not have an effect on the body and the person would fall through the floor, similar to trying to stand on a cloud.

What happens to the direction of normal force if the table, as in Fig. 3.7, is tilted (not horizontal), having an angle with the horizontal? The normal force is always perpendicular to the contact surface regardless of whether the contact surface is horizontal. Hence, normal force may be vertical or may not be vertical, depending on the orientation of the contact surface. Fig. 3.18 illustrates the normal force on a skier on a snow-covered hill.

Figure 3.18 Normal force on a skier.

CONCEPT QUESTION 3.18

Give an example in which the normal force is horizontal.

3.8.1.2: Force of Friction

Friction is another force that is all around us in our everyday lives. Friction can occur while two objects are in contact with each other. It is a contact surface force. The component of a surface force that is perpendicular to the surface is the normal force, as discussed above. The other component of a surface force that is parallel to the contact surface is the force of friction. The force of friction acts between two objects while their surfaces are in contact, opposing the motion of one object slipping along the other. The force of friction generally slows the motion of an object as its surface slides along the surface of the other object, or may even prevent a start of motion, keeping the object still on the surface. Thus friction is the resistive force opposing the relative motion of objects sliding along each other. So, most of the time, friction is a disadvantage in motion and undesirable. It is most often desirable to *reduce* the friction between two objects sliding across one another. Lubrication between two contacting surfaces is one way to reduce friction between them, such as using oil on a piston in a cylinder, and having synovial fluid squeezed through articular cartilage lining the joint in the knees or other leg joints, as illustrated in Fig 3.19.

Interestingly, friction is not always bad. It can also be useful and desirable. In fact, many things could not be done without friction in everyday life. Without friction it

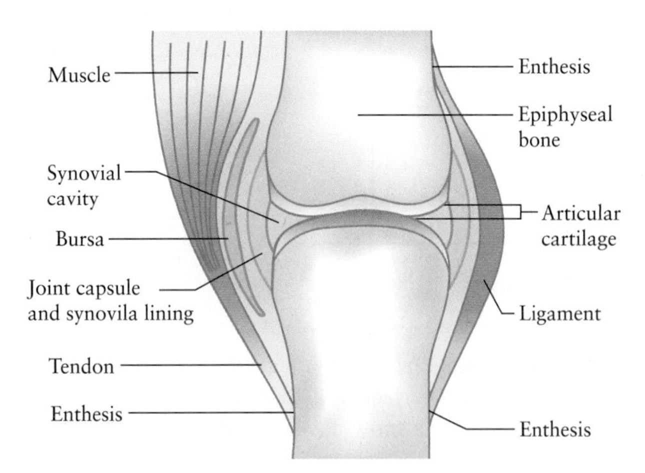

Figure 3.19 A knee joint with synovial fluid as a lubricant.

would be impossible to walk, to climb a stair, or even to sit on a chair. Without friction you could not write a note or erase it, clean a blackboard, or move your mouse to surf the Internet. The action of brakes slowing or stopping a moving object, the action of rubbing sandpaper on a surface, the uneven soles of our shoes, and the action of screws or bolts all depend on friction. Not only it is often helpful to have friction, sometimes we also want to increase it. During icy road conditions it is common to put sand on an icy driveway or road to increase traction between the tire surface and the road surface.

CONCEPT QUESTION 3.19

In walking, what force pushes you forward? (By starting to walk from a standing still position, what force is exerted on your foot?)

CONCEPT QUESTION 3.20

What happens if there is no friction between you and the ground, such as when walking on ice with slippery shoes?

In this section the force of friction between two solid (rigid) surfaces, which also is independent of velocity, will be discussed. Velocity-dependent resistance forces, such as the resistance an object experiences in motion through gas or liquid, will later be discussed separately.

What is the cause of friction? Friction is caused by the attractive electromagnetic force between atoms and molecules of contacting surfaces and the roughness and deformation of surfaces. As illustrated in Fig. 3.20, two contacting surfaces have some microscopic touch points that are much smaller than the apparent contact between them. Two surfaces that appear smooth to the naked eye might be uneven microscopically, with many points of contacts. At those contact points, atoms and molecules

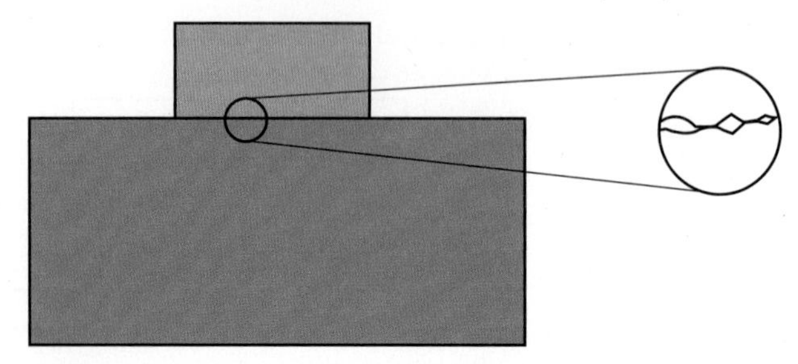

Figure 3.20 Microscopic contact points between two contacting surfaces.

of different objects get in such close proximity that intermolecular bonds are made between them. These atoms and molecules exert a strong attractive electromagnetic force on one another, forming so-called "cold-welds." As surfaces get rougher, their atoms and molecules get closer and, as a result, the force of friction gets larger. As surfaces become extremely smooth, the number of contact points and therefore the number of contact areas increase, forming stronger cold-welds, and the force of friction increases remarkably. Microscopically, any rigid surfaces, even those appearing very smooth, have bumps, hills, and valleys that get in the way of sliding. Also, some surfaces may deform slightly under pressure, which increases resistance to motion. All these parameters are involved in causing friction, but the exact details of the cause and microscopic origin of the force of friction are not completely understood. The microscopic cause and nature of force of friction are still under investigation. However, our physical model works well.

Although friction is a complicated force, there are some empirical models that describe many aspects of it. These models give some relations that are not fundamental, and instead are considered quantitative rules of thumb that can be used in everyday experiences. These relations were first investigated by Leonardo da Vinci 500 years ago. There are two kinds of friction between solids, static friction and kinetic friction, which are described by these models. Force of friction is shown by lowercase f.

3.8.1.2.1: Static friction

Static friction is the force that prevents objects from moving with respect to a contacting surface, and allows them to maintain their stationary status. Consider that you want to move a box across a floor. You push the box, applying a force horizontally, but the box does not budge. So there is another force on the box, a force of friction by the ground opposing your applied force. You may push harder, and still the box does not move, staying at rest. In any of these cases, the magnitude of the force of static friction is equal to the magnitude of your applied force:

$$f_s = F_{\text{applied}} \qquad [3.7]$$

You may get help from a friend and together push the box harder, and the box starts to move. Just before this point the force of static friction has its maximum magnitude. Experiments show that the maximum magnitude of the force of static friction between an object and the contacting surface is proportional to the magnitude of the normal force acting on the object by the surface:

$$f_{s,\text{Max}} = \mu_s N, \qquad [3.8]$$

in which μ_s is the coefficient of static friction, and $\vec{N}$ is the normal force. The coefficient μ_s depends on the materials of the two contacting objects. Its value can be as low as 0.1 for the synovial joint of a vertebrate and 1.0 for a rubber tire on dry concrete. The smaller the coefficient of static friction, the easier it is to set the object in motion. The direction of the force of static friction is the opposite direction of the supposed motion of the object if there were no friction. Note that, in any case, the magnitude of the force of static friction is always equal to the magnitude of the applied force. Static friction will be discussed again in more details in the next chapter.

CONCEPT QUESTION 3.21

The maximum force of friction between two surfaces is

(A) independent of the contact area.

(B) proportional to the normal force.

(C) both (A) and (B).

(D) none of the above.

EXAMPLE 3.6

A student wants to move his desk to another corner of his room. The desk weighs 320 N and the normal force from floor on the desk is equal to its weight. The coefficient of static friction between the desk and the floor is 0.5. The student applies 87 N to move it, but the desk does not budge. (a) What is the force of friction between the desk and the floor? (b) If his friend comes to help and together they apply 155 N on the desk, does the desk budge and move? How much is the force of friction now?

Solution

If the object is at rest, the static friction is equal to the applied force. If the applied force is slightly greater than the maximum static friction, then there is a net force on the object; it changes the object's velocity from zero to some magnitude and the object starts to move.

Solution to part (a): First we calculate the maximum magnitude of the force of static friction and then we

continued

compare it with the magnitude of the applied force. Using Eq. [3.8], we get

$$f_{s,\text{Max}} = \mu_s N = 0.5 \times 320 = 160 \text{ N}.$$

Therefore $f_s = F_{\text{App}} = 87\text{ N}$.

Solution to part (b): No it does not, since the magnitude of the applied force is less than the maximum magnitude of the force of static friction ($F_{\text{App}} = 155 \text{ N} < f_{s,\text{Max}} = 160 \text{ N}$). Static friction is equal to applied force

$$f_s = F_{\text{App}} = 155\text{ N}.$$

3.8.1.2.2: *Kinetic friction*

Kinetic friction is the force that opposes the motion of an object sliding across a contacting surface. Experiments show that the magnitude of kinetic friction is proportional to the magnitude of the normal force on the object exerted by the contacting surface.

$$f_k = \mu_k N, \qquad [3.9]$$

where μ_k is the coefficient of kinetic friction. The kinetic friction for an object is less than the maximum static friction for the same object on a surface. So, for two contacting surfaces

$$\mu_k < \mu_s.$$

Fig. 3.21 shows how the force of friction changes with respect to increasing applied forces. Kinetic friction will be discussed again in more detail in the next chapter.

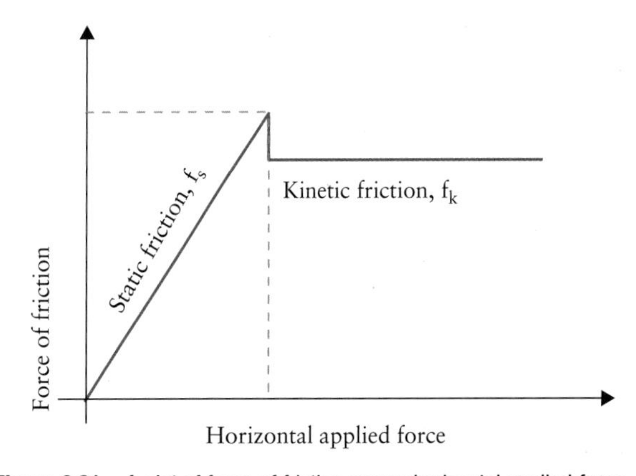

Figure 3.21 A plot of force of friction versus horizontal applied force.

CONCEPT QUESTION 3.22

What effect does the weight of a box have on the kinetic friction between the box and the floor on which it is moving?

CONCEPT QUESTION 3.23

A car is parked on a ramp that makes an angle θ with the ground. (a) What forces act on the car? (b) What force keeps it stationary on the ramp?

3.8.2: Tension

Tension is also known as the *pulling force*. An object can pull on another through the use of a string, cord, rope, chain, wire, tendon, cable, or other such object, as shown in Fig. 3.22. While pulling a box with a rope (ideal: massless and un-stretchable), the rope transmits the force to the box. So the rope is pulling the box, exerting a contact force on it. The force exerted by the rope on the box is called tension force, T, because the rope is under the tension. The direction of tension force is always along the rope and away from the box. The magnitude of the tension force along the rope is equal to the force on the rope. As discussed in Section 1.3, forces come in pairs. A pair of tension forces exists at any point along the rope, acting on each portion of the rope on either side of the point. As Fig. 3.23 illustrates, at a point along the rope, the forward portion of the rope pulls the backward portion of it, and vice versa.

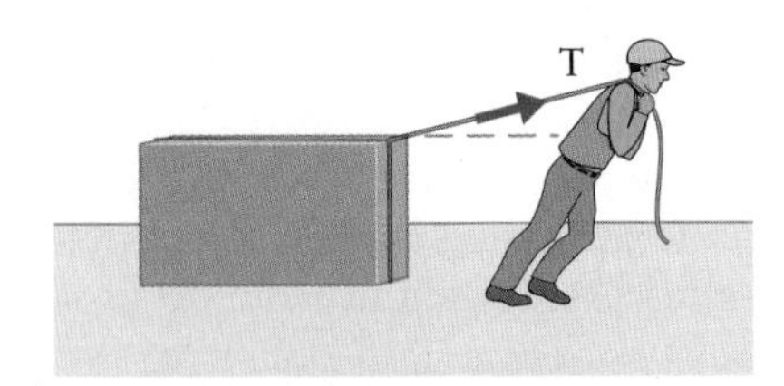

Figure 3.22 A box is pulled by a rope, which exerts a tension force on it.

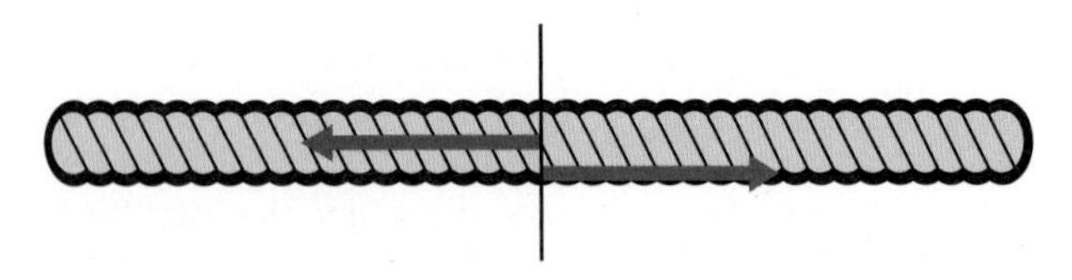

Figure 3.23 At each point on a rope, frontward pulls backward and vice versa.

In another example, the box is pulled by means of a pulley and a rope. Assume both the rope and the pulley are ideal (massless and frictionless; Fig. 3.24). The pulling force is transmitted to the box by the rope through the pulley, which changes only the direction, not the magnitude, of the force. This force acts on the box because of the tension force of the rope. This tension is equal to the applied force on the rope.

In the second example, the ball is hung by a chain from ceiling. Although gravity pulls the ball down, the ball remains at rest, hanging from the chain. Therefore the chain must exert a force on the ball, pulling the ball up, and opposing the force of gravity. This force is the tension force. The magnitude of tension will be the same at all points along the chain. Since, in reality, the chain has mass, the pulling force along the chain is different at different points of the chain, depending on the mass of the portion of chain below that point. The pulling forces at the top of the chain and at the bottom of it are different because of the relative masses of the chain. Note that the pulling force is not labelled a tension if the string's mass is included. We discuss this in Example 4.13.

Tensions occur often in physiological and kinesiological problems because we usually model tendons as massless strings, as illustrated in Fig. 3.25. The tension force will be visited again in more detail in the next chapter.

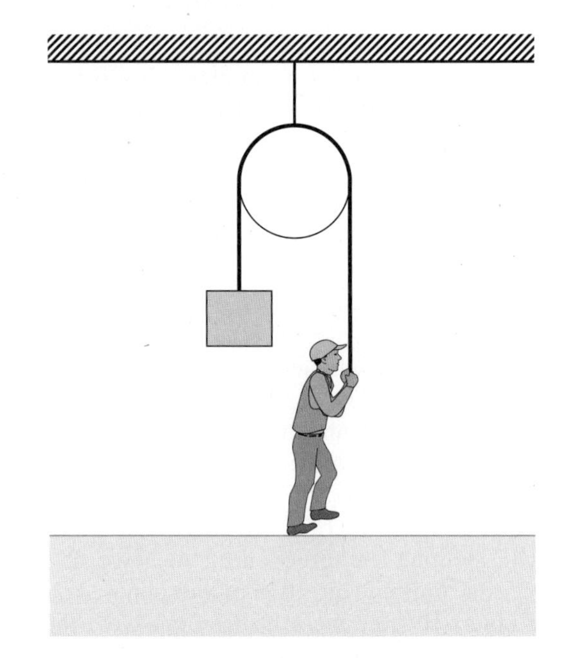

Figure 3.24 A box is pulled by a simple fixed pulley.

CONCEPT QUESTION 3.24

Two students are pulling opposite ends of a rope, each applying 100 N force. What is the tension in the rope?

(A) 0

(B) 50 N

(C) 100 N

(D) 200 N

(E) more information needed

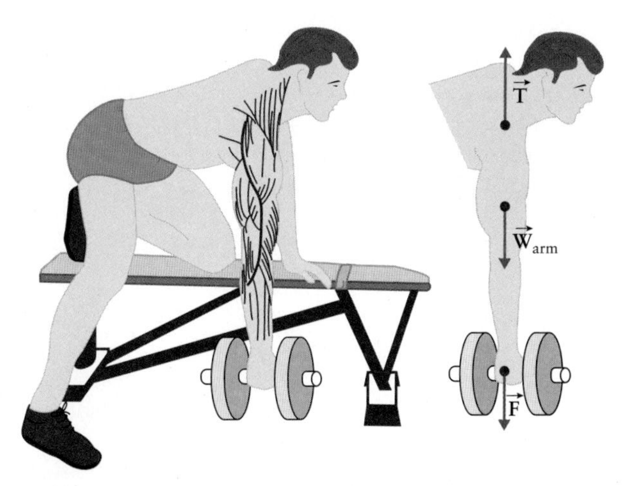

Figure 3.25 Forces also act between different parts of an organism. As an example, three major forces are highlighted as they act on the arm of a person intending to do one-arm dumbbell rows in a gym. The forces are indicated at the right: $\vec{F}$ is a downward-directed force due to the weight of the dumbbell, $\vec{W}_{arm}$ is the weight of the arm, and $\vec{T}$ is the force pulling the arm up due to the tension in muscles and ligaments connecting the trunk and the arm.

3.8.3: Spring Force

Any spring can be stretched or compressed. Take an unstretched spring, fix it at one end, and pull it at the other. To stretch it, you exert a force on the spring and pull it, and simultaneously the spring pulls you back. If you pull it harder, the spring pulls back harder. If you push the spring, it pushes you back in the same way as pulling you. The force the spring exerts on you, in pulling or pushing, is called the spring force. The magnitude of the spring force is approximately proportional to the length of extension or compression, which was first stated by Robert Hooke. Hooke's law says, as long as the applied force does not produce a permanent deformation, the change in spring's un-stretched/uncompressed length is proportional to the applied force. Mathematically, Hooke's law is given as

$$F_{App} = k\,x. \quad [3.10]$$

$\vec{F}_{App}$ is the applied force on the free end of the spring, x is the change in length of spring from its un-stretched (or uncompressed) length, and k is the spring constant. The SI unit for spring constant, k, is N/m. The spring force is given mathematically as

$$F_s = -k\,x. \quad [3.11]$$

The minus sign in Eq. [3.11] indicates that the spring force, $\vec{F}_s$, always opposes the change in the spring length and tends to take the spring back to its unstretched/uncompressed length. That is why the spring force, which is resisting the change in length, is also called *restoring force*. The restoring force of a spring, and hence Hooke's law, can be used for all elastic materials.

CONCEPT QUESTION 3.25

A spring balance with a linear scale can be used to measure mass because

(A) springs are strong.

(B) metal springs can be considered weightless.

(C) a spring force is linear in spring constant due to its strength.

(D) the spring force is linear in displacement through Hooke's law.

(E) none of the above.

3.8.4: Drag

So far, when we have described the interaction between two objects, the interaction between the object and the medium through which it was moving has been entirely disregarded. In the example of throwing and catching a ball, the air surrounding the ball and its interaction with the ball were ignored. What is the effect of the interaction between the ball and the air through which it moves? The medium, which is a fluid (a liquid or gas), has an interaction with the solid object that is moving through it, exerting a resistive force, $\vec{R}$, on it. This force of resistance in air is called **drag**, or specifically **air drag**, and in liquids it is called *viscous force*.

Thus drag is caused by the interaction of a solid object with a fluid (air). It is a contact force that requires contact between the fluid and the object. Drag is the resistive force that opposes the relative motion of an object through a fluid. The magnitude of drag depends on the relative speed of the object with respect to the medium. The direction of drag is always opposite to the velocity of the object. The magnitude of drag acting on an object that moves at relatively high speed through a fluid, such as air, is roughly proportional to the square of the object's speed. Mathematically drag is given by

$$R = \tfrac{1}{2}\, D \rho A v^2, \quad [3.12]$$

in which $\vec{R}$ is the drag force, D is the drag coefficient (a dimensionless parameter), ρ is the medium density (i.e. $\rho_{air} = 1.29\ \text{kg/m}^3$), A is the cross-section of the object, and v is its relative velocity with respect to the medium. This relation is not applicable for those objects that are very small (i.e. dust particles), move very fast, or move in liquid (i.e. water).

CONCEPT QUESTION 3.26

Why do birds fly in "V" formation, as illustrated in Fig. 3.26?

Figure 3.26 Snow geese are flying in a "V" shape.

EXAMPLE 3.7

A person standing in a wind that is blowing at 22 m/s experiences a drag force. Assuming the drag coefficient, *D*, is 1 and the person's area, *A*, is 0.8 m^2, estimate the magnitude of the drag force.

Solution

Using Eq. [3.12]

$$R = \tfrac{1}{2} D\rho A v^2,$$

$$R = \tfrac{1}{2}\,(1)(1.29\ \text{kg/m}^3)(0.8\ \text{kg/m}^3)(22\ \text{m/s})^2$$

$$= 249.7\ \text{N} = 2.5 \times 10^2\ \text{N},$$

$$R \approx 10^2\ \text{N}.$$

3.8.5: Viscous Force

The resistive force acting on an object moving through a liquid is called the **viscous force**. In simple terms, viscosity is a measure of how resistive the fluid is to flow. Viscous force is linearly proportional to the relative velocity of the object in the liquid. For the special case of a sphere moving through a liquid, the magnitude of the viscous force is given by Stokes' law as

$$F_{\text{vis}} = 6\pi\eta r v \qquad [3.13]$$

in which η is the viscosity of the liquid, r is the radius of the spherical object, and v is the relative velocity of object with respect to the liquid. This physical model is useful for cells in biology, as they can be considered approximately as spheres. Viscosity and the force of resistance of liquids to flow are discussed in Chapter 13.

EXAMPLE 3.8

What is the magnitude of the viscous force on a spherical dust particle of radius $r = 10^{-5}$ m in air? Assuming the dust particle is moving with constant velocity of $v = 2.4 \times 10^{-2}$ m/s. Take $\eta = 1.81 \times 10^{-5}$ Pa s:

$$(1\ \text{Pa} = 1\ \text{N/m}^2).$$

Solution

Using Eq. [3.13]:

$$F_{\text{vis}} = 6\pi\eta r v$$

$$F_{\text{vis}} = 6\pi(1.81 \times 10^{-5}\ \text{Pa s})(10^{-5}\ \text{m})(2.4 \times 10^{-2}\ \text{m/s})$$

$$= 8.2 \times 10^{-10}\ \text{N}.$$

3.9: Free Body Diagram

Before ending this chapter, an essential tool, the **free body diagram**, is introduced for finding the net force exerted on an object. A free body diagram is an outstanding conceptual tool and is very helpful in analyzing the forces and their accumulated effect on an object.

A free body diagram is a sketch of all forces on an object. It includes the object and arrows representing the applied forces on it. It shows all types of forces, both fundamental and convenience, exerted on the object. Keep in mind that only the forces exerted on the object are included in the free body diagram, not the forces that are exerted by the object on other objects or the surroundings. Free body diagrams will be used throughout this book.

To draw a free body diagram, you have to separate the object from its environment. For a system of more than one object, you have to draw a free body diagram for each object separately. You may represent the object by a square, or a circle, or even a dot; but the object is represented by a dot in a free body diagram. Replace each external, interacting object with an arrow representing the force exerted on the object at the contact point, and draw all external forces acting directly on the object. Therefore, the procedure for drawing a free body diagram can be outline as follows:

- Isolate the object by drawing a circle around it, separating it from other objects and its environment (ropes, pulley, etc.).
- Identify all forces acting on the object (contact and non-contact).
- Set up a coordinate system.
- Draw the object as a dot at the origin of the assigned coordinate system.
- Draw each force acting on the object as a vector arrow starting from the object and pointing in the direction that the force is applied.
- Draw every force, contact and contact-free, on the object.
- Do not draw forces that are not acting on the object.

A free body diagram is a very useful tool, helping us understand all forces acting on an object and analyzing net force (resultant force).

EXAMPLE 3.9

Draw a free body diagram of a book at rest on a table.

Solution

There are two forces acting on the book, gravity, $\vec{F}_g$, from Earth and the normal force, $\vec{N}$, (which will be discussed later) from the surface of the table preventing the book from falling. The free body diagram is shown in Fig. 3.27.

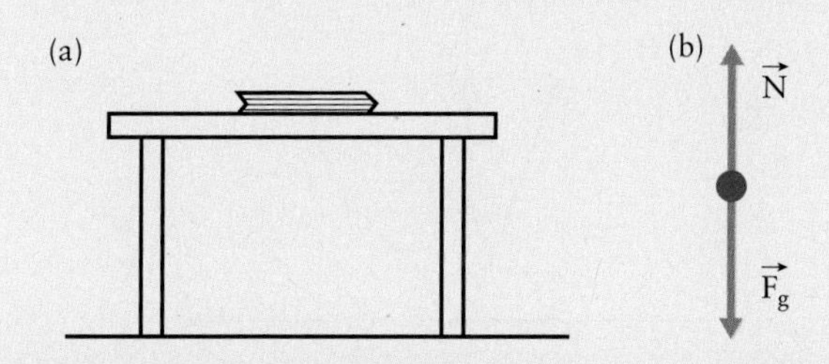

Figure 3.27 Free body diagram of a book on a table: (a) the book on the table, (b) its free body diagram.

EXAMPLE 3.10

Draw a free body diagram of a book pushed against the wall by your hand (shown in Fig. 28(a)).

Solution

Four forces can be identified acting on the book pushed against the wall, gravity from Earth, the pushing force on the book by you, the normal force from the wall perpendicular to the book, and the force of friction between the book and the wall. The free body diagram is shown in Fig. 3.28(b).

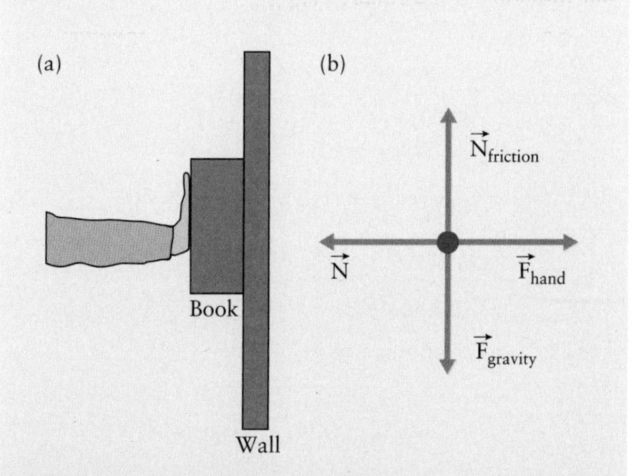

Figure 3.28 (a) A book is pushed against a wall, (b) free body diagram of a book pushed against a wall.

CONCEPT QUESTION 3.27

Some magnetic leaves are hung from ceiling attaching to a magnetic board, as illustrated in Fig. 3.29. Assume that $\vec{w}$ is the weight of one of the leaves, $\vec{M}$ is the magnetic force, and $\vec{N}$ is the normal force between the magnetic board and the leaf. Which of the diagrams shown in Fig. 3.30 best represents the free body diagram for one of the leaves?

Figure 3.29 Magnetic leaves attached to a magnetic board hang from a ceiling.

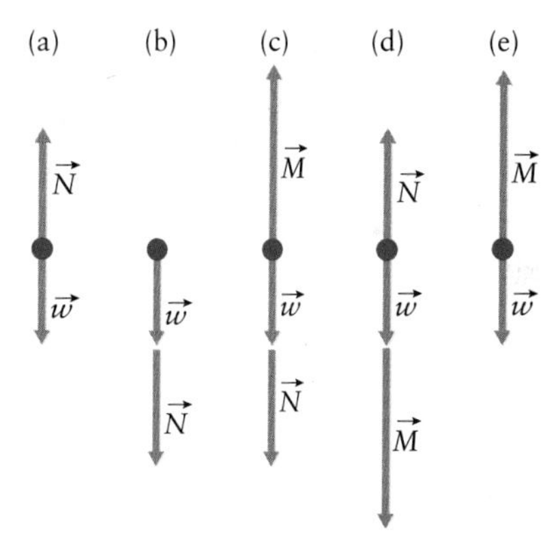

Figure 3.30 Free body diagrams of a magnetic leaf hanging from ceiling by a magnetic board.

3.10: Equilibrium

Force causes a change in velocity of an object. Consider a book on a table at rest, shown in Fig. 3.27(a) of Example 3.9. There is nothing changing its velocity, so there should be no forces acting on it. But there are two forces: weight from Earth and normal force from the table acting on the book, as illustrated in Fig. 3.27(b). So the net force on the book should be zero. The net force on the book is the resultant of the vector sum of the two forces, gravity and normal force, exerted on the book.

KEY POINT

If the net force that acts on an object is zero, the object stays at rest or moves without changing its velocity.

We say the object is in **equilibrium** or, specifically, in "*translational equilibrium.*" Mathematically, the condition for equilibrium is given as

$$\vec{F}_{net} = \sum_i \vec{F}_i = 0. \qquad [3.14]$$

In Eq. [3.14], Σ is the symbol for sum and "i" is an index that stands for different forces, such as weight, normal, friction, or tension, acting on the object.

Contact forces, particularly muscle forces, are often identified within our body. This is illustrated in Fig. 3.25 for the main forces acting on the arm of a person intending to do one-arm dumbbell rows in a gym. You can confirm the observations in Fig. 3.25 by holding this textbook with your arm relaxed beside your body (we used this self-test before). Notice that the book pulls your arm down, that your arm's own weight also pulls it down, and that your shoulder is pulling the arm upward. In the figure we label the weight of the arm $\vec{w}_{arm}$, the tension force the trapezius muscle in the shoulder exerts on the arm $\vec{T}$, and the force the book or the dumbbell exerts on the fist $\vec{F}$. Note that we did not call the latter force the weight of the book or the dumbbell because weight is a force acting on an object by Earth. In Fig. 3.25, in turn, the dumbbell exerts a normal force on the arm of the person; so $\vec{F}$ is a normal force. Here, the arm is in equilibrium, so the net force on the arm should be zero. Therefore

$$\vec{F}_{net} = \sum \vec{F}_i = \vec{T} + \vec{w}_{arm} + \vec{F} = 0\cdot$$

Since all these forces are in the y-direction, we can change the vector sum to an algebraic sum. Therefore

$$\vec{F}_{net,y} = \sum F_{i,y} = T - w_{arm} - F = 0\cdot$$

EXAMPLE 3.11

A picture frame is hung using two cords on a wall as illustrated in Fig. 3.31(a). If the mass of the frame is 200 g, find the tension in each cord.

Solution

First identify all forces acting on the frame so we can draw a free body diagram. There are three forces acting on the frame: weight $\vec{w}$, tension in the left cord, $\vec{T}_L$, and tension in the right cord, $\vec{T}_R$. Fig. 3.31(b) shows the free body diagram. Since the frame is at rest, the net force on it is zero. Using Eq. [3.14], we have

$$\vec{F}_{net} = \sum_i \vec{F}_i = 0.$$

continued

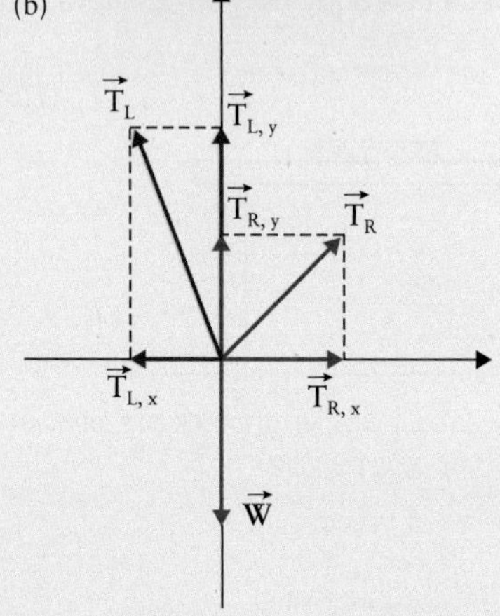

Figure 3.31 Free body diagram of a picture on the wall: (a) a picture frame is hung by two cords on a wall; (b) its free body diagram.

In the x- and y-directions, respectively, we have

$$x\text{-direction: } F_{net,x} = T_{R,x} - T_{L,x} = 0$$
$$T_{R,x} = T_{L,x}$$
$$T_R \cos(45°) = T_L \cos(68°)$$
$$T_R = 0.53 T_L$$

$$y\text{-direction: } F_{net,y} = T_{R,y} + T_{L,y} - w = 0$$
$$T_{R,y} + T_{L,y} = w$$
$$T_R \sin(45°) + T_L \sin(68°) = w = 0.2 \text{ kg}$$
$$T_L((\sin(68°) + 0.53\sin(45°))(0.2 \text{ kg})$$
$$(9.8 \text{ m/s}^2)$$
$$T_L(1.302) = 1.96 \text{ N}$$

$$T_L = 1.96 \text{ N}/1.302 = 1.51 \text{ N}$$

$$T_R = 0.53 \times 1.51 = 0.80 \text{ N}.$$

CONCEPT QUESTION 3.28

(a) Can the weight $\vec{w}$ in Eq. [3.4] be written as a vector? (b) The biceps tendon is attached to the radius, which is a bone in your lower arm. Does the biceps tendon exert a force on the radius? (c) Can the radius exert a force on the biceps tendon?

CONCEPT QUESTION 3.29

If you read Eq. [3.4] as a mathematical formula, the following statement is correct: to double m while holding w constant, we have to reduce g to one-half its original value. Using Eq. [3.4] to express the physical relation between mass, gravitational acceleration, and weight, is this statement still correct?

CONCEPT QUESTION 3.30

Fig. 3.32 shows an object as a white circle on an inclined surface. Six forces act on the object. Identify (a) the normal force and (b) the weight.

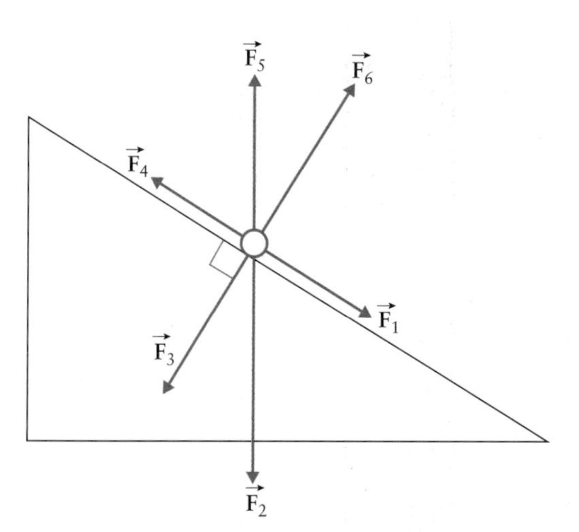

Figure 3.32 Object (dot) on an inclined surface; six forces act on the object in different directions.

3.11: Can Our Bodies Detect Forces?

In this section our awareness of the contact-free gravitational force and a typical contact force is surveyed. It is always instructive to start with a self-test. This allows us to locate the detection system in our bodies. We then describe the anatomical features of the detection system to later allow us to apply physical methods to quantify their function.

3.11.1: Detection of the Direction of Gravity

Close your eyes, stretch your arms, concentrate on your hands, and turn them to the side. The sensation in your hands has not changed despite the change of orientation. Thus, no gravity detector exists in your hands. Now, close your eyes again and lean your head to the side. This time you can even estimate the angle of tilt without a visual impression of your environment. The awareness of the direction of gravity resides in our heads. Note that the magnitude of the gravitational force during this self-test did not change; it is $w = mg$, in which m is the mass of the object, which our bodies use to detect this force. Thus, we detect a change in the direction of a force of constant magnitude.

The sensor to detect that direction is located in the maculae (plural of *macula*) of the *vestibular organ* in the inner ear. The overview of the inner ear in Fig. 2.26 identifies their location. How we use the utricular macula to determine the vertical direction during a sideways tilt of the head is illustrated in Fig. 3.33. The top sketch shows the macula in the upright position. It is built on supporting cells that house neurons, which are the main body of nerve cells. From these neurons emerge dendrites, the fine ends of nerves. The dendrites reach into a gelatin-like membrane above the supporting cells. This membrane is called the *otolithic membrane*, because it supports small calcite crystals that are called *otoliths*. The membrane has a density of about 1.0 g/cm^3 (close to the density of water) and supports $CaCO_3$ otoliths with a density of 3.0 g/cm^3, which is comparable to the density of rocks. If the head is turned, as shown in the bottom sketch of Fig. 3.33, the heavier otoliths pull the soft membrane in the direction of gravity, exerting a force on the dendrites. This force causes the dendrites to fire nerve impulses that travel to the brain (nerve impulse transport is discussed in Chapter 19).

3.11.2: Detection of the Weight of an Object

When you close your eyes and lay an object on your hand with the palm up, you notice at what position the object pushes on your skin. Note also that you sense the object continuously, even though it does not move. When you hold a different object you can tell which of the two objects is heavier. If you instead push the object against your hand from below, you also sense where the object touches you. This time, however, you do not measure the weight of the object but how hard it is pushed against your skin. Therefore, unlike our sense of gravity, the sensitivity for contact forces exists locally in the hand.

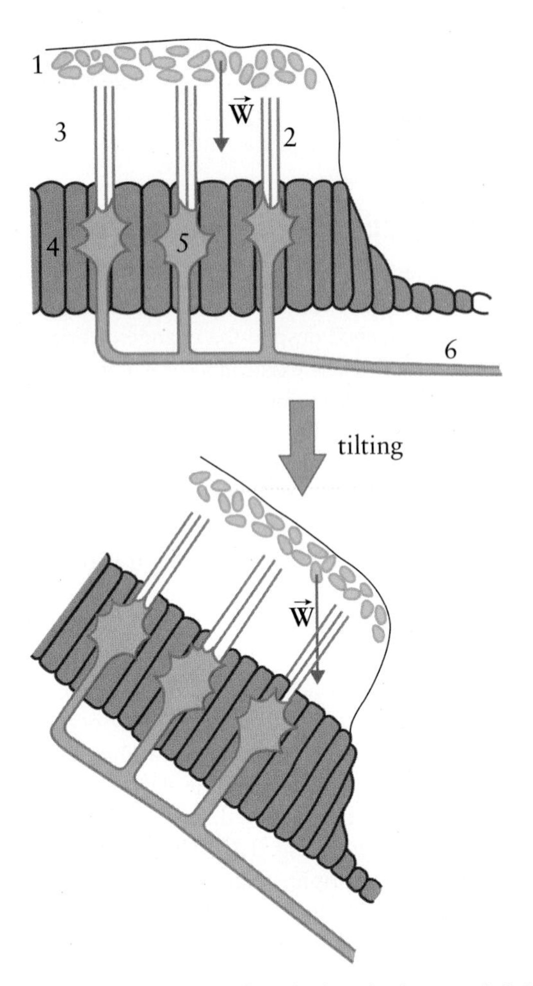

Figure 3.33 The mechanism of gravity detection in a macula is illustrated for a person tilting the head to one side. Above, the macula is shown in an upright position; below, the head is tilted. The components highlighted in the upper plot are the otoliths (1), the dendrites (2), which are embedded in the otolithic membrane (3), the supporting cells (4) containing the neurons (5), and the nerve to the brain (6).

Fig. 3.34 shows an overview of the near-surface structure of the palm, including the corpuscles typically contained in the skin. The outermost layer of the skin is called the *epidermis*. It varies in thickness between 30 μm and 4 mm. The next layer consists of 0.3- to 4-mm-thick connective tissue called *corium;* the fat cells below form the *subcutis*. Large numbers of various corpuscles are located in the shallower sections of the skin just below the epidermis, with other corpuscle types found deeper. We will discuss all these corpuscles in this textbook because they measure various physical parameters. The deep corpuscles (*Pacinian corpuscles*) are discussed in this chapter as acceleration detectors, and the beehive-shaped corpuscles (*Meissner's corpuscles*) are discussed in Chapter 2 as velocity detectors. Here we focus on the disc-shaped corpuscles that are sensitive to contact forces. These are called *Merkel's corpuscles* and are located just below the epidermis.

Merkel's corpuscles appear in high density in the palms of our hands and in the soles of our feet. The function of

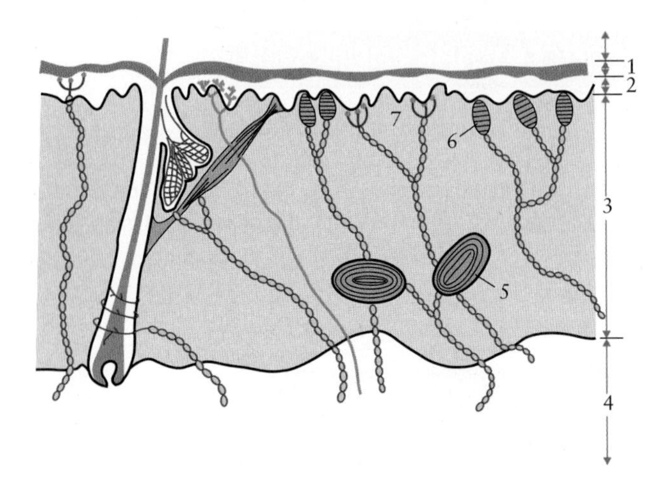

Figure 3.34 Cross-section through a section of skin with hair (left) and without hair (right). The skin is divided into four distinguishable layers: the epidermis (1 and 2) at the outer surface, and the corium (3) and the subcutis (4) forming its inner boundary. The skin contains a large number of receptor systems that measure a wide range of external parameters. Highlighted in the figure are four types of mechano-receptors, three corpuscle systems, and the hair for detecting mechanical stimuli. The Pacinian corpuscles (5), which measure the acceleration of the body, are located in the subcutis; Meissner's corpuscles (6), which measure the speed of incoming objects, and Merkel's corpuscles (7), which measure the weight of objects resting on the skin, lie in a shallow region below the epidermis.

Merkel's corpuscles can be illustrated most easily when we repeat our self-test placing objects of varying mass on the open hand with the palm up.

CASE STUDY 3.1

Magnetotactic bacteria live in the mud at the bottom of the sea where there is a low oxygen concentration just right for their needs. If somehow they were misplaced and brought up toward the surface, how would they find their way back to the right place at the bottom of the sea?

Answer: *Magnetotactic bacteria are found in a layer of water where it goes from having oxygen to not having oxygen. They are not found everywhere at this transition zone; they stay in a region where the water has exactly the right composition. They inhabit swampy water where the oxygen content in the water drops off sharply with increasing depth. Magnetotactic bacteria contain a row of magnetic crystals along with their long axis, giving them internal magnetic compass needles, as shown in Fig. 3.35. The effect of Earth's magnet on them is the same as that on compass needles, and they can use Earth's magnetic field to align themselves. If the bacteria stray too far above the preferred zone of habitat, they can swim back down the lines of Earth's magnetic field. Magnetotactic bacteria use their magnetic compass to tell them which way is down in a sea. Thus the bacteria find their way, going down, by aligning their internal magnetic compass with Earth's magnetic field, and swim along to the preferred*

continued

oxygen concentration. In the northern hemisphere they swim downward in the direction of Earth's magnetic field. Similarly, in the southern hemisphere they swim downward but opposite to Earth's magnetic field. The bacteria preferentially go away from the oxygen since oxygen is toxic for them. So they go toward the muddy, nutrient rich, right oxygen-content depths, and Earth's magnetic field direction points them in that direction.

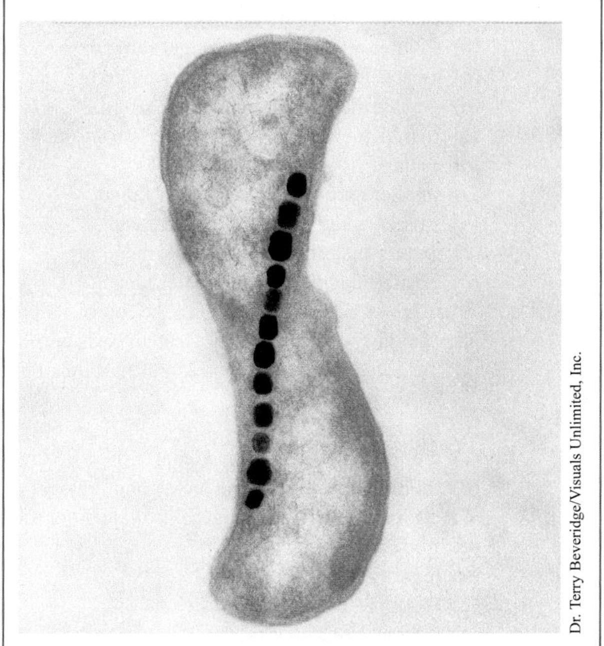

Figure 3.35 Magnetotactic bacteria have magnetic crystals inside, making them internally similar to magnetic compass needles.

Magnetotactic bacteria may not exist only on Earth. Examination of a 4.5-billion-year old Martian meteorite in 2000 disclosed the existence of magnetite crystals, an iron-rich compound. These Martian magnetite crystals are indistinguishable from those of the magnetotactic bacteria. Therefore, the existence of magnetite, as a biomarker, can be considered an indication of life on Mars, at least in the form of magnetotactic bacteria.

Since the performance of these bacterial magnets are much better than the performance of man-made magnets of the same size, magnetotactic bacteria are of significant interest in sciences and industries. The tiny magnets these bacteria make are far superior to those produced artificially. So, developing ways to use this magnetite crystal are under investigation.

SUMMARY

DEFINITIONS

- Forces represent the interaction of distinguishable objects.
- Fundamental forces:
  - Gravity: attractive force between two objects; its magnitude is proportional to the product of their masses and inversely proportional to the square of the distance between them
  - Electromagnetic
    - Electric: attractive or repulsive electric force between two charges; its magnitude is proportional to the product of the magnitude of their charges and inversely proportional to the square of the distance between them
    - Magnetic: attractive or repulsive force between two magnets or between two moving charges
  - Strong nuclear: attractive short-range force holding all protons and neutrons bonded in the atomic nucleus
  - Weak nuclear: short-range force involved in disintegration of certain radioactive nuclei, such as beta decay
- Convenience forces:
  - The weight $\vec{w}$ with magnitude $w = mg$ acts near Earth's surface with g the gravitational acceleration. The force is directed toward Earth's centre.
  - The normal force $\vec{N}$ is due to contact with a surface. It is directed perpendicular to the surface.
  - The force of friction $\vec{f}$ is due to contact with a surface. It is directed parallel to the surface. There are two forces of friction: static friction f_s and kinetic f_k friction.
  - The tension $\vec{T}$ is due to a taut, massless string. It is directed along the string.
  - Drag $\vec{R}$ is due to contact of a solid object with a fluid such as air. It is a resistive force, resisting the motion of a solid through air.
  - The viscose force $\vec{F}_{\text{vis}}$ is due to contact of a solid object with a liquid, such as water. It is resistive force acting on the object moving through a liquid.

UNITS

- Force: N (newton), $1\text{N} = 1 \text{ kg m/s}^2$
- Charge q: C (coulomb)
- Pressure p: Pa (pascal), $1 \text{ Pa} = 1 \text{ N/m}^2$

LAWS

- Newton's law of universal gravitation:

$$F = G\frac{m_1 m_2}{r^2}$$

- Coulomb's law:

$$F = k\frac{q_1 q_2}{r^2}$$

- For an object in mechanical equilibrium:

$$F_{\text{net}} = \sum_{i=1}^{n} F_i = 0$$

- If the forces are confined to the xy-plane:

$$F_{\text{net,x}} = \sum_i F_{ix} = 0$$

$$F_{\text{net,y}} = \sum_i F_{iy} = 0$$

MULTIPLE-CHOICE QUESTIONS

MC–3.1. Consider a planet that has half the mass and twice the radius of Earth. How does the acceleration of gravity on the surface of this planet compare to that on Earth?
(a) half that on Earth
(b) four times that on Earth
(c) twice that on Earth
(d) one-forth that on Earth
(e) one-eighth that on Earth

MC–3.2. Consider two planets with radius r_1 and r_2 are made from the same material, so have the same density. What is the ratio of the acceleration of gravity (g_1/g_2) at the surface of these two planets?
(a) r_2/r_1
(b) r_1/r_2
(c) $(r_2/r_1)^2$
(d) $(r_1/r_2)^2$
(e) $(r_2/r_1)^5$

MC–3.3. A negative electric charge
(a) interacts only with negative changes.
(b) interacts only with positive changes.
(c) interacts with both negative and positive changes.
(d) may interact with either negative or positive changes.

MC–3.4. Two charges of $+q$ are d cm apart. We replace one with a charge of $-q$. The magnitude of the force between them with respect to the initial case becomes
(a) zero.
(b) larger.
(c) smaller.
(d) stays the same.

MC–3.5. Two charges, 10 cm apart from each other, repel each other with a force of 3×10^{-7} N. If they were brought to 2 cm apart, the force between them becomes
(a) 1.2×10^{-7} N.
(b) 1.5×10^{-6} N.
(c) 7.2×10^{-6} N.
(d) 7.5×10^{-6} N.
(e) 2.3×10^{-5} N.

MC–3.6. Two charges, one negative and the other positive, are initially d cm apart. They are pulled away from each other and placed apart at a distance three times their initial distance. The force between them now is smaller by a factor of
(a) $\sqrt{3}$.
(b) 3.
(c) 9.
(d) 27.

MC–3.7. Two charges attract each other with a force $\vec{F}$. If the magnitude of one of them is doubled and the distance between them is doubled too, the magnitude of the force between them becomes
(a) $2F$.
(b) F.
(c) $F/2$.
(d) $F/4$.

MC–3.8. The strong nuclear force
(a) binds the orbital electrons to the atomic nucleus.
(b) is a long-range force.
(c) binds only identical particles.
(d) overcomes the repulsive force between protons in the nucleus.
(e) acts only on charged particles such as protons.

MC–3.9. Compared to electric force, strong nuclear forces have
(a) smaller magnitude and shorter range.
(b) smaller magnitude and longer range.
(c) greater magnitude and shorter range.
(d) greater magnitude and greater range.

MC–3.10. Compared to weak nuclear force, strong nuclear forces have
(a) smaller magnitude and shorter range.
(b) smaller magnitude and longer range.
(c) greater magnitude and shorter range.
(d) greater magnitude and greater range.

MC–3.11. What forces act on a football at the top of its path after having been kicked? Neglect air resistance.
(a) the force due to the horizontal motion of the football
(b) the force of gravity
(c) the force exerted by the kicker
(d) the force exerted by the kicker and gravity

MC–3.12. What forces act on a ball that rolls on the floor after being kicked?
(a) the force from the floor
(b) the force exerted by the kicker
(c) the force of gravity
(d) the force exerted by the kicker and gravity
(e) the force exerted by the floor and gravity
(f) the force exerted by the kicker, gravity, and the floor

MC–3.13. Which of the three forces shown to act on a man's arm in Fig. 3.25 are contact forces?
(a) all three forces
(b) $\vec{T}$ and $\vec{W}_{arm}$
(c) $\vec{T}$ and $\vec{F}$
(d) $\vec{W}_{arm}$ and $\vec{F}$
(e) only the force $\vec{F}$

MC–3.14. A block of mass m rests on an inclined plane that makes an angle of θ with the horizontal, as shown in Fig. 3.36. What is the static friction between the block and the inclined surface?
(a) $f_s \geq m\,g$
(b) $f_s \geq m\,g\cos\theta$
(c) $f_s = m\,g\sin\theta$
(d) $f_s = m\,g\cos\theta$
(e) zero because the plane is inclined

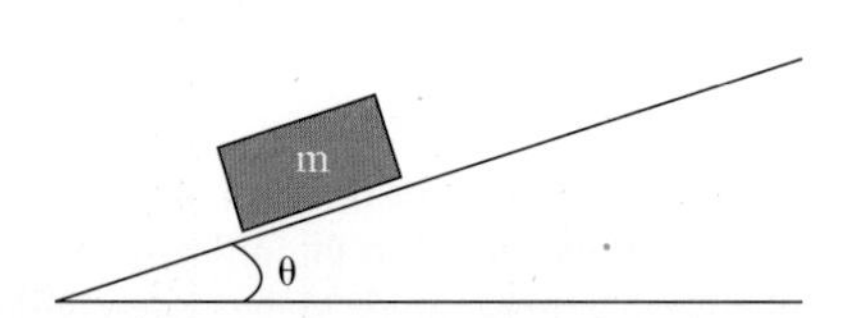

Figure 3.36

MC–3.15. A spring is stretched by a certain amount and then released to return to its original length. Then it is stretched twice that initial amount (without exceeding its elastic limit). Compared to the first stretching, how much force does the second one need?
(a) twice much force
(b) four times much force
(c) half as much force
(d) the same amount of force
(e) quarter as much force

MC–3.16. Suppose you drop two steel marbles into glycerine, a highly viscous liquid. The diameter of one marble is twice the diameter of the second one, and both marbles move with constant velocity. If the velocity of the larger marble is four times the velocity of the smaller one, how does the viscose force of the larger marble compare to the smaller one?
(a) twice the viscose force of the smaller one
(b) four times the viscose force of the smaller one
(c) half the viscose force of the smaller one
(d) the same as the viscose force of the smaller one
(e) quarter the viscose force of the smaller one

MC–3.17. The dumbbell in Fig. 3.25 has a weight, w_{dumb}. Why do we not include this force when discussing the equilibrium of the lower arm of the man in the figure?
(a) The force w_{dumb} acts in another direction than the listed forces and therefore has to be omitted.
(b) The force w_{dumb} has no interaction pair in the figure, and for that reason cannot be considered in the equilibrium equation.
(c) The force w_{dumb} does not act on the man's arm and has to be excluded when the equilibrium equation is applied to the arm.
(d) We have already included the weight of the arm, w_{arm}, and no way exists to include two different weights in Newton's laws.
(e) The dumbbell is not alive and therefore cannot exert a force on another object.

MC–3.18. A block of mass M is sliding along a frictionless inclined plane, as shown in Fig. 3.36. What is the normal force exerted on the block by the plane?
(a) $g \sin\theta$
(b) $Mg\sin\theta$
(c) $Mg\cos\theta$
(d) zero because the plane is frictionless

CONCEPTUAL QUESTIONS

Q–3.1. Fig. 3.37 shows five experimental arrangements. In part (A), the object is vertically attached to a string, in parts (B) and (C) the object is in a bowl-shaped structure, in part (D) it lies on a horizontal table, and in part (E) the object is held by a string on an inclined surface. (a) Identify all forces in each case separately. (b) Draw free body diagram for each one. (c) In which case is the object not in equilibrium?

Q–3.2. Is there a similarity between normal force and spring force? Are there any differences between the two forces?

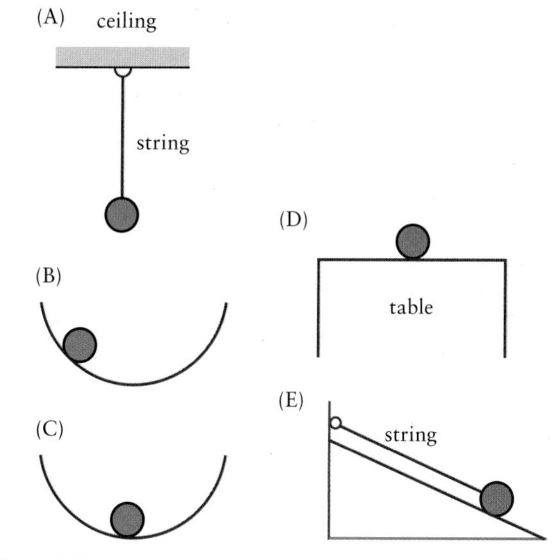

Figure 3.37

Q–3.3. Fig. 3.38 shows a free body diagram with three forces, a tension $\vec{T}$, a normal force $\vec{N}$, and a weight $\vec{W}$. For which of the five cases shown in Fig. 3.39 is this free body diagram correct?

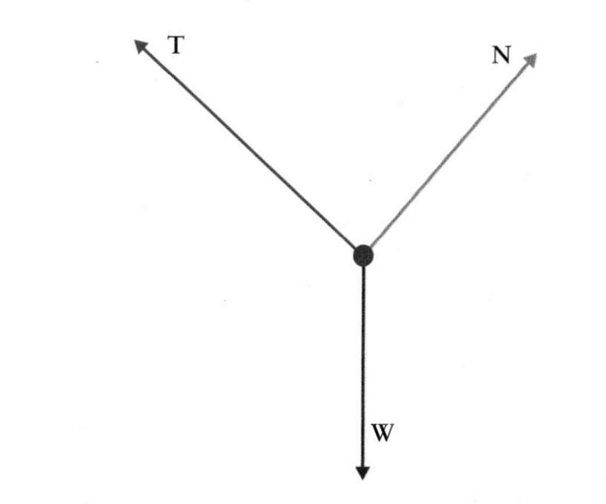

Figure 3.38

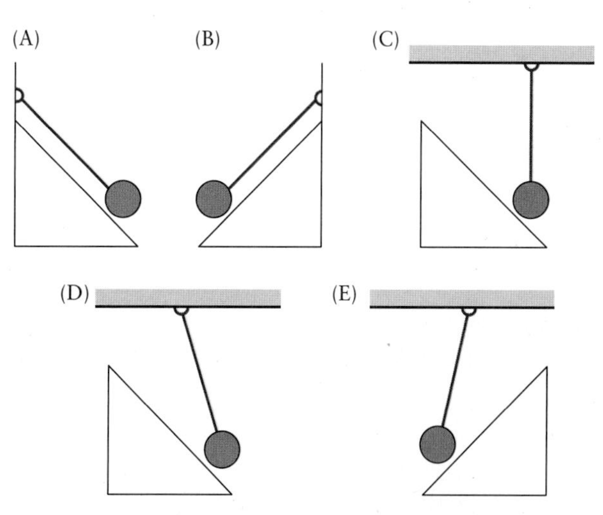

Figure 3.39

Q–3.4. The more forces act on a body part of interest, the more anatomical, physiological, and physical data are required for quantitative evaluation of a mechanical system. To deal with this, we often want to limit the number of forces that have to be taken into account. Under what circumstances can the weight of the body part be neglected? Can you suggest an example?

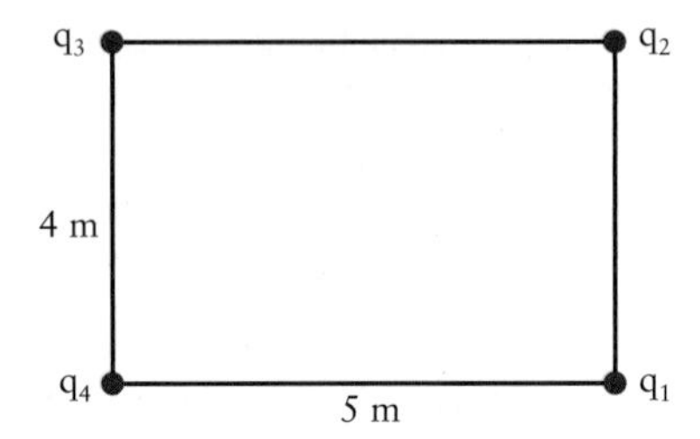

Figure 3.40

ANALYTICAL PROBLEMS

P–3.1. Find the force of gravity between two uniform spheres as they touch each other. Each sphere has a mass of $m = 15$ kg and a radius of $r = 0.5$ m. What is the force of gravity between them when they stand (surface to surface) 2 m from each other?

P–3.2. What is the acceleration of gravity 9800 km above the surface of Earth? Take the radius of Earth as $R_E = 6.37 \times 10^6$ m.

P–3.3. If the force of gravity between two identical spheres separated centre-to-centre by 2.00 m is 0.50 N, what is the mass of each sphere?

P–3.4. (a) What is the magnitude of the force of gravity between Earth and the Moon (take mass of Earth $M_E = 6.00 \times 10^{24}$ kg, mass of the Moon $M_M = 7.40 \times 10^{22}$ kg, and the distance between their centres $R_{EM} = 3.84 \times 10^8$ m)? (b) At what point between Earth and the Moon is the net force of gravity on a body by both Earth and the Moon exactly zero?

P–3.5. Two charged particles of $q_1 = +1.00\ \mu C$ and $q_2 = -1.00\ \mu C$ are separated by $l = 1.00$ m. Find the electric force each charge experiences.

P–3.6. How far apart are a proton and an electron if they experience an electric force of 1.00 N?

P–3.7. Compare the magnitude of the electric and gravitational force between (a) two protons; (b) two electrons; and (c) an electron and a proton. Give the results as the ratio of electric force to gravitational force (F_E/F_G). The charge and the mass for an electron are $q_e = -1.60 \times 10^{-19}$ C, $m_e = 9.11 \times 10^{-31}$ kg; and for a proton $q_P = 1.60 \times 10^{-19}$ C, $m_P = 1.67 \times 10^{-27}$ kg.

P–3.8. Three charges, $q_1 = +2.00$ nC, $q_2 = +2.00$ nC, and $q_3 = +4.00$ nC, are fixed at the corners of an equilateral triangle with a side of 4.00 cm. Find the magnitude and the direction of a net electric force on q_3.

P–3.9. If you change the sign of q_3 to negative in problem 3.8 ($q_3 = -4.00$ nC), what would be the net electric force (direction and magnitude) on q_3?

P–3.10. The singly charged sodium and chloride ions in crystals of table salt are separated by 2.82×10^{-10} m. Find the attractive electric force between these ions.

P–3.11. Four charges, $q_1 = +100\ \mu$C, $q_2 = +45\ \mu$C, $q_3 = -125\ \mu$C, and $q_4 = +25\ \mu$C, are fixed at the corners of a 4 m by 5 m rectangle, as illustrated in Fig. 3.40. What are the magnitude and the direction of the net force acting on q_1.

P–3.12. A 5.8-kg box is resting on an inclined surface 35° above the horizontal. Find the normal force exerted by the box on the inclined surface.

P–3.13. A standard man of mass 70 kg stands on a bathroom scale. What are (a) the standard man's weight and (b) the normal force acting on the standard man? (c) What does the standard man read off the scale if it is calibrated in unit N? Can you suggest reasons why that reading may deviate from the actual value? In particular, can you identify a reason why a balance scale might be more precise for a weight measurement?

P–3.14. A climber stands on the rocky face of a mountain that has a slope of 36° with respect to horizontal. If her mass is 64 kg and her boot has a static friction coefficient equal to 0.86, (a) find normal force on her and (b) the force of static friction between her boot and the rock. What is the maximum force of friction between her boot and the rock?

P–3.15. A 480-kg sea lion is resting on an inclined wooden surface 40° above horizontal, as illustrated in Fig. 3.41. The coefficient of static friction between the sea lion and the wooden surface is 0.96. Find (a) the normal force on the sea lion by the surface; (b) the magnitude of force of friction; and (c) the maximum force of friction between the sea lion and the wooden surface.

Jared Peterson/iStockphoto.com

Figure 3.41

P–3.16. A chandelier of mass 11 kg is hanging by a chain from the ceiling, as shown in Fig. 3.42. What is the tension force in the chain?

StudioSmart/Shutterstock

Figure 3.42

P–3.17. An 85-kg climber is secured by a rope hanging from a rock, as shown in Fig. 3.43. Find the tension in the rope.

Greg Epperson/Shutterstock

Figure 3.43

P–3.18. A 76-kg climber is using a rope to cross between two peaks of a mountain, as shown in Fig. 3.44. She pauses to rest near the right peak. Assume that the right-side rope and the left-side rope make angles of 18.5° and 11° with respect to horizontal respectively. Find the tension in the right- and left- side ropes.

Source: Archive Photos/Vintage Images/Getty Images

Figure 3.44

P–3.19. Two blocks, A and B, are pushed down a rough inclined ramp that makes an angle θ with horizontal. Draw a free body diagram for each box for each of the following circumstances: (a) block A is pushed down and block B is in front of it; (b) block A is pushed down and block B is sitting on it; (c) block B is pushed down and block A is in front of it. Assume the two boxes do not lose contact at any time.

P–3.20. Fig. 3.45 shows three blocks on a rough inclined ramp. Blocks A and B are connected by magnets on their faces and block C is sitting on block A. A force is applied on block A and pulls it up. Draw a free body diagram for each block.

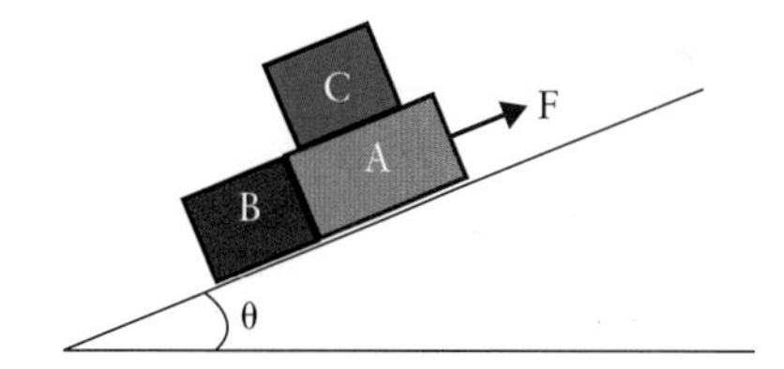

Figure 3.45

P–3.21. A box is lifted by a magnet suspended from the ceiling by a rope attached to the magnet, as illustrated in Fig. 3.46. Draw a free body diagram for the box and for the magnet.

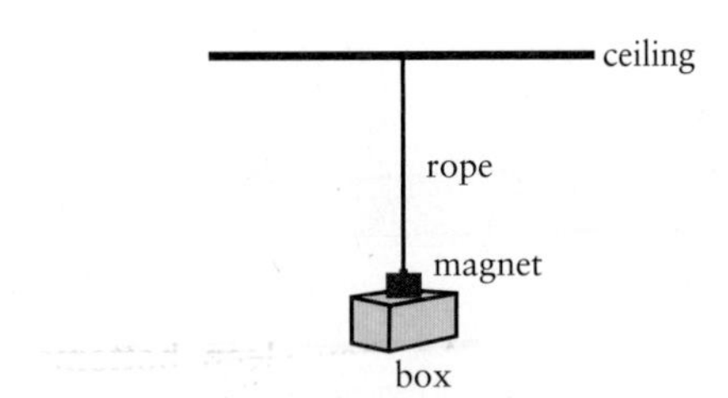

Figure 3.46

P–3.22. Fig. 3.47 shows a rock climber climbing up Devil's Tower in Wyoming. (a) Identify all forces acting on the climber. (b) Draw a free body diagram for him. (c) How does he support his weight?

Figure 3.47

P–3.23. Fig. 3.48 shows a standard man with mass of 70 kg intending to do *reverse curls* in a gym. The person holds the arms straight, using an overhand grip to hold the bar. If the mass of each arm is 4.6 kg and the mass of the bar is 100 kg, what is the tension in each of the shoulders? Consider the weight of the arm (see Table 4.1) and forces due to the weight of the bar.

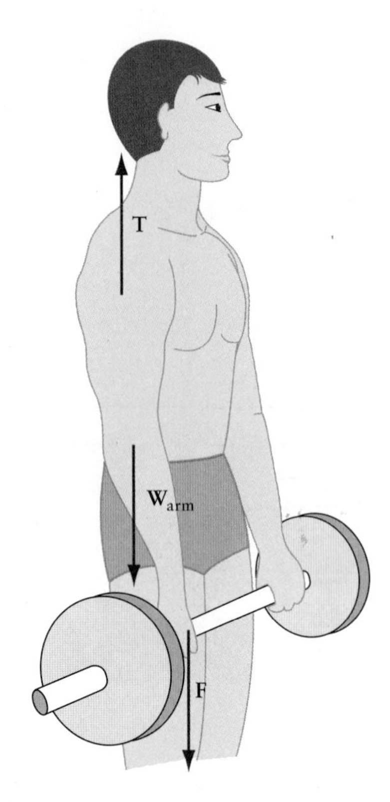

Figure 3.48

P–3.24. Fig. 3.49 shows a standard man intending to do *closed-grip lat pulldowns* in a gym. In this exercise, the person pulls the weight of the upper body (arms, head, and trunk) upward using a handle while the legs are wedged under a restraint pad. What is the magnitude of the force exerted by the handle on each of the standard man's hands?

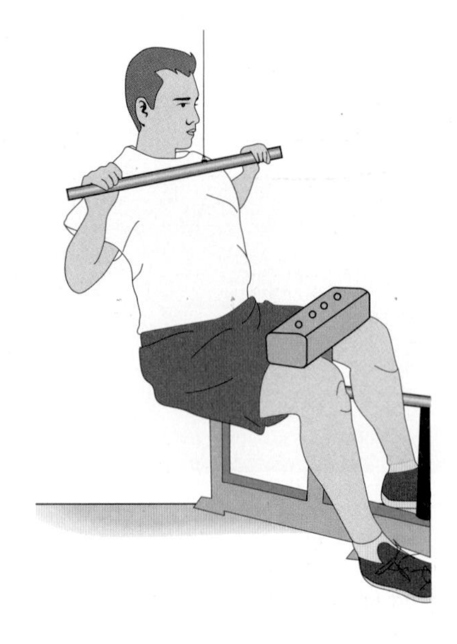

Figure 3.49

ANSWERS TO CONCEPT QUESTIONS

Concept Question 3.1: (a): In each direction, from you to the wall or from the wall to you, there are four interactions. (b): These interactions are between the ball and (1) your hand, (2) ground, (3) air, and (4) the wall.

Concept Question 3.2: (a): In each direction, from you up, or down to you, there are three interactions. (b): These interactions are between the ball and (1) your hand, (2) ground, and (3) air.

Concept Question 3.3: (a): Yes, force can change the magnitude of velocity. (b): Yes, force can change the direction of velocity. (c): Yes, force can change both the magnitude and direction of velocity.

Concept Question 3.4: (a): An athlete whirling a hammer, a ball on the end of string whirling, a radio satellite around Earth, an airplane in a loop. (b): Shooting a ball changes the magnitude of its velocity from zero to some magnitude; catching a ball; stopping a moving car.

Concept Question 3.5: While you exert a force on the ball (kicking it), the ball exerts a force back on you. Since the force makes the velocity of the object on which it is exerted change, the velocity of the foot is changed. So your foot slows down. You may not notice it at the point of impact, because you are holding and controlling your foot by exerting a force on it. In a case of kicking a heavy ball, your foot may bounce back a little. If you consider the foot acting as a pendulum, it may slow down or may bounce back depending how light or heavy is the ball.

Concept Question 3.6: This applied force will change the shape too. Since you apply force on the desk, the desk also will apply force back on you. Thus both your hand and the desk will change shape while the force is exerted on them. The change in your hand obviously is more noticeable than the desk.

Concept Question 3.7: (a): The gravitational force of the Moon, combined with Sun's, constantly pulls Earth's oceans in different directions, creating a tidal effect: high tide or low tide. (b): Yes, it has an effect almost similar to the tidal effect of the Moon, producing a tide in combination with the Moon's gravitational force.

Concept Question 3.8: The ratio in Example 3.4 is calculated as $F/w = 5.8 \times 10^{-15}$. The small ratio indicates that the presence of one bacterium does not affect the other one. The state of each bacterium does not change due to a gravitational effect of the other one.

Concept Question 3.9: Taking the mass of each person as 75 kg and assuming they are 1 m from each other, using Eqs. [3.1] and [3.4] gives

$$\frac{F}{w} = \frac{G\,mm/r^2}{mg} = \frac{Gm}{gr^2} \quad \frac{6.67 \times 10^{-11}}{9.8 \times (1)^2} = 6.8 \times 10^{-12}.$$

As seen in Concept Question 3.8, the ratio is a tiny value. This shows that the presence of one person next to another person does not affect either of them. The gravity between two small masses is small, thus gravity is a weak force but it has a remarkable effect for large masses such as planets.

Concept Question 3.10: As they sleep bottoms-up, infants keep their heads lower than their hearts. Due to the gravity pull supplying more blood, more oxygen goes to the brain, which is necessary for infants' brain growth.

Concept Question 3.11: Helping to supply more blood, and consequently more oxygen to the brain, is the influence of gravity pull.

Concept Question 3.12: Sitting on a chair for a long time may cause swelling in the feet due to accrual of excess fluid in the legs and feet. In this process, the influence of gravity pull has a remarkable effect. By propping the legs up, the effect of the influence of gravity is reversed, helping to regulate the flow of liquid in the lower limbs (body) and stopping the accumulation of liquid in the feet.

Concept Question 3.13: (a): The forces holding the molecules in the muscle sarcomeres are electromagnetic, acting as spring forces (tiny spring model between molecules), so the fundamental interaction in charge of muscle force is electromagnetic. (b): The interaction is electromagnetic, similar to the muscle force discussed in part (a).

Concept Question 3.14: The force holding a water strider upon the water surface is fundamentally electromagnetic in nature. A molecule of water is a strongly polar molecule, causing a strong attractive bond with adjacent water molecules. Thus, the molecules of water pull each other, causing the surface of water to act as an elastic film. This surface is strong and can be stretched, so can hold a water strider as shown in Fig. 3.16.

Concept Question 3.15: Earth is a huge magnet that affects a compass needle. So Earth's magnet exerts force on the compass needle, orienting it toward Earth's North Pole.

Concept Question 3.16: Using Eq. [3.6] we can calculate the repulsive electric force between two protons in the nucleus:

$$F = k\frac{q^2}{r^2} = \frac{(9 \times 10^9\,\mathrm{N\,m^2/C^2})(1.6 \times 10^{-19}\,\mathrm{C})^{92}}{(2.4 \times 10^{-15})^2} = 40\ \mathrm{N}.$$

Overcoming this huge repulsive electric force requires an enormous attractive nuclear force.

Concept Question 3.17: More weight in a book requires a higher normal force to hold it. The normal force will be as large as necessary to hold the heavy books until the table breaks.

Concept Question 3.18: A person leaning on a vertical wall, or pushing a book against a wall

Concept Question 3.19: Taking a step requires force since your body, and so your foot, gains velocity. You push on the ground, exerting a force on it; simultaneously, the ground exerts a force on you. So the ground pushes you forward. The force of the ground on you is a surface force parallel to the surface; it is the force of friction. So force of friction exerted on your foot by the ground pushes you forward.

Concept Question 3.20: There are roughly no microscopic points of hills and valley on the ice surface, so there is nothing to prevent the foot from sliding on the ice.

Concept Question 3.21: (C)

Concept Question 3.22: As discussed in Concept Question 3.17, more weight of the box requires a larger normal force from the floor on it. A larger normal force produces a larger force of friction.

Concept Question 3.23: (a): The forces are weight, normal force, and friction. (b): Force of friction (static) keeps the car on the ramp.

Concept Question 3.24: (C)

Concept Question 3.25: (D)

Concept Question 3.26: One reason why birds fly in a "V" shape is that this shape reduces the drag force on each bird compared to when it flies alone. The decrease in drag is due to wingtip vortices. Wingtip vortices are tubes of circulating air left behind a wing as it generates lift.

Concept Question 3.27: (C)

Concept Question 3.28: (a): Yes. The magnitude of the weight varies with the mass of the object. Its direction is fixed (always straight down toward the Earth's centre), but it has that direction nevertheless. A quantity with magnitude and direction is a vector, regardless of whether magnitude and direction can vary. (b): Not always. You can relax your muscle at its resting length. In that state, the tendon does not exert a force on the bone even though they are always connected with each other. (c): Yes. Recall that a force represents an interaction between two objects. The radius interacts with and affects the attached tendon. Imagine for a moment that the radius/biceps interaction did not exist while you stretch your arm. In this case, the biceps would not be stretched and later could no longer serve its purpose.

Concept Question 3.29: No. The gravitational acceleration g is constant for experiments done in a laboratory on Earth; thus, we cannot vary m while holding w constant in Eq. [3.5]. While this argument may appear trivial for the simple formula in Eq. [3.5], the distinction between the meaning of mathematical and physical formulas is fundamental to mastering physics. We will therefore reiterate this issue throughout the textbook.

Concept Question 3.30: (a): $\vec{F}_6$ is the normal force acting on the object in the direction normal to the underlying surface. Note that force $\vec{F}_3$ also acts perpendicular to the underlying surface, but cannot be exerted by that surface on account of its direction. For objects on a levelled surface, $\vec{F}_5$ represents the direction in which the normal force acts. However, on an inclined surface, the direction perpendicular to the surface is no longer vertical. (b): The weight is a force on an object that is always directed vertically downward, regardless of the orientation of surfaces in contact with the object. Thus, $\vec{F}_2$ is the weight. Force $\vec{F}_5$ also acts in the vertical direction; however, it cannot be caused by Earth because the gravitational force always attracts objects toward Earth's centre.

CHAPTER 4

Newton's Laws

In Chapter 2 we studied kinematics and learned how to describe motion mathematically. In describing motion, we did not get into the fundamental question of what causes motion. In Chapter 3 the concept of force was introduced as the cause of change of velocity. This chapter is centred on how a force affects the motion of an object on which it is exerted, and how it relates to the object's mass. Thus, a physical model that accurately formulates the laws governing mechanics is introduced.

Why does a car stop more easily on dry concrete than on ice? Why does your foot hurt less when you kick a ball than when you kick a brick wall? Why is it more difficult to move a heavy (massive) desk than a light chair across a floor? Why do some objects move faster than others, or some gain acceleration and others do not? How can an object moving in one direction be accelerated in another direction or even in the opposing direction? The answers to these and similar questions will help us to understand the physical phenomena in the world around us.

The key to answering these questions is the relation between force and mass. This is the subject of dynamics. In kinematics, the relation between quantities such as displacement, velocity, and acceleration are mathematically described. In dynamics, it is the relation between mass, force, and acceleration that are expressed mathematically.

In the seventeenth century, based on the work of his predecessors Galileo Galilei, Copernicus, and Kepler, Sir Isaac Newton developed a physical model that dealt with force and mass from his experimental studies on the motion of objects. He identified three laws that govern all forces. He published them in his "*Mathematical Principles of Natural Philosophy*" (briefly, "Principle") in 1686. These three laws shaped the foundation for our understanding of the effect of a force acting on an object.

Three laws, jointly called **Newton's laws of motion**, are adequate to describe a tremendous range of natural phenomena. Examples include planets orbiting the Sun, sending a shuttle to space, accelerating a car, the motion of red blood cells through the body, the infusion of an electrolyte solution into an artery. These laws of motion are as follow: (1) An object is in translational equilibrium if the forces that act on the object are balanced, resulting in no change of its velocity. (2) If the forces are unbalanced, the acceleration of the object is proportional to the magnitude of the net force acting on it, and is in the direction in which the net force acts. (3) Any two interacting objects exert equal but opposite forces upon each other; such forces are called an *interaction pair* or, in everyday language, *an action–reaction pair*. These laws are the basis of modern mechanics, which gives a simple and accurate explanation of a wide range of physical problems.

The first law also is called the law of inertia. Aristotle (384–322 B.C.) believed that "rest" is the "natural state" of objects and that force is required to maintain motion. At first it seemed to agree with everyday experience, but Galileo and Newton later declared that "constant velocity" is the "natural state" and a net force is required to change the object's velocity. So, "rest" often refers to a constant velocity with a particular value. Objects have an intrinsic tendency to keep their velocity constant. This is called **inertia**.

In the previous chapter, we discussed each of these three laws conceptually without naming them. These laws are presented in this chapter in more detail. Not only do we describe them quantitatively, we also show how they can be applied to everyday issues and problems.

4.1: Newton's Laws of Motion

Force is a means of quantitatively describing an interaction between two objects. The action of a force is change in the velocity of the object acted upon, as discussed in Chapter 3. So, force can be thought of as the **cause of acceleration**.

We start with a simple experiment that allows you to isolate the action of individual forces acting on an object on a horizontal surface. This is an experiment you can do yourself: You need a slippery surface, such as a frozen pond, and an object with a smooth surface, such as a hockey puck. Consider only horizontal forces acting on the object. There are other forces acting on the object, such as the force of gravity by Earth (its weight) and the normal force by the contact surface. The weight and the normal force are not of concern for this experiment because both act vertically. The only forces that act in the horizontal direction are the external force (such as pushing) and friction. By choosing an object with a smooth surface that sits on an icy surface, you can neglect friction.

Consider the following three experiments: (i) Place the puck on the ice surface. The puck, initially at rest, remains at rest if you do not push it. Similarly, if you leave your physics book on your desk, it will stay there as long as you leave it and do not touch it (Fig. 3.27 of Example 3.9). The situation here is the same. (ii) Next, set the puck in motion on the icy pond. It maintains the magnitude and direction of its speed (its velocity) once it is released and is no longer being touched. (iii) Last, push the puck, which is initially at rest, with two hands. You can do this with varying forces. You could even do it such that the puck remains at rest by pushing only very gently. Pushing the puck with two hands from opposite sides such that the puck remains at rest means that an object may remain at rest even though several forces act on it.

In all three experiments, there is nothing, no cause, to change the velocity. The last experiment deals with the concept of **net force** ($\vec{F}_{net}$), the vector summation of all forces acting on an object. For an object on which n forces ($\vec{F}_i$) act, the net force is given by:

$$\vec{F}_{net} = \sum_{i=1}^{n} \vec{F}_i . \qquad [4.1]$$

To simplify the calculations in this chapter, we limit our discussion to cases in which the forces acting on an object, and therefore its motion, occur in a two-dimensional plane. We define this as the xy-plane. In the xy-plane, the force vector can be decomposed into two orthogonal components: one parallel to the x-axis and the other parallel to the y-axis. In this case, two components of the net force are:

$$\begin{cases} F_{net,x} = \sum_i F_{i,x} & \text{in } x\text{-direction} \\ F_{net,y} = \sum_i F_{i,y} & \text{in } y\text{-direction.} \end{cases} \qquad [4.2]$$

The net force is a key concept in applying the laws of mechanics. **Free body diagrams**, first presented in Section 3.9, are a useful technique to identify and graphically combine all the forces that will contribute to the net force.

In each of these three experiments the net force on the puck is zero. So, *no force or zero net force acts on the puck while it is at rest or is moving with constant velocity.* To summarize the motion of the puck on the ice surface more quantitatively, we have:

$$\vec{F}_{net} = 0 \Rightarrow \begin{cases} \vec{v} = 0 \\ \vec{v} = \text{const} \end{cases} \Rightarrow \vec{a} = 0 . \qquad [4.3]$$

Consider two additional experiments: (iv) First, exert a constant single force on the hockey puck on the ice, e.g. using a hockey stick. In this case, the puck accelerates in the direction in which the force is exerted. So, the acting force causes acceleration in the same direction in which it is applied. (v) Second, push the puck with two sticks at the same time, but make sure that the two contact forces used differ in magnitude and direction. To make the experiment more general, do not use two forces that are directed exactly opposite to each other. In both cases, the object will accelerate in the same direction as that of the net applied force. A constant force applied on an object causes that object to move with a constant acceleration in the same direction as the applied force. This can be summarized quantitatively as follows:

$$\vec{F}_{net} \neq 0 \Rightarrow \vec{a} \neq 0 \Rightarrow \vec{v} \neq 0. \qquad [4.4]$$

Eq. [4.3] is identified as **Newton's first law**; Eq. [4.4] leads to Newton's second law.

4.2: Free Body Diagram, Revisited

A free body diagram is an essential graphical tool that is very helpful in finding the net force acting on an object and its subsequent effect.

KEY POINT

A free body diagram is a drawing that consists of all forces that act on the system of interest. No other physical features of the system are included, such as its velocity or acceleration. A coordinate system and labels for angles are often added.

A free body diagram illustrates an object as a point particle and demonstrates all forces acting on it. This is valid only if the non-rotational motion is concerned, which is the case in this chapter. To draw a free body diagram, you begin with a dot that represents the object of interest. Next, attach to the dot all the forces in your list of identified forces. The forces are drawn as arrows; be careful to draw them in the direction in which they act.

It is very important to note that a free body diagram must be drawn for each object in a system. For a system of more than one object, you must draw a free body diagram for each object separately. Therefore, a system of two objects has two free body diagrams, one for each object, and a system of three objects has three free body diagrams, and so on.

EXAMPLE 4.1

Draw the free body diagram for a heavy box at rest that is sitting on a rough inclined surface.

Solution

First, draw a coordinate system on an inclined surface. For the inclined surface, always draw the *x*-axis parallel to the inclined surface and the *y*-axis perpendicular to the inclined surface. Let $\vec{w}$ represents the weight of the box, $\vec{N}$ the normal force, and $\vec{f}$ the force of friction. Fig. 4.1 shows the free body diagram.

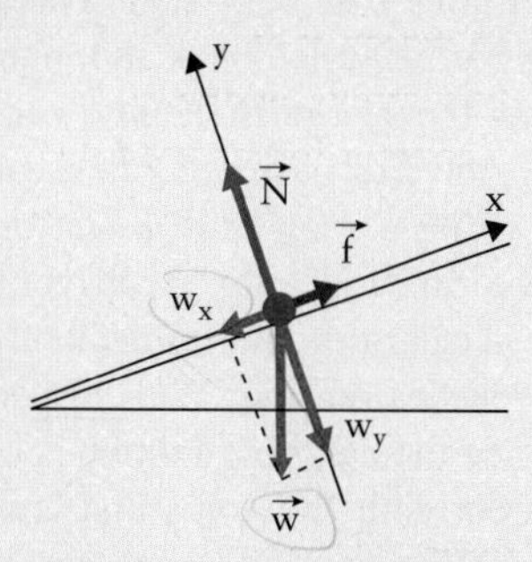

Figure 4.1 Free body diagram of a box on an inclined surface.

CONCEPT QUESTION 4.1

An iron man is lifted by a lift that has a magnetic hook, as shown in Fig. 4.2. Which one of the free body diagrams shown in Fig. 4.3 best represents forces on the iron man? Let $\vec{T}$ be the tension of the cable, $\vec{w}$ the weight of iron man, $\vec{M}$ the magnetic force, and $\vec{N}$ the normal force.

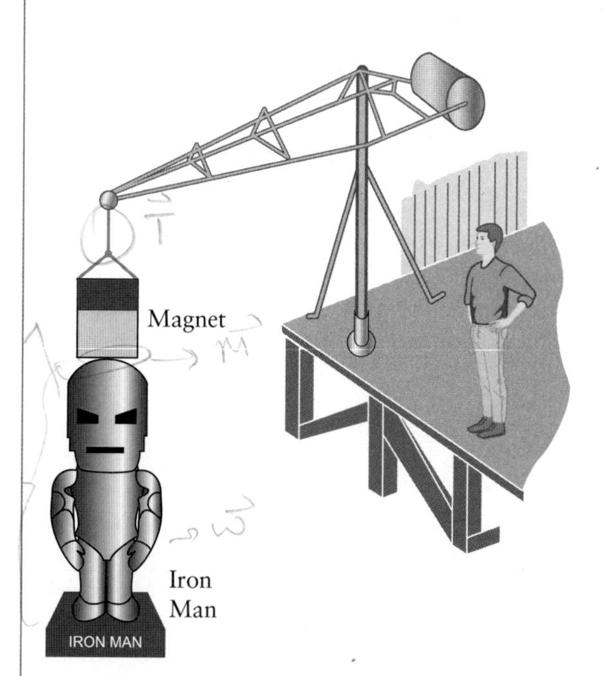

Figure 4.2 An iron man is pulled up by a lift.

continued

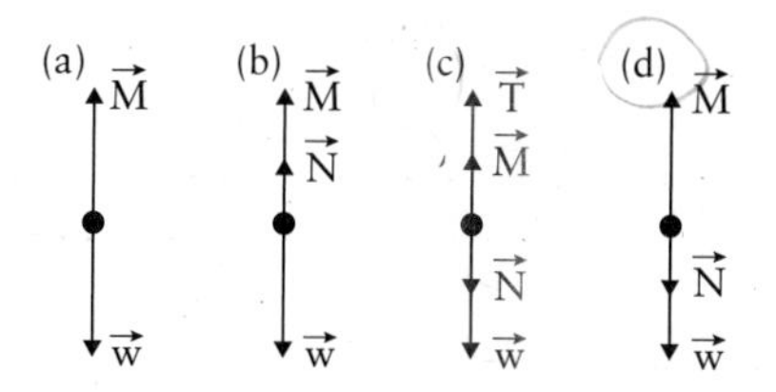

Figure 4.3 Proposed free body diagrams of the iron man.

EXAMPLE 4.2

Draw free body diagrams for a system of two blocks connected by a light cord, as shown in Fig. 4.4. Block 1 is moving on a rough surface of a table and block 2 is suspended at the side of the table.

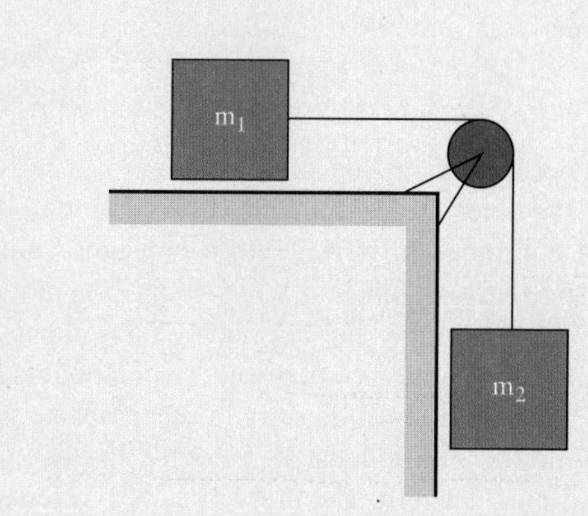

Figure 4.4 Two blocks, one on a rough surface and the other hung at the side, are connected by a light cord.

Solution

First, isolate each block and identify all forces acting on them. There are four forces acting on the first block: tension $\vec{T}$, weight $\vec{w}_1$, normal force $\vec{N}$ and force of friction $\vec{f}$. Two forces are acting on the second block: weight $\vec{w}_2$ and tension $\vec{T}$. Note that the cord is light and the tension on each block is the same. Figs. 4.5(a) and (b) show the free body diagrams for the two blocks.

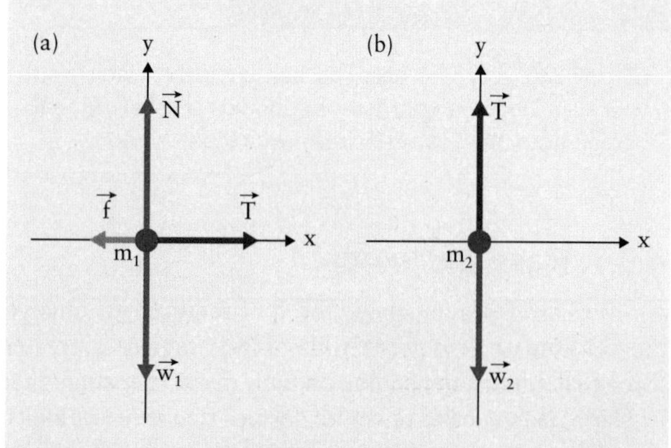

Figure 4.5 Free body diagram for the two blocks: (a) first block on the table, (b) second block hung at the table's side.

4.3: Newton's First Law

Newton's first law requires the net force acting on an object to be zero. This is written in component form for an object that is confined to the xy-plane as two simultaneous equations:

$$\vec{F}_{\text{net}} = \sum_{i=1}^{n} \vec{F}_i = 0 \quad \overset{\text{Components}}{\Rightarrow} \quad \begin{cases} F_{\text{net},x} = \sum_{i=1}^{n} F_{i,x} = 0 \\ F_{\text{net},y} = \sum_{i=1}^{n} F_{i,y} = 0. \end{cases} \qquad [4.5]$$

The first law can be stated in the following form:

KEY POINT

An object at rest or in motion with constant velocity remains in its state unless acted upon by a net force.

If the net force acting on an object is zero or no force acts on it, the object at rest remains at rest and the object in motion continues to move with a constant velocity (constant speed in a straight line, which is uniform motion). If in turn the object does not accelerate along the x-axis (y-axis), the sum of the x-components (y-components) of all forces acting on it must be zero. In other words, when an object moves with a constant velocity, the net force acting on it must be zero. The first law is also referred to as the law of inertia.

CONCEPT QUESTION 4.2

You are sitting next to a driver in a car travelling down a road; the driver suddenly hits the brakes. As a result, your body moves toward the front of the car. Why does this happen?

CONCEPT QUESTION 4.3

By shaking your wet hair, you can shed the water from your hair. Dogs can dry their wet bodies by shaking their bodies from head to tail. Explain why these work.

4.3.1: Mass and Inertia

As discussed previously, a force is required to change the velocity of an object. Some objects require a greater force than others, depending on their mass. It appears that massive objects have fewer tendencies to change velocity than lighter ones or, in other words, they have a greater tendency to remain at rest. The tendency of opposing a change in the velocity of an object is called **inertia**. In other words, inertia is the tendency of an object to remain in its state of motion. The inertia of an object is quantified by its mass; therefore, it is measured quantitatively by mass.

4.3.2: Inertial Reference Frame

Newton's first law appears not to be valid all the time. Let's re-examine Concept Question 4.1. You are sitting as a passenger next to a driver in a car. As long as the car is moving with constant speed along a straight line, you will remain stationary in your seat. You will not move toward the front or be pulled back and, more importantly, you will not feel the seat push against your back. This is consistent with the first law regardless of whether you observe it from the car or from a stationary reference point outside the car. Suddenly, the driver pushes the gas pedal. The car accelerates and immediately you feel that the chair-back pushes on you. With respect to the reference point outside the car, your motion is changed, which again is consistent with Newton's first law and also with his second law. With respect to the car, you are sitting stationary in your seat and you do not see any change in your motion. But you sense that a force is applied to your back. So, Newton's first law is not valid if the accelerated car is taken as a frame of reference. This means Newton's first law and also his second law do not hold for accelerated frames of reference.

The reference frame in which Newton's first law is valid is called an **inertial frame**. All reference frames that are stationary or have a uniform motion are inertial frames. In contrast, all accelerated frames of reference are *non-inertial*. We will deal with inertial frames of reference.

4.3.3: Equilibrium, Revisited

If the net force acting on an object is zero, the object is in equilibrium. Therefore, an object in equilibrium maintains its state of motion and does not change its velocity. Note that "equilibrium" does not mean the object must be at rest. Mathematically, Eq. 4.5 gives the condition for an object to be at equilibrium. In this case, the object is said to be in **translational equilibrium**:

$$\vec{F}_{\text{net}} = \sum_{i=1}^{n} \vec{F}_i = 0 \Rightarrow \text{translational equilibrium}. \qquad [4.6]$$

Keep in mind that the net force, $\vec{F}_{\text{net}}$, being equal to zero does not imply that any of the force $\vec{F}_i$ contributing to the net force in Eq. [4.6] must also be zero. Translational equilibrium can be **static equilibrium** or **dynamic equilibrium**. In static equilibrium, not only is the net force acting on the system zero, the system does not move and stays still. In dynamic equilibrium, the system moves with constant velocity and the net force acting on the system is zero.

CONCEPT QUESTION 4.4

For each of the six objects whose free body diagrams are shown in Fig. 4.6, the arrows indicate forces of equal magnitude that act in the directions shown. Which of the six objects are not in mechanical equilibrium?

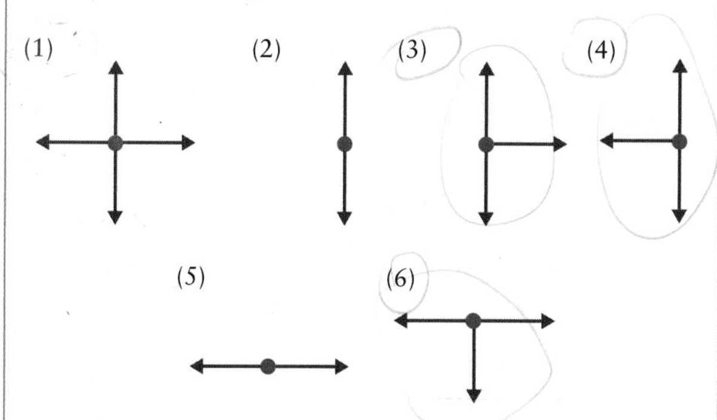

Figure 4.6 Six arrangements of equal-magnitude forces acting on an object.

CONCEPT QUESTION 4.5

Three forces are shown to act on a human leg in Fig. 4.7: the weight $\vec{w}$ of the leg, the normal force $\vec{N}$ acting at the sole, and the force $\vec{F}$ of the hip bone exerted on the head of the femur. Is the leg in equilibrium?

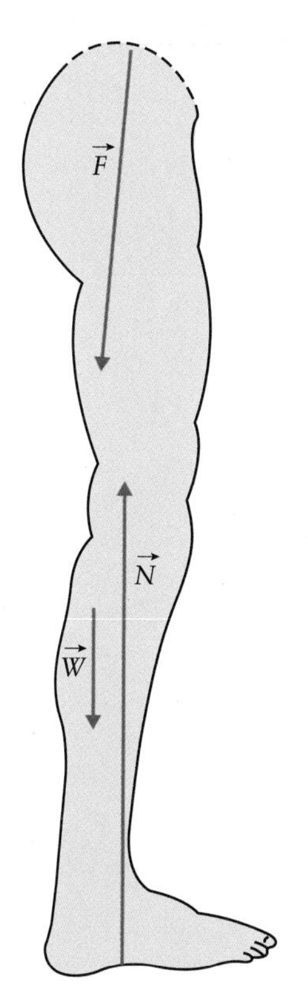

Figure 4.7 Human leg with three forces identified: its weight $\vec{w}$, the normal force $\vec{N}$, and a force $\vec{F}$ with which the upper body pushes the leg down.

CASE STUDY 4.1

Gliding Animals: Gliding is a mode of locomotion that some animals such as gliding fish, gliding lizards, gliding squirrels, gliding frogs, and gliding snakes employ to move through the air. How can these animals glide through the air?

Answer: *Gliding is a kind of controlled fall through the air in which air resistance plays a significant role. In gliding, air resistance slows the fall. More air resistance slows falling to the point that the fall becomes uniform and the net force on the glider becomes zero. When an animal leaps into the air, it falls due to gravity, but air resistance slows its fall. Air resistance depends on the falling speed and on the surface area of the glider. These animal gliders have special features (makeup, structure) that allow them to increase air resistance when gliding. For example, a gliding lizard has a gliding membrane (Fig. 4.8), a gliding squirrel has a large flap of skin as a gliding membrane between its front and hind legs, a gliding fish has unusually large pectoral fins. All these special features increase air resistance during a fall and allow the animal to glide.*

Figure 4.8 A gliding lizard.

EXAMPLE 4.3

A hawk is gliding at a constant velocity. What is the force due to air (lift force) on the hawk? Assume the hawk weighs 6.37 N.

Solution

First we identify the forces acting on the hawk and draw a free body diagram for the hawk. Since the hawk is gliding with constant velocity, the net force acting on the hawk is zero. So the net force on vertical direction is zero, and so the net force on horizontal. The two forces $\vec{F}_{\text{lift}}$ and $\vec{F}_g$ are the only forces acting on the hawk on the vertical direction. The thrust force in the forward direction and the drag force in the backward direction are horizontal, with

continued

zero net force. Fig. 4.9 shows the free body diagram of the hawk vertically.

Figure 4.9 A free body diagram of the hawk gliding with constant velocity.

The hawk is in dynamic equilibrium and the net force in both horizontal and vertical directions is zero. Using Eq. [4.5] in vertical direction gives:

$$\vec{F}_{net} = \vec{F}_{lift} + \vec{F}_g = 0$$

$$F_{lift} - F_g = 0$$

$$F_{lift} = F_g = 6.37 \text{ N}.$$

EXAMPLE 4.4

An object of mass m = 1.0 kg is attached to a taut string of length L = 2.0 cm. The object is held stationary by a horizontal external force $\vec{F}_{ext}$ at the position where the string makes an angle $\theta = 30°$ with the vertical, as shown in Fig. 4.10. (a) What is the magnitude of the tension T in the string? (b) What is the magnitude of the force $\vec{F}_{ext}$?

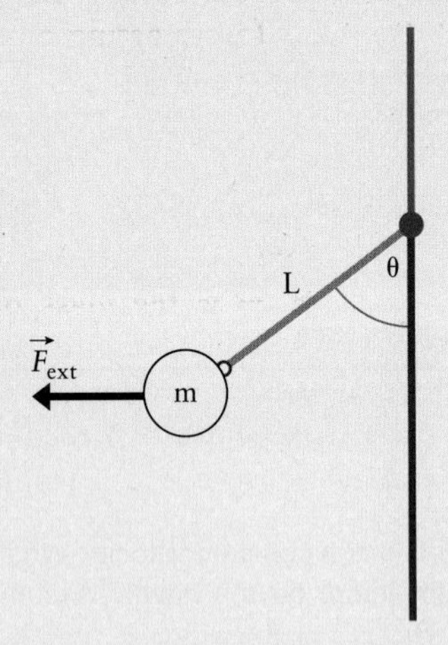

Figure 4.10 An object of mass m attached to a string of length L is pulled by an external force $\vec{F}_{ext}$ to the left, forming an angle θ with the vertical.

Solution

Draw a free body diagram, as shown in Fig. 4.11.

Using Eq. [4.5], gives:

$$x\text{-direction:} \quad -F_{ext} + T\sin\theta = 0$$

$$y\text{-direction:} \quad -w + T\cos\theta = 0$$

continued

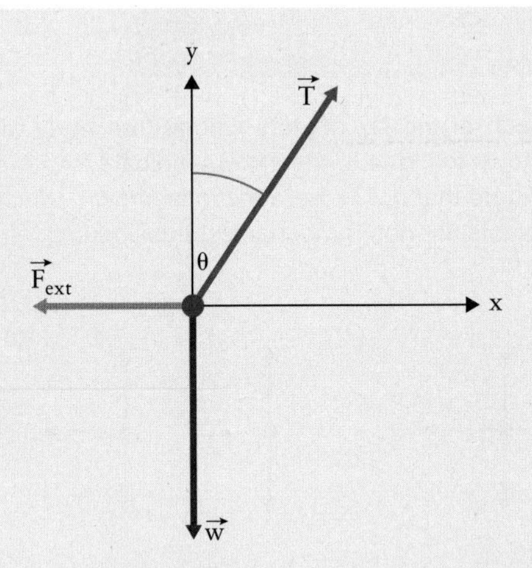

Figure 4.11 Free body diagram for the example shown in Fig. 4.10.

Solution to part (a): From the second formula above, we find:

$$T\cos\theta = w = mg.$$

With the given values of θ and m, we obtain:

$$T = \frac{m \cdot g}{\cos\theta} = \frac{(1.0 \text{ kg})\left(9.8\frac{\text{m}}{\text{s}^2}\right)}{\cos(30°)} = 11.3 \text{ N}.$$

Solution to part (b): Substitute magnitude of T into the first formula:

$$F_{ext} = T\sin\theta = m\frac{g}{\cos\theta}\sin\theta$$
$$= mg\tan\theta$$

which leads to:

$$F_{ext} = (1.0 \text{ kg})\left(9.8\frac{\text{m}}{\text{s}^2}\right)\tan(30°) = 5.7 \text{ N}.$$

EXAMPLE 4.5

An object of mass m = 3.0 kg is suspended from a massless string attached to two other massless strings. The right string is attached to the ceiling at an angle of $\theta = 45°$ with the horizontal, and the left string is attached horizontally to a vertical wall, as shown in Fig. 4.12. Find the tension in each string.

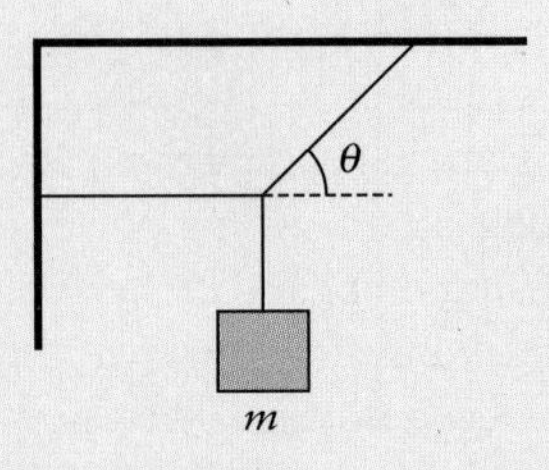

Figure 4.12 An object of mass m is suspended by a massless string connected to two strings.

continued

Solution

Identify all forces acting on the object. There are three forces, weight w, tension of the right string T_R, and tension of the left string T_L, acting on the object. Draw a free body diagram as shown in Fig. 4.13.

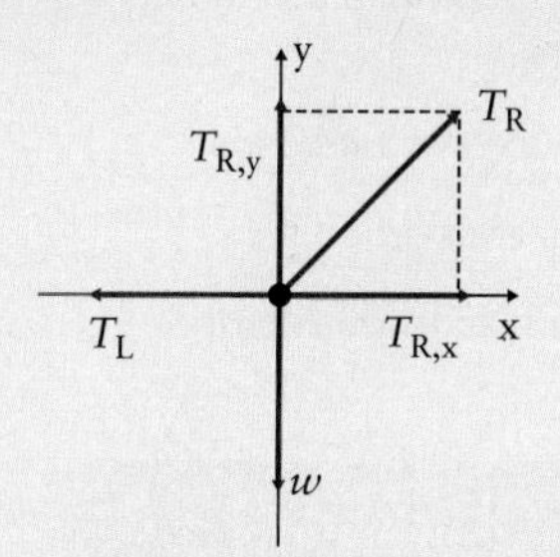

Figure 4.13 Free body diagram of an object of mass m with all forces acting on it.

The mass m is at equilibrium, so using Eq. [4.5] gives:

$$\text{in } x\text{-direction: } T_{R,x} - T_L = 0$$

$$\text{in } y\text{-direction: } T_{R,y} - w = 0.$$

From the second formula above, we find:

$$T_R \sin\theta - w = 0$$

$$T_R \sin\theta = w = mg.$$

With the given values of θ and m, we obtain:

$$T_R = \frac{mg}{\sin(45°)} = \frac{(3.0\text{ kg})(9.8\text{ m/s}^2)}{\sin(45°)} = 41.6\text{ N}.$$

From the first formula, we get:

$$T_R \cos\theta = T_L.$$

Substituting the magnitude of T_R into this formula gives us:

$$T_L = 41.6\cos(45°) = 29.4\text{ N}.$$

4.4: Newton's Second Law

Consider the result of the last two experiments (see (iv) and (v) on page 78) using the hockey puck on the ice pond as summarized in Eq. [4.4]. In both cases the net force acting on the puck is not zero. Therefore, in both cases the puck is not in translational equilibrium. Consider the same hockey puck sliding across an icy pond with no friction. Note that, after losing contact with your hand, the harder you push the puck, the faster it moves in the direction of the applied force. Keep in mind that, after losing contact with your hand, the puck moves with constant velocity on the ice surface since nothing is pushing it and it moves in the same direction in which it was pushed. So, achieving faster motion requires having a larger acceleration; a larger acceleration requires applying a larger force. By doubling or tripling the force, the acceleration of the puck is doubled or tripled; similarly, applying one half the force produces one half the acceleration. Thus, acceleration is directly proportional to and in the same direction of force. Repeat your experiment with two pucks stuck together, effectively doubling the mass. A force that produces a certain magnitude of acceleration for one puck produces one half the magnitude of acceleration for two attached pucks. So, twice the mass produces one half the acceleration. Thus, the magnitude of acceleration is inversely proportional to mass. These observations can be summarized as follows:

$$\begin{cases} \vec{a} \propto \vec{F} \\ |\vec{a}| \propto 1/m \end{cases} \qquad [4.7]$$

The relations between acceleration and force and mass in Eq. [4.7] can be joined as:

$$\vec{a} \propto \vec{F}/m.$$

This means the acceleration of an object is directly proportional to the net force applied to it, and inversely proportional to the object's mass. This leads to:

$$\vec{a} = \vec{F}/m. \qquad [4.8]$$

$$\vec{F} = m\vec{a}. \qquad [4.9]$$

Generally, acceleration of an object is produced by the resultant of several forces, the **net force**, acting on the object. So, for n forces acting on an object we have:

$$\vec{F}_{net} = \sum_{i=1}^{n} \vec{F}_i = m\vec{a}. \qquad [4.10]$$

In these equations, m is the mass of the object. Eq. [4.10] is called **Newton's second law**. It is often referred to as the **equation of motion**.

In the xy-plane, components are used and Newton's second law can be presented as two simultaneous equations:

$$\begin{cases} F_{net,x} = \sum_{i=1}^{n} F_{i,x} = m\,a_x \\ F_{net,y} = \sum_{i=1}^{n} F_{i,y} = ma_y \end{cases} \qquad [4.11]$$

Note that acceleration in the x-direction depends only on force components in the x-direction. Equally, acceleration in the y-direction depends only on force components in the y-direction. **Newton's second law** is stated as follows:

KEY POINT

If a net force is applied to an object of mass m it accelerates in the direction of the net force. The magnitude of the object's acceleration is directly proportional to the magnitude of the net force, and inversely proportional to the mass of the object.

4.4.1: Unit of Force

We introduced the unit of force, the newton (N), in Section 3.5. Now, based on Newton's second law we can identify it. Since $F = ma$, so the unit of force, a "newton" (N), is the unit of mass times the unit of acceleration:

$$\underbrace{F}_{\text{N}} = \underbrace{m}_{\text{kg}} \times \underbrace{a}_{\text{m/s}^2}$$

$$1\text{ N} = 1\text{ kg} \times 1\text{ m/s}^2$$

So, one **newton** is the force that, when exerted on a 1-kg mass, produces 1 m/s^2 acceleration:

$$1\text{ N} = 1\text{ kg m/s}^2.$$

EXAMPLE 4.6

A baby sea lion is pushed across a frozen pond by a constant horizontal force of 120 N. It gains an acceleration of 2 m/s^2. What is the baby sea lion's mass?

Solution

There are three forces acting on the baby sea lion, weight, the normal force by the pond, both vertical, and the horizontal applied force that causes acceleration. Using Eq. [4.10]:

$$F_{\text{net}} = ma$$

$$m = \frac{F}{a} = \frac{120\text{ N}}{2\text{ m/s}^2} = 60\text{ kg}.$$

CONCEPT QUESTION 4.6

The net force acting on an object is zero. The object can

(A) have zero acceleration

(B) move with constant velocity

(C) be at rest

(D) all of the above.

CONCEPT QUESTION 4.7

When an object moves with a constant acceleration, it means that

(A) its velocity always increases.

(B) its velocity sometimes increases and sometimes decreases.

(C) its mass always increases.

(D) a force is exerted on it.

(E) it always falls toward Earth.

CONCEPT QUESTION 4.8

A mule pulls a wagon at constant velocity. Which of the following statements is true?

(A) There must be no force on the wagon in the direction of its velocity.

(B) There must be a force on the wagon in the direction of its velocity.

(C) There must be a net force on the wagon in the direction of its velocity.

(D) There must be no net force on the wagon in the direction of its velocity.

(E) The mule must have acceleration.

EXAMPLE 4.7

A 1500-kg car is travelling at 24 m/s when its brakes fail. The driver manages to turn off the engine. How far will the car go before it comes to stop if the force of friction between the tire and the road is 4500 N.

Solution

There is only a force of friction, $\vec{F}_{\text{friction}} = \vec{f}$ on the car, which causes deceleration. Using Eq. [4.10] we can calculate the magnitude of deceleration of the car caused by force of friction:

$$F_{\text{net}} = ma.$$

Since friction opposes motion here:

$$f = -ma$$

$$a = -\frac{f}{m} = -\frac{4500\text{ N}}{1500\text{ kg}} = -3\text{ m/s}^2.$$

Now, we have initial velocity, final velocity (when it stops), and the magnitude of deceleration:

$$v^2 - v_0^2 = 2ax$$

$$(0)^2 - (24\text{ m/s})^2 = 2\,(-3\text{ m/s}^2)\,x$$

$$x = 96\text{ m}.$$

EXAMPLE 4.8

Fig. 4.14 shows an object of mass $m = 10.0$ kg held by a massless string on a frictionless inclined surface. (a) What is the tension in the string if $\theta = 35°$, and (b) what force does the surface exert on the object? (c) When the string is cut, the object accelerates down the inclined surface. What is its acceleration?

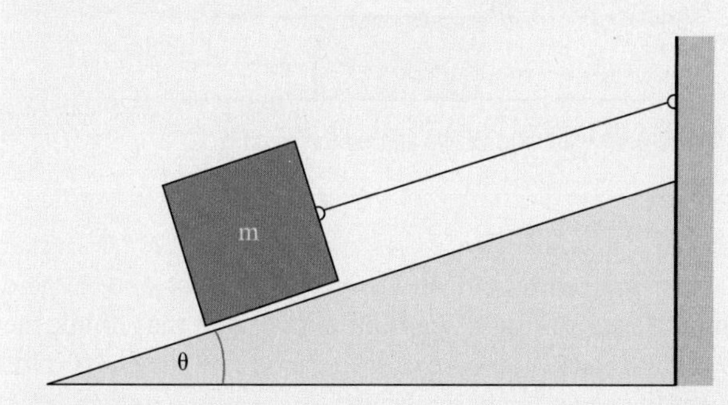

Figure 4.14 An object of mass m on an inclined frictionless surface is connected by a massless string to a vertical wall.

Solution

For parts (a) and (b), three forces, the tension $\vec{T}$ along the inclined surface, the weight $\vec{w}$ down, and the normal force $\vec{N}$ perpendicular to the surface, act on the object. For part (c) there is no tension. The free body diagram for parts (a) and (b) is shown in Fig. 4.15.

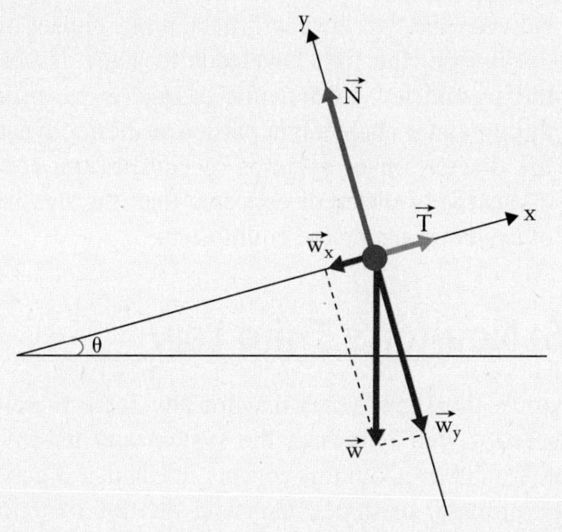

Figure 4.15 Free body diagram of an object on an inclined surface, hooked to the vertical wall.

Note that to draw the free body diagram, first the coordinate system is chosen with the x-axis parallel to the inclined surface and the y-axis perpendicular to it. Thus, $\vec{T}$ and $\vec{N}$ are parallel to the major axes x and y respectively. The components of the weight in this coordinate system are:

$$\begin{cases} w_x = mg\sin\theta \\ w_y = mg\cos\theta. \end{cases}$$

continued

Solutions to parts (a) and (b): The object is at rest. From the free body diagram in Fig. 4.15, and using Eq. [4.11], we find:

$$\text{in } x\text{-direction:} \quad T - w_x = 0$$

$$\text{in } y\text{-direction:} \quad N - w_y = 0.$$

Substituting for $\vec{w}_x$ and $\vec{w}_y$:

$$T - mg\sin\theta = 0$$

$$N - mg\cos\theta = 0.$$

From the first formula, we get:

$$T = (10 \text{ kg})(9.8 \text{ m/s}^2)\sin(35°) = 56 \text{ N}.$$

From the second formula, we obtain:

$$N = (10 \text{ kg})(9.8 \text{ m/s}^2)\cos(35°) = 80 \text{ N}.$$

Note that this example demonstrates you cannot assume $N = mg$ in all cases!

Solution to part (c): We have to redraw the free body diagram. By cutting the string, just two forces, $\vec{N}$ and $\vec{w}$ are acting on the object. So we use Fig. 4.15 but do not consider the tension of the string.

This time, because the two remaining forces, $\vec{N}$ and $\vec{w}$, do not balance each other, the object will accelerate. Thus, using Eq. [4.11] gives:

$$\text{in } x\text{-direction:} \quad -mg\sin\theta = -ma$$

$$\text{in } y\text{-direction:} \quad N - mg\cos\theta = 0.$$

Note that the acceleration component $\vec{a}_x = -\vec{a}$ because the object accelerates downhill while the x-axis is pointing uphill. Also, the second equation has not changed; this is an application of Newton's first law because the object does not move in the y-direction at any time. The first equation gives:

$$a = g\sin\theta,$$

which yields:

$$a = (9.8 \text{ m/s}^2)\sin(35°) = 5.6 \text{ m/s}^2.$$

The magnitude of the acceleration is $|\vec{a}| = 5.6$ m/s². The vector component along the x-axis is $\vec{a}_x = -5.6$ m/s².

EXAMPLE 4.9

A 60-kg skier is coming down a hill that has an incline of 35°. If the force of friction between his ski and the snow is 6 N, how fast does his velocity change per second?

Solution

There are two forces, weight $\vec{w}$ and the force of friction $\vec{f}$, on the skier. First draw a free body diagram for the skier, as shown in Fig. 4.16. To draw a free body diagram, the coordinate system is chosen with the x-axis parallel to the inclined surface and the y-axis perpendicular to it.

continued

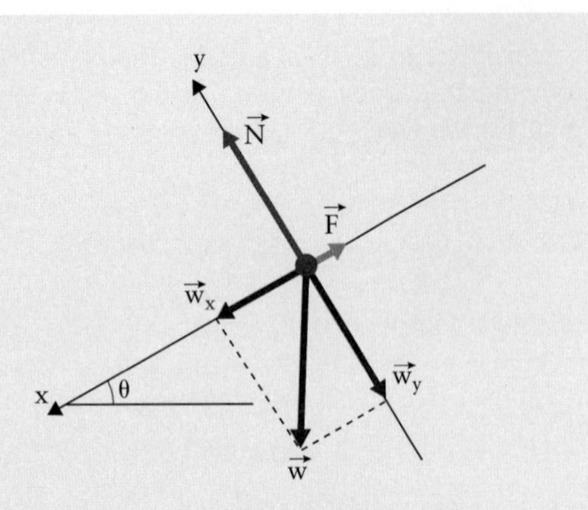

Figure 4.16 Free body diagram of the skier.

The components of the skier's weight in this coordinate system are:

$$\begin{cases} w_x = mg\sin\theta \\ w_y = mg\cos\theta. \end{cases}$$

Using Eq. [4.11], we can calculate the magnitude of acceleration (change of velocity per unit time) of the skier caused by the net force (algebraic sum of his weight, the *x*-component, and the force of friction) on him:

$$\text{in } x\text{-direction:} \quad w_x - f = m\,a$$

$$\text{in } y\text{-direction:} \quad N - w_y = 0.$$

Note that we took downhill as the positive *x*-axis, which means the *x*-component of weight and acceleration both are positive. The second equation shows that the skier does not move in the *y*-direction at any time. The first formula yields:

$$w\sin\theta - f = m\,a$$

$$a = (mg\sin\theta - f)/m = g\sin\theta - f/m$$

$$a = (9.8 \text{ m/s}^2)\sin(35°) - 6/60 = 5.5 \text{ m/s}^2.$$

The magnitude of acceleration is $|\vec{a}|$ = 5.6 m/s^2. The vector component along the *x*-axis is $\vec{a}_x$ = 5.5 m/s^2; it is positive because the object accelerates downhill, which is in the direction of the positive *x*-axis.

CONCEPT QUESTION 4.9

Two different forces are acting on the same object of mass $\vec{m}$ = 3.00 kg. One force, $\vec{F}_1$ = 2.31 N, is pulling the object under the angle θ = 42° above the horizontal, and the other force, $\vec{F}_2$ = 3.52 N, is pushing the object under an angle θ = 34° above the horizontal, as shown in Fig. 4.17. If the force of friction between the object and the floor is $\vec{f}$ = 1.22 N, the acceleration of the object is

(A) 1.94 m/s^2.

(B) 1.54 m/s^2.

continued

(C) 1.17 m/s^2.

(C) 1.14 m/s^2.

(E) 0.76 m/s^2.

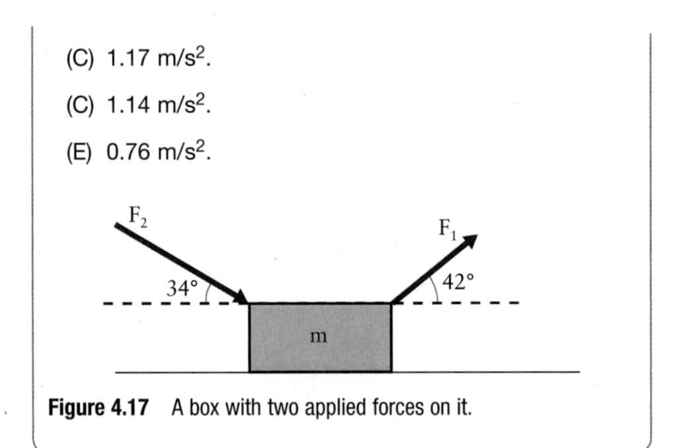

Figure 4.17 A box with two applied forces on it.

A final comment about Newton's first and second laws. Based on the discussion above, one may think the first law is a special case of the second law for $\vec{a} = 0$. This is not the case, for two reasons:

(1) In conceptual physics, the two laws are considered independent because of an aspect of the first law that we do not pursue further in this textbook. We define a frame of reference as a set of points or axes relative to which motion is described. The first law defines when a particular reference frame is an *inertial frame of reference*, in which an object that does not interact with other objects does not accelerate. The second and third laws then apply only in inertial frames of reference.

(2) For the applications in this book, another distinction between the two laws will have a big impact on our discussion: the first law leads to static issues and the second law to dynamic cases. As we progress through later chapters, in particular thermodynamics, the discussion of systems in equilibrium requires distinctively different concepts than the description of systems outside the equilibrium.

4.5: Newton's Third Law

Newton's third law states that for any force there is an "interaction pair" between the system and its environment. We talked about this concept in Chapter 3 generally as the property of forces. Consider the five experiments (i)–(v) with the puck on an ice pond in Section 4.1. In all those cases the puck applied a force on your hand while you applied a force on it simultaneously; regardless, it remained at rest, moved with constant velocity, or obtained acceleration. These two simultaneous forces, the force you applied on the puck and the force the puck applied on you, are equal in magnitude and opposite in direction.

Newton's third law is stated as follows:

KEY POINT

If an object A exerts a force on an object B, object B exerts a force equal in magnitude and opposite in direction on object A simultaneously.

Newton's third law can be written quantitatively as:

$$\vec{F}_{\text{on A by B}} = -\vec{F}_{\text{on B by A}}.$$

Writing the force exerted by object A on object B as $\vec{F}_{BA}$, and the force exerted by object B on object A as $\vec{F}_{AB}$, Newton's third law is written as:

$$\vec{F}_{AB} = -\vec{F}_{BA}. \qquad [4.12]$$

We call the two forces in Eq. [4.12] an **interaction pair** or, customarily, an **action–reaction pair**. Each force of the interaction pair is called the **interaction partner** of the other. It is very important to note that two interaction partners of an interaction pair (an action force and its reaction force) act on two different objects and never act on the same object! Another important point to notice is that the third law is applicable for all categories of forces and is not limited to mechanical forces.

EXAMPLE 4.10

Draw the free body diagrams for a system of two blocks in contact pushed to the right on a frictionless surface, as shown in Fig. 4.18.

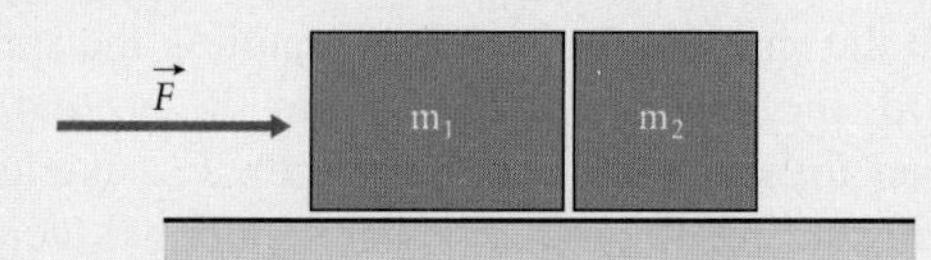

Figure 4.18 Two blocks in contact pushed to the right on a frictionless surface.

Solution

First, identify all forces acting on each block. There are four forces, applied force $\vec{F}$, weight $\vec{w}_1$, normal force $\vec{N}_1$, and contact force by the second block $\vec{F}_{c,12}$, all acting on the first block, and three forces, weight $\vec{w}_2$, normal force $\vec{N}_2$, and contact force by the first block $\vec{F}_{c,21}$, all acting on the second block. Note that the contact force on each block by the other one is an interaction pair, $\vec{F}_{c,12} = -\vec{F}_{c,21}$, and has the same magnitude, $|\vec{F}_{c,12}| = |\vec{F}_{c,21}|$, which is the application of the third law. Figs. 4.19(a) and (b) show the free body diagrams for the two blocks. In Fig. 4.19, the contact force is illustrated as $\vec{F}_c$ and the index 12 or 21 is dropped.

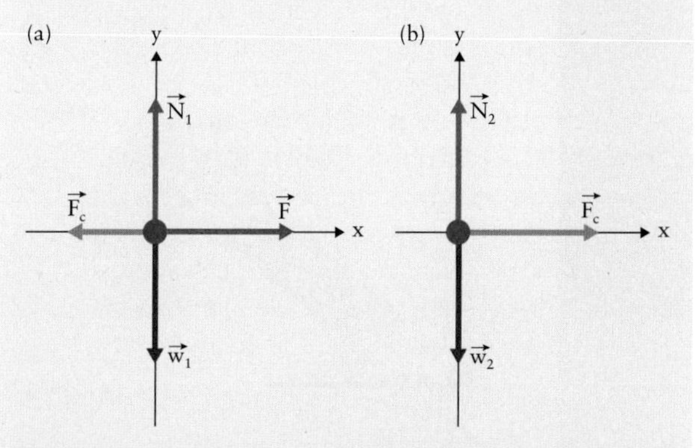

Figure 4.19 Free body diagram for the two blocks in contact: (a) first block on left, (b) second block on right.

CONCEPT QUESTION 4.10

Assume an astronaut's removable jetpack fails during his space walk. His jetpack does not work and he is stranded far away from his spaceship. How can the astronaut still use the failed jetpack and return to his ship?

CONCEPT QUESTION 4.11

Newton's third law states that a force as an interaction partner exists for each force, and that the two forces are equal but opposite in direction. Do we need some forces that have no interaction partner for an object to achieve acceleration, since each interaction pair adds up to a zero net force?

EXAMPLE 4.11

An Arctic explorer pushes two connected sleds, as shown in Fig. 4.20, moving with constant velocity. If each sled has a mass of 100 kg and the force of friction between first and second sleds and the snow are 80 N and 50 N respectively, find the normal (contact) force between the two sleds, $\vec{F}_c$, and the applied force, $\vec{F}_{app}$, on the first sled by the man.

Figure 4.20 An Arctic explorer pushes two sleds.

Solution

First, isolate each sled and identify all forces acting on each of them. There are four forces on sled two, the normal (contact) force $\vec{F}_c$ from sled one, weight $\vec{w}_2$, normal force $\vec{N}_2$ from the ground, and force of friction $\vec{f}_2$, and there are five forces on sled one, the normal force $\vec{F}_c$ from sled two, weight $\vec{w}_1$, the normal force $\vec{N}_1$ from the ground, applied force $\vec{F}_{app}$ from the explorer, and force of friction $\vec{f}_1$. Note that the contact force on each sled by the other one is an interaction (partner) force that has the same magnitude but is in opposite direction of the other partner. This is application of the third law. Figs 4.21(a) and (b) show the free body diagrams for the two sleds

continued

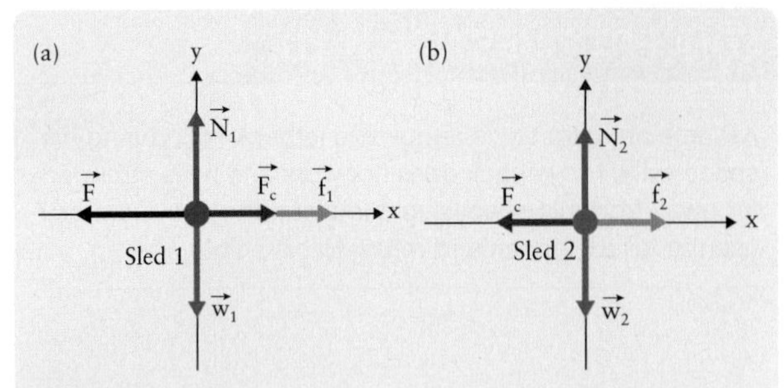

Figure 4.21 (a) Free body diagram of sled one; (b) free body diagram of sled two.

According to free body diagram of each sled, we can write the equation of motion for each of them. Note that the system, the explorer and the two sleds, are moving with constant velocity, so they do not have acceleration and the net force on each sled is zero. Using Eq. [4.10] gives:

$$\text{for sled 1:} \quad \vec{F}_{app} - F_c - f_1 = 0$$

$$\text{for sled 2:} \quad F_c - f_2 = 0$$

The second formula gives:

$$F_c = f_2 = 50 \text{ N}.$$

Substituting the magnitude of F_c in the first relation gives:

$$F_{app} = F_c + f_2 = 50 \text{ N} + 80 \text{ N} = 130 \text{ N}.$$

4.5.1: Newton's Third Law and Sprinters

A sprinter requires maximum forward acceleration during the initial stage of a sprint. This is achieved from a crouching position with the upper body of the sprinter well in front of the feet and pushing on the starting blocks, as shown in Fig. 4.22. In this position the sprinter pushes on the tilted starting blocks to gain the required acceleration. The first question that comes to mind is, how does the sprinter acquire his initial acceleration by pushing on the starting blocks? How does the force the sprinter applies on the starting blocks help him gain that initial acceleration?

This problem is one of the best examples that directly show an application of Newton's third law. According to the third law, in an interaction between two objects, two forces that are equal in magnitude and in opposite direction are acting on the objects at the same time. When the sprinter pushes on the blocks, the blocks simultaneously push the sprinter back with a force of the same magnitude but in the opposite direction. Therefore, the interaction partner of the force exerted by the sprinter on the starting blocks is the force responsible for causing the initial acceleration of the sprinter at the starting stage. These two forces, an interaction pair, are also shown in Fig. 4.22. A sprinter gains a large forward acceleration at the beginning of the sprint by pushing very hard on the starting blocks.

To calculate the sprinter's acceleration, a schematic of a sprinter shows the forces acting on the sprinter in Fig. 4.23. As Fig. 4.23 shows, we can identify three forces acting on the sprinter, the weight, $\vec{w}$, pulling the sprinter downward, the normal force, $\vec{N}$, from the ground, and the contact force (another normal force), $\vec{F}_N$, exerted by the angled starting blocks. Note that the normal (contact) force from the tilted starting blocks does not act vertically since the starting blocks are angled (tilted) and the normal force is perpendicular to its surface.

To draw a free body diagram, we choose x- and y-axes to be horizontal and vertical to the ground, respectively, because the horizontal component of the acceleration is of concern. Fig. 4.24 illustrates the free body diagram of the sprinter. The contact force is decomposed into two components, x and y. The y-component will contribute to balance

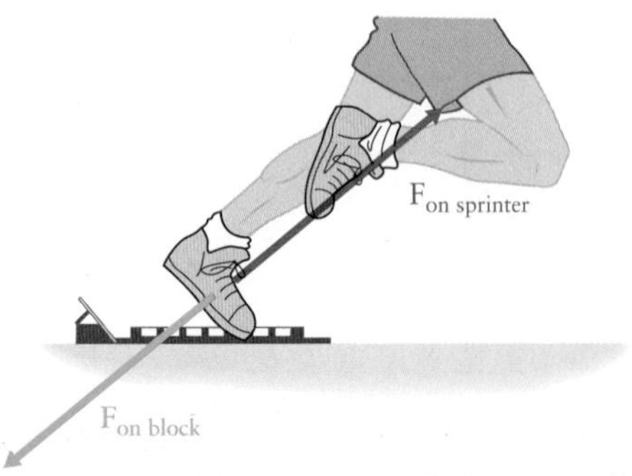

Figure 4.22 A sprinter pushing on the starting blocks in the initial stage of a sprint.

Figure 4.23 Schematic of all forces acting on a sprinter in the initial stage of a sprint.

the weight, and the x-component will cause acceleration. As Fig. 4.24 shows, the contact force by the starting blocks on the sprinter has a small vertical, y-, upward component but has a large horizontal, x-, forward component that is required to spring the sprinter into motion.

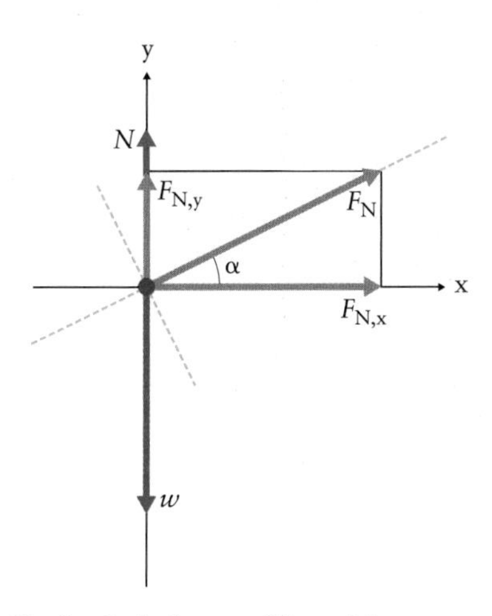

Figure 4.24 The free body diagram of the sprinter.

Note: The starting blocks must be tilted such that the vertical component of the force the starting blocks exert on the sprinter contributes in compensating for the weight of the sprinter. Otherwise, the sprinter would either fall to the ground or jump upward instead of accelerating forward.

The components of contact (normal) force by the starting block on the sprinter are:

$$\vec{F}_N \rightarrow \begin{cases} F_{N,x} = F_N \cos\alpha \\ F_{N,y} = F_N \sin\alpha. \end{cases}$$

Using Eq. [4.11] gives:

$$\begin{aligned} &\text{in } x\text{-direction:} \quad F_{N,x} = ma \\ &\text{in } y\text{-direction:} \quad F_{N,y} + N - w = 0. \end{aligned} \qquad [4.13]$$

Therefore:

$$F_N \cos\alpha = ma$$

$$F_N \sin\alpha + N = w,$$

in which the first formula yields the acceleration:

$$a = \frac{F_N \cos\alpha}{m}. \qquad [4.14]$$

EXAMPLE 4.12

If the sprinter has mass $m = 70$ kg, calculate the force the sprinter must apply on the starting blocks to achieve a horizontal acceleration of 15 m/s^2. Assume that the starting blocks are tilted such that the vertical component of the force the starting blocks exert on the sprinter compensates totally for the weight of the sprinter.

Solution

In this case, we identify two forces, the weight $\vec{w}$ and the contact (normal) force $\vec{F}_N$ exerted by the angled starting blocks, acting on the sprinter in the initial stage. We consider the normal force exerted on the sprinter by the ground, $\vec{N}$,is zero. We will use Fig. 4.24 for the free body diagram, disregarding normal force $\vec{N}$ exerted on the sprinter by the ground. Therefore, the vertical component of the contact force on the sprinter by the blocks compensates for the sprinter's weight.

Using Eq. [4.13] gives:

$$\text{in } x\text{-direction:} \quad F_{N,x} = ma,$$

$$\text{in } y\text{-direction:} \quad F_{N,y} - w = 0.$$

From these two formulas, we find the two components of the normal force:

$$F_{N,x} = (70 \text{ kg})\,(15 \text{ m/s}^2) = 1050 \text{ N}$$

$$F_{N,y} = mg = (70 \text{ kg})\,(9.8 \text{ m/s}^2) = 686 \text{ N}.$$

The two components are combined to calculate the magnitude of the normal force:

$$|F_N| = \sqrt{(F_{N,x}^2 + F_{N,y}^2)} = \sqrt{(1050)^2 + (686)^2} = 1255 \text{ N}.$$

The force exerted by the sprinter on the starting blocks is equal in magnitude but opposite in direction to the normal force exerted by the starting blocks on him, as we calculated. Therefore, the sprinter pushes with a force of 1255 N into the starting blocks.

EXAMPLE 4.13

A person's calf muscle pulls on the Achilles tendon attached to the heel bone. How are the various forces in the lower leg related when the person does push-ups? Fig. 4.25 shows a simplified sketch of the muscles and bones of a human body during a push-up. Note that the calf muscle connects to the heel via the Achilles tendon.

continued

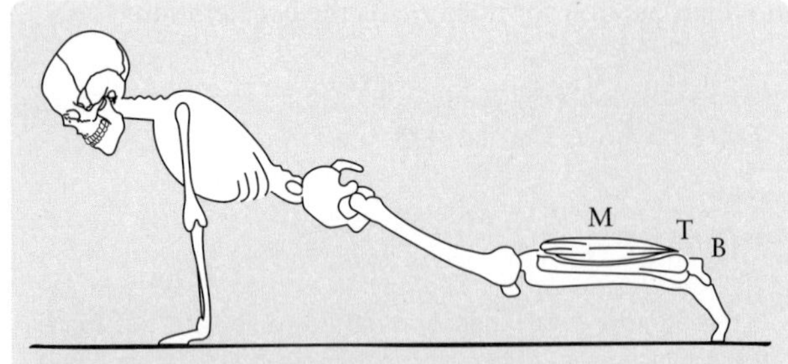

Figure 4.25 An anatomical sketch of a man doing push-ups. This side view highlights the relative position of the heel bone (B), the Achilles tendon (T), and the calf muscle (M). Note that the three elements are located along a horizontal line during push-ups.

Solution

Notice in Fig. 4.25 that the heel bone, the Achilles tendon, and the calf muscle are lined up horizontally when the body is in the position shown. This allows us to neglect weight as a force because we are interested only in the horizontal interactions. We label the calf muscle M, the Achilles tendon T, and the heel bone B, and we consider them as separate objects that interact with each other. We include the tendon since it may not be possible to assume that it is a massless string. In this problem, we need to consider only the forces the three parts of the leg exert on each other.

This problem is a typical application of the third law since several objects are involved.

We identify the horizontal interaction forces in each part. Then we draw free body diagrams for each part separately, as shown in Fig. 4.26. Note that the tendon is drawn as a bar. The muscle pulls the tendon to the left. At the same time, the muscle is pulled in the opposite direction by the tendon. At the other end, the tendon pulls on the bone, and at the same time the bone exerts an equal but opposite force on the tendon.

Figure 4.26 Free body diagram of the major horizontal forces acting on each of the three parts, the heel bone (B), the Achilles tendon (T), and the calf muscle (M).

Among these forces, shown in Fig. 4.26, are the following interaction (action–reaction) pairs:

$$\vec{F}_{TM} = -\vec{F}_{MT}$$

$$\vec{F}_{BT} = -\vec{F}_{TB}.$$

Note that $\vec{F}_{TB}$ and $\vec{F}_{TM}$ are not an interaction (action–reaction) pair because both forces act on the same object! Actually, these two forces need to be different for the muscle to succeed in moving the heel. In that case we write Newton's second law for the horizontal component of the net force on the tendon:

$$\text{horizontal direction: } F_{net} = F_{TB} - F_{TM} = m_T a,$$

continued

where the magnitude of $\vec{a}$ is equal to the horizontal acceleration component of the tendon. There are two applications for the above formula related to a tendon:

(a) For a given mass of the tendon, m_T, an acceleration can be determined. Note, however, that we need to identify two different forces, one at each end of the tendon.

(b) If we assume that the mass of the tendon is negligible, $m_T = 0$, the above relation simplifies to:

$$F_{TB} = F_{TM} = T.$$

This equation defines the magnitude of the tension $\vec{T}$. This is a simpler case since only one force is associated with the tendon.

4.6: Application of Newton's Laws, Convenience Forces Revisited

In this section, we apply Newton's laws to different cases whether the object moves with constant velocity or constant acceleration, or remains motionless in equilibrium. We revisit different convenience forces presented as example problems and apply Newton's laws to solve them. To solve problems involving Newton's laws, the following general steps are suggested:

- Identify all known and unknown variables in your problem.
- Sketch a neat, simple drawing of the system.
- Isolate the object of interest, and identify all forces acting on it.
- Choose an appropriate coordinate system with perpendicular axes in an inertial frame of reference. Generally, for inclined surface, choose coordinate axes parallel and perpendicular to the inclined surface.
- Draw a free body diagram for the object. Your drawing must be very neat and clear. For a system of several objects, draw a free body diagram for each object separately.
- Apply Newton's second law to each free body diagram. Use the component form.
- Solve the equations for the unknowns. You should have at least the same number of equations as unknowns.

4.6.1: Surface Forces

As it was discussed in Section 3.8.1, there are two surface forces: normal force and force of friction. Normal forces are perpendicular to the surface and forces of friction are parallel to the surface. We learned about static friction and kinetic friction forces in Chapter 3.

EXAMPLE 4.14

If the coefficient of kinetic friction between the baby sea lion of Example 4.6 and the frozen pond is 0.05, find its acceleration. The baby sea lion's mass is 60 kg and it is pushed by a constant horizontal force of 120 N.

Solution

There are four forces acting on the baby sea lion: weight, the normal force from the pond, the force of friction, and the applied force. Draw the free body diagram, as shown in Fig. 4.27.

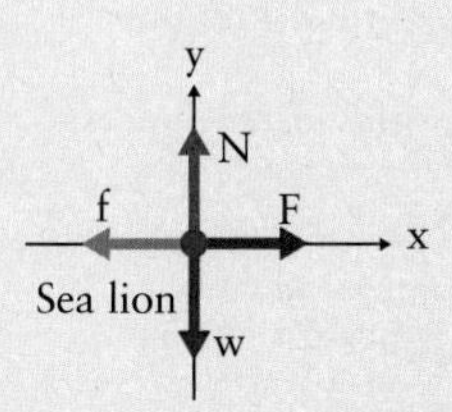

Figure 4.27 Free body diagram of the baby sea lion pushed by a horizontal force.

Using Eq. [4.11] gives:

$$\text{in } x\text{-direction:} \quad F - f = ma$$

$$\text{in } y\text{-direction:} \quad N - w = 0.$$

From the second equation, we find the normal force:

$$N = mg = (60\ \text{kg})(9.8\ \text{m/s}^2) = 588\ \text{N}.$$

Using Eq. [3.9] yields:

$$f_k = \mu_k N = (0.05)(588\ \text{N}) = 29.4\ \text{N}.$$

From the first equation, we find the acceleration:

$$a = (F - f)/m$$

$$a = (120\ \text{N} - 29.4\ \text{N})/60\ \text{kg} = 1.51\ \text{m/s}^2.$$

EXAMPLE 4.15

If a baby sea lion is pushed down a rough surface, inclined 21°, by a 120-N force parallel to the inclined surface, find its acceleration. The baby sea lion has a mass of 80 kg and the coefficient of kinetic friction with the surface is 0.45.

continued

Solution

There are four forces acting on the baby sea lion, weight, the normal force from the inclined surface, the force of friction, and the applied force. We select coordinates with the x-axis and y-axis parallel and perpendicular to the inclined surface respectively. Draw the free body diagram, as shown in Fig. 4.28.

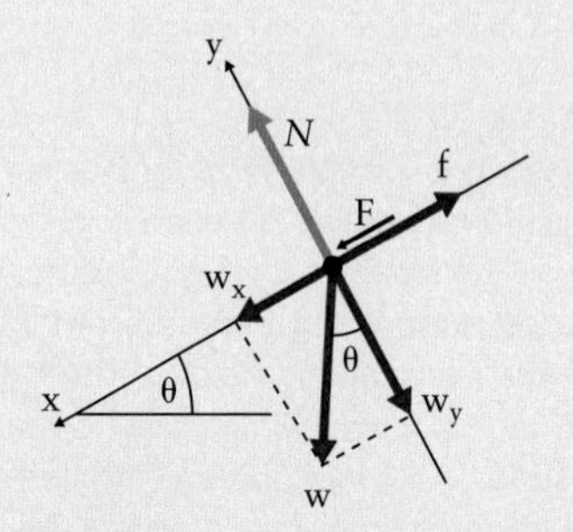

Figure 4.28 Free body diagram of the baby sea lion pushed on an inclined surface by a force parallel to the inclined surface.

The components of weight at the chosen coordinate are:

$$w \rightarrow \begin{cases} w_x = w \sin\theta \\ w_y = w \cos\theta. \end{cases}$$

Using Eq. [4.11] gives:

$$\text{in } x\text{-direction:} \quad F + w_x - f_k = ma$$

$$\text{in } y\text{-direction:} \quad N - w_y = 0.$$

From the second equation we obtain:

$$N = w_y = mg\cos\theta$$

Using Eq. [3.9] yields:

$$f_k = \mu_k N = \mu_k mg\cos\theta$$

$$f_k = 0.45(80\text{kg})(9.8\ \text{m/s}^2)\cos(21°) = 329.4\ \text{N}.$$

From the first equation we obtain:

$$a = (F + w_x - f_k)/m = (F + mg\sin\theta - f_k)/m$$

$$a = [(120\ \text{N}) + (80\ \text{kg})(9.8\ \text{m/s}^2)\sin(21°) - 329.4\ \text{N}]/80\text{kg}$$

$$a = 0.89\ \text{m/s}.$$

EXAMPLE 4.16

Two different boxes of masses $M = 125$ kg and $m = 45$ kg are sliding down a rough surface inclined $\theta = 35°$, as shown in Fig. 4.29. If the coefficients of kinetic friction between the big and small boxes and the inclined surface are 0.3 and 0.5 respectively, find the acceleration of the two boxes and the contact force between them.

continued

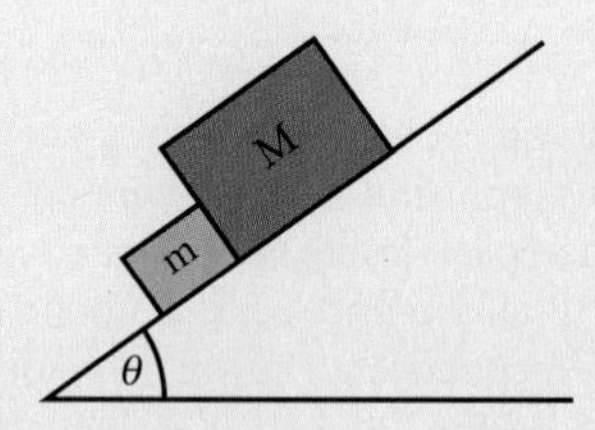

Figure 4.29 Free body diagram of two boxes sliding down a rough surface.

Solution

The system has two boxes. First isolate each box and then identify all forces acting on each one. There are four forces acting on the small box: weight $\vec{w}_S$, the normal force from the inclined surface $\vec{N}_S$, the force of friction $\vec{f}_S$, and the contact force from the big box $\vec{F}_C$. There are also four forces on the big box: weight $\vec{w}_B$, the normal force from the inclined surface $\vec{N}_B$, the force of friction $\vec{f}_B$, and the contact force from the small box $\vec{F}_C$. Note that the contact forces on each box by the other make an interaction pair and have the same magnitude, which is an application of the third law. We select coordinates with *x*-axis and *y*-axis parallel and perpendicular to the inclined surface respectively. Draw the free body diagram for each box separately, as shown in Figs. 4.30(a) and (b).

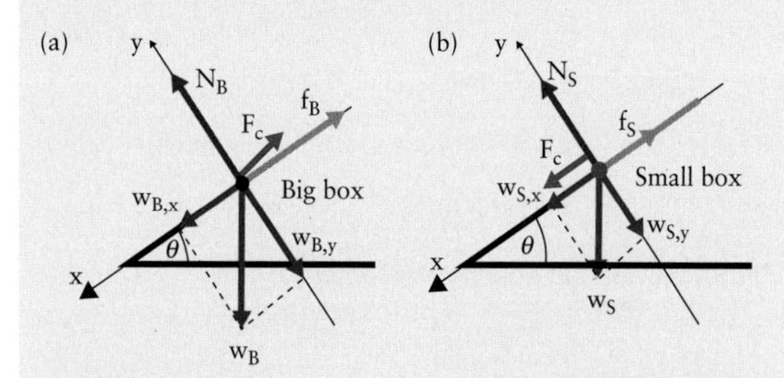

Figure 4.30 (a) Free body diagram of the big box on the inclined surface. (b) Free body diagram of the small box on the inclined surface.

According to free body diagram of each box, we can write an equation of motion for each box. The components of the weight of the small and big boxes at the chosen coordinates are:

$$w_S \rightarrow \begin{cases} w_{S,x} = w_S \sin\theta = mg\sin\theta \\ w_{S,y} = w_S \cos\theta = mg\cos\theta \end{cases}$$

$$w_B \rightarrow \begin{cases} w_{B,x} = w_B \sin\theta = Mg\sin\theta \\ w_{B,y} = w_B \cos\theta = Mg\cos\theta. \end{cases}$$

Using Eq. [4.11] along with Eq. [4.12] gives:

$$\text{for the small box} \begin{cases} \text{in } x\text{-direction:} & w_{S,x} + F_C - f_S = m\,a \\ \text{in } y\text{-direction:} & N_S - w_{S,y} = 0 \end{cases}$$

$$\text{for the big box} \begin{cases} \text{in } x\text{-direction:} & w_{B,x} - F_C - f_B = Ma \\ \text{in } y\text{-direction:} & N_B - w_{B,y} = 0. \end{cases}$$

From the second equation for each box we have:

$$N_S = w_{S,y} = mg\cos\theta = (45\text{ kg})\,(9.8\text{ m/s}^2)\cos(35°)$$
$$= 3.6\times10^2\text{ N}$$
$$N_B = w_{B,y} = Mg\cos\theta = (125\text{ kg})\,(9.8\text{ m/s}^2)\cos(35°)$$
$$= 1.0\times10^3\text{ N}.$$

So:

$$f_S = \mu_k N_S = (0.5)(3.6\times10^2) = 1.8\times10^2\text{ N}$$
$$f_B = \mu_{k,B} N_B = (0.3)(1.0\times10^3) = 3.0\times10^2\text{ N}.$$

From the first equation for each box we have:

$$\begin{cases} w_{S,x} + F_C - f_S = m\,a \\ w_{B,x} - F_C - f_B = Ma. \end{cases}$$

By adding these two equations, we get:

$$w_{S,x} + w_{B,x} - f_S - f_B = (m+M)a$$

$$a = \frac{w_{S,x} + w_{B,x} - f_S - f_B}{m+M} = \frac{mg\sin(35°) + Mg\sin(35°) - f_S - f_B}{m+M}$$

$$a = \frac{(45\text{ kg})(9.8\text{ m/s}^2)\sin(35°) + (125\text{ kg})(9.8\text{ m/s}^2)\sin(35°) - 1.8\times10^2\text{ N} - 3.0\times10^2\text{ N}}{45+125}$$

$$a = 2.7975\text{ m/s}^2 = 2.8\text{ m/s}^2.$$

Using either of the two first equations, we can find the contact force, $\vec{F}_C$, between the two boxes:

$$F_C = ma - w_{S,x} + f_S = ma - mg\sin(35°) + f_S$$

$$F_C = (45\text{ kg})(2.8\text{ m/s}^2) - (45\text{ kg})(9.8\text{ m/s}^2)\sin(35°) + 1.8\times10^2\text{ N} = 53\text{ N}.$$

4.6.2: Tension

EXAMPLE 4.17

A super man drags two heavy sleds uniformly, as shown in Fig. 4.31. If each sled has a mass of 500 kg and the coefficient of kinetic friction between each sled and the ground is $\mu_k = 0.4$, find the tensions in the rope between the two sleds, $\vec{T}_2$, and in the rope between sled one and the super man, $\vec{T}_1$, which has an angle of $\theta = 45°$ with respect to horizontal.

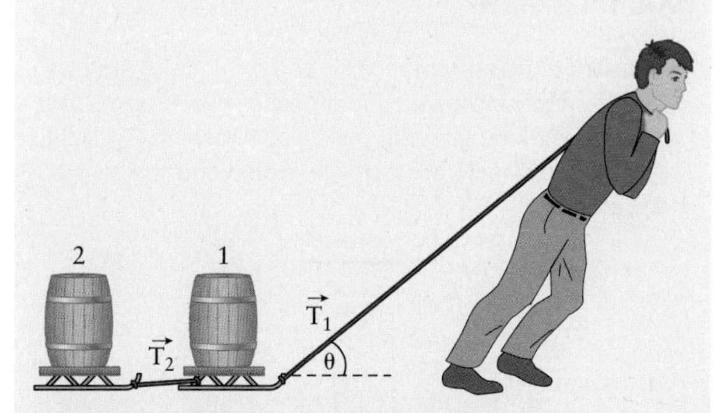

Figure 4.31 A superman drags two heavy sleds.

Solution

First, isolate each sled and identify all forces acting on each one. There are four forces on sled 2, tension $\vec{T}_2$, weight $\vec{w}_2$, normal force $\vec{N}_2$, and force of friction $\vec{f}_2$, and there are five forces on sled one, tension $\vec{T}_1$, weight $\vec{w}_1$, normal force $\vec{N}_1$, tension $\vec{T}_2$, and force of friction $\vec{f}_1$. Note that the contact forces on each block by the other one make an interaction pair and have the same magnitude, which is an application of the third law. Figs. 4.32(a) and (b) show the free body diagrams for the two blocks separately.

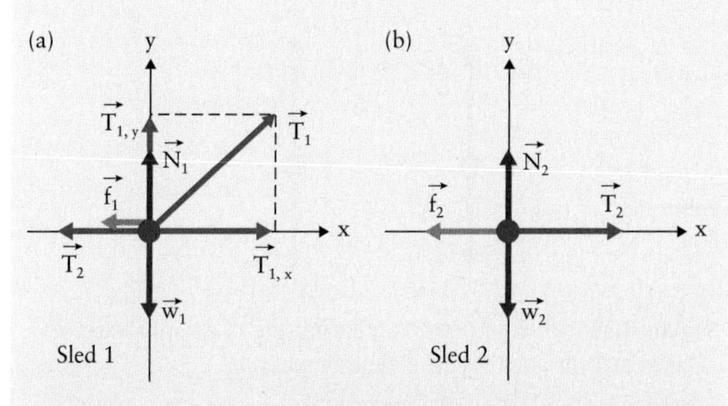

Figure 4.32 (a) Free body diagram for sled 1. (b) Free body diagram for sled 2.

According to free body diagram of each sled, we can write an equation of motion for each one. Note that the super man and the two sleds are moving uniformly, so they do not have acceleration and the net force on each sled is zero. Using Eq. [4.11] gives:

$$\text{for sled 1} \begin{cases} \text{in } x\text{-direction:} & T_{1,x} - T_2 - f_1 = 0 \\ \text{in } y\text{-direction:} & N_1 + T_{1,y} - m_1 g = 0 \end{cases}$$

$$\text{for sled 2} \begin{cases} \text{in } x\text{-direction:} & T_2 - f_2 = 0 \\ \text{in } y\text{-direction:} & N_2 - m_2 g = 0. \end{cases}$$

First we solve for sled 2. The fourth formula gives:

$$N_2 = m_2 g = (500 \text{ kg})(9.8 \text{ m/s}^2) = 4900 \text{ N}.$$

Therefore:

$$f_2 = \mu_k N_2 = (0.4)(4900) = 1960 \text{ N}.$$

The third formula gives:

$$T_2 = f_2 = 1960 \text{ N}.$$

For sled 1, the second formula gives:

$$N_1 = m_1 g - T_{1,x} = (500 \text{ kg})(9.8 \text{ m/s}^2) - T_1 \sin(45°)$$
$$= 4900 \text{ N} - 0.7\, T_1.$$

Therefore:

$$f_1 = \mu_k N_1 = m_1 g - T_{1,x} = (0.4)(4900 \text{ N} - 0.7\, T_1)$$
$$f_1 = 1960 \text{ N} - 0.28\, T_1.$$

Substituting f_1 and T_2 in the first formula gives:

$$T_1 \cos(45°) - 1960 \text{ N} - (1960 \text{ N} - 0.28 T_1) = 0$$
$$0.98\, T_1 - 3920 \text{ N} = 0$$
$$T_1 = 3920 \text{ N}/0.98$$
$$= 4000 \text{ N}.$$

EXAMPLE 4.18

A 15-kg wooden crate is being pulled up a 35° ramp by a rope, but the crate is held stationary by the force of static friction. The coefficient of static friction between the ramp and the wooden crate is $\mu_k = 0.65$. (a) What is the maximum tension in the rope? (b) What is the minimum tension in the rope?

Solution

In both cases, the crate is at equilibrium, so the net force on it should be zero. First, isolate the crate and identify all forces acting on it. There are four forces on the crate: tension $\vec{T}$ by the rope, weight $\vec{w}$, normal force $\vec{N}$ by the ramp, and force of friction $\vec{f}$ between the crate and the ramp. We select coordinates with *x*-axis and *y*-axis parallel and perpendicular to the inclined surface respectively.

continued

Draw the free body diagram. Fig. 4.33 shows the free body diagrams for the crate.

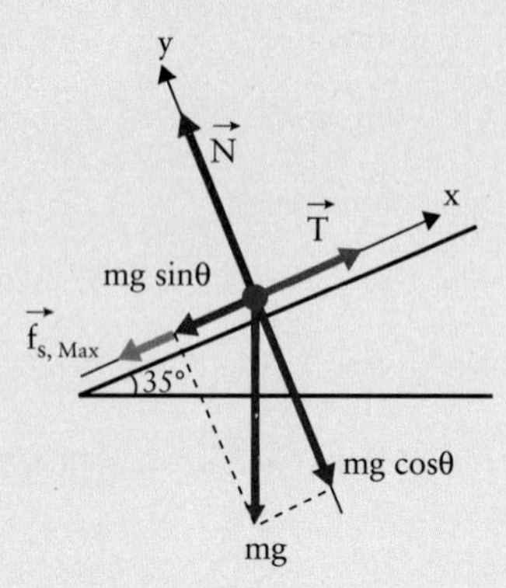

Figure 4.33 Free body diagram for the wooden crate pulled on the ramp.

In equilibrium, using Eq. [4.1] we have:

$$\text{in equilibrium: } F_{\text{net}} = \sum F_i = 0 \Rightarrow \begin{cases} \sum F_{i,x} = 0 \\ \sum F_{i,y} = 0. \end{cases}$$

Solution to part (a): According to the free body diagram, for the maximum tension we have:

$$\sum F_x = T - w\sin\theta - f_{\text{s,Max}} = 0$$

$$\sum F_x = N - w\cos\theta = 0$$

$$N = w\cos\theta = mg\cos\theta$$

$$f_{\text{s,Max}} = \mu_s N = \mu_s mg\cos\theta.$$

The first formula yields:

$$T_{\text{Max}} = w\sin\theta + f_{\text{s,Max}} = mg\sin\theta + f_{\text{s, Max}}.$$

Substituting $f_{\text{s, Max}}$ in above formula:

$$T_{\text{Max}} = mg\sin\theta + \mu_s mg\cos\theta$$

$$T_{\text{Max}} = mg(\sin\theta + \mu_s \cos\theta).$$

Substituting values for m, g, θ, and μ_s:

$$T_{\text{Max}} = (15 \text{ kg})\,(9.8 \text{ m/s}^2)\,[\sin(35°) + (0.56)\cos(35°)]$$

$$T_{\text{Max}} = 1.5 \times 10^2 \text{ N}.$$

Solution to part (b): According to the free body diagram, for minimum tension we get:

$$T_{\text{Min}} = mg \sin\theta$$

$$T_{\text{Min}} = (15 \text{ kg})\,(9.8 \text{ m/s}^2) \sin(35°)$$

$$T_{\text{Min}} = 84.3 \text{ N} = 8.4 \times 10^1 \text{ N}.$$

4.6.3: Ideal Pulley

A pulley can change the direction of an applied force. A *fixed* pulley has a fixed axle. It is used to change the direction of the force on a rope. A *movable* pulley has a free axle. It is used mostly to multiply forces. An ideal pulley is a massless and frictionless pulley. It does not exert any force on the cord; so the tension in the cord on both side of the pulley are the same. A real pulley with a negligible friction and small mass can be considered an ideal pulley.

EXAMPLE 4.19

Two crates of masses $m_1 = 8$ kg and $m_2 = 12$ kg are connected by a massless string that passes over a pulley that can be considered an ideal pulley, as shown in Fig. 4.34. Calculate the acceleration of the crates and the tension in the string.

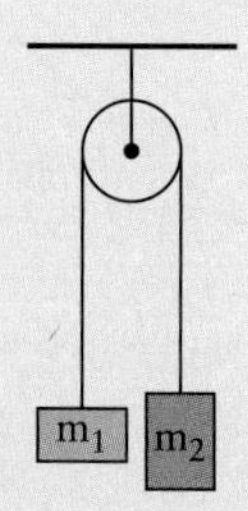

Figure 4.34 Two crates connected by a string passed through a pulley.

Solution

There are two crates in the system. Isolate each crate and identify all forces acting on each one. Two forces act on m_1, weight $\vec{w}_1$ and tension $\vec{T}$ in the string between the two masses; and two forces act on m_2, tension $\vec{T}$ in the string and weight $\vec{w}_2$. Figs. 4.35(a) and (b) show the free body diagrams for the two crates separately.

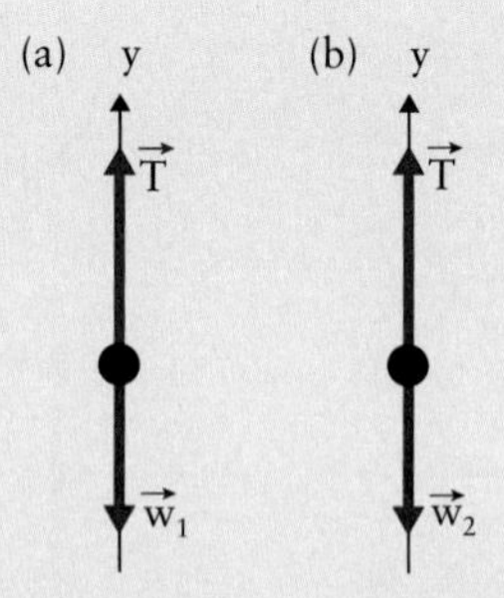

Figure 4.35 (a) Free body diagram for the crate with mass m_1. (b) Free body diagram for the crate with mass m_2.

According to the free body diagram of each crate, we can write an equation of motion for each of crate. Using Eq. [4.10] gives:

$$\text{for crate 1: } T - w_1 = m_1 a$$

$$\text{for crate 2: } w_2 - T = m_2 a.$$

continued

Adding these two equations gives:

$$-w_1 + w_2 = m_1 a + m_2 a$$

$$(m_2 - m_1)g = (m_1 + m_2)a$$

$$a = \left(\frac{m_2 - m_1}{m_2 + m_1}\right) g.$$

Substituting values for m_1, m_2, and g gives:

$$a = \left(\frac{12\text{ kg} - 8\text{ kg}}{12\text{ kg} + 8\text{ kg}}\right)(9.8\text{ m/s}^2) = 1.96\text{ m/s}^2.$$

Substituting $a = 1.96$ N in one of the first two equations will give *T*:

$$T = w_1 - m_1 a = m_1(g - a)$$

$$T = (8\text{ kg})(9.8\text{ m/s}^2 - 1.96\text{ m/s}^2) = 62.7\text{ N}.$$

EXAMPLE 4.20

A 600-kg whale is pulled upward with constant velocity by a rope and two pulleys, as shown in Fig. 4.36. What are the tensions in the three pieces of rope connecting the pulleys and holding the whale? Assume the pulley is ideal.

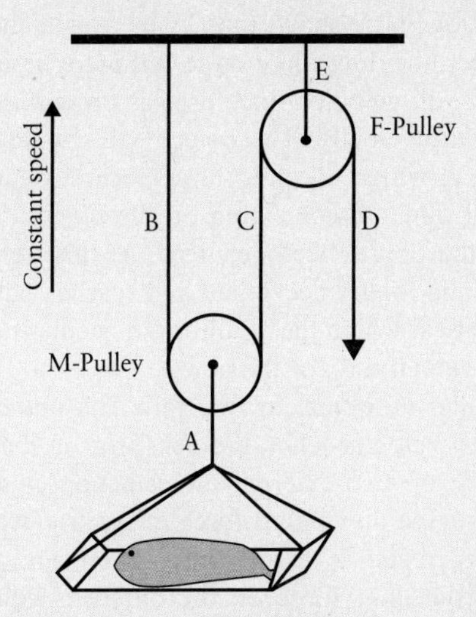

Figure 4.36 A rope-and-pulley system pulling up a whale using constant velocity pulleys.

Solution

Because the pulleys are assumed to be ideal, the tension force on the rope is the same on both sides of the pulley. The rope passes through both pulleys and has the same tension force everywhere, assuming the rope is massless: therefore $T_B = T_C = T_D = T$. The system has three components, the whale, the moveable pulley hauling the whale, and the fixed pulley attached to the ceiling. Because the whale and moveable pulley have constant velocity and the fixed pulley is at rest, the net forces on all of them are zero. Isolate each component and identify all forces acting on each one. Draw free body diagrams for each of these components separately. Figs. 4.37(a), (b), and (c) show three free body diagrams.

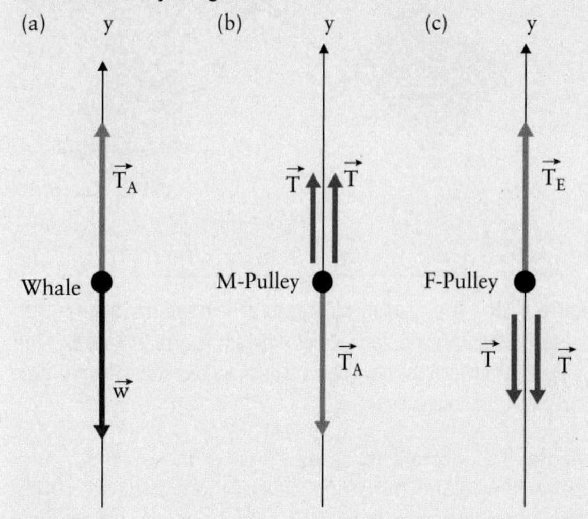

Figure 4.37 (a) Free body diagram of the whale. (b) Free body diagram of the moveable pulley. (c) Free body diagram of the fixed pulley.

According to the free body diagrams, and due to the fact that the motion is only in the *y*-direction, we can write an equation of motion for each component, in the *y*-direction, separately. Using Eq. [4.5] gives:

$$\text{for the whale:} \quad T_A - mg = 0$$

$$\text{for the moveable pulley:} \quad T + T - T_A = 0$$

$$\text{for the fixed pulley:} \quad T_E - T - T = 0.$$

Therefore:

$$T_A = mg = (600\text{ kg})(9.8\text{ m/s}^2) = 5880\text{ N}$$

$$2T = T_A$$

$$T = 5880/2 = 2940\text{ N}$$

$$T_E = 2T = 5880\text{ N}.$$

EXAMPLE 4.21

Two objects with $m_1 = 3.0$ kg and $m_2 = 2.0$ kg are connected by a massless, taut string that runs over a massless and frictionless pulley move on two frictionless inclined surfaces, as shown in Fig. 4.38. The inclined surface at the left is tilted at an angle $\theta = 45°$ with respect to the horizontal, and the inclined surface at the right at $\theta_2 = 60°$. (a) Find the magnitude of acceleration of the two objects, and (b) find the magnitude of tension on the taut string.

Solution

Isolate two objects and identify all forces acting on each of them separately. Three forces act on each object: its

continued

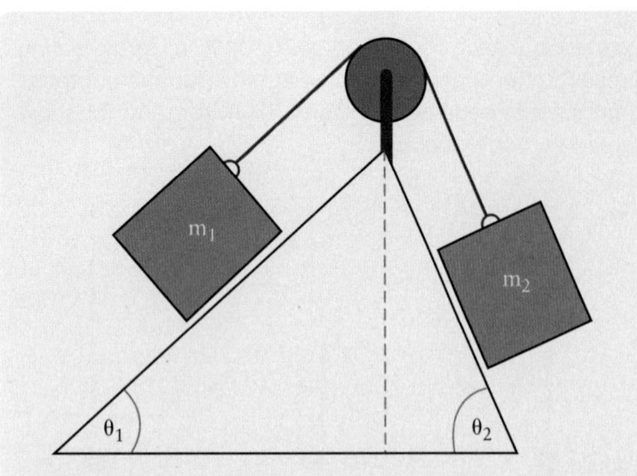

Figure 4.38 Two connected objects of masses m_1 and m_2 are placed on two inclined surfaces of different angles θ_1 and θ_2. The connection between the objects is a taut, massless string running over a massless and frictionless pulley.

weight, the normal force exerted by the surface, and the tension along the string. Select the coordinates with axes parallel and perpendicular to the inclined surfaces respectively. The two free body diagrams are shown in Figs. 4.39(a) and (b).

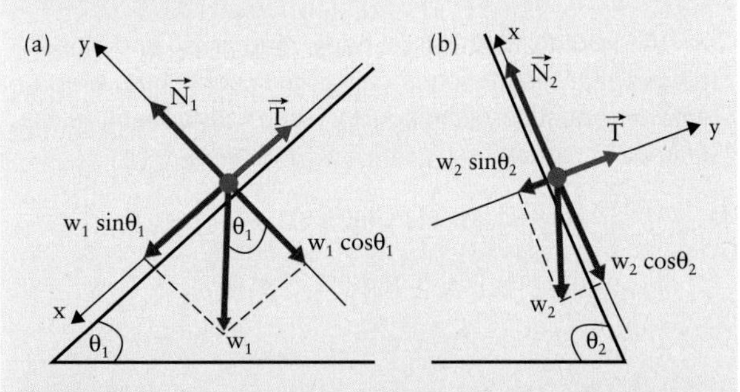

Figure 4.39 The two free body diagrams for the objects on two inclined surfaces. Note that we chose two different coordinate systems.

Since the objects are not moving in the *y*-direction, and the inclined surfaces are frictionless, we can write an equation of motion for each object in the *x*-direction, separately. There is no need to write an equation of motion in the *y*-direction. Using Eq. [4.11] gives:

for object of mass m_1, in x-direction: $m_1 g\sin(45°)-T=m_1 a$

for object of mass m_2, in y-direction: $T-m_2 g\sin(60°)=m_2 a$.

Eliminating T between these two equations by summing them, gives:

$$m_1 g\sin(45°) - m_2 g\sin(60°) = (m_1 + m_2)\, a$$

$$a = \frac{m_1 g\sin(45°) - m_2 g\sin(60°)}{m_1 + m_2}$$

$$a = \left(\frac{m_1 \sin(45°) - m_2 \sin(60°)}{m_1 + m_2}\right) g$$

$$a = \left(\frac{3\text{ kg}\sin(45°) - 2\text{ kg}\sin(60°)}{3\text{ kg} + 2\text{ kg}}\right)(9.8\text{ m/s}^2) = 0.76\text{ m/s}^2.$$

continued

Substitute the value of a in one of the first two equations and solve for T:

$$(3\text{ kg})(9.8\text{ m/s}^2)\sin(45°) - T = (3\text{ kg})(0.76\text{ m/s}^2)$$

$$T = (3\text{ kg})[(9.8\text{ m/s}^2)\sin(45°) - (0.76\text{ m/s}^2)] = 6.17\text{ N}.$$

4.6.4: Drag and Terminal Velocity

As discussed in Section 3.8.4, drag is the resistive force that opposes the relative motion of an object through air. The magnitude of drag depends on the relative speed of the object. It is either proportional to the speed of the object in air, $F_D \propto v$, which will be discussed in Chapter 12, or proportional to the square of the speed of the object, $F_D \propto v^2$, which was discussed in Section 3.8.4. Eq. [3.12] gives the drag force on a moving object mathematically, which we reintroduce again:

$$R = 1/2\, D\rho A v^2, \qquad [4.15]$$

where $\vec{R} = \vec{F}_D$ is the drag force, D is the drag coefficient, ρ is the medium (air) density, A is the object cross sectional area, and v is the relative velocity with respect to the medium.

Equation [4.15] shows that, by increasing the velocity of an object, the drag force on it will increase faster than the increase in velocity since drag is proportional to the square of the velocity. The object will eventually reach a velocity at which the drag force becomes equal to the sum of all other forces acting on the object; so the net force on the object becomes zero. At this velocity, the object can no longer accelerate and reaches equilibrium. The velocity at which the equilibrium occurs is called the **terminal velocity**, v_t, of the object.

Consider an object in free fall. The object reaches its terminal velocity when the net force on the object is zero. There are two external forces acting on the object in this case, the downward force of gravity, weight, and the upward resistive force, drag. Therefore, the magnitude of the drag force on the object is equal to the magnitude of its weight. We use Newton's first law to determine the terminal speed of an object falling toward Earth through air:

$$F_{net} = R - w = 0$$
$$R = w. \qquad [4.16]$$

Therefore, using Eq. [4.15] and Eq. [4.16] gives:

$$1/2\, D\rho A v_t^2 = mg$$
$$v_t = \sqrt{\frac{2mg}{D\rho A}}. \qquad [4.17]$$

Equation [4.15] can be rearranged in terms of v_t by using Eq. [4.17]. Substituting it in Eq. [4.15] gives:

$$R = mg\frac{v^2}{v_t^2} \qquad [4.18]$$

$$R = mg(v/v_t)^2.$$

EXAMPLE 4.22

A sphere with diameter of r = 17.5 cm and mass of m = 4.3 kg is dropped from an airplane. Take the drag coefficient of D = 0.5; what is its terminal velocity?

Solution

Take the air density ρ_{air} = 1.29 kg/m^3. Using Eq. [4.17] gives:

$$v_t = \sqrt{\frac{2(4.3\text{ kg})(9.8\text{ m/s}^2)}{(0.5)(1.29\text{ kg/m}^3)(\pi)(17.5\times10^{-2}/2)^2}}$$

$$v_t = \sqrt{\frac{84.28}{(1.55\times10^{-2})}} = 73.7\text{ m/s}.$$

EXAMPLE 4.23

Two same-size spheres, A and B, with masses m_A = 450 g and m_B = 1450 g are dropped from a very tall building. Find the ratio of their terminal speed.

Solution

Since the spheres are identical in size, we can assume that they have the same drag coefficient. So using Eq. [4.15], gives:

$$R = 1/2\, D\rho A v^2 = bv^2.$$

Substituting this equation into Eq. [4.16], gives:

$$bv_t^2 - mg = 0$$

$$v_t = \sqrt{mg/b}.$$

Therefore:

$$\frac{v_{t,A}}{v_{t,B}} = \frac{\sqrt{m_A g/b}}{\sqrt{m_B g/b}} = \sqrt{\frac{m_A}{m_B}}$$

$$\frac{v_{t,A}}{v_{t,B}} = \sqrt{\frac{450\text{ g}}{1450\text{ g}}} = 0.57.$$

This shows that the lighter sphere has a smaller terminal speed.

4.7: Weight and Apparent Weight

What is called weight of an object is really the force of gravity acting on the object. The weight of an object is proportional to its mass, m. So the greater the mass of an object, the greater is its weight. The weight of an object of mass m is given as:

$$\vec{w} = \vec{F}_g = m\vec{g}. \qquad [4.19]$$

As Eq. [4.19] shows, there is a difference between the mass and the weight of an object. Mass of an object is a scalar quantity that shows the amount of matter the object has. It depends on what an object is made of; thus it does not change by changing the location of the object. It stays the same if you move the object from one place to another.

The weight of an object is a vector quantity stating the strength of the force of gravity on the object. It may change with changing the location of the object, or changing the mass of the object. So it changes from one place to another place having different gravitational acceleration, $\vec{g}$. The weight of an object in Hamilton, Ontario, 75 m above sea level, is different from that at Mount Logan, Yukon, 5960 m above sea level.

To measure the weight of an object, we use scales, mostly spring scales like the usual bathroom scale. We place an object on, or hang an object from, the scale, which compresses its spring or stretches it, and shows the weight on the calibrated pointer. Therefore, what is measured as weight is the contact force between the scale and the object, which is equal to the spring force of the scale. We call this contact force **apparent weight**, $\vec{w}_{app}$, assuming that Earth is an inertial reference frame. Fig. 4.40(a) illustrates a spring scale without and with a brick resting on the scale, and Fig. 4.40(b) shows the free body diagrams of both the brick and the basket. You may convince yourself that the following relation is correct: $F_g = F_{Br-Ba} = F_{Ba-Br} = F_S$.

KEY POINT

Apparent weight is the weight that is measured by a contact force.

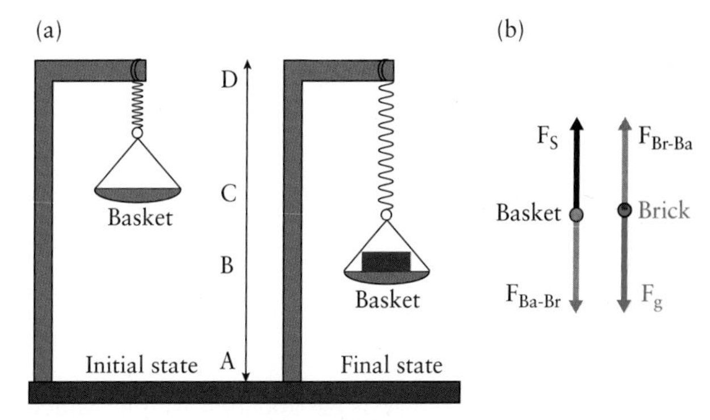

Figure 4.40 (a) A spring scale and a brick resting on it. (b) The free body diagrams of the brick and the scale's basket.

As discussed previously, the weight of an object can change by changing the mass or the location of the object. However, the apparent weight of an object changes whenever there is a change in the contact force between the object and the measuring device. For example, if weight is measured with a scale in an accelerating elevator, the reading will be different from that measured with the same scale moving uniformly or staying at rest. Apparent weight changes with acceleration.

Consider an object on a bathroom scale. In Fig. 4.41(a) the scale measures the apparent weight of the object. Fig. 4.41(b) shows its free body diagram. Newton's second law, Eq. [4.10], gives:

$$\vec{F}_{net} = \vec{N} + m\vec{g} = m\vec{a}, \qquad [4.20]$$

where $\vec{N}$ is the normal force "on the object by the scale." If the scale and the object on it do not have acceleration, then the above equation reduces to:

$$\vec{N} + m\vec{g} = 0.$$

The magnitude of $\vec{N}$ is the apparent weight:

$$|\vec{N}| = N = w_{app} \qquad [4.21]$$

$$w_{app} = N = mg = w. \qquad [4.22]$$

So the apparent weight is equal to weight.

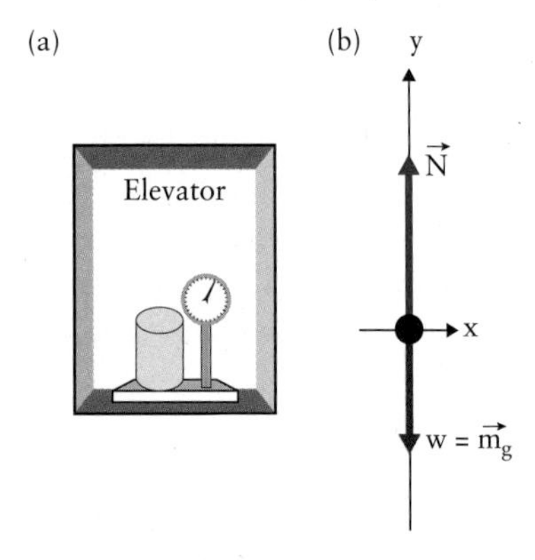

Figure 4.41 (a) An object resting on a scale in an elevator. (b) The free body diagram of an object resting on a scale in an elevator.

If the scale and the object resting on it accelerate in the y-direction, then Eq. [4.20] gives the apparent weight as:

$$\vec{N} = m\vec{a} - m\vec{g} \qquad [4.23]$$

Again, the magnitude of $\vec{N}$ is the apparent weight, which is not equal to the weight of the object:

$$w_{app} = |\vec{N}| = |m\vec{a} - m\vec{g}|. \qquad [4.24]$$

Thus, if the scale is taken into an elevator to measure the weight of an object, there will be different measurements for apparent weight in the elevator accelerating or decelerating, upward or downward.

EXAMPLE 4.24

A student of mass 60 kg stands on a scale in an elevator as shown in Fig. 4.42(a). How much is the scale reading in newtons, if the elevator is accelerating upward with $a = 0.6\ \text{m/s}^2$?

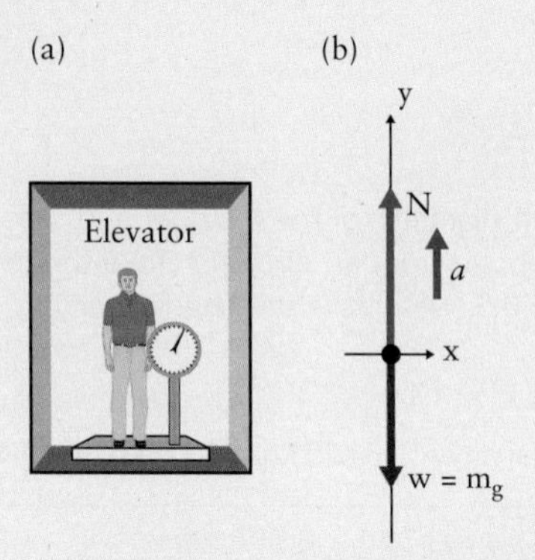

Figure 4.42 A student on a scale in elevator. (b) The free body diagram of the student on scale in an elevator accelerating upward.

Solution

Isolate the student and identify all forces acting on him. Draw the student's free body diagram as shown in Fig. 4.42(b). Using Newton's second law, Eq. [4.10] gives:

$$N - mg = ma$$

$$N = ma + mg = m(a+g)$$

$$w_{app} = N = (60\ \text{kg})(0.6\ \text{m/s}^2 + 9.8\ \text{m/s}^2) = 624\ \text{N}.$$

CONCEPT QUESTION 4.12

A student stands in an elevator, going from the first floor to the 15th floor. The elevator will go through three stages: it will first accelerate to reach a predetermined speed, then it will move uniformly for most of the lifting, then it will slow to a stop at the assigned floor, 15. The apparent weight of the student has

(A) largest magnitude when accelerating
(B) smallest magnitude when accelerating
(C) largest magnitude when decelerating
(D) smallest magnitude when moving uniformly
(E) smallest magnitude when decelerating
(F) largest magnitude when moving uniformly.

4.8: Physiological Applications of Newton's Laws

In this section, three examples highlight how the laws of mechanics are applied in kinesiology and physiology.

4.8.1: A Standard Man in the Gym

In physiology we are often concerned with humans in general, not with the features of a particular person. For this purpose, a **standard man** has been defined and the standard man's data, summarized in Table 4.1, are used in calculations.

TABLE 4.1

Standard man data. The percentage values indicate the fraction of the total body mass

General data	
Age	30 years
Height	173 cm
Mass Distribution	
Body mass M_{total}	70 kg
Mass of the trunk	48%
Muscle mass	43%
Mass of each leg	15%
Fat mass	14%
Bone mass	10%
Mass of the head	7%
Mass of each arm	6.5%
Brain mass	2.1%
Mass of both lungs	1.4%
Homeostasis data	
Surface area	1.85 m^2
Body core temperature*	310 K
Specific heat capacity	3.60 kJ/(kg K)
Respiratory data	
Total lung capacity	6.0 L
Tidal volume (lungs)	0.5 L
Breathing rate	15 breaths/min
Oxygen consumption	0.26 L/min
Carbon dioxide production	0.208 L/min
Cardiovascular data	
Blood volume	5.1 L
Cardiac output	5.0 L/min
Systolic blood pressure	16.0 kPa
Diastolic blood pressure	10.7 kPa
Heart rate	70 beats/min

*Skin surface temperatures vary with environmental temperature; see Fig. 10.5.

EXAMPLE 4.25

The standard man in Fig. 4.43 intends to do *concentration curls* in a gym. At the beginning of this exercise, the person holds with the left arm vertical a dumbbell of mass $M = 4$ kg. Calculate the ratio of the magnitude of the tension in the shoulder to the magnitude of the force that pulls the fist down.

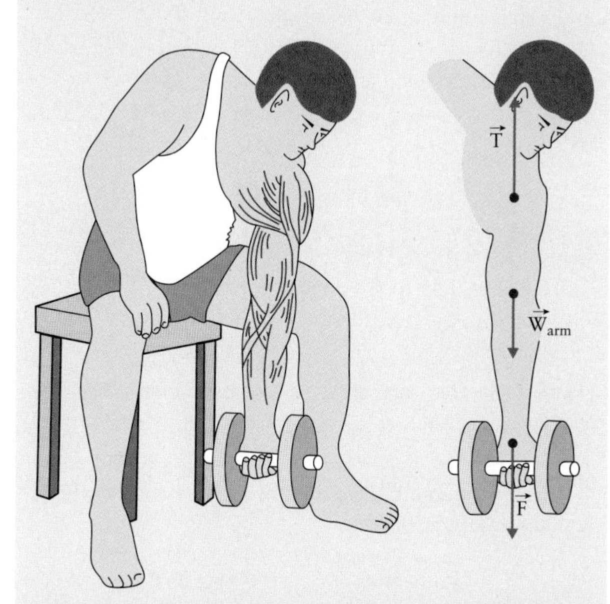

Figure 4.43 A standard man intends to do *concentration curls.* The right sketch indicates the major forces acting on the arm of the person: the tension $\vec{T}$ in the muscles and ligaments connecting the arm to the trunk, the weight of the arm, $\vec{w}_{arm}$, and the force, $\vec{F}$ acting on the fist as the interaction pair of the weight of the dumbbell.

Solution

Using Table 4.1, the mass of the left arm is 6.5% of the body mass of the standard man of 70 kg, i.e., 4.6 kg. The forces acting on the arm are the contact (normal) force between the fist and the dumbbell, $\vec{F}$, which is the force pulling down the fist, the weight of the arm, $\vec{w}_{arm}$, and the tension in the shoulder, $\vec{T}$. The person holds the dumbbell in the position shown, that is before doing an intended exercise with it. Thus, the arm and the dumbbell are at rest and the problem is an application of Newton's first law. The weight of the dumbbell and the force pulling the fist down are interaction pairs; so they are equal in magnitude.

Note: The tension in the shoulder is caused primarily in the tendon of the deltoid muscle and several ligaments that run across the interface between the trunk and the arm.

Note also that the question does not ask for a numerical value for one of the unknown parameters, but asks for a ratio instead. This type of question occurs frequently because such ratios help us to develop an intuitive idea of the magnitude of the effects we study.

Note: The forces related to the physiology in this problem are all given. You will find this to be typical for physiological or biological problems in this textbook because identifying the major forces acting in our bodies for a given exercise requires anatomical knowledge you may not yet have. Note that any one of these problems could be defined as a kinesiological research project, where identifying the major forces would consume most of the time invested in the project.

continued

Draw the free body diagrams for the arm and the dumbbell, as shown in Fig. 4.44.

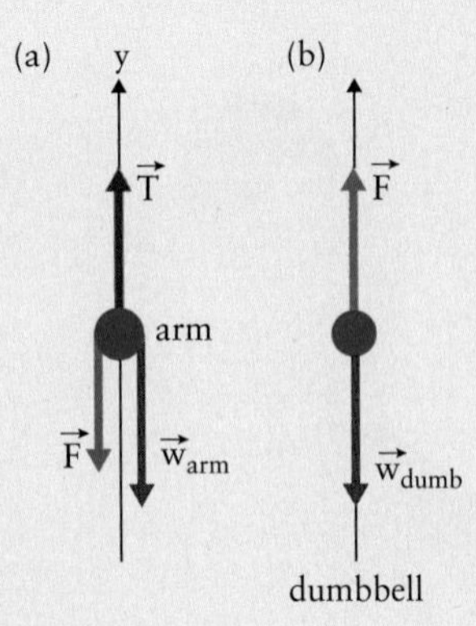

Figure 4.44 Free body diagrams for (a) the arm and (b) the dumbbell, separately.

Using the free body diagrams in Fig. 4.44, Newton's first law, Eq. [4.5], gives:

$$\text{for the arm:} \quad F_{net,y} = T - w_{arm} - F = 0$$

$$\text{for the dumbbell:} \quad F_{net,y} = F - w_{dumbb} = 0.$$

From the second equation, F is calculated:

$$F = w_{dumbb} = (m_{dumbb})\, g = 4\, g.$$

Substituting the values of F into the first formula gives:

$$T - w_{arm} - 4\, g = 0$$

$$T = w_{arm} + 4\, g.$$

Dividing these two equations gives:

$$\frac{T}{F} = \frac{w_{arm} + 4\, g}{4\, g} = \frac{w_{arm}}{4\, g} + 1$$

$$w_{arm} = m_{arm}\, g = 0.065\, m_{man}\, g = 0.065(70\text{kg})g = 4.6\, g.$$

Substituting the value for $\vec{w}_{arm}$ into the above ratio gives:

$$\frac{T}{F} = \frac{4.6\, g + 4\, g}{4\, g} = \frac{4.6\, g}{4\, g} + 1 = 2.15.$$

Ratio T/F shows that the tension in the shoulder is more than twice as large as the force pulling on the fist.

4.8.2: Gravity Detection in the Maculae

EXAMPLE 4.26

Determine the force acting perpendicular to the dendrites of each neuron in the *utricular macula* when it is tilted sideways by 30°. *Hint:* Fig. 3.33 shows the side view of the macula in the upright and the tilted positions. The mechanism of the macula was described earlier in Section 3.11.1 with Fig. 3.33.

Solution

Solution to part (a): Identify the system and the forces acting on it. Consider a single calcite *otolith* as the system, and the *otolithic membrane* of the macula as the environment. We cannot choose the dendrites directly as our system because we know too little about them.

Two forces act on the otolith in Fig. 3.33(upper): the otolith's weight $\vec{w}$ and a normal force $\vec{N}$ by the otolithic membrane.

Draw the free body diagram as shown in Fig. 4.45.

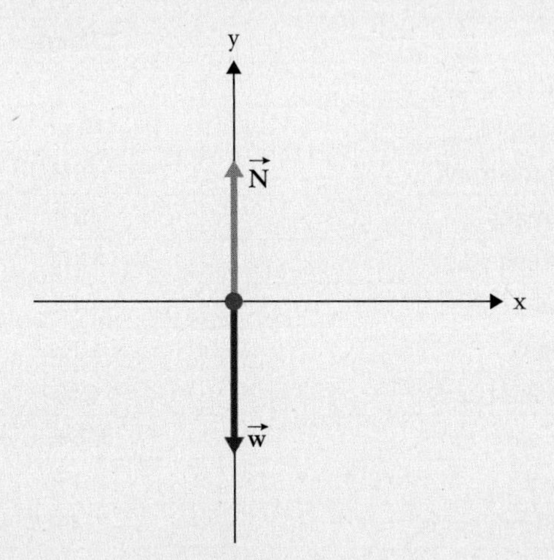

Figure 4.45 The free body diagram of a single otolith in the macula when the head is held upright.

The otolith is in mechanical equilibrium in both directions, so the net force on it is zero in both directions. Newton's first law in the y-direction gives:

$$\text{in } y\text{-direction: } N - w = 0$$

From this equation:

$$N = w = mg.$$

We do not know the magnitude of m, but there is information about the density and dimensions of the otolith. For calcite, $\rho = 3.0\ \text{g/cm}^3 = 3.0 \times 10^3\ \text{kg/m}$. From microscopic images of the macula we find that the length of an average otolith is about $l = 5\ \mu\text{m} = 5 \times 10^{-6}\ \text{m}$. Therefore:

$$V_{otolith} = (5 \times 10^{-6}\ \text{m})^3 = 1.25 \times 10^{-16}\ \text{m}^3$$

$$m = \rho\, V = (3.0 \times 10^3\ \text{kg/m}^3)(1.25 \times 10^{-16}\ \text{m}^3) = 3.75 \times 10^{-13}\ \text{kg}.$$

Substituting the magnitude of m into equation for N, gives:

$$N = (3.75 \times 10^{-13}\ \text{kg})\ (9.8\ \text{m/s}^2) = 3.7 \times 10^{-12}\ \text{N},$$

which gives the magnitude of the normal force for a single otolith.

Solution to part (b): Fig. 4.46(a) shows a simplified sketch of the otolith and the adjacent dendrites when

continued

the head is turned sideways by an angle θ. In addition to the otolith's weight, $\vec{w}$, and the normal force, $\vec{N}$, there is another force, a contact force along the interface between the otolithic membrane and the otolith, $\vec{F}_{gel}$. This contact force resists the slipping of the otolith, i.e., the gelatin pushes the otolith upward along the inclined surface. To have most forces coincide with the main axes, we chose y- and x-directions perpendicular and parallel to the surface plane of the membrane respectively.

The free-body diagram of a single otolith is shown in Fig. 4.46(b). The otolith is in mechanical equilibrium in both directions; so the net force on it is zero in both directions.

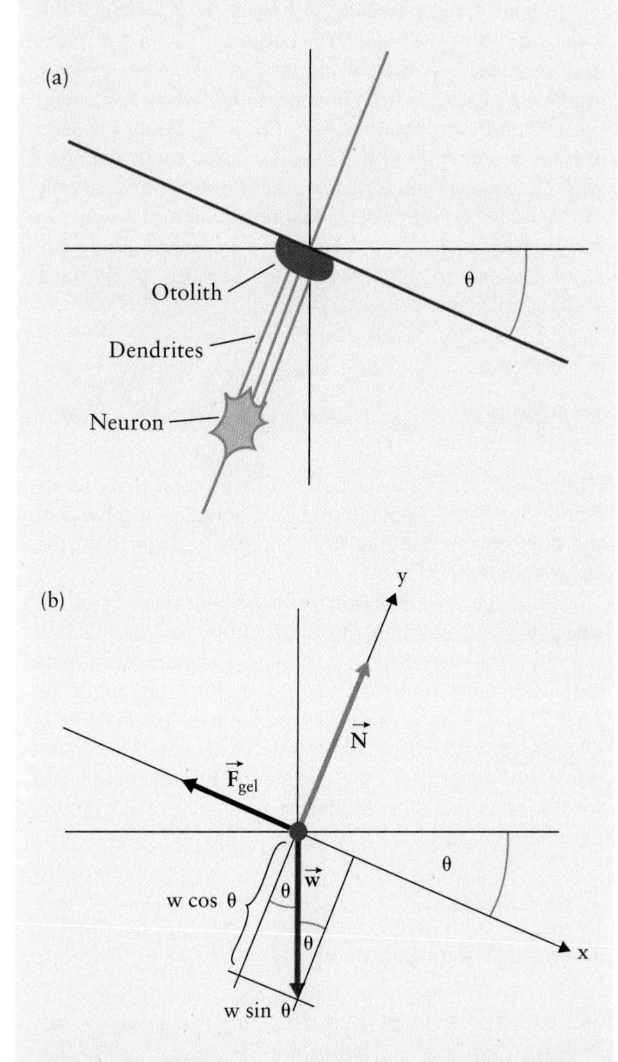

Figure 4.46 (a) Sketch of an otolith near the membrane surface of the utricular macula with the head tilted sideways by an angle θ. Dendrites near the otolith illustrate the geometry of the otolith–membrane–dendrite interaction. (b) The free body diagram of a single otolith as shown in part (a) with three forces acting on the system: its weight $\vec{W}$, the normal force $\vec{N}$, and a force parallel to the surface layer of the otolithic membrane $\vec{F}_{gel}$. The Cartesian coordinates are chosen with the x-axis parallel and the y-axis perpendicular to the membrane surface.

continued

Using Newton's first law in both the x- and the y-directions, gives:

$$\text{in } x\text{-direction:} \quad -F_{gel} + w\sin\theta = 0$$

$$\text{in } y\text{-direction:} \quad -w\cos\theta + N = 0$$

From equation (6) first equation, $\vec{F}_{gel}$ is calculated:

$$F_{gel} = w\sin\theta.$$

Substituting for the magnitude of m from part (a):

$$F_{gel} = (3.7\times10^{-12}\,\text{N})\sin(30°) = 1.85\times10^{-12}\,\text{N}.$$

The force F_{gel} calculated above is the force exerted by the otolithic membrane on a single otolith. However, due to the third law of mechanics, a force of the same magnitude is exerted by the otolith on the membrane, and in turn on the dendrites embedded in it. To determine the total lateral force on the dendrites of a single neuron, we multiply F_{gel} by the number of otoliths that interact with each neuron. This number is estimated from microscopic images to be about 15 otoliths per neuron. Since forces can be added, we estimate the magnitude of the lateral force on all dendrites of a single neuron to be $|F_{\text{Lateral on neuron}}| = 2.8 \times 10^{-11}$ N.

Table 4.2 allows us to compare this force with other typical forces. Note that the calculated force is comparably small due to the size of the objects involved.

Additional comments: The physical concepts we quantified in Example 4.25 explain why we are equipped with a utricular macula. The ability to detect the vertical direction allows us to judge the alignment of our head with the vertical; this information in turn allows our brain to coordinate our locomotion and correct the information received from other sensory components, such as the eyes. Indeed, due to the importance of our vision, a sideways tilt of the head leads to a horizontal realignment of the eyes.

TABLE 4.2

Some typical forces of physiological or biological relevance

Force	Magnitude N
Weight of a female blue whale (with a mass of up to 144 tonnes)	1.4×10^6
Weight of a male beluga whale	1.5×10^4
Typical contact force in a stressed joint (e.g., hip on head of femur)	1.7×10^3
Typical force exerted by a muscle on a large bone (e.g., abductor muscle on hip bone)	1.1×10^3
Weight of a standard man	6.9×10^2
Weight of an insect egg	3.5×10^{-6}
Weight of a single otolith in the macula of the vestibular organ	3.7×10^{-12}
Weight of a bacterium	2.0×10^{-18}

EXAMPLE 4.27

What horizontal force acts on the central dendrite in a *Pacinian corpuscle* when your hand, in pronation, accelerates with 90 m/s^2 horizontally? Note that this is a typical acceleration for a discus thrower just before releasing the discus.

Supplementary anatomical information: The term *pronation* describes the way the hand is held. Pronation means palm down and supination means palm up. In the current context, we conclude that the acceleration occurs parallel to the surface of the palm.

The corpuscles are located deep below the epidermis (see Fig. 3.34), with the central dendrite oriented either parallel to the skin surface or at about 45°. They are found primarily in the skin of the palms of our hands and the soles of our feet and serve as acceleration or vibration (cyclic acceleration) detectors.

Solution

We first identify the system and all the forces acting on it. Fig. 4.47(a) shows a schematic sketch of a Pacinian corpuscle. The *x*-axis is defined parallel and the *y*-axis

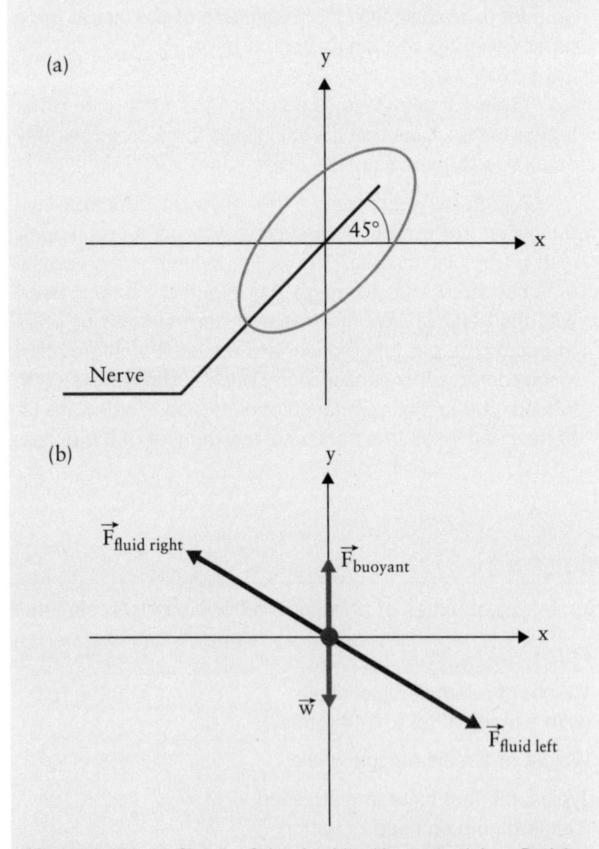

Figure 4.47 (a) Sketch of the dendrite (the system) in a Pacinian corpuscle tilted by 45° with respect to the palm. The *x*-coordinate is chosen parallel to the palm because it is the direction of the acceleration of the hand. (b) The corresponding free body diagram shows four forces acting on the dendrite: the weight and the buoyant force in the vertical direction, and two contact forces between the extracellular fluid and the dendrite surface.

continued

perpendicular to the palm surface. Four forces are identified:

- The weight $\vec{w}$ in the vertical direction downward.
- The buoyancy force, $\vec{F}_{buoyant}$, caused by the liquid in the vertical direction upward. The buoyant force is associated with the weight of the displaced fluid and is the reason why a diver can float underwater. This will be discussed in more detail in the Chapter 12.
- Two non-vertical contact forces that are due to the resting fluid pushing perpendicularly on the nerve from both sides: $\vec{F}_{fluid,left}$ is the force pushing the dendrite from the upper left side, and $\vec{F}_{fluid,right}$ is the force pushing the dendrite from the lower right side.

Figure 4.47(b) shows the free body diagram of the forces acting on the central dendrite. While the palm and thus the Pacinian corpuscle is at rest, a mechanical equilibrium exists in both directions, i.e., along the *x*- and *y*-axes, and Newton's first law, Eq. [4.5], could be used in both directions. But the question asks about the hand when it accelerates. Since the acceleration occurs only in the *x*-direction, a mechanical equilibrium still exists in the *y*-direction. So we use Newton second law, Eq. [4.10], in the *x*-direction and Newton's first law, Eq. [4.5], in the *y*-direction:

$$\text{in } x\text{-direction:} \quad F_{fluid,left,x} - F_{fluid,right,x} = ma_x$$

$$\text{in } y\text{-direction:} \quad F_{buoyant} - w - F_{fluid,left,y} + F_{fluid,right,y} = 0,$$

with $a_x = |a|$ since the acceleration occurs exclusively along the *x*-axis. According to first equation the force of the fluid on the dendrite from the left is larger than the force from the right.

To solve first equation we need the mass *m* of the dendrite. We estimate this mass from its volume and density. The dendrite in a typical Pacinian corpuscle extends slightly more than half the length of the corpuscle, $l \approx 0.33\ \text{mm}$. It is an unmyelinated nerve, as defined in Chapter18, with a typical diameter of 10 μm. We use the density of water ($\rho = 1.0\ \text{g/cm}^3$) as an approximate value for the nerve material. Modelling the nerve as a cylinder, its Volume, *V*, is:

$$V = \pi r^2 l = \pi(5\times10^{-6}\ \text{m})^2(3.3\times10^{-4}\ \text{m}) = 2.6\times10^{-14}\ \text{m}^3.$$

The mass of the nerve, *m*, is:

$$m = \rho V = (1\times10^3\ \text{kh/m}^3)(2.6\times10^{-14}\ \text{m}^3) = 2.6\times10^{-11}\ \text{kg}.$$

The difference in magnitude between the contact forces on the central dendrite in the first equation is:

$$\Delta F = F_{fluid,left,x} - F_{fluid,right,x} = ma_x$$

$$\Delta F = F_{fluid,left,x} - F_{fluid,right,x} = (2.6\times10^{-11}\ \text{kg})(90\ \text{m/s}^2)$$

$$\Delta F = F_{fluid,left,x} - F_{fluid,right,x} = 2.3\times10^{-9}\ \text{N}.$$

continued

What does this result mean? In order to accelerate a Pacinian corpuscle, the fluid must push the dendrite faster and faster to the left. In the same way the elastic capsule pushes the liquid, the skin pushes the Pacinian corpuscle, and the person moves the whole hand. Due to Newton's third law, the dendrite pushes in the opposite direction, i.e., toward the extracellular fluid that encloses it from the left. Since a fluid can flow, this force causes an evasion of the fluid. As this happens, the dendrite bends slightly to the left, triggering a signal in the nerve.

SUMMARY

DEFINITIONS

- Density: $\rho = m/V$, with m for mass and V for volume

UNITS

Force and force components: N (newton), with 1 N = 1 kg · m/s^2
Density: ρ: kg/m^3

LAWS

- Newton's first law (law of inertia) for an object in mechanical equilibrium:

$$\vec{F}_{\text{net}} = \sum_{i=1}^{n} \vec{F}_i = 0$$

- If the system's motion and the forces are confined to the xy-plane:

$$F_{\text{net},x} = \sum_{i=1}^{n} F_{i,x} = 0$$

$$F_{\text{net},y} = \sum_{i=1}^{n} F_{i,y} = 0$$

- Newton's second law (equation of motion) for an accelerating object of mass m:

$$\vec{F}_{\text{net}} = \sum_{i=1}^{n} \vec{F}_i = m\vec{a}$$

- If the system's motion and the forces are confined to the xy-plane:

$$F_{\text{net},x} = \sum_{i=1}^{n} F_{i,x} = ma_x$$

$$F_{\text{net},y} = \sum_{i=1}^{n} F_{i,y} = ma_y$$

- Newton's third law: When an object A acts on an object B with force $\vec{F}_{\text{BA}}$, then there is a force $\vec{F}_{\text{AB}}$ exerted by object B on object A with $\vec{F}_{\text{AB}} = -\vec{F}_{\text{BA}}$.

MULTIPLE-CHOICE QUESTIONS

MC–4.1. An object is in translational equilibrium if
(a) it is at rest.
(b) it moves with constant speed.
(c) the net force acting on it is zero.
(d) the net force acting on it is constant.
(e) only a single force is acting on it.

MC–4.2. Fig. 4.48 shows a round blue object on a table touching a green block. Which of the following six equations is the proper application of Newton's laws in the vertical direction describing the forces acting on the round object? Take $\vec{N}$ the normal force, $\vec{w}$ the weight, and $\vec{F}$ the contact force with the block.
(a) $\vec{F} + \vec{w} = 0$
(b) $\vec{F} - \vec{w} = 0$
(c) $\vec{N} + \vec{w} = 0$
(d) $\vec{N} - \vec{w} = 0$
(e) $\vec{N} - \vec{F} = 0$
(f) $\vec{N} + \vec{F} = 0$

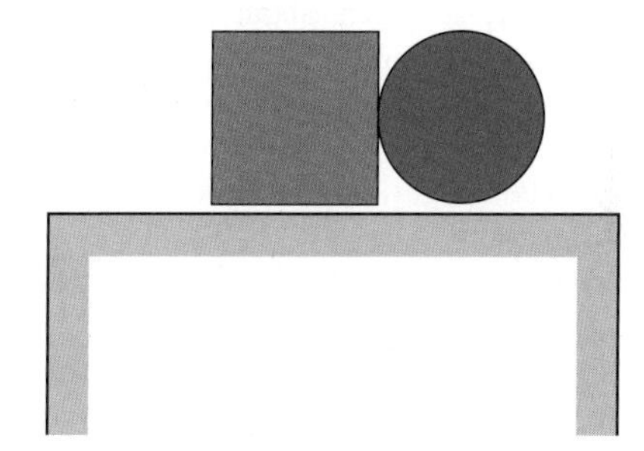

Figure 4.48

MC–4.3. Two forces act on a system, $\vec{F}_1$ $(F_1, 0)$ along the x-axis and $\vec{F}_2$ $(F_2, 0)$ along the y-axis. What is the minimum number of forces required in this case to establish mechanical equilibrium?
(a) 1
(b) 2
(c) 3
(d) 4
(e) 5

MC–4.4. An object of mass 92 g, which is initially at rest, attains a speed of 75.0 m/s in 0.028 s. What is the average net force acted on the object during this time interval?
(a) 1.2×10^2 N
(b) 2.5×10^2 N
(c) 2.8×10^2 N
(d) 4.9×10^2 N

MC–4.5. An object of mass m accelerates with acceleration a. How does the acceleration change if we double the mass of the object but keep the accelerating force unchanged?
(a) The acceleration remains unchanged.
(b) The magnitude of the acceleration doubles.
(c) The magnitude of the acceleration increases by a factor of 4.
(d) The magnitude of the acceleration is halved.
(e) Only the direction of the acceleration changes, but not its magnitude.

MC–4.6. When you lean heavily against a wall, you exert a force on it. This force is one member of an interaction pair. What is the other member of this pair?
(a) the force of friction between you and the wall
(b) the force of friction between your feet and the floor
(c) the normal force of the floor on you
(d) the force exerted by the wall on you
(e) the force exerted by you on the wall
(f) There is no other pair member to your force.

MC–4.7. A car while towing a trailer is accelerating on a level road. The force that the car exerts on the trailer is
(a) equal to the force that the trailer exerts on the car.
(b) equal to the force that the road exerts on the trailer.
(c) equal to the force that the car exerts on the road.
(d) greater than the force that the trailer exerts on the car.
(e) less than the force that the trailer exerts on the car.

MC–4.8. Due to Newton's third law we can make the following statement about the forces in Fig. 4.43. Which one is true?
(a) The arm exerts on the trunk a force that equals $-\vec{T}$.
(b) The weight $\vec{w}_{arm}$ is equal in magnitude to the tension $\vec{T}$.
(c) Forces $\vec{w}_{arm}$ and $\vec{T}$ are an action–reaction pair of forces.
(d) Forces $\vec{w}_{arm}$ and $\vec{F}$ are an action–reaction pair of forces.
(e) The weight of the arm has no reaction force.

MC–4.9. A big fish is hung on a spring scale that is hooked to the ceiling of an elevator. The scale shows the highest reading when the elevator
(a) moves downward with increasing speed.
(b) moves downward with decreasing speed.
(c) remains stationary.
(d) moves upward with decreasing speed.
(e) moves upward with constant speed.

MC–4.10. A small car is pushing a larger truck that has an engine problem. The truck is much heavier than the car. Which of the following statements is true?
(a) The car exerts a force on the truck, but the truck doesn't exert a force on the car.
(b) The car exerts a larger force on the truck than the truck exerts on the car.
(c) The car exerts the same amount of force on the truck as the truck exerts on the car.
(d) The truck exerts a larger force on the car than the car exerts on the truck.
(e) The truck exerts a force on the car, but the car doesn't exert a force on the truck.

MC–4.11. A 70.0 kg person stands on a bathroom scale in an elevator. What does the scale read, in kg, if the elevator is slowing down at a rate of 3.50 m/s^2. while descending?
(a) 49.5 kg
(b) 82.0 kg
(c) 83.7 kg
(d) 95.0 kg
(e) 111.0 kg

MC–4.12. A block at rest on a frictionless surface is acted on by two horizontal forces: $\vec{F}_1$ pushes it from left side and $\vec{F}_2$ $(\vec{F}_1 > \vec{F}_2)$ pulls it from right side. What additional horizontal force will keep the block at rest?
(a) $F = F_1 - F_2$ pushes the right side of the block.
(b) $F = F_1 - F_2$ pushes the left side of the block.
(c) $F = F_1 - F_2$ pulls the left side of the block.
(d) $F = F_1 - F_2$ pulls the right side of the block.
(e) $F = F_1 + F_2$ pulls the right side of the block.
(f) $F = F_1 + F_2$ pushes the left side of the block.
(g) $F = F_1 + F_2$ pushes the right side of the block.

MC–4.13. You want to move your desk to another corner of your room. You push it by a horizontal force of 275 N, but it doesn't budge. The 275 N force that you exert is one member of a pair of interaction forces; the other member of the pair is
(a) the frictional force exerted on the desk by the floor.
(b) the frictional force exerted on you by the floor.
(c) the force exerted on you by the desk.
(d) the normal force on you by the floor.
(e) the normal force on the desk by the floor.
(f) the force exerted by the desk on the floor.

MC–4.14. A box of mass m is pulled on a hard surface by a force $\vec{F}$ at angle of θ above horizontal. If the coefficient of kinetic friction between the block and the surface is μ_k, the acceleration of the block is
(a) $a = (F\cos\theta)/(\mu_k mg)$.
(b) $a = (F\cos\theta)/(\mu_k g)$.
(c) $a = (F\cos\theta)/m + (\mu_k mg)$.
(d) $a = F(\cos\theta + \mu_k \sin\theta)/m$.
(e) $a = F(\cos\theta + \mu_k \sin\theta)/m + (\mu_k mg)$.
(f) none of the above.

MC–4.15. A 69.1-kg skydiver reaches a terminal velocity of 30.0 m/s. What is the magnitude of the upward drag force on the skydiver due to air resistance?
(a) 9.80 N
(b) 339 N
(c) 452 N
(d) 677 N
(e) 1.02×10^3 N

MC–4.16. A horizontal force of 207 N is applied to a 63-kg box on a hard level floor. The coefficient of static friction and kinetic friction between the box and the floor are $\mu_s = 0.55$ and $\mu_k = 0.35$. What is the magnitude of the force of friction on the box?
(a) 207 N
(b) 216 N
(c) 340 N
(d) 556 N

MC–4.17. A person pushes a heavy book against the wall at an angle θ below horizontal. The book slides down the wall at a constant velocity. The coefficient of kinetic friction between the book and the wall is μ_k. Which one of the following answers best describes the magnitude of the force of kinetic friction between the book and the wall?
(a) mg
(b) $\mu_k F_{App} \sin\theta$
(c) $\mu_k mg$
(d) $\mu_k F_{App} \cos\theta$
(e) $\mu_k (F_{App} + mg)\sin\theta$
(f) $\mu_k (F_{App} + mg)\cos\theta$

MC–4.18. A sea lion of mass m is resting on a wooden plate that makes an angle of with the horizontal, as shown in Fig. 3.41. The static friction between the sea lion and the wooden plate is
(a) $f_s \geq mg$.
(b) $f_s \geq mg\cos\theta$.
(c) $f_s \geq mg\sin\theta$.
(d) $f_s = mg\cos\theta$.
(e) $f_s = mg\sin\theta$.

MC–4.19. In cases in which objects move through liquid at small speeds, the drag force is proportional to the speed v (linear relation). With the drag force being the only force acting in the horizontal direction on a fish that stops propelling its body forward, we expect that
(a) the fish slows down linearly with time, $v \propto t$ (t is time).
(b) the fish slows down until it reaches a finite terminal speed.
(c) the fish slows down but not with a linear relation between its speed and time.
(d) the fish comes instantaneously to rest.
(e) the fish slows down and is then pushed in the opposite direction by the drag force.

MC–4.20. Two identically shaped objects with same surface roughness and composition are placed on the same surface. One object is hollow and has mass m, the other object is solid and has mass $2m$. What is required to have both objects slide across the horizontal surface with the same frictional force?
(a) The hollow object must move twice as fast.
(b) The solid object must move twice as fast.
(c) The objects must move with the same speed across the surface.
(d) The hollow object must be pushed down with an additional force m g while it moves.
(e) The solid object must be pushed down with an additional force m g while it moves.

MC–4.21. Two objects are connected with a string. In the quantitative treatment of the problem we assume that the string exerts a tension force of magnitude T on each of the two objects. What has to be the case for this assumption to be valid?
(a) The string hangs loose between the objects.
(b) The string does not pass over a pulley.
(c) The string is massless.
(d) The two objects are not on a collision course with each other.

MC–4.22. The full title of Newton's seminal work on mechanics is *Philosophiae Naturalis Principia Mathematica* and it contains four statements labelled *lex prima, lex secunda, lex tertia,* and *lex quarta.* The second law, *lex secunda,* is formulated in a different way than we have used it in this chapter, but we will provide Newton's original formulation in Chapter 5. Newton's *lex quarta* states in translated form that "forces can be added like vectors." Why do we no longer identify this statement as Newton's fourth law?
(a) It is wrong—even Newton made errors once in a while.
(b) It is redundant—the information it contains isn't needed to establish the laws of mechanics.
(c) It has been combined with the first law in our modern formulation of Newton's laws.
(d) It has been combined with the second law in our modern formulation of Newton's laws.
(e) It has been combined with the first and second laws in our modern formulation of Newton's laws.

CONCEPTUAL QUESTIONS

Q–4.1. An object hangs in the middle of a rope whose ends are fixed at the same horizontal level. Can the rope stay perfectly horizontal? Why? Explain your answer.

Q–4.2. You are in a car, sitting next to the driver. If she suddenly pushes the gas pedal and accelerates rapidly, you feel a push backward against your seat. If she suddenly pushes the brake and comes to a stop, you feel a push forward toward the dashboard. Explain these situations according to the most suitable of Newton's laws.

Q–4.3. We study a person in a gym intending to do *seated rows.* In this exercise, the person sits on a bench facing the exercise equipment. The feet are placed against the foot stops. The person leans toward a pulley and holds the handle on a string. Then the person arches his back and pulls the handle until it touches the lower ribcage. The main muscles needed in this exercise are the deltoid muscle in the shoulder and the upper back muscles. When the arms are stretched out horizontally, the major forces acting on the arm include the tension in the deltoid/back muscles, $\vec{T}$, which acts in the direction away from the pulley; the force, $\vec{F}$, due to the shoulder bones pushing the upper arm bone against the pull of the tension force; the weight of the arm, $\vec{w}_{arm}$; and a force $\vec{F}_{handle}$ due to the mass of the handle. Which choice in Fig. 4.49 is the correct free-body diagram for the horizontal forces acting on the person's arm?

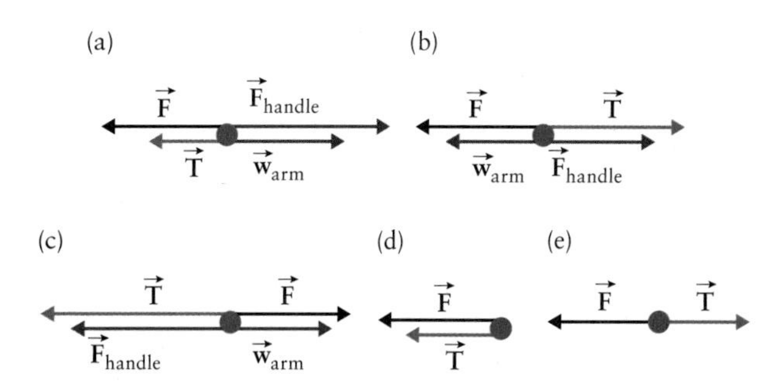

Figure 4.49

Q–4.4. A tablecloth can be removed from beneath dishes very quickly and the dishes on the tablecloth will not move. Explain why.

Q–4.5. Two equal forces are acting on four similar blocks sitting on a frictionless surface, as shown in Fig. 4.50. Which of these blocks may be at rest or moving uniformly?

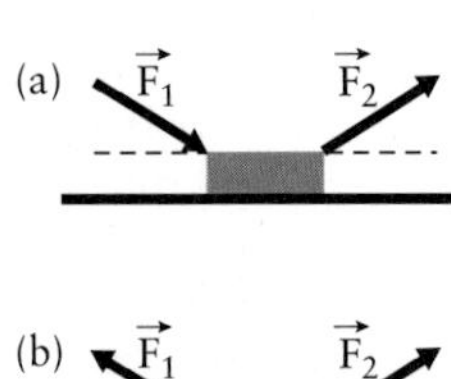

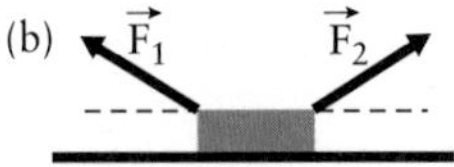

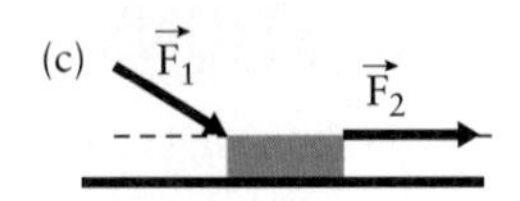

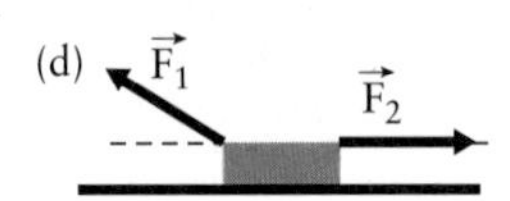

Figure 4.50

Q–4.6. On a snowy day you have to shovel the snow from the driveway. You scoop some snow from the driveway with your shovel and throw it to the side? How does the snow leave the shovel? Explain it according to the most appropriate of Newton's laws.

Q–4.7. (a) Only one force acts on an object. Can the object be at rest? (b) An object has zero acceleration. Does this mean that no forces act on it?

Q–4.8. Is it possible for an object to move if no net force acts on it?

Q–4.9. Is it possible that a net force on an object can accelerate the object in the direction opposite to the net force?

Q–4.10. Two forces $\vec{F}_1$ and $\vec{F}_2$ act on an object. It accelerates in a particular direction. Under what circumstances can you predict the direction of force $\vec{F}_1$ regardless of the specific value of force $\vec{F}_2$?

Q–4.11. If no forces other than the three forces shown act on the leg in Fig. 4.7, (a) what is the state of motion of the leg, and (b) in what direction would a fourth force have to act to establish mechanical equilibrium for the leg? Can you suggest the origin of such a force? (c) Identify the other member of the interaction pair to each of the three forces shown in Fig. 4.7.

Q–4.12. When a mule is pulling a cart from rest, he applies a force on it; the cart applies a force with the same magnitude and in opposite direction on the mule, according to Newton's third law. Explain how it is possible for the mule to make the cart begin to move.

Q–4.13. Two skaters of different masses push against each other on ice. (a) Which one will push harder? Explain your answer. (b) Which one will gain the larger velocity?

Q–4.14. We compare the terminal speed of a raindrop of mass 30 mg and an equally heavy ice pellet during a hailstorm. Noting the difference in density, with a value of $\rho = 1.0\ \text{g/cm}^3$ for liquid water and $\rho = 0.92\ \text{g/cm}^3$ for ice, (a) do both fall with the same terminal speed (since the density of the falling object is *not* a factor in Eq. [4.17]); and (b) if they fall with different terminal speeds, which one is faster?

Q–4.15. Why is the coefficient of kinetic friction smaller than the coefficient of static friction? (a) Give a reason based on the microscopic roughness of the interface, and (b) give a logical reason by looking at Eqs. [3.8] and [3.9] and thinking about what they imply.

Q–4.16. (a) When we refer to the shape of an animal as aerodynamic, what formula—and particularly what *term* in that formula—do we have in mind? (b) What is the underlying physics of an aerodynamic shape?

Q–4.17. Newton reported the following observation in *Principia*: Let's assume that two objects attract each other. Contrary to the third law of mechanics, however, we assume that object B is attracted by object A more strongly than object A is attracted by object B. To test this case further, we connect objects A and B with a stiff but massless string, forcing both objects to maintain a fixed distance. Since the force on object B is stronger than on object A, a net force acts on the combined object: it will accelerate without the action of an external force. Since this contradicts the first law of mechanics, we conclude that no violation of the third law is possible and that the third law is already included when we introduced the first law. Based on this argument, is the third law not required as a separate law?

Q–4.18. We live in a three-dimensional space. Consequently, each vestibular organ contains three orthogonal semicircular canals to measure the independent Cartesian components of the acceleration of the head. Why, then, do we have only two maculae per ear and not three?

ANALYTICAL PROBLEMS

P–4.1. One cubic centimetre (1 cm^3) of water has a mass of one gram (1 g). (a) Determine the mass of one cubic meter (1 m^3) of water. (b) Assume that a spherical bacterium consists of 98% water and has a diameter of (1.0 μm). Calculate the mass of its water content. (c) Modelling a fly as a water cylinder of 4 mm length and 1 mm radius, what is its mass? *Note:* Important formulas for volume and surface of two- and three-dimensional symmetric objects are listed in the Math Review section "Symmetric Objects" following Chapter 27.

P–4.2. (a) A 5.0-kg fish is held by a taut string. If the tension in the string is 49 N, what kind of motion does it have? (b) If the fish accelerated upward at a rate of 1.2 m/s^2, what would be the tension in the string?

P–4.3. An airplane is in level flight. Identify all forces acting on the plane and show them in a simple sketch. Is the airplane in equilibrium?

P–4.4. A 5.0-kg block is suspended by three taut strings as shown in Fig. 4.51. Find the tension in the strings.

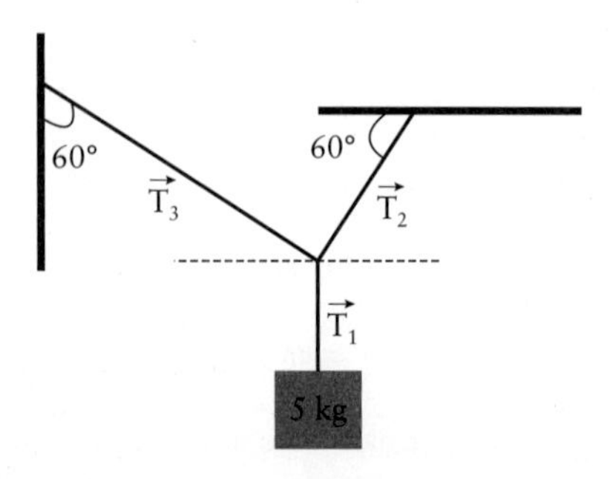

Figure 4.51

P–4.5. Large hawks, eagles, vultures, storks, the white pelican, and gulls are North American birds that sail on rising columns of warm air. This static soaring requires only 5% of the effort of flapping flight. The birds are essentially in a level flight, holding their wings steadily stretched. The weight of the bird is balanced by a vertical lift force, which is a force exerted by the air on the bird's wings. How large is the lift force for (a) a Franklin's gull (found in Alberta, Saskatchewan, and Manitoba) with an average mass of 280 g, and (b) an American white pelican (found in Western Canada) with an average mass of 7.0 kg?

P–4.6. Two horizontal forces, $\vec{F}_1$ and $\vec{F}_2$, are pulling an object of mass $m = 1.5$ kg from two opposite sides. $\vec{F}_1$ pulls to the right and $\vec{F}_2$ pulls to the left. The magnitude of $\vec{F}_1$ is $F_1 = 25$ N. The object moves strictly along the horizontal x-axis, which we chose as positive to the right. Find the magnitude of $\vec{F}_2$ if the object's horizontal acceleration is (a) $a = 10\ \text{m/s}^2$; (b) $a = 0\ \text{m/s}^2$, and (c) $a = -10\ \text{m/s}^2$.

P–4.7. Fig. 4.18 shows two objects, with masses $m_1 = 2.0$ kg and $m_2 = 1.0$ kg, in contact on a frictionless surface. A horizontal force $\vec{F}$ is applied to the object with mass m_1. (a) Calculate the magnitude of the contact force $\vec{F}_c$ between the two objects. (b) Find the magnitude of the force $\vec{F}_c$ between the two objects if the force $\vec{F}$ is instead applied to the object of mass m_2 but in the opposite direction.

P–4.8. Fig. 4.52 shows two blocks of masses m and m. The horizontal surface allows for frictionless motion. The string tied to the two blocks is massless and passes over a massless pulley that rotates without friction. (a) What resulting motion of the two blocks do you predict? If M = 3.0 kg and m = 2.0 kg, (b) find the magnitude of the acceleration of the sliding block, (c) find the magnitude of the acceleration of the hanging block, and (d) find the magnitude of the tension in the massless string.

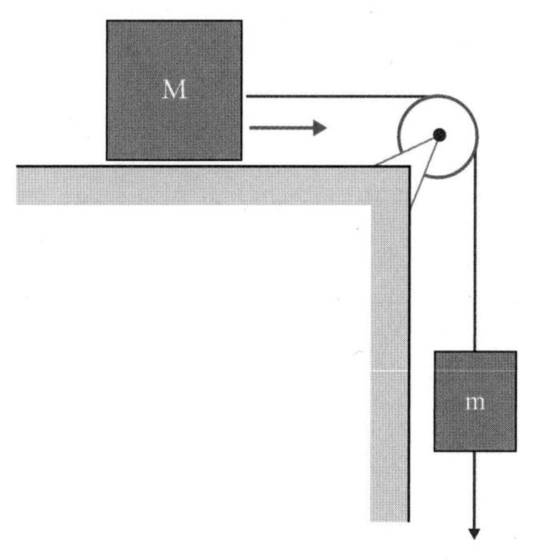

Figure 4.52

P–4.9. Fig. 4.53 shows two objects that are connected by a massless string. They are pulled along a frictionless surface by a horizontal external force. Using $F_{ext} = 50$ N, $m_1 = 10$ kg, and $m_2 = 10$ kg, calculate (a) the magnitude of the acceleration of the two objects, and (b) the magnitude of the tension $\vec{T}$ in the string.

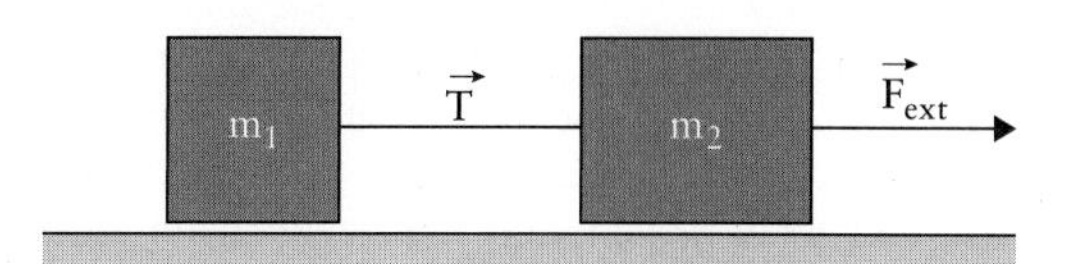

Figure 4.53

P–4.10. Fig. 4.54 illustrates the treatment of fractures of the humeral shaft with a cast on the upper arm bent by 100°–110° at the elbow joint, and an attached traction device. The traction device keeps the upper arm under tension to lower the risk of fracture dislocation due to muscle contraction. Similar traction devices can be applied for most fractures of the appendicular skeleton and for the cervical vertebrae after neck injuries. What is the range of masses for the traction device used if the muscle force it must compensate ranges from 5 N to 10 N?

Figure 4.54

P–4.11. Fig. 4.55 shows an object on a frictionless surface that forms an angle $\theta = 40°$ with the horizontal. The object is pushed by a horizontal external force $\vec{F}_{ext}$ such that it moves with constant speed. If the mass of the object is $m = 75$ kg, calculate (a) the magnitude of the external force $\vec{F}_{ext}$, and (b) the direction and the magnitude of the force exerted by the inclined surface on the object.

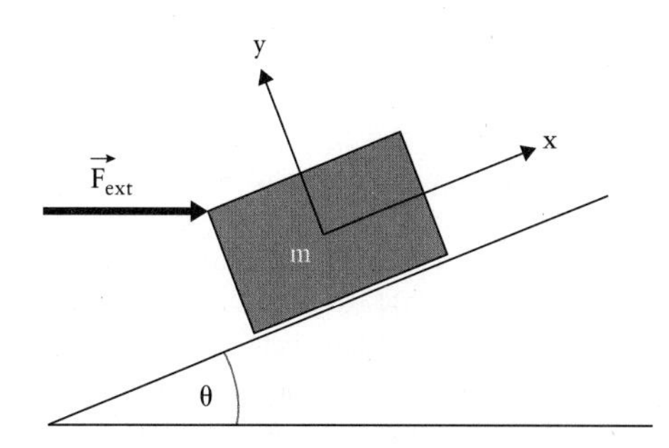

Figure 4.55

P–4.12. Fig. 4.56 shows a standard man hanging motionless from a horizontal bar. (a) Assume that the arms are stretched at 20° to the vertical on either side, as shown in the middle sketch. Find the magnitude of the force acting on each arm. (b) Assume that the arms are held at the two different angles shown in the bottom sketch. What are the magnitudes of the forces acting in each arm in this case? *Note:* The standard man data are given in Table 4.1.

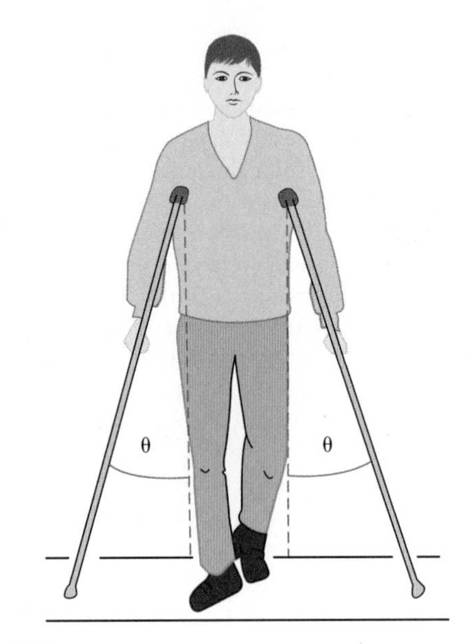

Figure 4.57

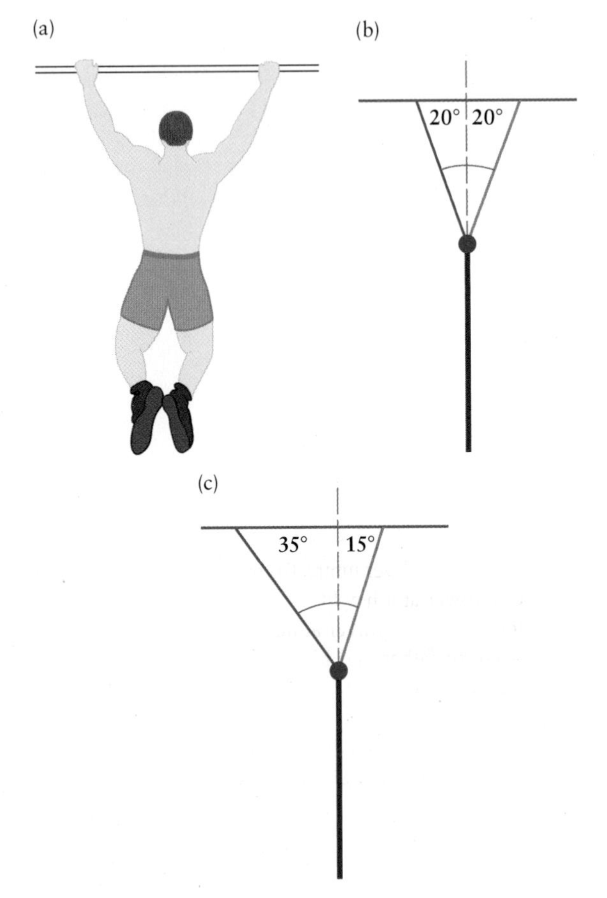

Figure 4.56

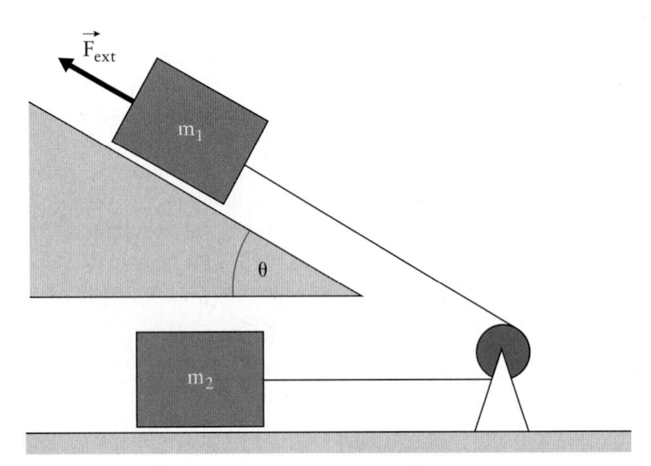

Figure 4.58

P–4.13. In Fig. 4.57 a standard man uses crutches. The crutches each make an angle $\theta = 25°$ with the vertical. Half the standard man's weight is supported by the crutches; the other half is supported by the normal forces acting on the soles of the feet. Assuming the standard man is motionless, find the magnitude of the force supported by each crutch.

P–4.14. Fig. 4.58 shows an object of mass $m_1 = 1.0$ kg on an inclined surface. The angle of the inclined surface is $\theta = 25°$ with the horizontal. The object m_1 is connected to a second object of mass $m_2 = 2.05$ kg on a horizontal surface below an overhang that is formed by the inclined surface. Further, an external force of magnitude $F_{ext} = 10$ N is exerted on the object with mass m_1. We observe both objects accelerate. Assuming the surfaces and the pulley are frictionless, and the pulley and the connecting string are massless, what is the tension in the string connecting the two objects?

P–4.15. An object of mass $m = 6.0$ kg accelerates at a rate of $2.0\ m/s^2$. (a) What is the magnitude of the net force acting on it? (b) If the same force is applied on an object with mass $M = 4.0$ kg, what is the magnitude of that object's acceleration?

P–4.16. A football punter accelerates a football from rest to a speed of 10 m/s during the time in which the shoe makes contact with the ball, typically $t = 0.2$ s. Using $m = 0.5$ kg as the mass of the football, what average force does the punter exert on the football?

P–4.17. The leg and cast in Fig. 4.59 have a mass of $w_{leg} = 22.5$ kg. Determine the mass of object 2, m_2, and the angle θ needed in order that there be no force acting on the hip joint by leg plus cast. Note that object 1 has a mass of $m_1 = 11$ kg and $\phi = 40°$.

P–4.18. Find the force applied to the patient's head by the traction device shown in Fig. 4.60.

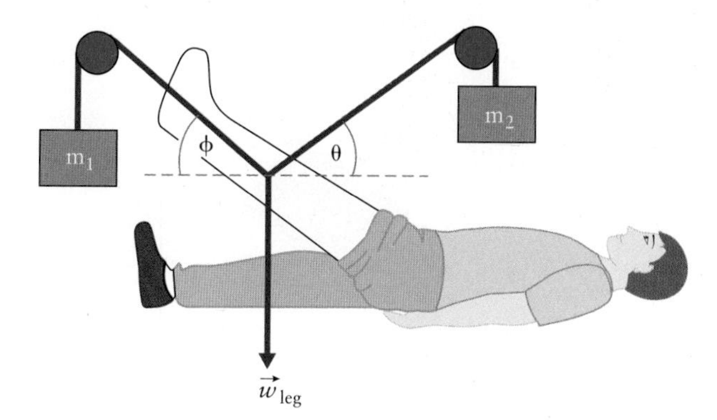

Figure 4.59

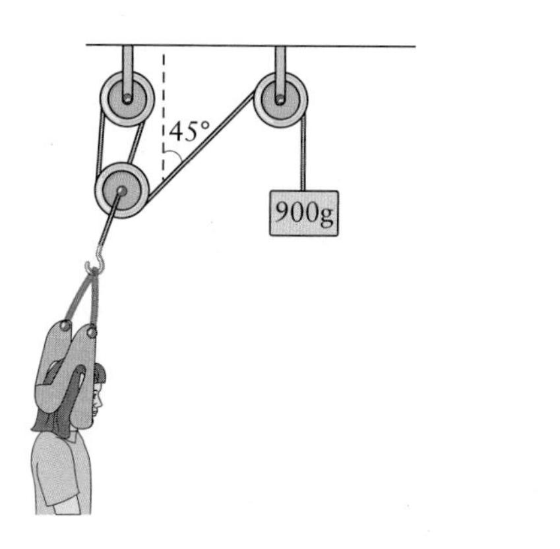

Figure 4.60

P–4.19. Fig. 4.61 shows a Russell traction apparatus for femoral fixation for a patient with a broken leg. (a) Find the magnitude of the total force, F_{ap}, applied to the leg by the traction apparatus assuming a 36 N weight is hanging from this apparatus, as shown. (b) What is the horizontal component of the traction force acting on the leg? (c) If the leg also weighs 36 N, what is the total magnitude of the resultant force of $F_{ap} + F_g$ on the leg? (d) What is the magnitude of the force, F_c, exerted on the femur by the leg?

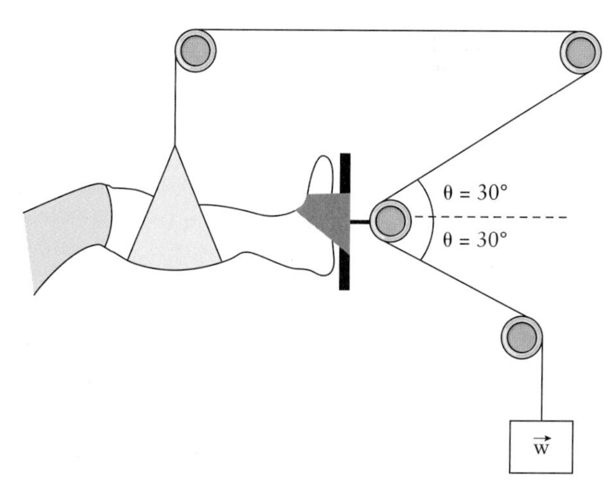

Figure 4.61

P–4.20. (a) A 56-kg climber is moving down a vertical cliff by pushing his feet against its surface and sliding down a rope, as shown in Fig. 4.62(a). Assuming the forces exerted by his feet are perpendicular to the face of the cliff, find the tension in the rope. Now, suppose the angle between the cliff face and the rope remains 12° but the cliff face makes an angle of 15° with the vertical, as shown in Fig. 4.62(b). If the force exerted by the climber's feet is again perpendicular to the cliff face, what is the tension in the rope now?

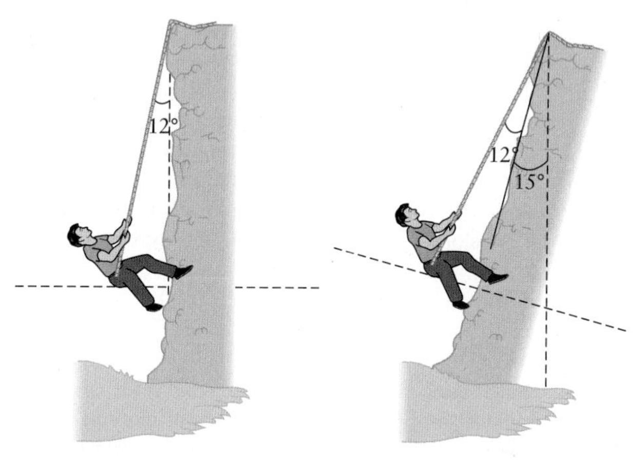

Figure 4.62

P–4.21. A 75-kg skydiver jumps from the top floor of Toronto's CN Tower at a height of $h = 447$ m. What is the drag force on the skydiver at his terminal velocity?

P–4.22. A 64-kg father and his 16-kg teenaged daughter are ice skating. They stand in front of each other face to face. The daughter pushes on her father such that he accelerates at 1.5 m/s^2 during the push. (a) What is the magnitude of the force the daughter applies on her father? (b) Does the daughter accelerate herself during the push? What is the daughter's acceleration?

P–4.23. At the instant of starting a race, a 63-kg sprinter exerted a force of 820 N on the starting blocks at an angle of 21° with respect to the ground. (a) What acceleration did the sprinter attain? With what speed did she leave the starting blocks if she exerted the force for 0.4 s.

P–4.24. (a) An object moves with an initial speed $v_0 = 10.0$ m/s on a horizontal surface. It slides for a distance of 20.0 m before it comes to rest. Determine the coefficient of kinetic friction between the object and the surface. (b) How long does the object move until it comes to rest? (c) With what speed does the object move after the same length of time if the surface is tilted to an angle of $\theta = 10°$ with respect to the horizontal and the object is sliding downhill?

P–4.25. Mary was driving on a country road at 80.0 km/h. After leaving a curve, she suddenly saw a deer standing in the middle of the road in front of her and she pressed the brakes. (a) If it took her 2 s to react, and the deer was 200 m away from her, what should be her acceleration be to avoid hitting the deer? (b) If she gained this deceleration from the road by the kinetic friction between the road and her sliding tires, what was the coefficient of the kinetic friction?

P–4.26. Fig. 4.4 shows two objects connected with a massless string. Object 1 has a mass $m_1 = 1.0$ kg and is placed on a horizontal surface with a coefficient of static friction $\mu_s = 0.35$ and a coefficient of kinetic friction $\mu_k = 0.25$. Object 2 has an unknown mass m_2. The connecting string passes over a massless and frictionless pulley. (a) Calculate the minimum mass m_2 that would allow the two objects to start moving when released from rest. (b) Once the two objects move with m_2 at the value calculated in part (a), what is the magnitude of the acceleration of the two objects?

P–4.27. A standard man rides a bicycle. Resistance against the forward motion is due to air resistance, friction of the tires on the road, and the metal-on-metal sliding in the lubricated axles. Combining the latter two effects as a velocity-independent friction force $\vec{f}$, determine the speed above which air resistance limits the cyclist's motion. $\vec{f}$ is determined from the fact that the cyclist rolls with constant speed downhill on a road that has a 1% slope. Use $\rho_{air} = 1.2$ kg/m^3 and a cross-sectional area $A = 0.5$ m^2 for the bicycle/standard man system in the direction of motion. Use $D = 0.5$ for the drag coefficient.

P–4.28. An object of mass m 20 kg is initially at rest on a horizontal surface. It requires a horizontal force $F = 75$ N to set it in motion. However, once in motion, only a horizontal force of 60 N is required to keep it moving with a constant speed. Find the coefficients of static and kinetic friction in this case.

P–4.29. In Fig. 4.4, assume $m_1 = 10$ kg and $m_2 = 4.0$ kg. The coefficient of static friction between the object on the table and the table surface is 0.5 and the coefficient of kinetic friction is 0.3. The system is set in motion with the object of mass m_2 moving downward. What is the acceleration of the object of mass m_1?

P–4.30. In Fig. 4.63, the coefficient of static friction is 0.3 between an object of mass $m = 3.0$ kg and a surface that is inclined by $\theta = 35°$ with the horizontal. What is the minimum magnitude of force $\vec{F}$ that must be applied to the object perpendicular to the inclined surface to prevent the object from sliding downward?

P–4.31. Two crates of masses $m_1 = 8$ kg and $m_2 = 12$ kg are connected by a massless string and placed on a rough surface. A force of $F = 96$ N is exerted on m_2, as illustrated in Fig. 4.53. The coefficient of kinetic friction between each crate and the surface is $\mu_k = 0.4$. What is the tension on the string between the two crates?

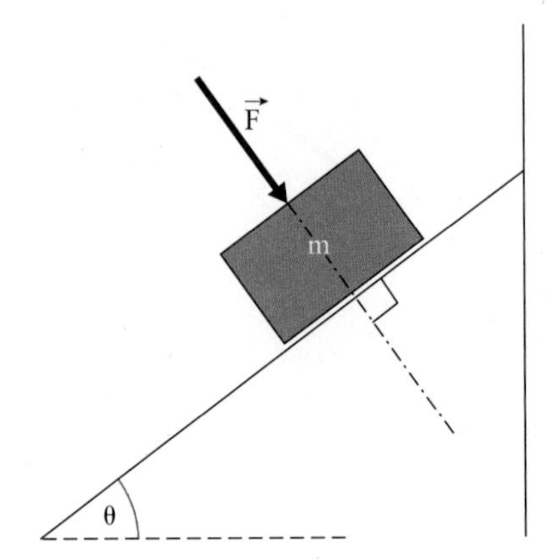

Figure 4.63

P–4.32. If the push on the baby sea lion of Example 4.14 is applied at the angle of 22° above the horizontal, what is its acceleration?

P–4.33. A box of mass 26 kg is hung by a thin string from the ceiling of an elevator. The string can hold up to 276.1 N and will break under larger tension force. How much can the maximum acceleration of the elevator be going upward so that the string does not break and continues to hold the box? Does the string hold the box if the elevator descends with the same magnitude of acceleration as it ascended?

P–4.34. A jellyfish floats motionless in water. At time $t = 0$ it expels water by contracting its bell, accelerating its body to a speed v_0. Determine the mathematical formula that describes the change of the jellyfish's speed until it returns to rest. How long does it take for the jellyfish to come to rest? *Note:* The drag force on the moving jellyfish is modelled by Stokes' law for drag force due to the motion of a sphere through a liquid: $\vec{F} = -kv$, where $k =$ const. This problem requires basic knowledge of differential equations. *Hint:* The jellyfish initially floats motionless because its weight is compensated by the buoyant force, which we introduce in Chapter 12. Thus, the drag force is the only unbalanced force acting on the jellyfish when in motion.

ANSWERS TO CONCEPT QUESTIONS

Concept Question 4.1: d

Concept Question 4.2: As a passenger, your body moves with the same velocity as the car while it is being driven. When the driver pushes the brakes, they apply stopping force on the car but not on your body. So, according to Newton's first law, your body will continue moving forward with the same velocity.

Concept Question 4.3: When you change your hair's direction of motion in shaking and apply force on it, the water droplets do not feel this force. According to Newton's first law, they continue to move, leaving your hair. The same thing happens when a dog shakes its body: the water leaves its body, according to Newton's first law.

Concept Question 4.4: Objects (3), (4), and (6). Mechanical equilibrium requires a balance of forces in every direction. We need to test only along the Cartesian coordinates, i.e., in Fig. 4.6 the horizontal x-axis and the vertical y-axis.

Concept Question 4.5: No. There is only one force, $\vec{F}$, that has a component in the horizontal direction. Thus, in the horizontal direction no balance of the force components exists as required by Eq. [4.5], so the leg is not in equilibrium.

Concept Question 4.6: D

Concept Question 4.7: D

Concept Question 4.8: D

Concept Question 4.9: D

Concept Question 4.10: According to Newton's third law, if the astronaut pushes the jet pack away from his spaceship, the jet pack pushes the astronaut back toward the spaceship.

Concept Question 4.11: This is a frequently stated misconception. Keep in mind that interaction pair (action and reaction) forces must act on different objects, but only forces that act on the object of interest determine the net force that is then used in Newton's first or second law. No forces exist without a reaction force.

Concept Question 4.12: A and E

CHAPTER 5

Centre of Mass and Linear Momentum

So far, we have limited our discussion to systems with two or three objects. In this chapter, we will develop an approach that will allow us to understand the motion of more complicated systems, such as those that a life scientist is more likely to encounter in nature. We approach such systems in several steps, first defining a new concept, the centre of mass, and then illustrating how the centre of mass allows us to simplify the description of physical processes for multi-object systems. We will see that the centre of mass is related to a fundamental physical parameter called linear momentum.

Linear momentum is a conserved quantity that can be used to model the behaviour of complex systems much more easily than we could by using only forces. However, the definition of an isolated system becomes very important when using the conservation of linear momentum, since linear momentum is only conserved in a system that is not being acted on by external forces. Although that sounds like a prohibitive restriction, we will see that conservation of linear momentum can be used to provide details of outcomes of two special types of collisions: elastic and perfectly inelastic. These two types of collisions describe opposite extremes and can be used to establish boundaries within which all collisions must take place.

5.1: Centre of Mass Definition

The **centre of mass** of a system of objects is a position in space mathematically described by a vector $\vec{r}_{\text{c.m.}}$:

$$\vec{r}_{\text{c.m.}} = \frac{1}{M}\sum_{i}^{n} m_i \vec{r}_i. \qquad [5.1]$$

In component form, this vector is given by $\vec{r}_{\text{c.m}} = (x_{\text{c.m.}}, y_{\text{c.m.}}, z_{\text{c.m.}})$, with:

$$\begin{aligned} x\text{-direction: } x_{\text{c.m.}} &= \frac{1}{M}\sum_{i=1}^{n} m_i x_i \\ y\text{-direction: } y_{\text{c.m.}} &= \frac{1}{M}\sum_{i=1}^{n} m_i y_i \\ z\text{-direction: } z_{\text{c.m.}} &= \frac{1}{M}\sum_{i=1}^{n} m_i z_i \end{aligned} \qquad [5.2]$$

where M represents the total mass of a system that consists of n objects of mass m_i at respective positions (x_i, y_i, z_i).

For systems that consist of a finite number of objects, it is straightforward to use Eq. [5.2] to calculate this position. This is illustrated in Case I below. Most objects, however, have a more complex distribution of mass. Eq. [5.2] is not suitable in these cases because the system would have to be broken down into a large number of small objects, and the sum in Eq. [5.2] would be extensive. Further, we may not know the structural details of the system well enough, and a division into small objects would be arbitrary. Two alternative approaches are available in these cases:

- We assume that all parts of the object are at fixed positions relative to each other and use a modified form of Eq. [5.2], as discussed in Case II; or
- We make no assumptions but use experimental methods or even "educated guesses," as illustrated in Case III. This includes all cases where the relative positions of the small objects that constitute the system vary as a function of time.

5.1.1: Case I: A System of Discrete Particles

In this textbook we frequently discuss objects that can be treated as if their volume is negligible. To simplify the discussion in the current section, where we also discuss extended objects, we will use the term particle to describe an object with negligible volume.

KEY POINT

A discrete particle is an object of negligible volume that is physically separated from all other particles in the system.

Most molecules consist of a small number of atoms that can be treated as n localized particles with given masses. The relative positions of the particles within a molecule are well defined. The simplest case is binary molecules, such as carbon monoxide. We consider this simple case first as it allows us to connect Eq. [5.2] with our well-developed intuition for the centre of mass in everyday situations.

We start by choosing the line through the two particles of a binary system as the x-axis, as illustrated in Fig. 5.1. This means that we need to consider only the first formula in Eq. [5.2]. For $n = 2$, that formula transforms into:

$$x_{\text{c.m.}} = \frac{x_1 m_1 + x_2 m_2}{m_1 + m_2} \qquad [5.3]$$

because the total mass is $M = m_1 + m_2$.

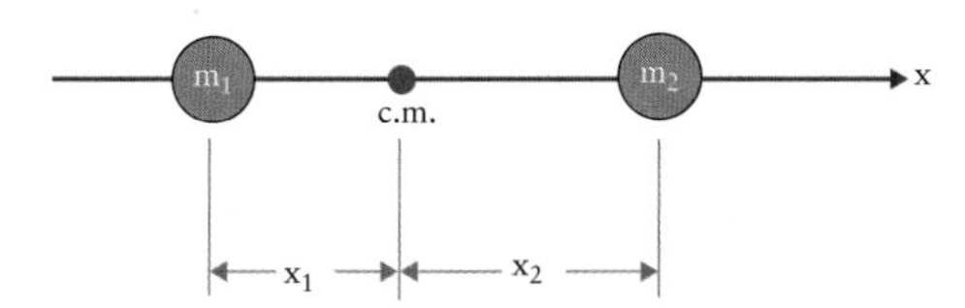

Figure 5.1 Illustration of the centre of mass for two objects positioned along the x-axis. Their masses are m_1 and m_2. The centre of mass is labelled c.m.

A frequently occurring case is a system with two particles of equal mass ($m_1 = m_2$), for example, molecules such as oxygen or nitrogen. In this special case, the centre of mass lies halfway between the two particles at the position $x_{\text{c.m.}} = (1/2)(x_1 + x_2)$. Another frequently occurring case is a system in which one particle has a mass much greater than the other particle, e.g., $m_1 >> m_2$. Examples include molecules such as HCl, HBr, and HI. This case leads to the centre of mass being at approximately the same position as the much heavier particle $x_{\text{c.m.}} \cong x_1$.

EXAMPLE 5.1

The four hydrogen atoms in the methane molecule CH_4 form a regular tetrahedron. Fig. 5.4 illustrates the positions of the hydrogen atoms with respect to a Cartesian coordinate system. The carbon atom is located at the centre of mass of the four hydrogen atoms. Assuming that all five atoms can be treated as particles, use the atomic positions shown in Fig. 5.4 to calculate the x-, y-, and z-coordinates of the position of the carbon atom.

Solution

The methane molecule is a three-dimensional structure. Thus, Eq. [5.2] is used to calculate the three Cartesian components of the position of the carbon atom in Fig. 5.4. In each case, the total mass of the four hydrogen atoms in the denominator is equal to $M = 4\ m_H$, where m_H is the mass of a single hydrogen atom. For the x-component of the centre of mass position, the figure contains two hydrogen atoms, H_B and H_D, with $x_i = l$. The other two hydrogen atoms, H_A and H_C, lead to zero contributions to the sum because their x-components are $x_i = 0$. Proceeding in the same fashion for the other components, we find:

$$x_{\text{c.m.}} = \frac{l m(H_B) + l\, m(H_D)}{4 m_H} = 0.5 l$$

$$y_{\text{c.m.}} = \frac{l m(H_B) + l\, m(H_C)}{4 m_H} = 0.5 l$$

$$z_{\text{c.m.}} = \frac{l m(H_C) + l\, m(H_D)}{4 m_H} = 0.5 l$$

continued

CASE STUDY 5.1

Why does the Fosbury flop allow an athlete to jump higher than the traditional straddle technique?

Supplementary information from kinesiology: In the straddle technique of high jumping, the athlete runs toward the bar from an angle and then leaps while facing the bar. The jump consists of swinging first one leg and then the other over the bar in a scissoring motion. At the highest point, the athlete's body is oriented parallel to the bar, facing downward.

In 1968, Dick Fosbury introduced a new technique called the Fosbury flop on his way to winning an Olympic gold medal in Mexico City. In this technique athletes turn as they leap, flinging their body backward over the bar with the back arched. Throughout the jump, the athlete's body is oriented perpendicular to the bar.

Answer: *For this example, we simplify the human body as a binary system of two particles of equal mass, as illustrated in Fig. 5.2(a). One particle is located at the upper chest, the other one in the pelvis area. The two particles are connected with a flexible, massless string of fixed length. The centre of mass is always positioned at the halfway point between the two particles.*

Fig. 5.2(b) illustrates the centre of mass of the athlete at different stages of a high jump using the straddle technique. The centre of mass of the athlete is briefly positioned above the bar because both the chest and the pelvis are located above the bar at the same time. The athlete must generate a sufficient force when leaping up to allow the centre of mass to reach this height.

Fig. 5.2(c) illustrates the athlete's body at various stages of a Fosbury flop. Note that the centre of mass always lies below the bar, allowing the athlete to pass the bar with a lesser force when leaping. A photograph of an athlete executing a Fosbury flop is shown in Fig. 5.3.

continued

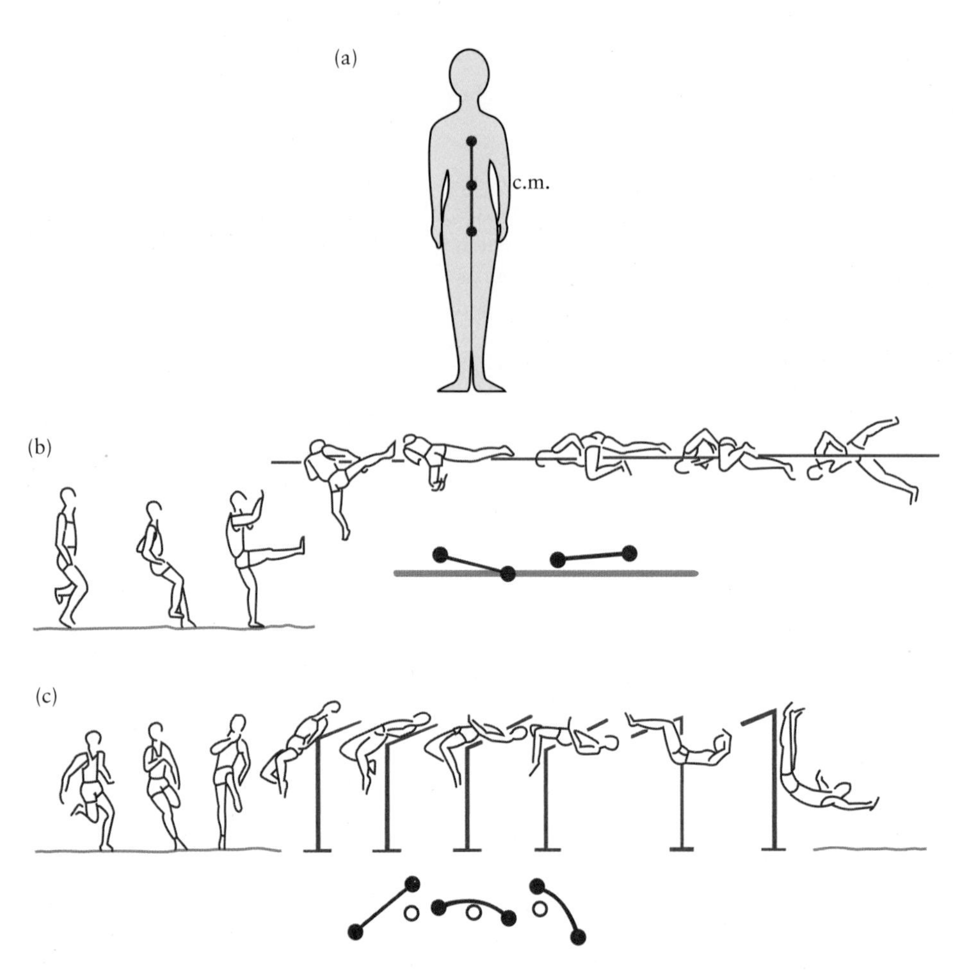

Figure 5.2 Comparison of the motion of the centre of mass of an athlete using the two main techniques for high jump. (a) For the comparison of the two techniques it is sufficient to simplify the human body as two equal point-like objects, located at the chest and at the pelvis. The two objects are connected with a massless string of fixed length. (b) For the straddle technique, the critical stage of the jump is highlighted, illustrating that both point-like objects pass over the bar at the same time. (c) For the Fosbury flop, three sequential frames of the jump show the two objects move across the bar, illustrating that the centre of mass of the athlete is never above the bar. Note that the centre of mass lies along a straight line between the two massive dots describing the athlete, i.e., not along their connection line if it is curved.

Figure 5.3 The Fosbury flop as executed by an athlete.

The carbon atom is located at the centre of the cube shown in Fig. 5.4.

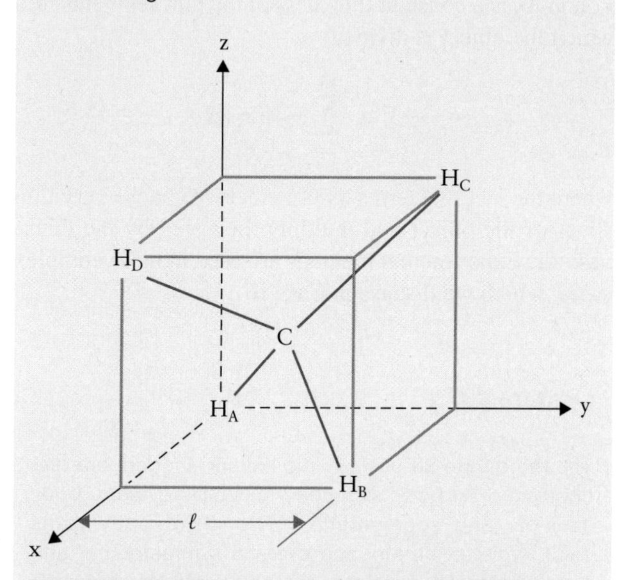

Figure 5.4 The geometry of the tetrahedral methane molecule, CH_4, is best described by placing the molecule in a square cube of side length ℓ in a Cartesian coordinate system. The four hydrogen atoms form four corners of the cube as shown. They are indistinguishable in a real molecule but have been labelled in the sketch with different indices for calculation purposes.

CONCEPT QUESTION 5.1

Figure 5.5 shows the positions of the hydrogen and oxygen atoms in a water molecule. If the mass of oxygen is 16 times larger than the mass of hydrogen, what is the distance between the centre of the oxygen atom and the centre of mass of the water molecule? (Note: 1 pm = 1×10^{-12} m)

(a) 5.99 pm

(b) 10.65 pm

(c) 85.19 pm

(d) 89.85 pm

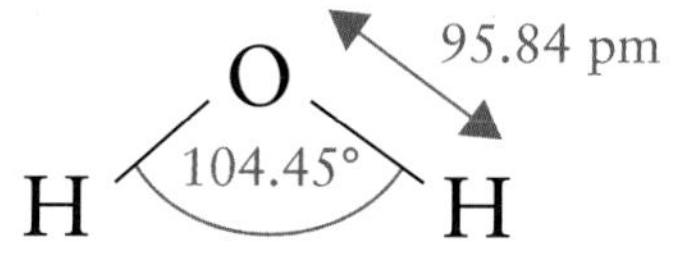

Figure 5.5 The relative positions of the hydrogen and oxygen atoms in a water molecule.

5.1.2: Case II: A Continuous Object

Objects in the real world are not particles. A continuous object is one that has a mass that is distributed over its whole volume. The centre of mass of such a continuous mass distribution can be approximated by modelling it as a system of a very large number of particles. Each of these particles is a small chunk of the continuous object. Smaller chunks provide a more precise approximation of the centre of mass. Writing Eq. [5.2] for a continuous object is greatly simplified if we assume that the position parameters $r_i = (x_i, y_i, z_i)$ for the particles in Eqs. [5.1] and [5.2] are time independent. This condition leads to the definition of a **rigid body**:

KEY POINT

The distance between any two points on a rigid body remains constant in time.

In order to understand how the distribution of mass affects the position of the centre of mass of a rigid body, let's consider what happens as we use an increasing number of slices of the object to predict the centre of mass. Fig 5.6(a) shows the cross-section of an irregularly shaped object that stretches from a position x_{min} to x_{max}. A best, first guess is that the centre of mass of a slice is at x-position in the middle of the slice. If we consider the object to be a single slice of thickness $\Delta x = x_{max} - x_{min}$ (i.e., the whole object), that means that the x-position of its centre of mass would be approximated as being at a point halfway between x_{min} and x_{max}.

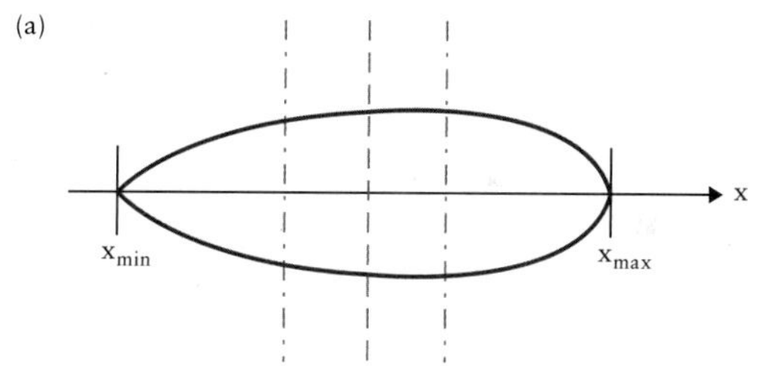

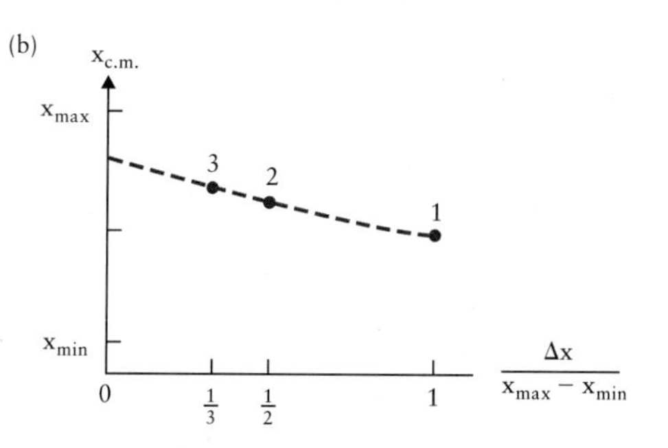

Figure 5.6 Method of finding the centre of mass on an irregularly shaped object. (a) For convenience of illustration we consider a two-dimensional object that we divide along the x-axis to find the $x_{c.m.}$-component. (b) The algorithm used is based on successively plotting estimated values for the centre of mass with decreasing segment lengths of the object, Δx. Shown are a first estimate based on the entire mass and the length of the object (1), a second estimate where the object is divided into two segments [dashed line in (a) and point 2 in (b)], and a third estimate with three segments [dash-dotted lines in (a) and point 3 in (b)]. Each estimate leads to a value closer to x_{max} for the particular object. The exact value follows from an extrapolation $\Delta x \to 0$.

That's not a very good approximation for an irregular object, but it can be made more precise by considering more slices. The vertical dashed line in Fig. 5.6(a) cuts the object in half, so the thickness of each slice is $\Delta x = \frac{1}{2}(x_{max} - x_{min})$. Once again, we will make the approximation that the centre of mass of each of these slices is the point halfway between the ends of the slice. You can tell by the shape of the object that there will be more mass on the right side of the object than on the left side. Looking at Eq. [5.3], we can see that the extra mass on the right side will move the centre of mass for the whole object toward x_{max}.

The estimate of the centre of mass position is further improved by dividing the system into three equally long segments (vertical dash-dotted lines). We find again the mass of each of these segments and their centre positions that serve as the positions assigned to the respective particle. Using the first formula in Eq. [5.2] yields a sum of three terms that is the third estimate for the x-component of the centre of mass:

$$\begin{aligned} x_{\text{c.m.}} &= \frac{1}{M}\sum_i x_i \Delta m_i \\ &= \frac{1}{M}(x_1 \Delta m_1 + x_2 \Delta m_2 + x_3 \Delta m_3) \end{aligned} \qquad [5.4]$$

We wrote the masses in Eq. [5.4] as Δm_i, noting that they become smaller with each successive division of the system. The position values x_i, in turn, remain finite numbers that lie equally spaced between the positions at the left and right ends of the system.

Fig. 5.6(b) shows the approximated values plotted against the relative size of the slice. If we repeat the last step iteratively, treating the system step by step as if composed of more and more, but increasingly shorter, sections, the estimated value of the x-component of the centre of mass approaches a value where the dashed line intersects the ordinate as the width Δx decreases toward zero. Mathematically, this approach is called an **extrapolation** because the final value of $x_{\text{c.m.}}$, at $\Delta x = 0$, lies to the left of all individual points we determined along the dashed line in Fig. 5.6(b). Thus, the precise value follows from extrapolating the successive estimates of the centre of mass with smaller and smaller segment lengths. Although Fig. 5.6 is shown specifically for the x-component of the object, the same technique can be used to determine the y- and z-components of the centre of mass of the object.

Taking this to the limit where the number of slices approaches infinity allows us to have a perfectly accurate approximation. We can write this in the following form:

$$x_{\text{c.m.}} = \frac{1}{M}\lim_{n\to\infty}\sum_{i=1}^{n} x_i \Delta m_i. \qquad [5.5]$$

Eq. [5.5] can be solved analytically for highly symmetric systems, and if the system has a uniform density ρ. In Eq. [5.5], we rewrite the formula using $\Delta m = \rho\, \Delta V = \rho\, A(x_i)\, \Delta x$, in which ΔV is the volume of each segment, $A(x_i)$ is the cross-sectional area of the object at position x_i, and Δx is a constant thickness of the thin segments into which the object is divided:

$$x_{\text{c.m.}} = \frac{\rho}{M}\lim_{n\to\infty}\sum_{i=1}^{n} x_i\, A(x_i)\,\Delta x, \qquad [5.6]$$

where the last two terms assume that we choose very thin slices of the object and multiply their area by the thickness Δx. Experimental methods are used in more complex cases, which we discuss in Case III.

EXAMPLE 5.2

The more than 30 phyla in the animal kingdom are distinguished by body symmetry, digestive system, body cavities, and segmentation. Two major body plans exist: radial symmetry and bilateral symmetry. For animals with radial symmetry, the body parts are arranged regularly around a central axis. Examples include starfish and jellyfish. Animals with bilateral symmetry have a mirror-plane passing through the body from the front to the back end, giving rise to similar left and right sides. Examples are arthropods, such as flying insects, and chordates, such as birds. What do you know about the position of the centre of mass for each of these body symmetries?

Solution

We inspect Eq. [5.6] to see how we benefit from symmetry in uniform systems when determining the centre of mass. Note that we discuss Eq. [5.6] in this answer to extract useful information without mathematically solving it. Assume that the system is symmetric to the centre point between x_{min} and x_{max}, i.e., the position $\frac{1}{2}(x_{min} + x_{max})$. The best way to see what consequences this symmetry has for Eq. [5.6] is to choose the centre position as the origin, i.e., $\frac{1}{2}(x_{min} + x_{max}) = 0$. This means that the cross-sectional area follows $A(x) = A(-x)$ due to the symmetry with respect to the origin along the x-axis. Since we multiply the cross-sectional area A by the position x_i in Eq. [5.6], we note that the sum of $x\,A(x)$ and $(-x)\,A(-x)$ is zero. Thus, we find $x_{\text{c.m.}} = 0$ for a symmetric object; the centre of mass lies at the spatial centre of the system.

For animals with radial symmetry, we choose the direction of the body axis as the z-direction. The symmetry then applies in both the x- and the y-directions. Using the argument made with Eq. [5.6], we find that the centre of mass lies on the body axis.

Based on the same argument, for animals with bilateral symmetry the centre of mass lies somewhere in the mirror-plane. We took this into account when we simplified the human body in Case Study 5.1: the two particles we used to model our body both lie on a central plane that divides our body into a left and a right side.

The centre of mass of a system is not always determined along the x-axis. Indeed, like any other position, the centre of mass has three vector components in three-dimensional space. We generalize Eq. [5.5] for numerical applications by including all three component formulas:

$$x_{\text{c.m.}} = \frac{1}{M} \lim_{n \to \infty} \sum_{i=1}^{n} x_i \Delta m_i$$
$$y_{\text{c.m.}} = \frac{1}{M} \lim_{n \to \infty} \sum_{i=1}^{n} y_i \Delta m_i \quad [5.7]$$
$$z_{\text{c.m.}} = \frac{1}{M} \lim_{n \to \infty} \sum_{i=1}^{n} z_i \Delta m_i$$

or, in vector notation:

$$\vec{r}_{\text{c.m.}} = \frac{1}{M} \lim_{n \to \infty} \sum_{i=1}^{n} \vec{r}_i \Delta m_i. \quad [5.8]$$

The centre of mass is easiest to find in highly symmetric systems such as cylinders, spheres, and rectangular prisms. Concept Question 5.1 illustrates that the centre of mass can be identified without a mathematical calculation if the object shows a high degree of symmetry. For example, the centre of mass of a uniform sphere lies at its centre; the centre of mass of a uniform bar lies halfway between the two ends of the bar.

5.1.3: Case III: Complex Systems

For an irregularly shaped rigid object, Eq. [5.8] is evaluated experimentally, as illustrated in Fig. 5.7. The object in the figure is a two-dimensional sheet of irregular shape. We pick two points, A and B, along the rim of the object and suspend the object from each of these points. The two vertical lines, which we draw on the object when it is in mechanical equilibrium in each case, intersect at the centre of mass of the object. The reason why this approach works is discussed further in a later chapter when we introduce *rotational equilibrium*.

The approach of Fig. 5.7 is often not feasible for biological systems. For example, we cannot cut down a tree just to find its centre of mass, nor can we model a tree as a simple geometric shape due to its low degree of symmetry. In these cases we place the centre of mass intuitively. If we consider the fact that Eq. [5.2] is essentially a mass-weighted average, the position with the most mass has the largest effect on the position of the centre of mass. This is shown in Fig. 5.8, which illustrates the correlation between the shape of coniferous trees and their usual place of growth. The tree in Fig. 5.8(a) has its centre of mass near the ground, allowing it to grow at isolated spots. Even gale storms will not topple this tree because the centre of mass is close to the ground. Because the trees shown in Fig. 5.8(b) have more mass concentrated at the top than the tree shown in Fig. 5.8(a), the centre of mass is further above the ground. Such trees grow in forests where the surrounding trees break the force of a storm. In turn, the tree must grow tall, with the majority of its needles near the top, as this is the only place where sufficient sunlight is available for photosynthesis.

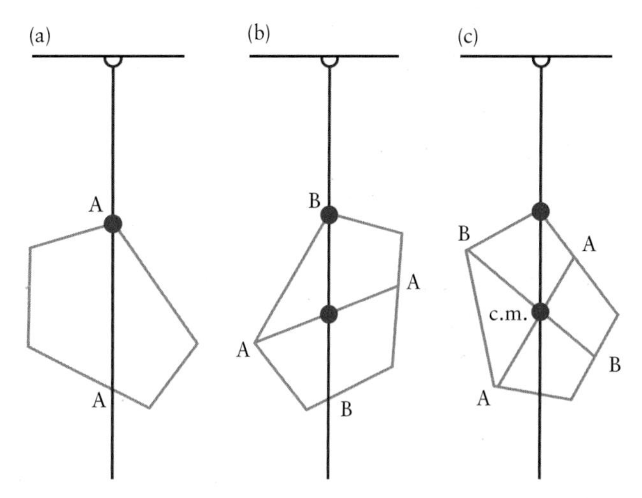

Figure 5.7 Construction of the centre of mass for an irregularly shaped two-dimensional object. (a) An arbitrary point, A, is chosen along the perimeter of the object. The object is then suspended at point A. When it is in mechanical equilibrium, a vertical line is drawn through point A and is labelled with two letter A's to indicate the vertex. (b) A different point, B, is chosen along the perimeter and the procedure is repeated. (c) The intersection of the two vertices A–A and B–B from parts (a) and (b) represents the centre of mass.

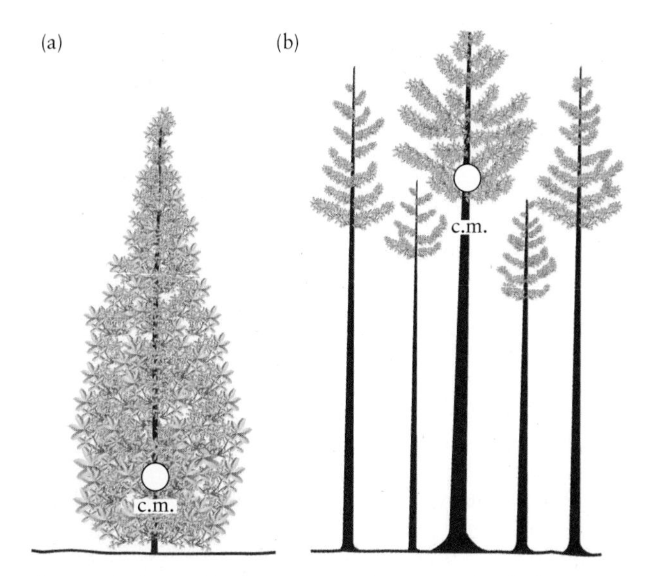

Figure 5.8 The centre of mass concept can be applied to the growth of coniferous trees. (a) The shape of isolated trees usually allows for a low centre of mass. (b) Trees in a forest often have a high-lying centre of mass. Note that we can guess the position of the centre of mass in these cases intuitively without requiring a precise calculation.

CONCEPT QUESTION 5.2

Rest a baseball bat on your finger as shown in Fig. 5.9. Move your finger to the right or left until the bat is perfectly balanced. Once the bat is balanced, which of the following statements must be true?

(a) The portion of the bat on the right side of your finger is equal in mass to the portion of the bat on the left side of your finger.

(b) The portion of the bat on the right side of your finger is equal in length to the portion of the bat on the left side of your finger.

(c) Both of the above.

(d) None of the above.

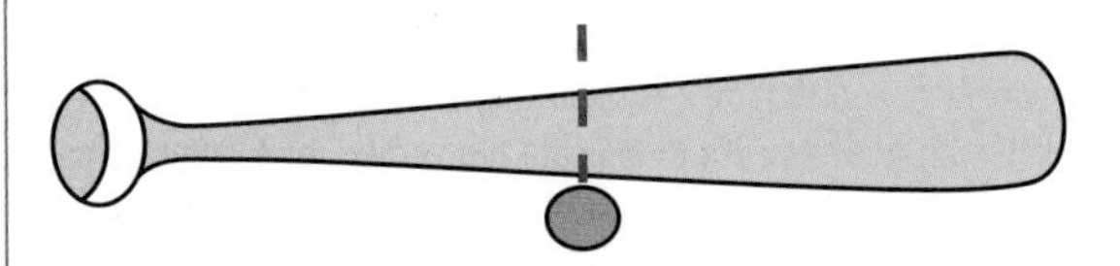

Figure 5.9 A baseball bat balanced at its centre of mass.

5.2: Motion of the Centre of Mass

The benefit of introducing the centre of mass becomes evident when we determine its variation with time. We know that the rate at which a position changes in time is a velocity. Thus, allowing both sides of Eq. [5.1] to vary with time, we find the velocity of the centre of mass of a system:

$$\frac{\Delta \vec{r}_{\text{c.m.}}}{\Delta t} = \frac{1}{M}\sum_i m_i \frac{\Delta \vec{r}_i}{\Delta t}, \qquad [5.9]$$

in which the masses M and m_i are time independent. Eq. [5.10] yields the velocity of the centre of mass of a system:

$$\vec{v}_{\text{c.m.}} = \frac{1}{M}\sum_i m_i \vec{v}_i \qquad [5.10]$$

Transferring the total mass M to the left-hand side of the equation yields:

$$M\vec{v}_{\text{c.m.}} = \sum_i m_i \vec{v}_i \qquad [5.11]$$

We can then use Eq. [5.11] to determine the acceleration of the centre of mass. The acceleration is the change in velocity with respect to time:

$$M\frac{\Delta \vec{v}_{\text{c.m.}}}{\Delta t} = \sum_i m_i \frac{\Delta \vec{v}_i}{\Delta t}. \qquad [5.12]$$

Rewriting Eq. [5.12] in terms of acceleration gives:

$$M\vec{a}_{\text{c.m.}} = \sum_i m_i \vec{a}_i = \vec{F}_{\text{net}}. \qquad [5.13]$$

The summation in Eq. [5.13] adds up all the forces acting on the individual particles that make up the system. Newton's second law tells us that the acceleration of an object can be predicted by the sum of all forces on that object. The particle model that we have been using to describe how objects move in response to forces actually describes a very particular particle: the centre of mass. Newton's second law describes how a net external force causes acceleration of an object's centre of mass.

Thus, we can state

KEY POINT

A net force on a system causes the centre of mass to accelerate as if the entire mass of the system was a point-like object located at the centre of mass.

In particular, when no external forces act on a system, the centre of mass of the system moves with constant speed along a straight line. Fig. 5.10 shows how we can use this fact to make sense out of complicated motion.

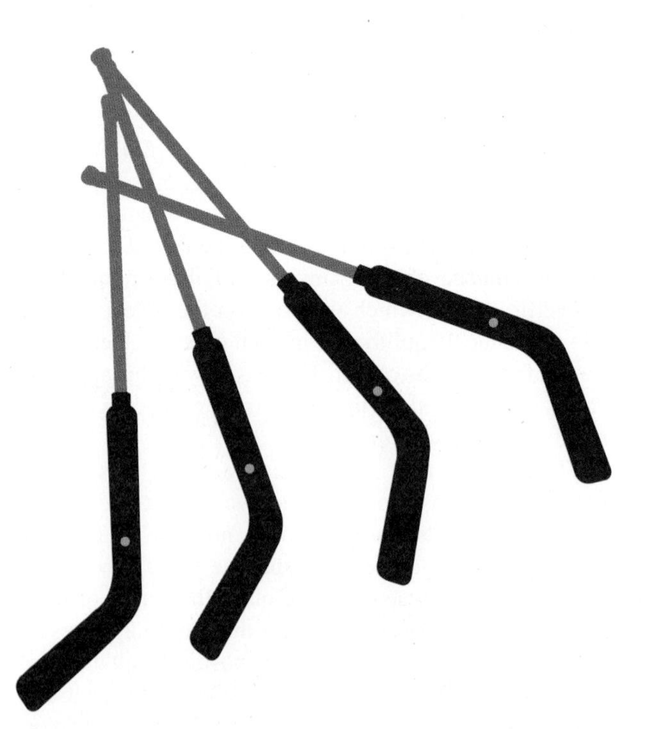

Figure 5.10 A hockey stick that is rotating about its centre of mass (the red dot) while its centre of mass moves at a constant speed along a straight line. Kicking a stick while it is on the ice will produce a similar effect since the force of friction between the stick and the ice is small.

CONCEPT QUESTION 5.3

In Case Study 5.1, we modelled the human body as a combination of two massive, point-like objects that are linked by a massless string. Comparing the straddle technique and the Fosbury flop in Fig. 5.2, what can be said about the maximum height of the centre of mass of the athlete for each of these techniques?

(a) The maximum height of the centre of mass is lower for the Fosbury flop.

(b) The maximum height of the centre of mass is lower for the straddle technique.

(c) The maximum height of the centre of mass is the same for both techniques.

5.3: Newton's Third Law and Linear Momentum

Let's consider two objects, 1 and 2, that interact with each other. Object 1 exerts a force $\vec{F}_{12}$ on object 2. If no other forces act on object 2, this force accelerates object 2 based on Newton's second law: $\vec{F}_{12} = m_2\,\vec{a}_2$. In turn, object 2 exerts a force $\vec{F}_{21}$ on object 1. If no other force acts on object 1, Newton's second law gives us the acceleration of object 1 as $\vec{F}_{21} = m_1\,\vec{a}_1$.

KEY POINT

We call the combined system of objects 1 and 2 an ***isolated system*** *if the paired forces $\vec{F}_{12}$ and $\vec{F}_{21}$ are the only forces acting between those objects:*

We know from Newton's third law that $\vec{F}_{12} = -\vec{F}_{21}$. Thus, we find for the combined system of two objects:

$$\vec{F}_{\text{net}} = \vec{F}_{12} + \vec{F}_{21} = 0 = m_1\vec{a}_1 + m_2\vec{a}_2. \qquad [5.14]$$

Each term on the right-hand side contains the mass and the acceleration of the object. We want to analyze Eq. [5.14] further thinking of the acceleration as the change of velocity of the object with respect to time, which we write as $\Delta\vec{v}/\Delta t$:

$$m_1\frac{\Delta\vec{v}_1}{\Delta t} + m_2\frac{\Delta\vec{v}_2}{\Delta t} = 0. \qquad [5.15]$$

If the two masses in Eq. [5.15] are time independent, we can combine them with the velocity without mathematically altering the equation. Thus, for the change over time of the product of mass and velocity, we can write:

$$\frac{\Delta(m_1\vec{v}_1)}{\Delta t} + \frac{\Delta(m_2\vec{v}_2)}{\Delta t} = 0, \qquad [5.16]$$

or, combining both terms in Eq. [5.16]:

$$\frac{\Delta(m_1\vec{v}_1 + m_2\vec{v}_2)}{\Delta t} = 0. \qquad [5.17]$$

If a term does not change with time, the term itself is constant:

$$m_1\vec{v}_1 + m_2\vec{v}_2 = \text{const.} \qquad [5.18]$$

Eq. [5.18] tells us the sum of the products of the mass and velocity of the objects in an isolated system is a constant. The only assumption that has been made so far is that the mass of the objects is fixed.

Remember that we saw that same product of the mass and the velocity when we were discussing the motion of the centre of mass. It turns out that the product of mass and velocity is a fundamental parameter we can use to describe motion. The name given to the product of mass and velocity is **linear momentum**, and we will use the symbol $\vec{p}$ to represent the linear momentum vector:

$$\vec{p} = m\vec{v}. \qquad [5.19]$$

The prefix "linear" is used to indicate that this parameter describes motion along a straight line. The standard unit of linear momentum is kilogram metres per second (kg m/s). No derived unit is introduced. Eq. [5.19] shows that linear momentum is a vector, and can be written in component form as:

$$\begin{aligned} x\text{-direction:}\quad & p_x = mv_x \\ y\text{-direction:}\quad & p_y = mv_y \\ z\text{-direction:}\quad & p_z = mv_z \end{aligned} \qquad [5.20]$$

Using the definition in Eq. [5.19], we can write Eq. [5.18] for the isolated system of two objects as:

$$\vec{p}_1 + \vec{p}_2 = \text{const.} \qquad [5.21]$$

The reasoning we have used can also be extended to describe an isolated system of more than two objects. To generalize the arguments, we define the total linear momentum, $\vec{p}_{\text{tot}}$, as the vector sum of the individual linear momentums $\vec{p}_i$, where i is the index identifying the individual objects that constitute the isolated system. This sum is constant if the multi-object system is isolated ($\vec{F}_{net} = 0$):

$$\vec{p}_{tot} = \sum_i \vec{p}_i = \text{const.} \qquad [5.22]$$

Eq. [5.22] states that the total linear momentum of an isolated system does not change with time.

KEY POINT

A quantity that does not change with time is ***conserved****, and the law that establishes this fact is called a conservation law.*

Thus, Eq. [5.22] establishes the **conservation of linear momentum** for an isolated system. Identifying an initial and a final state of interest, we write Eq. [5.23] in component form:

$$\begin{aligned} x\text{-direction:} \quad & \sum_i p_{i,x,\text{initial}} = \sum_i p_{i,x,\text{final}} \\ y\text{-direction:} \quad & \sum_i p_{i,y,\text{initial}} = \sum_i p_{i,y,\text{final}} \\ z\text{-direction:} \quad & \sum_i p_{i,z,\text{initial}} = \sum_i p_{i,z,\text{final}} \end{aligned} \qquad [5.23]$$

KEY POINT

The product of mass and velocity of an object defines the linear momentum. The total linear momentum of an isolated system of objects, i.e., a system on which no external forces act, is conserved.

CONCEPT QUESTION 5.4

A bus drives toward a bus stop to pick up a group of students. The bus comes to a complete stop, the students get on, and then the bus continues driving in the same direction as it was previously. If we take the system to be the students and the bus, is momentum conserved in this process?

(a) Yes, momentum is always conserved. It's the law.

(b) No, momentum is not conserved because the system is not isolated.

(c) Yes, momentum is conserved. The students move in a direction opposite to the motion of the bus.

(d) No, conservation of momentum doesn't apply to real things like buses.

One type of physical process for which linear momentum is particularly useful is **collisions**. Several different types of collisions between objects exist. The two most fundamental types are the *elastic collision* (the collision of two billiard balls comes close to this case) and the **perfectly inelastic collision** that we discuss in Example 5.3. Elastic collisions are more important for interactions at a microscopic level; however, we postpone their discussion to Chapter 7, when the energy concept is introduced, because the elastic collision is defined by the conservation of both linear momentum and energy.

EXAMPLE 5.3

We study the perfectly inelastic collision illustrated in Fig. 5.11: an object of mass m_1 hits with initial velocity $\vec{v}_{1i}$ a second object with mass m_2 and initial velocity $\vec{v}_{2i}$. Objects 1 and 2 merge in the collision, forming a combined object that moves with final velocity $\vec{v}_f$ after the collision. We assume in this and all other examples that the motion of the objects is confined to the *xy*-plane. Eq. [5.23] is written in this case as two component equations:

$$\begin{aligned} x\text{-direction:} \quad & m_1 v_{1xi} + m_2 v_{2xi} = (m_1 + m_2) v_{xf} \\ y\text{-direction:} \quad & m_1 v_{1yi} + m_2 v_{2yi} = (m_1 + m_2) v_{yf} \end{aligned} \qquad [5.24]$$

(a) Determine the speed $\vec{v}_f$. (b) What do you observe when $m_1 >> m_2$, or (c) when $m_1 << m_2$?

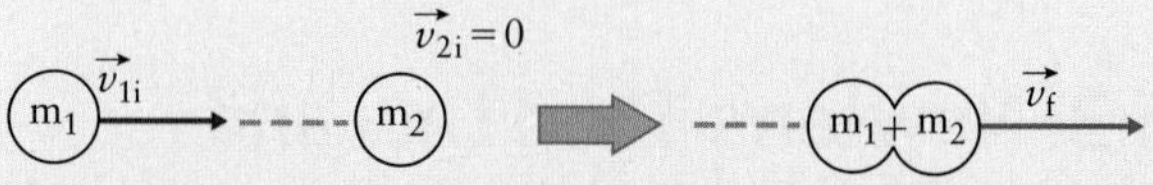

Figure 5.11 A specific case of an inelastic collision in which the two particles merge. Initially, the object with mass m_1 moves with velocity $\vec{v}_{1i}$ and the object with mass m_2 is at rest (seen at the left). The merged object moves with velocity $\vec{v}_f$ after the collision (seen at the right).

Solution

Solution to part (a): In the case of a perfectly inelastic collision with one object at rest, $\vec{v}_{1i}$ and $\vec{v}_f$ must be parallel; i.e., the entire motion occurs along a straight line, which we define as the *x*-axis. This means that $\vec{v}_{1i} = (v_{1i}, 0)$ and $\vec{v}_f = (v_f, 0)$; i.e., we need only the first formula of Eq. [5.24]:

$$m_1 v_{1i} = (m_1 + m_2) v_f \qquad [5.25]$$

where we used $v_{2i} = 0$ because the second object is initially at rest. The same final speed applies to both objects because they remain attached to each other after the collision. Eq. [5.25] is solved for the final speed:

$$v_f = \left(\frac{m_1}{m_1 + m_2} \right) v_{1i}. \qquad [5.26]$$

This is the final result because no specific values were given in part (a). Note that writing the result in the form of Eq. [5.26] is useful for discussing possible cases, as we illustrate with the two limiting cases in parts (b) and (c).

Solution to part (b): Eq. [5.26] illustrates that the final speed depends linearly on the initial speed of object 1 when the two mass values are given. More interesting for applications, though, is the dependence of the ratio v_f/v_{1i} on the two masses; $m_1 >> m_2$ implies that an object approaches and then collides with a much lighter object. Examples of a predator/prey system include a bat catching a slow-flying moth, a cougar ambushing an

continued

unsuspecting rabbit, or a dolphin feeding on a fish. We expect the effects predicted by Eq. [5.26] in these cases to be physiologically acceptable to the predator (i.e., the object of mass m_1) because they would otherwise not use a hunting method that is based on a perfectly inelastic collision. Mathematically, we find:

$$\frac{v_f}{v_{1i}} = \lim_{m_1 >> m_2} \left(\frac{m_1}{m_1 + m_2} \right) \cong 1 \cdot$$

Thus, the speed of the heavy object remains essentially unchanged; the predator can run full speed into the prey without a serious change of the linear momentum it has to accommodate. The prey is less lucky, as it is suddenly accelerated—which probably smashes a moth into the fangs of a bat before the bat has to close its bite. Indeed, the sudden change of linear momentum as endured by the prey is part of the tactics of a heavy ambush predator.

Solution to part (c): Next, we consider the case where the predator is lighter than the prey. Wolves are small compared to a moose they try to bring down; piranhas are small compared to a cow or a pig that falls into their South American river; and prehistoric men were dwarfed by mammoths. Let's see whether the same hunting practices discussed in part (b) make sense in these cases. We use $m_1 << m_2$ in Eq. [5.25]:

$$\frac{v_f}{v_{1i}} = \lim_{m_1 << m_2} \left(\frac{m_1}{m_1 + m_2} \right) \cong 0 \cdot$$

We find a negligible change in linear momentum for the prey (object 2), while a large change occurs for the predator (object 1). The predator, when stuck to the prey, would have to shake off the physical effects of a sudden change in speed, while the prey could address the nuisance of the attack in comparable comfort. Thus, other hunting methods have been developed to bring down heavy prey, such as hunting in packs. However, hunting prey in packs requires a well-developed brain in order to coordinate among the members of the pack. This is one reason why predatory dinosaurs were initially suspected to have been scavengers: only one 25-kg velociraptor was needed to enjoy feasting on a 180-kg protoceratops that had already dropped dead. We required fossil discoveries such as the famous 1924 find of a velociraptor and a protoceratops locked in battle in order to fully appreciate predatory dinosaurs as fierce and swift hunters.

EXAMPLE 5.4

Assume that the 180-kg protoceratops in the above-mentioned fossil find was standing on a slippery surface, such as an easily shifting sand layer, when the 25-kg velociraptor jumped on its back with a speed of 10 m/s. (a) With what speed would the battling pair have been sliding immediately after the impact? (b) In which direction would they have been sliding?

continued

Solution

Solution to part (a): We treat the surface as frictionless; i.e., the surface does not exert a horizontal force on objects. Thus, the interaction between the two dinosaurs in the horizontal direction can be treated as an interaction in an isolated system. Since the two dinosaur bodies are locked in battle after the velociraptor (object 1) jumps onto the protoceratops (object 2), the attack can be modelled as a perfectly inelastic collision, for which Eq. [5.25] applies:

$$m_1 v_{1i} = (25\ \text{kg})(10\ \text{m/s}) = (m_1 + m_2) v_f = (25\ \text{kg} + 180\ \text{kg}) v_f,$$

which yields:

$$v_f = \frac{25\ \text{kg}}{25\ \text{kg} + 180\ \text{kg}} (10\ \text{m/s}) = 1.2\ \text{m/s}\ .$$

This speed is small compared to the initial speed of the velociraptor; indeed, the velociraptor's impact on the protoceratops was likely not sufficient to cause the larger dinosaur to fall. This was not the velociraptor's intention anyway, as a falling protoceratops could well have crushed it to death. Instead, the success of the velociraptor's attack depended on positioning itself on the protoceratops such that it could bring its long, curved, blade-like claw on the second toe to devastating use, e.g., by slitting open the protoceratops' unprotected belly. This was no doubt on the mind of the velociraptor that battled with a protoceratops about 70 million years ago in Mongolia, when both were buried and eventually fossilized; its claw is still embedded in the larger dinosaur's belly section. Despite possession of this deadly weapon, velociraptors had developed large brains for a dinosaur, suggesting that they were able to coordinate with each other in a pack hunt.

Solution to part (b): The protoceratops was initially standing (i.e. motionless), so this can be modelled as a perfectly inelastic collision that occurs along a single axis. Thus, the direction of the incoming velociraptor determines the direction of motion for the battling pair.

CONCEPT QUESTION 5.5

A typical adult male bighorn sheep has a mass of 125 kg (females range from 50 to 90 kg but do not participate) and reaches maximum speeds of 50 km/h on flat terrain. If two adult male bighorn sheep were to run headfirst into each other, as illustrated in Fig. 5.12, what would be the combined linear momentum of both sheep immediately before and immediately after the collision?

(a) around 3500 kg m/s before and around 3500 kg m/s after

(b) around 3500 kg m/s before and around 0 kg m/s after

continued

(c) around 0 kg m/s before and around 3500 kg m/s after

(d) around 0 kg m/s before and around 0 kg m/s after.

© Bruce Coleman Inc./Alamy

Figure 5.12 Bighorn sheep.

EXAMPLE 5.5

Fig. 5.13(a) shows two objects of masses m_1 and m_2 travelling with velocities $\vec{v}_1$ and $\vec{v}_2$ toward a collision point (chosen at the origin). After a perfectly inelastic collision, the combined object travels with angle ϕ relative to the positive x-axis, as indicated in Fig. 5.13(b). (a) Calculate the x- and y-component formulas for the velocity of the combined object after the collision. (b) Choosing $m_1 = m_2$, $v_1 = v_2$, and $\theta = 45°$, calculate the angle ϕ.

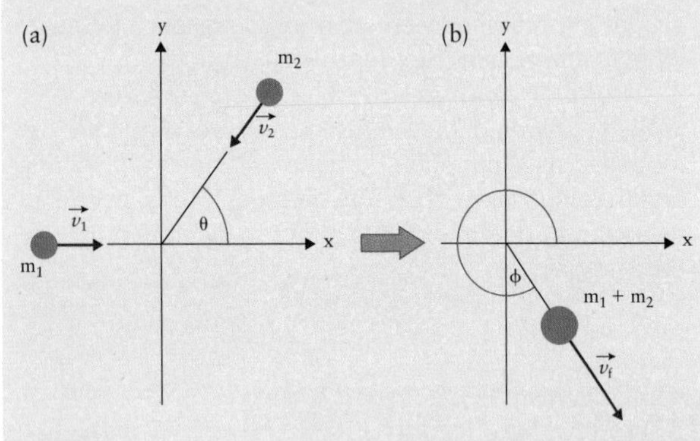

Figure 5.13 Inelastic collision of two moving objects in the xy-plane.

Solution

Solution to part (a): Inelastic collisions are governed by the conservation of momentum from Eq. [5.22]. We use the first two formulas from Eq. [5.23] because both objects in this example move in a common plane, which we define as the xy-plane. Fig. 5.13(a) indicates that object 1 travels along the x-axis; i.e., its velocity components are $(v_1, 0)$. Object 2 travels at an angle to the x-axis. Its velocity components are $(-v_2 \cos \theta, -v_2 \sin\theta)$. The velocity of the combined object after the collision is written as (v_{xf}, v_{yf}).

continued

Substituting these components in the first two formulas of Eq. [5.23], we find:

$$x\text{-direction:} \quad m_1 v_1 - m_2 v_2 \cos\theta = (m_1 + m_2) v_{xf}$$

$$y\text{-direction:} \quad -m_2 v_2 \sin\theta = (m_1 + m_2) v_{yf},$$

which, for the final velocity components, yields:

$$v_{xf} = \frac{m_1}{m_1 + m_2} v_1 - \frac{m_2}{m_1 + m_2} v_2 \cos\theta$$
$$v_{yf} = -\frac{m_2}{m_1 + m_2} v_2 \sin\theta. \qquad [5.27]$$

Solution to part (b): Eq. [5.27] is rewritten for the special case of $m_1 = m_2$, $v_1 = v_2 = v$, and $\phi = 45°$:

$$v_{xf} = \frac{v}{2}[1 - \cos(45°)]$$

$$v_{yf} = -\frac{v}{2}[\sin(45°)],$$

which, for the angle ϕ, yields:

$$\tan\varphi = \frac{v_{yf}}{v_{xf}} = \frac{-\sin(45°)}{1 - \cos(45°)} = -2.41.$$

This yields $\phi = -67.5° = 292.5°$.

5.4: Changes of Linear Momentum and Newton's Second Law

With the concept of momentum introduced, we can now remove the requirement of an isolated system. We consider first a system with a single object on which a non-zero external force acts. Newton's second law is rewritten with the linear momentum definition from Eq. [5.19]:

$$\vec{F}_{net} = m\vec{a} = m\frac{\Delta\vec{v}}{\Delta t} = \frac{\Delta(m\vec{v})}{\Delta t} = \frac{\Delta\vec{p}}{\Delta t}. \qquad [5.28]$$

Recall we discussed earlier that the inclusion of mass with velocity is mathematically correct as long as the mass of the object is time independent. Newton evaluated Eq. [5.28] carefully, and in *Principia* decided to write the second law of mechanics in the form $\vec{F}_{net} = \Delta\vec{p}/\Delta t$, and not in the form we used earlier, i.e., $\vec{F}_{net} = m\vec{a}$. Taking a fundamental view that physical laws are written such that a minimum of necessary assumptions are required, Newton chose the form connecting the linear momentum and the net force because he realized the assumption of a constant mass of the studied objects is not necessary:

KEY POINT

The laws of mechanics apply equally well to objects of variable mass.

We will see several examples below that use this fact to understand the strategies some sea animals use to escape their predators.

When Newton realized the laws of mechanics are not limited to objects of fixed masses, the formulation $\vec{F}_{net} = m\vec{a}$ for the second law became undesirable because it does not properly describe the motion of a system with varying mass. Thus, the most general formulation of Newton's second law reads

KEY POINT

The time rate of change of the linear momentum of a system is equal to the net force that acts on the system.

Eq. [5.28] is correct with respect to units; the force is given in unit kg m/s^2, while the linear momentum is given in unit kg m/s. The time rate of change of the linear momentum, $\Delta\vec{p}/\Delta t$, then, carries the unit kg m/s^2, i.e., the same unit as force. Eq. [5.28] is applied in component form in Example 5.6.

EXAMPLE 5.6

Squid have a streamlined shape, with flap-like fins that stabilize them in water. They either move slowly by rippling their fins or dart very fast forward using a kind of jet propulsion, squirting water rapidly out of the breathing tube (siphon). They use the latter movement to escape predators they spot early as a result of their good vision. What is the initial acceleration a squid can achieve when starting from rest? Use ρ = 1025 kg/m^3 for the density of seawater and 50 kg as the initial mass of the squid. The opening of the breathing tube has an area of 7 cm^2, and water is ejected at 15 m/s. *Note:* We ask about the initial acceleration only because this allows us to neglect (i) drag effects caused by moving through the water and (ii) the change of the mass of the squid as water is ejected.

Solution

The squid ejects water in a stream with a cross-sectional area of 7 cm^2 and a speed of 15 m/s. Multiplying these two quantities leads to a term of the form $\Delta V/\Delta t$, which is called **volume flow rate**, with unit m^3/s: $\Delta V/\Delta t$ = 0.0105 m^3/s = 10.5 L/s. Multiplying the volume flow rate by the density leads to a term of the form $\Delta m/\Delta t$, which is called the **mass flow rate**, with unit kg/s. The mass flow rate of ejected water can be found:

$$\frac{\Delta m}{\Delta t} = \rho\frac{\Delta V}{\Delta t} = \left(1025\frac{\text{kg}}{\text{m}^3}\right)\left(0.0105\frac{\text{m}^3}{\text{s}}\right) = 10.8\frac{\text{kg}}{\text{s}}. \qquad [5.29]$$

Note that this is only an initial rate. The squid cannot sustain that rate of expulsion of water. The result in Eq. [5.29] is then multiplied by the speed of the ejected water to

continued

obtain the rate of change of linear momentum of the water as it is brought from rest in the squid's body to ejection speed:

$$\frac{\Delta|\vec{p}_{\text{water}}|}{\Delta t} = \frac{\Delta m}{\Delta t}v = (10.8\text{ kg/s})(15\text{ m/s}) = 162\text{ kg m/s}^2.$$

Note that we wrote the rate of change of the linear momentum as the rate of change of the mass multiplied by the speed of the water. This is correct because that speed does not vary with time. Using Eq. [5.28], we note that 162 N is the net force acting on the water as it is ejected. Newton's third law tells us that the force of the squid on the water must have an equal and opposite paired force that is the force of the water on the squid. Thus, the water pushes the squid in the direction opposite to the ejected water with an acceleration of:

$$a_{\text{squid}} = \frac{F_{\text{net}}}{m_{\text{squid}}} = \frac{162\frac{\text{kg m}}{\text{s}^2}}{50\text{ kg}} = 3.2\frac{\text{m}}{\text{s}^2}$$

This acceleration will often be sufficient as it moves the squid a distance d = (1/2)(at^2) = 1.6 m within the first second of escape.

Note that we did not use a vector notation in this problem but treated the line along which the squid moves as the x-axis in Eq. [5.28]. This simplified the calculation because the force and the acceleration act directly along a straight line, so there were no y- or z-components.

EXAMPLE 5.7

Cuttlefish use rapidly elongating tentacles to snatch crabs and other prey. A pair of tentacles is hidden within the visible outer ring of arms. Despite the resistance of water, the cuttlefish's killing tentacles can accelerate with 25 g, i.e., 25 times the gravitational acceleration. What is the effect on the cuttlefish when it unleashes its tentacle? Does the targeted crab benefit from the backlash; i.e., does this effect increase the crab's chance of surviving the attack? Assume that the deadly tentacle represents 1/20th of the mass of the cuttlefish.

Solution

We use the conservation of linear momentum. The cuttlefish is initially motionless; i.e., its total linear momentum is zero. This must remain the case while the tentacle elongates forward. That means the rest of the cuttlefish must accelerate backward when the tentacle accelerates forward. This sounds good to the threatened crab because it means the cuttlefish moves away from the crab during the attack. But is the effect large enough to put the cuttlefish beyond its tentacles' reach before the crab is snatched? We expect the answer to be no, as otherwise the cuttlefish would have become extinct a long time ago.

continued

We use Eq. [5.28] to write the rate of change of the linear momentum for the cuttlefish's tentacle:

$$\frac{\Delta p_{\text{tentacles}}}{\Delta t} = m_{\text{tentacles}}\, a_{\text{tentacles}} = \frac{m_{\text{cuttlefish}}}{20} 25 \cdot g$$
$$= 1.25\, W_{\text{cuttlefish}}$$

where $W_{\text{cuttlefish}}$ represents the weight of the cuttlefish. Newton's third law tells us that a force of equal magnitude must act on the rest of the body of the cuttlefish (labelled m_{rest}):

$$1.25\, W_{\text{cuttlefish}} = m_{\text{rest}}\, a_{\text{rest}} = \frac{19}{20} \text{m}_{\text{cuttlefish}}\, a_{\text{rest}}$$

which yields $a_{\text{rest}} = 1.3\ g$. Receiving a backward acceleration larger than the gravitational acceleration isn't a negligible effect. Let's assume that a tentacle has a length L. It will be fully extended in time t, which we calculate from:

$$L = \frac{1}{2} a_{\text{tentacle}}\, t^2$$
$$t = \sqrt{\frac{2L}{25g}}$$

During this time, the rest of the cuttlefish is pushed backward a distance d:

$$d = \frac{1}{2} a_{\text{rest}}\, t^2 = \frac{1}{2}(1.3g)\frac{2L}{25g} = \left(\frac{1.3}{25}\right) L \cdot$$

The backward force causes the squid to move back a distance equal to about 5% of the tentacle's length. Good news for the cuttlefish. Not so good news for the crab.

SUMMARY

DEFINITIONS

- Linear momentum of a single particle (point-like object) in vector notation:

$$\vec{p} = m\vec{v}$$

- Linear momentum of a single particle in component form:

$$x\text{-direction:}\quad p_x = mv_x$$
$$y\text{-direction:}\quad p_y = mv_y$$
$$z\text{-direction:}\quad p_z = mv_z$$

- The position of the centre of mass (index c.m.) of a system of n objects with total mass M:

$$\text{r}_{\text{c.m.}} = \frac{1}{M}\sum_{i=1}^{n} m_i \text{r}_i$$

- Centre of mass of n objects of total mass M in component form:

$$x_{\text{c.m.}} = \frac{1}{M}\sum_{i=1}^{n} m_i x_i$$
$$y_{\text{c.m.}} = \frac{1}{M}\sum_{i=1}^{n} m_i y_i$$
$$z_{\text{c.m.}} = \frac{1}{M}\sum_{i=1}^{n} m_i z_i$$

- A rigid object is an extended object for which (i) the distance between any two parts of the object is fixed, and (ii) the angle between the lines connecting any three parts of the object is fixed.

UNITS

- Linear momentum $\vec{p}$: kg m/s

LAWS

- Conservation of linear momentum for an isolated multi-object system ($\vec{F}_{\text{net}} = 0$):

$$\vec{p}_{\text{tot}} = \sum_i \vec{p}_i = \text{const}$$

- Component form of linear momentum conservation in an isolated system:

$$x\text{-direction:}\quad \sum p_{\text{ix}} = \sum p_{\text{fx}}$$
$$y\text{-direction:}\quad \sum p_{\text{iy}} = \sum p_{\text{fy}}$$
$$z\text{-direction:}\quad \sum p_{\text{iz}} = \sum p_{\text{fz}}$$

- Linear momentum–net force relation for a single object (Newton's second law for a non-isolated system):

$$\vec{F}_{\text{net}} = m\vec{a} = m\frac{\Delta\vec{v}}{\Delta t} = \frac{\Delta(m\vec{v})}{\Delta t} = \frac{\Delta\vec{p}}{\Delta t}$$

MULTIPLE-CHOICE QUESTIONS

MC–5.1. An object of mass m moves with speed v. It collides head-on in a perfectly inelastic collision with an object of twice the mass but half the speed moving in the opposite direction. What is the speed of the combined object after the collision?
(a) $v_{\text{final}} = 0$
(b) $v_{\text{final}} = v/2$
(c) $v_{\text{final}} = 2\ v/3$
(d) $v_{\text{final}} = v$
(e) $v_{\text{final}} = 2\ v$

MC–5.2. You want to determine the mass of an object by exposing it to a perfectly inelastic collision with a test object. In this experiment, what is the minimum number of parameters you have to either measure or set?
(a) 1
(b) 2
(c) 3
(d) 4
(e) 5

MC–5.3. A 1.2-kg common raven just leaving a feeding platform at 0.2 m/s, and a 23-g Savannah sparrow in mid-flight at 10 m/s collide with a glass window (treat the collision as a perfectly inelastic collision). If the collisions between the birds and the window have approximately the same duration, which bird exerts the larger force on the window panel?
(a) the raven
(b) the sparrow
(c) They exert the same force.
(d) Neither exerts a force.

MC–5.4. We consider the water molecules in a beaker on a table as the system. We will later learn that the individual water molecules in the beaker move typically at several hundred metres per second (let v be the average molecular speed). What is the state of motion of the centre of mass of this system?
(a) It moves also with speed v like the individual molecules.
(b) It moves with speed v but, due to Newton's third law, in the opposite direction of the individual molecules.
(c) It moves with constant speed but that speed is different from v.
(d) It does not move; the centre of mass is at rest.
(e) It accelerates downward because gravity is the only force acting on the water molecules.

MC–5.5. A system consists of three objects. Initially, the centre of mass of the system moves with constant speed along a straight line. What happens if I exert an external force on only one of the three objects?
(a) Nothing; to change the state of motion I must exert the external force on all three objects.
(b) Nothing; I cannot exert a force on just a part of a system.
(c) Nothing; I need collisions among the three objects to affect the centre of mass.
(d) The object I interact with accelerates but the other two objects accelerate in the opposite direction, such that the centre of mass continues to move at constant speed.
(e) All three objects accelerate to allow the centre of mass to accelerate in the same manner as the object on which I exert the external force. This is necessary because, otherwise, the external force causes dismantling of the system, which is inconsistent with the definition of the system as three objects.
(f) None of the above. If you chose (f), do you have a better suggestion?

CONCEPTUAL QUESTIONS

Q–5.1. The kinetic gas theory is a microscopic model for gases. One of the assumptions in that model is that the gas particles move between collisions with constant speed along straight lines. For which part of a nitrogen molecule (N_2) should this condition apply exactly?

Q–5.2. A baseball player accidentally releases the bat in mid-swing. The bat sails toward the stands, spinning in the air. Neglect air resistance. (a) With the information given, can we determine any point on the baseball bat for which we can describe its motion after it left the player's hands? What motion does that point perform? (b) Assuming the baseball bat is made of solid wood (uniform density), where along the baseball bat would you expect the point in part (a) to lie?

Q–5.3. Stand perfectly still and then take a step forward. Before the step your momentum was zero and then it increased. Does this violate the conservation of momentum?

Q–5.4. If two objects collide and one was initially at rest, is it possible for both to be at rest after the collision?

Q–5.5. Two objects of equal mass m and speed $|\vec{v}_1| = |\vec{v}_2| = v$ collide in a perfectly inelastic collision at a right angle, as illustrated in Fig. 5.14. In what direction does the combined object move after the collision?

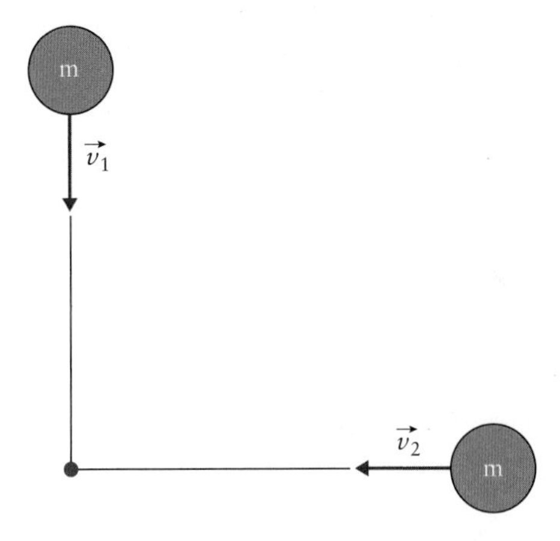

Figure 5.14 Two objects of equal mass collide perfectly at a 90° angle.

Q–5.6. In a sequence filmed with a high-speed camera, a northern goshawk is seen catching a moth in mid-flight without any sign of change of speed of the bird. Did the hawk find a way to violate the conservation of linear momentum?

ANALYTICAL PROBLEMS

P–5.1. An 80-kg man and his 20-kg daughter stand on opposite ends of a 4-m long wooden plank with a mass of 10 kg. (a) If the system is taken to be the man and the daughter, how far along the plank from the daughter is the centre of mass of that system? (b) If the system is taken to be the man, the daughter, and the wooden plank, how far along the plank from the daughter is the centre of mass of that system? (c) If you want to balance the wooden plank with the man and daughter on it, would you place the pivot directly under the centre of mass you found in part (a) or in part (b)?

P–5.2. You can arrange four rectangles such that they form an approximation of an isosceles triangle. The bottom rectangle represents the base of the triangle, with a width of 20 cm and a height of 5 cm. Just above that is another rectangle with a width of 15 cm and a height of 5 cm, followed by a rectangle with a width of 10 cm and a height of 5 cm. The top of the triangle is a square that is 5 cm in width and height. Arranged this way, the set of rectangles looks approximately like a triangle of base 20 cm and height 20 cm. (a) Where is the location of the centre of mass of this triangle as measured from the bottom corner? (b) A balsam fir tree with a uniform, triangular symmetry has a height of 2 m and a base of 2 m. Neglecting the trunk, estimate the height of the centre of mass of this tree using the same technique.

P–5.3. A group of seven Canada geese fly in a V-formation. Each goose is 1 m away from the next closest goose. There is an angle of 30° between the two arms of the V-formation. If we take the system to be all seven geese, what are the coordinates of the centre of mass of the system if the origin is taken to be the lead goose?

P–5.4. A uniform rectangular block of length 30 cm is placed so that its centre of mass is a distance of 5 cm from the edge of a table. Since its centre of mass is still over the table (i.e., not sticking out past the edge), the block is stable. An identical block is placed on top of that block. How far from the edge of the table can the centre of mass of the top block be before the blocks become unstable? Take the edge of the table to be the origin of your coordinate system, with negative values representing positions that are over the table, and positive values representing positions that are past the edge.

P–5.5. A system is made up of point masses P1 at position (−1 m, 5 m, 7 m), P2 at position (3 m, 3 m, 3 m), and P3 at position (9 m, −5 m, −2 m). Find the centre of mass of the system if (a) each of the point masses has the same mass, and (b) $m_1 = 2m_2 = 4m_3$.

P–5.6. In the ammonia molecule, NH_3, the three hydrogen atoms are located in a plane forming an equilateral triangle with side length a, as shown in Fig. 5.15. The nitrogen atom oscillates 24 billion times per second up and down along a line that intersects the plane of the hydrogen atoms at the centre of mass of the three hydrogen atoms (solid circles in Fig. 5.15). (a) Calculate the length a in Fig. 5.15 using for the N–H bond at length l = 0.1014 nm, and for the H–N–H-bond angle θ = 106.8°. (b) Calculate the distance between the centre of mass of the three hydrogen atoms and any one of the hydrogen atoms.

P–5.7. A water molecule consists of an oxygen atom and two hydrogen atoms. The bonds are O–H nm long and form an angle of 107°. Where is the molecule's centre of mass located? Consider the mass of the oxygen atom to be 16 times the mass of a hydrogen atom.

P–5.8. Table 5.1 shows the initial positions and velocities of four particles that make up an isolated system. (a) What are the (x, y)-coordinates and the velocity of the centre of mass of the system? (b) What are the (x, y)-coordinates and the velocity of the centre of mass of the system 10 s later? (c) If the particles are acted on by a uniform external force F = 10 N, what are the (x, y)-coordinates and the velocity of the centre of mass of the system 10 s later?

P–5.9. The mass of the Moon is 7×10^{22} kg, the mass of Earth is 6×10^{24} kg, and the mass of the Sun is 2×10^{30} kg. The distance from Earth to the Moon is 4×10^{8} m, the distance from Earth to the Sun is about 1×10^{11} m. (a) How far from Earth is the centre of mass of the Earth–Moon system? (b) How far from the Sun is the centre of mass of the Earth–Sun system? (c) What is the maximum amount the position of the centre of mass of the Earth–Sun system would change if you were to include the Moon in that system? (d) If the Earth–Moon–Sun system is an isolated system (which is a good approximation), can the centre of mass of that system move? If yes, what kind of motion is possible?

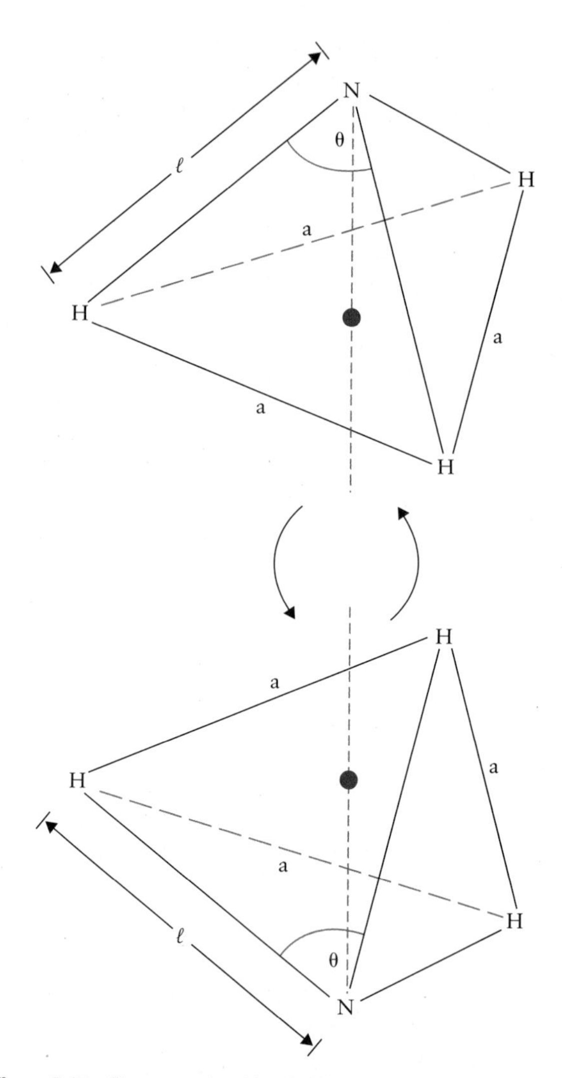

Figure 5.15 The ammonia molecule from Problem P–5.6.

TABLE 5.1

Data for problem 5.8

Particle [kg]	Mass [m]	position (x, y) [m/s]	Velocity (v_x, v_y)
1	3	(0, 0)	(3, −3)
2	10	(5, 3)	(2, 0)
3	4	(3, 2)	(0, −5)
4	2	(−3, −10)	(−4, 4)

P–5.10. A diver jumps off of a diving platform with an initial velocity of 3 m/s in the direction 10° to the vertical. The diver's centre of mass starts at a height of 11 m above the pool. What is the height of the diver's centre of mass 1 second after leaving the platform? If the diver is changing the relative positions of his legs and arms during the dive, can this question still be answered?

P–5.11. A student is standing on a very slippery surface that is flat and level (perfectly horizontal). She has a mass of 50 kg and a backpack with a mass of 10 kg. Consider the system to be her and her backpack. (a) If she throws the backpack horizontal and to her right with a speed of 2 m/s relative to her, what is the backpack's speed measured relative to the centre of mass of the system? (b) What is her speed measured relative to the centre of mass of the system? (c) If she throws the backpack straight up, is the total momentum of the system conserved? If not, define a system for which the total momentum is conserved.

P–5.12. Velvet worms use a cruel method of hunting: aiming at their prey for distances of up to one metre, they eject two jets of a sticky liquid from glands near their mouth. The liquid dries very fast, immobilizing the prey. The worm then injects digestive juices that cause internal liquefaction of the prey. Assume that the velvet worm's body has a mass of 10 g and that it ejects 0.1 g of liquid over a period of 0.5 seconds at an average speed of 5 m/s. What is the average force the velvet worm needs to hold itself steady while ejecting the liquid?

P–5.13. Polonium-210 (^{210}Po) is a radioactive isotope that has an atomic nucleus with 84 protons and 126 neutrons (together, 210 nucleons). It undergoes an α-decay with a half-life of 138.4 days, resulting in a stable lead-206 (^{206}Pb) isotope with a mass of 205.974 u. Assuming the polonium nucleus is at rest when it decays, an α-particle (^{4}He) with a speed of 1.6×10^7 m/s is emitted. The α-particle has a mass of 4.002 u. The atomic unit u is defined as $1\ \text{u} = 1.6605677 \times 10^{-27}$ kg. (a) In what direction does the daughter isotope ^{206}Pb move? (b) What is the speed of the ^{206}Pb leaving the decay zone? (c) Compare both speeds in this problem to the speed of light in a vacuum.

P–5.14. Two objects, each moving with the same speed but in opposite directions, collide head-on. After the collision, the smaller object of mass m = 100 g is embedded in the larger object of mass M = 500 g. Find an expression for the final speed in terms of the initial speed.

P–5.15. Two objects collide head-on. During the collision, the smaller object of mass m is embedded in the larger object of mass M. If the initial speed of the smaller object is twice the initial speed of the larger object, find an expression for the mass of the larger object in terms of the final speed of the combined object, the initial speed of the smaller object, and the mass of the smaller object.

P–5.16. Some birds of prey, such as the northern goshawk, hunt other birds in midair. Typically, the hawk spots the prey while soaring high above, then dives in for the kill with the deadly grip of its talons. They do not shy away from larger prey, such as ducks and crows. Using the change in linear momentum, we want to estimate the impact on the hawk when completing a successful kill by clawing into the prey, which we take to be an unsuspecting American crow. Use mass of crow m_C = 450 g, horizontal flight of the crow with speed v_C = 10 m/s, mass of hawk m_H = 900 g, flight path in the same plane as the crow but approaching from behind with v_H = 20 m/s and with an angle of θ = 70° with the horizontal, as illustrated in the side view of Fig. 5.16. (a) How fast does the hawk move immediately after catching the crow? (b) By what angle has the direction of motion of the hawk changed at the impact? *Note:* American crows take the danger posed by high-soaring hawks very seriously. They are often seen as a group attacking the predator to drive it out of their neighbourhood.

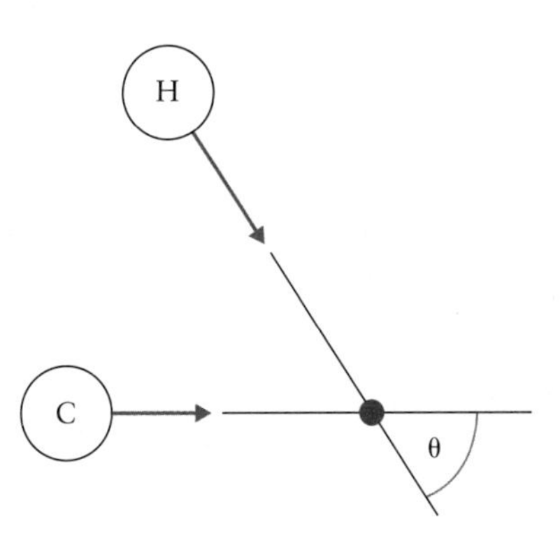

Figure 5.16 Flight path for crow and hawk.

P–5.17. The heart of the standard man ejects 5.0 L blood per minute into the aorta. We will calculate later that the average blood speed in the aorta is 22 cm/s. The density of blood is 1.06 g/cm^3. Take the mass of the heart to be 300 g. Use this data to estimate (a) the average net force acting on the heart, (b) the average acceleration of the heart as a result of this force, and (c) the average acceleration of the human body as a result of this force. (d) What assumptions do we make for the value in part (c) to have a physical meaning? (e) How accurate are the values we estimated in parts (a) and (b)?

P–5.18. A ballistocardiograph is an instrument that detects the recoil of the body caused by the change in momentum of the blood caused by the heart pumping. You can observe this at home if you stand very still on a bathroom scale (the analog spring-type, not a newer digital scale). You will notice that the reading will wobble in time with your heartbeat. This effect is small, but noticeable. If blood travels at around 1 m/s and the mass of blood in a 65-kg person is around 4 kg, what is the largest variation in apparent weight this effect could cause?

P–5.19. An object ejects 10% of its initial mass at time $t = 0$ along the negative x-axis with a speed of 1.0 m/s relative to the object. Shortly after, it ejects another 10% of its initial mass along the negative y-axis with the same speed relative to the object. (a) Assume the object started from rest and calculate the magnitude and direction of its final velocity. (b) If the object started at an initial speed $\vec{v}$= (2 m/s, 3 m/s), would the change in speed of the object be different than the change in speed in part (a)? (c) If the object ejected 20% of its initial mass at time $t = 0$ in the direction opposite to the direction found in part (a), what would be the magnitude and direction of its final velocity?

P–5.20. An object of mass $m = 3.0$ kg makes a perfectly inelastic collision with a second object that is initially at rest. The combined object moves after the collision with a speed equal to one-third of the object that was initially moving. What is the mass of the object that was initially at rest?

P–5.21. An object of mass $m = 8.0$ g is fired into a larger object of mass $M = 250$ g that was initially at rest at the edge of a table. The smaller object becomes embedded in the larger object (perfect inelastic collision) and the combined object lands on the floor a distance of 2 m away from the table. If the tabletop is 1 m above the floor, determine the initial speed of the smaller object.

ANSWERS TO CONCEPT QUESTIONS

Concept Question 5.1: (b). If you use symmetry, this can be calculated without using the angle. Be sure to divide by the total mass, not just the mass of oxygen.

Concept Question 5.2: (d). The balancing point is the centre of mass. Since the shape is not uniform, the fat end must be closer to the centre of mass, and the skinny end must be further away.

Concept Question 5.3: (a). In both techniques the weight is the only force acting on the athlete after the second foot leaves the ground, so the athlete's centre of mass follows a parabolic trajectory. In the case of the Fosbury flop this parabola peaks below the bar, while it must peak above the bar in the straddle technique.

Concept Question 5.4: (b). The ground is external to the system and it exerts a force of friction on the bus and the students. If the ground is considered part of the system, momentum would be conserved. That means the bus slowing down causes the length of the day to change. Thankfully, Earth is so massive compared to the bus that the effect is not noticeable.

Concept Question 5.5: (d). Linear momentum is a vector. Each sheep has approximately the same magnitude of linear momentum, but the directions are opposite so the sum is zero. During the collision, the force of sheep on sheep is much larger than any other force acting on the sheep, so it is safe to approximate this system as isolated, in which case, the total linear momentum does not change.

CHAPTER 6

Torque and Equilibrium

In this chapter we will examine the conditions under which an extended object can be considered to be in equilibrium. In Chapter 4 we studied Newton's laws and discussed the condition for an object to be in a "translational equilibrium," (assuming the object as a point-like particle). For an object in equilibrium, the particle model can hold if all external forces acting on the object pass through a single point. In cases where the applied forces on an object in equilibrium do not cross each other at a common point, the size and the shape of the object can no longer be overlooked, and the rigid object model should be used instead. In the latter case, the condition for translational equilibrium is no longer sufficient; another condition is needed to hold the object in equilibrium.

The **rigid object model** is a simplified model as it excludes vibrations and deformations; however, it allows us to focus on rotations, equilibrium, stability, and balance. Note that it makes no sense to define rotation for point-like particles that are zero-dimensional. When we start to consider the size and the shape of objects, more complex patterns of motion emerge. For a rigid body we ask, How effective is an applied force in causing or changing its rotational motion? Not only do the magnitude and direction of the force play a role, but the position at which the force is applied is also very important. The effectiveness (tendency) of a force in causing rotational motion of a rigid body is quantitatively measured by torque.

Consider the rotation of our lower arm about the elbow as a typical example: motion in a two-dimensional plane results when the rotation occurs about a fixed axis through the fulcrum (pivot) perpendicular to the plane of rotation. In a rotation of a rigid body with a fixed rotational axis, a force can be applied in two ways: the force can act along a line through the fulcrum, which does not lead to a rotation, or the force can act in any other direction and cause rotation to occur. In the latter case, there is an increasingly strong tendency to rotate as the magnitude of the force and/or the perpendicular distance from the axis of rotation increase. **Torque** can therefore be thought of as a measure of the strength of rotation. Actually, anytime you extend one of your fingers, raise your arm, or stretch your leg you exert a torque that makes these motions possible. To hold our body in balance while in motion, or while moving around from one place to another, is due to our ability to exert torques on our limbs.

The concept of equilibrium when applied to a rigid object extends Newton's first law (the first condition). While the net force on the object must be zero, the sum of the torques (the second condition) due to all forces acting on the object must also be zero. If the latter condition is not met, the object begins to spin faster and faster, or the rotating object slows down, as it gains or loses rotational acceleration. Stretching and holding your arm horizontally in line with your shoulder (Fig. 6.1) requires not only

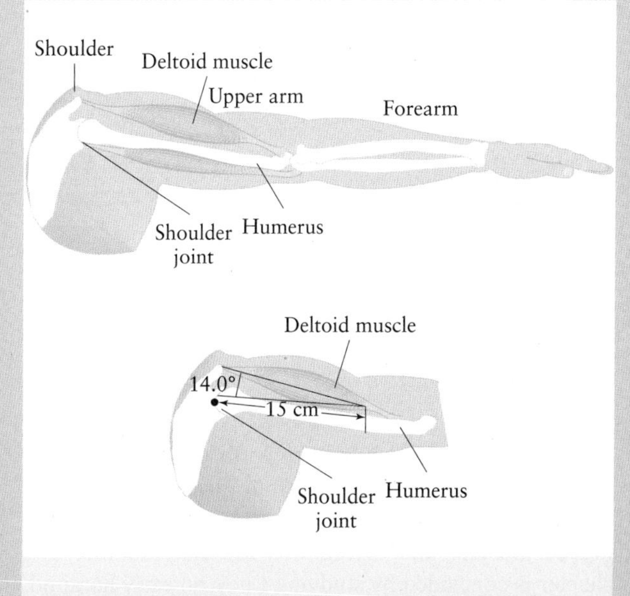

Figure 6.1 A stretched hand being held in horizontal position.

that the vector sum of all forces acting on your arm be zero, but also that the vector sum of all torques on it be zero.

The joints in our body are places where the translational and rotational motion and hence the equilibriums of our body are produced. The joints in our body all share the fundamental design illustrated in Fig. 6.2: two skeletal muscles are attached to the bones adjacent to the joint, one for clockwise and another for counter-clockwise rotation. These muscles exert torques to make motion. Since the tendons connecting these muscles to the bones are usually attached close to the joint, large forces are needed to achieve locomotion or equilibrium. These large forces are of interest in kinesiology and medicine: joints have a relatively high likelihood of failure during their lifetime, not only as a result of injury (tendon and ligament rupture) but also as a result of degeneration (arthritis).

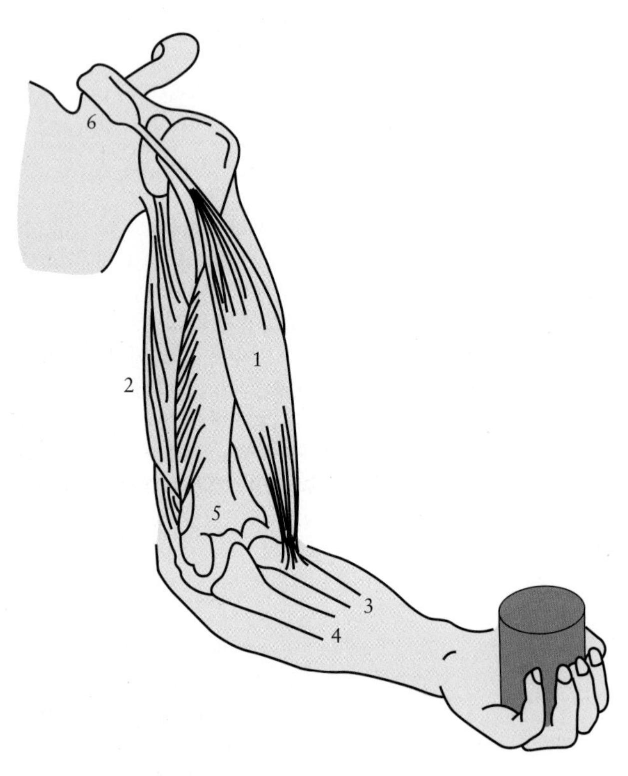

Figure 6.2 Arrangement of the biceps muscle (1) and the triceps muscle (2) in the upper arm. The biceps is connected to the scapula (6) and the radius (3), which runs parallel to the ulna (4) in the lower arm. The muscle actions of the biceps and triceps rotate the lower arm, relative to the humerus (5), at the elbow.

Finally, a case study will be performed to look at the conditions required to evolve from quadruped to biped. The requirements needed for effective and efficient bipedal posture and locomotion are also studied. The chapter is concluded by studying Lucy, an early hominid, from the physics perspective and by showing that she was a biped.

6.1: Force and Extended Object

Forces were identified in Chapters 3 and 4 as the cause of change in motion. But what determines how effective a force is in causing or changing the rotational motion of an object?

Figure 6.3(a) illustrates three concurrent forces (passing through a common point) applied to an object. These three forces do not set the object into rotation, and the object may be modelled as point-like. As long as the vector sum of the acting forces remains zero, the object continues to remains in equilibrium.

Figure 6.3(b) shows three forces that do not intersect at a common point (non-concurrent forces). Together, they can set the object into rotation and the object can no longer be modelled as point-like. Even when their vector sum becomes zero, the object may not remain in equilibrium. In this case, the object has to be modelled as a rigid object.

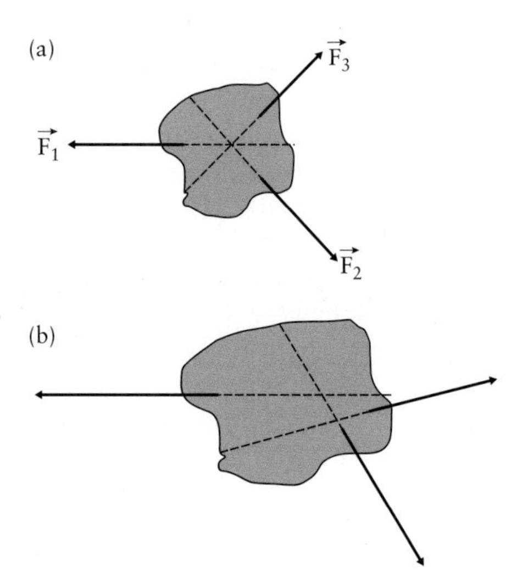

Figure 6.3 Concurrent forces and non-concurrent forces. (a) The line of actions of concurrent forces passes through a common point. (b) The line of actions of non-concurrent forces do not intersect at a common point.

6.1.1: Rigid Object as a Model for a Rotating System

For an extended object, the rigid object model is as follows:

- The distances between any two parts of the object are fixed.
- The angles between the lines connecting any three points of the object are also fixed.

To include rotations of a limb bone, two extensions to the approach taken so far are made: First, we note that modelling the bone as a point-like object is no longer valid because the position of the joint and the position at which the various muscles act upon it do not coincide;

this is illustrated in Fig. 6.2. In order to take this into account, we use the rigid object model. Second, for rotations, we have to distinguish between the effects of the forces that act along a line passing through the joint and those acting anywhere else on a line that passes through the pivot point (i.e., those that do not pass through the pivot point). In fact, we introduce the concept of torque in order to make that distinction. With these two concepts clarified, we will revisit Newton's first law and extend it in such a way to allow for the definition of equilibrium for a rigid object.

6.2: Torque

The tendency of a force to cause or change the rotational motion of an object is quantified by torque. Torque plays the same role in rotational motion as force plays in translational motion and is the cause of angular acceleration.

KEY POINT

Torque produces rotational acceleration, just as force produces translational acceleration.

Let's start with a simple experiment you can perform. Take two uniform rods or sticks, a short one and a long one. Make the short stick the base and hinge the long one in the middle of it, making a miniature seesaw as shown in Fig. 6.4(a). In equilibrium, the long rod should stand horizontally. Take three moveable masses, A, B, C, such that $m_A = m_B > m_C$. Perform the following simple experiments:

(i) Put the two masses A and B on each side of the rod at equal distances from the hinge (pivot point). As Fig. 6.4(b) shows, the rod with the two masses is balanced and stays in equilibrium horizontally.

(ii) Move one of the masses, say B, away from the pivot point so that it is farther from the centre than A, and then let it go. The rod will tilt toward B, as shown in Fig. 6.4(c). Only when these two equal masses sit at the same distance from the pivot will they stay balanced and not cause the rod to rotate.

(iii) Take mass B out and put on mass C at an equal distance to mass A from the pivot point. Now you have A and C on the rod at the same distance from the pivot. However, the rod will not stay in equilibrium horizontally and will tilt again, but this time toward A, as shown in Fig. 6.4(d).

(iv) Move mass C away from the pivot point. You will find a position for mass C at which the rod will stay in equilibrium horizontally, as shown in Fig. 6.4(e). Now the rod with two different masses, A and C, positioned at different distances from the pivot is balanced.

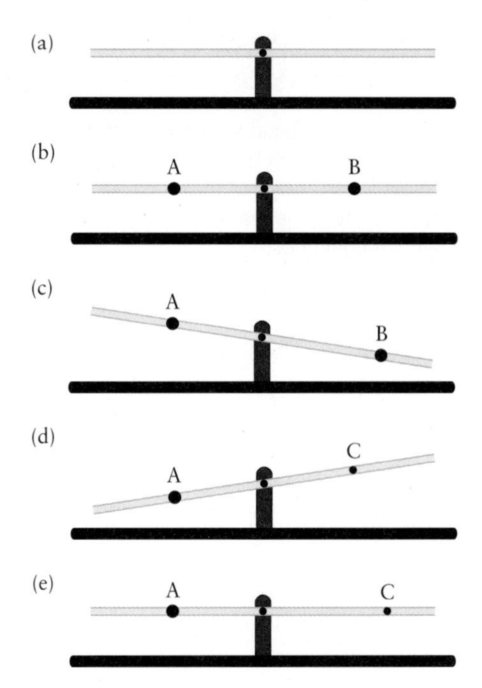

Figure 6.4 A miniature seesaw (a) in equilibrium with no masses attached to the horizontal rod, (b) with two equal masses at each side of the hinge at equal distance in equilibrium, (c) with two equal masses at each side of the hinge at different distances, (d) with two unequal masses at equal distances, (e) with two unequal masses at different distances in equilibrium.

In all four cases, the net force on the object, the horizontal rod, and the two masses is zero. But only in two cases does the rod stay in equilibrium horizontally and is in fact balanced. This will show that the magnitudes of the forces as well as their lines of action and the positions of their application together determine if the object is in equilibrium. As these experiments show, whenever the products of wl (the weight of the object, w, and the distance of the mass from the pivot, l) on both sides become equal, the object is balanced. Figs. 6.5(a) and (b) show the force diagrams for these two cases.

As shown, the rod and masses are balanced only if $w_A l_A = w_B l_B$ or $w_A l_A = w_C l_C$, where w_A, w_B, and w_C are the weights of masses A, B, and C, and l_A, l_B, and l_C are the distances of A, B, and C from the pivot point.

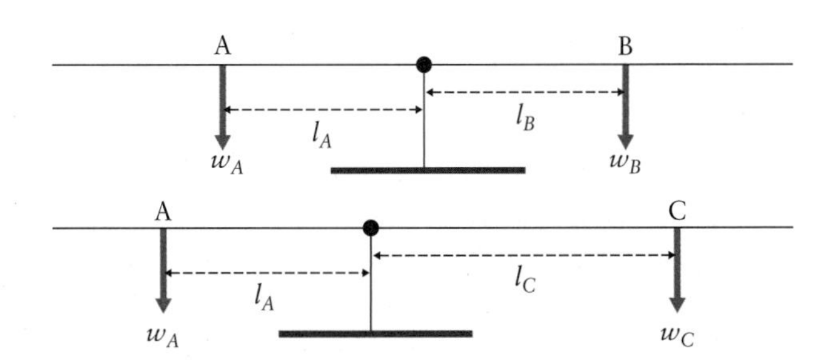

Figure 6.5 Force diagram (a) for two equal masses sitting on each side at equal distance from hinge in equilibrium; (b) for two unequal masses sitting at different distances but in equilibrium.

CONCEPT QUESTION 6.1

Fig. 6.6 shows an overhead view of a door on which five forces $\vec{a}$, $\vec{b}$, $\vec{c}$, $\vec{d}$, and $\vec{e}$ act horizontally. If you want to shut this door most easily, which of these forces will do the job most effectively? Assume that each of these force vectors has the same magnitude.

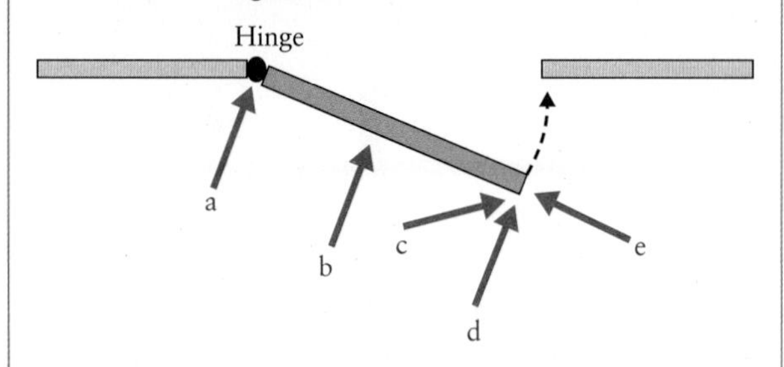

Figure 6.6 An overhead view of a door with five horizontally applied forces on it.

Figure 6.6 of Concept Question 6.1 shows that forces $\vec{a}$ and $\vec{e}$ have little effect in causing the door to close since their lines of action pass through the hinge. Therefore, if the line of action of a force passes through the pivot point, that force does not cause any rotation. Forces $\vec{b}$ and $\vec{d}$ have different effects because their perpendicular distances from the hinge are not equal. The effect of the force $\vec{c}$ is interesting: force $\vec{c}$ is applied to the door at the same point as force $\vec{d}$, but only its component normal to the door is what determines its effectiveness. The component of the force $\vec{c}$, which is parallel to the door, does not have any effect on closing the door, since its line of action passes through the hinge.

KEY POINT

The magnitude of torque is a measure of the strength of the rotation and is the product of the magnitude of the force and the perpendicular distance from its line of action to the fulcrum.

As was shown, not only the magnitude and the direction of a force are important in inducing a rotational motion, its line of action and the position of application are equally as important. The magnitude and the line of action together quantify the effectiveness of a force in causing or changing rotation.

Consider a force $\vec{F}$ acting on a bar that can rotate freely around a fixed axis, as shown in Fig. 6.7. If the force is applied on the bar at a distance r from the axis of rotation and at an angle θ, then the torque τ (Greek tau) is given as:

$$\tau = r F \sin\theta. \qquad [6.1]$$

The unit of torque is $\text{N} \cdot \text{m} = \text{kg} \cdot \text{m}^2/\text{s}^2$.

Note that torque is defined with respect to a specific pivot point. In fact, in all the examples we considered

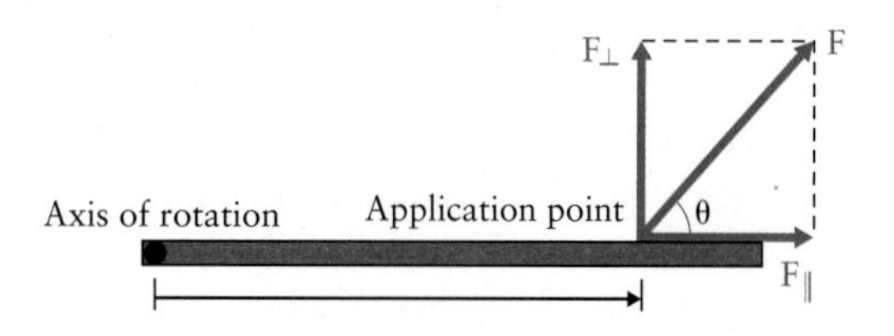

Figure 6.7 The force $\vec{F}$ acting on the bar rotating freely around the axis.

(the seesaw and the door), the pivot point (the axis of rotation) is assumed to be a fixed axis. In this chapter we will deal only with cases where this is so.

CONCEPT QUESTION 6.2

Fig. 6.8 shows different configurations where a force is applied on a uniform bar that can rotate freely around an axis. The dots in Fig. 6.8 represent the axes of rotation, and the force arrows all have the same length. Which of these configurations would produce the largest torque?

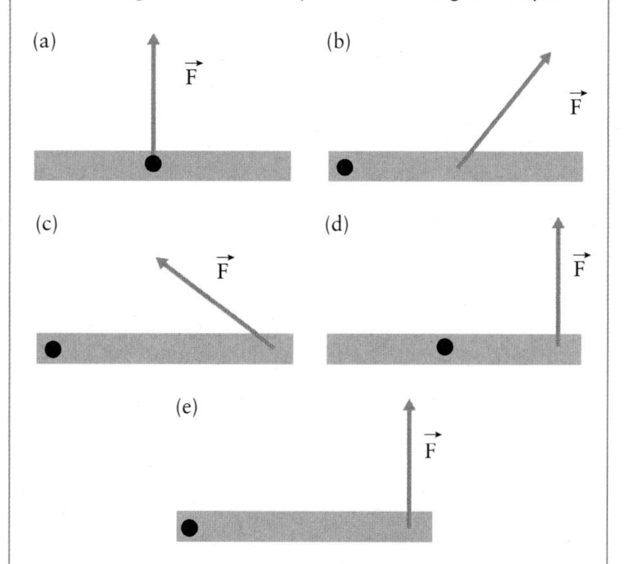

Figure 6.8 Different configurations of exerted force on a uniform bar.

In Eq. [6.1] both the force, $\vec{F}$, and the position, $\vec{r}$, are vectors. Similar to force and position, torque, $\vec{\tau}$, is also a vector. The torque vector acts along the axis of rotation; it does not act along the force line of action or along the position vector, and is perpendicular to both. Eq. [6.1] gives the magnitude of the torque. To find the direction of torque, let's go back to our miniature seesaw (Fig. 6.4). Take the seesaw without the moveable masses on it, shown in Fig. 6.4(a). The long rod stays horizontally in equilibrium. Now put mass A on the left side of the rod. The rod will lose its balance and tilt toward the left, rotating counter-clockwise. Now, take mass A from the left side of the rod and put it on the right side. The rod will tilt toward the right, rotating clockwise.

KEY POINT

If a force causes a counter-clockwise rotation, the torque τ is defined to be positive, $\tau > 0$. Conversely, if a force causes a clockwise rotation, the torque is a negative term, $\tau < 0$.

Figure 6.9 summarizes the sign conventions for a bar with the pivot at its centre. If either $\vec{F}_1$ or $\vec{F}_3$ acts on the bar, a counter-clockwise rotation about the pivot occurs. The corresponding torque terms, $\vec{\tau}_1$ and $\vec{\tau}_3$ are therefore positive, as indicated by a + sign in a small circle in Fig. 6.9. If, instead either $\vec{F}_2$ or $\vec{F}_4$ acts on the bar, the resulting rotation is clockwise with negative torque terms $\vec{\tau}_2$ and $\vec{\tau}_4$, as indicated by a minus sign in a small circle in Fig. 6.9.

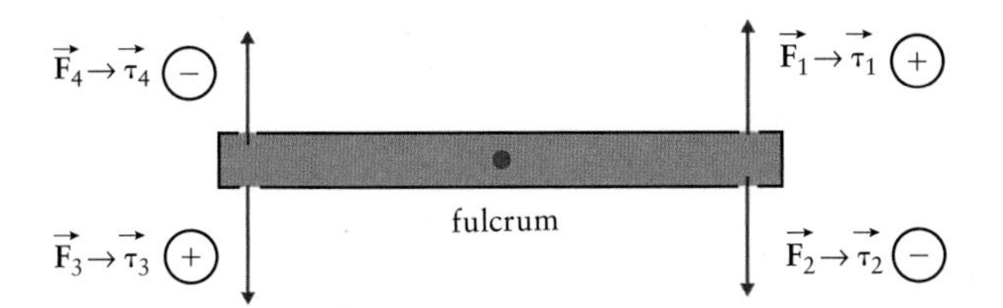

Figure 6.9 Conceptual sketch for the sign convention of the torque. A bar with the pivot at its centre is shown. Any force acting on the bar and causing a rotation can be thought of as one of the four forces shown in the sketch. For each force, the resulting torque is labelled with a corresponding sign in a circle, indicating the sign of the torque contribution.

EXAMPLE 6.1

Fig. 6.10 shows a bar that rotates freely over a fixed axis at one end. If a force of $F = 7.0$ N acts perpendicular to it at one end, where the length of the bar is $L = 1.7$ m, find the magnitude and the direction of the torque due to the force acting upon it.

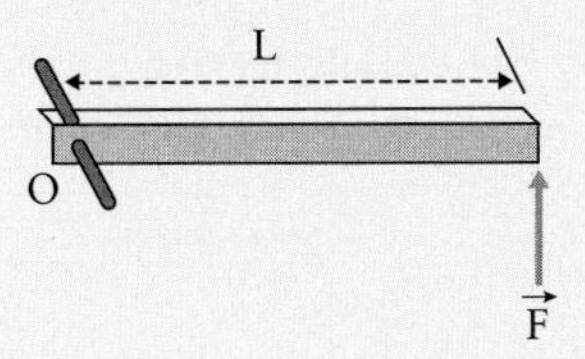

Figure 6.10 Force $\vec{F}$ acting on the bar at one end produces a torque with respect to the pivot point at the other end.

Solution

The known quantities are the magnitudes of force and the acting point distance. Using Eq. [6.1] to find the torque, and substituting the known quantities:

$$\tau = r F \sin\theta$$
$$\tau = (1.7 \text{ m})(7.0 \text{ N}) \sin(90°) = 11.9 \text{ Nm} \approx 12 \text{ Nm}.$$

The direction of the torque is pointing out of the page, causing a rotation in a counter-clockwise direction.

EXAMPLE 6.2

The bar in Example 6.1 is raised to an angle $\theta = 36°$, as shown in Fig. 6.11. Find the magnitude and direction of the torque acting on the bar due to a horizontal force with the same magnitude as the force of Example 6.1.

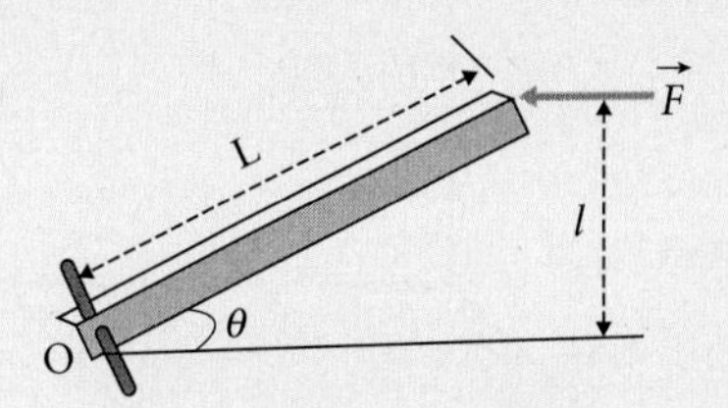

Figure 6.11 The force $\vec{F}$ acting on the bar raised by angle $\theta = 36°$ produces torque with respect to the pivot point.

Solution

As Fig. 6.11 shows, the angle between the force and the bar or, more accurately, between the force's line of action and the line connecting the application point to the pivot, is angle $\theta = 36°$. Using Eq. [6.1] gives:

$$\tau = r F \sin\theta$$
$$\tau = (1.7 \text{ m})(7.0 \text{ N}) \sin(36°) = 6.995 \text{ Nm} \approx 7 \text{ Nm}.$$

The direction of the torque is again pointing out of the page, indicating a counter-clockwise rotation.

6.2.1: Torque as a Vector

In general, torque is defined as the vector cross product of the position vector $\vec{r}$ and the force $\vec{F}$. When a force $\vec{F}$ acts on an object with a position vector $\vec{r}$ with respect to the pivot point, the resulting torque $\vec{\tau}$ with respect to the pivot point is given by:

$$\vec{\tau} = \vec{r} \times \vec{F}. \qquad [6.2]$$

Equation [6.2] adds a directional property to the torque: it is directed perpendicular to the plane of the position vector and the force. That is, it points along the direction of the axis of rotation, which is neither along the force nor along the position vector. Eq. [6.2] is illustrated schematically in Fig. 6.12 for the specific case in which the position vector $\vec{r}$ points in the x-direction and the force $\vec{F}$ is along the y-direction. The torque $\vec{\tau}$ will then point in the z-direction. As Fig. 6.12 indicates, $\vec{\tau}$ is perpendicular to the plane in which both $\vec{r}$ and $\vec{F}$ lie. Fig. 6.12 establishes the **right-hand rule** convention, which allows us to determine the direction of the third vector, $\vec{\tau}$, which is sought in Eq. [6.2]. As Fig. 6.13 shows, if you put the fingers of your right hand in the direction of $\vec{r}$, curl your fingers towards the direction of $\vec{F}$, then your outstretched right thumb will give the direction of $\vec{\tau}$.

The shaded area in Fig. 6.12 represents the product in Eq. [6.2]; i.e., this area is equal to the magnitude $|\vec{\tau}|$, which is given by $|\vec{\tau}| = \tau = F r \sin\theta$.

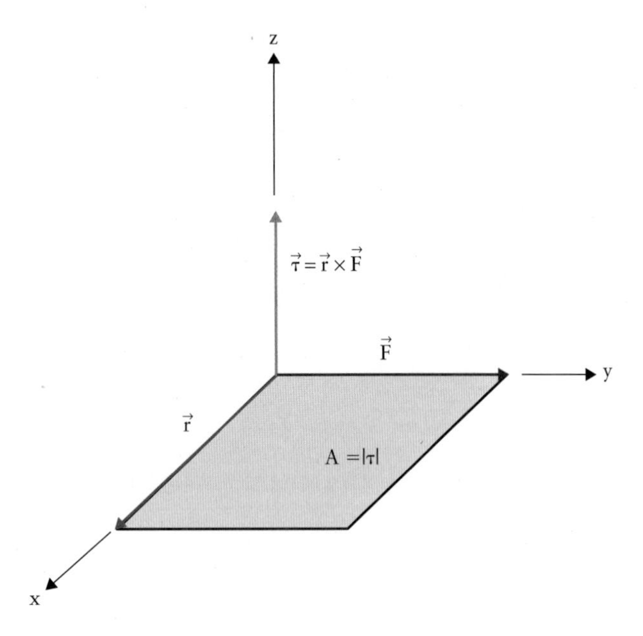

Figure 6.12 The torque vector, $\vec{\tau}$, is orthogonal to the plane defined by the force, $\vec{F}$, and the vector $\vec{r}$, which points from the fulcrum to the position on the rigid object on which $\vec{F}$ acts. The area defined by the vectors $\vec{F}$ and $\vec{r}$ equals the magnitude of the torque, denoted as $|\vec{\tau}|$. The directions of the three vectors are related by the right-hand rule.

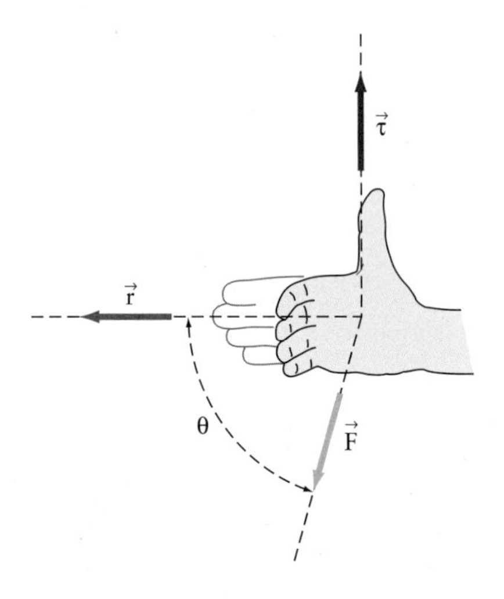

Figure 6.13 Curl your fingers from the line of $\vec{r}$ toward the line of $\vec{F}$, your thumb is now along the line of $\vec{\tau}$.

CONCEPT QUESTION 6.3

(a) Someone who reads this book too quickly is asked to recall Eq. [6.2] but cannot quite remember the order of vectors $\vec{r}$ and $\vec{F}$. What consequence will this have on the results? (b) Another reader writes down Eq. [6.2] with the correct order for the vectors $\vec{r}$ and $\vec{F}$, but uses the left hand to identify the direction of $\vec{\tau}$. What consequences does this error have?

6.2.2: Calculating Torque

We will not use vector cross products in this chapter any more. Since we are using a fixed axis, the direction of the torque can be assigned a + or − sign depending on whether a counter-clockwise or clockwise rotation is produced. Therefore,

- $\tau > 0$ if rotation is counter-clockwise, and
- $\tau < 0$ if rotation is clockwise.

Equation [6.1] can be used to obtain the magnitude of the torque. With Eq. [6.1] in mind, there are two ways to calculate the torque graphically, as we will see below.

6.2.2.1: Using the Components of the Force

Fig. 6.7 also shows the components of the force applied on the bar. The force is decomposed into two components: the perpendicular component and the parallel component. The perpendicular component of the force is perpendicular to the line passing through the pivot, and the parallel component of the force is parallel to the line. These two components of the force are given by:

$$F_{\perp} = F\sin\theta$$
$$F_{\parallel} = F\cos\theta$$

Rewriting Eq. [6.1] gives:

$$\tau = rF\sin\theta = r(F\sin\theta) = r\,F_{\perp}$$
$$\tau = r\,F_{\perp}. \qquad [6.3]$$

$F_{\perp}$ is the perpendicular component of the force $\vec{F}$, which does contribute to the torque. $F_{\parallel}$ is the parallel component of the force $\vec{F}$, which passes through the axis of rotation and therefore does not contribute to the torque.

Equation [6.3] shows that only the perpendicular component of the force produces a torque.

KEY POINT

Forces that act on the system along a line that passes through the pivot do not contribute to the net torque of the system.

EXAMPLE 6.3

Calculate the torques, τ_1 and τ_2 over the pivot, caused by the two forces, $F_1 = 30$ N and $F_2 = 20$ N acting on a bar as shown in Fig. 6.14(a). The bar is 2 m long and can rotate freely around a pivot point at the left end. F_1 is applied at an angle of 35° to the vertical in the middle of the bar and F_2 is applied at an angle of 15° to the horizontal at the end of the bar.

Solution

Draw the force diagram of the applied forces acting on the bar and the ⊥- and ||-components of the forces, as

continued

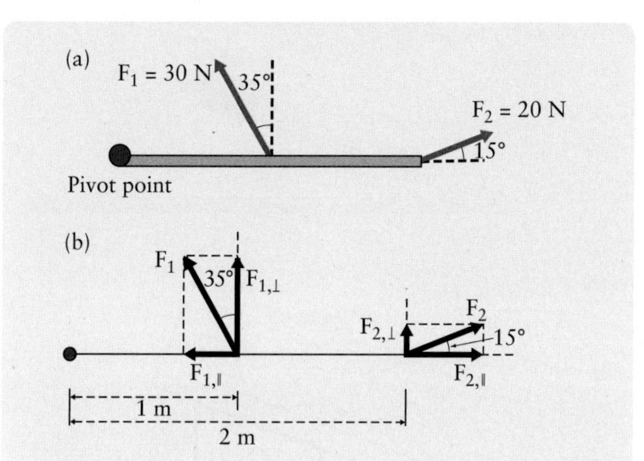

Figure 6.14 (a) The two forces acting on the bar. (b) The force diagram of the forces on the bar with the ⊥- and ||-components of the forces.

shown in Fig. 6.14(b). The components of the forces are given as:

$$\begin{cases} F_{1,\|} = -F_1 \sin(35°) = -(30\text{ N})\sin(35°) = -17.2\text{ N} \\ F_{1,\perp} = F_1 \cos(35°) = (30\text{ N})\cos(35°) = 24.6\text{ N}, \end{cases}$$

$$\begin{cases} F_{2,\|} = F_2 \cos(15°) = (20\text{ N})\cos(15°) = 19.3\text{ N} \\ F_{2,\perp} = F_2 \sin(15°) = (20\text{ N})\sin(15°) = 5.2\text{ N}. \end{cases}$$

Using Eq. [6.3] gives:

$$\tau_1 = +r_1 F_{1,\perp} = +(1\text{ m})(24.6\text{ N}) = +24.6\text{ Nm}$$

Direction of τ_1: out of the page, causes a counter-clockwise rotation:

$$\tau_2 = +r_2 F_{2,\perp} = +(2\text{ m})(5.2\text{ N}) = +10.4\text{ Nm}$$

Direction of τ_1: out of the page, causes a counter-clockwise rotation.

6.2.2.2: Using the Lever Arm

Instead of using the perpendicular component of the force, there is another way to calculate the torque that is often simpler to use and equally as effective as the first method. Consider the bar in Fig. 6.7. The torque produced by force $\vec{F}$ is given by Eq. [6.1]. Rearranging it gives:

$$\tau = rF\sin\theta = F(r\sin\theta).$$

As shown in Fig. 6.15, take:

$$r_\perp = r\sin\theta.$$

The magnitude of the torque is given by:

$$\tau = r_\perp F. \qquad [6.4]$$

The distance $r_\perp$ is called the *lever arm* (or *moment arm*). The lever arm is the perpendicular distance from the axis of rotation to the line of action of the force.

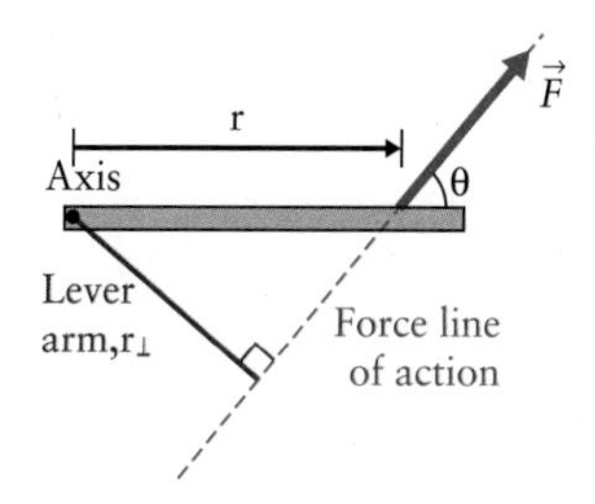

Figure 6.15 The lever arm of the force acting on the bar.

EXAMPLE 6.4

In Example 6.3, calculate the torques τ_1 and τ_2 on the bar as shown in Fig. 6.14(a), using the lever arm approach.

Solution

Figure 6.16 shows the forces on the bar of Example 6.4 and the level arm for each of the forces.

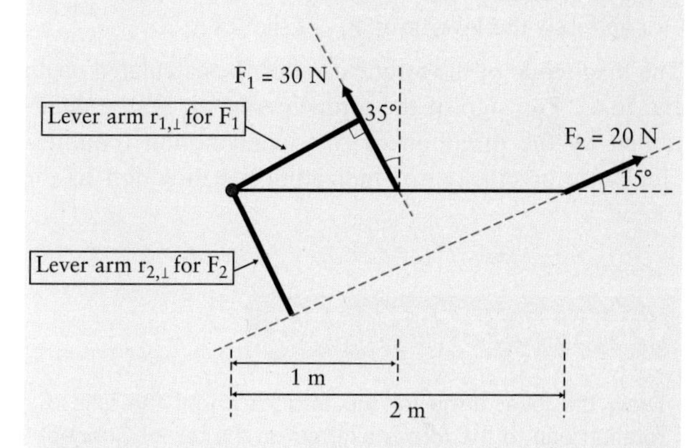

Figure 6.16 Lever arms of the two forces acting on the bar of Fig. 6.14(a).

The lever arms for two forces are:

$$r_{1,\perp} = (1\text{ m})\cos(35°) = 0.82\text{ m}$$
$$r_{2,\perp} = (2\text{ m})\sin(15°) = 0.52\text{ m}.$$

Using Eq. [6.4] gives:

$$\tau_1 = r_{1,\perp} F_1 = (0.82\text{ m})(30\text{ N}) = 24.6\text{ Nm}$$
$$\tau_2 = r_{2,\perp} F_2 = (0.52\text{ m})(20\text{ N}) = 10.4\text{ Nm}.$$

CONCEPT QUESTION 6.4

The forces $\vec{F}_1$, $\vec{F}_2$, $\vec{F}_3$, $\vec{F}_4$, and $\vec{F}_5$ are exerted on a uniform bar that can rotate freely around a fixed axis, as shown in Fig. 6.17. Which of these forces produce a positive torque and which produce a negative torque?

continued

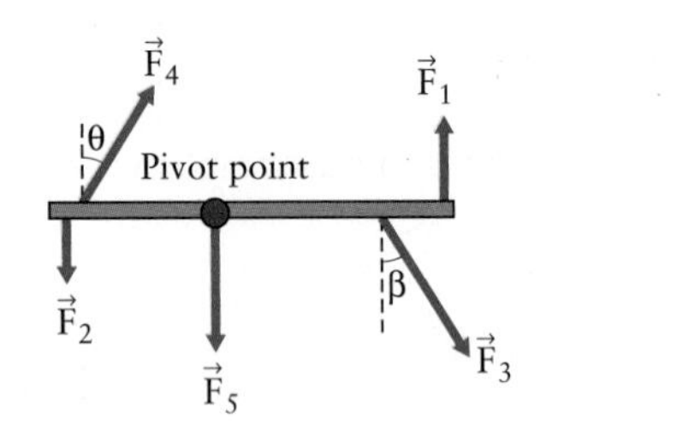

Figure 6.17 Five forces acting on the bar.

To draw the lever arm of a force, you may use the following steps:

(a) Draw an extension of the force passing through the application point. This line is called the force's *line of action* (the dashed line in Fig. 6.15).

(b) Draw a line from the axis of rotation perpendicular to the force's line of action. The distance along this perpendicular line from the axis point to the line of action is the lever arm, $r_{\perp}$, of the force.

The magnitude of the torque can then be calculated using Eq. [6.4]. The sign of the torque (+ *or* –), can be determined by the direction of rotational motion (counter-clockwise or clockwise) indicating the direction of the torque.

CONCEPT QUESTION 6.5

Draw the lever arms for the torques about the axis of rotation due to the forces exerted on the bar of Concept Question 6.4, as shown in Fig. 6.17.

EXAMPLE 6.5

Two uniform bars are hinged at their left ends. The upper bar is lifted at an angle 30.0° from horizontal with a 95.0 N weight attached to the end. The weight is located 86.0 cm from the hinge, as shown in Fig. 6.18(a). How large is the torque exerted on the bar by this weight about the hinge?

Solution

Draw the force diagram of the bar being lifted, with the force of the weight acting on it and its corresponding lever arm, as shown in Fig. 6.18(b). First, calculate the lever arm:

$$r_{\perp} = (86.0 \text{ cm}) \cos(30.0°) = 74.5 \text{ cm}.$$

Using Eq. [6.4] gives

$$\tau = r_{\perp} F = (74.5 \times 10^{-2} \text{ m})(95.0 \text{ N}) = 70.8 \text{ Nm}.$$

continued

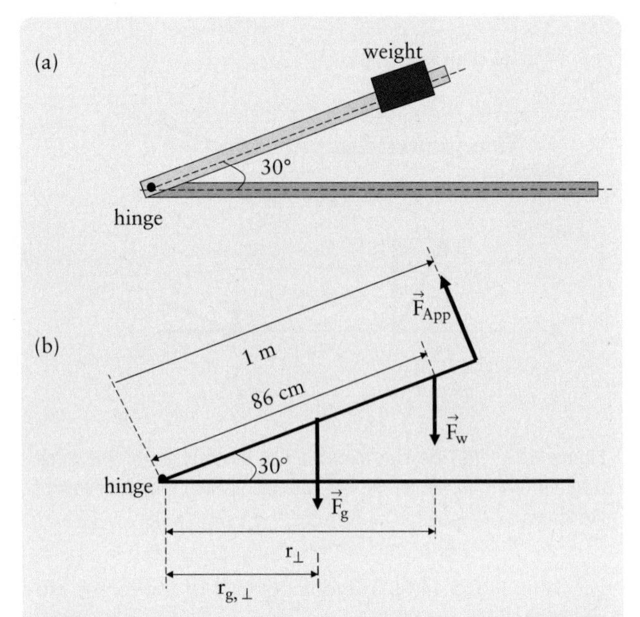

Figure 6.18 (a) Two bars hinged together at one end. The upper bar is lifted at an angle 30.0° from horizontal holding a weight of 95.0 N at the lifted end, 86 cm from the hinge. (b) Force diagram of the bar being lifted applied by the weight.

6.2.3: Net Torque

Consider the five forces acting on the freely rotating bar of Concept Question 6.4 about its axis, shown in Fig. 6.17. To determine the net torque (magnitude and direction) about the pivot due to all forces on the bar, we must calculate the torque (magnitude and direction) about the pivot due to each of the forces separately, and then add them algebraically. Therefore, the net torque on a system on which n forces are acting is given by:

$$\tau_{net} = \sum_{i=1}^{n} \tau_i, \tag{6.5}$$

where τ_i is the torque due to each force F_i and n is the total number of forces applied. Since the torques producing counter-clockwise (ccw) rotation are assigned a positive sign (+), those producing clockwise (cw) rotation are assigned a negative sign (–), Eq. [6.5] can be rewritten as:

$$\tau_{net} = \tau_{net,\ ccw} - \tau_{net,\ cw} \tag{6.6}$$

$$\tau_{net} = \sum_{i=1}^{n} \tau_{i,\ ccw} - \sum_{i=1}^{m} \tau_{i,\ cw} \tag{6.7}$$

Where *ccw* stands for counter-clockwise and *cw* for clockwise. These are simple scalar summations.

In general, the net torque acting on an object is the **vector sum** of all applied torques:

$$\vec{\tau}_{net} = \sum_{i=1}^{n} \vec{\tau}_i = \vec{\tau}_1 + \vec{\tau}_2 + \vec{\tau}_3 + \vec{\tau}_4 + \ldots \tag{6.8}$$

EXAMPLE 6.6

If the bar being lifted in Fig. 6.18(a) of Example 6.5 has a mass of 15.0 kg, what is the torque due to its weight with respect to the hinge? The bar being lifted has a length of 100 cm.

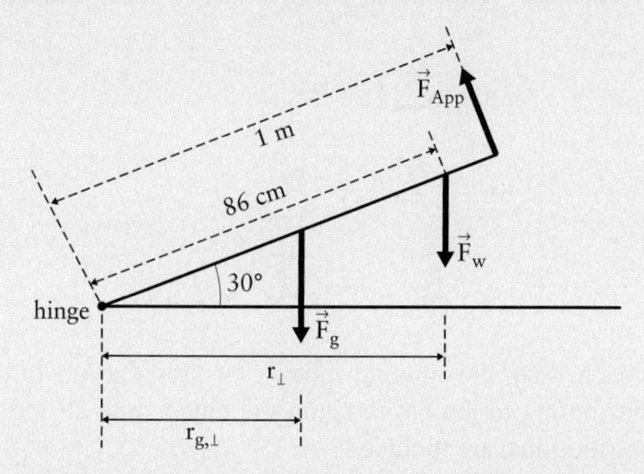

Figure 6.19 Force diagram of the bar being lifted acted on by all forces, the weight, the weight of the bar, and the applied force to hold it at the angle 30.0°.

Solution

Draw the force diagram with the lever arm for the bar as shown in Fig. 6.19. It shows the weight ($\vec{F}_g$) of the bar and its lever arm ($r_{g,\perp}$). The lever arm $r_{g,\perp}$ of F_g is:

$$r_{g,\perp} = (1.00/2 \text{ m}) \cos(30.0°) = 0.433 \text{ m}.$$

Using Eq. [6.4]:

$$\tau_g = r_{g,\perp} F_g = (0.433 \text{ m})(15.0 \text{ kg})(9.80 \text{ m/s}^2) = 63.6 \text{ Nm}.$$

EXAMPLE 6.7

If the lifted bar in Fig. 6.18(a) of Example 6.5 is held at the angle of 30.0° by a string attached to its end, it can be lifted by applying a force perpendicular to it. How large should the force be in order to keep the bar lifted at the angle of 30.0°. *Hint:* Assume the net torque on the bar being lifted is zero.

Solution

Figure 6.19 shows the force diagram of the bar being lifted and all forces acting on it. The force exerted by the string on the bar at its lifted edge is shown as $\vec{F}_{App}$, which is perpendicular to the bar. The two forces $\vec{F}_w$ and $\vec{F}_g$ produce torques that causes clockwise rotation, and $\vec{F}_{App}$ produces a torque causing counter-clockwise rotation. Using Eq. [6.7] and substituting the magnitude of torques produced by $\vec{F}_w$ and $\vec{F}_g$ from Examples 6.5 and 6.6, gives:

$$\tau_{net} = \sum_{i=1}^{n} \tau_{i,\,ccw} - \sum_{j=1}^{m} \tau_{j,cw} = 0$$

$$\tau_{net} = \tau_{F_{App}} - (\tau_{F_w} + \tau_{F_g})$$

$$\tau_{net} = [F_{App}\,(1 \text{ m})] - [(70.8 \text{ Nm}) + (63.6 \text{ Nm})] = 0$$

$$F_{App} = 134.4 \text{ N}.$$

CONCEPT QUESTION 6.6

Fig. 6.20 shows the human head pivoted on the vertebral column. The upper branch of the trapezius muscle connects the head to the shoulder to compensate with force $\vec{F}_{ext}$ for the action of the weight of the forehead, $\vec{w}$. Based on Fig. 6.20, which negative torque term will be used in the calculations?

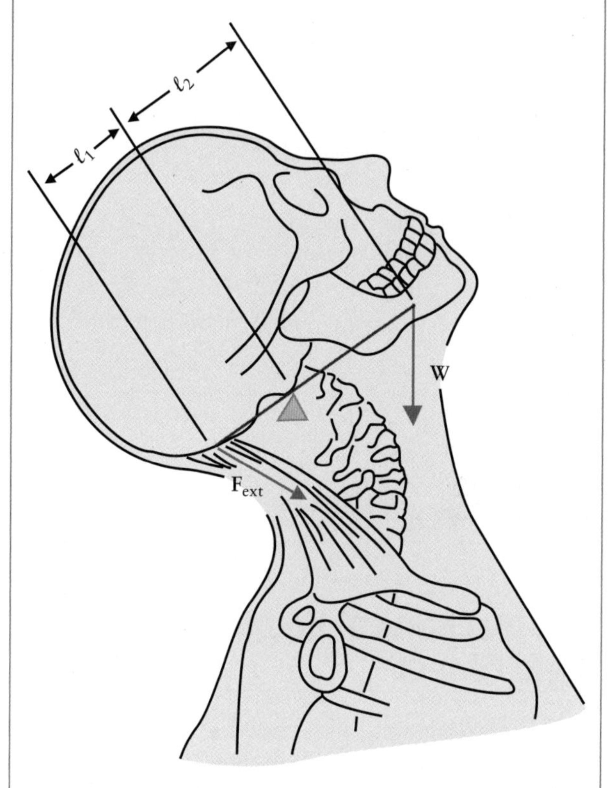

Figure 6.20 The head is pivoted on the vertebral column, defining the fulcrum (green triangle). The weight of the forehead pulls the head down in front of the fulcrum, while the upper branch of the trapezius muscle pulls the forehead up. To develop a mechanical model of the head we simplify it as a bar (red line). For calculations it is necessary to define the distances l_1 and l_2 from the fulcrum to the points at which the two forces act on the bar.

EXAMPLE 6.8

A board is pivoted at the left-hand side. Three forces, $\vec{F}_1$, $\vec{F}_2$, and $\vec{F}_3$, act on the board as shown in Fig. 6.21(a). Find the net torque acting on the board around the pivot point.

Solution

Draw the force diagram for the board as shown in Fig. 6.21(b). In this problem three forces and their distances from the pivot are known. Using Eqs. [6.7] and [6.3] gives:

$$\tau_{net} = \sum_{i=1}^{n} \tau_{i,\,ccw} - \sum_{j=1}^{m} \tau_{j,\,cw}$$
$$= (r_2 F_{2,y} + r_3 F_{3,y}) - (r_1 F_{1,y}).$$

continued

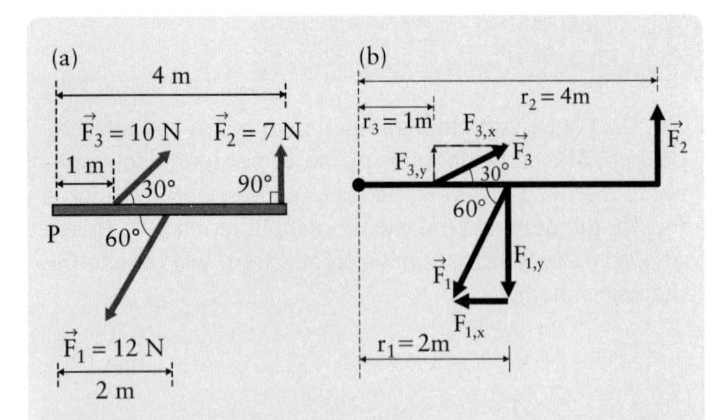

Figure 6.21 (a) The board pivoted at the left side, P, is acted on by three forces. (b) The force diagram of the board.

Calculating the perpendicular components of the forces $F_{i,y}$ and substituting into the above formula gives:

$$\begin{cases} F_{1,y} = F_1 \sin(60°) = (12 \text{ N})\sin(60°) = 10.39 \text{ N} \\ F_{2,y} = F_2 = 7 \text{ N} \\ F_{3,y} = F_3 \sin(30°) = (10 \text{ N})\sin(30°) = 5 \text{ N} \end{cases}$$

$$\tau_{\text{net}} = [(4 \text{ m})(7 \text{ N}) + (1 \text{ m})(5 \text{ N}) - (2 \text{ m})(10.39 \text{ N})$$
$$= +12.22 \text{ Nm}.$$

6.3: Mechanical Equilibrium for a Rigid Object

We can now introduce the concept of equilibrium for a rigid object and the conditions under which an object stays in equilibrium. Let's look back at our miniature seesaw experiments shown in Fig. 6.4. In all four cases discussed, the net force on the object (two masses and the rod holding them) is zero. But, in only two cases does the rod with two masses sitting on it stay in equilibrium horizontally, as shown in Fig. 6.5. In those two cases, the product of weight and the perpendicular distance from the hinge, which gives the torque, on one side of the rod becomes equal to that of the other side, preventing the rod from rotating. In both cases, both sides of the rod are acted on by torques having equal magnitude and opposite directions. So the equilibrium condition is a combination of the translational equilibrium for an object, as introduced in Chapter 4, and a condition that prevents the rigid object from rotating. Therefore, there are two conditions for the equilibrium of a rigid body:

(i) The net force acting on the object must be zero:

$$\vec{F}_{\text{net}} = 0. \qquad [6.9]$$

(ii) The net torque acting on the object about any pivot point must be zero:

$$\vec{\tau}_{\text{net}} = 0. \qquad [6.10]$$

In general, this would require three equations for torque, one for each of the three Cartesian components of the torque vector. Since we have limited our discussion to cases where the axis of rotation is fixed, say along the z-axis, only one torque component has to be considered, reducing the number of equations needed to provide rotational equilibrium. Therefore, the equilibrium condition for a rigid object with a fixed axis of rotation consists of a total of three equations:

$$\begin{cases} F_{\text{net},x} = \sum_i F_{i,x} = 0 \\ F_{\text{net},y} = \sum_i F_{i,y} = 0 \\ \tau_{\text{net}} = \sum_i \tau_{i,\text{ ccw}} - \sum_i \tau_{i,\text{ cw}} = 0 \end{cases} \qquad [6.11]$$

In each case, the running index i ensures that all forces contributing to the net force on the object, and all torque contributions, are included.

EXAMPLE 6.9

A light board (massless) 7 m in length is supported at its ends by two strings. An 85-kg man stands on the board 2.5 m from the right end, as shown in Fig. 22(a). What is the magnitude of the tension in the strings?

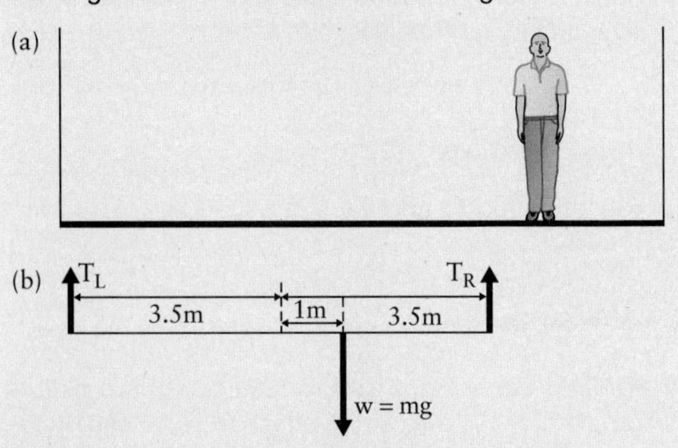

Figure 6.22 (a) A man standing on a light board supported by two strings. (b) The force diagram of the board.

Solution

Draw the force diagram for the board as shown in Fig. 6.22(b). In this problem three forces and their lever arms are known. The system (the board and the man) is in equilibrium and all forces acting on the board are balanced. So both conditions of equilibrium of a rigid body can be used. Three forces, T_R, T_L, w, are the forces acting on the board. Only w is known. We will use Eqs. [6.9] and [6.10]. Eq. [6.9] gives:

$$F_{\text{net}} = \sum_i F_i = T_L + T_R - w = 0.$$

This is only one equation with two unknowns. We have to use the second condition to calculate the net torque. We can choose any point as the pivot and calculate the net torque. Let's take one of two ends of the board as our pivot point. This way we eliminate one

continued

of the unknowns and make the solution simpler. Using Eq. [6.10], determining the torque about the left end gives:

$$\tau_{\text{net}}\big|_{\text{Left End}} = \sum_i \tau_i = \tau_{T_R} - \tau_w = 0$$
$$\tau_{\text{net}}\big|_{\text{Left End}} = +(7\text{ m})T_R - (4.5\text{ m})w = 0$$
$$(7\text{ m})T_R - (4.5\text{ m})(85\text{ kg})(9.8\text{ m/s}^2) = 0$$
$$T_R = (4.5\text{ m})(85\text{ kg})(9.8\text{ m/s}^2)/(7\text{ m}).$$

Substituting the magnitude of T_R into the first formula gives:

$$F_{\text{net}} = T_L + (535.5\text{ N}) - (85\text{ kg})(9.8\text{ m/s}^2) = 0$$
$$T_L + -(535.5\text{ N}) + (85\text{ kg})(9.8\text{ m/s}^2) = 297.5\text{ N}.$$

We can check our answers by calculating the net torque with respect to the right end:

$$\tau_{\text{net}}\big|_{\text{Left End}} = \sum_i \tau_i = \tau_w - \tau_{T_L} = 0$$
$$\tau_{\text{net}}\big|_{\text{Right End}} = (2.5\text{ m})(85\text{ kg})(9.8\text{ m/s}^2) - (7\text{ m})(297.5\text{ N})$$
$$\tau_{\text{net}}\big|_{\text{Right End}} = (2082.5\text{ N}\cdot\text{m}) - (2082.5\text{ N}\cdot\text{m}) = 0.$$

We have confirmed that the net torque with respect to the right end is zero, as expected.

EXAMPLE 6.10

Two children having masses of m_A = 20 kg and m_B = 40 kg are balanced on a seesaw board with a pivot at its centre. If child A sits at 1 m from the pivot, where should child B sit in order to balance child A? Assume the board is massless.

Solution

Draw the force diagram for the board, as shown in Fig. 6.22. Let's determine the net torque with respect to the pivot point. Using Eq. [9.10] gives

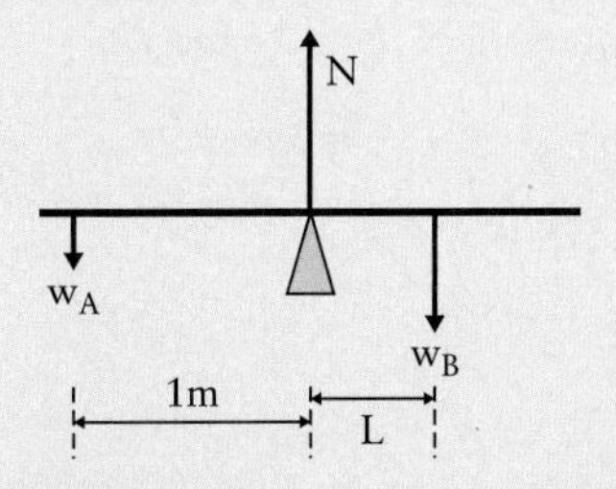

Figure 6.23 The force diagram of the board with two children on each sides of the pivot point:

$$\tau_{\text{net}}\big|_{\text{Pivot}} = \sum_i \tau_i = \tau_{w_A} - \tau_{w_B} = 0$$
$$\tau_{\text{net}}\big|_{\text{Pivot}} = (1\text{ m})(20\text{ kg})(9.8\text{ m/s}^2) - (L)(40\text{ kg})(9.8\text{ m/s}^2)$$
$$= 0$$
$$L = (1\text{ m})(20\text{ kg})(9.8\text{ m/s}^2)/(40\text{ kg})(9.8\text{ m/s}^2)$$
$$= 0.5\text{ m}.$$

CONCEPT QUESTION 6.7

Fig. 6.24 shows five ways a particular case of two forces acting on a bar at distance *d* from the fulcrum at the bar's centre can be labelled. Which of these cases leads to the rotational equilibrium condition *F*1sin *θ* – *F*2cos ϕ = 0 ?

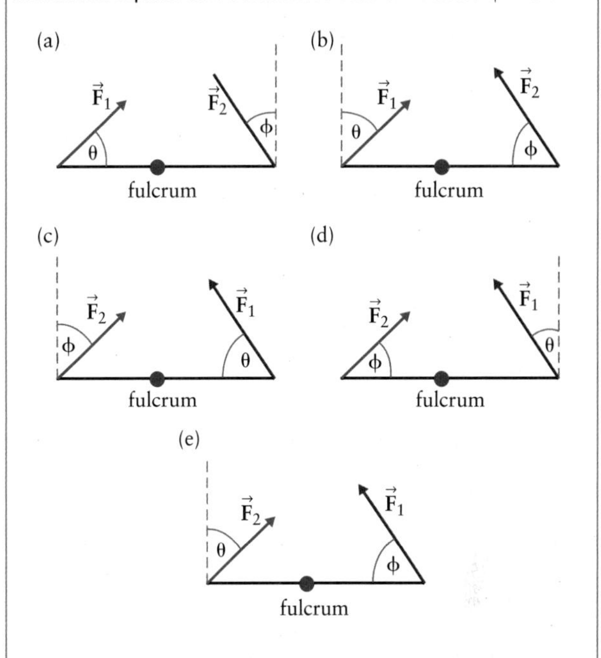

Figure 6.24 Five ways in which two forces, $\vec{F}_1$ and $\vec{F}_2$, may act on a bar with the fulcrum at its centre. Note also the variations in defining the angles *θ* and ϕ.

CONCEPT QUESTION 6.8

Fig. 6.25 shows five locations of the fulcrum along a red bar upon which five forces of any magnitude (blue arrows) act. Which of the five cases cannot be in rotational equilibrium?

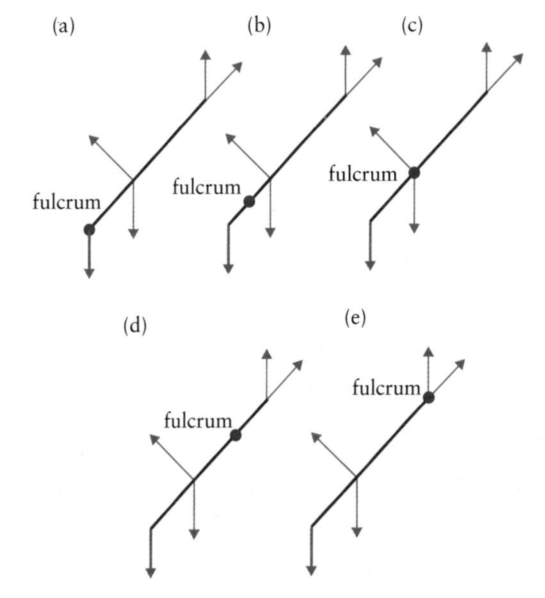

Figure 6.25 Five forces act on a red bar for which we consider five different fulcrum positions.

CONCEPT QUESTION 6.9

Figure 6.26(a) shows a standard man intending to do reverse pushdowns in the gym. In this exercise, the person stands facing the machine, holding the handle bar from below and flexing the elbows against the body. We identify the lower arm and hand as the system, with the ulna and hand bones defining the bar we use to address rotational equilibrium. Four forces act on the bar, as shown in Fig. 6.26(b): the force exerted by the triceps, $\vec{F}_1$, the force exerted by the humerus pushing into the elbow, $\vec{F}_2$, the weight of the system, $\vec{w}$, and the tension, $\vec{T}$, in the cable of the machine. Which of these forces can we neglect when addressing the rotational equilibrium in this case, and why?

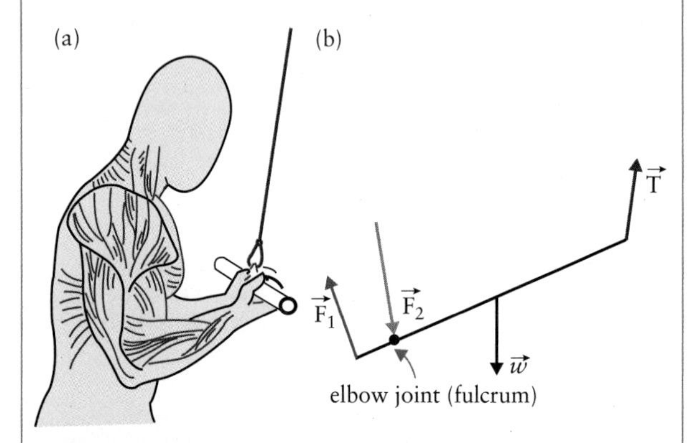

Figure 6.26 A standard man does reverse pushdowns in the gym. We can identify four major forces acting on the lower arm: The triceps muscle exerts the force $\vec{F}_1$, the humerus exerts force $\vec{F}_2$ at the elbow, $\vec{w}$ is the weight of the lower arm and hand, and $\vec{T}$ is the tension in the machine's string.

EXAMPLE 6.11

Figure 6.27 shows a uniform horizontal bar of mass m = 25 kg and length L = 4 m attached to a vertical wall by a point about which the bar can rotate. The bar's far end is held by a massless string that makes an angle of θ = 40° with the horizontal. An object of mass M = 50 kg is placed on the bar at a distance l = 1.2 m from the vertical wall. (a) Find the magnitude of the tension in the string and (b) find the horizontal and vertical components of the force exerted on the bar by the vertical wall.

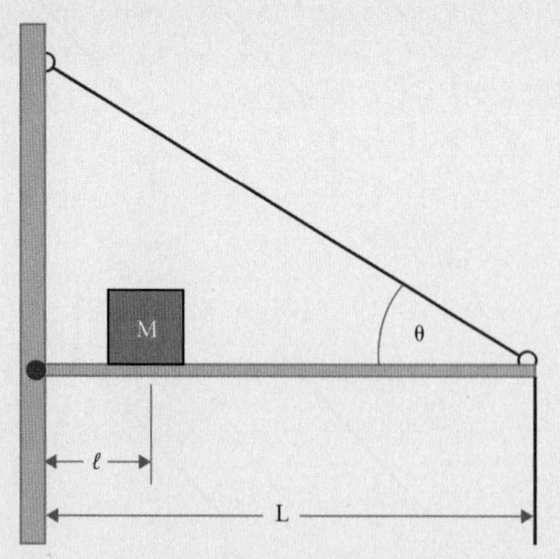

Figure 6.27 A uniform horizontal bar is attached to a wall at a point about which the bar may rotate. The bar's far end is supported by a massless string. An additional object is placed on the bar as shown.

continued

Solution

We identify all forces acting on the bar: the weight of the bar, $\vec{w}_b$, the contact force (normal) from the object of mass M on the bar, $\vec{F}$, the tension in the string, $\vec{T}$, and the contact force from the wall, $\vec{R}$. We can draw a force diagram for the bar, as shown in Fig. 6.28. We put the weight of the bar at its centre of mass. The magnitude of the normal force by the mass acting on the board is equal to the magnitude of the weight of the object, M.

$$F = Mg = (50\ \text{kg})(9.8\ \text{m/s}^2) = 490\ \text{N}.$$

Since the bar is in equilibrium, we use Eq. [6.11]. Take the contact point between the bar and the wall as the

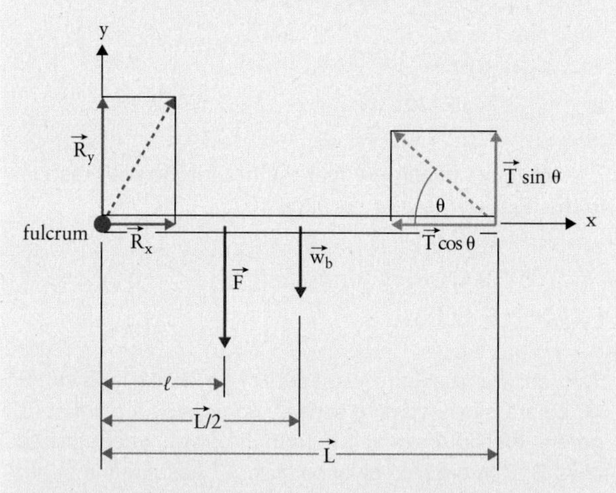

Figure 6.28 The force diagram of the bar presented in Fig. 6.27.

pivot point and determine the net torque on the board with respect to it. In this way, we eliminate the unknown force $\vec{R}$ from the torque equation so that we are only dealing with $\vec{T}$. First let's find the perpendicular and parallel components of $\vec{T}$ As Fig. 6.28 shows, we define the perpendicular and parallel components of $\vec{R}$ as R_x and R_y:

$$\begin{cases} T_x = T\cos\theta \\ T_y = T\sin\theta. \end{cases}$$

Using Eq. [6.11] gives:

$$\begin{cases} F_{\text{net},x} = R_x - T\cos\theta = 0 \\ F_{\text{net},y} = R_y + T\sin\theta - F - w_b = 0 \\ \tau_{\text{net}} = +L\,T\sin\theta - (L/2)w_b - lF = 0. \end{cases}$$

Isolating the magnitude of the tension $\vec{T}$ in the third formula gives:

$$T = ((L/2)mg + lF)/(L\sin\theta).$$

Substituting the magnitudes of L, m, F, and l into the above formula gives:

continued

$$T = ((2\ \text{m})(25\ \text{kg})(9.8\ \text{m/s}^2) + (1.2\ \text{m})(490\ \text{N}))/((4\ \text{m})\sin(40°))$$
$$T = 419\ \text{N}.$$

Now, substitute the magnitude of $\vec{T}$ into the first two formulas:

$$R_x = T\cos\theta = (419\ \text{N})\cos(40°) = 321\ \text{N}$$
$$R_y = F + w_b - T\cos\theta$$
$$= (490\ \text{N}) + (25\ \text{kg})(9.8\ \text{m/s}^2) - (419\ \text{N})\sin(40°)$$
$$= 466\ \text{N}.$$

Knowing R_x and R_y we can calculate the magnitude and direction of $\vec{R}$.

6.4: Classes of Levers and Physiological Applications

A lever is a rigid bar, such as a bone, that exerts a force on another object. Three important parameters of a lever are the "fulcrum" (or pivot), the "load," and the "effort" (the applied force). It has become common practice to distinguish three classes of lever systems in anatomy based on the relative positions of the fulcrum, the weight, and the applied force, like muscle force. These three types are described below.

A **class I lever system** is shown in Fig. 6.29(a). In this case, the fulcrum is between the load and the effort. Fig. 6.29(b) shows the head pivoted on the first cervical vertebra as an example; the fulcrum is indicated by a green triangle. In this type of lever system, the weight (load) acts on one side of the fulcrum and the force (effort) that balances the weight (applied by the muscle) acts on the opposite side of the fulcrum.

A **class II lever system** is shown in Fig. 6.30(a). In this case the fulcrum is at one end of the bar and the effort is exerted on the other end, with the load situated between them. Fig. 6.30(b) shows a foot as an example of a class II lever. The fulcrum is at the end and the weight (load) and the muscle force (effort) act on the

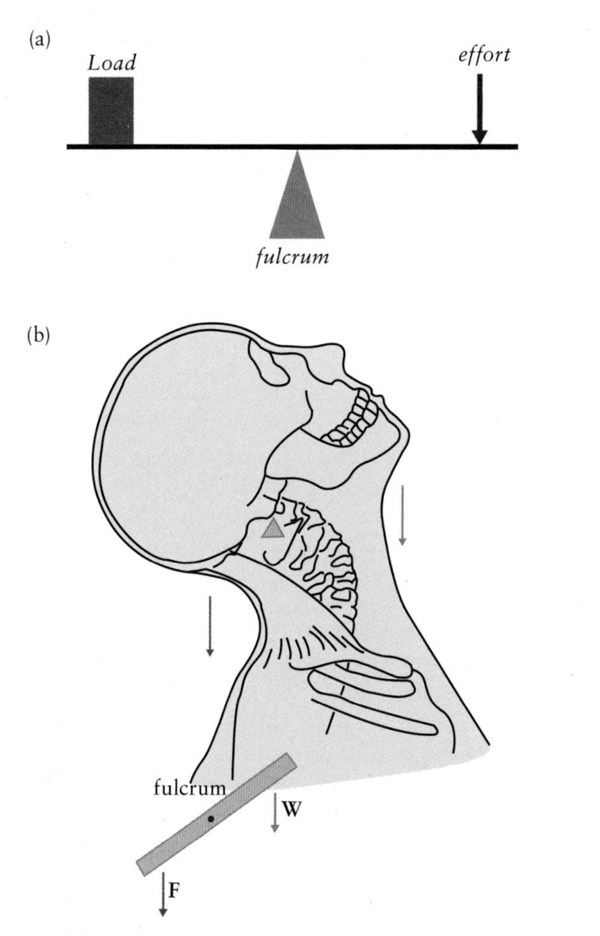

Figure 6.29 Class I lever system. The fulcrum is located between the points at which the weight and the muscle force act.

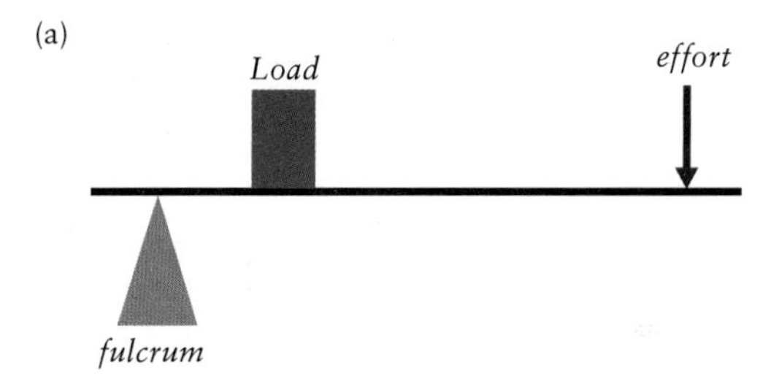

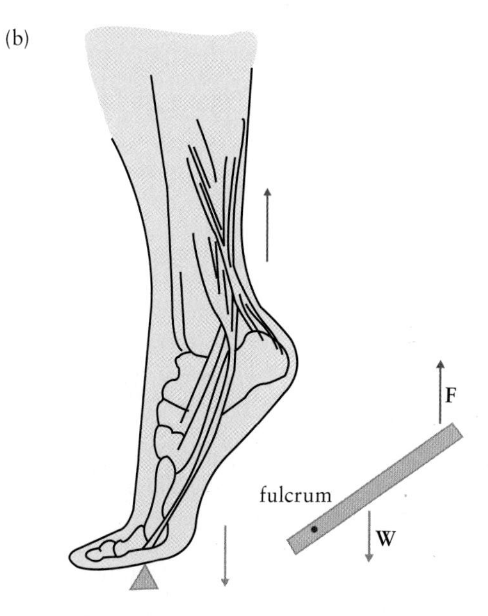

Figure 6.30 Class II lever system. The fulcrum is positioned near the end of the lever arm. The weight acts on the lever arm, closer to the fulcrum than the muscle force.

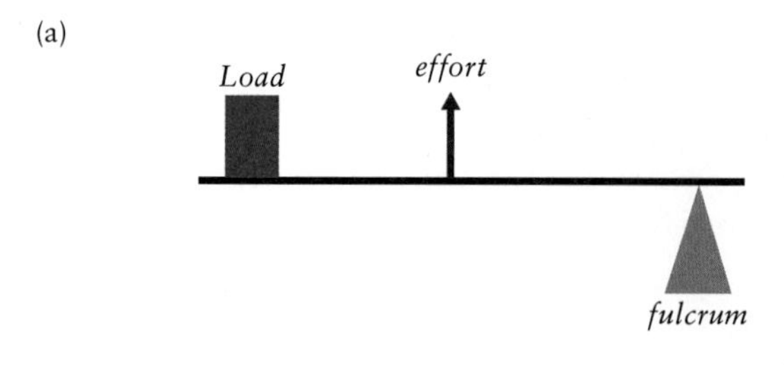

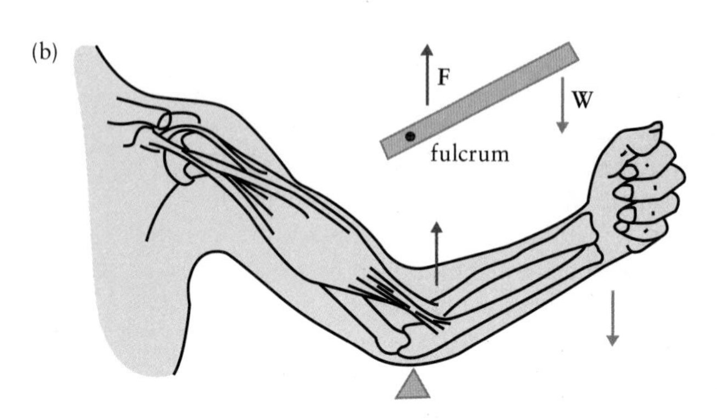

Figure 6.31 Class III lever system. The fulcrum is near the end of the lever arm. The muscle force acts closer to the fulcrum on the lever arm than the weight.

same side of the fulcrum. These forces must act in opposite directions to obtain a rotational equilibrium. Note that the weight acts at a point closer to the fulcrum than the muscle force.

A **class III lever system** is shown in Fig. 6.31(a). In this case, the fulcrum is at one end of the bar and the load is situated at the other end, with the effort acting on the bar between them. Fig. 6.31(b) illustrates a lower arm as an example. Note that the muscle force (effort) acts closer to the fulcrum than the weight (load).

EXAMPLE 6.12

The Achilles Tendon: Fig. 6.32(a) shows a standard man standing on a flat surface, and Fig. 6.32(b) shows a man standing backward on a diving board, intending to do a competitive backward dive into a pool. In Fig. 6.32(a), calculate (a) the magnitude of the two normal forces supporting the foot and (b) the magnitude of the force acting on the Achilles tendon. In Fig. 32(b), calculate (c) the magnitude of the force acting on the Achilles tendon for the motionless athlete on the diving board, and (d) the magnitude of $\vec{F}_{ext}$ and the angle θ.

In all cases, neglect the weight of the foot. Use $x_1 = 6.2\ \text{cm}$ and $x_2 = 12.3\ \text{cm}$, which are typical values for a shoe size of 11. Take the angle of the Achilles tendon with the vertical to be $\varphi = 8°$.

continued

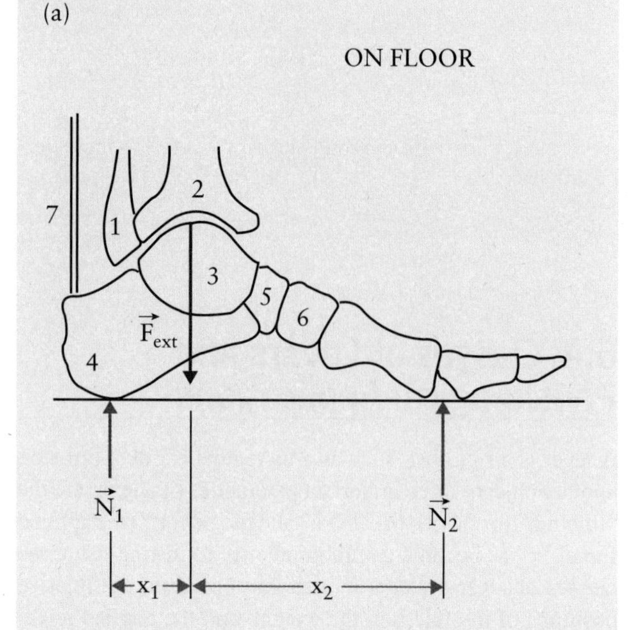

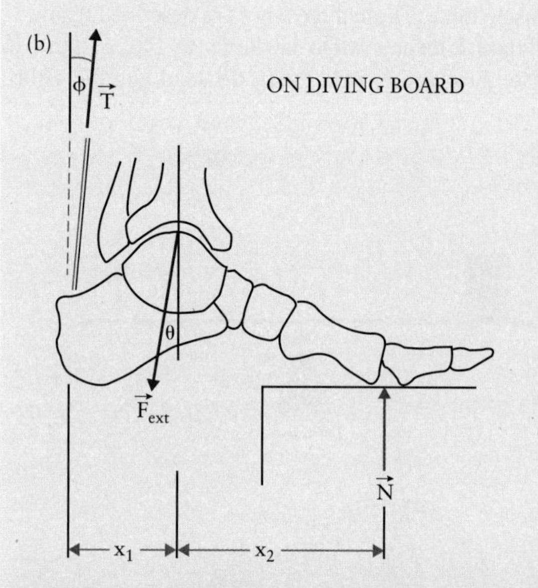

Figure 6.32 Anatomy of the foot of a standard man (a) resting on a flat surface, and (b) balancing backward on a diving board. Bones include the fibula (1), the tibia (2), the talus (3), the calcaneus (4), the navicular bone (5), and the medial cuneiform bone (6). Note that the Achilles tendon (7) is not vertical when balancing on the diving board. $\vec{F}_{ext}$, the force due to the upper body pushing into the foot, and the tension $\vec{T}$ in the Achilles tendon are very large forces. Note that $\vec{F}_{ext}$ is not vertical in the right sketch and is not equal to the weight of the upper body resting on the foot, since $\vec{T}$ pulls the foot toward the leg, beyond the effect of the weight.

Solution

Identify all forces acting on the foot in both cases. In Fig. 6.32(a), the forces are the normal forces from the floor, $\vec{N}_1$ and $\vec{N}_2$, and the force of the upper body on the foot $\vec{F}_{ext}$. In Fig. 6.32(b), the forces are the normal force from

continued

(a)

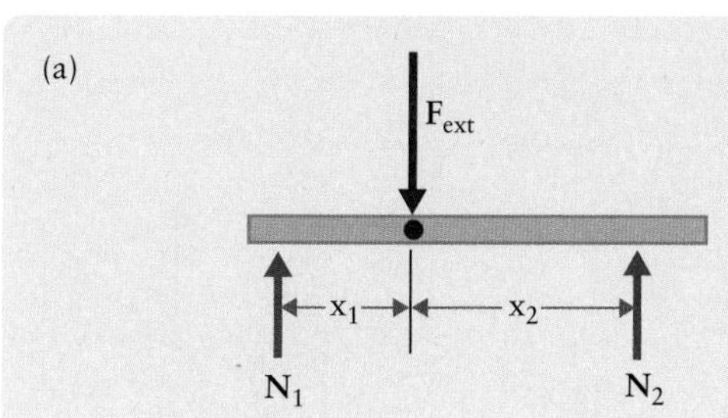

(b)

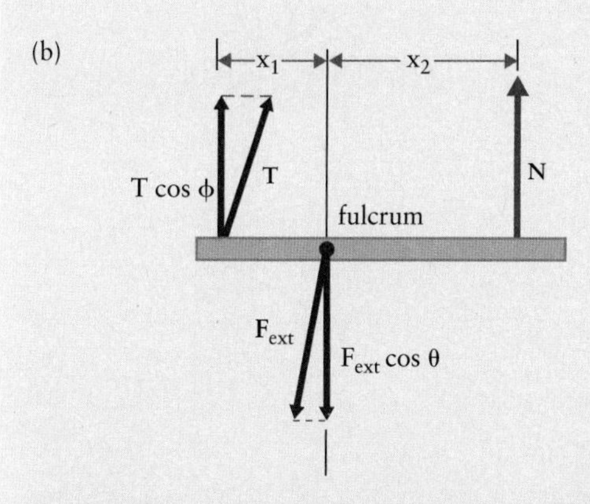

Figure 6.33 (a) The force diagram for the foot on the flat surface as shown in Fig. 6.32(a). Note that the two normal forces are drawn to different lengths and are labelled with different indexes because we cannot assume they are equal in magnitude. (b) The force diagram for the foot on the diving board.

the diving board, $\vec{N}$, the force of the upper body on the foot, $\vec{F}_{ext}$, and the tension in the Achilles tendon. Draw the force diagrams for both cases as shown in Figs. 6.33(a) and (b). For both cases we will use Eq. [6.11].

For the man standing on the floor, based on Fig. 6.33(a), Eq. [6.11] gives:

$$\begin{cases} F_{net,y} = N_1 + N_2 - F_{ext} = 0 \\ \tau_{net} = N_2 x_2 - N_1 x_1 = 0. \end{cases}$$

Take $F_{ext} = \frac{1}{2} w$, where w is the weight of the standard man. Note that this is a simplifying assumption, since the person's body weight may be distributed unequally. So:

$$F_{ext} = \frac{1}{2} w = \frac{1}{2}(70 \text{ kg})(9.8 \text{ m/s}^2) = 343 \text{ N}.$$

Substituting the known values into the above formulas gives:

$$\begin{cases} N_1 + N_2 = 343 \\ (12.3 \text{ cm}) N_2 = (6.2 \text{ cm}) N_1. \end{cases}$$

From the second formula:

$$\therefore N_2 = (6.2 \text{ cm}/12.3 \text{ cm}) N_1 = 0.504 N_1.$$

Substituting N_2 in first formula gives:

$$N_1 + 0.504 N_1 = 343 \text{ N}$$

$$N_1 = 343/1.504 = 228 \text{ N}$$

$$N_2 = 0.504 N_1 = 0.504(228 \text{ N}) = 115 \text{ N}.$$

For the man standing on the diving board, based on Fig. 6.33(b), Eq. [6.11] gives:

$$\begin{cases} F_{net,x} = T \sin\varphi - F_{ext} \sin\theta = 0 \\ F_{net,y} = T \cos\varphi + N - F_{ext} \cos\theta = 0 \\ \tau_{net} = N x_2 - T \cos\varphi \, x_1 = 0. \end{cases}$$

From the third formula, calculate the tension T:

$$T = ((w/2)\, x_2)/(x_1 \cos\phi) T = ((w/2)\, x_2)/(x_1 \cos\phi)$$

$$T = ((70 \text{ kg})(9.8 \text{ m/s}^2)(0.123 \text{ m})\, x_2)/(2\,(0.062 \text{ m}) \cos(8°))$$

$$T = 687 \text{ N}$$

From the first two formulas, we will get:

$$F_{ext} \sin\theta = T \sin\varphi$$

$$F_{ext} \cos\theta = T \cos\varphi + N$$

Also $N = w/2$.

Dividing the first by the second and using the result for T gives:

$$\tan\theta = \frac{T \sin\phi}{T \cos\phi + N} = \frac{T \sin\phi}{T \cos\phi + w/2}$$

$$\tan\theta = \frac{(687 \text{N}) \sin(8°)}{(687 \text{ N}) \cos(8°) + (70 \text{ kg})(9.8 \text{ m/s}^2)/2}$$

$$\tan\theta = 0.093$$

$$\theta = 5.3°.$$

From the first formula:

$$F_{ext} = \frac{T \sin\varphi}{\sin\theta}$$

$$F_{ext} = \frac{(687 \text{ N})\sin(8°)}{\sin(5.38°)} = 1035 \text{ N}.$$

EXAMPLE 6.13

The Mandible and the Masseter: Fig. 6.34(a) shows the masseter, which is one of the strongest muscles in the human body. It connects the mandible (the lower jaw bone) to the skull. The mandible is pivoted about a socket just in front of the ear. Three forces act on the jaw bone, as shown in Fig. 6.34(b): $\vec{F}_{ext}$ is the external force exerted by the chewed food, $\vec{T}$ is the tension in the masseter tendon, and $\vec{R}$ is the force exerted on the mandible by the skull. We make the simplifying assumption that these three forces act perpendicularly to the lower part of the mandible as shown. Use l_1 = 9 cm and l_2 = 5 cm, and an angle of 110° between the two parts of the mandible. Find (a) the magnitude of the tension $\vec{T}$ when the person bites down with a force of 40 N, and (b) the magnitude of the force $\vec{R}$ for the same bite.

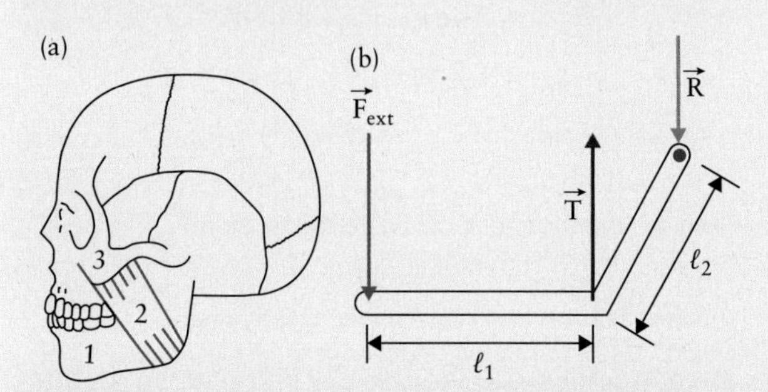

Figure 6.34 (a) Illustration of the attachment of the masseter (2) to the mandible (1) and the cheek bone (3). (b) Simplified arrangement of the forces acting on the mandible during chewing. The tension $\vec{T}$ is exerted by the masseter. The force $\vec{R}$ acts at the joint. The external force $\vec{F}_{ext}$ is due to the person chewing. The mandible consists of two straight parts of lengths l_1 and l_2; the angle between them is 110°.

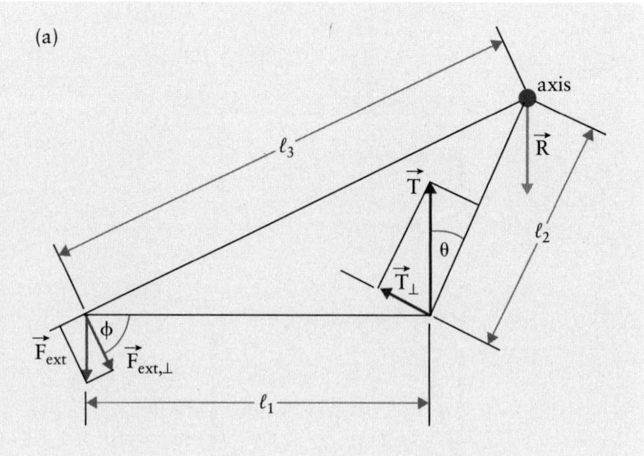

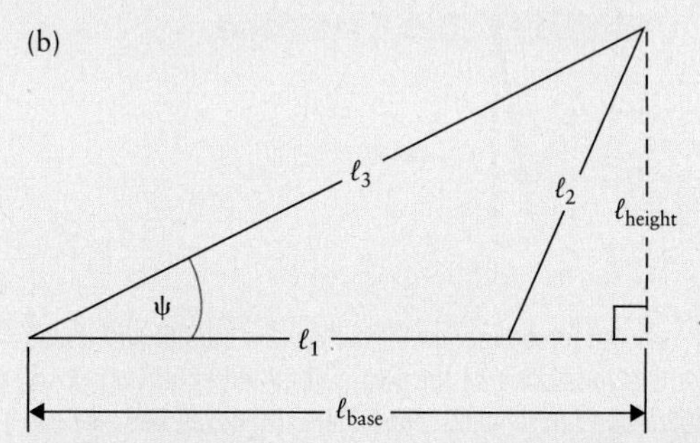

Figure 6.35 (a) The force diagram for the mandible system in Fig. 6.34. (b) Geometrical sketch illustrating how the angles and side lengths of the triangle in (a) are related to each other.

Solution

The mandible is the system; all relevant forces acting on it are shown in Fig. 6.34(b). The force diagram is shown in Fig. 6.35(a). Both the tension and the external force are divided into components parallel to and perpendicular to the line through the fulcrum. Of these components, the ones that are perpendicular to the line are needed for the balance-of-torque equation.

We chose the horizontal direction as the x-axis and the vertical direction as the y-axis in the diagram. Using Eq. [6.11] for equilibrium conditions together with Eq. [6.3] gives:

$$\text{in } y\text{-direction: } -F_{ext} + T - R = 0$$

$$\text{for torque: } +(F_{ext}\sin\varphi)\, l_3 - (T\sin\theta)\, l_2 = 0.$$

As Fig. 6.34(a) shows, $\vec{R}$ acts at the pivot. Isolating T from second formula gives:

$$T = F_{ext}\,(l_3\sin\varphi / l_2\sin\theta).$$

Since the angle between the two parts of the mandible is 110°:

$$\theta = 110° - 90° = 20°$$

Drawing Fig. 6.35(b) help us to determine ϕ and l_3. Using trigonometry gives:

$$l_{base} = l + l_2\cos(70°) = 10.7 \text{ cm}$$

$$l_{height} = l_2\sin(70°) = 4.7 \text{ cm}$$

$$\tan\psi = l_{height}/l_{base} = (4.7\text{ cm})/(10.7\text{ cm}) = 11.7\text{ cm}$$

$$\psi = \tan^{-1} 11.7 = 23.7°$$

$$\phi = 90° - \psi = 90° - 23.7° = 66.3°$$

$$l_3 = \sqrt{l_{base}^2 + l_{height}^2} = \sqrt{(10.7)^2 + (4.7)^2} = 11.7 \text{ cm.}$$

Substituting these values for ϕ and l_3 into the formula for calculating T gives:

$$T = (40\text{ N})\left(\frac{(11.7\text{ cm})\sin(66.3°)}{(5.0\text{ cm})\sin(20°)}\right) = 250\text{ N}$$

Substituting the magnitude of T into the first formula gives:

$$R = T - F_{ext} = (250\text{ N}) - (40\text{ N}) = 210\text{ N.}$$

Note: Calculations such as these in this example are useful when we attempt reconstructions of fossil skulls to assess physiological properties of an extinct species. Applying this approach to hominids from the past 5 million years provides a measure of strength of the masseter, which in turn is indicative of the diet of an individual.

continued

Calculating the required force for the same strength of bite allows us to judge whether the much more massive masseter of our ancestors like *Australopithecus robustus* represents a lesser adaptation due to the earlier stage of evolution, or whether it indicates a diet that required a lot of biting or chewing.

Another application of mechanical calculations for the mandible is the evolutionary adaptation to burrowing among *caecilians*, which are worm-like amphibians. Burrowing is a very demanding task and advanced caecilians, such as *Microcaecilia rabei*, have developed a second set of muscles to supplement the masseter in jaw-closing.

CONCEPT QUESTION 6.10

Figure 6.36 shows three bent-lever arms with the fulcrum indicated by a dot and a single force $\vec{F}$ acting at the far end of the lever arm. Which of the three systems (a) is in mechanical equilibrium, and (b) does not rotate as a result of the force?

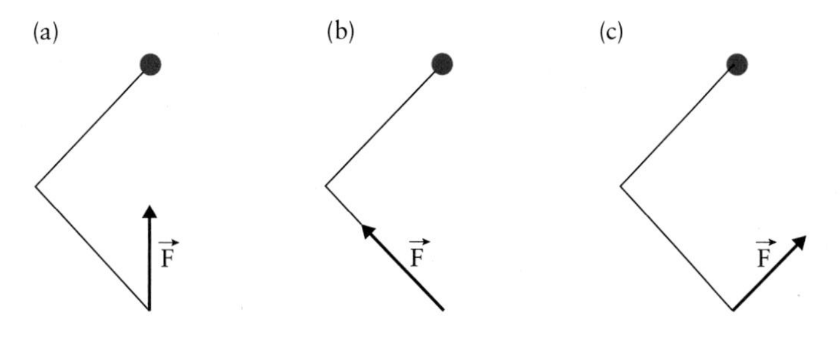

Figure 6.36 Three bent-lever arm systems with fulcrum (dot) and the action of a single force $\vec{F}$ at the opposite end.

EXAMPLE 6.14

The Hip Joint: Fig. 6.37 shows the anatomy and the main forces acting at the hip joint of a standard man (see data in Table 4.1) shifting from standing on both legs in Fig. 6.37(a) to balancing on one leg in Fig. 6.37(b). Calculate for the person balancing on one leg (a) the magnitude of the tension in the abductor muscle's tendon, $\vec{T}$, and (b) the components of the external force the pelvis exerts on the femur, $\vec{F}_{ext}$.

Solution

Consider the leg as the system. We identify four forces that act on the leg:

- the tension, $\vec{T}$, in the abductor muscle that acts on the great trochanter at an angle of $\theta = 65°$ to the positive *x*-axis, which runs horizontally to the right
- the external force, $\vec{F}_{ext}$, due to the acetabulum, which acts on the head of the femur, pushing in the direction of the hip joint
- the normal force $\vec{N}$ by the floor on the foot, $|N| = |w_{Person}|$
- the weight of the leg w_L, $|w_L| = 0.15\,|w_{Person}|$ (from Table 4.1).

The force diagram is shown in Fig. 6.37(c). The distances along the leg are given for an average person as l_1 = 8 cm, l_2 = 40 cm, and l_3 = 90 cm.

Based on Fig. 6.37(c) and using Eq. [6.11] gives:

$$\begin{cases} F_{net,x} = T\cos\theta - F_{ext,x} = 0 \\ F_{net,y} = T\sin\theta + N - F_{ext,y} - w_L = 0 \\ \tau_{net} = l_3 N\sin\psi - l_1 T\sin\theta - l_2 w_L \sin\psi = 0. \end{cases}$$

From third formula, we determine $\vec{T}$:

$$T = \frac{\sin\psi(l_3 - 0.5 l_2)}{l_2 \sin\theta}\, w$$

$$T = \frac{\sin(8°)(190\text{ cm} - (0.5)(40\text{ cm})}{(8\text{ cm})\sin(65°)}(70\text{ kg})(9.8\text{ m/s}^2)$$

$$= 1100\text{ N}.$$

From the first formula:

$$F_{ext,x} = T\cos\theta = (1100\text{ N})\cos(65°) = 465\text{ N}.$$

From the second formula:

$$F_{ext,y} = T\sin\theta + N - w_L = T\sin\theta + (w - 0.15\,w)$$

$$F_{ext,y} = T\sin\theta + 0.85\,w$$

$$= (1100\text{ N})\sin(65°) + 0.85(70\text{ kg})(9.8\text{ m/s}^2)$$

$$= 1580\text{ N}.$$

continued

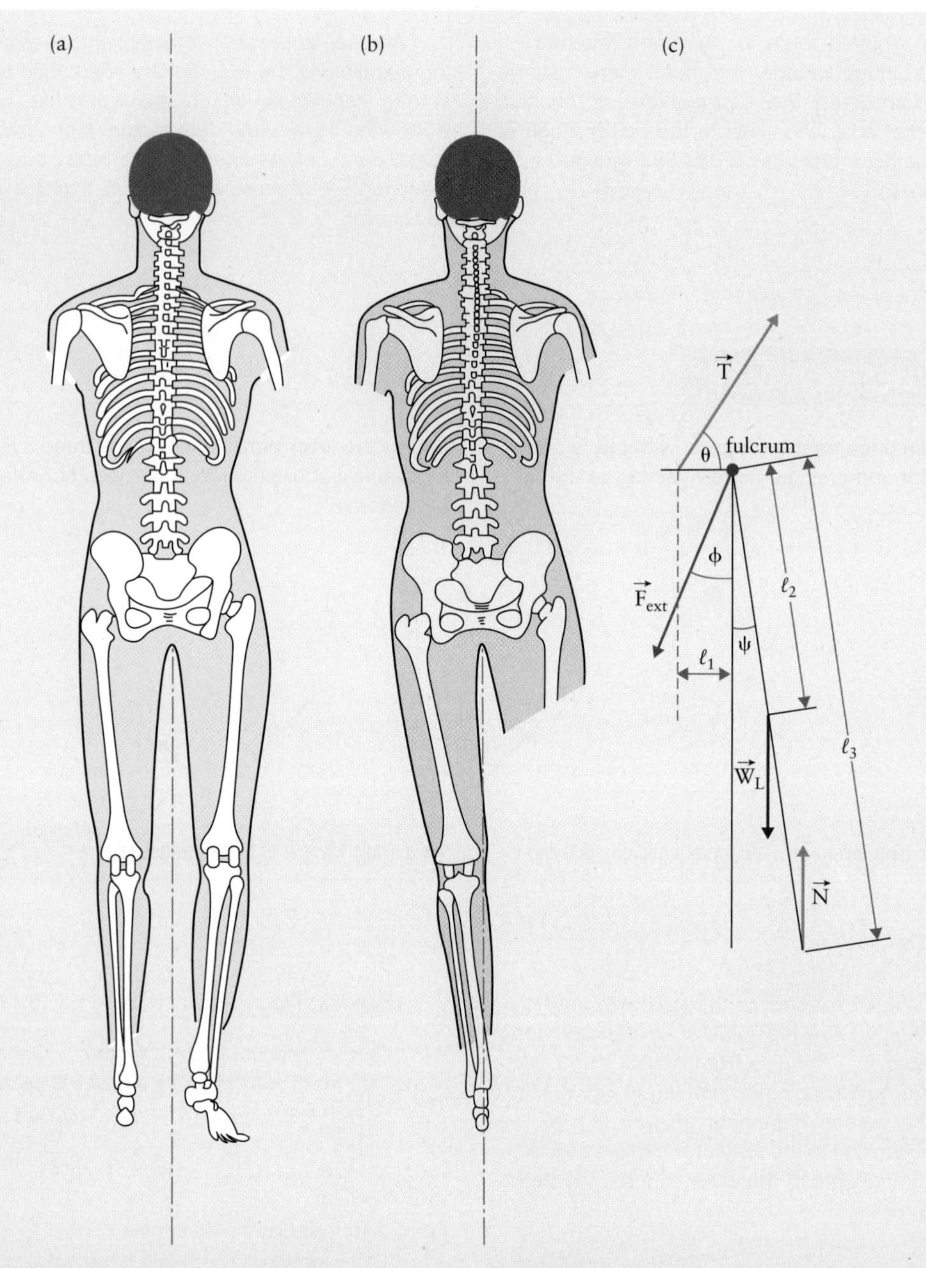

Figure 6.37 Posterior view of the hip joint (a) when the person stands on both feet and (b) when the person balances on one leg. (c) The arrangement of the various forces considered in the example: $\vec{T}$ is the tension in the abductor muscle, $\vec{F}_{ext}$ is the force of the upper body exerted through the acetabulum, $\vec{W}_L$ is the weight of the leg, and $\vec{N}$ is the normal force exerted by the floor.

Therefore:

$$F_{ext} = \sqrt{F_{ext,x}^2 + F_{ext,y}^2} = \sqrt{(465\ \text{N})^2 + (1580\ \text{N})^2} = 1645\ \text{N}$$

$$\tan\varphi = \frac{F_{ext,x}}{F_{ext,y}} = \frac{465\ \text{N}}{1580\ \text{N}} = 0.294$$

$$\varphi = \tan^{-1}0.294 = 16°.$$

Note:

$$\frac{T}{w} = \frac{1100\ \text{N}}{(70\ \text{kg})(9.8\ \text{m/s}^2)} = 1.6.$$

This means the abductor muscle must provide a force that is about 1.6 times the weight of the standard man, a large force that may strain tendons connecting the muscle to the pelvis or the great trochanter. The largest contribution to this force is due to the torque of the normal force, because the joint is far off the body's line of symmetry when the foot is vertically below the centre of mass. To circumvent such large forces, the person may use a cane on the opposite side, allowing the foot to be farther out, which greatly reduces the torque contribution due to the normal force.

The external force is also very large, with a value of about 2.4 times the weight of the person. Our anatomy takes great care to ensure that this external force does not contribute to the torque at the hip joint. This is done by moving the head of the femur deep into the acetabulum of the pelvis.

6.5: Since When Did Hominids Walk on Two Legs?

The three previous examples illustrate the use of mechanical concepts in human physiology and kinesiology. However, the same concepts also contribute to scientific research in other life sciences. As a special example, an interesting issue in human evolution is presented here.

The evolution of the primates began about 45 million years ago, when the monkeys of the Americas and the monkeys of Africa and Eurasia split. The African group split again about 32 million years ago into *cercopithecids* and *hominoids*. The *cercopithecids* include today's baboons, while the *hominoids* branched 4 more times between 22 and 7 million years ago, with gibbon, orangutan, gorilla, chimpanzee, and us as the contemporary representatives in each group. Of these, chimpanzees are our closest living relatives, with a DNA match exceeding 99%. The separation of their line occurred about half as long ago as the lines of fox and wolf split.

The evolution of the human branch during the past 5–7 million years is still not fully understood. The current section illustrates how physical reasoning in comparative anatomy and physiology can contribute to resolving new questions in this field.

We may ask whether humans began to walk upright to use tools, or whether walking upright is directly or indirectly linked to the increase in brain volume. We may ask what changes (morphology, physiology, and anatomy) are required for the basic biomechanical principles of bipedal locomotion with respect to quadrupedal locomotion. To answer these questions we inquire into the requirements for an effective and efficient bipedal posture and locomotion. The study is therefore based on comparison of three species:

- the chimpanzee (as a quadruped, knuckle walking), which moves on four legs on flat ground;
- modern *Homo sapiens* (human, as a biped, upright walking), which is the only living bipedal mammal; and
- Lucy, an *Australopithecus afarensis*, who lived 3.2 million years ago in what is today Ethiopia. Lucy was a female adult, 3.5 feet tall. Lucy's skeleton is shown in Fig. 6.38.

We first establish the major differences between chimpanzees and humans regarding the leg and the lower body anatomy. Next we use Lucy's fossil record to demonstrate that she was bipedal and, thus, that walking upright has been a feature of hominids for at least the past 3 million years. This is an important finding as it excludes the idea that the ability to walk upright was developed to allow humans to use tools with their hands. The use of tools actually emerged only about 2 million years ago, by which time the brain volume of hominids had sufficiently increased!

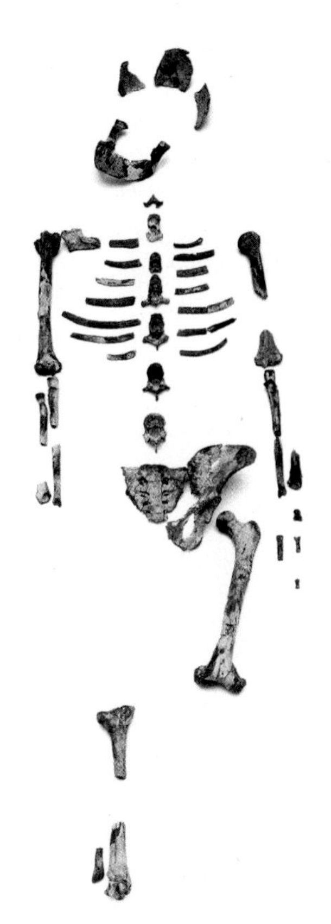

Figure 6.38 Lucy, 40% of an Australopithecus afarensis skeleton found in today's Ethiopia.

CASE STUDY 6.1

Bipedalism versus quadrupedalism: The most distinctive feature of humans is their ability to walk upright. The condition of bipedal locomotion is a derived state from a quadrupedal locomotion. Bipedalism in humans when compared with quadrupedalism in apes presents the following questions:

- Does walking upright need considerable change to the shape of the pelvis?
- Does walking upright need change in the position of centre of mass?
- How are the masses of bipeds and quadrupeds distributed?
- Does walking upright need considerable change to the function of the gluteal muscles?
- Does walking upright need considerable change to the function of the muscles in the leg (lower limb)?

Answer: *The bipedal human pelvis compared to that of quadrupedal apes shows many unique features necessary for bipedalism. Some important changes occurred in the shape of the pelvis, such as overall broadening, which was critical for the upright posture. These changes*

continued

provide a basin for the support of internal organs and facilitate our standing on two legs, but more significant changes in the shape of the pelvis and also in the function of some muscles attached to it were required for walking on two legs and maintaining balance.

Figure 6.39 shows the pelvis of a quadrupedal chimpanzee, with four-footed locomotion, and a bipedal human. The chimpanzee pelvis is a long and narrow plane shape. In contrast, the human pelvis is shaped more in the form of a basin (bowl) in order to support internal organs. Moreover, it is shorter and broader, thus stabilizing weight transmission in our walking steps.

(a)

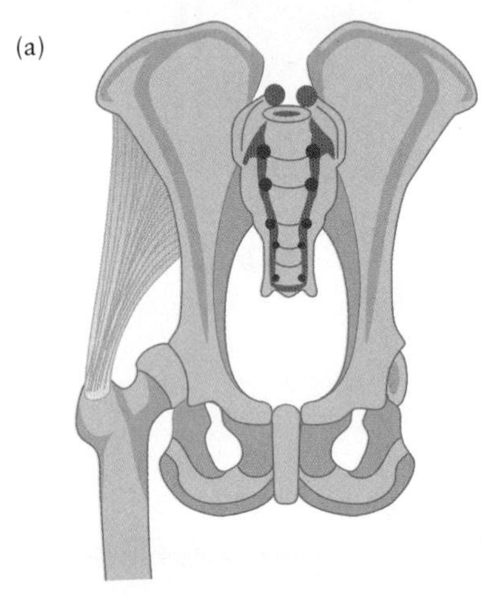

(b)

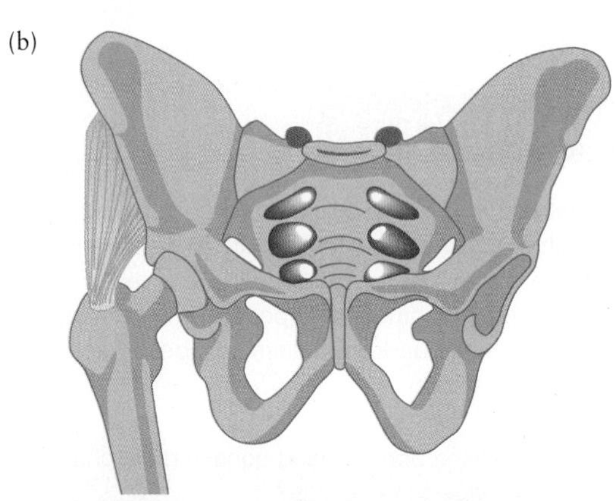

(c)

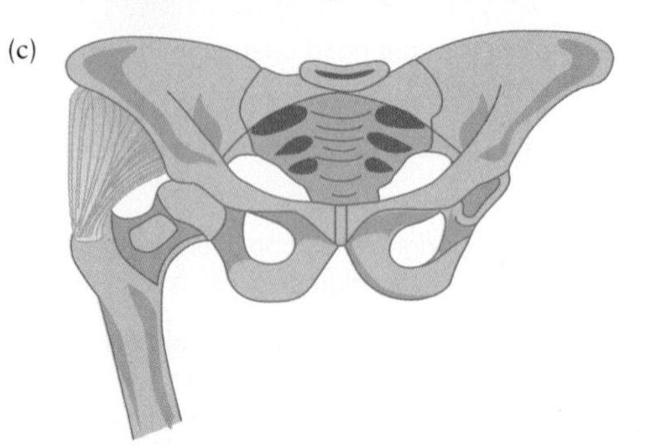

Figure 6.39 Comparison of the frontal views of pelvises of three primates: (a) the chimpanzee, with narrow plane shape; (b) the human, with basin shape; and (c) Lucy, a 3-million-year-old Australopithecus with a noticeably broader pelvis similar to that of the human.

The different positioning of centre of mass for a chimpanzee and a human is shown in Fig. 6.40. Most of the mass of the human lies along a vertical line passing through the pelvis; much of the mass of the chimp is not aligned with the pelvis. As Fig. 6.40 shows, the position of the weight line is changed too. Humans' weight line, a plumb line over which a significant weight load is carried, passes through the centre of mass.

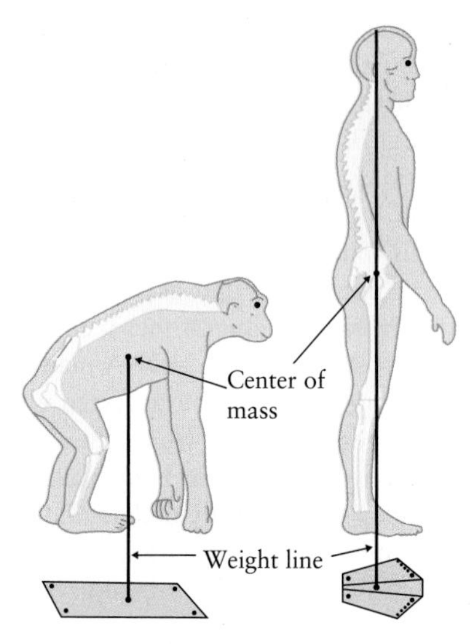

Figure 6.40 Position of centres of mass and weight lines in chimpanzees and humans.

Figure 6.41 compares the major muscles of the leg and pelvis of (a) a modern human and (b) a chimpanzee. Dominating the muscle arrangement of the chimpanzee's upper leg are the small and medium gluteal muscles (called the abductor muscles in humans). These muscles stretch the ape at the hip joint, efficiently accelerating its body forward since the centre of mass lies in front of the hind legs. This means that pushing into the ground under a small angle with the horizontal plane accelerates the chimpanzee in a manner similar to the starting sprinter (see Section 4.5.1 and Example 4.13). In this context it is also beneficial that the ape has a comparably long upper body, which shifts the centre of mass far ahead of the pelvis. The long upper body is the result of the shape of the pelvis bones, shown in Fig. 6.39(a), with the hip bone stretched upward, and a narrow sacrum.

As mentioned, the upright posture of humans is associated with significant changes to the shape of the pelvis bones and to the function of the muscles of the legs. Contracting the abductor muscle or the large gluteal muscle, which dominates the muscle arrangements of the human upper leg, would accelerate the upper body upward but not forward. Consequently, an entirely new mechanism for forward motion has developed, and the muscles causing acceleration in the leg of the chimpanzee serve new purposes in human legs. The abductor muscle stabilizes the upright posture in the sideways direction, as discussed in Example 6.14, and the gluteal muscle

continued

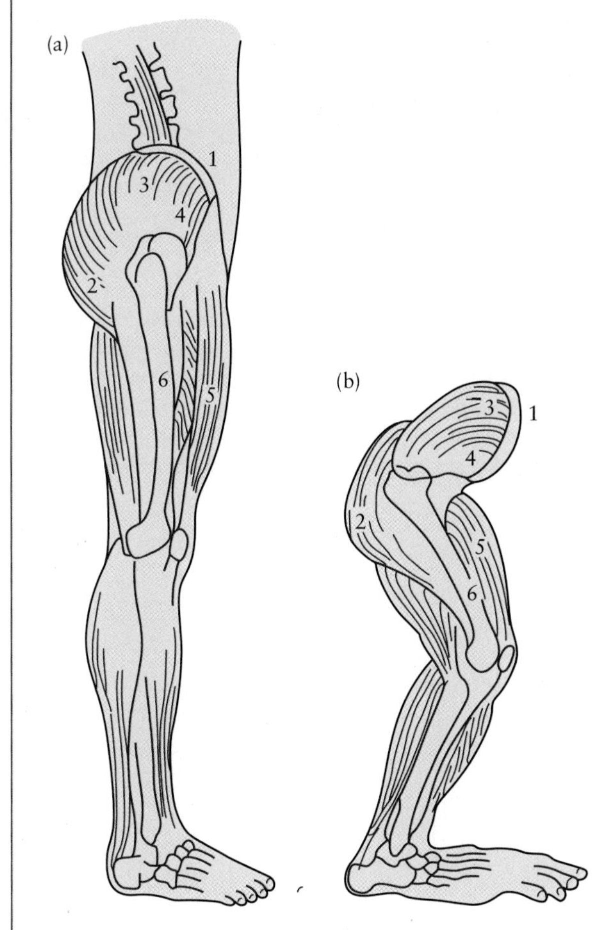

Figure 6.41 Comparison of muscles in the pelvis region and leg of (a) a modern human and (b) a chimpanzee. The figure emphasizes the difference in function of the upper leg and around the hip muscles. (a) The gluteal muscle (2) and abductor muscles (3,4) do not contribute to a forward acceleration for humans but are used to balance the upper body upright. This function is supported by a broader and flatter hip bone (1), lowering the centre of mass for the human body into the pelvis region and thus stabilizing the upright posture. (b) The three gluteal muscles [large gluteal muscle (2), medium gluteal muscle (3), and small gluteal muscle (4)] are the major means of acceleration of a chimpanzee, causing a large, mostly horizontal force on the body when stretched. Also labelled are the quadriceps femoris muscle (5) and the femur (6).

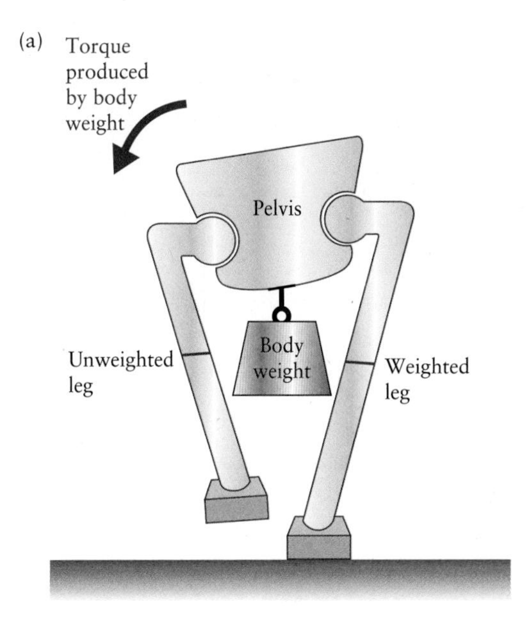

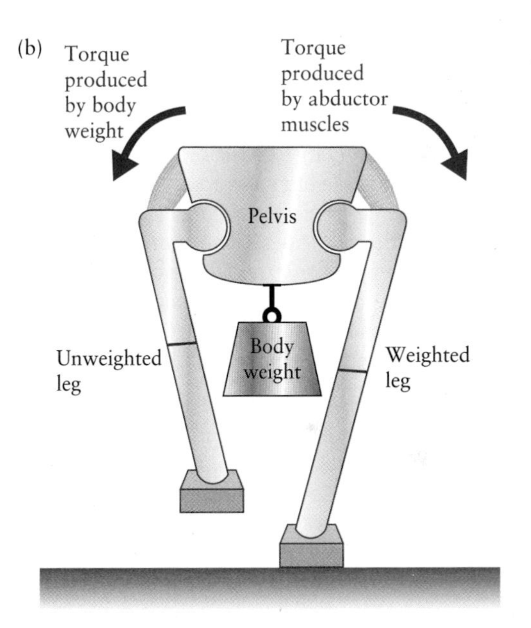

Figure 6.42 Comparison of pelvis positioning level without and with the gluteal muscles in human: (a) without abductors, (b) with abductors.

keeps our upper body from falling forward while walking. The gluteal muscles must resist hip tilting in various postures of standing, leaning, bending, carrying, walking, and running. In human walk, the pelvis does not drop dramatically on the non-supported side because the torque due to gravity is balanced by an equal and opposite torque due to the force applied by abductors, as illustrated in Fig. 6.42. A major function of the hip abductors is to maintain a level pelvis in an upright posture when walking.

While a human is walking or running, only one leg momentarily supports the upper body weight, which causes the hip to tend to fall on the side of the raised leg; in midstride with one leg off the ground, the gluteal muscles on the opposite side contract. You may verify this through a very simple test. Place your hand on your abductor and walk across your room very gently and slowly. While walking slowly you can notice that the contraction of the abductor muscle occurs opposite to the leg that is off the ground. As Fig. 6.43 illustrates, in walking and running, the weight of the upper body times the distance from its line of action to the centre of the hip joint is balanced by the abductor muscle force times its distance from the centre of the hip joint.

Therefore:

$$F_{Ab}\, l_{Ab} = w l_w .$$

continued

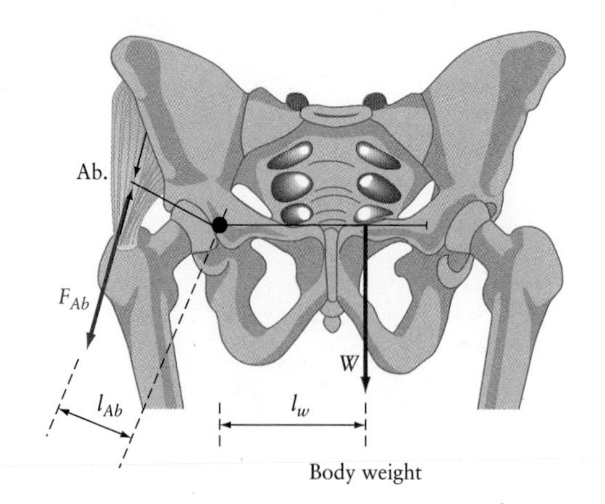

Figure 6.43 Torque configuration in pelvis.

So:

$$\tau_{\text{Ab|Hip}} = \tau_{\text{w|Hip}}.$$

The torque of the weight about the centre of the hip joint is balanced by the torque of force of the abductor muscle. Thus, to keep a bipedal balanced when the weight shifts to the right, the left abductor must pull it back to the right.

This focus on balance rather than rapid acceleration also resulted in evolutionary changes to the pelvis, as shown in Fig. 6.39(c). The wings of the hip bone turned inward and the sacrum widened to accommodate the intestines, which lowered the centre of mass. This stabilizes the upright posture because the pelvis plane contains the axes about which our body tilts forward and sideways.

CASE STUDY 6.2

Lucy: Was Lucy a bipedal?

Answer: *These differences between the quadrupedal chimpanzee and the bipedal modern human can now be used to study Lucy's fossil record. Although only about 40% of her skeleton was found (Fig. 6.38), this is possible since her pelvis was found almost completely intact [Fig. 6.39(b)]. The striking similarities between Lucy's hip bone and sacrum and those of modern humans indicate that she did indeed walk upright. This is further supported by marks on her pelvis that indicate where the tendons of the various leg muscles were once attached.*

Studying Lucy's pelvis more quantitatively, we surprisingly find her even better adapted to an upright posture than modern humans! This is illustrated in Fig. 6.44, where the front view of Lucy's pelvis and the pelvis of a modern human, as well as the respective abductor muscle arrangements, are overlapped with a balance-of-torque diagram (dashed line). In both cases, the fulcrum is located in the head of the femur, as discussed in Example 6.14. When the upper body is balanced on one leg, the abductor muscle must compensate the torque about the hip joint. The longer the lever arm of the upper body, $L_{c.m.}$, and the shorter the distance between the abductor muscle and the head of the femur, L_A, the greater the strain in the tendons of the abductor muscle. Measuring these lengths in Fig. 6.44, we find

$$\begin{cases} \left(\dfrac{L_{\text{c.m.}}}{L_{\text{A}}}\right)_{\text{Lucy}} = 2.1 \\ \left(\dfrac{L_{\text{c.m.}}}{L_{\text{A}}}\right)_{\text{human}} = 2.6. \end{cases}$$

The larger the ratio of ($L_{c.m.}/L_A$), the less favourable the lever arm arrangement of the pelvis region. Why, then, did the evolution of the genus Homo result in more poorly adapted individuals during the past 3 million years? A most likely explanation is based on a competing evolutionary process in modern humans: the development of a large brain. While Lucy was indeed well adapted to an upright posture, her pelvis had an elongated shape with an elliptic, narrow opening for the birth canal. She could never have

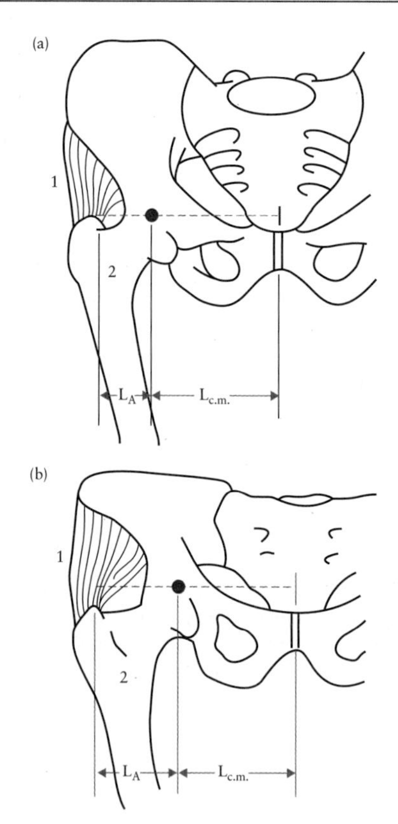

Figure 6.44 Comparison of the front view of the pelvis region with the abductor muscles (1) for (a) a modern human and (b) the *Australopithecine* Lucy. The *Australopithecines* were better adapted to upright walking as the head of their femur (2) is longer, providing a longer lever arm for the abductor muscle.

continued

given birth to the large-headed babies of modern humans. Nature had to compromise by reshaping the pelvis of modern humans to provide a rounder, larger birth canal at the expense of adaptation to the upright posture.

The form of locomotion known as bipedalism is the single most important difference between humans and apes. This is what places "Lucy" fairly within the hominid (human) family.

SUMMARY

DEFINITIONS

- If $\vec{F}$ is a force acting on a rigid object at distance $\vec{r}$ from the fixed axis of rotation, the torque $\vec{\tau}$ is defined as:

$$\vec{\tau} = \vec{r} \times \vec{F}.$$

- Magnitude of torque:

$$\tau = r\ F\ \sin\varphi,$$

where ϕ is the angle between the vectors $\vec{F}$ and $\vec{r}$.

- Magnitude of torque due to weight:

$$\tau_{\text{weight}} = r_{\text{c.m.}}\ mg\ \sin\varphi,$$

where $r_{\text{c.m.}}$ is the distance of the fulcrum from the centre of mass, M the mass of the system, and ϕ the angle between the weight vector and the lever arm.

- Sign convention for torque:

if the rotation is counter-clockwise, then $\tau > 0$;

if the rotation is clockwise, then $\tau < 0$.

UNITS

Torque τ with $\text{N} \cdot \text{m} = \text{kg} \cdot \text{m}^2/\text{s}^2$

LAWS

- The mechanical equilibrium for a rigid object with a fixed axis along the z-direction:

$$\begin{cases} \sum_i F_{i,x} = 0 \\ \sum_i F_{i,y} = 0 \\ \tau_{\text{net}} = \sum_i \tau_{i,\text{ ccw}} - \sum_i \tau_{i,\text{ cw}} = 0 \end{cases}$$

MULTIPLE-CHOICE QUESTIONS

MC–6.1. If the net torque on an object about a certain point is zero,
(a) it is zero about some other point.
(b) it is zero about all other points.
(c) it is not zero about any other point.
(d) the object must be in equilibrium.

MC–6.2. If an object is in equilibrium, the torque is computed about an axis that
(a) must pass through an end of the object.
(b) must pass through the centre of mass of the object.
(c) must intersect the line of action of at least one of the forces acting on the object.
(d) may be sited anyplace.

Note: In certain professional admission tests, some multiple-choice questions are grouped with a common text called a ***passage***. We illustrate this approach with the next four questions (Q.6.3–6.7) in this section.

Passage: We study a standard man bending forward to lift an object off the ground, as illustrated in Fig. 6.45(a). The fulcrum lies at the lower back, and the spinal column is the lever arm. We assume that the person is bending forward with the back horizontal and that the forearms are stretched downward. For this case draw the force diagram as shown in Fig. 6.45(b), with four forces acting on the spinal column:

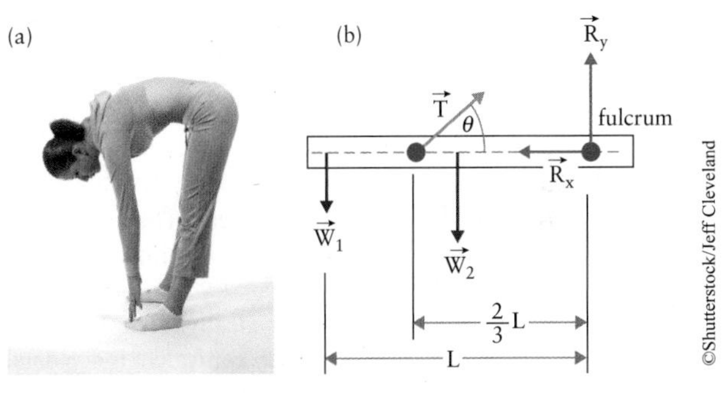

Figure 6.45 (a) A standard man bends down to lift an object off the ground. (b) The standard man's spine is shown as a bar with four major forces: the weight of the object and the arms, $\vec{w}_1$; the weight of the trunk and head, $\vec{w}_2$; the tension in the muscle that maintains the position of the back, $\vec{T}$; and the force pushing the spine up at the hip, $\vec{R}$.

- the weight of the object that is lifted and combines with the weight of the arms and hands as $\vec{w}_1$;
- the weight of the torso, $\vec{w}_2$, acting at the centre of mass located halfway between the two ends of the spinal column;
- the tension $\vec{T}$ exerted on the spinal column by the back muscle responsible for the lifting. We assume this force is applied at a distance $d = (2/3)L$ from the fulcrum at an angle θ above the horizontal.
- a compressive force, $\vec{R}$, exerted on the fulcrum in the lower back. Let β be the angle between $\vec{R}$ and the horizontal.

MC–6.3. An author reporting on this case may opt to exclude the force $\vec{R}$ from Fig. 6.45(b). Why would the author have done so?
(a) The author studies a problem of rotational equilibrium and anticipates that this force will not be required when writing the torque formula.
(b) Different from a free body diagram, we may eliminate up to all but one force from a diagram such as Fig. 6.45(b).
(c) The physiological knowledge about the magnitude of the force $\vec{R}$ and its angle β is uncertain; thus, omitting the force in Fig. 6.45(b) would eliminate a possible source of error.

(d) For the person in the posture shown in Fig. 6.45(a), the force $\vec{R}$ indeed does not exist; i.e., $R = 0$.
(e) Fig. 6.45(b) oversimplifies the problem: the force $\vec{R}$ acts in a direction out of the plane of the paper.

MC–6.4. Which of the following four formulas is the proper torque equation for the problem illustrated in Fig. 6.45? Note that the term "proper" includes the use of the generally accepted sign convention for torque as introduced in the textbook.
(a) $\tau_{net} = Lw_1 + (1/2)Lw_2 + (2/3)LT \sin\theta = 0$
(b) $\tau_{net} = Lw_1 + (1/2)Lw_2 - (2/3)LT \sin\theta = 0$
(c) $\tau_{net} = -Lw_1 - (1/2)Lw_2 + (2/3)LT \sin\theta = 0$
(d) $\tau_{net} = -Lw_1 + (1/2)Lw_2 - (2/3)LT \sin\theta = 0$

MC–6.5. If you answered the two previous questions choosing the same choice in Q.6.4, then skip this question. If you have chosen different answers for the two previous questions, then choose the statement below that best describes the consequences of your choices.
(a) We need two different formulas because the two cases are physically different.
(b) The two formulas we have chosen are mathematically equivalent (different due to the sign convention only). This is correct because the two cases are physically the same.
(c) The two cases should be the same, but the two formulas are indeed different. Thus, there must be something wrong with the introduced sign convention.
(d) The two cases differ physically. That the two formulas are mathematically equivalent is accidental.

MC–6.6. A steel band of a brace exerts an external force of magnitude $\vec{F}_{ext} = 30$ N on a tooth. The tooth is shown in Fig. 6.46, with point B a distance 1.0 cm above point A, which is the fulcrum of the tooth. The angle between the normal to the tooth and the external force is $\theta = 45°$. Which of the following values is closest to the torque on the root of the tooth about point A?

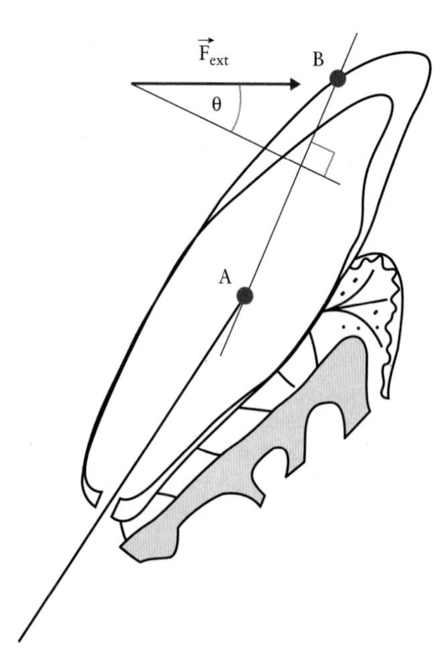

Figure 6.46 A horizontal force $\vec{F}_{ext}$ acts at point B on a tooth with the fulcrum at point A.

(a) 0.2 N · m
(b) 0.3 N · m
(c) 0.4 N · m
(d) 2.0 N · m
(e) 3.0 N · m

MC–6.7. The definition of torque contains the magnitude of the force $\vec{F}$ acting on a rigid object with a fixed axis, the magnitude of the vector $\vec{r}$ between the fulcrum and the point where the force is applied, and the angle ϕ between the force vector $\vec{F}$ and the position vector $\vec{r}$. Which of the following statements about torque is wrong?
(a) Torque is linearly proportional to the magnitude of the force $\vec{F}$.
(b) Torque is linearly proportional to the magnitude of the vector $\vec{r}$.
(c) Torque is linearly proportional to the angle ϕ.
(d) Torque can be positive or negative depending on the angle φ.
(e) The force $\vec{F}$ can be applied to the rigid object such that the resulting torque is zero.

MC–6.8. When a torque is acted on a rigid body, it always causes
(a) constant angular velocity.
(b) constant angular acceleration.
(c) rotational equilibrium.
(d) change in angular velocity.
(e) change in moment of inertia.

MC–6.9. A rod is 7 m long and is pivoted at a point 2.0 m from the left end. Both a force of magnitude 50 N at the left end and a force of magnitude 200 N at the right end act downward. At what distance from the pivot point must a third upward-directed force of magnitude 300 N be placed to establish rotational equilibrium? Neglect the weight of the rod.
(a) 1.0 m
(b) 2.0 m
(c) 3.0 m
(d) 4.0 m
(e) None of the first four answers is correct.

MC–6.10. If the weight of the rod in Q-6.9 is 200 N, at what distance should the third force be applied to have equilibrium?
(a) 1.0 m
(b) 2.0 m
(c) 3.0 m
(d) 4.0 m
(e) None of the first four answers is correct.

MC–6.11. Fig. 6.44 shows a comparison of the front view of the pelvis region with the abductor muscles for two bipedal species (a) and (b). The dot indicates the fulcrum located at the head of the femur. If we assume the same length of leg and weight of both individuals, but take into account that $L_A/L_{c.m.} = 0.38$ for individual (a) and $L_A/L_{c.m.} = 0.48$ for individual (b), then species (b) needs a greater strength (force) in the abductor muscle to balance its upper body on the shown leg.
(a) true
(b) false
(c) We cannot determine whether this is true or false.

MC–6.12. A horizontal trap door, hinged at one edge, is held open at an angle θ with respect to horizontal ground by a force. The door has length l and mass m.

The holding force applied to the edge opposite to the hinge side and perpendicular to the door is
(a) $mg/2\cos\theta$.
(b) $(mg/2)\cos\theta$.
(c) $(mg/2)\sin\theta$.
(d) $mg\cos\theta$.
(e) $2mg/\cos\theta$.

MC–6.13. A horizontal trap door, hinged at one edge, is held open at an angle $\theta = 70°$ with respect to horizontal ground by a force. The door is 1.2 m long and has a mass of 16 kg. The holding force applied to the edge opposite to the hinge side and perpendicular to the door is
(a) 157 N.
(b) 78 N.
(c) 54 N.
(d) 27 N.
(e) 13 N.

CONCEPTUAL QUESTIONS

Q–6.1. Assume Fig. 6.45(a) was instead drawn such that the standard man is shown from the opposite side, i.e., bending down toward the right. Correspondingly, Fig. 6.45(b) would be drawn with the left and right ends flipped horizontally. For this modified display, which of the four choices in Q-6.4 would now be the proper torque equation?

Q–6.2. Why can you not open a door by pushing on its hinge side?

Q–6.3. Fig. 6.47 shows the motion of the thorax during breathing. Air is pulled into the lungs and pushed out of the lungs by the active change of their volume associated with the change in the volume within the rib cage. Two sets of intercostal muscles allow for the increase and decrease of this volume. These muscles are shown in Fig. 6.48 and consist of the *intercostales interni muscles* (connecting points B' and C in the figure) and the *intercostales externi muscles* (connecting points B and C'). The fulcrum lies in the joint between the rib and the thoracic vertebrae. Determine from Fig. 6.48 which muscle contracts during inhalation (volume increase) and which muscle contracts during exhalation (volume decrease).

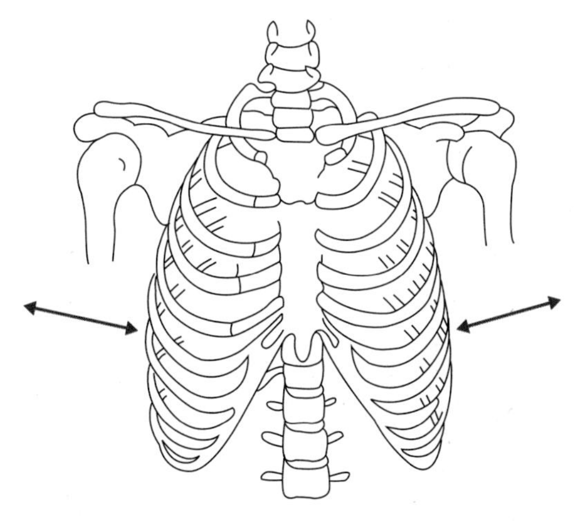

Figure 6.47 The thorax, with two arrows indicating the motion of the rib cage during breathing.

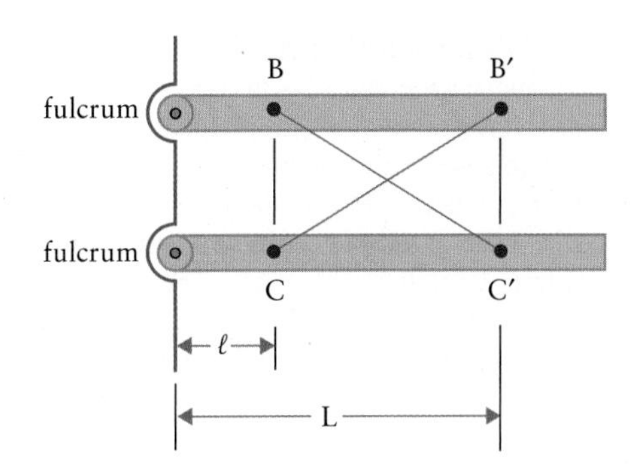

Figure 6.48 Muscle connections between two neighbouring ribs.

Q–6.4. (a) Give an example in which the net force on an object is zero, but the net torque is not zero. (b) Give an example in which the net torque on an object is zero, but the net force is not zero.

Q–6.5. What classes of lever systems are the pelvis and hip in Fig. 6.44 and the foot in Figs. 6.32(a) and (b)?

Q–6.6. Why does holding a long pole help a tightrope walker stay balanced?

Q–6.7. Fig. 6.49 shows a rack that will hold one wine bottle. What do we know about the centre of mass of the wine bottle and rack system if it is in rotational equilibrium, i.e., doesn't fall over?

© allOver photography/Alamy

Figure 6.49

Q–6.8. Why do we experience pain when we lift an object using the back (bend and lift) rather than the legs?

Q–6.9. What class of lever system is the outstretched arm and shoulder in Fig. 6.1?

Q–6.10. A flat square plywood sheet has a pivot at the mid-length of one of its edges and the axis of rotation is perpendicular to the sheet. How should a force be applied to the sheet, in the plane of the sheet, to produce the maximum torque?

ANALYTICAL PROBLEMS

P–6.1. If the torque required to loosen a nut has a magnitude of $\tau = 40$ N · m, what minimum force must be exerted at the end of a 30-cm long wrench?

P–6.2. A pendulum consists of an object of mass $m = 30$ kg that hangs at the end of a massless bar, a distance of 2.0 m from the pivot point. Calculate the magnitude of the torque due to gravity about the pivot point if the bar makes an angle of 5° with the vertical.

P–6.3. A steel band of a brace exerts an external force of magnitude $F_{ext} = 40$ N on a tooth. The tooth is shown in Fig. 6.46, with point B a distance 1.33 cm above point A, which is the fulcrum. The angle between the tooth and the external force is $\theta = 40°$. What is the torque on the root of the tooth about point A?

P–6.4. In Fig. 6.1, what is the force supplied by the deltoid muscle, 15 cm from the shoulder joint and acted upon at an angle of 14° with respect to the humerus, holding up the outstretched arm of weight 30 N? Assume the distance from centre of mass of the arm to the shoulder joint (pivot) is 24 cm.

P–6.5. In Fig. 6.21(a), find the lever arms of forces $\vec{F}_1$ and $\vec{F}_3$ for the torques about point P.

P–6.6. In Fig. 6.21(a), find the torques about point P due to forces $\vec{F}_1$ and $\vec{F}_3$ using their lever arms.

P–6.7. A standard man (see Table 4.1) stands on a scaffold supported by a vertical rope at each end. The scaffold has a mass of $m = 20.5$ kg and is 3.0 m long. What is the tension in each rope when the person stands 1.0 m from one end?

P–6.8. Fig. 6.50 shows a light of mass $m = 20$ kg supported at the end of a horizontal bar of negligible mass that is hinged to a pole. A cable at an angle of 30° with the bar helps to support the light. Find (a) the magnitude of the tension in the cable and (b) the horizontal and vertical force components exerted on the bar by the pole.

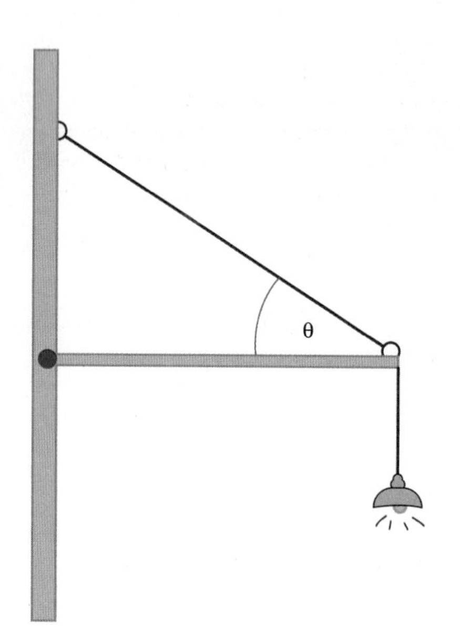

Figure 6.50

P–6.9. A ladder rests against a frictionless wall. It has a mass of 50 kg, a length of 5.0 m, and makes an angle of 60° with the ground. Find the horizontal and vertical force components exerted by the ground on the bottom of the ladder if a standard man stands on the ladder 4 m from the ground.

P–6.10. Fig. 6.51 shows a uniform boom of mass $m = 120$ kg supported by a cable perpendicular to the boom. The boom is hinged at the bottom and an object of mass $M = 200$ kg hangs from its top. Find the tension in the massless cable and the components of the force exerted on the boom at the hinge.

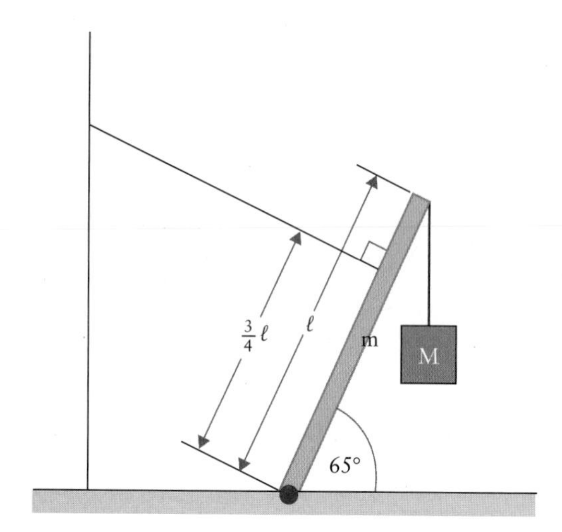

Figure 6.51

P–6.11. A standard man is doing push-ups, as shown in Fig. 6.52. The distances are $l_1 = 90$ cm and $l_2 = 55$ cm. (a) Calculate the vertical component of the normal force exerted by the floor on both hands, and (b) calculate the normal force exerted by the floor on both feet. Use Table 4.1 for the data of the standard man.

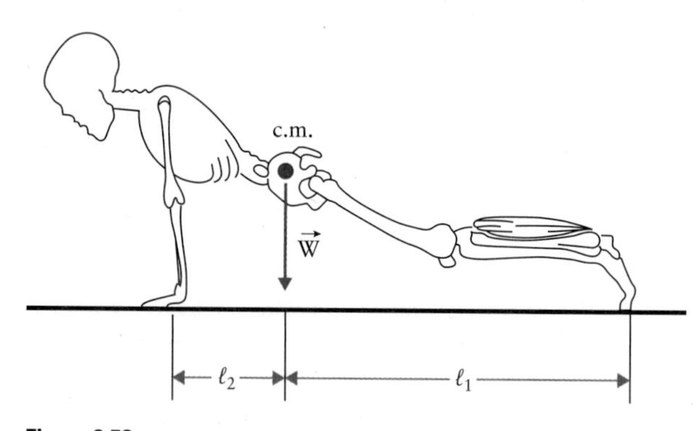

Figure 6.52

P–6.12. A standard man holds the upper arm vertical and the lower arm horizontal with an object of mass $M = 6$ kg resting on the hand, as illustrated in Fig. 6.2. The mass of the lower arm and hand is one-half the mass of the entire arm. As shown in Fig. 6.53, there are four forces acting on the lower arm: the external force $\vec{F}_{ext}$, exerted by the bones and ligaments of the upper arm at the elbow (fulcrum); the tension $\vec{T}$, exerted by the biceps; a force $\vec{w}$ due to the weight of the holding object; and the weight $\vec{w}_F$ of the lower arm. The points along the lower arm at which the forces act are $l_1 = 4$cm, $l_2 = 15$cm, and $l_3 = 40$cm, as shown in Fig. 6.53. (a) Calculate the vertical component of the force $\vec{F}_{ext}$, and (b) calculate the vertical component of the tension $\vec{T}$.

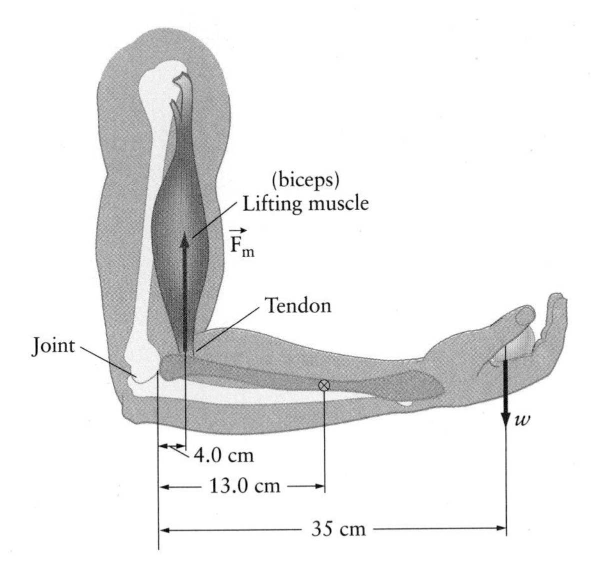

Figure 6.53

P–6.13. Fig. 6.53 illustrates a horizontal forearm held perpendicular to the upper arm. A 70-N weight is held in the hand. Neglect the weight of the forearm. (a) What is the torque about the joint due to the 70-N weight? (b) What is the torque due to the force on the forearm by the biceps, $\vec{T}$? (c) What is the magnitude of $\vec{T}$? Take the distances as $l_1 = 4$ cm, $l_2 = 13$ cm, and $l_3 = 35$ cm.

P–6.14. Repeat problem 6.11 assuming that the forearm weight is 20 N and its centre of mass is $l_2 = 13$ cm from the joint, as shown in Fig. 6.53.

P–6.15. An object of mass $M = 10$ kg is lifted by a standard man with the aid of a pulley, as shown in Fig. 6.54. The upper arm is held vertical and the lower arm has an angle of $\theta = 35°$ with the horizontal. The label c.m. marks the centre of mass of the lower arm. Consider the force due to the weight of the object M, the weight of the lower arm and hand, the tension due to the triceps muscle, and the force due to the humerus. For the lengths, we use the following values: $l_1 = 2$ cm, $l_2 = 15$ cm, and $l_3 = 40$ cm. The lower arm and the hand have a mass of 2.3 kg. (a) What is the magnitude of the vertical force exerted on the lower arm by the triceps muscle, and (b) what is the magnitude of the vertical force exerted on the lower arm by the humerus? *Hint:* The triceps muscle pulls vertically upward.

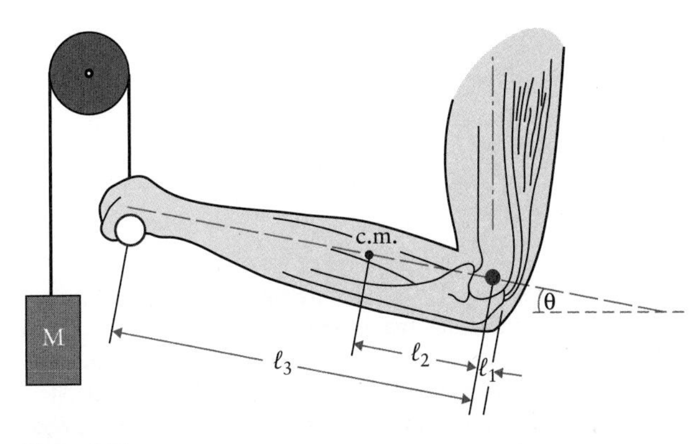

Figure 6.54

P–6.16. A standard man holds an object of mass $m = 2$ kg on the palm of the hand with the arm stretched, as shown in Fig. 6.55. (a) Use the torque equilibrium equation to determine the magnitude of the force $\vec{F}$ that is exerted by the biceps muscle, when $a = 35$ cm, $b = 5$ cm, and the angle $\theta = 80°$. Neglect the weight of the lower arm. (b) Is the assumption of a negligible mass of the lower arm in part (a) justified?

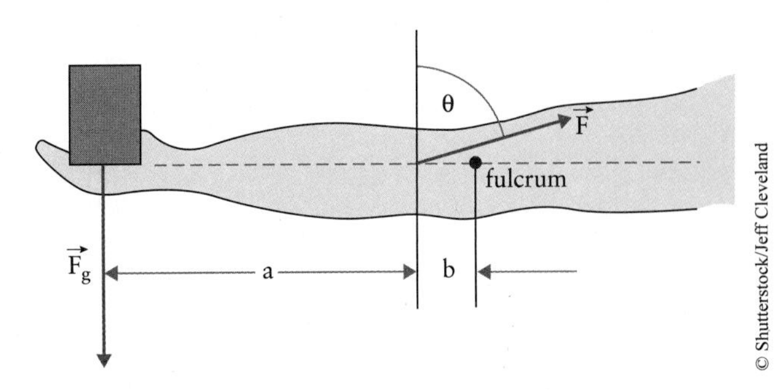

Figure 6.55

P–6.17. The quadriceps femoris muscle, shown as (1) in Fig. 6.56(a), is a muscle in the upper leg that serves a purpose analogous to the triceps in the upper arm. Its tendon (2) is attached to the upper end of the tibia (3), as shown in the figure. This muscle exerts the major force of the upper leg on the lower leg when it is stretched. Take the weight of the lower leg, w_L, the weight of the foot, F, and the tension T of the quadriceps femoris muscle as shown in Fig. 6.56(b). (a) Find the magnitude of the tension T when the tendon is at an angle of $\phi = 30°$ with the tibia using the torque equilibrium. Assume that the lower leg has a mass of 3 kg and the mass of the foot is 1.2 kg. The leg is extended at an angle of $\theta = 35°$ with the vertical, and the centre of mass of the lower leg is at its centre. The tendon attaches to the lower leg at a point 1/5 the way down the lower leg. (b) Establish whether the three forces shown in Fig. 6.56(b) are in mechanical equilibrium.

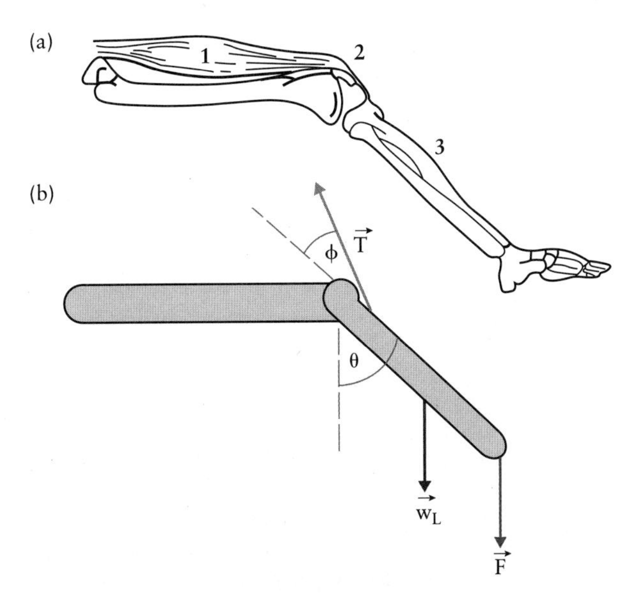

Figure 6.56

P–6.18. A standard man bends over as shown in Fig. 6.45(a) and lifts an object of mass $m = 15$ kg while keeping the back parallel with the floor. The muscle that attaches 2/3 the way up the spine maintains the position of the back. This muscle is called the back muscle or the *latissimus dorsi muscle*. The angle between the spine and the force $\vec{T}$ in this muscle is $\theta = 11°$. Use the force diagram in Fig. 6.45(b). The weight w_1 includes the object and the arms, and the weight w_2 includes the trunk and head of the standard man (use data from Table 4.1). (a) Find the magnitude of the tension $\vec{T}$ in the back muscle, and (b) find the x-component of the compressive force $\vec{R}$ in the spine.

P–6.19. A standard man holds the arm stretched out horizontally. The major forces acting on the arm are shown in Fig. 6.57: $\vec{F}_{ext}$ is the force acting into the shoulder joint, $\vec{T}$ is the tension of the *deltoid muscle*, and $\vec{w}$ is the weight of the arm. Use the standard man data from Table 4.1, take $\alpha = 15°$ for the angle, $l_1 = 13$ cm for the distance between the shoulder and the attachment point of the tendon of the *deltoid muscle*, and $l_2 = 35$ cm for the distance from the shoulder to the centre of mass of the arm. (a) Calculate the magnitude of the tension $\vec{T}$ in the tendon of the *deltoid muscle*, and (b) calculate the magnitude of the external force $\vec{F}_{ext}$ acting toward the shoulder.

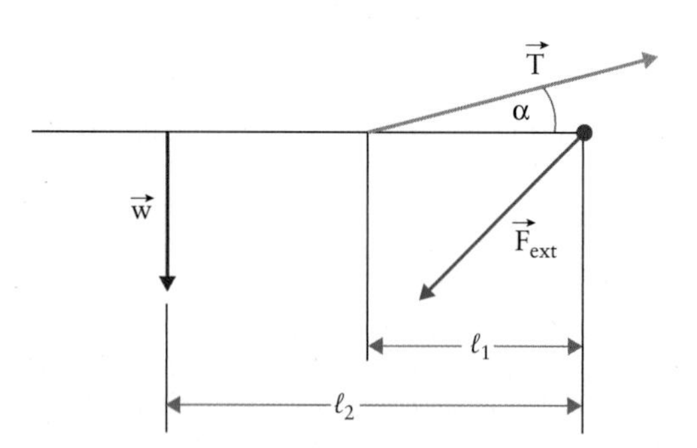

Figure 6.57

P–6.20. A standard man plays on the offence of a football team. Before a play, the centre, the guards, and the tackles bend the upper body forward, forming about an angle $\theta = 45°$ with the horizontal, then remain motionless until the play starts. Fig. 6.58 shows the corresponding force diagram for the standard man's back. Consider the weight of the head, $\vec{w}_H$, the arms $\vec{w}_A$, and the trunk $\vec{w}_T$ (see Table 4.1), and mass of typical helmet $m = 1.2$ kg. Calculate (a) the magnitude of the tension $\vec{T}$ in the back muscle and (b) the magnitude of the force $\vec{F}_B$ acting on the fifth lumbar vertebra (fulcrum). *Note:* The tension $\vec{T}$ forms an angle $\phi = 11°$ with the spinal column.

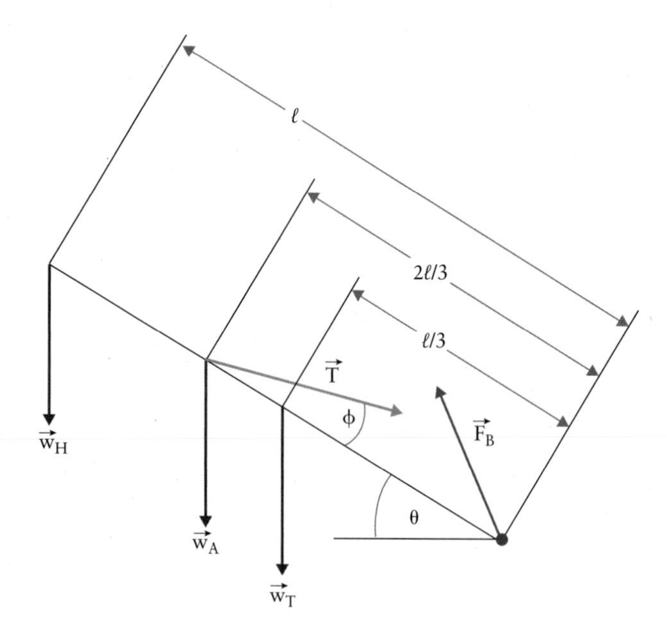

Figure 6.58

P–6.21. A standard man is suspended from a high bar. The sketch in Fig. 6.59 shows the lower arm in a side view with the person facing right. Four forces act on the lower arm: the external force, $\vec{F}_{ext}$, exerted by the high bar; the weight of the lower arm, $\vec{w}_{la}$; the tension in the biceps tendon, $\vec{T}$; and the force exerted by the humerus through the elbow, $\vec{F}_{elbow}$. Assume that the forces exerted by the bar on the left and right hand are each equal in magnitude to $\vec{F}_{ext}$ and directed parallel to each other. (a) Find the magnitude of the external force $\vec{F}_{ext}$; (b) find the magnitude of the tension $\vec{T}$; and (c) find the magnitude of $\vec{F}_{elbow}$ and its angle with the lower arm, ψ. *Hint:* Include the weight of the lower arm $\vec{w}_{la}$ with $m_{la} = 2.3$ kg, and assume that $\vec{T}$ and $\vec{F}_{elbow}$ are the only forces exerted on the lower arm by the upper arm. Also, use $l_1 = 40$ cm for the length from the hand to the elbow, $l_2 = 15$ cm for the length from the centre of mass of the

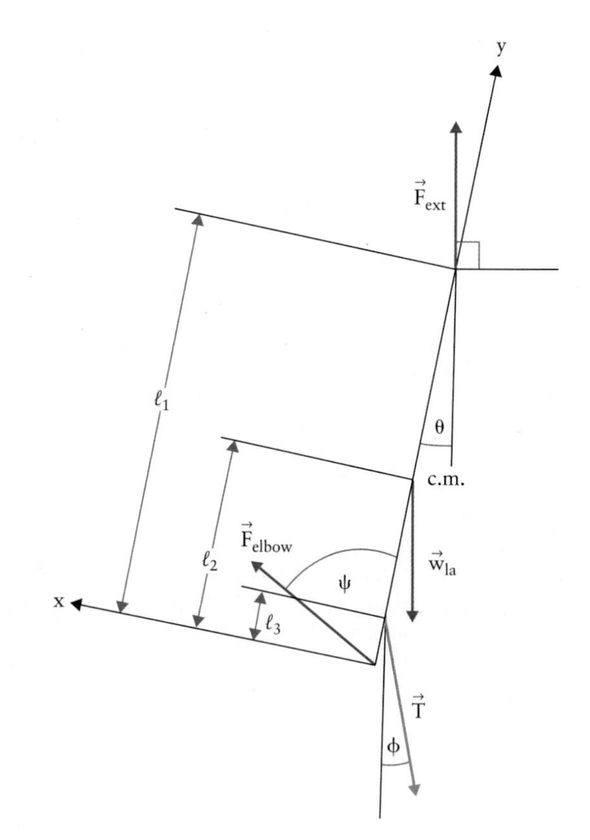

Figure 6.59

lower arm to the elbow, and $l_3 = 4$ cm for the distance of the attachment point of the biceps tendon to the elbow. The two angles are $\theta = \varphi = 8°$.

P–6.22. Fig. 6.60 shows a standard man wearing a cast supported by a sling that exerts a vertical force $\vec{F}$ on the lower arm. The distance between the shoulder joint and the elbow is $l_1 = 35$ cm, and the mass of the cast is 3 kg. The sling supports the lower arm at the centre of mass $l_2 = 15$ cm from the elbow. Use one-half the mass of the arm without a cast for the mass of the lower arm, and $\theta = 70°$ for the angle of the arm at the elbow. Calculate the magnitude of force $\vec{F}$. *Hint:* Other forces act on the arm, which we assume to act along a line through the shoulder joint. Thus, equating $\vec{F}$ with the weights of the upper and lower arm does not yield the correct result. Instead, the problem is solved with the balance-of-torque equation, in which the unknown forces at the shoulder joint are not included.

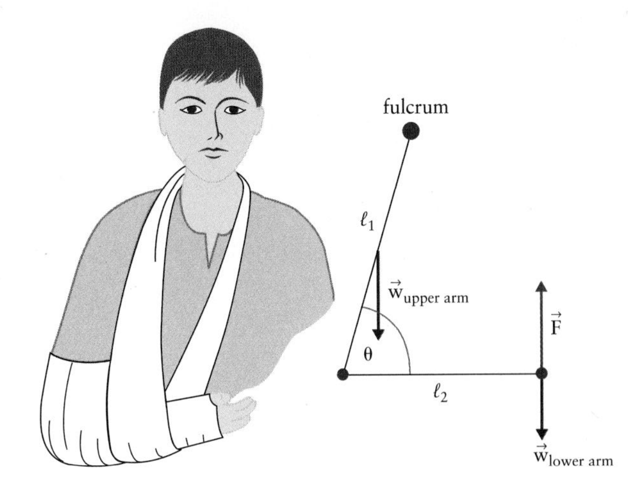

Figure 6.60

ANSWERS TO CONCEPT QUESTIONS

Concept Question 6.1: $\vec{d}$

Concept Question 6.2: e

Concept Question 6.3: (a): Using the right-hand rule with the wrong order of vectors yields a direction of the torque vector that is opposed to the correct direction. This means the person concludes that the rotation occurs in the opposite direction than is correct (clockwise versus counter-clockwise). (b): If the person works consistently with a left-hand rule, calculated results will be consistent with physical reality but the generally agreed upon conventions for torque have not been followed. When that person starts to communicate with others, hopeless confusion will result. Conventions are introduced in the scientific discussion only when needed to prevent such confusion; therefore, all conventions have to be followed consistently.

Concept Question 6.4: $\vec{F}_1$, $\vec{F}_2$ produce positive torque since they all cause counter-clockwise rotation. $\vec{F}_3$, $\vec{F}_4$ produce negative torque; they cause clockwise rotation.

Concept Question 6.5: Since $\vec{F}_1$ and $\vec{F}_2$ are perpendicular to the bar, the distance between them and the pivot point are their lever arms. For $\vec{F}_3$ and $\vec{F}_4$, just draw a line from the pivot point perpendicular to each of them.

Concept Question 6.6: Fig. 6.20 illustrates how we can reduce the current problem to a bar on which the two forces $\vec{F}_{ext}$ and $\vec{w}$ act: we identify the pivot point of the vertebral column as the fulcrum (green triangle). Next, we draw a bar to connect the fulcrum with the two points at which the relevant forces act (red line). In the case of the trapezius force, this is the point where the trapezius tendon is attached to the skull, and in the case of the weight of the forehead, this is the centre of mass of the forehead. We will discuss the role of the centre of mass of the rigid object in more detail following this Concept Question.

Now we use Fig. 6.9 to determine the resulting rotation if the action of either force were not balanced. The bar would spin in a counter-clockwise direction if responding to $\vec{F}_{ext}$, thus yielding a positive torque term $\tau_F > 0$. In turn, the clockwise rotation the weight would lead to a negative torque term, $\tau_w < 0$.

Concept Question 6.7: C.

Concept Question 6.8: C.

Concept Question 6.9: $\vec{F}_2$. This force acts into the elbow. Note that we do not neglect this force, arguing that it has a negligible magnitude (which it does not), because it leads to a torque contribution of zero due to r = 0 in Eq. [6.1].

Concept Question 6.10: (a): None. The mechanical equilibrium is defined by three conditions: a single force is always unbalanced and will cause acceleration. (b): Choice (A). Do the following self-test to convince yourself: Relax your facial muscles, open your mouth slightly, and push gently on your chin almost straight up. This corresponds to case (C) in Fig. 6.36 and results in your teeth being pushed together. Next, push gently in a horizontal direction toward your throat; this corresponds to case (B) in Fig. 6.36. As a result, your mouth will open. Last, push in the direction of your ear (roughly the location of the fulcrum of the mandible). No rotation will occur.

Part Two

ENERGY, BIOCHEMISTRY, AND TRANSPORT PHENOMENA

The second part of this book picks up on an idea we discussed earlier in the context of linear momentum: the question of what is conserved when we observe natural phenomena. This is an important question because it allows us to find the fundamental things "nature cares about." A quantity that is conserved has to be budgeted carefully because you cannot create or destroy it.

We start with a detailed discussion of the most important conserved quantity: energy. We will see how this opens the door into a new field of science, thermodynamics, which deals with heat and temperature, then chemistry, and ultimately biochemistry.

During the study of chemical processes in particular, the question will arise, how fast will a reaction occur? This brings us back to an issue we thought easily integrated in the discussion while studying mechanics: to get from a system in equilibrium to a dynamic system, we went from Newton's first law to his second law. Once we are beyond the realm of mechanics, this step is more complicated. We devote a separate chapter to this issue in which we introduce empirical dynamic properties, i.e., properties based on experimental observations. We find useful ways to describe heat conduction and diffusion, and generalize these concepts as transport phenomena.

By that point, we will have established two distinct lines of thought: how to characterize a system in equilibrium, and how to address dynamic changes. We round out the current part by illustrating how both lines of thought are applied to yet another new field in physics, the physics of fluids. Throughout this part we also highlight how all these insights apply to the life sciences, i.e., how they allow us to understand and quantify the foundations of chemistry, biology, and physiology.

CHAPTER 7

Energy and Its Conservation

Energy enables a system to do work. This definition of energy requires that we first quantify work. We limit our discussion at first to the same type of systems we studied in the previous chapters; the familiarity we established with Newtonian mechanics helps us in these cases to understand the new concepts. In this chapter, we will then reach the limits of Newtonian mechanics, setting the stage for the next group of chapters in which we expand our view to include thermal physics.

For an object, work is the product of a force applied to the object of interest and its resulting displacement. The sign convention attributes a positive value of work to work that is done on (received by) the object of interest.

Energy can be stored in a system in many ways. Each form of energy we introduce is quantified by allowing the system to undergo a specific change and then measuring the amount of work exchanged with the environment; in particular, we introduce kinetic energy, which is based on a change of the speed of the object of interest, and (gravitational) potential energy, which is based on its position relative to Earth's surface.

Energy obtains its central role in the physical sciences due to a unique feature: the energy of an isolated system is conserved. We approach this idea first in a more limited form by asking whether the sum of the kinetic and potential energies of an object is conserved. We find that such a restrictive conservation rule is very useful when it applies, but we also identify cases where it does not apply. This is illustrated with a comparison between inelastic and elastic collisions.

We introduced the concept of force to study the interaction of living organisms with objects in their environment. We saw that the magnitude and direction of forces can be predicted for a particular system, based on Newton's laws. The ability to exert forces is vital in locomotion, and control over forces is therefore an ever-present feature in our daily lives. The discussion of forces does not capture, however, every aspect of our interaction with objects in the environment. To determine the direction in which we want to proceed beyond Newtonian mechanics, which is the approach based on the concept of force we used in the first six chapters, we start with everyday observations.

7.1: Observations of Work and Energy

Figure 7.1 hints toward aspects we have so far omitted. Shown is a person on a bicycle ergometer in a clinical setting. The person operates the bicycle; this requires an effort as the person holds in motion an encased flywheel that is slowed with a preset level of friction. The clinical staff is not interested in the force the person exerts; they measure instead various parameters, such as blood pressure and blood oxygen content. This establishes the physiological changes the physical effort causes in the person's body.

We know from every day experience that exerting forces is not for free. When we exert a force, for example, to move a piece of furniture across a room, we refer to the effort as work. Such work we perceive

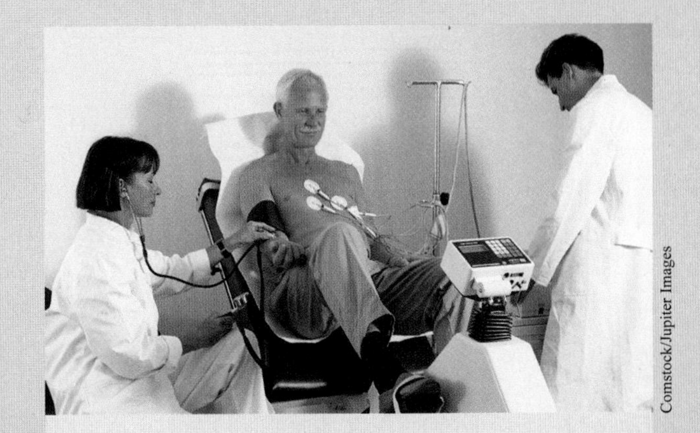
Comstock/Jupiter Images

Figure 7.1 Bicycle ergometer used in sports physiology and for exercise ECGs (electrocardiograms). The arrangement allows for continuous stationary recording of the person's vital parameters. The attending physician can vary the resistance of a frictional belt, slowing the encased flywheel.

as strenuous, indicating that it has an effect on our body; even after we stop pushing and the force exerted on the piece of furniture has been removed, we still feel fatigue, indicating that work has a lasting effect on us; i.e., we seem to have lost something in the effort. We feel the urge to eat and drink to recover what was lost while performing work.

What is lost when a muscle exerts a force, and what has to be recovered for the continuous operation of our muscles, can be pinpointed when we review the fundamental mechanism of muscle action: the sliding filament mechanism we refer to in Fig. 7.2 shows the sliding filament mechanism in greater detail, highlighting the important parts for the current chapter.

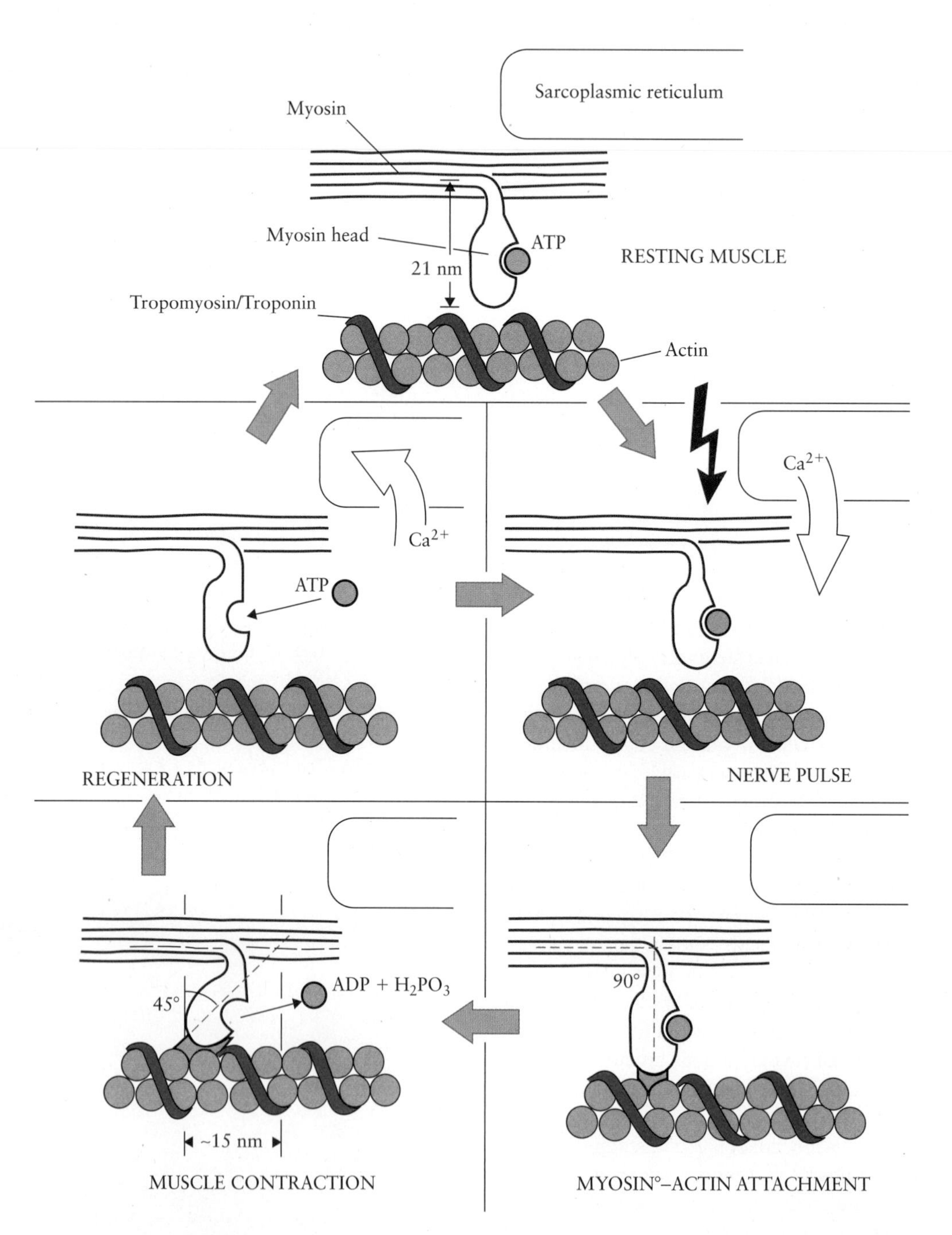

Figure 7.2 Illustration of the sliding filament model. A resting muscle is shown at the top with its major components: the sarcoplasmic reticulum at the upper right, the myosin filament with the myosin head charged with an ATP molecule below, and the intertwined actin and troponin/tropomyosin filaments at the bottom. When a nerve impulse arrives (indicated by a lightning bolt), a cycle of processes unfolds as shown in the four lower boxes. First, the nerve impulse triggers Ca^{2+} release from the sarcoplasmic reticulum. The calcium ions bond with troponin, deactivating the tropomyosin filament and allowing the myosin head to bond to the actin strand (bottom right box). When the myosin head is firmly attached to the actin filament it tilts to 45°. An ATP dissociation provides the energy for this step. Concurrent with the tilting of the myosin head, calcium is pumped back into the sarcoplasmic reticulum. This reactivates the troponin/tropomyosin, and the myosin–actin bond breaks. Further nerve impulses lead to a repetition of this cycle.

The top frame in Fig. 7.2 shows a resting muscle. A nerve signal triggers the release of Ca^{2+} ions from the sarcoplasmic reticulum. The calcium ions attach to troponin molecules. As a result, the tropomyosin protein strands that are coiled around the actin filament loosen. This allows the ends of the myosin filament to bond to the actin filaments. With the myosin head attached to the actin filament, a tilt from an angle of 90° to about 45° shortens the sarcomere. At the same time, the sarcoplasmic reticulum reabsorbs the calcium ions, causing the muscle fibre to regenerate as the actin–myosin bond is severed by the reactivated troponin. This cycle repeats for every new nerve impulse arriving at the muscle. Such impulses arrive at rates between 20 Hz and 100 Hz (hertz is a unit of frequency: 1 Hz = 1 s^{-1}; thus 20 Hz means 20 nerve impulses arrive per second). This leads to an appreciable contraction of the muscle in a short time.

If you focus on the step called *muscle contraction* you notice that it shows the splitting of a smaller molecule attached to the myosin head. This molecule is called ATP, which stands for *adenosine triphosphate*. The split of the ATP molecule to separate off a phosphate group leads to a new molecule called ADP, which stands for *adenosine diphosphate*. After the splitting, a replacement ATP molecule is reattached to the myosin head during the step called *regeneration*.

As noted in the macroscopic case of the person on the bicycle ergometer, generating a force by contracting a sarcomere does not come for free: an ATP molecule is sacrificed and a new ATP molecule has to be provided to allow the muscle to remain functional. We further see that the splitting of the ATP molecule happens exactly at the instant the actual effort during muscle contraction takes place, i.e., when the two Z-discs are pulled closer, shortening the sarcomere and thus the muscle. What we conclude from these observations guides us in the discussion of the current chapter: when we exert forces, an effort is made that we will define as **work**. Different from forces, which come and go, work done has a lasting effect on the person performing the work. This lasting effect we will define as a **change in energy**. Energy is therefore what we need to do work. We will investigate how work and changes in energy are linked, and this will lead us to the concept of energy conservation: energy cannot be created or eliminated like forces; it must be budgeted and accounted for.

To get an idea what energy is and how to deal with it quantitatively, it is not advisable to continue with the discussion of the microscopic processes in a muscle as a variety of energies—chemical energy, thermal energy, kinetic energy, and potential energy—are involved. In the current chapter we consider instead a limited number of energy forms that are related to the simplest possible interactions of objects. We have already discussed these interactions extensively in the previous chapters, enabling us to get a good idea of the various things energy can do. Once we have established the kinetic and potential energy, we will extend our discussion to non-mechanical systems, such as a muscle cell.

When we formulate the conservation law for energy, you will notice this does not mean energy is constant. There are many ways in which energy can change: it can be transferred from one object to another (or one system to another), or it can be converted from one form to another. Before defining quantitatively the various forms of energy and the many ways in which energy can affect a physical process, it is useful to consider a simple system to learn where to look to see energy in action. For this purpose we study a simple system we have already described in the context of Newton's laws. Fig. 7.3 shows an object on a frictionless table; that object is connected to a second object that hangs at the other end of a string. As before, we treat the string as massless and the pulley as rotating without friction. We know very well what will happen when we release this system from rest: the hanging object accelerates downward, pulling the object on the table along, as the overall length of the string does not change. Using Newton's laws in P–4.8, we calculated the acceleration of the two objects and found that it is less than what it would be for the hanging object if it were not attached to the string. In turn, the object on the table moves only because it is pulled toward the pulley by the hanging object. On its own, it is in mechanical equilibrium on the table and would not start to move.

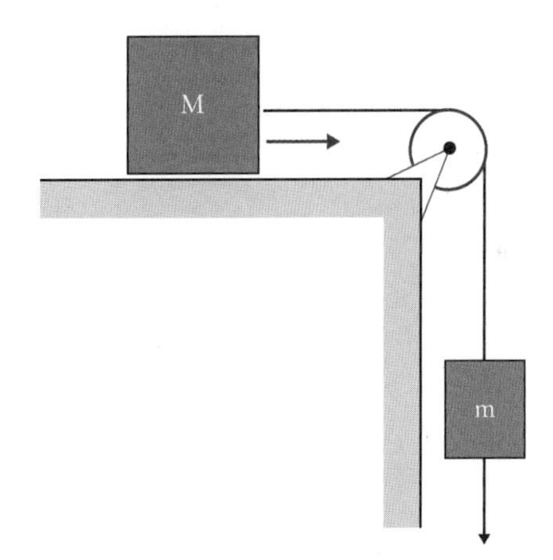

Figure 7.3 An object of mass M is placed on a frictionless table. It is connected to a second object of mass m by a massless string. The string runs over a massless pulley that rotates without friction. The weight of the object of mass m causes it to accelerate downward, accelerating the object of mass M at the same time toward the pulley.

EXAMPLE 7.1

For each of the two objects in Fig. 7.3, draw its height above ground and its speed as a function of time.

Solution

Inspecting Fig. 7.3 we notice that the object of mass M does not change its height above ground during its motion. This is shown in Fig. 7.4(a).

continued

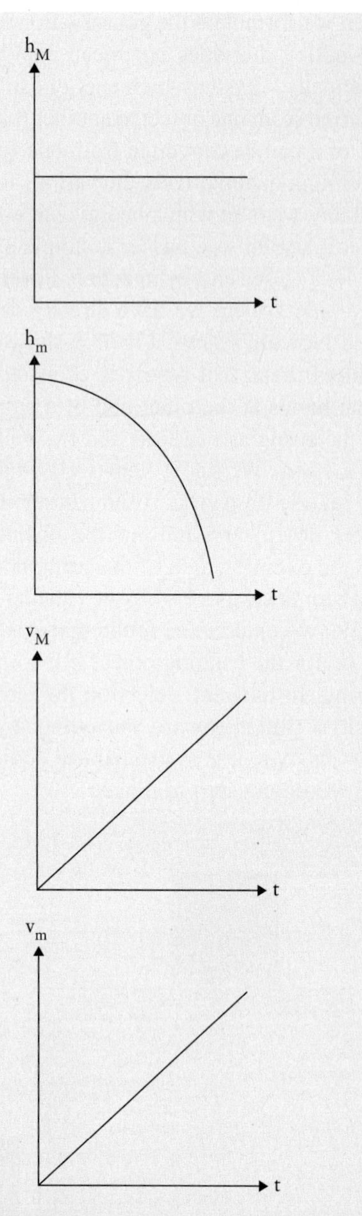

Figure 7.4 (a) The height above ground as a function of time for the object of mass M in Fig. 7.3. The object remains at a fixed distance from the ground as determined by the height of the table. (b) The same plot as (a) for the object of mass m. This object falls toward the ground with constant acceleration. (c) and (d) The speed of the two objects as a function of time is the same because both objects have the same magnitude of acceleration.

The solution to P–4.8(c) shows that the acceleration of the hanging object is constant. We found in this case the position as a function of time when the motion occurs in one dimension (along a straight line). If we substitute $v_{initial} = 0$, we find that the height is proportional to t^2: $h_m = h_0 - (a/2)\,t^2$, in which h_m is the height of the object of mass m as a function of time, h_0 is its height at time $t = 0$, and a is the acceleration we found in P–4.8. This dependence is shown in Fig. 7.4(b).

We found in P–4.8(b) and (c) that both objects accelerate with the same constant acceleration. Using $v_i = 0$ we find that the two plots for the speed as a function of time are identical and are linear in time. This is shown in Figs. 7.4(c) and (d).

A new way of interpreting the experiment in Fig. 7.3 is developed in the current chapter. We will introduce and quantitate two forms of energy that matter to this experiment: kinetic energy and potential energy. Kinetic energy is energy stored in the speed of an object: the faster it becomes the more energy is stored. Another way to store energy in a system is to move it vertically upward against the force of gravity. We will see that the higher an object is located relative to ground level, the more energy is stored in it as potential energy.

We quantify these forms of energy later in the chapter. At this time though we can already make interesting observations based on Fig. 7.3 that will guide us in the subsequent discussion. First we focus on the hanging object. It falls downward when the system is released from rest. As it falls, it picks up speed, as shown in Fig. 7.4. In the language of energy, we say it reduces its potential energy, which allows it to increase its kinetic energy. Thus, in a system that is not in mechanical equilibrium, we observe that an object can convert potential energy to kinetic energy (and vice versa). The interesting aspect of Fig. 7.3 is that we see energy being converted; this allows us to get a good handle on what otherwise might be perceived as an elusive concept.

We can see even more if we take a detailed look at what is happening with the falling object. Let's assume the object eventually hits the ground. We study its speed just before it hits the ground. If the object is released and it is connected to the object on the table, it travels faster to the ground and hits the ground with a greater speed than it did in the shown experiment. What does this mean? It means that, in both cases, the object reduced its potential energy by the same amount by the time it hit the ground but, since the amount of kinetic energy is determined by speed, it picked up less kinetic energy than when falling freely.

Now let's follow both objects carefully as the hanging object falls. We know from our discussion in previous chapters that the two objects interact because they act in a different manner than they would without the string connecting them. While the hanging object falls, the object on the table accelerates toward the pulley. Thus, the kinetic energy of the object on the table increases as its speed increases. At the same time, its potential energy remains unchanged, as it isn't coming closer to the ground while on the horizontal table (see Fig. 7.4). In comparison to the falling object, there can't be a debate about the sum of kinetic and potential energies for the object on the table: during the experiment its sum of the two energy forms increased. We know by now where this energy came from: the falling object transferred it through the string.

Thus, with these simple observations we can already make powerful statements about energy. Energy can be present in various forms in a system. If the system is not in mechanical equilibrium, energy can be converted between potential and kinetic energy. If two objects interact, energy can also be transferred between them. In the particular case of Fig. 7.3, we started with both objects at rest, then both picked up kinetic energy during the experiment at the expense of the potential energy of the hanging object.

7.2: Basic Concepts

From Fig. 7.3 we anticipate that a major challenge in the discussion of energy is the fact that several forms of energy exist and that we have to handle at least three properties of energy: the conservation of energy, the ability of energy to flow from place to place, and the ability of energy to convert back and forth between different energy forms. To avoid misconceptions in this discussion, we will start by paying close attention to the following:

- the distinction between equilibrium and non-equilibrium systems,
- the distinction between energy transfer and energy conversion, and
- the distinction between the properties of a system and the properties of the system's environment.

7.2.1: Equilibrium

We defined mechanical equilibrium using Newton's laws. When reading the earlier chapters you may have noticed that the distinction between Newton's first and second laws appeared somewhat artificial, as the second law appeared to include the first when we extended it to zero acceleration, $a = 0$. Why did we keep these two laws separate? The major reason is that both laws are at the foundation of different lines of thought:

- *Newton's first law focuses on the equilibrium state of a system.* The equilibrium concept links mechanics to thermodynamics and later to stationary fluids and electrostatics.
- *Newton's second law deals with dynamic changes of the state of a system.* The dynamics concept links mechanics to chemical kinetics and non-equilibrium thermodynamics processes, as well as fluid dynamics and the flow of electricity.

Mechanical equilibrium is distinguished as a special state for an object in which its mechanical properties do not change with time: the net force on the object and its acceleration are zero; its velocity and linear momentum are time independent. This observation characterizes equilibria in general:

KEY POINT

*A system is in **equilibrium** when all its essential physical parameters are time independent.*

The physical parameters in the definition of the equilibrium we consider to be essential vary with the physical concepts we study. For that reason, we must specify the context in which we refer to an equilibrium. This is done using a descriptive adjective with the term "equilibrium." Examples of equilibria used in this textbook include:

- the **mechanical equilibrium** of a point-like or extended rigid object in Chapters 4–6;
- **thermal equilibrium**, discussed in Chapter 8;
- **chemical equilibrium**, discussed in Chapter 10;
- **electrochemical equilibrium**, discussed in Chapter 19.

7.2.2: Energy Flow and Energy Conversion

Using Fig. 7.3 we illustrated above qualitatively that **energy can flow** from one place to another, and/or **energy can be converted** from one form into another. It is important that we distinguish these two features. We do this by using different terms for energy that flows and energy that may convert:

- The term **work** (and later also the term heat) refers to energy that is transferred in or out of a system; while
- the term **energy** with a descriptive adjective, for example *kinetic energy*, is used for energy contained within a system that may convert from one form to another. We label energy parameters always with a capital letter E, and identify the specific energy form with a subscript. For example, E_{kin} is written for kinetic energy.

Describing in precise scientific language how energy flows requires careful use of expressions adopted from everyday language. For work, we usually say it is *done on* or *done by* an object. These expressions refer to the direction of energy flow but do not express the fact that something is actually transferred from one object to another. We have chosen to use these standard notations in this book to allow the reader to relate easily to other literature. However, we will also use the terms *received* and *released* to emphasize the dynamics of energy flow between a system and its environment.

7.2.3: The System and Its Environment

When we studied the interaction of objects quantitatively with Newton's laws in Chapter 4, we always had to distinguish between the object of interest and other objects in its environment with which the object of interest interacts. This is important because we have to selectively identify forces that act on the object of interest. The distinction between the object of interest and other objects remains important when we study energy; therefore, a notation has to be chosen that generalizes this distinction. **System** is the term we use for the object or objects of interest, and we call the **environment** everything else that may interact with that system.

As we anticipate a careful budgeting of energy based on the observations we discussed at the beginning of the chapter, we have to be particularly careful with energy that flows between the system and its environment. To control the environment alongside the system, an additional boundary is drawn around the combination of the system and those parts of the environment that interact with the system. Everything that lies within this boundary is called an **isolated superstructure**.

To deal with the **system/environment interface**, we distinguish three types of systems based on their interactions with the environment, as sketched in Fig. 7.5.

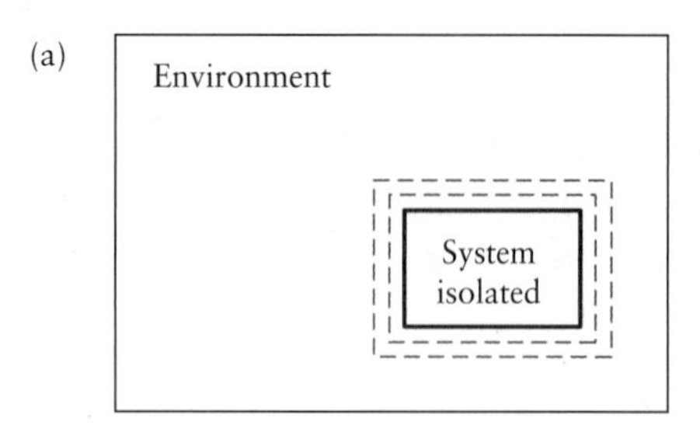

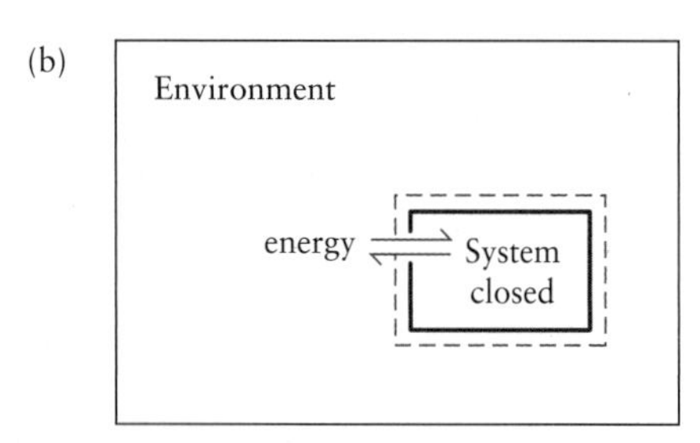

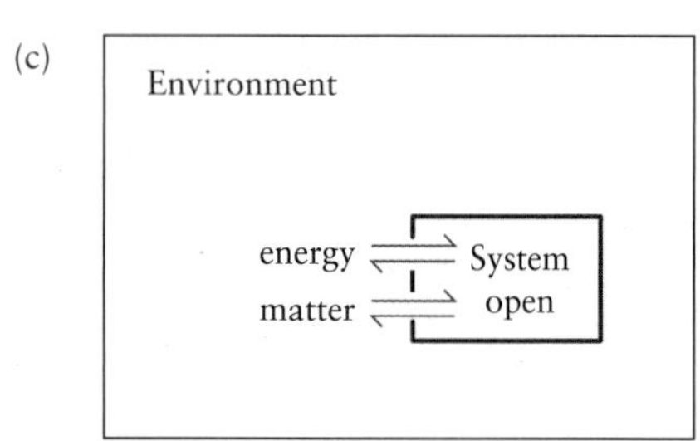

Figure 7.5 An isolated superstructure contains a system and all components that interact with that system in the environment. We distinguish three types of systems: (a) isolated systems do not interact with the environment, (b) closed systems exchange energy (for example in the form of work) with the environment but do not exchange matter, (c) open systems exchange both energy and matter with their environment. The isolating boundary that encloses the superstructure ensures the balance of energy and matter in the environment can be quantified.

7.2.3.1: Isolated Systems

Isolated systems [Fig. 7.5(a)] are systems that do not exchange energy or matter with their environment. We indicate this with two dashed lines surrounding the system to show that the surface of the system cannot be penetrated by energy or matter. Perfectly isolated systems are hard to establish experimentally but are of great relevance conceptually because they allow for the simplest mathematical formalism to quantify system parameters. Examples include the calorimeter, which is an instrument used to measure the energy content of chemical compounds, and closed insulated containers, which are used to store liquefied gases at low temperatures.

If the isolated system is in equilibrium, we know its essential parameters are time independent. Thus, for an isolated system in equilibrium, each type of energy will be shown to be separately constant. This represents the simplest situation we can study; therefore, isolated systems will be considered first whenever we enter a new field of inquiry.

Non-equilibrium situations are best conceptualized by studying the interaction of a system with its environment. We distinguish two ways in which this can occur: closed systems and open systems.

7.2.3.2: Closed Systems

Closed systems [Fig. 7.5(b)] are systems that interact with their environment in a limited fashion, exchanging energy but not transferring matter. This is indicated by a single dashed line surrounding the system to show that no matter can penetrate the surface of the system. Note that we now also need to identify the isolated superstructure; otherwise, we could not control the energy balance in the environment. A typical example is an organic chemistry experiment in which the reaction takes place in some glassware. Such a system is closed as long as no gas exchange with the external atmosphere occurs.

7.2.3.3: Open Systems

Open systems [Fig. 7.5(c)] are systems that exchange matter and energy with their environment. This is indicated by two double arrows at the left side of the system in the figure. Note that an exclusive exchange of matter without an exchange of energy is not possible because matter always carries energy. No dashed line surrounds the system because it can be penetrated by both energy and matter.

CONCEPT QUESTION 7.1

For which type of system do we not consider its environment?

(A) isolated system

(B) closed system

(C) open system

(D) closed and open systems

(E) all types of systems

7.3: Work for a Single Object

The discussion of energy starts with a definition of the concept of work for two reasons:

- There are many forms in which energy can be stored in a system, but there are only two forms in which it can flow in and out of a system: as work or as heat.
- Something that transfers out of a system is generally easier to capture and measure than something stored within a system. With this in mind we typically quantify energy based on the observed exchange of work, which must therefore be defined first. This is captured in the generalized definition of energy: *energy is that which enables a system to do work.*

7.3.1: Work Due to a Constant Force of Variable Direction

We base the definition of **work** on our everyday experience. Consider pushing this book across a table. We associate work with both the amount of force needed to move the book and the distance it is moved. Note that we do not consider all forces acting on the book but are concerned only with the specific force we exert on it. This distinguishes the discussion in this chapter from the discussion of Newton's laws, where we always had to consider the net force.

The simple experiment with the textbook allows us to make a first generalizing statement about work: work, W, is a function of the force, $\vec{F}$, that is applied to an object. It is further a function of the achieved displacement, $\Delta\vec{r}$, from an initial position, $\vec{r}_i$, to a final position, $\vec{r}_f$: $\Delta\vec{r} = \vec{r}_f - \vec{r}_i$. Thus work, or W, is a function of two vector parameters but work itself is not a quantity that depends on direction.

Let's confirm that we agree with these statements before continuing. This step is necessary because we made generalizing statements on the basis of just one observation. We know already that force and displacement are vectors. Let's vary the directions of the displacement and force vectors without varying their magnitude: push the textbook along the table in a different direction than before, but push it as far as you did in the first experiment. The effort you needed to invest is a measure of the work you did; that effort didn't change with the change in the direction. So, work is indeed not a vector. In mathematics, parameters that have no directional property are called **scalars**.

The descriptive statements above are not satisfactory though because we need to quantify work, and that requires that we know how it depends on force and displacement. To write work for this purpose in a more quantitative form, we repeat our experiment pushing the textbook across the table. Do it two more times, once pushing twice as far and once after stacking a few more books on top. You find from these two experiments that the work increases with increasing displacement distance and magnitude of the exerted force. We want to generalize these observations again, except this time we do it in the form of a formula for the work. We propose the following formula for work:

$$W = \vec{F} \cdot \Delta\vec{r} = \vec{F} \cdot (\vec{r}_f - \vec{r}_i). \qquad [7.1]$$

When we multiply two vectors and obtain a scalar, we call the underlying mathematical operation a scalar product. The mathematical aspects of the scalar product in Eq. [7.1] are reviewed in the section on "Vector Multiplication," which is in Math Review, after Chapter 27.

Again, Eq. [7.1] is a generalization of the specific experiment we performed. As always, we have to pause at this point and make sure the generalization is not in conflict with other experiments for which we want to use it. Note that Eq. [7.1] implies the following statements about work:

- The magnitude and direction of the force do not change during displacement. This is correct for the initial experiment we used to propose Eq. [7.1], but may not always be the case, as discussed in Case Study 7.1 below. To continue, we must therefore add an **assumption** to Eq. [7.1] for it to remain valid. The assumption is that the force in Eq. [7.1] is constant in direction and magnitude.
- Work depends linearly on both the force and the displacement. Pulling with twice the force for the same distance, or pulling with the same force for twice the distance, doubles the work. This is generally true for constant forces.
- Eq. [7.1] does not contain a constant offset term, e.g., a term W_0 added on the right-hand side. Such an offset term is not present because either $\vec{F} = 0$ or $\Delta\vec{r} = 0$ must result in $W = 0$. This means that if we do not exert a force, or do not accomplish a displacement, then no work has been done.
- No constant pre-factor is included with the dot product, which would have to be done in the form $c(\vec{F} \bullet \Delta\vec{r})$, with c a constant, because we use Eq. [7.1] as a **definition** for work.
- Eq. [7.1] allows us to identify the unit of work. Recall that the unit of force is newton (N) and the unit of displacement is metre (m). Thus the unit of work is $\text{N m} = \text{kg m}^2/\text{s}^2$. We define a derived SI unit for work, **joule** (J): 1 J is the work done when a force of 1 N moves an object a distance of 1 m, with force and displacement vectors parallel to each other.

CONCEPT QUESTION 7.2

In a first attempt, you lift a dumbbell in the gym by 1 m off the ground. Your trainer then doubles the weight of the dumbbell. In a second attempt, you now succeed to lift the dumbbell only by 50 cm. In which attempt did you do more work?

(A) attempt 1

(B) attempt 2

(C) The same work was done in both attempts.

For applications with constant forces, Eq. [7.1] is used in either one of two forms:

- If we know or are interested in the magnitudes of displacement, $|\Delta\vec{r}|$, and force, $|\vec{F}|$, and the angle between the two vectors, $\theta = \sphericalangle(\vec{F}, \vec{r})$, we use Eq. [7.1] in the form:

$$W = |\vec{F}|\,|\vec{r}|\cos\theta. \qquad [7.2]$$

- If we know or are interested in the components of the vectors in Eq. [7.1], we write it in component form:

$$W = F_x\,\Delta x + F_y\,\Delta y + F_z\,\Delta z, \qquad [7.3]$$

in which the last term on the right-hand side can be neglected if the physical action takes place in the xy-plane, i.e., when we use the vectors $\vec{F}$ and $\Delta\vec{r}$ to define the xy-plane.

CONCEPT QUESTION 7.3

You hold a heavy picture frame toward the wall while your friends decide whether they like it at the chosen location. They debate the issue lengthily and you get tired holding the picture. Finally, after more than a minute, you lower the frame. How much work did you do on the picture frame while holding it up?

CONCEPT QUESTION 7.4

Figure 7.6 shows three processes that unfold between an initial time t_i and a final time t_f. The first two processes are physically identical, but differ in the definition of the angle ϕ between the constant force acting on the object (green dot) and the displacement of length Δs. For which of the three cases is Eq. [7.2] suitable?

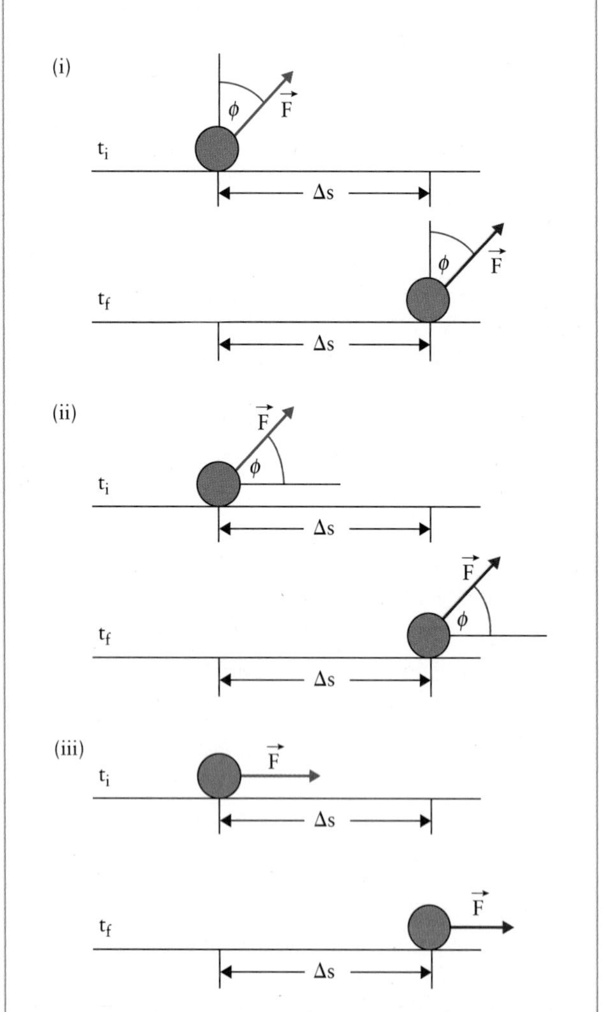

Figure 7.6 Three pairs of sketches showing an object at an initial (top) and final (bottom) instant. The displacement is labelled Δs. The force is not collinear with the displacement in the first two cases. Note that the angle labelled ϕ changes between case (i) and case (ii).

Power P is the rate at which work is done; the SI unit of power is **watt** (W, named for James Watt) with 1 W = 1 J/s. Since work isn't completed instantaneously, the rate at which it is done may vary with time. Therefore, we define an average power by dividing the entire work obtained or released, W, by the time interval it takes to complete the work, Δt

$$\Delta t: P = \frac{W}{\Delta t}.$$

Power is also related to the force $\vec{F}$ that causes the work W in turn:

$$P = \frac{W}{\Delta t} = \frac{\vec{F}\,\Delta\vec{r}}{\Delta t} = \vec{F}\,v.$$

In the last step of that equation, we assumed that the force is constant, i.e., that it does not change with time. Thus, power can be expressed as the dot product of a force acting on an object and the velocity of that object. A particular process does work at a higher rate (larger power) if it exerts a higher force or achieves a displacement in a shorter time, i.e., with a higher speed.

EXAMPLE 7.2

Fig. 7.7 shows an object on an inclined plane that forms a 20° angle with the horizontal. An external force $\vec{F}$ of magnitude 10 N is applied on the object. The force $\vec{F}$ acts in a direction that forms a 60° angle with the horizontal. As a result of the force $\vec{F}$, the object moves a distance of one metre along the inclined plane. What work is done by the source of force $\vec{F}$?

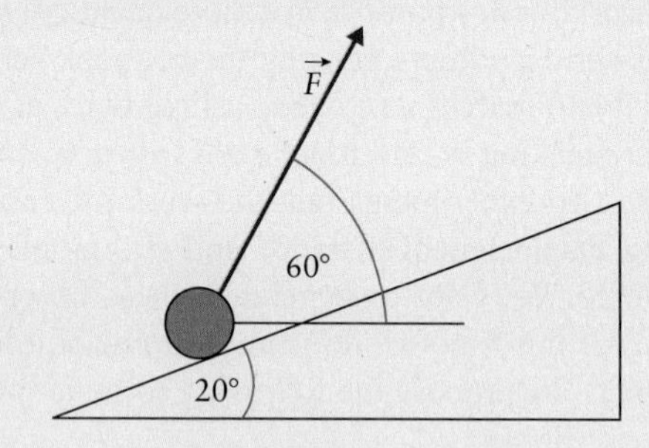

Figure 7.7 An object is pulled with a force $\vec{F}$ of given magnitude in a direction forming a 60° angle with the horizontal. As a result, the object is pulled a given distance along the underlying surface that forms a 20° angle with the horizontal.

Solution

We must first determine the angle between the displacement and the force because this angle is required in Eq. [7.2]. The angle θ is obtained from the data given in the problem text: $\theta = 60° - 20° = 40°$. Now we substitute the known data in Eq. [7.2]:

$$W = |\vec{F}|\,|\Delta\vec{r}|\cos\theta$$
$$= (10\text{ N})(1\text{ m})\cos(40°) = 7.7\text{ J}.$$

The work done by the source of force $\vec{F}$ is 7.7 J.

CASE STUDY 7.1

(a) Using Eq. [7.3], is it possible for work to have a negative value? (b) Is it possible that work is zero if a displacement is achieved in a given process?

Answer to Part (a): *Yes, for 90° < θ < 270° we find that cos θ < 0. This range of angles means that force and displacement vectors are directed in opposite directions when we project them onto a common axis. As always in physics, a change in the sign has ramifications beyond the mathematical fact. For work, this is explored below in the context of Eq. [7.4].*

Answer to Part (b): *Yes, if the displacement is perpendicular to the force exerted on the object. For example, I push an object against the wall but it is too heavy and slides down to the floor. In this case, I have not done any work on the object, nor has the object done work on me.*

In this context it is worthwhile to clarify a frequent misconception. Let's assume you stretch out your arm, place a paperclip on your hand and carry it from one side of the room to the other. Physically, you have not done work on the paperclip because you exert a vertical normal force on the object, but the displacement is horizontal. Yet, you will feel fatigue. The two observations are not contradictory; if you repeat the experiment without a paperclip, you feel the same fatigue. Work has been done, but it is work by your muscles on the arm.

Now someone may argue that the same argument that applies to the paperclip would apply to the arm, too: you hold it up, but carry it horizontally. To dispel this argument, you can indeed hold your arm up and not move at all. After the same time, you feel the same fatigue as you felt when walking with the arm lifted up. The fatigue is not related to the horizontal displacement. To see that in this case work is done on the arm, we have to look once more at the sliding filament model of the muscle. The weight of your arm pulls it down; continuous contraction of the muscle is needed to hold it up. The displacement in the process is the motion of the myosin heads along the actin filament; i.e., the displacement is in the axial direction of the muscle. The force a muscle exerts is also axial with its muscle fibres, thus displacement and force are collinear, which leads to a non-zero work.

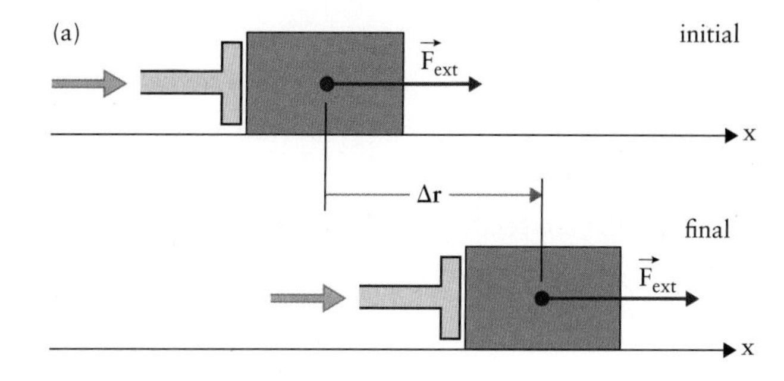

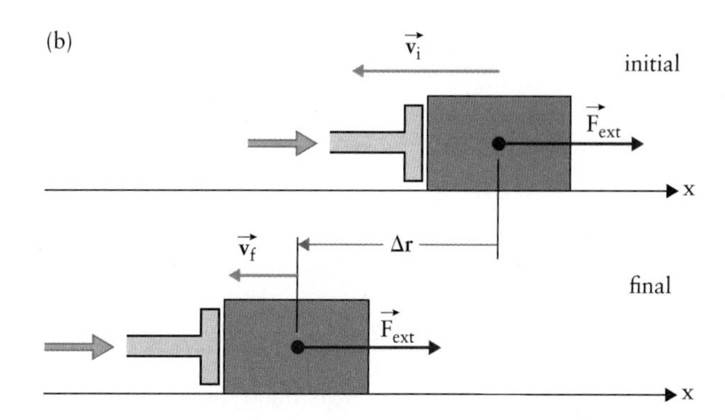

Figure 7.8 (a) Positive work is work done on the system: A piston is used to displace an object by a distance $\Delta\vec{r}$ along the x-axis. The experiment is conducted between an initial time (shown at top) and a final time (shown at bottom). To achieve the displacement, an external force $\vec{F}_{ext}$ is needed. We assume the external force acts collinear to the displacement. (b) Negative work is work done by the system: For the same arrangement, a displacement of the object in the negative x-direction is possible if, for example, the object had an initial negative velocity component v_i and is slowed down to v_f while moving toward the left. In this case the displacement and the external force act anti-parallel to each other.

7.3.2: Work Due to a Constant Force That Acts Parallel to the Displacement

Mathematically, Eq. [7.3] can be simplified in the special case where the force and the displacement vectors are parallel or anti-parallel:

$$W = \pm F\,\Delta r, \qquad [7.4]$$

where F is the magnitude of the force acting on the system and Δr is the length of the displacement vector. The ± symbol in Eq. [7.4] indicates that work can be positive or negative: if $\theta = 0°$ in Eq. [7.3]—i.e., force and displacement are parallel—we find $W > 0$; if $\theta = 180°$—i.e., force and displacement are anti-parallel—we find $W < 0$.

We illustrate this with two cases in Fig. 7.8. The figure explores the interaction of a piston pushing an object (green box) from the left. The piston exerts a constant force of magnitude F_{ext} on the object of interest, which is the green box. Note that the way the figure is drawn implicitly assumes that we choose the green box as the object of interest because the force $\vec{F}_{ext}$ is a force that acts on the green box. Fig. 7.8 generalizes our earlier experiment with the textbook and your hand.

- *Fig. 7.8(a):* The displacement occurs in the same direction as $\vec{F}_{ext}$ acts. The work is positive because $\vec{F}_{ext}$ and the displacement $\Delta\vec{r}$ point in the same direction regardless of our choice of coordinate system. We say in this case that **work is done on the system** by the environment (here, the piston).
- *Fig. 7.8(b):* Now the displacement occurs in the opposite direction of $\vec{F}_{ext}$. The piston still pushes against the green box; however, the box succeeds in pushing the piston to the left, causing a displacement to the left. The work is now negative because $\vec{F}_{ext}$ and $\Delta\vec{r}$ have opposite directions. The system does work on the environment.

KEY POINT

The mathematical sign of work is positive when work is DONE ON the system. The sign of work is negative when work is DONE BY the system.

This sign convention is easy to remember in the following way: identify yourself with the system; whatever you receive (here, work) is positive since you have more of it afterward; whatever you release is negative since you have less of it afterward.

Note that we return to Fig. 7.8 in Q–7.6 at the end of the chapter to discuss the case in which the piston is chosen as the object of interest instead of the green box.

EXAMPLE 7.3

(a) Calculate the work done on the object M in Fig. 7.3 if it moves 1.0 m toward the pulley. The objects in the figure have masses $M = 3.0$ kg and $m = 2.0$ kg. (b) Who or what has done the work you calculated? *Note:* In P–4.8(d) we found that the horizontal force acting on the object of mass M is $T = 11.8$ N.

Solution

Solution to part (a): The force acting on the object of mass M in Fig. 7.3 is constant and acts collinear with the displacement. Thus, we apply Eq. [7.4], which yields $W = +11.8$ J. Why did we find a positive value? We can argue in two ways. First, the force on the object and the displacement act in the same direction. If, for example, we return to the work calculation in Example 7.2, we would use $\theta = 0°$ for collinear force and displacement, and therefore find a positive value for work. Alternatively, we can argue that work is done on the object, or in other words, the object receives work. Based on the sign convention of Fig. 7.8, this again implies the work is positive.

Solution to part (b): What is the source of this work? It is done by the source of the force we used in the calculation, i.e., the tension force in Fig. 7.3. The tension is due to the object of mass m that pulls the string downward and thereby away from the object of mass M.

EXAMPLE 7.4

Determine the work done by a person when lowering an object of mass $m = 1.0$ kg by a distance of 1.0 m. The person applies a constant force $\vec{F}_{ext}$ such that the object moves with constant speed, i.e., without acceleration, as shown in Fig. 7.9.

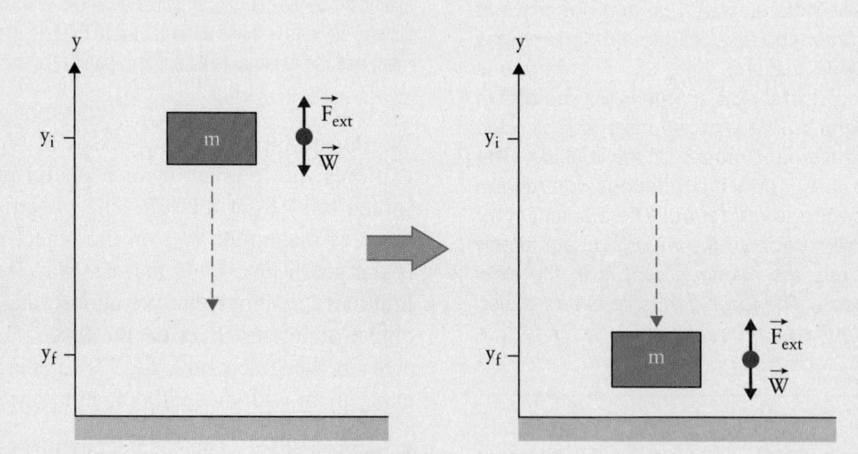

Figure 7.9 An object of mass m is lowered from an initial height y_i (shown at left) to a final height y_f (shown at right). The object moves from the initial to the final state with a constant speed if the magnitude of the external force $\vec{F}_{ext}$ is equal to the magnitude of the weight $\vec{W}$ of the object.

Solution

It is worthwhile to re-emphasize the shift of focus in this chapter: while we combined all forces acting on a system as a net force to apply Newton's laws, we are now focussing on a specific interaction between the system and its environment. In Fig. 7.9, the object (a rectangular box of mass m) is the system and the person (not shown) is one interacting component of the environment. Another component of the environment interacting with the system is Earth, as it exerts a gravitational pull on the object. Thus, the person, Earth, and the object together form an isolated superstructure with all interactions identified as forces.

Two free body diagrams are included in Fig. 7.9, each applies to the system in one of the two states we are interested in: the initial state at height y_i and the final state at height y_f. Note that mechanical equilibrium applies at both

continued

heights because the object does not accelerate; thus, the magnitude of the external force is equal to the weight of the object: $F_{ext} = m\,g$.

From the point of view of Newton's laws, no difference exists between the system at the initial height and the final height. However, based on our discussion in the current chapter, we can quantify as work the effort required to achieve displacement of the system. Based on Eq. [7.4], we expect that the work is negative because both external force and displacement are collinear but point in opposite directions. This is confirmed by using Eq. [7.3]: both the external force and the displacement have only y-components, with the external force $F_{ext,y} > 0$ and the displacement $\Delta y = (y_f - y_i) < 0$.

The work is quantified as:

$$W = F_{ext}\,\Delta y = F_{ext}(y_f - y_i)$$
$$= m\,g(y_f - y_i).$$

Substituting the values given in the problem text, we find:

$$W = (1.0\ \text{kg})\left(9.8\frac{\text{m}}{\text{s}^2}\right)(-1.0\ \text{m})$$
$$= -9.8\ \text{J}.$$

The negative result is interpreted with the sign convention of Fig. 7.8: the system has released work. Specifically, the person has received work from the system.

7.4: Energy

Even though a system may be capable of doing work, that work need not be done immediately. A system can store the capability to do work; what is stored is called energy.

KEY POINT

When a system interacts with its environment, work is transferred between the system and the environment. This work is equal to the amount of energy change in the system.

Energy can be stored in systems in many different ways. The following list includes the major forms of energy we discuss in this textbook:

- **Kinetic energy** is stored in the speed of an object.

  Example: Blood ejected by the heart carries a high kinetic energy into the aorta. This energy must be reduced in the aortic arch because blood has to flow more steadily through the remaining cardiovascular system.

- **Potential energy** is stored in the position of an object relative to other objects. We discuss:

  - *Gravitational potential energy,* which is the energy of an object due to its position relative to the surface of Earth.

    Example: The gravitational potential energy of blood varies when we study a standing person's feet, heart, or head. We will see in Chapter 12 that this has consequences for blood pressure.

  - *Electric potential energy,* which is the energy of an electrically charged object due to its position relative to other electric charges. This form of energy is introduced in Chapter 18.

    Example: Electric energy governs the signal transport in nerves, as discussed in Chapter 19.

  - *Elastic potential energy,* which is the energy due to the relative position of two objects that are connected through an elastic medium, most typically a spring. It is discussed in Chapter 14.

    Example: Elastic energy allows us to characterize molecular vibrations between the different parts of a chemical molecule.

- **Thermal energy** is stored in the irregular motion of particles. The temperature of the system is a measure of this energy. It is discussed in Chapter 9. Note that thermal energy flowing in or out of a system will be defined as heat, not work.

  Example: Thermal energy is generated and released in endotherms (warm-blooded animals) as they maintain a constant core body temperature.

- **Chemical energy** is stored in the relative position of atoms in a molecule (chemical bonds). It is discussed in Chapter 10.

  Example: Chemical energy is released during the splitting of ATP into ADP and a phosphate, as shown in the muscle contraction step of Fig. 7.2.

- **Latent heat** is stored in the phase of matter, such as the liquid, gaseous, or solid state.

  Example: Latent heat is dissipated during perspiration because perspiration is associated with the evaporation of water from the skin surface.

One consequence of the relation between work and the various forms of energy is that each form of energy has the same unit as work. Kinetic, gravitational potential, and elastic potential energies are usually defined for an object. We use sometimes the term mechanical energy for these forms of energy. With the current chapter still focusing on systems that can be discussed in the framework of Newton's laws (Newtonian mechanics), we introduce only the kinetic and the gravitational potential energy. Other forms of energy follow in later chapters.

7.4.1: Kinetic Energy

To define kinetic energy, we study a system for which its **speed** changes, but no other parameters. Such a system is shown in Fig. 7.10: an isolated object is allowed to interact for a well-defined time interval, Δt, with its environment. The interaction is based on a constant external force that acts along the horizontal direction of the object's motion, causing an acceleration of the system. The horizontal direction is labelled the x-axis. Newton's second law for a single external force, $\vec{F}_{ext} = m\,\vec{a}$, then leads to $\vec{a}$ = const during Δt. This acceleration changes the speed of the system from v_i to v_f.

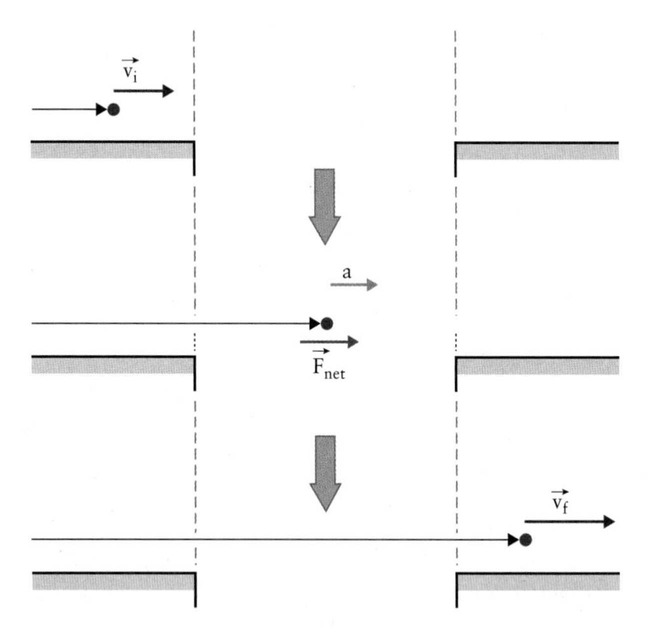

Figure 7.10 To quantify the kinetic energy of an object (blue dot) we compare two of its states that are each in mechanical equilibrium. The initial state is shown at the top with speed v_i and the final state is shown at the bottom v_f, with $v_f > v_i$; i.e., the two states differ in the speed of the object. The difference in speed is obtained by accelerating the object with a constant external force $\vec{F}_{net}$; this force acts on the object only while it passes from the left end to the right end of an interaction zone, shown in the middle section between the vertical dashed lines.

We use the acceleration step to quantify kinetic energy from the work done on the object. This work is $W = \vec{F}_{net} \bullet \Delta\vec{x}$. Fig. 7.10 illustrates which kinematic properties have to be included in the calculation: the speed of the object, its acceleration and displacement, but not the time. We use the third kinematic relation. Substituting the acceleration from that equation into Newton's equation of motion yields:

$$F = m\,a = \frac{m}{2\,\Delta x}(v_f^2 - v_i^2).$$

Work is done on the object during the time interval Δt, when the external force acts on the system. The work is obtained by multiplying both sides of this equation by the displacement, Δx:

$$W = F\,\Delta\,x = \frac{m}{2}(v_f^2 - v_i^2). \qquad [7.5]$$

The terms relating to the initial and final conditions can be separated on the right hand side:

$$W = \frac{m}{2}v_f^2 - \frac{m}{2}v_i^2. \qquad [7.6]$$

We notice an interesting mathematical feature with far-reaching physical consequences: the two terms on the right side have the same mathematical form. This means that both terms physically describe the same thing, except that

- the second term contains only system parameters at the initial instant, when the external force begins to act; and
- the first term contains only system parameters at the final instant, when the external force ceases to act.

Since both terms describe a feature of the system that is common at the initial and final states of a process, we create a new variable E_{kin}, called the kinetic energy:

$$E_{kin} = \frac{1}{2}mv^2. \qquad [7.7]$$

With this definition, we rewrite Eq. [7.6] in the form called the **work–kinetic energy theorem:**

$$W = E_{kin,f} - E_{kin,i}. \qquad [7.8]$$

KEY POINT

When work is done on an object causing exclusively a change in its speed, that work is equal to the change in kinetic energy of the object.

We conclude the following from Eqs. [7.5] to [7.8]: an object that does work on the environment becomes slower and its kinetic energy decreases by the amount of energy needed to do the work. If in turn work is done on the object, it accelerates and then stores a larger amount of kinetic energy. There is a maximum amount of work a moving object can do, which is equal to the kinetic energy it loses by slowing to rest, $v_f = 0$. We summarize these observations in the following form:

$$W > 0 \Leftrightarrow E_{kin,f} > E_{kin,i}$$

$$\Leftrightarrow \text{System receives work}$$

$$W < 0 \Leftrightarrow E_{kin,f} < E_{kin,i}$$

$$\Leftrightarrow \text{System releases work.}$$

CONCEPT QUESTION 7.5

Fig. 7.11 shows five plots of kinetic energy as a function of the velocity of an object. Which one is consistent with Eq. [7.7]?

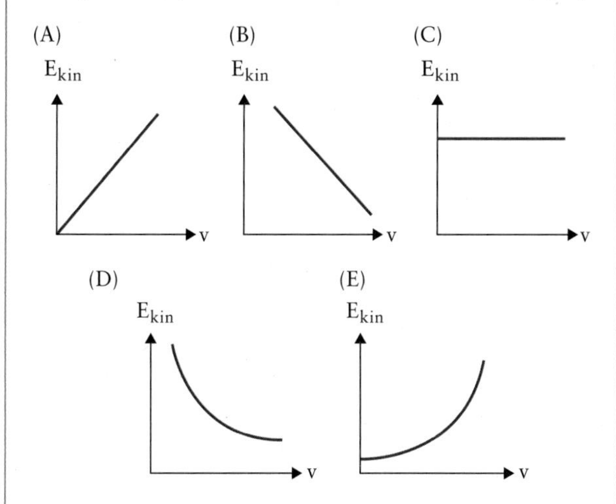

Figure 7.11 Five proposed graphs for kinetic energy as a function of velocity. The origin for each axis lies at the point where the vertical and horizontal axes intersect.

EXAMPLE 7.5

An object of mass $m = 0.5$ kg moves with initial speed $v_i = 5.0$ m/s, then interacts with its environment, releasing 5.0 J of work. What is the speed of the object just after the interaction?

Solution

We use Eq. [7.8]:

$$W = E_f - E_i$$
$$= \frac{1}{2} m v_f^2 - \frac{1}{2} m v_i^2.$$

Next, we isolate the final speed of the object:

$$\frac{1}{2} m v_f^2 = W + \frac{1}{2} m v_i^2.$$

After multiplication on both sides by $2/m$, the square root is taken:

$$v_{final} = \sqrt{\frac{2W}{m} + v_i^2}.$$

Applying the convention of Fig. 7.8, we note that work released by the system has to be entered as a negative value into the above equation, $W = -5.0$ J. This leads to:

$$v_f = \sqrt{\frac{2(-5.0 \text{ J})}{0.5 \text{ kg}} + \left(5.0 \frac{\text{m}}{\text{s}}\right)^2} = 2.23 \frac{\text{m}}{\text{s}}.$$

Note that it is important in the last step to understand whether work is done on or by the system because this determines the sign of W. Using $W = +5.0$ J instead of $W = -5.0$ J would lead to a wrong result.

EXAMPLE 7.6

Calculate for the system discussed in Example 7.3 (a) the change in kinetic energy for the object of mass m and (b) the change in the kinetic energy for the object of mass M. *Note:* Recall that the two masses in Fig. 7.3 are $M = 3.0$ kg and $m = 2.0$ kg. In P–4.8(c) we found for the acceleration of the hanging object $a = 3.9$ m/s^2. The displacement of the objects is given as 1.0 m in Example 7.3.

Solution

Solution to part (a): To determine the final speed of the hanging object, use $v_i = 0$ and $x - x_i = 1.0$ m. This yields $v_f = 2.8$ m/s. The change in the kinetic energy of the hanging object after moving 1.0 m is $\Delta E_{kin} = +7.8$ J.

Solution to part (b): It is easy to calculate now the change in the kinetic energy of the object on the table, as this object also moves 1.0 m with the same acceleration [see P–4.8(b)] and therefore has the same final speed as the hanging object. We find, due to the larger mass of 3.0 kg, $\Delta E_{kin} = +11.8$ J.

Note that the change in kinetic energy for the object on the table is equal to the work done on this object, as calculated in Example 7.3. Thus, the hanging object supplies in the form of work 11.8 J of its energy to the object on the table. During the same time, the hanging object increases its own kinetic energy by 7.8 J. Thus, we find an overall increase in kinetic energy of 19.6 J in the combined system.

CASE STUDY 7.2

Fig. 3.34 shows a sketch of a cross-section of the human skin containing hair at the left. In comparison to hair-free skin, no Meissner's corpuscles are present. Pacinian corpuscles are still found in the deeper layers, and Merkel's corpuscles are still present below the epidermis. Is the selective absence of Meissner's corpuscles explained by the hair replacing them as velocity detectors, as the physiological literature states?

Additional information: Unlike Meissner's corpuscles, hair can detect objects not yet in contact with the skin; i.e., they represent a rudimentary remote sensing device. Their mechanism is illustrated in Fig. 7.12. Shown is an object (illustrated by a piston) that moves parallel to the skin in near proximity to the skin surface. When the moving object pushes the hair to the side, a force is exerted on the root sheath. This force is sensed by the dendrites coiling around the root of the hair. As in the case of Meissner's corpuscles, the rate of triggered nerve impulses is a function of the speed of the piston.

continued

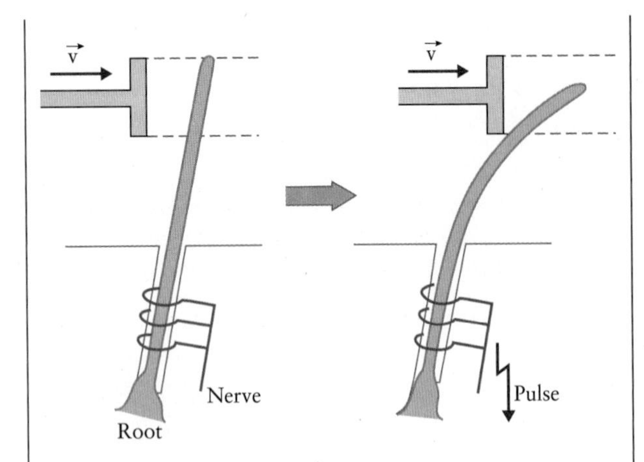

Figure 7.12 Illustration of the detection mechanism of hair. An object (illustrated as a piston) moves with velocity $\vec{v}$ parallel and in close proximity to the skin. The bending of the hair causes the nerves at the root to send a signal to the brain.

ANSWER: *The way Fig. 7.12 is drawn, the suggestion of hair as a velocity detector as opposed to a kinetic energy detector seems to gain the upper hand: velocity is a vector, kinetic energy is a scalar, and the figure seems to suggest that a directional dependence of the mechanism exists. However, if you close your eyes and let your left hand circle over your lower right arm (assuming that you have hair on the arm) you will notice that the direction of motion is not really detected by the single hair, but only by the sequence of neighbouring hairs being touched. If you focus on the shaft, you see that again elastic deformations take place that do not yield a simple relation with either speed or kinetic energy.*

We will return to this argument in the discussion of the kinetic gas theory in Chapter 8. There, a clear-cut distinction exists between a gas model that assumes relevance of the speed of gas particles versus relevance of the kinetic energy of gas particles, with the latter being consistent with the experimental observations.

7.4.2: Potential Energy

Next we study cases in which the work is stored in the relative position of objects. In the current chapter "relative position" refers specifically to the position of an object relative to Earth's surface. This potential energy is properly referred to as **gravitational potential energy** to distinguish it from other potential energy forms; however, the adjective "gravitational" is often dropped unless the various potential energies could be confused. Since we discuss only the gravitational potential energy in this chapter, the adjective "gravitational" is consistently omitted.

To quantify this form of energy, we have to develop a proper experimental arrangement. The system cannot be an isolated object, as used earlier when we defined kinetic energy in Fig. 7.10. Due to gravitational interaction, Earth becomes part of the environment. An isolated superstructure consisting of Earth as the environment and the object as the system is still insufficient because this isolated superstructure would allow for only one force, the force of gravity, to act on the object. This does not yield a mechanical equilibrium. We need to add a second force with its source in the environment such that the system is brought into mechanical equilibrium. A possible way to add this second force is to include a person holding the system. Note that we introduced this experiment previously in Fig. 7.9.

For the current discussion, the isolated superstructure consists of the object of interest as the system and Earth and a person as the system's environment. In this isolated superstructure, (at least) two different equilibrium states for the system can be established that enable us to define potential energy. The two equilibrium states we employ are the same ones we used in Fig. 7.9:

- the system of mass m held at rest at height y_i at an initial time, and
- the system held at rest at height y_f at a later, final time.

For the arguments we present here, we further choose the y-axis in Fig. 7.9 vertically upward; i.e., $y_f < y_i$.

The simplest way to conduct the experiment in Fig. 7.9 is to lower the system at constant speed from the initial to the final height. In this case, the system is always in mechanical equilibrium and we know that $F_{ext} - W = 0$; i.e., $F_{ext} = m\,g$ during the transfer. Thus, the work due to the external force—which by definition establishes the mechanical equilibrium for the system—is calculated as:

$$W = F_{ext}\,\Delta y = m\,g\,(y_f - y_i). \qquad [7.9]$$

The work for the process in Fig. 7.9 is negative because the expression in brackets, $y_f - y_i$, is negative while the external force is positive.

Postulating in analogy to Eq. [7.8] that work is a measure of the change of the potential energy of the system from the initial to the final equilibrium state, we write Eq. [7.9] in the form:

$$W = E_{pot,f} - E_{pot,i}. \qquad [7.10]$$

This is possible because the right-hand side of Eq. [7.9] consists of two mathematically identical terms. Comparing Eq. [7.10] with Eq. [7.9] then yields:

$$E_{pot,i} = m\,g\,y_i$$

$$E_{pot,f} = m\,g\,y_f$$

and we define the potential energy in the form:

$$E_{pot} = m\,g\,y. \qquad [7.11]$$

Note that reporting absolute potential energy values is meaningless because values calculated from Eq. [7.11] depend on the choice of origin for the y-axis. However, a physical quantity can never depend on an arbitrary mathematical construct such as choice of an origin. Thus, only differences between potential energies, ΔE_{pot}, play a role in physical processes.

CONCEPT QUESTION 7.6

A person states that an object of mass $m = 1$ kg has a potential energy of 0 J. Where is the object located relative to Earth's surface?

EXAMPLE 7.7

For the system discussed in Examples 7.3 and 7.6 calculate the change in potential energy for the hanging object of mass m. *Note:* Recall that the two masses in Fig. 7.3 are $M = 3.0$ kg and $m = 2.0$ kg. The displacement of the objects is given as 1.0 m in Example 7.3. In Example 7.6 we found an overall increase in kinetic energy in the combined system of 19.6 J.

Solution

We use Eqs. [7.9] and [7.10] in the form $\Delta E_{pot} = m\, g\, \Delta h$, with $\Delta h = -1.0$ m (a negative number because the final height is less than the initial height). This yields $\Delta E_{pot} = -19.6$ J. This change is negative because the falling object releases potential energy. Some of this energy is released as work and transferred to the object on the table; some is converted into kinetic energy. We note that the change in potential energy for the hanging object and the change in kinetic energy for the combined system are the same. We further note that the change in potential energy of the object on the table is zero because it moves horizontally.

A potential energy cannot be defined for all forces, leading in the literature to a distinction between **conservative forces** (with a potential energy) and **dissipative forces** (without a potential energy). We avoid a discussion of this issue by omitting dissipative forces until we have introduced thermal physics. At that point we are no longer trying to describe physical laws exclusively within the framework of Newton's laws, and the distinction between conservative and dissipative forces is no longer necessary.

7.5: Is Mechanical Energy Conserved?

The result of Examples 7.6 and 7.7 is intriguing: we found that the change of the total energy of the system, defined as the sum of the change of kinetic and potential energies of the two objects, is zero. Such a result might be accidental in a particular experiment but, alternatively, it may hint toward a more general underlying rule or law.

At this point, it would be easy to just state the proper form of the energy conservation law. However, we don't do this for two reasons:

- We can only understand the proper formulation of the energy conservation law after discussing further physical concepts that are presented in subsequent chapters.
- It is instructive to follow the chronological steps in the scientific discussion as they reveal how progress in the sciences is achieved in general.

This discussion started with observations based on experiments like the one in Fig. 7.3 during the latter part of the 1600s, and was completed only in the mid-1800s with the formulation of the first law of thermodynamics. It followed the following order of steps:

(1) The observations in a small number of experiments hint toward an underlying fundamental principle. In the current case we observed for Fig. 7.3 that the mechanical energy of a falling object is conserved.

(2) We formulate a generalization of these observations in the form of a hypothesis. The hypothesis in the current context is that mechanical energy is conserved in all possible experiments. We will state this in mathematical form in Eq. [7.13].

(3) The hypothesis is tested in further experiments and observations. In the particular case, we choose inelastic and elastic collisions later in the chapter to evaluate the validity of the hypothesis.

(4) If the hypothesis holds, it is eventually accepted as a natural law. However, a single, conclusive experiment that yields results inconsistent with the hypothesis will falsify the hypothesis. This happens often, as in our case when we studied inelastic collisions.

(5) This is then an interesting point in the scientific discovery process: the original idea can be abandoned, or it can be revised. Revisions are sought if the usefulness of the original idea has been established. In our case, the success of the energy conservation idea for elastic collisions merits further efforts.

(6) When pursued further, the hypothesis is reformulated to include the experimental results and observations that caused the falsification of the hypothesis in the first place.

In the current case we will ultimately find that the original hypothesis is too restrictive. Other forms of energy exist and cannot be kept separate from the mechanical forms of energy. Expanding our horizon beyond Newtonian mechanics in the next group of chapters will then allow us to formulate the broader energy conservation hypothesis that has now been accepted as the first law of thermodynamics.

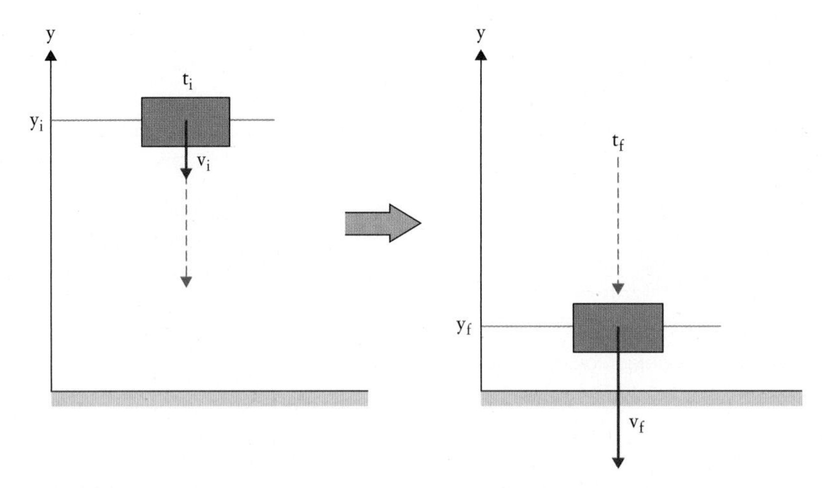

Figure 7.13 An object is released at time t_1 (shown at left) and falls toward Earth's surface. The final observation of the object is done at time t_2, when the object has not yet hit the ground (shown at right). Note that this experiment is similar to the one shown in Fig. 7.9. However, the person holding the object in Fig. 7.9 is no longer part of the isolated superstructure because no interaction between a person and the object is required.

7.5.1: An Object Falls Without Air Resistance: The Mechanical Energy Conservation Hypothesis

In equilibrium, both the kinetic and the potential energies are independently constant. In this section, we want to find out how these energy forms depend on time for a system that is not in mechanical equilibrium but that does not exchange work with its environment.

The specific experiment we study is shown in Fig. 7.13. An object is released at an initial height and drops. The final observation is taken at a height where the object is still above the point of impact on the ground. During the time that elapses while the object moves from its initial to its final height, $\Delta t = t_2 - t_1$, only the speed and the height above the ground change. This means we take a sufficient approach when considering only the kinetic energy and the potential energy. The system is the object; the isolated superstructure also contains Earth. No other object (such as a person) interacts with the system. For the system in this isolated superstructure, we quantify the energy changes during the process of falling.

We start with the relation between the parameters of the falling object. The displacement $y_f - y_i$ along the vertical y-direction in Fig. 7.13 and the y-component of the acceleration $a_y = -g$ are substituted in the third kinematics equation:

$$v_f^2 - v_i^2 = -2\,g\,(y_f - y_i).$$

We multiply this equation on both sides by the mass of the object and divide by 2:

$$\frac{1}{2}\,m\,v_f^2 - \frac{1}{2}\,m\,v_i^2 = -\,m\,g\,(y_f - y_i).$$

After separation of the terms related to the initial and final states respectively, this yields:

$$\frac{1}{2}\,m\,v_i^2 + m\,g\,y_i = \frac{1}{2}\,m\,v_f^2 + m\,g\,y_f.$$

We found that the sum of the potential and kinetic energies is the same at the initial and the final positions for the object in Fig. 7.13:

$$E_{kin,i} + E_{pot,i} = E_{kin,f} + E_{pot,f}. \qquad [7.12]$$

We generalize this result for any initial and final positions in two ways:

- The sum of kinetic and potential energy is constant at every position of the object or at any instant of the process in Fig. 7.13:

$$E_{kin} + E_{pot} = \text{const.} \qquad [7.13]$$

- The time rate of change of the sum of the kinetic and potential energy of the object is zero.

CONCEPT QUESTION 7.7

Figure 7.14 shows five graphs of energy versus vertical position for an object falling straight down from a given initial height. Which graph represents (a) the potential energy of the object and (b) its total mechanical (kinetic and potential) energy? Neglect air resistance.

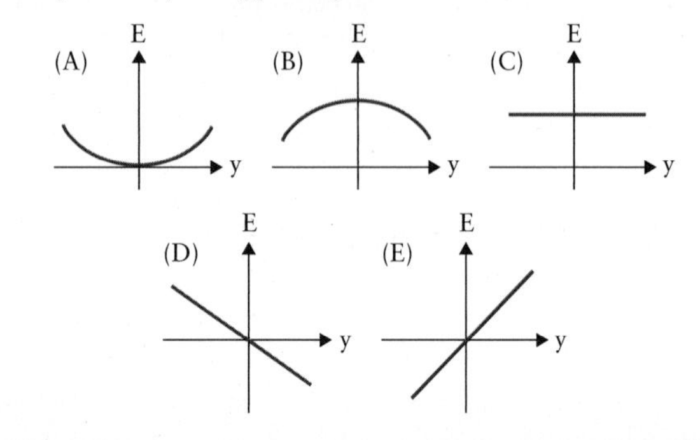

Figure 7.14 Five plots of energy E as a function of vertical position y.

EXAMPLE 7.8

An object of 0.2 kg moves initially at a speed of 30.0 m/s vertically upward. (a) What is the kinetic energy of the object at its highest point? (b) What is the potential energy of the object at its highest point? (c) How far is its highest point above the initial position of the object?

Solution

Solution to part (a): At the highest point, the downward-directed gravitational acceleration has eliminated the initially upward-directed speed component in the *y*-direction. Thus:

$$v_{\text{at top}} = 0.$$

This yields:

$$E_{\text{kin,at top}} = \frac{1}{2} m\, v_{\text{at top}}^2 = 0 \text{ J}.$$

Solution to part (b): We first calculate the total mechanical energy of the system. Since energy is conserved, we can do this at any instant of the experiment. For convenience we choose the initial instant, because we know all system data at that time. We further arbitrarily choose the initial height as the origin, $y_i = 0$; i.e., $E_{\text{pot,i}} = 0$. This is done without loss of generality and does not affect the result! As a consequence, the total energy of the object initially has only the kinetic energy component $E_{\text{kin,i}} = (1/2)m\, v^2 = 90$ J. This is, therefore, also the total energy of the system; in turn, the potential energy at the highest point is $m\, g\, y_{\text{at top}} = 90$ J because the kinetic energy at the highest point is zero.

Solution to part (c): We substitute the result from part (b) in Eq. [7.11] to obtain the maximum height $y_{\text{at top}}$:

$$y_{\text{at top}} = \frac{E_{\text{pot}}}{m \cdot g} = \frac{90 \text{ J}}{(0.2 \text{ kg})\left(9.8 \dfrac{\text{m}}{\text{s}^2}\right)} = 46 \text{ m}.$$

EXAMPLE 7.9

For the system discussed in Examples 7.3, 7.6, and 7.7, determine whether the mechanical energy is conserved for (a) the object on the frictionless table, (b) the hanging object, and (c) the combined system of the two objects together. (d) Give an explanation for the three results and propose a modification of the mechanical energy conservation hypothesis to take these findings into account.

Previous results: The experiment is shown in Fig. 7.3 with $M = 3.0$ kg and $m = 2.0$ kg. The displacement of the objects is given as 1.0 m in Example 7.3. In Example 7.6 we found an overall increase in kinetic energy of 19.6 J in the combined system, based on an increase in kinetic energy of 11.8 J for the object on the table and an increase in kinetic energy of 7.8 J for the hanging object. In Example 7.7 we found further a decrease in potential energy of 19.6 J for the hanging object, while the potential energy for the object on the table did not change during the experiment.

continued

Solution

Solution for part (a): Little calculation is needed in this example as the single terms have been determined already in the previous Examples 7.3, 7.6, and 7.7. Labelling the sum of kinetic and potential energies E_{mech}, we find for the object on the table $E_{\text{mech, M}} = E_{\text{kin, M}} + E_{\text{pot, M}} = +11.8 \text{ J} + 0 \text{ J} = +11.8 \text{ J} > 0$.

Solution for part (b): We repeat the same calculation for the hanging object: $E_{\text{mech, m}} = E_{\text{kin, m}} + E_{\text{pot, m}} = +7.8 \text{ J} - 19.6 \text{ J} = -11.8 \text{ J} < 0$. Note that, in Example 7.3, we found that the hanging object does work $W = 11.8$ J on the object on the table. Thus, the results of parts (a) and (b) indicate that a transfer of energy has occurred between the two objects, and this transfer occurred in the form of work.

Solution for part (c): Now we combine the calculations in parts (a) and (b) to find the total change in mechanical energy: $E_{\text{mech}} = E_{\text{mech, M}} + E_{\text{mech, m}} = E_{\text{kin, M}} + E_{\text{pot, M}} + E_{\text{kin, m}} + E_{\text{pot, m}} = +11.8 \text{ J} + 0 \text{ J} + 7.8 \text{ J} - 19.6 \text{ J} = 0 \text{ J}$. The mechanical energy of the combined system has not changed during the experiment; i.e., it was conserved.

Solution part (d): We find that the mechanical energy is not conserved for the object on the table or for the hanging object. This comes as no surprise as we have identified that energy is transferred from the hanging object to the object on the table. However, when we look at both objects together as a single system, the mechanical energy is conserved. Note that the combined system does not transfer energy to another object in its environment; in particular, it does not transfer energy to the table supporting the object of mass *M*. In turn, we note there is an interaction with Earth because we had to include the gravitational effect through the potential energy. If we define Earth, the object of mass *m*, the object of mass *M*, and the connecting massless string as our system, then we find the system to be isolated. Thus, we have to extend the mechanical energy conservation hypothesis to state: *In an isolated system, the sum of the mechanical forms of energy is conserved.* This is the hypothesis that is tested further.

7.5.2: Elastic Collisions and Energy Conservation

In section 7.4.1 and Example 7.9 the mechanical energy conservation hypothesis was formulated quantitatively. In the current section, we first use perfectly inelastic collisions, as originally discussed in Example 5.4, to falsify the hypothesis of mechanical energy conservation. We then introduce a new type of collision, called **elastic collisions,** to illustrate that there is a wide range of phenomena for which the conservation of mechanical energy is still a very useful concept.

7.5.2.1: Perfectly Inelastic Collisions Revisited

We showed in Chapter 5 that momentum is conserved for a system that undergoes a **perfectly inelastic collision.**

We assume that an object 1 with mass m_1 and speed v_{1i} collides with object 2 with mass m_2 that is initially at rest, as illustrated in Fig. 5.11. For this system we found in Chapter 5:

$$m_1 v_{1i} = (m_1 + m_2) v_f. \quad [7.14]$$

With no change in the potential energy during the collision, the conservation of mechanical energy would require $E_{kin,i} = E_{kin,f}$. This is our hypothesis. We can further write:

$$\text{(I)} \quad E_{kin,i} = \tfrac{1}{2} m_1 v_{1i}^2$$

$$\text{(II)} \quad E_{kin,f} = \tfrac{1}{2} (m_1 + m_2) v_f^2$$

We test the validity of the energy conservation hypothesis by rewriting Eq. [7.14] in the form:

$$v_f = \frac{m_1 v_{1i}}{(m_1 + m_2)}$$

and then substitute v_f from Formula (II):

$$E_{kin,f} = \tfrac{1}{2}(m_1 + m_2)\left(\frac{m_1}{m_1 + m_2} v_{1i}\right)^2 = \tfrac{1}{2}\frac{m_1^2}{m_1 + m_2} v_{1i}^2.$$

The final kinetic energy calculated this way is equal to that in Formula (I) only if:

$$\frac{m_1^2}{m_1 + m_2} = m_1 \Rightarrow \frac{m_1}{m_1 + m_2} = 1;$$

i.e., $m_2 = 0$, which is a physically meaningless case. Thus, we have dismissed the hypothesis and therefore conclude

KEY POINT

Perfectly inelastic collisions always violate the conservation of mechanical energy.

7.5.2.2: Elastic Collisions

To test whether the conservation of mechanical energy is at all applicable to collisions, we study a second type of collision in which the two colliding objects remain separated after the impact. An example is illustrated in Fig. 7.15. To treat this case without further assumptions is next to impossible. We therefore study this case from a different point of view, asking what predictions we can draw from Fig. 7.15 if both the conservation of momentum and the conservation of mechanical energy apply. We then compare the predictions with experimental results and draw conclusions on the usefulness of proceeding with the assumption that both conservation laws apply. We will find there are indeed collisions that can be described by these two assumptions. These collisions are called **elastic collisions** and are the subject of this section.

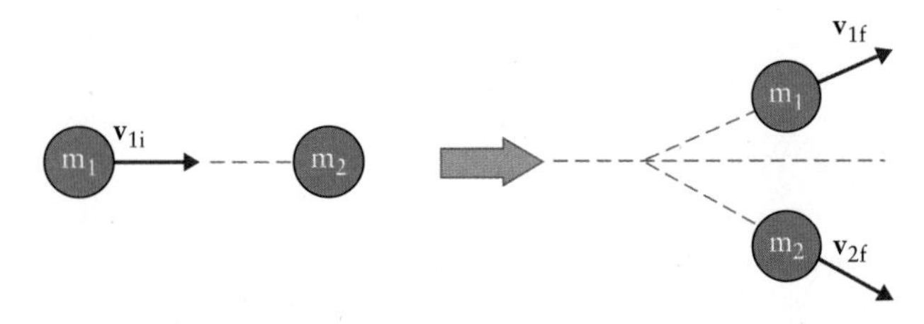

Figure 7.15 Collision of two objects of masses m_1 and m_2. Object 1 moves with velocity $\vec{v}_{1i}$ toward object 2, which is initially at rest. We define the *xy*-plane such that the two objects move in this plane after the collision at the right. If the collision is done horizontally, gravity-related effects can be omitted.

We chose a horizontal elastic collision so we can neglect the effects of gravity; in particular, we eliminate the need to include the potential energy as it does not change during horizontal motion. For a horizontal elastic collision we then quantify the two conservation assumptions in the form

Conservation of momentum (vector form):

$$m_1 \vec{v}_{1i} + m_2 \vec{v}_{2i} = m_1 \vec{v}_{1f} + m_2 \vec{v}_{2f}. \quad [7.15]$$

Conservation of mechanical energy (scalar equation):

$$\frac{m_1}{2} v_{1i}^2 + \frac{m_2}{2} v_{2i}^2 = \frac{m_1}{2} v_{1f}^2 + \frac{m_2}{2} v_{2f}^2. \quad [7.16]$$

We see right away that, together, Eqs. [7.15] and [7.16] would lead to extensive algebraic calculations because they consist of 4 formulas (3 Cartesian component formulas in the momentum conservation and 1 in the energy conservation), with a total of 12 velocity components and 4 corresponding speed terms. In applications, we can often reduce the number of parameters using additional conditions for the system. For example, in Fig. 7.15 we assume that object 2 is initially at rest. The figure allows us to further define the plane in which the two objects move after the collision as the *xy*-plane. Thus, starting with one object at rest significantly reduces the number of terms in Eqs. [7.15] and [7.16]:

Conservation of momentum in *x*-direction:

$$m_1 v_{1ix} = m_1 v_{1fx} + m_2 v_{2fx}; \quad [7.17a]$$

Conservation of momentum in *y*-direction:

$$m_1 v_{1iy} = m_1 v_{1fy} + m_2 v_{2fy}; \quad [7.17b]$$

Conservation of mechanical energy:

$$\frac{m_1}{2} v_{1i}^2 = \frac{m_1}{2} v_{1f}^2 + \frac{m_2}{2} v_{2f}^2. \quad [7.17c]$$

From the formulas in Eq. [7.17] we see that we now deal with six velocity components and three speed terms. Eq. [7.17] has been used successfully to explain a large number of molecular, atomic, and subatomic collisions, giving validity to the hypothesis that the conservation of energy is a useful concept in collisions. We use its predictions for example for molecular or atomic gas particles that are moving in a container in Example 7.10. Subatomic collisions are mostly of interest in astrophysics, nuclear physics, and high-energy physics; but the methods developed in these fields have found wider applications, for example in medical imaging and in nuclear medicine. For macroscopic processes, both perfectly inelastic and elastic collisions are idealizations. However, collisions of hard spheres, such as billiard balls, or a hard sphere with a rigid wall, such as a steel sphere on a glass plate, come close. Mathematically, perfectly inelastic and elastic collisions are the only two cases that can be solved without detailed further knowledge of the system.

EXAMPLE 7.10

In Chapter 8, we develop a mechanical model for a gas called the *kinetic gas theory*. In that model we are interested particularly in two aspects of elastic collisions: (a) the collision of two identical objects with $m_1 = m_2$, and (b) the collision of an object with a much heavier wall. To begin, we will discuss these cases for a one-dimensional elastic collision, because otherwise Eq. [7.17] leads to an extensive algebraic effort. Confining the motion to one dimension is achieved by assuming a head-on collision.

Solution

We first rewrite Eq. [7.17] for the one-dimensional case. The conservation of momentum simplifies to a single formula along the axis of motion:

$$m_1 v_{1i} = m_1 v_{1f} + m_2 v_{2f},$$

in which the final speed of the object of mass m_1 is v_{1f} and the final speed of the object of mass m_2 is v_{2f}. The conservation of energy remains unchanged from Eq. [7.17].

To combine the momentum and energy equations we first isolate v_{2f} in the momentum formula:

$$v_{2f} = \frac{m_1}{m_2}(v_{1i} - v_{1f}).$$

Then we substitute this formula in the energy formula:

$$m_1 v_{1i}^2 = m_1 v_{1f}^2 + m_2 \frac{m_1^2}{m_2^2}(v_{1i} - v_{1f})^2.$$

This is a quadratic equation. If you need a reminder how to solve such an equation, see the section on "Binomials and Quadratic Equations" in the Math Review section

continued

located after Chapter 27. Based on that approach, we sort the terms in the equation for the various v_{1f} factors:

$$v_{1f}^2\left[m_1\left(1+\frac{m_1}{m_2}\right)\right] - v_{1f}\left[2 v_{1i}\frac{m_1^2}{m_2}\right] + v_{1i}^2 m_1\left(\frac{m_1}{m_2} - 1\right) = 0. \quad [7.18]$$

This equation can be solved for v_{1f}. Instead of doing this, we use the conditions specified in the Example text to simplify yet again the algebraic effort.

Solution to part (a): A one-dimensional elastic collision of two objects of equal mass, $m_1 = m_2 = m$, leads to:

$$2\,m\,v_{1f}^2 - 2\,m\,v_{1f}\,v_{1i} = 0.$$

This equation has two solutions:

$$v_{1f} = 0$$

and:

$$v_{1f} = v_{1i}.$$

The second solution implies that the initially moving object passes through the object initially at rest, without interaction. This is physically not possible and therefore this solution is dismissed. The first solution implies that the moving object transfers its entire speed to the object at rest and comes to rest itself in turn. This solution is a suitable model for a head-on collision of equal gas particles.

Solution to part (b): A much heavier object hits a much lighter object at rest. An example is the collision between a fast car and a small, airborne pebble that bounces off the windshield without damaging it. In this case we use $m_1 >> m_2$. From Eq. [7.18] we find:

$$v_{1f}^2\frac{m_1^2}{m_2} - 2\,v_{1f}\,v_{1i}\frac{m_1^2}{m_2} + v_{1i}^2\frac{m_1^2}{m_2} = 0.$$

After being multiplied by m_2/m_1^2, this equation simplifies to:

$$(v_{1f} - v_{1i})^2 = 0.$$

This equation has only one solution:

$$v_{1f} = v_{1i}.$$

This result means that the heavier object pushes the lighter object out of its way without being affected.

Alternatively, a much lighter object hits a heavy object at rest. In this case, $m_1 << m_2$ and Eq. [7.18] simplifies to:

$$m_1 v_{1f}^2 - m_1 v_{1i}^2 = 0.$$

After division by m_1, this equation has two solutions:

$$v_{1f} = +v_{1i}$$

continued

and

$$v_{1f} = -v_{1i}.$$

In the case of the first solution, the incoming object passes through the heavy object without interaction, which is again physically impossible. In the case of the second solution, the incoming object is reflected with the same speed, moving after the collision in the direction opposite to where it came from. This case is suitable for the elastic collision of a gas particle with a container wall, as required in the kinetic gas theory.

SUMMARY

DEFINITIONS

- Isolated system: no exchange between system and environment
  - Closed system: no exchange of matter between system and environment, only exchange of work (and heat as discussed in Chapter 9).
  - Open system: exchange of matter and energy between system and environment
- Work W
  - force and displacement of object not collinear, force constant during displacement, dot product form:

$$W = \vec{F} \bullet \Delta\vec{r} = \vec{F} \bullet (\vec{r}_f - \vec{r}_i)$$

  - force and displacement of object not collinear, force constant during displacement, magnitude form:

$$W = |\vec{F}|\,|\vec{r}|\cos\theta$$

  - force and displacement of object not collinear, force constant during displacement, component form:

$$W = F_x\ \Delta_x + F_y\ \Delta_y + F_z\ \Delta z$$

  - force and displacement of object collinear, force constant during displacement:

$$W = \pm F\ \Delta r$$

  - sign convention for work: $W > 0$ when force and displacement occur in same direction, $W < 0$ when force and displacement occur in opposite directions
- Definition of energy E when only one form of energy changes in an experiment:

$$\Delta E = E_f - E_i = W$$

This is called the work–kinetic energy theorem for kinetic energy. It does not apply to thermal energy (see Chapter 9).

- Mechanical forms of energy:
  - kinetic energy: $E_{kin} = \frac{1}{2}\, m\, v^2$
  - potential energy: $E_{pot} = m\, g\, h$, in which h is the height of the object relative to a pre-set height $h = 0$

UNITS

- Work W, energy E: J = N m = kg m^2/s^2

LAWS

- Conservation of energy when applicable to a mechanical system:

$$E_{kin} + E_{pot} = \text{const}$$

MULTIPLE-CHOICE QUESTIONS

MC–7.1. Which is the standard unit for energy?
(a) N
(b) J/s
(c) N m
(d) Pa
(e) N/J

MC–7.2. Work is measured using the same unit as
(a) force.
(b) energy.
(c) torque.
(d) momentum.
(e) power.

MC–7.3. The change of work with time is
(a) energy.
(b) power.
(c) momentum.
(d) force.
(e) heat.

MC–7.4. A force of 5 N causes the displacement of an object by 3 m in the direction of the force. What work did the object do/has been done on the object?
(a) $W = +1.5$ J
(b) $W = -1.5$ J
(c) $W = +15.0$ J
(d) $W = -15.0$ J

MC–7.5. An object is pulled with a force $\vec{F}$ of magnitude 10 N upward by a distance of 3.0 m along an inclined plane that forms an angle of 30° with the horizontal. Which of the following choices comes closest to the result for the work done on the object by the force $\vec{F}$ if $\vec{F}$ is directed parallel to the surface of the inclined plane?
(a) $W = -15$ N
(b) $W = +15$ J
(c) $W = +26$ N
(d) $W = +26$ J
(e) $W = +30$ N
(f) $W = +30$ J

MC–7.6. A person pushes an object off a seat and then tries to lift it with an external force $\vec{F}$ upward into an overhead bin. However, the person is too weak and the object drops to the floor under its own weight $\vec{W}$, pulling the person's hands along a distance Δy. Which of the following statements is correct?
(a) Because the person did not succeed, no work is done in this experiment.
(b) The person has done work on the object; the absolute value of the work is $W = |\vec{F}|\ \Delta y$.
(c) The person has done work on the object; the absolute value of the work is $W = |\vec{W}|\ \Delta y$.

(d) The object has done work on the person; the absolute value of the work is $W = |\vec{W}|\,\Delta y$.
(e) The object has done work on the person; the absolute value of the work is $W = |\vec{F}|\,\Delta y$.

MC–7.7. When a system of interest does work on something else in the environment, the sign of the work is
(a) positive.
(b) negative.
(c) zero.
(d) either positive, zero, or negative, depending on the details of the process.

MC–7.8. Fig. 4.63 shows an object that slides a distance $d = 2$ m down along an inclined frictionless surface. A person pushes the object into the incline with the force $\vec{F}$. The direction of this force is shown by the black arrow; its magnitude is $F = 10$ N. What is the absolute value of the work done by the person on the object? Use $\theta = 30°$.
(a) 20 J or more
(b) more than 13 J, but less than 20 J
(c) more than 7 J, but less than 13 J
(d) less than 7 J

MC–7.9. How does the kinetic energy of an object change when its speed is reduced to 50% of its initial value?
(a) The kinetic energy remains unchanged.
(b) The kinetic energy becomes 50% of the initial value.
(c) The kinetic energy becomes 25% of the initial value.
(d) The kinetic energy doubles.
(e) The kinetic energy increases fourfold.

MC–7.10. When the ATP molecule splits, forming an ADP molecule, the following physical change happens.
(a) The ATP molecule takes up energy.
(b) The ATP molecule exerts a force.
(c) The ATP molecule does work.
(d) The ATP molecule releases energy.
(e) The ATP molecule exerts a torque.

MC–7.11. How does the potential energy of an object change when its speed is doubled? *Hint:* No further information is given about the system.
(a) The potential energy is halved.
(b) The potential energy becomes 1/4 of the initial value.
(c) The potential energy doubles.
(d) The potential energy increases by a factor of four.
(e) The information given is insufficient to choose one of the above answers.

MC–7.12. How does the potential energy of an object change when its speed is doubled? We further know that the experiment allows for the conservation of mechanical energy.
(a) The potential energy is halved.
(b) The potential energy becomes 1/4 of the initial value.
(c) The potential energy doubles.
(d) The potential energy increases by a factor of four.
(e) The information given is insufficient to choose one of the above answers.

MC–7.13. Labelling the kinetic energy E_{kin} and the gravitational potential energy E_{pot}, the conservation of energy can be written as $E_{kin} + E_{pot} = \text{const}$ if
(a) $E_{kin} = \text{const}$.
(b) $E_{pot} = \text{const}$.
(c) none of the energy of the system is converted into kinetic energy.
(d) all energy forms other than E_{kin} and E_{pot} are unchanged.

MC–7.14. In a mechanical experiment with an isolated object, only the values of E_{kin} and E_{pot} can vary. If the object accelerates from 5 m/s to 10 m/s, its potential energy has
(a) not changed.
(b) decreased by a factor of 2.
(c) decreased by a factor of 4.
(d) decreased by a factor we cannot determine from the problem as stated.
(e) increased.

MC–7.15. We consider a whale of 18 m length and mass $m = 4000$ kg. It leaps straight out of the water such that half its body is above the surface. If the entire upward surge is achieved by its speed at the instant when breaking the water surface, how fast was it going at that moment? Choose the closest answer.
(a) 5 m/s
(b) 15 m/s
(c) 20 m/s
(d) 200 m/s

MC–7.16. In pole vault, several forms of energy play a role: the kinetic energy of the runner, the elastic potential energy of the pole (discussed in Chapter 14), gravitational potential energy, and the internal energy of the vaulter, which is associated with muscles, tendons, and ligaments. For this question, however, we simplify the discussion by neglecting all but the kinetic and gravitational potential energy. We want to estimate what the highest possible pole vault is if the centre of mass of the athlete is 1.1 m above the ground and the maximum speed of approach is 11 m/s. Using conservation of energy to estimate this height, choose the closest value.
(a) 6.2 m
(b) 7.3 m
(c) 11.0 m
(d) 14.6 m

MC–7.17. What is your potential energy E_{pot} right now? If more than one answer or none apply, choose (d).
(a) $E_{pot} < 0$ J
(b) $E_{pot} = 0$ J
(c) $E_{pot} > 0$ J

MC–7.18. Which of the following is *not* a restriction on the applicability of the conservation of energy?
(a) The system of interest must be identified.
(b) The system of interest must be isolated.
(c) The system of interest cannot exchange matter with its environment.
(d) The system of interest cannot include a living organism.

MC-7.19. An object of mass $m = 7$ kg is thrown straight upwards with an initial speed of 15 m/s. What maximum height h_{max} does the object reach above the point of release?
(a) $h_{max} < 2.5$ m
(b) $2.5 \text{ m} \leq h_{max} < 5.0$ m
(c) $5.0 \text{ m} \leq h_{max} < 10.0$ m
(d) $10.0 \text{ m} \leq h_{max} < 20.0$ m
(e) $20.0 \text{ m} \leq h_{max}$

MC-7.20. An object of mass $m = 3.5$ kg moves with speed $v_i = 10.0$ m/s. Its kinetic energy is then increased by 85%. What is the final speed of the object?
(a) $35.0 \text{ km/h} \leq v_f < 45.0$ km/h
(b) $45.0 \text{ km/h} \leq v_f < 55.0$ km/h
(c) $55.0 \text{ km/h} \leq v_f < 65.0$ km/h
(d) $65.0 \text{ km/h} \leq v_f < 75.0$ km/h
(e) $v_f \geq 75.0$ km/h

MC-7.21. Fig. 4.4 shows two objects of different masses, m_1 and m_2, that are connected with a massless string that runs over a frictionless and massless pulley. We study the system from the instant the objects are released from rest until the hanging object has fallen a distance of 30 cm (and the object on the table has not yet hit the pulley). Which object has increased its kinetic energy by a larger amount?
(a) the object on the table
(b) the hanging object
(c) their change in kinetic energy is equal
(d) we cannot tell with the given information

MC-7.22. We study again the same system as in MC-7.21 based on Fig. 4.4. Which object has decreased its potential energy by a larger amount?
(a) the object on the table
(b) the hanging object
(c) their change in kinetic energy is equal
(d) we cannot tell with the given information

MC-7.23. We study again the same system as in MC-7.21 based on Fig. 4.4. Which object has changed its total energy by a larger amount?
(a) the object on the table
(b) the hanging object
(c) neither object because the total energy is conserved
(d) we cannot tell with the given information

MC-7.24. A basketball player passes the ball to another player such that it bounces once before reaching the second player's hands. At what instant is the kinetic energy of the ball the greatest?
(a) it is always the same, kinetic energy is conserved
(b) it is greatest just after leaving the hands of the first player
(c) just before touching the ground
(d) just before being caught by the second player

MC-7.25. Fig. 7.16 shows an object of mass $m = 5.0$ kg that is attached to a string of length 10 m. The object is initially held at an angle of $\theta = 45°$ below the horizontal. If we remove the table that provides the mechanical equilibrium in Fig. 7.16, with what speed will the object pass the point vertically below the hook holding the string at the ceiling?
(a) $0 \text{ m/s} \leq v < 3$ m/s
(b) $3 \text{ m/s} \leq v < 6$ m/s
(c) $6 \text{ m/s} \leq v < 9$ m/s
(d) $9 \text{ m/s} \leq v < 12$ m/s
(e) $v \geq 12$ m/s

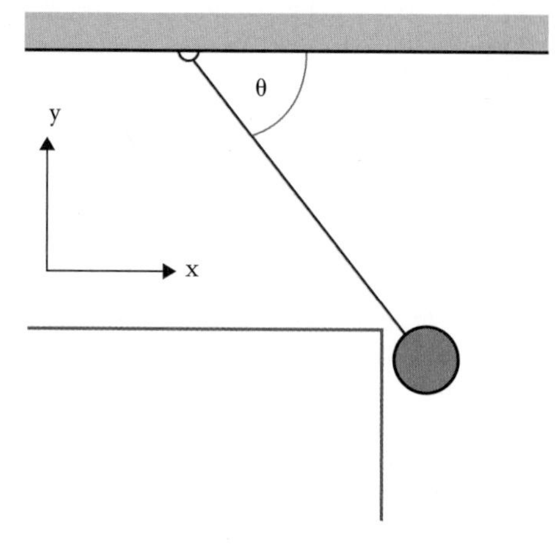

Figure 7.16

CONCEPTUAL QUESTIONS

Q-7.1. You want to push an object up a ramp onto a platform. Do you need to do less work if you lengthen the ramp, as this lowers the angle of the ramp with the horizontal?

Q-7.2. Can the kinetic energy of an object be negative?

Q-7.3. Can the (gravitational) potential energy be negative?

Q-7.4. If the speed of a particle is doubled, what happens to its kinetic energy?

Q-7.5. Attempt the experiment shown in Fig. 6.55. You will feel fatigue in your arm after a short while, indicating that you do work. What does work on what in this case?

Q-7.6. Discuss work done on or by the piston if we choose the piston in Fig. 7.8 as the system of interest.

ANALYTICAL PROBLEMS

P-7.1. A standard man uses a bicycle of mass 30 kg to travel at constant speed along a 5 km long road up a hill. The hilltop is 250 m above the point of departure and the standard man needs ½ hour for the trip. (a) How much work does the standard man do during the trip? (b) What is the power he invests? Assume the road has a constant slope of angle θ with the horizontal.

P-7.2. For the same trip as in P-7.1, a second standard man chooses to climb the hill along the shortest route, a trail of 1 km length. If this standard man uses the same power, will he reach the hilltop before, after, or together with the standard man in P-7.1?

P-7.3. A standard man climbs the stairs in a building. Assume that he reaches the fourth floor (16 m above the ground floor) in 15 seconds. How much work has the standard man done, and what was the power used for the climb?

P–7.4. (a) Estimate the work done and the power spent by a standard man performing 60 knee-bends in 10 seconds, during which the standard man lifts his body by 60 cm each time. This exercise is referred to as thigh flexion in strength training. Initially, the person stands upright with arms extended and then crouches until the thighs are horizontal. This is followed by extending the legs and straightening the torso back to the initial position. (b) Compare the result for the power with the non-standard unit of horse power. Does the result imply that a standard man's legs can outperform a horse? *Hint:* Neglect the downward motion, which in reality does contribute to the overall physiological work.

P–7.5. A standard man performs 12 chin-ups in one minute. In this exercise, the person raises himself while hanging by the hands until his chin is level with the support bar. Estimate the work and the power in this exercise, then compare with the result in P–7.4 to comment on the relative performance of our arms versus our legs.

P–7.6. A standard man excavates a hole of depth $d = 1.0$ m and area $A = 1.0\ \text{m}^2$ in soil of density $\rho = 2.0\ \text{g/cm}^3$. The work is completed in one hour. How much work has been done and what power was expanded during the excavation?

P–7.7. A weightlifter lifts a 40 kg set of weights from ground level to a position over the head. The vertical lift is 1.9 m. How much work does the weightlifter do? Assume that the weight is lifted at constant speed.

P–7.8. If a person lifts an 18.0 kg bucket from a well and does 5.50 kJ of work, how deep is the well? Assume the bucket is lifted at constant speed.

P–7.9. A shopper in a grocery store pushes a shopping cart with a force of 40 N directed at an angle of 30° below the horizontal. What is the work the shopper does on the cart for a horizontal distance of 10 m?

P–7.10. An object of mass $m = 2.80$ kg is pushed 1.90 m along a frictionless horizontal table by a constant force of magnitude $F = 18$ N, which is directed 30° below the horizontal. Determine the work done by (a) the force $\vec{F}$ and (b) the normal force exerted by the table.

P–7.11. (a) The highest headfirst dive is performed by professional divers near Acapulco, Mexico, from a height of 35 m (see Fig. 7.17). With what speed does a diver (standard man) enter the water if leaving the platform from rest? (b) Draw the speed, the kinetic energy, and the potential energy as a function of height for the diver. *Hint:* Neglect air resistance.

P–7.12. Two objects are connected by a massless string, as shown in Fig. 7.18. The pulley is massless and rotates without friction. The object of smaller mass $m = 1.2$ kg slides without friction on an inclined plane that makes an angle of $\theta = 35°$ with the horizontal. The mass of the larger object is given as $M = 2.5$ kg and it hangs on the string. If the two objects are released from rest with the string taut, what is their total kinetic energy when the object of mass M has fallen 30 cm?

Figure 7.17

P–7.13. Three objects with masses $m_1 = 5.0$ kg, $m_2 = 10.0$ kg, and $m_3 = 15.0$ kg are attached by massless strings over two frictionless pulleys, as shown in Fig. 7.19. The horizontal surface is frictionless and the system is released from rest. Using energy concepts, find the speed of m_3 after it has moved down 0.4 m.

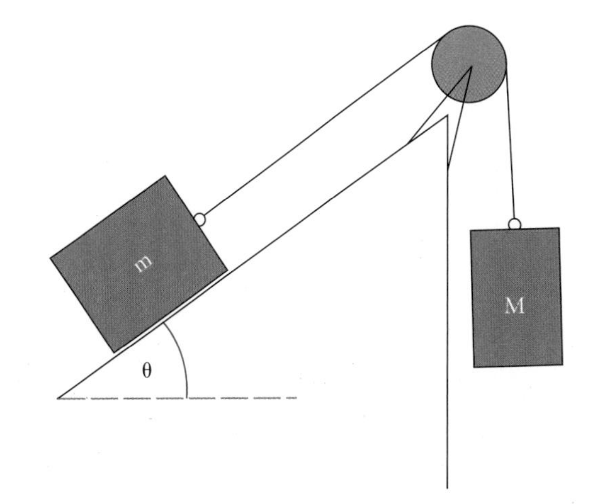

Figure 7.18

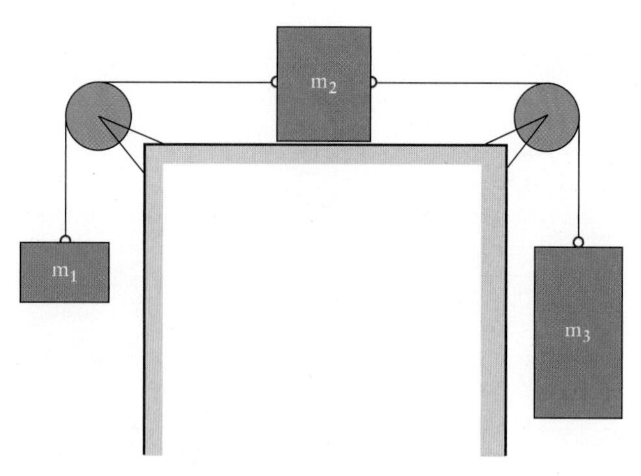

Figure 7.19

P–7.14. A pendulum consists of an object of mass $m = 1.5$ kg attached to a massless string of length $l = 3.0$ m. The object has a speed of 2.0 m/s when it passes through its lowest point. (a) If the potential energy is taken to be zero at the lowest point of the path of the object, what is the total mechanical energy of the system? (b) What is the speed of the object when the string is at 75° below the horizontal? (c) What is the greatest angle with the vertical that the string reaches during the motion of the object?

P–7.15. An outfielder in American baseball throws a baseball of mass 0.15 kg at a speed of 40 m/s at an initial angle of 30° with the horizontal. What is the kinetic energy of the baseball at its highest point of motion?

P–7.16. A standard man executes a pole vault. The approach is at 10 m/s and the speed moving above the bar is 1.0 m/s. Neglecting air resistance and energy absorbed by the pole, determine the maximum height of the bar.

P–7.17. A child and sled with a combined mass of 50 kg slide down a frictionless hill. If the sled starts from rest and has a speed of 3.0 m/s at the bottom, what is the height of the hill?

P–7.18. A bowling ball has a mass of $m_B = 7.0$ kg. It moves with a speed of 2.0 m/s. How fast must a Ping Pong ball of mass $m_P = 2.5$ g move so that both balls have the same kinetic energy?

P–7.19. An object of mass 0.5 kg has a speed of 2.5 m/s at position 1 and a kinetic energy of 10.0 J at position 2. Calculate (a) its kinetic energy at position 1, (b) its speed at position 2, and (c) the total work done on the object as it moves from position 1 to position 2.

P–7.20. A standard man does chin-ups in a gym. During the first 25 cm of the lift, each arm exerts an upward force of 350 N on the athlete's torso. If the upward movement starts from rest, what is the athlete's speed at this point? *Hint:* For the mass of the torso use the mass of the standard man minus the mass of the arms.

P–7.21. An object of mass $m = 1.8$ kg is attached to the ceiling by a 1.2 m long string. The height of the room is 2.8 m. What is the gravitational potential energy of the object relative to (a) the ceiling, (b) the floor, and (c) the equilibrium position of the object?

P–7.22. A child sits on a swing that is attached with ropes of length 2.5 m. The child has a mass of 30 kg and the seat 5 kg. Find the gravitational potential energy of the child relative to the lowest point of the swinging motion (a) when the ropes are horizontal, (b) when the ropes make an angle of 40° with the vertical, and (c) at the lowest point of the swinging motion.

P–7.23. We found $|\vec{F}_D| \propto v^2$ for the drag force in Chapter 3. Derive this velocity dependence from the power required to accelerate the displaced medium to the speed of the moving object, and then use Eq. [7.5] to determine the drag force.

P–7.24. A standard man is a base runner in baseball. The athlete begins sliding into second base while moving at a speed of 5.0 m/s. The slide is timed such that the athlete's speed is zero just as the base is reached. How much mechanical energy is lost due to friction acting on the athlete?

P–7.25. High-speed stroboscopic photographs show that the head of a golf club of mass 200 g is travelling at 55 m/s as it strikes a golf ball of mass of 46 g. At the collision, the golf ball is at rest on the tee. After the collision, the club head travels in the same direction at 40 m/s. Find the speed of the golf ball just after impact.

P–7.26. An object collides elastically head-on with an object twice its mass and initially at rest. Determine (a) the speed of each object after the collision and (b) the amount of kinetic energy transferred during the collision.

P–7.27. An object collides elastically with an identical object. Both objects moved before the collision with the same speed but in opposite directions along a common axis. Determine (a) the speed of each object after the collision and (b) the amount of kinetic energy transferred during the collision. (c) Compare your results with those for an elastic collision of an object with a stationary wall.

P–7.28. For an elastic head-on collision between two objects of arbitrary masses and initial speeds, (a) show that the following relation between the speeds of the objects holds:

$$v_{1i} + v_{1f} = v_{2i} + v_{2f}.$$

Hint: Start with Eqs. [7.15] and [7.16] for a head-on collision. Group all terms containing m_1 on the left side of each equation, and the terms containing m_2 on the right side. (b) Why is the formula stated in this problem often used instead of the combination of momentum and energy conservation formulas from which it is derived?

P–7.29. An object of mass $m_1 = 100$ g is attached to a string of length $L = 1.0$ m. The object is released from rest when the string forms an angle $\theta = 50°$ with the vertical, as shown in Fig. 7.20. When the object travels through the lowest point along its path, it hits, elastically, head-on an object of mass $m_2 = 50$ g. Calculate the final speed of object 2 if it was at rest before the collision. *Hint:* For a simplified calculation, use the equation discussed in problem P–7.28.

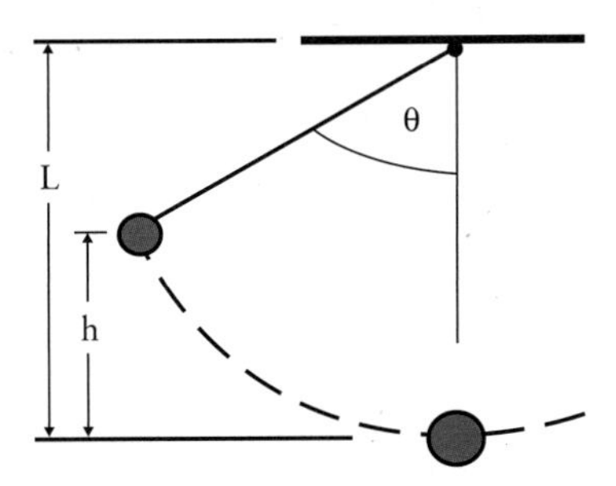

Figure 7.20

P–7.30. An object of mass $m_1 = 7.5$ g moves to the right at 25 cm/s. It makes an elastic head-on collision with a second object of mass $m_2 = 12.5$ g. The second object is at rest before the collision. Calculate (a) the speed of each object after the collision and (b) the fraction of the initial kinetic energy that is transferred to the second object.

P–7.31. An object of mass $m_1 = 7.5$ g moves to the right at 25 cm/s. It makes an elastic head-on collision with a second object of mass $m_2 = 12.5$ g. The second object moves initially to the left at 45 cm/s. Calculate the speed of each object after the collision.

P–7.32. A neutron generated in a nuclear reactor makes a head-on elastic collision with a carbon nucleus. (a) What fraction of the neutron's kinetic energy is transferred to the carbon atom? (b) If the neutron has an initial kinetic energy of 2.1×10^{-13} J, calculate the kinetic energy of both collision particles after the collision. *Hint:* Use $m_C = 12.0\, m_n$.

P–7.33. A bird is perched on a swing in a large cage in which it can fly. The bird has a weight of 0.6 N and the base of the swing has a weight of 1.6 N. Assume bird and swing are initially at rest. The bird then takes off in a horizontal direction at a speed of 2.5 m/s. If the base swings without friction about its pivot point, how high will the base of the swing rise above its original position?

ANSWERS TO CONCEPT QUESTIONS

Concept Question 7.1: (A)
Concept Question 7.2: (C)
Concept Question 7.3: None, because the displacement of the picture frame is zero.
Concept Question 7.4: Cases (ii) and (iii), but not case (i). For case (i) the angle is not defined between the force and displacement vectors. However, this is required to write Eq. [7.2] with the cosine term used. Case (iii) can be described by Eq. [7.2] since the case $\phi = 0°$ is included due to $\cos \phi = 1$.
Concept Question 7.5: None. The first three plots are linear, but we require a quadratic dependence to represent Eq. [7.7]. Plot (D) implies that the kinetic energy decreases with speed, which is inconsistent with Eq. [7.7]. Plot (E) is incorrect, as it shows a non-zero value for $v = 0$.
Concept Question 7.6: We can't tell. What we know is that the object is located at the origin of the y-axis as chosen by the person. However, that person may choose $y = 0$ on the surface of a table, or at any other height. Additional information is needed to render the given energy information useful.
Concept Question 7.7(a): The potential energy is given in Eq. [7.11], with $E_{pot} \propto y$. This corresponds to graph (E). Note that graph (D) corresponds to $E \propto -y$, which would be the correct choice if the y-axis were chosen to point downward.
Concept Question 7.7(b): The total mechanical energy is conserved for a falling object. We write this in the form $E_{tot} \neq f(y)$, which means the total energy is not a function of y. This corresponds to graph (C).

Richard Newstead/Getty Images

CHAPTER 8

Gases

Pursuing the theme of the previous chapter further, a more comprehensive study of energy leads us into the field of thermodynamics. As a first step toward thermal physics, a new model system has to be introduced because key parameters in thermodynamics cannot be studied with the tools developed in mechanics; in particular, pressure, volume, temperature, and the amount of matter in a system are of interest. We choose the ideal gas as our new model system because it allows us to investigate all the important relations between these system parameters, but is based at the same time on simple relations, with the ideal gas law the first milestone along that path. The ideal gas will later prove sufficient to introduce the two main laws of thermodynamics.

We discuss the parameters volume, pressure, and temperature first, based on applications in respiration. Then we establish how these parameters are related to each other, exploring the experimental observations of Boyle, Charles, and Avogadro. We find that the pressure of a given amount of an ideal gas is inversely proportional to its volume at constant temperature ($p \propto 1/V$), and that its volume is directly proportional to its temperature at constant pressure ($V \propto T$).

The ideal gas is further characterized microscopically with a mechanical model, called kinetic gas theory. In this model the gas is represented by a large number of particles (point-like objects) that collide elastically with the container walls and each other. The combination of the empirical ideal gas law and the microscopic model allows us to define and quantify the internal energy of the ideal gas. We find that the total energy of the ideal gas depends linearly on temperature, but interestingly, is independent of the volume and pressure of the gas.

Why do we bother discussing gases in a textbook about life sciences physics? There are two main reasons: first, there are specific gases that play a major role in human physiology and, second, gases are an ideal model system for all thermodynamic processes we need to discuss in this textbook.

Cellular metabolism requires oxygen, which is the chemically most active gas component in air. Thus, macroscopic amounts of air have to be brought into the organism via a process called respiration. In land-living animals, two internal organ systems cooperate to provide rapid access to sufficient levels of oxygen at the tissue level: oxygen is transported in gaseous form to the interface between the lungs and the blood system, and is then transported in dissolved form to the oxygen-consuming tissue. In turn, carbon dioxide, which we find as a component of exhaled air, is a by-product of cellular metabolism and is removed from the body in the reverse direction, first through the cardiovascular system, then through the respiratory system. What we learn in this and the following chapters about gases directly affects our understanding of the respiratory system and lays the foundation of many processes discussed in later chapters, even when we study non-gaseous systems, such as liquid solutions like blood.

To illustrate this argument more specifically for the supply of oxygen to organs in a human body, we give a brief overview of the many physical concepts contributing to oxygen supply in a four-step approach (all are discussed in detail in this textbook):

Step I: Oxygen reaches the lungs within 1 to 2 seconds as part of the inhaled air. The underlying physical processes include heating of a gas (to the core body temperature in the trachea), mixing of a gas (moisturizing to water vapour saturation in the trachea), turbulent fluid flow through the trachea and the bronchial tree, and gas interaction with a mechanical system (discussed as *compliance* in Physiology).

Step II: From the lungs, oxygen passes in less than 1 second to the red blood cells in the bloodstream. The underlying transport through the membrane is diffusion. Discussion of the driving force for the diffusive process requires use of the concept of partial pressure for gases and dilute solutions (Dalton's and Raoult's laws).

Step III: The cardiovascular system carries the oxygen in 30 seconds or less to the various organs. Blood flow is governed by a combination of several fluid flow laws, as it can be modelled alternatively as an ideal dynamic fluid or a Newtonian fluid, depending on the purpose of the discussion. Ultimately, blood is a mixed system and as such a non-Newtonian fluid, which may require more advanced concepts that include, for example, the transition between laminar and turbulent flow.

Step IV: In each organ, O_2 is distributed in the capillary bed to the cell tissue. It passes from the blood capillary to its final destination, again in about 1 second. The physical processes important at this stage are diffusion and osmosis.

This multitude of physical concepts necessitates a division of the discussion of gas processes into several chapters in the textbook. We have organized this material based on the major physical processes, following the typical order in introductory physics and physiology textbooks. Advanced texts in medicine, in which the authors assume familiarity with the basic processes, often divide the material by organ system or use case studies that pull together many processes in a common context, such as a particular disease.

8.1: The Basic Parameters of the Respiratory System at Rest

Before we discuss gas laws in this chapter, the parameters that appear in these laws are introduced individually. Each parameter is also discussed in a typical physiological context. This serves two purposes: to allow the reader to develop a sense of familiarity with the physical parameter in life science applications, and to point to interesting questions that will guide us through the ensuing material in thermodynamics.

As we now move beyond the realm of Newtonian mechanics, you will notice several changes in the way things are referred to. A major change is that we no longer use the terms object, or object of interest, and system interchangeably. Non-mechanical systems are always referred to as systems or systems of interest as the term object loses its meaning when, for example, thinking scientifically about the air you breathe.

8.1.1: Gas Parameter I: Volume

We are all familiar on a daily basis with the concept of volume. The volume of air outside your window is obviously of little practical use, but the volume of a well-defined amount of matter is of interest. When you buy milk, the price you are willing to pay depends on the volume of milk being sold to you. Also, when you dive below the surface of water, the volume of air you can hold in your lungs matters when deciding how long it is safe to stay below the surface.

The **spirometer** is the clinical instrument that allows us to measure gas volume in the lungs. This instrument is shown schematically in Fig. 8.1. The person breathes through a mouthpiece and a pipe into the instrument, which measures volume changes in the lungs in the form of the elevation of an inverted jar. This inverted jar floats between two beakers fixed to the bottom plate of the instrument. The two beakers hold water to seal the interior volume against outside air. The instrument is connected to a recording device (the cylinder drum at the left end of the instrument) that measures the total volume of the sealed air space, which is determined by the vertical motion of the inverted jar.

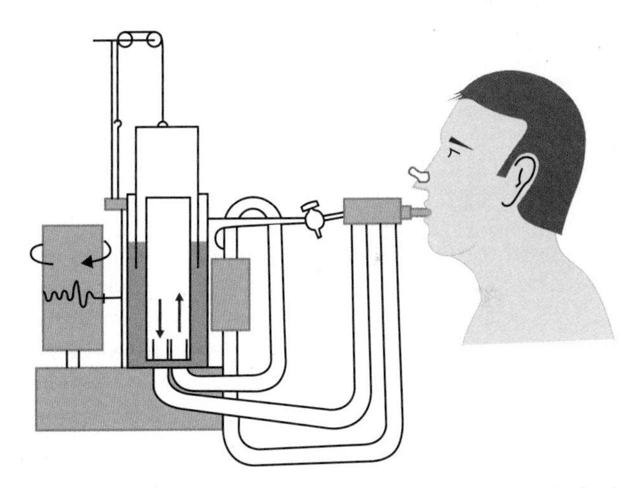

Figure 8.1 Sketch of a spirometer. The instrument consists of a fixed air volume, into which the patient breathes through a mouthpiece. The top part of the container is a freely moving inverted cylinder jar. The open end of the jar is immersed in water to seal off the air in the instrument. The changing amount of air in the sealed volume causes the jar to move up and down. This vertical motion is recorded on a plotter at the left.

The printout of a spirometer may look like Fig. 8.2. It allows us to define several components of the lung volume based on the typical breathing patterns of a healthy human. Fig. 8.2(a) shows the time variation of the volume during regular breathing. A person inhales and exhales about 0.5 L; this volume is called the **tidal volume** (2 in Fig. 8.2).

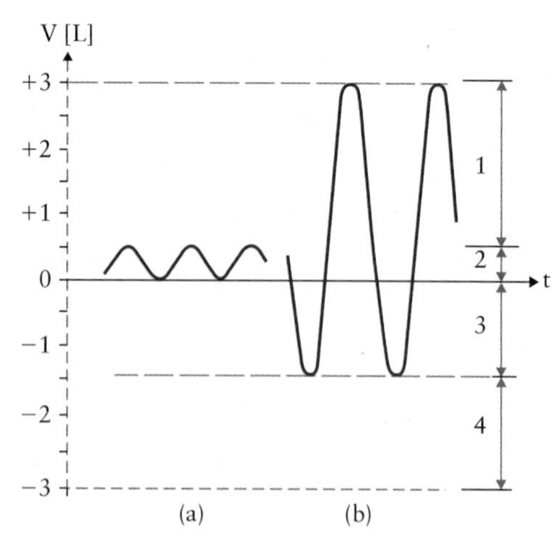

Figure 8.2 Typical data output of a spirometer. (a) A person breathing at rest, (b) a person breathing during heavy exercise. The data identify various contributions to the human lung volume: (1) the inspiratory reserve volume, (2) the tidal volume (regular breathing volume), and (3) the expiratory reserve volume. These three volumes combined are called the person's vital capacity. The instrument cannot measure the residual volume (4). Why are the two curves in (a) and (b) so different? We need this and the next chapter to answer this question! Note that the volume scale is relative to the volume after tidal expiration.

CONCEPT QUESTION 8.1

How does the non-standard unit litre (L) relate to the SI unit of length, metre (m)?

The spirometer measurement in Fig. 8.2 shows that a standard man breathing at rest (tidal breathing) inhales 0.5 L with every breath. Based on 15 inhalations per minute, this leads to an air exchange with the environment of 7.5 L per minute. During strenuous activity we breathe deeper, as shown in Fig. 8.2(b), with an **inspiratory reserve volume** (1) of 2.5 L and an **expiratory reserve volume** (3) of 1.5 L. The two reserve volumes allow for a short-term, additional air exchange when a larger oxygen intake is required. Even after a deep exhalation, the lungs are not empty. The remaining gas volume is called the **residual volume** (4) and is about 1.5 L. The residual volume cannot be measured with a spirometer, but must be determined indirectly by other methods. A typical approach is based on using an inert tracer gas, e.g., an air/helium mixture of known ratio. This method is illustrated in Example 8.1.

Combining the various volumes from Fig. 8.2 leads to a volume of 3.5 L after inhalation and 3.0 L after exhalation for the lungs of a standard man at rest (refer to Table 4.1 with standard man data).

EXAMPLE 8.1

A spirometer of volume $V_{spirometer}$ = 5.0 L is filled with air at atmospheric pressure and room temperature. A fraction of $f_{initial}$ = 10 vol% of that air has been replaced by helium gas [Fig. 8.3(a); the helium is indicated by red dots]. A standard man is connected to the spirometer via a mouthpiece. After a single inhalation and exhalation in Fig. 8.3(b), a valve in the mouthpiece is closed and the gas mixture in the spirometer is analyzed. The fraction of helium is found to be f_{final} = 6.25 vol%. Calculate the lung volume V_{lungs}.

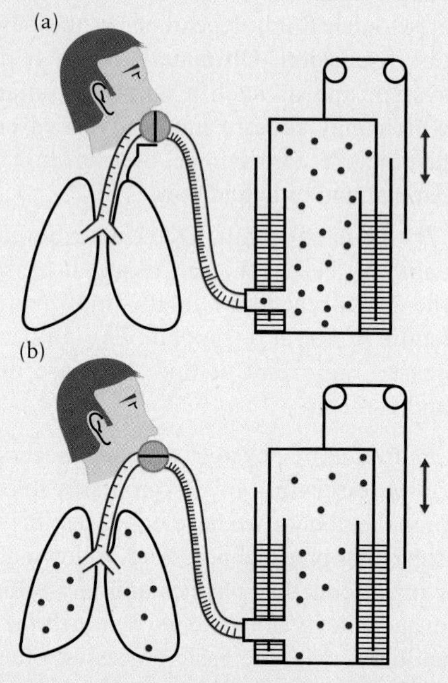

Figure 8.3 Experimental procedure to determine lung volume. (a) A known air/helium mixture fills a spirometer at atmospheric pressure. The helium component is illustrated by dots in the gas space. The test person is asked to inhale then exhale once, which brings us from frame (a) to frame (b). The breathing cycle leads to a uniform distribution of helium in the combined space of lungs and spirometer. The fraction of helium in the spirometer after the respiratory cycle is therefore a measure of the lung volume.

Additional information: When reporting fractions of matter, we use either volume percent (vol%) or weight percent (wt%) values. Gas and liquid mixtures are usually described in vol%. A given component with 5 vol% represents 5% of the total volume occupied by the system. Solid mixtures are often characterized by wt%, with 5 wt% referring to 5% of the total mass of the system.

Solution

A homogeneous mixture of the air in the spirometer and the gas in the lungs is established after breathing; i.e., the helium gas is diluted across the entire volume of lungs and spirometer. This is indicated by the uniform distribution of dots in the gas space in Fig. 8.3(b). To quantify the lung volume, we calculate the total volume of helium gas contained in the combined volume before and after the breathing cycle.

Before the standard man breathes, helium is exclusively in the spirometer:

$$V_{He} = f_i \, V_{spirometer} = 0.10\ (5.0\ \text{L}) = 0.5\ \text{L}.$$

continued

The spirometer volume occupied by helium gas is 0.5 L.

After completing the breathing cycle, the same amount of helium is distributed in the combined volume, with a uniform fraction f_f:

$$V_{He} = f_f(V_{spirometer} + V_{lungs}).$$

Since the total amount of helium has not changed during the experiment, the left-hand sides of the preceding two equations are equal. This leads to:

$$f_i V_{spirometer} = f_f(V_{spirometer} + V_{lungs}).$$

We isolate the volume of the lungs:

$$V_{lungs} = V_{spirometer}\left(\frac{f_i}{f_f} - 1\right).$$

Substituting the numerical values from the example text yields:

$$V_{lungs} = (5.0\ \text{L})\left(\frac{0.10}{0.0625} - 1\right) = 3.0\ \text{L}.$$

The volume in the lungs is therefore 3.0 L before inhalation.

Are calculations of parameters in the respiratory system indeed as simple as Example 8.1 makes us believe? Not quite. Before we accept any result, not just the values calculated in this example, the validity of the assumptions we made must be evaluated.

Which assumptions did we make? Intuitively you know we cannot simply neglect the pressure and temperature of the gas if it is initially different in the lungs and in the spirometer. At this point, though, we cannot proceed along this line of arguments; we have first to introduce pressure and temperature as parameters of a gas. Once we know how to include these two parameters, we will understand that, in Example 8.1, the pressure values initially in the lungs and in the spirometer are the same; however, the temperature is different (core body temperature in the lungs of 37°C and room temperature in the spirometer). Thus, a correction of the result has to take place; we will be able to make that correction in Example 8.4(a).

8.1.2: Gas Parameter II: Pressure

We start the discussion of pressure with the air in the room you are in. The measurement of the air pressure captures the effect of the air on any surface, including your skin. Were that pressure to decrease (as encountered by astronauts in outer space), a pressurized suit would be needed to protect you from the changed pressure; in turn, were the pressure to increase (as encountered during diving), other adverse effects can happen that you do not run into while sitting in your room.

The quantitative introduction of gas pressure requires us to specify more carefully the gas we are talking about. This is done in Fig. 8.4, which shows a gas in a container that is sealed off with a mobile piston. We assume the piston is an ideal piston, which means it can move without friction back and forth along the inside surface of the container. We arbitrarily choose the direction toward the right in Fig. 8.4 as our positive x-axis. When the piston moves in the positive x-direction the gas is compressed, and when the piston moves in the negative x-direction the gas expands. Thus, the piston provides for the means to interact with the gas. The source of the external force pushing the piston is part of the environment and is not shown in the figure.

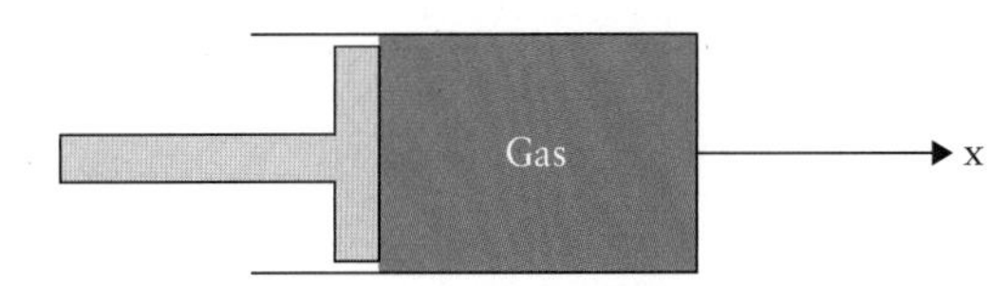

Figure 8.4 A piston seals a gas in an isolated container. The piston is mobile along the x-axis. The piston is assumed to move without friction.

From Fig. 8.4 we reason that the gas pressure depends on the force $\vec{F}$ exerted by the gas on the piston and the area A of the piston. You can do the following experiment yourself to specify the mathematical form of this dependence: use two syringes with different piston sizes. Keep the syringe closed. First push the piston of one syringe with variable force, then push the pistons of both syringes with the same force. Varying the two parameters leads to the sealed gas pushing back on the piston more strongly:

- when we increase the external force $\vec{F}_{ext}$ exerted on the piston, and
- when we decrease the piston area A at constant external force.

Thus, we suggest writing $p = F/A$. However, this is not correct! You can convince yourself if you contemplate pushing the piston in Fig. 8.4 at an angle. Only the component perpendicular to the piston surface, $F_\perp$, contributes to the pressure; the component parallel to the surface does not affect the piston. Thus, we write:

$$p = \frac{F_\perp}{A}. \quad [8.1]$$

Because we will use pressure often, we establish a derived unit, replacing the unit N/m^2 with the unit **pascal** (Pa), named for Blaise Pascal. Note that you frequently find pressures reported in kPa, particularly in Chemistry.

CASE STUDY 8.1

While Eq. [8.1] makes physical sense, you should wonder about its mathematical correctness. Can you suggest a way to write Eq. [8.1] mathematically correctly such that the force appears in the formula as a vector?

Answer: In Eq. [8.1], we calculate a scalar parameter (pressure) from a vector (force) and another scalar (area). The use of the normal component of force in Eq. [8.1] in particular confirms that the force $\vec{F}$ enters as a vector. To resolve this issue, let's study the area parameter more carefully. Think of the palm of your open hand as the surface you are interested in. Its area remains the same but its orientation changes as you rotate your hand. Thus, the area of a surface is a vector with magnitude and direction. We define the magnitude as A in unit m^2 and the direction with a vector of length 1 (called a unit vector) that is oriented perpendicular to the surface, $\vec{n}$, as illustrated in Fig. 8.5:

$$\vec{A} = A \bullet \vec{n}$$

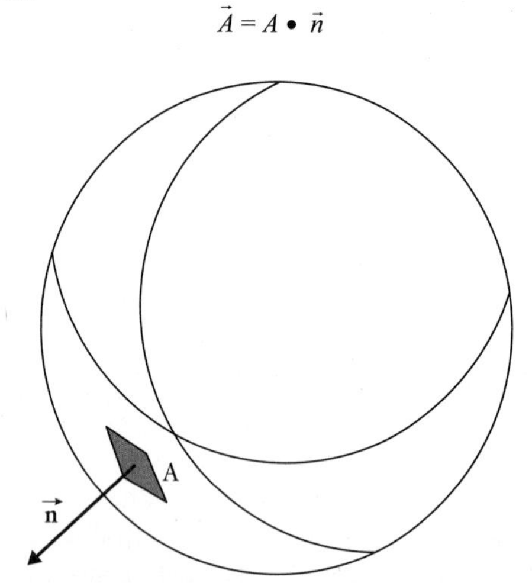

Figure 8.5 Representation of a surface segment in vector notation. The surface element is characterized by its area and by the direction of a vector of unit length that is oriented perpendicular to the surface. Note that a straight line that is normal to the surface of a sphere originates at the centre of the sphere.

Thus, the pressure is indeed the result of two vectors. Only one vector operation with two vectors results in a scalar, and this is the scalar product. Note that, in particular, division by a vector is not defined; i.e., we cannot write $p = \vec{F}/\vec{A}$! For this reasoning, refer to sections on "Vectors and Basic Vector Algebra," and "Vector Multiplication," which are Math Reviews located after Chapter 27. The proper way to write Eq. [8.1] in vector notation is therefore:

$$p = \frac{1}{A^2}\vec{F} \bullet \vec{A} = \frac{1}{A}\vec{F} \bullet \vec{n}. \qquad [8.2]$$

We often refer to atmospheric pressure in this textbook. This term is slightly ambiguous because the actual atmospheric pressure varies with varying weather patterns. Using the term **atmospheric pressure** in the context of an actual experiment, done at a particular time, refers to the pressure read off a barometer near the experimental set-up. In turn, when we refer to atmospheric in a general form, for example in a conceptual question in the textbook, then the term atmospheric pressure refers to an agreed upon fixed value, called the standard atmospheric pressure, with $p_{atm} = 101.3$ kPa.

CONCEPT QUESTION 8.2

Assume that no external force is applied to the piston in Fig. 8.4. Can you predict in which direction the piston accelerates?

EXAMPLE 8.2

In 1650, Otto von Guericke invented the vacuum pump. He demonstrated his invention at the Imperial Diet at Regensburg, Germany, in 1654. Historical sources report that he used two hollow bronze hemispheres 42 cm in diameter sealed together with a rubber gasket, as sketched in Fig. 8.6(a). Two teams of eight horses on each side could not pull the evacuated cavity apart. Do you believe the historical account?

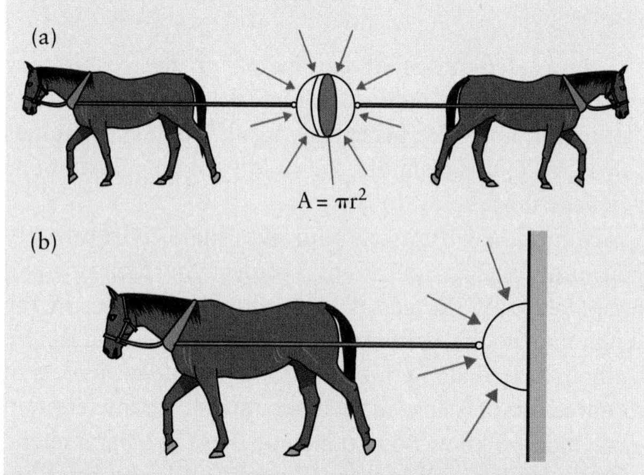

Figure 8.6 (a) Sketch of Otto von Guericke's experiment. Two teams of horses try to pull an evacuated cavity apart. The cross-sectional area of the cavity is $A = \pi r^2$ (green area) with r the radius. The arrows indicate the direction of the force exerted by the atmosphere on the surface of the cavity. (b) Equivalent approach with one-half the arrangement: one team of horses and half the cavity replaced by a smooth wall to which the cavity is sealed due to the vacuum.

Solution

No. Let's take the cavity as the system. Three forces act on it: (i) a tension caused by the team of horses on the right side, (ii) a tension caused by the team of horses on the left side, and (iii) a force due to the air pressure acting on the outer surface of the cavity. This force is not compensated by a force acting on the inner surface, since the cavity is evacuated. The arrows pointing toward the cavity in Fig. 8.6(a) illustrate the varying directions of the force due to the external air pressure. These arrows

continued

are all directed toward the centre of the sphere because this is the direction perpendicular to the cavity's surface, as required for Eq. [8.1].

First we simplify the problem using the symmetry of the experimental arrangement in Fig. 8.6(a). Of the two teams of horses, one is needed to keep the cavity in its place; were only one team present, the cavity would accelerate like a carriage. We can, however, replace one team, the attached rope, and one-half the cavity by a smooth wall, as shown in Fig. 8.6(b), because such a wall can provide for the same mechanical equilibrium. To quantify this simplified version of the problem, we have to identify the force the horses have to overcome to pull the half-cavity away from the wall.

The problem is not a simple application of the mechanics concepts we developed in earlier chapters since the force due to the external air pressure acts from variable directions across the extended half-cavity surface. How can we quantify the force against which the horses try to pull the half-cavity away from the wall? The proper approach is to find the horizontal component of the net force caused by the air pressure on the hemisphere. This is mathematically difficult. However, we can circumvent this problem with some straightforward reasoning: note that the wall in Fig. 8.6(b) forms the circular base of a hemisphere that is equivalent to the green cross-sectional area of the cavity in Fig. 8.6(a). All forces on the outer surface of the hemisphere must be equal to the force exerted by the wall because the cavity stays stationary after evacuation and before the horses pull. The net force acting on the flat circular area is much easier to calculate. If we label the radius of the sphere r, then $A = \pi r^2$ is the wall area covered by the cavity. Using for the atmospheric pressure $p_{atm} = 1.01 \times 10^5$ Pa, we find from Eq. [8.1] for the horizontal component of the net force due to the air pressure:

$$F_{air} = p_{atm}\,\pi r^2 = (1.01 \times 10^5 \text{Pa})\pi(0.21 \text{ m})^2$$
$$= 1.4 \times 10^4 \text{N}.$$

Eight horses can overcome that force. To see this, note that a force of 14 kN is equivalent to lifting an object with a mass of 14 kN/9.8 m/s^2 = 1400 kg, i.e., 175 kg per horse. von Guericke surely used fewer horses than the historical record claims.

8.1.3: Gas Parameter III: Temperature

In our daily lives we have a qualitative and, as we will see later, potentially misleading understanding what temperature means: we say something has a high temperature when it feels warm, and we say it has a low temperature when it feels cold.

In 1742, Anders Celsius was the first to propose a quantitative method to measure temperatures. He had observed that the height of a liquid in a hollow glass cylinder varies with temperature. Based on this observation he postulated that the expansion of common liquids, such as water, alcohol, and mercury (Hg; the only elemental metal that is liquid at room temperature), is linear with temperature. Choosing mercury, he then defined the temperature scale, which we call in his honour the Celsius-scale:

$$T = T_0 + \frac{1}{\alpha_{Hg}}\left(\frac{h - h_0}{h_0}\right)_{Hg}, \qquad [8.3]$$

in which T_0 is an arbitrary reference temperature, usually $T_0 = 0°C$ or the temperature at which the thermometer is filled with mercury. The term in the bracket on the right-hand side of Eq. [8.3] is the fraction of the height change of the mercury column during a temperature change; α_{Hg} is the **coefficient of linear expansion** of mercury, with a value of $\alpha_{Hg} = 1.82 \times 10^{-4}/°C$. This coefficient quantifies the proportionality between the length of the mercury column and the temperature. Both values T_0 and α_{Hg} are needed in order for us to use Eq. [8.3] to measure temperatures. These two parameters are determined by choosing two reference points. Celsius chose the melting point of ice, which he designated as 0 degrees, and the boiling point of water, which he designated as 100 degrees. In Celsius's honour, we use the unit °C for the values determined on the Celsius temperature scale.

CASE STUDY 8.2

Why did Celsius choose the melting and boiling temperature of water to define 0° and 100°?

Answer: *The two reference points Celsius chose are easily reproducible in an experiment, as illustrated in Fig. 8.7. Our system is a block of ice that is initially at –25°C. Heat is then supplied to the system continuously. As a result, the temperature increases steadily, except at the phase transition temperatures, when the heat supplied to the system is used to complete the phase transition of the material. This heat is called the* ***latent heat****. At 0°C we require the latent heat of melting and at 100°C we require the latent heat of evaporation.*

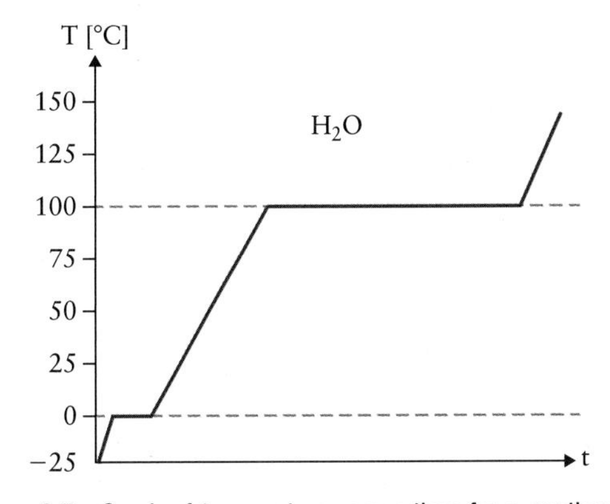

Figure 8.7 Graph of temperature versus time for a continuously heated beaker of water, starting with ice at –25°C. The temperature rises to 0°C, then remains unchanged while the ice melts. After the phase transition the temperature rises steadily to 100°C. The water begins to boil and the temperature remains constant until the entire amount of water has turned into vapour. After the second phase transition the vapour temperature rises steadily, shown up to about 140°C.

Note that the definition of temperature implies the existence of a thermal equilibrium. The **thermal equilibrium** is the subject of the **zeroth law of thermodynamics**.

KEY POINT

Zeroth law of thermodynamics: In a thermal equilibrium every part of the system has the same temperature. More specifically, if objects A and B are separately in thermal equilibrium with a third object (e.g., a thermometer), then objects A and B are in thermal equilibrium with each other.

Why is this statement a physical law, and why is it numbered in an unusual manner as the zeroth law? It is a law because it generalizes a few experimental observations and states that the observation applies universally. It is important because the possibility of measuring temperatures relies on it; were the law not correct, we would never be sure what the reading of a thermometer means for the investigated system (i.e., the parameter "temperature" would be rendered useless). It is numbered as the zeroth law because it was initially not recognized as a prerequisite for the first and second laws of thermodynamics (to be discuss later), which were named historically earlier.

In turn, if various parts of an object are at different temperatures we cannot define a meaningful single temperature for the object, and it is not in thermal equilibrium. If two systems in contact are not in thermal equilibrium, then we know from everyday experience that heat flows from the hotter to the colder system until their temperatures are equal. We discuss thermal non-equilibrium in Chapter 11.

Due to the complex nature of the liquid state of matter, the Celsius thermometer is scientifically unsatisfactory. However, the Celsius thermometer was sufficient to get studies on thermal physics started, and because of its simple design it continues to be used to measure moderate temperature changes, e.g., as a fever thermometer. We must return to the definition of temperature later for two reasons: to obtain a more precise definition to measure exact values, and to develop a better fundamental idea of what temperature physically tells us about the system.

8.2: Pressure-Volume Relations of the Air in the Lungs

In physiological studies of the respiratory system of endotherms, volume and pressure are the main variables, while the temperature remains usually constant. A more detailed look at the anatomy of the human chest indicates that we indeed need to measure two different pressure values. As illustrated in Fig. 8.8, each lung (1) is surrounded by a double-layered membrane called the **pleura** (2). The layer in contact with the lungs is called the visceral layer, and the outer layer is the parietal layer, which is attached to the inside of the rib cage (3). A small amount of fluid in the pleura allows frictionless movements of the lungs against the rib cage and prevents the harder ribs from puncturing the soft lung tissue. The pleura completely encloses the lungs, except for the entrance of blood vessels and the bronchus [primary bronchial branches (4) of the trachea (5)].

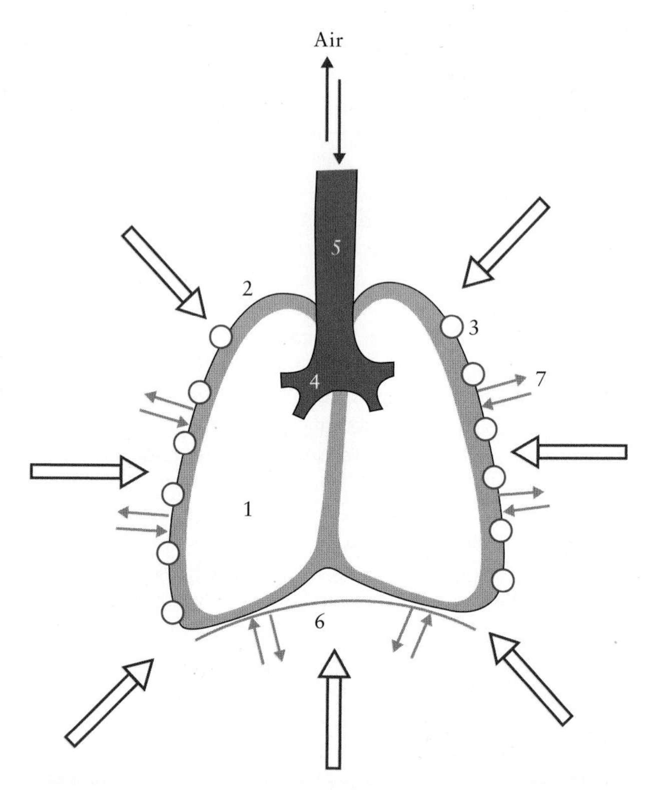

Figure 8.8 Functional sketch of the human thorax/lungs system shows the lungs (1); the double-layered membrane enclosing the lungs [pleura, (2)]; the rib cage (3); the trachea (5), which branches into the primary bronchi (4); the diaphragm and the abdominal muscle (6); and the intercostal muscles [(7), indicated through their action]. The large arrows indicate the effect of the external air pressure on the chest.

Breathing is achieved by several sets of muscles acting on the parietal layer: the diaphragm is a muscular layer that separates the chest cavity from the abdominal cavity. It is attached to the spine at the back, the sternum at the front, and the lower ribs along the side of the chest. The diaphragm pulls the lungs downward during inhalation and pushes them upward during exhalation. Muscles located between neighbouring ribs open the chest upward and sideways during inhalation and contract the rib cage during exhalation. These muscles are called intercostal muscles; their antagonistic action is discussed in Q–6.3.

Based on these physiological observations, we need to record the following two pressures during respiration: (i) the **pressure inside the lungs**, p_{lungs}, which is obtained from a pressure gauge in the spirometer tube on the mouth side of a valve, which allows the test person to be disconnected from the spirometer; and (ii) the **pressure in**

the pleura, which means specifically the pressure in the gap between the visceral and parietal layers of the pleura, p_{gap}. This pressure can be estimated with a non-intrusive pressure gauge lowered into the lower one-third of the esophagus, which is the muscular passage from the pharynx to the stomach.

With volume and pressure measured, and by applying the gas laws we will develop in the current chapter, the respiratory processes in the lungs can be understood. A brief overview of key observations will guide us through the development of the basic physics concepts. We saw in the previous chapter that work and energy are key parameters in characterizing physical and biological systems. When gases are involved, the work is determined from a p–V diagram of the process. Thus, representing respiration in a p–V diagram is the first step toward a quantitative discussion. Since the respiratory system is more complex than a simple piston-sealed container, respiration is not quite as straightforward as it sounds. We noted above that two relevant pressure values play a role; in addition, the experiment can be conducted in several different ways. This leads to a number of possible p–V diagrams for respiration. Of these, we present in the current chapter only the respiration curve at rest to motivate the discussion of its various features (Fig. 8.9). This p–V diagram contains three pressure curves:

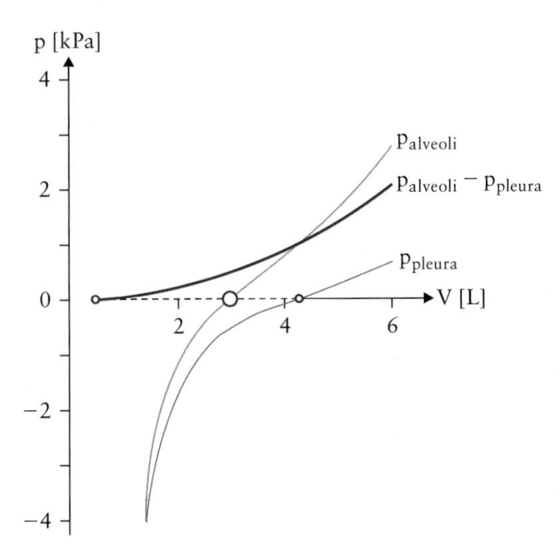

Figure 8.9 p–V diagram of the respiratory system of a person at rest, which means the breathing muscles are relaxed. The three curves show the alveolar pressure $p_{alveoli}$ (gauge pressure in the lungs; green curve), the pleural pressure p_{pleura} (gauge pressure in the pleural gap; blue curve), and the transmural pressure difference between lungs and thorax $p_{alveoli} - p_{pleura}$ (gauge pressure difference of lungs and pleural gap; red curve). To hold the lungs open, the transmural pressure difference is positive at all lung volumes. The pressure in the lungs exceeds atmospheric pressure for all volumes above 3.0 L (large open circle; lung volume after normal exhalation). The pressure in the pleural gap exceeds atmospheric pressure for lung volumes above 4.5 L. The horizontal dashed line indicates the consequences of a punctured lung: the lungs and the pleural gap collapse toward the small open circles; i.e., the lungs shrink to a volume of less than 1.0 L (at which the transmural pressure becomes zero) and the pleural gap widens toward the thorax (to a volume corresponding to a lung volume of 4.5 L).

- the curve labelled $p_{alveoli}$ represents the **gauge pressure inside the lungs,**
- the curve labelled p_{pleura} represents the **gauge pressure in the pleura**, and
- the curve labelled $p_{alveoli} - p_{pleura}$ represents the pressure difference between lungs and pleura, called the **transmural pressure** (*trans* means "across" and *murus* means "wall" in Latin).

Note that both pressures $p_{alveoli}$ and p_{pleura} are called **gauge pressures**.

KEY POINT

A gauge pressure is a pressure value relative to atmospheric pressure.

We can write relations between the various pressure terms we defined. Recall that we introduced as absolute pressure values above the pressure in the lungs, p_{lungs}, and the pressure in the gap between the visceral and the parietal layers of the pleura, p_{gap}. Thus, we can write:

$$p_{pleura} = p_{gap} - p_{atm}$$
$$p_{alveoli} = p_{lungs} - p_{atm}. \quad [8.4]$$

While absolute pressures are always positive, gauge pressures can be positive or negative. For example, the alveolar pressure is negative for lung volumes smaller than 3.0 L, as seen in Fig. 8.9. This means that the pressure inside the lungs is less than the atmospheric pressure for small lung volumes.

What does a p–V diagram like the one in Fig. 8.9 tell us? First we need to make sure we understand how the data in the figure were measured. To characterize the respiratory system at rest, the three curves in the figure were obtained as follows: the test person inhales a certain amount of air from the spirometer. Then the valve to the spirometer is closed and the test person relaxes his respiratory muscles. The pressure gauge on the mouthpiece records the alveolar pressure, $p_{alveoli}$, because the test person's epiglottis remains open.

The respiratory equilibrium is defined at a lung volume of 3.0 L because the alveolar gauge pressure is zero at that lung capacity, and therefore the pressure in the lungs is equal to atmospheric pressure (large open circle in Fig. 8.9). Nothing would happen if the test person removed the mouthpiece of the spirometer at this stage.

When the test person inhales a large amount of air (to a lung volume larger than 4.5 L), the figure shows that both the alveolar pressure and the pleural pressure become positive. This is possible for the pleural pressure because the lungs push the pleura outward against the ribcage. In turn, both the alveolar and the pleural pressures are negative when the test person exhales to a lung volume below 3.0 L. The transmural pressure difference between the lung and the pleura remains positive under all conditions because, otherwise, the lungs would collapse like a balloon from which the air has escaped. Note that we have drawn the curves in Fig. 8.9 between lung capacities of 1.5 L and 6.0 L: this range represents the maximum range of volume values accessible in breathing, as we saw in Fig. 8.2.

We fully appreciate p–V diagrams like Fig. 8.9 once we understand the relation between pressure and volume for gases in general. We build this understanding on a discussion of the basic laws that govern static gases. Once the gas laws are established, we can proceed in subsequent chapters to expand our discussion to dynamic breathing (as opposed to the respiratory system at rest), which will require us to also broaden our view of the ideal gas to its behaviour in basic thermodynamic processes.

8.3: The Empirical Gas Laws

We introduce the ideal gas as the model system of thermal physics, since it will turn out to be simple but still sufficient to describe all the important features. We will see that the versatility of this model stems from the fact that both a macroscopic and a microscopic approach to the ideal gas exist. The macroscopic model is empirical because it is based on experimental observations; it is discussed first. The microscopic model is an extension of the mechanical concepts we developed in mechanics and will be introduced later in the chapter.

We base the empirical gas model on two key experiments, the first one done by Robert Boyle in 1664 and the second one done by Jacques Alexandre Charles in 1787. Interpretation of the results of these two experiments then leads to a quantitative formulation of the ideal gas law. Studying Boyle's and Charles's experiments illustrates how the properties of gases were developed initially from an analogy to mechanical systems, and then extended beyond mechanics.

8.3.1: Boyle's Law

Boyle's experiment is illustrated in Fig. 8.10. A U-shaped glass tube has been filled with mercury in part (a) of the figure. The liquid metal reaches a mechanical equilibrium when its surfaces no longer move up or down. This mechanical equilibrium is indicated by a free body diagram for the mercury surface in the left column: the normal force due to the mercury beneath the surface pushes the surface up, and the weight of the air column above the metal pushes it down. In equilibrium, the mercury surfaces in both tubes are at the same height because the same air pressure, p_{atm}, acts on both sides.

In the next step, shown in Fig. 8.10(b), the left glass tube is sealed. Since this step as such does not change the air pressure in the sealed volume, no other changes

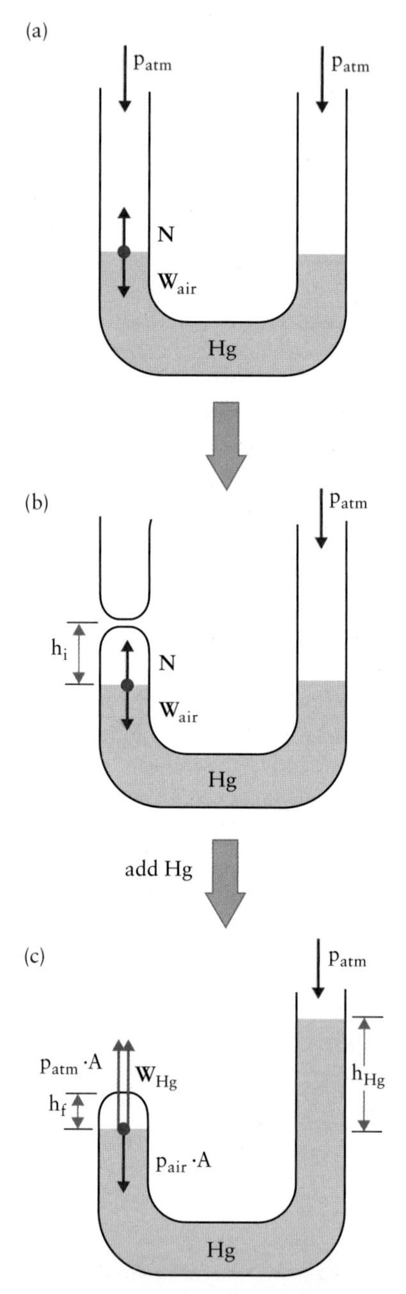

Figure 8.10 Boyle's experiment. (a) A U-shaped, hollow glass cylinder is filled with mercury and (b) the left glass column is sealed. (c) Adding mercury to the open column compresses the air in the sealed column. The free body diagram included at the left mercury surface in panel (b) indicates a mechanical equilibrium. This free body diagram changes in panel (c) due to the addition of mercury through the right column. The three forces that act on the mercury surface are balanced: the weight of the excess mercury and the air pressure above the open column push the left mercury surface upward, and the pressure of the sealed gas pushes the mercury surface downward. Vectors are labelled by their magnitude for clarity of the figure.

occur. Note that the sealed volume is now identified by the height of the air column above the mercury, $h_{i,air}$.

Boyle then added mercury through the open column, as illustrated in Fig. 8.10(c). The weight of the additional mercury pushes the mercury column down, which means it pushes the mercury surface upward in the left column. This upward push compresses the sealed gas—i.e., increases its pressure—until a new mechanical equilibrium is reached. Boyle measured the excess height of mercury between the two columns, h_{Hg}, and the height of the air in the sealed volume, $h_{f,air}$. When mercury was added repeatedly, he noticed that these measurements are related in the form:

$$h_{f,air} \propto \frac{1}{h_{Hg}}. \qquad [8.5]$$

To see what this equation implies, we replace both parameters with volume and pressure terms respectively:

- $h_{f,air}$ is related to the volume of air in the sealed space, V_{air}, by multiplication of the height by the cross-sectional area A of the glass tube:

$$V_{air} = h_{f,air}\, A \qquad [8.6]$$

- h_{Hg} is related to the pressure of the enclosed air in the following manner: We use Newton's first law for the mercury surface in the left column as the system. We identify three forces acting on this surface: (i) the weight of the excess mercury in the right column, $\vec{W}_{Hg}$, pushes the mercury below in the right column downward. Correspondingly, this force pushes the mercury in the horizontal part of the tube to the left, and pushes the mercury in the left column upward. Thus, $\vec{W}_{Hg}$ is in magnitude equal to a force pushing the left mercury surface toward the sealed air at the top of the left column. (ii) The open-air column above the right mercury surface pushes the right surface downward. It therefore contributes a second force pushing the left mercury surface upward, $p_{atm} A$. (iii) Acting in the opposite direction on the mercury surface is the gas pressure in the sealed-air volume. It exerts a force $p_{air} A$ downward. The condition of mechanical equilibrium for these three forces reads:

$$p_{atm} A + |\vec{W}_{Hg}| = p_{air} A.$$

In the weight term, the mass of mercury is replaced by the product of its volume and density:

$$|\vec{W}_{Hg}| = m_{Hg}\, g = \rho_{Hg} V_{Hg}\, g,$$

in which the mercury volume is $V_{Hg} = A\, h_{Hg}$. Substituting the last equation for the mercury weight into the previous one:

$$p_{atm} A + \rho_{Hg} A\, h_{Hg}\, g = p_{air} A.$$

Dividing both sides by the area A leads to:

$$\rho_{Hg} h_{Hg}\, g + p_{atm} = p_{air}.$$

Thus, the pressure in the sealed air volume, p_{air}, is proportional to the height of the excess mercury column, h_{Hg}. Note that $p_{air} \propto h_{Hg}$ is correct because p_{atm} in the last equation is constant.

Boyle's experimental result is therefore interpreted as:

$$V_{air} \propto \frac{1}{p_{air}}.$$

We generalize this result by omitting the subscripts:

$$V \propto \frac{1}{p} \Rightarrow p\, V = \text{const.} \qquad [8.7]$$

This relation is called **Boyle's law**. The dependence of the gas pressure on the gas volume in Boyle's law is shown in Fig. 8.11: the pressure and the volume are inversely proportional to each other in Eq. [8.7]. The product $p\, V$ has the standard unit Pa m^3. In the physico-chemical literature the equivalent unit of kPa L is often used alternatively. Boyle's law is applied in a second form when we compare an initial and a final state of the gas:

$$p_i V_i = p_f V_f. \qquad [8.8]$$

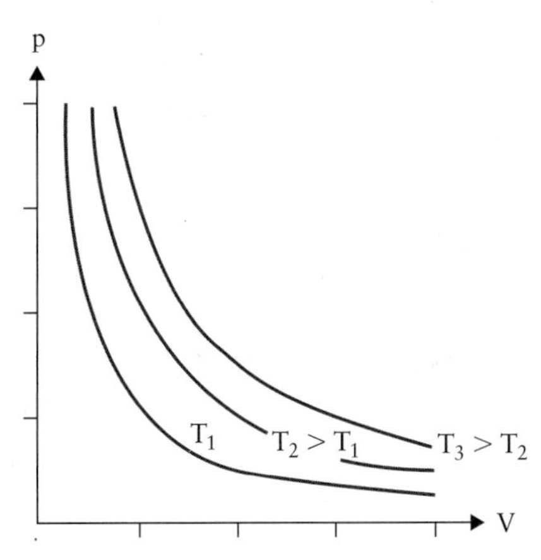

Figure 8.11 Boyle's data plotted as pressure versus volume for three different temperatures. Note that Boyle's law does not lead to a linear relation between pressure and volume but to an inverse relation of the form $p \propto 1/V$.

The constant term on the right-hand side of Eq. [8.7] is not a universal constant; it is only independent of the parameters p and V. In particular, a strong dependence on temperature is observed. Thus, Boyle's law applies only to *isothermal processes*, i.e., processes during which the temperature does not change. This is illustrated in Fig. 8.11 by displaying three separate curves for temperatures T_1, T_2, and T_3 where the temperatures increase from T_1 to T_3.

KEY POINT

Boyle's law: The product of pressure and volume of a gas at constant temperature is constant.

CONCEPT QUESTION 8.3

Which of the four graphs in Fig. 8.12 is consistent with Boyle's 1664 experiment that led to Boyle's law?

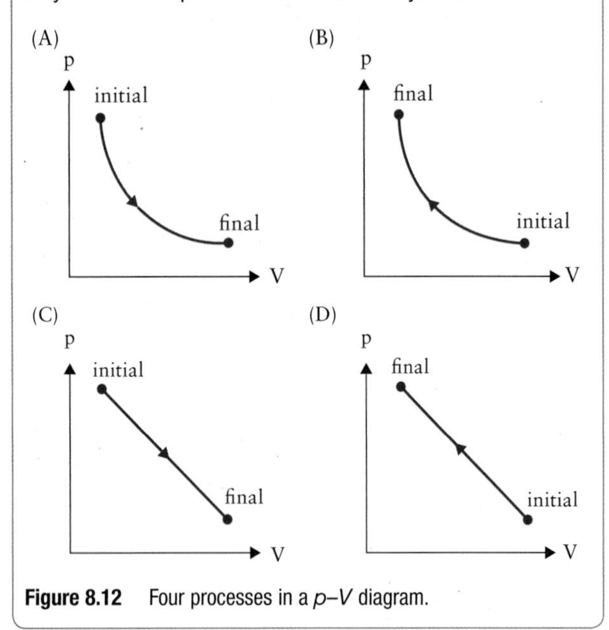

Figure 8.12 Four processes in a p–V diagram.

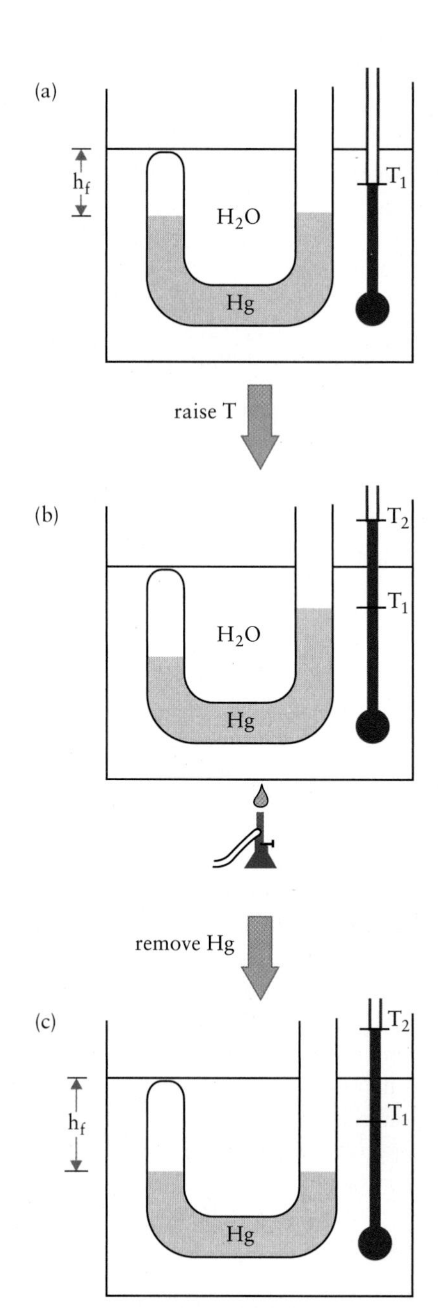

Figure 8.13 Charles's experiment. (a) A U-shaped, hollow glass cylinder is filled with mercury. (b) The left column is sealed and the arrangement is immersed in a bath at fixed temperature T_1. (c) Heating the bath to a higher temperature, T_2, causes the sealed air volume to increase. Mercury is then removed with a syringe to level the two mercury columns.

8.3.2: Charles's Law

We want to overcome the restriction to isothermal processes in Boyle's law to obtain a generally applicable law. Boyle could not do this, since Celsius invented the thermometer 78 years after Boyle's studies. Charles used the thermometer for studies of gases when he revisited Boyle's experiments in 1787. To investigate the temperature dependence of a gas, he modified the experiment, as shown in Fig. 8.13. After sealing an air volume of height h_i in the same way Boyle had done, the glass tube with the mercury was immersed in water at an initial temperature T_1 (e.g., room temperature), as shown in Fig. 8.13(a). Then the temperature of the arrangement was raised to a temperature T_2, with $T_2 > T_1$, and held at that temperature as indicated in Fig. 8.13(b). **Charles's experiment** showed that the volume of the sealed air increased, pushing some of the mercury to the column at the right side. In the last step, shown in Fig. 8.13(c), mercury was removed with a syringe until both mercury columns were levelled again. This meant that the pressure of the confined air column at the left returned to its initial value, since excess mercury no longer caused a compression. The final height of the air on the sealed side, h_f was recorded. Charles found:

$$h_f \propto T_{air},$$

in which T_{air} is the temperature of the water bath and everything immersed in it: $T_{air} = T_2$ (according to the zeroth law of thermodynamics).

To interpret his results, Charles used Eq. [8.6] to convert the height h_f in Fig. 8.13(c) to the air volume V_{air}. Generalizing again by omitting the subscripts, we formulate Charles's experimental results in the form:

$$V \propto T. \qquad [8.9]$$

You can easily convince yourself that a gas expands as a result of a temperature increase: thoroughly rinse with boiling water the outside of an open, empty plastic bottle. Then close the cap so that the bottle is airtight. Place the bottle in the refrigerator and let it cool down. It will crumple under the external air pressure as the cooling air inside requires less and less volume.

Fig. 8.14 shows a graphical representation of Charles's data. In his experiments, Charles was able to vary the temperature between 0°C and 300°C using an oil bath. Thus, we draw the results as solid lines between these two temperatures. The graph shows two experimental curves, one at atmospheric pressure p_1 = 1 atm and one at elevated pressure using a fixed excess height of mercury in the right glass tube of Fig. 8.14(c). When we extrapolate the two curves in Fig. 8.14 toward lower temperatures (shown as dashed lines), we find that the curves meet at a common point: $T = -273.15°C$ and $V = 0$. In 1848, William Thomson Baron Kelvin of Largs (Lord Kelvin) concluded that there is a physical meaning to this observation and that the temperature of −273.15°C is the lowest possible temperature we can achieve in any experiment. This allows us to eliminate negative temperature values by introducing the **Kelvin temperature scale**, which is calibrated with a new zero point at −273.15°C = 0 K:

$$T_{\text{Kelvin}} = T_{\text{Celsius}} + 273.15. \qquad [8.10]$$

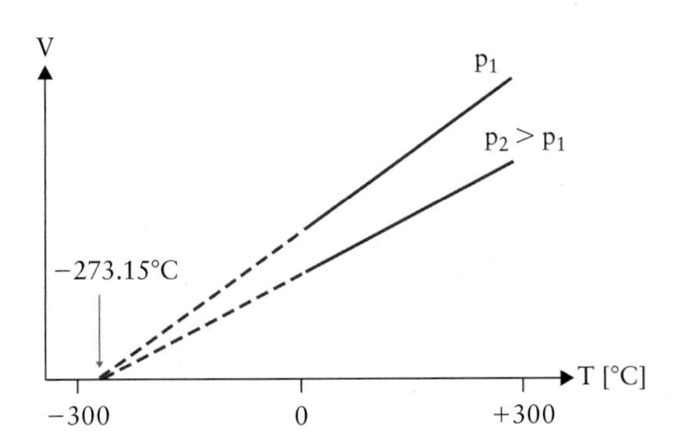

Figure 8.14 Charles's volume-versus-temperature data for two different pressures (combination of air pressure and excess mercury pressure). The solid parts of the lines indicate the temperature range experimentally accessible at Charles's time. The dashed parts of the lines are linear extrapolations of the experimental data.

The Kelvin scale is closely related to the Celsius scale because the difference between the melting and boiling temperatures of water is 100 degrees in both cases. The Kelvin scale, however, allows for simpler formulations of scientific laws than the Celsius scale. This is illustrated for **Charles's law**. If we quantify the experimental data of Fig. 8.14 for one of the given pressure values and with the temperature measured in degrees Celsius, we have to write:

$$V = V_0 + \text{const}\, T$$
$$\Rightarrow \frac{V}{T + 273.15} = \text{const},$$

in which the constant term V_0 requires an additional volume measurement at 0°C. For the temperature measured in kelvin, we find instead:

$$V = \text{const} \cdot T$$
$$\Rightarrow \frac{V}{T} = \text{const}. \qquad [8.11]$$

Comparing the last two equations, it is obvious that the formulation in Eq. [8.11] is preferable. As a consequence, all relations in this textbook that contain the temperature apply only if the temperature is written in unit kelvin. Charles's law, as stated in Eq. [8.11], is applied in a second form that compares an initial and a final state of a gas:

$$\frac{V_i}{T_i} = \frac{V_f}{T_f}.$$

From the existence of curves with different slopes in Fig. 8.14, we conclude that the constant factor in Eq. [8.11] is not a universal constant but varies with pressure; Charles's law applies only to experiments done at constant pressure. Such experiments are called *isobaric experiments*.

KEY POINT

Charles's law: The quotient of the volume and the temperature of a gas at a given pressure is constant.

8.3.3: Formulation of the Ideal Gas Law

In the next step, Charles's and Boyle's laws are combined to formulate the ideal gas law. We want to follow this step carefully for two reasons. First, we learn how two earlier but more restricted laws are combined to yield a law that is much more widely applicable—a prominent way that new insights emerge in the natural sciences. Second, the mathematical steps involved in the transition from Boyle's and Charles's laws to the ideal gas law illustrate how we can deal with **multi-variable functions**.

The derivation of the ideal gas law is illustrated in Figs. 8.15 and 8.16. We discuss these figures frame by frame. Fig. 8.15(a) shows the complete parameter space for a fixed amount of gas: from Charles's and Boyle's experiments we know that the three parameters pressure p, volume V, and temperature T have to be considered. If these three parameters were independent of each other, we would not need any further discussion: any value in a p–V–T diagram would describe an independent state of the gas. However, Boyle's and Charles's experiments show that p, V, and T cannot be chosen independently. Instead, we need a relation in the form:

$$V = f(p,T) \text{ or } p = f(V,T) \text{ or } T = f(p,V)\cdot$$

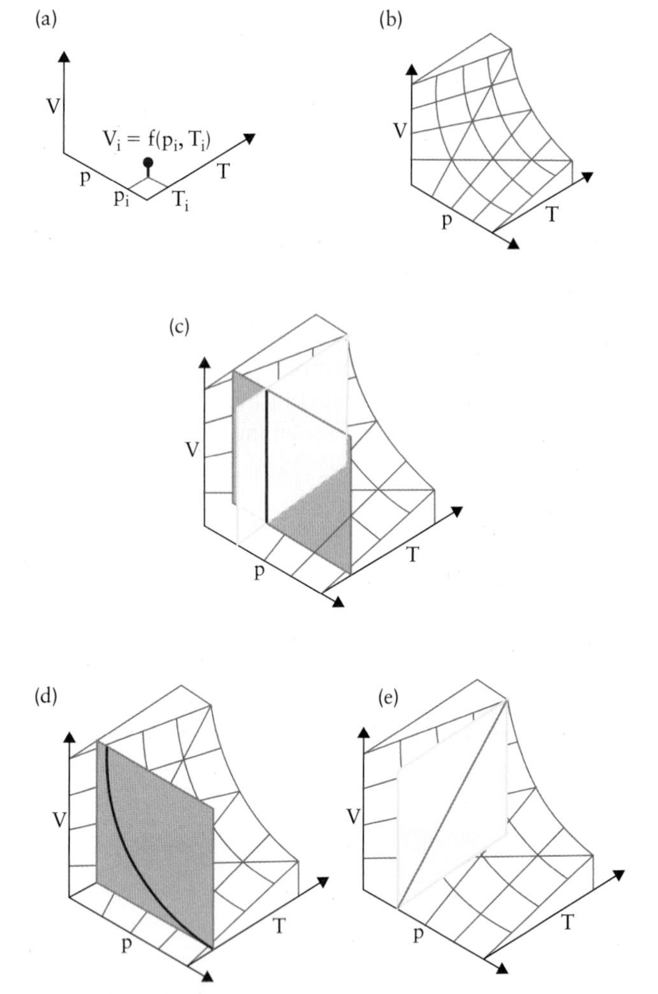

Figure 8.15 *p–V–T* diagram as a representation of the ideal gas. (a) The diagram is constructed by measuring the volume for a given amount of an ideal gas for each pair of independent parameters: p_i and T_i. The corresponding state of the system i has the volume V_i, which is entered into the graph. (b) The *p–V–T* diagram then combines the volume values for all parameter pairs (p, T), leading to a two-dimensional surface that represents all possible states of the ideal gas. (c) Vertical planes can be positioned in the diagram to represent all states for a given temperature (green plane) or a given pressure (yellow plane). These planes are perpendicular to each other. (d) The green plane for constant temperature intersects the plane of possible states of the system. The resulting curve shows all possible states of the gas at that fixed temperature (isothermal condition). (e) The yellow plane for constant pressure intersects the plane of possible states of the system. The resulting curve shows all possible states of the gas at that fixed pressure (isobaric condition).

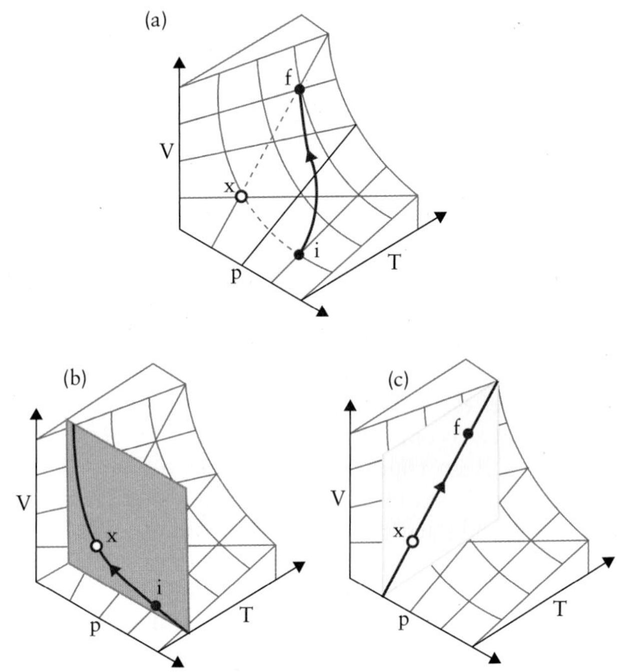

Figure 8.16 Formulating the ideal gas law with the use of a *p–V–T* diagram. (a) An arbitrary process (red line with arrow) guides the system from an initial state (labelled i) to a final state (labelled f). The dashed line shows an alternate possible path from the initial to the final state through an intermediate state (labelled *x*). The alternate path in part (a) consists of two steps we have discussed previously: (b) an isothermal process leads from the initial state to the intermediate state (Boyle's law), and (c) an isobaric process leads from the intermediate to the final state (Charles's law).

Which of the three forms, $V = f(p, T)$ or $p = f(V, T)$ or $T = f(p, V)$, is used depends on the specific purpose for which a given system is studied; in the remainder of this section, we use $V = f(p, T)$; i.e., we assume that the pressure and the temperature are independent variables and the volume is the dependent variable.

Mathematically, a function of two variables like $V = f(p, T)$ represents a two-dimensional surface in the three-dimensional *p–V–T* diagram. To illustrate this point, we consider Figs. 8.15(a) and (b). In Fig. 8.15(a) we find a particular volume, V_i, after choosing the pressure and temperature parameters p_i and T_i. This volume is found experimentally by measuring the volume for a fixed amount of gas once the desired pressure and temperature values are reached. Note that the subscript i indicates a particular state of the gas, which is characterized by the three parameter values V_i, p_i, and T_i. Later, we choose this state as the initial state for a process in which the state of the gas is changed.

The volume measurement in Fig. 8.15(a) can be repeated for other combinations of pressure and temperature. In each case a point in the *p–V–T* diagram is found that identifies the corresponding state of the gas. Instead of independent points, we combine them in Fig. 8.15(b) as a surface. The points on this surface represent all possible states of the gas; points that are not on the surface cannot be states of the gas. Finding this surface is important when we plan to discuss a process for the gas: such a process must start at a possible state on the surface and reach a final state that must again lie on it. The process must further move through a sequence of possible intermediate states, i.e., along a line that lies fully in the surface we found in Fig. 8.15(b).

We do not necessarily expect a simple mathematical formula to describe the surface in Fig. 8.15(b). However, we already know two quantitative statements that must hold: Boyle's law in Eq. [8.7] for constant temperature, and Charles's law in Eq. [8.11] for constant pressure. Requiring a constant temperature means we allow only states that lie on a vertical green plane as illustrated in Fig. 8.15(c); requiring a constant pressure leads to the yellow plane shown in the same figure. Note that these two planes are perpendicular to each other in the *p–V–T* diagram.

We study the green plane with $T = \text{const}$ in Fig. 8.15(d) first. The possible states of the gas in this plane are highlighted as a red line along the intersection with the surface we developed in Fig. 8.15(b). This curve follows the same relation as the curves shown in Fig. 8.12, each representing Boyle's law. Fig. 8.15(e) illustrates in the same fashion the yellow plane with $p = \text{const}$. The intersection of this plane with the surface of possible gas states leads to a straight red line, representing Charles's law in Fig. 8.14.

We now turn our attention to Fig. 8.16(a), in which we consider an *arbitrary process* of the gas between an initial state (labelled with subscript i in the figure) and a final state (with subscript f). We use this process to establish how the initial gas parameters V_i, p_i, T_i are related to V_f, p_f, T_f. We choose two perpendicular vertical planes such that the initial state lies in the constant-temperature plane and the final state lies in the constant-pressure plane. The intersection line of these two planes contains the intermediate state of the gas labelled x and highlighted with an open red circle.

Figs. 8.16(b) and (c) provide a new path from the initial to the final state of the gas through intermediate states that lie on either one of the two vertical planes. The first part of this process runs from the initial to the intermediate state in Fig. 8.16(b) as an isothermal process:

$$p_i, V_i, T_i \Rightarrow p_i, V_x, T_i.$$

The system reaches its final pressure while the temperature is still at the initial value. We apply Boyle's law to this process and isolate the intermediate volume V_x:

$$V_x = V_i \frac{p_i}{p_f}. \qquad [8.12]$$

The process from the intermediate to the final state, shown in Fig. 8.17(c), is then an isobaric process:

$$p_f, V_x, T_i \Rightarrow p_f, V_f, T_f.$$

In this equation we isolate the final volume using Charles's law:

$$V_f = V_x \frac{T_f}{T_i}. \qquad [8.13]$$

We eliminate the intermediate volume V_x by substituting Eq. [8.12] into Eq. [8.13]:

$$V_f = \left(V_i \frac{p_i}{p_f}\right)\frac{T_f}{T_i}.$$

Finally, the variables relating to the initial and final states are separated:

$$\frac{p_i V_i}{T_i} = \frac{p_f V_f}{T_f}. \qquad [8.14]$$

Note that Eq. [8.14] no longer contains any parameter of the intermediate state we chose in Fig. 8.16. The equation contains only initial and final parameters of the gas for the process of interest. Thus, Eq. [8.14] is valid independent of the path chosen between the initial and final states; i.e., the path in Fig. 8.16(a) must yield the same result as the path in frames (b) and (c) of the figure. Eq. [8.14] can be generalized in the form:

$$\frac{p V}{T} = \text{const.} \qquad [8.15]$$

Thus, Boyle's and Charles's laws represent special processes that are possible for an ideal gas, but Eq. [8.15] represents all possible processes for the ideal gas.

The constant in Eq. [8.15] depends only on the mass of the gas and is thus a more fundamental constant than the two constants that appear in Eqs. [8.7] and [8.11]. In 1811, Amedeo Avogadro noted that it is not the best approach to quantify the amount of gas as its mass (unit kg). If we do this, the constant in Eq. [8.15] depends on the chemical identity of the gas; i.e., it would become a materials constant. Instead, Avogadro expressed the amount of gas in unit mol, introducing the parameter n for the amount of gas, with $n = m/M$ in which m is the mass of the gas and M is its molar mass. For a given material, the value of the molar mass is obtained from the periodic table and is given in unit kg/mol; e.g., the molar mass of carbon is $M = 12.01$ g/mol $= 0.01201$ kg/mol, and the molar mass of methane (CH_4) is $M = 12.01 + 4 \times 1.008 = 16.04$ g/mol $= 0.01604$ kg/mol. This convention for the molar mass defines 1.0 mol as a macroscopic amount of matter. One mol of ideal gas contains 6.02×10^{23} atoms or molecules; this number is called the **Avogadro number**.

When the amount of gas in mol is separated from the constant in Eq. [8.15], we find const $= n\,R$, in which $R = 8.314$ J/(K mol), a fundamental constant called the **universal gas constant**. The universal gas constant is independent of any other parameter, including the chemical identity of the gas. Thus, the **ideal gas law** reads:

$$p V = n R T. \qquad [8.16]$$

KEY POINT

Ideal gas law: The product of the pressure and volume of a gas is proportional to the product of its amount, in unit mol, and temperature. The proportionality constant is the universal gas constant.

The observation that 1.0 mol always fills the same volume of $V = 22.414$ L for any gas at 0°C and at 1.0 atm pressure led to **Avogadro's hypothesis of the molecular nature of gases**. This was one of the earliest experimental hints of an atomic structure of matter; this aspect was, however, not fully explored until much later in the 19th century, as we will see later in the chapter.

CONCEPT QUESTION 8.4

Fig. 8.17 shows five processes by which a given amount of an ideal gas is brought from an initial to a final state. Which process occurs at constant temperature?

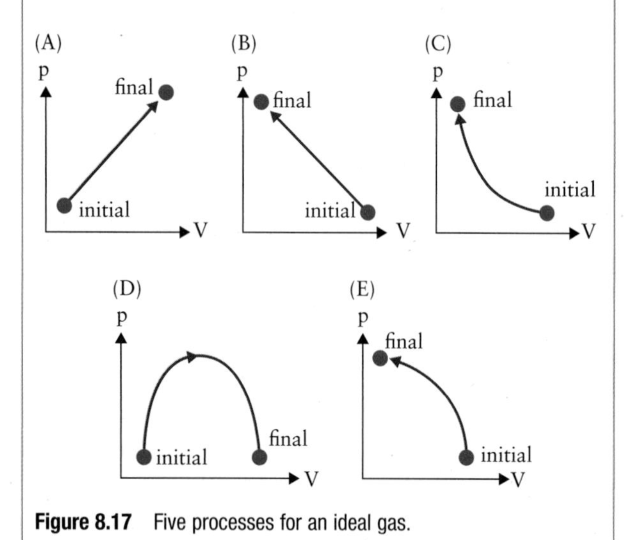

Figure 8.17 Five processes for an ideal gas.

CASE STUDY 8.3

The first manned free flight took place on November 21, 1783, in Paris, France. Two passengers travelled in a hot air balloon for 25 minutes to a maximum height of about 900 m. The balloon had a rigid envelope with a large opening at the bottom and did not carry a burner; the hot air was generated by a fire below the balloon before takeoff. Comparing the air in the balloon before the fire was started to the air in the balloon when the ropes were cut for takeoff, which of the parameters in the ideal gas equation were changed?

Answer: *Temperature and number of moles of air. The volume of the balloon remained unchanged, $V = V_0$, due to the rigid envelope. The pressure remained unchanged, $p = p_0$, due to the large opening in the envelope through which the air inside the envelope was heated. Thus, the ideal gas law reads:*

$$p_0 V_0 = n R T \Rightarrow n \propto \frac{1}{T};$$

continued

i.e., the molar amount of gas in a hot air balloon is reduced due to the increase in the gas temperature. Does this mean that doubling the gas temperature from 25°C to 50°C drives 50% of the gas out of the balloon envelope? No! This temperature increase is not a doubling of the temperature because it actually rises from 298 K to 323 K. The temperature increases in this case by only about 8%. Never use Celsius units when making quantitative physical statements about temperatures.

EXAMPLE 8.3

(a) Show that the ideal gas equation can be written as:

$$M = \frac{\rho}{p} R T, \qquad [8.17]$$

where ρ is the density and M is the molar mass. *Hint:* Use $\rho = m/V$ for the density definition and $n = m/M$ for the amount of gas in unit mol. (b) Consider an ideal gas at 20°C and 2.0 atm with a density of 2.5 kg/m^3. What is the molar mass of this gas? (c) Table 8.1 shows density measurements for CO_2 gas at 10°C as a function of pressure. Using Eq. [8.17] and graphical methods, determine the molar mass of CO_2.

TABLE 8.1

Density and pressure data pairs for carbon dioxide

Pressure (torr)	Density (g/cm^3)
515	0.0013
2065	0.0053
6185	0.0163

Solution

Solution to part (a): We start by substituting $n = m/M$ in the ideal gas law:

$$p V = \frac{m}{M} R T.$$

Next, we isolate the molar mass and group the parameters for the ratio m/V:

$$M = \frac{m}{V} \frac{R T}{p} = \rho \frac{R T}{p}.$$

This is the formula sought in the example text.

Solution to part (b): Using Eq. [8.17] with $T = 20°\text{C} = 293$ K and the data given in the example text, we find:

$$M = \frac{2.5 \frac{\text{kg}}{\text{m}^3}}{2.026 \times 10^5 \text{Pa}} \left(8.314 \frac{\text{J}}{\text{K mol}} \right) (293 \text{ K})$$

$$= 0.030 \frac{\text{kg}}{\text{mol}} = 30 \frac{\text{g}}{\text{mol}}.$$

Solution to part (c): The data given in Table 8.1 are plotted in Fig. 8.18 (red dots). Note that the pressure axis

continued

has been converted to the standard unit Pa. We analyze this plot to obtain the molar mass of carbon dioxide. First, we rewrite Eq. [8.17] in the form:

$$\rho = \frac{M}{R\,T}\,p, \qquad [8.18]$$

which shows that $\rho \propto p$ and explains why the data in Fig. 8.18 follow a straight line.

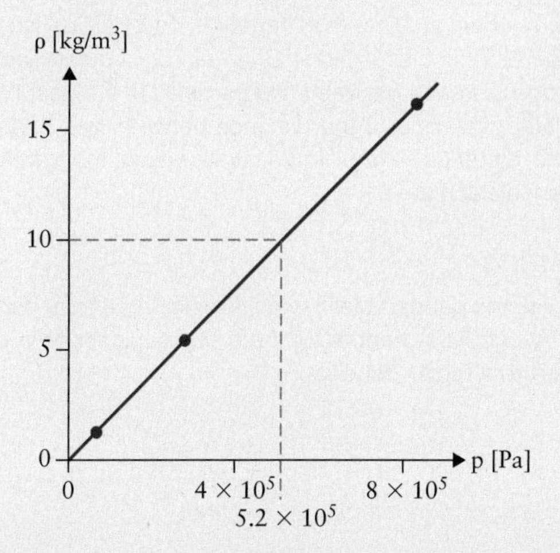

Figure 8.18 Plot of the density of carbon dioxide (CO_2) as a function of pressure.

We calculate the slope $\Delta\rho/\Delta p$ from Eq. [8.18]:

$$\frac{\Delta\rho}{\Delta p} = \frac{M}{R\,T}.$$

The slope contains the molar mass. Using the origin and the data point indicated by the dashed lines in Fig. 8.18, we find from the experimental data:

$$\frac{\Delta\rho}{\Delta p} = \frac{10.0\,\dfrac{\text{kg}}{\text{m}^3}}{5.2\times10^5\,\text{Pa}} = 1.92\times10^{-5}\,\frac{\text{s}^2}{\text{m}^2}.$$

A single data point is sufficient in this case because Eq. [8.18] requires that $\rho = 0$ at $p = 0$, which is consistent with the curve in Fig. 8.18 as it passes through the origin. Mathematically, we state that Eq. [8.18] is a linear function of the general type $y = a\,x + b$, with the constant b zero and constant $a = M/(R\,T)$.

Using the value from the previous equation and $T = 10°\text{C} = 283$ K, we calculate the molar mass:

$$M = R\,T\,\frac{\Delta\rho}{\Delta p}$$

$$= \left(8.314\,\frac{\text{J}}{\text{K}\cdot\text{mol}}\right)(283\ \text{K})\left(1.92\times10^{-5}\,\frac{\text{s}^2}{\text{m}^2}\right)$$

$$= 0.0452\,\frac{\text{kg}}{\text{mol}} = 45.2\,\frac{\text{g}}{\text{mol}}.$$

The molar mass is 45.2 g/mol.

EXAMPLE 8.4

Inhaled air reaches body temperature while travelling through the nasal cavities and the trachea, i.e., before it reaches the lungs. Is this required to prevent thermally expanding air from exerting excessive force in the lungs? Specifically, express the change in lung volume per inhalation if air at standard conditions ($T_{air} = 0°\text{C}$ and $p_{air} = 1$ atm) would reach the lungs. *Hint:* Assume the processes in the lungs occur under isobaric conditions.

Solution

We calculate first the temperature-related change in the tidal volume from the given physiological information:

$$V_{tidal} = \frac{7.5\,\dfrac{\text{L}}{\text{min}}}{15\,\dfrac{\text{inhalation}}{\text{min}}} = 0.5\,\frac{\text{L}}{\text{inhalation}}.$$

The calculation is done only for the tidal volume because the remaining lung volume of 3.0 L is filled with air that had been inhaled earlier and is therefore already at body temperature. Charles's law is used in this example because we assume the processes in the lungs are isobaric, i.e., done at constant pressure. Charles's law is written in the form:

$$\frac{V_{tidal}}{T_{outside}} = \frac{V_{in\ lungs}}{T_{in\ lungs}}.$$

We substitute the standard man data and the given external temperature in this equation:

$$V_{in\ lungs} = V_{tidal}\,\frac{T_{in\ lungs}}{T_{outside}} = (0.5\ \text{L})\,\frac{310\ \text{K}}{273\ \text{K}} = 0.57\ \text{L}.$$

Thus, 0.5 L dry air at 0°C becomes 0.57 L at core body temperature.

The calculated volume difference is now expressed as a fraction of the volume of the lungs after inhalation; i.e., from Fig. 8.2, $V_{inhaled} = 3.5$ L. The volume change $\Delta V = V_{in\ lungs} - V_{tidal} = 0.07$ L represents:

$$\frac{\Delta V}{V_{inhaled}} = \frac{0.07\ \text{L}}{3.5\ \text{L}} = 0.02,$$

i.e., a volume change of 2%. Accommodating a 2% volume expansion would not be a problem for the lungs; the reason for the warming of air before it enters the lungs must be a different one.

8.4: Mechanical Model of the Ideal Gas

The discussion of the ideal gas law in the previous section raises several new questions: Why is there a minimum temperature of −273.15°C below which matter cannot be cooled? Why is the gas defined in Eq. [8.16] called "ideal"? Which gases behave "ideally"? These questions are answered in this section as we obtain a deeper insight into the properties of the ideal gas by introducing a microscopic description. This approach was first proposed by Ludwig Boltzmann, James Clerk Maxwell, and Rudolf Clausius in the 1880s and is called the **kinetic gas theory**. Its most important application will be to allow us to quantify the energy contained in an ideal gas.

8.4.1: Four Postulates Define a Gas as a Mechanical System

The kinetic gas theory is based on a **mechanical model of the gas**. The gas itself is characterized by identical microscopic objects called **particles**. The term particle is a model for the atoms or molecules that make up a real gas. This term represents a simplification because we treat particles as *point-like objects* that can interact only in the most rudimentary manner with each other or the container walls. We exclude in particular vibrations and rotations, which are typical properties of molecules. Note that this is an additional restriction the empirically derived ideal gas law did not require:

KEY POINT

The empirical ideal gas law does not stipulate anything with respect to the microscopic structure of the gas. The kinetic gas theory stipulates that the gas consists of point-like objects.

As a result, the kinetic gas theory models primarily monatomic gases, such as noble gases. You therefore often find the term **monatomic** attached to the ideal gas law in the literature. If you want to extend the results of the kinetic gas theory to molecular gases, for example oxygen and nitrogen in air, corrections for their molecular nature have to be included. These corrections take into account vibrations and rotations of the molecules and are discussed in textbooks on physical chemistry. Unless otherwise stated, we omit these corrections in this textbook and limit the discussion to the monatomic ideal gas.

Four properties are postulated for these particles in a macroscopic container:

(i) *The individual volumes of the particles are negligible.* This is an acceptable assumption as long as the total volume of all particles in the container is small compared to the volume of the container itself.

(ii) *The gas consists of a very large number of identical particles.* This adds a further restriction on applicability: when we combine the first two postulates, we note that the actual size of the particles must also be much smaller than the inter-particle distance.

(iii) *The particles are in continuous random motion.* This means the motion of any particular particle is independent of the motion of all neighbouring particles (irregular motion).

(iv) *The only form of interaction between the particles or between particles and the container walls are elastic collisions.* Any other intermolecular interaction is neglected.

EXAMPLE 8.5

Test the validity of the first two postulates of the kinetic gas theory for air by estimating the ratio of the inter-particle distance to the particle radius.

For this test, first we estimate the size of the particles using liquid water of density $\rho = 1.0\ \mathrm{g/cm^3}$. Assume water molecules are spherical and fill the space in liquid water such that neighbouring particles touch each other. In the second step, we use the value $\rho = 1.2\ \mathrm{kg/m^3}$ for the density of air at 20°C (the dependence of air density on temperature is discussed in Chapter 15) to determine the average distance between two particles in the gas. In the last step, the ratio of the distance between gas particles to the distance in the liquid state (using the value for water) is calculated.

Solution

The volume per particle in water is calculated from density and Avogadro's number. In the first step, density is converted to a molar density:

$$\rho = \frac{m}{V} = \frac{n\,M}{V} \;\Rightarrow\; \frac{n}{V} = \frac{\rho}{M}.$$

With $M = 18$ g/mol for water, we find:

$$\left(\frac{n}{V}\right)_{H_2O} = \frac{1000\,\frac{\mathrm{kg}}{\mathrm{m^3}}}{0.018\,\frac{\mathrm{kg}}{\mathrm{mol}}} = 5.6\times10^{4}\,\frac{\mathrm{mol}}{\mathrm{m^3}}.$$

The inverse of this result, $(n/V)^{-1}$, is the volume per mol of water. With Avogadro's number, this is converted into volume per molecule:

$$\frac{1}{N_A}\left(\frac{n}{V}\right)^{-1}_{H_2O} = 3.0\times10^{-29}\,\mathrm{m^3}.$$

We estimate the diameter of the water molecule, d_{H_2O}, by drawing the third root from this value: $d_{H_2O} = 3.1 \times 10^{-10}$ m = 0.31 nm, in which the last value is given in unit nanometre (nm).

Repeating the same calculation for air yields the centre-to-centre distance between molecules in air. We find for the molar density:

$$\frac{n}{V} = \frac{1.2\,\frac{\mathrm{kg}}{\mathrm{m^3}}}{0.029\,\frac{\mathrm{kg}}{\mathrm{mol}}} = 40\,\frac{\mathrm{mol}}{\mathrm{m^3}},$$

in which we used $M = 29$ g/mol as an average molecular mass for air. This yields for the space occupied by a single molecule:

$$\frac{1}{N_A}\left(\frac{n}{V}\right)^{-1}_{air} = 4.2\times10^{-26}\,\mathrm{m^3},$$

which yields for the average distance between air molecules $d_{air} = 3.5 \times 10^{-9}$ m = 3.5 nm. Using the volume per molecule in liquid water as a typical particle size,

continued

we illustrate the implications of the above calculation in Fig. 8.19:

$$\frac{d_{air}}{d_{H_2O}} = \frac{3.5 \text{ nm}}{0.31 \text{ nm}} = 11.$$

The inter-particle distance in a gas is typically about 10 times larger than the size of the individual atoms or molecules.

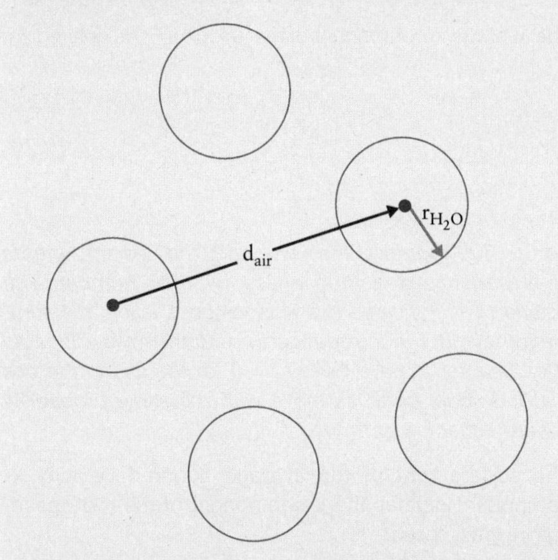

Figure 8.19 Five randomly chosen gas molecules in close proximity. The radius of the molecules is estimated from the intermolecular distance in liquid water and is labelled r_{H_2O}. The average distance between neighbouring air particles, d_{air}, is estimated from the density of air at 20°C.

This calculation resulted in a lower limit for the ratio because the diameter of an actual water molecule is smaller than the value we calculated earlier in the example.

8.4.2: Pressure Exerted by a Single Particle in a Box

We use the four postulates of the kinetic gas theory to develop a microscopic model of a gas. We begin with a single particle in a container, as shown in Fig. 8.20. The container is a cube of side length l and the particle initially has velocity $\vec{v}$.

The first parameter we calculate for this model is the pressure the particle causes in the container. We are interested in the pressure because it will allow us to relate this model to the ideal gas law. We determine the pressure from the force exerted by the particle on the container walls. To simplify the calculation, we focus initially on the yellow wall in Fig. 8.20. This wall is oriented perpendicular to the x-axis. We include therefore only the x-component of the net force because the definition of pressure, $p = F_{\perp}/A$, requires the force component

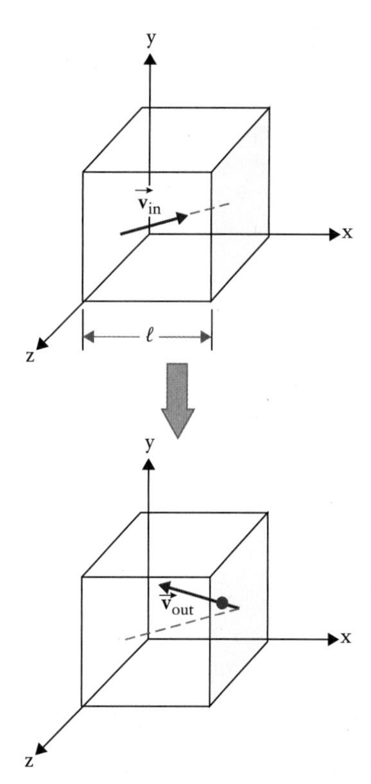

Figure 8.20 Model system for the kinetic gas theory. The system consists of a single particle that moves with velocity $\vec{v}$ in a cubic box of side length l. The interaction of the particle with the shaded wall is studied to relate the pressure and volume of the gas to its microscopic properties. The model assumes the particle collides with the wall elastically. From Example 7.11 we know that the velocity components of the particle perpendicular to the wall before and after an elastic collision are equal in magnitude but opposite in direction: $v_{x,in} = -v_{x,out}$.

perpendicular to the surface. We start with Newton's second law in the x-direction:

$$F_{net,x} = m\, a_x = m\frac{\Delta v_x}{\Delta t} = \frac{\Delta p_x}{\Delta t};$$

i.e., the force component on the yellow wall is equal to the change of the x-component of the momentum of the particle p_x with time. Note that we need to use the instantaneous velocity; i.e., we would have to consider the limit for $\Delta t \to 0$ if the velocity or momentum varied continuously with time. However, the momentum of the particle changes only at the instant it hits a wall. Thus, the change of the x-component of the momentum per second can be written as the product of two time-independent terms:

$$\frac{\Delta p_x}{\Delta t} = \Delta p_{x,\text{collision}} \frac{N_{\text{collision}}}{\Delta t}$$

$\Delta p_{x,\text{collision}}$ *is the change of the x-component of the momentum per collision.* For an elastic collision, we know from Example 7.10(b) that:

$$\Delta p_{x,\text{collision}} = 2\, m\, v_x.$$

The value is twice the momentum component of the incoming particle because it approaches with a positive velocity component, $+v_x$, and leaves the wall with a negative velocity component, $-v_x$.

$N_{collision}/\Delta t$ *is the number of collisions per second with the yellow wall.* We calculate this value from the x-component of the velocity of the particle, $v_x = 2\, l/\tau$, in which $2\, l$ is the distance the particle travels between two collisions with the yellow wall, and τ is the time of flight between two such collisions. Note that $1/\tau$ is the number of collisions per second:

$$\frac{1}{\tau} = \frac{N_{collision}}{\Delta t} = \frac{v_x}{2 \cdot l}.$$

We combine the last four equations to eliminate the momentum terms and collision frequency:

$$F_{net,x} = (2\, m\, v_x)\left(\frac{v_x}{2 \cdot l}\right) = \frac{m\, v_x^2}{l}.$$

To obtain the pressure from this equation, the force component perpendicular to the surface must be divided by the area on which the force acts. The surface is the yellow wall, which has an area $A = l^2$ in Fig. 8.21:

$$p = \frac{F_x}{l^2} = \frac{m\, v_x^2}{l^3} = \frac{m\, v_x^2}{V}, \qquad [8.19]$$

in which $l^3 = V$ is the volume of the cubic container in the figure.

8.4.3: Gas as a Very Large Number of Particles in a Box

To describe a gas we need not one but a very large number of particles in the box. As we add more and more particles, they start to hit each other. This is of no concern to our calculation because these collisions are elastic and therefore kinetic energy and linear momentum are conserved. Both are just redistributed among the particles, as discussed for their momentum in Example 7.10(a). This means that both the speed and the x-component of the velocity of individual particles vary. To avoid having to follow each of the particles individually, we use a particle with average properties, in particular a particle that travels at the root-mean-square speed.

Replacing v_x in Eq. [8.19] with $v_{x,rms}$ for N particles in the box yields:

$$p = N \frac{m \left\langle v_x^2 \right\rangle}{V}.$$

The pressure for N particles in a box differs from the pressure due to a single particle in two ways:

- the speed of an average particle is used, and
- the effect of a single particle is multiplied by the number of particles, N.

CASE STUDY 8.4

Identify the average speed of a particle in a gas that is suitable for the discussion in the kinetic gas theory.

Answer: *The average speed of a large number of particles can be determined in two ways, one based on momentum and one based on kinetic energy. We discuss both and then choose the more suitable one for the discussion of the kinetic gas theory.*

The average x-component of the velocity $\langle v_x \rangle$ is defined as:

$$\left\langle v_x \right\rangle = \frac{1}{N} \sum_{i=1}^{N} v_{x,i}. \qquad [8.20]$$

Eq. [8.20] requires that we add the x-components of N particles and then divide by their number. The notation $\langle \cdots \rangle$ indicates that an average is taken; the summation symbol is introduced in a Math Review located after Chapter 27. We find $\langle v_x \rangle = 0$ for the gas in the box that individual particles move with the same probability toward either left or right.

The square root of the average squared velocity x-component defines the x-component of the root-mean-square (rms) speed:

$$v_{x,rms} = \sqrt{\left\langle v_x^2 \right\rangle} = \sqrt{\frac{1}{N} \sum_{i=1}^{N} v_{x,i}^2}. \qquad [8.21]$$

In Eq. [8.21], we square the x-components of the velocities of all N particles, then add them together and divide by their number. The square root is taken to obtain a speed with unit m/s. The difference in this approach is that positive and negative values no longer offset each other, and we obtain $v_{x,rms} \neq 0$.

Two arguments lead to the choice of the root-mean-square speed instead of the average velocity for further discussion:

- The fact that the average velocity is zero does not yield a useful physical result, as its use would zero-out every subsequent formula.
- Eq. [8.19] contains the square of the velocity component v_x. Thus, it makes sense to use the average square of the x-component of the velocity, $\left\langle v_x^2 \right\rangle$, to replace the speed term in Eq. [8.19] for a box containing N particles.

Until now, we have focused on the yellow wall in Fig. 8.21. However, we expect pressure to be independent of any particular direction such as the x-direction because pressure is a scalar property. The only quantity in the last equation that is still direction-dependent is the speed term.

From our everyday experience, we know that no distinction exists between the x-, y-, and z-directions in a gas. Take the air in the room you are in as an example. You do not sense a higher air pressure from any particular direction on your skin. Thus the components of the root-mean-square speed must be equal in all directions: $v_{x,\mathrm{rms}} = v_{y,\mathrm{rms}} = v_{z,\mathrm{rms}}$, which yields:

$$\langle v_x^2 \rangle = \langle v_y^2 \rangle = \langle v_z^2 \rangle.$$

Applying the Pythagorean theorem in three dimensions yields:

$$\begin{aligned}\langle v^2 \rangle &= \langle v_x^2 \rangle + \langle v_y^2 \rangle + \langle v_z^2 \rangle \\ &= 3 \langle v_x^2 \rangle\end{aligned}$$

in which $\langle v^2 \rangle$ is the average of the squares of the speeds of the particles. Its square root defines the root-mean-square speed of the particles in the gas, v_{rms}. Substituting this result leads in the main quantitative prediction of the kinetic gas theory:

$$p\,V = \frac{1}{3} N\,m \langle v^2 \rangle. \qquad [8.22]$$

KEY POINT

Kinetic gas theory: The product of pressure and volume of a gas is proportional to two of its microscopic properties, the mass and the average of the squared speeds of the particles. It is also proportional to the number of particles in the container.

8.5: Energy Contained in the Ideal Gas

The result of the kinetic gas theory has one major consequence: it enables us to determine the energy of an ideal gas.

8.5.1: The Kinetic Energy of Gas Particles

We recognize the similarity of the term $m \langle v^2 \rangle$ on the right-hand side of Eq. [8.22] with the formula for the kinetic energy, which we introduced in Eq. [7.7]. Thus, $\frac{1}{2} m \langle v^2 \rangle$ is the kinetic energy of the average particle in the gas, and $N \frac{1}{2} m \langle v^2 \rangle$ is the kinetic energy of all N particles in the container. Eq. [8.22] relates the pressure and volume of an ideal gas to its kinetic energy:

$$p\,V = \frac{2}{3}\left(\frac{1}{2} N\,m \langle v^2 \rangle\right) = \frac{2}{3} E_{\mathrm{kin}}.$$

The product $p\,V$ for the gas is equivalent to two-thirds of the total kinetic energy stored in the gas.

8.5.2: The Internal Energy of the Ideal Gas

Eq. [8.22] is now compared with the empirical ideal gas law, $p\,V = n\,R\,T$. Because the product $p\,V$ is the same in both equations, we write:

$$\frac{2}{3} E_{\mathrm{kin}} = n\,R\,T. \qquad [8.23]$$

Recall that the kinetic gas theory defines the particle in the gas as a mechanical point-like object. This allows variations only in its kinetic and potential energies. The potential energy can be neglected because it does not play a role in the physical processes we observe. Otherwise, the gas would have to sink to the bottom of the container (sedimentation). This means the kinetic energy of the particles in the box is equal to their total energy, which is usually written as U: $E_{\mathrm{kin}} = U$. Replacing the kinetic energy with the internal energy of the gas in Eq. [8.23], we write:

$$U = \frac{3}{2} n\,R\,T. \qquad [8.24]$$

KEY POINT

The internal energy of an ideal gas of point-like particles depends only on the amount of the gas (in unit mol) and the temperature. It is independent of the pressure and the volume of the gas.

At this point we return briefly to the assumption of point-like particles we made at the start of the discussion of the kinetic gas theory. Recall that point-like particles don't allow for rotations or vibrations; i.e., only **translational motion** (which is motion along straight lines between particle–particle and particle–wall collisions) is allowed for the monatomic ideal gas. The motion of the particles has therefore three independent components along the basic Cartesian coordinates of space. We call these independent components of motion **degrees of freedom** of the particle. Thus, another way to read Eq. [8.24] is to state that one mol of an ideal gas has an internal energy of $\frac{1}{2} R\,T$ per degree of freedom.

This view points us toward the approach to correct Eq. [8.24] when modelling molecular gases: a molecule with N atoms has $3 \cdot N$ degrees of freedom, of which three are translational, three (or two for linear molecules) are rotational, and the remaining ones are vibrational. Thus, Eq. [8.24] maintains a physical purpose when we extend the discussion to a molecular ideal gas: it represents the total energy stored in the translational motion of the molecules. We will refer to this observation below when we discuss air as a molecular gas.

EXAMPLE 8.6

For the air in the lungs after a regular exhalation, calculate (a) the amount of gas in unit mol, and (b) its translational energy. *Hint:* For this calculation, we treat air as an ideal gas with molar mass $M = 29$ g/mol.

Solution

Solution to part (a): We know from the spirometer data in Fig. 8.2 that the gas volume of the lungs after exhalation is 3.0 L. Fig. 8.9 shows that the gas pressure in the lungs matches atmospheric pressure at $p_{atm} = 1.013 \times 10^5$ Pa after exhalation. The temperature of the gas in the lungs is at the human core body temperature of 310 K = 37°C. With these data, we use the ideal gas law to determine the amount of air in the lungs:

$$n = \frac{p\,V}{R\,T} = \frac{(1.013\times10^5\ \text{Pa})(3\times10^{-3}\ \text{m}^3)}{\left(8.314\,\frac{\text{J}}{\text{K}\cdot\text{mol}}\right)(310\ \text{K})} = 0.118\ \text{mol}.$$

Solution to part (b): Air consists primarily of nitrogen and oxygen molecules. Thus, Eq. [8.24] provides us with the total translational energy, which excludes vibrational and rotational contributions to the internal energy. From Eq. [8.24], we determine the translational energy of the gas in the lungs, assuming air is an ideal gas. With the data from the problem text and part (a), we find:

$$U = \frac{3}{2}\,n\,R\,T = \frac{3}{2}(0.118\ \text{mol})\left(8.314\,\frac{\text{J}}{\text{K}\cdot\text{mol}}\right)(310\ \text{K})$$
$$= 456\ \text{J}.$$

8.6: Implications of the Kinetic Gas Theory

The main ramifications of the result in Eq. [8.22] and the internal energy of a monatomic ideal gas in Eq. [8.24] are discussed in this section.

8.6.1: The Smallest Possible Temperature Is Zero Kelvin

Eq. [8.24] explains the existence of a minimum temperature, as postulated by Lord Kelvin, based on Fig. 8.14. At T = 0 K, the internal energy vanishes, $U = 0$ J. Thus, the entire kinetic energy is removed from the system at 0 K and v_{rms} = 0 m/s. With no motion left, the point-like particles would collapse and fill a zero volume, $V = 0\ \text{m}^3$. Thus, this prediction by Charles's law makes sense. In reality, however, atoms and molecules have a finite volume and the extrapolation of Charles's law cannot be made as indicated in Fig. 8.14. For that reason, the low-temperature ends of the lines in the figure are only dashed; we expect deviations from the ideal behaviour to be most notable at low temperatures.

8.6.2: The Root-Mean-Square Speed of Gas Particles

We introduced the **root-mean-square speed**, v_{rms}, for the particles in an ideal gas in Eq. [8.21]. We combine this definition with Eqs. [8.23] and [8.24] in the form:

$$v_{rms} = \sqrt{\langle v^2 \rangle} = \sqrt{\frac{3\,R\,T}{M}} = \sqrt{\frac{3\,k\,T}{m}}. \qquad [8.25]$$

The second-last term expresses the root-mean-square speed as a function of two macroscopic parameters, the gas constant and the molecular mass; the last term relates the root-mean-square speed to two microscopic parameters: the Boltzmann constant and the mass of the gas particles.

KEY POINT

The root-mean-square speed of a gas is proportional to the square root of the temperature and inversely proportional to the square root of the molar or molecular mass. It does not depend on the pressure or the volume of the gas.

EXAMPLE 8.7

(a) What is the speed of a typical nitrogen molecule at room temperature (T = 298 K)? (b) How much faster does a typical nitrogen molecule get when inhaled? *Hint:* Base your answers on the root-mean-square speed of the nitrogen component of air, which is treated as an ideal gas. Note that rotational and vibrational motions of the actual air molecules do not affect the root-mean-square speed of particles.

Solution

Solution to part (a): The molar mass of nitrogen is $M(N_2)$ = 28 g/mol. This quantity has to be converted into the standard unit kg/mol. We find:

$$v_{rms} = \sqrt{\frac{3\left(8.314\,\frac{\text{J}}{\text{K mol}}\right)(298\ \text{K})}{0.028\,\frac{\text{kg}}{\text{mol}}}} = 515\,\frac{\text{m}}{\text{s}}.$$

At first, this may appear to be a very high speed; 1850 km/h far exceeds the speed of most macroscopic objects we observe. It may also seem to be a high speed when you realize that this is the speed of billions and billions of particles hitting your skin right now. Can we put this result into perspective? The speed we found above can be related to the speed of sound because nitrogen is a major component of air and sound is carried by air. The speed of sound is about 330 m/s; thus, the speed of 515 m/s for the typical nitrogen molecule appears to be of the right order of magnitude. A detailed discussion of the speed of sound in gases is provided in Chapter 15.

continued

Solution to part (b): To solve this part, we could simply substitute the core body temperature into the previous equation. However, it is more useful to compare the two speeds:

$$\frac{v_{\text{rms}}(T_2 = 310\ \text{K})}{v_{\text{rms}}(T_1 = 298\ K)} = \frac{\sqrt{\dfrac{3\,R\,T_2}{M}}}{\sqrt{\dfrac{3\,R\,T_1}{M}}} = \sqrt{\frac{T_2}{T_1}}$$

$$= \sqrt{\frac{310\ \text{K}}{298\ \text{K}}} = \sqrt{1.04}$$

$$= 1.02.$$

The nitrogen molecules become 2% faster, leading to a root-mean-square speed of nitrogen in the lungs of 525 m/s.

8.7: Mixed Gases

8.7.1: Air Is a Gas Mixture

John Mayow recognized the **multi-component gas composition** of air in the 17th century, stating that not all the components of air are essential for living organisms. Air contains oxygen (first isolated by Joseph Priestley in 1774), which is its chemically most active component and which is essential for our metabolism. Antoine Lavoisier identified nitrogen in 1776 as the main component in air. He also described carbon dioxide as a by-product of respiration. Thus, air is an example for mixed gases and its role in respiration is the motivation to further pursue the phenomena that occur when gases are mixed.

The composition of air at the various stages of the breathing cycle of a healthy standard man is shown in Fig. 8.21. Fig. 8.21(a) shows the composition in vol%

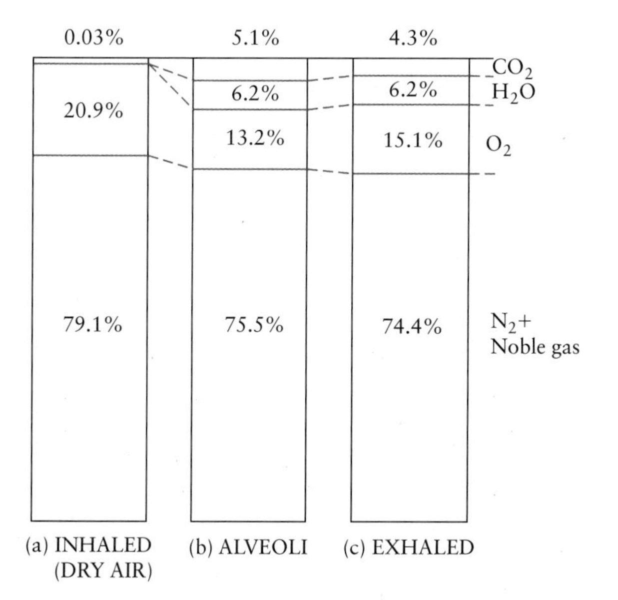

Figure 8.21 (a) The composition of air before inhaling, (b) the air in the alveoli, and (c) the composition of exhaled air. The increase of the water fraction is due to humidifying the air in the trachea to the saturation level.

of dry air, Fig. 8.21(b) identifies the gas composition in the lungs (alveolar space), and Fig. 8.21(c) shows the composition of exhaled air. Note that the unit vol% as a measure of the fraction of matter was introduced with Example 8.1.

EXAMPLE 8.8

(a) Determine Bohr's formula of the dead space in respiration. In this formula, the ratio of the dead space volume to the total volume is expressed as a function of the volume fraction of oxygen in the inhaled, exhaled, and alveolar gases. (b) Use the formula derived in part (a) to quantify the fraction of space in the human respiratory system that is dead space.

Additional information: The four volumes we defined in Fig. 8.2 are based on anatomical considerations. *Anatomy* deals with the structural makeup of an organism or its parts; *physiology* addresses the functions and activities of organisms with respect to the physical and chemical phenomena involved. This distinction therefore often leads to two ways of dividing a system: physiologists usually divide the air space into an active and a **dead space**. Air that remains in the dead space of the respiratory system after inhalation does not change, while air in the active space undergoes an oxygen–carbon dioxide exchange before leaving the body. Oxygen exchange takes place only in the alveoli at the end of the bronchial tree, deep in the lungs. Thus, we expect air from the dead space in the trachea and the mouth to leave first during exhalation, while the last part of the exhaled volume represents the gas from the active space.

Solution

Solution to part (a): In respiration studies we have to be careful when labelling parameters, as there are many in the calculations and they may well be similar to each other. In this example, we use non-abbreviated subscripts to prevent confusion, even though this is often not done in the literature because it requires more space.

We define three gas fractions for oxygen: f_{inhaled} = 0.209 is the fraction of oxygen in dry air as inhaled by the standard man [shown in Fig. 8.21(a)]; f_{alveoli} = 0.132 is the fraction of oxygen in the gas in the lungs [alveolar gas, Fig. 8.21(b)]; and f_{exhaled} = 0.151 is the fraction of oxygen in the exhaled gas [Fig. 8.21(c)]. We also define variables for the two physiological volumes: $V_{\text{dead space}}$ is the volume of the dead space and V_{total} is the total volume of the standard man's respiratory system filled with air, including the mouth and nose cavities, the trachea, and the lungs. The active space need not be labelled separately because it is equal to $V_{\text{total}} - V_{\text{dead space}}$.

We assume the standard man inhales and then exhales. The exhaled air can be seen as consisting of two fractions: one that is exhaled from the dead space and therefore with the same composition as the inhaled

continued

air, and one from the lungs, with the composition of the alveolar gas:

$$f_{exhaled} V_{total} = f_{inhaled} V_{dead\ space} + f_{alveoli}(V_{total} - V_{dead\ space}).$$

For Bohr's formula we need to isolate the two volumes and then determine their ratio. Separating the terms for the dead space and the total volume in the previous equation, we find:

$$(f_{exhaled} - f_{alveoli})V_{total} = (f_{inhaled} - f_{alveoli})V_{dead\ space}.$$

From this equation, we write the ratio of dead space to total volume:

$$\frac{V_{dead\ space}}{V_{total}} = \frac{f_{exhaled} - f_{alveoli}}{f_{inhaled} - f_{alveoli}}.$$

This is Bohr's formula for dead space in respiration.

Solution to part (b): This part requires only a straightforward substitution of the given values into the result of part (a). For a resting person, the total volume inhaled and exhaled is the tidal volume, $V_{total} = 0.5$ L. Thus:

$$V_{dead\ space} = V_{total}\frac{f_{exhaled} - f_{alveoli}}{f_{inhaled} - f_{alveoli}} = (0.5\ \text{L})\frac{0.151 - 0.132}{0.209 - 0.132} = 0.125\ \text{L};$$

i.e., about one-quarter of the inhaled air never moves beyond the physiological dead space.

Comparing the three panels in Fig. 8.21, we further note that the composition of air changes notably during respiration. The centre panel shows the composition of the air in the alveoli, which are the small hollow bubbles at the end of the bronchial tree where the gas exchange with the blood occurs. In the trachea, the air is humidified (saturated with water). In the lungs, the fraction of CO_2 has noticeably increased, mostly at the expense of the O_2 component. This is due to a diffusive exchange process, discussed in Chapter 11. The last panel then shows the composition of the exhaled air, which displays further changes due to the mixture of the gases in the active and dead spaces of the respiratory system during exhalation.

CASE STUDY 8.5

The values in Fig. 8.21 are given in vol%. Are the respective values written in wt% the same? If not, how is dry air composed in wt%? *Hint:* See the discussion of the terms vol% and wt% in Example 8.1.

Answer: *These values are not the same because the different components have different molecular mass. Volume percent (vol%) describes the fraction of space occupied by a component and is therefore given by the volume fraction occupied by particles (atoms or molecules) in the system. Weight percent (wt%) is the weight fraction of the system for one of its components. Mass and weight in this context can be used interchangeably as long as the experiment is done on Earth's surface.*

If we treat dry air as an ideal gas, 20.9 vol% oxygen means 0.209 mol of oxygen per mol of air. It contributes a mass of 0.209· M_{O_2} *to the mass of 1 mol of air. For simplicity, we restrict our considerations to the oxygen and nitrogen components of dry air and use* $M_{O_2} = 32$ *g/mol and* $M_{N_2} = 28$ *g/mol. Thus, the mass of 1 mol of air based on Fig. 8.21 is:*

$$M_{air} = n_{O_2} M_{O_2} + n_{N_2} M_{N_2} = 0.209\ \text{mol} \cdot 32\frac{\text{g}}{\text{mol}} + 0.791\ \text{mol} \cdot 28\frac{\text{g}}{\text{mol}} = 28.836\ \text{g},$$

and the weight contributions of its two components are:

$$f_{O_2} = \frac{m_{O_2}}{M_{air}} = \frac{n_{O_2} M_{O_2}}{M_{air}} = \frac{6.688\ \text{g}}{28.836\ \text{g}} = 23.2\ \text{wt\%}$$

$$f_{N_2} = \frac{m_{N_2}}{M_{air}} = \frac{n_{N_2} M_{N_2}}{M_{air}} = \frac{22.148\ \text{g}}{28.836\ \text{g}} = 76.8\ \text{wt\%}.$$

We made the assumption that dry air can be modelled as an ideal gas. Note that it is not sufficient that this assumption applies separately to oxygen and nitrogen. We further require that the gas mixture acts ideally; i.e., that oxygen and nitrogen molecules do not interact with each other. This additional condition is important in the derivation of Dalton's law.

8.7.2: Dalton's Law

To describe gas mixtures we begin with the definition of **partial pressure**. The partial pressure of a gas component is the pressure that would be measured if all other gas components were removed from the studied container. John Dalton studied a gas mixture of n components with partial pressures $p_1, p_2, \ldots, p_n$, and found in 1810 that the partial pressures add up to the total pressure measured for the mixture:

$$p_{total} = \sum_{i=1}^{n} p_i. \quad [8.26]$$

KEY POINT

Dalton's law *states that the total pressure of a gas is equal to the sum of the partial pressures of its components.*

Dalton's observations imply that the partial pressure of a gas component does not depend on any of the other components; i.e., intermolecular interactions are negligible. As in the case of the ideal gas law, a detailed experimental test shows that real gas mixtures do not act as ideal gases in Dalton's way. However, Dalton's law does apply to mixtures of ideal gases, as both the ideal gas law and Dalton's law have in common that we neglect inter-particle interactions. Thus, Dalton's law describes ideal gas mixtures. For these we can replace the pressure term in Eq. [8.26] by the ideal gas law applied to each gas component individually:

$$p_{total} = \frac{n_1 R T}{V} + \cdots + \frac{n_n R T}{V}.$$

This equation can be written in condensed notation using the summation symbol ($\sum$):

$$p_{total} = \frac{R T}{V} \sum_{i=1}^{n} n_i. \qquad [8.27]$$

Eq. [8.27] states that the amounts of the individual gases n_i in unit mol are added to obtain the total amount of the gas in an ideal gas mixture. This leads to the definition of the **molar fraction** x_i of the i-th component:

$$\frac{n_i}{\sum_i n_i} = \frac{n_i}{n_{total}} = x_i.$$

With this definition, Eq. [8.27] is rewritten for each single component of the gas mixture:

$$\frac{p_i}{p_{total}} = x_i. \qquad [8.28]$$

In this form, Dalton's law states that the ratio of the partial pressure of each component of a gas mixture to the total gas pressure is equal to the molar fraction of the same component. Historically, Dalton's discovery of the independent behaviour of gas components was one of the first significant indications that matter has a molecular structure.

Dalton's law can also be applied to gases dissolved in liquids if the solution can be modelled as an ideal solution; i.e., a solution in which interactions between solvent and solute molecules, and interactions among solute molecules can be neglected. The latter condition is often realizable if the solution is a **dilute solution**; i.e., only a small concentration of solute is mixed into the solvent.

CONCEPT QUESTION 8.5

Under what conditions does Dalton's law not apply to a gas mixture?

CASE STUDY 8.6

Using the gas composition in the lungs and Fig. 8.22, quantify the deviation from ideal behaviour at 0°C for the air components (a) oxygen and (b) carbon dioxide.

Supplementary information: Table 8.2 provides the partial gauge pressures of oxygen and carbon dioxide in the alveoli and in blood (pressures relative to air pressure) of a standard man. The higher partial gauge pressure of oxygen in the lungs causes oxygen to diffuse into the blood system. The opposite pressure difference pushes carbon dioxide from the venous blood into the lungs. Note that gas components always transfer from a space with higher partial pressure to a space with lower partial pressure. The pressure difference for CO_2 in Table 8.2 is lower since carbon dioxide penetrates the alveolar membrane more readily than oxygen and therefore does not need as high a driving force to pass through it. This phenomenon is discussed as a diffusion concept in Chapter 11.

TABLE 8.2

Partial pressures of CO_2 and O_2 in alveoli and venous blood

Gas component	Alveoli	Venous blood
O_2	13.33 kPa	5.33 kPa
CO_2	5.33 kPa	6.13 kPa

The values in the venous blood are obtained from Eq. [8.28] using the dissolved concentration of the gases.

Answer To Part (a): *The ideal gas value in Fig. 8.22 is p V = 22.418 L atm/mol. For oxygen at its partial pressure of p = 0.13 atm the value for p V is 22.415 L atm/mol; this corresponds to a deviation of slightly more than 0.01%.*

Answer To Part (b): *For CO_2 at partial pressure p = 0.05 atm, p V in Fig. 8.22 is 22.411 L atm/mol; this corresponds to a deviation of slightly more than 0.03%. Thus, both gases essentially act like ideal gases at 0°C. The same applies in the lungs at 37°C, as gases become more ideal at increased temperatures.*

8.7.3: Application of Dalton's Law: Diving

What is dangerous about diving while holding your breath? Specifically, we want to study the case in which a person hyperventilates first (to increase the fraction of oxygen in the lungs so that it is possible to dive longer) and then dives to a depth of 10 m, at which the total pressure in the lungs has doubled. We will see that the answer about the dangers in this diving attempt rest with (i) the development of the partial pressures of O_2 and CO_2 in the lungs, and (ii) the partial pressures of both gases in the venous blood in Table 8.2.

When a person is not breathing, the partial pressure of carbon dioxide in the blood increases because this gas cannot be removed through respiration. An increase in the partial pressure of CO_2 in the blood triggers the central chemoreceptor in the lowest section of the brain (medulla oblongata) to signal respiratory distress.

Fig. 8.22 illustrates the assumed diving attempt. The diver wants to extend the time to be spent under water and does this by initially hyperventilating (heavy breathing) to artificially decrease the carbon dioxide partial pressure in the blood. Hyperventilation allows the person to increase the partial pressure of oxygen in the alveoli, typically from 13.3 kPa to 15 kPa, as shown for the curve labelled O_2 in Fig. 8.22. The corresponding pressure scale is shown at the right side. The increase of the partial oxygen pressure is achieved by lowering the partial CO_2 pressure from 5.2 kPa to 3.5 kPa (upper solid curve, labelled CO_2). The figure also includes a sketch of the alveolus/blood capillary system below the graphs, which shows at each stage of the dive the direction of gas exchange through the alveolar surface. The arrows indicate the dominant direction of gas transfer. During hyperventilation, the partial pressure of nitrogen in the alveoli remains unchanged. Hyperventilation allows a healthy diver to stay below the surface for up to one minute.

Now the person dives to a depth of 10 m. The depth profile of the dive is shown in Fig. 8.22 above the graphs of the partial pressures, defining the horizontal axis of the figure. As the diver descends deeper and deeper below the surface, the pressure on the thorax increases significantly. This effect is due to Pascal's law (discussed in Chapter 12): the below-surface pressure of water doubles at 10 m below the surface. Due to Eq. [8.28], this two-fold increase of the external pressure on the thorax leads

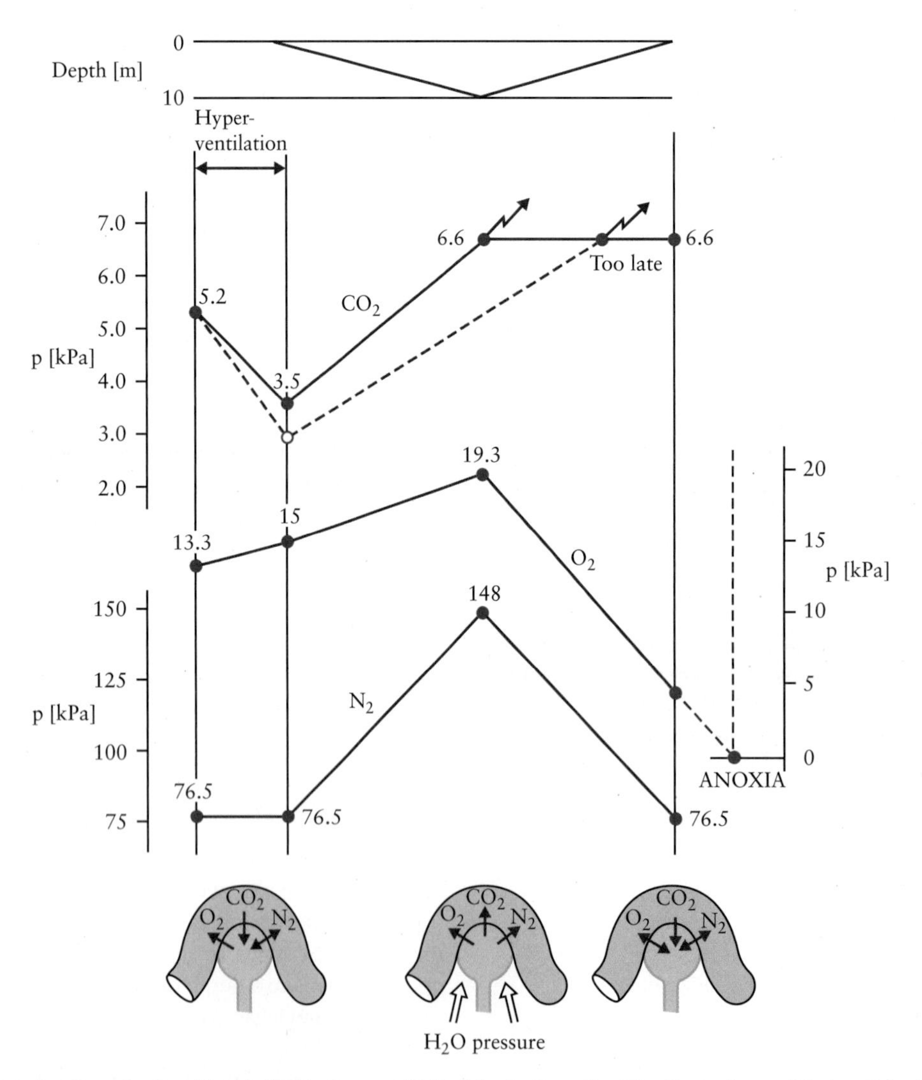

Figure 8.22 Sketch showing a dive to a 10-m depth after hyperventilation. The depth profile of the dive is shown at the top. Following the dive after the preceding hyperventilation, the changes of the partial pressures of nitrogen, oxygen, and carbon dioxide in the alveoli are recorded (solid lines). The dashed line corresponds to an even stronger initial hyperventilation, which causes the diver to resurface too late. In this case, anoxia-related unconsciousness may cause drowning. The curve for the partial pressure of CO_2 indicates that, at 6.6 kPa a signal is triggered, forcing the person to resurface to breathe. The pressure scale for the partial alveolar pressure of carbon dioxide is shown at the upper left, for oxygen at the right, and for nitrogen at the lower left.

initially to a doubling of each partial pressure in the lungs. However, as shown in the middle sketch at the bottom of Fig. 8.22, the increased pressure in the alveoli now pushes all gases into the blood system. CO_2 diffusion in particular is reversed when the CO_2 partial pressure in the lungs exceeds a value of 6.13 kPa, as shown in Table 8.2.

The validity of the changes in Fig. 8.22 for the CO_2 partial pressure in the alveoli and the blood capillaries is demonstrated with the following argument. When the external pressure on the human body rises, both the gas in the lungs and the blood in the arteries are subject to a pressure increase. The gas in the lungs responds like an ideal gas, allowing us to estimate the change in the concentration of CO_2 in the gas phase using the ideal gas law:

$$p_{CO_2} = \frac{n_{CO_2} R T}{V} = c_{CO_2} R T.$$

Thus, doubling the partial pressure doubles the concentration of the gas component.

On the other side of the membrane separating the alveolus and the blood capillary, the liquid blood can be treated as an incompressible fluid, for which a pressure increase does not lead to a volume change. Consequently, no change in the concentration of dissolved components in the fluid results. Thus, a doubling of the pressure on the human body inverts the direction of the concentration step across the alveolar membrane.

During the dive shown in Fig. 8.22, the oxygen and carbon dioxide partial pressures in the alveoli rise by less than a factor of 2: O_2 to only 19.3 kPa, since it is continuously consumed by the body, and CO_2 to 6.6 kPa. The partial pressure of CO_2 cannot rise higher since the blood system can take up a large amount of CO_2. The value of 6.6 kPa partial pressure of CO_2 triggers the central chemoreceptor, as described above; i.e., the diver now feels a strong urge to resurface.

A diver obeying this urge and approaching the surface from a depth of 10 m encounters a rapid decrease of the water pressure accompanied by noticeable changes in the partial pressures in the lungs. The nitrogen value comes down to its pre-dive value. The CO_2 partial pressure remains roughly constant, as the blood pushes into the alveoli all excess CO_2 that it had to absorb due to the high external pressure. At this stage the change of the oxygen partial pressure is critical. It decreases during the resurfacing for three reasons: (i) a drop of the total gas pressure in the lungs, (ii) a renewed ability of the blood to absorb oxygen to replace the leaving CO_2, and (iii) continued oxygen consumption by the body's metabolism. In the case shown as solid lines for CO_2 and O_2 in Fig. 8.22, the final partial pressure of O_2 in the alveoli is 4.2 kPa.

However, the dashed line for CO_2 illustrates the dangers associated with diving for a person misjudging the physical laws of nature: excessive hyperventilation may lower the initial partial pressure of CO_2 below the value considered above. Thus, the person feels comfortable below the surface for a longer time, particularly at a depth of 10 m, where the dangerous decrease of the partial pressure of oxygen in the lungs is not noticed because its partial pressure has increased due to the external water pressure. This time, following the dashed lines in Fig. 8.22, the signal to surface comes too late since the partial pressure of oxygen drops to 0.0 kPa during surfacing (anoxia). At that moment the diver loses consciousness and will drown if not rescued by others.

EXAMPLE 8.9

Standard diving gear allows diving to depths of about 60 m, since the gas cylinder automatically regulates the air supply so that the pressure of the inhaled air is always equal to the pressure of the surrounding water. Why is it dangerous to resurface from such a dive too quickly?

Solution

During the dive, the overall increase of the pressure leads also to an increase of the partial pressure of nitrogen in the lungs. This follows from Eq. [8.28] and a constant molar fraction of nitrogen:

$$p_{N_2} = x_{N_2} \, p_{total}.$$

Water pressure at a depth of 60 m increases sevenfold. Thus, this equation states that a sevenfold increase also exists in the partial pressure of nitrogen in the alveoli. This increase leads, in turn, to about a sevenfold increase of the partial pressure of N_2 in the blood because gas diffuses across the membrane between the alveolus and the blood capillary as a result of the increased partial pressure in the lungs.

When the diver resurfaces too quickly, the blood cannot push the dissolved nitrogen back into the alveoli fast enough because diffusion is a slow process. The lower external pressure closer to the surface then causes nitrogen to exceed its solubility in blood and form bubbles. This effect is called **embolism, Caisson disease**, or **diver's paralysis**.

Even if the diver surfaces slowly, danger still looms during deep sea diving. At great depths, the partial pressure of oxygen in the blood has also risen to dangerous levels. A prolonged increase of the partial pressure of O_2 increases the risk of **hyperoxia**, the condition caused by excess oxygen in the blood, acting toxic for partial pressures above 40 kPa. A partial pressure for O_2 of 70 kPa for several days, or 200 kPa for 3–6 hours, causes the alveolar surface to shrink irreversibly; an O_2 partial pressure of 220 kPa or higher results in cramping and unconsciousness. These conditions can occur when diving with compressed air/oxygen tanks at a depth of 100 m.

SUMMARY

DEFINITIONS

- Pressure
  - absolute value:

$$p = \frac{F_{\perp}}{A}$$

where the force acts perpendicular to the surface of area A.

  - in vector notation:

$$p = \frac{1}{A^2}\vec{F} \bullet \vec{A} = \frac{1}{A}\vec{F} \bullet \vec{n},$$

where $\vec{F}$ is the force acting on the surface of area $\vec{A} = A \bullet \vec{n}$ with $\vec{n}$ a vector of length 1 (unit vector) normal to the surface.

- Partial pressure of component i is the pressure of the i-th component with all other components removed from the container.
- Molar fraction x:

$$\frac{n_i}{\sum_i n_i} = \frac{n_i}{n_{total}} = x_i$$

UNITS

- Volume V: L (litre with 1.0 L = 1.0×10^{-3} m^3); m^3 (standard unit)
- Temperature T: K (standard unit), °C (frequently used non-standard unit)
- Pressure or gauge pressure p: Pa = N/m^2 (standard unit); mmHg, torr, atm, bar (frequently used non-standard units)

LAWS

- Ideal gas law:

$$p = V = n\,R\,T$$

with R the gas constant

- Internal energy of an ideal gas:
  - for 1 mol of gas (U in unit J/mol):

$$U = \frac{3}{2}R\,T = \frac{1}{2}M\langle v^2\rangle$$

  - per particle in the gas (ε in unit J/particle):

$$\varepsilon = \frac{3}{2}\frac{R}{N_A}T = \frac{3}{2}k\,T$$

N_A is the Avogadro number, k is the Boltzmann constant

- Root-mean-square speed of gas particles, $v_{rms} = (\langle v^2\rangle)^{1/2}$:

$$v_{rms} = \sqrt{\frac{3\,R\,T}{M}} = \sqrt{\frac{3\,k\,T}{m}},$$

m is the mass of a single gas particle

- Dalton's law for partial gas pressures:

$$p_{total} = \sum_{i=1}^{n} p_i$$

This is equivalent to:

$$\frac{p_i}{p_{total}} = x_i$$

MULTIPLE-CHOICE QUESTIONS

MC–8.1. Which of the following formulas describes the reading of a temperature T (in degrees Celsius) when measured with an expanding mercury column as proposed by Celsius? In these formulas, h is the height of the mercury column and H is the height of the mercury column at a chosen reference temperature. α and β are positive constants, with α the reference temperature at which the column height is $h = H$:

(a) $T = \alpha - \beta\left(\frac{h-H}{H}\right)$

(b) $T = \alpha + \beta\left(\frac{h-H}{H}\right)$

(c) $T = \alpha - \beta\left(\frac{h+H}{H}\right)$

(d) $T = \alpha + \beta\left(\frac{h+H}{H}\right)$

(e) $T = \alpha - \beta\left(\frac{h+H}{H}\right)^2$

MC–8.2. Which of the following reasons did not contribute to Celsius's choice of the melting point of water (at 0°C) and the boiling point of water (at 100°C) as the two reference points on his newly developed thermometer?

(a) reproducibility of a reference system at these two temperatures
(b) easy accessibility of water as the system providing the two reference temperatures
(c) technical simplicity of obtaining a water system at all three states of matter: solid, liquid, and vapour
(d) temperature independence of the heat capacity of water between 0°C and 100°C

MC–8.3. In Boyle's experiment, the volume of the sealed gas decreases as additional mercury is added to the open column. This effect is due to the fact that

(a) the volume of the gas decreases.
(b) pressure and volume are linearly proportional to each other.
(c) pressure and volume are proportional to each other.
(d) pressure and volume are inversely proportional to each other.
(e) pressure and volume are unrelated.

MC–8.4. Charles's law can be written as V/T = const if
(a) V is reported in non-standard unit cm^3.
(b) T is reported in non-standard unit degrees Celsius (°C).
(c) the pressure varies moderately.
(d) the gas is a noble gas.
(e) none of the above.

MC–8.5. We study the ideal gas equation. The gas constant can be given in different units; however, this unit is wrong:
(a) J/(K mol)
(b) (atm m^3)/(K mol)
(c) (Pa cm^2)/(K mol)
(d) cal/(K mol)
(e) None of the above; all are suitable for the gas constant R

MC–8.6. Evaluate the following four statements about an ideal gas in a closed container (n = const): (i) if the pressure doubles during an isothermal compression, the volume doubles as well; (ii) if the pressure and the volume of the gas double, the temperature of the gas must increase; (iii) if the volume doubles during an isobaric expansion, the temperature stays unchanged; (iv) if the temperature doubles, at least one of the parameters volume and pressure must increase. Which combination of these statements is correct?
(a) i and ii
(b) i and iv
(c) ii and iii
(d) ii and iv
(e) iii and iv

MC–8.7. In an ideal gas, the root-mean-square speed of the gas particles varies linearly with the following parameters (more than one may be correct):
(a) the gas temperature
(b) the molar mass of the gas
(c) the amount of gas in mol
(d) the gas pressure
(e) none of the above

MC–8.8. Rewrite the ideal gas law by combining n and V as ρ/M, in which ρ is the density of the gas (in unit kg/m^3) and M is the molar weight (in unit kg/mol). Which parameter in this rewritten ideal gas law *must* increase from inside to outside across the envelope of a *hot air balloon* for it to stay airborne?
(a) pressure p
(b) temperature T
(c) gas density ρ
(d) None or more than one of these parameters increases from inside to outside.

MC–8.9. Which parameter in the ideal gas law as written in MC–8.8 *must* increase from inside to outside across the envelope of a *helium-filled blimp* for it to stay airborne?
(a) pressure p
(b) temperature T
(c) gas density ρ
(d) None or more than one of these parameters increases from inside to outside.

MC–8.10. In the kinetic gas theory developed by Boltzmann, Maxwell, and Clausius, which of the following is a result, not an assumption made to develop the model?
(a) The gas consists of a very large number of particles with a combined volume that is negligible compared to the size of the container.
(b) The internal energy of an ideal gas depends linearly on temperature (in kelvin).
(c) The molecular size is much smaller than the interparticle distance.
(d) The molecules are in continuous random motion, travelling along straight lines while not colliding with other particles or the walls.
(e) Collisions with each other and the walls of the box are elastic, which excludes intermolecular interactions.

MC–8.11. One mol of hydrogen gas, which we treat as an ideal gas of molecular mass M = 2 g/mol, is held at 50°C. (A) Which of the following values is closest to the root-mean-square speed of the molecules in this gas?
(a) 2007 m/s
(b) 63.5 m/s
(c) 4 028 100 m/s
(d) 790 m/s
(e) 25.0 m/s

(B) Hydrogen in the gas is a molecule, H_2. Knowing this, why does it not alter the result? (C) If we now dissociate the gas to atomic hydrogen, $H_2 \rightarrow 2H$, will this alter the result?
(a) yes
(b) no
(c) We cannot predict this without doing the experiment.

MC–8.12. Which of the following statements about the temperature of a system is *not* correct?
(a) The temperature of a single gas particle is determined by its speed.
(b) The temperature measurement in Fig. 8.7 is correct during the transition from liquid water to water vapour because the continuous addition of heat does not lead to a temperature change.
(c) Temperature is a parameter that characterizes a system only when the system is in thermal equilibrium.
(d) Temperature measured with a Celsius thermometer is correct only when the expanding liquid in the thermometer and the system have the same temperature.
(e) The temperature of the human body is usually higher than the air temperature in the immediate environment. Therefore, the human body and the surrounding air are not in thermal equilibrium.

MC–8.13. The molar mass of He is 4×10^{-3} kg/mol and for water it is 18×10^{-3} kg/mol. What is the approximate ratio of the root-mean-square speed of He to that of water vapour molecules at room temperature?
(a) 2:1
(b) 1:2
(c) 4.5:1
(d) 1:4.5

MC–8.14. Which of the following assumptions is *not* made in developing the kinetic gas theory?
(a) The number of molecules is small.
(b) The molecules obey Newton's law of motion.
(c) Collisions between molecules are elastic.
(d) The gas is a pure substance, not a mixture.
(e) The average separation between molecules is large compared to their size.

MC–8.15. Suppose that, at some given instant, molecules hitting a container wall from inside collide not elastically but perfectly inelastically with the wall. How would the pressure change at that instant (and for a brief time afterward)?
(a) The pressure would be zero.
(b) The pressure would be halved.
(c) The pressure would be unchanged.
(d) The pressure would be doubled.

MC–8.16. We study two containers with helium ($M = 4$ g/mol, treated as an ideal gas). In container I, we have 5.0 L of gas with $p_I = 10$ atm at $T_I = -5°C$. In container II, we have 7.0 m^3 of gas with $p_I = 1200$ torr at $T_I = 263$ K. In which container does helium have the higher root-mean-square speed?
(a) container I
(b) container II
(c) v_{rms} is the same in both containers
(d) Too many parameters vary between the two containers to answer this with certainty.

MC–8.17. One container is filled with helium gas, another with argon gas. If both containers are at the same temperature, which molecules have the higher root-mean-square speed?
(a) helium atoms
(b) argon atoms
(c) Both have the same root-mean-square speed.
(d) This question can be decided only experimentally.

MC–8.18. If a gas has an internal energy $U = 0$ J, which of the following is *not* correct?
(a) None of the particles in the gas moves faster than the speed of sound.
(b) The kinetic energy of all gas particles is determined by their root-mean-square speed.
(c) The temperature of the gas is 0 K.
(d) The kinetic energy per particle is (3/2) $k\,T$.
(e) The gas cannot be a van der Waals gas.

MC–8.19. A standard man inhales a tidal volume of air at 20°C. When the air arrives in the lungs, it has a temperature of 37°C. Treating air as an ideal gas, the change in internal energy while travelling through the trachea is
(a) $\Delta U = 0$ J.
(b) about an 85% increase over the initial value.
(c) about a 5% increase over the initial value.
(d) unknown, because the amount of air is not specified.
(e) proportional to the change in the pressure of the air.

MC–8.20. We study 10 mol of an ideal gas at −10°C. What is the internal energy of that gas?
(a) $U < 0$ J
(b) $U = 0$ J
(c) $0\ \text{J} < U \leq 1$ kJ
(d) $1\ \text{kJ} < U \leq 100$ kJ
(e) $U > 100$ kJ

MC–8.21. The internal energy of an ideal gas increases when we increase which one of these parameters in an experiment?
(a) the gas pressure
(b) the gas temperature
(c) the gas volume
(d) the universal gas constant

MC–8.22. (A) How does the internal energy of an ideal gas change when I increase its pressure in an isothermal process?
(a) It increases.
(b) It stays unchanged.
(c) It decreases.
(d) The answer depends on the temperature of the gas.

(B) How does the internal energy of an ideal gas change when I expand its volume in an isothermal process?
(a) It increases.
(b) It stays unchanged.
(c) It decreases.
(d) The answer depends on the temperature of the gas.

CONCEPTUAL QUESTIONS

Q–8.1. Note that we have two temperature scales we often use: the scale introduced by Celsius and the scale introduced by Lord Kelvin, with $T\,(K) = T\,(°C) + 273.15$. In an astronomy class, the temperature at the core of a star is reported as 1.5×10^7 degrees. Does it make sense to ask the lecturer whether this is a temperature in units °C or K?

Q–8.2. When food has been cooked in a pressure cooker it is very important to cool the container with cold water before removing the lid. Why?

Q–8.3. Small air bubbles trapped between two sheets of plastic are often used to cushion breakable goods for shipping. Is this protection more effective on a warm or a cold day?

Q–8.4. Organisms in the deep sea are subjected to very high pressures, as we will discuss in the context of Pascal's law in Chapter 12. Why are these organisms destroyed when they are pulled up to the surface? *Note:* One animal living at those depths is the giant squid. Nobody has yet seen a live specimen of this species!

Q–8.5. Two identical cylinders at the same temperature contain the same gas. If cylinder A contains three times as much gas as cylinder B, what are the relative pressures of the two cylinders?

Q–8.6. (a) We consider 1.0 L of an ideal gas at standard conditions. Describe a state into which this gas cannot be transferred. (b) If you still need an ideal gas at the conditions you have chosen in part (a), how can you get it?

Q–8.7. Why is heat required to boil a liquid even though the molecules in the liquid and the vapour share the same temperature?

Q–8.8. Small planets like Mercury and Mars have very thin or no atmospheres. Why?

Q–8.9. Although the average speed of gas molecules in thermal equilibrium at a given temperature is always greater than zero, the average velocity is zero. Why?

Q–8.10. A child wants to pretend to have a fever. The child notices that air breathed onto an arm feels warmer than the arm itself and reasons that breathing on a thermometer should effectively drive up the mercury column. Will this trick deceive the parents?

Q–8.11. Which is lighter under otherwise equal conditions (i.e., same pressure, temperature, volume): humid air or dry air?

Q–8.12. In 1965, a French team led by Jacques-Yves Cousteau lived for 28 days in the deep-sea station *Conshelf III* at 108 m below the sea surface. They breathed an oxygen/helium mixture (called heliox) instead of air. (a) Would you agree with their claim that breathing this mixture is easier than breathing air? (b) Can you think of another reason why they used heliox instead of air? *Note:* One member of the team reported that, among other adverse effects, this exercise irritated his taste buds and he could no longer distinguish caviar from chicken. This, of course, is a disastrous effect for a Frenchman!

ANALYTICAL PROBLEMS

P–8.1. (a) Draw a graph for the volume of 1.0 mol of an ideal gas as a function of temperature in the range from 0 K to 400 K at constant gas pressures of first, 0.2 atm and second, 5 atm. (b) Draw a graph for the pressure of 1.0 mol of an ideal gas as a function of volume between 0 L and 20 L at constant temperatures of 150 K and 300 K.

P–8.2. A container of volume $V = 10.0\ \text{cm}^3$ is initially filled with air. The container is then evacuated at 20°C to a pressure of 5.0×10^{-6} mmHg. How many molecules are in the container after evacuation if we assume air is an ideal gas?

P–8.3. A container of volume $V = 400\ \text{cm}^3$ has a mass of 244.5500 g when evacuated. When the container is filled with air of pressure $p = 1$ atm at temperature $T = 20$°C, the mass of the system increases to 245.0307 g. Assuming air behaves like an ideal gas, calculate from these data the average molar mass of air.

P–8.4. 1.0 mol oxygen gas is initially at a pressure of 6.0 atm and a temperature of 300 K. (a) If the gas is heated at constant volume until the pressure has tripled, what will be the final temperature? (b) If the gas is heated such that both pressure and volume are doubled, what will be the final temperature?

P–8.5. The pressure of an ideal gas is reduced by 50%, resulting in a decrease in temperature to 75% of the initial value. Calculate the ratio of final to initial volumes of the gas.

P–8.6. An ideal gas is initially at temperature 300 K, volume $1.5\ \text{m}^3$, and pressure 2.0×10^4 Pa. What will be its final temperature if it is compressed to a volume of $0.7\ \text{m}^3$ and the final pressure is 8.0×10^4 Pa?

P–8.7. An ideal gas is confined in a container at a pressure of 10.0 atm and a temperature of 15°C. If 50% of the gas leaks from the container and the temperature of the remaining gas rises to 65°C, what is the final pressure in the container?

P–8.8. A container of fixed volume contains 0.4 mol of oxygen gas, which we treat as an ideal gas. Determine the mass of gas in unit kg that must be withdrawn in an isothermal manner from the container to lower the pressure from 40 atm to 25 atm.

P–8.9. A spherical weather balloon is designed to inflate to a maximum diameter of 40 m at its working altitude, where the air pressure is 0.3 atm and the temperature is 200 K. If the balloon is filled at atmospheric pressure and temperature 300 K, what is its radius at lift-off? Treat the gas as an ideal gas.

P–8.10. Use Avogadro's number to find the mass of a helium atom.

P–8.11. The temperature in the upper regions of the atmosphere of Venus is 240 K. (a) Find the root-mean-square speed of hydrogen molecules (H_2) and carbon dioxide (CO_2) in that region. (b) A result of planetary science is that a gas eventually is lost from a planet's atmosphere into outer space if its root-mean-square is one-sixth of the escape speed, which can be calculated from gravity. Using an escape speed of 10.3 km/s for Venus, does either of the two gases in part (a) escape from that planet?

P–8.12. A 10.0-L container holds 2.0×10^{-3} oxygen gas molecules (O_2). If the pressure of the gas is 300 mmHg, what are (a) the temperature and (b) the root-mean-square speed of the molecules? *Hint:* Treat oxygen as an ideal gas with a molecular mass of $M(O_2) =$ 32.0 g/mol.

P–8.13. Nitrogen is commercially available as a compressed gas contained in metal cylinders. (a) If a cylinder of 120 L is filled with N_2 to a pressure of 1.45×10^4 kPa at 20°C, how many mol of nitrogen does the cylinder contain? (b) If we open the valve on the cylinder and allow N_2 to escape, how many litres of nitrogen at $p = 1$ atm and 20°C would leave the cylinder? *Hint:* Treat nitrogen as an ideal gas.

P–8.14. Water boils at $p = 1$ atm and 100°C. Its density in the liquid state at the boiling point is $0.96\ \text{g/cm}^3$. Calculate the volume ratio of water vapour to liquid water at the boiling point. *Hint:* Model water vapour as an ideal gas. You may use a reference amount of water of 1 mol if this simplifies your calculations; however, return to the calculation afterward and see whether you need this value.

P–8.15. 2.0 mol of an ideal gas are sealed in a 10.0 L container at pressure $p = 5.0$ atm. What is the internal energy U of the gas (i.e., the total energy, not the internal energy per mol)?

P–8.16. A thermally isolated container is filled with 1.0 mol of an ideal gas at $T = 0$°C. The gas is then compressed from $1.0\ \text{m}^3$ to $0.8\ \text{m}^3$. What is the final internal energy U of the gas?

P–8.17. Calculate the pressure exerted by 25 g of nitrogen gas in a 1.0 L container at 298 K.

P–8.18. A *plethysmograph* is an airtight box that allows us to measure the volume change of a patient's body inside the box by recording the pressure in the box (Boyle's law). The residual volume in the human lungs is determined with a plethysmograph in the following manner: At the end of a normal exhalation through a mouthpiece, which is connected to the atmosphere outside the plethysmograph, the air pressure in the lungs equals the atmospheric pressure. A shutter then closes off the mouthpiece. The patient is requested to continue breathing against the closed shutter. During the next inhalation, the patient's chest enlarges, creating a new lung volume by decompression; i.e., the original volume V_{lungs} becomes $V_{lungs} + \Delta V$. At this point, two parameters are measured: ΔV is determined from the pressure change in the plethysmograph, and the final gas pressure in the lungs is measured between the shutter and the patient. For this method, determine a formula for the lung volume V_{lungs}.

P–8.19. We mix 0.25 g oxygen gas (O_2) and 1.5 g nitrogen gas (N_2) in a 2-L container at 20°C. Assuming both gases behave ideally, what are (a) the partial pressure of oxygen, (b) the partial pressure of nitrogen, and (c) the total pressure in the container? *Hint:* The molar masses are 32 g/mol for oxygen and 28 g/mol for nitrogen.

P–8.20. In an industrial accident, 10 L of oxygen gas and 10 L of hydrogen gas are mixed at 300°C and 1 atm pressure in a 20-L container. The mixture explodes but luckily remains confined to the original container. When the system has cooled down to 25°C, what is the pressure in the container? Treat both gases as ideal and neglect volume contribution due to liquids and solids. Neglect the vapour pressure of liquid water.

P–8.21. An evacuated glass bulb has a mass of 45.9214 g. Filled with dry air at 1 atm and 25°C, its mass is 46.0529 g. Filled with a mixture of methane (CH_4) and ethane (C_2H_6) its mass is 46.0141 g. Calculate the molar fraction of methane in the gas mixture.

P–8.22. N_2O_4 dissociates into 2 NO_2 at 45°C to the extent of 38%. Calculate the pressure that develops in a 20.0-L container filled initially with 1 mol N_2O_4 before it is heated to 45°C.

ANSWERS TO CONCEPT QUESTIONS

Concept Question 8.1: The unit L (litre) is a frequently used non-standard unit for volume. We convert with 1 L = 1×10^{-3} m^3; i.e., 1000 L = 1 m^3.

Concept Question 8.2: Not without further information. The gas exerts a pressure p on the inner surface of the piston. This results in a force of magnitude $F = p\,A$ pushing the piston to the left. However, the arrangement in the figure is in air. Thus, a second force acts on the piston toward the right, $F_{air} = p_{air} A$. The piston is in mechanical equilibrium if $p = p_{air}$; it will accelerate toward the left if $p > p_{air}$ and it will accelerate toward the right if $p < p_{air}$.

Concept Question 8.3: (B). Note that the relation of p and V is not linear. Also, Boyle *added* mercury; i.e., he recorded his results while increasingly compressing the sealed gas.

Concept Question 8.4: (C). We set the temperature constant in the ideal gas law, $T = T_0$. Since n, R, and T_0 on the right-hand side are all constant, they were combined in a single constant. Thus we have recovered Boyle's law from the ideal gas law at constant temperature. Boyle's law follows $p \propto 1/V$. It is a special case of the ideal gas law at constant temperature.

Concept Question 8.5: We used the ideal gas law for each gas component in Eq. [8.27]. This implies that two assumptions have been made: (1) the individual gas components are ideal gases and (2) the gas mixture is an ideal gas. This means the various gas components don't interact with each other except for elastic collisions. Based on the second assumption, a case that obviously is excluded is a chemically reacting gas mixture.

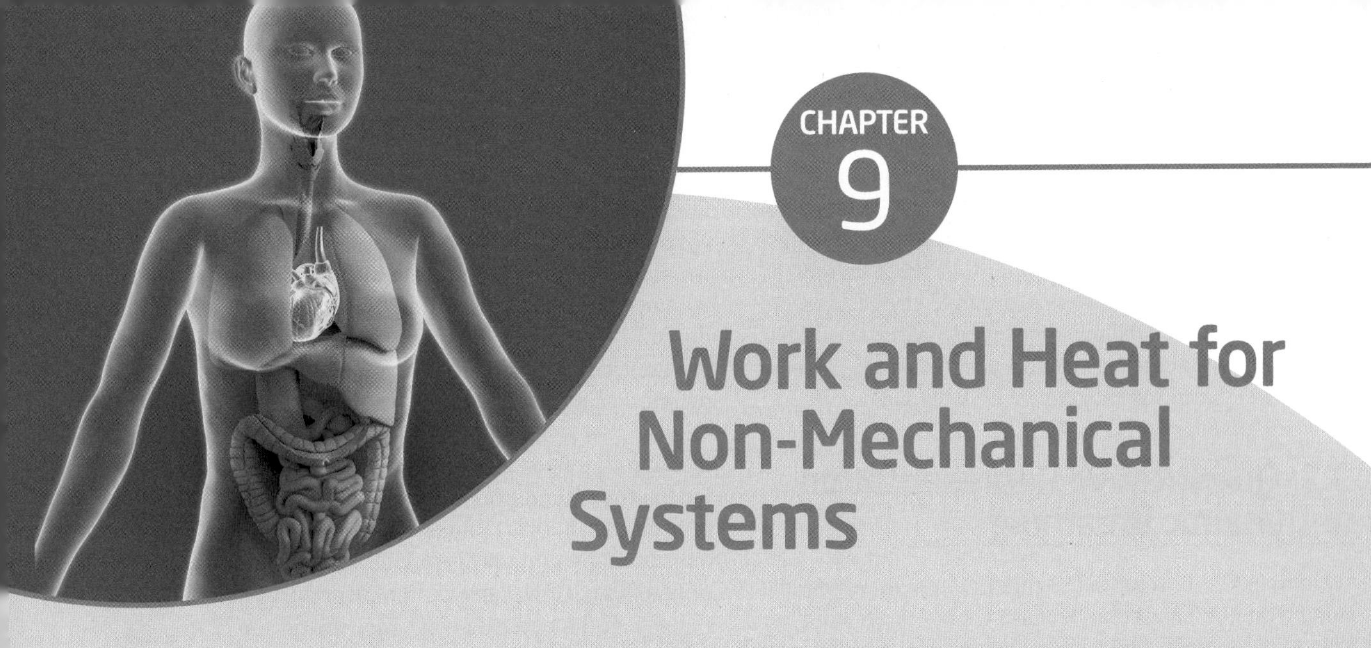

CHAPTER 9

Work and Heat for Non-Mechanical Systems

Living systems interact with their environment in three fundamental ways: (i) through the exchange of heat, (ii) through the performance of work, and (iii) through the exchange of matter. The first two interactions are quantified in this chapter for non-mechanical systems such as gases. We use in particular the definition of work to further pursue our discussion of the respiratory system; we discussed its static properties in the previous chapter.

We introduce work for a non-mechanical system as the negative product of the absolute gas pressure of the system and its volume change. This sign convention leads to a positive value of work when work is done on (received by) the system.

One method of energy exchange between a non-mechanical system and its environment is the uptake or release of heat. The corresponding form of energy contained in the system is called thermal energy. The thermal energy of a system is proportional to its temperature; heat exchange, then, is related to a temperature change in the system.

Heat and work allow us to formulate the concept of energy conservation for any closed system, leading to the first law of thermodynamics: the energy of an isolated system is constant, and the energy of a system in interaction with its environment varies exactly by the amount of work and heat exchanged. The consequences for the metabolism of animals are discussed for predatory dinosaurs.

The current chapter combines the discussions we left off at the end of Chapters 7 and 8. In Chapter 7 we found the concept of energy useful in discussing the properties of mechanical systems. An object can interact with another object in its environment through exchange of energy; the energy transfer occurs in the form of work. This discussion, however, did not result in finding a fundamental natural law at the end of Chapter 7 because we had taken a too restricted approach by limiting ourselves to mechanical processes.

In Chapter 8 we developed therefore a model system that allows the study of systems beyond the limitations of classical mechanics. The ideal gas law connects the gas parameters volume, pressure, and temperature and allows us to calculate the internal energy of the system as exclusively temperature dependent. We used the respiratory system as a physiological example for a non-mechanical system that includes static gas properties.

In this chapter we now continue to focus on non-mechanical systems (and the ideal gas in particular) to extend the concept of work beyond the limits of mechanics, and to introduce concurrently the concept of heat as a second form in which energy can be exchanged between a non-mechanical system and its environment. The dynamic breathing of the human respiratory system will serve as a major physiological example.

We used the respiratory system at rest to motivate the discussion of the gas system in the previous chapter. The starting point of the discussion was the development of a p–V diagram for the system in Fig. 8.9. Because the respiratory system is more complex than a simple piston-sealed container, both Fig. 8.9 and the ensuing discussion of respiration were extensive. Still, the concepts we developed for the ideal gas have proven to be a useful guide in characterizing air as it is cycled through the lungs.

A new concept we introduced in the previous chapter was the **gauge pressure** as the actual pressure of a system relative to atmospheric pressure. Fig. 8.9 then illustrated three gauge pressure curves relevant in respiration:

- the curve labelled $p_{alveoli}$ represents the gauge pressure inside the lungs,
- the curve labelled p_{pleura} represents the gauge pressure in the pleura, and
- the curve labelled $p_{alveoli} - p_{pleura}$ represents the pressure difference between lungs and pleura (transmural pressure).

These gauge pressures are derived from absolute pressure values that can be measured experimentally. The absolute pressure in the lungs is labelled p_{lungs}, which leads us to the alveolar gauge pressure, with $p_{alveoli} = p_{lungs} - p_{atm}$.

The absolute pressure in the gap between the visceral and the parietal layers is labelled p_{gap}, which leads us to the pleural gauge pressure, with $p_{pleura} = p_{gap} - p_{atm}$. Note that absolute pressures are always positive, but gauge pressures can be positive or negative. Fig. 8.9 provides an example for this distinction. A negative gauge pressure means the system is below atmospheric pressure and has the propensity to collapse, while a positive gauge pressure means the system has the propensity to expand.

9.1: Quantitative Representation of Dynamic Breathing

We now move beyond the respiratory system at rest and study dynamic breathing. Because we no longer hold our breath but operate with open air passageways, a different *p–V* diagram results; this is shown in Fig. 9.1. We neglect at first the dashed and dash-dotted cycles in the figure. The solid lines represent the respiratory system with very slow breathing, that is, a case in which the airflow resistance in the air passageways is negligible. Note that this diagram covers only the volume range from 3.0 L to 4.0 L because we are now interested in normal breathing with its tidal volume of 0.5 L (see the lung volume values defined in Fig. 8.2). The major difference between the *p–V* diagram in Fig. 9.1 and the diagram in Fig. 8.9 is that the alveolar pressure at all lung volumes is zero. This is so because the test person breathes with open air passageways; i.e., the lungs are open to the external air pressure. The curve for the transmural pressure difference, $p_{alveoli} - p_{pleura}$, is the same in both *p–V* diagrams. This indicates that the transmural pressure difference is independent of the type of breathing. Consequently, the pleural pressure varies in Fig. 9.1 in the opposite way from that in Fig. 8.9: p_{pleura} decreases with lung volume for slow breathing, while it increases with lung volume when the thorax is at rest and the respiratory muscles are relaxed.

Next we include the dynamic breathing we do continuously. The three basic pressure curves (solid lines in Fig. 9.1) remain unchanged, except that the alveolar and pleural pressures are modified between 3.0 L (exhaled) and 3.5 L (after inhalation of a tidal air volume). In regular breathing, air is inhaled and exhaled fast enough that flow resistance in the air passageways has to be taken into account. Viscous flow of air through a tube is discussed in Chapter 13, where we learn that a pressure difference along the tube is needed to achieve gas flow through the tube. Thus, the pressure in the lungs is smaller than atmospheric pressure during inhalation ($p_{alveoli} < 0$) and is larger than atmospheric pressure during exhalation ($p_{alveoli} > 0$). Dynamic breathing requires excess pressures of up to 100 Pa. Note again that dynamic breathing does not affect the transmural pressure difference in Fig. 9.1, as both alveolar and pleural pressures are modified in the same manner.

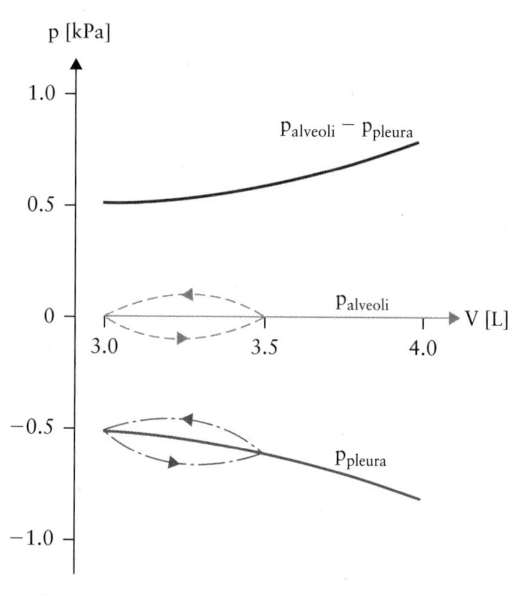

Figure 9.1 *p–V* diagram for a person breathing regularly. The solid curves result when a person breathes very slowly, that is, without airflow resistance in the airways. The figure shows the same three pressure curves we introduced for the respiratory curve at rest in Fig. 8.9, but the abscissa is limited to lung volumes between 3.0 L and 4.0 L. The pressure in the lungs (green curve) remains at atmospheric pressure because the lungs are open to the external atmosphere throughout the breathing cycle. The red curve for the transmural pressure difference is the same as in Fig. 8.9: the transmural pressure does not depend on the breathing process itself. The pleural pressure is negative across the regular breathing range (blue curve). This means the pleural gap has a propensity toward collapse during very slow breathing.

For tidal breathing, the dashed and dash-dotted curves show modifications between lung volumes of 3.0 L and 3.5 L: the pressure in the lungs is larger than atmospheric pressure during exhalation (upper dashed curve with arrow to the left) and smaller than atmospheric pressure during inhalation (lower dashed curve with arrow to the right). These pressure variations are needed to push or pull the air through the airways. These changes are mirrored in the pleural pressure (dash-dotted lines).

CONCEPT QUESTION 9.1

Which of the following statements about Fig. 9.1 is correct?

(A) The diagram does not allow us to calculate work done during respiration because the units kPa and L do not combine to the unit J.

(B) The absolute pressure in the lungs becomes negative during the breathing cycle.

(C) The transmural pressure increases faster during inhalation than it decreases during exhalation.

(D) The pressure in the pleural gap displays two different values at a given lung volume, one during inhalation and another one during exhalation.

What else can we learn from Fig. 9.1? Its major new feature, compared to Fig. 8.9, is the dynamic aspect of breathing: we no longer see the respiratory apparatus as static, but we perceive it as a process. The question then arises about the effort involved, that is, the amount of work per breath our body has to do. Can we calculate the work required for respiration from this figure? What role does the gas in the lungs play or, more precisely, how do the physical parameters of the inhaled air influence the physiological processes? To answer these questions, we need to further develop our understanding of the ideal gas.

9.2: Work on or by a Gas

We start the discussion of work with the model system shown in Fig. 9.2: an isolated container with an ideal gas includes a mobile ideal piston. Moving the piston from an initial to a final position allows us to vary the volume of the gas. The container/piston arrangement and its immediate environment form an isolated superstructure that allows us to account for work transfer, W, between the gas and the source of a force operating the piston. Note that the arrangement in Fig. 9.2 is sufficient to introduce the concept of work for a gas; later in the chapter we will then have to extend the system when we also include the exchange of heat based on Joule's experiments.

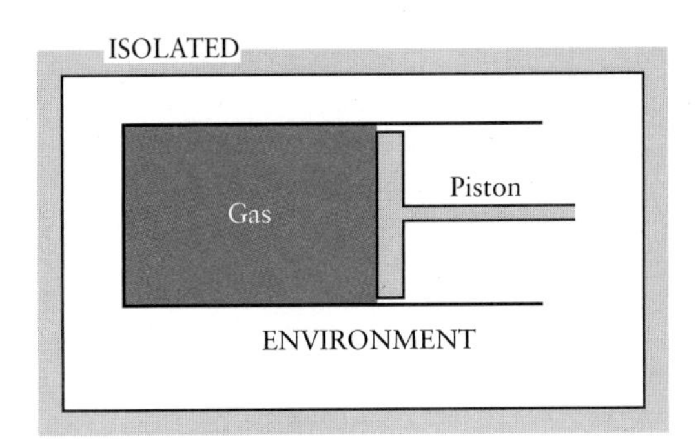

Figure 9.2 The model system to study the concept of work for a gas in a container. The container is confined by a mobile ideal piston. The piston allows us to vary the volume of the gas. Together, the box, the piston, and the box's immediate environment form an isolated superstructure.

We evaluate in Fig. 9.3 two ways in which the piston can move during a displacement Δx. Note that we also added the external force F_{ext}, which acts through the piston on the gas. As the two possibilities in the figure illustrate, it is important to understand that the external force is not the only force acting on the piston because the gas exerts a force in the opposite direction. However, we assume for all piston arrangements in this textbook that the external force acts parallel or anti-parallel to the displacement of the piston. Thus, vector notation is unnecessary and the work done on the system reads:

$$W = F_{ext}\,\Delta x = \frac{F_{ext}}{A}(A\,\Delta x). \qquad [9.1]$$

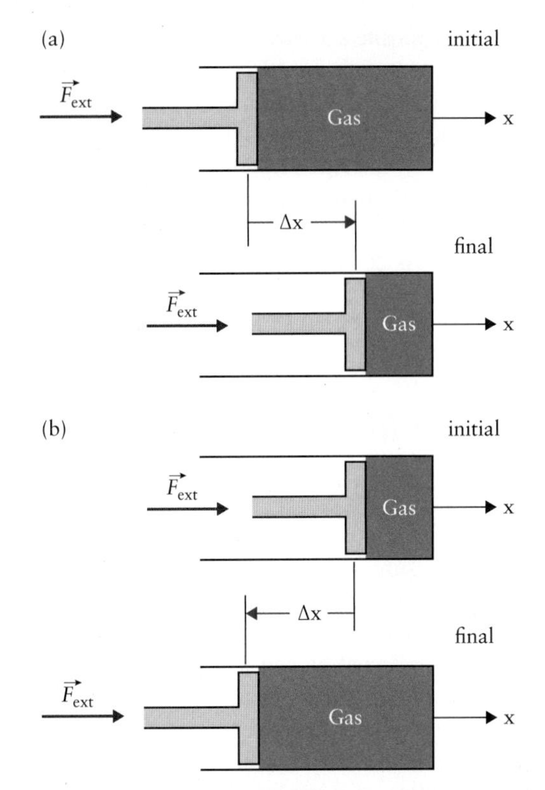

Figure 9.3 (a) A piston is pushed along the x-axis from an initial to a final state by an external force F_{ext}. This causes the piston to move a distance Δx toward the right, which leads to a compression of the gas. (b) Alternatively, the gas may expand and push the piston against an external force in the direction of the negative x-axis, i.e., toward the left.

We extended the basic formula for the work by a factor $A/A = 1$ in the last term. The area A in the numerator is combined with the displacement Δx to give the change in the volume: $\Delta V = A\ \Delta x$. The area A in the denominator is combined with the magnitude of the external force to give the pressure, $p = F_{ext}/A$. The work for a gas is then written in the form:

$$W = -p\,\Delta V. \qquad [9.2]$$

It is important to note the origin of the negative sign in this equation. As written in Eq. [9.1], the work is positive because the force F_{ext} acting on the gas and the displacement of the gas point in the same direction. You may think of the displacement of the gas as a shift to the right of its centre of mass. The subsequent multiplication with A/A in Eq. [9.1] has no effect on the sign. Proceeding from Eq. [9.1] to Eq. [9.2], we substitute the pressure p, which is the gas pressure when the system is in mechanical equilibrium. All absolute pressure values are positive. Studying Fig. 9.3(a) carefully though we note that the volume change introduces a negative sign because $A\ \ \Delta x$ in that figure is positive but ΔV is negative as the gas is compressed. The work for a **compression of a gas** is positive because work is done on the gas.

In Fig. 9.3(b), an **expansion of a gas** is studied. In this case, the external force and the displacement of the gas point in opposite directions and the work is negative. This is consistent with Eq. [9.2], in which p and ΔV are positive for Fig. 9.3(b). The work then is negative due to the additional negative sign on the right-hand side of Eq. [9.2].

In summary, the sign of the work for a gas system is determined by its volume change:

$$W > 0 \Leftrightarrow V_\mathrm{i} > V_\mathrm{f} \Leftrightarrow \text{gas compression}$$

$$W < 0 \Leftrightarrow V_\mathrm{i} < V_\mathrm{f} \Leftrightarrow \text{gas expansion.}$$

9.3: Work for Systems with Variable Pressure

The work definition for a gas in Eq. [9.2] has to be revised when the pressure varies during a compression or an expansion: In Eq. [9.2] a single value of p is assumed as the volume changes from its initial to a final value; i.e., the pressure must be constant during the expansion or compression. In this section we develop a method that allows us to determine the work when the pressure changes during the process; this method is based on a careful evaluation of what Eq. [9.2] implies mathematically in a p–V diagram.

9.3.1: Work as an Area in a *p*–*V* Diagram

We start with Fig. 9.4 with $p = p_0$, a constant pressure value for all volumes V, including the range from V_1 to V_2, which defines the volume change ΔV. The equation for the work in this case is $W = -p_0\ \Delta V$. Reading this formula as a geometric equation tells us that the purple area in Fig. 9.4 represents the absolute value of the work, because the area of a rectangle is given by the product of its height and width. In the figure, the width of the purple area is ΔV and the height is p_0. This observation can be generalized:

KEY POINT

The absolute value of work is the area under the curve of the pressure as a function of volume, p(V) in a p–V diagram. The area is taken between the initial and the final volumes of an expansion or a compression.

To obtain a correct value for the area A it is also important to ensure that the lower end of the area is taken at the absolute pressure $p = 0$. Note that we have to deal with the sign of the work separately; it is based on Fig. 9.3.

With these observations we can now quantify the work for a process during which the pressure varies. Such a case is illustrated in Fig. 9.5. The top part of the figure shows the p–V diagram of a gas expansion (red curve) from V_i to V_f. We write again $\Delta V = V_\mathrm{f} - V_\mathrm{i}$ for the volume change. This time we do not obtain a simple rectangular area under the curve like in Fig. 9.4 because the pressure increases (varies) during the volume increase. We have already seen that such processes exist, for example for the transmural pressure in Fig. 9.1.

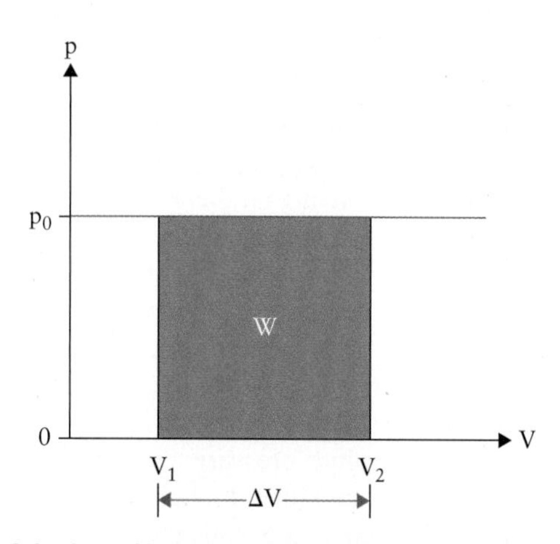

Figure 9.4 A graphical interpretation of the work for a gas that is represented in a p–V diagram. In the particular case the pressure is constant at value p_0, and the volume change is between volume V_1 and volume V_2 with $\Delta V = V_2 - V_1$. From Eq. [9.2] we find $W = -p_0\ \Delta V$, which corresponds to the purple area labelled A. Thus, work is quantified by the area A under the curve in a p–V diagram.

(a)

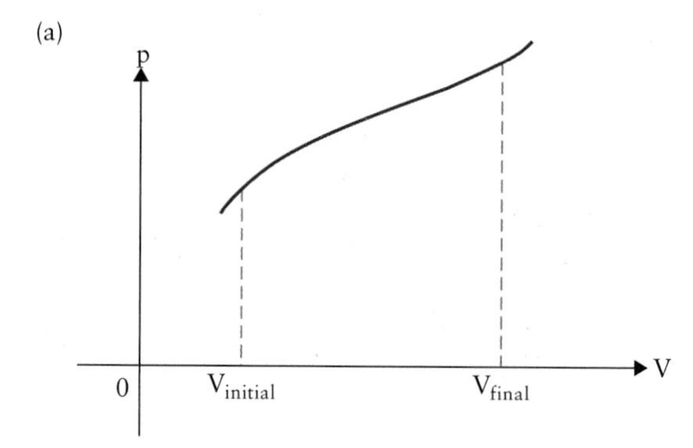

(b)

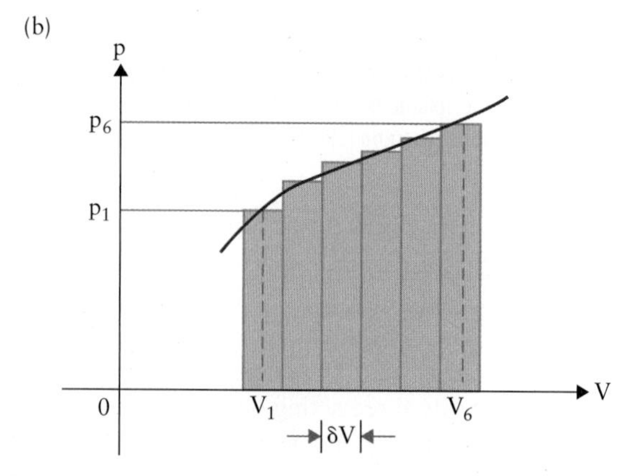

Figure 9.5 (a) A gas expansion from V_i to V_f with variable pressure. Mathematically, we write the pressure as a function of volume; i.e., $p = f(V)$. (b) Numerical approach to determining the work from part (a). The volume range V_i to V_f is successively divided into an increasing number of small intervals; shown is the case with six intervals of length $\delta V = (V_\mathrm{f} - V_\mathrm{i})/6$. The graph illustrates how the use of an increasing number of intervals allows us to estimate the actual work better and better.

9.3.2: Calculation of the Area in a *p*-*V* Diagram

Several mathematical methods exist that allow us to determine the area under the curve in Fig. 9.5(a). If the mathematical relation $p = f(V)$ is known, integral calculus allows us to obtain an analytical formula describing this area. In physiological applications, such as in Figs. 8.9 and 9.1, the curves are known only from connecting individual, experimentally measured data points. The numerical approach in these cases is illustrated in Fig. 9.5(b): the volume range of the process of interest from V_i to V_f is successively divided into an increasing number of shorter and shorter intervals δV. The area under the curve is then estimated by the summation of the individual, green-shaded areas. For example, the six rectangular areas in Fig. 9.5(b) are an estimate for the area under the curve in Fig. 9.5(a).

It is evident from Fig. 9.5 how this approach leads to a precise value for the area under the curve: we continuously divide the interval $V_f - V_i$ until we have an infinite number of individual intervals of length $\delta V = 0$. A good estimate for the area follows from a finite number of divisions of the volume range, combined with an extrapolation to an individual interval length of $\delta V = 0$. This method is a standard integration algorithm and is widely available as part of data analysis software packages.

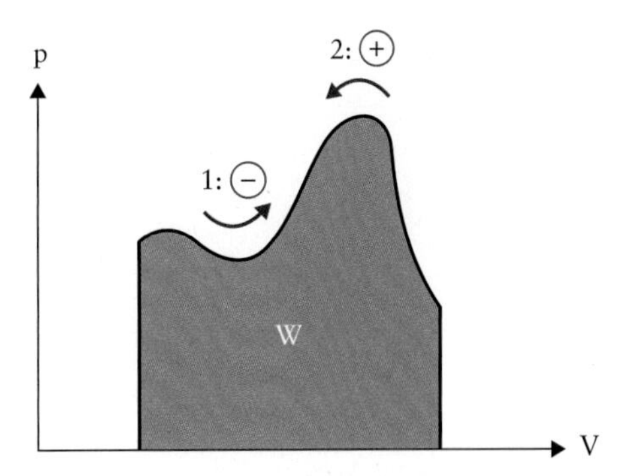

Figure 9.7 Sketch illustrating the two possible directions in which a process in a pressure versus volume diagram may occur. Path 1 is an expansion with a positive value for ΔV. Because pressure values are always positive, the sign in Eq. [9.2] leads to a negative value of work W. Path 2 is a compression with a negative value for ΔV. Combined with a positive pressure, this leads to a positive W value.

CONCEPT QUESTION 9.2

We consider the gas in Fig. 9.6 to be the system. In the process shown, is the work positive or negative?

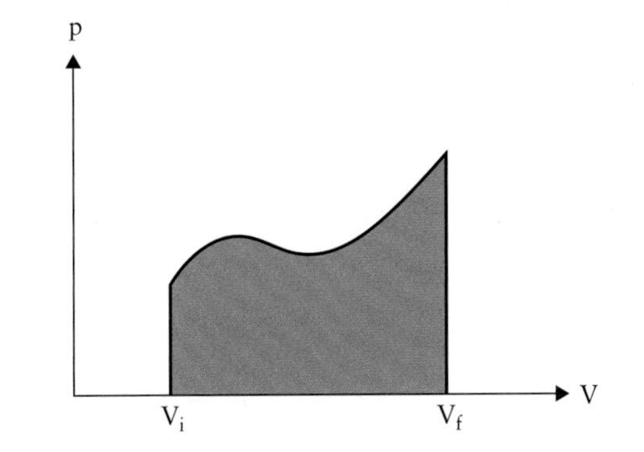

Figure 9.6 Sketch of a gas expansion with variable pressure.

CONCEPT QUESTION 9.3

Fig. 9. 8 shows three processes during which the volume of a gas expands from an initial volume V_i to a final volume V_f. Rank the work in each of the three cases, labelled (i), (ii), and (iii), starting with the highest value. *Hint:* –5 J < –2 J, but +5 J > +2 J.

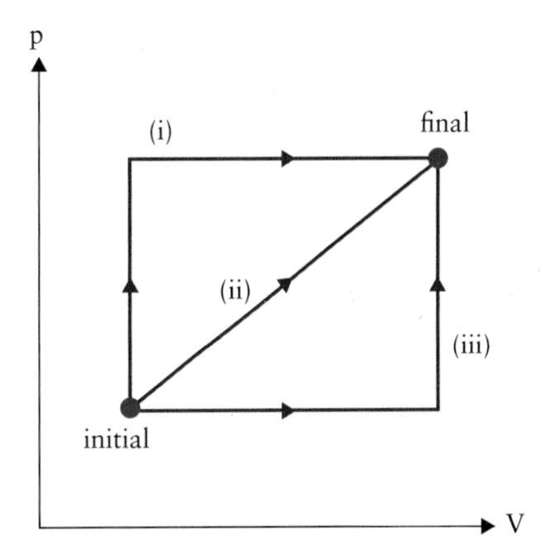

Figure 9.8 Three paths for an expansion of a gas system from an initial to a final volume. The paths are identified as (i), (ii), and (iii).

Fig. 9.7 is an alternative way to Fig. 9.3 and Eq. [9.2] to show the sign convention for gas expansions or compressions in a visual form. The red arrows for the two possible gas processes are accompanied by the respective sign of the work.

EXAMPLE 9.1

A gas is compressed from an initial volume of 8 m^3 to a final volume of 2 m^3. If the pressure depends on the volume as shown in Fig. 9.9, what is the work required to complete the compression?

Solution

We first determine the absolute value of the work. This value is obtained from the area under the curve in Fig. 9.9.

continued

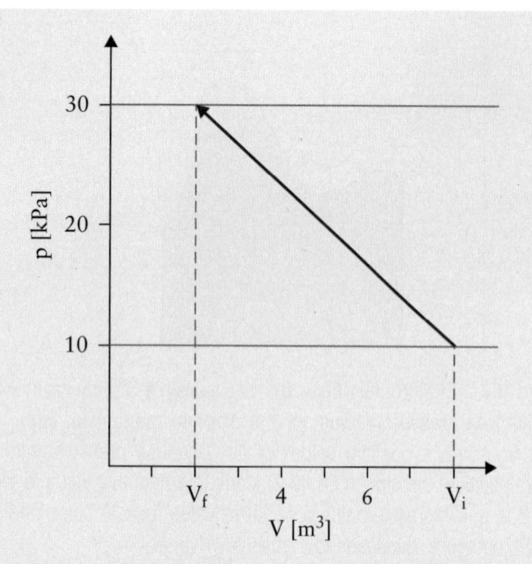

Figure 9.9 *p–V* diagram for a gas that is compressed from an initial volume of 8 m^3 to a final volume of 2 m^3. The pressure increases linearly during the compression: from 10 kPa to 30 kPa. This is an example of a system with varying pressure to which Eq. [9.2] cannot be applied.

Analyzing the area is straightforward because of the linear pressure increase. We add the area of the lower rectangle and the area of the top triangle:

$$A = (10\ \text{kPa})(8\ \text{m}^3 - 2\ \text{m}^3) + \frac{1}{2}(30\ \text{kPa} - 10\ \text{kPa})(8\ \text{m}^3 - 2\ \text{m}^3) = 1.2 \times 10^5\ \text{J}.$$

Note that the area of the lower rectangle must be included, because the area under the curve includes every part down to the line $p = 0$.

In the second step we determine the sign of the work. In a compression, the volume change is negative. Fig. 9.7 leads to a positive work $W > 0$. This means the gas receives $W = +1.2 \times 10^5$ J; in other words, this amount of work is done on the gas.

9.3.3: The Case of Human Respiration

EXAMPLE 9.2

Calculate the work for very slow breathing from the *p–V* diagram in Fig. 9.1 for (a) a single inhalation, (b) a single exhalation, and (c) a full breathing cycle. (d) The work done by muscles during breathing at rest is reported as 0.7 J/s in the physiological literature. How do you explain your results in parts (a) and (c) compared with this value? *Hint:* To simplify the calculations, approximate the pleural pressure curve in Fig. 9.1 as a straight line. This is shown in Fig. 9.10.

continued

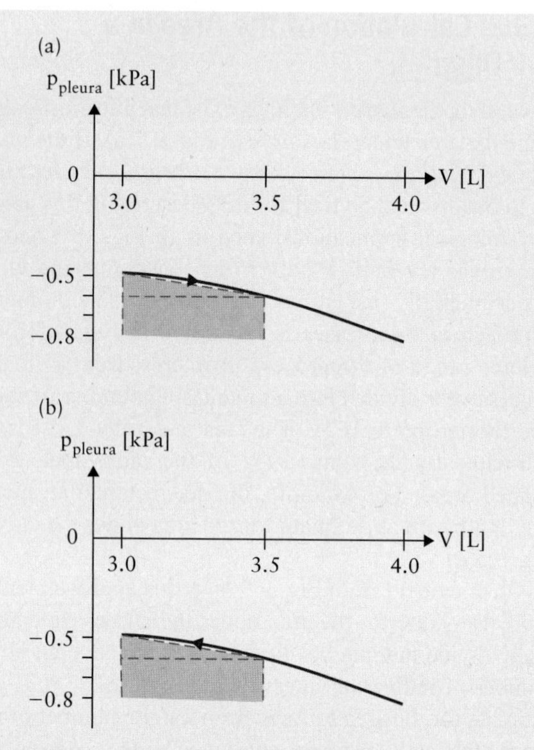

Figure 9.10 Geometric sketch of the area under the pleural pressure curve representing slow breathing between lung volumes of 3.0 L and 3.5 L (taken from Fig. 9.1).

Solution

Solution to part (a): We calculate the work required from the curve p_{pleura} versus lung volume in Fig. 9.1 because the active muscles act on the pleura and therefore most directly affect the pleural pressure. Fig. 9.10(a) shows the curve p_{pleura} versus lung volume for slow inhalation. Recall that the tidal volume is 0.5 L; i.e., a resting person inhales from a lung volume of $V_{exhaled}$ = 3.0 L to a lung volume of $V_{inhaled}$ = 3.5 L. Thus, the work required is the area under the curve between $V_{exhaled}$ and $V_{inhaled}$. This area is highlighted in grey in Fig. 9.10(a).

The area under the curve in Fig. 9.10(a) is divided into two parts: a triangle and a rectangle located below the triangle. The rectangle must be included because the work calculation requires the area under the curve down to absolute pressure zero. The area of the triangle is:

$$A_\Delta = \frac{1}{2}(0.5\ \text{L})(0.1\ \text{kPa}) = 0.025\ \text{J}.$$

To calculate the area of the rectangle, the absolute pressure at the horizontal dashed line is needed:

$$p_{\text{pleura, top of}} = p_{\text{atm}} - 0.6\ \text{kPa} = 1.013 \times 10^5\ \text{Pa} - 0.6 \times 10^3\ \text{Pa} = 1.007 \times 10^5\ \text{Pa}.$$

Thus, the rectangular area is:

$$A = (0.5\ \text{L})(1.007 \times 10^5\ \text{Pa}) = 50.35\ \text{J}.$$

continued

The work per inhalation is the sum of these two areas: $W_{inhale} = -50.375$ J.

The negative sign brings us back to the question of what we identify as the system and what we identify as the environment. Naively, we would have argued that the gas in the lungs is the system, as this gas expands from $V_{exhaled}$ to $V_{inhaled}$. However, the gas in the lungs does not do the work we calculated. Once inhaled, the only way the gas could do the work would be at the expense of its internal energy. This would require a cooling of the gas as discussed in Example 9.5. Cooling of the gas in the lungs is not possible. Instead the work originates in the muscles acting on the pleura. Thus, the environment receiving the work consists of the elastic tissues of the pleura and the lungs; the work is done by the active muscles (on behalf of the gas in the lungs). We will discuss this process in further detail in part (d).

Solution to part (b): Next we calculate the work associated with a single exhalation, as illustrated in Fig. 9.10(b). This figure again shows the curve p_{pleura} versus lung volume, with the work we want to calculate highlighted as the grey area under the curve between $V_{exhaled}$ and $V_{inhaled}$. The area under the curve in Fig. 9.10(b) is the same as in Fig. 9.10(a): the work per exhalation is $W_{exhale} = +50.375$ J. This value is positive because work is done on the system by its environment. Based on the same arguments we used in part (a), this result is interpreted in the following way: the work W_{exhale} originates from the elastic tissues of the pleura and the lungs; this work is done on the gas and the active muscles.

Solution to part (c): The work for a full breathing cycle is the sum of the work for a single inhalation and the work for a single exhalation. Using the two values we found in parts (a) and (b), we determine that:

$$W_{cycle} = -50.375 \text{ J} + 50.375 \text{ J} = 0 \text{ J}.$$

No net work has been done per cycle. From experience we know that this is not true; as the question in part (d) states, work is required during breathing. Thus, the assumptions we made to find this result must be revisited. We will do this by studying Fig. 9.1 for real tidal breathing in the next example. As we see in the current example, the assumption of very slow breathing oversimplifies the issue as it would imply that breathing can be done without effort.

Solution to part (d): Before we turn our attention to the actual breathing process, we can still learn more from the current example. In this example we assumed idealized, very slow breathing. For this we found in part (a) that the initial inhalation requires an effort that we quantified as slightly more than 50 J of work. Then we determined that the same amount of work is recovered during the subsequent exhalation, leading to no net effort for the entire cycle. This can be true only if the work done during the inhalation is indeed recovered by the system that did the work in the first step. Otherwise, per breathing cycle, work equivalent of 50 J has to be done and 50 J of energy would then be dissipated during the exhalation, leading to a net total work of –50 J per cycle.

continued

To clarify this issue, we begin with the actual, experimental value of the work done by the breathing muscles. This value is stated in the example text. Using 15 breathing cycles per minute for a resting person, 0.7 J/s corresponds to:

$$W_{cycle} = \frac{\left(0.7\frac{\text{J}}{\text{s}}\right)\left(60\frac{\text{s}}{\text{min}}\right)}{15\frac{\text{cycles}}{\text{min}}} = 2.8\frac{\text{J}}{\text{cycle}}.$$

This is only about 5% of the work we calculated as required for a single inhalation. Thus, in real breathing our muscles need to do only a small fraction of the work that has to be done during an inhalation. Consequently, something else must be doing the lion's share of that work. But what is that? Remember that we already ruled out the gas in the lungs.

The answer for the inhalation is the same as for the exhalation: the work is provided by the elastic action of stretched tissue, the lungs expanding and collapsing inside the pleura leading to an opposite collapsing and expanding of the pleural tissue. Recall that the dashed horizontal line in Fig. 8.9 indicates the response of pleura and lungs when the lungs are punctured: both tissues collapse, the lungs to a small volume and the pleura outward. Thus, the tissues are stretched in opposite directions. During inhalation the lungs expand and the pleura partially collapses. Then, during the following exhalation, the lungs partially collapse and the pleura expands. Thus, elastic energy is shifting back and forth between the lung tissue and the pleura tissue; only a small fraction of the work has to be done by the breathing muscles, accounting for energy losses during the energy transfer between lung tissue and pleural tissue.

EXAMPLE 9.3

Calculate the work for a regular breathing cycle due to air resistance during breathing. *Hint:* This work is given as the difference in work calculated for inhalation and exhalation in Fig. 9.1. To simplify the calculation, estimate this work from the four grey triangles highlighted in Fig. 9.11.

Solution

In the first step, the combined area of the four triangles in Fig. 9.1 is determined. Note that these triangles are identical, with each displaying a base of 0.25 L (e.g., the top left triangle has a base stretching from 3.0 L to 3.25 L), and a height of 0.1 kPa. Thus, their combined area is four times the area of each triangle:

$$A_{total} = 4\frac{1}{2}(0.25 \text{ L})(0.1 \text{ kPa}) = 0.05 \text{ J}.$$

continued

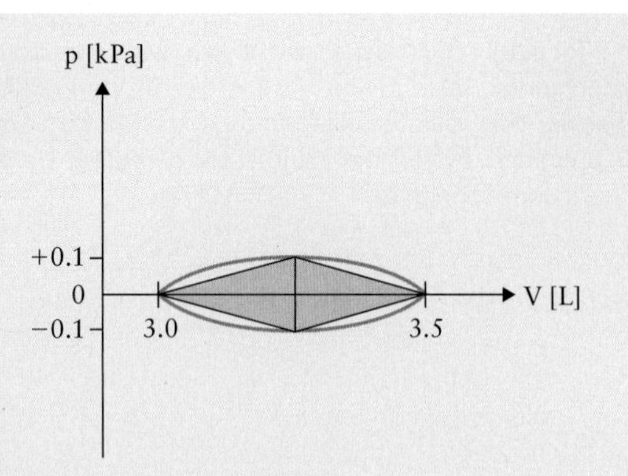

Figure 9.11 Geometric sketch of the area enclosed by the alveolar pressure (green line) during a tidal breathing cycle (based on Fig. 9.1).

The work in the cyclic process is positive; i.e., W_{cycle} = +0.05 J. Work has been done on the gas. This work is included in the work we calculated from experimental data as 2.8 J/cycle. We note that less than 2% of the respiratory work is required for forcing the air into and out of the lungs—the flow resistance of the gas is comparably small. A fish requires much more work to obtain the required flow of water through its gills because water is liquid and therefore has a much greater flow resistance than air.

EXAMPLE 9.4

Can the work required during inhalation (as calculated in Example 9.3) be provided by the internal energy of the air in the lungs? Compare your result also with Example 8.6, where we found that the total translational energy of the gas in the lungs of a standard man is U = 456 J.

Solution

Eq. [8.24] predicts the temperature change in the lungs if 50 J of energy—which is the energy for a single inhalation—were taken from the translational energy of the gas:

$$\Delta U = \frac{3}{2} n\, R(T_f - T_i).$$

This equation allows us to solve for the final temperature:

$$T_f = T_i + \frac{2\,\Delta U}{3\,n\,R}.$$

We substitute in this equation the change in internal energy of 50 J from Example 9.2 and n = 0.118 mol from Example 8.6 for the amount of gas in the lungs. With these values the right-hand side becomes:

$$T_f = 310\,\text{K} + \frac{2\,(-50\,\text{J})}{3\,(0.118\,\text{mol})\left(8.314\,\frac{\text{J}}{\text{K}\times\text{mol}}\right)},$$

which yields T_f = 276 K.

continued

Thus, the temperature in our lungs would sink close to the freezing point to accomplish just a single inhalation. Even though energy would flow fast into the lungs to maintain their temperature, during physical activities we are breathing at an accelerated rate and equilibration of the temperature would soon fall behind. Instead, the energy for the respiration work is provided in part by the various active components (muscles and diaphragm) and in part by the elastic action of the thorax.

The work of 50 J required in a single inhalation is slightly more than 10% of the translational energy of the gas in the lungs.

9.4: Heat and the First Law of Thermodynamics

The grand theme of Chapter 7 is the relationship between work and energy. We define energy as the capability of a system to do work. We then identified work as energy that flows in or out of a system and used this relation to quantify kinetic and gravitational potential energies by choosing a closed system in which only the energy form of interest varies. We will continue to use this approach throughout the remainder of the textbook as it is the best approach to introduce new energy forms, e.g., the electric potential energy in Chapter 18 and the elastic potential energy in Chapter 14.

The current section represents an exception to this approach. Focussing once more on the temperature of a non-mechanical system, we want to establish what we will call the thermal energy of a system. This is the energy that is stored in a system's temperature. Consistent with the approach summarized above, we will devise an experiment in which only the temperature of a system changes. However, unlike all other cases, we will not be able to identify the energy that flows into or out of the system as work. Due to its different nature, this energy flow is given the name **heat**.

Naively, one could argue that we should redefine the term work such that in includes heat. This would be possible if heat and work were just two forms of the same thing. However, this is not the case. Heat is distinct from work, as will become clear when we discuss the second law of thermodynamics in the next chapter.

There is also a difference in everyday use of the English language when we express the addition or removal of work versus heat from a system. For heat, terms such as *received* or *released* are usually used, which more clearly express both the transfer and direction of energy flow.

In the current section, we choose a generic non-mechanical system; the specific system of an ideal gas will then be considered in the following chapter where we introduce the fundamental processes of thermodynamics.

9.4.1: Joule's Experiments

Anders Celsius's invention of the thermometer made a proper definition of thermal energy possible. In 1798, Benjamin Thompson (Count Rumford) concluded from cannon drilling experiments that mechanical work leads to heating that in turn represents an increase in the thermal energy of the system. In 1842, Julius Robert von Mayer studied human blood in a tropical environment and found a correlation between the thermal energy and the work obtained from chemical energy. Herrmann von Helmholtz, who formulated the first law of thermodynamics, credited it to Mayer's work.

In 1843, James Prescott Joule defined both terms, thermal energy and heat, and connected them quantitatively to the concepts of work and mechanical energy. His arguments were based on two experiments shown in Fig. 9.12: the experiment in part (a) allows us to change exclusively the temperature in the system of interest, which is the water in a beaker. Thus, this experiment is used to define heat. The experiment in part (b) then relates the transfer of energy in the form of heat and work.

Joule used the first experiment to define heat, Q, in the following manner. Heat is provided from a combustible gas burner. We argue, even without further insight into the nature of heat, that burning twice the amount of the same gas will release twice the amount of heat. This variable amount of heat Joule added to a system that consisted of a beaker with water. The water temperature was measured with a thermometer. Joule showed that the system temperature is directly proportional to the heat added to the system, $Q \propto \Delta T$. Joule further noted that the heat needed to achieve a particular temperature increase is directly proportional to the amount of water in the beaker, $Q \propto m(H_2O)$. Finally, the heat needed for a given temperature increase of a fixed amount of liquid varied from liquid to liquid. This means that a materials-specific constant is needed. These observations allowed Joule to write the following **definition of heat**:

$$Q = c\, m\, \Delta T, \qquad [9.3]$$

in which the material constant c is called the **specific heat capacity**. Table 9.1 provides a list of values for c. The heat flowing into the beaker is then identified as the change in the thermal energy of the system:

$$Q = \Delta E_{\text{thermal}} = E_{\text{thermal,f}} - E_{\text{thermal,i}}. \qquad [9.4]$$

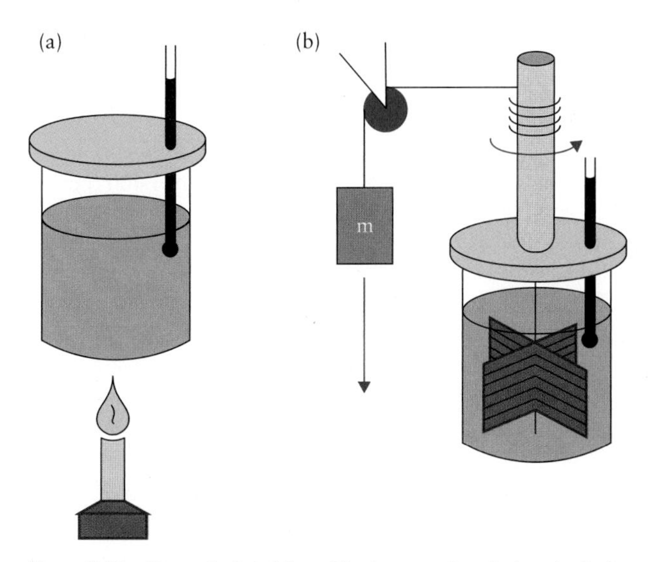

Figure 9.12 Conceptual sketches of the two experiments done by Joule. (a) In the first experiment heat is quantified by measuring the amount of gas burned. (b) The second experiment illustrates conceptually the equivalence of mechanical and thermal energy. The mechanical energy, which is released by the falling object at the left, is used to heat the water by operating a stirrer.

TABLE 9.1

Specific heat capacity values at atmospheric pressure

Material	Specific heat capacity (J/kg K)
Water	4186
Ice	2090
Steam	2010
Aluminium	900
Iron	448
Glass	837
Copper	387
Mercury	138
Gold	129

Historically, the unit **calorie** (cal) was introduced for heat, and the unit **Calorie** (Cal) for the energy content in food, with 1 Cal = 1000 cal. You should always convert these units to the SI unit joule. We note further that J/(kg K) is the unit of the specific heat capacity. *Heat capacity* is the proper term because a large heat capacity means the material can absorb more heat than a material with a small specific heat capacity before the temperature rises by a given amount.

For many applications, Eq. [9.3] is rewritten to replace the mass of the liquid by the amount of material in unit mol. For this step we define the number of moles n as:

$$n = \frac{m}{M},$$

in which m is the mass of the material and M is its molar mass. This leads to:

$$Q = C\, n\, \Delta T, \qquad [9.5]$$

with C the **molar heat capacity**, which is a material constant with unit J/(kg K). Values for both heat capacities are tabulated; when you find such values in the literature, always confirm the units used and then use either Eq. [9.3] or Eq. [9.5].

Eqs. [9.3] and [9.5] allow for an exchange of energy in two directions: heat can flow into or out of a system. This is consistent with our observation for work: a system can do work on a piston or the piston can do work on the system. Thus, we need to adhere to a strict sign convention for both work and heat.

KEY POINT

Any amount of heat flowing into the system and any work done on the system are positive, as they increase the total energy of the system. The opposite processes are negative, as they lead to a reduction of the total energy of the system.

When checking a calculation, always identify yourself with the system: whatever you receive is positive, whatever you give away is negative.

We want to emphasize once more the difference between heat and thermal energy in a physiological context. The temperature detectors in the bodies of endotherms that we discussed in Chapter 8 obviously are not needed in ectotherms, as their body temperature adjusts to the environmental temperature automatically. Thus, ectotherms are usually not able to measure the thermal energy content in their body. However, some ectotherms are able to measure heat arriving from the environment. They do this in a different manner and for a different purpose: vipers are nocturnal hunters that possess **heat detectors** between their eyes and nostrils to allow them to find their endothermic prey without visual contact.

9.4.2: General Form of the Conservation of Energy

Next, we examine Joule's second experiment in Fig. 9.12(b). The idea behind this experiment is that the object on the left falls, moving the stirrer through the water. Because the water resists the motion of the stirrer, the frictional interaction between the stirrer and the water heats the water. Note that Joule chose friction to achieve a non-mechanical energy conversion. He was motivated by the observation that friction is responsible for the sensation of warmth when you rub your hands against each other.

To allow us to quantify the conversion of potential to thermal energy, we need to ensure it is not converted into the kinetic energy of the falling object; i.e., we must ensure the object falls with a constant speed. Experiments such as the one sketched in Fig. 9.12(b) show that thermal energy and mechanical energy are equivalent. In Joule's case, the lost potential energy ΔE_{pot} is quantitatively converted into thermal energy $\Delta E_{thermal}$:

$$-m_{block}\, g\, \Delta y = c_{H_2O}\, m_{H_2O}\, \Delta T, \qquad [9.6]$$

which is generalized to:

$$E_{pot} + E_{thermal} = \text{const.} \qquad [9.7]$$

CASE STUDY 9.1

In using the experiment in Fig. 9.12(b) to prove the equivalence of thermal and mechanical energy, what should *not* happen or should *not* be done?

(A) The liquid will become warmer.

(B) The object suspended from the string will approach the ground.

(C) A liquid other than water should be used.

(D) The stirrer axis will spin faster and faster.

(E) The ambient pressure will rise above the normal air pressure.

Answer: *(D). The first two choices must happen for the experiment to work; the third and final choices are not excluded in Eq. [9.6] if we properly exchange the specific heat of water for that of the liquid used,* c_{liquid}*. In the case of (D), we convert potential into kinetic energy; i.e., we do not convert mechanical to thermal energy.*

EXAMPLE 9.5

(a) Calculate the work that an object with m = 400 g can do as a result of falling a distance of 3 m. (b) Assume that the object falls into 10 L of water in an isolated beaker. If the entire kinetic energy of the object is converted to thermal energy, by how much does the water temperature rise? *Hint:* For the specific heat capacity of water use the value from Table 9.1, and neglect the heat the object can absorb itself.

Solution

Solution to part (a): By dropping the object 3 m, its potential energy has been reduced by:

$$\Delta E_{pot} = m\, g(h_f - h_i)$$
$$= (0.4\ \text{kg})\left(9.8\frac{\text{m}}{\text{s}^2}\right)(-3.0\ \text{m}) = -11.8\ \text{J}.$$

continued

Solution to part (b): When the object is released and falling toward Earth, it is converting potential to kinetic energy. When it comes to rest in the isolated beaker its kinetic energy must convert to thermal energy, $\Delta E_{thermal} = +11.8$ J. This value is positive since the released mechanical energy is added as thermal energy to the system water. We use Eq. [9.3] in the form:

$$\Delta E_{\text{thermal}} = c_{H_2O}\, m_{H_2O}\, \Delta T$$

to determine the associated change in the water temperature, ΔT:

$$\Delta T = \frac{\Delta E_{\text{thermal}}}{\rho_{H_2O}\, V_{H_2O}\, c_{H_2O}},$$

in which ρ_{H_2O} is the density of water: $\rho_{H_2O} = 1.0$ g/cm^3. The value for c_{H_2O} in the denominator is taken from Table 9.1. This leads to:

$$\Delta T = \frac{11.8\text{ J}}{\left(1.0\frac{\text{kg}}{\text{L}}\right)(10\text{ L})\left(4.186\frac{\text{kJ}}{\text{Kkg}}\right)}$$

$$= 2.8 \times 10^{-4}\text{K}.$$

We find only a negligible temperature change!

9.4.3: Formulation of the First Law of Thermodynamics

Joule's experiment in Fig. 9.12(b) can be interpreted in two ways:

- We consider the falling object, the stirrer, the beaker, and the water together as an isolated system, or
- We consider the water in the beaker as the system and the stirrer and the falling object as the environment.

We use the above discussion of Fig. 9.12(b) to interpret the implications for the total energy of the isolated superstructure, for the conversion of energy between different forms, and for the flow of energy between system and environment.

9.4.3.1: A First View of Fig. 9.12(b)

We first consider all components in the figure as the system. In this case, Eq. [9.7] states that energy conservation applies also when non-mechanical energies, such as thermal energy, are involved. In 1847, Hermann von Helmholtz generalized this finding and formulated the **first law of thermodynamics for an isolated system**:

KEY POINT

Conservation of energy: the sum of all energy forms in an isolated system is constant.

The sum of all energy forms in the system is the **internal energy of the system**. As we introduced already in Chapter 8 for the ideal gas, the internal energy is labelled U. Thus, the energy conservation for an isolated system is written as:

$$\Delta U_{\text{isolated system}} = 0. \qquad [9.8]$$

This is a law since it predicts the outcome of future experiments. It even correctly predicted the outcome of experiments that Joule and Helmholtz could not have imagined in their days, such as Enrico Fermi's 1942 nuclear fission experiment that led to the development of modern nuclear power technology. The first law of thermodynamics is not limited to mechanical systems and therefore surpasses the range of applicability of Newton's laws.

9.4.3.2: A Second View of Fig. 9.12(b)

Now we consider only the water as the system. That means the left-hand side in Eq. [9.6] is a work term because it represents energy that flows into the system. We combine this observation with Fig. 9.12(a), in which heat is flowing into a system. In both cases the energy flowing leads to the same change in internal energy. This allows us to formulate the **first law of thermodynamics for a closed system**:

KEY POINT

The sum of all energy forms in a closed system changes by the amounts of heat and work that flow between system and environment.

This is written as:

$$\Delta U_{\text{closed system}} = Q + W. \qquad [9.9]$$

In this equation, neither the order in which the exchange of energy with the environment occurs nor its form (heat or work) matter for the change of the internal energy as the system goes from an initial to a final state. A system property that depends not on its detailed history but only on its current state is called a **variable of the state**. Thus, the internal energy is a variable of the state of the system, like temperature, pressure, and volume. The change of a variable of the state is zero for any sequence of processes that returns the system to its original state. Such processes are called cyclic processes and are the subject of the next chapter.

EXAMPLE 9.6

A standard man performs the standard ergometric test illustrated in Fig. 9.13. The test consists of a 6-m horizontal run followed by running up a staircase for at least 9 steps. For our calculations we focus only on the second part of the test, in which the person moves upward by $h = 1.05$ m from the third to the ninth step. For this part

continued

of the test the person requires an amount of 0.8 kcal of stored (food) energy. How much heat must the body of the person dissipate as the result of this part of the ergometric test?

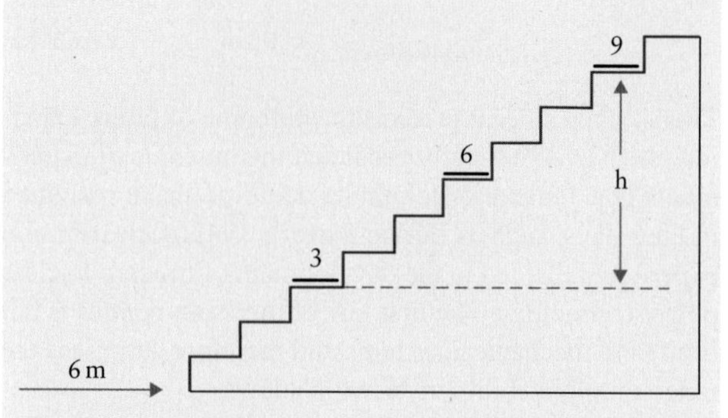

Figure 9.13 Standard ergometric test. The person must run 6 m toward a staircase and then move upward at least 9 stairs. Contact mats on the third, sixth, and ninth stairs record the motion. The height difference between the third and the ninth step is $h = 1.05$ m.

Supplementary physiological information: The step test in Fig. 9.13 is usually supplemented by contact mats on every third step to allow a measure of the time the patient needs to complete the test. It was considered the standard ergometric test until stationary devices such as bicycle ergometers or the treadmill came into use. A stationary bicycle ergometer is shown in Fig. 7.1. It is used for exercise ECGs in which several vital parameters are measured on the patient during the test. To test athletes for fitness, coaches use measurements of oxygen consumption, heart rate, and lactose levels in the blood of the athlete to predict performance in track and field events.

Solution

The problem is solved using conservation of energy for the person moving between the initial position on the third step and the final position on the ninth step. Treating the person as an isolated system in this experiment, the formula for the conservation of energy (Eq. [9.8]) can be written in the form:

$$\sum_i E_i = \text{const}$$
$$= E_{\text{pot}} + E_{\text{kin}} + E_{\text{thermal}} + E_{\text{chemical}}.$$

The food energy is labelled E_{chemical}. The kinetic energy in the equation can be neglected when assuming that the person's speed between the third and ninth step does not change. For the other three forms of energy, we use the last equation in the form:

$$E_{\text{pot,i}} + E_{\text{thermal,i}} + E_{\text{chemical,i}} = E_{\text{pot,f}} + E_{\text{thermal,f}} + E_{\text{chemical,f}},$$

which is written as:

$$\Delta E_{\text{pot}} + \Delta E_{\text{thermal}} + \Delta E_{\text{chemical}} = 0.$$

continued

Each term represents the difference between the final and initial amounts of energy in the system. The thermal energy difference, $\Delta E_{\text{thermal}}$, is sought in the problem; the other two terms are:

$$\Delta E_{\text{pot}} = m\,g(h_{\text{f}} - h_{\text{i}})$$
$$= (70\text{ kg})\left(9.8\frac{\text{m}}{\text{s}^2}\right)(1.05\text{ m}) = +720\text{ J}$$

and:

$$\Delta E_{\text{chemical}} = -800\text{ cal} = -3350\text{ J}.$$

The potential energy difference is positive because the final potential energy is higher than the initial potential energy of the body. The food energy is negative because the body stores this amount less at the end of the test. Substituting these values yields:

$$\Delta E_{\text{thermal}} = 3350\text{ J} - 720\text{ J} = +2630\text{ J}.$$

The thermal energy released in the process is about 2.6 kJ. It must be dissipated by the body in the form of heat; otherwise, as it would cause a dangerous temperature increase.

9.5: The Physics of the Respiratory System

We discussed intensively in the previous and the current chapter the physical parameters of the respiratory system. The key information on the system was presented in two p–V diagrams, Fig. 8.9 for the respiratory system at rest and in Fig. 9.1 for dynamic breathing of a person. In the current section we want to summarize what we have learned about the respiratory system. We do this by developing a physical model for respiration.

The best way to introduce this model is to start from the anatomical sketch of the thorax/lungs system in Fig. 8.8 and establish the important components in steps. We begin with the lungs. The figure indicates that no active muscle forces act directly on the lungs. The only direct interaction occurs with the air inside at pressure p_{alveoli} and with the pleura at pressure p_{pleura}. Recall that p_{alveoli} and p_{pleura} are gauge pressures, i.e., pressure values relative to atmospheric pressure. The suitable physical model we develop for the lungs is shown in the three parts of Fig. 9.14.

9.5.1: Fig. 9.14(a): The Lungs as a Balloon with a Piston

In the figure, the lungs are modelled by a balloon with a flexible outer membrane. The balloon has an internal pressure p_{alveoli} with an absolute outside pressure of atmospheric pressure. For such a balloon to remain inflated

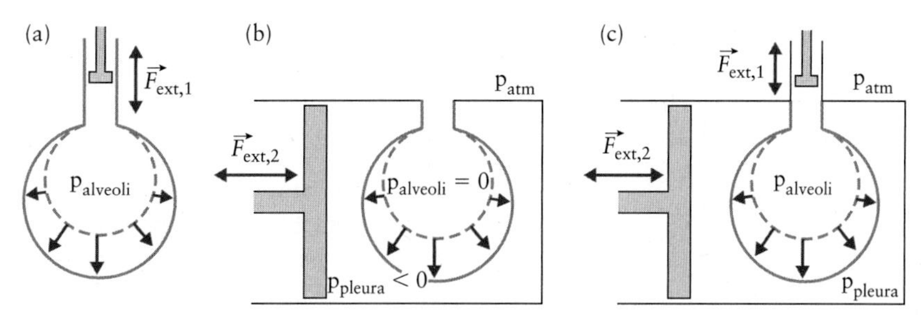

Figure 9.14 Three models used in the description of the respiratory system. (a) The lungs are represented by a balloon. The gas pressure in the balloon is varied by a piston (subscript 1). We define the gas pressure in the balloon as a gauge pressure (alveolar pressure $p_{alveoli}$). A pressure increase in the balloon causes an increase of its volume. Note that the pressure surrounding the balloon is atmospheric pressure. (b) The pleural gap is represented by a container that encloses the balloon. The balloon is open to the external atmospheric pressure. The container allows us to control the pressure surrounding the balloon (using the piston with subscript 2). The pressure surrounding the lungs is given as a gauge pressure, the pleural pressure p_{pleura}. (c) In the last model, two pistons are introduced: one to control the alveolar pressure and one to control the pleural pressure.

we require $p_{alveoli} > 0$. A piston is attached to the balloon, which allows us to vary an external force $\vec{F}_{ext,1}$, which enables us in turn to vary $p_{alveoli}$. From our experience in inflating children's balloons we know that the volume of the balloon increases with $p_{alveoli}$ and that, as the balloon becomes bigger, larger and larger pressure increases are needed to achieve a given size increase. This observation matches with the curve labelled $p_{alveoli} - p_{pleura}$ in Fig. 8.9, which is the transmural pressure difference from lungs to pleura, but it is not consistent with the curve labelled $p_{alveoli}$ in that figure. Thus, Fig. 9.14(a) is not a useful model for the lungs. Missing from the model are the pleura and the fact that the pleural pressure can differ from atmospheric pressure; i.e., $p_{pleura} \neq 0$ is possible.

9.5.2: Fig. 9.14(b): The Pleura as a Container with Piston Surrounding the Lungs

To improve the proposed model, the pleura is added as a container that encloses the balloon. This container is sealed by a second piston (shown at the left), with which we can exert a second external force, $\vec{F}_{ext,2}$. Moving the piston allows us to change p_{pleura}. In this first extension of our original model, the first piston has been removed and the balloon is open to the outside air. This means that $p_{alveoli} = 0$ and the balloon can be inflated only when $p_{pleura} < 0$.

Comparing the first two parts of Fig. 9.14 illustrates that it is neither one of the absolute pressures nor one of the gauge pressures that ultimately matters for the volume of the balloon: when we increase $p_{alveoli}$ in Fig. 9.14(a) and decrease p_{pleura} in Fig. 9.14(b) such that the balloon has the same volume in both cases, we find $+p_{alveoli}$ in part (a) equals $-p_{pleura}$ in part (b). Thus, the volume of the balloon depends on the pressure difference across the balloon's membrane. Note that this pressure difference is not a gauge pressure, as we will see when discussing the model in Fig. 9.14(c), where $p_{alveoli} \neq 0$ and $p_{pleura} \neq 0$.

9.5.3: Fig. 9.14(c): Container and Enclosed Balloon Are Sealed by a Piston

This model for the lung/pleura system is slightly more complex. The additional complexity comes from varying both $p_{alveoli}$ and p_{pleura}, which is possible because the model has two pistons, one attached to the balloon to exert the force $\vec{F}_{ext,1}$ and one attached to the container surrounding the balloon to exert the force $\vec{F}_{ext,2}$. It is important to introduce this model because we need both pistons when describing the various curves in Figs. 8.9 and 9.1: the model in Fig. 9.14(c) is used to characterize static properties of the respiratory system, i.e., when the test person is asked to inhale or exhale a certain amount of air, then stop (mouthpiece valve closed to spirometer) and relax the breathing muscles. The model in Fig. 9.14(b) is used when studying very slow (airflow resistance free) breathing with an open mouth. The open mouth guarantees that the pressure in the lungs is always equal to atmospheric pressure.

9.5.4: The Respiratory Curves at Rest

We first discuss in detail how the model in Fig. 9.14(c) explains the data shown in Fig. 8.9. The force $\vec{F}_{ext,1}$ is required to obtain $p_{alveoli} \neq 0$. The force $\vec{F}_{ext,1}$ can be exerted by two sources: (i) the streaming of air in or out of the lungs, and (ii) the propensity toward elastic collapse of the balloon-like lungs. When neither of these sources acts on the system, e.g., when you open your mouth and relax, then $p_{alveoli} = 0$ and the respiratory system is in mechanical equilibrium. In a simple self-experiment you can confirm that this state is reached after regular exhalation at a lung volume of 3.0 L: Concentrate on your chest as you breathe very slowly. Hold your breath after a regular inhalation and after a regular exhalation. You notice that you are most relaxed after exhalation because the forces on the lungs are balanced and do not need support through muscle action.

The airflow-related force contribution to $\vec{F}_{ext,1}$ is excluded in Fig. 8.9 because this p–V diagram represents a static view of the respiratory system. Indeed, airflow-related issues enter the discussion only when we discuss Fig. 9.1. This leaves the elastic force of the lungs—which is their propensity toward an elastic collapse—as the only force contributing to $\vec{F}_{ext,1}$ in Fig. 8.9.

The force $\vec{F}_{ext,2}$ is needed to allow variations of the pleural pressure, i.e., $p_{pleura} \neq 0$ for the pressure in the container surrounding the balloon. Again, two different sources can provide this force: (i) the active muscle forces exerted by the diaphragm and the intercostal muscles between the ribs, and (ii) the elastic propensity of the thorax to collapse (outward toward the rib cage).

The active muscle forces are not included in the force $\vec{F}_{ext,2}$ in Fig. 8.9 because the curves in that p–V diagram are drawn for a relaxed chest. Thus, the piston exerting the force $\vec{F}_{ext,2}$ in Fig. 9.14(c) addresses the tendency of the thorax to change toward its elastic equilibrium position. We conclude that both pistons in Fig. 9.14(c) model elastic forces when interpreting Fig. 8.9: one associated with the propensity toward elastic collapse of the balloon-like lungs, and the other associated with the propensity of the thorax to move toward its equilibrium position. For the healthy person, these two forces are not independent because a fixed amount of fluid in the pleural gap provides for strong adhesion between the visceral layer of the pleura on the lung surface and the parietal layer on the inside of the thorax. We can observe the action of both elastic forces independently for a patient with a punctured pleura. The injury allows air or fluids to enter the gap between lungs and thorax, and each relaxes to its equilibrium shape at zero gauge pressures $p_{alveoli}$ and p_{pleura} (along the dashed horizontal line in Fig. 8.9): the lungs collapse to a minimum volume of less than 1.0 L and the thorax widens to a size corresponding to a 4.5-L capacity of the lungs.

For the healthy person, the curve labelled $p_{alveoli} - p_{pleura}$ in Fig. 8.9 indicates the elastic response of the lungs to the transmural pressure difference; this pressure difference is always positive since otherwise the lungs could not stay inflated. In the range of normal breathing (breathing of the tidal volume between lung capacities of 3.0 L and 3.5 L), the transmural pressure difference varies between 500 Pa and 600 Pa. For lung volumes above 3.0 L, $p_{alveoli}$ is positive; i.e., upon opening the mouth, air will stream out of the lungs. In turn, for lung volumes below 3.0 L (after a deep exhalation), $p_{alveoli}$ is negative; i.e., opening the mouth will cause air to stream into the lungs.

The pleural pressure p_{pleura} always lies below the alveolar pressure $p_{alveoli}$, but crosses from negative to positive values at about 4.5-L lung capacity. At that point, the direction of stress in the thorax is inverted: for smaller volumes the thorax wants to expand, and for larger volumes it wants to contract. This is the reason why it is particularly hard for a test person to hold more than 4.5 L of air in the lungs.

9.5.5: Dynamic Breathing

In the next step toward understanding the actual processes during breathing we evaluate the solid lines in the p–V diagram in Fig. 9.1, for which the model in Fig. 9.14(b) applies. Note that the piston allowing for the force $\vec{F}_{ext,1}$ is not present in that figure. Instead, the balloon-like lungs are open to the external air pressure and $p_{alveoli} = 0$ at all lung volumes. We still neglect effects due to airflow resistance, for which the second piston would be required. In Fig. 9.14(b) the pressure p_{pleura} can still vary. The pleural pressure is changing now due to the action of both possible forces: the active muscle forces and the elastic force of the thorax. The need for an active muscle force is evident from Fig. 8.9: if $p_{alveoli} = 0$, and p_{pleura} were adjusted only by elastic forces in the thorax, then the only possible lung volume would be $V = 3.0$ L. However, we cannot breathe at a fixed lung volume. Thus, an active mechanism to vary the lung volume is required.

Breathing is modelled in Fig. 9.14(b) by the back-and-forth motion of the piston. During that motion the gauge pressure p_{pleura} changes and with it the pressure difference $p_{alveoli} - p_{pleura} = -p_{pleura}$ (because $p_{alveoli} = 0$). The variation of these pressure values corresponds to the different lung volume values shown by the solid lines in Fig. 9.1. We decrease the pleural pressure during very slow inhalation by expanding the thorax because the lungs expand faster. The effect is reversed during slow exhalation.

Note that the curve of the lung volume versus the pressure difference $p_{alveoli} - p_{pleura}$ is the same in Figs. 8.9 and 9.1: the volume of the lungs depends exclusively on the transmural pressure difference, as we have already concluded from comparing the first two parts of Figs. 9.14. We calculated in Example 9.2 the work done during very slow breathing using the p–V diagram in Fig. 9.1. We found from Fig. 9.10 that the work for a full breathing cycle is zero! This led us to dismiss the slow breathing model, as we know from experiments that a person requires 0.7 J/s or 2.8 J/cycle of work during tidal breathing. On the other side, we showed in Example 9.3 that this work is much less than the corresponding area under the breathing curve between 3.0 L and 3.5 L in Fig. 9.1.

The results in Examples 9.3 and 9.4 suggest that the breathing work is provided by the elastic action of stretched tissue. During inhalation the lungs expand and the pleura partially collapses. Then, during the following exhalation, the lungs partially collapse and the pleura expands. Elastic energy is shifting back and forth between the lung tissue and the pleural tissue; only a small fraction of the work has to be done by the breathing muscles, accounting for energy losses during the energy transfer between lung and pleural tissue.

The small fraction of energy needed to breathe is captured in the model shown in Fig. 9.14(c). The larger piston allows us to vary the pressure in the pleura surrounding the lungs, and the smaller piston allows us to vary the pressure inside the lungs. We need both abilities: the processes we just discussed for slow breathing continue to dominate the regular breathing process, including the alternating expansion and collapse of lungs and pleura to conserve the elastic energy. In addition, we need to account for the observation that air has to be pulled actively through the trachea into the lungs during inhalation, and that air has to be pushed actively out of the lungs during exhalation. The small piston in Fig. 9.14(c) allows for this to occur: during inhalation the pressure in the lungs is lowered by up to 100 Pa, and during exhalation the pressure in the lungs is increased by up to 100 Pa. This is illustrated in Fig. 9.15. These pressure differences are added to the slow breathing as dashed and dash-dotted lines in Fig. 9.1. Affected are only the pleural pressure and the pressure in the lungs; the transmural pressure remains unchanged because that pressure difference exclusively governs the size (volume) of the lungs. From Fig. 9.1 the fraction of work associated with the flow resistance of air in the trachea has been calculated for a full breathing cycle in Example 9.4.

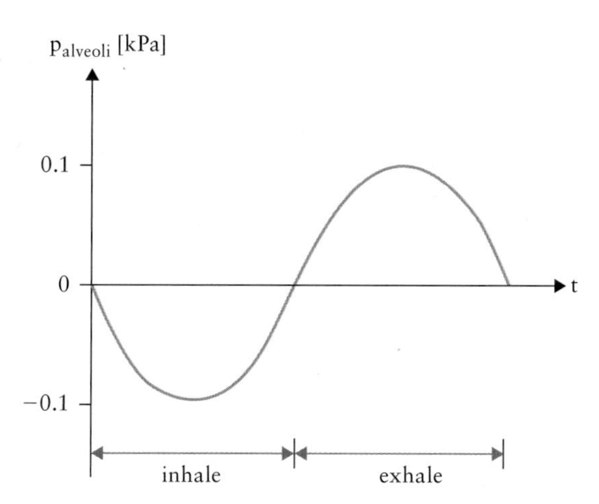

Figure 9.15 Alveolar pressure $p_{alveoli}$ as a function of time. The pressure in the lungs drops to 100 Pa below atmospheric pressure during inhalation, and exceeds the atmospheric pressure by up to 100 Pa during exhalation. These pressure variations allow the air to flow through the airways against the flow resistance.

SUMMARY

DEFINITIONS

- Work W:
  - for a gas in a container, pressure constant during volume change:

$$W = -p\,\Delta V$$

  - sign convention for pressure-related work (gas processes): when the gas undergoes an expansion, $V_i < V_f$, we use $W < 0$; when the gas undergoes a compression, $V_i > V_f$, we use $W > 0$.
- Heat Q
  - for a liquid or solid system of mass m given in unit kg:

$$Q = c\,m\,\Delta T,$$

    in which c is the specific heat capacity.
  - for a liquid or solid system of amount n in unit mol:

$$Q = C\,n\,\Delta T,$$

    in which C is the molar heat capacity.
  - sign convention for heat: heat flowing into the system is positive; heat leaving the system is negative.
- Energy E of a system in equilibrium:
  - for any type of energy except thermal energy:

$$\Delta E = E_f - E_i = W$$

    for kinetic energy this is called the work–kinetic energy theorem.
- for thermal energy:

$$\Delta E_{thermal} = E_{thermal,f} - E_{thermal,i} = Q$$

UNITS

- Work W, heat Q, energy E: J = N m = kg m^2/s^2
- Specific heat capacity c: J/(kg K)
- Molar heat capacity C: J/(mol K)

LAWS

- First law of thermodynamics for an isolated system (conservation of energy):

$$\Delta U_{isolated} = U_f - U_i = 0,$$

  in which U is the internal energy, i.e., the total energy of a system.
- First law of thermodynamics for a closed system:

$$\Delta U_{closed} = U_f - U_i = W + Q$$

MULTIPLE-CHOICE QUESTIONS

MC–9.1. Consider a generic system, mechanical or non-mechanical, that receives work. Which of the following *must* take place in this case?
(a) Its volume decreases.
(b) Its temperature increases.
(c) Its pressure increases.
(d) None of the above must take place.

MC–9.2. Air, initially at 100 kPa, is sealed in a container by a mobile piston of cross-sectional area 10.0 cm^2. Now we push the piston with an additional force of magnitude F = 100 N to compress the air. What is the final pressure p of the sealed air when the piston reaches mechanical equilibrium?
(a) $p = 2 \times 10^4$ Pa
(b) $p = 1.0001 \times 10^5$ Pa
(c) $p = 2 \times 10^5$ Pa
(d) $p = 1 \times 10^6$ Pa

MC-9.3. A gas sealed in a container by a mobile piston expands. Which of the following statements is correct?
(a) The piston does work on the gas.
(b) No work is done in the process.
(c) The gas does work on the piston.
(d) The gas and the piston exchange heat only.

MC-9.4. A gas is compressed from a volume of 5 L to a volume of 1 L with its pressure held constant at 4000 Pa. To achieve this compression, the following work is done on the gas.
(a) $W = 20$ kJ
(b) $W = 16$ kJ
(c) $W = 20$ J
(d) $W = 16$ J

MC-9.5. Fig. 9.16 shows a p–V diagram for a gas. Three processes (labelled a, b, and c) are investigated that bring the gas from an initial volume to a final volume. Which path yields the largest value for work? *Hint:* 5 J > 2 J, but −5 J < −2 J.
(a) path a
(b) path b
(c) path c
(d) At least two paths yield the same work.

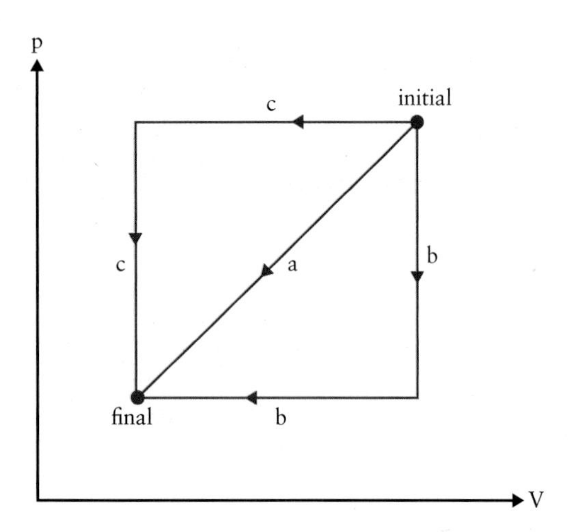

Figure 9.16 Three paths for the compression of a gas system from an initial to a smaller final volume. The paths are labelled a, b, and c.

MC-9.6. Fig. 9.17 shows a p–V diagram with the initial and final states of the system indicated. The system undergoes a process that follows the line shown in the diagram. During this process
(a) the system received a net transfer of work.
(b) the system released a net amount of work.
(c) no net transfer of work occurred.
(d) we cannot draw such conclusions from the shown figure.

MC-9.7. Fig. 10.18 shows a process we use to make an argument in Chapter 10; it is a cyclic process in a p–V diagram. A cyclic process is a process that repeats along the same path in a p–V diagram, for example, the heart cycle or the breathing cycle. In the shown process
(a) the system receives work per cycle.
(b) the system releases work per cycle.
(c) no net transfer of work occurs during a cycle.
(d) we cannot draw such conclusions from the shown figure.

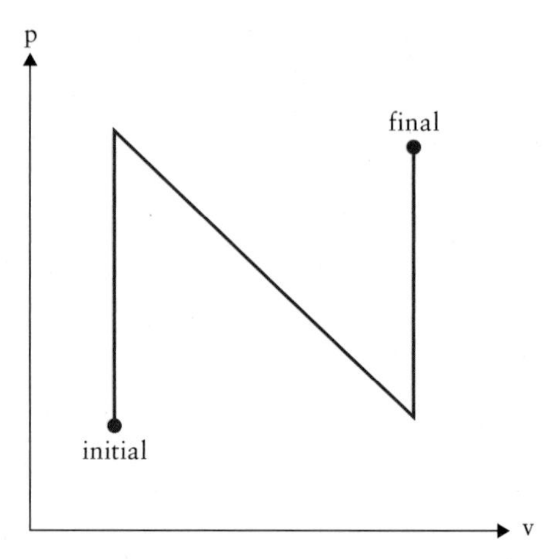

Figure 9.17

MC-9.8. This and the next question illustrate how concepts developed in one chapter will have an impact on problem solving in a later chapter. For this reason, we refer in this question to a figure from Chapter 10. Fig. 10.32 shows three steps, labelled I, II, and III, that form a cyclic process in a p–V diagram. Rank the work from largest to smallest for each individual step (Note: −5 J < −2 J, but + 2 J < + 5 J):
(a) I, II, III
(b) I, III, II
(c) II, I, III
(d) II, III, I
(e) III, I, II
(f) III, II, I

MC-9.9. This and the previous question illustrate how concepts developed in one chapter will have an impact on problem solving in a later chapter. For this reason, we refer in this question again to a figure from Chapter 10. Fig. 10.33 shows three steps forming a cyclic process in a p–V diagram: (I) from state A to state B, (II) from state B to state C, and (III) from state C back to state A. Rank the work from largest amount to smallest amount for each step. *Note:* −5 J < −2 J, but +2 J < + 5 J.
(a) I, II, III
(b) I, III, II
(c) II, I, III
(d) II, III, I
(e) III, I, II
(f) III, II, I

MC-9.10. An object of mass $m = 5$ kg is dropped from a height of 10 metres. Just before reaching the ground, the thermal energy of the object is (use $g = 10$ m/s^2)
(a) unchanged.
(b) $E = 50$ J.
(c) $E = 500$ J.
(d) $E = 5$ kJ.

MC-9.11. Which of the following things *cannot* happen in a closed system?
(a) Heat is transferred to the environment.
(b) Work is done on the system by an object in the environment.
(c) Matter flows into the system.

(d) The temperature of the system increases.
(e) The internal energy of the system remains unchanged.

MC-9.12. The specific heat for material A is greater than that for material B. If equal amounts of heat are added to both materials, the one reaching a higher temperature will be (assume that no phase transition occurs in either material)
(a) material A.
(b) material B.
(c) neither, they reach the same temperature.
(d) impossible to determine from the information provided.

MC-9.13. We double the amount of heat that is added to a closed system. If no phase transitions occur in the system and the system absorbs the heat fully into its thermal energy, we expect
(a) the temperature of the system to double.
(b) the temperature of the system to increase fourfold.
(c) the temperature of the system to be halved.
(d) the temperature of the system to decrease to one quarter.
(e) None of the values offered above is correct.

MC-9.14. We increase the gas pressure in a cylinder of fixed volume by heating the gas. The gas is the system. In this process,
(a) work is done on the gas.
(b) work is done by the gas.
(c) no work is done on the gas.
(d) whether work is done on the gas or by the gas depends on the temperature.

MC-9.15. The specific heat of ethyl alcohol is about 50% that of water. If equal amounts of alcohol and water in separate beakers are supplied with the same amount of heat, which liquid will show the larger increase in temperature?
(a) water
(b) ethyl alcohol
(c) Both show the same temperature increase.

MC-9.16. Fig. 9.18 shows the p–V relationship in the left ventricle of the human heart (heart cycle). The curve is traversed counter-clockwise with increasing time. The stroke volume is 100 mL − 35 mL = 65 L. The systolic pressure is 118 torr (equal to 15.7 kPa) and the diastolic pressure is 70 torr (equal to 9.3 kPa). The ventricular pressure drops to below the diastolic pressure, while the pressure in the arteries remains about 70 torr because the aortic valve has closed, preventing backflow. To simplify calculations, the dashed straight lines in the diagram allow us to replace the curved segments. Which of the following statements about the p–V diagram in Fig. 9.18 is correct?
(a) The diagram shows the data in standard units.
(b) The maximum pressure variation during the heart cycle exceeds the atmospheric pressure value.
(c) The work done in a single cycle can be determined from the area enclosed by the solid line in Fig. 9.18.
(d) The system is the blood in the ventricle. This system is an isolated system.

MC-9.17. Fig. 9.18 shows the p–V diagram for the left ventricle of the human heart (for details see previous question). The system is the blood.

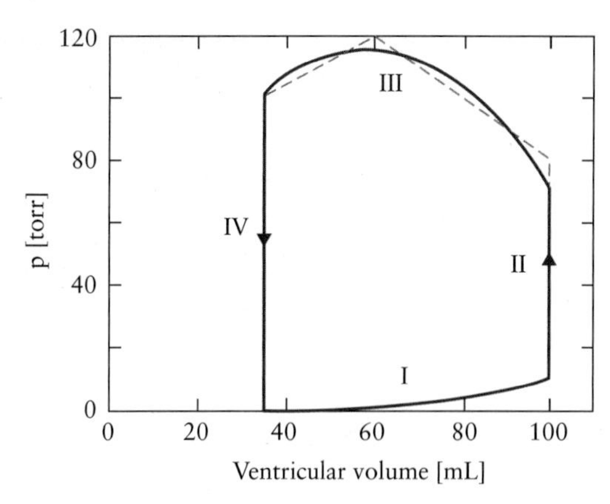

Figure 9.18 p–V diagram for the left ventricle of the human heart. The blood pressure initially increases slowly as blood flows into the ventricle from the left atrium (through the mitral valve, step I), but then jumps to about 75 torr during the contraction of the heart muscle (step II). This pressure causes the aortic valve to open. The blood pressure continues to rise, but the volume of the ventricular chamber decreases concurrently as blood is ejected from the heart (step III). The aortic valve closes when the muscle contraction is complete, leaving the ventricle at a fixed volume of 35 mL during step IV, until the mitral valve opens again for the next filling step.

(A) For the first two steps (steps I and II) the work on/by the blood is
(a) $W > 0$.
(b) $W = 0$.
(c) $W < 0$.
(d) indeterminate from the given figure.

(B) For the last two steps (steps III and IV) the work on/by the blood is
(a) $W > 0$.
(b) $W = 0$.
(c) $W < 0$.
(d) indeterminate from the given figure.

MC-9.18. We consider one breathing cycle, starting with an inhalation and then followed by an exhalation. The net work for such a cycle is
(a) positive ($W > 0$).
(b) zero ($W = 0$).
(c) negative ($W < 0$).
Note: The gas in the lungs is the system.

CONCEPTUAL QUESTIONS

Q–9.1. Fig. 9.8 shows a p–V diagram with three paths a gas can take from an initial to a final state. Rank the paths in decreasing order according to (a) the change of internal energy, ΔU, and (b) the amount of heat transfer, Q, between the system and the environment.

Q–9.2. During a stress test of the cardiovascular system, a patient walks and runs on a treadmill. (a) Is the energy dissipated by the patient equivalent to the energy dissipated by walking and running on the ground? (b) What would be the effect of tilting the treadmill upward?

Q–9.3. Early Europeans arriving in North America stored fruit and vegetables in underground cellars. In winter they also included a large open barrel with water. Why did they do this?

Q–9.4. Concrete has a higher specific heat than soil. Can this help us to explain why a city usually has a higher temperature than the surrounding countryside?

Q–9.5. What is wrong with stating that, of any two objects, the one with the higher temperature contains more thermal energy?

ANALYTICAL PROBLEMS

P–9.1. A gas expands from a volume of 1.0 L to a volume of 5.0 L, as shown in the *p–V* diagram of Fig. 9.19. How much work does the gas perform on the piston?

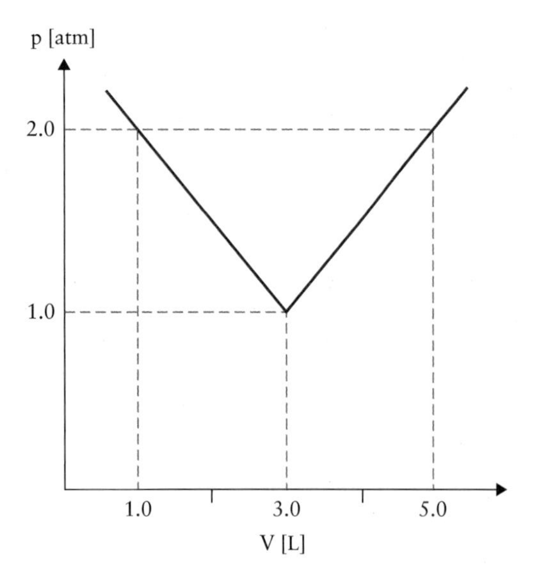

Figure 9.19

P–9.2. The density of solid aluminium at 20°C is $\rho = 2.70$ g/cm^3. When it is liquid at 660°C the density of the liquid state is decreased: 2.38 g/cm^3. How much work does a block of aluminium of mass 100 kg do if it is heated at $p = 1.0$ atm from 20°C to 660°C?

P–9.3. Calculate the work done by 1.0 mol of water when (a) it freezes at 1 atm and 0°C, and (b) when it boils at 1 atm and 100°C. (c) The latent heat is defined as the amount of energy required for a given amount of material to undergo a phase transition at the phase transition temperature, e.g., the latent heat of vaporization of water is the energy required to boil 1 mol of liquid water at 100°C to form water vapour. Compare the work values calculated in this problem with the respective latent heats of fusion and vaporization, with the values for water $E_{fusion} = 3.33 \times 10^5$ J/kg at 0°C and $E_{vaporization} = 2.26 \times 10^6$ J/kg at 100°C.

P–9.4. When people run, they dissipate about 0.6 J of mechanical energy per step per kilogram of body mass. If a standard man dissipates 80 J of energy per second while running, how fast is the person running? Assume the steps taken are 1.6 m long.

P–9.5. Table 9.2 shows the metabolic rate for given activities of the adult human body, and Table 9.3 gives the energy content of the three most important components of food. (a) How much energy is expended by a standard man who walks for one hour every morning? (b) If the body of the person consumes body fat reserves to produce this energy, how much mass will be lost per day?

TABLE 9.2

Metabolic rate for given activities

Activity	Metabolic rate [cal/(s kg)]
Sleeping	0.263
Sitting	0.358
Standing	0.621
Walking	1.0
Biking	1.81
Swimming	2.63
Running	4.3

TABLE 9.3

Food energy content

Food	Energy content (cal/g)
Carbohydrate	4100
Protein	4200
Fat	9300

P–9.6. A standard man climbs 10 m up a vertical rope. How much energy in calories is dissipated as heat in a single climb if 20% of the total energy required is used to do the work?

P–9.7. Assume that Joule's brewery horses each did P = 750 J/s of work per second (this corresponds roughly to the definition of horsepower). If Joule had four horses moving in a circle for one hour to operate a stirrer in a well-isolated container filled with 1 m^3 water at an initial temperature of 25°C, to what final value did the water temperature rise? Use Table 9.1 for the specific heat capacity of water.

P–9.8. When a raindrop of mass 30 milligrams (mg) hits the ground, by how much does the temperature increase if we assume its kinetic energy is completely converted into thermal energy? *Hint:* Use Table 9.1 for the specific heat capacity of water.

P–9.9. The energy extracted from burning sugar is used in the mitochondria to synthesize ATP from ADP. Considering glucose, which releases 675 kcal/mol during cellular respiration (formation of CO_2), what fraction of this energy is used in the formation of ATP if 38 molecules of ATP are formed for each molecule of glucose? Why is this value not 100%?

P–9.10. (a) How much chemical energy does an *E. coli* bacterium consume during its lifetime in the form of ATP (excluding the amount needed to replicate its DNA)? (b) Using the specific heat capacity of water, how hot would a thermally isolated *E. coli* get during its life if it formed at 37°C in a person's intestines? Note that about 60% of the consumed energy is released as thermal energy.

P–9.11. Water at the top of Niagara Falls has a temperature of +10.0°C. It falls a distance of 50 m. Assuming that all its potential energy is converted into thermal energy, calculate the temperature of the water at the bottom of the falls.

P–9.12. A piece of iron has a mass of 0.4 kg and is initially at 500°C. It is lowered into a beaker with 20 L of water at 22°C. What is the final equilibrium temperature? Neglect heat loss to the environment. The specific heat capacities of iron and water are found in Table 9.1.

P–9.13. Fig. 9.18 shows the p–V relationship in the left ventricle of the human heart. The curve is traversed counter-clockwise with increasing time. Using the stroke volume and pressure data from MC–9.16, determine graphically the amount of work done in a single cycle. *Hint:* Simplify the calculation by using the dashed straight lines in the p–V diagram instead of curved segments.

P–9.14. The air temperature above coastal areas is significantly affected by the large specific heat of water. Estimate the amount of air for which the temperature can rise one degree if 1 m^3 of water cools by one degree. The specific heat of air is 1.0 kJ/(kg K); for its density use 1.3 kg/m^3.

P–9.15. A standard man produces about 1×10^4 kJ of heat per day due to metabolic activity. (a) Assume the standard man is an isolated system with the heat capacity of water and estimate the temperature increase in his body in one day. (b) In reality, the standard man is an open system and the main mechanism of heat loss is evaporation of water. How much water would need to be evaporated per day to maintain a body temperature of 37°C? The amount of energy needed to evaporate 1 g of water is 2405 J (latent heat of vaporization).

P–9.16. 200 g of water at 10°C is contained in an aluminium container of mass 300 g. An additional 100 g of water at 100°C is added. What is the final equilibrium temperature if we treat the system's water and container as isolated? Use the heat capacity values from Table 9.1.

P–9.17. An aluminium calorimeter of mass 100 g contains 250 g of water. The system is in thermal equilibrium at +10°C. We place two blocks of metal in the water: one is a 50-g piece of copper with an initial temperature of +80°C; the second piece has a mass of 70 g and is initially at +100°C. The combined system reaches a final equilibrium temperature of +20°C. Calculate the specific heat capacity of the unknown second piece of metal.

P–9.18. Fig. 2.25(b) shows in double-logarithmic representation the nerve impulse rate P as a function of the speed of an approaching object for a Meissner's corpuscle. Using the power law relation $P = a\ v^b$, determine the constants a and b.

P–9.19. Fig. 9.20 shows the height h in mm versus mass m in kg (red line) and the active metabolic rate E in kcal/day versus height (blue line) for growing children. Determine the three exponents b in (a) $h = a_1\ m^{b_1}$ for $m < 15$ kg (curve I); (b) $h = a_2\ m^{b_2}$ for $m > 15$ kg (curve II); and (c) $E = a_3\ m^{b_3}$ for $m < 40$ kg (curve III). *For those interested:* (d) Find pictures of children and adults and compare the body proportions to see what causes the differences in the exponents.

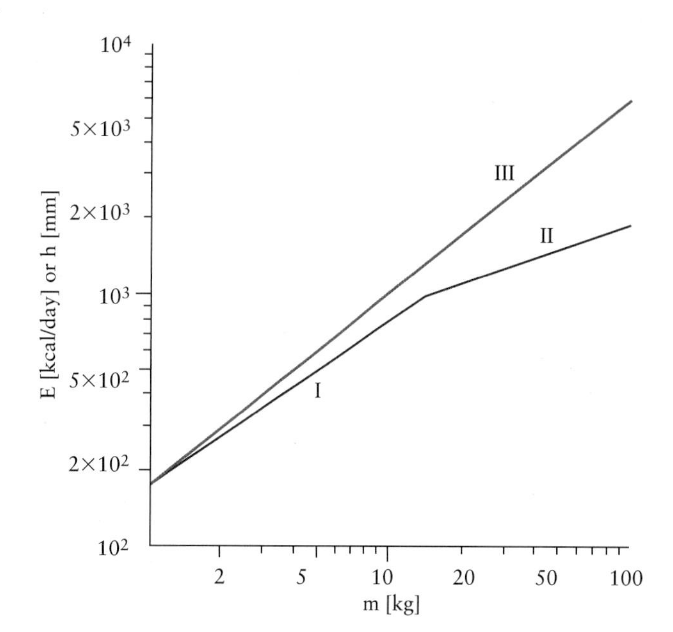

Figure 9.20

P–9.20. In the mid-Cretaceous (110–100 million years ago), dinosaurs lived near the poles, e.g., at 80°N for fossils in North Alaska and the Yukon, and at 80°S for fossils near Melbourne, Australia. The polar regions of the Cretaceous were densely forested, with only occasional light freezes in the winter, but non-hibernating ectotherms cannot tolerate prolonged periods without sunlight. Indeed, the most northern fossil find of a large ectotherm is a giant crocodile (*Phobosuchus*) at 55°N. The energy cost for long-distance migration across land is given in unit J/m as a function of the animal's body mass m in kg:

$$E_{\text{migration}} = 14\ m^{3/4}.$$

(a) Using this equation and

$$\text{ectotherm: } M = 20\ m^{3/4}$$

$$\text{endotherm: } M = 450\ m^{3/4},$$

which contain the metabolic rate in unit kJ/day and the mass in unit kg. Evaluate the hypothesis that ectothermic southern polar dinosaurs migrated annually between 80°S and 55°S latitudes for (I) Leaellynasaura, which was a 10-kg herbivore; (II) Dwarf Allosaurus, which was a 500-kg carnivore; and (III) Muttaburrasaurus, which was a 4-tonne herbivore. Note in this context that caribou in the Canadian North migrate 4000 km annually. (b) Using the energy consumption for migration and the potential energy, compare the benefits of living in plains versus mountainous terrain for small and large endotherms.

P–9.21. We considered a model in which an animal's metabolic requirements are determined by the loss of heat through the skin. Why does this lead to $b = 2/3$ as the exponent in $M \propto m^b$, with m the mass of the animal and M the metabolic rate?

P–9.22. Use the following data to determine the maximum rate at which a standard man can climb a mountain: Blood contains 15.5 wt% hemoglobin (with molecular weight 65 000 g/mol). Each hemoglobin molecule can carry four oxygen molecules. The heart pumps 115 cm^3/s blood of density 1.06 g/cm^3. Each oxygen molecule can oxidize one sugar unit (the chemical formula per sugar unit is CH_2O, which is an organic alcohol group) to CO_2 and H_2O; the oxidation of 1 g sugar yields about 17 kJ of energy, of which 25% can be used to do muscle work.

ANSWERS TO CONCEPT QUESTIONS

Concept Question 9.1: Option (A) is wrong because 1 kPa 1 L = 1 J. Option (B) is wrong because the figure shows gauge pressures. The maximum deviation from p_{atm} in Fig. 9.1 is 0.7 kPa, i.e., less than 0.7% p_{atm}. Option (C) is wrong because the top curve in Fig. 9.1, which shows the transmural pressure, has only one value at any given volume. Option (D) is correct as the figure shows. The pressure in the pleural gap varies even though it is not open to the lungs. The pressure in a closed container can be varied by squeezing from outside.

Concept Question 9.2: Fig. 9.6 shows a volume expansion: V_i is smaller than V_f. Based on Fig. 9.3 and Eq. [9.2] the work is negative, $W < 0$; i.e., the gas does work on the piston during an expansion.

Concept Question 9.3: Basing the answer on the area under the curve and Fig. 9.7, path 1 in Fig. 9.7 matches with the three expansions in Fig. 9.8. Thus, $0 > W_{(iii)} > W_{(ii)} > W_{(i)}$.

CHAPTER 10

Thermodynamics

Joule's basic experiments allowed us in the previous chapter to relate work, heat, and internal energy for isolated and closed systems. However, the formulations of the conservation of energy and the first law of thermodynamics remained very general because we did not yet connect them to the specific system of an ideal gas.

We identify four fundamental thermodynamic processes: the isochoric process (processes at constant volume), the isothermal process (processes at constant temperature), the isobaric process (processes at constant pressure), and the adiabatic process (processes that do not allow for exchange of heat between the system and the environment). All practical processes of interest can be derived from these four, including the important cyclic processes in respiration and blood circulation.

The first law of thermodynamics is insufficient to fully characterize systems that undergo dynamic processes such as chemical reactions. It distinguishes possible and impossible processes based on the conservation of energy, but it does not allow us to identify the spontaneous direction of a possible process. The second and third laws of thermodynamics fill this gap: the second law establishes entropy as a parameter of the state of a system that remains constant in an isolated system with reversible processes (processes that proceed exclusively via equilibrium states) but increases for spontaneous irreversible processes.

In a closed or open system at constant pressure, the Gibbs free energy, G, combines the enthalpy and entropy to predict the dynamics of the system: G remains constant if the process occurs between two states that are in equilibrium with each other, and G approaches a minimum value during a spontaneous irreversible process.

If the system is a liquid solution, we quantify its properties by studying the vapour phase, which is in thermodynamic equilibrium with the solution. The solution is called an ideal solution if no heat is released or required when mixing its components; thus, ideal solutions are conceptually equivalent to the ideal gas because intermolecular interactions are neglected in both systems. Raoult showed that the partial pressure of a component in the vapour phase is proportional to the molar fraction of the same component in the ideal solution with which the vapour phase is in equilibrium. The thermal equilibrium between the two phases is established when their respective Gibbs free energies are equal.

The term **bioenergetics** was created in 1912 to describe energy transformations and energy exchanges within and between living cells and their environment. It includes in particular the closely related concept of **metabolism**, which is the sum of all chemical changes in living cells by which energy is provided for vital processes.

10.1: Quantifying Metabolic Processes

10.1.1: Metabolism at the Molecular Level: The Role of ATP

We illustrated with Fig. 7.2 how a force is generated in muscle tissue by repetitively traversing the sliding filament mechanism. We return in the current chapter to this process to illustrate the connection between energy and chemical processes and to emphasize the prevalence of cyclic processes in the life sciences.

We start with the **adenosine triphosphate (ATP)** molecule in Fig. 10.1. It consists of three characteristic components, which are, from right to left, an adenine group, a pentose sugar group, and a chain of three phosphate groups. To see why this molecule carries energy we focus on the last of the three phosphate groups at the left. Recall that this molecule enables the tilting of the myosin head in Fig. 7.2 by splitting off the end-standing phosphate group. The release of the terminal phosphate group of the ATP molecule is a chemical reaction that results in the formation of an adenosine diphosphate (ADP) molecule:

$$ATP + ROH \rightarrow ADP + R-OPO(OH)_2.$$

ROH is an alcohol molecule in which R means *rest molecule*. It is the ROH molecule to which the energy in

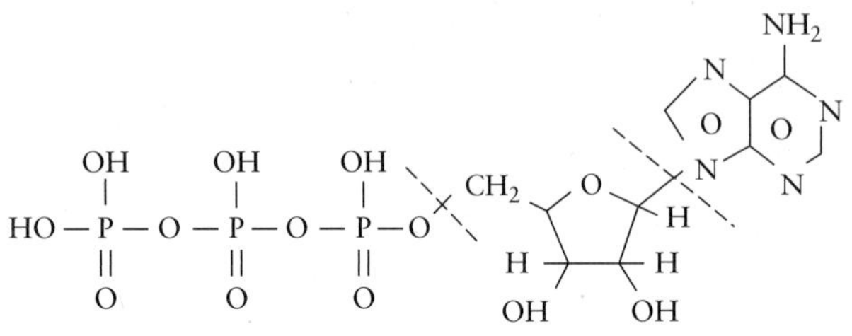

Figure 10.1 Chemical formula of adenosine triphosphate (ATP). The molecule consists of three components frequently found in biomolecules: a nitrogen-containing double-ring base (adenine, upper right part), a pentose sugar (ribose with an oxygen-containing ring at the centre), and three phosphate groups.

the **ATP hydrolysis** is transferred. The involvement of energy in chemical reactions like the one above is often highlighted in physical chemistry by adding to the reaction the amount of energy released or required per mol of the reactant of interest:

$$ATP \rightarrow ADP + H_3PO_4 \qquad -29\,\mathrm{kJ/mol},$$

which implies that the hydrolysis of ATP releases 29 kJ of energy per mol. This energy is sometimes referred to as **chemical energy**. It is usually not set free as heat but is utilized in a concurrent reaction that only becomes possible when receiving this energy.

This short description illustrates that we need to focus on the role of energy in processes, such as chemical reactions: when is energy required, when is it released, and where does it come from when a process cannot occur without it? We also look back at the first law of thermodynamics and ask what the 29 kJ/mol energy in the ATP hydrolysis means: is it released as work or heat, and are there restrictions on the use of energy stored in chemical bonds?

The hydrolysis of ATP also raises questions about the concept of energy conservation in a cyclic process. At any given time, the ATP concentration in muscle cells is relatively low and sufficient for only a few muscle contraction cycles. This means the muscle must quickly find energy from somewhere else or it will become disabled before much is achieved. Since energy cannot be created on a need basis, only two options exist: transport ATP molecules through the cell membrane into the muscle cell, or produce new ATP molecules. The first option is not feasible because ATP molecules are too big to pass the cell membrane easily. Thus, the cell must recycle its ADP molecules; i.e., it must run the ATP hydrolysis reaction backward. This process is called **phosphorylation**. A problem is associated with this approach: since the ATP hydrolysis releases 29 kJ/mol of energy, the ADP phosphorylation requires the same amount of energy. Thus, the cell needs energy to exert a force (muscle action), but it also needs energy to re-synthesize the energy agent it uses.

The cell deals with this apparent dilemma in the following fashion. The ADP molecule resulting during the muscle contraction step of Fig. 7.2 is removed and transported to a **mitochondrion** located within the muscle cell. Mitochondria are the source of ATP molecules; i.e., they are the intracellular power plants. The amount of energy that they have to provide to a cell varies vastly: a small microorganism may have just 10 mitochondria, while a human muscle cell requires on the order of 200 000 mitochondria to function properly. Fig. 10.2 shows a scanning electron

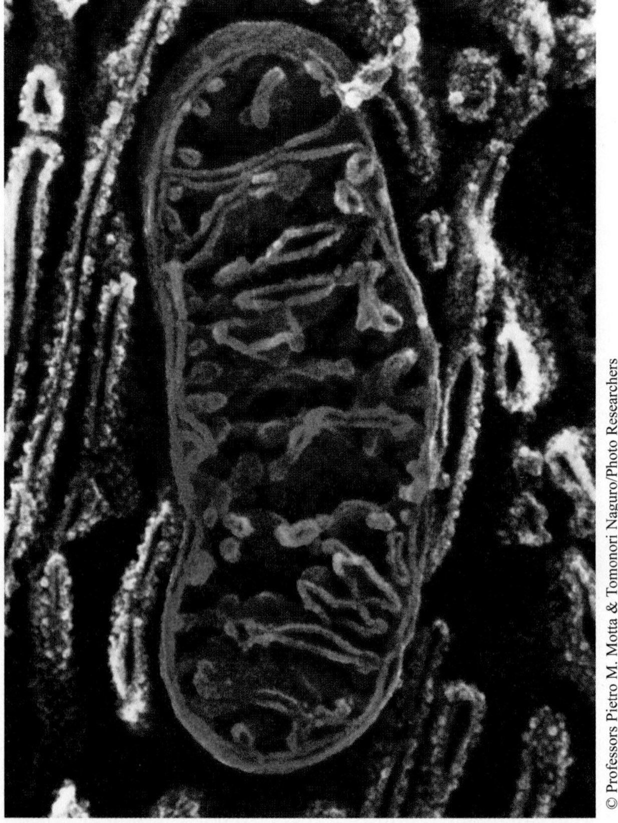

Figure 10.2 The mitochondrion is the power plant of the cell. This coloured high-resolution scanning electron micrograph (SEM) illustrates its internal structure. The mitochondrion is about 0.5 μm to 1.0 μm long and has two membranes, a smooth outer membrane and a folded inner membrane, called cristae. The folding increases the inner surface to allow for an increased rate of ADP phosphorylation to ATP. Mitochondria replicate independently during cell division. We believe therefore that they were originally bacteria that became incorporated into the eukaryotic cells by way of an intracellular symbiosis.

micrograph of a mitochondrion that is cut open to reveal its internal structure. Mitochondria absorb from the cytoplasm of the cell a compound called **pyruvic acid**, which is a primary product of the chemical breakdown of food. In a series of chemical reactions, single hydrogen atoms are isolated from the pyruvic acid. These hydrogen atoms combine with oxygen to form water. The energy released in this reaction is used in the phosphorylation of ADP.

Does the recycling of ADP molecules in the mitochondria solve the muscle cell's energy problem? Only for a short period: for the first 50 to 100 contraction cycles, which corresponds at most to a few seconds of muscle activity, the mitochondria have stored enough of an energy-rich molecule called keratin phosphate to recover ATP from ADP molecules. Thereafter, for up to one minute, glycogen is used instead of keratin phosphate. The chemical reactions of both compounds have short response times because they are **anaerobic reactions**. Such reactions do not need the supply of oxygen from outside the cell and are called *lactic acid fermentations* because of the product formed in the process. Fig. 10.3 shows the typical energy output per second for an average healthy adult on a bicycle. The time axis of the figure is a logarithmic scale to highlight short times. As Fig. 10.3 illustrates, the energy output decreases fast initially, but after about one minute continues at a lower rate. This marks the transition to the **aerobic phase**, in which the muscle cells use oxygen acquired through the cell membrane.

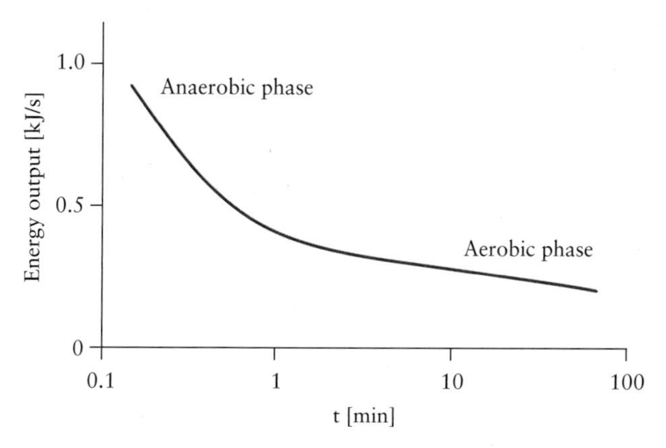

Figure 10.3 Energy dissipated per second by an average healthy adult on a bicycle. The logarithmic time scale highlights the change at shorter times during the anaerobic phase. The loss of power is slowed in the aerobic phase. The data representation of this figure shows in particular the transition from the anaerobic to the aerobic phase after about one minute.

Figure 10.3 illustrates how energy considerations at the level of individual chemical reactions relate to the metabolism of an organism. In the next section we consider a cell as an intermediate level to further highlight this connection.

10.1.2: Metabolism at the Cellular Level: *Escherichia coli* (*E. coli*)

E. coli is a **bacterium** responsible for several illnesses such as peritonitis, appendicitis, sepsis, sinusitis, otitis, diarrhea, and some forms of meningitis. *E. coli* are small with a volume of 2.25×10^{-18} m^3 and a mass of about 2.5×10^{-15} kg. Because they divide into two daughter bacteria every 20 minutes (the life of a bacterium is a cyclic process), their need for energy from ATP molecules is immense, as illustrated in Table 10.1: excluding the ATP molecules needed to replicate the bacterium's DNA, each *E. coli* produces more than 14 000 biomolecules per second. To accomplish this it must consume the energy of more than 2.3 million ATP molecules! Since *E. coli* bacteria do not photosynthesize, this energy must be absorbed as food from their environment.

To appreciate the details of the discussion of metabolic processes in the literature, the governing principles of energy conservation, energy flow, energy conversion, heat generation, and maximum achievable work must be established. We have touched on these issues frequently in the previous three chapters; in the current chapter we want to place thermodynamics on a firm quantitative footing through the introduction of basic thermodynamic principles using the ideal gas as a sufficiently simple model system.

TABLE 10.1

Metabolism of *Escherichia coli*

Compound	Weight fraction (%)	Molecules	Molecules/s	ATP/s
Protein	70	1.7×10^6	1400	2.1×10^6
Fat	10	1.5×10^7	12 500	8.8×10^4
Polysaccharides	5	4.0×10^4	32	6.5×10^4
RNA	10	1.5×10^4	12	7.5×10^4

The values in the first column indicate the fraction of the total weight due to each type of compound; these values don't add up to 100% because DNA contributes another 5%. The last two columns provide the number of each type of molecule synthesized per second and the number of ATP molecules needed for the synthesis per second.

10.2: Basic Thermodynamic Processes

Figure 10.4 illustrates the model system we will use in the current chapter: a closed container with an ideal gas includes a mobile ideal piston that allows us to exchange work, W, with the gas. The double arrow indicates that work can be done both on and by the gas. Secondly, a heat source is present (called reservoir) that allows us to exchange heat, Q, with the gas. Again, the double arrow indicates that heat can be released or taken up by the gas.

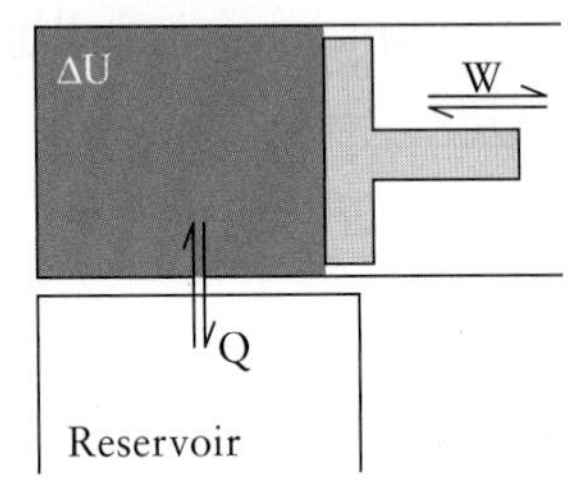

Figure 10.4 Model system for the discussion of the basic processes in thermodynamics. An ideal gas is sealed in a container (closed system with internal energy U). Work W is exchanged with the environment via motion of a frictionless piston. Heat Q is exchanged with a heat reservoir in the environment. The heat reservoir is ideal; i.e., its temperature remains unchanged during heat transfer.

In the discussion of the first law of thermodynamics for a closed system at the end of the previous chapter, we found that the change in the internal energy of the gas ΔU is written in the form:

$$\Delta U = U_{\mathrm{f}} - U_{\mathrm{i}} = Q + W. \qquad [10.1]$$

The ideal piston in Fig. 10.4 allows for the work exchange in Eq. [10.1], and the ideal heat source in Fig. 10.4 allows for the exchange of heat. An ideal heat source is a new concept we introduce in this chapter. When we refer to this concept we think of it as being one of two idealizations of practical devices you know from your chemistry and physics laboratories:

- A simple Bunsen burner is depicted when we allow for heating of the system with a source at a temperature higher than that of the system. This leads to what we define later in the chapter as an irreversible heating under thermal non-equilibrium conditions.
- An oil bath (or heating jacket) allows heat to flow into or out of the system when brought into thermal contact with the gas under equilibrium conditions, i.e., with equal temperatures of the system and the bath. The oil bath becomes an ideal heat reservoir when we assume that it maintains a given temperature even during heat exchange.

What are the **fundamental thermodynamic processes** the system in Fig. 10.4 allows us to introduce? Fundamental for a thermodynamic process can mean several things: (i) the process should be important enough that we benefit from studying it, but (ii) it should also be sufficiently simple that its properties are easily applied in a wide range of practical cases. A good candidate for the latter requirement is a process that is sufficiently easy to analyze mathematically. Eq. [10.1] suggests that the mathematical treatment of a process is minimized when any one of the three terms in that formula is zero. Alternatively, we expect more, simple calculations for processes where at least one major system parameter in the ideal gas equation does not change. The following four sections identify such processes.

While you read through these sections, carefully observe at what point in each case we require the specific properties of the ideal gas: formulas derived *before* that point apply to gases in general; formulas derived *after* that point apply *only* to the ideal gas. When we use results from the kinetic gas theory, in particular the formula describing the internal energy of the ideal gas, we assume the gas is monatomic (i.e., molecular vibrations and rotations don't contribute to its internal energy).

10.2.1: Isochoric Processes

An **isochoric process** is a process that takes place at constant volume; such a process is illustrated in Fig. 10.5. The top frames in the figure show the system before (initial) and after (final) an **isochoric heating**. We expect the temperature and the internal energy of the system to increase in this process. The sketch in the large dashed circle at the bottom shows what happens during the process: a propane flame provides heat to the gas. Isochoric processes

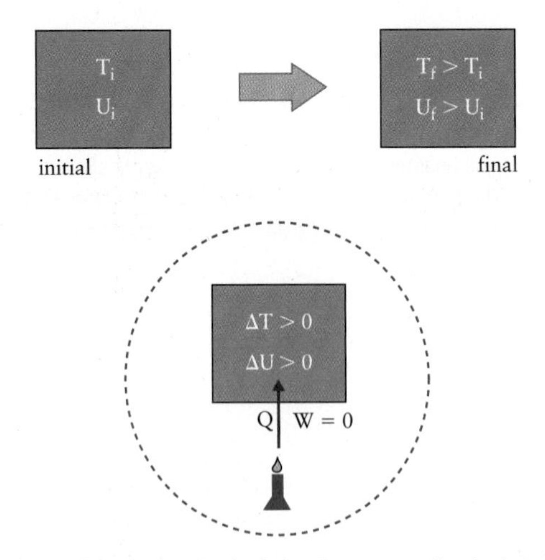

Figure 10.5 Sketch of an isochoric heating process. Isochoric processes are processes at constant volume. The top two frames show the system before and after the heating process and indicate the temperature and internal energy of the system. The frame in the large dashed circle illustrates the heating process itself. The container has fixed walls. Heat is supplied to the system from a hotter source, e.g., a Bunsen burner. The work done in the process is zero and the internal energy increases with the temperature. Note that the heat source must have a temperature of at least T_{f}.

require rigid containers such as steel vessels. Rigid confinement is unusual in living systems; thus, isochoric conditions usually are not applicable in life sciences systems. In physics, such conditions are typically found in high-pressure or high-vacuum experiments. High-pressure experiments are used to simulate conditions below Earth's crust (geophysics), and vacuum experiments are often used in surface and thin film physics.

The work for the isochoric process is $W = 0$ because the volume does not change and therefore the area under the curve between V_i and V_f is zero. From the first law of thermodynamics it follows that $\Delta U = Q$: heat taken up or released in an isochoric process is completely taken up from or released into the internal energy of the gas. The heat exchange is quantified with Joule's definition, $Q = n\ C_V\,\Delta T$, in which n is the amount of gas in mol, ΔT is the temperature change, and the index V indicates that **C_V is the molar heat capacity of a gas at constant volume**.

When comparing the formula for the heat discussed above and the formula we derived based on Joule's experiment in Fig. 9.12, we note that the specific or molar heat capacities were previously not specified for constant volume. The difference is that we used in Joule's experiment a liquid system (for example, water), while we now focus on gases. As we will see below in the discussion of processes that occur at constant pressure, constant volume and constant pressure conditions lead to different properties for gas systems, while this distinction leads to no differences for liquid or solid systems.

Summary: Isochoric process of a gas in general:

$$W = 0$$
$$Q = n\,C_V(T_f - T_i) \qquad [10.2]$$
$$\Delta U = Q.$$

We combine the last two formulas in Eq. [10.2] to eliminate the heat: $\Delta U = n\ C_V\ \Delta T$. This relation defines a constant rate of change of the internal energy with temperature: $\Delta U/\Delta T = n\ C_V$.

KEY POINT

The molar heat capacity C_V is the change of the internal energy with temperature at constant volume for 1 mol of an ideal gas.

Now we focus on the specific case of an ideal gas. Fig. 10.6 shows the p–V diagram for an isochoric heating process. The two red lines in the figure represent the p–V relation for an ideal gas for two different temperatures T_{low} and T_{high}. The green arrow indicates the specific path the system takes in the p–V diagram during the process in Fig. 10.5: both pressure and temperature increase but the volume remains unchanged.

For the ideal gas, we have a second formula addressing the change of the internal energy, which was derived as Eq. [8.24] in the kinetic gas theory: $\Delta U = (3/2)\ n\ R\ \Delta T$. Comparing Eq. [8.24] with $\Delta U = n\ C_V\ \Delta T$ provides us

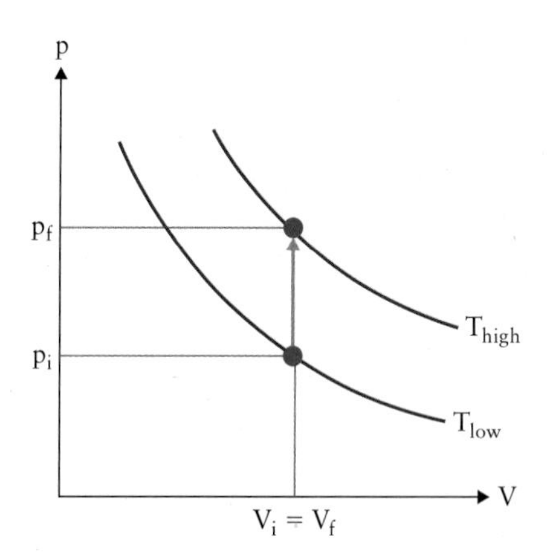

Figure 10.6 p–V diagram for isochoric heating of an ideal gas. The green arrow indicates the change of the temperature from T_{low} to T_{high} during the process. The arrow is vertical, as this corresponds to a constant volume process with $V_i = V_f$.

with a value for the molar heat capacity of an ideal gas at constant volume:

$$C_V = \frac{3}{2}R. \qquad [10.3]$$

Summary: Isochoric process of an ideal gas:

$$W = 0$$
$$\Delta U = \frac{3}{2}n\ R(T_f - T_i). \qquad [10.4]$$
$$Q = \Delta U.$$

Note that Eq. [8.24] applies to ideal gases regardless of whether the process is isochoric; however, $W = 0$ and $Q = \Delta U$ apply only for an isochoric process.

From $Q = \Delta U$ we conclude that an ideal gas system absorbs heat from the environment to increase its internal energy during isochoric heating. During isochoric cooling—the reverse process—heat is removed from the system, which leads to a lowering of its internal energy.

CONCEPT QUESTION 10.1

Figure 10.7 shows a T–V diagram (not a p–V diagram!) with four processes for an ideal gas. Which of the processes is an isochoric process?

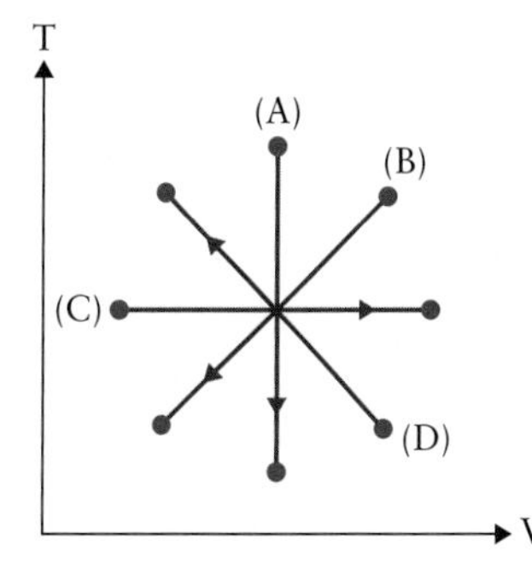

Figure 10.7 T–V diagram with four processes for an ideal gas.

10.2.2: Isothermal Processes

Isothermal processes are processes that take place at **constant temperature**. An isothermal process is illustrated in Fig. 10.8. The top frames show the system before (initial) and after (final) an isothermal expansion. The frame in the large dashed circle at the bottom of the figure shows what happens during an isothermal expansion: heat flows into the system and work is done by the system on a piston. Boyle's experiment is an example of an isothermal process. We cannot specify this process further without specifying the system.

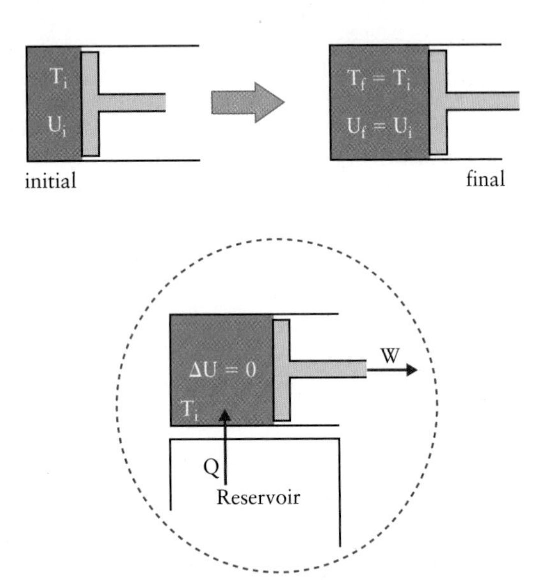

Figure 10.8 Sketch of an isothermal expansion. Isothermal processes are processes at constant temperature. The top two frames show the system before and after the expansion, indicating the temperature and internal energy of the system. The frame in the large dashed circle illustrates the expansion process itself. Heat Q is absorbed by the gas from a heat reservoir, which is at the same temperature as the system. This heat does not change the temperature or the internal energy of the gas, but transfers through the system and is released as work to the piston.

We study an ideal gas using a p–V diagram. Fig. 10.9 shows this p–V diagram for an isothermal expansion in which the pressure decreases from p_i to p_f and the volume increases from V_i to V_f.

For an isothermal process the kinetic gas theory (Eq. [8.24]) states:

$$\Delta U = \frac{3}{2} n\, R\, \Delta T = 0. \qquad [10.5]$$

Thus, the first law of thermodynamics connects heat and work of an ideal gas for an isothermal process in the form $Q + W = 0$ or $Q = -W$. The work is determined from the area under the curve in Fig. 10.9. This area can be quantified analytically; we state without derivation that:

$$W = -n\, R\, T \ln\left(\frac{V_f}{V_i}\right), \qquad [10.6]$$

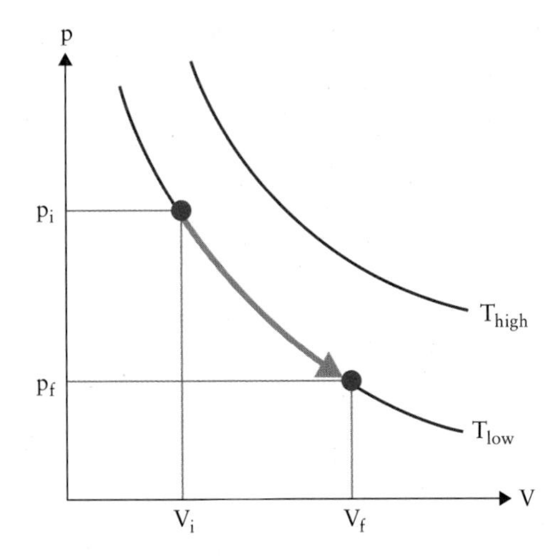

Figure 10.9 p–V diagram of an isothermal expansion of an ideal gas. The green arrow shows an isothermal expansion at temperature T_{low} from an initial volume V_i to a final volume V_f.

in which the notion ln(···) represents the natural logarithm. The logarithm function is reviewed in section "Powers and Logarithms," which is a Math Review located after Chapter 27.

Problem P–10.11 discusses Eq. [10.6] further and shows how the work can be estimated reasonably precisely from a p–V diagram. Eq. [10.6] is consistent with our basic observation that work is positive for a compression ($V_f < V_i$ leads to a negative value of the logarithm term) and negative for an expansion ($V_f > V_i$ leads to a positive value of the logarithm term). The latter case is shown in Fig. 10.8. This means that the system releases work to the piston. The energy to do this work does not come from the internal energy of the gas, as the internal energy remains unchanged in an isothermal process. Instead, the energy is passed through the system and originates in a heat reservoir that is in thermal contact with the system.

Summary: Isothermal process of an ideal gas:

$$W = -n\, R\, T \ln\left(\frac{V_f}{V_i}\right)$$
$$Q = -W \qquad [10.7]$$
$$\Delta U = 0.$$

CASE STUDY 10.1

Someone wants to report the result in Eq. [10.6] but writes:

$$W = n\, R\, T \ln\left(\frac{V_i}{V_f}\right).$$

Do you accept this formula?

Answer: *Yes. You have to familiarize yourself with the basic properties of the logarithm function, because*

continued

it occurs frequently in physical applications of the life sciences. Note that we already used ln a − ln b = ln(a/b). Here we use ln(a/b) = −ln(b/a) because:

$$\ln\left(\frac{a}{b}\right) = \ln a - \ln b = -(\ln b - \ln a)$$
$$= -\ln\left(\frac{b}{a}\right).$$

EXAMPLE 10.1

Find the work for an isothermal expansion between p_i = 1.0 atm and p_f = 0.1 atm for 10 mol of an ideal gas at temperature 0°C.

Solution

Because the expansion is done isothermally, we use Eq. [10.7]. Boyle's law in the form $p_i\ V_i = p_f\ V_f$ is used to substitute the volume for pressure terms as independent variables:

$$W = -n\,R\,T\ln\left(\frac{V_f}{V_i}\right)$$
$$= -n\,R\,T\ln\left(\frac{p_i}{p_f}\right).$$

Substituting the given values, we find:

$$W = -(10\text{ mol})\left(8.314\frac{\text{J}}{\text{K}\cdot\text{mol}}\right)(273\text{ K})\ln\left(\frac{1.0}{0.1}\right)$$
$$= -52.3\text{ kJ}.$$

This work is negative, which indicates that the gas does work on the environment, e.g., by pushing a piston.

10.2.3: Isobaric Processes

Isobaric processes are processes that take place at constant pressure. An isobaric process is illustrated in Fig. 10.10. The top frames show the system before and after an isobaric expansion. The frame in the large dashed circle at the bottom of the figure shows what happens during an isobaric expansion: heat from a heat source flows into the system and work is done by the system on a piston. In addition, the internal energy of the gas increases. We discussed Charles's experiment in Fig. 8.14, which is an example of an isobaric process.

The work associated with an isobaric process is:

$$W = -p(V_f - V_i). \quad [10.8]$$

This equation is correct because the pressure p is constant. The calculation of other quantities for isobaric processes is, however, slightly more complicated because none of the variables in the first law of thermodynamics vanish in this case, even for the ideal gas: $\Delta U \neq 0$ and $Q \neq 0$ since $\Delta T \neq 0$. Still, isobaric processes are important because any experiment conducted in a container that is open to air (i.e., most experiments in chemistry, and all in biochemistry and molecular biology) takes place at constant pressure.

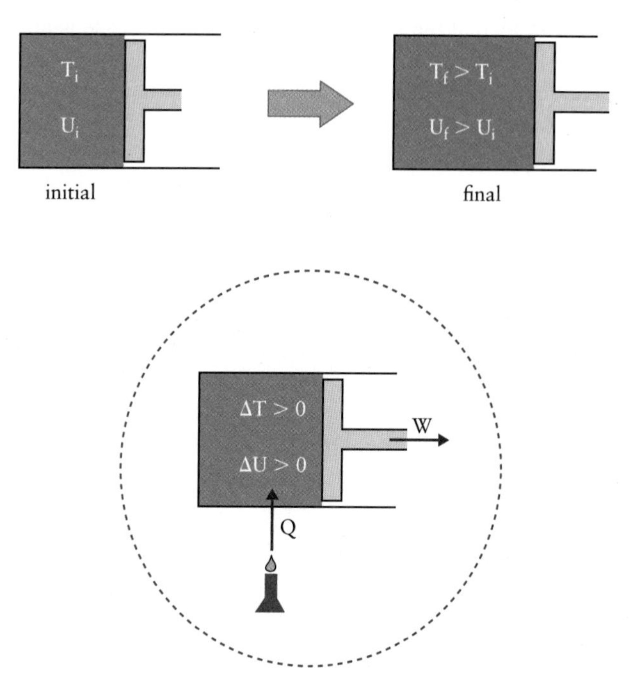

Figure 10.10 Sketch of an isobaric expansion. Isobaric processes are processes at constant pressure. The top two frames show the system before and after the expansion, indicating the temperature and internal energy of the system. The frame in the large dashed circle illustrates the expansion process itself. Heat Q is absorbed by the gas from a Bunsen burner with a flame at least as hot as the highest temperature of the gas system. The added heat does increase the temperature and the internal energy of the gas, even though concurrently the gas does work during the expansion.

Physical chemists in the 19th century chose a particular approach to address this case. Since this approach is now universally accepted, we follow their reasoning. When you read through the earlier discussion of the isochoric process you may have noted that it remained mathematically simple because the work term in $\Delta U = Q + W$ was zero, and consequently we were able to eliminate the heat in favour of the change of the internal energy, which is a variable of the state, $\Delta U = Q$. As we switch from the case of constant volume to constant pressure, we can no longer neglect the work term, as shown in Eq. [10.8]. To recover the mathematical simplicity of the isochoric case we "absorb" the work term in the internal energy change, introducing a new variable H, which we call **enthalpy**:

$$\begin{aligned} &\Delta H = Q \quad \text{for} \quad p = \text{const} \\ \text{equivalent to:}\ &\Delta U = Q \quad \text{for} \quad V = \text{const}. \end{aligned} \quad [10.9]$$

This allows us in particular to rewrite Joule's definition of heat:

$$\Delta H = Q = n\,C_p\,\Delta T \quad \text{for} \quad p = \text{const}$$
$$\text{equivalent to:} \quad \Delta U = n\,C_V\,\Delta T \quad \text{for} \quad V = \text{const}, \qquad [10.10]$$

with C_p the **molar heat capacity at constant pressure**. This means that the simplicity of the isochoric case is not limited to that case, with its vanishing work term. It can be extended to the important isobaric case if we replace the internal energy U with the enthalpy H, and the molar heat capacity for a gas at constant volume with the molar heat capacity at constant pressure. Before we accept this approach we must confirm two properties of the newly introduced functions: (i) the enthalpy must be a function of the state of the gas, and (ii) C_p must be a materials constant like C_V.

Both points are tested by relating the new parameters H and C_p to the respective parameters in the isochoric case. We start with enthalpy. Using the work term for the isobaric case, Eq. [10.8], we write for the first law:

$$\Delta U = Q - p\,\Delta V \Rightarrow Q = \Delta U + p\,\Delta V.$$

The right-hand side equals the enthalpy change for the isobaric case, as stated in Eq. [10.9]:

$$\Delta H = \Delta U + p\,\Delta V. \qquad [10.11]$$

Since U, p, and V are all variables of the state, so is the enthalpy. The **enthalpy** $H(U, p, V)$ is found from Eq. [10.11]:

$$H = U + p\,V. \qquad [10.12]$$

Note that we have not used the ideal gas law to this point. Thus, the enthalpy is a useful system parameter for all gases and even for liquid and solid materials, for which $H = U$, because volume changes are usually negligible for non-gaseous matter.

Summary: Isobaric process of a gas in general:

$$W = -p(V_f - V_i)$$
$$Q = n\,C_p(T_f - T_i) \qquad [10.13]$$
$$\Delta H = Q.$$

We now study the specific case of an ideal gas. Fig. 10.11 shows the p–V diagram for an ideal gas undergoing an isobaric process. The specific process shown in the figure is an isobaric expansion, in which the volume increases from V_i to V_f and the temperature increases from T_{low} to T_{high}.

For the ideal gas the two molar heat capacities can be related to each other. In Eq. [10.11] we substitute ΔH and ΔU from Eq. [10.10] and use the ideal gas law to replace $p\;\Delta V$ with $n\;R\;\Delta T$:

$$n\,C_p\,\Delta T = n\,C_V\,\Delta T + n\,R\,\Delta T,$$

which yields:

$$C_p = C_V + R = \frac{5}{2}R. \qquad [10.14]$$

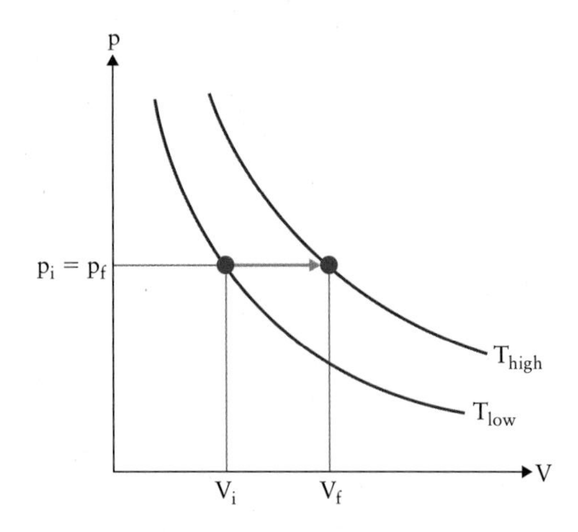

Figure 10.11 *p–V* diagram of an isobaric expansion of an ideal gas. The green arrow indicates an expansion from an initial volume V_i to a final volume V_f. The temperature increases concurrently from T_{low} to T_{high}. The arrow is horizontal because this ensures $p_i = p_f$.

Thus, C_p is a materials constant in the same fashion as the molar heat capacity at constant volume. We emphasize again that we distinguish two molar heat capacities for a gaseous system, C_V and C_p, but do not need to make this distinction when studying solids or liquids.

CONCEPT QUESTION 10.2

Figure 10.12 shows a process for a gas in a *V–T* diagram. The initial and final stages of the gas are indicated. Which of the following statements is correct?

(A) The process is an isobaric compression of an ideal gas.

(B) The process as shown is physically impossible.

(C) The process is an isothermal process.

(D) The process is an isobaric expansion of an ideal gas.

(E) The process is an isochoric process.

(F) Too little information is given.

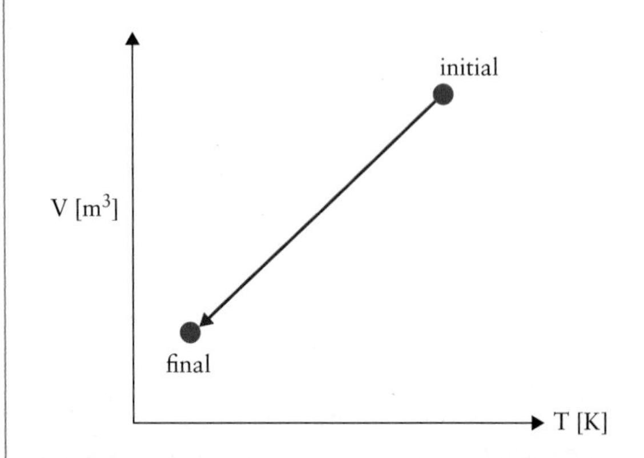

Figure 10.12 *V–T* diagram for a gas with the initial and final states indicated.

10.2.4: Adiabatic Processes

So far we have found that the first law of thermodynamics is easiest to apply when one of its variables is zero: for isothermal processes of the ideal gas $\Delta U = 0$, and for all isochoric processes $W = 0$. This leads us to wonder whether processes also exist with $Q = 0$, and whether they are limited to the ideal gas. Processes with no heat exchange with the environment do exist and are called **adiabatic processes**. For example, in Chapters 15 and 16 we will treat the density vibrations in air that carry sound as adiabatic changes. Fig. 10.13 is a sketch of an adiabatic expansion.

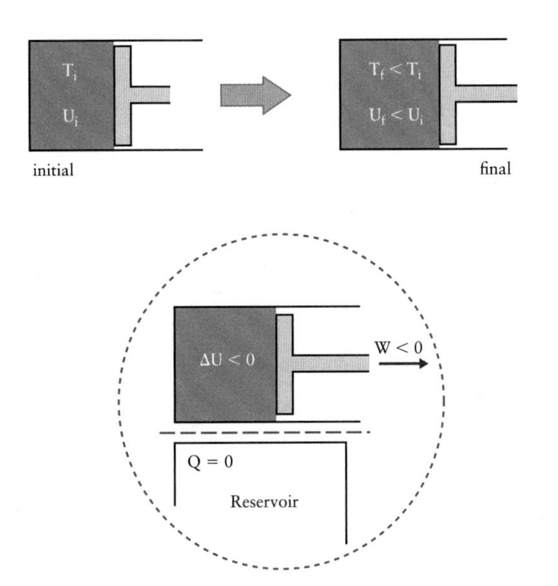

Figure 10.13 Sketch of an adiabatic expansion. In an adiabatic process the system does not exchange heat with the environment. The top two frames show the system before and after the expansion, indicating the temperature and internal energy of the system. The frame in the large dashed circle illustrates the expansion process itself. The dashed line between the reservoir and the system indicates they are isolated from each other.

The lower part of Fig. 10.13 illustrates system changes during the adiabatic expansion. Note that an adiabatic process requires that all heat reservoirs are disconnected from the system; this is indicated by the dashed line between the heat reservoir and the gas container.

For adiabatic processes, we determine from the first law of thermodynamics that:

$$\Delta U = W,$$

which applies to all systems. Thus, the energy for pushing the piston during the adiabatic expansion in the bottom part of Fig. 10.13 ($W < 0$) must be taken out of the internal energy of the system ($\Delta U < 0$).

Quantitative ramifications of the relation $\Delta U = W$ are discussed specifically for the ideal gas. We use Eq. [8.24] from the kinetic gas theory for the change of its internal energy:

$$\Delta U = \frac{3}{2} n\, R\, \Delta T. \qquad [10.15]$$

The volume and the temperature for the initial and final states of an adiabatic process are related in the form (derivation based on Eq. [10.15], but not shown here):

$$V_i\, T_i^{3/2} = V_f\, T_f^{3/2}.$$

It turns out this equation is also useful when describing gas systems other than the (monatomic) ideal gas if we replace the exponent 3/2 by C_V/R, as taken from Eq. [10.3]:

$$V_i\, T_i^{C_V/R} = V_f\, T_f^{C_V/R}. \qquad [10.16]$$

This formula is called **Poisson's equation**. Alternatively, we can replace temperature in Eq. [10.16] by using the ideal gas law in the form $T = p\ V/(n\ R)$. We further use $C_V = C_p - R$ for the relation between the two molar heat capacities for the ideal gas. This leads to a second formulation for the adiabatic process in the form:

$$p_i\, V_i^{\kappa} = p_f\, V_f^{\kappa}. \qquad [10.17]$$

$\kappa = C_p/C_V$ is called the **adiabatic coefficient**, with a value $\kappa = 5/3$ for the ideal gas. Even though we used specific results that are only applicable to the ideal gas to proceed from Eq. [10.16] to Eq. [10.17], the latter equation can be applied generally when using an experimental value for the adiabatic coefficient, κ. For example, $\kappa = 1.4$ for molecular nitrogen (N_2).

Fig. 10.14 compares an isothermal expansion and an adiabatic expansion from 10 atm to 1.0 atm pressure for an ideal gas initially at 20°C. The comparison shows that the adiabatic expansion in volume is smaller but that a significant drop in temperature occurs: from 293 K to 116.6 K. Since $\kappa > 1$, the solid adiabatic curve in Fig. 10.14 is steeper than the dashed isothermal curve, which follows $p\ V = \text{const.}$

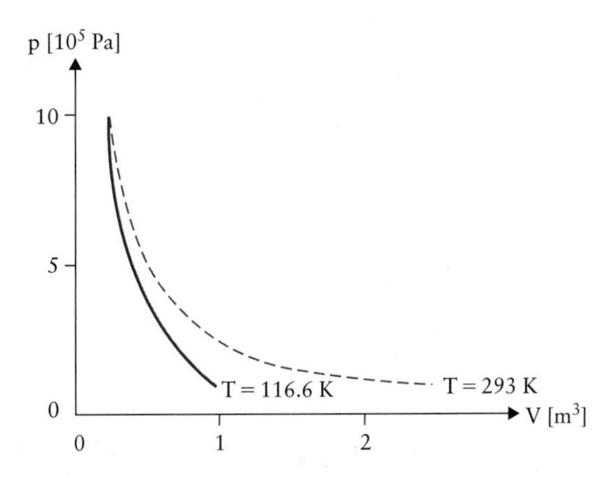

Figure 10.14 Comparison of an adiabatic expansion (solid line) and an isothermal expansion (dashed line) for 1 mol of an ideal gas in a p–V diagram. Both processes start from the same state of the gas, with $p = 1.0 \times 10^6$ Pa and $T = 293$ K. The final pressure is $p = 1.0 \times 10^5$ Pa. During the isothermal expansion, heat from a heat reservoir maintains the temperature of the ideal gas, while its temperature drops to 116.6 K during the adiabatic expansion, which occurs at the expense of the internal energy.

The difference between the two curves in Fig. 10.14 results from the differences in the physical processes of both expansions. In the isothermal case, the work done by the gas during the expansion is supplied by a heat reservoir, allowing the internal energy and therefore the temperature to remain constant. In the adiabatic case, the same amount of energy is required, but it is removed from the internal energy of the gas, which consequently cools down.

Summary: Adiabatic process of an ideal gas:

$$W = \Delta U$$
$$Q = 0 \qquad [10.18]$$
$$\Delta U = n \cdot C_V (T_f - T_i).$$

EXAMPLE 10.2

One mol of an ideal gas undergoes an adiabatic expansion from an initial volume of $V_i = 1.0\ \text{m}^3$ to twice that volume. The initial temperature is $T_i = 270$ K. (a) What is the initial pressure and (b) what is the final pressure?

Solution

Solution to part (a): The initial pressure is obtained from the ideal gas law:

$$p_i = \frac{n R T_i}{V_i} = \frac{(1.0\ \text{mol})\left(8.31 \frac{\text{J}}{\text{K mol}}\right)(270\ \text{K})}{1.0\ \text{m}^3} = 2245\ \text{Pa}.$$

Solution to part (b): We use Poisson's equation with $C_V/R = 3/2$ to determine the final temperature of the gas:

$$T_f^{3/2} = \left(\frac{V_i}{V_f}\right) T_i^{3/2}$$

Bringing both sides to the power of 2/3 leads to:

$$T_f = \left(\frac{V_i}{V_f}\right)^{2/3} T_i.$$

We substitute the given values:

$$T_f = \left(\frac{1}{2}\right)^{2/3} (270\ \text{K}) = 170\ \text{K}.$$

In the next step, the ideal gas law is used to determine the final pressure:

$$p_f = \frac{n R T_f}{V_f} = \frac{(1.0\ \text{mol})\left(8.31 \frac{\text{J}}{\text{K mol}}\right)(170\ \text{K})}{2.0\ \text{m}^3} = 707\ \text{Pa}.$$

10.3: Cyclic Processes

10.3.1: Role of Cyclic Processes in Physiology

A major application of the thermodynamic processes discussed in this chapter are **cyclic processes**. These processes begin and end at the same point in a *p–V* diagram, which means they start and end at the same state of the system. Cyclic processes are important because such processes can be continuously repeated. A general form of a cyclic process is shown in Fig. 10.15(a). Two applications in physiology and biology are highlighted in the same figure, respiration in Fig. 10.15(b) and blood circulation in Fig. 10.15(c).

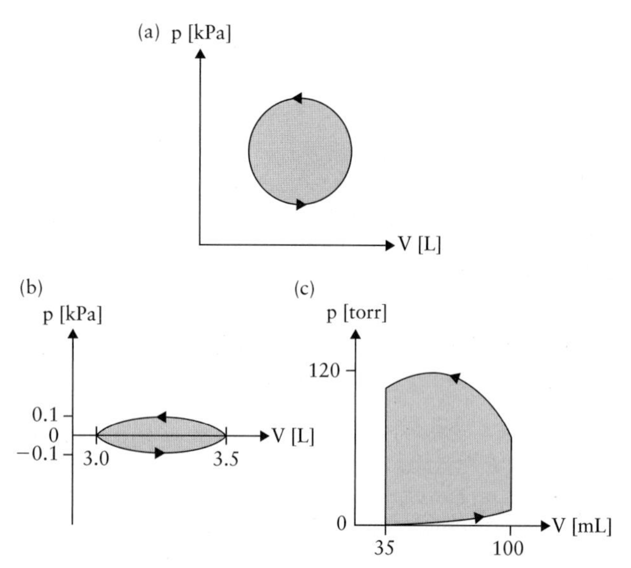

Figure 10.15 Conceptual sketch and key applications of cyclic processes. (a) A cyclic process is a process that returns to its initial state, following a closed curve in a *p–V* diagram. The grey area enclosed by the curve indicates the amount of work required or released per cycle. In the case shown the work is positive as the system receives work from the environment. (b) Respiration is a cyclic process discussed in detail in this textbook. The curve shows the alveolar pressure during regular breathing (compare Fig. 9.1). (c) The repetitive action of the heart. The system in this cycle is the blood. It is discussed in MC–9.16 and P–9.13 (compare Fig. 9.18).

One cyclic process is of particular importance for the development of thermal physics and serves as a reference process for all other cyclic processes: the Carnot cycle.

10.3.2: The Carnot Process

In 1824, Nicolas Léonard Sadi Carnot studied the cyclic process shown in Fig. 10.16. The system is an ideal gas sealed in a container with an ideal piston. This piston and two ideal heat reservoirs, one at a higher temperature

T_{high} and one at a lower temperature T_{low}, are part of the environment. The superstructure, consisting of the system and the environment, is not in thermal equilibrium since the two heat reservoirs are at different temperatures. However, we will see how Carnot's careful choice of the steps of the cyclic process enables us to describe the cycle with the formulas we have derived so far.

We start with the gas at temperature T_{high}, the pressure at p_1, and the volume at V_1. The cyclic process consists of four steps:

I. *Isothermal expansion:* The system is in thermal contact with the heat reservoir at temperature T_{high}. The dashed line between the heat reservoir at temperature T_{low} and the system indicates that the low-temperature heat reservoir is currently isolated from the system. The pressure and volume at the end of the first step are p_2 and V_2. Heat $Q_1 = Q_{high}$ is required to maintain the temperature of the system. The gas does work on the piston.

II. *Adiabatic expansion:* The high-temperature heat reservoir is disconnected from the system. The gas continues to expand, but now adiabatically, to a pressure p_3 and a volume V_3. During the adiabatic expansion the temperature decreases from T_{high} to T_{low}. At the end of this step, the gas has expanded to its largest volume and lowest pressure. The gas continues to do work on the piston.

III. *Isothermal compression:* The low-temperature heat reservoir at temperature T_{low} is brought into thermal contact with the system. The gas is compressed to pressure p_4 and volume V_4. Heat $Q_3 = Q_{low}$ is

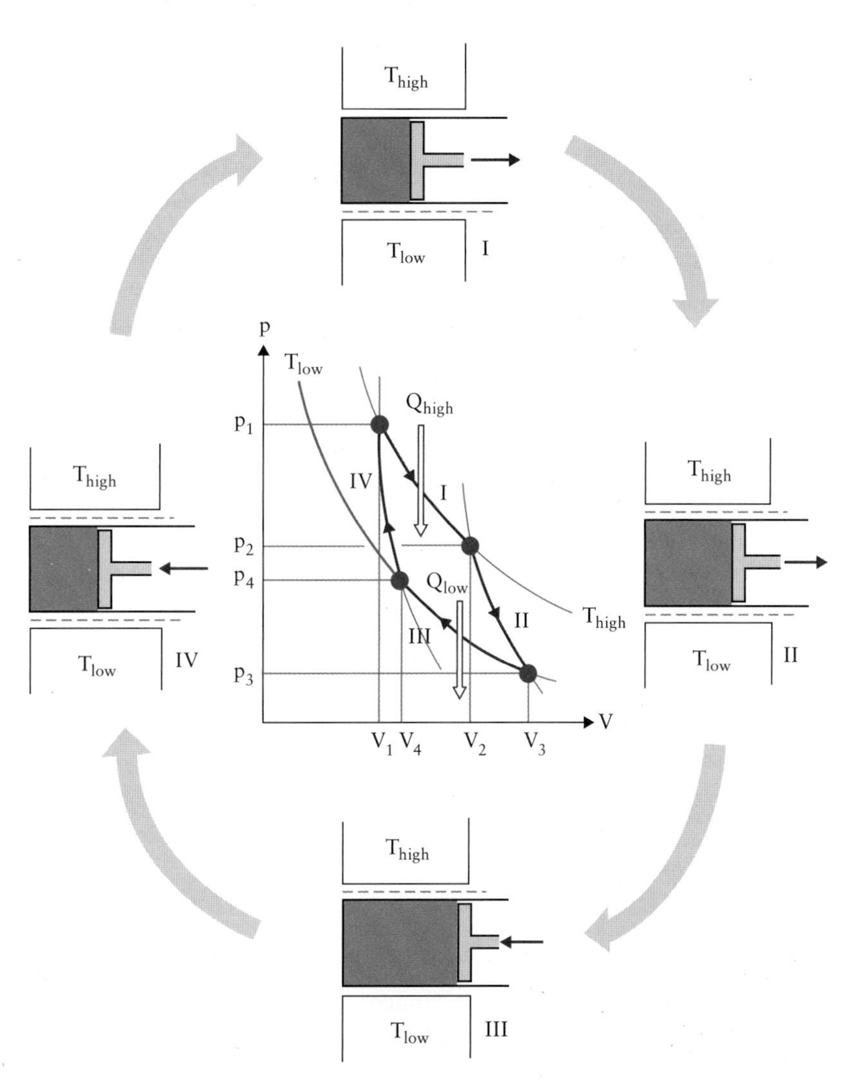

Figure 10.16 Sketch and p–V diagram for the Carnot process. An ideal gas is sealed in a chamber by an ideal piston. Two ideal heat reservoirs at temperatures T_{high} and T_{low} are part of the environment. The cycle is divided into four steps: (I) an isothermal expansion, (II) an adiabatic expansion, (III) an isothermal compression, and (IV) an adiabatic compression. The Carnot process is a cyclic process because the system returns at the end of the four steps to its initial state. The p–V diagram is shown at the centre. The two isothermal curves are extended to both sides as thin blue lines. Heat is taken up from the high-temperature heat reservoir in step I and released to the low-temperature heat reservoir in step III.

released to the reservoir in order to maintain the temperature of the system. The piston does work on the gas.

IV. *Adiabatic compression:* When the pressure reaches p_4 and the volume reaches V_4, the heat reservoir is disconnected from the system. The gas is still compressed, but adiabatically, which raises the temperature of the gas and brings the system back to its initial state at temperature T_{high}, pressure p_1, and volume V_1. In this last step the piston has again done work on the gas.

Then the process repeats, forming a continuous sequence of four-step cycles. The p–V diagram for the Carnot process is shown at the centre of Fig. 10.16. The figure is based on two isothermal lines at temperatures T_{high} and T_{low}. The four steps of the Carnot process are labelled with roman numerals I, II, III, and IV. The flow of heat is also indicated, with Q_{high} received by the gas to maintain its temperature during the isothermal expansion, and Q_{low} released by the system to keep its temperature from rising above T_{low} during the isothermal compression.

CONCEPT QUESTION 10.3

Carnot wanted to study a cyclic process. Why did he not simply combine step I (an isothermal expansion) with step III (an isothermal compression)?

We have already quantified heat, work, and internal energy for each of the single steps in the Carnot cycle. The relevant energy terms are summarized in Table 10.2 for 1.0 mol of an ideal gas as the system.

Using Table 10.2, we determine the work done by the gas, the heat exchange with the heat reservoirs, and the change of the internal energy for one full cycle. Noting that $W_2 = -W_4$ from the table, the total work is calculated as:

$$\sum_{i=1}^{4} W_i = -R\,T_{\text{high}} \cdot \ln\left(\frac{V_2}{V_1}\right) - R\,T_{\text{low}}\,\ln\left(\frac{V_4}{V_3}\right).$$

To simplify this formula, we notice that the four volume terms are not independent of each other. After completing the first two steps, the isothermal compression has to lead to a specific value V_4, which connects back to V_1 in the final adiabatic step. To quantify the relation between the four volume terms we first write Eq. [10.16] for each of the two adiabatic steps in the Carnot process (step II and step IV):

$$\text{Step II:}\quad T_{\text{high}}^{C_V/R}\,V_2 = T_{\text{low}}^{C_V/R}\,V$$

$$\text{Step IV:}\quad T_{\text{high}}^{C_V/R}\,V_1 = T_{\text{low}}^{C_V/R}\,V.$$

The two formulas in this equation are divided by each other. This yields:

$$\frac{V_2}{V_1} = \frac{V_3}{V_4}.$$

We substitute this result into the sum of the work terms for the Carnot process:

$$W_{\text{total}} = -R(T_{\text{high}} - T_{\text{low}})\ln\left(\frac{V_2}{V_1}\right).$$

Next, we determine the heat exchange with the heat reservoirs using Table 10.2:

$$\sum_{i=1}^{4} Q_i = Q_{\text{high}} + Q_{\text{low}} = -W_1 - W_3 = -W_{\text{total}}.$$

From the two results for total work and total heat, the total change of internal energy for the Carnot cycle is calculated:

$$\sum_{i=1}^{4} \Delta U_i = 0.$$

This last result is not surprising, since the gas returns to its initial state after each cycle of the Carnot process. The internal energy is a function of the state and thus must have the same value whenever the gas returns to the same state. Comparing the last three equations, we note once more the difference between a variable of state, such as the internal energy U, and quantities that depend on the history of the system, such as work W and heat Q: only the variable of state must return to its original value after completion of a cyclic process.

TABLE 10.2

Summary of work, heat, and internal energy change per mol during each step of the Carnot cycle

Process	Isothermal expansion	Adiabatic expansion	Isothermal compression	Adiabatic compression
Work W	$-R\ T_{\text{high}}\ \ln(V_2/V_1)$	ΔU_2	$-R\ T_{\text{low}}\ \ln(V_4/V_3)$	ΔU_4
Heat Q	$-W_1$	0	$-W_3$	0
Internal energy change ΔU	0	$C_V\,(T_{\text{low}} - T_{\text{high}})$	0	$C_V\,(T_{\text{high}} - T_{\text{low}})$

Summary: Carnot process per cycle with an ideal gas:

$$W = -R(T_{high} - T_{low})\ln\left(\frac{V_2}{V_1}\right)$$
$$Q = -W \qquad [10.19]$$
$$\Delta U = 0.$$

10.4: Reversibility

To develop a deeper understanding of thermodynamics, a closer look at physical processes is necessary, where we no longer limit our interest to a comparison of the initial and final states but follow the system parameters along the path between these states. A fundamental distinction of processes follows when we study an arbitrary Process I that leads from an initial state A to a final state B, and its reverse Process II from state B to state A. Process I is called **reversible** if all associated parameter changes of the system and its environment are completely reversed during Process II. This is possible only if Process I travels from state A to state B through a continuous sequence of equilibrium states. The biological and physiological processes discussed throughout this textbook are not reversible; we therefore call them **irreversible**.

CASE STUDY 10.2

(a) You put a room-temperature glass of water in a microwave oven and heat it to 50°C. Then you place it back on the table and let it cool down to room temperature. Neglecting a slightly increased rate of evaporation, did you perform a reversible process when heating the water in the microwave oven? (b) If your answer is no, can you name another process you can perform that is perfectly reversible? If you can't name such a process, why do we introduce the concept?

Answer to Part (a): *The process is irreversible. The fact that you recovered the system in its original state (a room-temperature glass of water on your table) does not prove reversibility of the heating process because the cooling process has not restored all the parameters of the environment of your system: electric energy available to do work before the experiment has been lost as thermal energy to the air and the table beneath the glass.*

Answer to Part (b): *There are no reversible processes you can actually perform. We have introduced idealized concepts before in this textbook with no match in the real world, such as point-like particles and the ideal gas. This is justified in each case because such idealizations allow us to model a physical situation or process under simplified conditions, which minimizes the mathematical complexity while preserving the key physical properties. In the current case, idealized reversible processes can be described with the equilibrium thermodynamics concepts we introduced earlier.*

Reversible processes and all physical laws we have introduced so far in this textbook are **time invariant**, which means their predictions remain unchanged whether time moves from the past to the future or whether it were to move from the future toward the past. The first law of thermodynamics serves as an example. For a closed system it is given in Eq. [10.1]. Reversing the time in this equation means that the initial state of the system becomes the final state, and vice versa. Thus, the term $U_f - U_i$ becomes $U_i - U_f$, which is the same numerical value but carries the opposite sign. On the right-hand side of Eq. [10.1] the heat term changes under time reversal from Q to $-Q$ (the heat flows in the opposite direction) and the work term changes from W to $-W$ (work is undone). Thus, Eq. [10.1] looks exactly the same; we cannot distinguish the direction of time (past to future or future to past) using that law.

CASE STUDY 10.3

Assume that you watch a movie that is run in reverse. Parts of this movie will look funny, such as people running backward. What sequence of the movie would appear equally acceptable whether run forward or in reverse?

Answer: *Static scenes, such as the camera sweeping across a room to show the setting prior to any action taking place. In this part of the movie, no time-dependent processes occur; i.e., all parameters are constant.*

We choose in Fig. 10.17 a thermal expansion of an ideal gas to quantify the differences between reversible and irreversible processes. The reversible expansion is illustrated in Fig. 10.17(a): heat transferred from the heat reservoir is transported through the system and deposited as work in the piston. The bottom frame indicates that this process is reversible: each successive state of the system is an equilibrium state along an isothermal curve in the p–V diagram. The process is done slowly enough to allow the pressure in the gas to remain uniform throughout the expanding gas. We can further prove the reversibility by following the expansion with an isothermal compression in which all parameter changes are exactly reversed; e.g., work is done by the piston and returns therefore as energy to the gas while an equivalent amount of heat flows from the gas to the heat reservoir. At the end of the reversed process the gas and its environment are in exactly the same state as they were before the expansion started. This is an idealization, as we assume both an ideal heat reservoir with a fixed temperature and an ideal piston moving without friction.

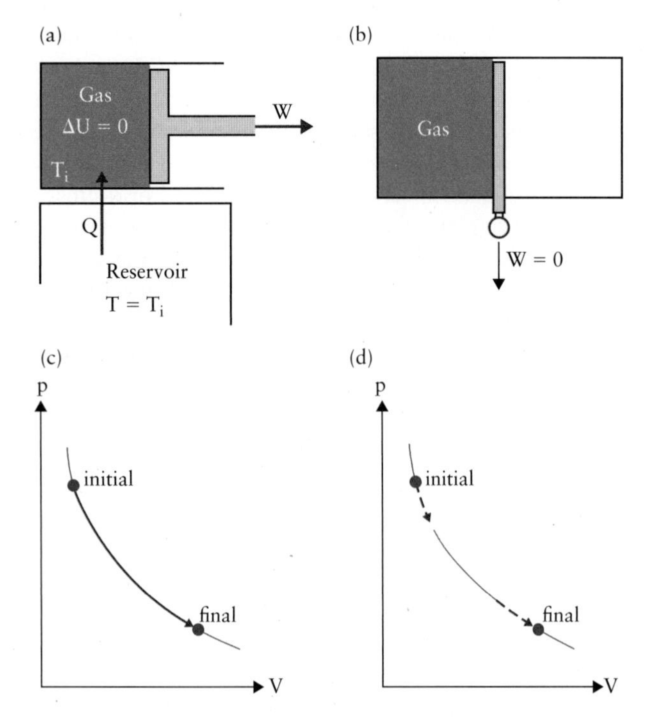

Figure 10.17 Comparison of a reversible and an irreversible isothermal expansion. (a) and (c) In the reversible case, heat flows to the gas from a heat reservoir at the same temperature as the gas. The piston moves without friction, representing the work released by the system. The process follows a continuous line in a p–V diagram. (b) and (d) In the irreversible case, a shutter is removed, allowing the gas to expand into a part of the container that was previously evacuated. No work is done in this case, and re-inserting the shutter does not recover the initial state of the system. The process cannot be represented in a p–V diagram because it is irreversible.

Figure 10.17(b) shows an irreversible isothermal expansion. The gas is initially sealed in a fraction of the container at the left, separated by a shutter from the remaining space in which vacuum conditions apply. When the shutter is allowed to drop down, the gas expands isothermally to fill the whole container. The p–V diagram below the sketch shows that the initial and final states of the gas are the same as for the isothermal expansion in Fig. 10.17(a), but that the process between these states cannot be depicted in this diagram because every point in the p–V diagram represents an equilibrium state of the gas. Thus, the irreversible process does not follow the isothermal curve, as shown in Fig. 10.17(a); the gas parameters vary in an uncontrolled fashion between the initial and final equilibrium states. Pushing the shutter back in an attempt to reverse the process does not recover the initial state of the system, as the gas remains expanded. Even if we use other means to force the gas back into its initial state, permanent changes in the environment have taken place that can no longer be reversed. The major difference between the two processes is that in Fig. 10.17(b) we allowed a change that suddenly converted the initial equilibrium state into a non-equilibrium state.

EXAMPLE 10.3

Calculate the work done for the reversible and irreversible isothermal expansions in Fig. 10.17.

Solution

We determined the work for the reversible process in Fig. 10.17(a) already earlier in the chapter (see Eq. [10.6]). We found that the reversible isothermal expansion allowed us to obtain work from the system. In turn, the irreversible process in Fig. 10.17(b) does not release any useful work. Thus, we summarize:

$$W_{\text{reversible}} = -n\,R\,T\,\ln\left(\frac{V_f}{V_i}\right)$$

$$W_{\text{irreversible}} = 0.$$

The result for the work of the irreversible process is generalized as $W_{\text{irreversible}} < W_{\text{reversible}}$.

We conclude from Fig. 10.17 and the last equation that work and heat exchange cannot be determined for a process by just specifying the initial and final states. In addition, a detailed knowledge of the history of the system is needed, e.g., whether it undergoes reversible or irreversible processes.

CASE STUDY 10.4

Consider Figs. 10.5 and 10.6. Are these two figures drawn appropriately based on our discussion of reversibility?

Answer: *No. The continuous vertical line in Fig. 10.6 implies that we conduct the isochoric heating in a reversible fashion because the figure shows equilibrium states all the way through. However, the illustration of the isochoric heating in the bottom frame of Fig. 10.5 uses a propane flame (Bunsen burner) for heating. For this approach to yield a final state at T_{high}, the propane flame must be hotter than the gas. Creating contact between a system and a component of its environment that are at different temperatures means that the isolated superstructure is not in thermal equilibrium. Processes that involve non-equilibrium states are not reversible and therefore do not proceed through equilibrium states.*

We make two comments in defence of Figs. 10.5 and 10.6: (i) Conducting an isochoric process as reversible or irreversible does not make a difference as both approaches yield $W = 0$. (ii) Conducting the experiment in Fig. 10.5 in a reversible manner would be cumbersome. The way to do it would be to use an infinite number of heat reservoirs with infinitesimally small temperature differences ΔT (all of them part of the system's environment) and bringing one after the other into contact with the system to raise its temperature ever so slightly. If you insist on that level of precision, a corresponding revision of Fig. 10.11 for the isobaric process should also be made.

10.5: The Second Law of Thermodynamics

With the definition of reversibility, we are now in a position to go beyond the first law of thermodynamics. Remember that the first law compares equilibrium states of a system. Thus, we were able to formulate the conservation of energy concept without details of the process by which the system transfers from the initial to the final state. We cheated a bit in this respect when we calculated work and heat transfer because we made the assumption of a reversible process without stating it. The second law of thermodynamics now builds on those calculations but focuses on the history of the actual process, i.e., the heat and work terms that occur during the process. The starting point is the **Carnot process** because

- it is reversible; i.e., we were able to quantify it using the first law of thermodynamics, but
- due to its use of two heat reservoirs at different temperatures, it contains fundamental features about heat and work that have not yet been exploited completely.

10.5.1: Formulation of the Second Law of Thermodynamics

To characterize the Carnot process an efficiency coefficient is introduced. The motivation stems from an economic interpretation of the Carnot process: the heat input in step I represents an investment, while the total work output represents a gain. Thus, we define the **efficiency coefficient** η for a cyclic process as the ratio of the net work to the heat input per cycle, with the heat input given by the heat that is taken up by the system from the high-temperature heat reservoir:

$$\eta = \frac{|\text{net work}|}{\text{heat input}} = \frac{|W_1 + W_3|}{Q_{\text{high}}}. \qquad [10.20]$$

The absolute value of the work is included in Eq. [10.20] to obtain η as a positive number for a system that does work. The work and heat terms in Eq. [10.20] can be substituted from Table 10.2:

$$\eta = \frac{T_{\text{high}} - T_{\text{low}}}{T_{\text{high}}}. \qquad [10.21]$$

Thus, the efficiency coefficient η is always smaller than unity because it is technically and conceptually impossible to operate a cycle with the low-temperature heat reservoir at $T_{\text{low}} = 0$ K. The work obtained from a Carnot process never matches the energy invested into the system in the form of heat.

In turn, we found earlier in the chapter that a maximum amount of work is obtained for reversible processes. This statement can be extended to cyclic processes: **reversible cyclic processes** are those with the greatest efficiency coefficient. In particular, the efficiency coefficient calculated for the Carnot process in Eq. [10.21] is the maximum possible efficiency coefficient for any cyclical process because the Carnot process is done in a reversible fashion. Real cyclic processes are never as ideal as the Carnot process; i.e., the net work is always less than the net work calculated for the Carnot process.

The efficiency coefficient satisfies the interest of engineers in determining the cost-efficiency of engines. For example, a 19th century steam engine operated with steam at 120°C and its cooling water at 20°C. Thus, the maximum efficiency coefficient predicted by Eq. [10.21] is:

$$\eta = \frac{(393\text{ K}) - (293\text{ K})}{393\text{ K}} = 0.25.$$

The efficiency coefficient of a 19th century steam engine cannot exceed 25%. At that time, no actual steam engine had an efficiency coefficient larger than 10%! Natural cyclic processes, such as the ATP cycle in muscle cells, come close to the efficiency of the Carnot process as written in Eq. [10.20]—man-made processes usually fall significantly short.

Because $\eta < 1$, we note that thermal energy is different from all other energy forms because it cannot be converted completely into work. Following Carnot's calculations, it became evident that his findings are applicable to more than the specific cyclic process he studied. Rudolf Clausius and Lord Kelvin generalized in 1850 the implications of the Carnot process to formulate the second law of thermodynamics. In the form as stated by Lord Kelvin, it reads

KEY POINT

In a cyclic process it is impossible to take heat from a reservoir and change it into work without releasing a fraction of the heat to a second reservoir at lower temperature.

CASE STUDY 10.5

The reversibility of the Carnot process means that it could be operated in the opposite direction, i.e., with an isothermal compression while in contact with the high-temperature heat reservoir, and with an isothermal expansion while in contact with the low-temperature heat reservoir. (a) Identify a process for which this inverse Carnot process is an idealization? (b) Try to formulate the second law of thermodynamics in analogy to Lord Kelvin's statement but based on the inverse Carnot process.

Answer to part (a): *Comparing with Fig. 10.16, we find that the inverse Carnot process removes heat from the low-temperature heat reservoir and deposits heat in the high-temperature heat reservoir. For this process, net*

continued

work has to be done on the system; this work is converted into heat, which also ends up in the high-temperature heat reservoir. If we allow the low-temperature heat reservoir to change its temperature, then a lowering of its temperature occurs. If we allow the low-temperature heat reservoir to pick up heat from its environment, then the inverse Carnot process is suitable to maintain a low temperature of that environment. Thus, the inverse Carnot process is a model of a ***refrigeration*** *device.*

Answer to Part (b): *In a cyclic process it is impossible to exclusively transfer heat from a low-temperature heat reservoir to one at a higher temperature. This is the formulation of the second law as introduced by Clausius.*

10.5.2: Definition of Entropy

For applications, we need to find a way to express the second law of thermodynamics quantitatively. To do this, we re-examine the efficiency coefficient η of the Carnot process. For this we start with Eq. [10.21]. The two work terms in the numerator can be exchanged for the two corresponding heat terms of the isothermal steps in Table 10.2:

$$\eta = \frac{|W_{\text{net}}|}{Q_{\text{high}}} = \frac{Q_{\text{low}} + Q_{\text{high}}}{Q_{\text{high}}} = \frac{T_{\text{high}} + T_{\text{low}}}{T_{\text{high}}}.$$

Next, we sort the terms based on the subscripts "high" and "low"; i.e., we separate the terms associated with temperature based on the high- and low-temperature heat reservoirs respectively:

$$\frac{Q_{\text{low}}}{T_{\text{low}}} + \frac{Q_{\text{high}}}{T_{\text{high}}} = 0.$$

This is an interesting result because it states that the sum of the ratios of heat to temperature for a cyclic process is zero. Before we discuss the implications further, we want to confirm that the result in this equation applies to any reversible cyclic process, not just the Carnot process for which it was derived. This is shown in Fig. 10.18 for an arbitrary cyclic process in a p–V diagram. By drawing a continuous line in a p–V diagram we assume the cycle operates reversibly. For such a process, a net of closely spaced isothermal and adiabatic lines can be superimposed on the cycle as shown. The more shallow lines are isotherms and the steeper lines are adiabates, like those drawn in Fig. 10.14. This net of lines divides the arbitrary cycle into a large number of small Carnot processes, each with a very small temperature difference between the high- and low-temperature heat reservoirs. One such Carnot cycle created by the net of thin lines is highlighted near the centre for illustration. Following each small Carnot cycle we notice that its four contributing steps are shared with neighbouring Carnot cycles, except for the segments that run along the outer line, which constitutes the large arbitrary cycle. Thus, the sum over all Q/T terms for all Carnot cycles in Fig. 10.18 provides a value corresponding to the last equation for the arbitrary reversible cycle:

$$\sum_i \frac{Q_i}{T_i} = 0. \qquad [10.22]$$

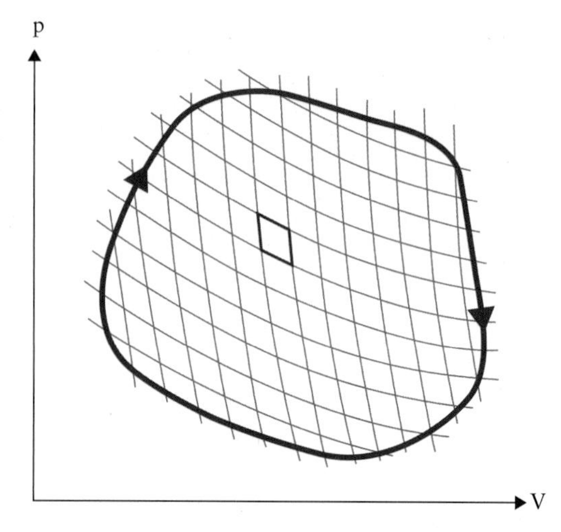

Figure 10.18 A p–V diagram for an arbitrary cyclic process. To quantify the properties of such a process, it is divided into a very large number of Carnot processes, which are operated in adjacent cycles with very small temperature differences between the two heat reservoirs (thin lines). One such Carnot process is highlighted near the centre of the figure. The approach taken in the figure allows us to quantify the entropy change in the process.

We find that the sum of all quotients of heat transfer and temperature during a cyclic process is zero. The sum in Eq. [10.22] behaves in the same way as changes of the internal energy: whenever the system returns to its initial state, these values return to the same value they had before the cycle started. Thus, the physical property underlying Eq. [10.22] has a defined value for a given state of the system, in the same way a defined value of internal energy is attributed to each possible equilibrium state of a system. Any parameter that has a defined value for each equilibrium state of a system—i.e., a value independent of the history of the system—is called a **variable of the state** of the system. Thus, the ratio Q over T is a variable of the state, while we know from our earlier discussions of the first law of thermodynamics that Q separately is not a variable of the state. To fully describe a system, such variables of the state have to be measured. A complete description of the state of a system therefore includes temperature, volume, pressure, and the amount of material in unit mol or kg, as well as the internal energy and the newly introduced ratio of heat transfer and temperature. With this importance attached to the new quantity it was given a name, **entropy** S with unit J/K. The name *entropy* is taken from Classical Greek and means "change" or "transformation."

The quantitative definition of entropy is developed from Eq. [10.22], which represents the change of the entropy for a cyclic process. If an arbitrary process does not return to the initial state, then an entropy change ΔS occurs, where ΔS is the difference in entropy between its initial and final states:

$$\Delta S = S_f - S_i = \sum_i \frac{Q_i}{T_i}. \qquad [10.23]$$

This definition is useful as it provides a method to measure entropy: The system is brought into a "standard" initial state, for which the entropy is known; then the state of the system is changed to the final state, for which the entropy value is sought. Measuring the heat transfer at each temperature during the change of state allows us to quantify the change in entropy between the two states. This procedure is discussed in detail later in this chapter.

EXAMPLE 10.4

Two mol of air is initially confined to the left side of the container shown in Fig. 10.19. Assume that the air volume is doubled when the valve is opened and that the air temperature is held constant. (a) Is the expansion process reversible or irreversible? If it is irreversible, what is the corresponding reversible process? (b) Show that the entropy change for this process is:

$$\Delta S = n\,R \ln\left(\frac{V_f}{V_i}\right).$$

(c) Quantify ΔS for the case of Fig. 10.19. (d) Does the formula in part (b) depend on whether the process is reversible or irreversible?

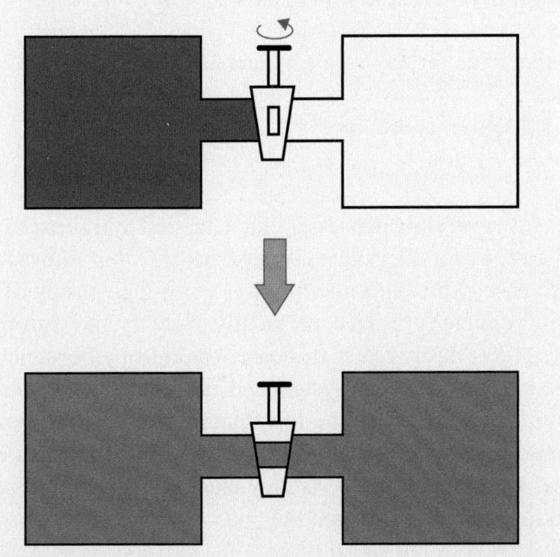

Figure 10.19 Expansion of 2 mol of air, which is initially confined to the half-space at the left. The expansion becomes possible when a valve between the two half-chambers is opened.

continued

Solution

Solution to part (a): The process is irreversible since it cannot be reversed by returning the valve to its original position. The corresponding reversible process is an air expansion with a mobile frictionless piston [Fig. 10.17(a)].

Solution to part (b): The heat added to an ideal gas when it expands isothermally was derived in Eq. [10.7]. To quantify the process, we have to assume air is an ideal gas; however, we note that the irreversibility of the process has no bearing on this question, as discussed in part (d). Thus:

$$Q = n\,R\,T \ln\left(\frac{V_f}{V_i}\right),$$

and with the definition of the entropy in Eq. [10.23], we find for the entropy change for an isothermal process:

$$\Delta S = \frac{Q}{T} = n\,R \ln\left(\frac{V_f}{V_i}\right).$$

Solution to part (c): We use $n = 2$ mol with a ratio of final to initial volume of $V_f/V_i = 2$:

$$\Delta S = (2.0\ \text{mol})\left(8.314\frac{\text{J}}{\text{K mol}}\right)\ln 2$$

$$= +11.5\frac{\text{J}}{\text{K}}.$$

Solution to part (d): When switching from a reversible to an irreversible process, the amount of heat exchange is not affected, but the maximum work (as calculated for the reversible process) is not obtainable in the irreversible process. The calculation in part (b) does not include a work term and therefore is not affected by the manner in which the process is conducted.

CONCEPT QUESTION 10.4

Figure 9.1 shows the p–V diagram for dynamic breathing. (a) Is dynamic breathing a reversible process? (b) What is the entropy change per cycle based on Fig. 9.1?

The following three sections focus on key properties of entropy that we will use later in the textbook and that you often will find references to in biophysical or biochemical studies.

10.5.3: Entropy and Work

The Carnot process showed that entropy is associated with the fraction of heat that cannot be utilized as work during a cyclic process, i.e., the heat lost to a low-temperature part of the environment. This is one interpretation of the entropy concept. We will expand on this interpretation when we introduce the Gibbs free energy later in this chapter: the total energy of a system can be diminished by the entropy-related useless energy to determine the energy available from the process to do work (to which the term *free energy* then refers).

10.5.4: Entropy and Reversibility

We want to compare the values of the change of entropy ΔS for reversible and irreversible spontaneous processes, where **spontaneous process** refers to a process that progresses on its own. The explosive formation of water from oxygen and hydrogen gases is such a spontaneous process, while the opposite process, the splitting of water to obtain hydrogen and oxygen gases, is not spontaneous as it never happens without external effort.

Fig. 10.20 shows idealized reversible and irreversible processes for equilibrating the temperature of two identical objects (1 and 2) that differ only in temperatures T_1 and T_2, with $T_1 > T_2$. The top sketch shows the reversible approach, and the bottom sketch shows an irreversible approach.

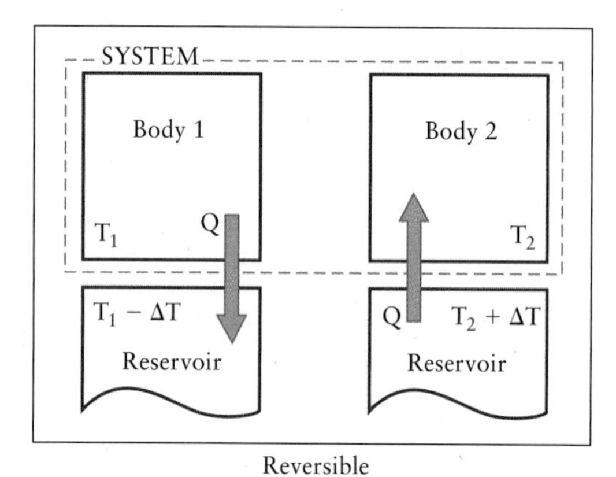

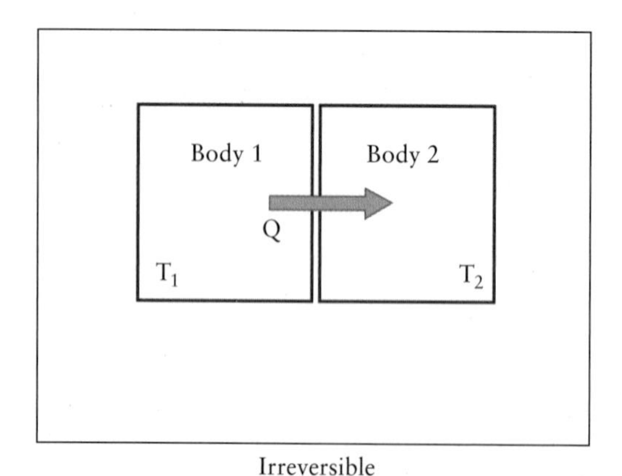

Figure 10.20 Comparison between a reversible (top) and an irreversible (bottom) experiment to equilibrate the temperature of two objects B1 and B2 with initial $\Delta T > 0$. For the irreversible case the two objects are brought into direct thermal contact. For the reversible process heat reservoirs R1 and R2 have to be used to adjust the temperature of each part of the system in very small steps until the temperature difference is eliminated.

In the simpler, irreversible case, the two objects are brought into direct thermal contact. As a result, an amount of heat Q flows from the hotter to the colder body until both have the same temperature. This process is irreversible as no work is extracted and stored in the isolated superstructure to reverse the process, i.e., to re-establish different temperatures for both objects after the temperature equilibrium has been established.

The reversible approach is more complex. The two objects are not brought into direct contact; instead, each is brought into thermal contact with a heat reservoir that is minimally warmer in the case of object 2 and minimally colder in the case of object 1. The slight temperature differences of the objects and the respective heat reservoirs are labelled $\Delta T/2$. We assume that the temperature difference $\Delta T/2$ is the same in both cases. Heat exchange with the heat reservoirs occurs. For simplicity, we assume the initial temperatures of the two objects also differ only by $T_1 - T_2 = \Delta T$. Otherwise, a very large number of heat reservoirs would have to be lined up for each object, with each one having a temperature smaller or higher than the previous heat reservoir by a difference of $\Delta T/2$. Even though such a reversible approach is obviously not practical for achieving a notable temperature change, we consider it for the sake of argument.

We carefully distinguish the *system,* which consists of the two objects 1 and 2, and the *environment,* which consists of the two heat reservoirs 1 and 2 within the isolated superstructure of Fig. 10.20. For the respective changes of entropy in the reversible case, we find from Eq. [10.23]:

$$\Delta S_{\text{system}} = \Delta S_{B1} + \Delta S_{B2} = -\frac{Q}{T_1} + \frac{Q}{T_2} > 0$$

$$\Delta S_{\text{environment}} = \Delta S_{R1} + \Delta S_{R2} = \frac{Q}{T_1} - \frac{Q}{T_2} < 0.$$

This yields for the isolated superstructure:

$$\Delta S_{\text{isolated superstructure}} = \Delta S_{\text{system}} + \Delta S_{\text{environment}} = 0 \qquad [10.24]$$

A **reversible process** is characterized by a zero change in entropy for the isolated superstructure. This allows for a decrease in the entropy of the system if the entropy of the environment is raised concurrently. Note that no entropy is created or destroyed in this case, but entropy flows across the interface between system and environment. The observations we made for the top frame of Fig. 10.20 characterize reversible processes in general—including systems that are in equilibrium, as they undergo no processes: the entropy remains constant, $\Delta S = 0$.

KEY POINT

$\Delta S = 0$ applies to isolated superstructures with reversible processes or systems that are continuously in an equilibrium state.

Next, we study the irreversible process shown in the bottom frame of Fig. 10.20. In this case, the isolated superstructure contains only the system, which again consists of the two objects 1 and 2. No heat reservoirs are needed as no heat exchange with the environment occurs. The entropy change of the system is the same as in the reversible case because the same amount of heat flows out of object 1 and into object 2. However, no entropy change occurs in the environment as no interaction with the environment takes place:

$$\Delta S_{\text{system}} = -\frac{Q}{T_1} + \frac{Q}{T_2} > 0$$

$$\Delta S_{\text{environment}} = 0.$$

This yields for the isolated superstructure:

$$\Delta S_{\text{isolated superstructure}} = \Delta S_{\text{system}} + \Delta S_{\text{environment}} > 0. \quad [10.25]$$

This result is different from the result of the reversible case in Eq. [10.24]: in a **spontaneous irreversible process** the entropy of an isolated system increases. Entropy is created in the system! Interpreting the results in Eqs. [10.24] and [10.25], we note that an isolated system not in equilibrium will undergo spontaneous irreversible processes that increase its entropy. Such processes cease when the system reaches equilibrium. At that point the entropy does not increase any further.

KEY POINT

The entropy reaches a maximum value for a system in equilibrium: S_{eq} *= maximum.*

CONCEPT QUESTION 10.5

We did not identify a heat transfer to the gas in the irreversible isothermal expansion in Fig. 10.17(b). Does the entropy of the gas change?

10.5.5: Entropy and Order

Ludwig Boltzmann showed that entropy is proportional to the number of accessible microscopic states of a system. While the number of accessible states is usually complicated to determine, we can use the concept qualitatively to judge whether processes are associated with an increase in entropy. Fig. 10.21 shows five processes, each associated with an increase of entropy when followed from the left to the right box:

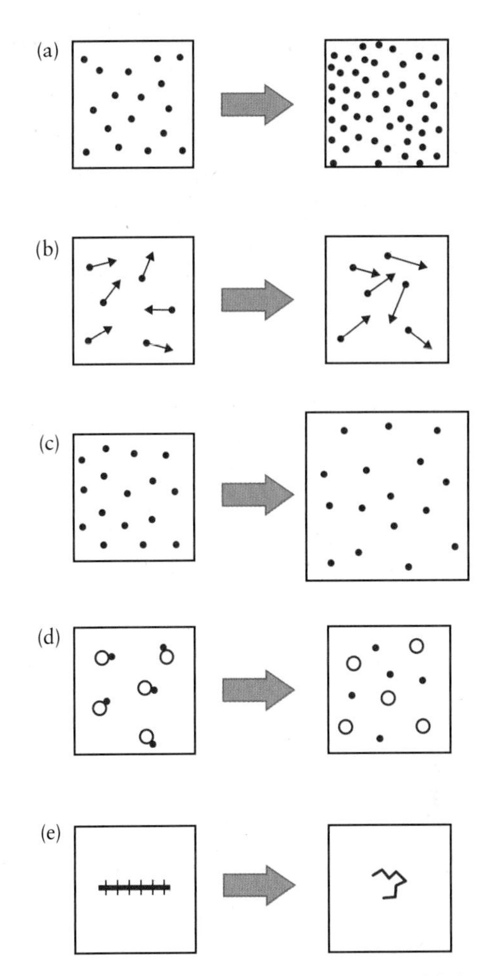

Figure 10.21 Processes with entropy increase: (a) increase of particle number; (b) increase of internal energy; (c) isothermal expansion; (d) dissociation of molecules; and (e) polymer relaxation.

(a) Adding particles (i.e., an open system with material influx) increases the entropy since each of the previously present particles has as many accessible states as before, but the new particles contribute additional states for the combined system.

(b) Adding energy to a closed system leads to an increase of the entropy. If the system is a gas, the molecules all gain speed; i.e., the range of accessible velocity components becomes larger. Note that no low-speed components are lost since, for example, a very fast particle can move mostly in the xy-plane and have still a small z-component of velocity.

(c) Increasing the volume can be done in two ways: by holding the internal energy constant (isothermal process) or by isolating the system (adiabatic process). An isothermal expansion increases the entropy since after the expansion the particles can resume all their previous positions but have access to additional positions with the number of accessible velocity components unchanged. Adiabatic expansions obey the entropy equation $\Delta S = 0$ since no heat exchange occurs, $Q = 0$. Therefore, adiabatic processes are also called **isentropic**. Why does the volume increase in this case not lead to an entropy increase? During an adiabatic expansion, each particle gains new accessible positions, but as the temperature drops sharply it loses the same number of accessible states linked to its velocity.

(d) Dissociation of molecules. If there are N molecules initially, then the system has $6N$ free parameters (3 spatial and 3 velocity components per particle). In addition, the molecule may possess various states of rotation and vibration. Dissociation is favoured because it doubles the number of free parameters to $12N$ because the number of independent particles doubles, while the loss of vibrational and rotational states is much less.

(e) The last example is a polymer consisting of a large number of repetitive monomer units. If these units are connected with chemical bonds that can rotate, e.g., C–C single bonds, then the relaxation of the polymer represents an increase of entropy. This is due to the fact that a fully stretched polymer has only one relative orientation for adjacent monomers, while a very large number of relative positions of twisted repetitive units is possible for the relaxed polymer.

What the processes in Fig. 10.21 have in common is that the **degree of disorder** of the system increases as the entropy increases. Thus, entropy can be understood as a measure of the disorder in a system. Again, the second law does not prohibit a system from becoming ordered (e.g., crystals growing from solution or patterns developing in biological systems) unless it is an isolated system. If the system is open or closed, the total entropy or disorder of the isolated superstructure must increase in spontaneous processes but the entropy or disorder of the system itself can decrease.

CASE STUDY 10.6

What is the most likely shape in which large biomolecules might be found in the human body?

Answer: *Most polymers resume a folded equilibrium shape. Polymer relaxation is favoured by the entropy argument made with Fig. 10.21(e). We will see later in this chapter that the internal energy also plays a role because the system is not isolated; this contribution then favours a particular relaxed structure over random relaxation, i.e., supports the formation of the biochemically active structure. Neither entropy nor energy contributions favour a stretched polymer.*

CASE STUDY 10.7

(a) Sugar crystals form when a supersaturated sugar solution evaporates slowly. Crystals are a more ordered form of matter than a solution. Why does this observation not violate the second law of thermodynamics? (b) Stanley Miller achieved abiotic synthesis of organic compounds under "Early Earth" conditions in his famous 1953 experiment (Fig. 10.22 shows Miller and the apparatus). In a sealed

continued

container, electrical discharges in an atmosphere that consisted of water vapour, hydrogen, ammonia, and methane gases produced complex amino acids. Is this result consistent with the second law of thermodynamics?

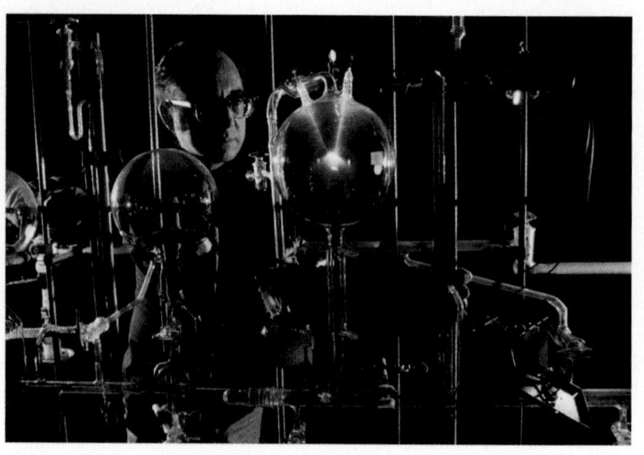

Figure 10.22 Stanley Miller with the original equipment used in his 1953 experiments, which gave credence to the idea that organic molecules were created by conditions of early Earth's atmosphere.

Answer to Part (a): *The system in this experiment is an open system that continuously loses solvent molecules to its environment. The second law of thermodynamics restricts ordering only for isolated systems.*

Answer to Part (b): *Miller's experiment was done in a closed system (no exchange of matter with the environment). However, energy had to be supplied to the system to generate the electric discharges. Thus, again, the second law of thermodynamics does not restrict the complexity of compounds in Miller's arrangement since it was not an isolated system. If the battery generating the discharges had been integrated with the arrangement to allow the combined system to be isolated, Miller's experiments would still have produced highly ordered molecules, at the expense of a net entropy increase due to irreversible processes in the discharging battery.*

10.6: Chemical Thermodynamics: An Overview

Chemical reactions are non-equilibrium processes. However, it is possible to quantify some aspects of chemical reactions using equilibrium thermodynamics concepts. For example, chemists often are interested in whether a chemical reaction is possible as proposed, or which parameter ranges of temperature, pressure, and concentrations favour the reaction.

We need both the internal energy and the entropy to judge whether a chemical reaction is possible. The internal energy allows us to separate chemical reactions into **endothermic reactions**, which are reactions that require the supply of heat, and **exothermic reactions**, which are reactions that release heat. In turn, the entropy allows us to separate **spontaneous reactions** from reactions that have to be forced externally.

10.6.1: Internal Energy and Enthalpy

Changes of the internal energy are measured with a **calorimeter**. In a calorimeter we study processes within an isolated superstructure. The instrument includes a heat reservoir that measures the amount of heat released from the system. This is a direct measurement of the change of the internal energy because it is equal to the heat exchange of the system at constant volume.

Fig. 10.23 shows a calorimeter that is set up specifically to study oxidation (combustion) processes at constant volume. The heat reservoir consists of a water-filled chamber surrounding the system; its temperature change is directly proportional to the amount of heat released.

Most chemical processes, including biochemical processes, occur at constant pressure, i.e., at atmospheric pressure. For isobaric processes, we saw earlier that the enthalpy is better suited than the internal energy to describe the heat exchange, as it accounts properly for a possible volume change during the process.

CASE STUDY 10.8

Calorimeter arrangements like the one in Fig. 10.23 have been used in chemistry since the late 1700s. For example, in 1780 Antoine de Lavoisier and Pierre-Simon Laplace placed a guinea pig in the calorimeter and measured the heat dissipated by the animal as a function of the amount of oxygen it consumed. What idea did they test with this experiment?

Answer: *They suggested respiration is a form of combustion within the body of the animal, comparable to the combustion of carbon:* $C + O_2 \rightarrow CO_2$*. Comparing changes in the internal energies allowed them to quantify their model.*

The volume change leads to a work term of the form $p_{atm}\,\Delta V$; this work is done *against* (expansion) or *by* (compression) the external atmosphere. The heat exchange at constant pressure defines the enthalpy change ΔH (see Eq. [10.10]).

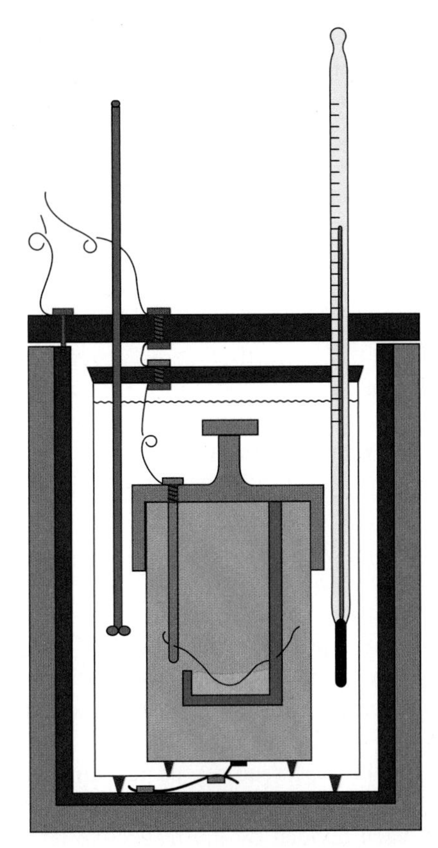

Figure 10.23 Sketch of a calorimeter used to determine the amount of heat released in combustion at constant volume. The probe material is placed in an inner steel container with a high-pressure oxygen atmosphere. The reaction is electrically ignited. The steel container is immersed in a water bath in which the temperature change is measured after the reaction is completed. The water bath in turn is isolated from the environment to prevent measurement errors due to heat loss through the outer wall.

EXAMPLE 10.5

We study the thermal decomposition of 1 mol $CaCO_3$ into CaO and CO_2 at 900°C. A final CO_2 gas pressure of 1 atm is reached in this process. The system absorbs 175.7 kJ of heat. What is the change of the internal energy for 1 mol of $CaCO_3$? *Hint:* Neglect volume changes of the solid components CaO and $CaCO_3$, and treat CO_2 as an ideal gas.

Solution

In the process $CaCO_3 \rightarrow CaO + CO_2$, only CO_2 is a gas. Its initial volume is zero, $V_i = 0$. After the reaction is completed, the product $p\ V$ for CO_2 is:

$$pV = nRT = (1.0\,\text{mol})\left(8.314\frac{\text{J}}{\text{K mol}}\right)(1173\,\text{K}) = 9750\,\text{J}.$$

We calculated the term $p\ V$ for CO_2 from the ideal gas law, using $p = p_{atm}$ and T = 900°C = 1173 K. n = 1 mol because 1 mol of CO_2 is formed for every mol of $CaCO_3$.

The internal energy is calculated from the measured enthalpy using Eq. [10.11]. We note that the result in Example 10.5 equals the term $p\,\Delta V$ in Eq. [10.11]. Further, we know from the heat required for this reaction that ΔH = +175.7 kJ. Thus:

$$\Delta U = 175.7\ \text{kJ} - 9.75\ \text{kJ} = +166.0\ \text{kJ}.$$

This means 175.7 kJ is supplied to the system. Of this energy, 9.75 kJ is required for the volume expansion of the gaseous product and the balance of +166 kJ represents the increase of the internal energy of the system.

Absolute values of the internal energy and enthalpy of a system are physically meaningless values, like an absolute value of the gravitational potential energy. We use this fact to freely choose a reference point for these energies: the internal energy or enthalpy of an elementary system in its most stable form at standard conditions (standard conditions are defined in chemistry and physics as 25°C and 1 atm) is set equal to zero: $H = 0$ J. For any other system, we determine the enthalpy relative to this reference state. For example, $H = 0$ J for molecular oxygen O_2 at standard conditions (25°C, 1 atm), while $H = +247.3$ kJ/mol for atomic oxygen (O) under the same conditions. Of particular use in chemistry is the **standard enthalpy of formation** of a compound, which is the energy needed to chemically form 1 mol of the compound from the elements at standard conditions.

In chemical and biological processes, the energy stored in a molecule (e.g., a sugar molecule) may be used in two ways. First, it can be turned completely into heat. This is the case when a human body responds to undercooling by shivering. For such cases, the knowledge of the change in internal energy in the reaction is sufficient (more precisely, the change of the enthalpy, since biological processes occur under isobaric rather than isochoric conditions). Table 10.3 compares enthalpy values for the combustion of the main components of food and some other organic compounds.

TABLE 10.3

Combustion heat at 25°C for reactions of various compounds to CO_2 and H_2O

Compound	ΔH (kJ/g)
Fat	−38.9
Carbohydrates	−17.2
Protein*	−17.2
C_nH_{2n+2}, $1 \le n \le 7$ (methane to heptane)	−55.7 to −48.1
Ethanol	−29.7
Benzene	−42.3
Acetic acid	−14.5

* ΔH for reaction to urea since urea is the final combustion product of proteins in the human body.

Alternatively, the stored energy can be used as work. However, only part of the energy listed in Table 10.3 results in work, as illustrated in Example 9.6. Here the distinction between useful and useless energy matters, as the latter increases only the temperature of the body (as in the shivering case); it does not contribute to the work of muscle contraction. The enthalpy is not sufficient to describe such a process; the change in entropy needs to be taken into account.

10.6.2: Standard Entropy for Chemical Reactions

The enthalpy of formation of a chemical compound tells us nothing about the spontaneity of the process. The reason for this limitation lies in the fact that the internal energy and enthalpy are system properties obtained from the first law of thermodynamics. The conservation of energy does not reveal the direction in which a process evolves. We need to include one more parameter to completely describe a process: the entropy. Entropy characterizes the state of a system and allows us to distinguish spontaneous from non-spontaneous processes. For the entropy, we have established

$\Delta S > 0$: *irreversible process; can occur spontaneously*

$\Delta S = 0$: *reversible process; the system is continuously in equilibrium with its environment*

$\Delta S < 0$: *process that requires a significant increase in the environmental entropy such that the entropy change of the superstructure is* $\Delta S_{superstructure} \ge 0$.

Example 10.4 illustrates how entropy changes are calculated for physical processes. We are equally interested in the change of the entropy for chemical reactions, as these values indicate whether the chemical reaction is spontaneous. **Standard entropy** values at 25°C and 1 atm are tabulated for a wide range of chemical compounds. It is interesting to note that these values are obtained in a different manner than the standard enthalpy of formation. In the case of the entropy, no compound or element has a zero value at the standard state. This is due to the physical meaning of entropy as it relates to the degree of disorder in the system. Therefore, all absolute values of entropy are positive, or zero for a perfectly ordered system. Walther Nernst stated that perfectly ordered systems exist only at $T = 0$ K because the thermal energy of a system at $T > 0$ K causes at least some disorder. He therefore formulated the **third law of thermodynamics**:

KEY POINT

The entropy of a perfect crystal of a chemical element or compound at $T = 0$ K is zero, $S = 0$ J/K.

Thus, the standard entropy at 25°C and 1 atm for 1 mol of any compound has a well-defined value. We find this value from Eq. [10.23], with the initial state at $T = 0$ K and the final state at $T = 25$°C. The right-hand side of Eq. [10.23] requires us to sum over all heat exchange contributions from 0 K to 298 K at atmospheric pressure, each divided by the respective temperature at which it occurs. These heat contributions, in turn, are written for 1 mol in the form $Q = C_p \ \Delta T$. Thus, we can determine the standard entropy from the area under the curve of C_p/T versus T. An example is shown in Fig. 10.24, which illustrates the C_p/T-versus-T curve for N_2 from 0 K to 298 K. From the area under the curve we find $S^0_{N_2} = +191.2$ J/(K mol).

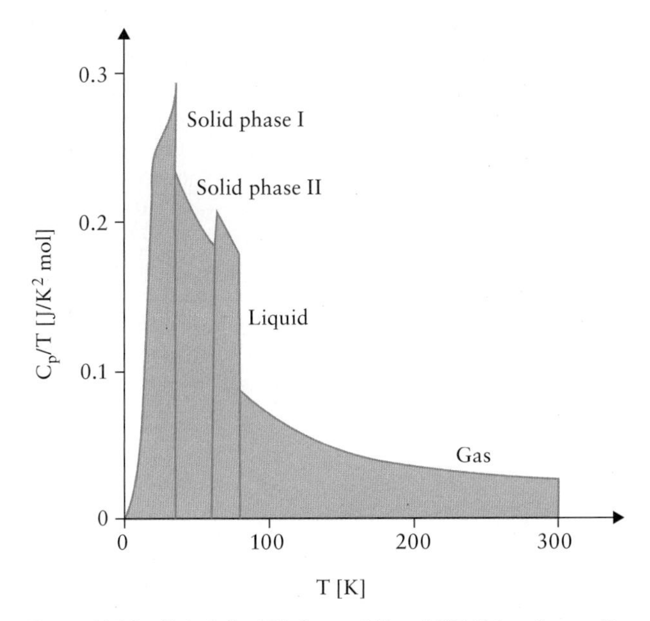

Figure 10.24 Plot of C_p/T between 0 K and 298 K for nitrogen N_2. The area under the curve is the standard entropy for the compound $S^0_{N_2}$. In the temperature interval of interest, nitrogen passes through two solid phases, melts at 63.14 K, and evaporates at 77.32 K. These phase transitions are noted in the figure as discontinuities of the heat-capacity function. At each phase transition, a term for the latent heat divided by the transition temperature must be added.

10.6.3: Gibbs Free Energy

With the discussion of entropy changes in the previous section we have now enough information to predict the outcome of any process we are interested in. However, for practical purposes, the entropy has a disadvantage in that it requires a combined study of the system and its environment. As the statement for a process with $\Delta S < 0$ indicates, such processes can be achieved if appropriate processes occur concurrently in the environment. It would be more convenient to find a thermodynamic system parameter that allows us to make statements about the system and its propensity with respect to a certain process without considerations of the environment.

Initially, it was suggested to use the enthalpy for this purpose. However, spontaneous processes exist that are endothermic. Thus, even though most systems have a propensity toward a low energy state, this is not the sole driving force for chemical processes. Josiah Willard Gibbs resolved this issue in 1875 by introducing the **Gibbs free energy**:

$$G = H - T\,S$$
$$\Delta G = \Delta H - T\,\Delta S, \qquad [10.26]$$

in which the second formula is used for the change of the state of a system or for a chemical reaction. The product $T\ S$ represents the amount of heat lost during the process. Thus, the Gibbs free energy represents the maximum obtainable work. The obtainable work is a maximum since a reversible process is required to actually gain this amount of work. Eq. [10.26] allows us to summarize the properties of the Gibbs free energy for various processes:

KEY POINT

All natural phenomena are governed first by the propensity of the system to lower its internal energy or enthalpy, and second by the propensity of the system to increase its entropy.

$\Delta G < 0$: a spontaneous process

$\Delta G = 0$: a process in which the initial and final states coexist in equilibrium

$\Delta G > 0$: a process that occurs spontaneously in the reverse direction.

Whether we study the Gibbs free energy, the entropy, or the enthalpy in a given process depends on its details. For example, a process in an isolated system is fully described by studying the change in entropy because the internal energy of an isolated system is constant (first law of thermodynamics). On the other hand, a combination of isothermal and isobaric processes requires us to calculate the Gibbs free energy, as both the enthalpy and the entropy may vary during the process. For most known organic and inorganic chemical processes, values of the standard Gibbs free energy are tabulated or can be calculated from tabulated values for the standard enthalpy and standard entropy.

The concept of maximum obtainable work as related to the Gibbs free energy further allows us to define the **chemical equilibrium**. If a system is in chemical equilibrium then the Gibbs free energy G resumes a minimum value. If G is not a minimum, then the system spontaneously decreases its Gibbs free energy to the minimum value. If the process reducing the Gibbs free energy is reversible, then a maximum of work is obtained.

The parameters of state of a system, p, V, T, n, U, H, S, and G, are all linked, with well-defined relations. We do not discuss these relations in the textbook; they are found in the physical chemistry literature when needed. The exception is the dependence of the Gibbs free energy on pressure for a process in which temperature is held constant:

$$\Delta G = n\,R\,T \ln\left(\frac{p_f}{p_i}\right). \qquad [10.27]$$

This relation is used below to define a chemical equilibrium between a solution and its vapour phase, and in Chapter 19 to discuss the electrochemical equilibrium across a nerve membrane.

EXAMPLE 10.6

An instructive example of the role of the Gibbs free energy is the comparison of the formation of carbonates of group II elements from aqueous solutions. Table 10.4 provides the standard enthalpy and standard entropy for four carbonate formation reactions of the type:

$$\left[X^{2+}\right]_{aq} + \left[CO_3^{\ 2-}\right]_{aq} \rightarrow \left[XCO_3\right]_{solid}.$$

The subscript "aq" denotes that the ions are in solution, where they are stabilized by a hydration shell, which is an interaction between the ion and the water dipoles, as discussed in Chapter 17. What predictions can we make about the four processes?

TABLE 10.4

Standard enthalpy and entropy data for four carbonate formation reactions in aqueous solution under standard conditions (25°C and 1 atm)

Element X	ΔH^0 (kJ/mol)	$T \cdot \Delta S^0$ (kJ/mol)
Mg	+25.1	+71.1
Ca	+12.3	+59.0
Sr	+3.3	+55.6
Ba	−4.2	+46.0

Solution

We determine the change of the Gibbs free energy for each of the four processes in Table 10.4. Using $\Delta G = \Delta H - T \cdot \Delta S$ we find values ranging from $\Delta G = -46.0$ kJ/mol to $\Delta G = -50.2$ kJ/mol. Thus, all values are negative and similar: the four reactions occur spontaneously when the metal and carbonate ions meet in a solution.

However, the first three reactions have a positive ΔH^0 value. This means that they are endothermic processes; i.e., the chemical reaction requires heat. This case is illustrated by the red curve in Fig. 10.25, which shows the enthalpy of a chemical reaction at constant pressure as a function of time. The reactants A and B form a transition state AB*, which leads to products C and D, which are either energetically less stable (case 1) or more stable (case 2) than the reactants. In case 1 the reaction is endothermic; in case 2 the reaction is exothermic.

Comparing the two columns in Table 10.4 shows that all four reactions are driven by the entropy term. The reason is the existence of a hydration shell surrounding the ions. The hydration shell must be removed when forming a solid precipitate. The elimination of the hydration shell allows the involved water molecules to move more freely; this represents a higher degree of disorder in the system.

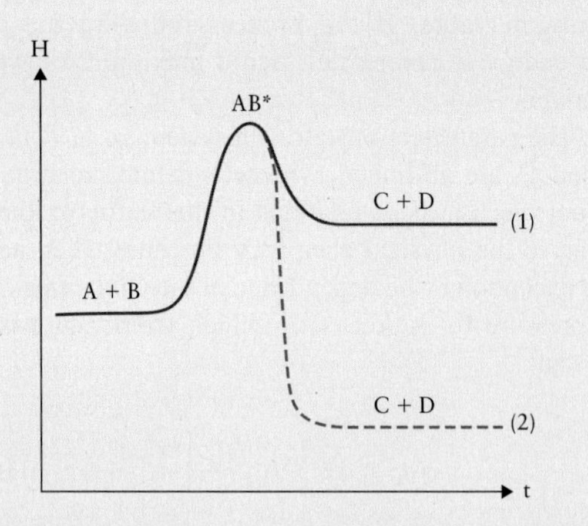

Figure 10.25 Enthalpy as a function of time for a chemical reaction. The reactants A and B form a transition state AB*, which then decomposes into the products C and D. Case 1 shows an endothermic process and case 2 an exothermic process.

continued

CONCEPT QUESTION 10.6

If mixing of components is always associated with an increase of entropy, why do water and oil not mix in Fig. 10.26?

Charles D. Winters/Science Photo Library/Custom Medical Stock

Figure 10.26 Oil and water in a beaker.

10.7: Liquid Solutions: Raoult's Law

As we turn our attention from gases to liquids, Dalton's law is no longer sufficient because it assumes ideal gas behaviour for each component of the mixture, which requires that intermolecular interactions be neglected. This is an inappropriate assumption for a **solution** because intermolecular interactions are required to cause the gas to condense to become a liquid in the first place. Thus, we need other tools to describe liquid mixtures such as blood, extracellular fluids, and the cytoplasm in the cell.

A liquid in a beaker has a surface that separates the **condensed phase** from the **vapour phase**. It is useful to develop a description of liquid solutions starting at that interface. We know that the solution is in mechanical and chemical equilibrium with the vapour at the surface. It would have to accelerate upward or downward were there no mechanical equilibrium, and the amount of vapour would have to increase or decrease if there were no chemical equilibrium. This allows us to utilize the gas laws we have already established to characterize the vapour phase and thereby the condensed phase indirectly.

An important experimental observation is the fact that in equilibrium a well-defined, non-zero vapour pressure is present above a liquid at any temperature. The pressure of this vapour is called the **saturation vapour pressure**. If a gas comes into contact with a liquid and is undersaturated, evaporation occurs until saturation is reached. This is the mechanism by which the dry air we inhale is saturated with water vapour in the trachea before it reaches the alveoli in the lungs.

With the concept of a vapour pressure above a liquid introduced, we can define and study ideal solutions.

KEY POINT

*An **ideal solution** is a solution for which the enthalpy of the system is the same for the sum of the separated components and for the mixed system.*

This means no heat is absorbed or released while mixing the components of an ideal solution, except for the energy needed to adjust the volume (due to the $p\ V$ term in the definition of the enthalpy H). This definition utilizes the same concept used in Dalton's definition of partial pressures in gas mixtures: ideal means that interactions between the various molecular species in the mixed state are negligible.

François Raoult made the following experimental observation for such a system, called **Raoult's law**:

$$p_i = x_i\, p_i^0. \qquad [10.28]$$

KEY POINT

Raoult's law states that the partial pressure p_i of the i-th component in the vapour phase above a solution is proportional to the molar fraction x_i of the i-th component in the solution. The proportionality factor is the vapour pressure of the i-th component above a pure liquid of the same component (p_i^0).

This law is important for the description of solutions as it allows us to predict their properties from measurements in the vapour phase. The vapour pressure curves for an ideal solution of two components A and B are shown in Fig. 10.27. The vertical axis is the partial pressure of component A or B, and the horizontal axis is the molar fraction of component B, ranging from $x_B = 0$ for pure liquid A to $x_B = 1$ for pure liquid B.

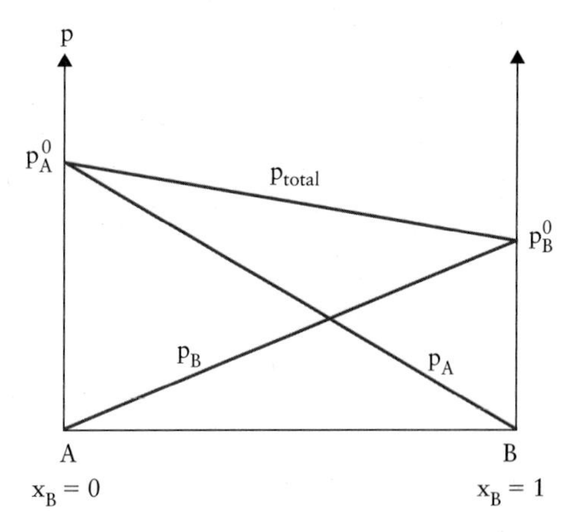

Figure 10.27 Sketch of a two-phase system illustrating Raoult's law. The partial pressures of the two components in the gas phase are linearly proportional to the molar fractions of the components in solution.

Figure 10.28 shows experimentally measured partial and total vapour pressure data for two real systems. Toluene and benzene, shown in Fig. 10.28(a), behave ideally since the data follow Raoult's law across the entire molar fraction range. This indicates that, effectively, no interaction occurs between the benzene and toluene molecules; in particular, no interaction that differs from the interactions between either a pair of toluene molecules

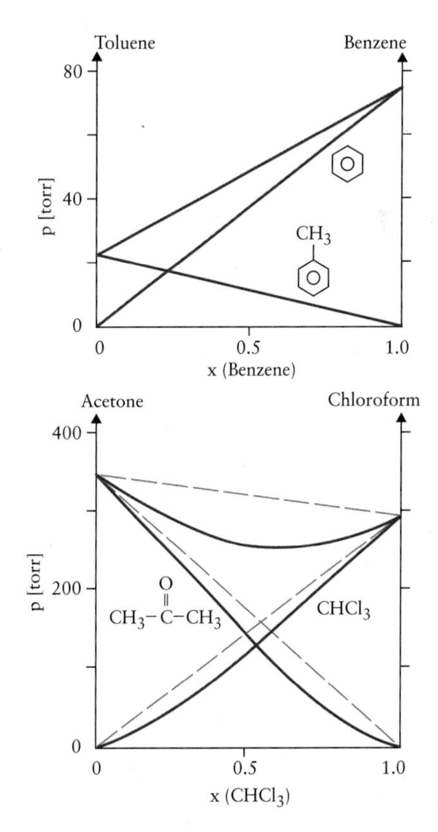

Figure 10.28 Experimental partial pressure curves over a liquid solution for (a) toluene and benzene, a nearly ideal solution that follows Raoult's law, and (b) acetone and chloroform, a non-ideal solution that follows Raoult's law only in the very dilute limit, i.e., near $x = 1$ and $x = 0$.

or a pair of benzene molecules. Many other systems behave non-ideally. As an example, a system consisting of acetone and chloroform is shown in Fig. 10.28(b). The vapour pressure curves predicted by Raoult's law are shown as dashed lines. The solid lines are the experimental data and they clearly deviate from Raoult's law.

CASE STUDY 10.9

The system acetone/chloroform in Fig. 10.28 does not obey Raoult's law (dashed lines). Under what conditions would you still use Raoult's law for this system?

Answer: *The figure confirms that ideal behaviour is still a good approximation if the system is dilute (lim $x_i \to 0$ or lim $x_i \to 1$). Note that the solid curves and the dashed curves coincide reasonably well near x($CHCl_3$) = 0 and x($CHCl_3$) = 1.*

Sometimes Raoult's law is confused with Henry's law. **Henry's law** states that the solubility of a gas in a liquid is proportional to its pressure above the surface of the liquid. Henry's law therefore allows us to *control* the composition of a solution by adjusting parameters in the gas phase. Raoult's law in turn allows us to *measure* the composition of a solution using the gas phase.

SUMMARY

DEFINITIONS

- Entropy: change of entropy with the state of a system:

$$\Delta S = S_f - S_i = \sum_i \frac{Q_i}{T_i}$$

- Gibbs free energy:

$$G = H - TS$$

UNITS

- Entropy S: J/K
- Gibbs free energy G: J

LAWS

- Heat Q for:
  - isochoric process for an ideal gas:

$$Q = C_V\, n\, \Delta T$$

    C_V in unit J/(mol K) is the molar heat capacity at constant volume, $C_V = 3\,R/2$ for the (monoatomic) ideal gas.
  - isobaric process for an ideal gas:

$$Q = C_p\, n\, \Delta T$$

    C_p in unit J/(mol K) is the molar heat capacity at constant pressure: $C_p = 5\,R/2$ for the (monoatomic) ideal gas.
- Work: for an ideal gas with frictionless piston
  - isochoric process (V = const):

$$W = 0$$

  - isothermal process (T = const):

$$W = -n\,RT\,\ln\left(\frac{V_f}{V_i}\right)$$

  - isobaric process (p = const):

$$W = -p\,(V_f - V_i)$$

- Change of the internal energy for a cyclic process:

$$\Delta U_{cycle} = 0$$

- Second law of thermodynamics: In a cyclic process it is impossible to take heat from a reservoir and change it into work without releasing a fraction of the heat to a second reservoir at lower temperature.
- Third law of thermodynamics: The entropy of a perfect crystal of a chemical element/compound at $T = 0$ K is zero; $S = 0$ J/K.
- Entropy change for
  - reversible process: $\Delta S = 0$
  - spontaneous irreversible process (system and environment): $\Delta S > 0$
- Gibbs free energy change for
  - the transition between coexisting system components in equilibrium: $\Delta G = 0$
  - spontaneous processes: $\Delta G < 0$
- Raoult's law for ideal solutions:

$$p_i = x_i p_i^0,$$

  where p_i is the partial pressure in the vapour phase, x_i is the molar fraction, and p_i^0 is the vapour pressure of the i-th component.

MULTIPLE-CHOICE QUESTIONS

MC–10.1. We study the process shown in Fig. 10.29, which is a plot of pressure versus temperature following the relation $p \propto T$. Which of the following five statements is true?
(a) The process in Fig. 10.29 is an isobaric compression of an ideal gas.
(b) The process in Fig. 10.29 is an isothermal expansion of an ideal gas.
(c) The process in Fig. 10.29 is an isobaric expansion of an ideal gas.
(d) The process in Fig. 10.29 is an isothermal compression of an ideal gas.
(e) The ideal gas must be heated during the process shown in Fig. 10.29.

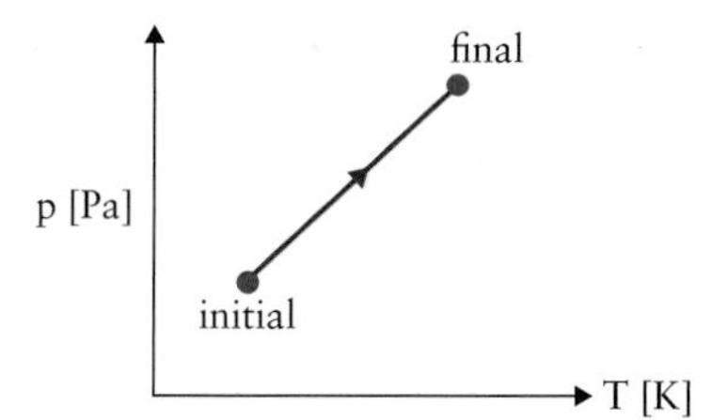

Figure 10.29 Graph of the pressure versus the temperature for a gas that follows the relation $p \propto T$.

MC–10.2. We plot the following processes for 1 mol of an ideal gas in a p–V diagram. Which one leads to a linear plot, i.e., a plot that can be described by the linear formula $p = a + b\,V$?
(a) an isothermal expansion
(b) an isobaric heating
(c) an adiabatic cooling

MC–10.3. In which of the following processes does the volume of the system not change?
(a) isothermal expansion
(b) isothermal compression
(c) isobaric heating
(d) isobaric cooling
(e) isochoric heating

MC–10.4. A process is an isochoric process when
(a) the temperature remains constant and the pressure changes.
(b) no work is done.
(c) no heat is exchanged with the environment.
(d) the internal energy remains constant.
(e) the pressure remains constant and the temperature changes.

MC–10.5. We consider 1 mol of an ideal gas under isothermal conditions. If the pressure is doubled,
(a) the volume remains unchanged.
(b) the volume doubles as well.
(c) the volume is halved.
(d) the volume increases by a factor of 4.
(e) the volume decreases to 25% of its original value.

MC–10.6. A process is called adiabatic if
(a) the temperature remains constant.
(b) no work is done.
(c) no heat is exchanged with the environment.
(d) the internal energy remains constant.
(e) the volume remains constant.

MC–10.7. If a process starts at pressure p_0 and volume V_0 and leads to a doubling of the volume, which of the following processes involves the most work?
(a) an adiabatic process
(b) an isothermal process
(c) an isobaric process
(d) No answer is always correct.

MC–10.8. 2.0 mol of an ideal gas is maintained at constant volume in a 4-L container. If 100 J of heat is added to the gas, what is the change in its internal energy? Choose the closest value.
(a) zero
(b) 50 J
(c) 67 J
(d) 100 J

MC–10.9. 1 mol of an ideal gas is expanded from initial volume V_i to a final volume V_f, in each case starting at pressure p_0.
(A) For which type of reversible process does the gas do the most work?
(a) adiabatic expansion
(b) isothermal expansion
(c) isobaric expansion
(d) cannot be determined without further information
(B) For which type of irreversible process does the gas do the most work?
(a) adiabatic expansion
(b) isothermal expansion
(c) isobaric expansion
(d) cannot be determined without further information

MC–10.10. Which of the following statements about the entropy of a system is *not* correct.
(a) It is a measure of the degree of order/disorder in the system.
(b) It is a measure of the fraction of heat taken up by the system that is turned into work during a reversible cyclic process.
(c) It determines whether an isolated system is in equilibrium or may undergo spontaneous processes.
(d) It can be combined with the temperature and the enthalpy to determine the Gibbs free energy of the system.
(e) The entropy difference between two states of the system is equal to the amount of heat exchanged with the environment during a process leading from one state to the other.

MC–10.11. Which of the following processes is associated with an increase in entropy?
(a) a reversible expansion of a gas within an isolated superstructure
(b) an adiabatic expansion of a gas to twice its volume
(c) a complete Carnot cycle for an ideal gas
(d) melting of a cube of ice in a beaker at room temperature
(e) freezing of a litre of liquid water in a cold room at 5°C

MC–10.12. A typical 1800s steam engine operated with steam of 125°C. Room-temperature air served as the low-temperature heat reservoir ($T = 25$°C). What was the maximum efficiency coefficient η of that machine? (Choose the closest value.)
(a) 100%
(b) 80%
(c) 34%
(d) 25%
(e) 10%

MC–10.13. A cyclic Carnot process operates with a high-temperature heat reservoir at 500 K and a low-temperature heat reservoir at 300 K. To what temperature must the low-temperature heat reservoir be brought to increase the efficiency coefficient of the Carnot machine by a factor of 1.5? (Choose the closest value.)
(a) 200 K
(b) 250 K

(c) 300 K
(d) 350 K
(e) 500 K

MC–10.14. A machine operates within an isolated superstructure. Its operation causes an entropy increase of $\Delta S = 5$ J/K. What change ΔS^* is required in the machine's environment within the isolated superstructure to identify the machine as operating reversibly?
(a) $\Delta S^* = 0$ J/K
(b) $\Delta S^* = +5$ J/K
(c) $\Delta S^* = -5$ J/K
(d) No value exists to answer the question.
(e) The value of ΔS^* is different from those listed above.

MC–10.15. Which of the following processes has a negative change in entropy?
(a) increase in the number of gas particles in a box
(b) increase of the internal energy of the particles in a box
(c) isothermal expansion
(d) formation of molecules from atoms, e.g., $2H \rightarrow H_2$
(e) relaxation of a stretched polymer

MC–10.16. A piece of ice at temperature 0°C and 1.0 atm pressure has a mass of 1.0 kg. It then completely melts to water. What is its change in entropy? (The latent heat of freezing of water is given as 3.34×10^5 J/kg.)
(a) 3340 J/K
(b) 2170 J/K
(c) 613 J/K
(d) 1220 J/K

CONCEPTUAL QUESTIONS

Q–10.1. The internal energy of an ideal gas in an isothermal process does not change. Assume that the gas does work W during this process. How much energy was transferred as heat?

Q–10.2. A standard man performs a strenuous exercise, e.g., lifting a weight or riding a bicycle. The standard man does work in this exercise and dissipates heat. Would the first law of thermodynamics not require that the internal energy, and therefore the temperature, of the standard man decrease?

Q–10.3. Study once more the multiple-choice question MC–10.7. Why is "isochoric process" *not* a sensible choice to be offered as an answer?

Q–10.4. If you chose one of the three processes offered as answers in MC–10.9(a), explain why the other two are still reversible despite the definition that a reversible process is the process that involves the most work.

Q–10.5. We use the arrangement of Fig. 10.19 but fill one chamber with oxygen and one chamber with nitrogen. (a) When you open the valve, what process occurs? (b) Propose a modification to this experiment where the same process can be done reversibly.

Q–10.6. Marcellin Berthelot measured a large number of standard enthalpy of formation values. He justified that effort in an 1878 publication by saying "every chemical process in an isolated system tends toward the products that release the most heat"; i.e., Berthelot believed his measurements reveal the **chemical affinity** of the reactants. Do you agree?

Q–10.7. We want to show that no engine (super-engine) can be more efficient than the Carnot process. For this we consider Fig. 10.30, which shows a proposed super-engine generating work and a reverse Carnot process restoring the heat extracted by the super-engine. Both engines operate between the same two heat reservoirs. (a) Is this process possible with respect to the first law of thermodynamics? (b) Is this process possible with respect to the second law of thermodynamics?

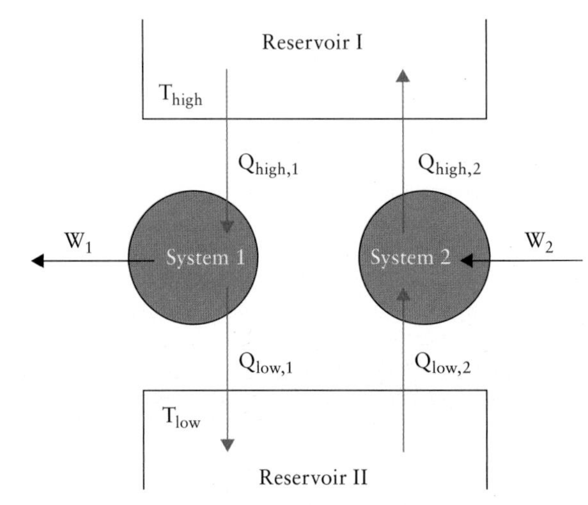

Figure 10.30 Two cyclic processes operating between two heat reservoirs. System 1 is a super-engine and system 2 is a Carnot process that operates in the reverse direction.

Q–10.8. Is it possible to build a heat engine that causes no thermal pollution?

Q–10.9. A thermodynamic process occurs in which the entropy of a system changes by −8.0 J/K. Based on the second law of thermodynamics, what can you state about the entropy change in the environment?

Q–10.10. The thermodynamic equilibrium between two states of a single component always lies entirely on one or the other side; e.g., at temperatures below the freezing point all water molecules are part of the solid state, and above the freezing point all water molecules are part of the liquid state. This situation is different for chemical reactions, where even for the most complete reactions a finite concentration of the reactants remains. What explains this difference?

Q–10.11. Two gas containers are each thermally isolated. They are separated by a valve. Container I contains 1.5 L of ideal gas I at 0°C, and container II contains 2.5 L of ideal gas II at 25°C. The pressures in the two containers are initially the same. at $p_i = 1$ atm. Assume the mixed gas is also ideal. (a) Is this mixing process reversible? (b) If not, how can the mixing be modified so that it is done reversibly?

ANALYTICAL PROBLEMS

P–10.1. 1.0 mol of an ideal gas isothermally expands from an initial pressure of 20 atm to a final pressure of 5 atm. Calculate separately for two temperatures, 0°C and 25°C, (a) the work done by the gas, (b) the change of internal energy of the gas, and (c) the amount of heat taken from the environment.

P–10.2. (a) How much heat is needed to increase the temperature of 100 g argon gas (Ar) in a 1-m^3 container from −10°C to +10°C? Treat Ar as an ideal gas and use $M(\text{Ar}) = 39.95$ g/mol. (b) By how many percent does $\langle v^2 \rangle$ of the argon atoms in the gas increase in this process? *Hint:* Part (b) does not ask for v_{rms}.

P–10.3. Show that Poisson's equation leads to $p \; V^{\kappa} = \text{const}$ when using the operations specified in the text.

P–10.4. 1.0 mol of an ideal gas that starts at 1.0 atm and 25°C does 1.0 kJ of work during an adiabatic expansion. (a) What is the final temperature of the gas? (b) What is the final volume of the gas?

P–10.5. In the text we stated that $C_p = C_V + R$ holds for an ideal gas. Derive this result for an isobaric expansion of 1.0 mol of ideal gas without using the enthalpy concept. For this, start with the work in Eq. [10.8] and the change of internal energy in Eq. [8.24] for the expansion in Fig. 10.11. Then use the first law of thermodynamics and the definition

$$Q = n \cdot C_p \cdot \Delta T,$$

which applies for an isobaric process.

P–10.6. A container with $V = 0.4$ m^3 holds 3.0 mol of argon gas at 303 K. What is the internal energy of this gas if we assume it behaves like an ideal gas?

P–10.7. Sketch a p–V diagram for the following processes: (a) An ideal gas expands at constant pressure p_1 from volume V_1 to volume V_2, and is then kept at constant volume while the pressure is reduced to p_2. (b) An ideal gas is reduced in pressure from p_1 to p_2 while its volume is held constant at V_1. It is then expanded at constant pressure to a final volume V_2. (c) In which process is more work done?

P–10.8. An ideal gas is in a container at pressure 1.5 atm and volume 4.0 m^3. What is the work done by/on the gas if (a) it expands at constant pressure to twice its initial volume, or (b) it is compressed at constant pressure to one-quarter of its initial volume?

P–10.9. 1.0 mol of an ideal gas is initially at 0°C. It undergoes an isobaric expansion at $p = 1.0$ atm to four times its initial volume. (a) Calculate the final temperature of the gas, T_f. (b) Calculate the work done by/on the gas during the expansion.

P–10.10. An ideal gas undergoes a process in which its pressure is doubled from p_0 to $2 \; p_0$ while the volume doubles from V_0 to $2 \; V_0$. The pressure increases linearly with the volume in this process. How much heat does the gas exchange with its environment during this process?

P–10.11. 1.0 mol of an ideal gas expands isothermally from an initial volume of 20 L to a final volume of 40 L at $T = 300$ K. (a) Determine the work done by the gas during this process from Eq. [10.6] (Fig. 10.31(a)). (b) Compare the result in part (a) with an estimate of the same work by connecting the initial and the final states in the p–V diagram linearly, as shown in Fig. 10.31(b). (c) Repeat part (b) by estimating the work based on Fig. 10.31(c).

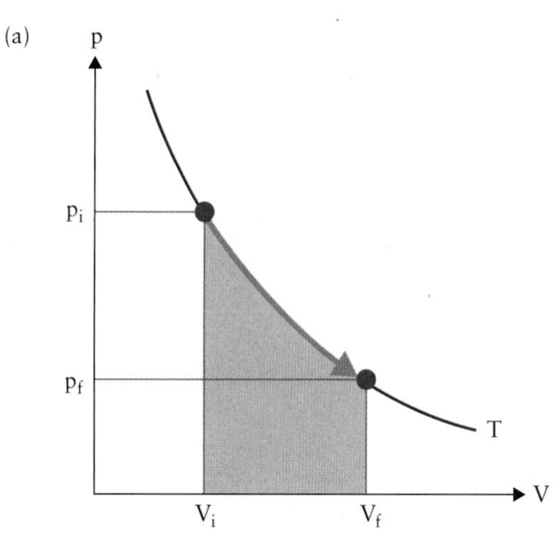

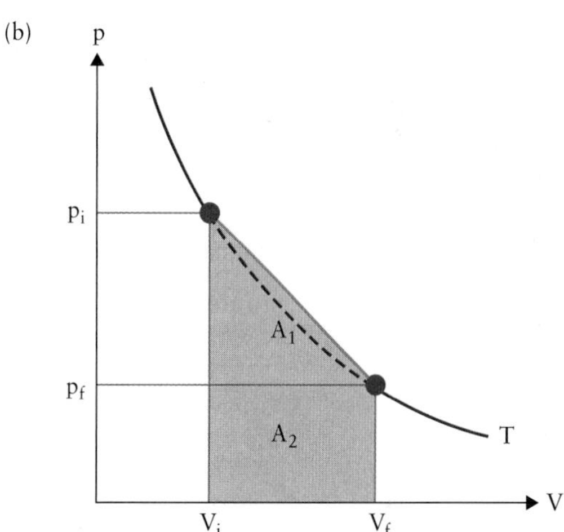

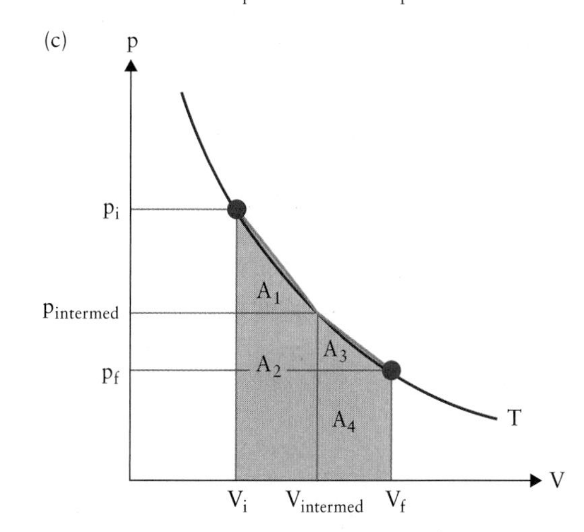

Figure 10.31

P–10.12. Compare the efficiency coefficient for a Carnot machine operating between a low-temperature heat reservoir at room temperature (25°C) and a high-temperature heat reservoir at the boiling point of water at two different pressures: (a) 5 atm with $T_{boil} = 152°C$, and (b) 100 atm with $T_{boil} = 312°C$.

P–10.13. The cyclic process in Fig. 10.32 consists of (i) an isothermal expansion, (ii) an isochoric cooling, and (iii) an adiabatic compression. If the process is done with $n = 1.0$ mol of an ideal gas, what are (a) the total work done by the gas, (b) the heat exchanged with the environment, and (c) the change of the internal energy for one cycle? (d) Sketch the cyclic process of Fig. 10.32 as p–T, V–T, and U–T diagrams.

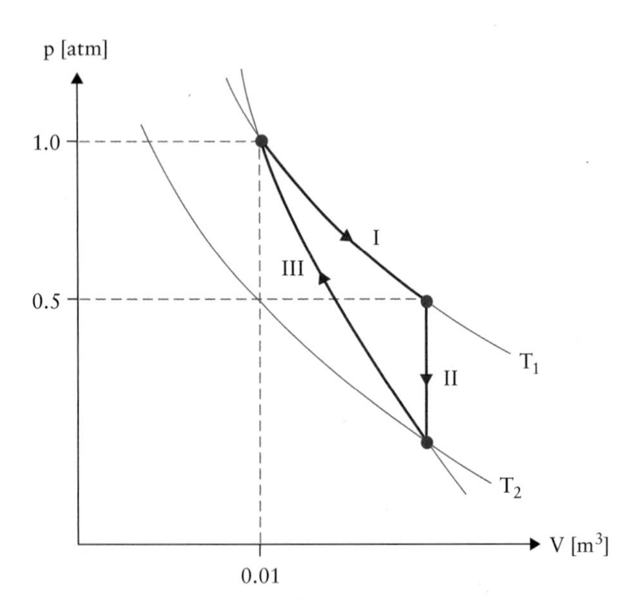

Figure 10.32

P–10.14. An ideal gas is taken through a cyclic process as shown in Fig. 10.33. (a) Find the net heat transferred between the system and its environment during one complete cycle. (b) What is the net heat transfer if the cycle is reversed; i.e., it follows the path ACBA?

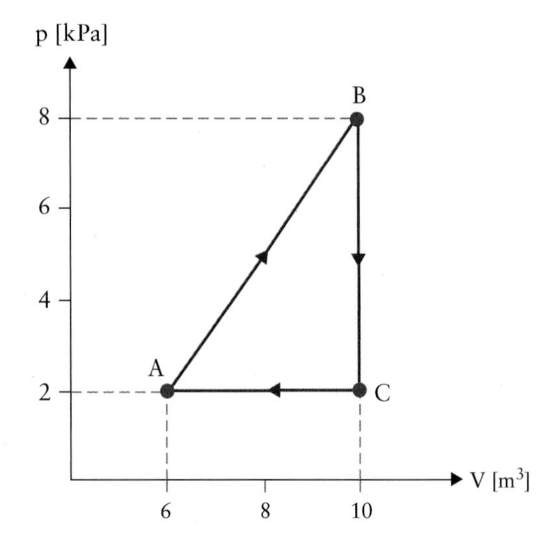

Figure 10.33

P–10.15. For any system other than an ideal (monatomic) gas, we need to use tabulated values for the heat capacity. These values usually depend on the temperature. As an example, the molar heat capacity of solid and liquid water is listed as a function of temperature in Table 10.5.

TABLE 10.5

Molar heat capacity of liquid and solid water

State	Temperature	C (J mol^{-1} K^{-1})
Solid	−34°C	33.30
Solid	−2.2°C	37.78
Liquid	0°C	75.86
Liquid	25°C	75.23
Liquid	100°C	75.90

In this problem, we want to bring 1 kg water, initially at −10°C, to a final temperature of +10°C. What is the change in enthalpy for this process if the latent heat of melting is 6.0 kJ/mol? *Hint:* Use the heat capacity value at −2.2°C from Table 10.5 for ice, and use the value at 0°C for liquid water. The molecular mass of water is $M = 18$ g/mol.

P–10.16. Standard enthalpy of formation values are tabulated and are usually labelled ΔH_f^0, in which the subscript f stands for *formation* and the superscript 0 indicates the standard state. Most standard enthalpy of formation values cannot be obtained experimentally. For example, it is impossible to mix carbon, oxygen, hydrogen, and nitrogen and hope to ignite the mixture to obtain the essential amino acid isoleucine. In such cases the **theorem of Hess** is applied, which states that the enthalpy of formation is independent of the actual reaction by which the product is formed (because it is a variable of the state of the system). Thus, we can combine the enthalpies of several chemical reactions to obtain a value for a particular process we want to study.

To illustrate the use of the theorem of Hess, determine the standard enthalpy of formation of CO. This value cannot be obtained experimentally, as oxidation of carbon always leads to the formation of (some) CO_2. However, the following processes yield experimentally accessible values:

$$C + O_2 \rightarrow CO_2 \quad \Delta H_f^0 = -393.5\frac{\text{kJ}}{\text{mol}}$$

$$CO + \frac{1}{2}O_2 \rightarrow CO_2 \quad \Delta H_f^0 = -283.0\frac{\text{kJ}}{\text{mol}}.$$

P–10.17. A Carnot process is operated with 1.0 mol of an ideal gas of heat capacity $C_V = 3\ R/2$. The pressure of the gas is 20 atm and the temperature is 500 K in the most compressed state. From there, an isothermal expansion leads to a pressure of 2.0 atm. The lower process temperature is 250 K. (a) Calculate for each step of this Carnot process the work and heat exchange with the environment. (b) What is the efficiency coefficient of this machine? (c) Draw this Carnot process as a p–V diagram, then sketch it as a p–T diagram, a V–T diagram, a U–T diagram, and an S–T diagram.

P–10.18. (a) We consider a reversible isothermal expansion of 1 mol of an ideal gas. Calculate the entropy change for a pressure decrease from 10 atm to 1 atm at 0°C. (b) What is the entropy change in the environment within the isolated superstructure? (c) What is the entropy change if the expansion is done adiabatically instead?

P–10.19. Calculate the entropy change during melting of 1.0 mol benzene. The melting point of benzene at atmospheric pressure is $T_m = 288.6$ K and its latent heat of melting is 30 kcal/kg.

P–10.20. Calculate the entropy of evaporation for 1 mol of the elements and compounds listed in Table 10.6.

TABLE 10.6

Boiling point $T_{boiling}$ and latent heat of evaporation for various materials

Material	$T_{boiling}$ (°C)	ΔH (kcal/mol)
Argon (Ar)	87.5	1.88
CCl_4	349.9	7.17
C_6H_6 (benzene)	353.3	7.35
Mercury (Hg)	629.8	15.50

P–10.21. We place 150 g ice at 0°C (latent heat of melting 1430 cal/mol) in a calorimeter with 250 g water at 80°C. Use 18.0 cal/(K · mol) for the molar heat capacity of liquid water and assume this value is temperature independent. (a) What is the final temperature of the water? (b) If the process is done reversibly, what is the entropy change of the combined ice/water system? (c) What is the entropy change in the environment for the reversible process? (d) What is the entropy change if the process is done irreversibly in an isolated beaker? *Hint:* Use the following dependence of the entropy on the temperature for a process in which the pressure is held constant:

$$\Delta S = n\,C_p \ln\left(\frac{T_f}{T_i}\right).$$

P–10.22. Calculate the entropy of 1 mol oxygen gas (O_2) at atmospheric pressure and $T = 50$°C. Use for the molar heat capacity of oxygen its standard value at 25°C; i.e., $C_p = 29.4$ J/(K mol).

P–10.23. Determine graphically the standard entropy of silver from the data given in Table 10.7.

P–10.24. We consider two phases of solid carbon: diamond and graphite, with $S^0_{diamond} = 2.44$ J/(K · mol) and $S^0_{graphite} = 2.3 \cdot S^0_{diamond}$. Which of the two carbon modifications is more stable if we establish a thermal equilibrium between them in an isolated system?

P–10.25. Assume that we found for a given chemical reaction $\Delta H = -100$ kJ and $\Delta S = -200$ J/K. Neglect the temperature dependence of these two values. What is the Gibbs free energy for the reaction (a) at room temperature and (b) at 800°C?

P–10.26. Calculate ΔS and ΔG for the evaporation of 1 mol water at $T = 100$°C and $p = 1$ atm. The latent heat of evaporation of water is 9.7 kcal/mol.

P–10.27. Two gas containers are each thermally isolated. They are separated by a valve. Container I contains 1.5 L of ideal gas I at 0°C, and container II contains 2.5 L of ideal gas II at 25°C. The pressures in the two containers are initially the same at $p_i = 1$ atm. Assume that the mixed gas is also ideal. (a) Express the amount of gas in each container in unit mol. (b) What are the final pressure and temperature after the valve has been opened? (c) What are the molar fraction and the partial pressure of ideal gas I? (d) What is the entropy change during mixing?

P–10.28. Use Raoult's law to derive the **vapour pressure depression** for an ideal dilute solution. *Hint:* Consider a system with two components, solvent A and solute B, with $x_A >> x_B$.

P–10.29. A kilogram of ethylene (C_2H_4) is compressed from 1.0 L to 0.1 L at a constant temperature of 27.5°C. Calculate the work that must be expended, assuming the gas behaves like an ideal gas.

TABLE 10.7

Molar heat capacity C_p for silver (Ag) at various temperatures

T (°C)	C_p [cal/(K · mol)]	T (°C)	C_p [cal/(K · mol)]	T (°C)	C_p [cal/(K · mol)]
−263	0.07	−243	1.14	−223	2.78
−203	3.90	−183	4.57	−163	5.01
−143	5.29	−123	5.49	−103	5.64
−83	5.76	−63	5.84	−43	5.91
−23	5.98	−3	6.05	+17	6.08

ANSWERS TO CONCEPT QUESTIONS

Concept Question 10.1: (A)

Concept Question 10.2: (A) if the gas is an ideal gas, otherwise (F). Note that the sketch is a *V–T* diagram. The ideal gas law leads to a linear relation between volume and temperature at constant pressure p_0: $p_0 \; V = n \; R \; T$; i.e., $V \propto T$. The arrow of the process in Fig. 10.12 shows that the process leads from a larger to a smaller volume; i.e., the process is a compression. The van der Waals equation (Eq. [8.28]) shows that (F) is correct if we don't specify the gas as ideal.

Concept Question 10.3: While such a process would return the system to its initial state, no net work would be achievable per cycle. You see this by drawing this simplified cycle in a *p–V* diagram: the area underneath the expansion step is the same as underneath the compression step.

Concept Question 10.4(a): Dynamic breathing is a real physiological process and is therefore irreversible. However, Fig. 9.1 shows continuous lines for a full breathing cycle, which implies that Fig. 9.1 illustrates an idealized reversible version of breathing.

Concept Question 10.4(b): Example 10.4 shows that the reversibility of a process does not affect the change of entropy of the system, as no work term is included in the calculation. Thus, the entropy change for a cyclic process, whether it is reversible or irreversible, is the same and is zero because entropy is a variable of state. If the cyclic process is irreversible it will cause an increase of entropy *in the environment*.

Concept Question 10.5: Don't try to answer by looking at Fig. 10.17(b); the formulas we introduced require that we study an equivalent reversible process. In the reversible process of Fig. 10.17(a), energy is required to push the piston; this energy is recovered as heat from a heat reservoir. The initial and final equilibrium states in Figs. 10.17(a) and (b) are identical, thus the entropy change in both cases is the same whether the process is done reversibly or irreversibly.

Concept Question 10.6: Mixing is governed by the change of the Gibbs free energy and not the entropy. In the case of oil and water, the enthalpy of mixing exceeds the gain from entropy, which is about 4 J/(K · mol). ΔH is large to provide for the formation of an increasing surface between oil and water droplets. Surface energy is discussed in Chapter 12. The result is different for mixing water and rock salt, where energy is gained when water molecules form a hydration shell around salt ions. Hydration shell formation is discussed in Chapter 17.

CHAPTER 11

Transport of Energy and Matter

Membranes separate two systems. A semipermeable membrane enables two adjacent systems to interact through exchange of heat and/or matter. The exchange occurs if the two systems differ in one or more essential parameters: if they differ in temperature, heat flows toward the colder system (heat conduction), and if they differ in the concentration of a chemical component, matter flows toward the more dilute system (diffusion). The transport across the membrane is proportional to the cross-sectional area of the membrane and is inversely proportional to its length. The proportionality constants between heat flow and temperature difference (called the thermal conductivity coefficient), and between matter flow and concentration difference (called the diffusion coefficient) are materials constants.

Diffusion coefficients are strongly temperature dependent. This is explained by a microscopic model of individual particles hopping between energetically favoured adjacent sites in the matrix. In the hopping process an activation energy must be overcome, utilizing the thermal energy of the diffusing particle (Arrhenius model). Each particle jumps randomly to any of the available adjacent sites, including the site from which it came in the previous jump. Consequently, the particle traverses in N jumps of length a a total distance less than Na. Einstein found that the total distance, defined as the diffusion length, is proportional to $(Dt)^{1/2}$ with D the diffusion coefficient and t the diffusion time.

In our discussion of the basic laws of thermodynamics we often referred to closed systems that exchange heat and work with their environment. We also noted that biological systems are usually open systems, i.e., systems that further exchange matter with their environment. We then studied the underlying exchange processes in more detail and distinguished between idealized processes, which we used further to establish the basic laws of thermodynamics, and real systems, which display irreversible interactions with the environment. Despite this observation we pursued the discussion of thermodynamics first, as it compares initial and final equilibrium states and thus governs the propensity of processes toward one or the other state.

In the current chapter we want now to turn our attention more closely to the dynamic processes in which heat or matter are exchanged between system and environment. This will require a different approach, and we need to focus on a new model system to allow us to develop the concepts that govern unfolding irreversible processes. The best way to identify this new model system is to carefully study how nature deals with establishing and maintaining non-equilibrium conditions.

11.1: Membranes in Living Organisms

At the macroscopic level, we look at any living organism. While alive, it is per definition in a non-equilibrium state with its environment, as the living body contains many molecules you don't find outside, and, for mammals and birds, is at a different temperature than the environment. The way to manage this non-equilibrium is with a barrier, for example the human skin, often combined with fur, scales, or feathers for insulation.

At the microscopic level two living systems come to mind: bacteria, which have to maintain a direct separation with the environment, and eukaryotic cells of larger organisms, like ourselves. To manage the interactions with their environments, these microscopic systems have developed **biological membranes**. These membranes were the focus of biological studies for most of the 19th century. A heated debate concerned first the presence of membranes encapsulating the cells (the so-called **plasma membrane**), then the membranes encapsulating the nucleus.

Particularly puzzling to biologists was that membranes not only serve as passive envelopes for biological units, but also are actively involved in the dynamic processes occurring in our body, e.g., serving as a filter in the kidneys (this will be described in Chapter 13), or regulating the metabolism of a cell using enzymes embedded in the membrane to gather various chemical molecules from the environment and dispose of them back into the environment.

Biological membranes come in many different forms. One chemically simpler example—but at the same time one of the great success stories of the evolutionary process—is the eggshell. Consisting mainly of calcium carbonate, it establishes a non-equilibrium between the egg inside and the environment outside. The eggshell allows air to infiltrate but does not allow water to escape. This feature allowed *cotylosaurs* about 300 million years ago to move to dry land for their entire life cycle. Their amphibian ancestors had to spend at least part of their life in water since their eggs would desiccate on dry land. *Cotylosaurs* in turn were able to move far away from the coastline and eventually became the ancestor of all reptiles, birds, and mammals. You can verify the air exchange across the eggshell in a simple experiment. Place an egg in hot water. Air bubbles will form all over the eggshell. This is due to the thermal expansion of air inside the egg: the expansion pushes some of the air out.

We illustrate the complexity of biological membranes for the plasma membrane of a eukaryotic cell (e.g., a human cell).

Fig. 11.1 shows a schematic sketch of a human cell membrane (plasma membrane), which separates the cytoplasm inside from the interstitium outside. Membranes are an example of organic molecules organized into a higher level of complexity. The aggregation of phospholipids in an aqueous solution is driven by the molecule's *amphipathic* character: one end of the molecule is *hydrophobic* (water repelling) and the other end is *hydrophilic* (water attracting); the hydrophilic ends will reach into the aqueous solution, while the hydrophobic ends will merge together to exclude water. The result is a bilayer of phospholipids with the hydrophilic ends directed outward. These double-layer structures self-assemble into three-dimensional spherical forms called *protobionts,* which can maintain an internal aqueous environment that is different from the external environment.

The human cell is far more complex than a simple phospholipid bilayer. Various protein molecules are associated with the membrane, some stretching across the membrane layer and some attached only extrinsically. These proteins play a role in specific transport processes across the membrane.

With the current chapter we follow nature's lead in trying to address non-equilibria. Throughout this textbook, you will learn that a step forward in our scientific understanding requires that we have a firm comprehension of the previous knowledge, as its conclusions remain true and the new ideas must be consistent with what we already know. At the same time, nature must be observed from a different angle, with new questions asked; otherwise we won't make any progress. The current chapter is a great example of this approach. We spent quite some time in the previous chapters to establish a solid footing in thermodynamics, which allows us to compare equilibrium states of a system. Now we want to tackle non-equilibrium processes. Our observations of nature tell us that using barriers, or membranes, is an effective way to manage non-equilibrium processes. Thus, we develop a new model system, which we call a physical membrane, and then develop experiments that allow us to mimic the natural processes occurring at actual biological barriers. The new laws we develop from these experiments will bring us a big step closer to understanding nature, as the many examples in the current chapter will show.

Hybrid Medical Animation/Science Photo Library

Figure 11.1 Fluid mosaic model of a human cell membrane. The lipid bilayer (green) consists of chain-like macromolecules with hydrophobic and hydrophilic ends. The hydrophilic ends are directed toward the external aqueous milieu. Proteins can be embedded in the membrane or pass completely through it (purple). They are able to diffuse laterally on or within the membrane. Protein mobility is the reason the term fluid (versus rigid) is used to label this membrane model. Inside the cell, the Golgi apparatus (lower right) produces lysosomes (blue spheres).

11.2: A New Model System: Physical Membranes as an Idealized Concept

Throughout the sciences we depend on suitable models when developing new concepts. Membranes prove to be a simple but powerful model for introducing time-dependent, non-mechanical processes. To be able to quantify these processes, a simplified membrane is introduced. This model is called a **physical membrane**. It is a uniform barrier that is characterized by a very limited number of variables, primarily its width and a homogeneous chemical composition.

KEY POINT

A physical membrane is a barrier separating two uniform systems. Each system is in equilibrium. At least one parameter varies between the two systems, establishing a non-equilibrium across the membrane.

The parameter varying across the membrane is either a physical parameter, such as pressure or temperature, or a chemical parameter, such as the concentration of molecules in solution. Often, more than one parameter varies across real membranes. In Chapter 18, for example, nerve membranes are discussed that separate different concentrations of positive and negative ions, leading to a combination of electric and chemical effects. Before studying such combinations of effects, however, we need to develop the basic concepts of interactions across a membrane. For our discussion we need to specify two properties of membranes: permeability and width.

11.2.1: Permeability

The major physical property of a membrane is the degree of interaction it allows between the two systems it is separating. This degree of interaction is characterized by the permeability of the membrane.

A membrane can be **impermeable**, which means it is completely blocking interaction across the membrane. This case is of limited interest as it creates isolated systems on either side, each fully described by the concepts we already discussed in the context of thermal physics. The opposite extreme is a membrane that is **fully permeable**; i.e., everything transfers freely in both directions across the membrane. Again, such a membrane is not particularly interesting as it does not alter the interactions within the system.

The type of membrane used to develop the concepts of this chapter is, therefore, the **semipermeable membrane**. In the biological literature, **selective permeability** is often used as an alternate notion. A semipermeable membrane allows some system components to pass but blocks others. Fig. 11.2 illustrates the most important system components in transport processes: energy in the form of heat, and matter in the form of atoms, ions, or molecules. The simplest type of semipermeable membrane is a membrane that allows only heat to pass and blocks matter. Such a membrane is used to introduce heat conduction. We later proceed to membranes that block only some chemical components while allowing others to pass. Semipermeable membranes of this type are used to develop the concept of diffusion.

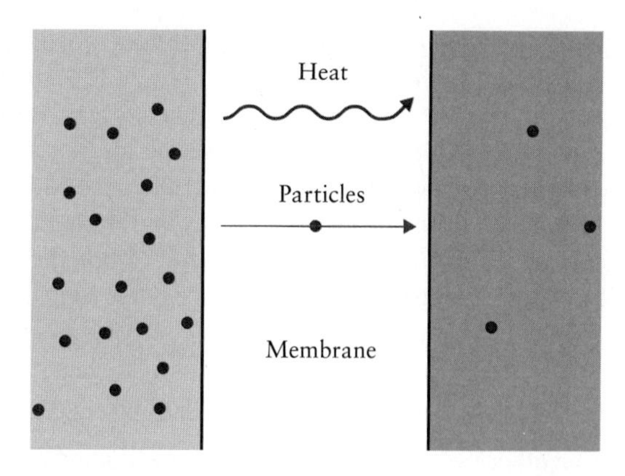

Figure 11.2 Sketch of the major transport mechanisms operating across a physical membrane (white area at centre): particles and/or energy may be transported. When energy flows independently of a material transport it is in the form of heat.

11.2.2: Membrane Width

The thickness of a membrane also plays a role in the extent of interactions allowed between the two adjacent systems. If it is too thick, a semipermeable membrane may be impermeable in practice. We distinguish conceptually two types of membranes: a **zero-width membrane** and a **finite-width membrane**. If we want to study the consequences of interactions across the membrane focussing on the two adjacent systems, a zero-width membrane allows us to neglect the mechanisms of transport through the membrane. In other cases we want to focus on the transport mechanisms in the membrane itself. In these cases, a membrane of finite width has to be taken into account. The variation of a system parameter across a finite-width membrane requires the introduction of the gradient as a new mathematical concept in this chapter.

The physical membrane is obviously a good description for physical barriers such as windows. Biological membranes are usually more complex. In our body, membranes are neither uniform in thickness nor chemically homogeneous, and they often actively participate in the physiological processes rather than form a passive barrier. The concepts we develop for the physical membrane still apply to biological membranes and illustrate the role of basic transport processes in physiology.

The physical membrane is sufficient to discuss the time-dependent transport of physical or chemical properties. These phenomena are grouped together under the term **transport phenomena** and include heat conduction, diffusion, viscosity, and electric current.

- *Heat conduction* is the flow of energy to eliminate temperature differences. We assume the temperature difference occurs across a membrane that is impermeable to matter but allows transfer of energy in the form of heat. An example is the continuous heat loss of our body through the skin at moderate temperatures. Heat conduction is discussed first in this chapter.

- *Diffusion* is the process by which molecules move from a region of higher concentration to one of lower concentration. The concentration difference occurs across a membrane that is semipermeable, permitting some chemical components to pass while others are blocked or significantly slowed down. Two important examples are (a) the transfer of oxygen and carbon dioxide between capillaries and the gas space in the lungs, and (b) the resorption of blood plasma components in the renal tubes of the kidney as discussed in Chapter 13. Diffusion is discussed in a later part of this chapter.

The other two transport phenomena are studied in different sections of this textbook, viscosity in Chapter 13 and electric currents in Chapter 19.

11.3: Heat Conduction

11.3.1: Fourier's Law

In 1822, Jean Baptiste Fourier set up an experiment suitable for investigating the flow of heat. The experiment is shown in Fig. 11.3(a) and in conceptualized form in Fig. 11.3(b). He quantified heat conduction in a rod connecting two heat reservoirs at different temperatures. For easy control of their temperatures, one is held at 0°C with an ice–water mixture, while the other is held at 100°C by boiling water. The rod is well insulated so that all heat transported through the rod is transferred from the high-temperature heat reservoir to the reservoir of lower temperature. The cylindrical rod has length L and a cross-sectional area A. Fourier found an empirical relation for the heat flow by varying every parameter in Fig. 11.3. Increasing the temperature difference between the two heat reservoirs and increasing the cross-sectional area of the rod increased the rate of heat flow toward the low-temperature heat reservoir. Decreasing the length of the rod also increased the heat flow rate, indicating that the length of the rod is inversely proportional to that rate. Defining the flow of heat per time interval as Q/t with unit J/s, Fourier wrote:

$$\frac{Q}{t} = \lambda A \frac{T_{high} - T_{low}}{L}. \quad [11.1]$$

Note that area A is not the surface of the membrane. Only the cross-sectional area perpendicular to the transport direction of heat enters Fourier's law. The proportionality constant λ is a materials constant because it depends on the composition of the rod; λ is called the **thermal conductivity coefficient** and has unit J/(m s K). Table 11.1 lists

(a)

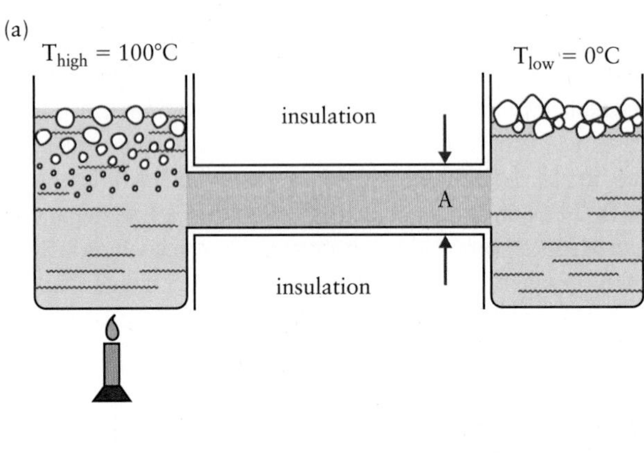

(b)

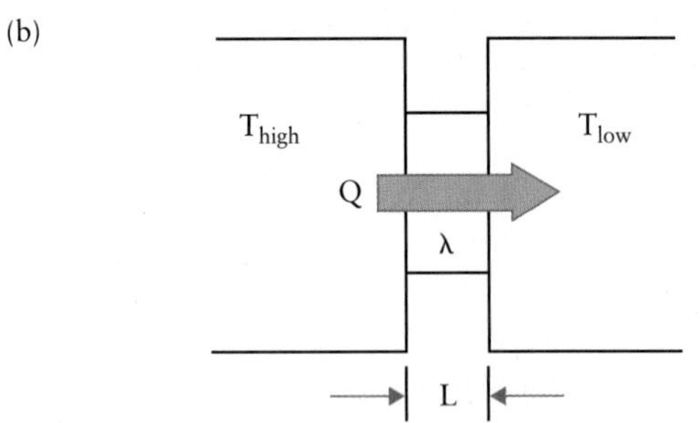

Figure 11.3 Fourier's experiment of heat conduction. (a) Experimental set-up: A steady-state heat flow across a rod of length L and cross-sectional area A is achieved by providing for a thermal contact to heat reservoirs at T_{low} = 0°C and T_{high} = 100°C at the two ends of the rod. The areas above and below the rod indicate thermal insulation of the rod to prevent lateral heat loss. (b) Conceptual sketch: Heat Q flows from a high-temperature heat reservoir through the system with thermal conductivity λ to a low-temperature heat reservoir.

TABLE 11.1

Thermal conductivity coefficients at room temperature

Material	Thermal conductivity λ ($J \cdot m^{-1} \cdot s^{-1} \cdot K^{-1}$)
Solid metals and alloys	
Silver (Ag)	420
Copper (Cu)	390
Gold (Au)	310
Aluminum (Al)	230
Iron (Fe)	80
Steel	50
Non-metallic solids	
Ice	1.6
Quartz glass (SiO_2)	1.4
Window glass	0.8
Fat	0.24
Rubber	0.2
Wood	0.12–0.04
Felt, silk	0.04
Liquids	
Mercury (Hg)	8.3
Water (H_2O)	0.6
Ethanol (C_2H_5OH)	0.18
Gases	
Air	0.026

Note that the coefficients vary by less than 5 orders of magnitude. Compare this with the variations of diffusion coefficients in Table 11.5 and electric resistivities in Table 19.1.

these coefficients for a range of materials. All values of λ are positive; i.e., heat always flows from an area of higher temperature to one with lower temperature.

Although Eq. [11.1] is written as if the thermal conductivity coefficient is temperature independent, the values in Table 11.1 are given specifically at room temperature. It has been found experimentally that the thermal conductivity of most materials varies with temperature, e.g., by more than 20% for water and air between 0°C and 100°C. We neglect such temperature dependences in this textbook.

The term $\Delta T/L$ represents the temperature step along the length L of the rod. If the rod is oriented along the x-axis this term can be written as the change of the temperature with position along the rod, which is the slope $\Delta T/\Delta x$, in which Δx is an interval along the rod. The term $\Delta T/\Delta x$ is called a **gradient**. The gradient itself can be a function of the position x, requiring more complicated mathematical expressions than Eq. [11.1]. To keep the current discussion simple, we limit ourselves to cases in which the gradient has a constant value.

CONCEPT QUESTION 11.1

Which of the following statements is true?

(A) Gold conducts heat better than copper.

(B) Under otherwise equal conditions, about two times more heat transfers through gold than through iron per second.

(C) If I conduct Fourier's experiment with a steel rod and an aluminum rod of the same shape, then I must use a longer rod for aluminum to obtain the same rate of heat transfer between beakers of freezing and boiling water.

(D) The thermal conductivity in Table 11.1 is not given in SI units.

(E) Table 11.1 shows that the thermal conductivity is a materials-independent constant.

11.3.2: Heat Loss of the Human Body

When we touch an object, it is its thermal conductivity—and not its actual temperature—that affects our impression of warmth or coldness. Conduct the following self-experiment: touch a piece of metal and a piece of wood that are both at room temperature. The metal feels cold and the wood feels warm. Since our hand in this experiment is warmer than the objects we touch, heat flows from our body into the object. The greater this flow of heat, the colder we perceive the object to be.

The same heat loss occurs continuously from the skin to the surrounding air. The air is not usually perceived as cold because the heat conductivity of gases is low (see Table 11.1), and thus the heat loss in room-temperature air is not large enough to be felt. Still, the thermal non-equilibrium between the body temperature of 37.0 ± 0.5°C and the air temperature is reflected in a gradual temperature drop in our bodies toward the surface. This is illustrated in Fig. 11.4 for two environmental

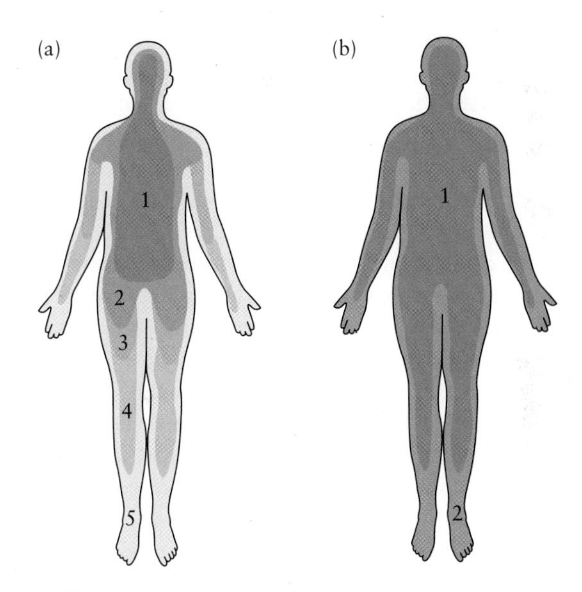

Figure 11.4 The temperature profile for a human body at ambient temperatures of (a) 20°C and (b) 35°C. Both sketches show a cross-section through the centre of the body. The different shades represent different temperatures. They are numbered as follows: (1) 37°C; (2) 36°C; (3) 34°C; (4) 31°C; and (5) 28°C.

CASE STUDY 11.1

In Fourier's experiment, heat is irreversibly transferred from a high- to a low-temperature heat reservoir. Can you suggest a reversible approach?

Answer: *We have discussed the idea of heat transfer between objects of different temperature before; in particular, reversible and irreversible conduction of these experiments was studied in Fig. 10.20. An elegant way to answer this question is to refer back to the Carnot process in Fig. 10.16 and consider only the first three steps: isothermal expansion, adiabatic expansion, and isothermal compression. During the isothermal reversible expansion in step I, heat Q is taken from the high-temperature heat reservoir and is deposited as work in the piston. The adiabatic expansion then adjusts reversibly the temperature of the system to that of the low-temperature heat reservoir. The third step is an isothermal reversible compression in which the work stored in the piston is turned into heat that is deposited in the low-temperature heat reservoir. In the Carnot process we stop the third process such that a subsequent adiabatic compression closes a cyclic process. In the current case the compression in the third step has to be done to a smaller volume to deposit the same amount of heat in the low-temperature heat reservoir as was removed from the high-temperature heat reservoir.*

TABLE 11.2

Mechanisms of heat loss from the human body as a function of the ambient temperature

Temperature of environment	Total heat loss	Fraction of evaporation	Fraction of convection	Fraction of radiation
20°C	63 J/(m^2 s)	13%	26%	61%
30°C	38 J/(m^2 s)	27%	27%	46%
36°C	43 J/(m^2 s)	100%	—	—

The total loss is given in the second column in unit J/(m^2 s). The third, fourth, and fifth columns show how this amount is distributed among perspiration, convection, and radiation.

temperatures. Fig. 11.4(a) shows a cross-sectional temperature profile in the body for an ambient temperature of 20°C, and Fig. 11.4(b) illustrates the same profile for an ambient temperature of 35°C. As the figure shows, the temperature profile is not a single temperature step from 37°C to the environmental temperature. Instead, the temperature decreases by as much as 9 degrees in the limbs in a room-temperature environment.

Three primary processes facilitate the continuous loss of heat to the environment: perspiration (evaporation of water from the skin), convection (heat carried away by air passing across the skin), and radiation. Table 11.2 summarizes the relative contributions to the heat loss of the human body at various environmental temperature conditions (dry air). Note that neither convection nor radiation contribute at 36°C and above.

Heat conduction is an effective heat loss process when swimming in water but contributes little in air. **Convection**, on the other hand, which is caused by the turbulent flow of air across the skin, can significantly enhance the heat loss in air. It is for this reason that Bedouins in northern Africa wear black or dark blue robes; convection contributes more effectively to the cooling of the skin beneath dark cloth than beneath cloth of brighter colours.

Heat loss by **perspiration** is based on the phase transition of water from liquid to vapour on the skin. Sweat glands bring liquid water to the skin. The evaporation of water requires energy. This amount of energy is called the latent heat of evaporation. It is supplied from the body's thermal energy. The vapour leaves the skin, carrying the latent heat into the environment. This heat transfer is very effective, corresponding to a loss of 2428 kJ per litre of water evaporated. However, perspiration is effective only in dry air. If the air is humid (saturated with water vapour), temperatures of about 33°C become unbearable.

Heat loss by **radiation** is a totally different process for which we discuss the physical principles in Chapter 21. Radiative energy flow is not carried by the medium air like heat in heat conduction or convection. You notice this when you hold the palm of your hand facing the Sun and then turn it 90°. The Sun causes the sensation of warmth because heat from the Sun reaches the hand. This heat flow must be independent of a medium to transport the heat, because most of the distance between the Sun and your hand passes through vacuum. The experiment with the hand also illustrates that heat transport by radiation works well when a cooler surface lies in the line of sight of a hotter surface. Thus, a cold wall contributes to loss of heat by radiation even if the air in the room is warm.

Example 11.1

(a) Calculate the steady rate at which a standard man (see data in Table 4.1) in winter clothing loses body heat. Assume the clothing is 1.5 cm thick, the average surface temperature of the standard man is 34°C, and the temperature of the outer surface of the clothing is 0°C. For the thermal conductivity of the clothing, use the value for felt in Table 11.1. (b) How does the answer to part (a) change if the clothes of the standard man get soaked with water?

Solution

Solution to part (a): We use Fourier's law as given in Eq. [11.1]. Substituting A = 1.85 m^2 from Table 4.1, L = 0.015 m, the temperature difference of ΔT = 34°C − 0°C = 34 K, and the thermal conductivity for felt λ = 0.04 J/(ms K), we find:

$$\frac{Q}{t} = \frac{\lambda A \Delta T}{L} = \frac{\left(0.04\frac{\text{J}}{\text{msK}}\right)(1.85\ \text{m}^2)(34\text{K})}{0.015\ \text{m}} = 168\frac{\text{J}}{\text{s}}.$$

Solution to part (b): In this part we are asked to derive an answer relative to the answer in part (a). Instead of repeating the calculation as shown in the first part, we divide the respective left- and right-hand sides of Fourier's law for the "dry" and the "wet" cases:

$$\frac{(Q/t)_{\text{dry}}}{(Q/t)_{\text{wet}}} = \frac{\dfrac{\lambda_{\text{dry}} A (T_{\text{high}} - T_{\text{low}})}{L}}{\dfrac{\lambda_{\text{wet}} A (T_{\text{high}} - T_{\text{low}})}{L}} = \frac{\lambda_{\text{dry}}}{\lambda_{\text{wet}}},$$

continued

Using the thermal conductivity of water from Table 11.1 for the soaked clothes, this equation yields:

$$\left(\frac{Q}{t}\right)_{\text{wet}} = \left(\frac{Q}{t}\right)_{\text{dry}} \frac{\lambda_{\text{wet}}}{\lambda_{\text{dry}}}$$

$$= \left(168\frac{\text{J}}{\text{s}}\right) \frac{0.6\dfrac{\text{J}}{\text{msK}}}{0.04\dfrac{\text{J}}{\text{msK}}} = 2520\frac{\text{J}}{\text{s}}.$$

Thus, the person loses heat 15 times faster when the clothes are wet! An interesting case in this context: Fritjof Nansen, a famous Norwegian polar explorer, tried to reach the North Pole with a second Norwegian in March of 1895. After leaving their ship at 84.4°N they failed in their quest for the Pole, drifting away from it faster than they could travel. They reached as far north as 86.14°N, which was a record at the time. Later they had to struggle through breaking ice on their way back. At one point all their supplies, which were stored in small boats, drifted away and Nansen had to swim after them. Before doing so, he stripped off his clothes. The calculation above demonstrates that this was an intelligent decision despite the freezing temperatures. It took Nansen only a few minutes to salvage the boats, but his colleague had to warm him up for several hours. Luckily, Nansen survived the incident. Had he jumped into the water with his warm clothes, they would not have protected him from the frigid water but would have dragged him down once soaked.

11.3.3: Examples from Biology

Example 11.2

Whales are mammals that live in the sea. Their bodies generate heat at the rate given for endotherms burning their food. Quantitatively, this is expressed as the metabolic rate M in kJ/day, with $M = 450\ m^{3/4}$, where m is the mass in kg. Like humans, they must maintain a core body temperature of about 37°C. Some of these whales spend part or all of the year in the frigid Arctic Ocean near the edge of the pack ice (water temperature 0°C). These include large whales such as bowhead whales of 100 tonnes body mass, and small whales such as the narwhal at about 1.5 tonnes. Their only protection in the chilly water is a thick fat layer. Calculate the minimum thickness of that fat layer for these whales as a function of body size. *Hint:* Use a spherical shape for the whale's body.

Solution

Fig. 11.5 shows the simplified model used for the whale in this problem: the spherical animal has an inner body of radius R and a fat layer of uniform thickness l. The minimum thickness of the fat layer is determined by balancing the heat loss and the conversion of food to thermal energy (metabolic rate) in the whale's body.

continued

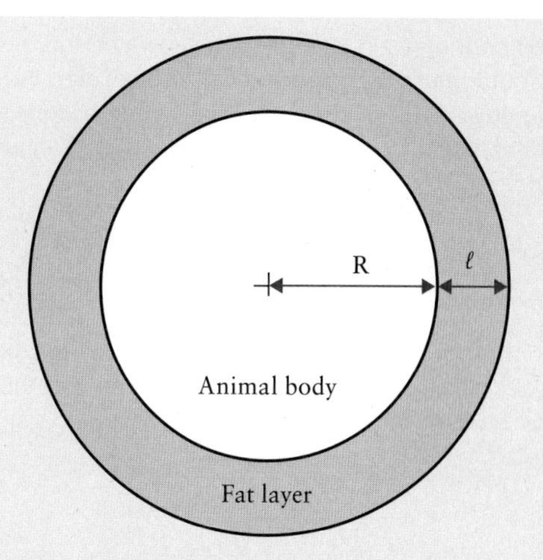

Figure 11.5 Simplified model of a whale consisting of an inner body of radius R, covered by a fat layer of uniform thickness l.

For the heat loss, we use Fourier's law with the thermal conductivity coefficient of fat from Table 11.1. The higher temperature in Fourier's law is the core body temperature and the lower temperature is the temperature of the seawater; i.e., $\Delta T = 37$ K. The area across which the heat flows is the surface of the inner body of the whale, $A = 4\pi R^2$ (which is the surface of a sphere of radius R). Thus:

$$\left(\frac{Q}{t}\right)_{\text{loss}} = \lambda 4\pi R^2 \frac{\Delta T}{l}$$

$$= \left(110\frac{\text{J}}{\text{ms}}\right)\frac{R^2}{l},$$

in which we left the radius of the whale, R, and the fat layer thickness, l, as variables and combined all other parameters to a single pre-factor of 110 J/(ms). Thus, the heat loss is expressed as a function of the radius of the whale and the thickness of the fat layer. We keep these two parameters variable because we want to find their relation for whales of any size.

To offset the heat loss, the whale's metabolism converts food energy into thermal energy as given in the Example text. We rewrite that equation in standard units, defining the metabolic rate as $(Q/t)_{\text{gain}}$ in unit J/s, and expressing it as a function of the mass m in unit kg:

$$\left(\frac{Q}{t}\right)_{\text{gain}} = 5.2\, m^{3/4}.$$

To relate again to the radius R of the whale, the mass of the animal is rewritten as its density and the volume of a sphere of radius R. This leads to:

$$\left(\frac{Q}{t}\right)_{\text{gain}} = 5.2\left(\rho\frac{4}{3}\pi R^3\right)^{3/4}$$

$$= 5.2\left(\frac{4}{3}\pi\rho\right)^{3/4} R^{9/4}.$$

continued

Since the whale is floating in seawater, the density of its body must be close to that of seawater. We use the value ρ = 1020 kg/m^3 for the average density of seawater. With Q/t in unit J/s and the radius R in unit m we find:

$$\left(\frac{Q}{t}\right)_{\text{gain}} = 2750\, R^{9/4}.$$

For calculating the minimum fat layer thickness, the balance between the whale's heat loss and thermal energy gain must be found; i.e., heat loss and gain are set equal:

$$\left(\frac{Q}{t}\right)_{\text{loss}} = \left(\frac{Q}{t}\right)_{\text{gain}}.$$

Substituting the equations we found above:

$$110\frac{R^2}{l} = 2750\, R^{9/4}.$$

With both length l and radius R given in units m, this leads to:

$$l = \frac{0.04}{R^{1/4}}.$$

This equation relates the fat layer thickness to the radius of the whale. The bigger the animal (i.e., the larger R) the thinner a sufficient fat layer (this is called the **bulk effect**). An animal of radius R = 0.1 m = 10 cm (rat-sized) needs a fat layer thickness of about l = 7 cm; for an animal of radius R = 1 m (sea lion-sized) a fat layer of 4 cm is needed; and for R = 10 m (whale-sized) a fat layer of 2.5 cm is sufficient. Obviously, a polar rat would be a clumsy creature and therefore does not exist. Sea lions and whales have no problem developing a sufficient fat layer that does not hinder them in their daily lives.

Indeed, real whales typically have fat layer thicknesses (blubber) of about 0.5 m; in the case of the right whale, as much as 40% of its body mass is blubber. Comparing with our calculation above, this means the whale's fat layer is significantly oversized. This is due to the fact that the fat layer thickness cannot be adjusted on a short time scale, e.g., to accommodate day–night temperature changes or water temperature variations caused by currents. Thus, whales require another mechanism of temperature regulation. Nature's solution to the problem includes an oversized fat layer and a secondary mechanism to dissipate excess thermal energy through the fat layer-free tail. Thermal energy is carried by blood flow to the tail, where it is brought close to the skin (perfusion). The fluke has no fat layer and therefore allows for effective heat conduction into the surrounding water. The effectiveness of this mechanism is illustrated in Fig. 11.6 for a dolphin's fluke. The figure shows the temperature profile of the tail from an infrared photograph

continued

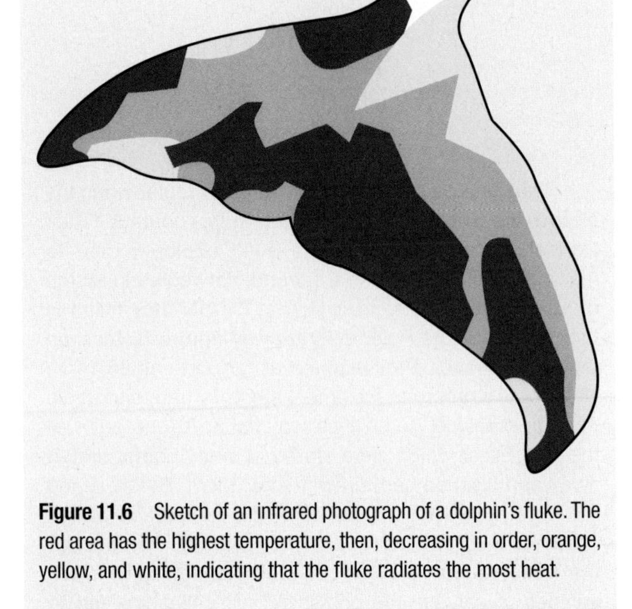

Figure 11.6 Sketch of an infrared photograph of a dolphin's fluke. The red area has the highest temperature, then, decreasing in order, orange, yellow, and white, indicating that the fluke radiates the most heat.

(**thermography**; Fig. 11.7 shows this equipment used in a kinesiology laboratory). The various temperatures are shown in different shades: red is hottest and white is coldest. The figure clearly illustrates that the fluke is used not only for motion but also serves as the body's temperature-control device.

Whaling was a major branch of the fishing industry in the mid-1800s when Herman Melville wrote the famous adventure story *Moby Dick*. At that time whales were little-understood creatures, and the described

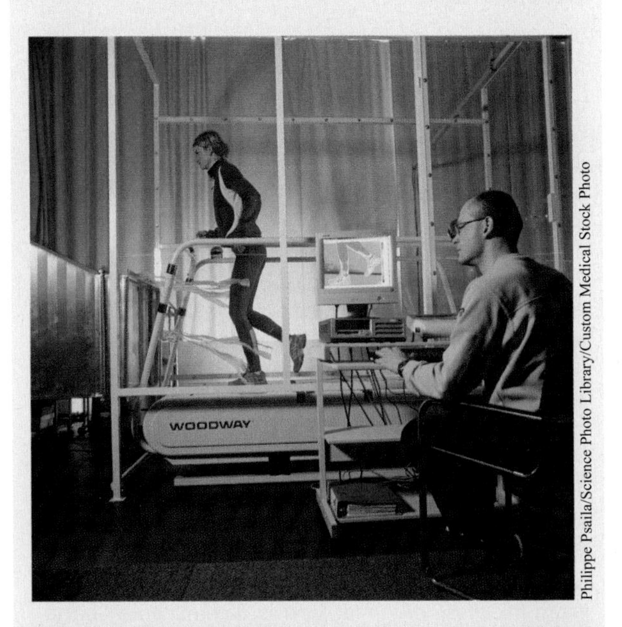

Philippe Psaila/Science Photo Library/Custom Medical Stock Photo

Figure 11.7 Thermogram taken of the legs and shoes of an athlete on a treadmill. A wind machine is replicating the conditions during running.

continued

mechanism of heat dissipation was unknown. Thus, whalers often reported their amazement when they found temperatures in whale carcasses to be as high as 60°C and the whale meat half-cooked. This is due to the fact that the heart of a dying whale stops immediately, interrupting the blood flow to the fluke. However, thermal-energy-generating metabolic processes continue for a short time after death.

Is there an actual case in nature to which the direct balance of heat loss and gain applies? The family *Megapodiidae* contains 22 bird species that are found in Australasia, including the Australian brush-turkey, shown in Fig. 11.8. They have in common a unique nesting and incubation technique: they build a large mound of forest litter in which they bury their eggs after a strong rainfall. Heat in the mound is generated by decomposition (putrefaction). The parent birds must return frequently to the mound to adjust its thickness because ambient temperatures vary but the incubation temperature must remain constant at $T_{incubation}$ = 35°C to 36°C. A calculation of the dependence of the radius of the mound on the ambient temperature yields (the complete formula is derived in P–11.11):

$$R \propto \sqrt{T_{incubation} - T_{ambient}}.$$

With day/night temperature variations moderate in the habitat of these birds, and the square-root dependence in the above equation reducing the work to alter the size of the mounds, the brush-turkeys manage their task of balancing heat loss and heat generation well.

© Martin Harvey/Alamy

Figure 11.8 Australian brush-turkey.

CASE STUDY 11.2

Not every problem in the sciences requires a full calculation; often, so-called back-of-the-envelope estimates are sufficient. These include cases where the actual formula for the physical process is derived through an educated guess. Educated guesses are frequently based on dimensionality considerations. The following is an instructive example:

Food that has a high concentration of fat has to be heated longer during **pasteurization**. For a while this observation was explained as follows: In fatty food, bacteria cover themselves with a fat layer that, for practical purposes, cannot be much thicker than the size of the bacteria. Heat cannot penetrate this layer as readily because heat conductivity is much lower in fat than in water, effectively protecting the bacteria against thermal destruction. Do you accept this explanation?

Hint: Determine the **thermal relaxation time** τ, which we define as the time it takes for a sphere of radius R to adjust its initial temperature $T_{initial}$ to the ambient temperature $T_{ambient}$ after immersion. *Note:* An analytical derivation of τ is provided in P–11.12.

Answer: *If heat conduction governs this process, the physical formula for the relaxation time should contain the radius R of the sphere, the density of the sphere ρ, the specific heat capacity of the sphere c, the initial temperature difference ΔT, and the heat conductivity coefficient of fat λ. Table 11.3 shows the units of these parameters in terms of the fundamental SI units.*

continued

TABLE 11.3

Dimensional analysis for thermal relaxation time in pasteurization

	Radius R	Density ρ	Heat capacity c	Thermal conductivity λ	Temperature K	Relaxation time τ
Length (m)	1	−3	2	1	0	0
Mass (kg)	0	1	0	1	0	0
Time (s)	0	0	−2	−3	0	1
Temperature (K)	0	0	−1	−1	1	0

We seek a combination of the five parameters R, ρ, ΔT, c, and λ to describe the parameter τ with unit s. We start with a combination of c and λ. Since none of R, ΔT, or ρ contains the unit s, the combination of c and λ must result in a unit s. This is obtained with the ratio c/λ. This term also carries a unit kg^{-1}. Multiplication by the density compensates for this unit: $c\rho/\lambda$ has the unit s/m^2. Now we introduce a factor R^2 in the numerator to eliminate the unit m in the denominator. Thus, we propose for the thermal relaxation time:

$$\tau \propto \frac{R^2 \rho c}{\lambda}.$$

Note that we did not need the initial temperature difference $\Delta T = T_{ambient} - T_{initial}$ in this formula. This makes sense because a larger temperature difference would require more heat to flow to the colder sphere, but the larger temperature difference would drive this heat into the sphere faster.

Assuming that above relation for the thermal relaxation time is a reasonable estimate (it is indeed correct to within a factor of 3, as shown in P–11.12), we quantify the various parameters, modelling the bacterium as a spherical droplet of water. The radius of a bacterium is about 10 μm, ρ_{water} = 1000 kg/m^3, c_{water} = 4210 J/(kg K) at 100°C, and λ_{fat} = 0.24 J/(m s K) from Table 11.1. Substituting these values in the previous equation yields:

$$\tau = \frac{R^2 \rho c}{\lambda}$$

$$= \frac{(1\times10^{-5}\,\mathrm{m})^2\left(1000\frac{\mathrm{kg}}{\mathrm{m}^3}\right)\left(4210\frac{\mathrm{J}}{\mathrm{kg\,K}}\right)}{0.24\frac{\mathrm{J}}{\mathrm{msK}}}$$

$$= 1.8\times10^{-3}\,\mathrm{s}.$$

Thus, the bacterium with the assumed fat layer thermally equilibrates within a few milliseconds with its environment. The idea that the fat layer is a thermal protective layer has to be dismissed.

11.3.4: Lord Kelvin's Age of Earth

Example 11.3

After publishing *The Origin of Species by Means of Natural Selection* in 1859, Charles Darwin had to endure some malicious criticism. On the other hand, there were only very few serious objections. The most credible one came from Lord Kelvin, who estimated the age of Earth to be 400 million years or less, and the age of the Sun to be 100 million years or less. He later even corrected these numbers downward to as little as 20 million years.

We consider Lord Kelvin's argument for the **age of Earth**. It is based on the following data: From underground mining we know that the temperature below the surface increases by 3 K per 100 metres. This is called the **geothermal effect**. Assuming Earth started as molten rock called the **Proto-Earth**, with a uniform temperature of T = 3000°C, how long did it take to reach the current state? For the density of rock use ρ = 3 g/cm^3, for the heat capacity of rock c = 1470 J/(kg K), and for its thermal conductivity coefficient λ = 1.7 J/(m s K).

continued

Solution

The current temperature profile of Earth, based on the geothermal effect, is shown in Fig. 11.9. At the surface (the radius of Earth is taken as 6400 km), the temperature is about 0°C (not 0 K, due to the atmosphere). Based on the geothermal effect, it rises to 3000°C at 100 km depth, i.e., 6300 km from the centre of Earth.

For our quantitative calculations, we consider a rectangular segment of 100 km depth and of surface area A = 1 m^2, as illustrated in Fig. 11.10. The heat loss of this segment is determined using Fourier's law as given in Eq. [11.1]. We write it in the form:

$$\frac{Q/t}{A} = \lambda \frac{\Delta T}{l}.$$

Substituting the given values yields:

$$\frac{Q/t}{A} = \left(1.7\frac{\mathrm{J}}{\mathrm{smK}}\right)\left(0.03\frac{\mathrm{K}}{\mathrm{m}}\right) = 0.05\frac{\mathrm{J}}{\mathrm{sm}^2}.$$

continued

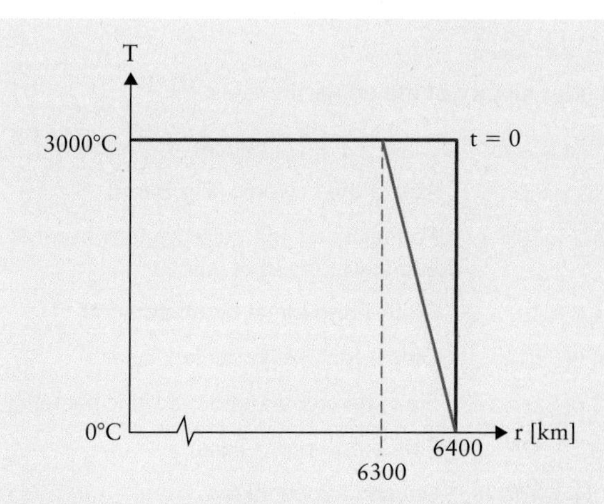

Figure 11.9 Temperature profile of Earth based on the geothermal effect (blue line). With the temperature increasing by one degree for each 30 m, the temperature increases for about 100 km until the temperature of molten rock (3000°C) is reached. Using 6400 km for the radius of Earth, molten rock is reached at $r = 6300$ km. The red line, labelled $t = 0$, indicates the temperature profile of the Proto-Earth, which was a liquid sphere throughout due to the *great bombardment* with space debris during the time of the early solar system.

The term $(Q/t)/A$ is the rate of loss of heat per time interval through the area A. The term $\Delta T/l$ is the **geothermal temperature gradient**.

Next, using Joule's definition of heat we calculate the total heat lost since the times of the liquid Proto-Earth:

$$Q = mc_{\text{rock}}\Delta T = \rho V c_{\text{rock}}\Delta T,$$

in which the mass of the rock slab of Fig. 11.10 has been rewritten as its density and volume. The volume can further be rewritten as the surface area A times the depth l, leading to:

$$Q = \rho A l c_{\text{rock}}\Delta T.$$

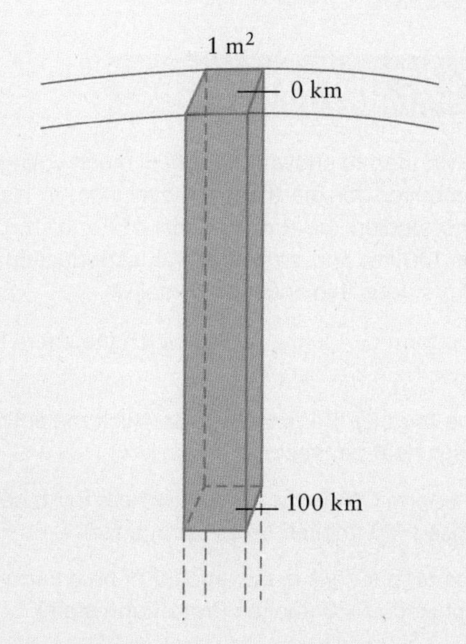

Figure 11.10 Segment of Earth's crust used for the calculation of the rate of heat loss.

continued

We divide by the area to express the heat per unit surface area:

$$\frac{Q}{A} = \rho l c_{\text{rock}}\Delta T,$$

where Q/A is the total heat lost through the surface area A. Evaluating this equation must be done carefully since the temperature change, ΔT, varies with depth as shown in the lower part of Fig. 11.11. The figure indicates that, due to the direct proportionality between heat and temperature in the previous equation, the total loss of heat corresponds to the triangle enclosed by the dashed lines and the line indicated with vertical arrows. Thus, the total amount of heat lost through the surface area A is found from Fig. 11.11 graphically. The area of the triangle equals half the area of the rectangle with a temperature drop of 3000 degrees throughout. Thus, we substitute for ΔT:

$$\frac{Q}{A} = \frac{1}{2}\left(3000\frac{\text{kg}}{\text{m}^3}\right)(1.0\times10^5\,\text{m})$$
$$\times\left(1470\frac{\text{J}}{\text{kg K}}\right)(3000\,\text{K})$$
$$= 6.6\times10^{14}\,\frac{\text{J}}{\text{m}^2},$$

in which 3000 K is the difference between 0°C and 3000°C.

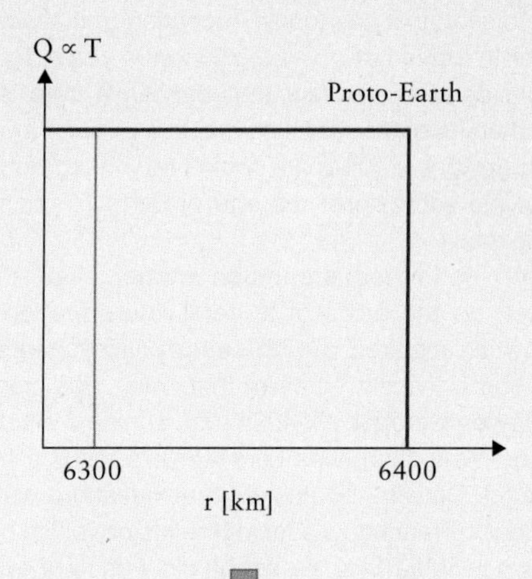

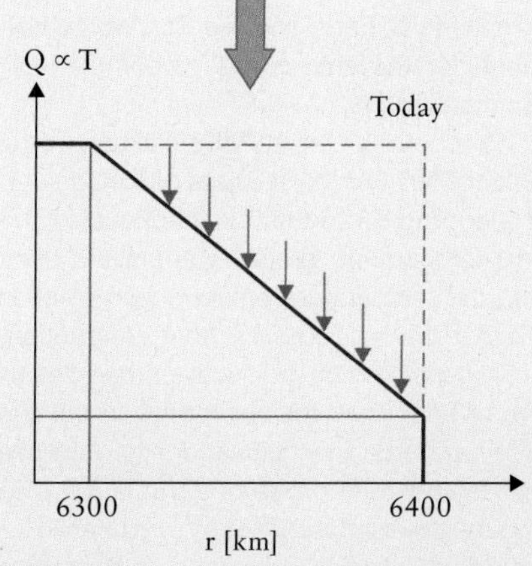

Figure 11.11 Illustration of the total amount of heat, Q, lost between the times of the Proto-Earth (top) and the current state (bottom).

continued

Lord Kelvin then divided the total energy lost from the 100-km-deep segment by the rate of loss through area A:

$$t_{\text{Earth}} = \frac{\frac{Q}{A}}{\frac{Q/t}{A}} = \frac{6.6\times10^{14}\,\frac{\text{J}}{\text{m}^2}}{0.05\,\frac{\text{J}}{\text{sm}^2}}$$

$$= 1.3\times10^{16}\,\text{s} = 4\times10^{8}\ \text{years}.$$

This is the predicted age of Earth since cooling began, $t_{\text{Earth}} = 400$ million years.

Lord Kelvin tried to confirm this estimate by comparing it with an estimate of the age of the Sun. The modern view of the solar system is that the Sun and the Proto-Earth formed at about the same time. Thus, since the Sun is still operating, Lord Kelvin was able to calculate an upper limit to the age of the Sun by determining how long the fuel of the Sun would last.

His first estimate of the Sun's age was based on the assumption that the Sun operates with chemical energy, i.e., energy obtained by burning energy-rich compounds like coal or hydrogen. This would allow the Sun to operate for only 1500 to 5000 years, depending on the type of fuel. A better estimate followed when Lord Kelvin added the energy released during a gravitational collapse of the Sun. This obviously requires that the Sun started with a much larger radius than it has today. Accepting that assumption, Lord Kelvin arrived at a value of 2×10^7 years. By taking into consideration the fact that the Sun's core is much denser than its outer shell, he was able to push the age of the Sun up to 1×10^8 years, which he judged sufficiently close to his estimate of the age of Earth to confirm the previous result.

Both estimates are much longer than the age predicted on the basis of a literal interpretation of the Bible, which inspired a 17th-century vice-chancellor of Cambridge University to claim that "man was created by the Trinity on October 23, 4004 B.C., at nine o'clock in the morning." Why, then, was Lord Kelvin's result a problem for Charles Darwin? To answer that question, we take a look at our current knowledge of the history of life on Earth as shown in Table 11.4: Given the slow pace of evolution, Lord Kelvin's age of Earth and the Sun would not provide enough time for the emergence of complex organisms such as human beings.

Why were both of Lord Kelvin's estimates wrong by at least a factor of 10? For the estimate of the age of Earth, his model of a cooling Proto-Earth is inadequate. The current geothermal temperature profile is actually a steady-state profile; i.e., the temperature profile does not change with time. To offset the heat loss by heat conduction, heat is continuously generated by radioactive processes that occur in the core of Earth, and the convection in the liquid outer core, which converts gravitational energy into heat. Thus, the current temperature profile of Earth is not the result of the cooling mechanism assumed by Lord Kelvin.

For the Sun, Lord Kelvin's model of heat generation is incorrect. Neither chemical energy nor gravitational energy contributes significantly; rather, nuclear fusion in the core

continued

TABLE 11.4

A brief history of life on earth

Years ago	Event
13×10^9	Age of the universe (**Big Bang**)
4.7×10^9	Formation of the solar system from an interstellar cloud of gas
4.6×10^9	Proto-Earth (great bombardment)
4.03×10^9	Oldest rock (Yellowknife, Canada)
$3.6–3.8 \times 10^9$	First prokaryotes (stromatolithic bacteria)
2.5×10^9	First eukaryotes (algae)
1.7×10^9	Oxygen atmosphere
1.0×10^9	Sexual reproduction
6.7×10^8	Multicellular animal fossils found at many places on Earth
5.8×10^8	Animals with shells and skeleton
4.8×10^8	Plants expand from sea to land
4.2×10^8	Animals expand from sea to land
2.4×10^8	First mammals
$1–4 \times 10^8$	*Kelvin's age of Earth and Sun*
65×10^6	End of dinosaurs
4.0×10^6	Early hominids (*Australopithecus*)
2.5×10^6	Genus *Homo*
0.125×10^6	Modern *Homo sapiens*

of the Sun generates the heat the Sun radiates. In Lord Kelvin's defence, it should be noted that in both cases the underlying physics had not yet been discovered at the time he criticized Charles Darwin's ideas.

CONCEPT QUESTION 11.2

Lord Kelvin tried to show that Earth is much younger than Darwin required for the theory of evolution. In his calculations he studied the rock material of Earth's crust to a depth of 100 km and arrived at the last equation above. What terms does this equation contain?

(A) The term Q/A is the rate at which the entire Earth loses heat per second.

(B) The term $(Q/t)/A$ is the rate at which the entire Earth loses heat per second.

(C) The term Q/A is the amount of heat Earth has lost since $t = 0$ (molten Proto-Earth state).

(D) The term $(Q/t)/A$ is the amount of heat Earth has lost since $t = 0$ (molten Proto-Earth state).

(E) The term Q/A is the amount of heat Earth has lost since $t = 0$ (molten Proto-Earth state) through each square metre of its surface.

11.4: Diffusion

All membranes have in common the ability to maintain a non-equilibrium between the systems they separate. In addition, biological membranes participate actively in the exchange processes between the two systems they separate. For this role, cell membranes command a range of transport processes as illustrated in Fig. 11.12:

- *Passive diffusion*, sketched as an arrow through the uniform membrane material, allows small ions and molecules to penetrate the membrane in the direction of an existing concentration step. In the case of diffusion, the flow of material is in the direction from the higher to the lower concentration.
- *Facilitated diffusion* is an enhanced diffusion of specific molecules, supported by proteins. This type of diffusion is indicated in the figure by an arrow with the label "protein." These proteins are embedded in the cell membrane, as shown in Fig. 11.1. The diffusion process is the same as in passive diffusion. In biology, proteins that facilitate diffusion are called **gated channels** for the motion of the diffusing species.

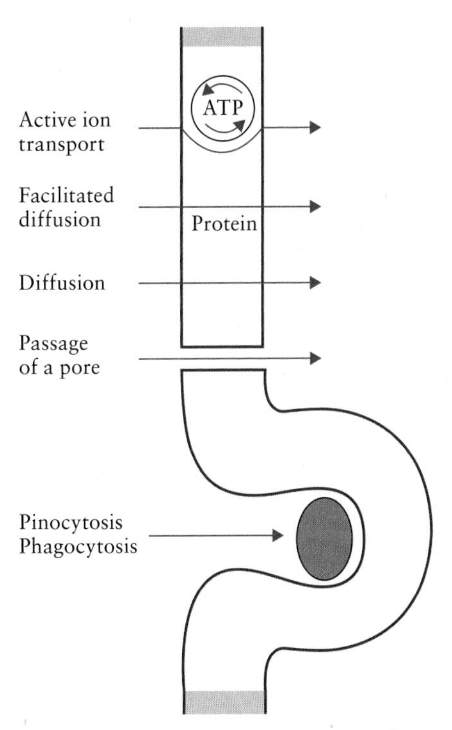

Figure 11.12 Conceptual sketch illustrating the main mechanisms of material transport across biological membranes. Diffusion along a concentration gradient is discussed in this chapter. Facilitated diffusion allows for chemically specific transport and is described by the same mechanisms we introduce for passive diffusion. Active ion transport against a concentration gradient is discussed in Chapter 19 for the potassium and sodium ion transport across nerve membranes. Passage of pores in a membrane is discussed in Chapter 13 as a fluid flow phenomenon. Pinocytosis and phagocytosis are more complex processes that include shape changes of the membrane.

- *Active ion transport*. Proteins and enzymes are also capable of transporting (mostly smaller) ions across the membrane against a concentration step. This process is called active ion transport and requires energy, typically provided by ATP molecules, as indicated in Fig. 11.12. Important examples include the transport of potassium and sodium across the membrane of nerve cells, as discussed in Chapter 19, and proton pumps in plants, fungi, and bacteria to remove H^+ from the cell.
- *Fluid flow*. Small-sized molecules pass through pores in the membrane. An example is the passage of blood plasma components through the basement membrane in the kidneys. This process is not considered a diffusion process in the context of this chapter but is treated as a flow process in Chapter 13.
- *Pinocytosis*. Very large molecules or bacteria are transported across the membrane in processes called **pinocytosis** (for molecules) or **phagocytosis** (for bacteria). These processes are complex encapsulations and are not included in the current discussion.

11.4.1: Fick's Law

In 1855, the physician Adolf Fick described the transport of matter across a membrane in analogy to Fourier's law of heat transport. Fick observed empirically that the rate of a gas passing through a membrane of contact area A and width l is proportional to A and the density difference on both sides of the membrane, $\rho_{high} - \rho_{low}$. The transport rate of a given component i is also inversely proportional to the width of the membrane:

$$\frac{m_i}{t} = DA\frac{\rho_{i,high} - \rho_{i,low}}{l}. \qquad [11.2]$$

This is **Fick's law**. The proportionality factor D is called the **diffusion coefficient** and has SI unit m^2/s. The index i in Eq. [11.2] indicates that Fick's law applies to components of a mixed phase as well. In biological systems, mixed systems occur frequently as membranes separate liquid solutions, such as cytoplasm from the extracellular fluid; or membranes separate gaseous and liquid solutions, such as blood and air in the lungs. For such systems the mass of the i-th component, m_i, is usually rewritten as the amount of the component n_i in unit mol:

$$n_i(\text{mol}) = \frac{m_i(\text{kg})}{M\left(\frac{\text{kg}}{\text{mol}}\right)}.$$

Note that the units of the variables are given in brackets. Further, the density ρ_i is rewritten as **concentration** c_i:

$$c_i\left(\frac{\text{mol}}{\text{m}^3}\right) = \frac{\rho_i\left(\frac{\text{kg}}{\text{m}^3}\right)}{M\left(\frac{\text{kg}}{\text{mol}}\right)},$$

in which we included again the units of the variables in brackets. Using these definitions, Fick's law of diffusion can be given in a second form:

$$j_i = \frac{n_i}{t} = DA\frac{c_{i,\text{high}} - c_{i,\text{low}}}{l}, \qquad [11.3]$$

in which j_i is the **material flux** (the amount of matter that continuously passes a given location, in unit mol/s). It is equal to the amount of component i, n_i, crossing the membrane from the side of higher concentration to the side of lower concentration during the time interval t.

The term $\Delta c/l$ is the concentration step from one side of the membrane to the other. This term can be generalized as the change of the concentration with position across the membrane, i.e., a **concentration gradient** $\Delta c/\Delta x$ when the direction perpendicular to the membrane is the x-axis. The concept of a concentration gradient is more useful when we do not want to refer to a particular membrane thickness. For example, concentration gradients of morphogens are thought to define the body axes in embryos (**gradient hypothesis in embryology**). In this case you are interested in the local change of a concentration, not a concentration step across the entire embryo. As with temperature gradients, concentration gradients need not be constant but can be a function of the position. In the current chapter, however, we continue to confine our discussion to constant gradients.

In analogy to Fourier's law, Fick's law of diffusion applies only when transport occurs in a steady state, which means no essential parameter of the system varies with time. In particular, the concentration of component i on both sides of the membrane must be constant, and the profile of the diffusing component across the membrane must not change with time. A steady state therefore excludes any initial, transient period after the experiment has started. During the early transient period the concentration profile across the membrane varies, eventually approaching the steady-state profile. The steady state is an important state because the system will develop toward this state if it cannot develop toward an equilibrium state. As an example, the concentration difference across the membrane is always maintained in a human cell, preventing the cell from approaching a chemical equilibrium.

Since material flows continuously from one side to the other in a steady-state diffusion process, independent processes are required to ensure that the concentration of component i does not decrease on the higher-concentration side of the membrane (continuous supply), and that it does not increase on the lower-concentration side of the membrane (continuous consumption). Thus, whenever a concentration gradient is maintained across a semipermeable biological membrane, additional physiological processes are involved. Examples include

- *passive chemical processes*, such as the buffer effect regulating the acidity (pH value) of blood by using the chemical reaction $CO_2 + H_2O \rightleftarrows HCO_3^- + H^+$;
- *active biosynthesis* of components such as immunoglobulin (antibodies in blood); or
- *active transport* across the membrane against the concentration gradient, as discussed in Chapter 19 for nerve cells.

Example 11.4

We consider sucrose diffusing along a 10-cm-long tube filled with water. The cross-sectional area of the tube is 6.0 cm^2. The diffusion coefficient is 5.0×10^{-10} m^2/s and a total amount of sucrose of 8.0×10^{-14} kg is transported in a steady state along the tube in 15 s. What is the difference in the density levels of sucrose at the two ends of the tube?

Solution

We use Fick's law in the form given in Eq. [11.2]. This equation is rewritten to isolate the unknown density difference:

$$\Delta\rho = \left(\frac{m}{t}\right)\frac{l}{DA}.$$

Now we substitute the given data:

$$\Delta\rho = \frac{(8\times10^{-14}\,\text{kg})\,(0.1\,\text{m})}{(15\ \text{s})\left(5\times10^{-10}\,\frac{\text{m}^2}{\text{s}}\right)(6\times10^{-4}\text{m}^2)} = 1.8\times10^{-3}\,\frac{\text{kg}}{\text{m}^3}.$$

CASE STUDY 11.3

Anoxia is a lack of oxygen supply to cells; hypoxia is a deficiency of oxygen supply. Which parameters in Fick's law play a possible role in anoxia or hypoxia?

ANSWER: *We use Eq. [11.3]. The oxygen flux j_{O_2} into the receiving tissue is lowered when*

- *the concentration step is too small because the blood oxygen concentration $c_{high} = c_{blood}$ is too low;*
- *the concentration gradient is too low because the membrane thickness l is too large; or*
- *the diffusion coefficient is lowered.*

Too low oxygen levels in the blood can be the result of a diminished uptake of oxygen (at high altitudes; too high pressure on the thorax while diving with a snorkel that is too long; sleep apnea), a too low oxygen capacity of the blood (too low erythrocyte count; hemoglobin deficiency, e.g., due to iron deficiency, sickle cell anemia, or carbon monoxide poisoning), or a diminished flow of blood

continued

through the blood capillaries (too low blood pressure, e.g., during heart failure or serious injury; local circulatory disorder due to embolism or thrombosis). Diffusion paths get too long during irregular tissue growth. The diffusion coefficient cannot be altered directly; however, alterations in the cells can diminish the rate of oxygen uptake, e.g., hydrocyanic acid poisoning, in which the oxygen reaction with food in the mitochondria is inhibited.

CASE STUDY 11.4

Each blood capillary is capable of supplying oxygen to a surrounding tissue cylinder with a radius of about 20 μm, called a Krogh cylinder. Which factors affect this radius?

ANSWER: *We start with Eq. [11.3]. The blood oxygen concentration or its partial pressure in the blood, and the oxygen permeability of the tissue, which is defined in physiology as the diffusion coefficient per length of membrane tissue, D/l, govern the oxygen flux per unit area of the capillary wall. Further, the oxygen consumption rate of the receiving tissue is a limiting factor.*

11.4.2: Temperature Dependence of Diffusion

Table 11.5 lists several diffusion coefficients D for biologically relevant systems at 20°C. The data indicate some interesting trends; for example, that the same molecule (e.g., oxygen) diffuses faster in less dense media, i.e., fastest in gases and slowest in solid tissue. The data also show that bigger particles diffuse more slowly in a given medium; for example, sucrose diffuses faster than hemoglobin in water.

When wondering about such observations and the fact that they seem not to be predicted by Eq. [11.2], we need to keep in mind that Fick's law is a **phenomenological law**. This means that it is an adequate description of experimental observations of the relation between the flux of matter across a membrane and the corresponding concentration change, but it is not derived from the fundamental laws of physics, e.g., the three laws of thermodynamics or Newton's three laws of mechanics. Empirical laws usually contain constants, such as the diffusion coefficient in Fick's law, for which a fundamental origin is not revealed by the empirical law itself. Consequently, although we know the diffusion coefficients in Table 11.5, we cannot predict values for other diffusion coefficients based on our understanding of nature in general, nor can we explain the temperature dependence of D in spite of our intuition that it should be significant.

To understand the temperature dependence of diffusion, a microscopic look at the membrane is necessary. Since diffusion is based on the motion of single atoms or molecules, the membrane model has to be developed at the atomic length scale. We used this approach before, when we developed the kinetic gas theory as an atomic scale model in Chapter 8 to link temperature and the root-mean-square speed of particles.

A microscopic model for the matrix in which diffusion is observed is illustrated in the top panels of Fig. 11.13: it consists of a regular array of atoms or molecules with fixed relative positions (open red circles). This is a good model for crystalline solids, but it is also suitable to describe diffusion in solids without long-range order, e.g., amorphous solids, liquids, most biological systems.

This microscopic model of diffusion was first introduced quantitatively by Svante August Arrhenius in 1889. Diffusion results when the foreign particle, shown as the solid dot in Fig. 11.13, travels from one open site in the matrix to the next by passing through a zone in which the matrix particles are more crowded. The plot in the lower panel of Fig. 11.13 shows the potential energy of

TABLE 11.5

Biologically relevant diffusion coefficients at $T = 20$°C

System	Diffusion coefficient
Oxygen (O_2) in air	6.4×10^{-5} m²/s
Oxygen (O_2) in water	1×10^{-9} m²/s
Oxygen (O_2) in tissue	1×10^{-11} m²/s
Water in water	2.4×10^{-9} m²/s
Sucrose in water	5×10^{-10} m²/s
Hemoglobin in water	7×10^{-11} m²/s
Tobacco mosaic virus in water	5×10^{-12} m²/s

The value for water self-diffusion applies at $T = 25$°C.

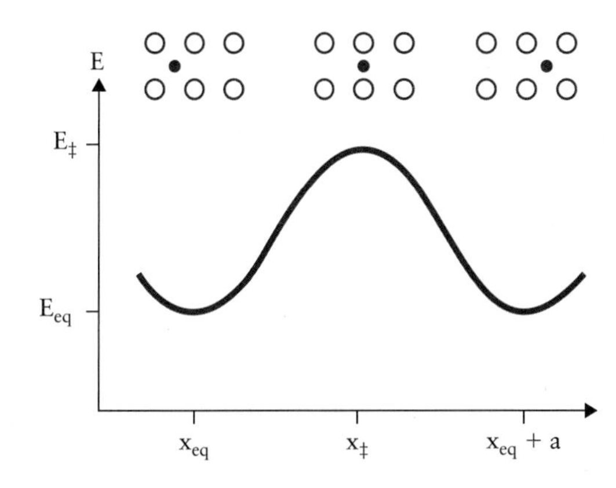

Figure 11.13 Top panel: three sequential sketches for the microscopic diffusion mechanism in an ordered matrix. The matrix particles are indicated by open circles. The diffusing foreign particle (solid dot) hops from one wide-open site within the matrix to another, passing through a zone in which the matrix particles are located in a denser array. The wide-open sites are equilibrium sites for the foreign particle in which it spends most of the time. Bottom panel: the corresponding plot of the potential energy of the foreign atom as it moves from its initial site at x_{eq} to the final site of the jump at $x_{eq} + a$. At the equilibrium sites the energy of the atom is E_{eq}. The atom has a maximum energy $E_{\ddagger}$ in the transition state at position $x_{\ddagger}$.

the foreign atom as a function of its position. Sites that provide more space for the diffusing particle are associated with a smaller potential energy. Such sites are shown in the first and last panels at the top of the figure and correspond to positions x_{eq} and $x_{eq} + a$, where a is a typical spacing of adjacent open sites for a given matrix. Since the potential energy of the diffusing particle has a minimum value at x_{eq}, we call this an equilibrium position and label the corresponding energy E_{eq}.

The foreign particles are in thermal equilibrium with the matrix, which means they will not have a sufficiently high total energy to move through the matrix like gas particles move in the kinetic gas theory model. Instead, the total energy of the particle is much smaller than the energy barrier between positions x_{eq} and $x_{eq} + a$ in Fig. 11.13, and the particle is confined to the equilibrium position for most of the time.

When an atom moves, it must pass through an area of higher potential energy. The state where the foreign atom reaches the maximum potential energy is called the **transition state** and its potential energy is labelled $E_{\ddagger}$. $\Delta E = E_{\ddagger} - E_{eq}$ is then called the **activation energy**, since this energy difference must be provided to the atom in order for it to be able to cross into a neighbouring equilibrium site. How can particles ever pass the energy barrier when the transition-state energy exceeds the total energy of the diffusing particle? The atom jumps successfully when its kinetic energy is higher than the activation energy at the instant it attempts to jump. For atoms at a given temperature we determine the fraction of foreign atoms that have enough kinetic energy to pass the barrier, e.g., using the results of the kinetic gas theory in Chapter 8. In almost all cases these will be a small fraction of the particles present in the matrix. This fraction increases with temperature; thus, more particles have enough energy to overcome the activation energy barrier if the system is at a higher temperature. We expect therefore that diffusion coefficients are larger at higher temperatures.

Based on the model in Fig. 11.13, Arrhenius predicted the following temperature dependence of the diffusion coefficient:

$$D = D_0 e^{-\Delta E/(kT)} = D_0 \exp\left(-\frac{\Delta E}{kT}\right), \qquad [11.4]$$

in which k is the Boltzmann constant and D_0 is called a pre-exponential factor with unit m^2/s. The letter e indicates the exponential function, which is also often written as $\exp(\cdots)$. Thus, the diffusion coefficient depends on the temperature in a non-linear fashion.

Arrhenius thought about how to best represent experimental diffusion data in graphic form. Due to the exponential function on the right-hand side of Eq. [11.4], an attempt to plot diffusion data with the temperature T as the abscissa (x-axis) and the corresponding diffusion coefficient D as the ordinate (y-axis) results in steep curves that would be useless to a reader who wants to use the data for conceptual arguments or for further quantitative analysis.

A first step in improving the usefulness of the graphs is to follow the arguments made in the section on Graph Analysis Methods, which is a Math Review located after Chapter 27. There we outline that a logarithmic plot is most suitable for data that follow an exponential function. Following the outlined procedure we take the logarithm on both sides of Eq. [11.4]:

$$\ln(D) = \ln(D_0) - \frac{\Delta E}{kT}. \qquad [11.5]$$

When writing Arrhenius formula in this form, we refer to it as **linearized**. This term stems from the fact that we obtain a straight-line graph if we plot $\ln(D)$ on the ordinate and $1/T$ on the abscissa because Eq. [11.5] follows the general form of a linear equation; i.e., it is of the form $y = a + bx$, with $y = \ln(D)$ the dependent variable and $x = 1/T$ the independent variable. In Eq. [11.5], $\ln(D_0)$ is the constant offset term a and $\Delta E/k$ is the slope b. In honour of Arrhenius work, we call such a plot an **Arrhenius plot**. If you compare with the section Graph Analysis Methods in the Math Review further, you note that we can alternatively use a logarithmic scale for the ordinate and then plot the diffusion coefficient D directly. We follow this approach in Fig. 11.14 for the diffusion coefficients in silicon. Note that the abscissa in that figure is $1000/T$, not $1/T$.

Fig. 11.14 tests Arrhenius model because it displays the diffusion data for a wide range of elements in silicon. If Arrhenius model and his prediction in Eq. [11.4] hold, all data must follow straight lines in the figure. Close inspection shows that this is indeed the case. The crystalline silicon matrix is chosen for this test because by far the best-established diffusion data exist for that system. Note that the figure also shows the temperature as a scale above the panel.

Once we are convinced that Arrhenius model correctly describes diffusion, further conclusions can be drawn from Fig. 11.14. We see that copper has the highest diffusion coefficient of the displayed elements: it is the fastest-diffusing element in silicon. In turn, silicon self-diffusion has the smallest shown diffusion coefficient; i.e., silicon diffuses more slowly than any other element in silicon. The wide range of diffusion coefficients in the silicon system is due partially to the existence of two different diffusion mechanisms. Fast diffusers like copper move in the manner indicated in Fig. 11.13, which is called **interstitial diffusion**. Particles that diffuse more slowly, i.e., those located below the dashed line in Fig. 11.14, diffuse **substitutionally**: they move from lattice site to lattice site displacing silicon atoms. Arrhenius model applies to both types of diffusion.

Eq. [11.4] allows us also to distinguish two contributing factors in the diffusion coefficient: the diffusion coefficient decreases as the ratio of the activation energy, ΔE, to the term kT increases; kT is an energy term related to the temperature of the matrix. Therefore, diffusion coefficients become larger at higher temperatures.

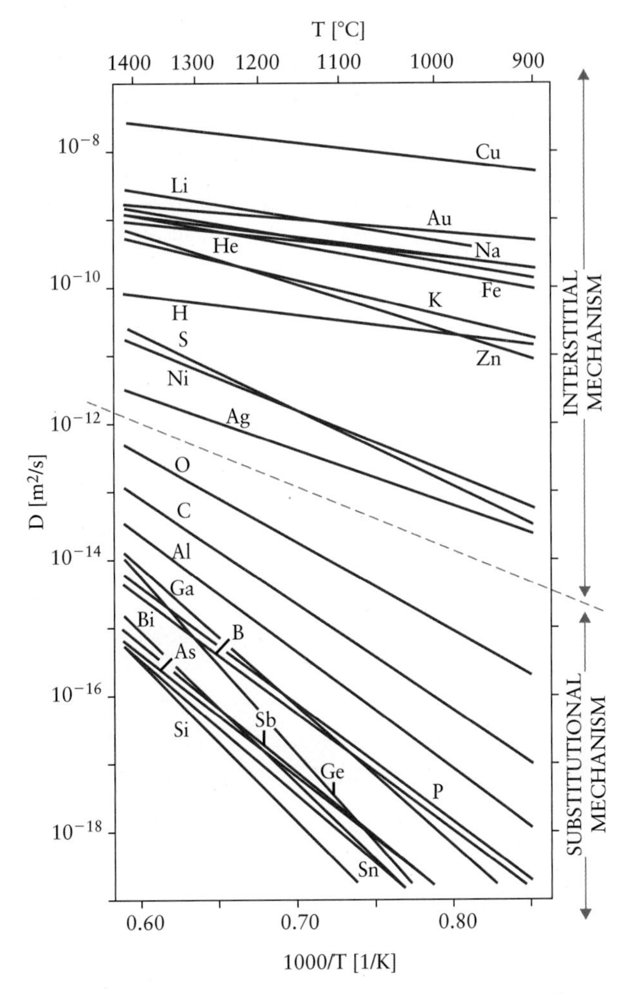

Figure 11.14 Arrhenius plot of diffusion coefficients (in unit m²/s) for various elements in silicon. Temperatures are given in unit °C at the top and inversely as 1000/*T* in unit 1/K at the bottom. Elements above the dashed line follow an interstitial diffusion mechanism, and elements below a substitutional diffusion mechanism. Both mechanisms are consistent with Arrhenius equation for the diffusion coefficient as a function of temperature.

On the other side, the diffusion coefficient also contains the pre-exponential factor D_0: D_0 is associated with the frequency of attempts by the particle to overcome the activation energy barrier and is therefore related to its vibration frequency.

CONCEPT QUESTION 11.3

In Arrhenius model, the activation energy depends on

(A) the temperature.

(B) the pre-exponential factor D_0.

(C) the concentration of the diffusing atoms.

(D) the concentration gradient (or concentration step) of the diffusing atoms across the matrix.

(E) none of the above.

EXAMPLE 11.5

We want to determine from Fig. 11.14 the activation energy ΔE and the pre-exponential factor D_0 for (a) the diffusion of germanium (Ge) in silicon, and (b) the diffusion of copper (Cu) in silicon. *Hint:* If you find the figure too small to read two pairs of ordinate and abscissa values, ln(*D*) and 1/*T*, for each of the two elements, use the values given in Table 11.6 for Ge and in Table 11.7 for Cu.

TABLE 11.6

Parameter sets for Ge diffusion in Si

ln (*D*)	1/*T* (1/K)
−44.74	8.0×10^{-4}
−32.80	6.0×10^{-4}

TABLE 11.7

Parameter sets for Cu diffusion in Si

ln (*D*)	1/*T* (1/K)
−19.17	8.5×10^{-4}
−17.71	6.0×10^{-4}

Solution

The approach to analyzing logarithmic plots is discussed in detail in the section on Graph Analysis Methods, which is a Math Review located following Chapter 27. The simplest way to analyze a logarithmic plot such as Fig. 11.14 is to first replace the *D*-axis (ordinate) by a linear ln(*D*)-axis. This is achieved by replacing each value along the axis by the logarithm of that number; e.g., the value 10^{-14} m²/s is assigned a new value of −32.24 and 10^{-8} m²/s is assigned a new value of −18.42. The new ordinate is linear with an increment of 2.303 for each decade of the original axis.

Solution to part (a): From this modified plot, two pairs of ordinate and abscissa are selected in Table 11.6 to determine the unknown parameters in Eq. [11.5]. We then substitute each data pair from the table into Eq. [11.5]:

$$\text{(I)}\quad -44.74 = \ln(D_0) - \frac{\Delta E}{k} 8.0 \times 10^{-4}$$

$$\text{(II)}\quad -32.80 = \ln(D_0) - \frac{\Delta E}{k} 6.0 \times 10^{-4}.$$

The two formulas are subtracted from each other to solve for the unknown variable $\Delta E/k$. We find:

$$\frac{\Delta E}{k} = 59\ 700\ \text{K}.$$

The activation energy is isolated by multiplying by the Boltzmann constant *k*:

$$\Delta E = 8.24 \times 10^{-19}\ \text{J}.$$

Activation energies for diffusion are often reported in another energy unit, the electron volt (eV). The conversion is 1 eV = 1.6×10^{-19} J (for the physical motivation of

continued

this unit, see section 18.3. With this conversion we find that the activation energy for diffusion of Ge in Si can be expressed as 5.1 eV. In comparison, the average thermal energy of a particle at room temperature is $kT = 0.025$ eV, i.e., 0.5% of the energy needed for a Ge atom to overcome the diffusion barrier between neighbouring sites in silicon.

A value for $\ln(D_0)$ is obtained by substituting $\Delta E = 8.24 \times 10^{-19}$ J into either one of the two formulas we derived from Table 11.7. This leads to $D_0 = 2 \times 10^1$ m^2/s. The pre-exponential factor is discussed further in Example 11.6.

Solution to part (b): This part is solved in an analogous manner to part (a). Table 11.7 shows the data pairs chosen from the curve labelled copper (Cu). The data pairs in Table 11.7 are again substituted into Eq. [11.5]:

$$\text{(I)} \quad -19.17 = \ln(D_0) - \frac{\Delta E}{k} 8.5 \times 10^{-4}$$

$$\text{(II)} \quad -17.71 = \ln(D_0) - \frac{\Delta E}{k} 6.0 \times 10^{-4}.$$

This leads to $\Delta E = 8.06 \times 10^{-20}$ J = 0.5 eV. This is an energy barrier that is an order of magnitude smaller than the energy barrier in the case of germanium diffusion. We further find $D_0 = 6.8 \times 10^{-7}$ m^2/s.

11.4.3: Diffusion Length

Albert Einstein further developed the model laid out by Arrhenius, considering particles that make not one but many jumps. For each jump a particle selects randomly from its neighbouring equilibrium sites. This means, for example, that two consecutive jumps can bring the atom back to its initial position. Thus, the atom does not move a total distance of Na after N jumps of length a. From a statistical analysis of this problem, Einstein found instead a non-linear formula for the **diffusion length**. The diffusion length Λ at a given temperature is defined as the average distance a diffusing particle moves during a time period t (Einstein's formula for the diffusion length):

$$\Lambda = \sqrt{2Dt}, \qquad [11.6]$$

in which D is the diffusion coefficient.

CONCEPT QUESTION 11.4

If we double the membrane width of the membrane between the alveolar air space and an adjacent capillary from 2 μm to 4 μm, oxygen diffusion across the membrane at body temperature of 37°C will require

(A) the same time.

(B) half the previous time.

(C) double the previous time.

(D) one-quarter of the previous time.

(E) four times as long.

EXAMPLE 11.6

Fig. 11.15 shows the alveolar sacs at the end of the bronchial tree. The alveoli are in direct contact with blood capillaries. In the alveoli, the inhaled air gets in closest contact to the red blood cells, which carry oxygen. Fig. 11.16 highlights the transport process barrier between the gas space of an individual alveolus and the adjacent blood capillary. The narrow capillary forces the erythrocyte to deform in order to tightly squeeze through. The membrane width is 1–2 μm. An erythrocyte passes through the contact zone in about 0.75 s. (a) How long do oxygen molecules need to diffuse from the gas phase to a passing erythrocyte? (b) How thick may the membrane tissue between alveolus and capillary become before the oxygen transfer is significantly reduced? This occurs when the oxygen cannot diffuse to the erythrocyte during the time period an erythrocyte passes the alveolus.

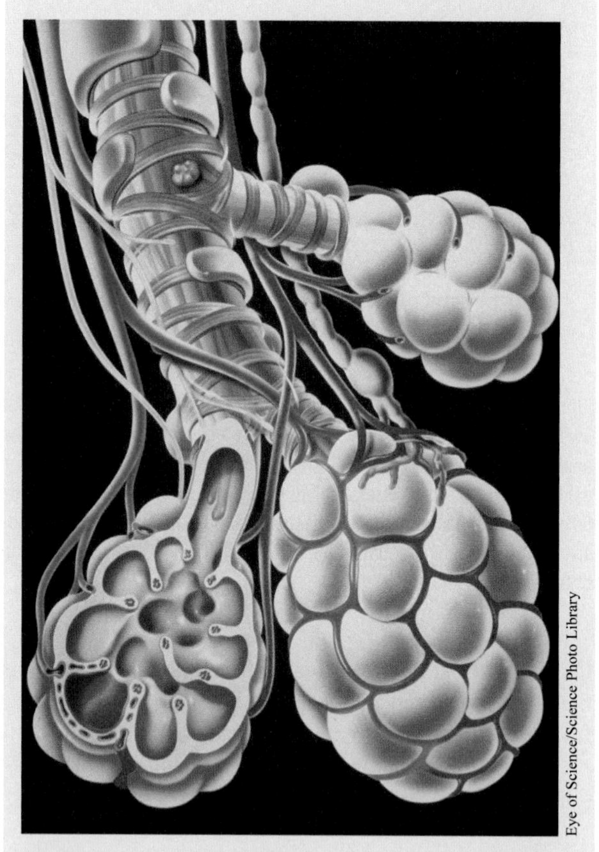

Figure 11.15 At the end of each bronchial branch in the lungs, small sacs of about 0.3 mm diameter form an interface between the inhaled air and blood capillaries. These sacs are called alveoli. Oxygen transfer from air to blood occurs across the alveolar membrane.

Solution

Solution to part (a): We use Eq. [11.6] with the diffusion coefficient for oxygen in tissue taken from Table 11.5. The diffusion length is the maximum distance in healthy

continued

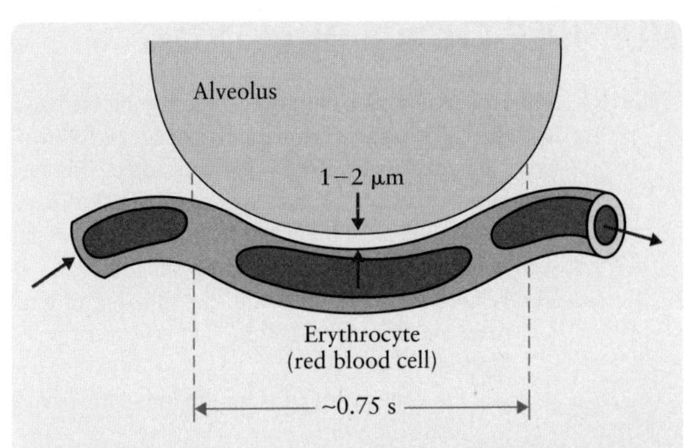

Figure 11.16 Sketch of a blood capillary in contact with an alveolus across a 1- to 2-μm-wide membrane. An erythrocyte, which is about 7.5 μm wide and 1 μm to 2 μm thick, passes through the contact area in 0.75 s. A healthy red blood cell can deform to squeeze through the capillary because it does not contain a nucleus.

tissue, which is 2 μm. Eq. [11.6] is rewritten to isolate the time for diffusion:

$$t = \frac{\Lambda^2}{2D} = \frac{(2\times10^{-6}\text{m})^2}{2\cdot1\times10^{-11}\,\frac{\text{m}^2}{\text{s}}} = 0.2\text{ s.}$$

Oxygen diffusion occurs without a problem across the membrane between alveolus and blood capillary in the allotted time.

Solution to part (b): We use the same diffusion coefficient as before and $t = 0.75$ s. This time, Eq. [11.6] is used as written since the diffusion length is sought:

$$\begin{aligned}\Lambda &= \sqrt{2Dt} \\ &= \sqrt{2\left(1\times10^{-11}\,\frac{\text{m}^2}{\text{s}}\right)(0.75\text{ s})} \\ &= 3.9\times10^{-6}\text{m.}\end{aligned}$$

Doubling of the membrane width between alveolus and blood capillary from 2 μm to 4 μm already shifts the diffusion barrier into a physiologically dangerous range. The medical term for alveolar membrane thickness increase is *pneumonosis.* It is usually caused by atypical pneumonia (which is caused by viruses) or pulmonary congestion. The resulting drop in the oxygen concentration in the arterial blood is referred to as *hypoxemia.*

(a)

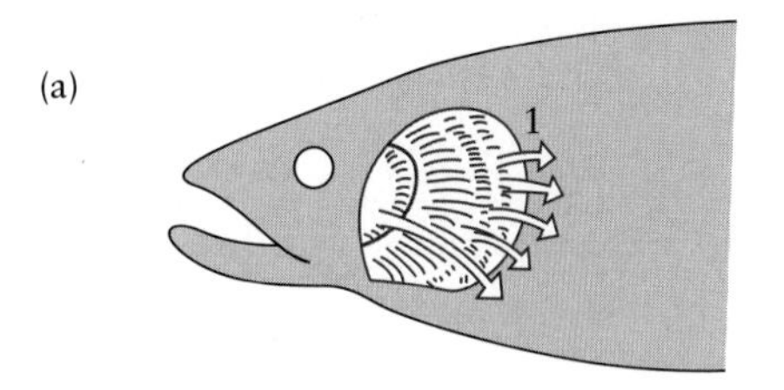

(b)

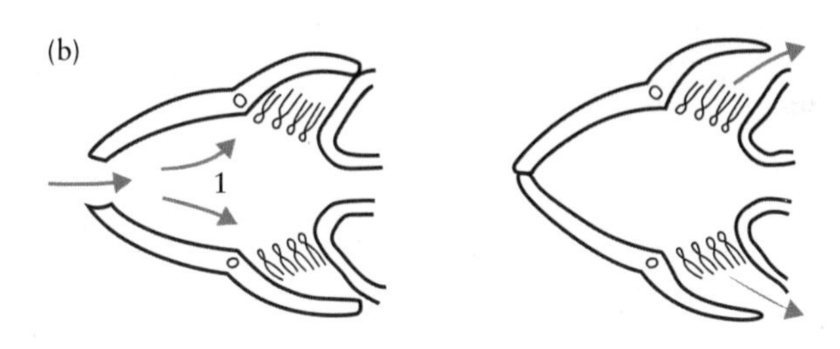

(c)

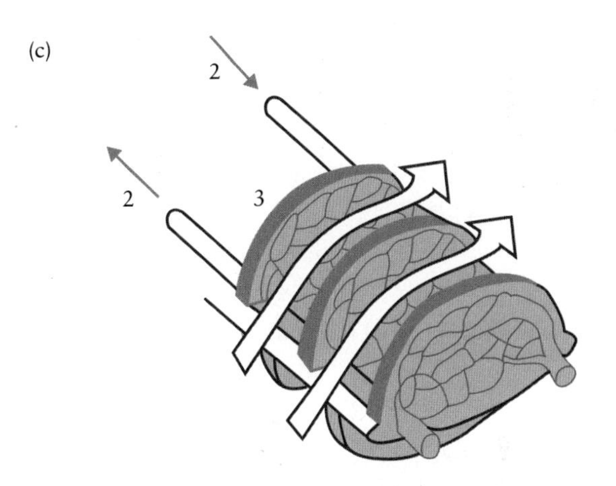

Figure 11.17 Illustration of the function of fish gills. Fish require a continuous flow of water over their gill arches (1). These arches contain blood vessels (2) that branch into capillaries absorbing oxygen from the water in the lamellae (3) of the gill filaments. To maximize the oxygen harvesting, blood flow and water flow are arranged in a countercurrent pattern, with oxygen-poor blood entering the lamellae upstream of the water flow and oxygen-rich blood leaving the lamellae downstream.

We compare the respiratory effort of fish, as illustrated in Fig. 11.17, and that of humans, as discussed in Example 11.6. Breathing in air has the obvious advantage of access to a more abundant source of oxygen. Water in most fish habitats contains about 4–8 mL of O_2 per litre, and each litre of air contains 210 mL O_2. Because oxygen and CO_2 diffuse much faster in air than in water, ventilation of the internal surfaces of the alveoli for us is a much lesser concern. Indeed, a resting fish must actively pump water through its gills to not suffer from oxygen deprivation.

However, a problem with air breathing had to be addressed before life on land became possible: the humidity of air changes often, which would lead to varying degrees of water vapour inhalation. The inner surfaces of the alveoli would change (dry versus moist), causing significant variations in the physiological effectiveness of the respiration process. As a solution, the respiratory system of land-living animals always operates with completely water-saturated air. The humidity of the air is controlled by a turbulent flow of the inhaled air through the trachea, which has a moist inner surface. In this process moisture is picked up to the saturation level before the air reaches the lungs.

SUMMARY

DEFINITIONS

- Amount in mol and concentration of the i-th component in solution:

$$n_i = \frac{m_i}{M}$$

$$c_i = \frac{\rho_i}{M}$$

UNITS

- Molar mass M: kg/mol
- Amount of matter n: mol
- Concentration c: mol/m^3
- Density ρ: kg/m^3

LAWS

- Fourier's law of steady-state heat conduction:

$$\frac{Q}{t} = \lambda A \frac{T_{high} - T_{low}}{l}$$

where λ in unit J/(msK) is the thermal conductivity coefficient, l is the length of the membrane, and A is its cross-sectional area.

- Fick's law of steady-state diffusion for the transport of
- an amount of material in unit kg:

$$\frac{m_i}{t} = DA \frac{\rho_{i,high} - \rho_{i,low}}{l},$$

where D in unit m^2/s is the diffusion coefficient, and ρ is the density;

- an amount of material in unit mol:

$$j_i = \frac{n_i}{t} = DA \frac{c_{i,high} - c_{i,low}}{l},$$

where j_i in unit mol/s is the material flux of the i-th component.

- Temperature dependence of the diffusion coefficient:

$$D = D_0 e^{-\Delta E/(kT)} = D_0 \exp\left(-\frac{\Delta E}{kT}\right),$$

where k is the Boltzmann constant and ΔE is the activation energy. This is equivalent to:

$$\ln(D) = \ln(D_0) - \frac{\Delta E}{kT}$$

- Einstein's formula for the diffusion length Λ:

$$\Lambda = \sqrt{2Dt}.$$

MULTIPLE-CHOICE QUESTIONS

MC–11.1. Some admission tests for professional schools are structured such that a leadoff paragraph (passage) is followed by several questions related to the same topic. This concept question serves as an example. The leadoff paragraph would be an abbreviated form of the text in this section to this point, then three questions follow:

(A) What did Fourier measure as a function of temperature difference, length, and cross-section of the rod?
- (a) the heat removed from the low-temperature heat reservoir
- (b) the heat transfer through the rod per second
- (c) the heat deposited in the high-temperature heat reservoir
- (d) the work done on the low-temperature heat reservoir by dropping ice pellets into it
- (e) the internal energy of the water in the low-temperature heat reservoir

(B) In order to increase the effect Fourier observed, he could have done which of the following?
- (a) increase the length of the rod
- (b) decrease the diameter of the rod (assumed to be of cylindrical shape)
- (c) increase the temperature of the low-temperature heat reservoir
- (d) decrease the length of the rod
- (e) decrease the temperature of the high-temperature heat reservoir

(C) When Fourier changed the rod from aluminum to iron, which of the following happened?
- (a) He had to increase the temperature of the low-temperature heat reservoir to obtain the same heat flow rate as before.
- (b) He had to increase the temperature of the high-temperature heat reservoir to obtain the same heat flow rate as before.
- (c) His observations did not change.
- (d) He could not do the experiment because it doesn't work with iron rods.
- (e) He had to lengthen the rod to obtain the same heat flow rate as before. *Hint:* Use Table 11.1.

MC–11.2. The material of the rod in Fourier's experiment is changed such that its thermal conductivity decreases by 20%. What change allows us to best re-establish the previous flow rate of heat (Q/t)?
- (a) decreasing the length of the rod by 10%
- (b) increasing the length of the rod by 10%
- (c) increasing the diameter of the rod by more than 20%
- (d) increasing the diameter of the rod by 20%
- (e) increasing the diameter of the rod by less than 20%

MC–11.3. Use the thermal conductivity coefficient for window glass from Table 11.1, and for wood use λ_{wood} = 0.08 J/(msK). What happens if you replace a broken window by a wood panel of one-quarter the thickness of the glass?
- (a) The window opening with the wood panel is thermally, better insulated.

(b) The window opening with the wood panel is about as well insulated as before.
(c) The window opening with the wood panel allows an increased amount of heat to escape the room.
(d) The stated data do not allow us to draw any of the previous three conclusions.

MC–11.4. When studying the heat loss of a sphere of radius R we use which of the following terms for the surface area A through which heat flows in Fourier's law?
(a) $A = \pi R$
(b) $A = \pi R^2$
(c) $A = 4 \pi R^2$
(d) $A = 4 \pi R^3$
(e) $A = 4 \pi R^3/3$

MC–11.5. Which of these changes in Fourier's experiment causes a doubling of the flow rate of heat?
(a) doubling the diameter of the rod between the two heat reservoirs
(b) doubling the length of the rod connecting the two heat reservoirs
(c) cutting in half the length of the rod connecting the two heat reservoirs
(d) doubling the temperature of the high-temperature heat reservoir
(e) doubling the temperature of the low-temperature heat reservoir

MC–11.6. The geothermal effect states that the temperature below Earth's surface increases by one degree Celsius for every 30 metres in depth. Assuming a surface temperature of 0°C, which of the following statements is true? The temperature rises to 1000 K at what depth?
(a) 8 km
(b) 15 km
(c) 22 km
(d) 30 km
(e) 100 km

MC–11.7. Heat conduction and diffusion are called transport phenomena. What exactly is transported in heat conduction?
(a) temperature
(b) thermal energy
(c) heat
(d) internal energy
(e) entropy

MC–11.8. Fig. 11.14 shows the temperature dependence of the diffusion coefficient for various elements in silicon. Which of the following statements is consistent with that data?
(a) The activation energy for copper (Cu) is smaller than the activation energy for aluminum (Al).
(b) The diffusion coefficient is linearly proportional to the temperature.
(c) The diffusion coefficient is inversely proportional to the temperature.
(d) The diffusion coefficient is a materials-independent constant.
(e) The activation energies for elements at the upper end of the graph are higher than for those at the lower end of the graph.

MC–11.9. Use the diffusion coefficients for sucrose and tobacco mosaic virus in water from Table 11.5. To obtain the same rate of mass transfer for equal concentrations of sucrose and the virus on both sides of a water-filled cylindrical tube of fixed tube radius, the tube length has to be reduced for the virus by a factor of (choose the closest value)
(a) 3.2.
(b) 10.
(c) 32.
(d) 100.
(e) 320.

MC–11.10. When an erythrocyte passes through a blood capillary adjacent to an alveolus, oxygen diffusion occurs because
(a) a small temperature difference is established between both sides of the membrane.
(b) the moving erythrocyte causes a drag effect, pulling the oxygen through the membrane.
(c) a concentration step for oxygen is established with the lower concentration at the erythrocyte's side.
(d) there is no concentration gradient, but oxygen moves randomly in the membrane and ends up in the erythrocyte by chance.

MC–11.11. In two separate experiments the following two diffusion coefficients of a contaminant in tissue are found: experiment I: $D = 7 \times 10^{-11}$ m²/s, and experiment II: $D = 7 \times 10^{-9}$ cm²/s. Your laboratory head suggests further experiments to check whether the two contaminants are the same. What do you do?
(a) Conduct the suggested experiments because the two diffusion coefficients are essentially the same.
(b) Repeat the previous experiments to see whether the new data are still so close to each other.
(c) Ask another researcher to confirm the group head's conclusion because it isn't that easy to compare the two given diffusion coefficients.
(d) Reject the laboratory head's suggestion and proceed with the conclusion that the two contaminants are different.

MC–11.12. If time is the dependent variable, in which of the following forms is Einstein's diffusion equation written correctly?

(a) $t = \dfrac{\Lambda^2}{2D}$ (b) $t = \dfrac{\Lambda}{2D^2}$

(c) $t = \sqrt{\dfrac{\Lambda^2}{2D}}$ (d) $t = \sqrt{\dfrac{\Lambda}{2D^2}}$

(e) $t = \dfrac{\Lambda}{2D}$ (f) $t = \left(\dfrac{\Lambda}{2D}\right)^{1/2}$

MC–11.13. The diffusion length of a molecule in a solution is $\Lambda = 1$ cm after the experiment is run for $t = 1$ hour. When will the diffusion length double?
(a) after an additional hour
(b) after an additional 2 hours
(c) after an additional 3 hours
(d) after an additional 4 hours

MC–11.14. The diffusion length for a given system does *not* depend on which of the following parameters?
(a) the temperature
(b) the pre-exponential factor D_0 of the diffusion system
(c) the diffusion coefficient
(d) the thickness of the sample in which the diffusion is observed
(e) the time duration of the experiment

CONCEPTUAL QUESTIONS

Q–11.1. A tile floor may feel uncomfortably cold to your bare feet, but a carpeted floor feels warm. Why?

Q–11.2. The column of mercury in a thermometer initially descends slightly before rising when the instrument is placed in a hot liquid. Why?

Q–11.3. Summers in sunny southern California can be very hot and very dry. If one were to jump in an outdoor swimming pool, one would realize that the water would be too warm to induce any refreshment. However, as soon as one gets out of the pool and lies in the sun to dry off, one would quickly feel cold for a while and would have goosebumps even on a very hot day indeed. How can one explain this phenomenon?

Q–11.4. The metallic lid of a glass jar is stuck. Given the fact that the coefficient of thermal expansion of the glass is less than that of the metal lid, explain what would happen if the jar and the lid are heated.

Q–11.5. A chair with a metal base and a wooden desk are located in a room at temperature T. What would you feel if you were to touch the metal base of the chair and the surface of the desk?

ANALYTICAL PROBLEMS

P–11.1. We quantify Fourier's experiment, shown in Fig. 11.3, for a cylindrical copper rod of a length 1.2 m and a cross-sectional area 4.8 cm^2. The rod is insulated to prevent heat loss through its surface. A temperature difference of 100 K is maintained between the ends. Find the rate at which heat is conducted through the rod.

P–11.2. We focus on the geothermal effect in a slightly different manner than in Example 11.3. We know that the average rate at which heat is conducted through the surface of the ground in North America is 55 mJ/(s m^2). Assuming a surface temperature in summer of 25°C, what is the temperature at a depth of 30 km (near the base of Earth's crust, which together with the brittle upper portion of the mantle forms the 100-km-deep lithosphere)? *Hint:* Ignore the heat generated by the presence of radioactive elements and use 1.7 J/(m s K) for the average thermal conductivity of the near-surface rocks. Start with Fourier's law.

P–11.3. For poor heat conductors the thermal resistance R has been introduced. The thermal resistance of a piece of material of thermal conductivity λ and thickness l is defined as:

$$R = \frac{l}{\lambda}$$

(a) Show that this equation allows us to rewrite Fourier's law in the form:

$$\frac{Q}{t} = A\frac{T_{high} - T_{low}}{R},$$

in which A is the cross-sectional area of the piece of material. (b) What is the SI unit of the thermal resistance R?

P–11.4. In a table you find for the thermal conductivity of quartz glass at 0°C a value of $\lambda = 3.4 \times 10^{-3}$ cal/(cm s K). (a) Is this value consistent with the value given in Table 11.1? (b) What is the thermal resistance of a quartz glass sheet of thickness 1.0 cm?

P–11.5. In a cookbook, a poultry thawing chart states that a 10-kg whole turkey takes four days to defrost in a refrigerator. Estimate how long it would take to defrost a 2-tonne Siberian mammoth from the same initial temperature in an industrial refrigeration hall. *Hint:* Treat both animals as spherically shaped and use the same approach we applied in Examples 11.2 and 11.3.

P–11.6. We want to measure the thermal conductivity of an unknown insulator material. For this we use the following set-up: A 4-mm-thick plate of the unknown material is placed between two iron plates of thickness 3 cm each. All three plates are 15 cm by 15 cm in size. The upper iron plate is heated to 350 K and the lower iron plate is kept at 290 K. Once a stationary temperature profile has developed across the insulator, the heater is removed from the upper iron plate. We observe that the temperature of the upper iron plate drops by 2.5 K after 100 seconds. Neglecting any loss of heat to the environment, what is the thermal conductivity coefficient l for the unknown insulator material? *Hint:* Approach this problem in the same manner as we solved Example 11.3. The density of iron is $\rho = 7.9$ g/cm^3 and its specific heat capacity is given in Table 9.1.

P–11.7. Fig. 11.18 shows a block that consists of two materials with different thicknesses l_1 and l_2 and different thermal conductivities λ_1 and λ_2. The temperatures of the outer surfaces of the block are T_{high} and T_{low}, as shown in the figure. Each face of the block has a cross-sectional area A. (a) Show that the formula:

$$\frac{Q}{t} = \frac{A(T_{high} - T_{low})}{(l_1/\lambda_1)+(l_2/\lambda_2)}$$

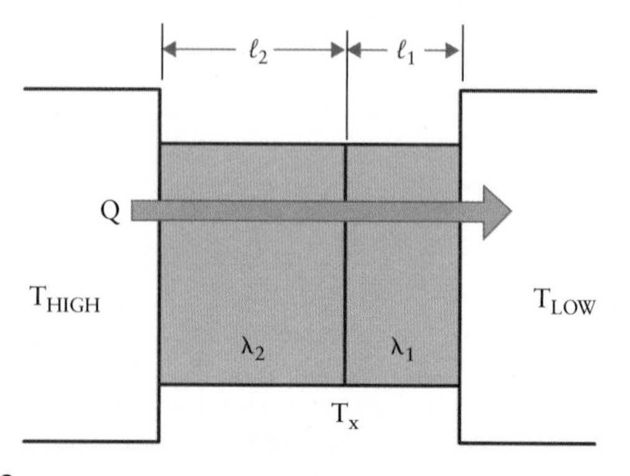

Figure 11.18

correctly expresses the steady-state rate of heat transfer. *Hint:* In the steady state, the heat transfer through any part of the block must be equal to the heat transfer through the other part of the block. Introduce a temperature T_x at the interface of the two parts as shown in Fig. 11.18, and then express the rate of heat transfer for each part of the block separately. (b) Rewrite the above equation using the thermal resistance from P–11.3, which introduces R_1 and R_2 as the thermal resistances for the two parts of the block. By comparing the result with the second equation given in P–11.3, determine how thermal resistances are combined for materials in sequence.

P–11.8. Show that the temperature T_x at the interface of the block in Fig. 11.19 is given by:

$$T_x = \frac{R_1 T_{\text{high}} + R_2 T_{\text{low}}}{R_1 + R_2}.$$

P–11.9. We want to compare combinations of insulator materials in the case shown in Fig. 11.19. The figure shows a two-layer system with a total of four square pieces. Each piece has an area A, thus the two systems each cover an area of 2 A. The two arrangements in Fig. 11.19 differ in the order of the two materials (labelled 1 and 2). Which arrangement, (a) or (b), allows for a greater heat flow?

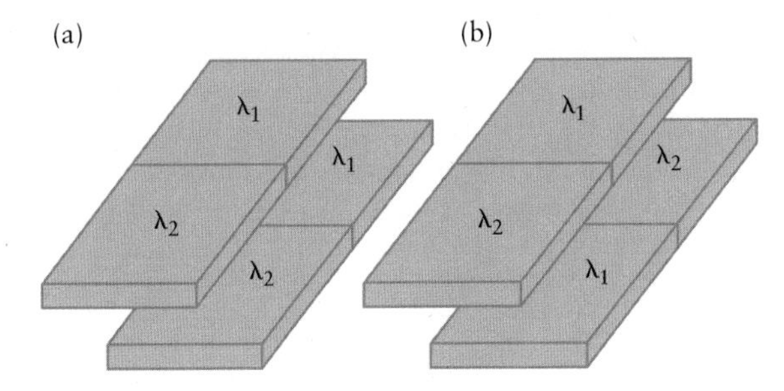

Figure 11.19

P–11.10. Two identical rectangular rods of metal are welded together as shown in Fig. 11.20(a), and 1 J of heat is conducted in a steady-state process through the combined rod in 1 minute. How long would it take for the same amount of heat to be conducted through the rods if they were welded end to end, as shown in Fig. 11.20(b)?

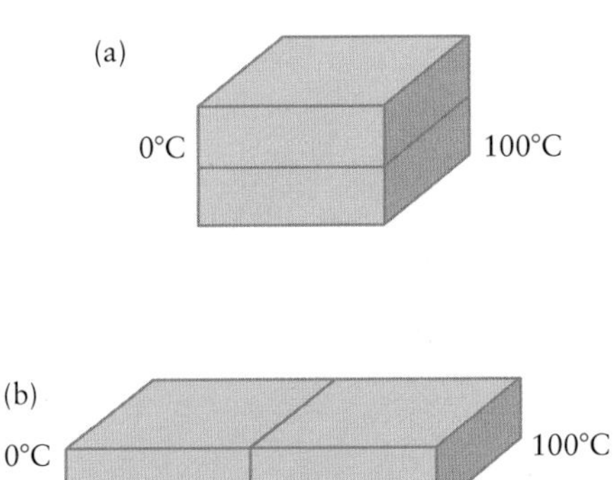

Figure 11.20

P–11.11. In the text we discussed the Australian brush-turkey's approach to nesting. We noted that it must maintain a radius R of a decomposing forest litter mound in which its eggs incubate according to $T \propto \sqrt{T_{\text{incubation}} - T_{\text{ambient}}}$. Confirm that this equation properly describes the relation between mound radius and its temperature profile.

P–11.12. In Concept Question 11.2, we used dimensional analysis to estimate the thermal relaxation time for a spherical bacterium in a fatty fluid. We derived an equation based on these arguments. Confirm this equation based on the total amount of heat needed to bring the bacterium up to ambient temperature using a reasonable estimate of the temperature gradient in the bacterium during the process.

P–11.13. Heat loss via convection occurs only when heat is carried by a moving fluid. For example, when heating water in a beaker from below, the increase of the water temperature at the bottom leads to a decrease of the water density and causes the warmer water to rise due to buoyancy. The rising water carries excess heat to the surface. (a) Compare bare skin to skin covered with clothes. Why is the heat loss of the body significantly reduced when wearing clothes? (b) At temperate lakes and ponds it is often observed that algae bloom for a short period during spring and autumn. Consider Fig. 11.21, which shows the stratification during summer (top) as well as the seasonal turnover in spring and autumn (bottom). How can the convection-driven turnover cause algal blooms?

P–11.14. A Styrofoam box with surface area 0.8 m^2 and wall thickness of 20 mm has a temperature of +5°C on its inner surface and +25°C on the outer surface. Calculate the thermal conductivity of Styrofoam if it takes 480 minutes for a 5.0-kg piece of ice to melt in the box. The latent heat of freezing of water is 3.33×10^5 J/kg.

P–11.15. Determine the diffusion coefficient for glycerine in H_2O using the following observations: glycerine diffuses along a horizontal, water-filled column that has a cross-sectional area of 2.0 cm^2. The density step across $\Delta\rho / l$ is 3.0×10^{-2} kg/m^4 and the steady-state diffusion rate is 5.7×10^{-15} kg/s.

P–11.16. We want to test a statement we made in Chapter 8: carbon dioxide diffuses more easily than oxygen across the membrane between the alveoli and the blood capillaries. To show this, calculate the ratio of the diffusion coefficients of CO_2 and O_2 in tissue at 37°C. *Hint:* Start with Eq. [11.3]. Rewrite the concentration difference as a pressure difference using the ideal gas law. Apply this equation for both gases separately. For the pressure differences across the membranes in the lungs use the values $\Delta p(CO_2) = 0.8$ kPa and $\Delta p(O_2) = 8.0$ kPa. The amount of both gases diffusing across the interface alveoli/capillaries can be determined from the data given in Example 8.4.

P–11.17. Why can bacteria rely on passive diffusion for their oxygen supply but human beings cannot? *Hint:* Using Eq. [11.6], calculate (a) the time it takes for oxygen to diffuse from the interface with the environment to the centre of a bacterium of radius $r = 1.0$ μm, and

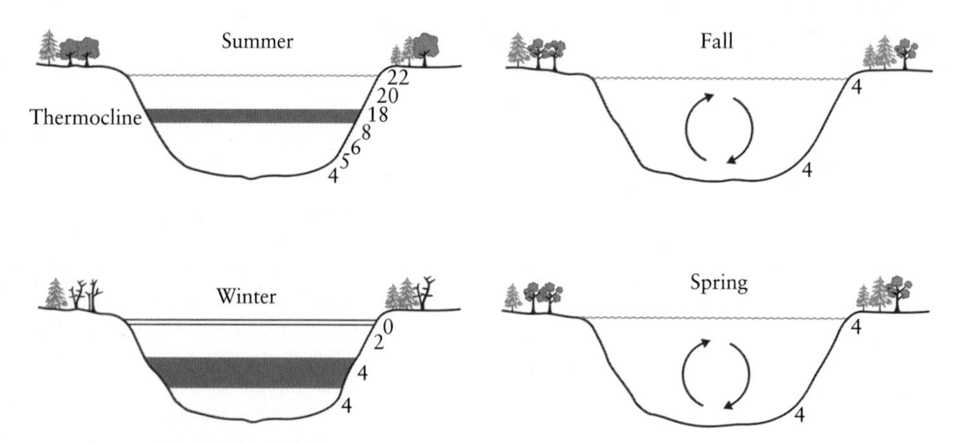

Figure 11.21 Lakes and ponds in temperate climates usually stratify by temperature and water density in winter and summer. In the summer, the warmer water is above the thermocline (water zone between 8°C and 18°C), with temperatures below the thermocline near 4°C (temperature of water at its highest density). In the winter, the water above the thermocline is cooler (between 0°C and 4°C) than the water below. The seasonal turnover occurs biannually as denser water sinks to the bottom of the lake.

(b) the time it takes for oxygen to diffuse from the external air to an organ 10 cm below human skin. *Note:* For an upper limit use the diffusion coefficient of oxygen in water, and for a lower limit use the diffusion coefficient of oxygen in tissue from Table 11.5. These two values give you a good approximation since humans consist of roughly 10 L extracellular fluid and 30 L cells. (c) *If you are interested:* Why can many relatively large invertebrates such as hydras survive without a cardiovascular system?

P–11.18. We want to determine the relation between the diffusion coefficient and the molecular mass of macromolecules. Use a double-logarithmic plot of the data listed in Table 11.8 to determine the coefficients a and b in:

$$D = aM^b.$$

TABLE 11.8

Diffusion coefficients in aqueous solution at 20°C and molecular mass for various biomolecules and viruses

Protein	D (m²/s)	M (kg/mol)
Tobacco mosaic virus	5.3×10^{-12}	31 000
Urease	3.5×10^{-11}	470
Catalase	4.1×10^{-11}	250
Hemoglobin	6.3×10^{-11}	67
Insulin	8.2×10^{-11}	41

P–11.19. (a) How far does a tobacco mosaic virus move in water at 20°C in 1 hour? (b) Using the ratio of the diffusion coefficients for oxygen and carbon dioxide in tissue from P–11.16, what is the ratio of diffusion lengths for these molecules in tissue at 20°C?

P–11.20. Fig. 11.22 shows a block that consists of two materials with different thicknesses, l_1 and l_2, and different diffusion coefficients for oxygen, D_1 and D_2. The oxygen concentrations at the outer surfaces of the block are $c_{Ox,1}$ and $c_{Ox,2}$, as shown in the figure. Each face of the block has a cross-sectional area A. Determine a formula for the steady-state mass transport through the block in analogy to P–11.7.

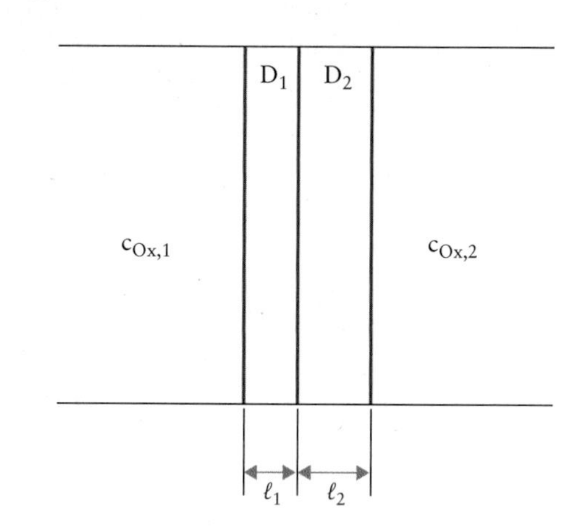

Figure 11.22

ANSWERS TO CONCEPT QUESTIONS

Concept Question 11.1: (C)

Concept Question 11.2: (E)

Concept Question 11.3: (E). The activation energy measures the energy barrier for a single hop of a diffusing particle from one possible site to an adjacent one. The temperature plays a role as part of the thermal energy available to the particle to overcome the barrier, but the barrier itself is predetermined by the arrangement of the matrix atoms.

Concept Question 11.4: (E) based on Eq. [11.6].

Dorling Kindersley/Getty Images

CHAPTER 12

Static Fluids

The term *fluid* includes liquids and gases. Fluids are deformable, which allows them to evolve toward a mechanical equilibrium in a given space. When intermolecular forces dominate, the fluid is found in the liquid state. When the thermal energy of the system dominates, the fluid is found in the gaseous state. Liquids occupy a well-defined volume, forming a surface when the size of the accessible space exceeds the liquid's volume.

The ideal fluid model assumes the fluid is incompressible. If such a fluid is in mechanical equilibrium it is called an ideal stationary fluid. If the ideal stationary fluid is in the condensed liquid state, Pascal's law states that the pressure increases linearly with depth below the surface. The mechanical equilibrium condition requires the pressure at the liquid's surface to be equal to the pressure in the adjacent gas space.

Fluids support immersed objects with a buoyant force. This force counteracts the weight of the object and is proportional to the volume of the object and the density of the displaced fluid.

The surface of a condensed ideal stationary fluid has properties that are distinct from those of the bulk material. This is due to an excess amount of energy required to form a surface, called surface tension or surface energy. Surface energy causes a pressure difference across a bubble or droplet surface that is inversely proportional to its radius. This relation is known as Laplace's law.

Fluid surfaces facing a substance other than air or vacuum are called interfaces. Interfacial energies are conceptually similar to surface energies and are related to the wetting properties at the interface. A consequence of interfacial interactions is capillarity, which is the action by which the surface of a liquid is elevated or depressed in a tube based on its surface energy and the tube/liquid interfacial energy.

At this point we have reached an important milestone in the discussion of physics: we followed thermal physics, which Albert Einstein once identified as its most unshakeable pillar, the entire distance from simple equilibrium to complex pattern formation in living systems. We have developed many powerful concepts along the way, but we have done it with an incredibly simple model system: the ideal gas.

Now it is time to shift gears and use these tools to develop the properties of those physiological systems that play important roles in our body, e.g., liquids such as water or blood, elastic materials such as tissues, vibrating fluids such as the perilymph in the inner ear, electrically active solutions at the surfaces of nerves, and optical materials such as lenses in our eyes or rods and cones in the retina.

What will guide us in the discussions of these systems in the following chapters are the tools we have successfully developed using the ideal gas: Starting with equilibrium properties and the first law of thermodynamics, we will establish dynamic properties by seeking linear near-equilibrium phenomena first and more complex nonlinear properties last.

The current chapter is one of four chapters that focus on the most important biological system: **water**. We approach water from four different directions:

- the macroscopic properties of stationary water as an equilibrium system in the current chapter,
- the macroscopic phenomena of flowing water as a dynamic non-equilibrium system in Chapter 13,
- the microscopic structure of the water molecule as an electric dipole in Chapter 17, and
- water as a solvent and main constituent of mixed phases such as blood. We discussed this aspect already in Chapter 10; it will be useful to recapitulate that chapter after reading Chapters 12, 13, and 17.

Life on Earth began in water and remained there exclusively for more than three billion years. Even life on dry land maintains close ties to water. The importance of water is evident from its abundance throughout the human body. The fraction of our body mass made up by water is about 75% for a baby, decreases to about 60% for young adults, and is as low as 50% for seniors. Two-thirds of the water in an adult's body is located in the cells, with the remaining one-third in extracellular fluids, including blood plasma.

12.1: Model System: The Ideal Stationary Fluid

We discuss the macroscopic properties of water in this and the next chapter. However, we don't focus exclusively on water but study a more general model system that includes many of the properties of liquid water. This model system is called a **fluid**. Fluids are systems that yield to any force that attempts to alter their shape, causing the system to flow until it reaches a mechanical equilibrium in which the fluid then conforms to the shape of the container. Based on this definition, the term *fluid* refers to both liquids and gases, but distinguishes them from solids, which remain unchanged when placed in containers of different sizes.

There are differences between liquids and gases that we have to take into account. The molecules in a liquid are in a condensed state; i.e., they maintain a fixed intermolecular distance. If the liquid is brought into a container with a volume larger than the volume the liquid occupies, the liquid forms a surface. In contrast, gases adjust their intermolecular distance and fill any provided space uniformly. Thus, gases have no natural surface.

We turn our attention first to fluids in mechanical equilibrium; in that state we call the fluid **stationary**. Remember that we defined the term *equilibrium* such that it refers to the state in which all essential parameters of the system are time independent. What we mean by "essential parameters" in the current context is illustrated using a glass filled with water. Next time you drink, look at the liquid before you touch the glass. The liquid is in equilibrium: no obvious changes occur while you observe the system. However, as in a gas, a tremendous amount of motion of the particles exists at the molecular level. Thus, the equilibrium of a macroscopic fluid does not include as an essential parameter the microscopic motion of the fluid molecules. The properties essential for an **ideal stationary fluid** are these:

- *The ideal stationary fluid is incompressible.* This means that both the volume of the fluid, V, and its density, $\rho = m/V$, are constant; i.e., they particularly do not depend on pressure. This is a good approximation for liquids but does not apply to gases. For this reason we have to retain the ideal gas model as introduced in the previous chapters.
- *The ideal stationary fluid is deformable* under the influence of forces and seeks a mechanical equilibrium. Only when the mechanical equilibrium is established does the fluid become stationary. This applies equally to liquids and gases. This condition is obviously very useful because we already know a great deal about the mechanical equilibrium.

Note that we did not include a condition for the type of interactions between the fluid molecules or with the container walls. In particular, the limitation to elastic collisions we used for the ideal gas is not used, as it would exclude concepts such as surface tension and capillarity, which we introduce later in the chapter.

12.2: Pressure in an Ideal Stationary Fluid

We are well aware of pressure variations in the atmosphere. These are most notable after take-off in an airplane. Still, pressure variations in liquids are much more profound. Many diving-related accidents and the need for lifeguards in all public pools are clear signs that we are exposed to unexpectedly strong effects once we immerse our bodies in water. But how can we express this strong effect quantitatively? The idea goes back to Blaise Pascal and the year 1653, i.e., 34 years before Sir Isaac Newton published his foundation of mechanics.

12.2.1: Pascal's Law

Instead of following Pascal's original reasoning, we utilize the mechanical equilibrium introduced in Chapter 4 to study an ideal stationary fluid. The approach is illustrated in Fig. 12.1. The pressure dependence on depth in the fluid is established by selecting a small fluid element at a certain depth, as highlighted in Fig. 12.1 as a small rectangular prism. This fluid element must be small compared to the size of the beaker, but must still

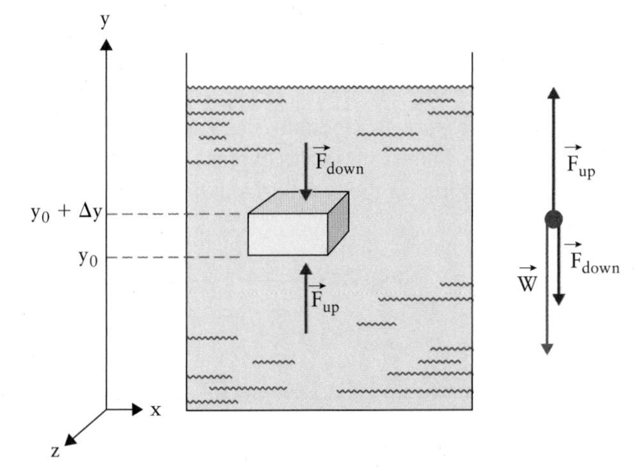

Figure 12.1 A small rectangular prism of fluid is identified in a beaker. Three forces act on this element: two contact forces due to the remaining fluid, and the weight of the fluid element. Note that the element is at rest in an ideal stationary fluid. The corresponding free body diagram for the fluid element is shown at the right.

contain a macroscopic amount of the fluid. The rectangular prism has a horizontal surface of area A and a height $\Delta y = y_{up} - y_{down}$, which we choose to be a small length because the pressure in the fluid is expected to vary in the vertical direction. The volume of the fluid element is then $V = A\Delta y$.

The sketch at the right side of Fig. 12.1 shows the vertical forces of the free body diagram for this fluid element. Horizontal forces act on it, but cancel each other. The three vertical forces acting on the fluid element include

- the weight of the fluid element, $\vec{W}$, which is directed downward;
- the contact force due to the fluid below the element, $\vec{F}_{up}$, which pushes the fluid element upward; and
- the contact force due to the fluid above the element, $\vec{F}_{down}$, which pushes the fluid element downward.

The fluid element neither rises nor sinks because it is in mechanical equilibrium. Newton's first law applies in the vertical direction:

$$F_{net,y} = 0 = F_{up} - W - F_{down}.$$

We use this equation to find a relation between pressure and depth. All three forces are rewritten: the weight to show density and volume, and the two contact forces to show the related pressure terms. Because density is mass divided by volume, the mass of the fluid element is given as:

$$m = \rho V = \rho A \Delta y,$$

where A is the horizontal surface area of the water element and ρ is the density of the fluid. Thus, the weight of the fluid element in Fig. 12.1 is:

$$W = mg = \rho g A \Delta y.$$

The two contact forces contributing to the net force in the vertical direction are replaced by the respective pressure terms because the rectangular prism is an extended object. For this, we first note that the fluid element stretches vertically from y_0 to $y_0 + \Delta y$. At the position y_0 the pressure is labelled p and at $y_0 + \Delta y$ it is labelled $p + \Delta p$. This allows us to express the magnitudes of the forces acting on the two horizontal surfaces of the fluid element:

$$|\vec{F}_{up}| = pA$$

$$|\vec{F}_{down}| = (p + \Delta p)A.$$

Using this equation and the formula for the weight, we rewrite Newton's law in the form:

$$pA - (p + \Delta p)A - \rho g A \Delta y = 0.$$

After combining the first two terms and then dividing by A, we obtain:

$$\Delta p = -\rho g \Delta y.$$

Note that this equation applies regardless of whether the density is constant or varies with depth [$\rho = f(y)$]. We restrict the further discussion to cases with constant density for two reasons: (i) the ideal stationary fluid is incompressible, and (ii) the more general case requires calculus methods to derive the dependence of the pressure on depth. We will test the applicability of the assumption of a constant density for every system we discuss below.

When ρ = const, the previous equation applies for any depth difference Δy. Thus, choosing two arbitrary depths y_1 and y_2 with respective pressures p_1 and p_2, the previous equation is written in the form of **Pascal's law**:

$$p_2 - p_1 = -\rho g(y_2 - y_1). \qquad [12.1]$$

KEY POINT

Pascal's law states that the difference between the pressures at two different positions in a fluid of constant density is proportional to the vertical distance between these two positions. The proportionality factor is the product of the density of the fluid and the gravitational acceleration.

Eq. [12.1] is the first of two formulations we introduce for Pascal's law. It is used in this general form when the surface of the fluid cannot be identified and thus cannot be used as a reference point. An important example is the blood in the cardiovascular system because it is a closed system with no identifiable surface of blood toward air.

12.2.2: Pressure in Liquids with a Visible Surface

In systems with an identifiable surface of the fluid, e.g., for water in a glass, index 1 in Eq. [12.1] is chosen to refer to the surface of the liquid. Therefore, we set $y_1 = 0$ and $p_1 = p_{atm}$. The atmospheric pressure is the proper value for the pressure of the fluid surface since it is in mechanical equilibrium. The force pushing the surface upward equals the force caused by the air pressure pushing downward. Note that we used this argument before when we studied Boyle's experiment in Chapter 8.

It is more convenient in this case to define the position axis downward, i.e., to define the depth below the surface as a positive distance. This changes the negative sign on the right-hand side of Eq. [12.1] into a positive sign. Writing $y_2 = d$, with d the depth below the surface of the fluid in unit m, we get:

$$p = p_{atm} + \rho g d. \qquad [12.2]$$

This equation is a second, frequently used formulation of Pascal's law. It expresses the pressure at depth d below the surface as a function of the pressure at the surface and the weight of the water column above depth d.

The following comments on the two formulations of Pascal's law are useful:

- Pascal's law does not apply to the fluid **air**; i.e., Eq. [12.1] does not describe pressure variations in the atmosphere. This is illustrated by trying to calculate the height of the upper end (surface) of the atmosphere by substituting for the height $y_2 = y_{max}$, which is the maximum height of the atmosphere:

$$(p_{y_{max}} - p_{ground}) = -\rho g(y_{max} - y_{ground}).$$

At the maximum height the pressure drops to a value of $p_{y_{max}} = 0$ atm; i.e., y_{max} is the height at which the vacuum of outer space would begin. The ground-level values are $y_{ground} = 0$ m and the pressure $p_{ground} = p_{atm}$, with $\rho = 1.2\ kg/m^3$ for the density of air at sea level. We find from the previous equation:

$$y_{max} = \frac{1.01 \times 10^5 \text{Pa}}{\left(1.2 \frac{\text{kg}}{\text{m}^3}\right)\left(9.8 \frac{\text{m}}{\text{s}^2}\right)} = 8614 \text{ m}.$$

The assumption of a constant density throughout the atmosphere is clearly inadequate, as the atmosphere would terminate at 8614 m height—234 m below the peak of Mount Everest! Thus, Pascal's law does not apply to gases because gases are compressible and their density depends on pressure. The dependence of the gas density on the pressure was discussed in detail for carbon dioxide in Example 8.3.

- Pascal's law does not contain any information about the shape of the container. Thus, regardless of the shape of the container, the pressure increases below the surface and results in a fixed value at a given depth. This is illustrated in Fig. 12.2, in which the fluid surface is located at the same height in each column above the connected bottom tube. Note that deviations from this observation occur for fluid containers with tiny diameters. We discuss these capillarity effects later in this chapter.

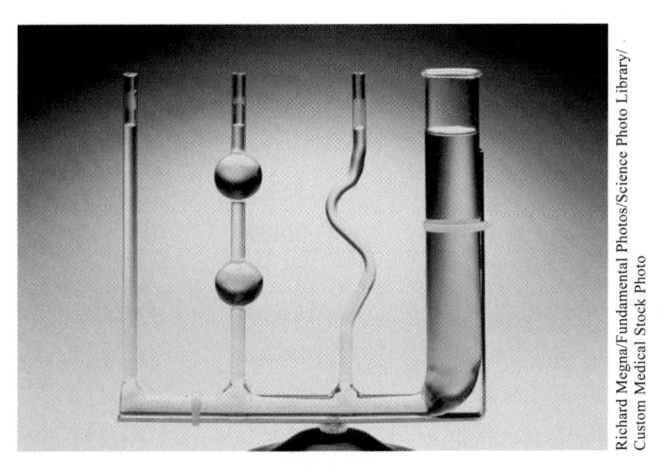

Richard Megna/Fundamental Photos/Science Photo Library/ Custom Medical Stock Photo

Figure 12.2 Experimental illustration of the result of Pascal's law (Eq. [12.2]) that the pressure at various depths in a fluid does not depend on the shape of the fluid container.

- Pressure data are often given in non-standard units. Blood pressure, for example, is usually recorded in unit mmHg. For calculations it is advisable to convert such units to standard units. The standard unit of pressure is Pa. Some pressure data, such as blood pressure, are reported relative to air pressure. A pressure value relative to air pressure is called a **gauge pressure**:

$$p_{gauge} = p_{absolute} - p_{air}, \quad [12.3]$$

which may have either a positive or a negative value. In particular, the term $\rho g d$ in Eq. [12.2] represents a gauge pressure.

EXAMPLE 12.1

What is the pressure 10 m below the surface in a lake?

Solution

The density of fresh water has a value of $\rho = 1.0\ g/cm^3 = 1.0\ kg/L = 1 \times 10^3\ kg/m^3$. We convert $p_{atm} = 1.0$ atm to $p_{atm} = 1.01 \times 10^5$ Pa, because we want to do the calculation in standard units. From Eq. [12.2], at 10 m depth we get:

$$p_{10\text{ m}} = 1.01 \times 10^5 \text{Pa} + \left(1000 \frac{\text{kg}}{\text{m}^3}\right)\left(9.8 \frac{\text{m}}{\text{s}^2}\right)(10 \text{ m})$$
$$= 1.99 \times 10^5 \text{Pa}.$$

We use the inverse pressure conversion from unit Pa to unit atm to find that this result is equivalent to $p_{10\text{ m}} = 1.97$ atm; i.e., the pressure below the water surface rises fast, doubling at just 10 m in depth. In Fig. 12.3, the result of this example is used to illustrate the pressure in water as a function of depth. The fast pressure increase is a critical issue for diving, as we noted in Example 8.9.

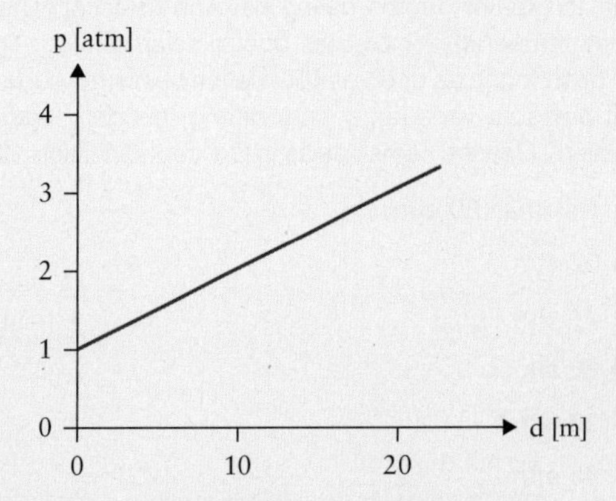

Figure 12.3 The water pressure as a function of depth.

CONCEPT QUESTION 12.1

The nautilus in Fig. 12.4 rises to a depth of 60 metres below the surface of the Pacific Ocean at night to feed on plankton. What is the water pressure at that depth using the rule from Example 12.1 that the pressure in water roughly increases by 1 atm per 10 metres depth?

(A) 3 atm
(B) 4 atm
(C) 5 atm
(D) 6 atm
(E) 7 atm

Figure 12.4 The nautilus is a relative of the long-extinct spiral-shelled ammonites. Its shell (16–27 cm in diameter) not only protects the soft-bodied mollusc but also provides the animal with perfect control over its buoyancy. The mollusc inhabits the last chamber inside the shell, while the other chambers (up to 30 for an adult animal) are filled with a mixture of air and seawater to adjust its overall density. The nautilus's shell is coiled, calcareous, and lined with mother-of-pearl for pressure resistance (preventing implosion to a depth of 800 m).

CONCEPT QUESTION 12.2

The nautilus in Fig. 12.4 is a temperature-sensitive animal, diving to greater depths during daytime to escape near-surface temperature increases due to solar heating. They have been found as deep as 400 metres below the surface. What pressure variation occurs during this daily vertical migration? Use the same rule as in Concept Question 12.1.

(A) less than 30 atm
(B) 32 atm
(C) 34 atm
(D) 36 atm
(E) 38 atm
(F) 40 atm
(G) more than 40 atm

12.2.3: Blood Pressure

We distinguish high- and low-pressure sections of the cardiovascular system because the **blood pressure** varies significantly, as illustrated in Fig. 12.5. The high-pressure part includes the aorta, the arteries and arterioles, and the capillaries of the systemic circulation. The blood pressure in the arteries varies between 10.7 kPa (**diastolic pressure**, equal to 80 mmHg) and 16.0 kPa (**systolic pressure**, equal to 120 mmHg). The low-pressure part includes the veins and the pulmonary circulation; in this circulation the pressure varies only between 1.3 kPa and 3.3 kPa (10–25 mmHg). Since all blood pressure values in Fig. 12.5 are positive gauge pressure values, we conclude that the pressure in our cardiovascular system exceeds the ambient air pressure everywhere. This is, however, only correct while lying down; negative gauge pressures can occur while standing, as outlined in Example 12.2. The pressure in the high-pressure part also changes with age, as illustrated in Table 12.1.

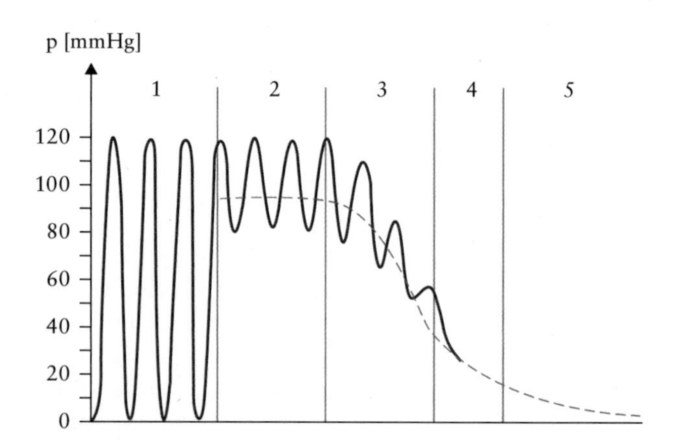

Figure 12.5 Blood pressure variations along the cardiovascular system: (1) left ventricle of heart, (2) large arteries, (3) arterioles, (4) capillaries, (5) venules and veins. Note that the pressure values are gauge pressures, i.e., values given relative to the atmospheric pressure.

TABLE 12.1

Blood pressure as a function of age

	Normal blood pressure	
Age	**(mmHg)**	**(kPa)**
Newborn†	60–80	8.0–10.7
Baby†	80–90	10.7–12.0
Up to 10 years‡	90/60	12.0/8.0
10–30 years‡	110/75	14.7/10.0
30–40 years‡	125/85	16.7/11.3
40–60 years‡	140/90	18.7/12.0
>60 years‡	150/90	20.0/12.0

†Systolic blood pressure only.
‡Systolic/diastolic blood pressure.

EXAMPLE 12.2

Relative to the blood pressure in supine position, calculate the additional blood pressure difference between the brain and the feet in a standing standard man. Use ρ = 1.06 g/cm^3 for the density of blood.

Supplementary physiological information: The term *supine position* specifies that the person is lying down, as shown in Fig. 12.6(a). The blood pressure in supine position is quantified in Fig. 12.5, with a maximum variation of about 15% of the atmospheric pressure value. When the person stands, as shown in Fig. 12.6(b), an additional difference between the blood pressure in the feet and in the brain is due to the extra column of blood that rests on the blood in the feet. Recall that a standard man is 173 cm tall, as noted in Table 4.1.

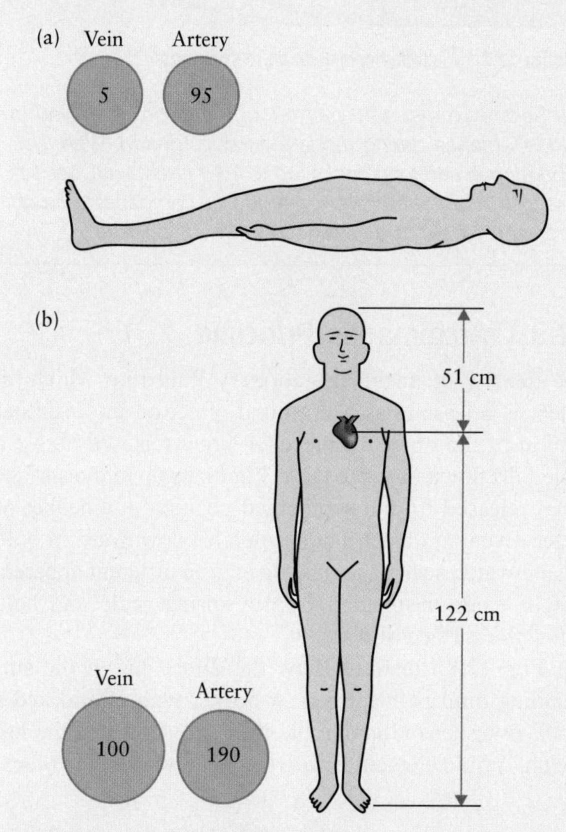

Figure 12.6 Blood pressures and blood vessel sizes in the feet of a person (a) in supine position and (b) standing upright. The vessel sizes are indicated by the areas of the circles shown. The numbers in the vessels are the respective blood pressures in unit mmHg.

Solution

To quantify the additional difference for the standing standard man, we use Pascal's law in the general form as given in Eq. [12.1] (because no blood surface exists):

$$p_{\text{brain}} - p_{\text{feet}} = -\rho_{\text{blood}} g (y_{\text{brain}} - y_{\text{feet}}).$$

continued

This equation is rewritten with $\Delta p = p_{\text{feet}} - p_{\text{brain}}$ for the pressure difference, and $\Delta h = y_{\text{brain}} - y_{\text{feet}}$ for the height of the person. The choice to write Δp in this form eliminates the extra minus sign on the right-hand side of the last equation:

$$\Delta p = \rho_{\text{blood}} g \Delta h.$$

Substituting the given values yields:

$$\Delta p = \left(1.06\times10^3 \frac{\text{kg}}{\text{m}^3}\right)\left(9.8\frac{\text{m}}{\text{s}^2}\right)(1.73\ \text{m})$$
$$= 1.80\times10^4\,\text{Pa} = 18.0\ \text{kPa}.$$

This difference is about 20% of the atmospheric pressure; i.e., it is of the same order of magnitude as the pressure variations within the cardiovascular system in supine position.

For physiological applications, it is more useful to refer to pressures that are measured relative to the pressure at the height of the heart. For a standing person of 1.73 m height, the heart is at a height of 1.22 m and the arterial and venous pressures in the feet are increased relative to the pressures at the height of the heart by:

$$\Delta p = \rho g \Delta h$$
$$= \left(1.06\times10^3 \frac{\text{kg}}{\text{m}^3}\right)\left(9.8\frac{\text{m}}{\text{s}^2}\right)(1.22\ \text{m})$$
$$= 12.7\ \text{kPa} = 95\ \text{mmHg}.$$

The pressure is increased to an average arterial value of 190 mmHg and an average venous value of 100 mmHg. This is illustrated in Fig. 12.6. The figure shows two pairs of circles that indicate the relative sizes of veins and arteries in the feet. The numbers in the circles refer to blood pressures in the respective vessel in unit mmHg. In the scalp, the pressures decrease for a standing person by:

$$\Delta p = \left(1.06\times10^3 \frac{\text{kg}}{\text{m}^3}\right)\left(9.8\frac{\text{m}}{\text{s}^2}\right)(-0.51\ \text{m})$$
$$= -5.3\ \text{kPa} = -40\ \text{mmHg}.$$

The average arterial pressure drops to a value of 55 mmHg and the average venous pressure becomes –35 mmHg. This low venous value does not lead to the closing of the veins in the skull, though, since the blood vessels in the brain are surrounded by cerebrospinal fluid and the pressure in that fluid also drops by a corresponding amount relative to the extracellular fluid in the chest when the person is standing upright.

CASE STUDY 12.1

For which animals does the effect discussed in Example 12.2 pose the greatest challenge?

Answer: *In large land animals, this effect can be much more profound than in humans. The additional pressure required when blood must be pushed above the level of the heart can be generated only with the four-chambered heart of mammals. Of these, the pumping challenge is greatest for animals with long necks. A standing giraffe needs to pump blood as much as 2.5 m above the heart to the brain. That requires significantly more of an additional blood pressure in the left ventricle; the normal systolic pressure at the heart of a giraffe is therefore more than 250 mmHg. Such a systolic pressure would be extremely dangerous for humans. Special valves and a feedback mechanism reduce cardiac output when the giraffe bends its neck down to drink (see Fig. 12.7). In this position the brain is suddenly almost 2 m below the heart and would otherwise be exposed to a tremendous blood pressure due to the changed height difference.*

Another group of animals that had similar issues to deal with were the large quadrupedal dinosaurs, such as Seismosaurus, a 40-metre-long herbivore that lived in North America during the late Jurassic period. With their upright gait and long necks (up to 10 metres long), cardiovascular adaptations were needed to compensate blood pressure variations as a function of body posture.

Figure 12.7 A giraffe bending down to drink water.

CONCEPT QUESTION 12.3

An oceanographer chooses to report water pressure values below the ocean surface as gauge pressures. (a) What values do his reports contain when referring to surface water? (b) When are negative values reported?

12.3: Buoyancy

Ideal stationary fluids display a range of physical properties that are important for biological and physiological applications. We discuss buoyancy, surface tension, and capillarity in the remainder of this chapter.

12.3.1: Archimedes' Principle

The density of an ideal stationary fluid into which an object is released has a profound effect on the resulting motion of the object: a piece of wood released above a table falls down onto the table, but floats up to the surface when released from a submerged position in a beaker of water. Even an object that accelerates downward in both air and water, such as a rock, displays a different apparent weight when suspended from a spring scale and held below the surface of water.

Fig. 12.8 illustrates how the effect due to the surrounding fluid is quantified. A beaker with a fluid and a block B suspended above the fluid are shown in the left sketch. A fluid element F with the same shape and volume

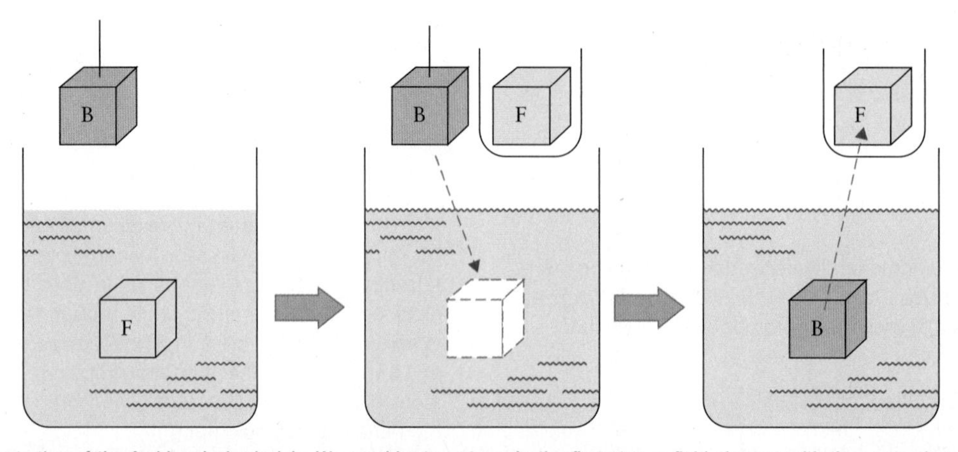

Figure 12.8 Illustration of the Archimedes' principle. We consider two steps: In the first step, a fluid element with the same shape as the object is removed. In the second step, the object is placed in the void created in step 1.

as the block B is identified below the fluid surface. For the fluid element F, mechanical equilibrium conditions apply in the same manner as discussed in Fig. 12.1. Choosing the positive y-axis to be directed upward, the mechanical equilibrium condition for the fluid element F is based on the vertical force components as:

$$F_{\text{up}} - F_{\text{down}} - W_F = 0.$$

$\vec{F}_{\text{up}}$ is the contact force due to the fluid below the chosen fluid element, $\vec{F}_{\text{down}}$ is the contact force due to the fluid above the chosen fluid element, and $\vec{W}_F$ is the weight of the chosen fluid element itself. In the first step, we assume that the fluid element F is removed from the beaker and placed in a small container, leaving an empty bubble of equal size in the beaker. We can simply think of this bubble to be air-filled with the mass of air in the bubble negligible. This bubble is not in mechanical equilibrium since its weight is significantly reduced compared to the removed fluid, but the two contact forces remain unaltered as the fluid around the bubble, which exerts these forces, has not changed. Newton's second law for the bubble reads:

$$F_{\text{up}} - F_{\text{down}} = W_F > 0.$$

As a result, an acceleration of the bubble to the surface is observed. Before this happens, the bubble is replaced by block B with weight $\vec{W}_{\text{B}}$. This leads to the following net force component acting on the block in the vertical direction:

$$F_{\text{net}} = F_{\text{up}} - F_{\text{down}} - W_{\text{B}} = W_F - W_{\text{B}}. \qquad [12.4]$$

Eq. [12.4] illustrates which terms we need to study to describe the submerged block in Fig. 12.8: its weight $\vec{W}_{\text{B}}$ and the weight of the displaced fluid $\vec{W}_F$. Three cases are distinguished for Eq. [12.4]:

- $F_{net} > 0$: The weight of the block is less than the weight of the displaced fluid, and the block B rises to the surface. Examples of this case are ice or wood in water. The wood fibres are denser than the displaced water, but wood contains a large fraction of enclosed air. Ice has a lower density than liquid water due to its peculiar structure in the solid state at atmospheric pressure.
- $F_{net} = 0$: The block floats at its current depth below the fluid surface. The weight of the block equals the weight of the displaced fluid. This case describes a fish, or a submerged diver with breathing equipment.
- $F_{net} < 0$: The weight of the block is larger than the weight of the displaced fluid and the block sinks to the bottom of the container.

These findings were first established by Archimedes in antiquity. Based on Eq. [12.4], the principle is usually stated in the following form:

KEY POINT

Archimedes' principle: *When an object is immersed in a fluid, the fluid exerts an upward force on the object equal to the weight of the fluid displaced by the object.*

We define the buoyant force as equal in magnitude to the weight of the displaced fluid. Thus, the magnitude of the **buoyant force** is:

$$F_{\text{buoyant}} = \rho_{\text{fluid}} V_{\text{object}} g, \qquad [12.5]$$

in which the density is the value for the displaced fluid and the volume is the value for the immersed object. The buoyant force is always directed upward, i.e., in the direction opposite to gravity. Comparing Eq. [12.5] to Eq. [12.4], we emphasize that F_{buoyant} replaces W_F and not F_{net}: the net force can in particular be zero, but the buoyant force is always non-zero unless the object of interest is placed in a vacuum.

Mechanical equilibrium applies in one of the three cases; the system accelerates toward a new equilibrium in the other two cases. In the case where $F_{\text{net}} > 0$, which means the buoyant force exceeds the weight, the object floats to the surface of the fluid. When a sufficient fraction of the object rises above the fluid surface, the object's volume that displaces fluid is reduced and the weight of the displaced fluid becomes equal to the weight of the block. Note that, for an object floating at the surface of a fluid, only the fraction of the volume below the surface of the fluid enters Eq. [12.5]. In the case where $F_{\text{net}} < 0$, the object sinks. At the bottom of the beaker, a mechanical equilibrium is reached due to an additional normal force acting upward on the object.

12.3.2: Applications of Buoyancy

Buoyant forces are observed in many contexts. In Example 4.27, we calculated the force on the central dendrite in a Pacinian corpuscle (shown in Fig. 4.47). That example illustrated how the buoyant force is taken into account in a mechanical problem involving a floating object. Buoyancy can also be illustrated in simple experiments you can do at home. For example, place a freshly cut piece of *lemon peel* in a bottle filled with water. Close the bottle and exert variable pressure on the plastic cap with your thumb. The lemon peel rises or sinks depending on the exerted pressure. The lemon peel floats in the first place due to very small air bubbles in its porous structure. When you squeeze the cap, variations in the water pressure occur. The size of the air bubbles varies significantly with pressure changes because air as a gas is highly compressible. As a result, the volume of the peel varies in Eq. [12.5].

EXAMPLE 12.3

In 1936, the zeppelin LZ-129 *Hindenburg* successfully completed 17 round trips across the Atlantic Ocean. During the winter, 10 passenger cabins were added, accommodating up to 22 additional passengers. On May 6, 1937, on its first flight to the United States, the *Hindenburg* exploded at Lakehurst Naval Air Station in New Jersey. Did the increased load due to additional passengers contribute to this famous accident?

Supplementary historical information: The *Hindenburg* was the largest aircraft ever to fly. At 245 metres in length it was longer than three Boeing 747s and almost as long as the *Titanic*. It had a capacity of 50 passengers (72 after the upgrade) and 61 crew. Its 16 gas cells had a combined volume of 210 000 m^3 and were initially designed to be filled with helium gas (density ρ_{He} = 0.16 kg/m^3). When helium remained unavailable due to an embargo, the Deutsche Zeppelin-Reederei opted for hydrogen gas instead (density ρ_{H_2} = 0.08 kg/m^3). The *Hindenburg* had a mass of 130 tonnes without passengers, crew, and cargo.

With these data, calculate (a) the maximum payload (in kg) that the *Hindenburg* could carry as designed (i.e., with a helium filling). Then calculate (b) by how much the payload had increased with a hydrogen filling. *If you are interested:* (c) find out what factors led to the disaster at Lakehurst in 1937.

Solution

Buoyancy also occurs in gases such as air. In the case of hot air balloons, lighter hot air displaces heavier cold air in the balloon's envelope. Hot-air balloon operation is discussed in P–12.10. Alternatively, gases that are lighter than air can be employed. This is illustrated with the current example.

Solution to part (a): We use Eq. [12.4] with F_{net} = 0 to determine the maximum payload; i.e., we find the total mass of the zeppelin at which its weight and its buoyant force are balanced. The buoyant force is obtained from Eq. [12.5]. The weight has three contributions: the weight of the empty zeppelin, the weight of the payload (passengers, crew, and cargo), and the weight of the gas in the 16 gas cells:

$$F_{net} = 0 = \rho_{air} V_{gas\ cells}\, g - W_{ship} - W_{payload} - \rho_{He} V_{gas\ cells}\, g.$$

We divide this equation by the gravitational acceleration g, switching the various weight terms to the respective masses. We use for the air density at 1 atm a conservative (lower) estimate. Note that ρ_{air} = 1.29 kg/m^3 at 0°C, and ρ_{air} = 1.21 kg/m^3 at 20°C. Thus, the value at 20°C is substituted in the last equation. For the payload of the *Hindenburg*'s original design, this yields:

$$m_{payload} = (\rho_{air} - \rho_{He}) V_{gas\ cells} - m_{ship} = \left(1.05 \frac{kg}{m^3}\right)(2.1\times10^5 m^3) - 1.3\times10^5 kg.$$

continued

i.e., $m_{payload} = 9.6 \times 10^4$ kg = 96 tonnes. Note that this result corresponds to more than 600 standard men (each with an additional 70 kg luggage). There were only 50 spaces for passengers on the *Hindenburg,* showing that it was never intended for modern mass tourism. During the two days of travel—at a maximum speed of 135 km/h—the passengers enjoyed luxury comparable to that on a modern cruise ship, including dining halls, reading rooms, a lounge, a music salon with piano, and even a smoking room!

Solution to part (b): Exchanging helium for hydrogen in the equation we found in part (a) yields a payload of 107 tons, i.e., an increase of about 11%. An additional 11-tonne lifting capacity is plenty for 22 additional passengers. Thus, they were not to blame for the accident. Indeed, prior to the trip to Lakehurst, the modified *Hindenburg* had already completed a round trip to Brazil to start off the 1937 season. On the doomed flight it carried only 36 passengers.

Solution to part (c): The generally accepted explanation is the static spark theory, which requires concepts we discuss in Chapter 17. The outer envelope of the airship was electrically isolated from its aluminium frame. When the *Hindenburg* flew through a weather front, both its outer skin and the mooring lines became wet, turning them into electrically conducting surfaces. The wet skin then accumulated electric charges, causing a significant electric potential relative to the ground. The mooring lines were anchored on the aluminium frame. When they were lowered to the ground the aluminium frame became electrically grounded. The high potential difference at close proximity between skin and metal frame probably caused sparking. Sparks wouldn't ignite hydrogen in the gas tanks; however, the *Hindenburg* was venting some hydrogen in preparation for landing. Hydrogen mixed with oxygen in the air formed a combustible mixture.

CASE STUDY 12.2

Greek mythology reports the story of the Argonauts. Under Jason's leadership, they travelled to Aeëtes, king of Colchis, in their quest for the Golden Fleece. On the way they were forced to throw significant amounts of gold overboard to prevent their ship from sinking as they passed from the Mediterranean Sea into the Black Sea, near modern-day Istanbul in Turkey. Can this part of the myth be true?

Answer: *A significant difference between the salt concentrations of these two bodies of water exists. The more southern Mediterranean Sea loses more water through evaporation. It receives fresh water from only four major rivers: Po (Italy), Rhône (France), Ebro (Spain), and Nile (Egypt). Thus, the salt content in the Mediterranean Sea is high, at 38 g/L NaCl, which corresponds to a density of* ρ = *1.028 g/cm*^{3}*. The Black Sea is smaller than the Mediterranean Sea, but receives fresh water from three major rivers: Danube (Romania), Dnieper (Ukraine), and Don (Russia). Thus, its salt content is lower, at 16 g/L NaCl and its density is* ρ = *1.014 g/cm*^{3}*.*

continued

The Archimedes' principle predicts that the Mediterranean Sea can float 1.4% more weight than the Black Sea. Thus, an overloaded ship must reduce its cargo when travelling north through Istanbul Bogazi (Bosporus). Cargo ships travelling across the Atlantic and continuing along the St. Lawrence River to Montreal are affected in the same way.

Average seawater in Earth's oceans contains 25.5 g/L NaCl and has a density of $\rho = 1.02$ g/cm^3. With the molar mass of sodium $M = 23$ g/mol and chlorine $M = 35.5$ g/mol, the NaCl salt concentration of average seawater corresponds to 435 mmol/L. The extracellular fluid in our body has a NaCl concentration of 165 mmol/L; i.e., the NaCl concentration in the interstitium is 40% that of seawater.

The effect discussed above can be verified in a simple experiment. In your kitchen, fill a jar about halfway with water and dissolve a large amount of table salt in it. Then add more salt-free water, pouring it into the jar carefully such that the two liquids don't mix. If you place a chicken egg in the water, it will sink to the interface between the two liquids. The reason for this is the same effect we just discussed for ships on the high seas: the egg displaces a fixed volume of water equal to its own volume; the weight of the displaced tap water is less than the weight of the egg, but the weight of the same volume of salt water is greater than the weight of the egg.

12.3.3: Buoyancy in Physiology

EXAMPLE 12.4

The human brain is immersed in cerebrospinal fluid of density 1.007 g/cm^3. This density is slightly less than the average density of the brain, which is 1.04 g/cm^3. Thus, most of the weight of the brain is supported by the buoyant force of the surrounding fluid. What fraction of the weight of the brain is not supported by this force?

Solution

This problem is an application of the Archimedes' principle. The magnitude of the weight of the brain is:

$$W = \rho_{\text{brain}} V_{\text{brain}} g.$$

The buoyant force, in turn, for the brain fully immersed in the cerebrospinal fluid follows from Eq. [12.5]:

$$F_{\text{buoyant}} = \rho_{\text{cerebrospinal fluid}} V_{\text{brain}} g.$$

We know that these two forces do not balance each other because the brain is connected through the medulla oblongata to the spinal cord, which exerts an additional force on the brain. To determine the fraction of the weight of the brain that is not balanced by the buoyant force, we calculate:

continued

$$\frac{W - F_{\text{buoyant}}}{W} = \frac{\rho_{\text{brain}} - \rho_{\text{cerebrospinal fluid}}}{\rho_{\text{brain}}} = \frac{\left(1.04\frac{\text{g}}{\text{cm}^3}\right) - \left(1.007\frac{\text{g}}{\text{cm}^3}\right)}{1.04\frac{\text{g}}{\text{cm}^3}} = 0.032.$$

Just 3.2% of the brain's weight is not balanced by the cerebrospinal fluid, requiring only a small force to be exerted by the spinal cord on the brain. You can calculate the actual brain mass not supported using $m_{\text{brain}} = 1.5$ kg for the standard man (from Table 4.1). This yields 50 g.

Buoyancy is used by many marine animals to float, rise, or sink in seawater without effort. A famous example is **spirula**, shown in Fig. 12.9. Spirula is a cephalopod that was thought to have died out 50 million years ago but was then discovered by a scientific expedition of *H.M.S. Challenger* during its voyage around the world in 1873–1876. The animal swims head down because it has a buoyant, gas-filled shell at its posterior end.

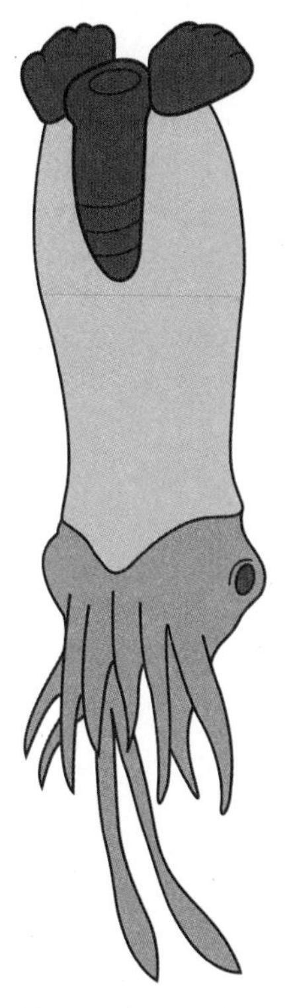

Figure 12.9 As described in the *Report on the Scientific Results of the Voyage of the H.M.S.* Challenger *During the Years 1873–1876*, one of the more remarkable events during the long voyage to chart the world's oceans was the discovery of the spirula, a cephalopod that was thought from fossil evidence to have died out 50 million years ago. The spirula has a squid-like body between 3.5 cm and 4.5 cm long. It has eight arms and two longer tentacles. The most distinctive feature is the buoyancy chamber, which is an internal shell that keeps the animal's body vertical.

Most bony fish possess a swim bladder to control buoyancy. The swim bladder is an air sac that allows gas exchange with the fish's blood. This gas exchange leads to a variation in size of the swim bladder, which in turn adjusts the density of the fish. Bony fish can therefore conserve energy by remaining motionless at a chosen depth in the water. Sharks don't have swim bladders and must swim all their life to prevent their body from sinking.

An interesting living fossil is the **nautilus** (see Fig. 12.4), a relative of the spiral-shelled ammonites. The nautilus's shell not only protects the soft-bodied mollusc but also provides the animal with perfect control over its buoyancy. The mollusc inhabits only the last of a spiralling series of chambers inside the shell. Using osmosis it regulates the seawater fraction in its inner chambers, which contain a mixture of air and seawater, to adjust its overall density. Decreasing the density allows the nautilus to rise during its nightly migration from the depths of the Pacific Ocean to the surface.

12.3.4: Sedimentation

Differences in the buoyant force for different components in a solution allow for the separation of heterogeneous systems. This process is called **sedimentation** and is sketched in Fig. 12.10(a). For the buoyant force in Eq. [12.5] you find two variables that may differ between different suspended particles in a solution: volume and/or density. Thus buoyancy is suitable to separate chemically identical particles on the basis of their shape and size, or to separate chemically different particles based on their density. Variations in shape and size are often arbitrary for suspended particles in a solution, and using buoyancy for their separation is less effective than filtering with varying pore size sieves. Consequently, in chemistry sedimentation is often referred to as separation due to density differences. An example is given by *tailings*. Tailings are suspensions of crushed gold-carrying ores in water; gold has a density of 19.3 g/cm^3, and sedimentation takes place rapidly because the remaining rock material has only a density of 2.5 − 3.0 g/cm^3.

Separation at the end of the sedimentation process depends on the system: for a suspension of a solid component in a liquid, the process is called **decantation** (in which the cleared solution is poured out); for a mixture of immiscible fluids a separating funnel is used [see Fig. 12.10(b)].

The sedimentation process itself is based on two physical principles:

- *Buoyancy:* The vertical net force component on a suspended particle is determined from Eq. [12.5] as:

$$F_{\text{net}} = F_{\text{buoyant}} - W = (\rho_{\text{fluid}} - \rho_{\text{particle}})V_{\text{particle}}g.$$

  The net force acts downwards in the solution if the density of the particle is greater than that of the solvent. Based on Newton's second law, this will lead to an acceleration downward.

- *Friction:* The solvent will resist the downward motion of the particle by exerting a drag force on the particle. Friction was discussed in Chapter 3. We distinguish frictional effects based on their dependence on the speed of the object moving through the medium: we observe either $f \propto v$ or $f \propto v^2$. The linear case applies to objects sinking slowly through a solution.

Once the net force due to weight, buoyant force, and drag force is zero, the particle will move downward at a

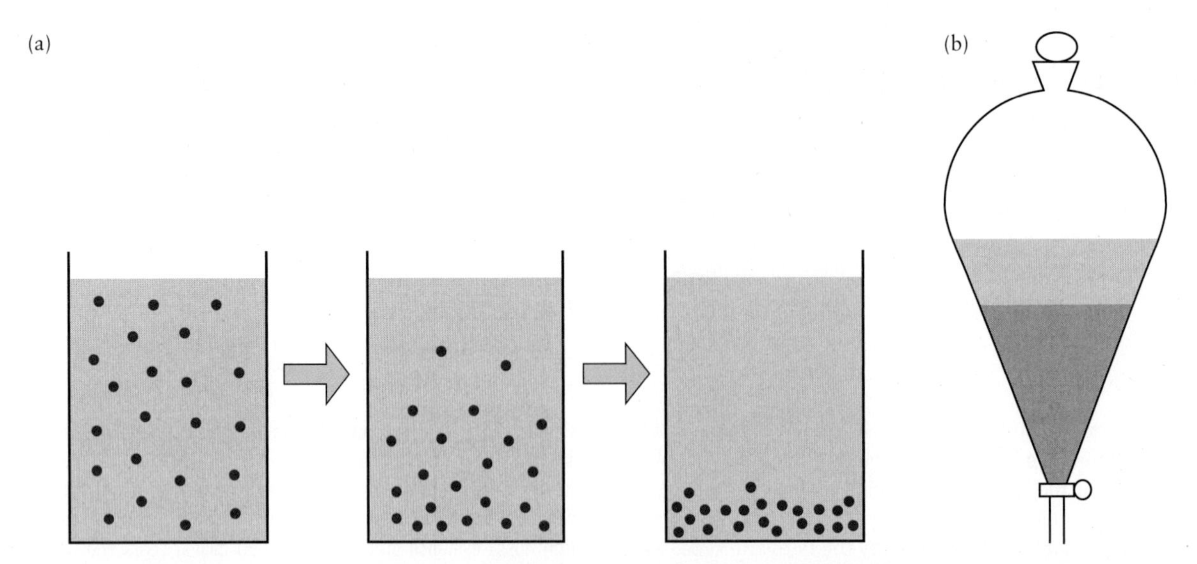

Figure 12.10 (a) Sedimentation of a solid suspension. If the solid particles in the solution have a density greater than that of the liquid phase, buoyancy causes the solid phase to collect at the bottom of the beaker. (b) If the mixture consists of two immiscible liquids, a separating funnel is used to separate the components after sedimentation.

constant speed, called the **terminal speed**. The mechanical equilibrium condition that leads to a constant (terminal) speed is:

$$F_{net} = F_{buoyant} + f - W = 0,$$

in which the buoyant force and the frictional force act upward and the weight of the object acts downward. Substituting the buoyant force from Eq. [12.5] and $f = kv$, with k a constant for the object's friction, we find:

$$F_{net} = \frac{\rho_{fluid}}{\rho_{object}} mg + kv_{terminal} - W = 0,$$

in which m is the mass of the object. This equation allows us to write the terminal speed $v_{terminal}$:

$$v_{terminal} = \frac{mg}{k}\left(1 - \frac{\rho_{fluid}}{\rho_{object}}\right). \qquad [12.6]$$

This speed is called the **sedimentation rate**.

In 1908, Jean Perrin pointed out that sedimentation cannot be used not only as a chemical separation process but also to measure the molecular mass of macromolecules, because $M = N_A m$ with N_A the Avogadro number. Sedimentation rates are small for most biological samples of interest; e.g., a blood cell sinks in blood plasma at a rate of about 5 cm per hour. Biomolecules are even slower, as they are smaller than a cell. These rates therefore limit the speed with which biological samples can be analyzed. We note in particular two problems with Eq. [12.6]:

- The frictional coefficient k is unknown (it can be determined analytically only for a smooth spherical particle, which is not a suitable model for most particles of interest).
- Only very large molecular masses provide sufficient sedimentation velocities to be measured under the microscope because the gravitational pull is weak.

The second problem is addressed by replacing the gravitational acceleration $g = 9.8\ \mathrm{m/s^2}$ with a centrifugal acceleration (see Chapter 2). For a centrifuge the gravitational acceleration g in Eq. [12.6] is replaced by a radial acceleration that is caused by the rapid rotation of the solution in a test tube (as illustrated in Fig. 12.11). The radial acceleration is $a_{radial} = \omega^2 r$, with ω the angular frequency of the centrifuge and r the distance of the sample from the rotation axis. For a centrifuge, Eq. [12.6] is then written in the form:

$$v_{terminal} = \frac{m\omega^2 r}{k}\left(1 - \frac{\rho_{fluid}}{\rho_{object}}\right). \qquad [12.7]$$

The sedimentation rate is significantly increased because a typical laboratory centrifuge, such as the one shown in Fig. 12.11, can reach centrifugal accelerations of $3 \times 10^5\ g$.

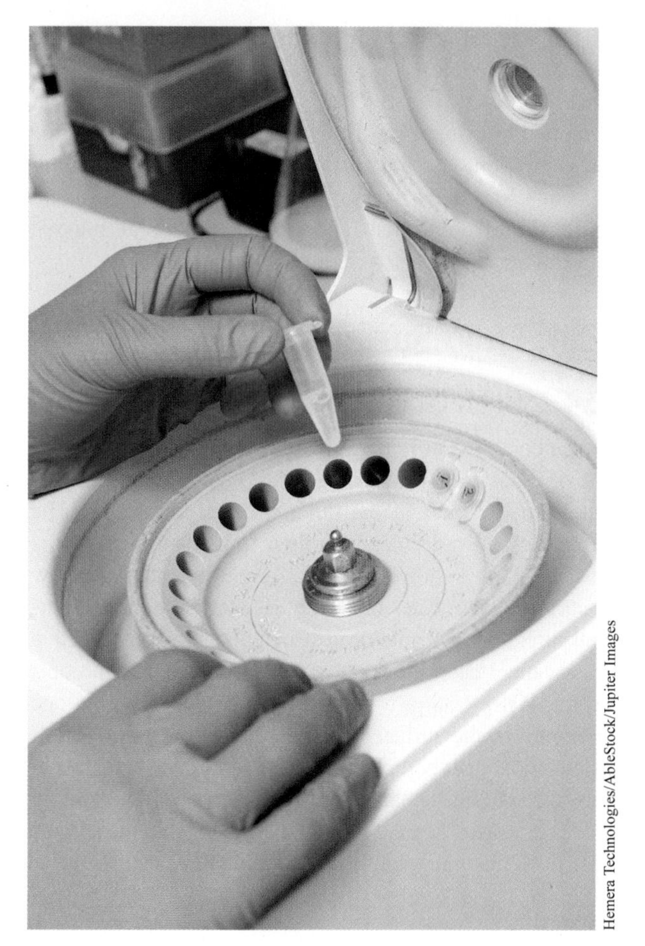

Hemera Technologies/AbleStock/Jupiter Images

Figure 12.11 A laboratory centrifuge.

EXAMPLE 12.5

A typical centrifuge spins with an angular frequency of 5250 rad/s. The top end of an 8-cm-long test tube in this centrifuge is 5 cm from the axis of rotation. Find the effective value of the radial acceleration as a multiple of the gravitational acceleration g at the bottom of the test tube.

Solution

We use $a_{radial} = \omega^2 r$ with $r = 8\ \mathrm{cm} + 5\ \mathrm{cm} = 13\ \mathrm{cm}$:

$$\frac{a_{radial}}{g} = \frac{\omega^2 r_{bottom}}{g} = \frac{\left(5250\,\frac{\mathrm{rad}}{\mathrm{s}}\right)^2 (0.13\ \mathrm{m})}{9.8\ \frac{\mathrm{m}}{\mathrm{s}^2}} = 3.7 \times 10^5.$$

The test tubes must be carefully supported in the centrifuge to withstand this tremendous acceleration!

For precision measurements of molecular weights of macromolecules, this leaves us to address the frictional coefficient k in Eq. [12.7]. In 1905 Albert Einstein developed a formula for diffusion that contains the diffusion coefficient and the frictional coefficient. Combining Einstein's formula with Eq. [12.7] allows us to eliminate the frictional coefficient and obtain molecular masses from the measurement of sedimentation velocities.

12.4: Fluid Surfaces

12.4.1: Surface Energy

Surface energy is a fluid property associated with the presence of a surface toward air. Surfaces have properties distinct from the bulk of the fluid, as the symmetry of the interactions of molecules with their immediate neighbours is broken. This is shown in Fig. 12.12. Sufficiently far below the surface, each water molecule has attractive interactions in all directions with its nearest neighbours. The attractive forces cancel each other out and the molecule is in equilibrium. When such a water molecule comes to within one nanometre of the surface, the sphere of neighbouring molecules is no longer complete and a net force acts on the water molecule, pulling it back from the surface. This resulting force is greatest when the water molecule reaches the surface because, at this point, half its neighbours have vanished.

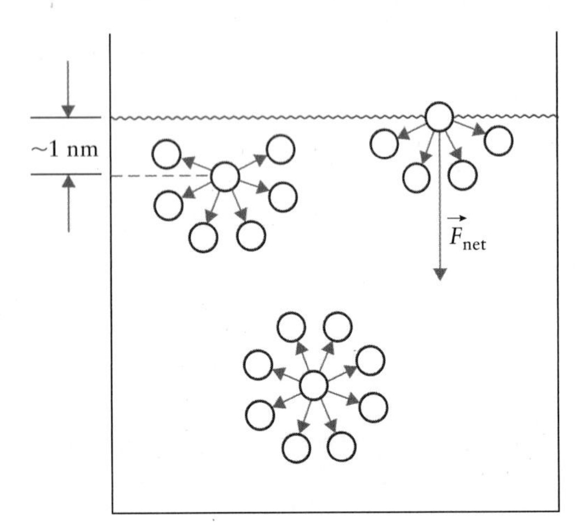

Figure 12.12 Three water molecules, shown as open circles, are illustrated with their respective neighbours. Any molecule that is 1 nm or farther below the surface has a symmetric cloud of other water molecules around it. Only molecules at the surface encounter a net force downward due to an incomplete cloud of attracting neighbours.

The resulting force pulls molecules away from the surface. However, molecules cannot leave the surface toward the bulk as the surface area cannot shrink. Since the molecules in the surface layer have to be brought to the surface against the net force shown in Fig. 12.12, there is an energy associated with the formation of a surface. We define the surface energy σ:

$$\sigma = \frac{\Delta E}{\Delta A}, \qquad [12.8]$$

in which ΔE is the energy needed to increase the surface of a fluid by an area ΔA. Thus, σ is the energy required to form 1 m^2 of new surface and carries the unit J/m^2.

12.4.2: Surface Tension

There is a second way to look at σ. For this approach, Eq. [12.8] is rewritten as:

$$W = \Delta E = \sigma \Delta A,$$

in which W is the work required to increase the surface by ΔA. As you recall from Chapter 7, work is connected to a force by $W = Fd$, in which d is the displacement vector, attributed to the action of the force F. We use the device shown in Fig. 12.13 to derive a relationship between the force F exerted on a surface and the surface energy σ. The device consists of a thin film, such as a soap film, spanned by a fixed U-shaped wire and a mobile straight wire. Attached to the straight wire is a handle that can be pulled with force $\vec{F}$ to enlarge the surface enclosed by the wire frame. We consider an increase in area A from $A = l_x l_y$ to $A = l_x(l_y + \Delta l_y)$. The work needed to enlarge the area is $W = F\Delta l_y$. Note that the force and the displacement are parallel and vector notation is no longer needed. The resulting change in area is $\Delta A = l_x \Delta l_y$. These terms are substituted in the work formula:

$$\sigma \Delta A = \sigma l_x \Delta l_y = W = F \Delta l_y.$$

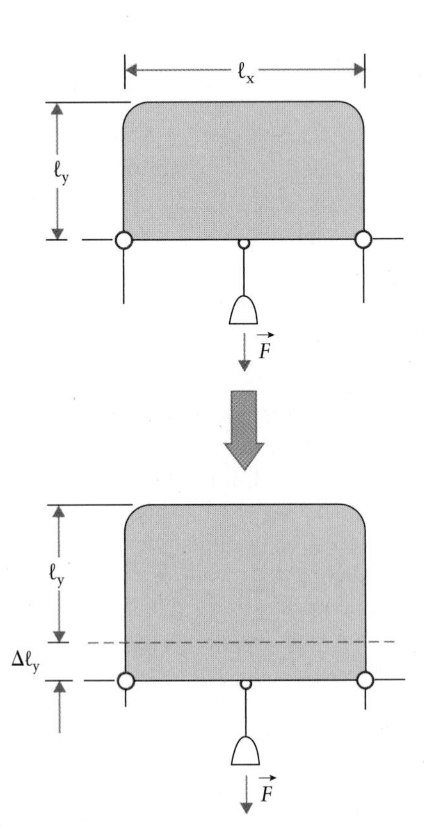

Figure 12.13 Sketch of a device to measure surface tension in the form of the force $\vec{F}$ acting per unit length on a film. The U-shaped wire and the mobile bar enclose the film. The mobile bar has length l_x and in the lower frame is pulled from the dashed line down to the solid line by a distance Δl_y.

Division by Δl_y yields:

$$\sigma = \frac{F}{l_x}, \qquad [12.9]$$

in which the right-hand side has the unit N/m, which is therefore also the unit of σ. σ represents in Eq. [12.9] a tangential force needed to increase the surface per unit length. For this reason, σ is also called **surface tension.** This double interpretation of σ as energy to increase a surface or force to stretch a film is possible because the units for an energy per surface area and for a force per unit length are the same, J/m^2 = N/m. Table 12.2 provides surface tensions for several fluids.

TABLE 12.2

Surface tensions for various fluids at 20°C

Fluid	Surface tension σ (N/m)
Acetone	0.0252
Benzene	0.0288
Chloroform	0.0275
Cyclohexane	0.02495
Ethanol	0.0221
Mercury	0.4254
Methanol	0.0227
Toluene	0.0284
Water	0.073
Whole blood (at 37°C)	0.058

Fig. 12.13 illustrates one practical way in which the surface of a fluid can be increased. This approach is particularly suitable for soap films due to their high cohesiveness. Surface tensions of other fluids such as water cannot be measured this way. For such fluids an alternative set-up exists in which the fluid surface area is initially reduced by bringing a flat solid surface of known area into contact with it. The surface tension is then determined from the force needed to lift the solid surface off the fluid.

12.5: Bubbles and Droplets

Two key consequences of the surface tension concept will be discussed in this section: the equilibrium shape of bubbles and droplets, and the pressure in small bubbles, droplets, and cylindrical tubes.

We first study a free system—i.e., a fluid that is not confined by any external surface such as a container wall. An example is a raindrop. Since work has to be done to increase the surface of a liquid, the drop minimizes its surface to achieve its energetically most favourable shape. The equilibrium shape of the raindrop is the geometrical figure that has the least surface for a fixed volume; this is the sphere. If a drop is not spherical, it releases energy while reshaping toward a sphere. When the drop is spherical, external energy has to be provided to change its shape.

If we introduce additional boundary conditions—e.g., a droplet sitting on a flat surface—the shape with a minimum surface becomes more complex; however, some simpler cases can be noted. For any flat frame, as in Fig. 12.13, the minimum surface for a film is a flat layer, not a bent structure. For droplets on inert surfaces, such as water droplets on glass, a circular interface forms and the droplet assumes a partially spherical shape with a well-defined contact angle.

A physiological example is the 300 to 400 million alveoli at the end of the bronchial tree in the lungs. Each alveolus is a partial sphere with a diameter of about 0.3 mm placed on top of a circular bronchial tube. However, before we can discuss physiological consequences of the bubble shape of the alveoli we need to find how the pressure in a bubble depends on its radius.

12.5.1: Pressure in a Bubble: Laplace's Law

Use a children's soap solution from a toy store. Blow a few soap bubbles and observe them. Their spherical surface is stretched smoothly, suggesting that it is held open by the air inside. Even though the soap film forming these **bubbles** is less than a micrometre thick, it is able to retain compressed air inside. To illustrate that indeed an enhanced pressure exists inside a bubble, we need a traditional pipe or a pipe-like bubble blower from a toy store. Catch a soap bubble on the pipe bowl while covering the mouthpiece with your finger, as shown in Fig. 12.14. Then hold the mouthpiece close to the flame of a burning candle and remove your finger. The candle flame is blown to the side while the bubble shrinks and vanishes into the pipe bowl.

Figure 12.14 A soap bubble forms on the bowl of a toy pipe.

We want to quantify by how much the pressure inside a bubble exceeds atmospheric pressure. Fig. 12.15(a) shows a nearly square-shaped segment selected on the spherical surface of a bubble. In Fig. 12.15(b) a coordinate system is chosen for this segment. The z-axis is perpendicular to its surface. The x- and y-axes are parallel to the edges of the segment. They are chosen to intersect with the z-axis

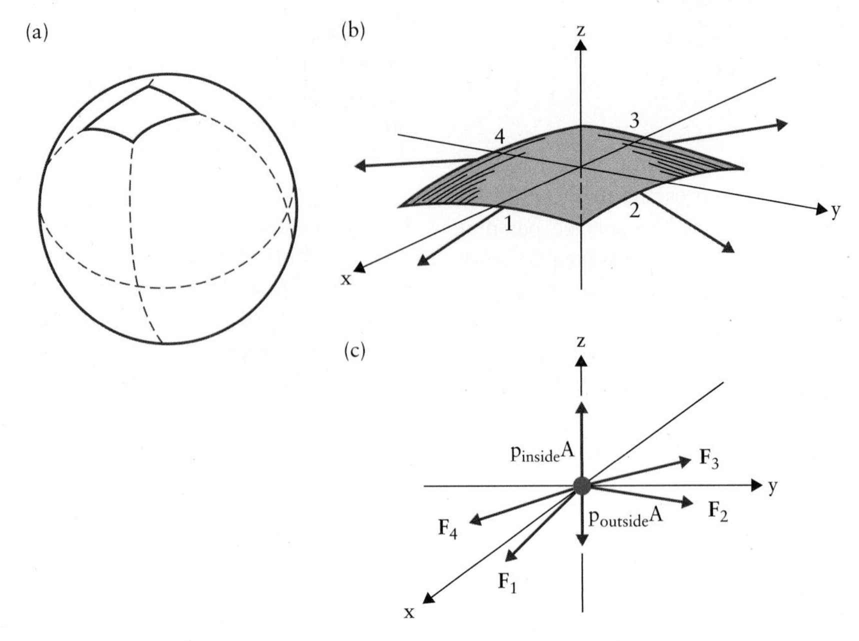

Figure 12.15 (a) A bubble with a highlighted nearly square area close to the bubble's upper pole. (b) The same segment of the bubble is shown with sides 1 and 3 aligned in the *yz*-plane and sides 2 and 4 aligned in the *xz*-plane. (c) The corresponding free body diagram for the bubble segment contains four contact forces that are due to the tangential pull of the remaining bubble membrane, and two contact forces due to the gas pressure acting perpendicular to the segment's inner and outer surfaces. For simplicity, forces are labelled by their magnitude.

at the point where the *z*-axis intercepts the bubble surface. Thus, edges 1 and 3 of the segment lie in the *yz*-plane and edges 2 and 4 lie in the *xz*-plane. The four arrows in the figure indicate the directions in which the remaining bubble surface pulls to stretch the segment open. Since the segment bends downward from the origin, these four forces are all directed below the horizontal.

Fig. 12.15(c) shows the free body diagram for the bubble segment. The four forces discussed above are identified as $\vec{F}_i$, with $i = 1$ to 4. Due to the symmetry of the surface segment in Fig. 12.15(b), the following relations among the components of the four forces $\vec{F}_i$ apply:

$$\text{x-direction} \quad F_{1,x} = -F_{3,x} \quad F_{2,x} = F_{4,x} = 0$$

$$\text{y-direction} \quad F_{2,y} = -F_{4,y} \quad F_{1,y} = F_{3,y} = 0$$

$$\text{z-direction} \quad F_{1,z} = F_{2,z} = F_{3,z} = F_{4,z}.$$

The force components in the *x*- and *y*-directions (first two lines) compensate each other. Only the *z*-components yield a non-vanishing net force.

If these four forces were the only forces acting on the bubble segment, then the segment would not be in mechanical equilibrium in the *z*-direction, as a net force downward would result, which would cause the segment to accelerate. Thus, for a bubble in mechanical equilibrium, additional forces must be present and be taken into account. These are contact forces due to air inside the bubble with pressure p_{inside}, and air outside the bubble with pressure $p_{outside}$. To express a force, each pressure is multiplied by A, where A is the area of the bubble segment shown in the figure. Newton's first law then describes the mechanical equilibrium of the bubble segment in the *z*-direction:

$$\sum_{i=1}^{6} F_{i,z} = 0 = 4F_{1,z} + p_{inside}A - p_{outside}A.$$

Eq. [12.9] is used to quantify $F_{1,z}$ in this equation further. If l is the length of each edge of the segment, i.e., $l^2 = A$, then $|\vec{F}_1| = 2\sigma l$. The factor 2 in this relation is due to the fact that the bubble has an inner and an outer surface. We use Fig. 12.16 to determine the *z*-component of this force. In the figure, r is the radius of the bubble and φ is the opening angle of the segment of length $l/2$.

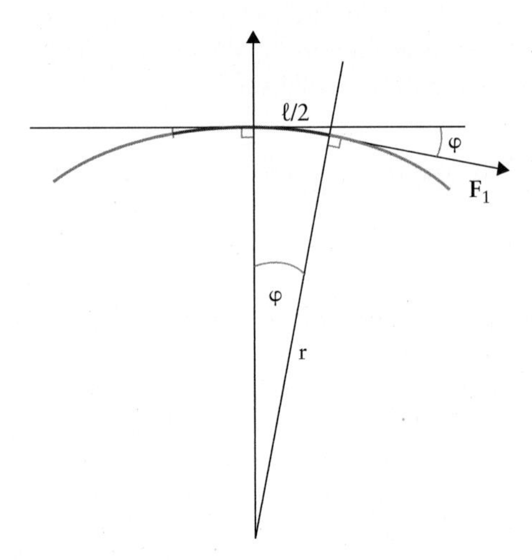

Figure 12.16 Geometric sketch of the *z*-components of the forces shown in Fig. 12.15. The red line represents the side view of the square segment highlighted in Fig. 12.15.

The red line in Fig. 12.16 corresponds to a side view of the segment shown in Fig. 12.15(b). Applying trigonometric relations to express sin φ in Fig. 12.16, we find:

$$\sin\,\varphi = \frac{l/2}{r}$$

$$\sin\,\varphi = \frac{F_{1,z}}{|\vec{F}_1|}.$$

The last equation contains two relations because the angle φ occurs twice in Fig. 12.16. We equate the right-hand sides of these two formulas to express $F_{1,z}$:

$$F_{1,z} = 2\sigma l \frac{l/2}{r} = \sigma \frac{l^2}{r} = \frac{\sigma A}{r},$$

in which we replaced $l^2 = A$. Substituting the z-component of the force from this equation into the mechanical equilibrium condition, we find:

$$\sum_{i=1}^{6} F_{i,z} = 0 = -\frac{4\sigma A}{r} + A\Delta p,$$

which leads to:

$$\Delta p = \frac{4\sigma}{r}.$$

Eq. [12.10] is **Laplace's law**. $\Delta p = p_{\text{inside}} - p_{\text{outside}}$ is called the **transmural pressure**. Thus, the internal pressure in a small bubble is found to be larger than the external pressure. This pressure difference maintains the curved surface of the soap bubbles you observed earlier.

12.5.2: Systems for Which Laplace's Law Is Applied in Biology

Formulas equivalent to Eq. [12.10] can also be derived for homogeneous droplets or other curved shapes of hollow or homogeneous liquids. The derivation of such formulas follows the same steps we took with Fig. 12.15 to derive Eq. [12.10]. Two factors vary between the different cases:

- Droplets and homogeneous cylinders display only one curved surface, while bubbles and hollow tubes have two curved surfaces, an outer and an inner surface.
- Cylindrically symmetric systems, such as hollow and homogeneous tubes, display a finite curvature in only one direction across their surface, while spherically symmetric systems, such as bubbles and droplets, have a finite curvature in two perpendicular directions across their surface. When drawing a figure equivalent to Fig. 12.15 for a cylinder only one curved edge of the surface segment will be present.

Either of these changes eliminates a factor of 2: for homogeneous systems a factor of 2 is eliminated from the magnitude of the force F_1, i.e., $|F_1| = \sigma l$; and for cylindrically symmetric systems only two instead of four z-components occur in Fig. 12.15. These factors carry through the calculation to Eq. [12.10]. Thus, we find three formulations for Laplace's law:

bubble—

$$\Delta p = p_{\text{inside}} - p_{\text{outside}} = \frac{4\sigma}{r}; \qquad [12.10]$$

droplet or hollow cylinder—

$$\Delta p = p_{\text{inside}} - p_{\text{outside}} = \frac{2\sigma}{r}; \qquad [12.11]$$

solid cylinder—

$$\Delta p = p_{\text{inside}} - p_{\text{outside}} = \frac{\sigma}{r}. \qquad [12.12]$$

KEY POINT

Laplace's law states that the pressure difference between the inside and the outside of a fluid with a curved surface is inversely proportional to the radius of curvature of the curved surface. This means that a smaller bubble, droplet, or cylinder has a larger pressure difference Δp.

Each of Eqs. [12.10] to [12.12] has physiological applications. We highlight blood vessels and alveoli in the lungs as examples in the remainder of this section. The formula for the homogeneous cylinder applies to blood vessels: their elastic tissue must be capable of sustaining the pressure difference in Laplace's law. We compare blood capillaries and small arterioles to illustrate the consequences. The transmural pressure values between blood in a vessel and the surrounding tissue are essentially identical to the values in Fig. 12.5. From that figure we find that similar pressure differences apply for the smallest arterioles (near the right end of the interval labelled 3 in Fig. 12.5) and the capillaries (labelled 4 in Fig. 12.5). This leads to:

$$\Delta p_{\text{arteriole}} = \Delta p_{\text{capillary}} \Rightarrow \frac{\sigma_{\text{arteriole}}}{r_{\text{arteriole}}} = \frac{\sigma_{\text{capillary}}}{r_{\text{capillary}}}.$$

Therefore, the surface tension for a capillary of small radius must be smaller than the surface tension of an arteriole with a larger radius:

$$r_{\text{arteriole}} > r_{\text{capillary}} \Rightarrow \sigma_{\text{arteriole}} > \sigma_{\text{capillary}}.$$

This equation is physiologically important because it allows the walls of the capillaries to be thinner; this in turn improves the efficiency of diffusive transport of small ions and oxygen from the blood into the surrounding tissue.

CONCEPT QUESTION 12.4

If we use a surfactant that reduces to 50% the surface tension of a bubble membrane, we obtain the same transmural pressure if

(A) the radius is increased by a factor of 4.

(B) the radius is doubled.

(C) the radius remains unaffected.

(D) the radius is reduced to one-half the previous value.

(E) the radius is reduced to one-quarter the previous value.

CASE STUDY 12.3

A lung has collapsed when the alveoli have retracted so that their membranes stretch almost flat across the ends of the bronchial tubes, as shown in the second sketch from the left in Fig. 12.17. What happens when we try to inflate a collapsed lung?

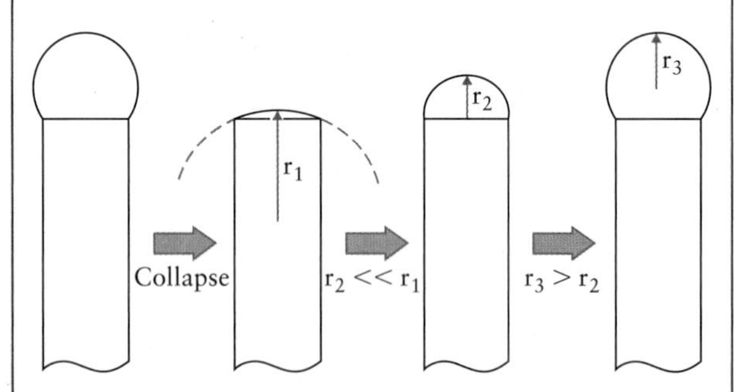

Figure 12.17 Left panel: healthy alveolus sealing the end of a bronchial tube. Three right panels: stages of the inflation of a collapsed lung. Initially (second panel) the lungs resist inflation until the alveoli reach an intermediate, smallest radius (third panel). Thereafter, the completion of the inflation does not require further pressurizing (last panel).

Answer: *When the alveoli are collapsed, the corresponding radius of the alveolar surface, r_1, is large and the transmural pressure in the alveolus is low due to the inverse relation between radius and pressure difference in Laplace's law. When trying to inflate such a collapsed lung, a significant resistance must be overcome. Initially, the tissue resists the inflating of the lung, because pushing the alveoli out of the bronchial tubes significantly reduces their radii, $r_1 >> r_2$. Thus, pressurizing the lung has initially very little effect until an external pressure has been reached that provides the transmural pressure needed for alveoli of radius r_2. When this radius is reached the external pressure must be lowered suddenly because inflating the alveoli from r_2 to r_3 in Fig. 12.17 requires a decreasing pressure based on Laplace's law and $r_2 < r_3$. Thanks to the great elasticity of human tissues the risk of over-pressurizing and rupturing the lungs during medically controlled inflation is not high.*

EXAMPLE 12.6

We study two alveoli in competition at adjacent bronchial tubes, as shown in Fig. 12.18. The actual size of any two adjacent alveoli will usually differ slightly. How can nature address this problem?

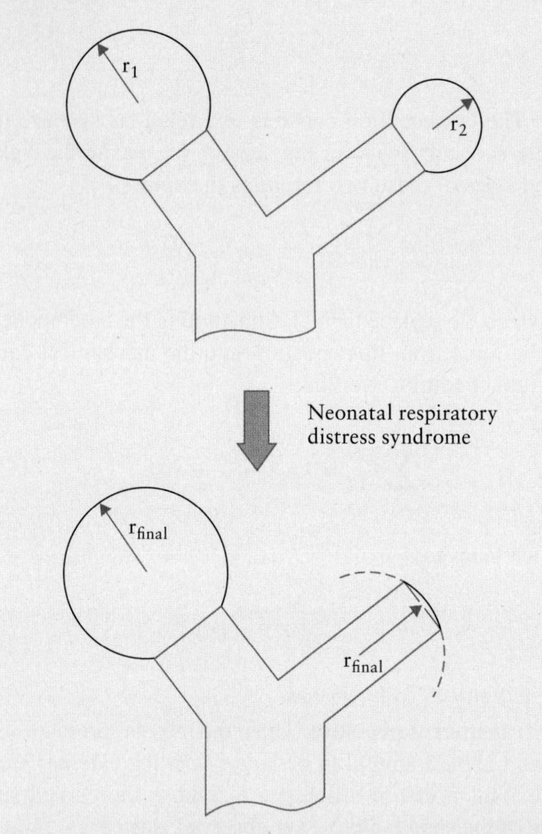

Figure 12.18 Sketch of two alveoli of different radii located near a bronchial branching point. The smaller alveolus requires a larger pressure difference; therefore, it should collapse in favour of the bigger alveolus. After collapsing, both alveoli have the same radius r_{final}, but the previously smaller alveolus is essentially completely retracted into the bronchial tube. A healthy lung has means of preventing this effect using pulmonary surfactants to modify the surface tension in the alveoli. However, prematurely born babies lack these surfactants, and the collapse of small alveoli is observed. This neonatal respiratory distress syndrome was in the past a major cause of death in prematurely born babies.

Solution

When using Laplace's law quantitatively, we have to carefully study the experimental situation because the law contains a numerical factor that varies from situation to situation (compare Eqs. [12.10] to [12.12]). For the alveoli in the lungs we choose Eq. [12.11] because they are spherical but have only one open surface inside, with the outside immersed in extracellular fluid. This is an inverse droplet configuration.

Based on Fig. 12.18 we consider two adjacent alveoli with different radii, $r_1 > r_2$. Since the external pressure $p_{outside}$ in the surrounding tissue is the same for both

continued

alveoli, the corresponding internal pressures of the two alveoli would have to vary due to Laplace's law:

$$r_1 > r_2 \Rightarrow \Delta p_1 = \frac{2\sigma}{r_1} < \frac{2\sigma}{r_2} = \Delta p_2.$$

However, the internal pressures cannot be different for adjacent alveoli because their air spaces are connected through the bronchial tree. This means that either one of the two alveoli cannot be in mechanical equilibrium for a given pulmonary pressure. In practice, the air pressure in the lungs would not sufficiently pressurize the smaller alveoli and they would collapse. We cannot allow this to happen, since the air exchange in the lungs is proportional to the contact area of the alveoli with the adjacent blood capillaries; i.e., small alveoli contribute more effectively than larger ones.

To prevent this problem, alveolar cells produce **pulmonary surfactants** that wet the alveolar surface to counterbalance the radius effect in Laplace's law by particularly lowering the surface tension σ for smaller alveoli. Pulmonary surfactant is formed starting late in fetal life. For this reason, it is often the case in premature births that insufficient pulmonary surfactant is present, which causes *neonatal respiratory distress syndrome*. For a baby with this syndrome the lungs are stiff, with some alveoli collapsed and others likely to be filled with fluid. Immediate medical attention is required at a neonatal hospital unit, as shown in Fig. 12.19.

By adding some soap to a bottle filled with water, you can easily demonstrate that the competitive effect between small and large bubbles discussed in this example does indeed exist. When water is poured out of the bottle, air streams in and bubbles form. You observe that the smaller bubbles arch into larger, more stable bubbles.

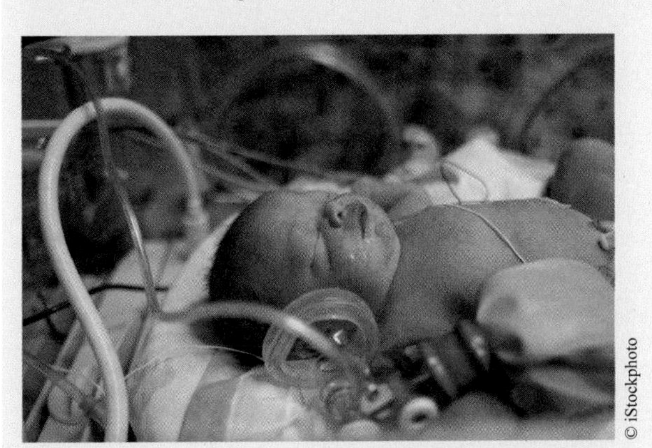

Figure 12.19 A newborn baby in a neonatal hospital unit.

EXAMPLE 12.7

An average alveolus in the human lungs has a radius of about 50 μm. (a) What is the predicted transmural pressure if we model the extracellular fluid outside the alveolus as water with the surface tension value from Table 12.2? (b) Do you recall the actual value from our earlier discussions in this textbook?

continued

Solution

Solution to part (a): We use Eq. [12.11] for the same reason outlined in Example 12.5. Substituting the given radius and σ from Table 12.2, we get:

$$\Delta p = \frac{2\sigma}{r} = \frac{2\left(0.073\frac{\text{N}}{\text{m}}\right)}{5.0\times10^{-5}\text{m}} = 2900 \text{ Pa},$$

i.e., a value close to 3 kPa.

Solution to part (b): The transmural pressure in the lungs is defined as the difference between the alveolar and pleural pressures, $p_{\text{alveoli}} - p_{\text{pleura}}$. This pressure is included in Figs. 8.9 and 9.1, where we found that this value never exceeds 2 kPa and for regular breathing lies in the range of 0.5 to 0.7 kPa. Thus, pulmonary surfactants cause the transmural pressure to be lowered from the value in part (a) by a factor of 4 to 6.

12.6: Capillarity

Capillarity is an effect closely related to surface tension. It is illustrated in Fig. 12.20, which compares the results of two experiments. A small, hollow glass tube is immersed in water (left sketch) or liquid mercury (right sketch). It is observed that the water rises in the tube and its surface bends up slightly at the water/glass/air interface. In turn, the mercury level in the tube is lower than the surface of the surrounding mercury and the mercury bends downward at the mercury/glass/air interface. This indicates that both the surface tension of the liquid

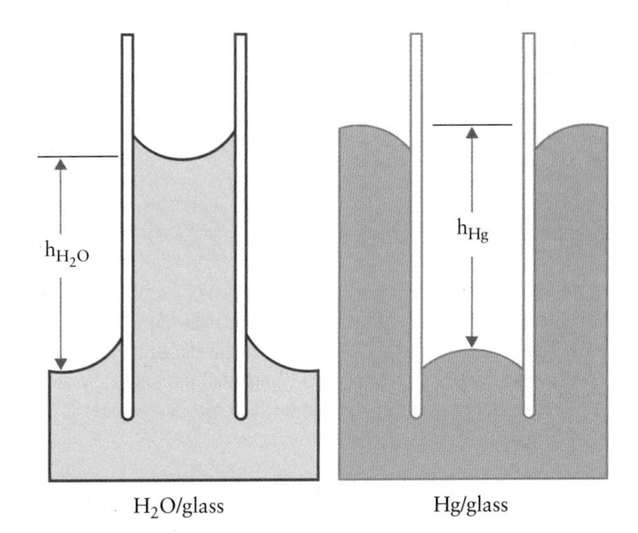

Figure 12.20 Comparison between the wetting of a hollow glass tube immersed in water (left panel) and immersed in mercury (right panel). The two final equilibrium states shown indicate that water wets glass, which leads to an upward force on the water in the tube. This effect is called capillarity (capillary action). Mercury does not wet glass, leading to a mercury level in the tube below the mercury surface in the remaining beaker.

and the interfacial tension at the water/glass or mercury/glass interfaces play a role in this experiment. Surface tension values are always positive since forming a surface toward air requires energy to reduce the number of attractive neighbouring molecules in the liquid, as indicated in Fig. 12.12. Interfacial tensions can be positive or negative since a solid or liquid surface contains a similar density of molecules for interaction as the studied liquid. The attractive interaction with equal molecules is replaced by a possibly stronger attractive interaction with the molecules on the other side of the interface. This is the case for water and glass. Consequently, water wets glass because forming this interface requires less energy than forming a surface with air. Mercury, in turn, does not wet glass, as formation of the glass/mercury interface requires a large amount of energy.

A second experiment illustrating the effect of capillarity is shown in Fig. 12.21. Two tubes with different inner diameters are submerged in water. The height of the water level in a tube is greater when the tube diameter is smaller. This is due to the total energy required in each tube to form surfaces toward air and interfaces along the glass walls. Capillarity is an essential part of many life processes. In taller animals blood is pumped through arteries and veins but capillarity is important in the smallest blood vessels, which are therefore called capillaries. We address this issue in part (b) of Example 12.8.

Andrew Lambert Photography/Science Photo Library/Custom Medical Stock Photo

Figure 12.21 The capillarity effect of water. Two glass tubes of varying diameter are submerged into water. The tube with the smaller diameter at left shows the greater capillarity effect because the interaction between water and the glass surface contributes more strongly to the water column's mechanical equilibrium, which allows the water to rise to higher levels.

EXAMPLE 12.8

(a) Using Fig. 12.22(a) for a liquid that has risen in a capillary to height h_{liquid} and forms a contact angle θ with the inner surface of the capillary of radius r, show that

continued

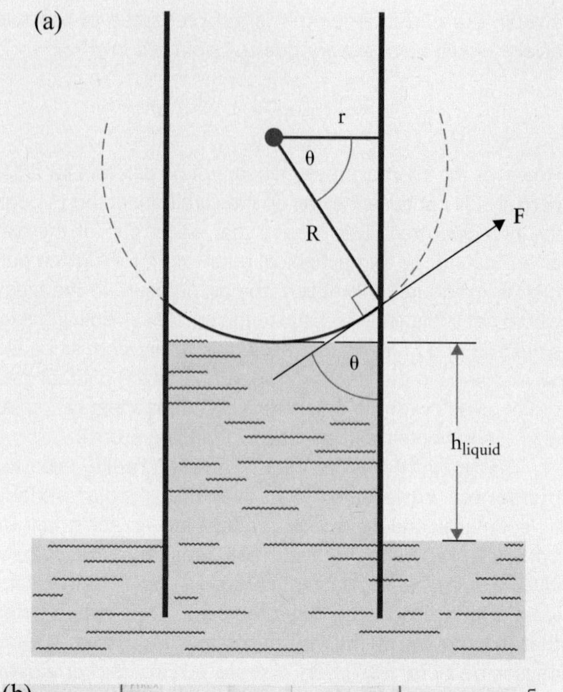

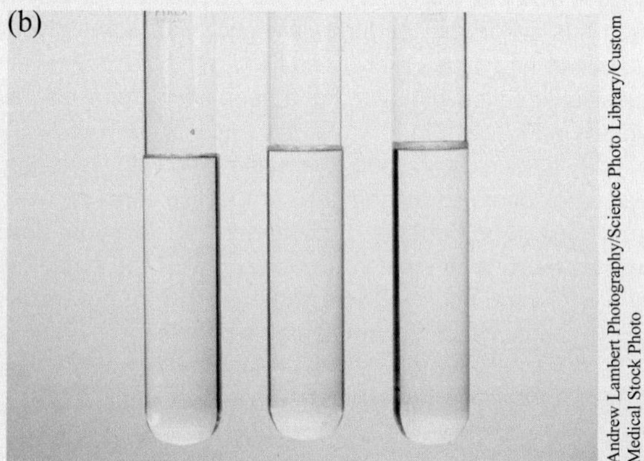

Andrew Lambert Photography/Science Photo Library/Custom Medical Stock Photo

Figure 12.22 (a) Sketch of a liquid column rising in a narrow capillary of radius r to a height h_{liquid}. The curvature of the meniscus of the liquid is defined by the radius of curvature, R, which leads to a contact angle θ of the liquid with the inner capillary wall. The force related to the surface tension is indicated. (b) Comparison of the water menisci of distilled water (left), tap water (centre), and seawater (right).

the height of the liquid column in the capillary is given by **Jurin's law**:

$$h_{liquid} = \frac{2\sigma_{liquid}}{\rho_{liquid} g} \frac{\cos\theta}{r}. \qquad [12.13]$$

(b) The value of the surface tension for whole blood is given in Table 12.2. Its density is 1.06 g/cm^3. Calculate the maximum height to which blood can rise in a capillary that has a diameter of 4.5 μm.

Solution

Solution to part (a): In Fig. 12.22(a) you notice that the meniscus of the liquid surface is not flat but curved like an inverse sphere. This is due to the capillarity effect in

continued

the tube and is quantified with Laplace's law. We use Eq. [12.11] with the proportionality factor 2 for the droplet because there is only one interface between liquid and air. The pressure difference between the liquid at the meniscus in the tube and the outside air is:

$$p_{\text{liquid surface}} - p_{\text{atm}} = -\frac{2\sigma_{\text{liquid}}}{R},$$

in which R is the radius of curvature of the meniscus, as shown in Fig. 12.22(a). This equation has a negative sign on the right-hand side since the meniscus is curved upward (as opposed to the curvature of the bubbles in Figs. 12.17 and 12.18). The radius of curvature is related to the inner radius of the capillary and the contact angle θ, as shown in Fig. 12.22(a):

$$R = \frac{r}{\cos\theta}.$$

Pascal's law enables us to write a second condition for the same pressure difference since we know that the surface pressure of water and the air pressure are the same at the water/air interface outside the tube, i.e., at a height h_{liquid} below the meniscus in the capillary in Fig. 12.22(a). Thus, at the meniscus in the tube we find:

$$p_{\text{liquid surface}} - p_{\text{atm}} = -\rho_{\text{liquid}} g h_{\text{liquid}}.$$

We combine the last three equations to eliminate the gauge pressure of the liquid surface and the radius R:

$$-\frac{2\sigma_{\text{liquid}}}{r/\cos\theta} = -\rho_{\text{liquid}} g h_{\text{liquid}},$$

which is rearranged as Jurin's law.

Fig. 12.22(b) shows that this effect can be displayed for water with different concentrations of salt: the three test tubes contain distilled water (left), tap water (centre), and seawater (right). Distilled (i.e., highly purified) water has a small meniscus height and seawater, with its high ion concentration, has the highest meniscus. Dissolved salts increase the surface tension of seawater.

Solution to part (b): This part of the example is an application of Jurin's law. First, we need to interpret the term "maximum height" in the example text. Note that we do not identify the contact angle θ. The range of possible contact angles between 0° and 180° allows for the range of cosine values $-1 \le \cos\theta \le +1$ in Jurin's law. Thus, a maximum height is reached when $\cos\theta = 1$:

$$h_{\text{max,blood}} = \frac{2\sigma_{\text{blood}}}{r\rho_{\text{blood}} g}.$$

We substitute the values given in the example text and Table 12.2, and obtain:

$$h_{\text{max,blood}} = \frac{2\left(5.8\times10^{-2}\,\frac{\text{N}}{\text{m}}\right)}{\left(1.06\times10^{3}\,\frac{\text{kg}}{\text{m}^3}\right)\left(9.8\,\frac{\text{m}}{\text{s}^2}\right)(2.25\times10^{-6}\,\text{m})}$$

$$= 5.0\ \text{m}.$$

Significant amounts of water are transported upward in trees: a full-grown birch evaporates about 350 litres of water from its leaves per summer day. It is still a matter of research to what extent the effect shown in Fig. 12.21 plays a role in the transport of water and sap in tall trees. In order to transport water by capillarity into the canopies of even the tallest trees (e.g., eucalyptus trees of 150 m height), capillary diameters of smaller than 0.1 μm would be required. However, the xylem fibres, in which water has been shown to rise in trees, typically have diameters of 20 μm to 300 μm. Xylem tissue of a leaf midrib is shown in a coloured scanning electron micrograph in Fig. 12.23. In the cross-section shown, the midvein runs through the centre. The layer surrounding it is large mesophyll cells, followed by a thin outer layer of epidermal cells. The midvein contains the larger, water-carrying xylem tubes bundled at the centre and surrounded by a ring of smaller phloem tubes that carry nutrients.

Clearly, other effects play a role in the transport of water from the ground to the leaves. We study the wood anemone as an example. The wood anemone is a flower that stands erect and opens only when the sun shines. When a cloud covers the sun or when we create an artificial shade for the flower, the flower closes and the stem bends down. This is a sign of reduced water pressure in the stem. When the shade is removed, the flower resumes its erect posture as the water pressure in the stem increases again. In the anemone, the water pressure is regulated by chemical reactions that operate only under sunny and warm conditions.

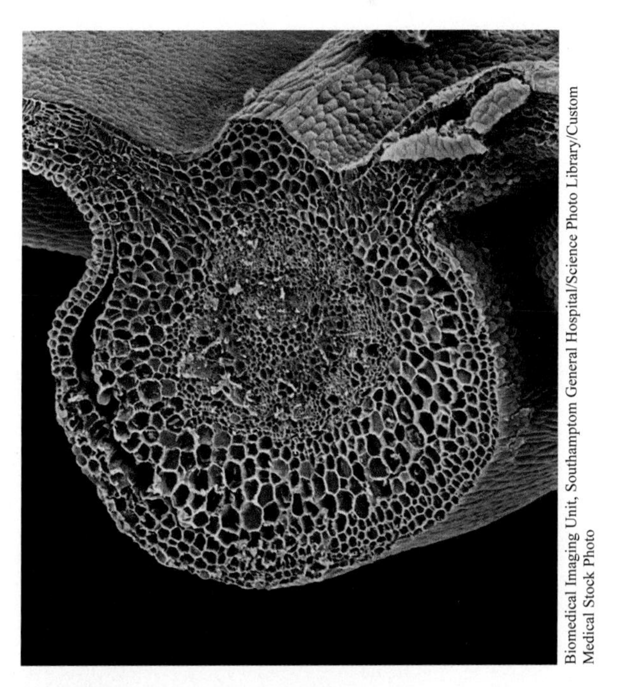

Figure 12.23 Leaf midrib as seen with coloured scanning electron microscopy (SEM). This is a cross-section through the midrib, the continuation of a leaf's stem along the centre of the leaf. The main vein, called the midvein, runs through the centre of the midrib.

SUMMARY

DEFINITIONS

- Atmospheric pressure: $p_{atm} = 1.01 \times 10^5$ Pa
- Gauge pressure: a pressure value relative to air pressure:

$$p_{gauge} = p_{absolute} - p_{air}$$

- Buoyant force:

$$F_{buoyant} = \rho_{fluid} V_{object} g$$

- Surface energy σ is the energy required to form an area of 1 m^2 of new surface:

$$\sigma = \frac{\Delta E}{\Delta A}$$

- Surface tension (equivalent to surface energy):

$$\sigma = \frac{F}{l_x}$$

l_x is the length along which the force acts tangentially to the surface

UNITS

- Surface tension σ: J/m^2 = N/m

LAWS

- Pascal's law (for fluid without identifiable surface):

$$p_2 - p_1 = -\rho g(y_2 - y_1),$$

in which index 1 and index 2 refer to two arbitrarily chosen vertical positions in the fluid

- Pascal's law (depth d measured from surface downward):

$$p = p_{atm} + \rho g d$$

- The Archimedes' principle (for the vertical components of the forces):

$$F_{net} = F_{buoyant} - W_{object}$$

- Laplace's formula for pressure difference across the surface for (r is radius)
- hollow bubble: $\Delta p = 4\sigma/r$
- homogeneous droplet or hollow cylinder: $\Delta p = 2\sigma/r$
- homogeneous cylinder: $\Delta p = \sigma/r$
- Jurin's law for capillarity:

$$h_{liquid} = \frac{2\sigma_{liquid}}{\rho_{liquid} g} \frac{\cos\theta}{r},$$

in which r is radius, and θ is contact angle of fluid with capillary wall

MULTIPLE-CHOICE QUESTIONS

MC–12.1. Which law is used to quantify the pressure at the bottom of a lake?
(a) Arrhenius law
(b) Pascal's law
(c) Newton's first law
(d) Laplace's law

MC–12.2. Which of the following pressures can be negative?
(a) the transmural pressure in the lungs
(b) the blood pressure in supine position
(c) the alveolar pressure
(d) the air pressure
(e) the water pressure below the surface of a lake

MC–12.3. The vertical distance from feet to heart of a standing individual is 1.2 m. Using for the density of blood $\rho = 1.06$ g/cm^3, find the difference in blood pressure between the two levels (heart and feet). Choose the closest value.
(a) 1270 Pa
(b) 1.06 kPa
(c) 12.5 kPa
(d) 1.0 atm

MC–12.4. Pascal's law is applied to an unknown liquid in an open container. We observe that the pressure is 1.05 atm at 10 cm below the surface. At what depth in the liquid is the pressure 1.2 atm?
(a) 0.2 m
(b) 0.4 m
(c) 0.8 m
(d) 1.2 m
(e) 2.5 m

MC–12.5. Under what circumstances can air *not* be modelled as an ideal stationary fluid?
(a) when applying Pascal's law
(b) when determining the buoyant force acting on an object
(c) when applying Laplace's law
(d) when describing the pressure in a soap bubble

MC–12.6. Exchanging one fluid for another in a given beaker, the density of the beaker's content increases if
(a) the fluid mass decreases per unit volume.
(b) the fluid volume decreases per unit mass.
(c) both fluid mass and volume double.
(d) both fluid mass and volume are halved.

MC–12.7. We consider a motionless shark sinking slowly to the bottom of a lagoon. Which of the choices in Fig. 12.24 represents the proper free body diagram for the shark of weight $\vec{W}$? *Hint:* The arrows are not drawn proportional to magnitude of the respective force.

MC–12.8. Most fish can float in water. What must be true for such fish?
(a) Their density is larger than that of water.
(b) Their density is equal to that of water.
(c) Their density is smaller than that of water.
(d) No conclusions about the fish's density can be drawn.

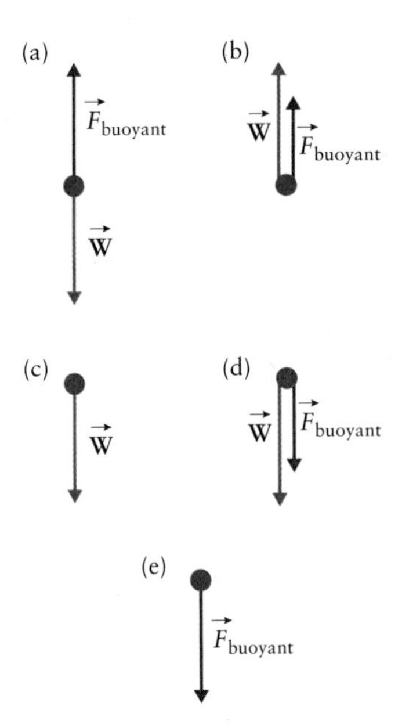

Figure 12.24

MC–12.9. Siamese fighting fish (shown in Fig. 12.25) have an unusual way of caring for their young. Males prepare a home for their future offspring by taking gulps of air and blowing saliva-coated bubbles that collect at the water's surface as a glistening froth. When a female arrives ready to mate, she swims under the bubble-nest where the male embraces her and fertilizes her eggs. Then, picking up the eggs in his mouth, he spits them, one by one, into the bubbles. The role of father continues for the male Siamese fighting fish as he conscientiously watches over the developing eggs. Any that slip from a bubble and start to sink are carefully retrieved and spat back into a bubble. Based on these observations, what statement can we make about the Siamese fighting fish's fertilized eggs?
(a) They are heavy.
(b) They have a small volume.
(c) Their density is larger than that of water.
(d) Their density is equal to that of water.
(e) Their density is smaller than that of water.

Figure 12.25 A Siamese fighting fish.

MC–12.10. Consider the *Hindenburg* zeppelin, which exploded in 1937 at Lakehurst, New Jersey. At the time of the accident, its tanks were filled with hydrogen gas. Had they been filled with the following gas, the airship could have carried more passengers.
(a) helium
(b) methane
(c) carbon monoxide
(d) nitrogen
(e) no existing gas could carry more passengers when used as a filling of the tanks

MC–12.11. The food label on a package of cheese starts with the line: "Nutrition facts per 3 cm dice (30 g)." If we take a piece of that cheese and throw it into seawater of density 1025 kg/m^3, what happens?
(a) The piece sinks.
(b) The piece floats at the surface; i.e., part of the piece rises above the water surface.
(c) The piece floats fully immersed in the water, i.e., with its entire volume below the water surface.
(d) What happens depends on the size of the piece we throw into the water.
(e) What happens depends on the mass of the piece we throw into the water.

MC–12.12. We study a solid steel sphere completely immersed in water (which we treat as an ideal stationary fluid). We use the variable d to represent the depth below the water surface. Which sketch in Fig. 12.26 shows the dependence of the magnitude of the buoyant force acting on the solid sphere as a function of depth?

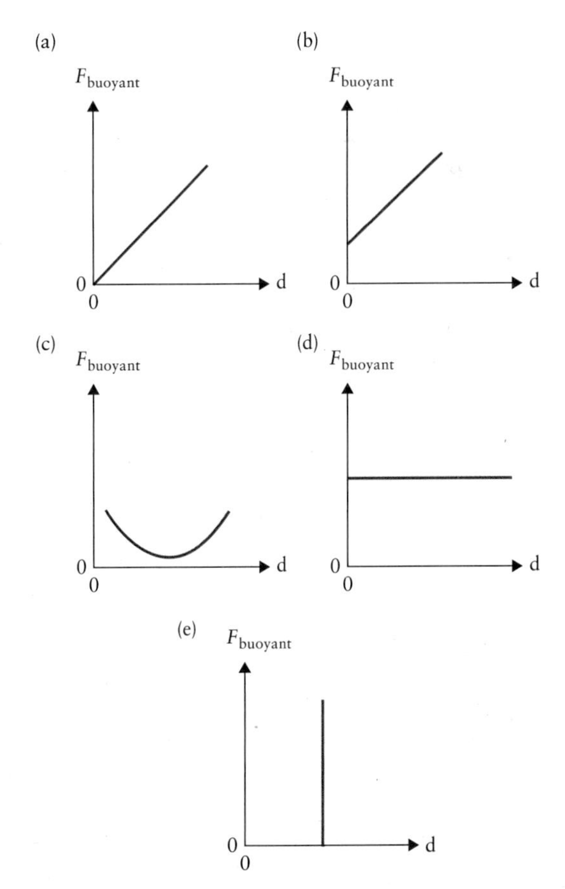

Figure 12.26

MC–12.13. The magnitude of the buoyant force acting on a seagull in level flight is equal to
(a) the density of the bird.
(b) the mass of the bird.
(c) the volume of the bird.
(d) the weight of the bird.
(e) the weight of the air displaced by the bird.

MC–12.14. We study Fig. 12.27. Part (a) shows a sphere with a radius $r = 10$ cm and density $\rho_a = 1.8$ g/cm^3 suspended in water. Part (b) shows a wooden sphere of diameter $d = 10$ cm (density $\rho_b = 0.95$ g/cm^3) anchored under water with a string. Which of the four free body diagrams shown in Fig. 12.28 correspond to the two cases in Fig. 12.27? *Note:* $\vec{T}$ is tension, $\vec{W}$ is weight, and $\vec{F}_{\text{buoyant}}$ is buoyant force.
(a) The free body diagram in sketch Fig. 12.28(a) belongs to the case in Fig. 12.27(a), and the free body diagram in sketch Fig. 12.28(b) belongs to the case in Fig. 12.27(b).
(b) The free body diagram in sketch Fig. 12.28(b) belongs to the case in Fig. 12.27(a), and the free body diagram in sketch Fig. 12.28(c) belongs to the case in Fig. 12.27(b).
(c) The free body diagram in sketch Fig. 12.28(c) belongs to the case in Fig. 12.27(a), and the free body diagram in sketch Fig. 12.28(d) belongs to the case in Fig. 12.27(b).
(d) The free body diagram in sketch Fig. 12.28(c) belongs to the case in Fig. 12.27(a), and the free body diagram in sketch Fig. 12.28(a) belongs to the case in Fig. 12.27(b).
(e) The free body diagram in sketch Fig. 12.28(b) belongs to the case in Fig. 12.27(b), and the free body diagram in sketch Fig. 12.28(c) belongs to the case in Fig. 12.27(a).

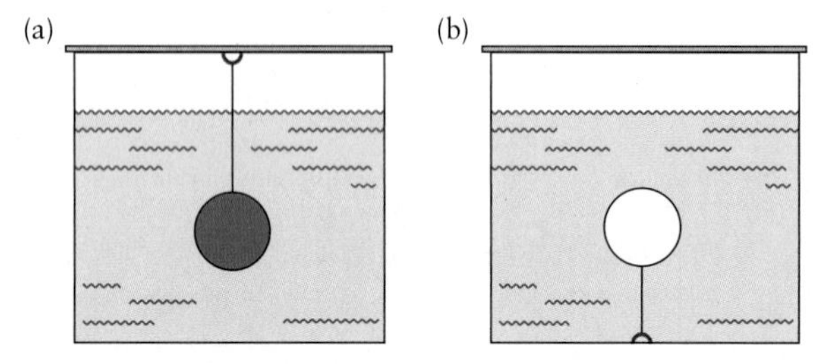

Figure 12.27 (a) A sphere of radius $r = 10$ cm and density $\rho_a = 1.8$ g/cm^3 is suspended in water. (b) A wooden sphere of diameter $d = 10$ cm and density $\rho_b = 0.95$ g/cm^3 is held under water by a string.

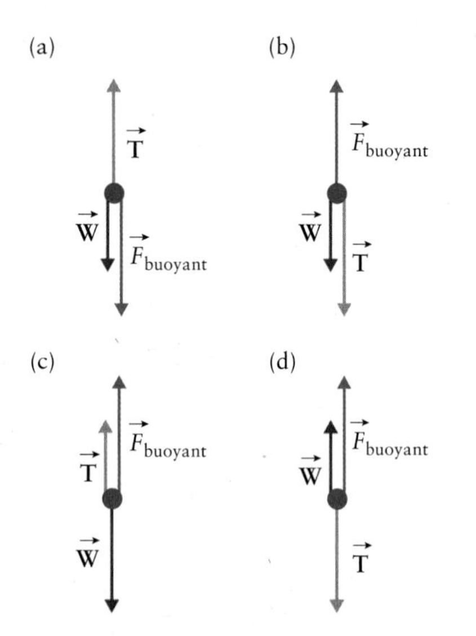

Figure 12.28

MC–12.15. Which is the standard unit of surface tension?
(a) J
(b) Pa
(c) (kg m)/s^2
(d) kg/s^2
(e) kg/(ms^2)

MC–12.16. Laplace's law describes the pressure in an alveolus in the lungs in the form $p_{\text{inside}} - p_{\text{outside}} = 2\sigma/r$, in which σ is the surface tension and r is the radius of curvature of the alveolus. In healthy alveoli a surfactant is used to reduce the surface tension by coating parts of the surface. Which of the following statements is false?
(a) The surfactant particularly must coat areas with a large radius of curvature.
(b) The surfactant particularly must coat areas with a small radius of curvature.
(c) A surfactant does not change the pressure in the bubble.

MC–12.17. Fig. 12.17 shows in three steps how a lung collapses and then is re-inflated under medical observation. We call the collapse "step 1," the step from alveolar radius r_1 to r_2 "step 2," and the final step from alveolar radius r_2 to r_3 "step 3." In which step must a health practitioner exert the greatest pressure from outside through the mouth?
(a) during step 1
(b) during step 2
(c) during step 3
(d) after step 3 to keep the lung from collapsing again

MC–12.18. Which law is used to quantify the pressure in a soap bubble?
(a) Jurin's law
(b) Pascal's law
(c) Newton's third law
(d) Laplace's law

MC–12.19. You study a large and a small soap bubble. In which of the two is the air pressure higher?
(a) neither; it's the same
(b) in the larger bubble
(c) in the smaller bubble
(d) depends on the common outside pressure
(e) impossible to predict

MC–12.20. A water strider is an insect that can walk on water. Looking at one of the six legs of the water strider resting on the water surface, which of the following is the case for the surface underneath the insect's foot?
(a) It is perfectly flat.
(b) It is bent upward due to the attractive force of the foot.
(c) It is curved downward because the foot simulates an increased pressure from above the water surface.
(d) It splits and the lower end of the foot dangles below the surface.

MC–12.21. In Fig. 12.12 a force is shown acting on a molecule in the water surface. Our knowledge of Newton's laws tells us that
(a) the water molecule accelerates downward until it hits the bottom of the beaker.
(b) the water molecule accelerates downward until it has left the surface.
(c) at least one more force must act on the molecule to keep it at the surface.
(d) the water molecule is ejected into the gas space above due to the reaction force to force $\vec{F}$.

CONCEPTUAL QUESTIONS

Q–12.1. Vapour bubbles in a beaker of boiling water get larger as they approach the surface. Why?

Q–12.2. Fig. 12.29(a) shows a beaker on a scale. The beaker is filled to the rim with water. It is then taken from the scale and a piece of wood is lowered into the beaker. The beaker with wood is then placed back on the scale, as shown in Fig. 12.29(b). How has the reading of the scale changed from part (a) to part (b)?

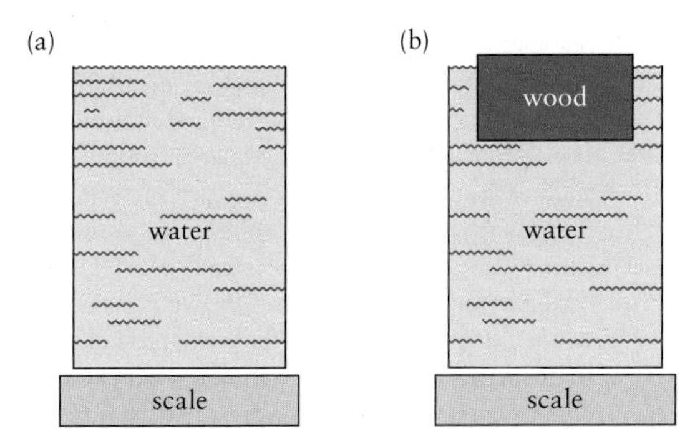

Figure 12.29 (a) A beaker filled to the rim with water is placed on a scale. (b) A piece of wood is lowered into the beaker and the beaker is placed again on the scale.

Q–12.3. A few lizards have developed specialized feet that allow them to climb smooth surfaces, such as windows, without the aid of claws. Examples include several gecko species, which can even walk across the ceiling in tropical homes. (a) How can they hold on to such a smooth surface? (b) How can they walk at a reasonable pace up the window glass surface?

Q–12.4. We introduced Pascal's law in two forms in Eqs. [12.1] and [12.2]. Under what circumstances can only the form in Eq. [12.1] be used?

ANALYTICAL PROBLEMS

P–12.1. The sphere is the shape with the smallest surface for a given volume. To prove this statement properly requires variational analysis. Here we want to confirm this result only for a selection of highly symmetric shapes by calculating the ratio of surface to volume. Find these ratios for each of these shapes.
(a) sphere
(b) cylinder
(c) cube
(d) pyramid
(e) tetrahedron
(f) cone

Does the statement hold for these six shapes? For some of the required data see the section on Symmetric Objects, which is a Math Review located after Chapter 27.

P–12.2. A diver accustomed to standard snorkel tubing of length 25 cm tries a self-made tube of length 7 m. During the attempt, what is the pressure difference between the external pressure on the diver's chest and the air pressure in the diver's lungs? *For those interested:* What happens to the diver as a result of the attempt?

P–12.3. A scuba diver takes a deep breath from an air-filled tank at depth d and then abandons the tank. During the subsequent ascent to the surface the diver fails to exhale. When reaching the surface, the pressure difference between the external pressure and the pressure in the lungs is 76 torr. At what depth did the diver abandon the tank? *For those interested:* What potentially lethal danger does the diver face?

P–12.4. What is the pressure increase in the fluid in a syringe when a force of 50 N is applied to its circular piston, which has a radius of 1.25 cm?

P–12.5. What minimum gauge pressure is needed to suck water up a straw to a height of 10 cm? Recall that the gauge pressure is defined as the pressure relative to atmospheric pressure, $p_{gauge} = p - p_{atm}$.

P–12.6. Collapsible plastic bags are used in hospitals for infusions. We want to use such a bag to infuse an electrolyte solution into the artery of a patient. For this we mount the bag at a height h above the arm of the patient, as shown in Fig. 12.30. Assuming the average gauge pressure in the artery is 13.3 kPa and the density of the electrolyte solution is 1.03 g/cm^3, what is the minimum height h at which the infusion would work?

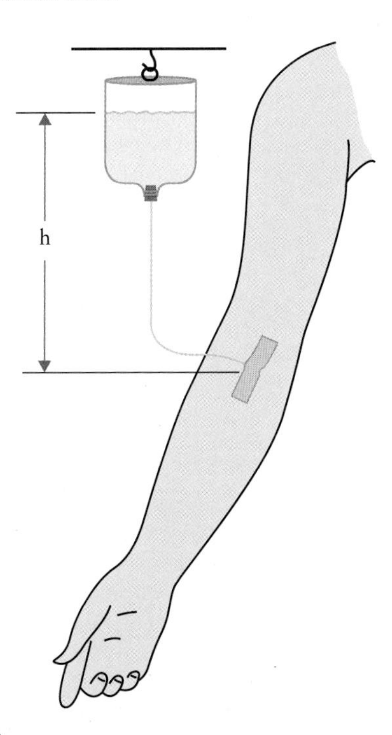

Figure 12.30

P–12.7. Water is pumped to the top of the 365-m-tall Empire State Building in New York City. What gauge pressure is needed in the water line at the base of the building to achieve this?

P–12.8. The U-shaped glass tube in Fig. 12.31 contains two liquids in mechanical equilibrium: water of density $\rho_w = 1.0$ kg/L and an unknown liquid of density ρ_l. The unknown liquid is in the left tube, floating on top of the water with a clearly visible interface. Use $h_1 =$ 150 mm and $h_2 = 15$ mm with the heights as labelled in Fig. 12.31. What is the density ρ_l?

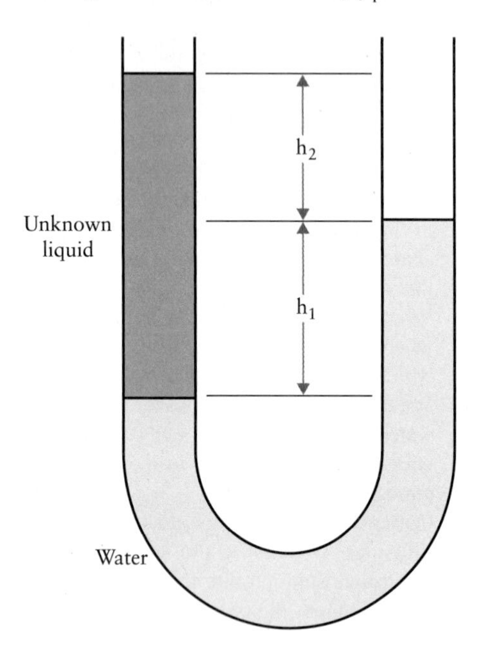

Figure 12.31

P–12.9. The density of ice is $\rho_{ice} = 920$ kg/m^3 and the average density of seawater is $\rho_w = 1.025$ g/cm^3. What fraction of the total volume of an iceberg is exposed?

P–12.10. We consider a hot-air balloon of mass 250 kg (basket and envelope). The spherical envelope of the balloon has a diameter of 16 m when fully inflated. To what temperature must the enclosed air be heated for the balloon to carry four standard men? Assume the surrounding air is at 20°C and is treated as an ideal gas. Use 29 g/mol for the molar mass of air.

P–12.11. (a) Fig. 12.27(b) shows a wooden sphere with a diameter of $d = 10$ cm (density $\rho = 0.9$ g/cm^3) held under water by a string. What is the tension in the string? (b) Fig. 12.27(a) shows a sphere of radius $r = 10$ cm and density of $\rho = 2.0$ g/cm^3 suspended in water. What is the tension in the string? *Note:* Draw the free body diagram in each case.

P–12.12. An object is placed in a hemispherical pod such that the pod just floats without sinking in a fluid of density 1.50 g/cm^3. If the pod has a radius of 5.0 cm, what is the combined mass of the pod and the object placed in it?

P–12.13. A small boat is 4.5 m wide and 6.5 m long. When a cargo is placed on the boat the boat sinks an additional 4.5 cm into the water. What is the mass of the cargo?

P–12.14. A particular object weighs 350 N in air and 250 N when immersed in alcohol of density 0.7 g/cm^3. What are (a) the volume and (b) the density of the object?

P–12.15. An object has a mass of 35 kg when measured in air. It is then tied to a string and immersed in water. The tension in the string is now 280 N. Later, when immersed in oil the tension becomes 300 N. Find (a) the density of the object and (b) the density of the oil.

P–12.16. In order to lift a wire ring of radius 1.75 cm from the surface of a container of blood plasma, a vertical force of 1.61×10^{-2} N greater than the weight of the ring is required. Calculate the surface tension of blood plasma from this information.

P–12.17. A biology lab staining solution has a surface tension of $\sigma = 0.088$ N/m and a density 1.035 g/cm^3. What must the diameter of a capillary tube be so that this solution will rise to a height of 5.0 cm? Assume a contact angle of 0°.

P–12.18. Water is transported upward in plants through xylem tissue, which consists of cells of 1 mm length and a species-dependent diameter between 40 μm and 400 μm. The xylem cells are attached to each other to form a channel. To what maximum height can water rise in these xylem channels due to capillarity? *For those interested:* Confirm this result with a simple experiment: Cut and split the stem of a flower with white petals (e.g., a dahlia or a carnation) and place one half the stem in a glass with dilute red ink and the other half in a glass with dilute blue ink. After several hours the flower will be half red and half blue.

P–12.19. Fig. 3.16 shows how surface tension supports insects such as water striders on the water surface. Assume that an insect's foot is spherical, as shown in Fig. 12.32, and that the insect stands with all its six feet on the water. Each foot presses the water surface down, while the surface tension of the water produces upward forces to restore the flat shape of the water surface. A characteristic profile of the water surface results, as shown in the figure. The mass of the insect is 15 mg and the diameter of the insect's foot is 250 μm. Find the angle θ as indicated in Fig. 12.32(b). *Hint:* The definition of the surface tension provides for a tangential force along the depressed surface of the water, shown as force $\vec{F}$ in the figure. The surface tension of water is given in Table 12.2.

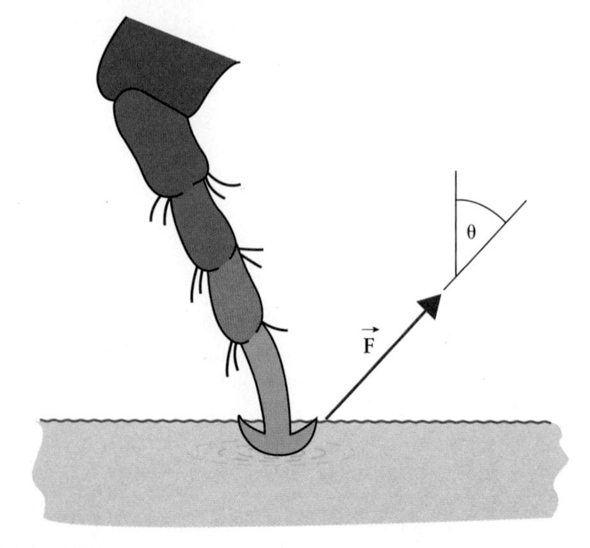

Figure 12.32

ANSWERS TO CONCEPT QUESTIONS

Concept Question 12.1: (E). Start at 1 atm at the surface then add 6 atm to get to 60 m depth.

Concept Question 12.2: (C). At 400 m depth, the pressure is 41 atm. The variation in pressure is 41 atm − 7 atm = 34 atm.

Concept Question 12.3(a): The gauge pressure in seawater near the surface is close to 0 Pa because the water surface is defined by the mechanical equilibrium between air pushing the surface down and water pushing the surface up.

Concept Question 12.3(b): No negative values will be reported because the water pressure never drops below atmospheric pressure.

Concept Question 12.4: (D) based on Laplace's law.

CHAPTER 13

Fluid Flow

A fluid that is not in mechanical equilibrium will flow. Different aspects of flow are described in this chapter, with two models introduced for dynamic fluids: the ideal dynamic fluid and the Newtonian fluid. Both are idealized, as we assume the fluid is incompressible and turbulence free. The flow under these conditions is called laminar flow.

In the ideal dynamic fluid, molecular interactions are limited to elastic collisions. This yields frictionless motion of the fluid at stationary walls. Two laws determine the properties of the resulting flow: the equation of continuity is an expression of the conservation of fluid mass, and Bernoulli's law represents the conservation of energy. These laws predict that the flow through a tapering tube accelerates, and that the pressure in the fluid decreases with increasing speed.

In Newtonian fluid interaction, the container walls become important as fluid molecules transfer energy in inelastic collisions. The equation of continuity continues to apply, but Bernoulli's law is no longer sufficient to describe the pressure in the fluid. We develop the concept of viscosity to take into account flow resistance. In a Newtonian fluid two forces are present in the direction of the flow: a forward-acting force based on a pressure difference along the tube, and a resistance force that depends on the viscosity of the fluid. In steady state a parabolic velocity distribution results in a cylindrical tube and the volume flow rate is proportional to the fourth power of the radius of the tube (Poiseuille's law).

When the properties of a flowing fluid violate the assumptions made for a Newtonian fluid non-Newtonian behaviour is observed, most notably as deviations from Poiseuille's law. Examples include turbulent flow at high fluid speeds and velocity-dependent interactions within the fluid if it is a mixed phase.

Two physiological systems in the human body feature fluid flow: the respiratory system as air streams back and forth into the lungs, and the cardiovascular system as blood flows through the pulmonary and systemic circulations in sequence. We were able to discuss key properties of the respiratory system when developing the gas laws; for the cardiovascular system we need the fluid model introduced in the previous chapter.

13.1: Basic Issues in Blood Flow

Our survey of the most important features of the cardiovascular system starts with Fig. 13.1, in which we combined an anatomical overview with quantifying its prime physiological function:

- In the **systemic circulation**, blood is pumped out of the left ventricle of the heart to the capillaries in the organs throughout the body, where it delivers oxygen and nutrients. Loaded with carbon dioxide, a by-product of cellular respiration, the blood then returns to the heart. From the right atrium it proceeds into the right ventricle, entering
- the **pulmonary circulation**. From the right ventricle blood is pumped to the capillary bed in the lungs, where carbon dioxide is exchanged for oxygen. From the lungs, oxygenated blood then returns to the left atrium to complete a full cycle.

In the systemic circulation, organs are arranged in parallel to allow the body to prioritize oxygen supply based on the vital relevance of the organ and the current metabolic demand. For the major organs, Fig. 13.1 shows the fraction of the received blood flow, I/I_0. $I = \Delta V/\Delta t$ is the volume flow rate through an organ, and I_0 is the total volume flow rate of the systemic circulation. The term *volume flow rate* refers to a fluid volume ΔV that flows through a vessel per time interval Δt. I_0 can be determined from the amount of blood ejected into the aorta per heart beat, which is 70 mL. With the repetition rate of 70 heart beats per minute, a total volume flow rate of $I_0 = 5$ L/min results.

Fig. 13.1 also shows the fraction of oxygen consumed in each organ as a percentage value (labelled O_2). The rate of oxygen uptake in the pulmonary circulation is 0.25 L/min.

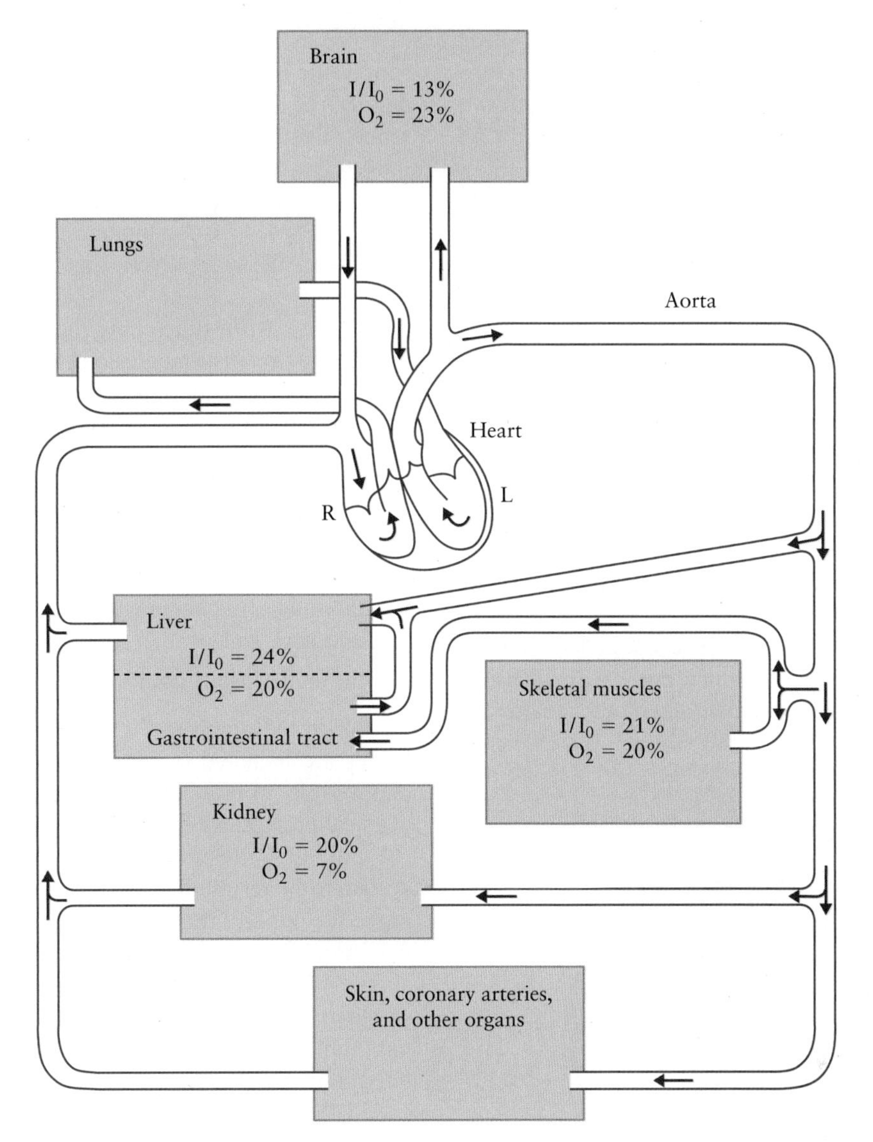

Figure 13.1 Quantitative overview of the flow diagram of the human cardiovascular system. Listed for each major organ are its fraction of the volume flow rate, I/I_0, and its fraction of oxygen consumed (labelled O_2).

Sufficient blood supply to the brain has the highest priority, as the brain is very sensitive to oxygen deficiency (hypoxia). In turn, the brain is efficient in retrieving oxygen from the blood: it requires only 13% of the blood flow to consume 23% of the total oxygen. The same priority is given to the heart muscle, which receives about 5% of the blood, because survival requires continuous blood circulation.

A high priority is further given to the kidneys, which receive 20% to 25% of the blood even though their share is less than 0.5% of the body mass. This preferred supply is due to the filtration function of the kidneys, as discussed in an example later in the chapter. Supply to the skeletal muscles and the gastrointestinal tract with the liver varies significantly with demand. While physically active, up to two-thirds of the blood flow is distributed to the skeletal muscles. While digesting, a similarly high fraction of blood reaches the gastrointestinal tract. Thus, one shouldn't force both organ systems to work at the same time, e.g., by eating just before physical exercise!

Fig. 13.1 points to some of the physical concepts and parameters we need to discuss to understand the physics of the cardiovascular system. We saw that blood circulates in two ways: it passes the systemic and the pulmonary circulations in series, but it flows through several organ systems in parallel. These two flow patterns have to be distinguished. To quantify blood flow along a single vessel, we have to study the dependence of the volume flow rate on blood velocity, blood pressure, and the cross-sectional area of the blood vessel. To correlate the total volume flow rate and the size and strength of the heart, the origin of flow resistance and its consequences for blood flow in various vessels have to be established. We further noted that blood supply can vary; an active regulation process is required for this. Two mechanisms have been proposed for the control of blood flow through organs: (i) a smooth muscle layer on the arterioles near the capillary bed; when these muscles contract, the blood vessel is constricted and blood flow is reduced; and (ii) rings of smooth muscles, called

precapillary sphincters, which are a means of closing capillaries near the afferent arteriole. As a result, blood flow bypasses the capillary bed and reaches a nearby venule directly from the arteriole via a single channel that always remains open.

Blood flow at bifurcations depends on the flow resistance of the two downstream tube systems. Due to this pivotal role of flow resistance, we devote an entire section to the underlying physics based on the concept of viscosity.

The total blood volume of a standard man is 5.1 L (about 8% of the total body mass; see Table 4.1). Fig. 13.2(a) shows the volume distribution in the cardiovascular system. At any time, at least 80% of the blood is in veins, the right ventricle, and the pulmonary circulation. This part of the cardiovascular system is called the low-pressure system since the blood pressure is no greater than 2 kPa. Note that all blood pressure data are given relative to atmospheric pressure, which is about 101.3 kPa. The main purpose of the low-pressure system is the storage of blood. This part of the system can accommodate as much as 98% of the blood in the human body, e.g., when the total blood volume is increased during blood transfusion.

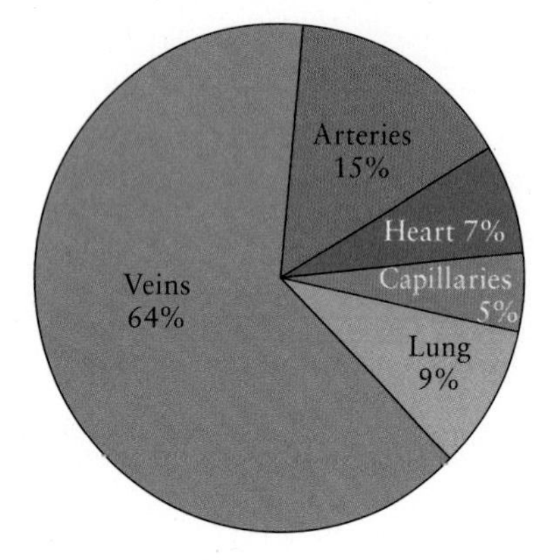

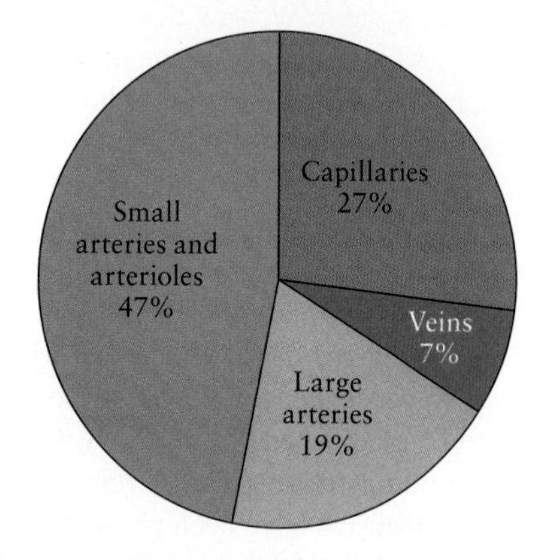

Figure 13.2 (a) Blood volume distribution and (b) the contribution to the flow resistance for the main components of the systemic circulation. While the veins accommodate a major fraction of the blood, it is the small arteries and arterioles that accommodate the largest fraction of the flow resistance.

The main purpose of the high-pressure system is to supply the organs with oxygen. This requires a speedy delivery of blood to the capillaries, leading to the occurrence of more than 90% of the total flow resistance in the systemic circulation, as shown in Fig. 13.2(b). Note that the figure shows that the arterioles, and not the fragile capillaries, accommodate the biggest fraction of this effect. We will explain this observation later in the chapter.

Fig. 13.3 illustrates how the respective volumes in Fig. 13.2(a) are distributed along the systemic circulation. The numbers at the top indicate how many blood vessels of each type are found in a standard man. They vary from one aorta to 5 billion systemic capillaries. Below these numbers are three sketches for

- the individual diameter of a vessel of a given type in unit cm,
- the total cross-section for all vessels of this type in unit cm^2, and
- the total capacity of all vessels of this type in unit cm^3.

Since the graph in Fig. 13.3 is not drawn to scale, values are included to permit its quantitative use. Note that we refer to Fig. 13.3 frequently when quantifying the physical properties of the cardiovascular system.

Fig. 13.4 illustrates that the human cardiovascular system is the result of an extensive evolutionary development; it compares the generalized circulatory schemes of fish, amphibians, and mammals. Fish have a relatively simple two-chambered heart that provides blood to a single circulation. As life moved to dry land, the cardiovascular system became more complex. Amphibians have a three-chambered heart with two circulations, called the pulmocutaneous and the systemic systems. The pulmocutaneous system delivers blood to the skin and to the lungs for oxygen uptake. The systemic system delivers blood under high pressure to the systemic organs. Mammals have a **four-chambered heart** that represents, essentially, the two pumps of a double circulation system.

But how do we know that the anatomical changes from (a) to (c) in Fig. 13.4 that occurred in the past 420 million years are an example of evolutionary progress? We can judge this by how effectively each system is optimized within the framework of applicable physical laws; therefore, let's now discuss these laws.

13.2: Flow of an Ideal Dynamic Fluid

The description of flow cannot be based on the stationary fluid model we introduced in the previous chapter, because it is a dynamic process that requires us to drop the requirement of a mechanical equilibrium. To develop a suitable model, in this chapter we ultimately modify two properties of the stationary fluid model. In this section, we introduce the ideal dynamic fluid model, which

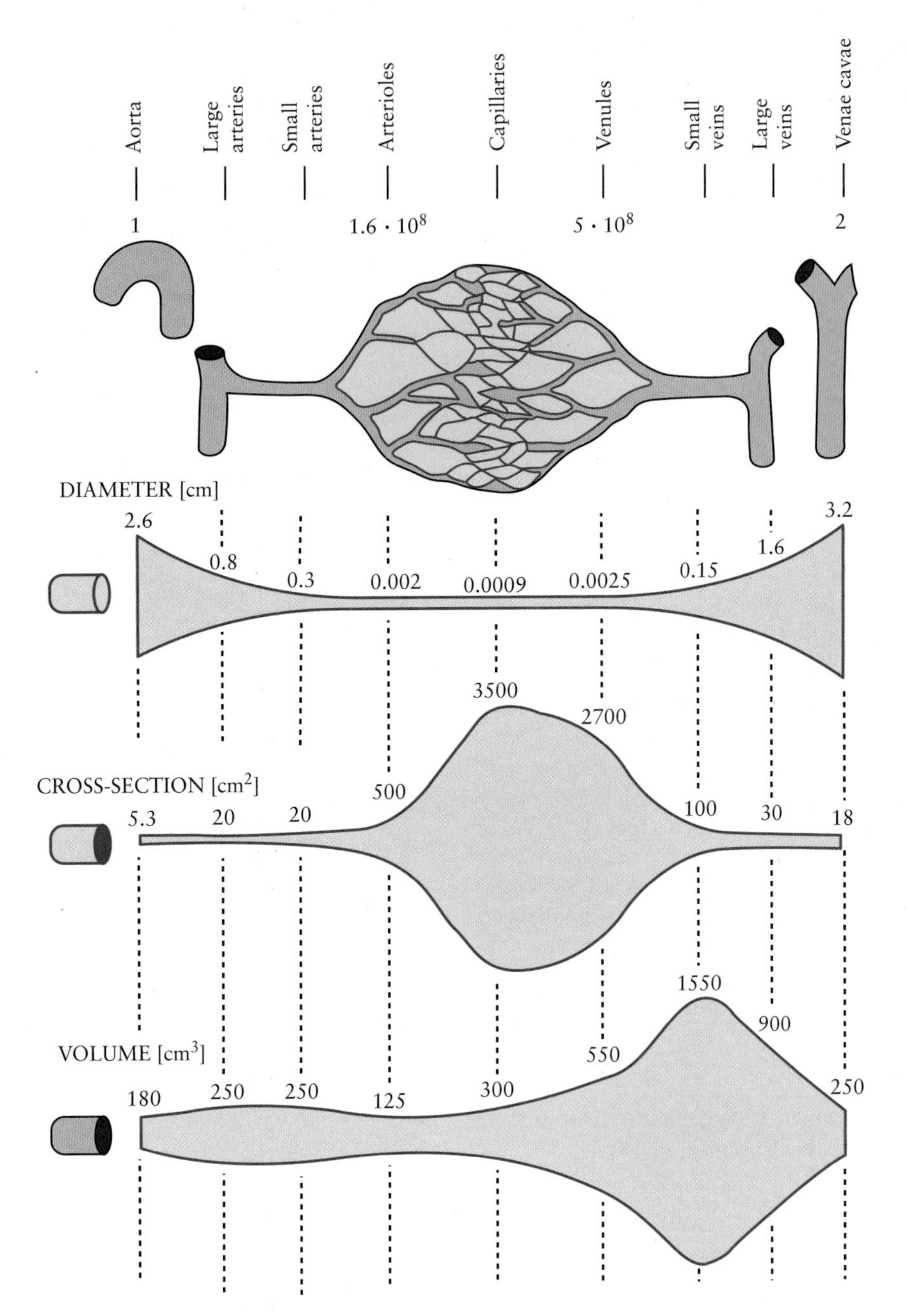

Figure 13.3 Anatomical data for various types of vessels of the systemic circulation. The top row identifies the type of blood vessel. The second row gives numbers of blood vessels for some types. For each type, the *outer* diameter of a single vessel, the *outer* cross-sectional area, and the volume of all vessels of this type are illustrated. The numerical values are given in the indicated units.

requires the fewest modifications to address the mechanical non-equilibrium of the fluid during flow. Later, then, we develop the Newtonian fluid model, which further allows an interaction between the fluid and the confining container walls.

13.2.1: Ideal Dynamic Fluid Model

The ideal stationary fluid is incompressible and deformable. Both of these properties are retained in the dynamic case. The ideal stationary fluid further fills any given container such that the fluid is in mechanical equilibrium. This condition has to be replaced: flow occurs specifically because the fluid has not yet achieved this mechanical equilibrium.

However, this alone does not define the ideal dynamic fluid, because additional effects can occur when a fluid flows. Thus, we have to make further assumptions that address the inclusion or exclusion of such phenomena. The **ideal dynamic fluid** is called ideal because it excludes the largest number possible of phenomena we observe in real fluids. Specifically, we require that

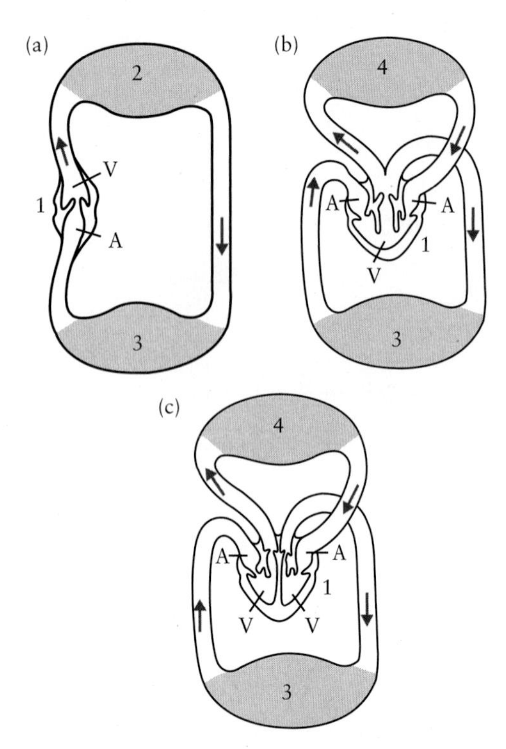

Figure 13.4 Comparison of the circulatory systems of various vertebrates. (a) Fish have a two-chambered heart (1) with a single circulation. Gills (2) and capillary beds in the rest of the body (3) are shaded. (b) Amphibians have a three-chambered heart with two circulatory systems. These are called the pulmocutaneous system and the systemic system. The systemic system is a high-pressure system. The lungs (4) are shaded. (c) Mammals have a four-chambered heart with two circulations: the systemic and the pulmonary circulations. The heart completely separates oxygen-rich blood from the lungs and oxygen-depleted blood emerging from the systemic circulation.

(i) **no turbulences** (which are a departure from smooth flow as specified below) occur during flow,

(ii) **no sound waves** (which require density fluctuations) develop in the flowing fluid, and

(iii) **no friction** occurs with walls or other objects adjacent to the flowing fluid that move at a different speed than the fluid. This requires that the interaction of the fluid particles is limited to **elastic collisions**.

What do these three assumptions imply? The first assumption rules out consideration of the actual flow of air through the trachea, which is usually turbulent. Including turbulence is mathematically challenging and in this textbook we touch the issue only briefly later in this chapter, when we introduce an empirical threshold speed above which flow becomes turbulent.

The second assumption is automatically applicable if the fluid is incompressible, since sound waves are the result of localized compressions in the fluid. However, sound propagation through a fluid is extremely important in physiology as it leads to acoustics. Thus, we will eventually drop this assumption (in Chapter 15).

Fluid flow that satisfies the first two assumptions is called **laminar flow** and is illustrated in Fig. 13.5. The figure shows an amount of fluid entering the field of view through the green area at the left side. If we imagine the fluid divided into small fluid segments, then we can follow each segment as a function of time. Its path is called a **flow line** and can be envisaged to lie fully within an envelope we call a **flow tube**, as indicated for one flow line in the lower part of Fig. 13.5.

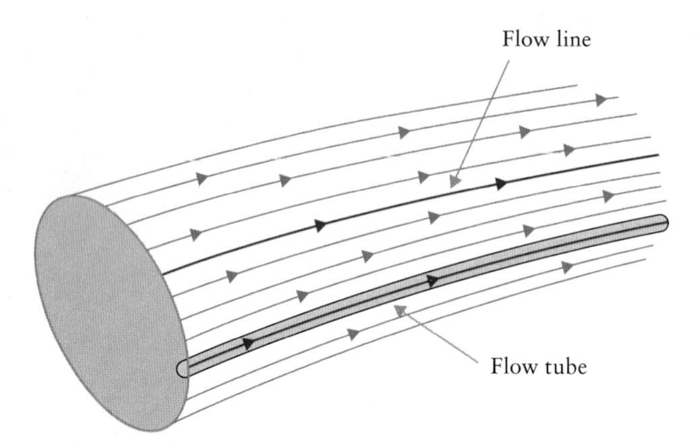

Figure 13.5 Sketch of laminar flow through an arbitrarily chosen area. The path taken by each fluid segment is drawn as a flow line (highlighted in red). One of the lower flow lines in the sketch is enclosed in a flow tube (grey cylinder). Flow lines cannot cross each other during laminar flow. Laminar flow characterizes ideal dynamic fluids and Newtonian fluids.

KEY POINT

Laminar flow is established when (i) flow lines in the fluid never cross each other, and (ii) flow tubes never penetrate each other.

Note that laminar flow and the ideal dynamic fluid are therefore not synonymous concepts. For an ideal dynamic fluid, flow must always be laminar because it shares the same assumptions with the laminar flow concept. However, laminar flow can occur for fluids that violate the third assumption for the ideal dynamic fluid. For example, we will still discuss laminar flow after we include flow resistance in the second half of this chapter.

The third assumption for the ideal dynamic fluid has to be discussed a bit further as it can easily lead to misconceptions. First, it differs notably from the other two conditions in that it addresses objects beyond the fluid. This condition has to be formulated somewhat vaguely because we want to develop widely applicable laws that are not confined to particular containers. For example, we want to apply these laws to the flow of air past the wing of a bird, or to a stream of water falling from a water tap toward the kitchen sink. Deriving laws in this chapter is easier when referring to a fluid-confining container or tube; however, the existence of the container or its shape may not matter to the laws' applicability.

In the easiest case, the ideal dynamic fluid is an ideal gas. In this case, the third assumption is equivalent to the respective assumption in the kinetic gas theory we introduced in Chapter 8. Elastic collisions were studied in

Chapter 7, including results for a collision with a wall. This interaction is shown in Fig. 13.6: the component of the velocity of the fluid particle parallel to the wall (||-direction in Fig. 13.6) remains unchanged, while the velocity component perpendicular to the wall changes its sign (⊥-direction in Fig. 13.6). In the kinetic gas theory, a non-zero component parallel to the wall allows for collisions with variable angles; for a flowing ideal gas the component parallel to the wall also contains the velocity of the collective flow. Since this velocity component does not change as a result of an elastic collision, no additional effect occurs due to the particle–wall interaction that is not already covered in the discussion of the ideal gas. This means wall interactions are frictionless and can be neglected in the discussion of flow of an ideal gas. Unfortunately, though, we saw in the previous chapter that the ideal gas is not a good example for a fluid because it is compressible.

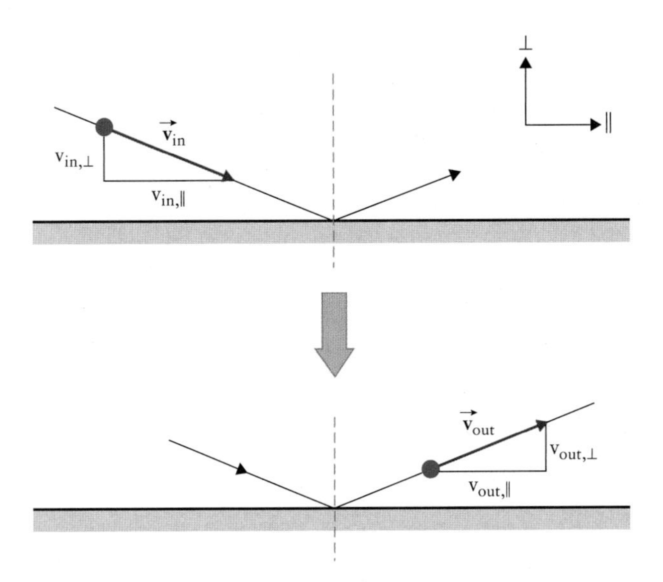

Figure 13.6 Elastic collisions of a fluid particle with a stationary wall. Its velocity component parallel to the wall (||) remains unchanged; its velocity component perpendicular to the wall changes its sign (⊥) but not its magnitude.

If the ideal dynamic fluid is a liquid, neglecting inelastic collisions is problematic: in a dense liquid, the fluid particles interact extensively with each other and the walls. We exclude inelastic collisions in the current section regardless, but will be forced to abandon this assumption when we introduce Newtonian fluids.

13.2.2: Equation of Continuity

We now establish the laws governing flow of an ideal dynamic fluid. We start with Fig. 13.7, which shows a fluid flowing from left to right through a tube of varying cross-sectional area. The cross-sectional area of the tube changes from A_1 to A_2 with $A_1 > A_2$. From experience, we know that fluid is neither created nor lost along the tube. Thus, during any given time interval, the same mass of fluid that enters through cross-section A_1 (grey area in upper sketch) must leave through cross-section A_2 (grey area in lower sketch):

$$\frac{\Delta m_1}{\Delta t} = \frac{\Delta m_2}{\Delta t}. \qquad [13.1]$$

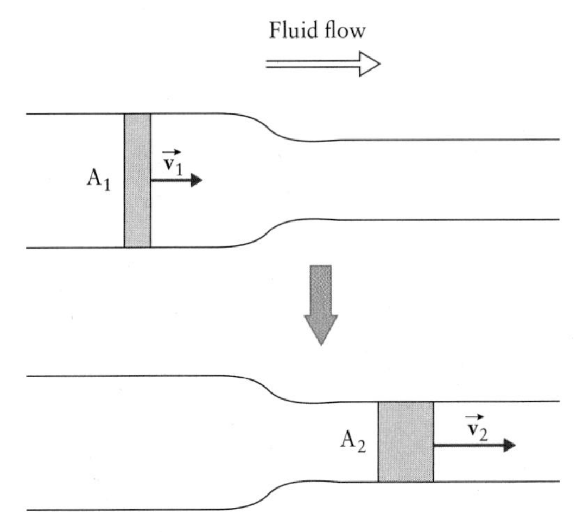

Figure 13.7 Model of a tapering tube that we use to derive the equation of continuity. Fluid flow occurs from left to right. (Top) The fluid segment is initially represented by the grey area at the left with cross-sectional area A_1 and fluid speed v_1. (Bottom) Later, the fluid segment is positioned in the grey area at the right with cross-sectional area A_2 and flow speed v_2.

Since the fluid is incompressible, the **conservation of fluid mass** is equivalent to a conservation of fluid volume. Using ρ for the density of the fluid, Eq. [13.1] is written in the form:

$$\rho \frac{\Delta V_1}{\Delta t} = \rho \frac{\Delta V_2}{\Delta t}.$$

The term $\Delta m/\Delta t$ in Eq. [13.1] is called the **mass flow rate**, and $\Delta V/\Delta t$ is called the **volume flow rate**.

The volume of a fluid segment in Fig. 13.7 can be written as the product of the cross-sectional area of the tube and the segment's length:

$$\rho\, A_1 \frac{\Delta l_1}{\Delta t} = \rho\, A_2 \frac{\Delta l_2}{\Delta t}.$$

$\Delta l/\Delta t$ is now interpreted as a displacement of the fluid along the tube per time interval Δt. This is equivalent to the speed of the fluid in the tube. With $|\vec{v}| = \Delta l/\Delta t$ we find:

$$A_1 |\vec{v}_1| = A_2 |\vec{v}_2|. \qquad [13.2]$$

Because Eq. [13.2] applies between any two points 1 and 2 along the tube, we can write it in a generalized form:

$$A\, |\vec{v}| = \text{const.} \qquad [13.3]$$

This is the **equation of continuity**. Note that the reasoning that led to Eq. [13.3] also establishes a second useful equation for the volume flow rate:

$$\frac{\Delta V}{\Delta t} = A\,|\vec{v}|, \qquad [13.4]$$

which applies anywhere along the tube, including at the cross-sectional areas with indexes 1 and 2 in Fig. 13.7.

KEY POINT

The equation of continuity is an expression of the conservation of mass or the conservation of volume of an incompressible fluid. It states that the volume flow rate is constant along a tube. The fluid flows faster when it passes through a section of the tube with a smaller cross-section.

Let's check this law carefully. We still need to establish whether it applies to laminar flow and/or to an ideal dynamic fluid. The two differ in that the former does not include an assumption about the interactions within the fluid and with the container wall. If you read once more through the derivation of the equation of continuity, you note that the type of interaction plays no role: the mass of the fluid is conserved regardless. Thus, the equation of continuity applies to laminar flow whether the fluid is an ideal dynamic fluid or not.

CONCEPT QUESTION 13.1

A tube widens from a cross-sectional area A_1 to a cross-sectional area $A_2 = 3\,A_1$. As a result the speed of an ideal dynamic fluid in the tube changes from v_1 to

(A) $v_2 = v_1$.

(B) $v_2 = v_1/3$.

(C) $v_2 = 3\,v_1$.

(D) $v_2 = v_1/9$.

(E) $v_2 = 9\,v_1$.

CASE STUDY 13.1

We study the steady flow of water from a water tap, e.g., in your kitchen sink. The jet of water (A) broadens as it falls; (B) narrows as it falls; (C) does not change its cross-sectional shape; (D) slows before hitting the bottom of the sink. *Hint:* Neglect effects that could lead to the break-up of a continuous flow into droplet formation.

Answer: *(B). The equation of continuity applies even though the flow does not occur in a tube. The presence of the wall of a tube makes no difference for an ideal dynamic fluid since the third condition neglects interactions with the wall. From faucet to sink the water accelerates due to gravity. With the speed of the fluid increasing and the volume flow rate constant, the cross-section of the fluid must diminish.*

EXAMPLE 13.1

The heart of a standard man (Table 4.1) pumps 5 litres of blood per minute into the aorta. (a) What is the volume flow rate in the cardiovascular system? (b) What is the speed of blood in the aorta? (c) If we assume that the blood passes through all systemic capillaries in our body in series, how fast would it have to flow through each capillary? Use data from Fig. 13.3. Would this result make sense? (d) What is the speed of blood in a capillary if we instead assume that the blood flows in parallel through the systemic capillaries in the human body?

Solution

Solution to part (a): The amount of blood flowing through the aorta per minute corresponds to a volume flow rate of:

$$\left(\frac{\Delta V}{\Delta t}\right)_{\text{aorta}} = \frac{5.0\ \text{L}}{60\ \text{s}} = 8.3\times 10^{-5}\,\frac{\text{m}^3}{\text{s}}.$$

Solution to part (b): The diameter of the aorta is given in Fig. 13.4 as $d = 2.6$ cm, which leads to an outer cross-sectional area $A = \pi(d/2)^2 = 5.3\ \text{cm}^2$ (also shown in the figure). The inner diameter defines the *lumen*, which is the open volume inside a blood vessel. To calculate the inner diameter of the blood vessel, the wall thickness has to be taken into account. The fraction of the total diameter attributed to the blood vessel wall lies between 15% and 20%; thus, a typical inner diameter of the aorta is $d_{\text{aorta}} = 2.2$ cm. This leads to the cross-sectional area of the lumen:

$$A_{\text{aorta}} = \pi\left(\frac{d_{\text{aorta}}}{2}\right)^2 = \pi\left(\frac{2.2\ \text{cm}}{2}\right)^2$$
$$= 3.8\times 10^{-4}\,\text{m}^2.$$

The speed of the blood in the aorta is obtained from its inner cross-sectional area and the volume passing per second. Using Eq. [13.4], we obtain:

$$|\vec{v}_{\text{aorta}}| = \frac{\left(\dfrac{\Delta V_{\text{aorta}}}{\Delta t}\right)}{A_{\text{aorta}}}$$
$$= \frac{8.3\times 10^{-5}\,\dfrac{\text{m}^3}{\text{s}}}{3.8\times 10^{-4}\,\text{m}^2} = 0.22\,\frac{\text{m}}{\text{s}}.$$

This is a frequently used result: blood flows through the aorta at an *average speed* of about 20 cm/s.

Solution to part (c): Let's assume blood passes through each single systemic capillary at the rate found in part (a). For the outer diameter of a capillary we use 9 μm from Fig. 13.4. This value leads to an inner diameter of $d_{\text{capillary}} =$ 7 μm (capillary wall thickness is about 1 μm). The cross-sectional area of the capillary is:

$$A_{\text{capillary}} = \pi\left(\frac{d_{\text{capillary}}}{2}\right)^2 = \pi\left(\frac{7\times 10^{-6}\,\text{m}}{2}\right)^2$$
$$= 3.8\times 10^{-11}\,\text{m}^2.$$

continued

We use the equation of continuity to derive the speed for blood in the capillary, $|\vec{v}_{\text{capillary}}|$:

$$|\vec{v}_{\text{capillary}}| = \frac{A_{\text{aorta}}|\vec{v}_{\text{aorta}}|}{A_{\text{capillary}}}$$

$$= \frac{8.3\times10^{-5}\,\frac{\text{m}^3}{\text{s}}}{3.8\times10^{-11}\,\text{m}^2} = 2200\frac{\text{km}}{\text{s}}.$$

Even if the capillaries could sustain blood rushing through at such speed, it would no longer be possible to exchange oxygen and nutrients with the surrounding tissue; i.e., the physiological purpose of the cardiovascular system would be lost.

Solution to part (d): A slow flow of blood in the systemic capillaries is achieved by arranging them parallel to each other with a combined cross-section that is larger than the cross-section of the aorta. Fig. 13.4 suggests that this is the case with 3500 cm^2 for the capillaries, compared to 5.3 cm^2 for the aorta. The equation of continuity allows us to determine the actual speed of blood in the capillaries once we have corrected the cross-sectional areas from Fig. 13.4 to represent the lumen. A correction factor k is defined as the ratio of the lumen cross-sectional area to the outer cross-sectional area and is quantified with the inner and outer diameter for a typical capillary:

$$k = \frac{A_{\text{lumen}}}{A_{\text{outer}}} = \frac{(7\mu\text{m})^2}{(9\mu\text{m})^2} = 0.6.$$

The two diameters are squared because the area is proportional to the square of the radius. With this factor we obtain for the cross-sectional area of the lumen of the capillaries $A_{\text{capillary}} = 0.6\ 3500\ \text{cm}^2 = 2100\ \text{cm}^2$. Thus:

$$|\vec{v}_{\text{capillary}}| = \frac{\left(\frac{\Delta V}{\Delta t}\right)_{\text{aorta}}}{A_{\text{capillary}}}$$

$$= \frac{8.3\times10^{-5}\,\frac{\text{m}^3}{\text{s}}}{0.21\text{m}^2} = 4\times10^{-4}\,\frac{\text{m}}{\text{s}}.$$

Again, this is a frequently used value: blood flows very slowly through the capillaries, at less than 1 mm/s.

13.2.3: Bernoulli's Law

The changes we observed in the speed of a flowing fluid due to changes in the cross-section of the tube also lead to changes in the pressure in the fluid. This can be illustrated experimentally with an instrument called a **Venturi meter**, which is shown in Fig. 13.8. The instrument consists of a tube with a constriction zone at its centre. A W-shaped tube connects at the left and right to the wide sections of the main tube; in the middle it is open to the constriction zone. The W-shaped tube is partially filled with a liquid to indicate the pressure in the section of the horizontal tube above each column. Initially, while the fluid in the main horizontal tube is at rest, the liquid in the W-shaped tube rises to the same level in all three columns, indicating that the pressure is the same. When the fluid flows through the main tube, a pressure difference is observed: the liquid in the middle column rises highest, indicating that the pressure in the constriction zone is lower than in the wider sections. Thus, the speed of the fluid, which is higher in the constricted section due to the equation of continuity, is inversely related to the pressure.

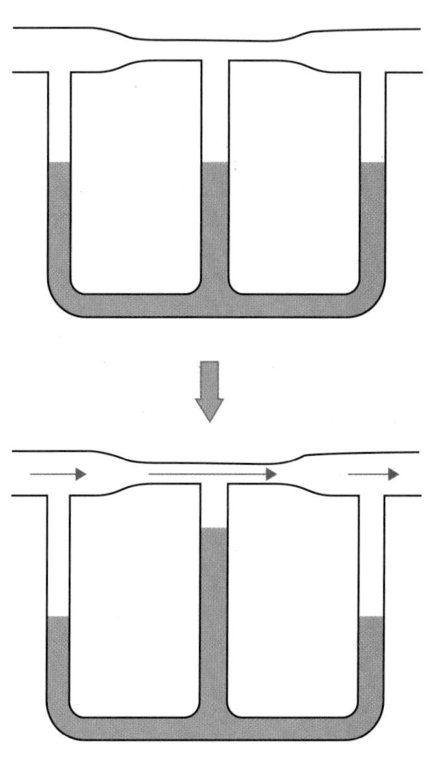

Figure 13.8 The Venturi meter is an instrument to measure the speed of a fluid in a horizontal tube. (Top) It indicates the same pressure in every section of the tube while the fluid is at rest. (Bottom) When the fluid flows as indicated by the arrows in the tube, pressure variations become evident: the pressure is higher where the speed of the fluid is slower.

Daniel Bernoulli quantified this observation starting from Fig. 13.9. Shown is a horizontal tube that tapers from cross-section A_1 to A_2. We assign a pressure p_1 to the wider section of the tube and p_2 to the narrower section. With these definitions, the grey fluid segment is studied. We want to determine its kinetic energy at an initial and a final instant and then relate the change in the kinetic energy to the work needed to move it into the constricted section. We specifically choose the initial time (index 1) when the fluid segment occupies the grey volume shown in the top part of Fig. 13.9. At that instant the volume of the fluid segment is $\Delta V = A_1\,\Delta x_1$. The final instant is shown in the bottom part of Fig. 13.9 (index 2). Now its volume is $\Delta V = A_2\,\Delta x_2$.

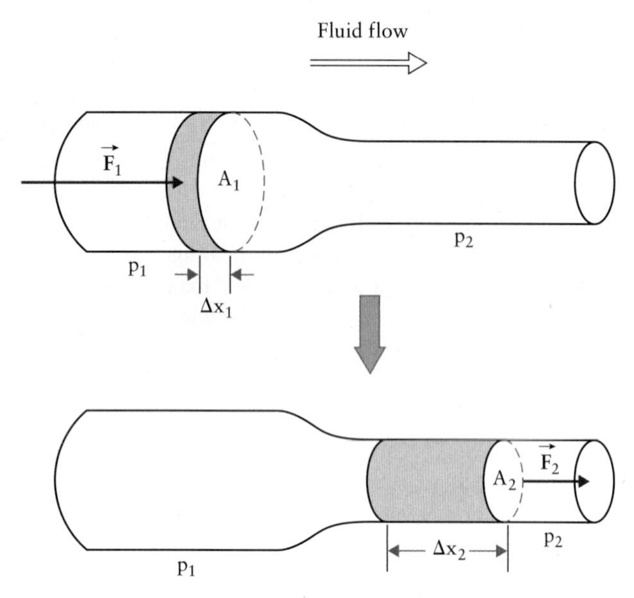

Figure 13.9 A sketch defining the parameters needed to derive Bernoulli's law. We study a fluid segment (grey) that initially occupies the volume $A_1\,\Delta x_1$ (top) and later the volume $A_2\,\Delta x_2$ (bottom). Fluid flow in the sketch occurs from left to right through a tapering tube. The fluid pressure varies from p_1 to p_2 at the constriction. The change in the speed of the fluid causes a change in the kinetic energy that is accounted for by a work term associated with the transfer of fluid into the constriction.

The cross-sectional area and the speed of the fluid are related by the equation of continuity. A force $\vec{F}_1$ has to be applied to accelerate the fluid segment through the tube. The change in the kinetic energy of the fluid segment is:

$$\Delta E_{\text{kin}} = \frac{1}{2}\Delta m\, v_2^2 - \frac{1}{2}\Delta m\, v_1^2,$$

in which Δm is the mass of the fluid segment. With ρ the density of the fluid, we rewrite the mass as $\Delta m = \rho\,\Delta V$:

$$\Delta E_{\text{kin}} = \frac{1}{2}\rho(v_2^2 - v_1^2)\Delta V.$$

Since the tube becomes narrower, the speed must increase and thus the kinetic energy of the fluid segment increases. To achieve this increase in kinetic energy, work must be done on the fluid segment. This work is required to transfer the fluid segment from its initial to its final position. In a gedanken experiment, we can split the work into two contributions: removing the fluid segment in the part of the tube with pressure p_1, and adding the fluid segment to the part of the tube with pressure p_2. Quantitatively, this means that the volume of the segment is changed from ΔV to 0 in the top part of Fig. 13.9, and, concurrently, its volume is changed from 0 to ΔV in the bottom part. The work is:

$$W = -p_2\,\Delta V - p_1(-\Delta V)$$
$$= -(p_2 - p_1)\Delta V.$$

The conservation of energy for the fluid segment as a closed system requires that $\Delta U = W$, because no heat exchange takes place. The only form of energy that changes in Fig. 13.9 is the kinetic energy, thus $\Delta U = \Delta E_{\text{kin}}$:

$$-(p_2 - p_1)\Delta V = \frac{1}{2}\rho(v_2^2 - v_1^2)\Delta V.$$

We separate all the terms related to positions 1 and 2 in the equation and divide by ΔV:

$$p_1 + \frac{1}{2}\rho\, v_1^2 = p_2 + \frac{1}{2}\rho\, v_2^2. \qquad [13.5]$$

Eq. [13.5] applies at any position along the tube:

$$p + \frac{1}{2}\rho\, v^2 = \text{const.} \qquad [13.6]$$

This is **Bernoulli's law**.

KEY POINT

Bernoulli's law is an expression of the conservation of energy for a closed system. It states that an increase in the speed of an ideal dynamic fluid in a tube is accompanied by a drop in its pressure.

Bernoulli's law indeed applies only to the ideal dynamic fluid. The third condition limiting the interactions of the fluid particles with the container wall to elastic collisions is necessary to apply the conservation of energy in the form used to derive Bernoulli's law. When we revise this condition in the next section, we must revisit the calculation of the pressure in the flowing fluid.

Bernoulli's law is sometimes reported in the physics literature with an additional term that accommodates possible changes in the height of the tube. We do not introduce that term because Example 12.2 provides an alternative approach to include height differences via Pascal's law.

CONCEPT QUESTION 13.2

A blood vessel of radius r splits into two smaller vessels, each with radius $r/4$. If the speed of the blood in the large vessel is v_{large}, what is the speed of the blood in each of the smaller vessels (v_{small})? Treat blood as an ideal dynamic fluid.

(A) $v_{\text{small}} = 8\, v_{\text{large}}$

(B) $v_{\text{small}} = 4\, v_{\text{large}}$

(C) $v_{\text{small}} = v_{\text{large}}$

(D) $v_{\text{small}} = v_{\text{large}}/4$

(E) $v_{\text{small}} = v_{\text{large}}/8$

CASE STUDY 13.2

(a) In a person with advanced arteriosclerosis (artery constriction due to accumulated plaque on the inner walls, as shown in Fig. 13.10), Bernoulli's effect produces a symptom called vascular flutter. To maintain a constant volume flow rate in this situation, the blood must travel faster than normal through the constriction. At a sufficiently high blood speed, the artery collapses and immediately reopens, leading to a repeated temporary interruption of the blood flow that can be heard with a stethoscope. Why does vascular flutter occur? (b) An aneurysm is a weakened spot of an artery where the artery walls balloon outward; Fig. 13.11 shows an aneurysm of the aorta. Blood flows more slowly through this region, resulting in an increase in pressure at the aneurysm relative to the pressure in adjacent sections of the artery. This condition is dangerous because the increased pressure can cause the artery to rupture (see Chapter 14 for a detailed discussion of aneurysms). What slows blood flow in an aneurysm?

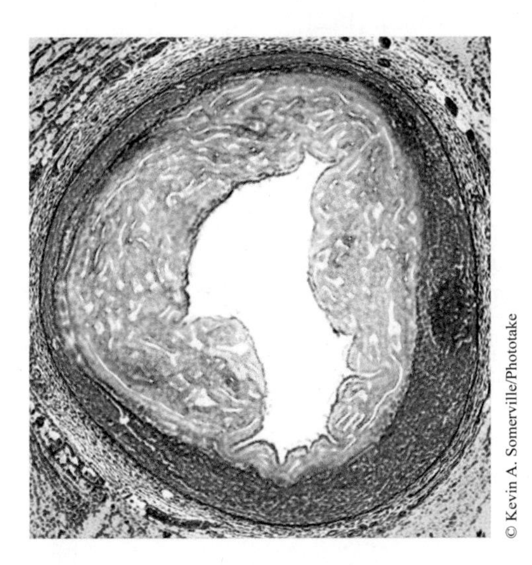

Figure 13.10 Advanced arteriosclerosis is an artery constriction due to accumulated plaque on the inner vessel walls. Shown is a coronary artery cross-section with atherosclerotic plaque (yellow) in the lumen.

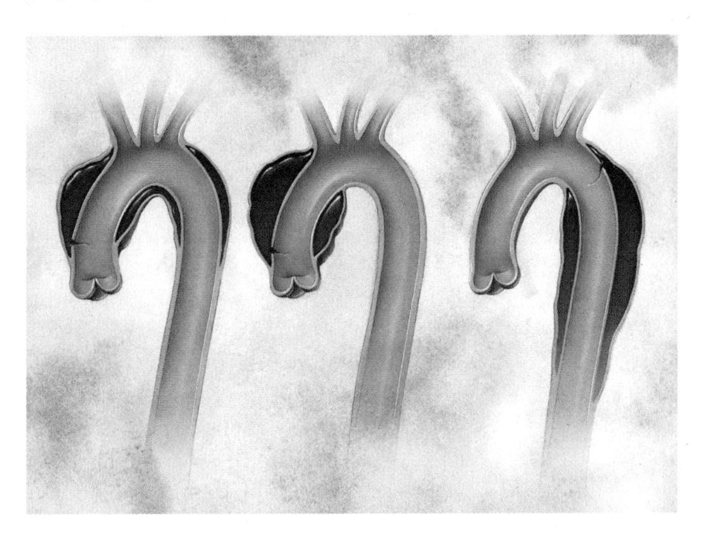

Figure 13.11 An artist's rendering of an aneurysm of the aorta. The inner layer of the aorta wall (pink) has ruptured. Blood (dark red) has pooled in the fissure of the wall, producing the visible bulge. Surgical repair of the aorta is required.

Answer to Part (a): *The artery collapses since the high speed of the blood inside the vessel lowers the pressure in the bloodstream relative to the pressure in the stationary extracellular fluid. This is due to Bernoulli's law: a high value for the speed, v, leads to a low value of the pressure, p. Once the pressure difference is large enough to close the artery, the blood flow stops momentarily. When this happens the blood upstream from the clogged vessel causes a pressure increase that is sufficient to reopen the artery. The closing and reopening of the artery then continues in a cyclic manner.*

Answer to Part (b): *The cross-sectional area of a blood vessel and the speed of blood in the vessel are related by the equation of continuity. Thus, the speed of blood flow decreases in a blood vessel when its cross-section increases in an aneurysm.*

EXAMPLE 13.2

Blood flows smoothly through the aorta as its cross-section tapers to 75% of its initial value, similar to the case illustrated in Fig. 13.9. What is the pressure difference Δp between the wide and the narrow sections? *Hint:* Data we used before and require for this calculation are the volume flow rate in the aorta $\Delta V/\Delta t = 83\ \text{cm}^3/\text{s}$, the density of blood $\rho = 1.06\ \text{g/cm}^3$, and the lumen cross-section of the aorta $A_{aorta} = 3.8\ \text{cm}^2$.

Solution

We start with Bernoulli's law. Let index 1 in Eq. [13.5] refer to the wide section and index 2 to the narrow section of the aorta. From the equation of continuity we know that the flow is faster in the narrow section; i.e., $v_2 > v_1$. Inserting this inequality in Eq. [13.5] leads to $p_1 > p_2$. Thus, we predict that the pressure drops from section 1 to section 2. This must be taken into account when writing the pressure difference in the form $\Delta p = p_1 - p_2$. In this form, $\Delta p > 0$. Note that we could have chosen to define Δp as $p_2 - p_1$, in which case Δp would be a negative value.

Eq. [13.4] allows us to quantify the two speeds, v_1 and v_2, since we know how the two cross-sectional areas are related: $A_2 = 0.75\ A_1$. Thus:

$$v_1 = \frac{\Delta V/\Delta t}{A_1}$$

and:

$$v_2 = \frac{\Delta V/\Delta t}{A_2} = \frac{4\Delta V/\Delta t}{3A_1}.$$

Substituting the last two equations into Eq. [13.5], we find:

$$\Delta p = \frac{1}{2}\rho\left(\frac{16(\Delta V/\Delta t)^2}{9\ A_1^2} - \frac{(\Delta V/\Delta t)^2}{A_1^2}\right),$$

which leads to:

$$\Delta p = \frac{7\ \rho(\Delta V/\Delta t)^2}{18\ A_1^2}.$$

The given values are substituted next:

$$\Delta p = \frac{7\left(1060\frac{\text{kg}}{\text{m}^3}\right)\left(8.3\times10^{-5}\frac{\text{m}^3}{\text{s}}\right)^2}{18\,(3.8\times10^{-4}\text{m}^2)^2}$$

$$= 20\ \text{Pa}.$$

13.3: Flow of a Newtonian Fluid

Up to this point, fluid flow was discussed for the ideal dynamic fluid. We established two laws: the equation of continuity and Bernoulli's law. We now test how closely predictions based on these laws correlate with experimental observations. The experiment we use is illustrated in Fig. 13.12: a liquid flows through a horizontal tube from left to right (direction of arrow). Three smaller, vertical columns are placed at different positions along the tube. The height to which the liquid rises in each of these columns depends on the pressure in the flowing liquid below. If that liquid can be modelled as an ideal dynamic fluid, all three columns have to be equally high, as shown in Fig. 13.12(a). This prediction results from the laws of the previous section: the equation of continuity states that the fluid speed does not change for a constant cross-section of the tube, then Bernoulli's law states that the pressure in the fluid does not change either.

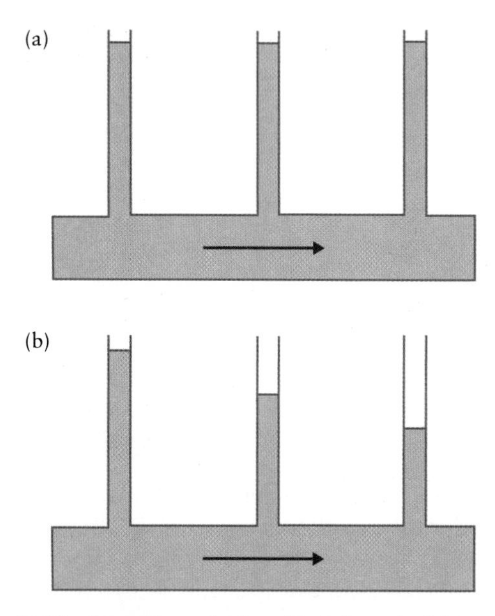

Figure 13.12 Comparison of (a) an ideal dynamic fluid and (b) a Newtonian fluid flowing through a horizontal tube. Flow resistance leads to a pressure drop along the tube, as indicated by the lower column height of the fluid above the tube at the right in part (b).

The actual experimental result for a real liquid is shown in Fig. 13.12(b). The farther the liquid progresses along the tube, the shorter the vertical columns. Thus, the experimental result differs fundamentally from the prediction in Fig. 13.12(a): the liquid speed and/or the pressure along the tube must change; i.e., the equation of continuity and/or Bernoulli's law must be modified. The equation of continuity applies as long as fluid flow is laminar: the fluid speed cannot change because fluid would have to accumulate in or vanish from the tube. Therefore, an approach beyond Bernoulli's law is needed to quantify the fluid pressure.

13.3.1: Newtonian Fluid Model

The Newtonian fluid model is developed to correctly describe the observation in Fig. 13.12(b). It is derived from the ideal dynamic fluid model by removing the assumption that was identified as too restrictive: a dense fluid such as a liquid cannot travel past a solid wall without extensive interactions. We noted these interactions already for the ideal stationary fluid in the last chapter, when we attributed the capillarity effect to a significant interface energy term.

How do we formulate a condition to replace the elastic collision restriction we used in the previous section? In the natural sciences we usually proceed by conducting experiments. In this particular case, the macroscopic observation of viscosity provides a promising approach even though it is not primarily a phenomenon describing the interaction between a fluid and a container wall.

13.3.2: Viscosity

Viscosity is an interaction between neighbouring layers of a moving fluid in a direction perpendicular to the flow lines. This is illustrated in Fig. 13.13, in which two parallel fluid layers of area A are highlighted. Let's assume the lower layer is at rest. This could be, for example, due to close proximity to the resting walls of the tube. The upper layer, a distance Δy away from the lower layer, moves with velocity $\Delta \vec{v}$ toward the right. The moving layer encounters a resistance force $\vec{R}$ that tries to slow it down. To maintain a constant velocity, Newton's first law requires the presence of a second force, $\vec{F}_{ext}$, with which the upper layer of fluid is pushed forward. Thus, if a fluid encounters flow resistance, an external force must be applied to push the fluid through the tube. The magnitude of this external force is found empirically by submerging two parallel test plates into a resting fluid and moving one plate relative to the other, as Fig. 13.13 illustrates. From such experiments the force $\vec{F}_{ext}$ is found to be proportional to (i) the area A of the fluid layers that face each other and (ii) the difference in speed of these layers, $|\Delta \vec{v}|$. The force is also inversely proportional to (iii) the distance Δy between the two layers:

$$F_{ext} = \eta A \frac{|\Delta \vec{v}|}{\Delta y}. \qquad [13.7]$$

Since the required external force varies further from fluid to fluid, with smaller forces typically needed in gases and larger forces in liquids, a materials constant is introduced called the **viscosity coefficient** η. Based on Eq. [13.7], the unit of η is N s/m^2. Table 13.1 lists viscosity coefficients for several fluids.

TABLE 13.1

Viscosity coefficients of various fluids

Fluid	Viscosity coefficient η (N s/m²)	Temperature (°C)
Gases		
N_2	1.78×10^{-5}	25
O_2	2.08×10^{-5}	25
Air	1.71×10^{-5}	0
H_2	9.0×10^{-6}	25
H_2	8.4×10^{-6}	0
H_2O	9.8×10^{-6}	25
Liquids		
H_2O	1.79×10^{-3}	0
H_2O	1.01×10^{-3}	20
H_2O	2.8×10^{-4}	100
(Whole) blood	$2.3 - 2.7 \times 10^{-3}$	37
Blood serum	$1.6 - 2.2 \times 10^{-3}$	20
Ethanol	1.19×10^{-3}	20
Glycerine	1.5	20

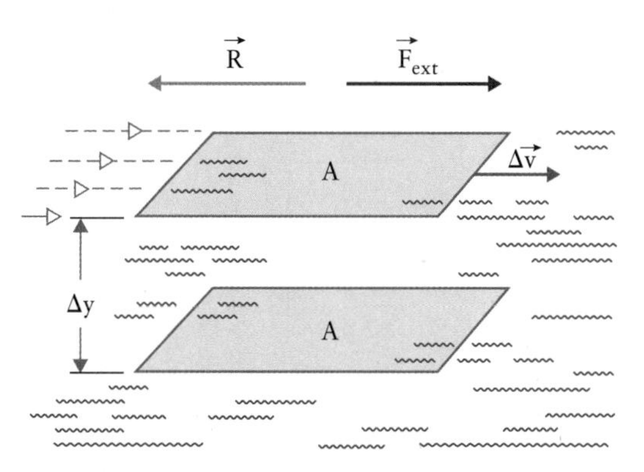

Figure 13.13 Experimental set-up to measure viscosity coefficients. Two parallel plates are immersed in a fluid at a distance Δy from each other. The lower plate is held at rest and the upper plate is pushed by an external force $\vec{F}_{ext}$. The resistance force of the fluid $\vec{R}$ balances the external force, leading to a constant velocity $\Delta \vec{v}$ of the plate toward the right).

Note that the third column in Table 13.1 gives the temperature at which the reported values apply, implying that these values change with temperature. Viscosity is one of the transport phenomena like diffusion and heat conduction, which we discussed in Chapter 11. A microscopic model is needed in addition to the phenomenological law of Eq. [13.7] to describe the temperature dependence of the viscosity coefficient.

Eq. [13.7] establishes that viscosity is a dynamic effect that requires a velocity gradient perpendicular to the direction of the flow lines in the fluid, $\Delta \vec{v}/\Delta y$. This means in turn that viscosity does not play a role in stationary fluids ($\vec{v} = 0$) or in ideal dynamic fluids ($\Delta \vec{v}/\Delta y = 0$), even though the viscosity coefficient for all physiologically relevant fluids is $\eta > 0$. This means, further, that there is no fluid, not even the ideal gas, that doesn't behave as a Newtonian fluid once a velocity gradient is introduced.

KEY POINT

In a Newtonian fluid, the inelastic interaction with the fluid-confining walls causes velocity gradients. Viscosity replaces the assumption of elastic collisions required between the ideal dynamic fluid and its confining walls. In Newtonian fluids the flow is laminar and viscous.

EXAMPLE 13.3

A 1.0-mm-thick coating of glycerine is placed between two microscope slides of width 2 cm and length 7 cm each. Find the force required to move the microscope slides at a constant speed of 10 cm/s relative to each other. The viscosity coefficient of glycerine is found in Table 13.1.

Solution

This problem is solved with Eq. [13.7]. Each of the terms on the right-hand side of the equation is given in the example text. The area is:

$$A = 0.02 \text{ m} 0.07 \text{ m} = 1.4 \times 10^{-3} \text{m}^2.$$

Note that A isn't twice this value for the two faces of a slide because only the cross-sectional area enters Eq. [13.7]. The difference in speed is $\Delta v = 0.1$ m/s and the coating thickness is the distance $\Delta y = 1.0 \times 10^{-3}$ m. Using $\eta = 1.5$ N s/m^2, we find:

$$F = \frac{\left(1.5 \frac{\text{N s}}{\text{m}^2}\right)(1.4 \times 10^{-3} \text{m}^2)\left(0.1 \frac{\text{m}}{\text{s}}\right)}{1.0 \times 10^{-3} \text{m}}$$
$$= 0.21 \text{ N}.$$

This is a notable force, given the rather wide 1-mm separation between the slides. Imagine you reduce their separation to 1 μm. Now a force of $F = 210$ N is required; i.e., a mass of more than 20 kg has to be suspended from one of the slides to achieve the stated motion. This phenomenon was used in the development of adhesive tape.

13.3.3: Fluid Velocity Profile in a Cylindrical Vessel

In a Newtonian fluid the inelastic interaction of fluid molecules with a stationary wall causes velocity gradients perpendicular to the flow lines. This leads to a non-uniform velocity profile across the fluid in the direction perpendicular to the stationary wall. The actual velocity profile depends on the shape of the stationary wall; we confine the discussion in this section to cylindrical tubes, which are physiologically important because they include blood vessels.

Eq. [13.7] is used to quantify the inclusion of inelastic collisions of the fluid particles with the stationary wall. Eq. [13.7] can be used only when the fluid flow has reached a steady state; thus, steady state is an additional assumption for quantitative predictions for the Newtonian fluid. This assumption is reasonable for blood flow in the cardiovascular system. A more general approach would require us to replace Eq. [13.7] with a formula that includes transient fluid accelerations. Recall that we excluded transient behaviour in transport phenomena before, e.g., when using Fick's law for diffusion.

In steady-state flow, two forces act on the fluid in the tube. In the direction of the motion of the fluid there is a force due to a pressure difference along the tube. This force pushes the fluid through the tube. The viscosity of the fluid causes a resistance force acting in the direction opposite to the direction of motion of the fluid. This force tries to slow the moving fluid. The mechanical equilibrium between these two forces varies with the position in the tube: near the stationary wall, viscosity dominates and the fluid flows slowly (with the speed vanishing directly at the wall), whereas toward the centre of the tube the force pushing the fluid dominates and the fluid moves comparably fast. The **velocity profile** is written quantitatively in the form:

$$v = \frac{r_{\text{tube}}^2 - r^2}{4\eta} \frac{\Delta p}{l}, \qquad [13.8]$$

in which r_{tube} is the radius of the tube and Δp is the pressure difference along a segment of the tube of length l. η is the viscosity coefficient of the fluid. The term $\Delta p/l$ is a constant pressure gradient along the tube. Eq. [13.8] provides a parabolic velocity distribution, as shown in Fig. 13.14. This velocity profile can be demonstrated with a slow-flowing fluid such as honey. If you open a jar of honey at room temperature and turn it upside down, the honey travels fastest at the centre of the jar, while honey near the glass surface won't flow out.

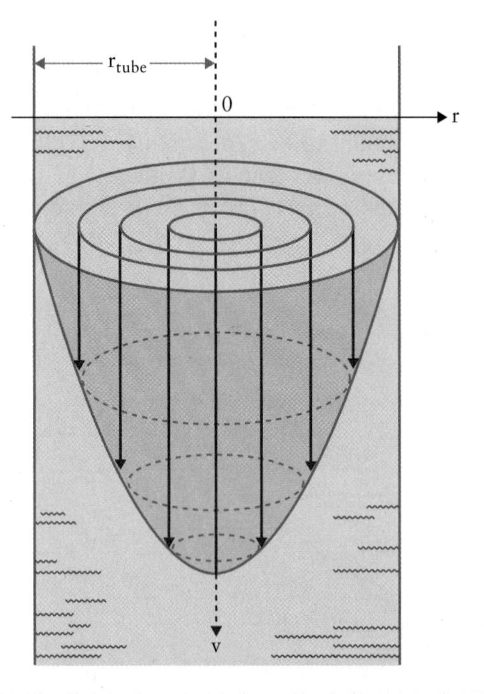

Figure 13.14 The steady-state velocity profile of a Newtonian fluid flowing through a cylindrical tube. The highest speed is reached at the centre of the tube (longest red arrows), while the fluid layer directly in contact with the wall does not move. Note that the figure shows a physical sketch of the system, which includes the walls of the tube, and a diagram with the velocity axis pointing downward and the position axis pointing to the right.

CASE STUDY 13.3

In hospitals and in the food industry, workers are required to wash their hands frequently and extensively. Why are elaborate hand-washing procedures required in these environments?

Answer: *Eq. [13.8] shows that fluid flow near a stationary surface vanishes ($v \to 0$ for $r \to r_{tube}$). Thus, allowing water to flow past your hands does not wash away pathogens and toxins that are attached to the skin, where the water flows slowly if at all. Only extensive rubbing with soap may loosen these. This always applies; however, hospitals are more directly concerned because of the larger number of pathogens and toxins their employees come in contact with.*

13.3.4: Poiseuille's Law

Jean Leonard Poiseuille used Eq. [13.8] to determine the volume flow rate through a cylindrical tube. We motivate his result with a simplified argument. Eq. [13.8] shows that an average velocity of the fluid is proportional to the square of the radius of the tube, r_{tube}^2, and the pressure gradient along the tube, $\Delta p/l$. It is also inversely proportional to the viscosity coefficient of the fluid, η. We can substitute this average velocity in the equation of continuity, which states that the volume flow rate $\Delta V/\Delta t$ is equal to the (average) speed of the fluid and the cross-sectional area of the tube, A, with $A \propto r_{\text{tube}}^2$. Thus, the volume flow rate must be proportional to the pressure gradient and inversely proportional to the viscosity coefficient. It also is proportional to the fourth power of the radius of the tube, r_{tube}^4:

$$\frac{\Delta V}{\Delta t} = \frac{\pi}{8\,\eta} r_{\text{tube}}^4 \frac{\Delta p}{l}. \qquad [13.9]$$

This is **Poiseuille's law**, with the proportionality factor $\pi/8$ that applies specifically to a cylindrical tube.

KEY POINT

Poiseuille's law states that the volume flow rate of a Newtonian fluid through a cylindrical tube is proportional to the fourth power of the radius of the tube.

Thus, a narrower tube reduces the flow severely; e.g., when the diameter of a tube is reduced by a factor of 2, the flow through the tube is diminished by a factor of 16!

Eq. [13.9] can be generalized for arbitrarily shaped containers in the form of **Ohm's law**:

$$\Delta p = R\frac{\Delta V}{\Delta t}, \qquad [13.10]$$

where R is the **flow resistance**, with unit Pa s/m^3.

KEY POINT

Ohm's law states that the volume flow rate of a Newtonian fluid is proportional to the pressure difference along the tube, and that the proportionality constant is the flow resistance.

This relates to the everyday use of the word resistance since, if the resistance is high, a large pressure difference leads to only a small volume flow rate. In our discussion of electric currents in Chapter 19 we will compare Eq. [13.10] to one of the laws of electricity, which is also called Ohm's law. We will see then that both laws are conceptually the same, except that we study viscous flow of fluids in the current chapter and then the flow of charges in a conductor.

For a cylindrical tube, the flow resistance is defined by Eq. [13.9]. It is directly proportional to the viscosity coefficient η of the fluid:

$$R = \frac{8\,l}{\pi\, r_{\text{tube}}^4}\eta. \qquad [13.11]$$

It is important to note that Poiseuille's law cannot be extrapolated to the case $\eta = 0$. In particular, Eqs. [13.8] and [13.9] do not predict an infinite velocity or an infinite volume flow rate in this case. This interpretation would be inconsistent with many other laws of physics. Why is that so? Essentially, the answer is that $\eta > 0$ is an assumption in the derivation of Eq. [13.8]: we used mechanical equilibrium between a force pushing the fluid forward and a force holding it back. If $\eta = 0$, this equilibrium requires that the pressure gradient along the tube is also zero. Thus, substituting the conditions for an ideal dynamic fluid into Eqs. [13.8] and [13.9] leads to a division of zero by zero, which is mathematically undefined.

CONCEPT QUESTION 13.3

A Newtonian fluid is forced through a tube to obtain a certain volume flow rate (experiment 1). If the same fluid is then forced through a tube of the same cross-sectional area with double the length (experiment 2), how has the pressure difference Δp_2 along the tube changed from the previous value Δp_1 if we observe the same volume flow rate?

(A) $\Delta p_2 = \Delta p_1$

(B) $\Delta p_2 = 2\ \Delta p_1$

(C) $\Delta p_2 = 4\ \Delta p_1$

(D) $\Delta p_2 = 8\ \Delta p_1$

(E) $\Delta p_2 = 16\ \Delta p_1$

EXAMPLE 13.4

What is the pressure gradient (the drop of pressure per length unit) in the aorta? Assume that blood flows as a Newtonian fluid. The viscosity coefficient of blood is from Table 13.1 as $\eta_{blood} = 2.5 \times 10^{-3}$ N s/m^2. Note this example differs from Example 13.2 because no change in the aortic cross-section is assumed.

Solution

The volume flow rate in the aorta is $\Delta V/\Delta t = 8.3 \times 10^{-5}$ m^3/s (see Example 13.1). The aorta is cylinder-shaped with an inner diameter of 2.2 cm. We apply Poiseuille's law to obtain the pressure gradient $\Delta p/l$:

$$\frac{\Delta p}{l} = \frac{\Delta V}{\Delta t}\frac{8\eta}{\pi}\frac{1}{r_{tube}^4}.$$

This yields:

$$\frac{\Delta p}{l} = \frac{8\left(8.3\times10^{-5}\frac{\text{m}^3}{\text{s}}\right)\left(2.5\times10^{-3}\frac{\text{Ns}}{\text{m}^2}\right)}{\pi(1.1\times10^{-2}\text{m})^4}$$

$$= 36\frac{\text{Pa}}{\text{m}}.$$

13.3.5: Newtonian Fluid Flow with Variable Tube Size

When ideal dynamic fluids flow through a tube of variable size, we see that the combination of the equation of continuity and Bernoulli's law is sufficient to determine all fluid parameters at any point along the tube. The fluid pressure in particular varies because of variations in the speed of the fluid. The associated changes in kinetic energy are accounted for with work done against the pressure in the fluid.

Once we move from the ideal dynamic fluid to the Newtonian fluid, that is, once we include interactions of the fluid molecules with each other and the stationary tube walls, an additional effect on the pressure in the fluid has to be included: to overcome viscous flow resistance to sustain steady-state flow, a pressure gradient is required along the tube that leads to a decreasing pressure toward downstream.

In the next step we return with the Newtonian fluid to the case of variable tube size. More specifically, we identify which role the equation of continuity, Bernoulli's law, and Poiseuille's law play in this case.

- Fig. 13.15(b) illustrates why the equation of continuity must also apply to Newtonian fluids in tubes with variable tube size: with or without flow resistance, no place exists for either excess fluid to collect or fluid to disappear along the tube. Thus, all equations from the beginning of this chapter up to Eq. [13.7] apply to Newtonian fluids.
- Fig. 13.15 illustrates the effect of Bernoulli's law in a Newtonian fluid for a variable tube (while neglecting at this point changes to Poiseuille's law due to the constriction). Part (a) of the figure shows the laminar flow through a tube of fixed diameter. A constant pressure gradient applies as predicted by Poiseuille's law. Fig. 13.15(b) highlights the changes that occur if the tube diameter varies locally: Bernoulli's law allows us to determine, at any given point along the tube, the pressure relative to the value at the same location if the tube diameter had not changed (dashed line). Thus, Poiseuille's law governs the actual flow through the tube, while Bernoulli's law allows us to correct the pressure locally due to tube diameter variations.

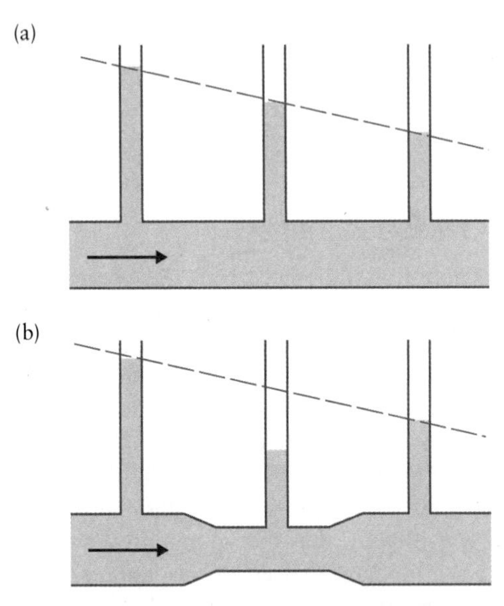

Figure 13.15 Role of Bernoulli's law for Newtonian fluids. (a) Laminar flow through a tube of fixed diameter. (b) The tube diameter varies. Bernoulli's law predicts the pressure relative to the value at the same location if the tube diameter had not changed (dashed line). Note that pressure corrections to Poiseuille's law caused by the constriction are not included.

Why is that so? The physics we discussed when deriving Bernoulli's law was correct. We neglected only the wall interaction, which we later found to be substantial enough that it cannot be neglected in practical cases. However, we can argue that the pressure gradient in a blood vessel is small enough that its effect over short distances can be neglected. Thus, if a blood vessel changes its diameter along a distance of a few centimetres or millimetres, (e.g., at a vasoconstriction), the pressure variation predicted by Bernoulli's law will be the dominant effect we will observe.

- Further to the effects that we established already for the ideal dynamic fluid, the formula for the flow resistance (Eq. [13.11]) also suggests that the flow resistance changes as the tube diameter varies. Thus, we cannot use Poiseuille's law in the form of Eq. [13.9] as it does not allow for a variable tube radius along the studied tube length. Instead, we need to segment the tube into sections of fixed radius, and then combine the individual flow resistances to an overall flow resistance for the variable tube. This approach was taken by Robert Gustav Kirchhoff in 1845 and led to two new laws, one for a single tube of varying radius and one for branching tubes.

13.3.6: Kirchhoff's Laws

Kirchhoff's laws describe how flow resistances have to be combined in cases in which a Newtonian fluid flows through tubes in series or parallel, e.g., when blood passes through the various sections of the cardiovascular system. Two laws are to be formulated, one for **vessels in series** (e.g., the aorta and an artery leading to the liver) and one for **parallel vessels** (e.g., a bed of capillaries between an arteriole and a venule in the liver).

Kirchhoff derived these laws originally not for fluids but for flowing electric charges. In the physics literature, therefore, you find them primarily applied in electricity. However, in physiology they are more important in fluid flow.

13.3.6.1: Blood Vessels in Series

Let's assume that a given amount of blood passes a vessel 1 with a given flow resistance R_1, and then passes a vessel 2 with a flow resistance R_2. We assume further that the blood vessel does not branch between vessel 1 and vessel 2.

The volume flow rate of each of the two vessels obeys Ohm's law, as given in Eq. [13.10]. In other words, the respective drop in blood pressure along the vessel is equal to the product of the flow resistance in the vessel and the volume flow rate of blood through the vessel:

$$\Delta p_1 = R_1 \frac{\Delta V}{\Delta t}$$

$$\Delta p_2 = R_2 \frac{\Delta V}{\Delta t}.$$

Note that the volume flow rate is the same in both vessels because the volume of blood is conserved for an incompressible fluid. And it does not change because no branching points exist between the two vessels in the combined system.

In addition to studying each vessel separately, we can also describe the combined system with Eq. [13.10]:

$$\Delta p = R_{\text{equivalent}} \frac{\Delta V}{\Delta t},$$

in which we introduce an equivalent flow resistance, $R_{\text{equivalent}}$, which must be a combination of the two individual flow resistances. The term $\Delta V/\Delta t$ in the last equation is equal to the same terms in the two previous equations. We can therefore combine the last three equations to relate the equivalent flow resistance to the individual flow resistances by recognizing that $\Delta p = \Delta p_1 + \Delta p_2$; i.e., that the pressure drop along the first vessel and the pressure drop along the second vessel combine to give the total pressure drop along the two vessels in series. This leads to:

$$R_{\text{equivalent}} \frac{\Delta V}{\Delta t} = (R_1 + R_2) \frac{\Delta V}{\Delta t};$$

i.e., for n resistances in series we write:

$$R_{\text{equivalent}} = \sum_{i=1}^{n} R_i. \qquad [13.12]$$

This is **Kirchhoff's law for serial flow resistances.**

KEY POINT

Kirchhoff's law for serial flow resistances states that flow resistances in series are added to obtain the equivalent flow resistance.

13.3.6.2: Blood Vessels in Parallel

In the second case we study a blood vessel that branches into two vessels, a vessel with flow resistance R_1 and a vessel with flow resistance R_2. Downstream, the two vessels recombine into a single vessel. No further branching occurs.

In this case, the pressure drop along the two separated vessels must be the same because there must be a well-defined pressure value in the vessel before branching and because there can be only one particular pressure value after the vessels merge. Thus, we write Eq. [13.10] for the two parallel vessels in the form:

$$\Delta p = R_1 \left(\frac{\Delta V}{\Delta t}\right)_1$$

$$\Delta p = R_2 \left(\frac{\Delta V}{\Delta t}\right)_2.$$

These two equations show that the fraction of the blood passing through each of the two vessels depends on their respective flow resistances. We can alternatively study the two vessels as a combined system. For this we apply Eq. [13.10] to the combined system by assigning an equivalent flow resistance, $R_{\text{equivalent}}$:

$$\Delta p = R_{\text{equivalent}} \frac{\Delta V}{\Delta t}.$$

The last three equations are combined to determine the dependence of the equivalent flow resistance on the two individual flow resistances. First we note that the total amount of blood flowing into the branching point per time, $\Delta V/\Delta t$, must be equal to the sum of the amounts of blood passing through the two vessels during the same time interval; i.e:

$$\frac{\Delta V}{\Delta t} = \left(\frac{\Delta V}{\Delta t}\right)_1 + \left(\frac{\Delta V}{\Delta t}\right)_2.$$

Were this not so, either blood would have to accumulate in vessels 1 and 2, or blood would have to accumulate in the upstream vessel. Combining the last four equations, we find:

$$\frac{\Delta V}{\Delta t} = \frac{\Delta p}{R_{\text{equivalent}}} = \left(\frac{\Delta V}{\Delta t}\right)_1 + \left(\frac{\Delta V}{\Delta t}\right)_2$$
$$= \frac{\Delta p}{R_1} + \frac{\Delta p}{R_2}.$$

From this equation we conclude that:

$$\frac{1}{R_{\text{equivalent}}} = \frac{1}{R_1} + \frac{1}{R_2},$$

which can be generalized for n flow resistances in parallel:

$$\frac{1}{R_{\text{equivalent}}} = \sum_{i=1}^{n} \frac{1}{R_i}. \quad [13.13]$$

This is **Kirchhoff's law for parallel flow resistances**.

KEY POINT

Kirchhoff's law for parallel flow resistances states that flow resistances in parallel are added inversely to obtain the equivalent flow resistance.

EXAMPLE 13.5

Fig. 13.16 shows a cylindrical blood vessel A of radius r_1, in which blood flows from point P_1 to point P_2, with the two points a distance l_1 apart. At point P_2 the blood vessel splits into three cylindrical vessels, one of which we label vessel B. Each of these parallel vessels has a radius $r_2 = (1/2)r_1$. The three vessels merge at a distance $l_2 = (1/2)l_1$ downstream from point P_2; this position we define as point P_3. (a) What fraction of the blood volume passing through vessel A is passing through vessel B? (b) What fraction of the drop in pressure between points P_1 and P_3 is occurring in vessel A?

continued

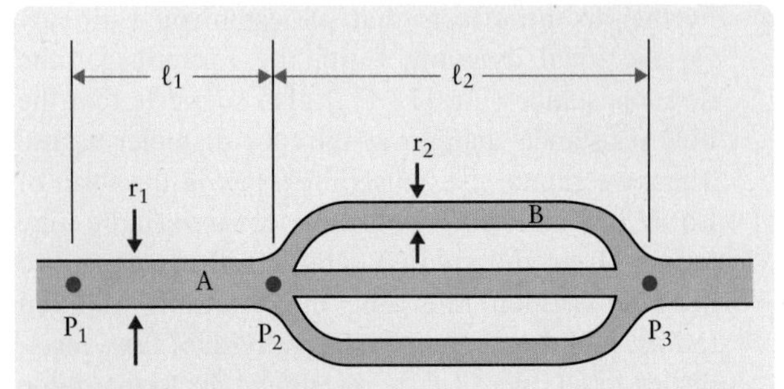

Figure 13.16 Example for Kirchhoff's laws. Shown is a single blood vessel A that splits into three capillaries of type B. The capillaries reunite to form a single vessel at point P_3. Note the various geometric data shown in the figure. Blood is modelled as a Newtonian fluid in this example.

Solution

Solution to part (a): The volume flow rates at points P_1 and P_3 are equal. Using Eq. [13.11] for the flow resistance in a cylindrical vessel, we note further that the three parallel vessels have the same flow resistance because they have the same geometrical parameters. Using Eq. [13.10] we find, therefore, that the volume flow rate in each of the three parallel vessels must be the same. With this information we calculate the ratio of the volume flow rate in vessel A and in vessel B:

$$\frac{\left(\frac{\Delta V}{\Delta t}\right)_B}{\left(\frac{\Delta V}{\Delta t}\right)_A} = \frac{\frac{1}{3}\left(\frac{\Delta V}{\Delta t}\right)_A}{\left(\frac{\Delta V}{\Delta t}\right)_A} = \frac{1}{3}.$$

Solution to part (b): Using Eq. [13.10] for both the entire system in Fig. 13.16 and for vessel A, we write:

$$\Delta p_{P_1P_3} = R_{P_1P_3}\left(\frac{\Delta V}{\Delta t}\right)$$
$$\Delta p_A = R_A\left(\frac{\Delta V}{\Delta t}\right).$$

These two formulas are then combined:

$$\frac{\Delta p_A}{\Delta p_{P_1P_3}} = \frac{R_A}{R_{P_1P_3}}. \quad [13.14]$$

We obtain the ratio of the pressure drops on the left-hand side of Eq. [13.14] once the ratio of the equivalent flow resistances for the system in Fig. 13.16 has been determined. These equivalent flow resistances are calculated in three steps:

(i) we calculate R_A and R_B,

(ii) we combine the three parallel vessels and calculate their equivalent flow resistance, and

continued

13.3.7: Newtonian Fluids in Physiology

13.3.7.1: Flow Resistance in the Human Cardiovascular System

(iii) the equivalent flow resistance for the three parallel vessels and the flow resistance for vessel A are combined to obtain the overall equivalent flow resistance.

Step (i): The flow resistances for sections A and B are calculated from Eq. [13.11]. We obtain:

$$R_A = \frac{8\, l_1\, \eta}{\pi\, r_1^4}$$

and:

$$R_B = \frac{8 l_2 \eta}{\pi r_2^4}$$

$$= \frac{8 \frac{1}{2} l_1 \eta}{\pi \left(\frac{1}{2} r_1\right)^4} = 8\, R_A.$$

Step (ii): We combine the contributions of the three parallel vessels to an equivalent flow resistance for the part of Fig. 13.16 that lies between points P_2 and P_3. For this we use Kirchhoff's law for parallel vessels:

$$\frac{1}{R_{P_2P_3}} = \frac{1}{R_B} + \frac{1}{R_B} + \frac{1}{R_B} = \frac{3}{R_B}.$$

Step (iii): In the last step we combine the flow resistances for the section between points P_1 and P_2 (vessel A with R_A) and the section between points P_2 and P_3. The equivalent flow resistance for the entire system in Fig. 13.19 is obtained from Kirchhoff's law for vessels in series:

$$R_{P_1P_3} = R_A + R_{P_2P_3} = R_A + \frac{R_B}{3}$$

$$= R_A + \frac{8\, R_A}{3} = \frac{11 R_A}{3}.$$

in which we used the result from step (i) to replace R_B. By substituting the last equation in Eq. [13.14], we find:

$$\frac{\Delta p_A}{\Delta p_{P_1P_3}} = \frac{R_A}{R_{P_1P_3}} = \frac{R_A}{\frac{11 R_A}{3}} = \frac{3}{11}.$$

Thus, only 27% of the drop in pressure occurs in vessel A.

EXAMPLE 13.6

In Fig. 13.2(b), we emphasized that 50% of the flow resistance in the systemic circulation is caused in the arterioles (small arteries), a higher fraction than in either the aorta or the capillaries. Using an average value for the viscosity of blood from Table 13.1 and data from Fig. 13.3, confirm that the total flow resistance is indeed greatest in the arterioles. Treat blood as a Newtonian fluid and blood vessels as cylinder-shaped. First calculate the flow resistance for a single vessel, then calculate the flow resistance for all vessels of the same type by using Kirchhoff's laws.

Solution

We start with the aorta. The average length $\langle l \rangle$ of any vessel of a given type is determined from Fig. 13.3 by dividing the total volume of the particular type of vessel by its total cross-sectional area. For the aorta, this leads to:

$$\langle l \rangle = \frac{V}{A} = \frac{180\ \text{cm}^3}{5.3\ \text{cm}^2} = 34\ \text{cm}.$$

The radius of the tube, r_{tube}, is half of the inner diameter of the aorta, $r_{tube} = 1.1$ cm. The flow resistance then follows from Eq. [13.11]:

$$R_{aorta} = \frac{8(0.34\ \text{m})\left(2.5\times10^{-3}\ \frac{\text{N s}}{\text{m}^2}\right)}{\pi(0.011\ \text{m})^4}$$

$$= 1.5\times10^5\ \frac{\text{Pa}\cdot\text{s}}{\text{m}^3}.$$

Since there is only a single aorta in our body, the summation parameter n is one ($n = 1$) in Kirchhoff's laws; i.e., Eq. [13.13] yields:

$$R_{total} = R_{aorta}.$$

The calculations of the corresponding values for the other blood vessel types proceed in the same manner. For the capillaries, we have $n = 5 \times 10^9$; i.e., Eq. [13.13] yields:

$$\frac{1}{R_{total}} = \sum_i \frac{1}{R_{capillary}} = \frac{5\times10^9}{R_{capillary}}$$

$$\Rightarrow R_{total} = \frac{R_{capillary}}{5\times10^9}.$$

The results for all vessel types are summarized in Table 13.2. We learn from this table that two factors cause

continued

the arterioles to dominate the flow resistance in our body compared to the smaller capillaries:

- the arterioles are significantly longer than the capillaries, and
- there are 20 times fewer arterioles than capillaries.

TABLE 13.2

Length, radius, individual flow resistance, and collective flow resistance for arterioles, aorta, and capillaries

Vessel type	Length (m)	Radius (m)	R_{single} (Pa s/m^3)	R_{total} (Pa s/m^3)
Arterioles	2.5×10^{-3}	8.0×10^{-6}	3.9×10^{15}	2.4×10^{7}
Aorta	3.4×10^{-1}	1.1×10^{-2}	1.5×10^{5}	1.5×10^{5}
Capillaries	8.5×10^{-4}	3.5×10^{-6}	3.6×10^{16}	7.2×10^{6}

The arterioles contribute most to the flow resistance in the systemic circulation.

What consequence does this large flow resistance of the arterioles have for the systemic circulation? To answer, we study the pressure–current relation in Ohm's law. The current $\Delta V/\Delta t$ is constant throughout the systemic circulation, since otherwise blood would collect or disappear somewhere. Thus, a section with a large flow resistance R must have a large pressure difference Δp: the blood pressure drops significantly in the arterioles. Therefore, shifting the flow resistance to the arterioles protects the capillaries, in which a significant pressure drop is undesirable due to the thinner and more fragile nature of these blood vessels.

13.3.7.2: Filtration in the Kidneys

EXAMPLE 13.7

(a) Quantify the volume flow rate through a single pore in the basement membrane in the kidneys, using for the pressure difference across the membrane $\Delta p = 1.3$ kPa and a viscosity coefficient of $\eta = 1.4 \times 10^{-3}$ N s/m^2 (this value lies between the values of blood serum and water). (b) How many pores are needed in the kidneys and how many pores are needed per nephron?

Supplementary anatomical information: The *kidneys* serve two purposes: to regulate the total water volume and the pH (acidity) of the blood, and to filter the end products of metabolism, especially urea and uric acid, out of the blood. Both purposes are accomplished in the functional unit of the kidneys, called the *nephron*.

A kidney contains about 1.2 million nephrons. An overview of the nephron is shown in Fig. 13.17. The filtration process in the nephron is a two-step process. An arteriole branches into the *glomerulus*, which is embedded in *Bowman's capsule*. The glomerulus filters the blood by holding back only proteins and blood cells. About 180 L/day of filtrate

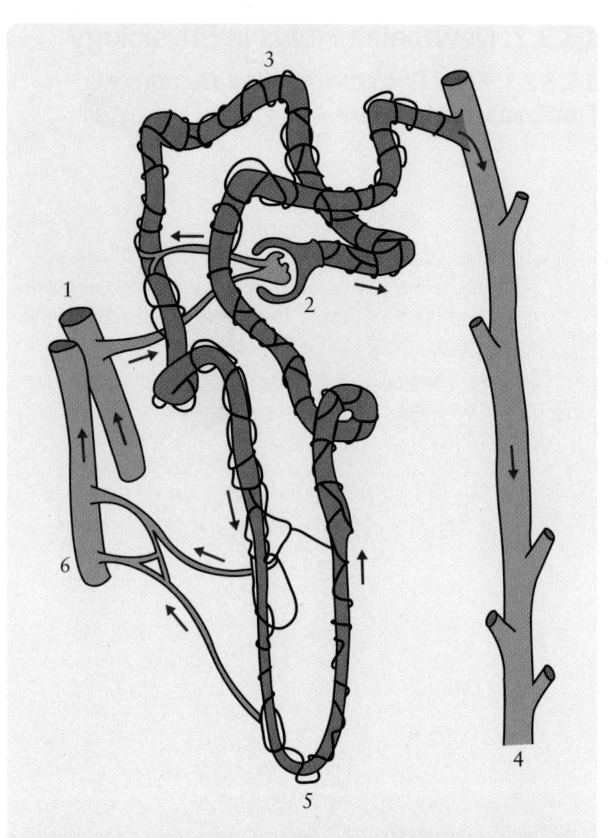

Figure 13.17 Overview of a nephron, showing the afferent arteriole (1), the glomerulus in Bowman's capsule (2), the renal tube (3), the urinary tract collection tube (4), the loop of Henle (5), and the efferent arteriole (6).

reach the renal tube. There, more than 99% of the fluid is resorbed into the circulatory system. The remainder reaches the collecting tube and is eliminated from the body, leading to excretion of about 1.5 L/day in the form of urine.

A detailed view of the glomerulus in Bowman's capsule is shown in Fig. 13.18, which is a sketch that shows, from left to right, two different magnifications. In part (a), the supplying

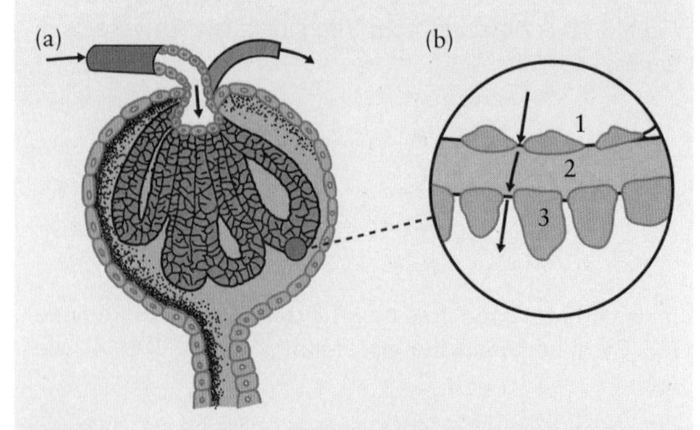

Figure 13.18 Detailed sketch of Bowman's capsule. (a) Blood supply to the capsule is shown at the top and the renal tube for filtrate removal at the bottom. (b) A cross-section of the wall inside Bowman's capsule is highlighted (green dot) with the capillary membrane (1), the basement membrane (2), and the podocytes (3). In this sketch blood is at the top and urine is at the bottom.

continued

arteriole is visible at the top left. It leads to tangled loops of capillaries, resembling a skein of wool, that are embedded in a capsule and finally leave as a blood vessel at the top right. The renal tubule, which collects the filtrate, is shown at the bottom where it leaves Bowman's capsule, which therefore serves as the primary fluid collection container.

Fig. 13.18(b) shows a cross-section through the capillary inside Bowman's capsule. The capillary membrane (1) is porous, but enclosed by the **basement membrane** (2). The basement membrane is embraced by podocytes (3)—cells with arm-like extensions—leaving slits open for fluid flow. The pores in the capillary membrane on the blood side are typically 20 nm in diameter. The basement membrane at the centre is 50 nm to 80 nm thick and contains pores of 12 nm diameter. Thus, the pores in the basement membrane determine the volume flow rate.

The actual value for the pressure drop across the basement membrane varies between zero and 1.3 kPa to allow the body to regulate the flow using variations in the blood plasma pressure.

Solution

We use r_{tube} = 6 nm for the radius of the pores in the basement membrane; Δp = 1.3 kPa is the maximum pressure difference across the membrane; 50 nm for the length of the pore, which is equivalent to the thickness of the basement membrane; and $\eta = 1.4 \times 10^{-3}$ N s/m^2 for the viscosity coefficient. With these values we find from Poiseuille's law the volume flow rate of a single pore:

$$\frac{\Delta V}{\Delta t} = \frac{\pi(6\times10^{-9}\,\text{m})^4(1.3\times10^3\,\text{Pa})}{8\left(1.4\times10^{-3}\,\dfrac{\text{N}\cdot\text{s}}{\text{m}^2}\right)(5\times10^{-8}\,\text{m})}$$

$$= 9.5\times10^{-21}\,\frac{\text{m}^3}{\text{s}}.$$

To handle the daily filtration of 180 L, the two kidneys must have more than $N = 2 \times 10^{14}$ pores. This number is obtained from the result in the last equation:

$$\left(\frac{\Delta V}{\Delta t}\right)_{\text{kidneys}} = \left(\frac{\Delta V}{\Delta t}\right)_{\text{pore}} N,$$

which leads to:

$$N = \frac{180\,\dfrac{\text{L}}{\text{day}}}{9.5\times10^{-21}\,\dfrac{\text{m}^3}{\text{s}}} = 2.2\times10^{14}.$$

The number of pores per nephron, the unit shown in Fig. 13.17, is obtained by dividing N by the number of nephrons in our kidneys, which is about 2.4 million nephrons. Thus, we need roughly 9×10^7 pores per nephron, which is a number close to 100 million. We see that the physiological performance of the kidneys on a macroscopic scale is based on the physical properties of a tremendous number of microscopic functional units.

13.4: Special Topics in Fluid Flow

13.4.1: Beyond Laminar Flow: Turbulence and Convection

When local velocity gradients in the fluid become too big, turbulences occur. Turbulent flow is a superposition of laminar flow (discussed above) and vortex formation/vortex motion. The different flow patterns for laminar flow and turbulent flow around a solid cylinder are shown in Fig. 13.19. In the turbulent case, flow lines are not continuous but terminate or start in vortices.

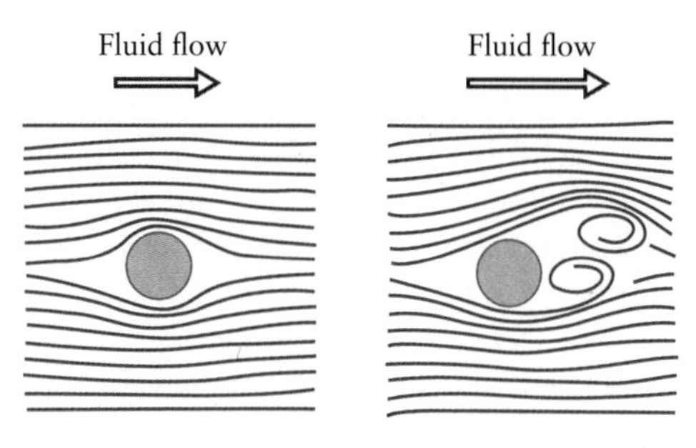

Figure 13.19 Two flow patterns: laminar flow (left) and turbulent flow (right) around a solid cylinder (blue dot) immersed in fluid. Note the vortex formation for turbulent flow.

Turbulent flow has a major effect on the volume flow rate. This is illustrated in Fig. 13.20, which shows the volume flow rate as a function of pressure difference along a given tube. At low pressure differences Ohm's law applies; i.e., the volume flow rate is proportional to the pressure difference. The vertical dashed line indicates the pressure difference at which the flow undergoes the transition from laminar to turbulent flow. At larger pressure differences Ohm's law no longer applies, and increasing the pressure difference to obtain a volume flow rate increase beyond the transition is ineffective. Thus, once the flow has become turbulent, no significant increase of the volume flow rate can be achieved.

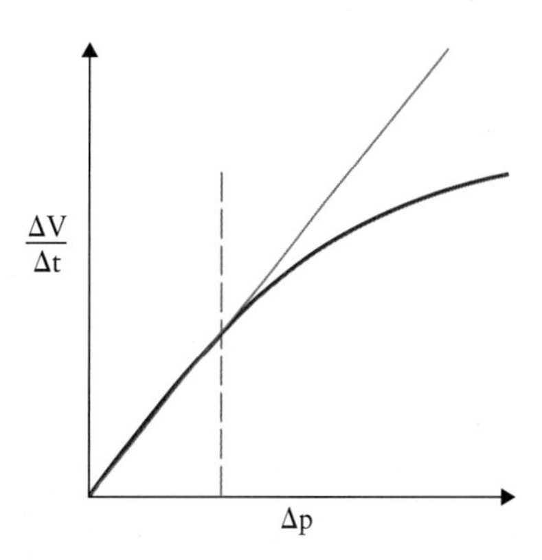

Figure 13.20 Transition of the volume flow rate as a function of pressure difference along a tube (red curve) from laminar to turbulent flow (dashed threshold line). Turbulent flow is characterized by minimum or negligible volume flow increments for increasing pressure differences.

A semi-empirical threshold number is introduced to determine whether flow is laminar or turbulent. It is called the **Reynolds number**, *Re*, in honour of Osborne Reynolds, who proposed it in 1883. The present discussion is based on flow in a cylindrical tube. For that tube geometry, the Reynolds number is given as:

$$Re = \frac{\rho \langle v \rangle d}{\eta}, \qquad [13.15]$$

in which $\langle v \rangle$ is a typical speed of the fluid, d is the diameter of the tube, η is the viscosity coefficient, and ρ is the density of the fluid. *Re* is a dimensionless parameter. In a cylindrical tube, laminar flow is predicted for Reynolds numbers $Re \leq 2000$ and turbulent flow for $Re \geq 2000$.

The Reynolds number is useful when we discuss systems in which turbulence control is required. For example, turbulence suppression is a design criterion for birds' wings in addition to providing lift and thrust to overcome air drag. Wings are primarily shaped such that air above the wing travels faster than that below. Due to Bernoulli's law, a lower air pressure results above the wing. The net effect is a lift force sufficient to compensate the effect of gravity on the bird. The faster the bird flies through the air the stronger the lift force, and flapping of the wings is no longer necessary. Energy conservation is, however, only one reason why large birds, such as most birds of prey, use a flap-and-glide flight pattern. Avoiding flapping or flapping slowly further addresses turbulence: air is a Newtonian fluid in which a transition from laminar to turbulent flow occurs at high relative speeds of air and wing surface, i.e., when large velocity gradients are involved. Ceasing wing motion therefore minimizes the occurrence of turbulences. Smaller birds, particularly finches and woodpeckers, use another approach to minimize flow resistance related to slowing of their flight: they rise on one or two wing beats, then fold their wings to the body and dart through the air, eliminating turbulent air motion past their bodies at high speed. These birds can be identified by their undulating flight pattern because they need to re-establish lift through another few wing beats after several metres to avoid crashing to the ground.

Birds show a good sense for turbulence in many ways, even when they are not airborne. When frigid winds blow along the seashore in winter, you can see seagulls on the beach all facing in the same direction. As illustrated in Fig. 13.21, the birds align their streamlined bodies such that they offer the least resistance to the oncoming breeze. This leads to a laminar flow of air around their bodies, avoiding the ruffling of their feathers due to turbulences that would expose their body to the low temperatures and possibly cause hypothermia.

EXAMPLE 13.8

Determine the Reynolds number of the following three systems: (a) water flow in a creek, (b) airflow through the trachea, and (c) blood flow in the aorta.

Solution

Solution to part (a): In most creeks of 1.0 m width, water flows at speeds of 1.0 m/s to 10.0 m/s. The density of water is 1.0×10^3 kg/m^3 and its viscosity coefficient is 1.8×10^{-3} N s/m^2 (see Table 13.1). This leads to a Reynolds number of $1.0 \times 10^6 \leq Re \leq 1.0 \times 10^7$; i.e., the flow in a creek is always turbulent.

continued

Figure 13.21 When cold winds blow across the shore, seagulls align their bodies with the wind to allow a laminar airflow. Minimizing turbulences is essential for these animals as the air vortices would ruffle some of their feathers and allow the body to excessively lose heat.

Solution to part (b): We assume 15 inhalations per minute of 0.5 L each. For the volume flow rate of air through the trachea, this yields:

$$\frac{\Delta V}{\Delta t} = 2\,(15\ \text{min}^{-1})(0.5\ \text{L})$$
$$= 2.5 \times 10^{-4}\,\frac{\text{m}^3}{\text{s}}.$$

An additional factor of 2 is introduced since each inhalation is followed by an exhalation, doubling the volume flow through the trachea per breath. Using the diameter of the trachea as d = 1 cm, the average speed of air is determined from the equation of continuity:

$$v = \frac{1}{A}\frac{\Delta V}{\Delta t} = \frac{2.5 \times 10^{-4}\,\frac{\text{m}^3}{\text{s}}}{\pi(5 \times 10^{-3}\,\text{m})^2} = 3.2\,\frac{\text{m}}{\text{s}}\,.$$

Using the density of air as ρ = 1.2 kg/m^3 and the viscosity coefficient as $\eta = 2 \times 10^{-5}$ N s/m^2 (Table 13.1), we find Re = 1900, i.e., a value near the threshold to turbulent flow. The actual flow is turbulent because the inner trachea surface is not smooth. Turbulent flow is desired because the inhaled air must be moistened in the trachea; moistening occurs when dry air is brought in contact with the moist trachea wall. This contact is more efficient for turbulent flow. Fig. 13.20 then shows why airflow in the trachea has a Reynolds number close to the laminar-to-turbulent transition. Once flow is turbulent, little gain in the volume flow rate is achieved by increasing pressure gradients along the tube. Thus, operating far into the turbulent regime would unnecessarily increase the physical work required for breathing.

Solution to part (c): The average speed of blood in the aorta is 20 cm/s (see Example 13.1). Using d = 2.2 cm for the inner diameter of the aorta, $\eta = 2.5 \times 10^{-3}$ N s/m^2 for the viscosity coefficient of blood, and $\rho = 1.06 \times 10^3$ kg/m^3 for the density of blood, we find the Reynolds number is Re = 1900, i.e., again a value close to the transition laminar to turbulent. If you keep in mind that blood flow into the aorta is pulsatile, with peak velocities in the 1 – 2 m/s range, turbulent flow seems to be favoured. However, turbulent flow is particularly undesirable in blood vessels since it greatly diminishes the volume flow rate for a given pressure gradient. Nature again maximizes the efficiency of the aortic blood flow based on Fig. 13.20; it pushes the volume flow rate to the greatest possible value for laminar flow, then develops a way to compensate for the peak velocities at which blood flow would have to be turbulent: immediately beyond the heart the aorta arches 180°, which allows it to buffer the rushing blood with the *Windkessel effect.* We discuss this effect in Chapter 14, because the elastic response of the aortic wall plays a key role in it.

Pathological vasoconstriction (vessels becoming narrower due to illness) may cause turbulent blood flow.

continued

Based on the equation of continuity, the blood speed $\langle v \rangle$ increases in this case because the heart still pumps the same amount of blood through the aorta. This leads to a potentially dangerous increase in the Reynolds number even though the diameter of the blood vessel, d, is reduced. To illustrate the net effect, we use the equation of continuity to determine the dependence of the blood speed on the vessel diameter:

$$|\vec{v}|A = |\vec{v}|\pi\left(\frac{d}{2}\right)^2 = \text{const} \quad \Rightarrow \quad |\vec{v}| \propto \frac{1}{d^2}.$$

Thus, even though the diameter of the blood vessel is reduced in the case of pathological vasoconstriction, the overall effect on the Reynolds number is an increase:

$$Re = \frac{\rho\,\langle v \rangle d}{\eta} \propto \frac{1}{d}.$$

Let's focus specifically on the aorta. Peak blood velocities in the aorta can also increase due to a defective aortic valve. Fig. 13.22 shows the peak velocity of blood ejected from the heart as a function of time for a particular patient. The speed of blood is measured by Doppler ultrasound, a diagnostic tool we discuss in Chapter 16. This patient's peak blood flow velocity increased steadily during a three-year observation period, approaching a threshold at 5 m/s, which is considered clinically the maximum tolerable value. This value exceeds the value we used to calculate the Reynolds number for this system by a factor of 25, illustrating the added tolerance due to the elasticity of the blood vessel walls. Data such as those shown in Fig. 13.22 allow the medical team to plan a heart operation and prepare the patient for the subsequent treatment over a considerable time span.

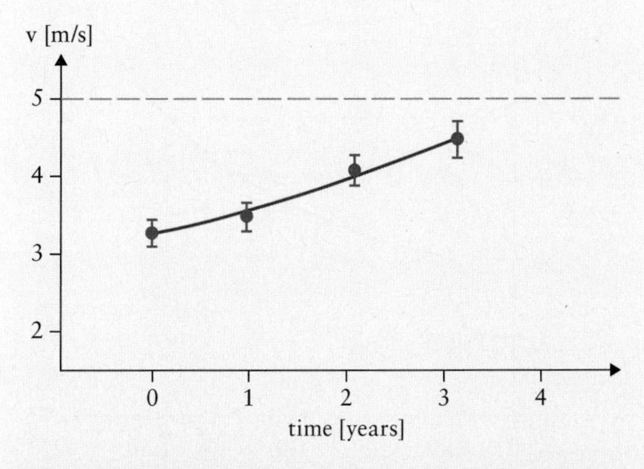

Figure 13.22 Peak velocity of blood ejected from the heart as a function of time for a particular patient, as measured by Doppler ultrasound. The patient's peak blood flow velocity increased steadily during a three-year observation period, approaching a threshold at 5 m/s. Once this threshold is reached, open-heart surgery is required to address a defective heart valve.

continued

13.4.2: Beyond Newtonian Fluids: The Viscosity of Blood

We have so far treated blood as a Newtonian fluid. However, its heterogeneous composition leads to novel properties that we cannot explain with the model developed for the Newtonian fluid. We illustrate this point for the viscosity coefficient of blood η, as listed in Table 13.1, and the flow resistance R, which we defined in Eq. [13.11]. As that equation showed, both parameters are closely related in the form $R \propto \eta$. Earlier in this chapter we defined the viscosity coefficient as a materials constant, which depends only on the temperature. As no dependence on other macroscopic parameters was identified, blood viscosity in the cardiovascular system of an endothermic species should be constant. The flow resistance R in turn depends on geometric factors, such as the tube length and the tube radius, and the viscosity coefficient η. Thus, for blood flow in a particular blood vessel of an endotherm, the flow resistance should be constant as well. Physiological observations illustrate, however, that the viscosity of blood and its flow resistance in a given blood vessel depend strongly on two additional factors:

- *The hematocrit value* is the volume fraction of blood cells in blood. A higher hematocrit value leads to a higher viscosity. This is illustrated in Fig. 13.23, which shows the viscosity coefficient of blood (relative to the viscosity coefficient of water) as a function of the hematocrit value. With an average hematocrit value of 46 for males and 41 for females we note that the blood of males is more viscous.

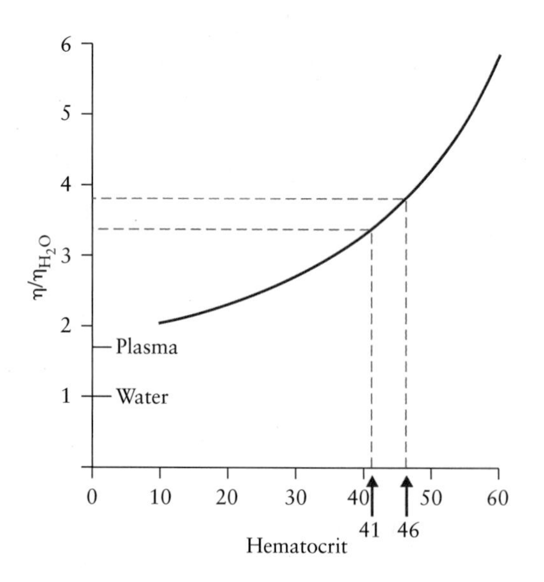

Figure 13.23 The viscosity coefficient for whole blood relative to the viscosity coefficient of water as a function of the hematocrit value, which is a parameter measuring the volume fraction of blood cells in whole blood. The average hematocrit value for males is 46 and for females 41. Thus, males have, on average, blood of a higher viscosity, as indicated by the dashed lines.

- *The flow velocity of the blood:* The viscosity coefficient is inversely proportional to the flow velocity. This is illustrated in Fig. 13.24, which shows the volume flow rate of a Newtonian fluid (1) and the volume flow rate of blood (2) as a function of the pressure difference along a vessel. The viscosity coefficient affects the flow resistance; it is constant in the case of a Newtonian fluid (blue curve in Fig. 13.24) but varies as a function of pressure difference for blood (red curve).

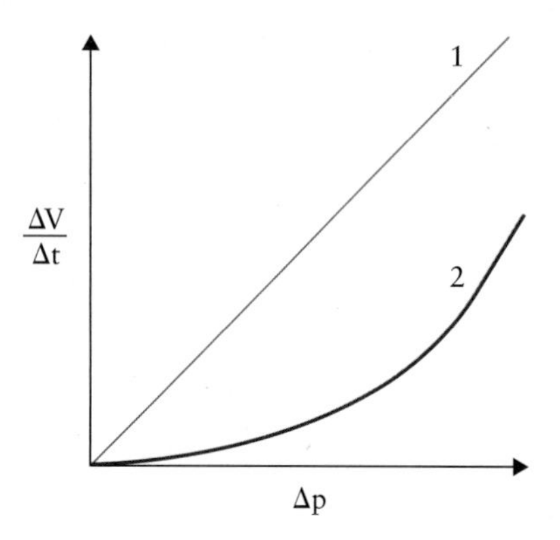

Figure 13.24 Pumping is required to force Newtonian and non-Newtonian fluids (e.g., water and blood) at various rates of flow through a straight tube. As the intended volume flow rate increases, the pressure the pump must produce increases as well. Note that for Newtonian fluids (1) the slope of the line relating flow rate and pressure is constant. However, for blood (2) the flow resistance is very high at low flow rates but approaches the value for Newtonian fluids at higher flow rates.

As a consequence of Fig. 13.24, blood flow cannot be allowed to fall below a minimum speed, as a dangerous feedback-loop effect may occur. We illustrate this feedback loop in the context of an anaphylactic shock. An *anaphylactic shock* is an allergic reaction of the body's immune system in response to a second contact with an antigen to which the body has become sensitized. An example is a severe peanut allergy. The initial reaction of the body is a histamine release that leads to a peripheral vasodilation, i.e., blood vessels widening. The increased cross-sectional area of the blood vessels causes the blood flow to slow down, as predicted by the equation of continuity. Due to the non-Newtonian behaviour of blood, the slower flow leads to an increase in the viscosity. This causes further slowing of the blood flow, which again results in a further increase in the viscosity. In the end, this feedback loop leads to the cessation of the volume flow rate, $\Delta V/\Delta t = 0$, a state called *stasis*.

But why does the viscosity change with the flow velocity? The answer lies in the fact that blood is heterogeneous. Normally, blood cells are well-immersed in the blood plasma due to the fact that they are nucleus free, contain a low-viscosity cytoplasm, and have a highly flexible cell membrane. This allows blood to behave like a low-viscosity emulsion (mixture of two liquids). When blood flows slowly, however, aggregation of the red blood cells (erythrocytes) occurs in a process called *nummulation*. The red blood cells form a structure resembling a stack of coins. Nummulation creates a highly viscous suspension (mixture of solid in liquid).

SUMMARY

DEFINITIONS

- Laminar flow: flow tubes (around flow lines) are not created in the flow; they do not intersect or vanish in the flow.
- Ideal dynamic fluid: a fluid whose flow is laminar. The fluid molecules interact only through elastic collisions with confining walls (frictionless motion).
- Newtonian fluid: a fluid whose flow is laminar. The fluid molecules interact inelastically with confining walls, causing a velocity gradient in the fluid.
- Viscosity is a property of resistance to flow in a fluid. It is quantified by the force needed to move a plate at constant velocity $\Delta\vec{v}$ parallel to a plate at rest at distance Δy, both immersed in a fluid:

$$F_{ext} = \eta A \frac{|\Delta\vec{v}|}{\Delta y},$$

where A is the cross-sectional area of the plates, η is the viscosity coefficient, $|\Delta\vec{v}|/\Delta y$ is a velocity gradient.

- Reynolds number for the transition from laminar to turbulent flow in a cylindrical tube ($Re < 2000$ is laminar; $Re > 2000$ is turbulent):

$$Re = \frac{\rho\langle v\rangle d}{\eta},$$

where $\langle v\rangle$ is the average speed of the fluid, d is the diameter of the tube, η is the viscosity coefficient, and ρ is the density of the fluid.

UNITS

- Volume flow rate $\Delta V/\Delta t$: m^3/s
- Mass flow rate $\Delta m/\Delta t$: kg/s
- Viscosity coefficient η: N s/m^2
- Flow resistance R: Pa s/m^3

LAWS

- Equation of continuity (fluid mass conservation) for laminar flow:

$$\frac{\Delta V}{\Delta t} = A|\vec{v}| = \text{const},$$

where $\Delta V/\Delta t$ is the volume flow rate, A is the cross-sectional area of the tube, and $|\vec{v}|$ is the speed of the fluid.

- Bernoulli's law for an ideal dynamic fluid in a horizontal tube:

$$p + \frac{1}{2}\rho v^2 = \text{const},$$

where p is the pressure in the fluid.

- Poiseuille's law for a Newtonian fluid in a cylindrical tube:

$$\frac{\Delta V}{\Delta t} = \frac{\pi}{8\eta} r_{tube}^4 \frac{\Delta p}{l},$$

where r_{tube} is the radius of the tube, and Δp is the pressure difference along the length l of the tube.

- Ohm's law for a Newtonian fluid:

$$\Delta p = R\frac{\Delta V}{\Delta t},$$

where R is the flow resistance. The flow resistance in a cylindrical tube is:

$$R = \frac{8l}{\pi r_{tube}^4}\eta.$$

MULTIPLE-CHOICE QUESTIONS

MC–13.1. The diameter of a tube increases from d_1 to $d_2 = 2\,d_1$. As a result, the volume flow rate of laminar flow changes to
(a) $\Delta V_2/\Delta t = \Delta V_1/\Delta t$.
(b) $\Delta V_2/\Delta t = (1/2)\Delta V_1/\Delta t$.
(c) $\Delta V_2/\Delta t = 2\,\Delta V_1/\Delta t$.
(d) $\Delta V_2/\Delta t = (1/4)\Delta V_1/\Delta t$.
(e) $\Delta V_2/\Delta t = 4\,\Delta V_1/\Delta t$.

MC–13.2. Fig. 13.25 shows a cylindrical tube of changing diameter with an ideal dynamic fluid (blue) flowing toward the right with initial speed v. The vertical columns are connected to the main tube. Which of the five choices shows the proper elevations of the fluid in each of the three vertical columns?

MC–13.3. The volume flow rate and the mass flow rate in laminar flow are
(a) the same.
(b) proportional to each other.
(c) inversely proportional to each other.
(d) unrelated.
(e) related in a non-linear manner.

MC–13.4. An artery has ballooned at one location outward (to a larger cross-section) due to an aneurysm. Which of the following statements is correct for blood flowing through this broadened section? Treat blood as an ideal dynamic fluid.
(a) Blood will rush faster through the broadened section due to the equation of continuity.
(b) The equation of continuity predicts an increase of the blood pressure in the broadened section.

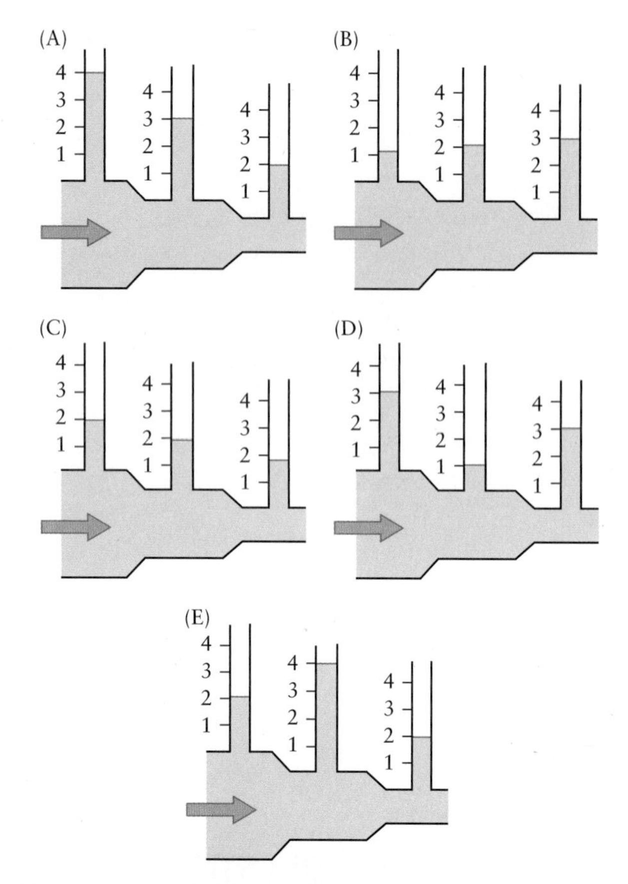

Figure 13.25 An ideal dynamic fluid (blue) flows through a cylindrical tube with initial speed $\vec{v}$. Vertical columns are connected to the main tube to measure the fluid pressure.

(c) Bernoulli's law predicts that the blood pressure in the broadened section is lower than in an adjacent section of the blood vessel, causing the blood vessel to temporarily collapse.
(d) Bernoulli's law predicts that the blood pressure in the broadened section is higher than in an adjacent section of the blood vessel.

MC–13.5. The equation of continuity is an expression of
(a) the conservation of mass.
(b) the conservation of total energy.
(c) the conservation of kinetic energy.
(d) the conservation of velocity.

MC–13.6. Bernoulli's law contains a term $\frac{1}{2}\rho v^2$, which was derived from the kinetic energy of the ideal dynamic fluid. What units does this term carry?
(a) J
(b) N
(c) m/s
(d) m/s^2
(e) Pa

MC–13.7. Which law connects the speed of an ideal dynamic fluid to its pressure?
(a) Pascal's law
(b) Newton's second law
(c) Bernoulli's law
(d) Laplace's law

MC–13.8. Bernoulli's law is an expression of
(a) the conservation of mass.
(b) the conservation of kinetic energy.
(c) the conservation of total energy.
(d) the conservation of velocity.
(e) the conservation of momentum.

MC–13.9. The conservation of mass leads to the following law we use to describe laminar flow in a fluid.
(a) Bernoulli's law
(b) Pascal's law
(c) equation of continuity
(d) Poiseuille's law
(e) Ohm's law

MC–13.10. Do the following experiment as shown in Fig. 13.26: Push a pin through the centre of a thin sheet of cardboard. Locate the tip of the pin in the central hole of a thread spool from below. Hold the cardboard from below and start to blow through the hole. The cardboard will not drop to the floor when you release it. Which law explains this effect?
(a) Poiseuille's law
(b) Ohm's law
(c) equation of continuity
(d) Bernoulli's law
(e) Pascal's law

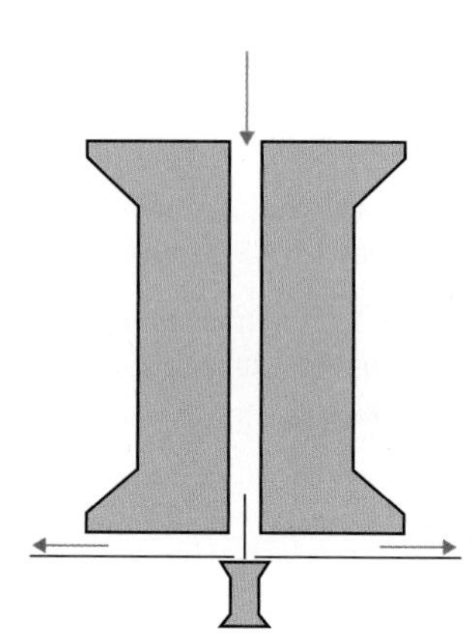

Figure 13.26 A pin is placed at the centre of a thin cardboard sheet. The pin's tip is then placed in the central hole of a thread spool. Hold the cardboard then start blowing through the hole. Release the cardboard.

MC–13.11. Do the following experiment: Hold two sheets of paper parallel to each other at a distance of about 1 cm to 3 cm. Blow gently between the two sheets from their edges. The two sheets will be pulled together. Which law explains this observation?
(a) Poiseuille's law
(b) Ohm's law
(c) equation of continuity
(d) Bernoulli's law
(e) Pascal's law

MC–13.12. What additional information is needed to calculate the mass flow rate from the product of the cross-sectional area and the fluid speed in the equation of continuity?
(a) none
(b) length of tube

(c) flow resistance
(d) density
(e) viscosity coefficient

MC–13.13. An artery is partially clogged by a deposit on its inner wall, as shown in Fig. 13.27. Which of the following statements best describes the processes that occur when blood rushes through this constriction? Treat blood as a Newtonian fluid.
(a) Blood will rush faster through the constriction due to the equation of continuity, causing additional wear and tear on the nearby blood vessel walls.
(b) Bernoulli's law and the equation of continuity predict a variation of the blood pressure in the constricted zone, but the blood vessel walls prevent any adverse effect due to this pressure variation.
(c) The blood pressure in the constriction zone is lower than in the adjacent blood vessel, causing the blood vessel to temporarily collapse at the constriction (vascular flutter).
(c) The blood pressure in the constriction zone is higher than in the adjacent blood vessel, causing a ballooning effect of the blood vessel at the constriction (aneurysm).

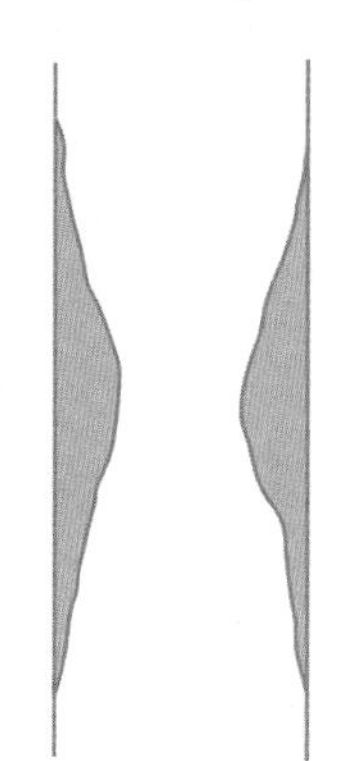

Figure 13.27 Cross-sectional view of a partially clogged artery.

MC–13.14. Use the viscosity coefficients from Table 13.1 for water at 20°C and for (whole) blood. With these values, a model for blood flow through the aorta is developed. We start with measuring the volume flow rate of water through an appropriately sized tube, using for the relevant parameters (such as pressure difference) physiological data for a standard man. Then we repeat the experiment with the same parameters and the same tube, but using whole blood. How will the volume flow rate change?
(a) It will be the same in both experiments.
(b) It will increase due to Poiseuille's law.
(c) It will increase due to Pascal's law.
(d) It will decrease due to Poiseuille's law.
(e) It will decrease due to Bernoulli's law.

MC–13.15. Newtonian fluids include which of the following assumptions that is excluded in ideal dynamic fluids?
(a) incompressible fluid
(b) fluid molecules interact inelastically
(c) the flow is laminar
(d) flow tubes do not intersect
(e) flow lines do not vanish

MC–13.16. Which of the following statements is wrong?
(a) Poiseuille's law applies as derived only to laminar flow.
(b) Poiseuille's law applies as derived only to ideal dynamic fluids.
(c) Poiseuille's law applies as derived only to incompressible fluids.
(d) Poiseuille's law and the equation of continuity can be used together for the same system.
(e) Poiseuille's law applies as derived only to Newtonian fluids.

MC–13.17. Fig. 13.28 shows a cylindrical tube through which a Newtonian fluid flows. Which velocity profile $\vec{v}$ (x), with x the axis to the right, best describes the actual velocity profile of the fluid?

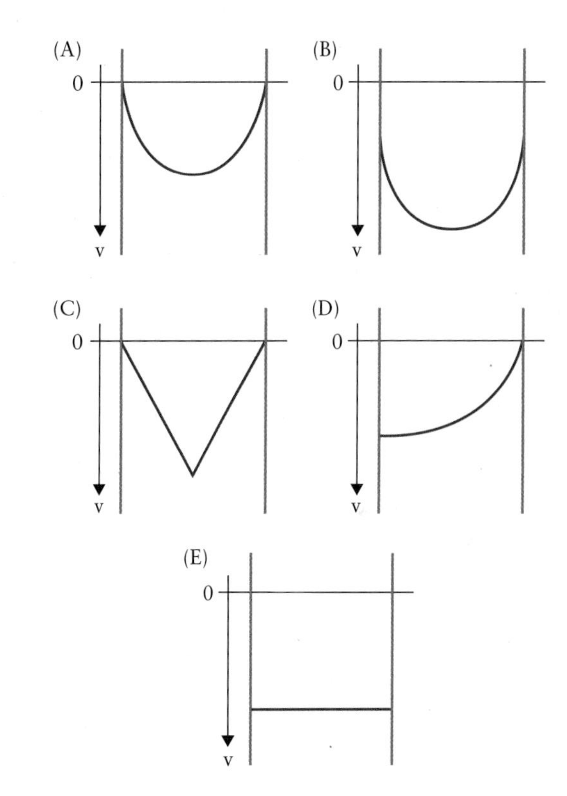

Figure 13.28 A Newtonian fluid passes through a cylindrical tube. Various velocity profiles $\vec{v}$ (x) are proposed (red curves), with x the axis to the right.

CONCEPTUAL QUESTIONS

Q–13.1. North American prairie dogs live in underground burrows with several exits. They usually build a mound over one exit, which causes a draft past that hole. How does this arrangement allow for ventilation of the burrow with the air stagnant above all other exits?

Q–13.2. In chemistry laboratories, moderate vacuum conditions are obtained in an experimental set-up when connecting the sealed apparatus to an aspirator. How does this instrument produce suction? *Note:* An aspirator is

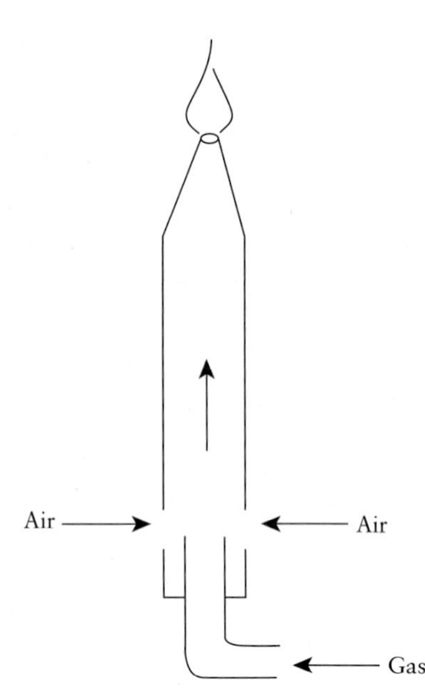

Figure 13.29 Design principle of the Bunsen burner used in chemistry laboratories.

a device consisting of a T-shaped tube with tap water running through vertically and the side tube connected to the system you want to evacuate.

Q–13.3. The design concept of a Bunsen burner is illustrated in Fig. 13.29. Which physical principle is employed to provide a sufficient mixing of the gas with air for effective combustion?

Q–13.4. Tornadoes and hurricanes can lift roofs off houses. A standard recommendation in affected areas is to keep windows open when a storm approaches. What happens to the roof, and why would open windows help?

ANALYTICAL PROBLEMS

P–13.1. What is the net upward force on an airplane wing of area $A = 20\ \text{m}^2$ if air streams at 300 m/s across its top and at 280 m/s past the bottom? Note that this airplane moves at subsonic speed with respect to the speed of sound (called Mach 1), about 330 m/s.

P–13.2. The instrument shown in Fig. 13.8 (Venturi meter) is used to measure the flow speed v of a fluid in a tube of cross-sectional area A. This is done by integrating the instrument into the tube, with the entry and exit cross-sectional areas identical to the primary tube. Between the entry and exit points, the fluid flows through a narrow constriction of cross-sectional area a. At the constriction the speed of the fluid is v_{con}. A manometer tube, connecting the wider and narrower portions of the main tube, shows a difference Δh in the liquid levels in its two arms.

(a) Using Bernoulli's law and the equation of continuity, show that:

$$v = \sqrt{\frac{2a^2 g \Delta h}{A^2 - a^2} \frac{\rho_{liquid}}{\rho_{fluid}}}.$$

(b) What is the volume flow rate $\Delta V/\Delta t$ if we use water for the fluid in the tube? The tube diameter is 0.8 m, the diameter of the constriction is 20 cm, and the pressure difference is 15 kPa.

P–13.3. A large water-containing tank is open to air. It has a small hole 16 m below the water surface through which water leaks at a rate of 2.5 L/min. Determine (a) the speed of the water that is ejected from the hole and (b) the diameter of the hole.

P–13.4. An ideal dynamic fluid flows through a tapering tube. Upstream, the tube has a cross-sectional area of 10 cm^2; the fluid pressure is 120 kPa, its density is $\rho = 1.65\ \text{g/cm}^3$, and the flow speed is 2.75 m/s. In the downstream section the cross-sectional area is 2.5 cm^2. Calculate in the downstream section (a) the fluid density, (b) the fluid flow speed, and (c) the fluid pressure.

P–13.5. Water is supplied to a building through a pipe of radius $R = 3.0$ cm. In the building, a faucet tap of radius $r = 1.0$ cm is located 2.0 m above the entering pipe. When the faucet is fully open, it allows us to fill a 25-L bucket in 0.5 minutes. (a) With what speed does the water leave the faucet? (b) What is the gauge pressure in the pipe entering the building? Assume that no other faucets are opened during the experiment.

P–13.6. Fig. 13.30 shows a tube A with radius R that splits into two equal tubes B and C with radii $r = R/3$. (a) For an ideal dynamic fluid in this system of tubes, what is the ratio of the fluid speed in tubes A and B, v_A/v_B? (b) What is the pressure difference from the bifurcation point to the point of merger at the beginning of tube D?

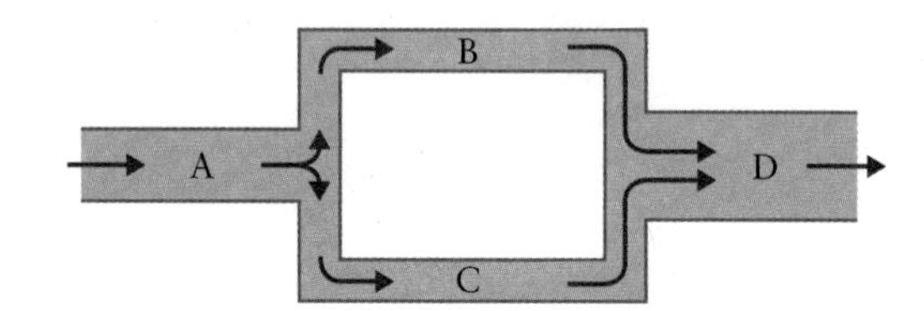

Figure 13.30

P–13.7. Fig. 13.31 shows a siphon. Flow in this device must be initiated with suction but it then proceeds on its own. (a) Show that water emerges from the open end at a speed of $v = (2\ g\ h)^{1/2}$. (b) For what range of y-values will this device work?

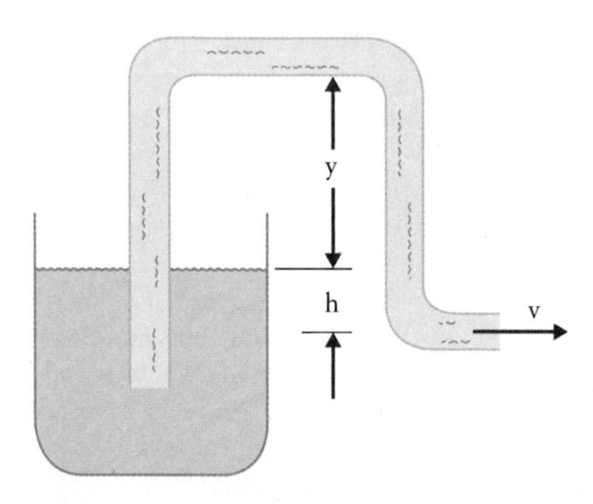

Figure 13.31

P–13.8. Fig. 13.32 shows a horizontal tube with a constriction and two open, vertical columns. The inner radius of the larger sections of the horizontal tube is 1.25 cm. Water passes through the tube at a rate of 0.18 L/s. If $h_1 = 10$ cm and $h_2 = 5$ cm, what is the inner radius at the constriction?

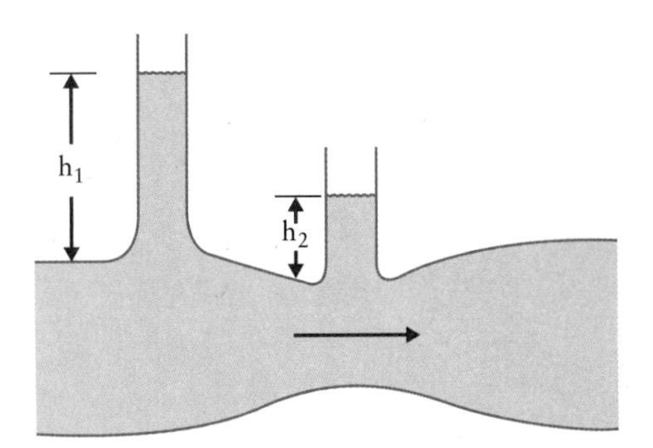

Figure 13.32

P–13.9. A beaker has a hole of radius $r = 1.75$ mm near its bottom, from which water is ejected as shown in Fig. 13.33. Calculate the height h of the water in the beaker if $h_1 = 1.0$ m and $h_2 = 0.6$ m.

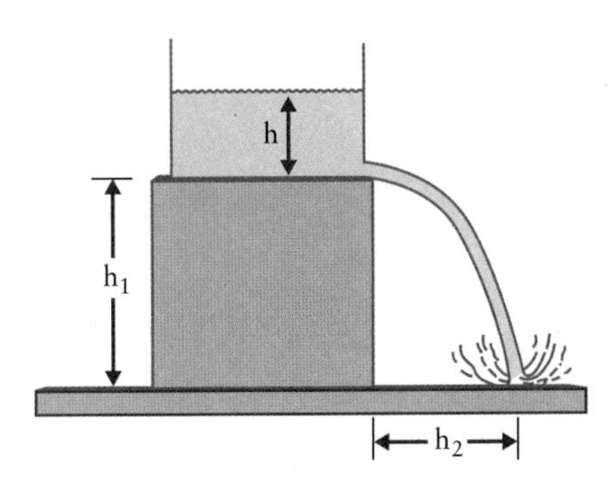

Figure 13.33

P–13.10. During level flight, air flows over the top of a bird's wing of area A with speed v_{top} and past the underside of the wing with speed v_{below}. Show that Bernoulli's law predicts that the magnitude F_{lift} of the upward lift-force on the wing is given by:

$$F_{lift} = \frac{\rho A}{2}(v_{top}^2 - v_{below}^2),$$

with ρ the density of the air.

P–13.11. Air moves through the human trachea at 3 m/s during inhalation. Assume that a constriction in the bronchus exists at which the speed doubles. Treating air as an ideal dynamic fluid, calculate the pressure in the constriction.

P–13.12. Water is pumped into a storage tank from a well delivering 140 L/min through a pipe of 6.0 cm^2 cross-sectional area. What is the average velocity of the water in the pipe as it is pumped from the well?

P–13.13. A liquid of density 1.5 g/cm^3 flows through two horizontal sections of tubing joined end to end. In the first tube the cross-sectional area is 8.0 cm^2, the flow speed is 250 cm/s, and the pressure is 1.1×10^5 Pa. In the second section the cross-sectional area is 3.0 cm^2. Calculate the smaller section's (a) flow speed and (b) pressure.

P–13.14. Confirm the data shown in Table 13.2 for the average length, radius, and individual and total flow resistance in (a) arterioles, (b) the aorta, and (c) capillaries. *Hint:* Use $\eta = 2.5 \times 10^{-3}$ N s/m^2 as an average value for the blood viscosity coefficient from Table 13.1 at 37°C.

P–13.15. The hypodermic syringe in Fig. 13.34 contains water. The barrel of the syringe has a cross-sectional area $A_1 = 30$ mm^2. The pressure is 1.0 atm everywhere while no force is exerted on the plunger. When a force $\vec{F}_{ext}$ of magnitude 2.0 N is exerted on the plunger, the water squirts from the needle. Determine the water's flow speed through the needle, v_2. Assume that the pressure in the needle remains at a value of $p_2 =$ 1.0 atm and that the syringe is held horizontal. The final speed of the water in the barrel is negligible.

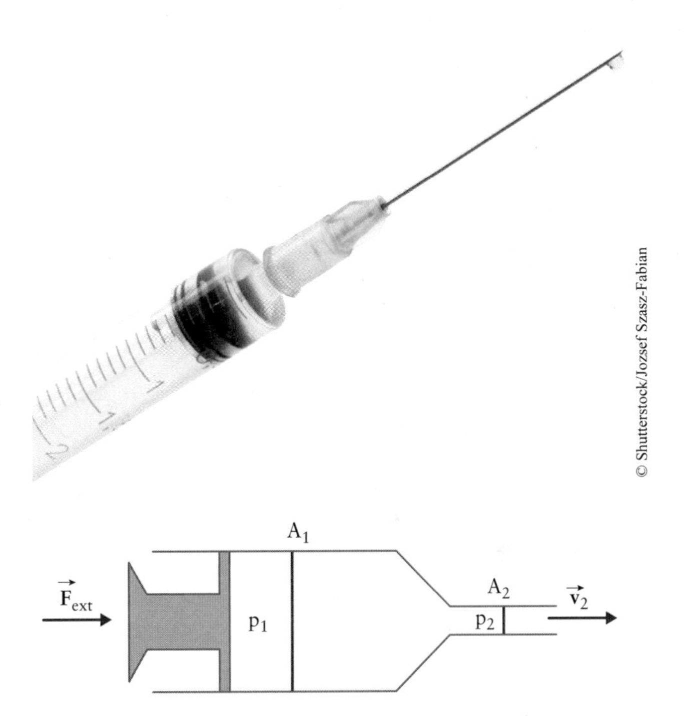

Figure 13.34

P–13.16. A hypodermic needle is 4.0 cm long and has an inner diameter of 0.25 mm. What excess pressure is required along the needle so that the flow rate of water through it is 1.0 g/s? Use the viscosity coefficient of water at 20°C from Table 13.1.

P–13.17. A horizontal tube of radius $r = 5.0$ mm and length $l =$ 50 m carries oil ($\eta = 0.12$ N s/m^2). At the end of the tube the flow rate is 85 cm^3/s and the pressure is $p =$ 1.0 atm. What is the gauge pressure at the beginning of the tube?

P–13.18. A patient is to be injected with 0.5 L of an electrolyte solution over half an hour. Assuming that the solution is elevated by 1.0 m above the arm and the needle is 2.5 cm long, what inner radius should the needle have? Use water parameters for the solution and assume the pressure in the patient's vein is atmospheric pressure.

P–13.19. We study flow of a Newtonian fluid through two different tubes (index 1 and index 2). The pressure differences between the two ends of the tubes, Δp, are the same for both tubes, $\Delta p_1 = \Delta p_2$. The tubes differ in radius and length: length of tube 1 is $l_1 = 2$ m, length of tube 2 is $l_2 = 1$ m, radius of tube 1 is $r_1 = 2$ cm, and radius of tube 2 is $r_2 = 1$ cm. Calculate the ratio of volume flow rates through the two tubes.

P–13.20. Fig. 13.30 shows a tube A with radius R that splits into two equal tubes B and C with radii $r = R/3$. (a) If we substitute the ideal dynamic fluid for a Newtonian fluid with given viscosity coefficient, how do the results in P–13.6 change? What additional parameter must be measured for tubes B and C? (b) Sketch the pressure in the fluid from left to right along tubes A, B, and D.

ANSWERS TO CONCEPT QUESTIONS

Concept Question 13.1: (B). Eq. [13.2] reads in this case $v_1/v_2 = A_2/A_1 = 3A_1/A_1 = 3$. Thus, $v_1 = 3v_2$, or $v_2 = v_1/3$.

Concept Question 13.2: (A)

Concept Question 13.3: (B). We use Poiseuille's law twice to write the ratio for the volume flow rates in both experiments (note that the radius of the tube is the same in both cases): $(\Delta V/\Delta t)_1 : (\Delta V/\Delta t)_2 = (\Delta p_1\, l_2)/(\Delta p_2\, l_1)$. Now we substitute $(\Delta V/\Delta t)_1 = (\Delta V/\Delta t)_2$ and $l_2 = 2\, l_1$. This yields $\Delta p_2 = 2\, \Delta p_1$.

Part Three

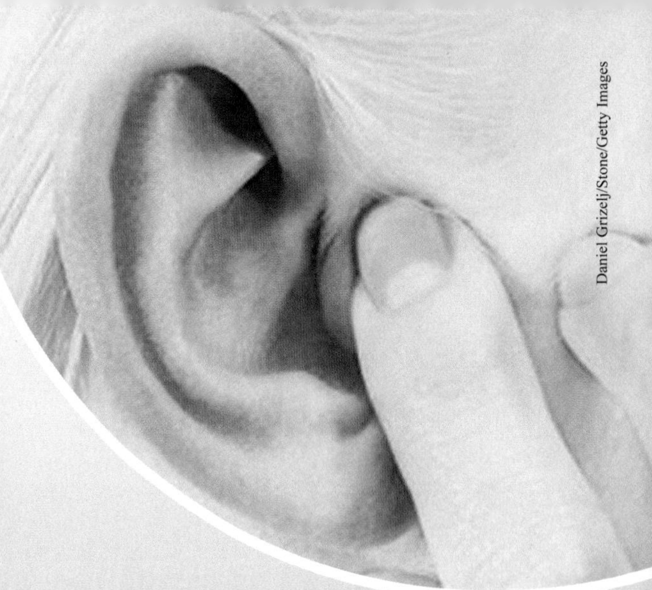

VIBRATIONS, ACOUSTICS, AND HEARING

New physical concepts are often discovered when we try to explain everyday phenomena with previously acquired knowledge and find it insufficient. The third part of this textbook illustrates such a case: before we move on to study a new fundamental force in Part 4, we look at vibrations. These we can satisfactorily describe with the methods we developed in the first part of the book. But then we note that the vibration of a string, let's say in a piano or on a guitar, allows us to hear a sound. We explore the connection of sound and vibration, and end up opening an entirely new field of study: the generation and propagation of waves. In the current part we will find the waves to be carried by the medium (air, water, or within the tissues of our body), but we will see in the next part that waves are a much more general phenomenon than what we find with sound waves; electromagnetic waves can even travel through empty space.

We find again that new physical phenomena we discuss in this part are immediately applicable to explaining phenomena observed in the other life sciences. Physics lies at the root of all natural phenomena, so it is not surprising to find that the chemical method of infrared spectroscopy or the physiological phenomenon of hearing can be explained with the methods introduced in this part. But the relevance of physics goes further: we see also that the understanding of its concepts allows us to develop new methods of medical diagnosis, such as ultrasound imaging, which is discussed at the end of this part. This is just one example, with others to follow in later parts, where physics allows us not only to understand and explain phenomena beyond the limits of its realm, but also allows us to develop new techniques to be applied in distinct areas, such as medicine. This relevance makes Physics an integral part of the Life Sciences, often captured by labelling the field as Biophysics.

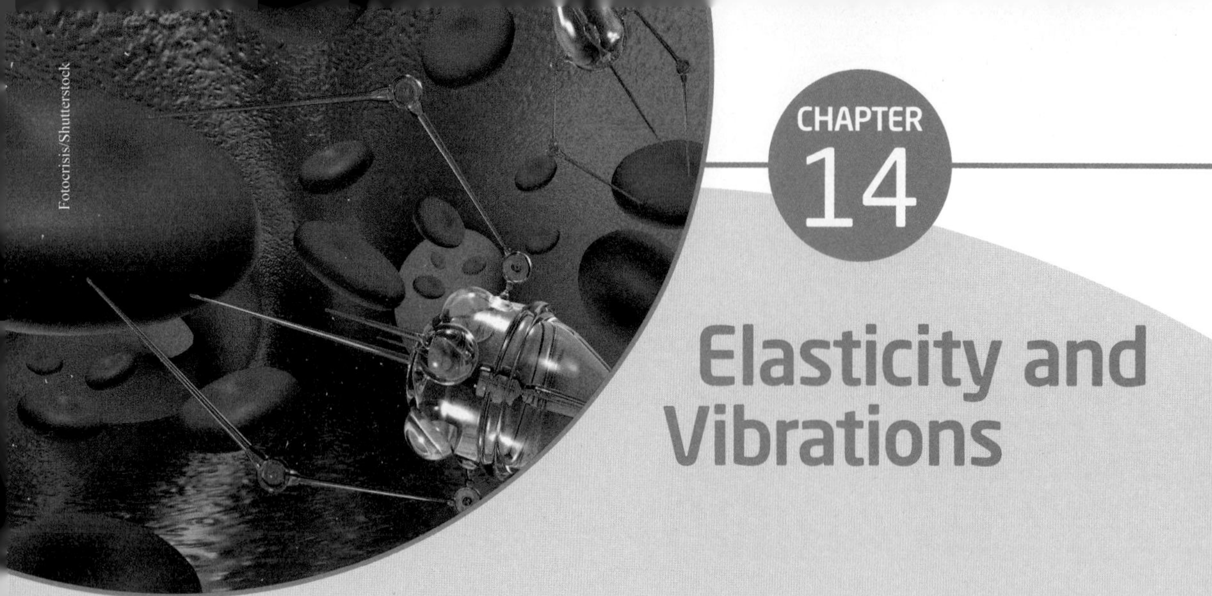

CHAPTER 14

Elasticity and Vibrations

The rigid-object model we introduced in Chapters 5 and 6 is insufficient to describe many fundamental properties of extended objects, such as stretching, twisting, deformation, and rupture. In the current chapter, two more general models are introduced: the elastic and the plastic object. An elastic object responds to a stress (a force acting on the surface of an extended object) with a strain (a change in size or shape of the object) that is linearly proportional to the stress. This linear relation is called Hooke's law. Examples include elongation in response to tensile stress, twisting as a result of shear stress, and compression due to hydraulic stress. Deformation of objects is called plastic when the stress exceeds the range of applicability of Hooke's law.

The most important property of elastic systems is their dynamic behaviour when released from a mechanical non-equilibrium state: the system undergoes sinusoidal vibrations about its equilibrium state in response to a restoring force that is linear in the displacement. The elastic potential energy is introduced as a new form of potential energy for such systems. It depends quadratically on the displacement from the equilibrium position.

We continue in this chapter where we left off with the discussion of rigid objects in Chapter 6. In that chapter we found that the rigid object is a useful model when we want to describe rotations. It was an important extension of the previously introduced model of a point-like object because it allowed us to describe and quantify the basic processes of vertebrate locomotion.

Like any other physical model, the rigid-object model contains restrictions. The rigid object cannot change its shape or volume, vibrate, bend, or rupture. If any of these processes is important for life science applications, then we need to extend the rigid-object model once more. The objective of the current chapter is first to establish that these processes are relevant and then to introduce two new models that go beyond the rigid object: the elastic and the plastic object.

Mechanical processes in our body beyond rotation are easily established in a self-test. Push your right index finger into the skin of your lower left arm. You feel a localized force acting on your arm and you observe a deformation of the skin; the stronger you push, the further the skin is displaced. The skin re-shapes into its original form when you withdraw your finger. The reversible deformation of the skin is an example of an elastic response to an external force. Plastic response occurs when the forces are larger. For example, bone fractures are the result of excessive forces acting on our skeleton.

The scientific relevance of these observations stems from the fact that corresponding microscopic processes exist in which the system responds elastically or plastically to external forces. We illustrate this with the intended function of muscles (elastic response) and deviations for torn muscles (plastic response). In Fig. 7.2, we introduced the sarcomere as the unit that can actively contract when the myosin filaments slide toward the two terminal Z-discs along the actin filaments. The sarcomere can also be stretched passively by the action of an antagonistic muscle. Fig. 14.1 shows a simplified graph of the dependence of the contractile (active) force a sarcomere can exert after being stretched to a given length l. The length l of the sarcomere is defined as the distance between two adjacent Z-discs. The force is displayed on the abscissa as a value relative to the maximum force the sarcomere can exert. The maximum force is available when a sarcomere is at its resting length.

The graph allows us to highlight some important features of the active muscle force: as the sarcomere length varies, the overlap of the myosin and actin filaments varies as well. As shown in the insets along the curve, the variation of overlap of actin and myosin filaments causes a loss of muscle force when operating far from the resting length. When the sarcomere shortens to less than 1.65 μm in length, the myosin filaments touch the two adjacent Z-discs. This provides a physical limitation, as muscle contraction by more than 20% to 25% is not possible.

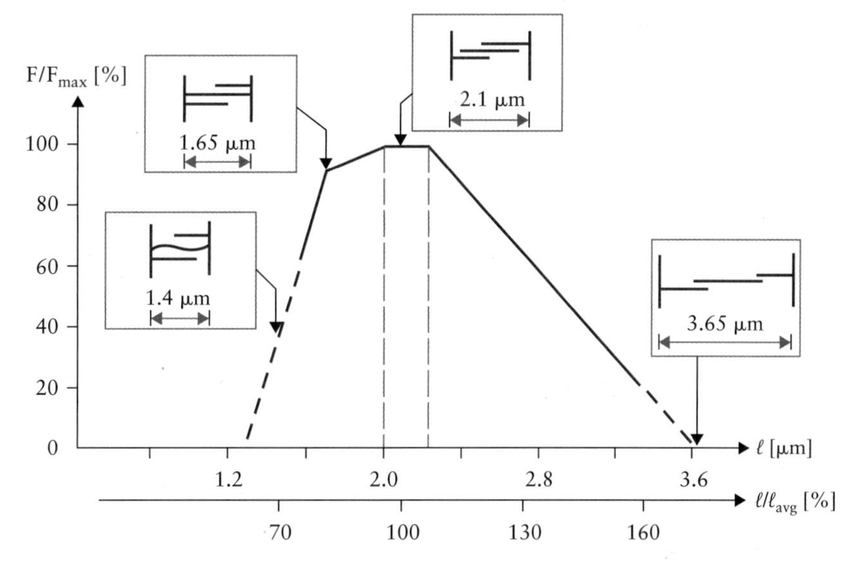

Figure 14.1 The active muscle force is shown as a function of the sarcomere length. The force is given as a fraction of the maximum force a muscle can exert when it is at its average length $l_{average} = 2.1\ \mu m$. The sarcomere length is given as an absolute value l in unit μm (top abscissa) and as a length relative to the resting length $l/l_{average}$ in % (bottom abscissa). For several specific sarcomere lengths the overlap of the myosin and actin filaments is illustrated in the inset boxes.

When the sarcomere in turn is passively stretched to a length of 2.8 μm and more (i.e., when the muscle is stretched to 130% or more of its resting length), its ability to contract is significantly reduced as the overlap of the actin and myosin filaments is reduced. If there were no counteracting process, this would be dangerous for the muscle as its own ability to withstand the stretching through contraction is weakened.

The curves in Fig. 14.2(a) illustrate how the muscle is protected against overstretching. The dash-dotted curve shows again the active force the muscle can exert when stretched to various lengths (shown relative to the resting length). This curve corresponds to the curve in Fig. 14.1. Note that you find both Figs. 14.1 and 14.2 in the physiological literature; Fig. 14.2 is a plot of the actual forces, and Fig. 14.1 is used for conceptual discussions of muscle action. Comparing Figs. 14.1 and 14.2 we note the benefit of the simplified graph: the straight-line segments in Fig. 14.1 are easier to quantify than the curved, dashed-dotted function in Fig. 14.2. Whether it is sufficient to use Fig. 14.1 or whether it is necessary to use Fig. 14.2 depends on the specific purpose or argument we want to make.

Fig. 14.2 shows a second curve applicable to the sarcomere. The dashed curve is the passive stretching force, i.e., the force by which the muscle tissue resists being stretched. This passive stretching force protects the muscle against being extended to a length from which it can no longer contract by its own action. The net force a muscle applies to oppose being stretched is shown in Fig. 14.2 as a red line. This net force is the sum of the active and passive

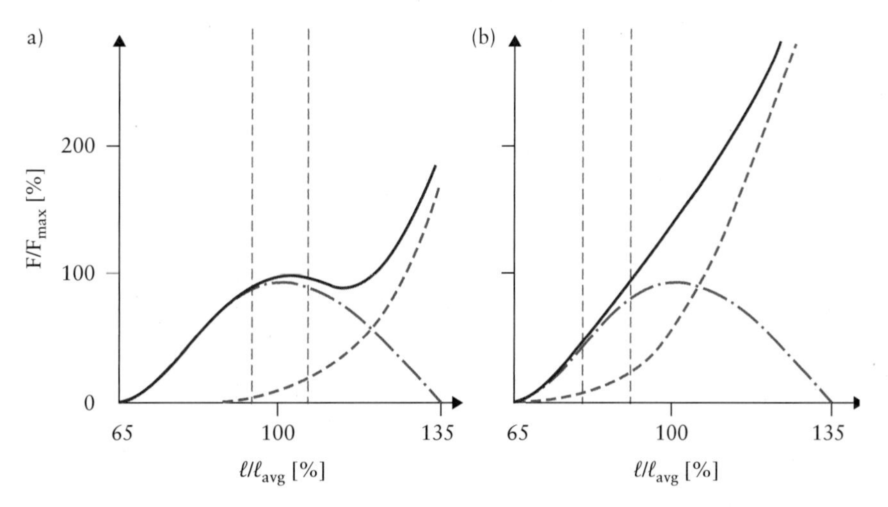

Figure 14.2 Active (dash-dotted) and passive (dashed) components of the muscle force for (a) skeletal muscles, and (b) the heart muscle. The figures show the force, relative to the maximum force. It is plotted as a function of the muscle length, relative to the muscle length at the point where the muscle exerts a maximum active force. The red lines show the total force, which is the sum of the active force and the passive stretching force. The two vertical dashed lines indicate the normal operating range of the muscle. The dash-dotted curve in part (a) is equivalent to the curve shown in Fig. 14.1.

forces. Muscles cannot be overstretched, because the passive force limits the stretching at large values of length l.

Fig. 14.2(b) illustrates that the same observations apply to muscles with other functions than skeletal muscles. For the heart muscle the passive force counteracting its stretching is even more dominant than in the case of a skeletal muscle. The total force acting on the tissue (red curve) is again a combination of the active force (dash-dotted curve) and the passive stretching force (dashed curve).

Figs. 14.1 and 14.2 illustrate two points. First, a muscle is not a rigid object, as it can contract or be stretched. Second, when a muscle is stretched beyond its average length, a force counteracts against further stretching. This force increases with increasing length of the muscle; i.e., it acts to restore the original length of the muscle. Thus, it is necessary to extend the rigid-object model to allow for deformations as a result of external forces. The ability of biological materials to be deformed and the response to the deformation are important to prevent damage to tissues. The response to a deformation is a force that aims at restoring the resting state of the tissue. We call this the **elastic behaviour** of tissue. However, external forces may exceed a threshold beyond which the restoring force can no longer prevent tissue damage. Once this threshold is passed, **permanent plastic deformations** occur that require medical attention.

CONCEPT QUESTION 14.1

In Chapters 3 to 6 we discussed mechanical concepts using the assumption that forces are position independent. (a) Can we use this assumption for the force shown in Fig. 14.1? (b) Can we use it for the forces shown in Fig. 14.2?

Elastic behaviour of tissues can lead to vibration. Note that vibration did not result in our earlier self-experiment involving pushing the skin on the lower arm. Two reasons prevent the skin from vibrating: it is too soft, and it is too well connected to the tissues below. However, the human body contains other membranes that can vibrate: the vocal cords to create our voice, and the eardrum to couple sound into the middle and inner ear. Vibration is discussed later in this chapter.

14.1: Elasticity

The quantitative description of elastic behaviour starts with a more specific look at the connection between an external force and the resulting deformation. We define **stress** as a force exerted per unit area of surface on an extended object. The stress leads to a **strain**, which is the relative change in the size of the object. We distinguish three types of deformation based on the type of strain: a change in the length of an object is called stretching, a change in an angle of the extended object is called twisting, and a change in the volume of an object is called a compression or an expansion. Each of these types of strain is caused by a different form of stress. Tensile stress leads to stretching, shearing stress leads to twisting, and hydraulic stress causes expansion or compression.

14.1.1: Tensile Stress and Stretching

Stretching of an extended object is illustrated in Fig. 14.3. Stretching in the vertical direction is achieved by two opposite forces, $\vec{F}_1$ and $\vec{F}_2$. The volume of the object, and therefore its density, are conserved in this process since the width and depth of the object are reduced simultaneously with the vertical stretching. This process is possible only for solids and liquids.

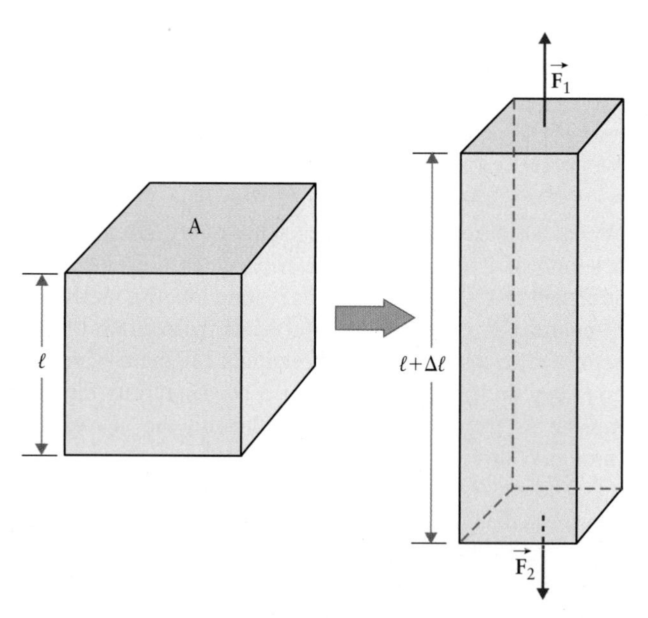

Figure 14.3 Sketch of a stretching deformation of an extended object, chosen to be initially a cube. Two equal but opposite forces pull on opposing surfaces of area A. The stress $|\vec{F}|/A$ leads to a strain in the form of a length increase from l to $l + \Delta l$.

We quantify the terms stress and strain for Fig. 14.3. **Strain** is the relative change in length, expressed as $\Delta l/l$ with $\Delta l = l_{\text{final}} - l_{\text{initial}}$. Thus, strain is a dimensionless quantity. The stress leading to this stretching is called tensile stress and is given by F/A with F the magnitude of force $\vec{F}_1$. A is the area of the object surface on which force $\vec{F}_1$ acts. F/A has unit Pa, the same as the unit of pressure. The two forces in Fig. 14.3 are related by Newton's first law due to the condition of mechanical equilibrium: $\vec{F}_2 = -\vec{F}_1$. Note that this is not an application of Newton's third law, as both forces act on the same object!

We expect an elongation of an object with a tensile stress. The object stretches to a certain length at which the restoring force within the object balances the external forces, establishing a new mechanical equilibrium for the system. We call the response of the object **elastic** if the strain and stress are proportional to each other, i.e., when the length change of the object increases linearly with the stress:

$$\frac{F}{A} = Y\frac{\Delta l}{l}. \quad [14.1]$$

The proportionality factor Y is called **Young's modulus** (named for Thomas Young), with unit Pa. Since this formula is used frequently in the description of mechanical properties of solids and liquids, new variables are introduced: σ for the stress, with $\sigma = F/A$; and ε for the strain, with $\varepsilon = \Delta l/l$. With these variables, Eq. [14.1] is written as:

$$\sigma = Y\,\varepsilon. \quad [14.2]$$

A large value of Young's modulus, e.g., for steel, means that even a large force acting on a piece of material leads to only a small length increase. In turn, small values of Young's modulus, e.g., for a blood vessel, mean that even small forces cause a large length variation. We call materials with large Young's moduli **strong materials** and materials with small Young's moduli **soft materials**. Table 14.1 lists several values for Young's moduli, primarily comparing biological materials.

TABLE 14.1

Young's modulus for various materials

Material	Y (Pa)
Steel	2×10^{11}
Douglas fir wood	1.3×10^{10}
Compact bone (e.g., femur)	$1–2 \times 10^{10}$
Teeth	$7 \times 10^{9}–1.5 \times 10^{10}$
Cartilage	$1–4 \times 10^{7}$
Tendon	$2 \times 10^{7}–1 \times 10^{6}$
Rubber	$7 \times 10^{6}–2 \times 10^{7}$
Blood vessels	$1.2–4.0 \times 10^{5}$

CASE STUDY 14.1

What conditions apply to Young's modulus Y when we say that the stress depends linearly on the strain in Eq. [14.2]?

Answer: *The answer is not simply that Eq. [14.2] is linear, because the strain occurs with an exponent 1 on the right-hand side of the equation; i.e.,* $\sigma = Y \cdot \varepsilon^n$ *with* $n = 1$*. This is a necessary condition because* $n \neq 1$ *means that an explicit non-linearity exists.* $n = 1$ *is not sufficient, however, because Eq. [14.2] does not rule out an implicit dependence of* Y *on the strain,* $Y(\varepsilon)$*.*

In turn, requiring that Y *be a constant is too severe a restriction. Indeed,* Y *does depend on the temperature and is therefore only a materials constant. Thus, the proper way to ensure the linearity of Eq. [14.2] is to require* $Y \neq f(\varepsilon, \sigma)$*, i.e., that the Young modulus does not depend on the strain or stress.*

Extensive stretching usually doesn't obey Eq. [14.2]. If we still use Eq. [14.2], then Young's modulus must become strain-dependent. Thus, an **elastic limit** for the strain is introduced that describes the strain threshold beyond which Eq. [14.2] is no longer applicable. For a larger strain the object encounters permanent plastic deformations. The elastic limit is discussed later in the next section.

14.1.2: Shearing Stress and Twisting

A different type of deformation results in the case shown in Fig. 14.4: two equal but opposite forces, $\vec{F}_1$ and $\vec{F}_2$, are applied tangentially to two opposite faces of an extended object; i.e., the force vectors lie parallel to the surfaces they act on, as shown in the figure. To maintain mechanical equilibrium, i.e., to avoid a torque acting on the object, additional forces must be applied to hold the object stationary. This type of deformation can occur only in solids, because liquids and gases cannot sustain tangential forces. The two forces shown in Fig. 14.4 lead to a deformation of a rectangular prism, which we can quantify either by the angle θ or by the ratio of the displacement Δx to the width of the object l:

$$\tan\theta = \frac{\Delta x}{l}.$$

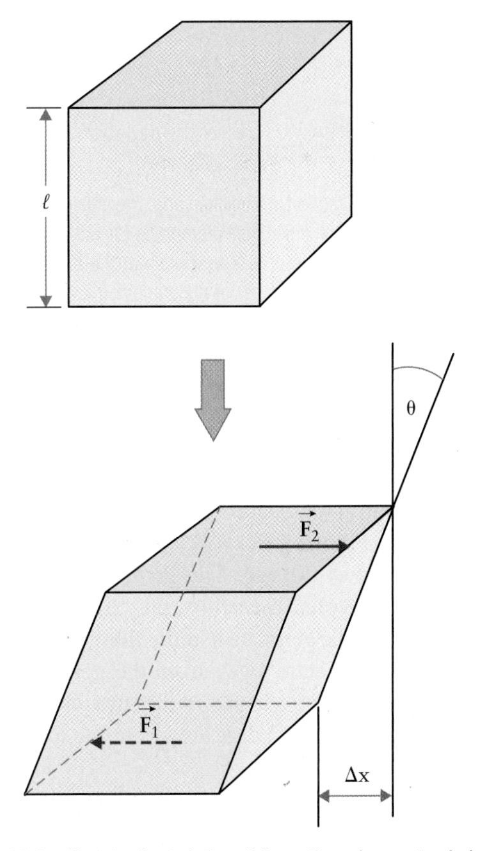

Figure 14.4 Sketch of a twisting deformation of an extended object, chosen to be initially a cube. Two tangential forces acting on opposing surfaces cause a change in the angle θ. This change is expressed as a function of the lengths Δx and l.

In analogy to the stretching case, a linear relation between stress and strain is observed for small values of Δx:

$$\frac{F}{A} = G\frac{\Delta x}{l}. \quad [14.3]$$

The proportionality factor G is called the **shear modulus**, with unit Pa. A large value of G means that even a large tangential force leads to only a small twisting. An example is steel, which has a value of $G = 8 \times 10^{10}$ Pa. In turn, a material with a small G value is easily twisted by small forces. An example is cartilage with $G = 2.5 \times 10^{7}$ Pa.

An interesting application of the twisting motion is illustrated in Fig. 14.5 for the *chicken egg*. The egg yolk (5) is connected to the germ disk (7), from which the chicken develops. It is beneficial for the developing chicken to have the germ disk directed upward at all times because the warmth of the body of the breeding hen flows into the egg from above. Rotation of the egg in the nest should not allow the germ disk to turn to the side or facing downward.

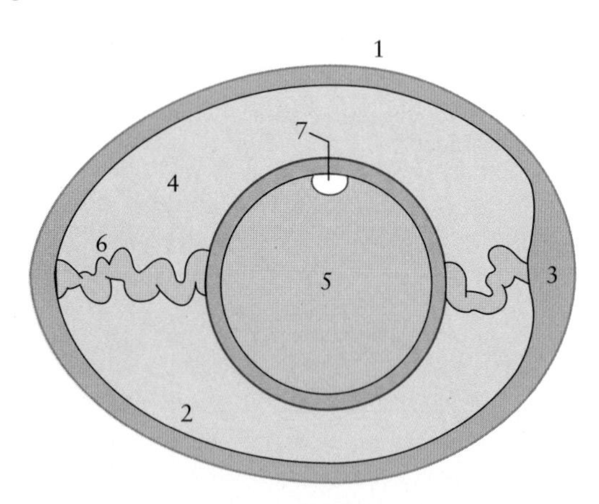

Figure 14.5 Twisting motion in a chicken egg. The yolk (5) and the germ disk (7) are attached to the inner shell membrane (2) with a pair of elastic cords (6). These allow the germ disk to remain at the top of the yolk at all times because the centre of mass of the combined yolk/germ disk lies off-centre away from the germ disk. This allows for a small net torque due to gravity when the germ disk is rotated sideways. Other components of the egg are: (1) shell, (3) air sac, and (4) egg white.

To stabilize the germ disk in the upward position, the yolk is connected to the inside of the eggshell (2) as shown. The spiral-shaped cords (6) have a very small shear modulus, allowing the yolk to turn easily about the longitudinal axis of the egg. The germ disk has a lower density than the yolk. Therefore, the centre of mass of the suspended yolk/germ disk unit, floating in the egg white (4), lies off-centre away from the germ disk. This provides a small net torque due to the gravitational force, enough to rotate the germ disk always to the top.

14.1.3: Hydraulic Stress and Compression

A third type of deformation is possible for solids, liquids, and confined gases: a compression or expansion is obtained by applying a force of magnitude F perpendicular to the surface of an extended object from all sides, as illustrated in Fig. 14.6. In this case the stress not only carries the unit of pressure, Pa, but physically is a pressure. The strain is defined as the relative volume change, $\Delta V/V$.

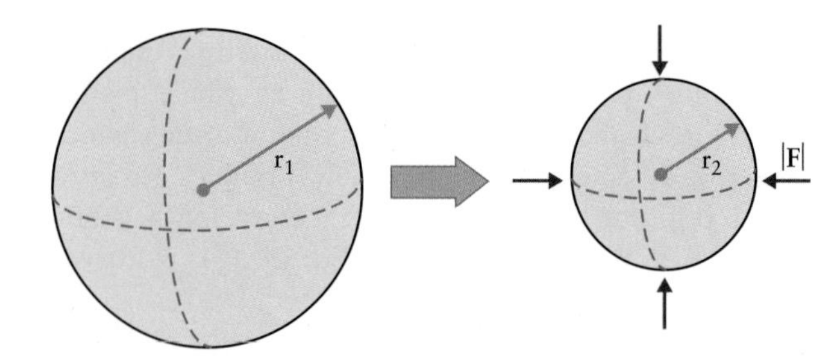

Figure 14.6 Sketch of a compression of a spherical extended object due to external forces of uniform magnitude F. The volume change is based on the radius change of the sphere.

An elastic deformation is given when the stress and strain are related linearly to each other:

$$p = -B\frac{\Delta V}{V}. \quad [14.4]$$

The negative sign is introduced in Eq. [14.4] to ensure that the coefficient B is a positive number. It is needed since an increasing external pressure leads to a decreasing volume. The coefficient B is a materials constant that we call **bulk modulus**. It has the unit Pa. A typical value for the bulk modulus of a solid material is $B = 1.6 \times 10^{11}$ Pa, which is the value for steel. Liquids have only slightly lower values; e.g., for water we find $B = 2.1 \times 10^{9}$ Pa. Gases in turn have significantly lower values as they are easily compressed. It is important to distinguish that the bulk modulus describes the change in volume of a material (with unit m^3) while Young's modulus describes the change in length of a material (with unit m).

The bulk modulus is used to define the **compressibility** of a material. The compressibility is given by $1/B$; i.e., a material with a large bulk modulus has a small compressibility.

CONCEPT QUESTION 14.2

Using Eq. [14.4], which statement is wrong?

(A) The relative volume change $\Delta V/V$ does not depend linearly on pressure due to the negative sign on the right-hand side.

(B) The compressibility is a materials constant.

(C) The compressibility of a material is not linearly related to its bulk modulus.

(D) The absolute volume change of an object, ΔV, alone does not reveal anything about the elastic properties of the object.

(E) Eq. [14.4] is not a linear relation if the compressibility depends on the absolute pressure.

14.2: Plastic Deformations

14.2.1: General Concept

The three elastic relations we introduced in Eqs. [14.1], [14.3], and [14.4] can be generalized in the form of Eq. [14.2]. An elastic deformation is reversible because the object resumes its original shape when the stress is removed.

We know from experience that Eq. [14.2] does not always hold. Only minor forces have to be applied to deform play dough permanently. This deformation is irreversible because the strain does not return to zero when the stress on the material is removed.

Stronger materials also respond to a large stress in an irreversible fashion, e.g., when a bone breaks. Fig. 14.7 shows the entire range of the stress–strain relation for a medium-strength steel. For low stress values a linear relation is observed (this elastic part of the curve is extended in a dashed line). As the stress is increased beyond the linear regime, microscopic structural alterations in the steel take place leading to a significant strain increase, often occurring in sudden bursts. In this regime a permanent **plastic deformation** of the material takes place. Eventually the material can no longer withstand the stress, passing through its **ultimate strength point**. Beyond this point the material approaches the rupture point quickly, without need for a further increase of the stress.

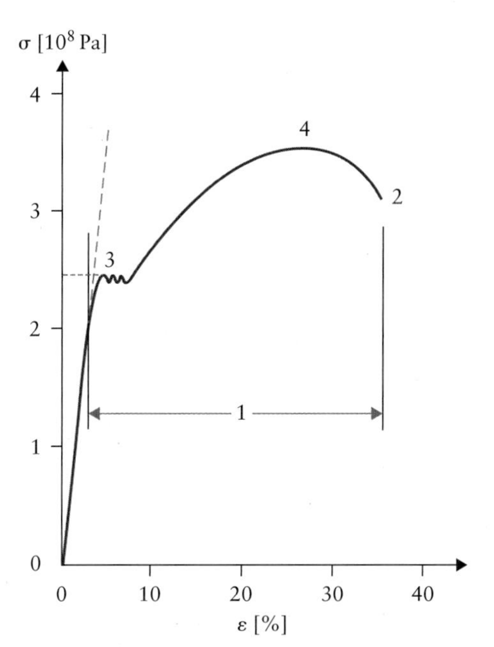

Figure 14.7 Stress–strain relation of medium-strength steel. The stress is linearly proportional to the strain for strain values up to about 3%. This linear regime is indicated by the dashed straight line. The plastic deformation regime (1) is more complex. Sudden bursts of strain increase indicate that structural changes occur within the material (3). The maximum strength is reached at a strain of about 28% (4). The ultimate strength is reached near a strain value of 40%, at which value the material ruptures (2).

KEY POINT

A material is called elastic when it responds to external forces (stress) with a linear deformation (strain). Elastic deformations are reversible. A material is called plastic when it responds to a stress in a non-linear fashion. Plastic deformations are irreversible.

CASE STUDY 14.2

Fig. 14.8 shows the stress–strain relation of compact bone. (a) Fig. 14.7 for steel shows only positive values for stress and strain. Does it make sense for Fig. 14.8 to show also negative values? (b) The red curve in Fig. 14.8 ends both at the lower left and at the upper right. Is there a physical meaning to this?

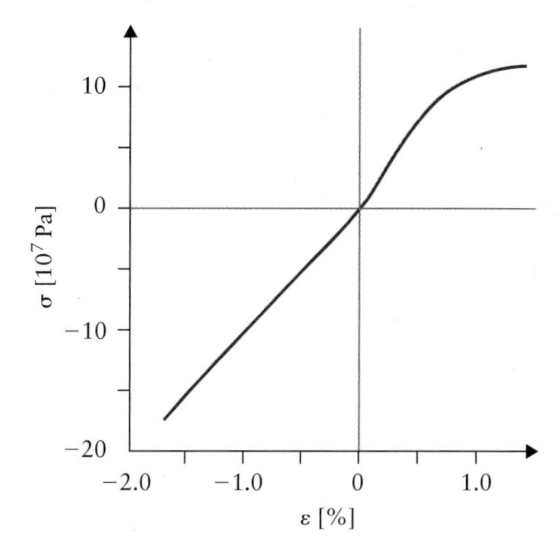

Figure 14.8 Stress–strain relation for compact bone. The graph includes both tensile stress, when the bone is stretched, and compressive stress, when the bone is compressed. The two branches of the curve differ slightly, but compact bone reaches its rupture point very quickly for either stress at strains of less than ±2%.

Answer to Part (a): *Yes. Positive and negative strain values mean physically different alterations, with positive strains representing stretching and negative strains a uniaxial compression. The corresponding stress intervals are called tensile stress (positive branch) and compressive stress (negative branch).*

Answer to Part (b): *The curve ends in these points because the bone has then reached its ultimate strength points, in the same fashion as discussed for steel in Fig. 14.7. When the stress exceeds the corresponding values the bone gets crushed or splinters.*

EXAMPLE 14.1

We use again Fig. 14.8 for the stress–strain relation of compact bone. In an adult male the femur has a minimum diameter of about 3 cm. At what force along its axial direction will the femur break?

continued

Solution

We read the maximum tensile and compressive stress from Fig. 14.8. Rupturing occurs where the curves end at each side. For bone this occurs at strains below 2%, which means that bones safely can be stretched only very little. The maximum compressive stress is -1.7×10^8 Pa, and the maximum tensile stress is 1.2×10^8 Pa. The corresponding forces are obtained from the definition of the stress, $\sigma = F/A$, when rewritten as $F = \sigma \cdot A$. In this formula the cross-sectional area of the bone is identified as the area onto which the force acts. The area A is calculated by assuming a cylindrical shape of the bone with a circular cross-section of radius r. With $r = 1.5$ cm, we find $A = r^2 \pi = 7.0 \times 10^{-4}$ m^2. The magnitude of the compressive force then follows as:

$$F_{\text{compressive}} = (1.7\times10^{8}\,\text{Pa})(7.0\times10^{-4}\,\text{m}^2)$$
$$= 1.2\times10^{5}\,\text{N}.$$

Note that we used the absolute value of the maximum compressive stress because we seek only the magnitude of the compression force. The magnitude of the tensile force is found in the same manner:

$$F_{\text{tensile}} = (1.2\times10^{8}\,\text{Pa})(7.0\times10^{-4}\,\text{m}^2)$$
$$= 8.4\times10^{4}\,\text{N}.$$

To obtain an idea of what these values mean, we convert the tensile force into a corresponding weight (using $F = m\,g$): the femur can withstand the pull of a mass of up to 8.5 tons (8600 kg). Although this is a large mass the corresponding forces unfortunately do occur, even for only a fraction of a second—e.g., during a fall onto a hard surface.

Some systems don't display a linear stress–strain regime at all. The most prominent case is blood vessel tissue. Fig. 14.9 shows a light micrograph of a cross-section through a muscular artery. The thick wall is required to carry blood under pressure. It consists of three layers: the innermost *tunica intima* surrounds the lumen (white). It consists of a fine elastic sheet (dark purple, convoluted), which we call the *internal elastic lamina.* The centre layer, the thick *tunica media,* is mainly composed of smooth muscle fibres (purple) interspersed with a few connective collagen fibres (blue). The external *tunica adventitia* (blue) consists mainly of connective tissue.

Figure 14.9 The three layers of a blood vessel wall shown in cross-section. The innermost layer is elastic thanks to a fine elastic sheet (dark purple, convoluted). Collagen strengthens the centre layer.

Fig. 14.10 shows the stress–strain curve for the tissue material of large arteries. Blood vessel tissue contains two main structural materials, elastin and collagen. Their elastic properties combine, leading to a continuously curved dependence of the stress on the strain. Contractile smooth muscle cells are a third component of blood vessel walls, but do not contribute to their mechanical properties.

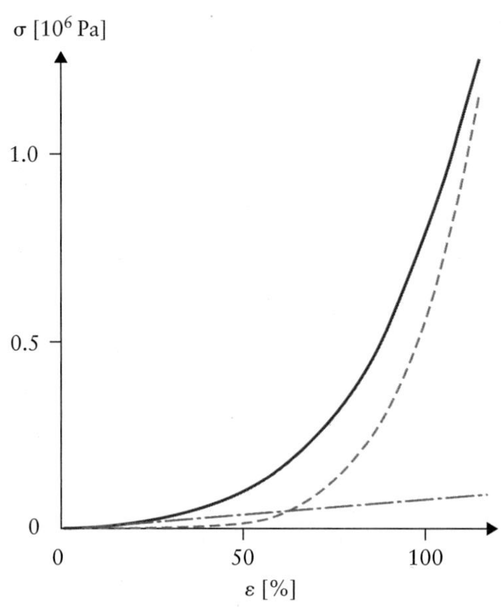

Figure 14.10 Stress–strain relation for blood vessel tissue. Note the absence of a linear regime. Blood vessels contain two components that play a role in its deformation properties: elastin, which has a small bulk modulus (dash-dotted curve), and collagen, with a bulk modulus that continuously increases with strain (dashed curve). The two contributions combine to the red curve, which then represents the stress response of the actual blood vessel.

Elastin is a protein that determines the deformation of an artery at small stress values, but plays a negligible role at large stress values due to its small bulk modulus. Collagen fibrils dominate when the strain increases beyond 25%. The collagen is slack while the artery is narrow, but stiffens notably when the artery widens. This effect can be visualized with a dish of spaghetti. If you use two forks to pull spaghetti apart, you need little force until they are untangled and straightened. To pull beyond the original length of the spaghetti strands, a larger force is needed.

We consider two physiological applications of this type of response to deforming forces in arteries: the Windkessel effect of the aorta and the development of aneurysms.

14.2.2: Windkessel Effect of the Aorta

The ability of a blood vessel to deform under stress is used in the aorta to buffer blood flow variations that result from the pulsatile operation of the heart. The Windkessel effect is illustrated in Fig. 14.11. The aorta exits the heart upwards, followed by a bend of almost 180°. This bend must accommodate the largest fraction of the rush of blood ejected by the heart during each heart beat. When blood rushes out of the heart, the aortic blood pressure increases, as shown at the centre of Fig. 14.11. The pressure increases at stages 1 and 3 represent an increased stress on the aorta wall. The aorta wall responds, as predicted by Fig. 14.10, with a widening. An enlarged aorta lumen accommodates the blood volume increase while moderating the blood pressure increase.

As blood now starts to flow from the bend its pressure drops. The resulting decrease in strain in the aorta wall contributes a forward push to the blood to overcome its flow resistance, as illustrated in stages 2 and 4 in Fig. 14.11. Note that blood cannot flow back to the heart at this point because the heart valve has closed.

The elasticity of blood vessel tissue reduces with age, causing the aorta to stiffen. This process is called **arteriosclerosis**. The age-related change in the elasticity leads also to changes of the average blood pressure as indicated in Table 12.1.

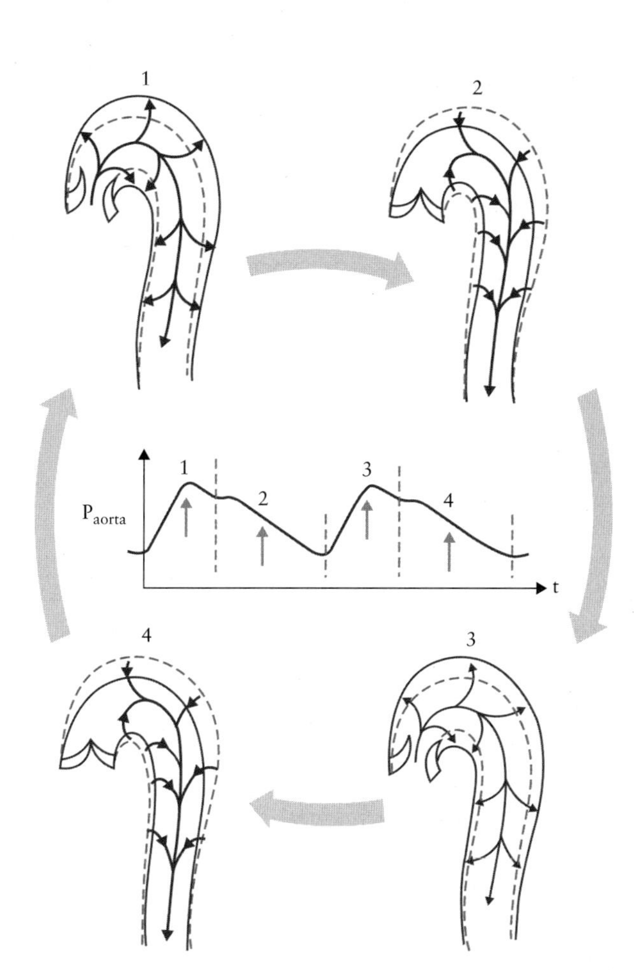

Figure 14.11 The Windkessel effect of the aorta. Shown at the centre is a plot of the blood pressure in the aorta, p_{aorta}, as a function of time during two heart beats. When blood is ejected from the heart (frames 1 and 3) the blood vessel tissue responds to the pressure increase (stress) by widening (strain). As the heart valve closes (at the left end of the aortic bend), we observe a lowered pressure (frames 2 and 4). This reduction of the stress leads to a contraction of the vessel (strain reduction), which pushes the blood downstream.

14.2.3: Aneurysms

The stress–strain curve shown in Fig. 14.10 applies to healthy blood vessel tissue. If a blood vessel is weakened, the tissue responds to the same blood pressure with a larger strain, as shown by the blue curve in Fig. 14.12.

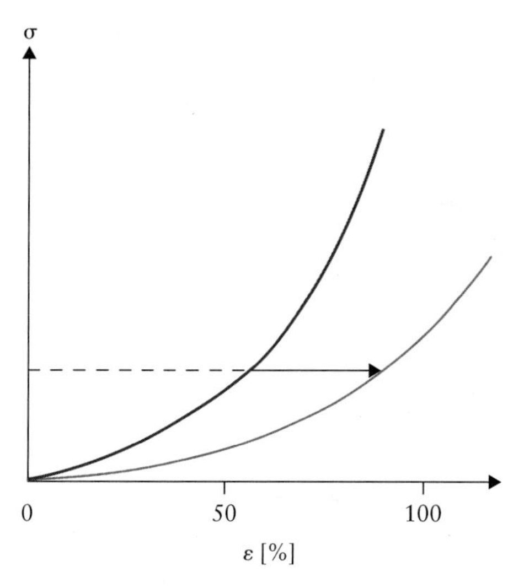

Figure 14.12 Stress–strain curves for healthy (red curve) and weakened blood vessel tissue (blue curve). The healthy tissue curve corresponds to the red curve shown in Fig. 14.10. The horizontal arrow indicates the excess strain for the weakened tissue in comparison to adjacent healthy tissue at a particular stress.

We can predict the development of an aneurysm—i.e., the ballooning of the blood vessel at such a weakened spot in the cardiovascular system—by combining Fig. 14.12 with the fluid flow concepts of Chapter 13.

The horizontal line in Fig. 14.12 indicates a given stress as exerted by a typical blood pressure on the blood vessel tissue. A healthy blood vessel opens as a result of this stress to its typical lumen value (red curve in Fig. 14.12). At a weakened spot, a larger strain means that the blood vessel widens locally further (blue curve). The equation of continuity then causes the blood flow to slow down in the weakened section due to its larger cross-section. This slowing causes a local increase in the blood pressure as described by Bernoulli's law. With the blood pressure rising locally, an increasing stress on the blood vessel occurs. We see from Fig. 14.12 that this leads yet again to a larger strain, i.e., a further widening of the vessel. This is called a positive feedback loop, which further enhances the size of the blood vessel and therefore has a negative effect on the affected patient.

Blood vessels can widen significantly due to this process. The widened section of the blood vessel is then called an **aneurysm**. Such aneurysms may eventually rupture, releasing blood into the adjacent tissue. Depending on where this happens the result may be fatal, e.g., when occurring in the brain.

14.3: Hooke's Law

The stress–strain curves of most materials are similar to the curves shown in Figs. 14.7 and 14.8 in that there is a linear regime for small strain values where the stress σ and the strain ε are related in the form $\sigma \propto \varepsilon$. The linear relation between stress and strain in this regime is called **Hooke's law**. It is named after Robert Hooke, who first wrote this relation in the late 17th century.

KEY POINT

Hooke's law applies in the elastic regime of a material and states that the stress and the strain of the material are linearly proportional.

Confining our discussion in this chapter to systems in the regime where Hooke's law is applicable, we study vibrations of the system as a new type of motion resulting from this law. All three types of deformation introduced at the beginning of the chapter lead to important vibrations:

- *The vibration of an ideal gas confined by a piston:* The vibration is the result of compressions and expansions under hydraulic stress.
- *The vibration of an object attached to a spring:* The vibration is the result of a stretching or compression of a spring under uniaxial (tensile or compressive) stress.
- *Pendulum motion of an object attached to a massless string:* This case can be discussed in analogy to the motion of an object attached to a spring.

In this chapter we establish the specific form of Hooke's law specifically for an object attached to a spring. From Newton's laws we know that an isolated object moves along a straight line when no forces act on it. We now attach such an object to a spring, as illustrated in Fig. 14.13. This establishes a continuous interaction with the object, which we want to introduce quantitatively as a new force.

A spring is a device that causes an object to move toward its equilibrium position in response to a displacement. Before quantifying the spring force further, we note that a horizontal arrangement is chosen in Fig. 14.13. An object could also be attached to a spring in a vertical arrangement. In that case, as seen with the problem sets with this chapter, the gravitational force needs to

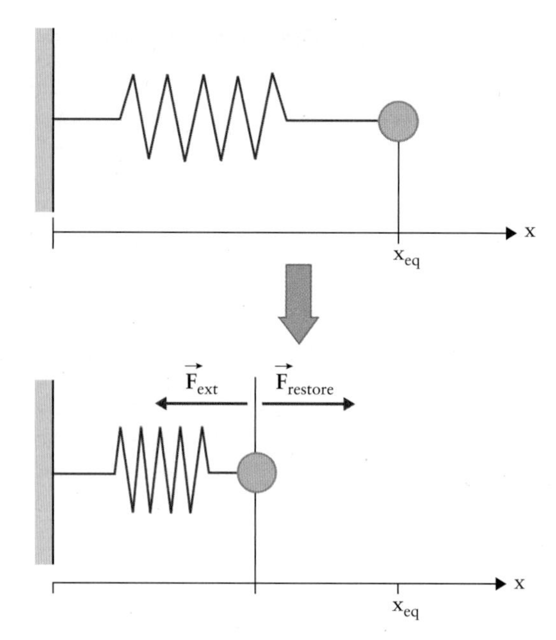

Figure 14.13 An object is attached to a horizontal spring, which in turn is attached to a rigid wall at the left. The equilibrium position of the object is at x_{eq}; at this position no external forces act on it. With an external force applied toward the left, the object is moved to a new equilibrium position at which the external force and the restoring force, exerted by the spring, are balanced.

be included in the discussion as an additional force. We avoid this in Fig. 14.13, as the gravitational force plays a negligible role at the microscopic-length scales to which we ultimately want to apply the spring force concept.

We emphasize the choice of a horizontal arrangement, and the choice of the horizontal x-axis, before proceeding with the discussion. In the discussion of forces that act on a system attached to a spring, the direction of these forces are as important as the magnitude of the forces to predict the outcome of an experiment. In particular, Newton's laws cannot be applied when we only know the magnitude of a force, we need the complete information about the force vector; i.e., we need to include the directional information. By choosing a horizontal x-axis and allowing only for horizontal forces, the formalism is simplified as we can represent each force by its magnitude and a sign that notes the direction in which the force acts relative to the positive x-axis. Thus, a single equation will suffice to apply Newton's laws.

Before the object is released, it is held in mechanical equilibrium by an external force at the position shown in the figure. For a sufficiently small displacement $\Delta x = x - x_{eq}$, the external force is linearly proportional to the displacement:

$$F_{ext} = k(x - x_{eq}). \qquad [14.5]$$

This is Hooke's law for an object attached to a horizontal spring. The parameter k is called the **spring constant** and has unit N/m. It is derived from Eq. [14.1] by combining the cross-sectional area A, the length l, and Young's modulus in a single constant: $k = A\,Y/l$. A large spring constant means therefore a stiffer spring.

The mechanical equilibrium in Fig. 14.13 is due to a force exerted on the object by the spring. We call this force the **elastic spring force**:

$$F_{\text{elast}} = -k(x - x_{\text{eq}}). \qquad [14.6]$$

$\vec{F}_{\text{elast}}$ is a **restoring force** as it causes the object to accelerate toward the equilibrium position once the external force is removed. The restoring character of the force is represented by the negative sign in Eq. [14.6]: for positions $x > x_{\text{eq}}$ the force $\vec{F}_{\text{elast}}$ is negative; i.e., the force pulls the object back to smaller values of x. For positions $x < x_{\text{eq}}$ the force $\vec{F}_{\text{elast}}$ is positive, pushing the object toward larger values of x.

The spring described by Eq. [14.6] is called an **ideal spring**, because real springs need not follow this law. In general, springs do not follow Hooke's law if the displacement is too large, i.e., too far from the equilibrium position. Most springs, however, are well described by Hooke's law near the equilibrium position, and this is the reason why applications of Eq. [14.6] are widespread—including the description of intramolecular forces as discussed in section 14.4.4.

KEY POINT

An object on an ideal spring that undergoes a small displacement from its equilibrium position is a valid case for Hooke's law.

14.4: Vibrations

In this textbook, we will discuss the resulting motion for two systems that follow Hooke's law; i.e., systems that have a restoring force that is linearly proportional to a displacement:

- We start the discussion with a single object attached to a horizontal ideal spring. This case is important as it can be applied to molecular vibrations.
- The case of a mobile piston sealing an ideal gas is pursued in the next chapter and leads to sound generation and detection.

14.4.1: Elastic Potential Energy

Newton's second law of mechanics applies to an object that is attached to a spring. We quantify this law for cases where Hooke's law applies, like in Fig. 14.13. Choosing the horizontal axis as the x-axis and with the equilibrium position written as x_{eq}, Newton's second law reads:

$$-k(x - x_{\text{eq}}) = ma. \qquad [14.7]$$

The right-hand side is the net force that acts on the object. This force is the restoring force as this is the only force that acts on the object in the horizontal direction. In Eq. [14.7], it is written as the magnitude combined with a minus sign to indicate the direction in which the force acts.

Deriving the formula for the position of the object as a function of time from Eq. [14.7] is mathematically more complicated than solving previously discussed cases of Newton's second law. The reason is that Eq. [14.7] contains both the position x and the acceleration a. The acceleration is the change of velocity with time, and in turn the velocity is the change of position with time. Thus, we cannot simply isolate the position in Eq. [14.7] as the independent variable.

The energy concept allows for an alternative way to approach Eq. [14.7]. In a first step toward solving Eq. [14.7], we determine the potential energy of the object on the spring. We call this energy the **elastic potential energy**, or just the **elastic energy**, to distinguish it from the gravitational potential energy we introduced in Chapter 7. Both are potential energies because they are energy forms dependent on the relative position of objects —here, the object relative to its equilibrium position on the spring.

Following the same reasoning as in Chapter 7, the potential energy is derived from the work required to displace the object. Let's assume that we move the object from its equilibrium position, for which we choose $x_{\text{eq}} = 0$, to a final position x_{final}, holding it at that point in a mechanical equilibrium with an external force $\vec{F}_{\text{ext}}$. From the discussion of Hooke's law for an object attached to a spring, we know that $\vec{F}_{\text{ext}} = -\vec{F}_{\text{elast}}$. We recall that the work is the area under the curve of the force as a function of position, as illustrated in Fig. 14.14 for the object on a spring. With the force linearly proportional to the displacement in Hooke's law (Eq. [14.6]), the area under the curve between positions $x = 0$ and $x = x_{\text{final}}$ is a triangle with area:

$$\begin{aligned} W &= \frac{1}{2} F_{\text{ext}}\, x_{\text{final}} = \frac{1}{2}(k\, x_{\text{final}})x_{\text{final}} \\ &= \frac{1}{2} k\, x_{\text{final}}^2. \end{aligned} \qquad [14.8]$$

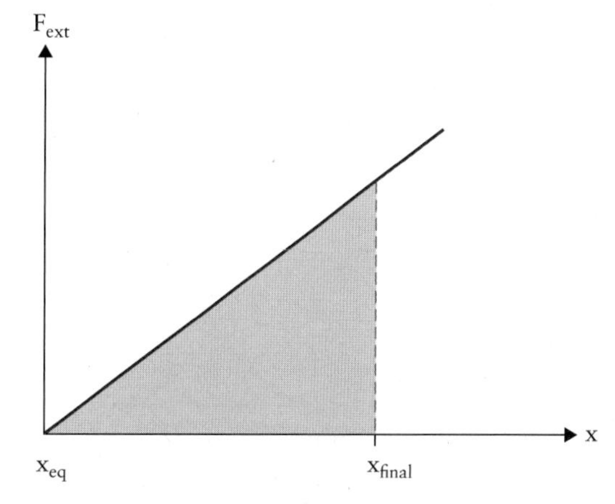

Figure 14.14 The external force as a function of displacement for an object attached to a spring. The work is given by the area under the curve. The elastic potential energy is related to this work because the external force is the force that establishes the mechanical equilibrium for the system.

Thus, we define the elastic potential energy of an object attached to a spring:

$$E_{elast} = \frac{1}{2} k x^2 \qquad [14.9]$$

for $x_{eq} = 0$, i.e., when the equilibrium position of the spring is chosen as the origin, or:

$$E_{elast} = \frac{1}{2} k (x - x_{eq})^2 \qquad [14.10]$$

for $x_{eq} \neq 0$—i.e., when the origin is chosen arbitrarily.

KEY POINT

A system with a linear restoring force (Hooke's law) has an elastic potential energy that is proportional to the square of the displacement from the equilibrium position of the system.

Fig. 14.15 compares the potential energies, shown as solid lines, (a) for gravity acting on an object and (b) for an object attached to a spring, i.e., an object on which an elastic force acts. The parabolic shape of the elastic potential energy in Fig. 14.15(b) results from Eq. [14.9] and defines a minimum in the elastic energy at the equilibrium position. No well-defined point of minimum potential energy exists in the case of gravity in Fig. 14.15(a) because the gravitational force is given as the weight $W = m\,g$ with the direction always pointing downward.

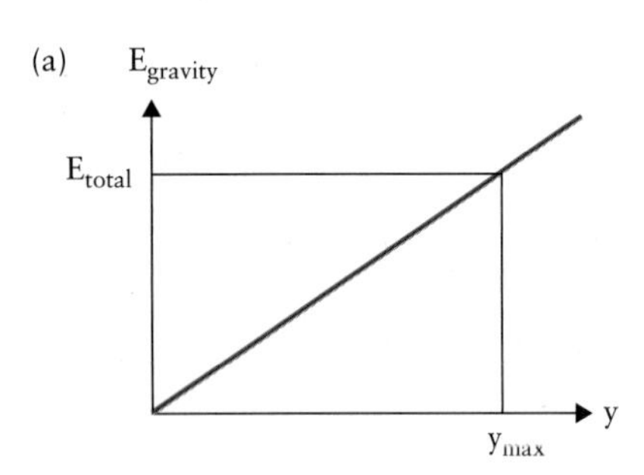

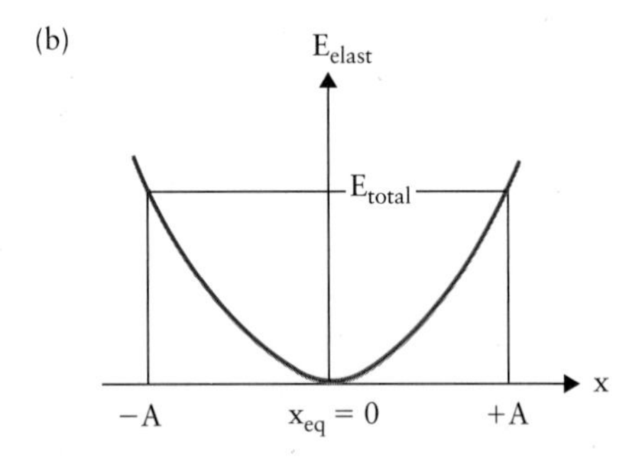

Figure 14.15 Comparison between the potential energy (a) due to gravity and (b) due to an elastic spring force. The potential energy is shown as the red curve. The horizontal lines indicate an arbitrarily chosen total energy for the system. The curve for the elastic spring force allows us to define the amplitude *A* for the motion of an object.

With the definition of the elastic potential energy in Eq. [14.9], the conservation of energy for an object on a horizontal spring is written as:

$$E_{total} = E_{kin} + E_{elast} = \frac{1}{2} m v^2 + \frac{1}{2} k x^2, \qquad [14.11]$$

where we have chosen the x-axis such that $x_{eq} = 0$.

The interpretation of Eq. [14.11] is best done using Fig. 14.15. The two horizontal lines in the figure represent a particular total energy for the system. The difference between the total energy (black line) and the potential energy (red line) represents the kinetic energy at each position along the x-axis. Obviously, positions for which the potential energy would exceed the total energy are not allowed. The gravitational system in Fig. 14.15(a) is not confined; it can move toward the left indefinitely. An object that travels in the graph toward the right (physically travelling upward) will turn around at the point where the kinetic energy becomes zero, but then never returns to this point. In contrast, the system in Fig. 14.15(b) is confined; i.e., an object moves back and forth between the two points at which its kinetic energy is zero.

CONCEPT QUESTION 14.3

Fig. 14.16 shows the elastic potential energy of an object attached to a horizontal spring. The object slides on a table without friction. At which position does it reach its greatest total energy?

(A) at the position labelled (i)

(B) at both positions (i) and (iv)

(C) at the position labelled (iii)

(D) at the position labelled (ii)

(E) None of the four answers (A) to (D) is correct.

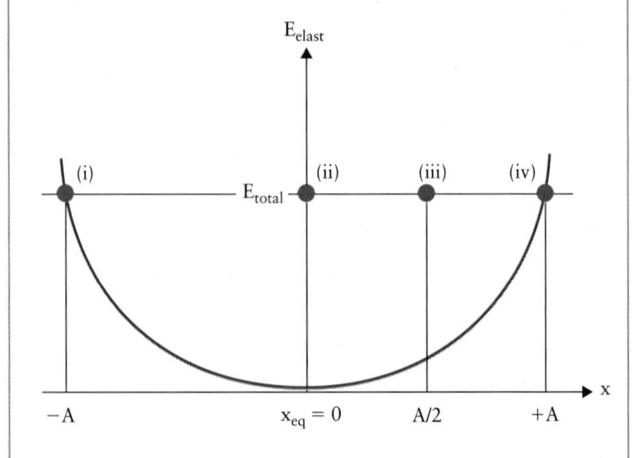

Figure 14.16 The elastic potential energy (red curve) of an object attached to a horizontal spring. The object slides on a table without friction.

CONCEPT QUESTION 14.4

At the position labelled (iii) in Fig. 14.16, the object is halfway between its greatest displacement and its equilibrium position. Why does it not store 50% of its total energy in each of the kinetic and elastic energies at this point?

EXAMPLE 14.2

We consider the spring in Fig. 14.17. The top panel shows the case when the spring is relaxed and the bottom panel when the spring is stretched by a distance $d = 0.15$ m. To pull the spring this distance, a total work of 20 J is needed (typical workout equipment). What force do you need to hold the stretched spring at that point?

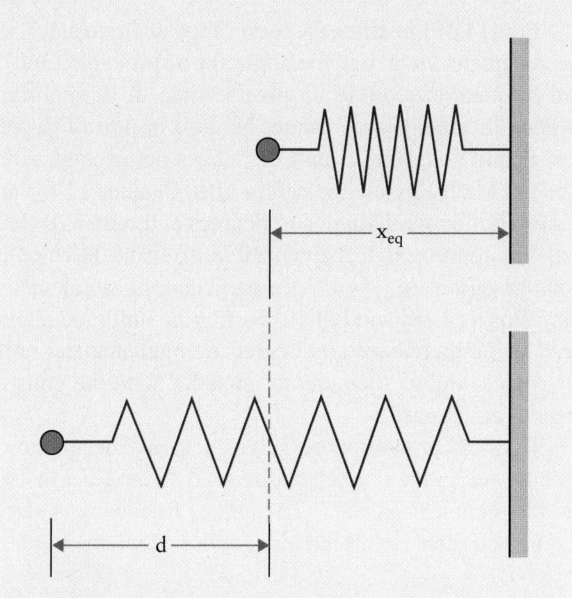

Figure 14.17 Sketches of a relaxed and a stretched horizontal spring. The spring in the bottom panel is elongated by distance *d*.

Solution

We use $x_i = x_{eq} = 0$ and $x_f = d$. The work needed to pull the spring to the final position is given by

$$W = \frac{1}{2}k(x_f^2 - x_i^2) = \frac{k}{2}d^2.$$

In an intermediate step, this equation allows us to calculate the spring constant k:

$$k = \frac{2W}{d^2} = \frac{2 \bullet 20\ \text{J}}{(0.15\ \text{m})^2} = 1780\frac{\text{N}}{\text{m}}.$$

continued

The spring constant is then used in Hooke's law to determine the elastic force:

$$F_{elast} = |k(x - x_{eq})| = |k\,d| = \left(1780\frac{\text{N}}{\text{m}}\right)(0.15\ \text{m}) = 267\ \text{N}.$$

This force is directed to the right in the lower panel of Fig. 14.17. Thus, an external force $F_{ext} = F_{elast} = 267$ N toward the left in Fig. 14.17 is needed to hold the spring in its stretched position.

14.4.2: Parameters of Motion: Amplitude and Maximum Speed

The amplitude of the motion of an object attached to a spring is defined with Fig. 14.15.

KEY POINT

The amplitude A is the maximum displacement of a vibrating object.

When the object has moved to the amplitude its kinetic energy becomes zero, $E_{kin} = 0$, and $E_{total} = E_{elast}$:

$$E_{total} = \frac{1}{2}kA^2, \qquad [14.12]$$

which yields for the amplitude:

$$A = \sqrt{\frac{2E_{total}}{k}}. \qquad [14.13]$$

We can further determine the maximum speed of an object attached to a spring by studying the instant when $x = x_{eq} = 0$. At that point the elastic energy is zero, $E_{elast} = 0$, and the conservation of energy in Eq. [14.11] leads to:

$$E_{total} = E_{kin} = \frac{1}{2}m\,v_{max}^2, \qquad [14.14]$$

which yields for the maximum speed of the object:

$$v_{max} = \sqrt{\frac{2\,E_{total}}{m}}. \qquad [14.15]$$

A general relation between the amplitude and the maximum speed of an object attached to a spring is then derived from Eqs. [14.13] and [14.15]:

$$E_{total} = \frac{1}{2}k\,A^2 = \frac{1}{2}m\,v_{max}^2,$$

which yields:

$$v_{max} = \sqrt{\frac{k}{m}}A. \qquad [14.16]$$

With k and m given for a particular system, we see that the maximum speed depends linearly on the amplitude of the motion.

EXAMPLE 14.3

An object of mass 1.0 kg is attached to a spring with spring constant $k = 1000$ N/m. If the object is displaced by 30 cm from its equilibrium position and is released, with what speed will it pass through the equilibrium position?

Solution

Using Eq. [14.13], we first calculate the total energy of the object on the spring:

$$E_{\text{total}} = \frac{1}{2}kA^2$$
$$= \frac{1}{2}\left(1000\frac{\text{N}}{\text{m}}\right)(0.3\ \text{m})^2 = 45\ \text{J}.$$

Using Eq. [14.15], we find the maximum speed of the object:

$$v_{\max} = \sqrt{\frac{2E_{\text{total}}}{m}} = \sqrt{\frac{245\ \text{J}}{1.0\ \text{kg}}} = 9.5\frac{\text{m}}{\text{s}}.$$

You find the same result in a single step by substituting the given values in Eq. [14.16].

This completes the discussion of the energy concept for the motion of an object attached to a spring. In the next section we continue to seek an equation that describes the position of the object during its motion as a function of time, i.e., a formula of the type $x = f(t)$.

14.4.3: Harmonic Motion

We start with Eq. [14.11], which contains the position and the velocity, with the velocity a change of the position with time. It is mathematically simpler than Eq. [14.7] since it does not contain the acceleration. However, Eq. [14.11] is still not simple to solve in the form $x = f(t)$ because it is a non-linear equation. Therefore, we don't discuss the required mathematical operations but just note the result. The equation for the position of an object on a spring as a function of time is:

$$x(t) = A\cos\left(\sqrt{\frac{k}{m}}t\right) = A\cos(\omega t). \qquad [14.17]$$

This formula describes a cyclic motion called **simple harmonic motion**. In simple harmonic motion, an object oscillates about a point of mechanical equilibrium. Eq. [14.17] is plotted in Fig. 14.18. In the last formula of Eq. [14.17] a new parameter is introduced, ω (lower-case Greek omega), which is called the **angular frequency** and has unit 1/s:

$$\omega = \sqrt{\frac{k}{m}}. \qquad [14.18]$$

Eq. [14.18] indicates that the angular frequency contains the physical parameters of the system, i.e., its mass and the spring constant. ω also governs the motion of the object as a function of time since it carries an inverse time unit. We illustrate this by following an object through a full cycle of its motion, i.e., between two instants in Fig. 14.18 where the argument of the cosine function has the same value—for example, cos(0) and cos(2 π). Any two such instants differ by $\omega t = 2\pi$, which we define in Fig. 14.18 as the **period** T. Thus, $2\pi = \omega T$, or:

$$\omega = \frac{2\pi}{T}. \qquad [14.19]$$

Eq. [14.19] justifies the term "angular frequency" for the parameter ω: 2 π is the angle for a full cycle (360°), and frequency refers to an inverse time. It is important to keep in mind that ω cannot be used in unit of degree per second (°/s), but in unit of radians per second, rad/s (see the Math Review located in after Chapter 27 for the conversion between angles in degrees and radians). Note that this represents a major source of errors in calculations based on Eq. [14.17]. In particular, ω as calculated from Eqs. [14.18] and [14.19] carries the unit rad/s; however, the units radians and degree are mathematical units for angles, and as such are not included with the units in physics equations.

Further, ω cannot be referred to as "frequency," because we define frequency differently from Eq. [14.19]: the **frequency** f is $f = 1/T$. The unit of frequency is hertz, Hz, named after Heinrich Hertz, with 1 Hz = 1 s^{-1}.

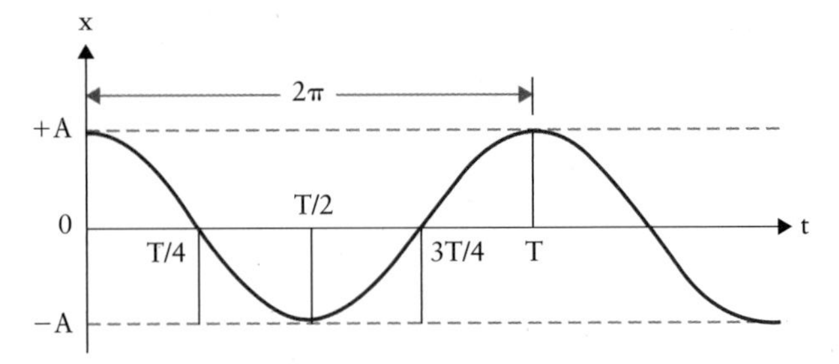

Figure 14.18 Position as a function of time, $x = f(t)$, for an object attached to a horizontal spring (harmonic oscillator).

EXAMPLE 14.4

We study a simplified model for the tympanic membrane in the human ear (eardrum): a circular membrane that can undergo a harmonic motion like a trampoline sheet.

continued

Assume a mass of 20 mg and a vibration frequency of 3.0 kHz. Calculate (a) the spring constant of the eardrum and (b) the angular frequency and the period of the vibration.

Solution

Solution to part (a): We note that the frequency, not the angular frequency, is given in the example text. We combine Eqs. [14.18] and [14.19] to relate the frequency to the spring constant:

$$f = \frac{\omega}{2\pi} = \frac{1}{2\pi}\sqrt{\frac{k}{m}},$$

which yields:

$$k = 4\pi^2 f^2 m.$$

Substituting the given values, we find:

$$k = 4\pi^2 (3000\ \text{Hz})^2 (20 \times 10^{-5}\ \text{kg}) = 71\,000 \frac{\text{N}}{\text{m}}.$$

Comparing this result with other spring constants we calculated earlier, we note that this value represents a very stiff membrane.

Solution to part (b): The angular frequency ω is obtained from Eq. [14.19]:

$$\omega = 2\pi f = 2\pi(3000\ \text{Hz}) = 19\,000 \frac{\text{rad}}{\text{s}}.$$

Eq. [14.19] also relates the period and the frequency:

$$T = \frac{1}{f} = \frac{1}{3000\ \text{Hz}} = 3.3 \times 10^{-4}\,\text{s} = 0.33\ \text{ms}.$$

This is a very rapid vibration. The large value is necessary to allow acoustic coupling to the sound frequencies we want to hear.

CONCEPT QUESTION 14.5

The amplitude of a system moving in simple harmonic motion is increased by a factor of 3. Determine the change in (a) the total energy, (b) the maximum speed, (c) the period, and (d) the angular frequency.

14.4.4: Application in Chemistry: Chemical Bonds in Molecules

The relative motion of atoms in a molecule is an important application of simple harmonic motion. This is illustrated for the HCl molecule and a NaCl bond in a rock salt crystal. HCl is a binary molecule in which the chlorine atom is about 35 times heavier than the hydrogen atom. This allows us to simplify the model for the HCl molecule as shown in Fig. 14.19, where the chlorine atom is considered to be immobile—like the wall to which the spring is attached on the right side in Fig. 14.17. The hydrogen atom is modelled as an object attached to the Cl atom by an ideal spring. In comparison, the description of a chemical bond with atoms of similar masses, e.g., NaCl, requires a formalism allowing both atoms to vibrate simultaneously relative to their common centre of mass.

Figure 14.19 Simplified model of the HCl molecule. The chlorine atom is considered immobile. The spring allows the hydrogen atom to undergo simple harmonic oscillation.

We test whether the model proposed in Fig. 14.19 is adequate to describe real chemical bonds. For this we analyze the interaction between two neighbouring atoms or ions. Their interaction has two contributions, as illustrated in Fig. 14.20 for NaCl:

- an attractive component shown at negative energies. In the case of an electrostatic interaction, the attractive component is the electrostatic potential energy:

$$E_{\text{attract}} = -\frac{1}{4\pi\varepsilon_0}\frac{e^2}{r},$$

which we derive as Coulomb's law in Chapter 17. ε_0 is the permittivity of vacuum and e is the elementary charge. If this attraction were the only interaction in the Na–Cl system, the sodium atom would crash into the chlorine atom as if it were swallowed up by a black hole.

- This doesn't happen, because the attraction is shielded when the two ions penetrate each other and negative electrons and positive nuclei come respectively close to each other. This interaction leads to a repulsive term, shown in Fig. 14.19 as a red curve at positive energy values. Max Born determined a semi-empirical formula for the repulsive contribution:

$$E_{\text{repulsive}} = b e^{-ar},$$

in which a and b are constants that have to be determined experimentally for each chemical bond. The exponential function in Born's formula is steeper than the $1/r$ dependence for the electrostatic attraction. Therefore, the repulsive term dominates at shorter distances between the two ions and the attractive term dominates at longer distances.

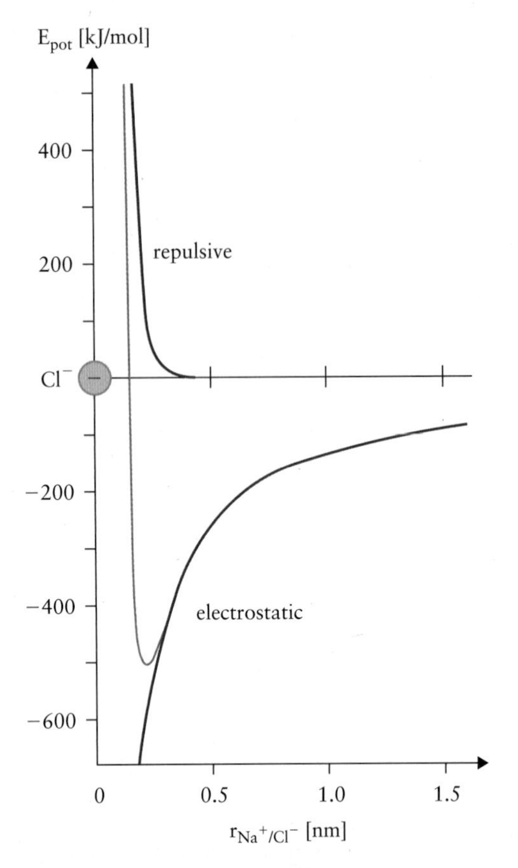

Figure 14.20 Potential energy as a function of distance between Na^+ and Cl^- ions in rock salt (NaCl). The electrostatic attraction between the ions (red curve at negative energies) is overcompensated by a repulsive term (red curve at positive energies) at closer proximity. The combined potential energy curve for the Na–Cl bond is shown as an asymmetric blue curve.

Both contributions were combined by Heitler and London in 1927, describing the complete potential function as shown as the blue line in Fig. 14.20 for an NaCl bond, and as shown as the red line in Fig. 14.21 for an HCl molecule. In the case of HCl, a minimum energy is reached when the H atom is separated by a distance r_0 of about 0.13 nm from the Cl atom.

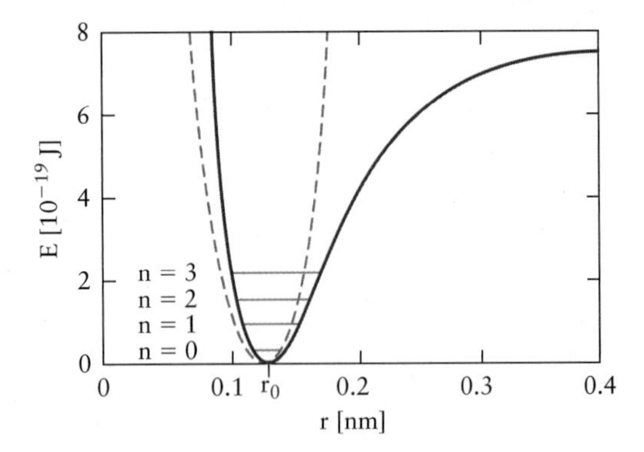

Figure 14.21 Intramolecular potential energy curve (red line) and harmonic oscillator model (dashed blue curve) as a function of the separation distance between the atoms in an HCl molecule. Horizontal lines represent the allowed total energy levels of the molecule.

The combination of the potential energies for the electrostatic interaction and Born's repulsion doesn't result in the same potential energy curve as for an object on an ideal spring. This is highlighted in Fig. 14.21, where the blue line is the curve for an object attached to an ideal spring while the actual potential energy curve for the HCl molecule is shown as the red curve. Recall that the spring model in Fig. 14.19 predicts a potential energy of the form $E_{elast} = (1/2)k(x - x_{eq})^2$. The disagreement between the two potential energy curves is particularly large at high total energies.

Do we therefore dismiss the harmonic oscillator as a model for a molecular bond? Not necessarily: at typical temperatures, such as room temperature or temperatures that molecules reach during most chemical reactions in solutions, almost all of the HCl molecules have a total energy corresponding to the ground state, which is indicated in Fig. 14.21 by the horizontal line labelled $n = 0$. Note that the energy of this ground state is higher than the minimum of the potential energy curve for quantum-mechanical reasons (Heisenberg's uncertainty relation). Only a negligible fraction of the HCl molecules have an energy corresponding to one of the excited states of the molecule (labelled $n = 1, 2,\ldots$). The reason for this small fraction of molecules being at an energy other than the ground state is due to a peculiar result of quantum-mechanics: the molecules cannot have any other total energy than the ones indicated by the horizontal lines in Fig. 14.21. Thus, for a molecule to reach an excited state from the ground state, a significantly higher amount of energy is needed than is available as a result of the ambient thermal energy, which is of the order of $k\,T \approx 4 \times 10^{-21}$ J at room temperature (k is the Boltzmann constant).

Remember that the hydrogen atom in Fig. 14.21 cannot travel beyond the two points at which its potential energy is equal to its total energy. The range accessible to the atom coincides with the range in which the potential energy of the harmonic oscillator (dashed blue line) and the actual potential energy curve (red line) match quite well.

KEY POINT

The simple harmonic oscillator is a good model for chemical molecules in their ground state.

EXAMPLE 14.5

We quantify key properties of HCl molecules using the simple harmonic oscillator model. The mass of a hydrogen atom is $m = 1.67 \times 10^{-27}$ kg and the spring constant of the molecule is $k = 484$ N/m, which is a value obtained from spectroscopic data listed in Table 14.2. Calculate (a) the angular frequency, (b) the frequency, and (c) the period for the vibration of the hydrogen atom in the HCl molecule.

continued

TABLE 14.2

Spring constants for various chemical bonds

Bond	Molecule	k (N/m)
H–Cl	HCl	484
H–O	H_2O	780
H–C	CH_3R	470–500
C–C		450–560
C≡C		950–990
C=C		1560–1700
N–N		350–550
C–O		500–580
C=O		1180–1340

R stands for *rest*, i.e., an organic extension of the functional group.

Solution

Solution to part (a): We use Eq. [14.18] to determine the angular frequency:

$$\omega = \sqrt{\frac{k}{m}} = 5.38 \times 10^{14} \frac{\text{rad}}{\text{s}}.$$

Solution to part (b): Substituting the value for the angular frequency from part (a) in Eq. [14.19], we find for the frequency:

$$f = \frac{\omega}{2\pi} = 8.6 \times 10^{13} \text{ Hz}.$$

Solution to part (c): Frequency and period are inversely related:

$$T = \frac{1}{f} = 1.2 \times 10^{-14} \text{s}.$$

The hydrogen atom in the HCl molecule vibrates with an extremely short period.

The short period of vibration of molecules is very important for their chemical properties. The hydrogen atom moves away from the chlorine atom once during each vibration cycle. If another molecule is close to the HCl molecule, this represents the opportunity for the hydrogen atom to engage in a chemical reaction with a neighbouring molecule. Therefore, f is the frequency with which the hydrogen atom tries to escape from the molecule by moving toward the outer limit of the potential energy barrier in Fig. 14.21. While no neighbouring molecule is present, a dissociation of the HCl molecule does not take place (which would require the hydrogen atom to move all the way beyond the right end of Fig. 14.21). However, when a neighbouring molecule is close enough for a strong interaction, an escape attempt may result in a regrouping of atoms, i.e., in the successful completion of a chemical reaction. Since the period of closest proximity between molecules in a reaction volume is short, a high frequency of attempts is vital to obtain an appreciable rate at which chemical reactions take place. Due to the relevance of the vibrational frequency for chemical kinetics, some f values are summarized in Table 14.3.

TABLE 14.3

Vibration frequencies for various chemical bonds in organic molecules

Bond	f (Hz)
H–O	1.05×10^{14}–1.11×10^{14}
H–N	9.9×10^{13}–1.05×10^{14}
H–C	8.64×10^{13}–9.09×10^{13}
C=C	4.8×10^{13}–5.04×10^{13}
C=C	6.6×10^{13}–6.78×10^{13}
C=O	4.98×10^{13}–5.61×10^{13}

EXAMPLE 14.6

Read the total energy of the HCl molecule in the ground state from Fig. 14.21 and calculate the amplitude and the maximum speed of the hydrogen atom in the molecule.

Solution

The total energy of the molecule in the ground state is $E_{total} = 2.87 \times 10^{-20}$ J. With this value we use Eq. [14.13] to obtain the amplitude of the vibrating hydrogen atom:

$$A = \sqrt{\frac{2E_{total}}{k}} = 1.09 \times 10^{-11} \text{m}$$

The amplitude is 0.011 nm, or about 10% of the distance to the chlorine atom. From Eq. [14.19] we obtain the maximum speed of the vibrating hydrogen atom (use data from Example 14.6):

$$v_{max} = \sqrt{\frac{k}{m}} A = 5860 \frac{\text{m}}{\text{s}}.$$

SUMMARY

DEFINITIONS

- Stress: $\sigma = F/A$
- Strain: $\varepsilon = \Delta l/l$
- Amplitude A is the maximum displacement during a vibration.
- Period T is the time to complete a full cycle during a vibration.

- Frequency: $f = 1/T$
- Angular frequency: $\omega = 2 \cdot \pi \cdot f = 2 \cdot \pi/T$

UNITS

- Stress σ: N/m^2 = Pa
- Strain ε: no units
- Spring constant k: N/m
- Amplitude A: m
- Period T: s
- Frequency f: 1/s = Hz
- Angular frequency ω: rad/s

LAWS

- Elastic deformation (Hooke's law)
- for tensile stress (stretching):

$$\sigma = Y\varepsilon,$$

where Y is Young's modulus.

- for hydraulic stress (for volume compression):

$$p = -B\frac{\Delta V}{V}$$

where B is bulk modulus.

- Elastic force along the axis of an elastic spring (x-axis):

$$F_{\text{elast}} = -k(x - x_{\text{eq}}),$$

where k is the spring constant, and x_{eq} is the equilibrium position of the spring.

- Elastic energy for an object attached to a spring
  - when we set $x_{\text{eq}} = 0$:

$$E_{\text{elast}} = \frac{1}{2}kx^2$$

  - when $x_{\text{eq}} \neq 0$ is chosen:

$$E_{\text{elast}} = \frac{1}{2}k(x - x_{\text{eq}})^2$$

- Simple harmonic oscillation:

$$x(t) = A\cos(\omega t),$$

where ω is angular frequency, which is given for an object attached to a spring as:

$$\omega = \sqrt{\frac{k}{m}}$$

MULTIPLE-CHOICE QUESTIONS

MC–14.1. The dash-dotted curve in Fig. 14.2(a) is the active stretching force for a human muscle. This curve shows the force (as a fraction of the maximum force in %) a muscle can exert at a given length. For example: the maximum force can be exerted when the muscle is at its resting length, but only about 50% of that force can be exerted when the muscle has contracted to 80% of its resting length. The dashed curve in Fig. 14.2(a) shows the passive stretching force for a human muscle. This curve indicates the external force required to stretch the muscle at the given length. For example, to stretch a muscle when it is about 5% longer than its resting length requires about double the force as stretching the same muscle just beyond its resting length. Which of the following statements is true?
(a) Hooke's law is suitable to describe the passive stretching of a muscle near its resting length.
(b) Hooke's law is suitable to describe the active muscle force for a given muscle near its resting length.
(c) Fig. 14.2(a) does not allow us to verify statements such as those made in (a) and (b).
(d) Both curves in Fig. 14.2(a) imply that muscle tissue has no elastic regime near the resting length.
(e) Fig. 14.2(a) implies that a muscle released after stretching should undergo simple harmonic motion.

MC–14.2. The solid curve in Fig. 14.10 shows the stress–strain relation for blood vessel tissue. Up to which stress value does the strain in the blood vessel respond linearly to the exerted stress? Choose the closest value.
(a) The stress–strain relation is not linear for any stress.
(b) $\varepsilon = 100\%$
(c) $\varepsilon = 50\%$
(d) $\sigma = 0.5 \times 10^6$ Pa
(e) $\sigma = 1.0 \times 10^6$ Pa

MC–14.3. The elastic behaviour of the blood vessel in Fig. 14.10 is the result of two contributing components in the tissue: elastin (dash-dotted curve) and collagen, with a bulk modulus increasing with strain (dashed curve). Based on Fig. 14.10, which of the following statements is wrong?
(a) The elastic properties of elastin can be described by Hooke's law.
(b) Collagen shows a non-linear stress–strain behaviour.
(c) Up to a strain of about 60%, elastin dominates the elastic response of the system.
(d) The actual blood vessel wall shows non-linear stress–strain behaviour due to its collagen component.
(e) All four statements are correct.

MC–14.4. Fig. 14.22 shows the stress–strain curve for compact bone from Fig. 14.8. It further includes four points (i) to (iv) along this curve. In which interval is the response of the compact bone not fully elastic?
(a) from the origin to point (iii)
(b) from the origin to point (iv)
(c) from the origin to point (i)
(d) from point (i) to point (iii)
(e) Its response is fully elastic in all four intervals.

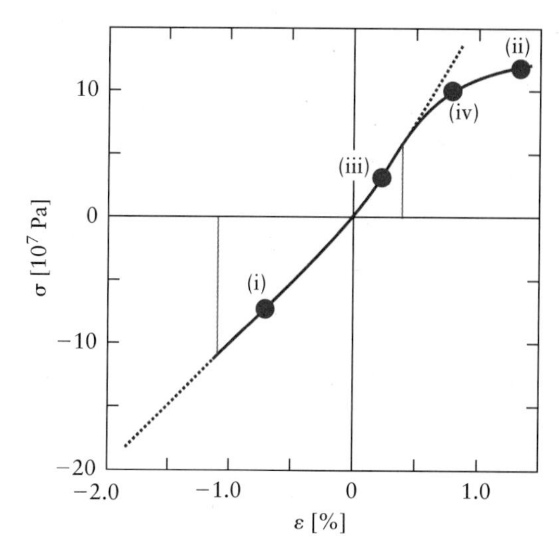

Figure 14.22

MC–14.5. What do we imply when we say that a force acting on an object is an elastic force?

(a) We imply that we hold the object in a mechanical non-equilibrium position.
(b) We imply that the object moves fast in the vicinity of the equilibrium position.
(c) We imply that the elastic energy exceeds 1.0 J.
(d) We imply that the total energy of the object is conserved.
(e) We imply that the strain is proportional to the stress for the system we study.

MC–14.6. We consider an object that is attached to a horizontal spring and moves without friction on a flat surface. The object is displaced by a given distance such that the spring is stretched relative to its equilibrium length. The external force is then removed (i.e., the object is released). Immediately after its release, which statement is *correct* regarding the object?

(a) The object is in mechanical equilibrium.
(b) The object moves with its maximum speed.
(c) The object moves away from the equilibrium position.
(d) The object remains at rest.
(e) The object decreases its elastic potential energy.

MC–14.7. We define the positive x-direction as the direction in which an object stretches a spring (k = 100 N/m) during a simple harmonic motion. At the instant the object moves through its equilibrium position after having been released at the amplitude point, $x = +A = +0.1$ m, what is the elastic force acting on the object?

(a) $F_{elast} = -100$ N
(b) $F_{elast} = -10$ N
(c) $F_{elast} = 0$ N
(d) $F_{elast} = +10$ N
(e) $F_{elast} = +100$ N

MC–14.8. An object is attached to a horizontal spring and moves along the x-axis. It is initially displaced to the positive amplitude point, $x = +A$. At that point its elastic potential energy is E_1. The object is then released. At what point along its path will the total energy of the system be smaller than E_1?

(a) at no point
(b) at all points other than the point $x = -A$
(c) at all points with $x < 0$
(d) at all points where the kinetic energy exceeds $(1/2)E_1$
(e) at all points where the object's speed is positive, $v > 0$

MC–14.9. An object is attached to a horizontal spring. It is initially displaced by a distance Δx from the equilibrium position and then released from rest. The object passes the equilibrium position with speed v. From this speed we can calculate the initial displacement Δx using which formula?:

(a) $\Delta x = \sqrt{k}\, v$
(b) $\Delta x = \sqrt{m}\, v$
(c) $\Delta x = \sqrt{\frac{m}{k}}\, v$
(d) $\Delta x = \sqrt{\frac{k}{m}}\, v$
(e) $\Delta x = \sqrt{mk}\, v$

MC–14.10. An object is attached to a horizontal spring that is oriented along the x-axis. The object is initially located at the equilibrium position of the spring and is at rest. We then hit the object such that it moves along the x-axis, stretching the spring. At what point of its motion along the x-axis does the object reach its highest kinetic energy?

(a) at the equilibrium position
(b) at the positive amplitude position
(c) midway between the amplitude and equilibrium positions
(d) at the negative amplitude position
(e) The kinetic energy is the same at all positions.

MC–14.11. An object is attached to a horizontal spring that is oriented along the x-axis. The object is initially located at the equilibrium position of the spring and is at rest. We then hit the object such that it moves along the x-axis, stretching the spring. At what point of its motion along the x-axis does the object reach its lowest kinetic energy?

(a) at the equilibrium position
(b) at the positive amplitude position
(c) midway between the amplitude and equilibrium positions
(d) at the initial position
(e) The kinetic energy is the same at all positions.

MC–14.12. An object attached to a spring has how many mechanical equilibrium positions?

(a) zero
(b) one
(c) two
(d) a finite number larger than two
(e) an infinite number

MC–14.13. We study an object that is attached to a spring and performs a harmonic oscillation on a frictionless horizontal surface. Which parameter set does *not* allow us to calculate the angular frequency of the motion?
(a) mass of object and spring constant
(b) frequency of the motion
(c) period of the motion
(d) mass of the object and amplitude of the motion

MC–14.14. Why is the force acting on an object attached to a spring called a restoring force?
(a) It restores the system to the state in which it was just prior to the release of the object.
(b) It acts in the direction of the equilibrium position.
(c) It causes a repetitive motion.
(d) The force always acts in the negative x-direction.
(e) The system restores the initial state of elastic potential energy (a special type of energy conservation).

MC–14.15. An object is attached to a horizontal spring and moves along the x-axis. It is initially displaced to the positive amplitude point, $x = +A$. At that point its elastic potential energy is E_1. Next the object is moved to the opposite amplitude point at $x = -A$. Now its elastic potential energy is E_2. The following relation holds for E_1 and E_2:
(a) $E_2 = E_1$
(b) $E_2 = -E_1$
(c) $E_2 = 2\,E_1$
(d) $E_2 = -2\,E_1$

MC–14.16. An object is attached to a horizontal spring. It is initially displaced by a distance Δx from the equilibrium position and then released from rest. When the object passes the equilibrium position, what will its speed be?:

(a) $v = \dfrac{\Delta x}{\sqrt{k}}$
(b) $v = \dfrac{\Delta x}{\sqrt{m}}$
(c) $v = \dfrac{\Delta x}{\sqrt{\dfrac{m}{k}}}$
(d) $v = \dfrac{\Delta x}{\sqrt{\dfrac{k}{m}}}$
(e) $v = \dfrac{\Delta x}{\sqrt{m\,k}}$

MC–14.17. We replace an object on a spring with one that has four times the mass. How does the frequency of the system change?
(a) by a factor of 0.25
(b) by a factor of 0.5
(c) It remains unchanged.
(d) by a factor of 2
(e) by a factor of 4

CONCEPTUAL QUESTIONS

Q–14.1. An object is attached to a horizontal spring and undergoes simple harmonic motion with an amplitude A. Does the total energy of the system change if the mass of the object is doubled but the amplitude remains unchanged? In particular, are the potential and/or kinetic energies at a given point along the object's motion affected?

Q–14.2. Does the acceleration of a simple harmonic oscillator remain constant during its motion? Is it ever zero?

Q–14.3. What is the total distance travelled by an object attached to a spring during a full period of its simple harmonic motion? Use A for its amplitude.

Q–14.4. Determine whether the following vectors can point in the same direction during a simple harmonic motion: (a) displacement and velocity, (b) velocity and acceleration, and (c) displacement and acceleration.

ANALYTICAL PROBLEMS

P–14.1. For the graph in Fig. 14.1, express the force (in % of the maximum force) as a mathematical function of the sarcomere length l (in μm) for the linear segments in the intervals (a) 2.2 μm $\le l \le$ 3.2 μm, (b) 2.0 μm $\le l \le$ 2.2 μm, (c) 1.4 μm $\le l \le$ 1.65 μm.

P–14.2. Assume a leg contains a 50-cm-long bone with an average cross-sectional area of 3 cm^2. By how much does the bone shorten when the entire body weight of the person (use 700 N) is supported by the leg? Use for Young's modulus of the bone $Y = 1.8 \times 10^{10}$ Pa.

P–14.3. Determine an upper limit of the maximum height of building construction on Earth. This limit is due to the maximum stress in the building material prior to rupture. For steel of density $\rho = 7.9$ g/cm^3 the maximum stress is $\sigma = 2.0 \times 10^8$ Pa. Note that the pressure in the steel at the ground level may not exceed the maximum stress.

P–14.4. If the ultimate strength for a particular steel is reached at 5.0×10^8 Pa, determine the minimum diameter of a wire made of this steel that can support a standard man (see Table 4.1).

P–14.5. The four tires of a car are inflated to a gauge pressure of 2.0×10^5 Pa. Each tire is in contact with the ground with an area $A = 240$ cm^2. Determine the weight of the car.

P–14.6. An object has a mass $m = 0.7$ kg. It is attached to a spring that has spring constant of $k = 80$ N/m. At time $t = 0$ the object is pulled to a distance of 10 cm from its equilibrium position (which you may choose conveniently at $x = 0$). The surface on which the object moves is frictionless. (a) What force does the spring exert on the object just before it is released? (b) What are the angular frequency, the frequency, and the period of the oscillation? (c) What is the amplitude of the oscillation? (d) What is the maximum speed of the object?

P–14.7. (a) What is the total energy of the system in P–14.6? (b) What is the elastic potential energy of this system when the object is halfway between the equilibrium position and its turning point?

P–14.8. An object undergoes a simple harmonic motion. During that motion the object needs 0.4 s to reach one point of zero velocity from the previous such point. If the distance between those points is 50 cm, calculate (a) the period, (b) the frequency, and (c) the amplitude of the motion.

P–14.9. An object has a mass of 250 g. It undergoes a simple harmonic motion. The amplitude of that motion is 10 cm and the period is 0.5 s. (a) What is the spring constant (assuming that the spring obeys Hooke's law)? (b) What is the maximum magnitude of the force that acts on the object?

P–14.10. An object is attached to an ideal spring. It undergoes a simple harmonic motion with a total energy of E = 1.0 J. The amplitude of the motion is 15.0 cm and the maximum speed of the object is 1.2 m/s. Find (a) the spring constant, (b) the mass of the object, and (c) the frequency of the oscillation.

P–14.11. An object of mass m = 1.0 kg starts from rest and slides a distance d = 10 cm down a frictionless inclined plane of angle θ = 50° with the horizontal, as shown in Fig. 14.23. It then attaches to the end of a relaxed spring that has a spring constant k = 100 N/m. By what length does the object compress the spring by the time it comes (momentarily) to rest?

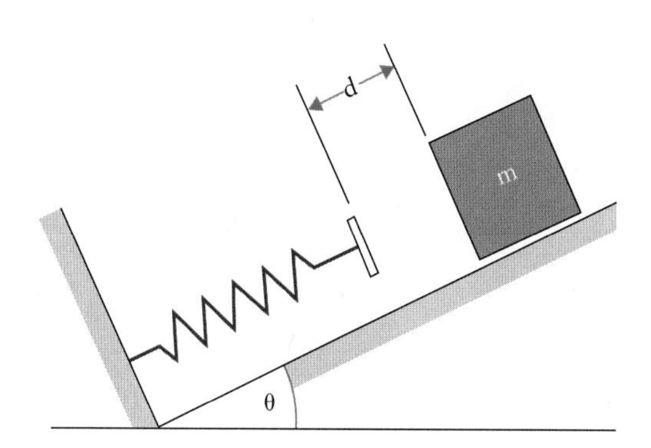

Figure 14.23

P–14.12. An object of mass m = 250 g is placed on a vertical spring with k = 5.0 kN/m. It is pushed down, compressing the spring by 10 cm. When the object is released, it leaves the spring and continues to travel upward. What maximum height above the release point does the object reach? Neglect air resistance.

P–14.13. In Fig. 14.23, the object of mass m = 2.0 kg is brought into contact with the spring. The object and spring are at rest with the spring compressed by 10 cm. If θ = 30°, what is the spring constant?

P–14.14. In Fig. 14.24, a spring with k = 140 N/m rests vertically on the bottom of a beaker filled with water. A block of wood of mass m = 500 g and density ρ = 0.65 g/cm^3 is attached to the spring. The system reaches a new static equilibrium, shown in the figure at right, after the spring is elongated by a length ΔL. Calculate ΔL. (This problem requires concepts from Chapter 12.)

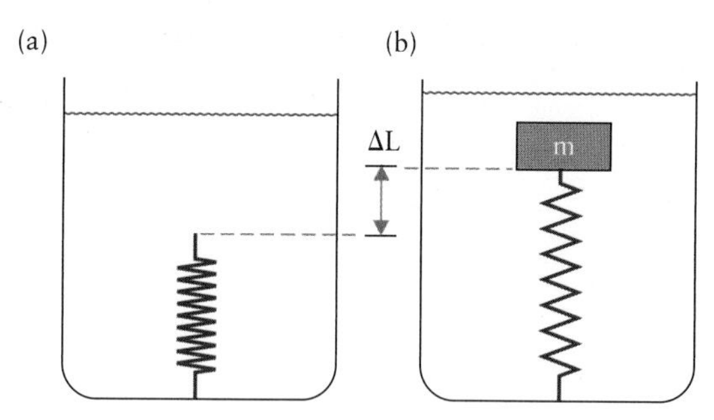

Figure 14.24

P–14.15. An object of mass m = 0.3 kg is attached to a horizontal spring. Its position varies with time as x = (0.25 m)cos(0.4 · π · τ). Find (a) the amplitude of its motion, (b) the spring constant, (c) the position at t = 0.3 s, and (d) the speed of the object at t = 0.3 s.

P–14.16. A spring with k = 30 N/m is stretched by a distance of 20 cm from its equilibrium position. How much work must be done to stretch it an additional 10 cm?

P–14.17. Fig. 14.25(a) shows a fly and Fig. 14.25(b) shows a simplified model for the insect's moving wings during flight. The wing is pivoted about the outer chitin capsule (on red arrowhead). The end of the wing lies l_1 = 0.5 mm inside the insect's body and moves up and down by 0.3 mm. We use an effective spring constant of k = 0.74 N/m for the elastic tissue in the insect's body surrounding the end of the wing, and an effective mass of 0.3 mg for the wing. The wing motion is described as a vibration of the end of the wing attached to a spring (elastic tissue). (a) With what frequency do the wings of the insect flap during flight? (b) What is the maximum speed of the inner end of the wing? (c) What is the maximum speed of the outer tip of the wing if the wing is treated as a rigid object? Use l_2 = 1.4 cm.

P–14.18. The vibration frequencies of atoms in solids at room temperature are about 10^{13} Hz, similar to the values shown in Table 14.3. Using a simplified model for a solid in which the atoms are connected by ideal springs, we want to study how a single atom in a piece of copper vibrates with this frequency relative to surrounding atoms that are at rest. (a) Calculate the (effective) spring constant, using M_{Cu} = 63.55 g/mol. (b) What is the ratio of the (effective) spring constant of a gold atom in a piece of gold and the copper atom in part (a)?

P–14.19. A pogo stick [see Fig. 14.26(a)] is a toy that stores energy in a spring with typical spring constant k = 25 000 N/m. Fig. 14.26(b) shows a child at three different instants when playing with a pogo stick. At position d_1 = −10 cm [panel (i)], the spring compression is at a maximum and the child is momentarily at rest. At position d = 0 [panel (ii)], the spring is relaxed and the child is moving upward. At position d_2 [panel (iii)], the child reaches the highest point of the jump and is momentarily at rest. Assume that the combined mass of the child and the pogo stick

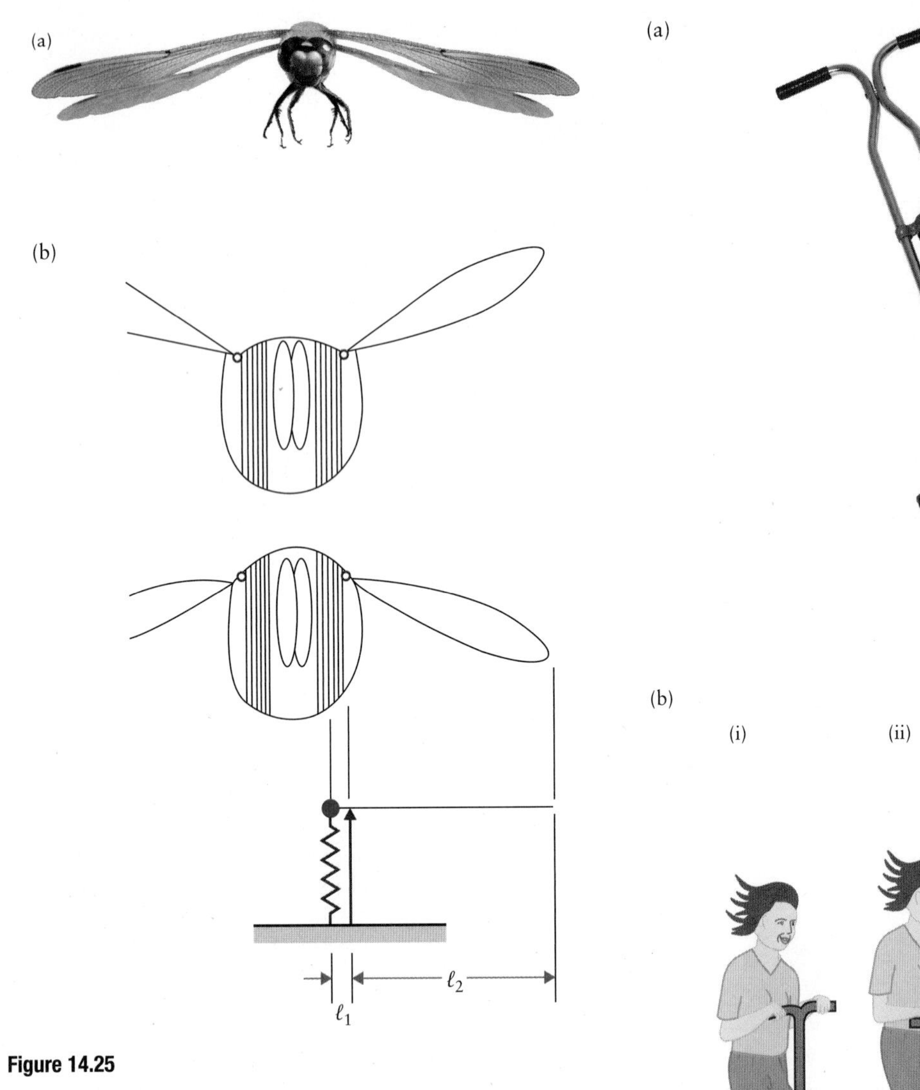

Figure 14.25

is 25 kg. (a) Calculate the total energy of the system if we choose the gravitational potential energy to be zero at $d = 0$. (b) Calculate d_2. (c) Calculate the speed of the child at $d = 0$. (d) Calculate the acceleration of the child at d_1. (e) *For those interested:* Should any of these results be a matter of concern to the parents?

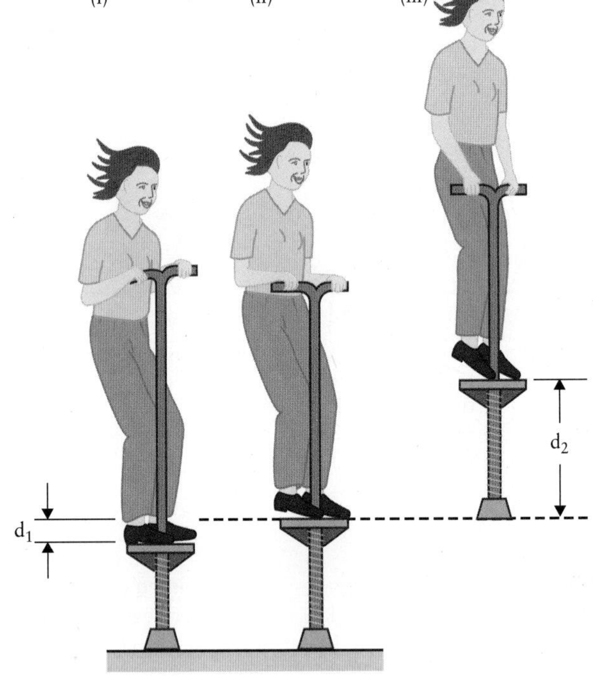

Figure 14.26

ANSWERS TO CONCEPT QUESTIONS

Concept Question 14.1(a): The force shown in Fig. 14.1 is only position independent when the sarcomere length lies in the interval 2.0 μm to 2.2 μm.

Concept Question 14.1(b): The forces shown in Fig. 14.2 are position dependent throughout the accessible range. Note that Fig. 14.1 is a simplification of the dash-dotted curve in Fig. 14.2(a). Thus, the answer to part (a) is correct only because of our choosing to simplify in that figure. In practice, forces in muscles cannot be treated as position independent, not even close to their resting length.

Concept Question 14.2: (A). In a linear relation the slope can be positive or negative.

Concept Question 14.3: (E). Note that "none of the above" would not be correct either, because the total energy of the system—i.e., the sum of its elastic and kinetic energy—is conserved. Thus, it has the same total energy value at all four positions identified in Fig. 14.17.

Concept Question 14.4: The elastic energy in Eq. [14.9] is not linear in the displacement of the object because $E \propto x^2$. The red curve in Fig. 14.16 therefore cannot be a straight line as in the gravitational case in Fig. 14.14(a). When the displacement from the equilibrium position doubles, the elastic energy increases by a factor of 4.

Concept Question 14.5(a): The total energy increases by a factor $3^2 = 9$ due to Eq. [14.12].

Concept Question 14.5(b): The maximum speed is linear with the amplitude due to Eq. [14.14]. Thus, it increases by a factor of 3.

Concept Question 14.5(c) and (d): The angular frequency depends only on the system parameters mass and spring constant in Eq. [14.18]; thus it does not change. The period and the frequency can be calculated from the angular frequency without using the amplitude; therefore these two parameters are also unaffected.

CHAPTER 15

Sound I

A mechanical deformation caused in a medium is not confined to its source but travels as a perturbation at a given speed away from the source. This perturbation is called a wave; acoustic waves (sound) are the physiologically important example considered in this chapter. Sound is a longitudinal wave, which is a wave for which the propagation direction is collinear with the direction in which the deformation oscillates (e.g., air elements vibrate back and forth in the direction in which the sound travels).

The perturbation in a gas that carries sound is an oscillating variation of the density and pressure of small gas elements. The displacement and motion of the gas element represent the total energy contained in the wave. This energy travels with the speed of sound. The product of energy and speed defines the sound intensity. The intensity of a point sound source attenuates with the inverse square of the distance from the source.

In confined media, such as closed or half-closed tubes, the wave is described by a sinusoidal function if it is caused by an elastic vibration obeying Hooke's law; such waves are called harmonic waves. When harmonic waves reflect off the closed end of the tube and are superimposed on themselves, standing waves emerge if a half-wavelength is a multiple of the length of the tube. Such standing waves are called harmonics. Harmonics result from an external excitation that couples into the confined system resonantly. An example is the vibration of the vocal cords causing the human voice. The inverse process causes the vibrations of the eardrum when sound enters the outer ear.

Psychophysics is the term invented more than 120 years ago by Gustav Fechner to describe the physics and physiology of the human senses. This term combines the analytic physics approach with the more interpretative concepts of psychology. Despite a solid foundation in fundamental physics, this field often defies the basic assumption of objectivity. Whether we hear Sergei Rachmaninoff's Concerto for Piano and Orchestra No. 2 played by the New York Philharmonic Orchestra under Leonard Bernstein, see Claude Monet's 1875 painting "La Japonaise" at the Boston Museum of Art, taste a Chateaubriand with sauce Béarnaise, or smell the freshness of a crisp Canadian winter morning, we have to understand that the beauty of these things does not exist anywhere but in our mind. It is the interpretation that our brain attaches to the tremendous flow of stimuli arriving from the environment that we perceive as reality, while the physical and chemical reality is a much more profane maze of electromagnetic waves, acoustic waves, and chemical compounds that cause reactions in our mucous membranes.

We focus in this chapter on the acoustic waves to establish the basic physics of waves and the processes and limitations of sound perception. We motivate that discussion first with a discussion of the hearing of dolphins.

15.1: The Hearing of Dolphins

Biological research often compares different species to extract common, and thus more generally applicable principles. This is in particular a useful first approach to something as familiar as hearing because we otherwise easily lose the uniqueness of our acoustic abilities out of sight. We approach hearing therefore in this chapter motivation with a look at dolphins. What do they hear? How do they hear under water? Which concepts from Physics do we need to answer these questions and explain our own hearing?

We know dolphins can hear due to observations of their vocalizations in the wild or during tests with animals in captivity. Hydrophones are used to quantify what they hear. A hydrophone records sounds in the form of frequency spectra, sound intensities and sound amplitudes. For dolphins, these data allow us to distinguish an almost constantly emitted clicking sound (about 300 sounds per second) and whistle sounds. With a proper combination of acoustic receivers we can also establish that the clicking sounds are focussed in the forward direction, like the headlights of a car.

These acoustic data are supplemented by anatomical data on dolphins. Some important components are highlighted in Fig. 15.1. The **oily melon** is a unique feature that acts as an acoustic lens. Whales use this device as a powerful weapon, emitting sound shockwaves that stun squids and render them defenceless. Dolphins share an inner ear and a middle ear with other mammals; however, the outer ear is absent. For the following discussion we keep in mind that the outer ear of humans consists of the external pinna and the auditory canal, which collect sound waves and channel them to the eardrum. Instead, dolphins have a unique fatty organ that connects the rear end of the lower jawbone to the middle ear section.

Animal behavioural studies demonstrate that the sounds of dolphins are not just simple species or mate identification patterns. Particularly the whistle sounds are used to communicate with other dolphins of the same species, to express alarm, sexual excitement, and likely a range of other emotions. The clicking sounds, in turn, are used to navigate in the physical terrain and relative to other dolphins in the pod, and to detect fish, squid, and shrimp for food.

As we know from our own species, the ability to communicate requires two attributes in addition to a variable vocalization: good hearing and a large brain to process the information. Fig. 1.9 demonstrates how we quantify brain size. The figure is a double-logarithmic plot of the brain mass versus the body mass of average individuals of various species. Larger animals have larger brains, with the brain mass of the elephant and the blue whale exceeding our brain mass. This trend is indicated in the figure by a straight line (power law). More interestingly, we note that some species deviate from the general trend in that they possess particularly large brains for their body size. These species are found far above the solid line in the figure: humans and dolphins are distinguished in that respect. This reinforces our earlier conclusion that dolphins display a well-developed ability to hear.

With no vocal tract and no outer ear, how can acoustic signals play an important role in the dolphin's life? The answer is that there are other ways to communicate and hear than we are familiar with: the whistle sounds originate from a deeper range in the dolphin's larynx, but the clicking sounds are generated by moving air in and out of air sacs near the blowhole. This sound is then focused forward by the oily melon. Focussing the clicking sound allows the dolphin to increase the range of the initial sound energy, which is necessary because sound energy attenuates rapidly when it travels away from a point source in all directions, an issue we discuss in the next chapter for ultrasound.

The clicking sounds are useful for echolocation because sound waves reflect off an object in water. The received echo allows the dolphin to locate the reflecting object and determine the object's direction of motion and speed based on the Doppler effect, which we will also discuss in the next chapter. Even though dolphins have eyesight comparable to that of cats, the echolocation system is necessary at depths greater than 70 metres, beyond which sunlight cannot penetrate seawater. Dolphins often dive to such depths; they have been observed as deep as 300 metres below the sea surface. Such dives don't take the dolphin long because they can swim at sustained speeds of 30 km/h.

For a mammal to hear, an external sound has to cause an excitation of sound-sensitive cells in the inner ear. There are two ways in which this can occur:

- External sounds enter the outer ear through the auditory canal and set the eardrum in vibration. The ossicles of the middle ear amplify these vibrations

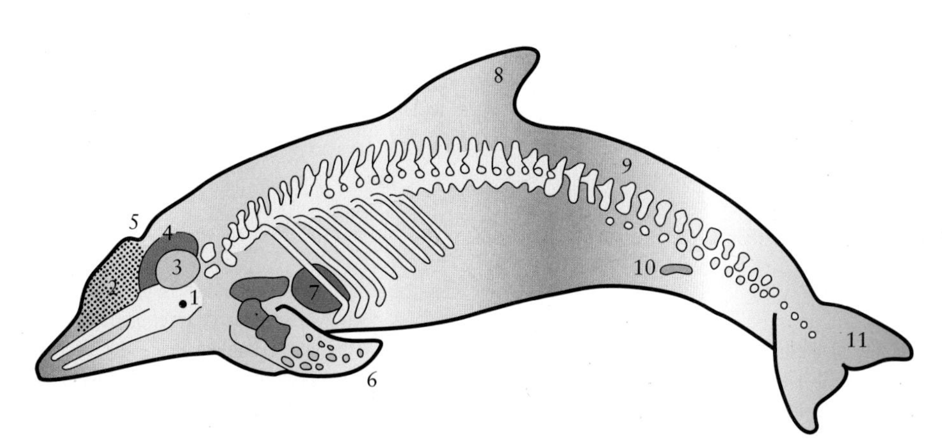

Figure 15.1 Dolphins can vocalize, hear, and interpret a wide range of clicking and whistle sounds. The clicking sounds are focussed in the forward direction by the oily melon (2), which is located in the forehead in front of the blowhole (5). The dolphin's ear (1) consists of an inner ear and a middle ear, but lacks the outer ear of land mammals. Arriving sound is transmitted to the middle ear by bone conduction in the skull (4). Dolphins need large brains (3) to interpret the information contained in the sounds from other dolphins and the echoes of their own clicking sounds. Other anatomical features highlighted in the figure include the (6) flipper, (7) heart, (8) dorsal fin, (9) spinal column, (10) pelvis, and (11) fluke.

(in humans by a factor 30, as discussed in section 15.6.2) and transmit it to the oval window, which separates the inner ear from the middle ear. The sound then propagates through the liquid-filled medium of the inner ear and excites sound-sensitive cells. This process is quantitatively discussed in section 15.6.3 for our own hearing.

- Alternatively, external sound waves cause vibrations of the skull bones surrounding the ear. The whistle sounds of other dolphins and the echo of the clicking sounds are incident sound waves that are received by the lower jawbone of a dolphin and are then transmitted to the middle ear via the fatty organ. These vibrations directly stimulate the sound-sensitive cells of the inner ear. This process is called **bone conduction**.

Both mechanisms contribute to human hearing, even though we are not consciously aware of bone conduction (which is the main mechanism by which we hear ourselves speak). In audiology, bone conduction is applied by two tests to distinguish hearing impairments caused by diseases of the middle ear versus diseases of the inner ear. The anatomy of the human middle ear is shown in Fig. 15.2. Chronic hearing impairments of the middle ear are caused by *otitis media* (bacterial infection of the middle ear) or by *otosclerosis*. Otitis media leads to malformed tissue in the middle ear, which negatively affects the mobility of the eardrum and the ossicles. Otosclerosis affects 1% of the adult population and leads to an abnormal amount of spongy bone deposition between the stapes and the oval window, which immobilizes the stapes. Both diseases lead to a loss of sound transmission from the eardrum to the oval window. Hearing impairments of the inner ear are usually associated with damaged organs of Corti; i.e., sound is not properly processed into signals sent to the brain by the sound-sensitive cells.

The two audiological tests are illustrated in Fig. 15.3. The first test, in Fig. 15.3(a), is the **Weber test**: An A_1 tuning fork, vibrating at 55 Hz, touches the top centre of the patient's head. A healthy patient locates the source of the sound at the proper central position. A patient with a disease of the middle ear locates the source of the sound

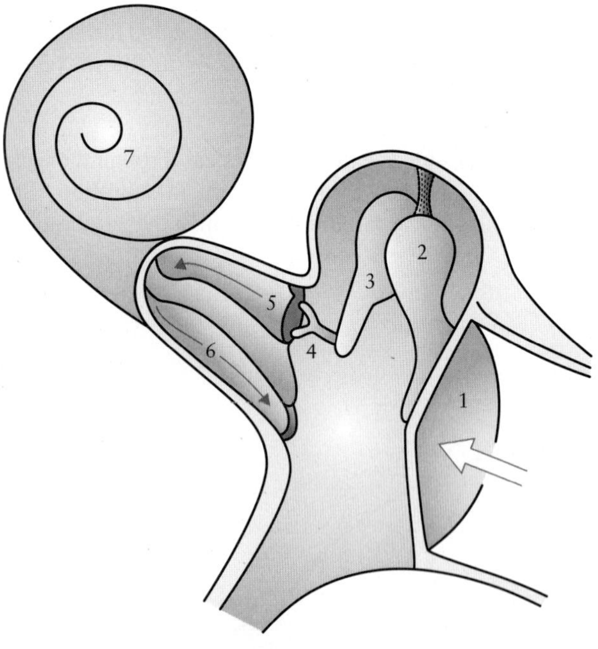

Figure 15.2 The anatomy of the human ear represents 300 million years of evolutionary adaptation to hearing in air. The arriving sound (open arrow) causes the eardrum (1) to vibrate. This vibration is mechanically transmitted to the oval window (5) of the cochlea (7; solid arrows). The sound is amplified by a factor of 30 due to the arrangement of the three ossicles: the hammer (2), the anvil (3), and the stirrup (4). The middle ear converts a sound wave in air into a sound wave in the fluid (perilymph; 6) of the inner ear.

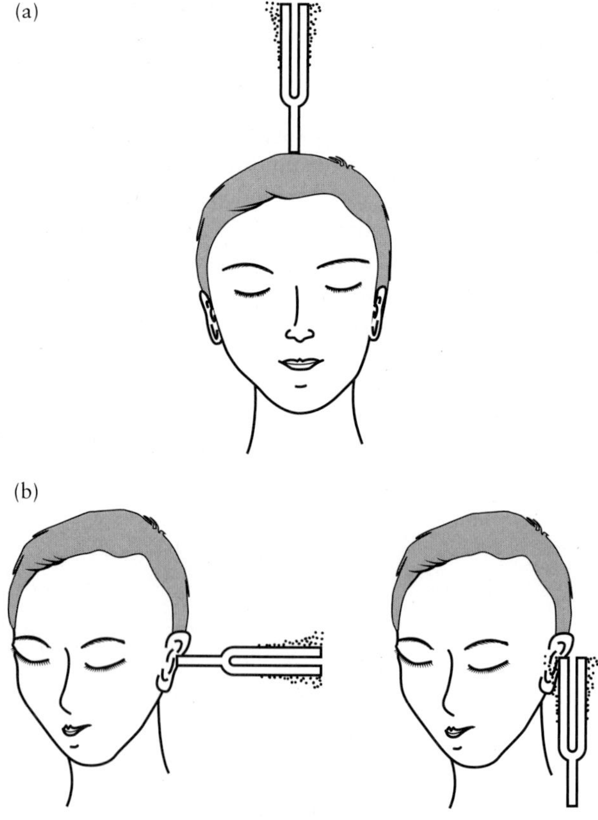

Figure 15.3 Bone conduction is employed in two audiological tests that allow us to distinguish between middle ear and inner ear diseases. (a) In the Weber test, a vibrating tuning fork touches the top of the patient's head. The diagnosis of the health practitioner depends on where the patient perceives the sound source. (b) In the Rinne test, the vibrating tuning fork is first brought into contact with the *mastoid process* behind the auricle until the patient can no longer hear the sound. Then the tuning fork is moved in front of the auricle. The health practitioner's diagnosis depends on whether the patient hears the sound after the tuning fork has been moved.

near the ailing ear because these patients rely stronger on bone conduction on the ailing side. A patient with a hearing impairment of the inner ear locates the source of the sound near the healthy ear since the ailing side does not receive a strong signal either way.

The second test in Fig. 15.3(b) is the **Rinne test**. In this test, a vibrating A_1 tuning fork is first brought in contact with the mastoid process behind the auricle. When the patient doesn't hear the diminishing sound any longer, the tuning fork is brought in front of the auricle. A healthy person or a person with a hearing impairment in the inner ear now hears the sound again since bone conduction in humans is less effective than hearing a sound, which is transmitted through air to the eardrum. A patient with a hearing impairment in the middle ear, however, does not pick up the sound of the moved tuning fork. This is due to the ability to circumvent the middle ear mechanism with bone conduction but not with sound arriving at the eardrums.

In a modern audiological setting, these tests have been replaced by instruments that generate sounds across a wide range of frequencies with tunable sound intensity. For each frequency, the intensity threshold below which the patient can no longer hear the sound is recorded. Fig. 15.4 shows four plots of hearing intensity loss in unit decibel (dB) versus the sound frequency. Two curves are shown in each case: the solid line is sound conduction by air (done with headphones) and the dashed line is bone conduction of the sound (done similarly to in the Rinne test). The figure shows that the three shown cases of hearing impairment can easily be distinguished from each other and normal hearing.

Bone conduction is more important for hearing under water. Adjustments in the lower jaw of dolphins and the development of the fatty organ to connect the lower jaw to the middle ear are improvements in the ability to hear by bone conduction.

The discussion in this section has revealed a great number of physical parameters, units, and basic concepts that we need to establish before we can fully appreciate the description of hearing. In this chapter and the next we introduce these concepts. We then return to the physical aspects of hearing in the latter part of the current chapter, and supplement the discussion with the medical application of ultrasound in Chapter 16.

15.2: Piston-Confined Gas Under Hydraulic Stress

We do not need to start with the discussion of sound generation and sound detection from the beginning. The discussion of vibrations in Chapter 14 has laid a good foundation for discussing sound waves in this chapter because

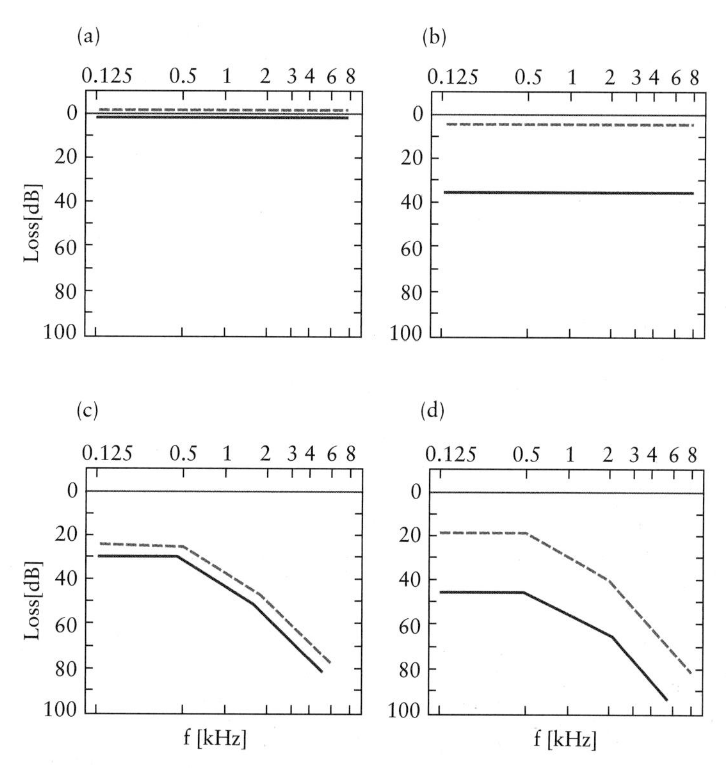

Figure 15.4 Four audiometric measurements leading to different diagnoses for human hearing impairments. Audiometers use electroacoustic sound generators that emit single frequency sounds, allowing us to determine the frequency dependence of the intensity threshold of hearing by a patient. Each audiogram shows a curve for bone conduction hearing (dashed curve) and hearing of sounds travelling through air (solid curve, which is located in all four plots below the dashed line): (a) normal hearing; (b) hearing impairment of the middle ear, e.g. due to otosclerosis; (c) hearing impairment of the inner ear, e.g. due to old age; (d) a combination of hearing impairment of the middle and the inner ear.

sound is generated by mechanical vibrations. However, we cannot progress straight from the conclusions in Chapter 14 because in that chapter we focused on vibrations of an object attached to a string. Sound is generated in a medium, for example in air. Thus, we start the discussion of sound by returning in this first section to the basic discussion of vibrations, but we introduce the concept alternatively to Chapter 14 for a piston that confines a gas in a closed container. As in Chapter 14, we use this model system first to demonstrate that a restoring force acts on the piston if moved away from its equilibrium position by an external force. The restoring force is shown to obey Hooke's law, which then allows us to apply all subsequent conclusions we drew in Chapter 14 for the object attached to a spring. This includes the harmonic motion of the piston when the external force is removed.

Fig. 15.5 illustrates an ideal gas that is confined in a container with a mobile piston at its right end. Choosing a horizontal arrangement with the x-axis toward the right allows us to neglect gravitational effects on gas or piston. The piston is at its mechanical equilibrium position x_{eq} when the pressure of the ideal gas inside is equal to the atmospheric pressure outside. The origin in Fig. 15.5 is chosen at the fixed left wall of the cylindrical container.

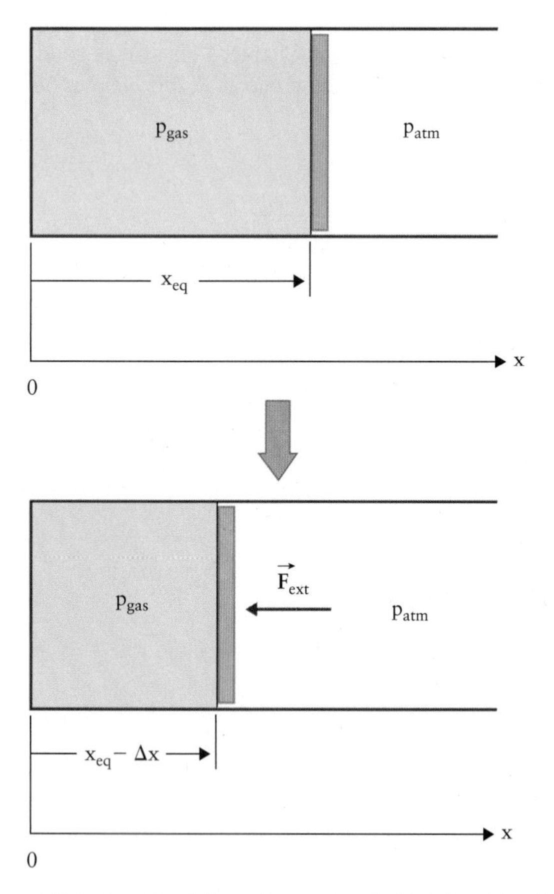

Figure 15.5 A mobile piston seals a gas in a horizontal container. The piston is in mechanical equilibrium at position x_{eq} when the gas pressure p_{gas} is equal to the external atmospheric pressure p_{atm}. An external force allows us to move the piston to a new position at $x_{eq} - \Delta x$. In this graph, the origin of the x-axis is chosen at the left end of the gas container.

We displace the piston by a small distance Δx to a new position at $x = x_{eq} - \Delta x$. To achieve this displacement, an external force of magnitude F_{ext} is applied toward the left; i.e., the x-component of the external force is negative, $F_x = -F_{ext} < 0$. A new mechanical equilibrium, governed by Newton's first law, is established when we hold the piston at this position. The mechanical equilibrium then allows us to determine the specific form of Hooke's law for the piston-confined ideal gas. Defining the area of the piston as A, we find:

$$\sum_i F_{i,x} = -p_{atm}A - F_{ext} + p_{gas}A = 0.$$

Sorting the terms in this equation yields:

$$F_{ext} = A(p_{gas} - p_{atm}).$$

This equation allows for positive and negative values for F_{ext}, which is inconsistent with the definition of F_{ext} as the magnitude of a vector. What does this imply? It means that Newton's first law above must be written with $\pm F_{ext}$ where $+F_{ext}$ applies when $p_{gas} < p_{atm}$. This hints toward the restoring effect of the pressure difference. We avoid carrying two formulations of Newton's law by rewriting the last equation in the form:

$$F_{ext} = A\left|p_{gas} - p_{atm}\right|,$$

where we no longer carry directional information.

We assume isothermal conditions and use the ideal gas law to rewrite the pressure in this equation as $p = n\ R\ T/V$. We note that the gas pressure at the equilibrium position of the piston is equal to the atmospheric pressure, $p_{atm} = n\ R\ T/V_{eq}$. Thus:

$$F_{ext} = nRTA\left|\frac{1}{V_{gas}} - \frac{1}{V_{eq}}\right|.$$

The volume terms are expressed as the product of the cross-sectional area of the cylindrical piston, A, and the length of the container between the left wall and the piston:

$$F_{ext} = nRTA\left|\frac{1}{A(x_{eq} - \Delta x)} - \frac{1}{Ax_{eq}}\right|.$$

Cancelling the cross-sectional area A in numerator and denominator simplifies this equation to:

$$F_{ext} = nRT \left| \frac{1}{x_{eq} - \Delta x} - \frac{1}{x_{eq}} \right|.$$

The term in the parentheses can be simplified for small piston displacements; i.e., when we assume $\Delta x << x_{eq}$:

$$\lim_{\Delta x \ll x_{eq}} \left| \frac{1}{x_{eq} - \Delta x} - \frac{1}{x_{eq}} \right| = \lim_{\Delta x \ll x_{eq}} \left| \frac{x_{eq} - (x_{eq} - \Delta x)}{x_{eq}(x_{eq} - \Delta x)} \right|$$

$$= \frac{|\Delta x|}{x_{eq}^2}.$$

Inserting the last in the previous equation then leads to:

$$F_{ext} = \frac{nRT}{x_{eq}^2} |\Delta x|.$$

Dividing both sides by A yields a stress term F/A on the left-hand side and converts one of the factors x_{eq} in the denominator on the right-hand side into the equilibrium volume of the gas, $V_{eq} = A\, x_{eq}$. Further multiplying the right-hand side with a factor A/A converts the other factor x_{eq} in the denominator into V_{eq} and converts the Δx term in the numerator into a volume change, ΔV:

$$\frac{F_{ext}}{A} = \frac{nRT}{V_{eq}} \frac{|\Delta V|}{V_{eq}}.$$

We can compare this equation with the experimental observations in Fig. 15.5. We note that an increasing external force causes a decreasing volume in Fig. 15.5. This is in turn consistent with Hooke's law; i.e., $F_{ext}/A \propto |\Delta V|$ means that the force is linear in the volume change. The directional interpretation complements this analysis in demonstrating the restoring nature of the force.

Comparing this equation with Hooke's law for a spring in Eq. [14.5] yields then for the bulk modulus of the ideal gas:

$$B_{ideal\ gas} = \frac{nRT}{V_{eq}} = p_{eq}. \qquad [15.1]$$

The bulk modulus for an ideal gas under isothermal conditions is equal to its equilibrium pressure.

KEY POINT

A small displacement of a piston sealing an ideal isothermal gas in a container is a valid case of Hooke's law.

CASE STUDY 15.1

Assume that the external force $\vec{F}_{ext}$ is suddenly removed in the lower panel of Fig. 15.5. Describe qualitatively what happens next.

Answer: *The system discussed in Fig. 15.5 is in mechanical equilibrium since the external force is balanced by an equal but opposite force exerted by the gas on the piston. If we remove the external force by releasing the piston, a mechanical non-equilibrium situation is created, i.e., a case governed by Newton's second law. The unbalanced force acting on the piston is the force exerted by the gas on the piston. This is a restoring force since it points in the direction toward the equilibrium position of the piston.*

The unbalanced restoring force accelerates the piston initially toward the right in Fig. 15.5. As the piston moves, it reaches the equilibrium position. At that instant no force acts on the piston; i.e., it no longer accelerates. However, the inertia of the piston prevents it from suddenly coming to rest. Thus, the piston moves farther to the right. Once the piston has moved beyond the equilibrium position, the confined ideal gas has a pressure lower than the external atmospheric pressure. Therefore, a restoring force acts on the piston, pulling it back toward the equilibrium position; i.e., a force acts toward the left. This force slows the piston down to rest, which occurs at $x_{eq} + \Delta x$ *if the piston is moving without friction in the cylindrical container. Thereafter, the piston continues to move back and forth.*

15.3: Waves in an Unconfined Medium

Since the particles in any homogeneous medium (e.g., solids, liquids, and gases) interact with each other, local vibrations of atoms or molecules around their equilibrium position affect other particles in the close vicinity. These interactions lead to a propagating wave.

15.3.1: Physical Properties of Waves: Speed, Wavelength, and Frequency

An easy way to illustrate the connection between a local vibration and the resulting wave is to attach a rope to a wall and swing its free end up and down. The formation of a wave along the rope is sketched in Fig. 15.6. Assume that the rope is a taut string with one end attached to a rigid wall and the other end held horizontally with your hand. At time $t = 0$, you start to move your hand up and down so that it oscillates vertically with amplitude A. At time $t = T/4$, your hand has completed a quarter period and has reached position $+A$. Since the rope is continuous, its parts adjacent to your hand must have been pulled up as well, as illustrated in the second frame of Fig. 15.6. Later,

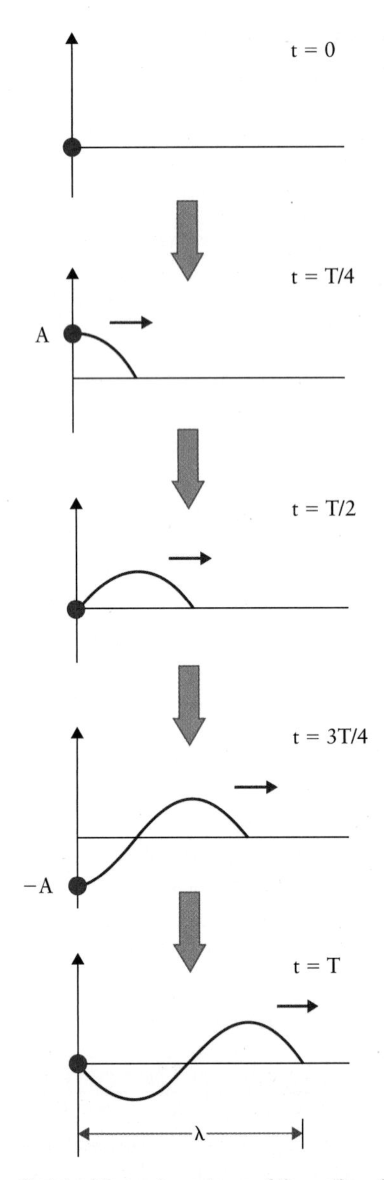

Figure 15.6 Sketch of the various stages of the motion of a rope when the free end at the origin oscillates up and down with period T. Once a full period of the vibration is completed (bottom panel), the rope is displaced to a distance λ, called the wavelength. Note that the rope is assumed to be elastic and therefore stretches to maintain the left end at the same horizontal position, in the same fashion as a string on a violin.

at time $t = T/2$, your hand has returned to its original position. However, since the rope has been following your hand, it is not stretched horizontally as it was initially but contains a bulge, as shown in the third frame of Fig. 15.6. In the next two frames your hand completes another half cycle by moving to the position $-A$ and back. Again, the adjacent part of the rope follows and a downward bulge results. The initial, upward-directed bulge did not disappear but has moved farther to the right. Thus, at time $t = T$, the local oscillation of your hand has been transformed into a full sinusoidal shape of the rope, reaching to a point at distance λ from your hand. The rope now describes a wave with λ the **wavelength**, which we combine with the **period** T to calculate the speed at which the perturbation caused by your hand propagates along the rope:

$$v_{\text{wave}} = \frac{\lambda}{T} = \lambda f, \qquad [15.2]$$

in which f is the **frequency,** which is related to the period as $f = 1/T$.

KEY POINT

The speed of a wave is equal to the product of its wavelength and frequency.

CONCEPT QUESTION 15.1

In air at room temperature, the wavelength of a sound increases as a result of which of the following changes in parameters?

(A) a frequency increase

(B) a frequency decrease

(C) a speed of sound increase

(D) a speed of sound decrease

EXAMPLE 15.1

(a) The musical note C_4 (middle C) on a piano is caused by a string vibrating with a frequency of 261.6 Hz. The vibrating string interacts with the adjacent air, causing a sound wave to propagate away from the piano with a wavelength of 1.31 m. What is the speed of sound in air? (b) An FM station broadcasts at $f = 100$ MHz with radio waves of wavelength $\lambda = 3$ m. Find the speed of the radio wave.

Solution

Solution to part (a): The speed of waves is given by Eq. [15.2]:

$$v_{\text{wave}} = \lambda f = (1.31 \text{ m})(261.6 \text{ Hz})$$
$$= 342.7 \frac{\text{m}}{\text{s}}.$$

The speed of a sound wave at room temperature is about 340 m/s.

Solution to part (b): We again use Eq. [15.2]:

$$v_{\text{wave}} = \lambda f = (3 \text{ m})(100 \text{ MHz})$$
$$= 3 \times 10^8 \frac{\text{m}}{\text{s}}.$$

The speed of a radio wave is significantly higher than the speed of sound.

EXAMPLE 15.2

Bats use ultrasound echolocation to detect small insects in flight. For this to work, the wavelength used by bats must be smaller than or equal to the size of their prey. Bats therefore use frequencies of about 80 kHz. Dolphins and porpoises also use ultrasound echolocation for hunting. (a) If a dolphin's prey were as small as the insects eaten by bats, what frequency would dolphins have to use? (b) Dolphins actually use frequencies up to 225 kHz. How much bigger is their smallest prey compared to the insects that bats hunt? *Hint:* Use c_{air} = 340 m/s for the speed of sound in air and use c_{sea} = 1530 m/s for the speed of sound in seawater.

Solution

Solution to part (a): We use Eq. [15.2], written once for the medium air and once for the medium seawater. In both cases, the wavelength is specified by the smallest length required for each predator, $\lambda = L_{insect}$ for the bat and $\lambda = L_{squid}$ for the dolphin (assuming that squid is the smallest prey a dolphin hunts):

$$\begin{aligned} &\text{bat in air:} && c_{air} = L_{insect} f_{bat} \\ &\text{dolphin in sea:} && c_{water} = L_{squid} f_{dolphin}. \end{aligned}$$

In part (a) we assume that $L_{insect} = L_{squid}$. This leads to the frequency needed by a dolphin to detect a squid the size of an insect:

$$\begin{aligned} f_{dolphin} &= \frac{c_{water}}{c_{air}} f_{bat} \\ &= \frac{1530 \text{ m/s}}{340 \text{ m/s}} (80 \times 10^3 \text{ Hz}) \\ &= 360 \text{ kHz}. \end{aligned}$$

Solution to part (b): The result in part (a) shows that dolphins cannot hunt prey as small as the insects that are hunted by bats. To answer part (b), we allow for the size of a squid and the size of an insect to differ, using the two formulas at the beginning of part (a) to calculate the ratio L_{squid}/L_{insect}:

$$\begin{aligned} \frac{L_{squid}}{L_{insect}} &= \frac{c_{water}}{c_{air}} \frac{f_{bat}}{f_{dolphin}} \\ &= \frac{1530 \text{ m/s}}{340 \text{ m/s}} \frac{80 \times 10^3 \text{ Hz}}{225 \times 10^3 \text{ Hz}} = 1.6. \end{aligned}$$

Thus, among those hunting in three-dimensional space, dolphins are probably the most versatile. Given the typical size of an insect hunted by bats, with a size just over 4 mm, the much larger dolphin can detect objects as small as 7 mm in size under water. The smallest adult squid are 2 cm to 3 cm long. In addition, the dolphin can detect objects of that size over a much longer distance because sound absorption in water is much less than in air.

15.3.2: Longitudinal and Transverse Waves

It is worthwhile to analyze the results of Example 15.1 in more detail. The wave propagation speed in part (a) is consistent with the range of speeds we calculated for air molecules at room temperature using the kinetic gas theory in Chapter 8. Thus, we proposed that air is the **medium** that carries sound waves. We will see that the vibration of a sound source creates a periodic variation in the density of the adjacent air, and this density perturbation travels through the medium, superimposed on the random motion of the gas particles. The speed of individual air molecules limits the propagation speed of sound, as they need to collide with gas molecules in neighbouring gas elements.

Accepting this reasoning requires us to identify a different medium for the propagation of radio waves as studied in part (b) of Example 15.1. We recognize the speed found in that case to be the speed of light. Thus, we expect that radio waves and light are related. Radio waves and light are electromagnetic waves, which we discuss in Chapter 21. Electromagnetic waves do not require a wave-carrying medium as illustrated by the fact that they can be transmitted through the vacuum of outer space (the field of radio astronomy analyzes radio signals from astronomical objects).

Waves are not just distinguished by different wavelengths, frequencies, and wave propagation speeds. We illustrate another major difference by looking once more at Fig. 15.6. Note that the vibration that causes the wave is based on an up–down motion of the left end of the rope. The propagating wave in turn moves toward the right, i.e., in a direction perpendicular to the vibration. We call such a wave a **transverse wave**. Important examples of this type of wave are radio and light waves.

Other wave forms, in particular the sound waves that interest us in this chapter, differ from the transverse wave type. This is illustrated with the five sketches in Fig. 15.7 showing a gas in which an acoustic wave is generated. In the first frame at time $t = 0$ the gas shown to the right of the piston has a uniform density. The equilibrium position of the piston is indicated by a vertical dashed line. The gas density is indicated graphically as the density of vertical lines in the gas. At time $t = 0$ the piston is set into motion toward the right, as indicated by an arrow. At $t > 0$ the piston will undergo a harmonic oscillation, as introduced in the previous section. In the current context our focus is now on the air in front of the piston.

The four lower frames in Fig. 15.7 are equivalent to frames (b) to (e) in Fig. 15.6 for the formation of a wave on a rope. When the piston reaches its amplitude toward the right at $t = T/4$, a locally compressed pocket of air is created in front of the piston (indicated by a high density of vertical lines representing the gas). The gas molecules in this pocket of increased density have also been given an additional velocity component toward the right. When the piston then moves back to its equilibrium position (reached at time $t = T/2$), the gas molecules in front of the

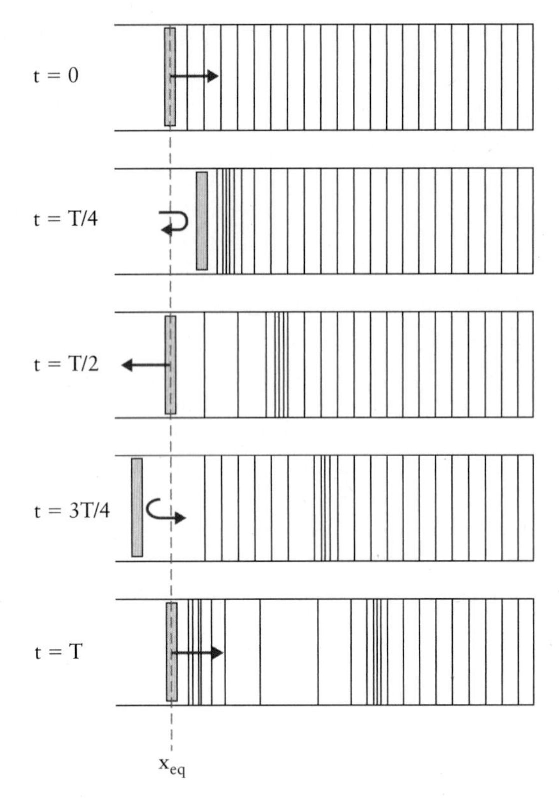

Figure 15.7 Five snapshots illustrating the development of an acoustic wave in a gas-filled container due to the vibrational motion of a piston about its equilibrium position x_{eq}. The piston reaches the amplitude point toward the right at time $T/4$ and the amplitude point toward the left at time 3 $T/4$. When the piston moves toward the right the air pocket adjacent to the piston is compressed (indicated by an increased density of vertical lines). When the piston moves toward the left the air pocket adjacent to the piston is expanded (indicated by a decreased density of vertical lines).

piston are pulled back toward the left, causing a pocket of decreased air density (indicated by a low density of vertical lines).

By moving beyond the equilibrium position to the amplitude point at the left (reached at time $t = 3\ T/4$), the piston increases the volume of the gas pocket directly in front of it at the right. Thus, the volume with decreased air density is further enlarged. In the meantime, the initially created zone of increased air density has travelled toward the right and remains present in the system (still illustrated as a pocket of denser vertical lines). Note that it is the density variations that travel, not the individual molecules in the gas. The absolute density of the gas is very high and thus individual gas molecules encounter frequent collisions that essentially keep them near their original position.

In the last frame of Fig. 15.7 we see that the piston, turning toward the right after $t = 3\ T/4$, compresses the air in front of it once more while completing a full period of motion. Thus, at the final time $t = T$, a second pocket of increased air density is generated that subsequently travels toward the right following the initial high-density pocket at a constant distance.

What distinguishes the case in Fig. 15.7 from the case in Fig. 15.6 is the fact that the direction of the vibration of the piston and the direction of the propagating sound wave are collinear. Such waves are called **longitudinal waves**.

KEY POINT

The direction of the exciting oscillation and the direction of the propagating wave are collinear for longitudinal waves and perpendicular for transverse waves.

Important differences exist between longitudinal and transverse waves. We will discuss one such difference in detail when introducing the linear polarization of light. Acoustic waves do not have this feature, because longitudinal waves cannot be polarized.

CASE STUDY 15.2

Identify examples of (a) transverse waves and (b) longitudinal waves. Use a literature or Internet search for this question.

Answer to Part (a): *Electromagnetic waves, including visible light, ultraviolet and infrared light, microwaves, and radio waves.* ***Seismic waves*** *caused by an earthquake contain both a longitudinal and a transverse component. They travel with different speeds in Earth's crust and are recorded in sequence by a seismograph. The relative delay allows earth scientists to determine the epicentre of the earthquake.*

Answer to Part (b): *Sound waves, including audible sound and ultrasound for medical applications. Note that we don't list surface waves of water because the medium (small water elements) moves in a circular fashion.*

15.3.3: Longitudinal Waves in a Gas

Now we want to establish a quantitative description of wave propagation in a gas. This step is based on Fig. 15.8, in which we study the speed and the pressure variations of a gas element. The notation $p(x)$ means $p = f(x)$; i.e.,

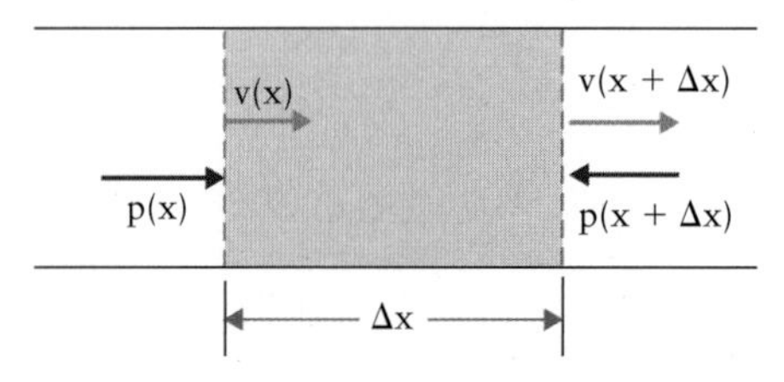

Figure 15.8 A small gas element is shown in a cylindrical tube with a one-dimensional sound wave generated by a vibrating piston, which is somewhere to the left of the grey gas element. The vertical dashed lines define the gas element's length Δx. We distinguish the gas pressure and the speed of the gas at both interfaces confining the gas element.

the pressure is a function of position along the x-axis, and $v(x)$ means $v = f^*(x)$; i.e., the gas element's speed is also a function of position. We use the asterisk to note that the two functions $f(x)$ and $f^*(x)$ are not assumed to be the same. A small gas element of length Δx is highlighted in grey in Fig. 15.8. It lies between two dashed vertical lines. To simplify its motion, we assume the gas is contained in a cylindrical tube that is aligned with the x-axis. The gas to the left of the element causes a local pressure $p(x)$ at position x and the gas to the right causes a local pressure $p(x + \Delta x)$ at position $x + \Delta x$. The small gas element is not in mechanical equilibrium if we assume that somewhere to the left a piston vibrates, as in Fig. 15.7. This means that the two local pressure values differ and therefore cause an acceleration of the small gas element. Newton's second law relates the pressure change along the x-axis to the change of gas element speed with time, i.e., its acceleration.

The speeds of the two interfaces that separate the gas element from the gas outside in Fig. 15.8 may vary as well, with values $v(x)$ and $v(x + \Delta x)$. This difference leads to a change in the volume of the small gas element. A volume change causes a change in pressure based on the ideal gas law. Thus, the change in speed along the x-axis causes a change in pressure.

When these two arguments are combined a **wave equation** follows, i.e., an equation that relates the time and position dependence of the parameters of a wave to each other. The specific dependence of the displacement, the pressure variations, and the density variations on position and time (called the **wave function**) obviously cannot be specified based on Fig. 15.8 alone, as the choice of a particular vibration that causes the wave will lead to different wave forms. Thus, additional information is needed to specify the wave function. We study the most important case, in which the vibration of the piston is a simple harmonic oscillation, in the next section.

KEY POINT

Pressure differences in a gas cause the acceleration of small gas elements. Their velocity differences in turn cause pressure differences. Both effects combine to create a wave function that relates the time and position dependences of gas pressure variations or small gas element displacements.

A few general properties of waves apply regardless of the specific form of vibration causing the wave. Here, we discuss two of these issues for a wave that may result from an experiment like the one shown in Fig. 15.8: the dependence of position and time for pressure variations, and the speed of sound.

15.3.3.1: Interdependence of Position and Time in a Wave

We define the **displacement** of a small gas element in Fig. 15.8 as D, with $D = x - x_{eq}$. For a sound wave the displacement varies from point to point in space and also in time. Thus, the most general mathematical form for the displacement is $D = f(x, t)$. This means the displacement is a function that contains two independent variables, time and position. However, writing $D = f(x, t)$ is too general because the variables x and t are not independent from each other. To see this, we study an arbitrary wave that travels along the x-axis at two different times, t_1 and t_2. You can do this by dropping a pebble in a pond and observing the evolving pattern of water surface waves moving away from the impact point.

At the earlier time t_1 the wave function has a particular profile as a function of the position, for example, the profile shown in the top sketch of Fig. 15.9. Studying the wave at a later time t_2 we find the case illustrated in the lower sketch of Fig. 15.9. The wave function differs at the later time only in that the wave pattern has moved along the x-axis. This is highlighted in Fig. 15.9 by identifying the position of a specific feature of the wave, which is initially at position x_1 and later at position x_2. The wave function is in both cases identical if the appropriate combination of position shift and elapsed time between an initial and a final snapshot is taken into account:

$$x_1 - ct_1 = x_2 - ct_2;$$

i.e., the position shift from x_1 to x_2 is linearly related to the elapsed time, from t_1 to t_2, with a prefactor c for the time. This prefactor must have the unit of speed m/s for the equation to be physically correct. Since both sides of the equation have the same mathematical form, we conclude that the displacement of the gas element D depends on position and time in the form:

$$D = f(x - ct), \qquad [15.3]$$

which states that the displacement depends on a single independent variable, which is $x - ct$. This variable is

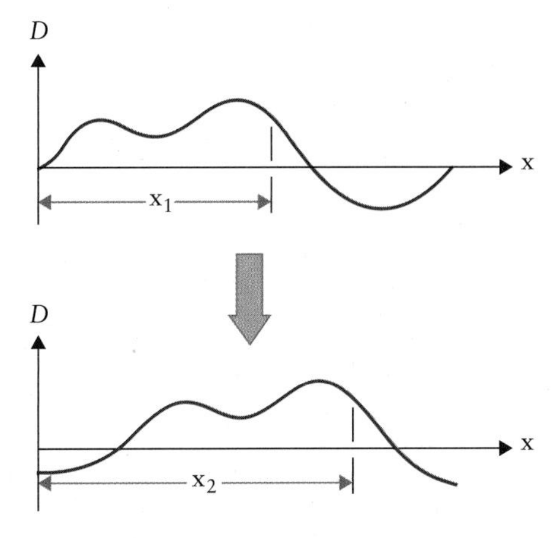

Figure 15.9 A displacement of a gas element from its equilibrium position, $D = x - x_{eq}$, due to a perturbation moving through the medium. The figure shows the wave twice, at time t_1 (top) and at time t_2 (bottom). During the time interval $\Delta t = t_2 - t_1$ the wave travels a distance $\Delta x = x_2 - x_1$ toward the right. The speed of the wave is $c = \Delta x/\Delta t$.

called the **phase of the wave**; the constant c is the speed of sound.

15.3.3.2: Speed of Sound

We can quantify the speed of sound from Eq. [15.3] by selecting for the wave two arbitrary instances, 1 and 2:

$$c = \frac{x_2 - x_1}{t_2 - t_1}.$$

c is specifically called the **phase velocity** of the wave. This quantity should not be mixed up with the speed of a particular gas particle or the root-mean-square speed introduced in Chapter 8.

Sound propagation is not limited to gaseous media. Fig. 15.8 illustrates that the occurrence of a travelling sound wave in a gas is due to the compressibility of the gas. In Chapter 14 we determined that real solids and liquids are also compressible. Thus, wave phenomena occur in these two states of matter as well. Compressibility in solids is directional, as we discussed at the beginning of Chapter 14. Liquids and gases are fluids in which compression must occur from all directions. Their compressibility is governed by the bulk modulus B. The speed of sound then follows from the bulk modulus and the density of the medium:

$$c_{\text{fluid}} = \sqrt{\frac{B}{\rho}}. \qquad [15.4]$$

KEY POINT

Wave equations describe the wave as a function of position and time. In a sound wave, position and time dependences are linked through the speed of sound, which depends on the elastic properties and the density of the medium.

The speed of sound in a gas can also be derived from Eq. [15.4]. Note that we calculated the bulk modulus for an isothermal ideal gas in Eq. [15.1]. We cannot substitute that result in Eq. [15.4] because Eq. [15.1] is derived with the assumption of a constant temperature (T = const). For a gas expansion to occur isothermally, it must be done slowly. This does not apply to the fast vibrations of air in a travelling sound wave. The sound vibrations of air have to be described by an adiabatic compression since the time during which the compression occurs is too short to allow a heat exchange with the adjacent air. Using an adiabatic process to calculate the bulk modulus of an ideal gas leads to **Laplace's equation** for the speed of sound:

$$c_{\text{sound in gas}} = \sqrt{\frac{\kappa p}{\rho}}, \qquad [15.5]$$

in which $\kappa = C_p/C_V$ is the adiabatic coefficient, with C_p the molar heat capacity of the gas at constant pressure and C_V at constant volume. Table 15.1 summarizes values for the speed of sound in various solids, liquids, and gases.

TABLE 15.1

Speed of sound in various materials

Material	Speed of sound (m/s)	Temperature (K)
Gases		
Air	331	273
Air	343	293
Air	386	373
Liquids		
Water	1400	273
Water	1490	298
Seawater (3.5% salt)	1530	298
Solids and soft matter		
Steel	5940	
Granite	6000	
Human body tissue	1540	310
Vulcanized rubber	55	

CASE STUDY 15.3

The speed of sound in a gas is limited by the speed of the gas particles in the gas. Why then can a solid carry sound even though its atoms or molecules remain at the same location within the lattice?

Answer: *Gas molecules are not chemically or physically bound to each other and interact only through collisions. Thus, a physical change of the parameters for a given gas particle can reach other gas particles only when the initial particle has travelled into their vicinity and undergone a collision. In a solid, all particles are chemically or physically bound to each other. When a physical parameter in one neighbourhood changes, neighbouring atoms or molecules are affected very fast due to the strong interactions along these bonds. Thus, the speed of sound in a solid is usually much higher than in a gas.*

EXAMPLE 15.3

The adiabatic coefficient for air under normal conditions is $\kappa = 1.4$. If we use for the density of air at 0°C the value $\rho = 1.293\ \text{kg/m}^3$, we can confirm the speed of sound at 0°C and air pressure 1.0 atm, as shown in Table 15.1.

Solution

Substituting the given values into Eq. [15.5], we find:

$$c_{\text{air}} = \sqrt{\kappa\frac{p}{\rho}} = \sqrt{1.4\frac{1.013\times10^5\ \text{Pa}}{1.293\frac{\text{kg}}{\text{m}^3}}}$$

$$= 331.3\frac{\text{m}}{\text{s}}.$$

As Table 15.1 shows, the speed of sound in gases and liquids depends strongly on the temperature. We quantify this for gases using the ideal gas law. We know from Charles's law that the ideal gas expands linearly when the temperature is increased. Charles's law is rewritten with the reference volume V_0 at 0°C: $V = V_0(1 + \alpha T)$. In this formula, α is the **linear expansion coefficient** for an ideal gas, with $\alpha = 1/273.15\ \text{K}^{-1}$. This means that the gas expands by a fraction of 1/273.15 per degree of temperature increase. With this volume formula, the definition of the gas density, $\rho = m/V$, is written in the form:

$$\rho = \frac{\rho_0}{1+\alpha T}, \qquad [15.6]$$

in which ρ_0 is the reference density at 0°C temperature. Eq. [15.6] is substituted into Eq. [15.5] to determine the temperature dependence of the speed of sound at constant gas pressure:

$$c = c_0\sqrt{1+\alpha T}, \qquad [15.7]$$

in which c_0 is the reference speed at 0°C temperature.

EXAMPLE 15.4

Using the data for air at 0°C from Table 15.1, find at $p = 1$ atm and $T = 20$°C (room temperature) (a) the density of air and (b) the speed of sound in air. *Hint:* Use an ideal-gas approximation for the linear expansion coefficient for air.

Solution

Solution to part (a): We calculate the density of air at room temperature from Eq. [15.6] first, using a temperature change of 20 K:

$$\rho = \frac{1.293\frac{\text{kg}}{\text{m}^3}}{1+\left(\frac{1}{273.15\ \text{K}}\right)(20\ \text{K})} = 1.205\frac{\text{kg}}{\text{m}^3}.$$

Solution to part (b): Eq. [15.7] allows us to calculate the speed of sound at 20°C:

$$c = \left(331.3\frac{\text{m}}{\text{s}}\right)\sqrt{1+\frac{20\ \text{K}}{273.15\ \text{K}}} = 343.2\frac{\text{m}}{\text{s}}.$$

CASE STUDY 15.4

In Chapter 8 we stated that the density of an ideal gas can be written in the form $\rho = Mp/(RT)$. Is this formula consistent with Eq. [15.6]?

Answer: *The difference between the two equations is that the temperature in Eq. [15.6] is given in degrees Celsius, while it is given in unit kelvin in the formula from Chapter 8. To test consistency, we convert the Celsius temperature in Eq. [15.6] into the unit kelvin with T(K) = T(°C) + 273.15 K. Multiplying by the coefficient α yields $\alpha T(\text{K}) = \alpha T(°\text{C}) + 1$. Substituting this term in Eq. [15.6] then leads to:*

$$\rho = \frac{\rho_0}{\alpha T}.$$

The formula from Chapter 8 and this equation are equivalent if:

$$\frac{\rho_0}{\alpha} = \frac{M}{R}p,$$

which shows that ρ_0 and the coefficient α are materials dependent (due to the dependence on M) and have a different pressure dependence. Both of these conclusions are consistent with what we expect: the density at 0°C is proportional to p and the coefficient α is independent of pressure.

15.3.4: Harmonic Waves

In our discussion of vibrations we identified a unique role for the harmonic vibrations because they are the result of a mechanical system that obeys Hooke's law with a linear restoring force.

KEY POINT

The waves caused by harmonic vibrations are called harmonic waves. This name was chosen since sinusoidal functions describe these waves.

With harmonic vibrations causing the wave, the general wave function in Eq. [15.3] takes a specific form for the displacement of the gas element D:

$$D = A\sin\left(2\pi f\left(t-\frac{x}{c}\right)\right) = A\sin\left(2\pi\left(ft-\frac{x}{\lambda}\right)\right). \qquad [15.8]$$

in which A is the amplitude, f the frequency, λ the wavelength, and c the speed of sound. The negative sign in the argument of the sine function is due to Eq. [15.3]. Note that we have inverted the argument of Eq. [15.3]: $x - ct$ is now written as $ct - x$. This follows standard notation in the literature and is mathematically equivalent because of the symmetry of the sine function, $\sin(-\alpha) = -\sin\alpha$.

Eq. [15.8] applies for a wave travelling in the x-direction (plane wave). Other interesting waveforms include spherical waves travelling outward from a point source. For spherical waves the position parameter in the wave equation is the radius r, replacing x in Eq. [15.8].

The harmonic wave function is illustrated in Fig. 15.10 at two different times. The top curve shows the harmonic wave function at time $t = 0$ and the bottom curve shows it at time $t = T/4$ when the wave has travelled a quarter of a wavelength; i.e., the wave function has shifted by $\lambda/4$.

The harmonic wave function is often written in a slightly different form in the literature, introducing the **angular frequency** $\omega = 2\pi f$ and the **wave number** $\kappa = 2\pi/\lambda$. With these two parameters, Eq. [15.8] reads:

$$D = A\sin(\omega t - \kappa x). \qquad [15.9]$$

Fig. 15.8 indicates that the pressure in the gas element varies in the same fashion as the displacement of the gas element. This remains true for harmonic waves, leading to another way in which this wave function can be written:

$$\Delta p = \Delta p_{max}\cos(\omega t - \kappa x), \qquad [15.10]$$

in which the cosine function is used to ensure that maximum displacement and pressure maximums do not occur at the respectively same time and position. The maximum pressure variation, Δp_{max}, is linearly related to the maximum displacement (amplitude A) in Eq. [15.9]:

$$\Delta p_{max} = (c\rho\omega)A, \qquad [15.11]$$

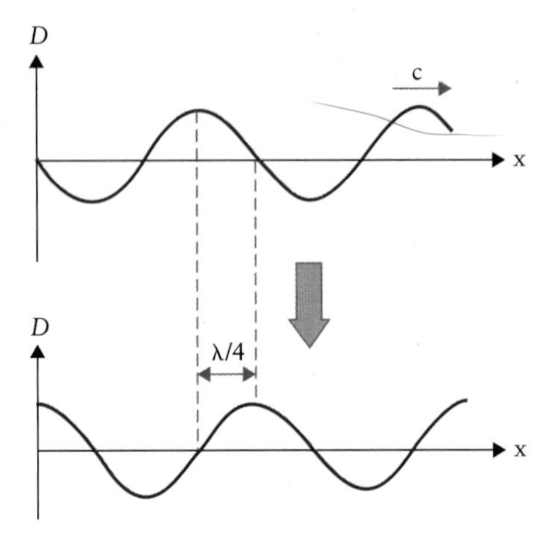

Figure 15.10 Two harmonic waves. The top shows the harmonic wave at time $t = 0$ and the bottom shows it a quarter of a period later ($t = T/4$). During this time the wave has travelled toward the right by a quarter of a wavelength, $\lambda/4$ (dashed interval).

in which c is the speed of sound and ρ is the density of the medium. It is always useful to check the units for new equations, particularly when we do not derive them in the text. In Eq. [15.11], the units on the right-hand side are m/s for the speed of sound, kg/m^3 for the density of the medium, rad/s for the angular frequency, and m for the amplitude. Neglecting the mathematical unit rad, these units combine to kg/(ms^2) = N/m^2 = Pa, i.e., the unit of pressure. This unit is consistent with the term on the left-hand side of Eq. [15.11].

EXAMPLE 15.5

The equation of a wave travelling along the x-axis is given as:

$$D = (2.0\text{ cm})\sin\{(1.0\text{ s}^{-1})t - (1.5\text{ cm}^{-1})x\},$$

in which D is the displacement, given in unit cm. Determine (a) the amplitude, (b) the wavelength, (c) the angular frequency, (d) the period, (e) the frequency, and (f) the travelling speed of the wave. (g) Draw two sketches for the wave: D versus x at time $t = 0$, and D versus t at the position $x = 0$.

Solution

Solution to part (a): We compare the specific wave equation in this example with the generic harmonic wave equation in Eq. [15.8]. The amplitude is the prefactor of the *sine* function, i.e., $A = 2.0$ cm. The units of A and D must be the same since the argument of the sine term cannot carry a physical unit.

Solution to part (b): The wave number is read from the given wave equation in the example text as $\kappa = 1.5\text{ cm}^{-1}$. With the definition of the wave number, $\kappa = 2\pi/\lambda$, we find the wavelength:

$$\lambda = \frac{2\pi}{\kappa} = \frac{2\pi}{1.5\text{ cm}^{-1}} = 4.2\text{ cm}.$$

Solution to part (c): The angular frequency is also read directly from the given wave equation as $\omega = 1.0$ rad/s.

Solution to part (d): The period follows from ω:

$$T = \frac{2\pi}{\omega} = \frac{2\pi}{1.0\frac{\text{rad}}{\text{s}}} = 6.3\text{ s}.$$

Solution to part (e): The frequency is inversely proportional to the period:

$$f = \frac{1}{T} = \frac{1}{6.3\text{ s}} = 0.16\text{ Hz}.$$

continued

Solution to part (f): The travelling speed of the wave follows from Eq. [15.2]:

$$v_{wave} = \lambda f = (0.042\ \text{m})(0.16\ \text{Hz})$$
$$= 6.7 \times 10^{-3}\ \frac{\text{m}}{\text{s}} = 0.67 \frac{\text{cm}}{\text{s}}.$$

Because of the negative sign in the argument of the sine function in the wave function of this example, the wave travels with 0.67 cm/s in the positive x-direction.

Solution to part (g): The two sketches are shown in Fig. 15.11. Note that the following trigonometric identity holds: sin(–1.5x) = –sin(1.5x).

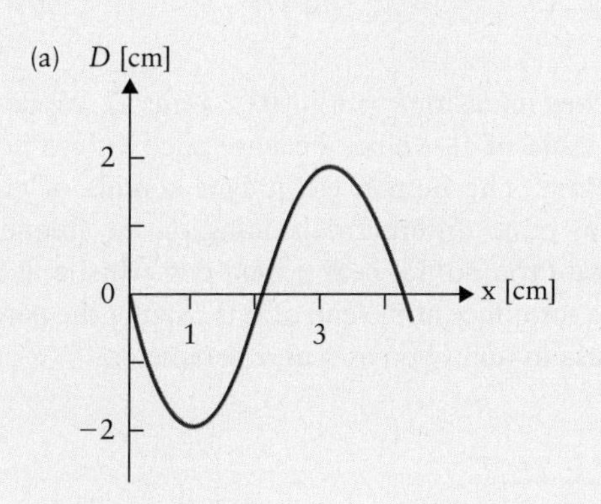

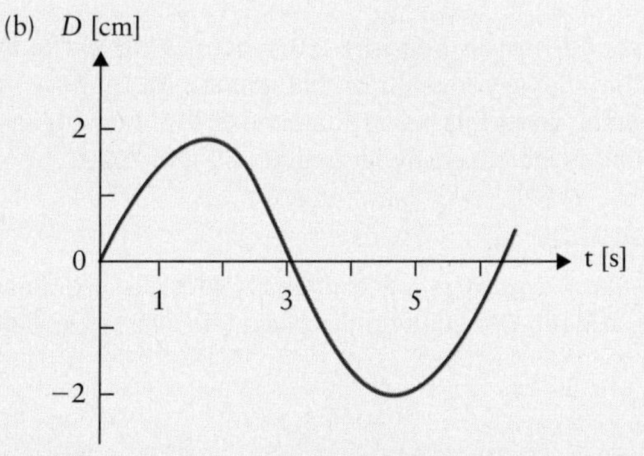

Figure 15.11 Displacement function for a given wave. (a) The displacement as a function of position along the x-axis at time $t = 0$; (b) the displacement as a function of time at the origin $x = 0$.

EXAMPLE 15.6

The maximum pressure variation that is acceptable for the human ear is about 30 Pa. What is the amplitude of a gas element in this case at 20°C when the sound source emits a frequency of 3.0 kHz, which is a frequency in the range where the ear is most sensitive?

Solution

We use the air density we calculated in Example 15.4. From Eq. [15.11], we find:

$$A = \frac{\Delta p_{max}}{c\rho\omega}$$
$$= \frac{30\ \text{Pa}}{\left(343.2 \frac{\text{m}}{\text{s}}\right)\left(1.205 \frac{\text{kg}}{\text{m}^3}\right) 2\pi\,(3000\ \text{Hz})}$$
$$= 3.85 \times 10^{-6}\ \text{m};$$

i.e., the maximum amplitude tolerated is 3.85 μm. Note that both the pressure variations and the amplitude are comparably small values. The maximum pressure variation used in this example corresponds to less than 0.03% of the normal air pressure (1 atm), and the amplitude of 3.85 μm is a distance we cannot see with the naked eye (the size range of typical bacteria is 1 μm to 2 μm).

15.3.5: Sound Intensity

Intensity is a measure of the amount of energy transported by a wave per time interval Δt through a plane of unit area that is placed perpendicular to the wave's propagation direction. To evaluate the intensity of a sound wave, we first have to establish the total energy carried in a wave. The total energy is the sum of the kinetic and potential energies of the local vibration of its small gas elements:

$$E_{total} = \frac{1}{2} m v^2 + \frac{1}{2} k D^2, \qquad [15.12]$$

in which v is the speed at which the gas element moves when displaced by D. We quantify the total energy in analogy to the calculation we used in Chapter 14 to determine the total energy of a vibrating system. We found that the total energy can be expressed as the elastic potential energy at the point of maximum displacement (amplitude) or, alternatively, as the kinetic energy at the maximum speed, i.e., when the object passes through its equilibrium position. The same applies to a vibrating gas element. We use the instant when the gas element passes through its equilibrium position and express the total energy as the kinetic energy at that instant. From Chapter 14 we know that $v_{max} = \omega A$ and therefore can write the kinetic energy as:

$$E_{kin,max} = \frac{1}{2} m\omega^2 A^2,$$

where m is the mass of the vibrating gas element. We want to circumvent the need to identify this mass, because the gas element size is arbitrary. We therefore replace m with the density of the gas, ρ, multiplied by the volume of the gas element, V:

$$E_{\text{total}} = \frac{1}{2}\rho V A^2 \omega^2.$$

By dividing both sides by the arbitrary volume of the gas element, we introduce the **energy density** $\varepsilon_{\text{total}}$ with unit J/m^3:

$$\frac{E_{\text{total}}}{V} = \varepsilon_{\text{total}} = \frac{1}{2}\rho A^2 \omega^2. \qquad [15.13]$$

Introducing the energy density relieves us therefore of the problem of specifying what we mean by a small gas element: the right-hand side of Eq. [15.13] no longer contains arbitrary parameters such as V.

KEY POINT

The energy density of a sound wave is proportional to the square of the amplitude, $\varepsilon_{total} \propto A^2$.

The energy density travels with speed c in a medium carrying sound. Thus, the **intensity** I is given as:

$$I = c\varepsilon_{\text{total}} = \frac{1}{2} c\rho A^2 \omega^2, \qquad [15.14]$$

with the unit for the intensity $J/(m^2 s)$.

KEY POINT

The intensity of a wave is the amount of energy passing through a plane of unit area perpendicular to the propagation direction of the wave. It is proportional to the square of the amplitude.

Eq. [15.11] allows us to relate the intensity to the pressure variation: the intensity of the sound wave is proportional to the square of the maximum pressure variation; i.e., $I \propto (\Delta p_{\text{max}})^2$. This result is useful when we quantify the sound intensity as judged by the human ear. This relation also allows us to convert sound intensities to sound pressure variations, which is important since both quantities are used in the literature.

The most commonly used unit system for sound pressure variations and intensity is based on a logarithmic scale, because both parameters vary widely—e.g., between a whisper and a running jet engine—by about seven orders of magnitude (a factor of 10^7). The pressure variation in a sound wave is defined as **sound pressure level (SPL)**:

$$\text{SPL} = 20 \log_{10} \frac{p}{p_0}, \qquad [15.15]$$

with p_0 a constant chosen as $p_0 = 2 \times 10^{-5}$ Pa. In Eq. [15.15] the notation $\log_{10}$ indicates the logarithm with the base 10 (not the natural logarithm with the base e). The reference pressure p_0 is chosen near the faintest detectable sound for the human ear. The unit of SPL is called **decibel (dB)**. The prefactor 20 is chosen arbitrarily, except that a factor of 2 is included to accommodate the difference to the **intensity level (IL)**. The relation $I \propto \Delta p^2$ leads to $2 \log_{10}(p/p_0) = \log_{10}(I/I_0)$. Thus:

$$\text{IL} = 10 \log_{10} \frac{I}{I_0}, \qquad [15.16]$$

with reference intensity $I_0 = 1 \times 10^{-12}$ $J/(m^2 s)$. We do not provide a table of IL values because such values could be misleading. The human ear judges sounds of equal IL values as quite different depending on the frequency of the sound (you do not hear a loud dog whistle at all). Instead, we introduce at the end of this chapter the parameter loudness in unit phon as a new parameter.

EXAMPLE 15.7

A sound has an intensity 5.0×10^{-7} $J/(m^2 s)$. (a) What is the intensity level IL of this sound? (b) By how much decibel does this value increase if the intensity of the sound is increased by factors of 100 and 1000?

Solution

Solution to part (a): The intensity level IL is defined in Eq. [15.16]. Substituting the given intensity value yields:

$$\text{IL} = 10 \log_{10} \left(\frac{5.0\times10^{-7} \frac{J}{m^2 s}}{1.0\times10^{-12} \frac{J}{m^2 s}} \right)$$
$$= 57 \text{ dB}.$$

Solution to part (b): We look again at Eq. [15.16], and write it for the difference of two sound intensities I_1 and I_2:

$$\Delta\text{IL} = 10 \log_{10}\left(\frac{I_2}{I_0}\right) - 10 \log_{10}\left(\frac{I_1}{I_0}\right) = 10 \log_{10}\left(\frac{I_2}{I_1}\right).$$

Substituting $I_2 = 100\, I_1$ yields a difference in intensity level of $\Delta\text{IL} = 20$ dB; i.e., when the sound intensity increases by a factor of 100 it is reported as an increase of 20 dB. For $I_2 = 1000\, I_1$ in turn we find $\Delta\text{IL} = 30$ dB.

CASE STUDY 15.5

How does the sound intensity change with distance from a point sound source?

Answer: *Fig. 15.12 shows a point sound source. Two concentric spherical surfaces with areas A_1 and A_2 are drawn with the point-like source at the centre. We know that the same amount of energy must flow through each of these surfaces per time unit to satisfy energy conservation, $\Delta E/\Delta t$ = const. Sound would have to pile up or disappear between the two spheres were this not the case. Thus, for Fig. 15.12 we specifically write:*

$$\left(\frac{\Delta E}{\Delta t}\right)_{A_1} = \left(\frac{\Delta E}{\Delta t}\right)_{A_2} \qquad [15.17]$$

In Eq. [15.20] we replace the time change of the total energy with the intensity. Starting from the first formula in Eq. [15.14], we write:

$$I = c\varepsilon_{\text{total}} = c\frac{\Delta E}{V} = \frac{\Delta r}{\Delta t}\frac{\Delta E}{A\Delta r} = \frac{\Delta E/\Delta t}{A},$$

in which the speed of sound is expressed as the change of the radial position with time, $c = \Delta r/\Delta t$, and in which the volume containing the sound energy is calculated as the area multiplied by the width, $V = A\,\Delta r$. This volume is the volume of a spherical shell of thickness Δr. Thus, the intensity is the rate of change of the total energy per unit area. Substituting this in Eq. [15.20] yields:

$$I_1 A_1 = I_2 A_2,$$

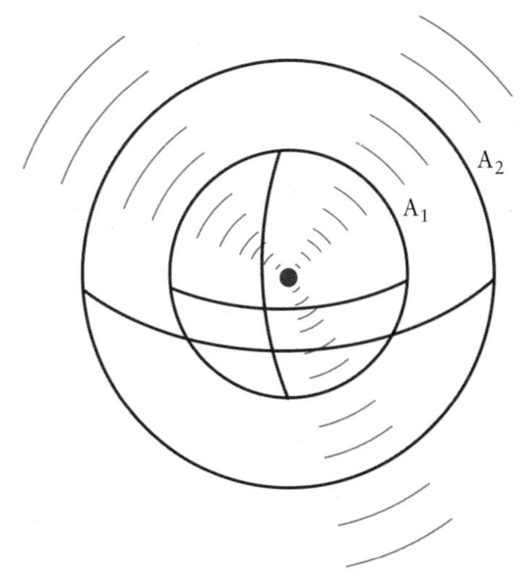

Figure 15.12 A point sound source (blue dot) shown at the centre of two concentric spherical surfaces with areas A_1 and A_2. The sound intensity per unit area, travelling through the two surfaces, diminishes as their areas increase.

which provides us with the intensity ratio:

$$\frac{I_2}{I_1} = \frac{A_1}{A_2} = \frac{4\pi r_1^2}{4\pi r_2^2}. \qquad [15.18]$$

We note that $I \propto 1/r^2$; i.e., the intensity diminishes in proportion to the square of the distance from the sound source.

The reduction of sound intensity in Case Study 15.5 is a purely geometric effect. The sound travelling through a medium can further diminish due to energy loss to the medium when the vibration of the gas elements is not perfectly harmonic. The maximum sound pressure level produced by an animal is 188 dB for the low-frequency moans of the blue whale. These sounds can be heard by other whales for hundreds of kilometres.

15.4: Waves in a Confined Medium

The previous two sections allow us to understand how sound travels from a sound source through air. However, the concepts introduced are not sufficient for us to understand how the human voice operates, or how the human ear detects sound. The most important issue not included up to now is a spatial confinement of the sound wave. Sound waves are confined during the hearing process, as illustrated by the anatomic overview of the ear in Fig. 15.13. The auditory canal (1) resembles a cylindrical tube, allowing one-dimensional waves to travel inside. However, the auditory canal is a half-closed tube and sound cannot travel farther than the eardrum, which separates the outer ear from the middle ear. The current section focuses on waves that are confined in either closed or half-closed tubes. Like in the previous sections, we assume that the absorption of sound by the medium is negligible.

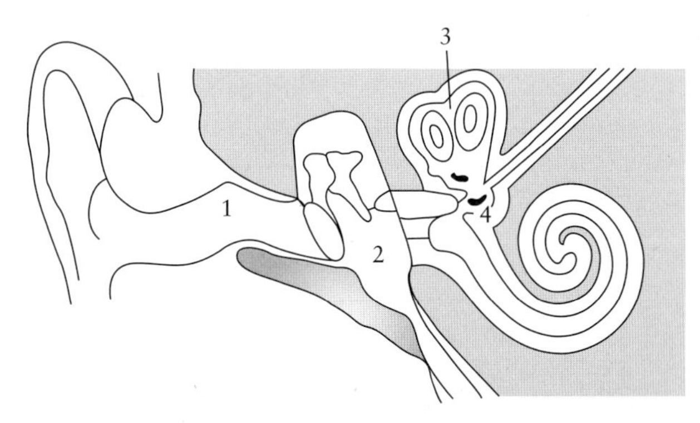

Figure 15.13 Overview of the human ear. We can distinguish three main sections of the ear: the outer ear with the auditory canal (1) ending at the eardrum; the middle ear with the three ossicles: hammer, anvil, and stirrup (from left; 2); and the inner ear with the vestibular organ. The vestibular organ includes the semicircular canals (3) and the maculae (4), both discussed in earlier chapters.

The study of a longitudinal wave in a confined space begins with a simple model system: a cylindrical tube filled with an ideal gas and closed at both ends. We use **closed tubes** to introduce the most important features of confined waves, such as standing waves, harmonics, and the concept of resonance.

15.4.1: Standing Waves

The biggest difference between waves in open space and waves in a confined medium is the presence of reflected waves. The easiest case is the closed tube, where waves reflect at both ends and travel back and forth inside the tube. We treat a wave and the corresponding reflected wave as two separate waves. This is necessary because each of these waves is described by a different wave function. Thus, the first issue to tackle in a discussion that leads toward wave phenomena in a closed tube is the superposition of two waves. This discussion is actually simplified when we choose a wave and its reflection because several wave parameters are identical for both waves: the wavelength or wave number and the angular frequency do not change at the time of reflection.

To see what happens with more than one wave present, consider two waves travelling in a tube in the same direction with the same wavelength and angular frequency. To study how these waves interact, their respective wave functions have to be written in a more complete form than that given in Eq. [15.9]:

$$D_1 = A\sin(\omega t - \kappa x + \phi_1)$$
$$D_2 = A\sin(\omega - \kappa x + \phi_2). \quad [15.19]$$

Why does the inclusion of the terms ϕ_1 and ϕ_2 in the argument make the wave function more general? To see this, assume that we study the waves at the particular position $x = 0$. Fig. 15.14 shows the waves with $\phi_1 = 0$ (top) and $\phi_2 = \pi/2$ (bottom). We see that the term ϕ

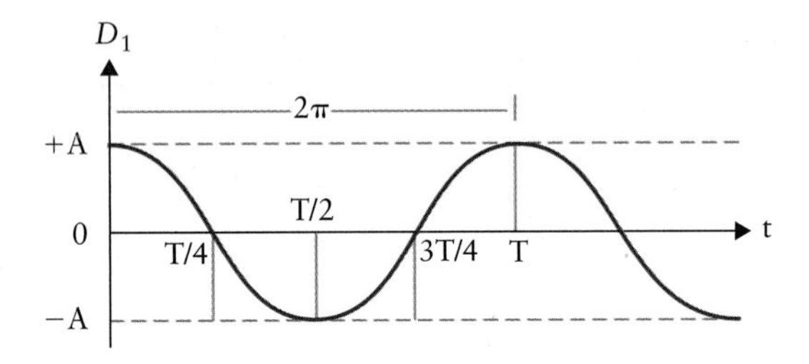

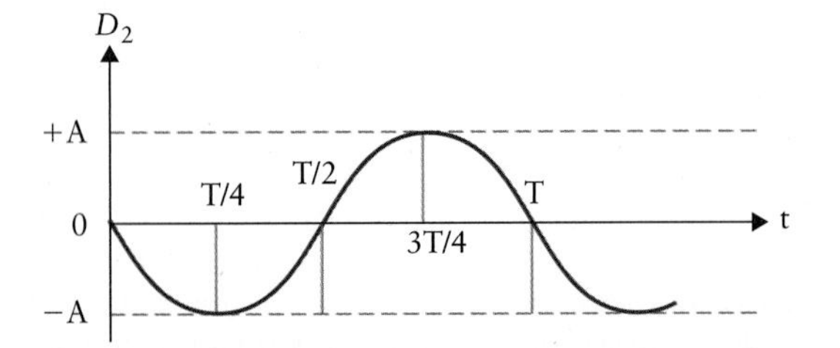

Figure 15.14 Two sinusoidal waves at $x = 0$ that differ by a phase shift of $\Delta\phi = \pi/2$.

in Eq. [15.19] is a phase shift because the two waves are different only in that the lower wave is shifted to the left by a time difference of $\Delta t = T/4$. We refer to this observation in Eq. [15.19] by introducing $\Delta\phi = \phi_2 - \phi_1$, which we call a **phase angle difference** between the two waves.

As long as the amplitudes are not too large, the principle of **additive superposition** of waves is valid. This simply means that the superposition of the two waves is the sum of their respective displacements at each point along the tube. This leads to two cases of particular interest for the waves in Eq. [15.19]:

- When the two waves are shifted relative to each other by half a wavelength—i.e., when the phase angle difference is $\phi_2 - \phi_1 = (2\ n + 1)\pi$, with n an integer number—**destructive superposition** occurs and the resulting wave has a vanishing amplitude, $A_{sup} = 0$. This is illustrated in Fig. 15.15(a), where the two initial waves are indicated by a dashed and a dash-dotted line. The superpositioned wave is coincident with the x-axis (no wave).
- When the two waves are shifted by a multiple of a full wavelength, which corresponds to $\phi_2 - \phi_1 = (2\ n)\ \pi$, this called **constructive superposition** and is shown in Fig. 15.15(b). Again, the dashed and dash-dotted curves are the original waves, drawn beside each other only to allow them both to be seen. The superposition of both waves yields a wave with double the amplitude (solid red curve).

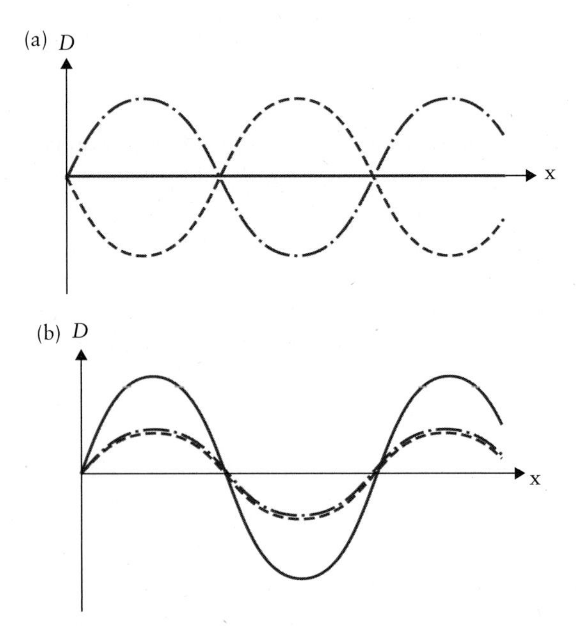

Figure 15.15 (a) Destructive superposition of two harmonic waves with same angular frequency, wave number, and amplitude. This occurs for a half-wavelength phase shift between the two waves. The two original waves are shown as dashed and dash-dotted curves. The resulting wave coincides with the x-axis. (b) Constructive superposition of two harmonic waves with same angular frequency, wave number, and amplitude. This occurs for a phase shift equal to a multiple of a full wavelength between the two waves. The two original waves are shown as dashed and dash-dotted curves. The resulting wave is shown as a solid red curve with double the amplitude.

Harmonic waves travelling back and forth in a closed tube are a special application of the superposition principle since the reflection at the end of the tube automatically guarantees that both waves have the same amplitude (because of the absence of sound absorption), angular frequency, and wave number because the initial and the reflected wave are caused by the same vibration. In this case, the two waves are written as:

$$D_1 = A\sin(\omega t - \kappa x)$$
$$D_2 = -A\sin(\omega t + \kappa x). \qquad [15.20]$$

Note that the reflected wave carries an additional negative sign. This results because the wave has a phase shift of $\Delta\phi = \pi$ when reflected off a wall. Fig. 15.16 illustrates how this arrangement can lead to a **standing wave**. Fig. 15.16(a) shows a sound source at the right end of an enclosed tube filled with a gas. The sound waves moving back and forth coincide such that the amplitude of the local vibration of a small gas element becomes time independent. A standing wave is generated if the distance between the sound source and the wall of the tube at the left end is a multiple of half a wavelength.

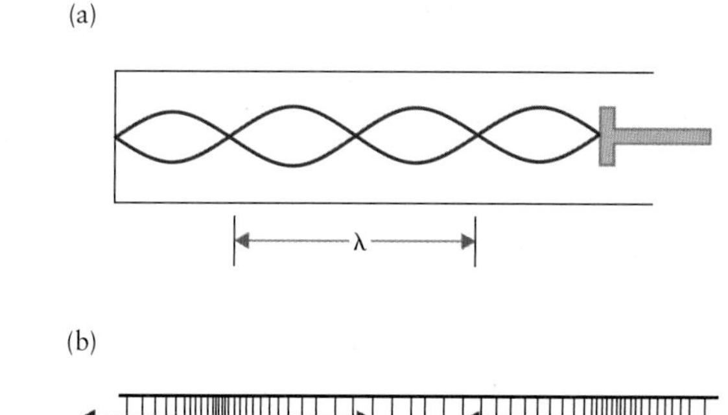

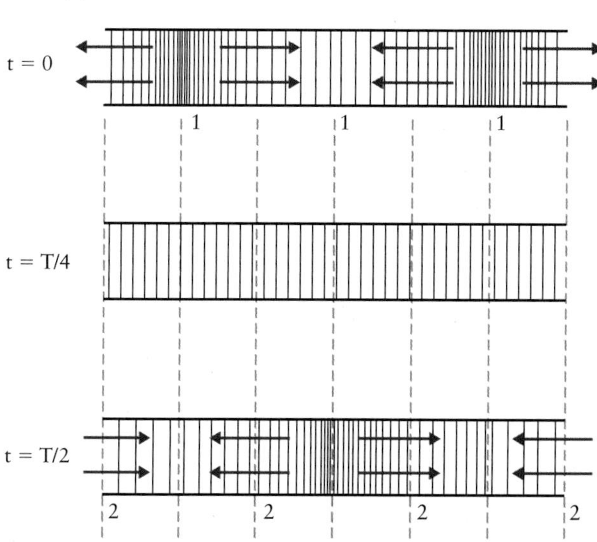

Figure 15.16 Standing wave in a closed, air-filled tube. (a) A mobile piston varies the length of the tube. A standing wave occurs when the distance between the piston and the end of the tube is a multiple of a half-wavelength of the sound wave. (b) The air density profile in the tube at three different times. Note that a standing wave does not travel toward the left or the right. Amplitude nodes (1) and pressure or density nodes (2) remain stationary.

We can also derive this result mathematically from Eq. [15.20]. Additive superposition of the two waves leads to

$$D_{superposition} = A\,[\sin(\omega t - \kappa x) - \sin(\omega t + \kappa x)].$$

We use the trigonometric relation $\sin(\alpha - \beta) - \sin(\alpha + \beta) = -2\cos\alpha\sin\beta$, to simplify this equation:

$$D_{superposition} = -\{2A\cos(\omega t)\}\sin(\kappa x). \qquad [15.21]$$

This is a wave with a fixed wavelength of $\lambda = 2\pi/\kappa$ and an amplitude that varies with time as given in the curly braces in the equation. The maximum amplitude is $2A$.

We want to interpret Eq. [15.21] with a microscopic picture. Fig. 15.16(b) shows a section of the tube and the respective motion of the gas elements. In a standing wave, certain points have a zero amplitude all the time. These points are labelled 2 in the three snapshots of the gas in the tube, i.e., at times $t = 0$, $t = T/4$, and $t = T/2$. They are called **pressure nodes**. Between two pressure nodes the gas pressure and density oscillate between an increased value, indicated by a higher density of lines, and a decreased value, indicated by a lesser density of lines.

The points labelled 1 in Fig. 15.16(b) are **velocity nodes** since the gas element at those points does not move (motion of gas is indicated by arrows). Velocity nodes are alternatively called amplitude nodes since they identify gas elements with $A = 0$. Such a node must be located at the end of the tube in Fig. 15.16(a), since the gas cannot move into the wall. Thus, no pressure node occurs at the end of the tube. This is important in the ear because pressure variations cause the eardrum to vibrate and thereby transmit sound to the middle ear.

KEY POINT

A standing wave forms in a closed tube when the wavelength of the sound and the distance to the end wall in the tube allow for a velocity node at that wall.

CONCEPT QUESTION 15.2

What is the meaning of "standing" in the term standing wave?

(A) The amplitude is time independent.

(B) The frequency is time independent.

(C) The location of density nodes is time independent.

(D) The wavelength is time independent.

EXAMPLE 15.8

Consider again the wave travelling along the *x*-axis given in Example 15.5:

$$D = (2.0\ \text{cm})\sin\{(1.0\ \text{s}^{-1})t - (1.5\ \text{cm}^{-1})x\},$$

in which *D* is the displacement, given in unit cm. Assume this wave reflects off a rigid surface. (a) What is the equation for the reflected wave? (b) What is the equation of a resulting standing wave? (c) Sketch the standing wave at $t = 0$ and $t = T/2$, with *T* being the period.

Solution

Solution to part (a): The reflected wave has exactly the same properties as the incoming wave, except that (i) it travels in the opposite direction and (ii) it is phase-shifted by half a wavelength (i.e., a phase shift of π radians). This is achieved with the wave equation in this example by (i) switching the sign of the position-dependent term and by (ii) switching the sign of the amplitude term in the same fashion as in Eq. [15.20]:

$$D_{\text{reflect}} = -(2.0\ \text{cm})\sin\{(1.0\ \text{s}^{-1})t + (1.5\ \text{cm}^{-1})x\}.$$

Solution to part (b): The standing wave is the superposition of the wave given in the example text and the one that resulted in part (a). We need not repeat the mathematical steps leading to the resulting standing wave because the calculation has already been done from Eq. [15.20] to Eq. [15.21]. Inserting the specific values from the two wave equations in this example, we find:

$$D_{\text{superposition}} = -\{(4.0\ \text{cm})\cos(1.0\text{s}^{-1})t\} \times \sin(1.5\ \text{cm}^{-1})x.$$

Solution to part (c): The two sketches are shown in Fig. 15.17, with the solid curve for $t = 0$ and the dashed curve for $t = T/2$.

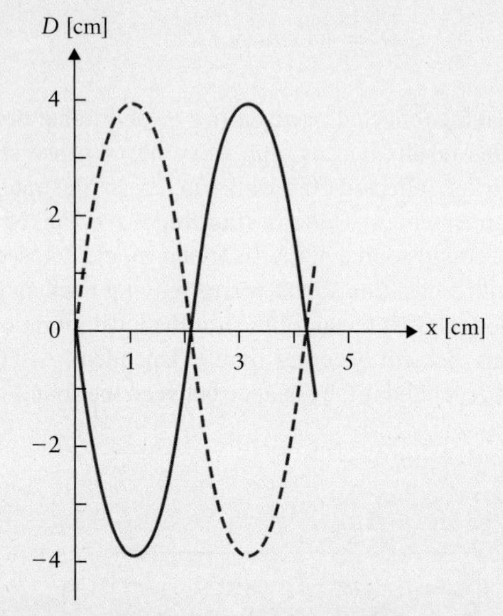

Figure 15.17 Sketch of a standing wave. The solid curve represents it at time $t = 0$; the dashed curve represents the same wave at time $t = T/2$, i.e., half a period later.

15.4.2: Harmonics

In free space, waves may occur with any combination of wavelengths and angular frequencies consistent with the speed of sound in the medium. In a confined space, only certain waves, the standing waves, can be sustained. Their selection rule is based on the size and type of the confining space: for longitudinal waves in a closed tube a multiple of half-wavelengths must fit into the space of the tube. However, any integer number of half-wavelengths is acceptable, and thus several standing waves with different wavelengths can form in the same closed tube.

Fig. 15.18 shows the three longitudinal waves with the longest wavelengths between the ends of the tube. The standing wave in part (a) is called the **first harmonic**, which has the lowest frequency that the tube can support. The standing waves in parts (b) and (c) are the second and third harmonics respectively. Note that the harmonic of *n*-th order has two amplitude nodes at the fixed ends of the tube, and $n - 1$ amplitude nodes between the ends at equal distances. The term *harmonic* is derived from the use of the term in music. It is applied in the current context despite the fact that we are not studying a string

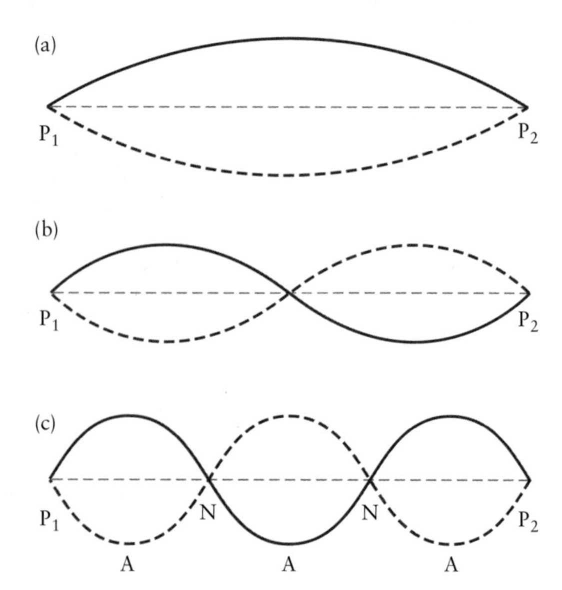

Figure 15.18 (a) The first, (b) second, and (c) third harmonics for a standing longitudinal wave in a closed tube between P_1 and P_2. N identifies amplitude nodes and A anti-nodes. The vertical deviation from the horizontal line is a measure of the longitudinal displacement of the gas element.

on a guitar because a string on a musical instrument must be set into transverse motion.

Table 15.2 lists the wavelengths and frequencies for the various longitudinal harmonics for a one-dimensional closed tube. The frequencies and wavelengths are related to each other by the speed of sound, $c = \lambda_n f_n$. As the table illustrates, higher harmonics have shorter wavelengths and higher frequencies but the same wave speed.

TABLE 15.2

Wavelengths and frequencies for various harmonics of air in a tube closed at both ends. *n* is an integer number.

Mode	Wavelength	Frequency
1st harmonic	$\lambda_1 = 2\,L$	$f_1 = c/(2\,L)$
2nd harmonic	$\lambda_2 = L$	$f_2 = c/L = 2\,f_1$
3rd harmonic	$\lambda_3 = 2\,L/3$	$f_3 = 3\,c/(2\,L) = 3\,f_1$
⋮	⋮	⋮
n-th harmonic	$\lambda_n = 2\,L/n$	$f_n = n\,c/(2\,L) = n\,f_1$

For applications, we have to supplement the discussion of closed tubes with a second system in which a confined gas can sustain a longitudinal wave: a **half-open tube**. The half-open tube is, e.g., a model for the outer ear in Fig. 15.13. The waves sustained in a half-open tube differ from the harmonic waves of the closed tube because the air can move back and forth at the open end of the tube. Indeed, a half-open tube has an amplitude anti-node at its open end; i.e., a standing wave forms such that a pressure node instead of an amplitude node lies in the open end. This is illustrated in Fig. 15.19, where

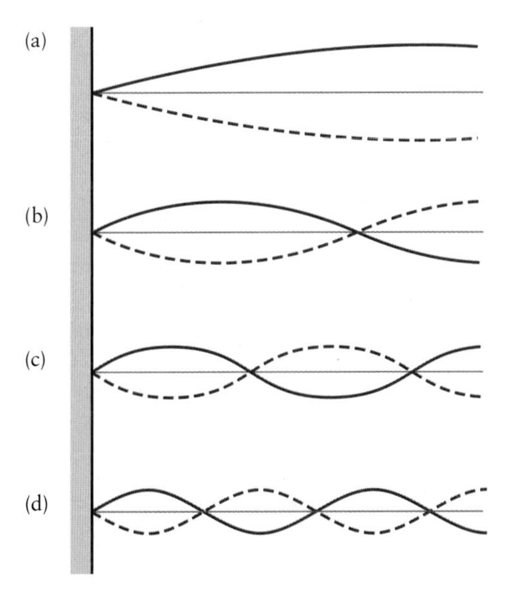

Figure 15.19 (a) The first harmonic and (b–d) the corresponding higher harmonics in a tube open at one end. Note that an amplitude anti-node (a pressure node) forms at the open end (right end of red curves). The vertical deviation from the horizontal line in the figure indicates the longitudinal displacement of a gas element in the tube.

the open end of the tube lies at the right end of the plot. Fig. 15.19(a) shows the first harmonic, and Figs. 15.19(b–d) illustrate the three next-lowest harmonics. For the half-closed tube the frequencies of the allowed harmonics are given by:

$$f = \frac{nc}{4L} \quad \text{with } n = 1, 3, 5, 7, \ldots ; \qquad [15.22]$$

i.e., only odd-numbered harmonics are possible.

KEY POINT

Closed and half-closed tubes can sustain standing waves, which we call harmonics.

CASE STUDY 15.6

Fig. 15.20(a) shows a blue whale and Fig. 15.20(b) shows the intensity spectrum of its moan. (a) What model would you use to describe the technique of sound generation by the blue whale: an air column closed at both ends, or a half-open air column? (b) What is the first harmonic of the blue whale? What is the highest harmonic shown in Fig. 15.20?

Biological information: Blue whales generate two types of sound: the low-frequency moans shown in Fig. 15.20 and the high-frequency click sounds, which are in the range between 21 kHz and 31 kHz. The moans occur in the range of 12.5 Hz to 200 Hz, with the greatest intensities between 20 Hz and 32 Hz. The sound duration lies usually between 15 seconds and 40 seconds; however, with more careful measurements more detailed features can be identified than are revealed in Fig. 15.20, including amplitude modulations at 0.26-second repetition cycles and, later during the sound, with 0.13-second repetition cycles.

Answer to Part (a): *Table 15.2 and Fig. 15.18 show that the frequencies of the various harmonics of a closed tube are all equally spaced:*

$$\Delta f = f_{n+1} - f_n = \frac{c}{2L} = \text{const},$$

which applies for all values of n ≥ 1. This frequency spacing is equal to the absolute frequency of the first harmonic:

$$\text{closed:} \quad f_1 = \Delta f.$$

On the other hand, Eq. [15.22] and Fig. 15.19 show that the frequencies of the various harmonics of a half-closed tube, while also spaced as for the closed system, are twice as wide-spaced as the value of the first harmonic:

$$\text{half-closed:} \quad 2f_1 = \Delta f.$$

continued

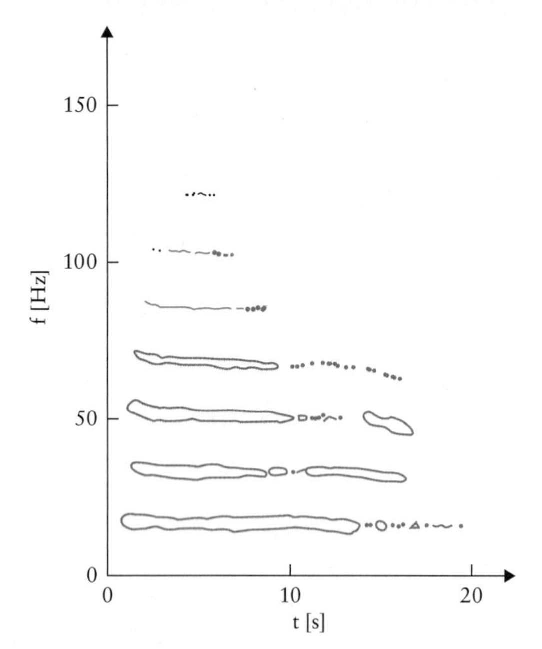

Figure 15.20 (a) Blue whale. (b) Spectrum of the sound intensity emitted by a blue whale. Bands represent frequency ranges of high sound intensity. These low-frequency moans last for 15 to 40 seconds.

The frequencies shown in Fig. 15.20 are 20 Hz, 34 Hz, 51 Hz, 68 Hz, 85 Hz, 102 Hz, and 119 Hz. Thus, comparing the first harmonic with the frequency separation of higher harmonics we conclude that a closed-tube model describes the moans of the blue whale best.

Answer to Part (b): *The first harmonic is the lowest frequency emitted. In the case of the moan of the blue whale shown in Fig. 15.20, this frequency is 20 Hz. Every higher-frequency band shown indicates the next-higher harmonic according to Table 15.2. Thus, the highest harmonic shown at 119 Hz is the seventh harmonic.*

Higher harmonics play an essential role in the human voice. Human voices are characterized by the number and relative amplitude (or intensity) of higher harmonics generated. The first harmonic in a human voice is defined as the **pitch** and the higher harmonics define the **timbre**.

15.5: Resonance

In the previous two sections we deviated from our earlier approach and derived the concepts of sound in a confined space from the wave model and not from the vibrations that cause the wave. For the description of resonances, in turn, it is useful to return to our original approach and link the vibration of an object to the sound phenomenon.

If you consider the human voice, or, more simply, a musical instrument such as a flute, you notice that neither generates a sound unless a mechanical excitation is applied. In the case of the human voice, you sense the effort involved in speaking. In both cases, the voice and the flute, the external excitation is usually not a vibration with a frequency perfectly matching the frequency of a standing wave in the adjacent air column. This is obvious for the flute when you quickly blow into the mouthpiece. In this case, you obviously do not provide a harmonic vibration at all. But even if you tried to vibrate the piston in Fig. 15.16, it is unlikely that you would do so with the right frequency for a standing wave. Still, the flute responds to the blowing with a sound, representing the first and several higher harmonics characteristic of the length of the flute's barrel. In the same fashion, the human voice has a characteristic frequency pattern.

To understand the response of a flute or of the vocal tract, we need to study two phenomena:

- the sound amplitude in a closed or half-closed tube in response to an external excitation, and
- the reason why a sudden excitation as much as a harmonic excitation causes harmonic waves.

We start with Fig. 15.16 and assume an external force is used to move the piston harmonically back and forth with a maximum force of magnitude F_{max} and an angular frequency ω_{ext}:

$$F_{ext,x} = F_{max} \cos(\omega_{ext} t).$$

Note that we need not consider the vector character of the force, as the entire experiment is done one-dimensionally along the x-axis. We will find that such an externally caused vibration can lead to high sound intensities under certain conditions.

We use this external force to extend Newton's second law for a vibrating object, as written in Eq. [14.7], to describe the piston displacement in response to the external force:

$$-kx + F_{max} \cos(\omega_{ext} t) = ma_x.$$

The second term on the left-hand side is a newly introduced force acting on the piston. Solving this equation of motion to write the position as a function of time, $x = f(t)$, is not particularly easy; instead, we choose an approach based on an experiment: an object is attached to a horizontal spring. A taut string is also attached to the object, which allows us to exert the external force by moving the string continuously back and forth. Depending on the frequency with which we move the string, the object's response will be somewhere between little and very strong. We express the response to the motion of the string as the amplitude of the vibration of the object.

The observed amplitude is illustrated in Fig. 15.21. If you move the string very slowly back and forth, the object follows the motion with exactly the same amplitude as the external motion. We define this amplitude as A_0. The plot shows values near $A/A_0 = 1$ for angular frequencies near $\omega_{ext} = 0$, where ω_{ext} describes the angular frequency of the motion of the externally driven string. In this case, the motion of the object is not altered by the spring to which it is attached.

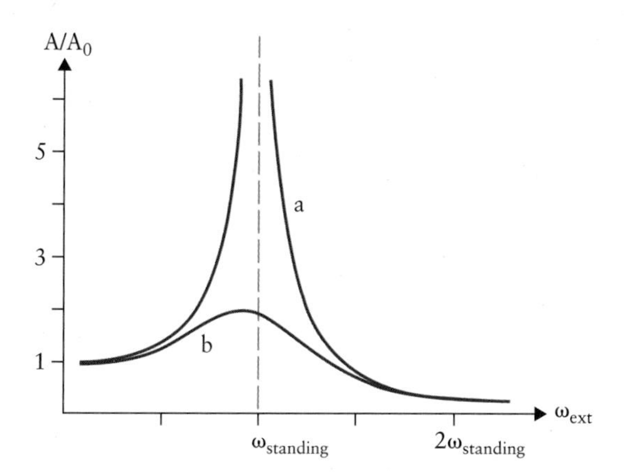

Figure 15.21 The amplitude in a resonant system, shown as a value relative to the amplitude A_0 for small external angular frequencies, ω_{ext}. The system follows the sinusoidal variation of the external force when $\omega_{ext} \ll \omega_{standing}$, with $\omega_{standing}$ the natural angular frequency of the system. When ω_{ext} reaches $\omega_{standing}$, an amplification of the amplitude occurs, which is called a resonant amplification. Curve (a) applies when damping is negligible, and curve (b) when moderate damping occurs.

As we increase the frequency with which we move the string back and forth, we get closer and closer to the natural angular frequency of the object on the spring, which is its angular frequency without an external force:

$$\omega_{natural} = \sqrt{\frac{k}{m}}. \qquad [15.23]$$

Equally, if we consider the case of a piston moving back and forth at the end of an air column, moving the piston faster means that we are getting closer to the angular frequency of the first harmonic of the standing wave forming in the closed gas tube, $\omega_{standing}$. As Fig. 15.21 shows, this leads to an increasing amplitude because the external push transfers energy to the vibrating object (or piston) more and more effectively.

Following curve (a) in Fig. 15.21, an infinite amplitude occurs when the natural frequency of the object or the first harmonic of a closed tube is reached. This is called a **resonance**. An infinite amplitude is not a physically possible result. Real systems follow curves like the one shown in Fig. 15.21(b), with a finite amplitude at the resonance. The change from curve (a) to curve (b) is called **damping**. Damping is the result of non-ideal behaviour, for example friction between the edge of the piston and the inside wall of the tube. In an energy-based picture, damping represents the absorption of mechanical energy and its conversion into non-mechanical energy forms, primarily thermal energy. Curve (b) assumes a moderate damping that results in a finite amplitude oscillation at the resonance. The shock absorbers in your car, for example, have a significantly stronger damping—you do not want increased vertical amplitude for the car's cabin when you are driving over potholes!

The resonance is reached when the external angular frequency equals the natural angular frequency or the angular frequency of the first harmonic of the system. Fig. 15.21 shows that the confined system responds to an external excitation with an enhanced amplitude in the vicinity of the resonance, and therefore with an enhanced sound intensity, since we saw earlier that the intensity is proportional to the square of the amplitude. This phenomenon is called **resonant amplification**.

If we increase the frequency of the external vibration beyond the resonance frequency, we observe a decrease in the amplitude of the object. When the external frequency is much larger than the natural frequency of the object, the amplitude of the object ceases and the object does not respond at all to the external force; this is due to the inertia of the system. In order to follow the motion of the string, the object must constantly accelerate back or forth. Every object requires some time for that: the heavier the object, the slower the response to a force. You can redirect a tennis ball in an instant; redirecting a shot put thrown to you takes longer. With the change of direction of the external force occurring faster and faster, eventually the inertia of the object no longer allows enough time for the object to follow.

The curve labelled (a) in Fig. 15.21 is described quantitatively in the form:

$$A = \frac{F_{max}}{m(\omega_{standing}^2 - \omega_{ext}^2)}, \qquad [15.24]$$

in which F_{max} is the magnitude of the maximum force exerted on the system of mass m, ω_{ext} is the angular frequency of the external force, and $\omega_{standing}$ is the angular frequency of the first harmonic of the system. The difference of quadratic terms in the denominator causes the steep increase close to the resonance. This formula is useful for systems with negligible damping.

KEY POINT

A system is in resonance when an external harmonic excitation causes it to respond with a maximum amplitude. This occurs at the first harmonic of a closed or half-closed tube.

15.6: Hearing

15.6.1: The Outer Ear

The human ear is an extremely sensitive sound detection system. It can analyze sound intensities ranging over 12 orders of magnitude and is able to distinguish frequencies between 16 Hz and 20 kHz.

To achieve this performance, all three components of the ear shown in Fig. 15.13—the outer ear, the middle ear, and the inner ear—are essential. The outer ear is a half-closed tube with a first harmonic that generates a resonant amplification across most of the range of audible frequencies. The mechanical mechanism in the middle ear circumvents a major loss in intensity that would occur if the external medium air were directly coupled to the inner ear's fluid. The inner ear provides frequency analysis that is then encoded in electric signals sent to the brain.

EXAMPLE 15.9

The outer ear consists of the auditory canal and ends at the eardrum. The auditory canal is about 2.5 cm long. What is the first harmonic of the human auditory canal?

Solution

We use Eq. [15.22] for a half-open tube to calculate the first harmonic. We find for $n = 1$:

$$f_1 = \frac{c}{4L} = \frac{343\frac{\text{m}}{\text{s}}}{4(2.5\times10^{-2}\text{m})} 3.4 \text{ kHz}.$$

This value is close to the frequency at which the ear reaches its greatest sensitivity.

Due to the width of the peak of the resonance curves in Fig. 15.21 (the range where $A/A_0 > 1$), an appreciable range of frequencies around the first harmonic is amplified. This range and the amplification factor depend on the damping of the resonance curve. The outer ear amplifies the arriving sound by about a factor of two; i.e., the curve in Fig. 15.21(b) is a good representation of the resonance behaviour of the auditory canal. We benefit from the resonant amplification of the outer ear in the frequency interval from about 2 kHz to 7 kHz.

CASE STUDY 15.7

Prior to the development of modern hearing aids, people with age-related hearing loss used funnel or conical ear trumpets, long tubes they held to their outer ear. These devices improved hearing because of two effects: the sound intensity reaching the eardrum was increased as the result of the large opening of the ear trumpet, and the frequency range of resonant sound amplification in the outer ear was shifted to lower frequencies, which individuals who are hearing impaired usually lose last. Explain the mechanism of the second effect.

Answer: *The argument is based on the result of Example 15.10 and Fig. 15.21(b). Eq. [15.22] shows that the first harmonic is inversely proportional to the length of the half-open tube, L. Extending the outer ear with the ear trumpet causes a lower-resonance frequency for the combined tube. Fig. 15.21(b) illustrates that the largest resonant sound amplification is obtained near the first harmonic of the tube. Thus, lower frequencies are heard better with an ear trumpet.*

Ear trumpets work well and dominated the hearing aid market from the mid-20th century until just a few decades ago. Electronic hearing aid manufacturers were always aware of this low-tech, low-budget competition: a major issue in advertising the more expensive state-of-the-art aids is their virtual invisibility.

15.6.2: The Middle Ear

The middle ear transports the sound signal from the eardrum to the oval window. The anatomical set-up is shown in Fig. 15.2. The vibrations of the oval window (5) cause waves in the **perilymph** (6). Perilymph is a fluid similar to highly filtered blood plasma, comparable to the extracellular fluid. Like other body fluids, perilymph is an electrolyte with a concentration of 14 mmol/L Na^+, but it has only a low concentration of proteins.

The middle ear is physiologically necessary because the sound travels toward a medium of significantly altered density (air to perilymph). It consists of the hammer (2), which is attached to the eardrum (1); the anvil (3); and the stirrup (4), which is attached to the oval window of the cochlea (7).

To understand the purpose of the middle ear, assume for a moment that it did not exist. In this case the eardrum and the oval window would be the same membrane and the sound would have to be coupled across this membrane from air to perilymph. An incoming wave would be partially reflected and partially transmitted. We quantify this situation in the next chapter because it is particularly important in ultrasound imaging. We will find that the efficiency of sound transmission across the membrane for the simplified middle ear would be negligibly small.

So, how does the middle ear circumvent this problem? It transports the vibration of the eardrum as a mechanical vibration to the oval window. This eliminates the intensity loss referred to above. Beyond that, the middle ear provides a moderate amplification of the vibration due to the specific design of the interacting mechanical components. This amplification has two components: a force amplification and a pressure amplification. We calculate each component separately and then combine them to arrive at the overall sound amplification of the middle ear.

15.6.2.1: Force Amplification in the Middle Ear

Fig. 15.22 shows the anatomy of the middle ear of humans. Sound transmission from the eardrum (4) to the oval window (5) is accomplished by **three ossicles**. The ligaments in Fig. 15.23 identify their mechanical mobility. The *ligamentum mallei superius* (1) and the *ligamentum incudis superius* (2) hold the hammer and the anvil in position. The *ligamentum mallei laterale* (3) is responsible for allowing a rotation of the hammer. We identify this last ligament as the fulcrum. Fig. 15.23 is a sketch of the hammer as a mechanical lever-arm system. The eardrum can exert a force $\vec{F}_1$ and the anvil can exert a force $\vec{F}_2$ on the lever arm at distances r_1 and r_2 from the fulcrum respectively. The hammer is in mechanical equilibrium when a torque equilibrium is established, as defined in Chapter 6:

$$\sum_i \tau_i = r_1 F_1 - r_2 F_2 = 0,$$

which yields:

$$F_2 = \frac{r_1}{r_2} F_1.$$

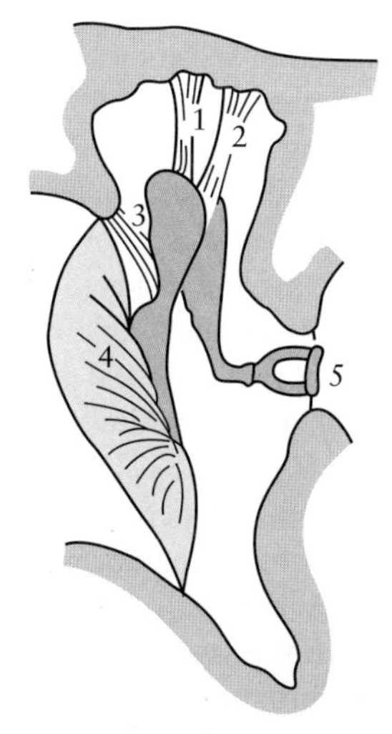

Figure 15.22 Human middle ear anatomy, highlighting the ligaments that stabilize the three ossicles between the eardrum (4) and the oval window (5). *Ligamentum mallei superius* (1) and *Ligamentum incudis superius* (2) hold the hammer and the anvil in position. *Ligamentum mallei laterale* (3) acts as a fulcrum for the rotation of the hammer.

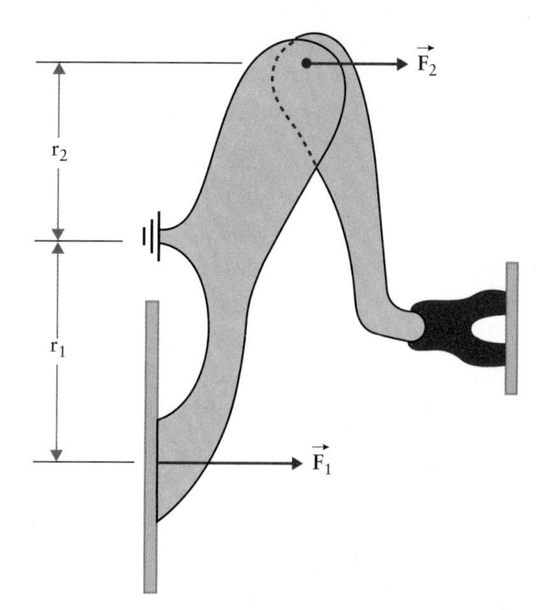

Figure 15.23 Mechanical arrangement of the three ossicles in the middle ear. The hammer is a lever-arm system with a fulcrum at $r_1 = 1.5\ r_2$. In mechanical equilibrium, the torques due to the forces exerted by the eardrum ($\vec{F}_1$) and the anvil ($\vec{F}_2$) must be equal. The stirrup (dark red) connects the anvil to the oval window (green).

With the fulcrum located above the halfway point of the lever arm, i.e., $r_1 = 1.5\ r_2$, we find that the force acting on the oval window is about 1.5 times the force exerted by the eardrum.

15.6.2.2: Pressure Amplification in the Middle Ear

A pressure amplification is the result of the difference in area of eardrum and oval window. The pressure on the eardrum equals the force acting on the eardrum divided by its area. Equally, the pressure at the oval window equals the force on the oval window divided by its area. Anatomically, we find an area for the eardrum of 65 mm^2 (of which, however, as little as 45 mm^2 might be mechanically active) and an area of 3.2 mm^2 for the oval window. Thus, the pressure amplification accounts for a factor of 15 to 20.

Combining force and pressure amplifications, we get:

$$\Delta p_{\text{oval}} = \frac{F_{\text{oval}}}{A_{\text{oval}}} = \frac{1.5\ F_{\text{eardrum}}}{A_{\text{oval}}} = 1.5 \frac{A_{\text{eardrum}} \Delta p_{\text{air}}}{A_{\text{oval}}},$$

which yields:

$$\frac{\Delta p_{\text{oval}}}{\Delta p_{\text{air}}} = 1.5 \frac{65\ \text{mm}^2}{3.2\ \text{mm}^2} = 30.$$

The middle ear provides an amplification of the pressure difference arriving at the eardrum by a factor of 30 instead of diminishing the signal to less than 1% as expected for a single membrane separating outer and inner ear! Note that the factor of 30 is derived neglecting a damping loss across the middle ear. This damping loss means that the total energy transferred from the auditory canal to the inner ear is reduced; however, focussing the energy transfer onto the small area of the oval window means the intensity, which is energy transfer per unit area, is still enhanced.

15.6.3: The Inner Ear

The acoustic components of the inner ear are located in the **cochlea**, shown in Fig. 15.13 in its characteristic curled shape for mammals. This shape is the origin of its name, which means *snail* in Latin. The cochlear cross-section is shown in Fig. 15.24, highlighting its three separate channels. The **vestibular chamber** (top cavity) starts at the oval window and runs along the cochlea to its far end, called the apex or the helicotrema. There the vestibular chamber is open to the **tympanic chamber** (bottom cavity), which runs back to the round window at the end of the inner ear. Both chambers are filled with perilymph and are separated by the **basilar membrane**. The third channel, called the **cochlear duct** (at the centre left), contains the organ of Corti (named for Alfonso de Corti). The cochlear duct is separated from the other two channels by the basilar membrane and by Reissner's membrane (named for Ernst Reißner) and is filled with endolymph, a liquid solution similar to the perilymph but that contains 145 mmol/L K^+. This difference in ionic concentrations leads to an electric potential of the endolymph relative to the perilymph of +80 mV. In the present discussion we neglect the related electric phenomena.

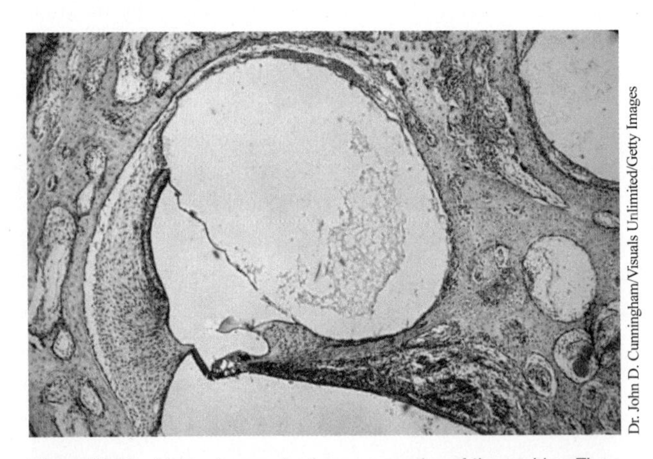

Figure 15.24 Light micrograph of a cross-section of the cochlea. Three liquid-filled channels are shown: the vestibular chamber (top centre) and the tympanic chamber (bottom), both containing perilymph; and the cochlear duct (centre left), containing endolymph. The tympanic chamber on one side and the vestibular chamber and the cochlear duct on the other side are separated by the basilar membrane. The vestibular chamber and the cochlear duct are separated by Reissner's membrane. The organ of Corti is supported by the basilar membrane in the cochlear duct. It consists of hair cells and the tectorial membrane.

Dendrites emerge from the organ of Corti and run through the basilar membrane toward the brain. The **organ of Corti** is highlighted in Fig. 15.25. The basilar membrane (horizontal section at the bottom) carries an array of support cells in which one internal auditory and three external (at left) hair cells are embedded. The auditory hair cells can be seen as they each carry about 80 hair-like extensions, called stereovilli (just below the plate-like structure that extends from the right). The stereovilli extend into the cochlear duct, which forms a narrow gap between the basilar membrane and the tectorial membrane above the auditory hair cells.

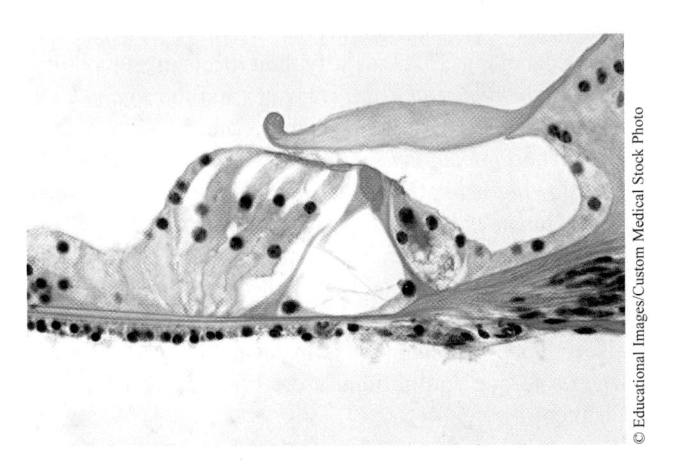

Figure 15.25 Detail of the organ of Corti. The basilar membrane (at bottom) carries a layer of supportive cells in which one internal and three external auditory hair cells are embedded. The auditory hair cells have hair-like extensions (stereovilli) that extend into a gap between the basilar membrane and the tectorial membrane, which extends from the upper right.

These components allow for sound detection and sound frequency analysis in the following fashion. The inner ear represents a closed tube (both confining windows are elastic membranes). However, an excitation at the oval window cannot form resonances along this tube because that would limit the frequencies we hear to a set of harmonics, as discussed in section 15.4.2. Instead, the excitation at the oval window leads to a one-dimensional travelling wave, similar to the single bulge you form on a rope stretched between your hand and the wall if you briefly swing your hand up and down. Such a **travelling wave** is sketched in Fig. 15.26. The figure illustrates the amplitude of a wave in the perilymph as a function of position along the vestibular chamber; however, the figure is not correct in that the travelling wave is a longitudinal and not a transverse wave as the graphic implies. The two arrows indicate the initial excitation at the oval window and the later mechanical response at the round window below. For each sound frequency, a specific point exists, between the oval window and the apex, where the travelling wave amplifies in the perilymph and causes the basilar membrane to vibrate. This vibrational excitation is

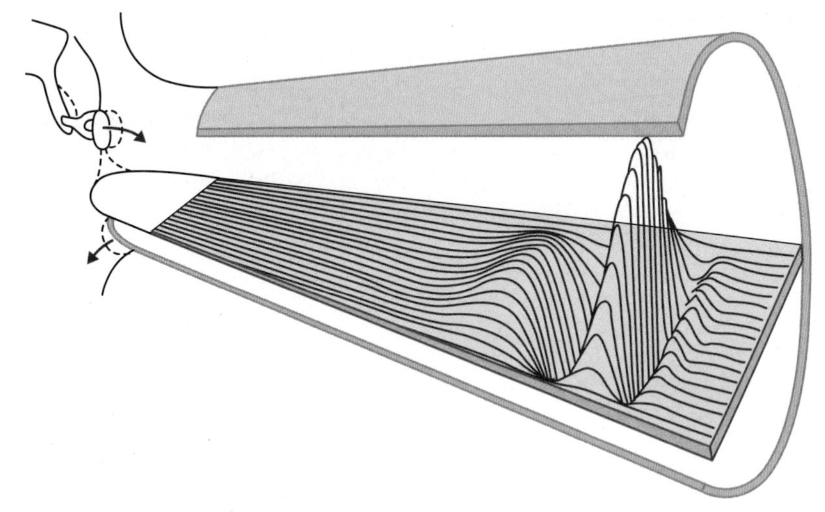

Figure 15.26 A travelling wave in the vestibular chamber, caused by a vibration of the stirrup at the oval window (upper arrow). The basilar membrane vibrates at a point determined by the frequency of the travelling wave. This vibration causes the wave to be transferred to the perilymph in the tympanic chamber, in which it travels toward the round window (lower arrow).

controlled by the stiffness of the basilar membrane, which reduces by a factor of 10^4 along its entire length. This is caused by a thickness variation of the basilar membrane. It starts with a thickness of 0.04 mm at the oval window and ends with a thickness of 0.5 mm at the apex.

Fig. 15.27 illustrates the position dependence of the point at which a certain frequency causes the basilar membrane to vibrate. Fig. 15.27(a) shows a conceptual sketch in which the cochlea is shown as if stretched out. The oval window is excited by the stirrup (1). The wave then travels along the perilymph (2). The frequency properties are shown in Fig. 15.27(b). For higher frequencies, e.g., 10 kHz, the wave causes the basilar membrane (3) to vibrate close to the oval window. This causes the wave to transfer to the tympanic chamber (4), where it travels back to the round window (6). At an intermediate frequency, e.g., 500 Hz, the wave travels to an intermediate point along the basilar membrane, and at low frequencies, e.g., 50 Hz, the wave travels all the way to the apex.

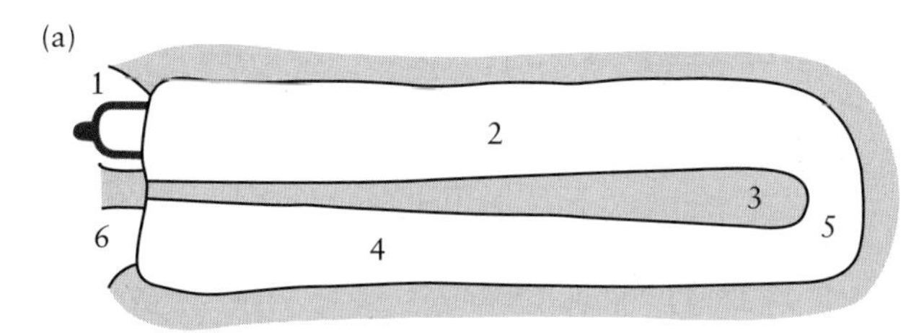

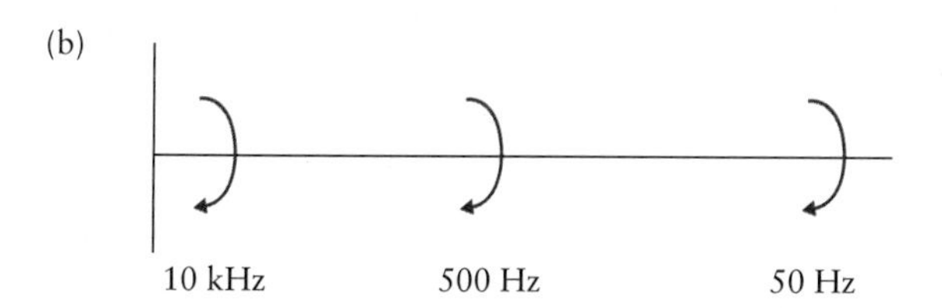

Figure 15.27 Schematic sketch of the path taken in the inner ear by a wave of specific frequency. (a) Overview of the cochlea with the stirrup and the oval window (1), the vestibular chamber (2), the basilar membrane (3) with its varying thickness, the tympanic chamber (4), the apex (5), and the round window (6). (b) Waves of higher frequencies travel a shorter distance along the cochlea.

How does the resonance of the basilar membrane cause a signal to the brain? The vibration of the basilar membrane causes the tectorial membrane in the organ of Corti to vibrate in synchrony. The mechanism by which this leads to an excitation of the dendrites in the organ of Corti is illustrated in Fig. 15.28, which shows the two membranes in their equilibrium position in the upper sketch and when moved upward by an angle ϕ in the lower sketch. The narrowness of the gap between the two membranes causes the stereovilli to bend sideways during the vibration. Only the internal auditory hair cell is connected to dendrites; thus the stereovilli of the internal auditory cell must be bent for a signal to be sent to the brain.

What is reported to the brain? Two components of the vibrational motion of the basilar membrane are encoded in the sequence of the nerve impulses:

- **Tonotopic mapping (frequency-to-place mapping).** The response of a particular dendrite identifies the position along the cochlea where the resonance in the basilar membrane has occurred. If the travelling wave contains different frequencies, they are separately and detected concurrently by different dendrites along the cochlea, pretty much in the same fashion that a Fourier analysis is developed mathematically. This mechanism was already proposed by Georg Ohm and Hermann von Helmholtz in the 19th century.
- **Temporal coding of frequencies.** The frequency at which the dendrites in the organ of Corti send their signals to the brain is synchronized with the actual frequency of the sound wave.

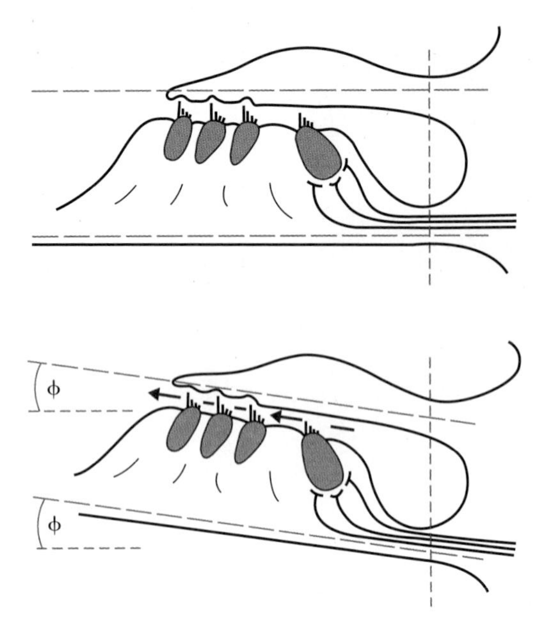

Figure 15.28 Mechanism of dendrite excitation in the organ of Corti. The basilar membrane and the tectorial membrane vibrate in synchrony in response to a travelling wave. The upper sketch shows the equilibrium position and the lower sketch shows the position at the instant when both membranes are moved upward by an angle ϕ. The stereovilli of the auditory hair cells are bent sideways in this process due to the narrowness of the gap between the two membranes. A signal is sent to the brain only from the internal auditory hair cell.

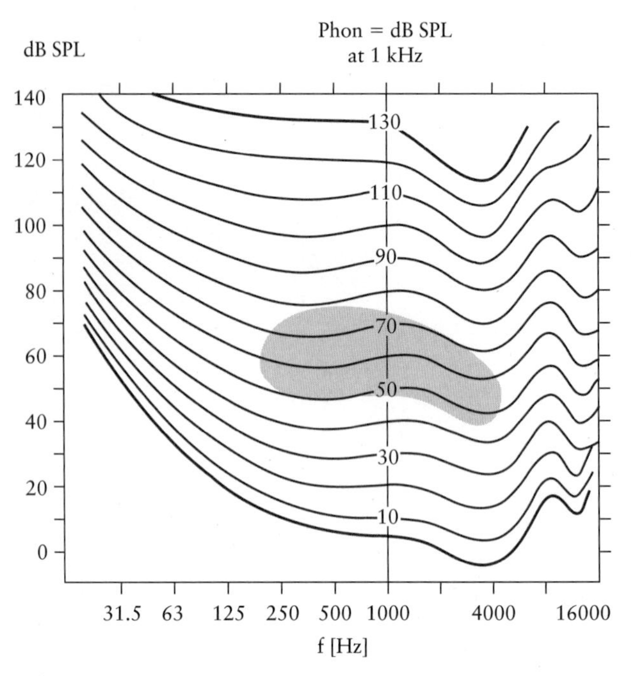

Figure 15.29 The hearing range of the human ear, shown as a function of the frequency of the sound (f; abscissa) and as a function of the sound pressure level (SPL; ordinate). Each line in the plot represents sounds that are judged to be equally loud. The lowest curve is the acoustic reflex threshold and the highest curve is the pain threshold. The grey area corresponds to the normal range of conversations. The loudness in unit phon is equal to the intensity level (IL) or sound pressure level (SPL) in unit dB at a sound frequency of 1 kHz.

The human ear is extremely sensitive to sounds ranging from 16 Hz to about 20 kHz (dogs hear up to 40 kHz, while elephants communicate subsonically as these frequencies carry much farther in air). However, the sensitivity of human hearing varies across this frequency interval, primarily as the result of the resonance properties we discussed earlier for the outer ear. This is illustrated quantitatively in Fig. 15.29. The figure shows the frequency range from about 10 Hz to 16 kHz and sound pressure levels from 0 dB to 140 dB (for the definition of the sound pressure level scale see Eq. [15.15]). Each curve in the plot represents the pressure levels as a function of frequency that a person judges to be equally loud. The thicker line at the bottom corresponds to the **normal acoustic reflex threshold**, and the thicker line at the top corresponds to the **pain threshold**. The range of normal conversation is shown as a grey area near the centre of the plot

A maximum sensitivity near 3 kHz, in agreement with Example 15.10, is clearly demonstrated. As frequencies are lower or higher, the sensitivity diminishes until it ceases at the lower- or higher-frequency limit of the ear. The variations in perceived **loudness** as a function of frequency render the physical scales of intensity level (IL; see Eq. [15.16]) and sound pressure level (SPL; see Eq. [15.15]) less useful, because we all disagree with the idea that a 100-dB sound at 40 Hz is equally as loud as a 100-dB sound at 3000 Hz. For this reason, a new parameter for loudness is introduced based on Fig. 15.29 and is recorded in unit **phon**. The convention is to set the decibel scale and the phon scale equal at a sound frequency of 1 kHz: 100 dB = 100 phon at 1 kHz. Loudness values deviate from SPL or IL values at all other frequencies. Table 15.3 provides several examples for sounds, with loudness reported in unit phon.

TABLE 15.3

Sound perception in unit phon

Loudness (phon)	Example
4	threshold of normal hearing
20	rustle of leaves
40	whispering, talking in a low voice
60	normal conversation
80	city traffic noise
100	industrial plant
110	comfort limit
120	thunder
130	pain threshold
140	jet engine

SUMMARY

DEFINITIONS

- Displacement D in a one-dimensional harmonic wave (called wave function):

$$D = A\sin(\omega t - \kappa x),$$

where κ is the wave number, with $\kappa = 2\,\pi/\lambda$, and λ is the wavelength

- Speed of wave: $c = \lambda f$, with f the frequency
- Energy density ε of a sound wave:

$$\varepsilon_{\text{total}} = \frac{1}{2}\rho A^2\omega^2,$$

where A is the amplitude, ρ is the density, and ω is the angular frequency

- Intensity I of a sound wave, i.e., the energy passing an area A per time unit Δt:

$$I = c\varepsilon_{\text{total}} = \frac{1}{2}c\rho A^2\omega^2,$$

where c is the speed of sound

- Sound pressure level SPL (in unit dB):

$$\text{SPL} = 20\log_{10}\frac{p}{p_0} \quad \text{with } p_0 = 2\times10^{-5}\,\text{Pa}$$

- Intensity level (IL; in unit dB):

$$\text{IL} = 10\log_{10}\frac{I}{I_0} \quad \text{with } I_0 = 1\times10^{-12}\,\frac{\text{J}}{\text{m}^2\text{s}}$$

UNITS

- Wave number κ: m^{-1}
- Wavelength λ: m
- Frequency f: Hz
- Energy density ε: J/m^3
- Intensity of a sound wave I: $\text{J/(m}^2\text{ s)}$
- Sound pressure level (SPL) and intensity level (IL): dB

LAWS

- Speed of waves
  - in fluids:

$$c = \sqrt{\frac{B}{\rho}},$$

with B the bulk modulus and ρ the density of the medium

  - in air, assuming adiabatic pressure variations (Laplace's equation):

$$c = \sqrt{\kappa\frac{p}{\rho}},$$

with κ the adiabatic coefficient, p the pressure, and ρ the density of the medium

- Standing waves for a reflected wave:

$$D_{\text{superposition}} = -\{2A\cos(\omega t)\}\sin(\kappa x),$$

where $D_{\text{superposition}}$ is the displacement in the wave that results from the superposition

- Harmonics
  - for a closed tube for n-th harmonic:
    –wavelength: $\lambda_n = 2\,L/n$
    –frequency: $f_n = n\,c/(2\,L) = n\,f_1$
  - for a half-open tube:

$$f_n = \frac{nc}{4L} \text{ with } n = 1, 3, 5, 7, \ldots$$

- Amplitude for resonant coupling of a system with negligible damping:

$$A = \frac{F_{\max}}{m\left(\omega^2_{\text{standing}} - \omega^2_{\text{ext}}\right)},$$

where $F_{\max}$ is the maximum of the periodic external force applied to the system, ω_{ext} is the angular frequency of the external force, and ω_{standing} is the angular frequency of the first harmonic (standing wave)

MULTIPLE-CHOICE QUESTIONS

MC–15.1. The frequency of a sound wave has which of the following units?
(a) s
(b) 1/s
(c) m/s
(d) s^2
(e) $1/\text{s}^2$

MC–15.2. We compare two sound waves in air at room temperature. Wave II has twice the frequency of wave I. Which of the following relation holds between their speeds of sound?
(a) $c_\text{I} = c_\text{II}$
(b) $c_\text{I} > c_\text{II}$
(c) $c_\text{I} < c_\text{II}$
(d) Such a conclusion cannot be drawn with the given information.

MC–15.3. We compare two sound waves in air at room temperature. Wave II has twice the frequency of wave I. Which of the following relation holds between their wavelengths?
(a) $\lambda_\text{I} = \lambda_\text{II}$
(b) $\lambda_\text{I} > \lambda_\text{II}$
(c) $\lambda_\text{I} < \lambda_\text{II}$
(d) Such a conclusion cannot be drawn with the given information.

MC–15.4. Waves are typically characterized by frequency, angular frequency, period, wavelength, and amplitude. Which of these parameters are related to each other in a linear fashion? *Note:* More than one answer may apply.
(a) period and frequency
(b) period and angular frequency
(c) frequency and angular frequency
(d) wavelength and period
(e) amplitude and frequency

MC–15.5. The distance between a crest of a sinusoidal water wave and the next trough is 2 m. If the frequency of the water wave is 2 Hz, what is its speed?
(a) 8 m/s
(b) 4 m/s
(c) 2 m/s
(d) 1 m/s
(e) Not enough information is given to determine the wave speed.

MC–15.6. A sound source I generates sound with twice the frequency of sound source II. Compared to the speed of sound of source I, the speed of sound of source II is
(a) twice as fast.
(b) half as fast.
(c) four times as fast.
(d) one-fourth as fast.
(e) the same.

MC–15.7. If you perceive a point-like source of sound as too loud, you should move away from the source. This is because of the following relation between the sound intensity and the distance from the source.
(a) Intensity is independent of distance.
(b) Intensity increases linearly with distance.
(c) Intensity increases non-linearly with distance.
(d) Intensity decreases linearly with distance.
(e) Intensity decreases non-linearly with distance.

MC–15.8. The intensity level (IL) of a sound is reported in unit decibel (dB). How does IL change if we increase a sound intensity by a factor of 10?
(a) It remains unchanged.
(b) It increases by 1 dB to 2 dB.
(c) It increases by 2 dB to 20 dB.
(d) It increases by 20 dB to 200 dB.
(e) It decreases.

MC–15.9. Doubling the rate at which a sound source emits energy at a single frequency leads to the following increase of the sound intensity level:
(a) 0.5 dB.
(b) 2.0 dB.
(c) 3.0 dB.
(d) 20 dB.
(e) 100 dB.

MC–15.10. A tube is initially filled with air (use c_{air} = 340 m/s), then with water (c_{water} = 1500 m/s). How does the frequency of the first harmonic change for the tube?
(a) It doesn't change.
(b) It increases.
(c) It decreases.

MC–15.11. Which of the following does a sound wave transmitted to the inner ear form in the perilymph?
(a) a standing wave
(b) the first and second harmonic
(c) a travelling wave
(d) no wave at all

CONCEPTUAL QUESTIONS

Q–15.1. We frequently modelled tendons as massless strings. Why is this not a useful model when describing waves on strings?

Q–15.2. Having established that a sound wave corresponds to pressure fluctuations in the medium, what can you conclude about the direction in which the pressure fluctuations travel?

Q–15.3. Describe the motion of the particles that make up the medium through which travels a sound wave.

Q–15.4. What kind of information can one deduce from the amplitude of a wave?

Q–15.5. How can an object generate sound waves?

ANALYTICAL PROBLEMS

P–15.1. A wave with frequency 5.0 Hz and amplitude 40 mm moves in the positive x-direction with speed 6.5 m/s. What are (a) the wavelength, (b) the period, and (c) the angular frequency? (d) Write a formula for the wave.

P–15.2. The best way to measure the compressibility of liquids or solids is to measure the speed of sound in the material. If such a measurement for water yields c = 1.4 km/s (which is about four times the value in air!), what is the compressibility of water?

P–15.3. The range of frequencies heard by the healthy human ear stretches from about 16 Hz to 20 kHz. What are the corresponding wavelengths of sound waves at these frequencies?

P–15.4. Bats can detect small insects that are about equal in size to the wavelength of the sound the bat makes with its echolocation system. A bat emits a chirp at a frequency of 60 kHz. Using the speed of sound in air as 340 m/s, what is the smallest insect this bat can detect?

P–15.5. Fig. 15.30 shows a 25-Hz wave travelling in the x-direction. Calculate (a) its amplitude, (b) its wavelength, (c) its period, and (d) its wave speed. Use L_1 = 18 cm and L_2 = 10 cm.

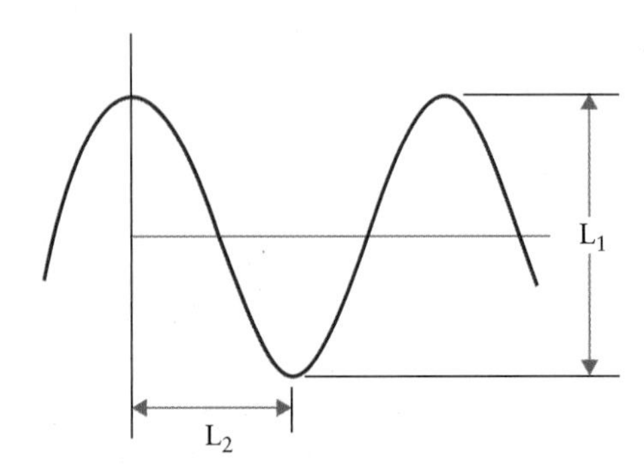

Figure 15.30

P–15.6. An FM radio station broadcasts at 88 MHz. Determine for the radio waves (a) their period and (b) their wavelength.

P–15.7. A piano emits sounds in the range of 28 Hz to 4200 Hz. Find the range of wavelengths at room temperature for this instrument.

P–15.8. A sound wave has a frequency of 700 Hz and a wavelength of 0.5 m. What is the temperature of the air in which this sound wave travels?

P–15.9. A person hears an echo 3.0 seconds after emitting a sound. In air of 22°C, how far away is the sound-reflecting wall?

P–15.10. A supersonic jet travels at 3.0 Mach, i.e., at three times the speed of sound. It cruises at 20000 m above ground. We choose $t = 0$ when the plane passes directly overhead of an observer, as shown in Fig. 15.31(a). (a) At what time t will the observer hear the plane? (b) What distance Δx has the plane travelled by that time?

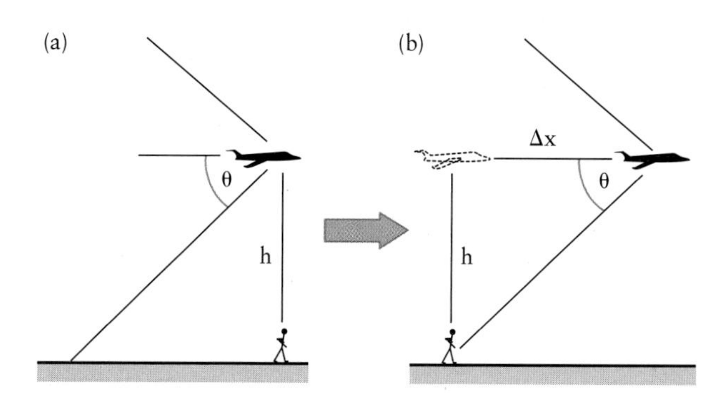

Figure 15.31

P–15.11. The only supersonic jet ever used for commercial air travel was the Concorde. It travelled at 1.5 Mach. What was its angle θ, as defined in Fig. 15.31(b), between the direction of propagation of its shock wave and the direction of flight?

P–15.12. The sound intensity of 1.0×10^{-12} J/(m^2 s) is the threshold of hearing for humans. What is the amplitude of the motion of the air molecules? Use $c = 340$ m/s and 1.2 kg/m^3 as the density of air.

P–15.13. (a) A microphone has an area of 5 cm^2. It receives during a 4.0-s time period a sound energy of 2.0×10^{-11} J. What is the intensity of the sound? (b) Using the sound intensity from part (a), what is the variation in pressure in the sound wave, Δp? Use $T = 293$ K and $\rho_{air} = 1.2$ kg/m^3.

P–15.14. A jet airplane has an intensity of 100 J/(m^2 s) when heard at a distance of 30 m. (a) What is the maximum sound intensity heard by a person on the ground when the airplane cruises 10000 m above the ground? (b) What is the intensity level (IL) heard?

P–15.15. A certain sound has an intensity that is four times the intensity of a reference sound at the same frequency. (a) What is the difference in the intensity level of the two sounds? (b) If the reference sound causes a sound perception of 60 phon, what is the sound perception value of the more intense sound?

P–15.16. An underwater microphone is used to record sounds emitted by porpoises. The minimum intensity level the instrument can record is 10 dB. Assuming a porpoise emits sound at a rate of 0.05 J/s, what is the maximum distance at which the animal will still be recorded? Neglect sound absorption in water and treat the porpoise as a point sound source.

P–15.17. Two sound waves have intensities of $I_1 = 100$ J/(m^2 s) and $I_2 = 200$ J/(m^2 s). By how many decibels do the two sounds differ in intensity level?

P–15.18. A standard man shouting loudly produces a 70-dB sound at a distance of 5 m. At what rate does the person emit sound energy? Express the result in J/s.

P–15.19. A hypothesis says the upper limit in frequency a human ear can hear can be determined by the diameter of the eardrum, which should have approximately the same diameter as the wavelength at the upper limit. If we use this hypothesis, what would be the radius of the eardrum for a person able to hear frequencies up to 18.5 kHz?

P–15.20. If we model the human auditory canal as a tube that is closed at one end and that resonates at a fundamental frequency of 3000 Hz, what is the length of the canal? Use normal body temperature for the air in the canal.

ANSWERS TO CONCEPT QUESTIONS

Concept Question 15.1: (B). Eq. [15.2] shows that wavelength and frequency are inversely proportional to each other for a given speed of sound. The speed of sound does not vary with wavelength or frequency; it is a parameter that depends exclusively on the properties of the sound-carrying medium.

Concept Question 15.2: (C). All four statements are correct, but only answer (C) specifically characterizes standing waves. The amplitude is time independent for non-attenuating waves; the frequency is time independent for a wave of given energy density; and the wavelength is time independent in a uniform medium.

CHAPTER 16

Sound II

Beer's law describes the absorption of sound intensity as an exponential function of the distance of a wave from the sound source. The absorption coefficient depends on the medium and the frequency of the sound.

Waves reflect at interfaces. They travel in a different direction after they pass an interface at an angle (law of refraction). At most interfaces, both reflection and transmission occur; however, the transmitted intensity is greatly reduced when the two media adjacent to an interface differ significantly in their density and/or speed of sound. The combined factor of density and speed of sound is therefore defined as the acoustic impedance.

In ultrasound imaging, a transducer generates a sound that travels through the skin into the tissue along a line that is called a ray. It reflects at various interfaces and is detected as an echo when it returns to the transducer. Several parameters have to be considered for ultrasound imaging applications. Sound absorption limits the depth to which tissue can be sampled. The echo delay time plays a role in the total time the acquisition of an image will take. It determines the maximum pulse repetition, which in turn determines the frame rate of imaging when combined with the number of lines that constitute the image. The depth resolution (i.e., the independent observation of two interfaces at different depths) depends inversely on the frequency.

The Doppler effect is the change of received frequency when either the source or the receiver move relative to the sound-carrying medium. In Doppler ultrasound diagnosis, the speed of blood is measured for the echo received from individual erythrocytes. The distribution of speeds allows the health practitioner for example to determine blood vessel stenosis that results in turbulent blood flow near the constriction.

Ultrasound applications in medicine differ from sound perception in physiology, as discussed in the previous chapter, not only in the relevant frequency range. The discussion of audiology focussed on waves in confined media and their resonance phenomena. Concepts important in the medical use of ultrasound focus more on absorption of a travelling wave, and sound transmission and reflection at interfaces. The current chapter starts therefore with the introduction of additional fundamental concepts.

16.1: Sound Absorption

The sound intensity reduction in Case Study 15.5 is a geometric effect. The sound travelling through a medium can further diminish due to energy loss to the medium when the gas pocket vibrations are not perfectly harmonic. This effect is called **sound absorption** because sound energy is absorbed by the medium. For example, if the medium is air the absorption is caused by vibrations of air molecules that are not perfectly adiabatic. Thermal energy loss occurs, which slightly heats up the air through which the sound travels.

For a quantitative description of this effect, we study a planar wave travelling in a one-dimensional gas column that is aligned with the x-axis. This eliminates the geometric sound intensity loss we discussed in Case Study 15.5, allowing us to focus exclusively on the absorption effect.

August Beer observed that the rate of loss of sound intensity along the x-axis is proportional to the intensity itself at every point along the axis. We want to follow Beer's further reasoning carefully as he formulated this observation in equation form.

First, we cannot write for the rate of loss of intensity along the axis $\Delta I/\Delta x$. If we were to do this, we would average the change of the intensity over a finite interval Δx. This would yield a correct value for the rate of loss at every position along the x-axis only if that rate were constant. However, Beer's experimental observation states that the rate of loss varies along the axis with the changing value of the intensity. We have to write dI/dx for the rate of intensity loss instead. Mathematically, this is a calculus-based term and is called the derivative—in particular the derivative of the intensity function with the position along the x-axis. For a review of basic calculus methods see the section on Calculus in the Math Review section at the end of the book. Physically, dI/dx is a function that represents the local rate of intensity change at any particular position along the x-axis. A derivative d/dx of a function is called a gradient of the function. The gradient of the intensity is the quantity that is consistent with Beer's experimental observations: the rate of intensity loss at any given position x_0 along the x-axis is proportional to the intensity at the same position. This leads to **Beer's law**:

$$\frac{dI}{dx}\Big|_{x=x_0} = -\beta\, I(x_0) \qquad [16.1]$$

We note several features of this law:

- Eq. [16.1] contains a minus sign. This sign is needed because the gradient term represents a loss of intensity; i.e., it is a negative value while the intensity on the right-hand side is always positive (no negative intensity values exist). If the minus sign were not included, Eq. [16.1] would imply that the intensity gradient along the x-axis is positive; i.e., the intensity would increase with distance from the sound source.
- Eq. [16.1] is written in equation form, not just as a proportionality between the gradient and the intensity. For this, Beer included a proportionality factor β. The coefficient β is called the absorption coefficient with unit m^{-1}. You can confirm the unit of the absorption coefficient from Eq. [16.1]. Were this coefficient a materials constant, a table of values would be given. However, we discuss below that the **absorption coefficient** in this form depends also on the frequency of the sound; a new coefficient will therefore be introduced to replace the absorption coefficient and tabulated values will be given then.
- Mathematically, Eq. [16.1] is a linear differential equation. Physical phenomena are often governed by differential equation; for example the kinematics equations we discussed in Chapter 2 are differential equations as they relate a function (position) with its derivative to time (velocity) and second derivate to time (acceleration). In Beer's law, we gave a linear differential equation because the intensity on the right side is a linear term; i.e., it has an exponent of 1.

The solution of a differential equation provides a function written in explicit form, e.g. in the case of Beer's law we expect a function $I = f(x)$. We find this solution for Eq. [16.1] in the form:

$$I = I_{incident}\, e^{-\beta x}. \qquad [16.2]$$

Eq. [16.2] is a mathematical solution to the differential equation in Eq. [16.1] since the exponential function is the only function for which the function and the slope of the function are the same (except for a constant factor). In Eq. [16.2], $I_{incident}$ is the intensity emitted by the source, i.e., the intensity at $x = 0$.

KEY POINT

Sound absorption is governed by Beer's law, which states that the loss of intensity per unit length of medium, dI/dx, is proportional to the absolute intensity of the sound wave.

CASE STUDY 16.1

Where does the factor $I_{incident}$ in Eq. [16.2] come from mathematically?

Answer: *Eq. [16.2] is the result of an integration of Eq. [16.1]. To illustrate this, we first separate the variables in Eq. [16.1]:*

$$\frac{dI}{I} = -\beta dx.$$

Both differential terms lead to respective functions when we integrate:

$$\int_{I_{incident}}^{I} \frac{I'}{dI'} = \ln I - \ln I_{incident} = -\beta \int_0^x dx' = -\beta x,$$

in which a prime was added to the original functions to avoid confusing them with the variable upper boundary of the integral. The lower boundary is at the location of the source. This last equation yields Eq. [16.2]. Thus, $I_{incident}$ is an integration constant that will always occur unless the function passes through the origin.

CASE STUDY 16.2

Beer's law in differential equation form (Eq. [16.1]) and in integrated form (Eq. [16.2]) is mathematically similar to the radioactive decay law; e.g., both laws lead to exponential functions in the integrated form. What is conceptually the biggest difference between these two cases?

Answer: *The independent variable of the radioactive decay law is time, while it is position for Beer's law. Time derivatives describe rates, while position derivatives describe gradients. Even though some mathematical operations look similar, the two types of problems lead to very different discussions as Beer's law has a static, time-independent character.*

We illustrate the implications of Eq. [16.2] further by rewriting it with a new parameter, the **absorption length** $x_{absorption} = 1/\beta$:

$$\frac{I}{I_{incident}} = \exp\left(-\frac{x}{x_{absorption}}\right). \qquad [16.3]$$

At the origin, which is the location of the sound source, the intensity is per definition $I = I_{incident}$. When we move away from the source to a distance equal to the absorption length, i.e., to $x = x_{absorption}$, then Eq. [16.3] yields $I/I_{incident} = 1/e \cong 0.37$. Moving further to a distance twice the absorption length, $x = 2\,x_{absorption}$, we find $I/I_{incident} = 1/e^2 \cong 0.14$ and finally, at a distance of three times the absorption length, $x = 3\,x_{absorption}$, we obtain $I/I_{incident} = 1/e^3 \cong 0.05$. Thus, the sound intensity drops to about 1/3 of the initial intensity at the absorption length and drops to 5% at three times the absorption length. The absorption length is therefore a good measure of the distance at which the sound intensity is significantly reduced due to absorption.

In the literature, a second length is introduced to describe the distance from the source at which the attenuation is 50% of the initial intensity. This length is called the **half-value thickness** (*HVT*).

The **sound attenuation coefficient** is a materials constant because the absorption coefficient is roughly proportional to the frequency, thus α and β differ by a factor f. To quantify their relation more specifically, we write the formula that defines α:

$$-\Delta \mathrm{IL} = \alpha f x, \qquad [16.4]$$

in which x is the depth in the medium and ΔIL is the change in the intensity level at position x relative to the position of the source. ΔIL carries an additional negative sign to obtain positive α-values for attenuation rather than intensity increases. Introducing the coefficient α is useful because the attenuation case corresponds to applications in ultrasound imaging. Values used for α in ultrasound applications are listed in Table 16.1. Eq. [16.4] shows that sound absorption increases at higher frequencies. Thus, sound absorption is a critical issue in ultrasound applications because ultrasound cannot penetrate tissue far due to Eq. [16.4].

TABLE 16.1

Sound absorption at higher frequencies

	Material	α (dB cm^{-1} MHz^{-1})
Fluids		
	Water	0.0002
	Blood	0.18
Tissues		
	Smooth muscle	0.2–0.6
	Soft tissue	0.3–0.8
	Fat	0.5–1.8
	Tendon	0.9–1.1
	Bone	13–26
Organs		
	Brain	0.3–0.5
	Liver	0.4–0.7
	Lungs	40

KEY POINT

The absorption coefficient β depends on the medium (materials parameter), but also on the frequency of the wave. It is therefore not tabulated. Instead, we introduce the frequency-independent sound attenuation coefficient α with unit dB/(cm MHz) as a materials constant.

EXAMPLE 16.1

(a) Determine the conversion coefficient K between the absorption coefficient β defined in Eq. [16.2] and the attenuation coefficient α in Eq. [16.4]:

$$\beta = K\,\alpha f.$$

(b) What is the reduction in intensity level IL when the intensity of a sound is cut in half? What is the half-value thickness *HVT* for soft tissue at (c) 2 MHz and (d) 10 MHz?

Solution

Solution to part (a): We start with Eq. [16.2] and rewrite the left side of the equation to obtain ΔIL:

$$\log_{10}\frac{I}{I_0} - \log_{10}\frac{I_{incident}}{I_0} = \log_{10}\frac{I}{I_{incident}} = \log_{10} e^{-\beta x},$$

in which I_0 is an arbitrary reference intensity as defined in Eq. [15.16]. We also note that $\log_{10}(1) = 0$. An additional factor of 10 must be included due to the definition of IL in Eq. [15.16]. Thus, this equation leads to:

$$\Delta \mathrm{IL} = 10\log_{10}\frac{I}{I_{incident}} = -4.343\,\beta x,$$

in which we used $\log_{10}e = 0.4343$. Next we compare this result with Eq. [16.4]. Note that x in the last equation has unit m because m^{-1} was introduced as the unit of β, but x in Eq. [16.4] has unit cm. Thus, we first rewrite the last equation for x in unit cm and then substitute it in Eq. [16.4]:

$$0.0434\,\beta = \alpha f.$$

Using the definition of the conversion coefficient K in the example text we find:

$$K = \frac{\beta}{\alpha f} = 23.0.$$

The numerical value results when the denominator αf is replaced by the same product in the previous equation.

Solution to part (b): We substitute $I = I_0/2$ in Eq. [15.16]:

$$\mathrm{IL} = 10\log_{10}\frac{0.5\,I_0}{I_0} = -3.01.$$

Solution to part (c): We use for soft tissue an attenuation coefficient of $\alpha = 0.5$ dB cm^{-1} MHz^{-1}, which

continued

is an average value based on Table 16.1. We noted in part (b) that a sound loss of −3 dB occurs from the surface of a tissue sample to *HVT*. Thus, we find from Eq. [16.4]:

$$HVT = \frac{\Delta IL}{\alpha f} = \frac{3.0 \text{ dB}}{\left(0.5\frac{\text{dB}}{\text{mHz cm}}\right)(2.0\,MHz)} = 3.0 \text{ cm}.$$

Solution to part (d): We use the same calculation as in part (c). This yields *HVT* = 6 mm at 10 MHz, which is a significant attenuation at a short distance into the tissue.

We saw in the previous chapter that the sound intensity is proportional to the square of the amplitude of the gas vibrations in the gas column; i.e., $I \propto A^2$. Thus, we can extend Eq. [15.9] for the displacement by using Eq. [16.2] for an absorbing medium in the form:

$$D = A\,e^{-\gamma x}\sin(\omega\,t - k\,x), \qquad [16.5]$$

in which γ is the decay coefficient for the sound amplitude, which is closely related to the absorption coefficient β. This function is shown in Fig. 16.1.

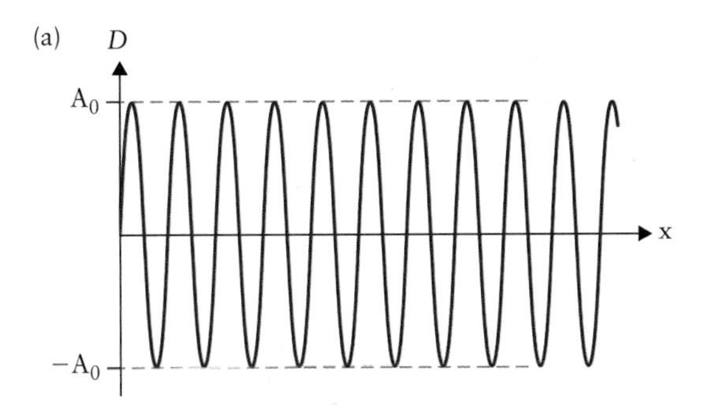

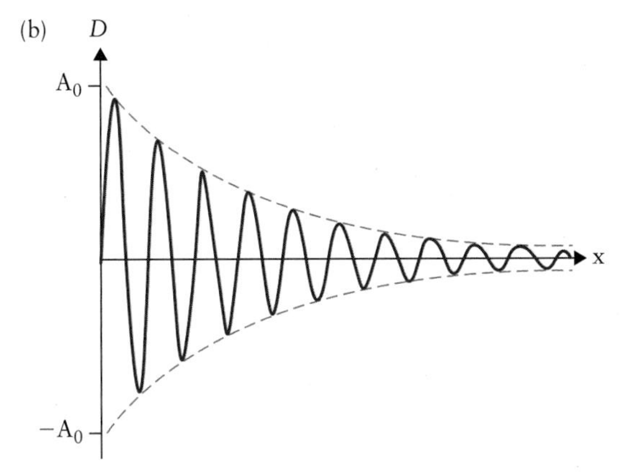

Figure 16.1 Free and damped planar wave. The free wave has a constant amplitude. In the damped case (bottom) the amplitude decreases exponentially from its initial value (following the envelope function that is included as a solid line).

16.2: Reflection and Transmission of Waves at Flat Interfaces

In ultrasound imaging we obtain images of organs and tissues within the human body. These images show anatomical features, such as the position of the surface of a specific organ. Such features can be visualized because ultrasound waves sent into the human body are reflected at interfaces between different tissues. The reflected wave (called an **echo**) is detected with the same device that emitted the initial ultrasound signal (called a **transducer**). We are interested in both reflection and transmission because the transducer signal has to pass the transducer–skin interface for imaging applications.

KEY POINT

Three physical principles contribute to the ultrasound image: (I) the relative directions of incident, reflected, and transmitted sound waves; (II) the relative intensities of the three waves; and (III) the tissue-specific attenuation of these waves.

We have addressed the attenuation in the previous section; in the current section we focus on wave directions and then on wave intensities.

16.2.1: Relative Directions of Waves: Huygens' Principle and Snell's Law

The propagation of a wave, including planar waves, can be constructed with Huygens' principle. Christiaan Huygens introduced in the late 1600s the concept of an elementary spherical point source. He showed that we could describe any wavefront at a later time by allowing a spherical wave to emerge from each point of the wavefront at an earlier instant. We use this principle to establish quantitative relations for reflection and transmission of sound waves at a flat interface.

16.2.1.1: Law of Reflection

Wave reflection is a familiar phenomena for light and sound waves, for example light reflection from the surface of water or echo formation in mountains. We use Fig. 16.2 to relate the angles between incident and reflected sound waves. The figure shows a planar wavefront approaching a flat interface at an angle α_i. The interface is defined by the points A_1, A_2, and A_3, with A_2 half way between A_1 and A_3. We choose the initial time $t = 0$ at the instant when the wavefront has reached point A_1. This wavefront is shown in the top part of Fig. 16.2 as a straight line labelled W.

Using Huygens' principle we construct the location of the wavefront after reflection in the bottom part of Fig. 16.2. The figure shows the system at the time instant when the incident wavefront has reached point A_3. The time elapsed between the top frame and the bottom frame in Fig. 16.2 is $\Delta t = R/c$ in which R is the distance from B_3 to A_3, and c is the speed of sound.

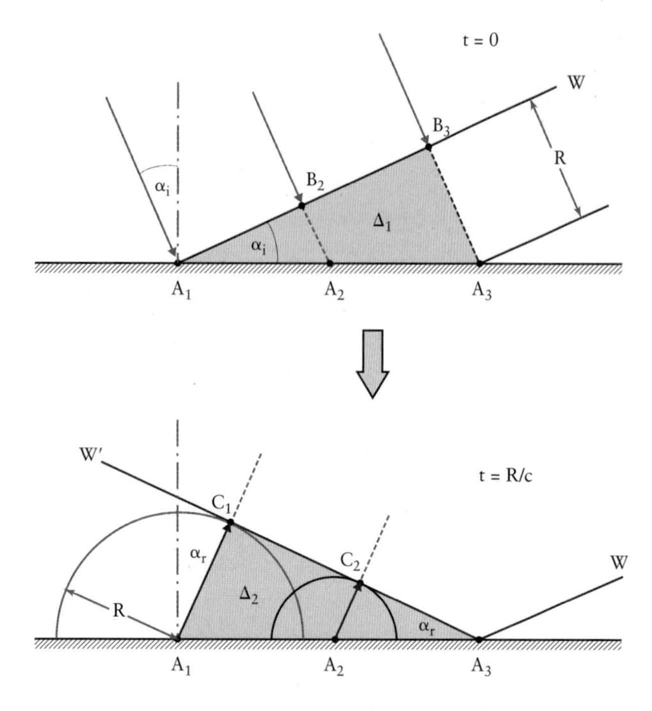

Figure 16.2 Geometric sketch for the law of reflection on a flat interface based on Huygens' principle. A planar wavefront approaches the interface under an angle α. At time $t = 0$ (top sketch) it reaches point A_1 at the interface. In the bottom sketch spherical waves are drawn from points A_1 and A_2 forming the reflected wavefront W'. Note that the initial wavefront W has just reached point A_3 in the bottom sketch, which applies at time $t = R/c$ with R a distance defined in the top sketch and c the speed of sound.

The elementary spherical wave emerging from point A_1 in the bottom part of the figure has the same radius R. The elementary wave emerging from point A_2 has radius $R/2$ because the incident wavefront reached point A_2 after time $\Delta t/2$ and it travelled a distance $R/2$ in the remaining time. Note that we connect the elementary waves in the bottom of the figure according to Huygens' principle; this yields the wavefront W'.

The construction in Fig. 16.2 allows us to use geometric arguments to derive the law of reflection for a sound wave at a flat interface. The law of reflection connects α_i, which is the angle of the incident wavefront with the interface, and α_r, which is the angle of the reflected wavefront with the interface, as illustrated in the bottom of Fig. 16.2. For a quantitative result, we compare the two triangles Δ_1 and Δ_2 in the figure. We note the following three features:

- The sides from A_1 to A_3 are equal in both triangles.
- The sides from A_1 to C_1 and A_3 to B_3 are equally long because sound travels same distances in the same time.
- The angle $A_1C_1A_3$ is 90° because the tangent to a circle is perpendicular to the radius. The angle $A_1B_3A_3$ is also 90° because the wavefront is per definition perpendicular to the wave propagation direction.

These three conditions establish that the two shaded triangles in Fig. 16.2 are the same; this leads to the **law of reflection**:

$$a_i = a_r. \qquad [16.6]$$

KEY POINT

The law of reflection states that the angle of an incident wave with a flat surface is the same as the angle of reflection.

16.2.1.2: Law of Refraction (Snell's Law)

Sound waves are not completely reflected at most surfaces but penetrate the surface and travel in the new medium. Fig. 16.3 allows us to relate the angle of incidence with the interface to the angle of the transmitted wave; the resulting law is called the law of refraction. The figure shows two different media, labelled I and II. From Table 15.1 we expect the speed of sound to vary between the two media with values c_I and c_{II} respectively. Fig. 16.3 illustrates the case $c_I > c_{II}$.

The top sketch of Fig. 16.3 is identical to the top sketch in Fig. 16.2: it shows the instant $t = 0$ when the incident wavefront has reached point A_1. The wavefront arrives at an angle α_i with the interface. It will reach

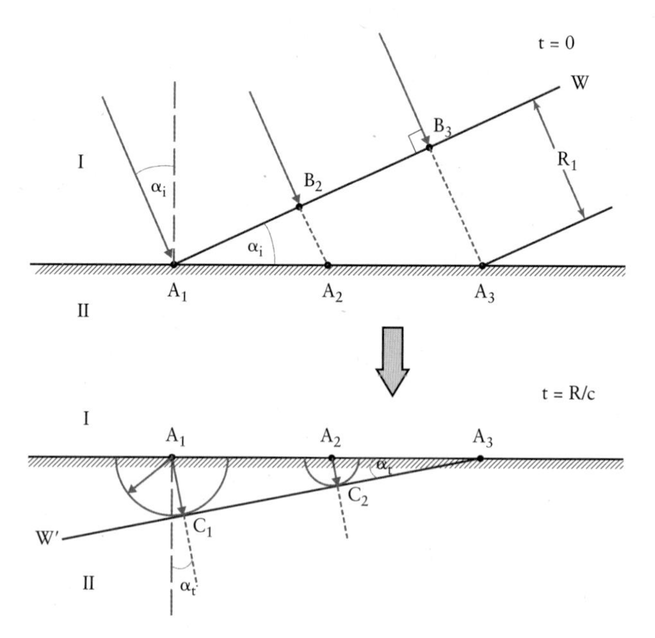

Figure 16.3 Geometric sketch for the law of refraction at a flat interface based on Huygens' principle. The top sketch shows an incoming wavefront W at an angle α_i. At $t = 0$ the wave has just reached point A_1 in the interface. The bottom sketch shows the refracted wavefront W′ at time $t = R/c_I$, with c_I the speed of sound in medium I. At that time, the wavefront W has just reached point A_3 in the interface. Huygens' spherical waves have been drawn at points A_1 and A_2 to illustrate the formation of the wavefront W'.

point A_3 a time $\Delta t = R_I/c_I$ later. This later instant is illustrated in the bottom sketch of Fig. 16.3, which shows two elementary spherical waves, one emerging from point A_1 and one from point A_2. Both elementary waves are in medium II and travel therefore slower; the wave from point A_1 has reached a distance of $R_{II} = c_{II}\,\Delta t$ and the elementary wave from point A_2 has travelled half as far as it emerged only at time $\Delta t/2$. The following relations are obtained geometrically for the angles of the incident and transmitted wavefronts:

$$\sin\alpha_i = \frac{R_I}{A_1A_3}$$

$$\sin\alpha_t = \frac{R_{II}}{A_1A_3}.$$

These two formulas are combined by eliminating the identical length A_1A_3 and substituting the product of speed of sound and Δt for the two radii:

$$\frac{\sin\alpha_i}{\sin\alpha_t} = \frac{R_I}{R_{II}} = \frac{c_I}{c_{II}}. \qquad [16.7]$$

This is called **Snell's law** for Willebrord Snell, who discovered it in 1621. It states that the sine of the angle of a wavefront with a flat interface is proportional to the speed of sound in the respective medium.

16.2.2: Relative Intensities of Waves: Acoustic Impedance

We consider again a flat interface between different media, for example the interface between two different tissues in ultrasound imaging. Fig. 16.4 shows schematically the acoustic interaction zone (dashed box) at a flat

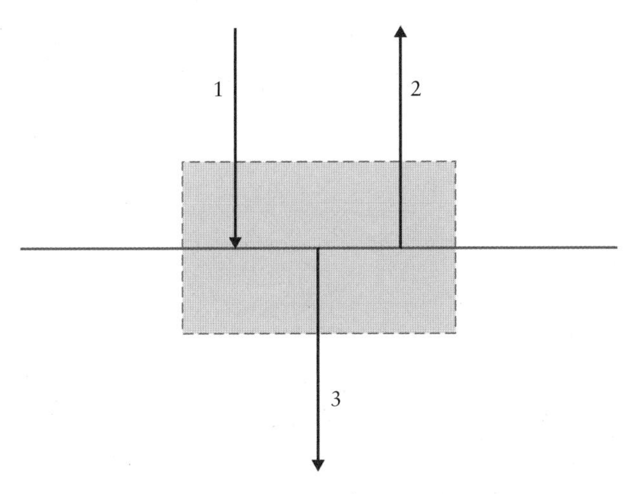

Figure 16.4 Conceptual sketch of a wave passing through an interface. At the interface section (dashed box) three wave components have to be considered: (1) an incoming wave, (2) a reflected wave, and (3) a transmitted wave.

interface. At this interface an incident wave (1) is partially reflected (2) and partially transmitted (3). To quantify sound reflection and sound transmission at the interface we determine the ratios of the reflected and transmitted wave intensities to the incident intensity respectively. Two conditions must apply in the dashed zone:

- To satisfy the continuity of physical properties across the interface, the sum of the maximum gas pocket speed of vibration in the incident wave, v_i, and the maximum gas pocket speed in the reflected wave, v_r, must match the maximum gas pocket speed in the transmitted wave, v_t:

$$v_i + v_r = v_t. \qquad [16.8]$$

- The energy passing through the dashed zone in Fig. 16.4 must be conserved. The energy density transported in the incident wave is given in Eq. [15.14]. In that equation we rewrite the angular frequency ω of the gas pocket as its maximum velocity: $v_{max} = \omega\,A$ with A the amplitude of the wave. Thus:

$$\frac{1}{2}\rho_1\,c_1\,v_i^2 = \frac{1}{2}\rho_1\,c_1\,v_r^2 + \frac{1}{2}\rho_2\,c_2\,v_t^2, \qquad [16.9]$$

in which the index 1 represents the medium carrying the incident wave and the index 2 represents the medium carrying the transmitted wave.
We use Eq. [16.8] to eliminate v_r in Eq. [16.9]:

$$\begin{aligned}\rho_1\,c_1\,v_i^2 &= \rho_1\,c_1(v_t - v_i)^2 \\ &\quad + \rho_2\,c_2\,v_t^2 \\ &= \rho_1\,c_1\,v_t^2 \\ &\quad - 2\,\rho_1\,c_1\,v_t\,v_i \\ &\quad + \rho_1\,c_1\,v_i^2 + \rho_2 c_2\,v_t^2.\end{aligned}$$

Note that the third term in the last line cancels with the term on the left-hand side of the equation. We divide the last equation by v_t on both sides to obtain:

$$2\,\rho_1\,c_1\,v_i = (\rho_1\,c_1 + \rho_2\,c_2)v_t.$$

This yields a relation between the transmitted and the incoming speeds of the wave:

$$v_t = \frac{2\,\rho_1\,c_1}{\rho_1\,c_1 + \rho_2\,c_2}v_i.$$

This equation illustrates that the various gas pocket velocities in sound propagation are connected to each other with terms containing the product of the speed of sound and the density of the medium. Such terms are called **acoustic impedance** Z:

$$Z = \rho\, c. \qquad [16.10]$$

Z has unit kg m^{-2} s^{-1}, which is usually called **rayl**. Table 16.2 shows that impedances vary widely, with gases at the low end of the range with values between 1×10^2 and 1×10^3 rayl, and condensed matter at the upper end with values of 1×10^6 to 1×10^7 rayl. Inserting Eq. [16.10] in the formula for v_t yields:

$$v_t = \frac{2\, z_1}{z_1 + z_2} v_i.$$

TABLE 16.2

Acoustic impedance values for various physiologically important materials

Material	Z
Air at 20°C	414 rayl
Fat at 37°C	1.33×10^6 rayl
Water at 20°C	1.48×10^6 rayl
Muscle at 37°C	1.66×10^6 rayl
Bone	6.73×10^6 rayl

Next we substitute this equation in the equation for the intensity of a wave, given in Eq. [15.14]. The ratio of transmitted to incident intensity follows:

$$T_I = \frac{I_t}{I_i} = \frac{\frac{1}{2} c_2\, \rho_2\, v_t^2}{\frac{1}{2} c_1\, \rho_1\, v_i^2} = \frac{c_2 \rho_2}{c_1 \rho_1}\left(\frac{2\, \rho_1\, c_1}{\rho_1\, c_1 + \rho_2\, c_2}\right)^2,$$

which yields:

$$T_I = \frac{4\, \rho_1\, \rho_2\, c_1\, c_2}{(\rho_1\, c_1 + \rho_2\, c_2)^2} = \frac{4\, Z_1\, Z_2}{(Z_1 + Z_2)^2} \qquad [16.11]$$

The ratio I_t/I_i is called the **intensity transmission** T_I.

We further determine the **intensity reflection** R_I, which is defined as $R_I = I_r/I_i$. A fast way to quantify R_I starts from substituting the formula for v_t in Eq. [16.9] and sorting the terms that contain v_i and v_r:

$$v_r^2 = \left(1 - \frac{4\, \rho_1\, \rho_2\, c_1\, c_2}{(\rho_1\, c_1 + \rho_2\, c_2)^2}\right) v_i^2.$$

The ratio of the reflected intensity to incident intensity is then calculated as:

$$R_I = \frac{I_r}{I_i} = \frac{\frac{1}{2} c_1\, \rho_1\, v_r^2}{\frac{1}{2} \cdot c_1\, \rho_1\, v_i^2},$$

which yields:

$$R_I = \left(\frac{\rho_2\, c_2 - \rho_1\, c_1}{\rho_1\, c_1 - \rho_2\, c_2}\right)^2 = \left(\frac{Z_2 - Z_1}{Z_1 + Z_2}\right)^2. \qquad [16.12]$$

KEY POINT

Intensity transmission and intensity reflection of sound at an interface are governed by the acoustic impedances of the two adjacent media. The acoustic impedance is the product of the density and the speed of sound in a given material.

CASE STUDY 16.3

Why do land-living mammals have a middle ear that separates the outer ear's eardrum from the oval window at the entry of the cochlea?

Hypothesis: We answer the question by assuming the middle ear does not exist and propose that the ear then can no longer serve its purpose. Once proven, this hypothesis allows us to conclude that the middle ear is necessary because the sound travels through a significant difference in density from air to perilymph (for a description of the ear anatomy see Chapter 15).

Answer: *The alternative anatomical set-up we assume in this Case Study is obtained by removing the middle ear such that the eardrum and the oval window collapse into a single membrane. Sound must then couple across this membrane from the medium air of the outer ear to the denser medium perilymph of the inner ear. The hypothetical interface is equivalent to the wave reflection model we introduced in Fig. 16.4. We are interested in the transmission across the interface. The ratio of the transmitted to the incident sound intensity is taken from Eq. [16.11]:*

$$T_I = \frac{4\, \rho_{air}\, \rho_{perilymph}\, c_{air}\, c_{perilymph}}{(\rho_{air}\, c_{air} + \rho_{perilymph}\, c_{perilymph})^2} = 4 \frac{Z_{air}\, Z_{perilymph}}{(Z_{air} + Z_{perilymph})^2},$$

in which ρ is the density, c is the speed of sound, and Z is the acoustic impedance as defined in Eq. [16.10]. We identify medium 1, which carries the incident wave, as air and medium 2 as perilymph. Note that this equation is symmetric with respect to the two media adjacent to the membrane; the same result applies if the wave is incident from the perilymph side.

continued

The last equation can be simplified in the case of a significant difference between the two media, i.e., when $\rho_{perilymph}\, c_{perilymph} >> \rho_{air}\, c_{air}$ *which means also* $Z_{perilymph} >> Z_{air}$ *In this case we get:*

$$T_I = 4\frac{\rho_{air}\, c_{air}}{\rho_{perilymph}\, c_{perilymph}} = 4\frac{Z_{air}}{Z_{perilymph}}.$$

Before we apply this result, we confirm the condition $Z_{perilymph} >> Z_{air}$ *for the human ear with the impedance values given in Table 16.2. Then we calculate the ratio of sound intensities between air and water:*

$$T_I = \frac{4.414\ \text{rayl}}{1.48\times10^6\,\text{rayl}} = 1.1\times10^{-3}.$$

This value confirms a more general observation: sound transfer between media of significantly different density is very ineffective. In particular, this means that nature could not construct our ear with a single membrane between the outer and inner ear, as a sound intensity transfer of only about 0.1% would leave us essentially deaf.

EXAMPLE 16.2

For an ultrasound wave with an incident intensity $I_{incident} = 5.0 \times 10^{-2}$ J cm^{-2} s^{-1} = 50 mW/cm^2 calculate at a flat interface between fat and muscle tissue (a) the intensity reflection R_I and (b) the intensity transmission T_I. *Note:* Ultrasound applications in medicine use typically sound intensities between 10 mW/cm^2 and 100 mW/cm^2.

Solution

Solution to part (a): We use $Z_{fat} = 1.33 \times 10^6$ rayl and $Z_{muscle} = 1.66 \times 10^6$ rayl from Table 16.2. With Eq. [16.12] we find:

$$R_I = \left(\frac{Z_2 - Z_1}{Z_1 + Z_2}\right)^2 = \left(\frac{1.66-1.33}{1.66+1.33}\right)^2 = 0.012 = 1.2\%.$$

Thus, only 1.2% of the incident intensity is reflected, which in this particular case corresponds to 0.6 mW/cm^2.

Solution to part (b): We use the same *Z* values as in part (a). From Eq. [16.11] we find:

$$T_I = \frac{4Z_1Z_2}{(Z_1+Z_2)^2} = \frac{4\,1.66\,1.33}{(1.66+1.33)^2} = 0.988 = 98.8\%.$$

Note that the sum of intensity reflection and intensity transmission must be 100%; thus the last result could have been obtained without calculation once the result to part (a) was found. 98.8% of the incident intensity passes through the fat–muscle interface; for ultrasound imaging this means little contrast can be expected from an interface between two similar tissues.

16.3: The Ultrasound Image

In this section we simulate an actual ultrasound imaging application as illustrated in Fig. 16.5. The ultrasound signal is emitted from a transducer at the left. We assume that ultrasound of frequency 2.0 MHz and intensity of 50 mW/cm^2 is used. The interface to the skin is adjusted to the acoustic impedance of fat with a jelly-like fluid (shaded area). The tissue system consists of a fat layer of acoustic impedance Z_1 and thickness $l_1 = 2.0$ cm, followed by muscle tissue of acoustic impedance Z_2. We want to explore the various properties of the ultrasound echo we receive from the fat–muscle interface labelled In$_1$ (where we use the abbreviation In for interface).

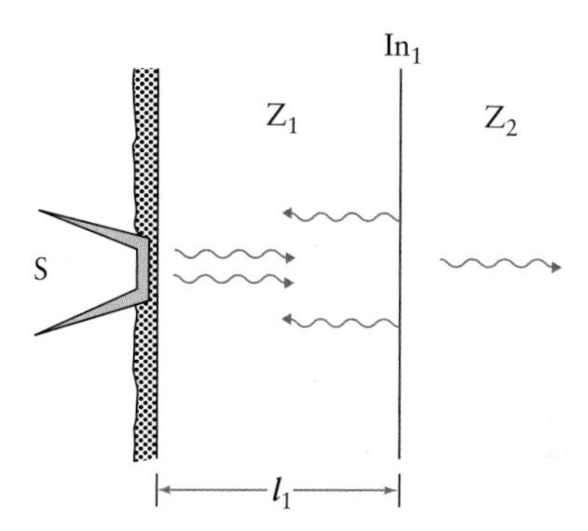

Figure 16.5 Ultrasound measurement with an internal interface In$_1$ at a distance l_1 below the skin. The interface separates two media with acoustic impedances Z_1 and Z_2. A gel is used to provide perfect acoustic coupling of the transducer at the transducer–skin interface.

Part of this application has been addressed in Example 16.2. From that example we know that only 1.2% of the incident sound intensity is reflected at the fat–muscle interface. If we neglect sound absorption in the fat layer, this corresponds to an echo intensity of 0.6 mW/cm^2.

The current section extends on those results by including sound absorption and **echo delay time**. Both effects play an important role in choosing the various parameters for the imaging application, and allow us to put limits to what can be imaged. Since we have introduced the relevant physical concepts already, we present the discussion in the form of two examples.

16.3.1: Sound Absorption in Ultrasound Imaging

EXAMPLE 16.3

We use an ultrasound signal with frequency *f* = 2.0 MHz and an incident intensity of $I_{incident} = 5.0 \times 10^{-2}$ J cm^{-2} s^{-1} = 50 mW/cm^2. (a) How much intensity does the signal lose before it reaches the fat–muscle interface In$_1$ in Fig. 16.5,

continued

l_1 = 2 cm below the skin surface? (b) What is the total intensity loss for the recorded echo?

Solution

Solution to part (a): Typical sound attenuation coefficient values for fat are found in Table 16.1. We use a value of $\alpha_{fat} = 1.0$ dB cm^{-1} MHz^{-1}. We calculate from this value the intensity loss (ΔIL in unit dB) at 2 cm depth of a fat layer, using Eq. [16.4]:

$$\Delta\text{IL} = -\left(1.0\frac{\text{dB}}{\text{MHz cm}}\right)(2.0\ \text{MHz})(2.0\ \text{cm}) = -4.0\ \text{dB}.$$

Thus, the signal loses 4 dB from the skin surface to the fat–muscle interface.

Solution to part (b): Two steps follow the process in part (a) before the echo can be recorded: the signal has to be reflected and then it has to travel back to the skin surface, where the transducer will act as a detector.

The intensity loss in reflection at the interface is calculated in Example 16.2 as 1.2%. The corresponding intensity loss in dB is:

$$\Delta\text{IL} = 10\log_{10}\frac{I_{reflected}}{I_{incident}} = 10\log_{10}R_I = 10\log_{10}0.012$$
$$= -19.2\ \text{dB}.$$

The intensity loss for the echo travelling back from the fat–muscle interface to the skin surface is again 4 dB because the previous equation applies equally for this step. The convenient property of ΔIL values is that we can add losses in subsequent processes; i.e., the total loss in the current example is:

$$\Delta\text{IL}_{total} = \Delta\text{IL}_{travel\ in} + \Delta\text{IL}_{reflection} + \Delta\text{IL}_{travel\ out}$$

Numerically this corresponds to:

$$\Delta\text{IL}_{total} = -4.0\ \text{dB} - 19.2\ \text{dB} - 4.0\ \text{dB} = -27.2\ \text{dB}$$

We can express this result as an absolute intensity value by inverting Eq. [15.16]:

$$-27.2\ \text{dB} = 10\log_{10}\frac{I_\text{f}}{I_\text{i}},$$

which leads to:

$$I_\text{f} = 10^{-2.72}I_\text{i}.$$

Substituting the given incident intensity we find:

$$I_\text{f} = 10^{-2.72}\left(50.0\frac{\text{mW}}{\text{cm}^2}\right) = 0.095\frac{\text{mW}}{\text{cm}^2}.$$

Such a significant decrease in sound intensity sets an upper limit to how deep a certain interface can be located to be imaged. In the case of the use of clinical ultrasound methods operating in the range between 2 MHz and 10 MHz, typical sampling depths of 3 cm to 15 cm result for a maximum acceptable sound loss of 60 dB.

CASE STUDY 16.4

Why is the intensity loss an issue in ultrasound imaging given that blue whales can communicate over many hundred kilometres in water?

Answer: *The intensity level loss is a function of frequency; the moans of the blue whale occur at just a few hertz.*

16.3.2: Echo Delay

EXAMPLE 16.4

(a) For the sample in Fig. 16.5, how long does an incident ultrasound signal need to return as an echo from interface In_1 to the detector? (b) Assuming we do not know the thickness of the fat layer, what is l_1 if the echo arrives with a delay of 42.3 μs?

Solution

Solution to part (a): The basic definition of speed as the distance divided by the time is sufficient to solve the first part of the problem. We note that the distance the signal travels until it reaches the detector is 2 l_1, consisting of a distance l_1 from the transducer to the fat–muscle interface and a distance l_1 from In_1 to the detector. With the speed of sound in fat from Table 15.1, c_{fat} = 1450 m/s, we find:

$$t_\text{echo} = \frac{2l_1}{c_{fat}} = \frac{2(2\ \text{cm})}{1450\ \frac{\text{m}}{\text{s}}} = 2.76\times10^{-5}s = 27.6\ \mu\text{s}.$$

This time difference can be resolved with modern electronics. A short t_echo is beneficial for imaging because the number of signals processed determines the resolution of the image.

Solution to part (b): We need to find l_1 in part (b). From the definition of the speed of sound in fat we find:

$$l_1 = \frac{1}{2}\frac{c_{fat}}{t_{echo}} = \frac{1}{2}\frac{1450\frac{\text{m}}{\text{s}}}{4.23\times10^{-5}s} = 3.05\ \text{cm}.$$

16.3.3: Depth Resolution and Maximum Imaging Depth

The depth resolution is defined by studying two buried interfaces at different depths. Fig. 16.6 is an extension of Fig. 16.5 to include a finite width of the buried tissue layer with acoustic impedance Z_2. We choose this layer to have a thickness l_2 with ultrasound wave reflection at the interface between the first two layers, In_1, and at the interface of the two deeper layers, In_2.

We further note that the ultrasound signal is not continuous but is a short pulse. The signal pulse length L_signal

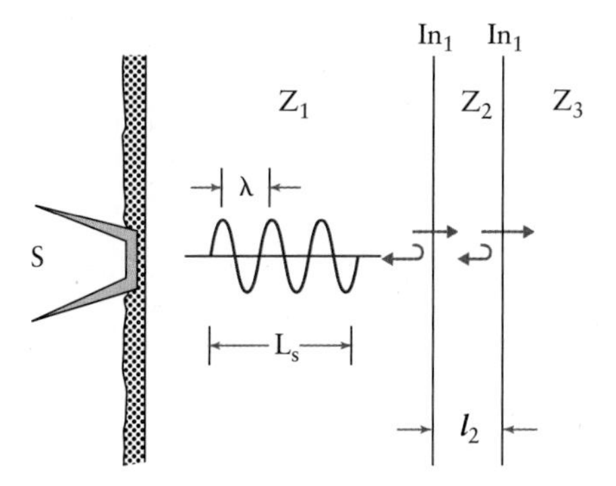

Figure 16.6 Ultrasound measurement with two parallel internal interfaces In_1 and In_2 that are separated by a distance l_2. The interfaces separate media with different acoustic impedances Z_1 to Z_3. A gel is used to provide perfect acoustic coupling of the transducer at the transducer–skin interface.

is best characterized by the wavelength λ and the number of complete wave cycles, *n*:

$$L_{signal} = n\lambda.$$

Fig. 16.6 shows the particular case with $n = 3$; *n* depends on the set-up of the transducer, but $n = 3$ is a typical number for most ultrasound applications.

The minimum thickness l_2 that we can resolve is determined by the condition that no overlap of wave reflection from In_1 and In_2 should occur. This means that the wave transmitted through In_1 must have passed that interface completely before its reflection from In_2 passes the first interface on the way back out:

$$L_{signal} \le 2\,l_2.$$

We combine the last two equations to express the minimum resolvable thickness l_2:

$$l_2 = \frac{L_{signal}}{2} = \frac{n\lambda}{2} = \frac{nc}{2f}, \qquad [16.13]$$

in which we replaced the wavelength λ with c/f where f is the frequency of the ultrasound wave. This result shows that the spatial resolution of a buried feature is inversely proportional to the frequency of the used ultrasound.

EXAMPLE 16.5

(a) Assuming three complete wave cycles per signal and an average speed of sound in tissue of c = 1540 m/s, what difference in spatial resolution occurs for ultrasound of 2 MHz and 10 MHz? (b) Does this mean that we should always use the highest available frequency for ultrasound imaging?

continued

Solution

Solution to part (a): We use Eq. [16.13] in both cases with $n = 3$. This yields minimum l_2 values of 1.15 mm and 0.23 mm respectively.

Solution to part (b): The answer is no, and the reason is found in Eq. [16.4]. That equation illustrates that the sound attenuation in tissue is proportional to the frequency of the sound. Thus, a higher frequency allows for greater spatial resolution but confines the maximum depth at which imaging is possible. For typical ultrasound applications with incident sound intensities in the range from 10 mW/cm^2 to 100 mW/cm^2, a maximum acceptable sound intensity loss in imaging is about 60 dB. In fat tissue with α = 1.0 dB/(cm MHz), 60 dB loss is reached after a 2-MHz signal has travelled 30 cm; this means the deepest structures we can image lie 15 cm below the skin (because the signal travels in and out). At 10 MHz, the deepest structure can only be 3 cm below the skin! This illustrates that we deal with a trade-off between the spatial resolution and maximum imaging depth in ultrasound imaging.

16.3.4: Pulse Repetition and Image Frame Rate

In the previous section we addressed critical length scales, such as spatial resolution and maximum imaging depth. The current section in turn addresses critical time scales in ultrasound imaging applications.

Ultrasound imaging uses the same device, called a transducer, both for signal generation and echo detection. The transducer contains a ceramic element that converts electric energy into mechanical vibrations (**piezoelectric effect**). The piezoelectric effect is applied both ways: To generate ultrasound, an electric field is used to alter the crystalline structure of the ceramic element to cause vibrations. As a detector, the arriving ultrasound is used to generate an electric field in the crystal that varies in phase with the vibrations caused by the arriving sound.

As a result, after generating a signal, the ceramic element must be available for sound detection long enough to receive sound that reached the deepest structures we can image. How deep these structures are located is determined by the applied frequency and Eq. [16.4].

Considering Examples 16.5 and 16.6 we note that the key properties in ultrasound imaging are interconnected: when using higher ultrasound frequencies, we gain higher spatial resolution and faster repetition rates, but have to accept shallow sampling depths; when using lower ultrasound frequencies, we can image to greater depths, but must allow for a slower repetition rate and a lower spatial resolution.

But does the signal repetition time really matter? It plays a role because an ultrasound image is formed from many single echoes. Thus, a slower repetition rate forces a longer imaging session. To quantify this argument, we

EXAMPLE 16.6

We use a 5 MHz ultrasound transducer. What shortest **pulse repetition time** applies to this device based on the definition of pulse repetition in Fig. 16.7? *Hint:* Use a maximum allowable sound intensity loss of 60 dB, a sound attenuation of $\alpha = 0.5$ dB/(cm MHz), and a speed of sound of 1540 m/s.

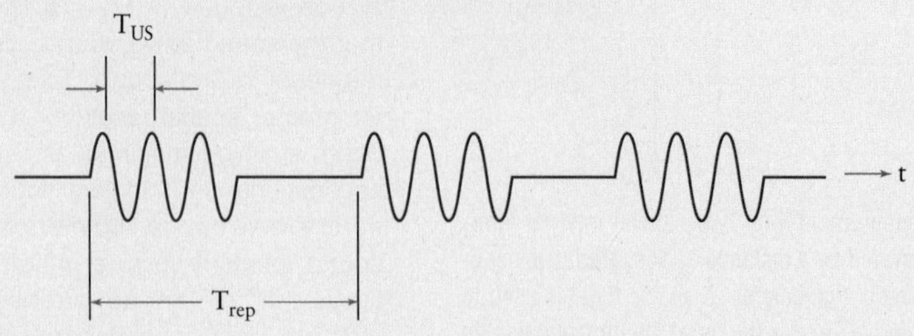

Figure 16.7 Time pattern of ultrasound signals used in medical imaging: a short ultrasound pulse T_{US} is followed by a dead time (combined time length T_{rep} is the pulse repetition time) during which the transducer operates as a receiver.

Solution

We use Eq. [16.4] to determine the maximum depth of a feature we can image with a 5-MHz ultrasound wave. Note that the maximum depth is ½ the maximum distance a signal can travel before its sound intensity loss is 60 dB:

$$d_{max} = \frac{1}{2}\frac{60 \text{ dB}}{\left(0.5\frac{\text{dB}}{\text{MHz}\cdot\text{cm}}\right)(5.0 \text{ MHz})} = 12 \text{ cm}.$$

Next we use the definition of speed to determine the time the sound signal needs to travel 12 cm in and 12 cm out of the tissue sample:

$$t_{echo,max} = \frac{2d_{max}}{c_{sound}} = \frac{24 \text{ cm}}{1540 \text{ m/s}} = 1.56\times10^{-4} s = 156\ \mu\text{s}.$$

This corresponds to a repetition frequency of $1/t_{echo,\ max}$ = 6410 Hz = 6.41 kHz.

start with the actual procedure by which an ultrasound image is formed. For this we focus on the so-called **b-mode** of imaging, which leads to two-dimensional images. The **a-mode** is a one-dimensional method and is rarely used in modern medicine; the **m-mode** is used to measure the speed of movement (hence the letter "m" to label the mode) and is discussed in the context of the Doppler effect later on.

In the b-mode, the transducer sends signals in a wide range of directions, as illustrated in Fig. 16.8. Each direction is called a **line** and requires the transducer to be switched to the receiving mode for the full repetition time calculated in Example 16.6. These signals are transmitted in a particular order, which then correlates with the position in the image. A typical image consists of 100 to 200 lines. The **frame rate** is defined as the inverse of the product of the number of lines per image and the repetition time (which applies per line). For example, for the 5-MHz ultrasound transducer of Example 16.6, a 200-line image is obtained in 156 μs/line × 200 lines/image = 31 ms (milliseconds). This corresponds to a frame rate of 1/(31 ms/image) ≈ 32 images/s. We could increase the frame rate by lowering the number of lines used to form an image. This would, however, reduce the sideways resolution in the image.

Fig. 16.8 also explains the well-known crescent-shape of ultrasound images. The maximum imaging depth d_{max}, which is 12 cm in Example 16.6, applies for each line in the direction of the signal emission.

Table 16.3 summarizes the main parameters that are relevant for ultrasound imaging at the two ends of the usual operation range, 2 MHz and 10 MHz.

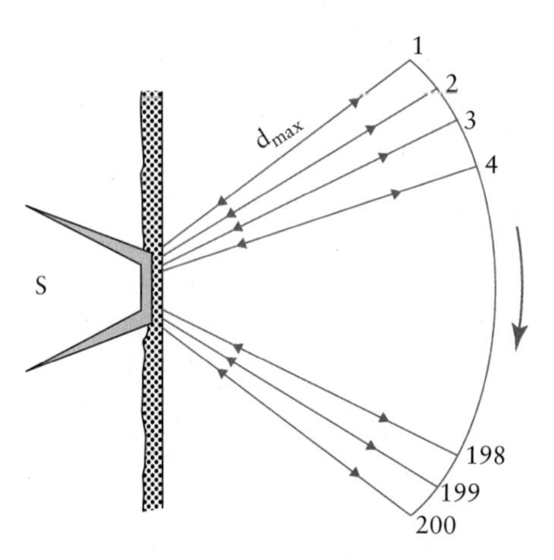

Figure 16.8 The maximum imaging depth d_{max} is independent of the angle of the transmitted signal/returning echo, causing the well-known pie-slice shape of ultrasound images. Each numbered ray corresponds to a single echo signal analyzed by the device.

TABLE 16.3

Critical parameters for ultrasound imaging at 2 MHz and 10 MHz

Parameter	Formula	High frequency case	Low frequency case	Notes
Frequency		10 MHz	2 MHz	
Depth resolution	$l_{res} \propto n/f$	0.3 mm	1–2 mm	$n = 3$, $c = 1540$ m/s
Maximum sampling depth	$d_{max} \propto 1/(\alpha f)$	3 cm	15 cm	$I = 10$–100 mW/cm^2 $\equiv 60$ dB loss $\alpha_{fat} = 1$ dB/(cm MHz)
Pulse repetition time	$t_{echo} \propto d_{max}$	40 μs	200 μs	
Frame rate	$T_{frame} \propto 1/t_{echo}$	125 frames/s	25 frames/s	image of 200 lines
Application	blood vessel	fetus		

16.4: Doppler Ultrasound: The Use of the Doppler Effect in Medicine

An altered sound frequency is detected when the sound source or the receiver moves relative to the medium. The underlying physical phenomenon is called the Doppler effect (named for Christian Doppler). Two cases are distinguished:

- the sound source at rest and the receiver moving with speed $v_{receiver}$, and
- the sound source moving with speed v_{source} while the receiver is at rest. The latter case is well-known from police car sirens when they approach us or move away from us.

16.4.1: Sound Detection with Moving Receiver and Stationary Source

Fig. 16.9 shows a source (S) emitting waves that travel outward. The lines in the figure indicate subsequent crests of the sound wave; thus, between every two lines lies a full wavelength, and the lines are separated by equal distances as long as the frequency of the source does not change. When the receiver (R) moves toward the source, the receiver records more waves per time unit than if it was at rest. Correspondingly, when the receiver moves away from the sound source, a lesser number of waves is recorded per time unit.

We quantify this effect for a source of frequency f_0. We consider a time interval Δt during which a receiver at rest records $f_0\,\Delta t$ wavelength cycles. If the receiver moves with speed $v_{receiver}$, it travels during time interval Δt a distance of $v_{receiver}\,\Delta t$. Within this distance, $v_{receiver}\,\Delta t/\lambda$ wavelength cycles are counted. Thus, the total number of wavelength cycles is the sum of $f_0\,\Delta t$ and $v_{receiver}\,\Delta t/\lambda$, and the frequency observed by the receiver is obtained from this number after dividing by Δt:

$$f_{receiver} = f_0 \pm \frac{v_{receiver}}{\lambda} = f_0\left(1 \pm \frac{v_{receiver}}{c}\right). \quad [16.14]$$

The $\pm$ sign in Eq. [16.14] has been introduced to describe both possibilities: the receiver moving toward (+) or away (−) from the source.

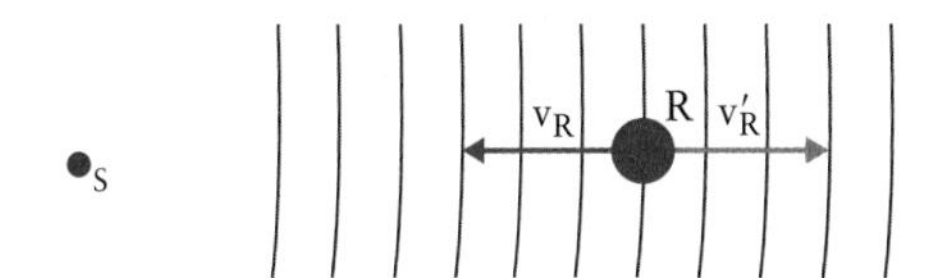

Figure 16.9 Sketch of the motion of a receiver R moving with speed $v_{receiver}$ toward the sound source S or moving with speed $v'_{receiver}$ away from the sound source S. The nearly vertical lines indicate the waves emitted by the source; the distance between the lines corresponds to the wavelength.

16.4.2: Sound Detection with Moving Source and Stationary Receiver

Fig. 16.10 shows a source that emits waves of wavelength λ. The concurrent motion of the source leads to an apparent wavelength λ_{source}. If the speed of the source is v_{source}, we calculate a recorded wavelength of $\lambda_{source} = \lambda_0 \pm v_{source}/f_0$ if the source moves straight toward or straight away from the receiver. The recorded frequency is then calculated from the speed of sound $c = \lambda_{source}\, f_{source}$:

$$f_{source} = \frac{c}{\lambda_{source}} = \frac{c}{\lambda_0 \pm \frac{v_{source}}{f_0}} = f_0 \frac{c}{c \pm v_{source}},$$

which is usually written in the form:

$$f_{source} = f_0 \frac{1}{1 \pm \frac{v_{source}}{c}} \quad [16.15]$$

The (+) sign applies when the source moves away from the receiver and the (−) sign applies when the source moves toward the receiver.

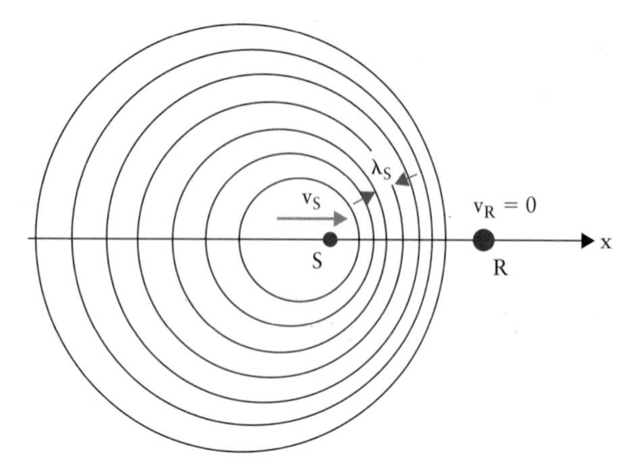

Figure 16.10 Sketch of the motion of a sound source S moving with speed v_{source} toward a receiver R at rest, $v_{receiver} = 0$. The large circles represent spherical sound waves emitted from the point sound source. The motion of the source along the x-axis leads to a change in the wavelength λ_{source} of the emitted sound as received by the receiver.

16.4.3: Doppler Ultrasound Diagnosis

A combination of both effects leads to the m-mode of ultrasound imaging, usually referred to as **Doppler ultrasound**. In this technique we measure the speed of moving components in the human body, such as erythrocytes in blood vessels. The principle is illustrated in Fig. 16.11. A standard transducer (shown at right) is brought into air-free contact with skin. A typical transducer frequency in this technique lies between 2 MHz and 8 MHz. At the same time, the transducer serves as a receiver. The rate of switching between the two functions is 1 kHz. Three processes are combined in the Doppler ultrasound technique:

- The sound wave emitted by the resting transducer is received by the moving blood cell (receiver). The blood cell moves with speed v, e.g. in Fig. 16.11 toward the transducer, but possibly at an angle for which further corrections are needed that we do not discuss here.
- At the same instant it receives the ultrasound, the erythrocyte becomes a passive source by reflecting the sound wave (echo).
- The transducer receives the sound wave emitted from a moving source.

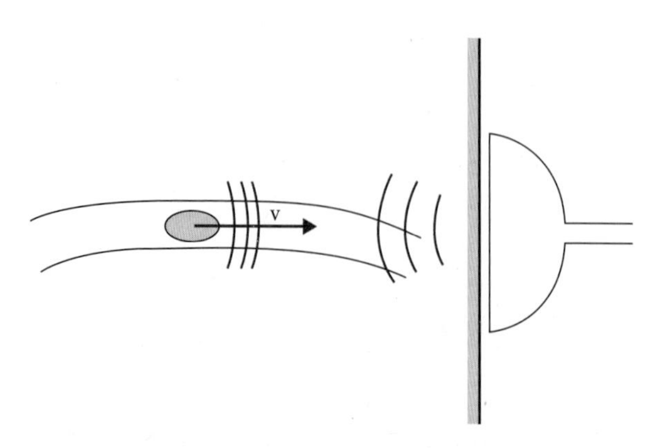

Figure 16.11 Sketch of the Doppler ultrasound method. A transducer at right, in contact with the skin, sends an ultrasound signal, which is reflected by an object moving with speed v, e.g. an erythrocyte.

To describe the overall combined effect of both cases of the Doppler effect discussed above. We write for the first step in which the moving blood cell is the receiver:

$$f_{erythrocyte} = f_0\left(1+\frac{v}{c}\right).$$

The sign is chosen positive because the receiver moves toward the sound source in Fig. 16.11. For the third step, in which the moving blood cell is the sound source, we write:

$$f_{transducer} = f_{erythrocyte}\left(\frac{1}{1-v/c}\right).$$

This time, the sign is chosen negative because the source moves toward the receiver. Note that the velocity v has no subscript because it is the same velocity of the erythrocyte in both cases. The last two equations lead to:

$$f_{transducer} = f_0\left(\frac{1+v/c}{1-v/c}\right). \qquad [16.16]$$

Eq. [16.16] also confirms that the measured frequency would remain f_0 if both receiver and source moved with the same velocity in the same direction. In this case, the signs in the brackets of the previous two equations would be the same and the bracket in Eq. [16.16] would become 1.

The difference between the frequency emitted and received by the transducer is called the **Doppler shift** Δf. Eq. [16.16] can be simplified in the limiting case that the speed of the blood cell is much slower than the speed of sound in the tissue, $v << c$. Since this is always the case, the Doppler shift can be shown to be linearly dependent on the speed of the blood cell:

$$\Delta f = f_{transducer} - f_0 = f_0\left(\frac{1+v/c}{1-v/c}-1\right)$$
$$= f_0\left(\frac{1+v/c-(1-v/c)}{1-v/c}\right),$$

which simplifies to:

$$\Delta f = 2f_0\frac{v}{c} \; for \; v \ll c. \qquad [16.17]$$

CASE STUDY 16.5

How does a Doppler ultrasound measurement change for a blood vessel with a stenosis, i.e., with a constriction?

ANSWER: *We first describe the result of a Doppler ultrasound measurement for a healthy blood vessel. The average speed of the erythrocytes in the aorta is 0.16 m/s, while a typical speed of sound in body tissue is 1540 m/s. Thus, the assumption v << c is satisfied and Eq. [16.17] applies. Assuming a transducer frequency of 5 MHz, an average Doppler shift of about 1 kHz is expected:*

$$\Delta f = \frac{2 f_0 v}{c} = \frac{2(5 \times 10^6 s^{-1})(0.16 \frac{m}{s})}{1540 \frac{m}{s}} = 1040\ \text{Hz} = 1.04\ \text{kHz}.$$

Since the blood pressure varies, variations in the speed of the erythrocytes during the pumping cycle of the heart are expected. This is illustrated in Fig. 16.12, in which the Doppler shift is plotted as a function of time.

For a stenosis, a different blood speed pattern is expected. This is illustrated in Fig. 16.13. A constriction in a blood vessel causes the flow speed to increase and even to reverse due to turbulence. A range of different erythrocyte speeds exists in a blood vessel section near a stenosis, thus a broadening of the speed-versus-time curves of a Doppler ultrasound measurement is observed.

(a)

v

t

t_A

t_B

(b)

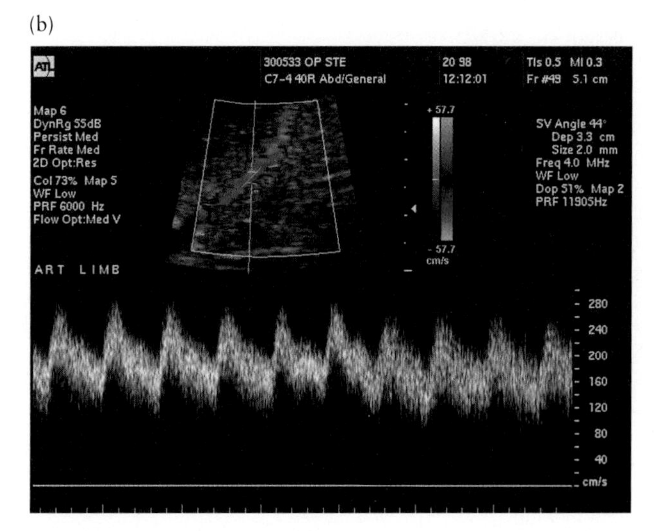

Alain Pol, ISM/Science Photo Library/Custom Medical Stock Photo

Figure 16.12 (a) Typical Doppler shift pattern for erythrocytes in an artery. The Doppler shift is converted to a speed of the blood using Eq. [16.17]. The periodic speed variation of blood cells between times t_A and t_B is due to the rhythmic action of the heart. (b) The same data as recorded with a clinical ultrasound set-up.

(a)

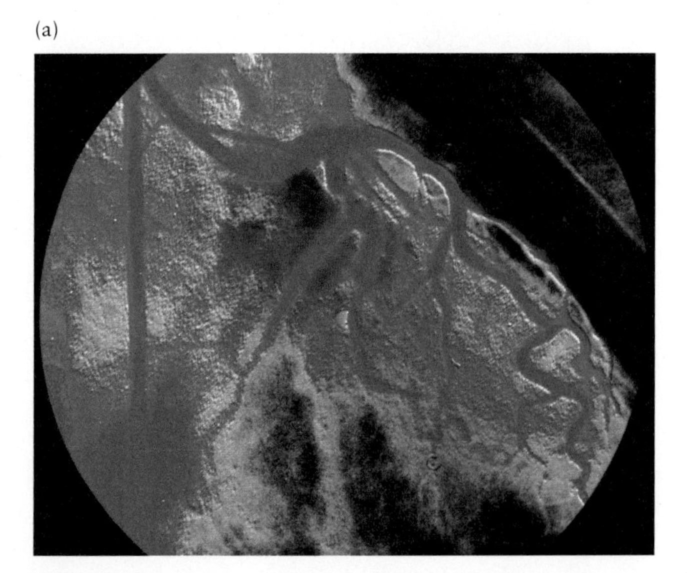

(b)

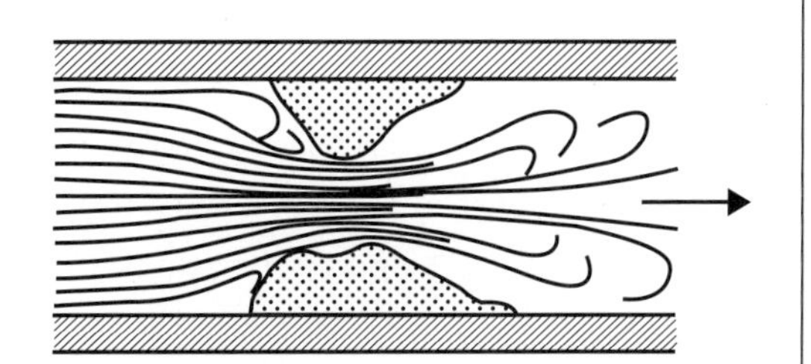

Alain Pol, ISM/Science Photo Library/Custom Medical Stock Photo

Figure 16.13 (a) Coloured angiogram of the heart with a blood vessel obstruction (stenosis). The stenosis section occurs in the circumflex coronary artery: find the narrowed section immediately above the inverted U-shaped artery in the centre. (b) Illustration of the origin of a broadening of the Doppler shift for blood cells passing through a blood vessel with a stenosis.

SUMMARY

DEFINITIONS

- Acoustic impedance Z:

$$Z = \rho \; c$$

with ρ the density and c the speed of sound in the medium

- Intensity transmission T_I is the ratio I_t/I_i for sound reaching an interface
- Intensity reflection R_I is the ratio I_r/I_i for sound reaching an interface

UNITS

- Acoustic impedance Z: kg m^{-2} s^{-1} = rayl

LAWS

- Sound absorption:
  - for intensity (Beer's law):

$$I = I_0 \, e^{-\beta x} = I_0 \, e^{-\frac{x}{x_{abs}}}$$

I_0 is the source intensity at $x = 0$, β is absorption coefficient, x_{abs} is the absorption length

  - for amplitude:

$$D = A_0 \, e^{-\alpha x} \sin(\omega \cdot t - kx)$$

D is the gas pocket displacement and α is the decay coefficient for the sound amplitude

- Intensity transmission T_I at an interface between medium 1 and medium 2:

$$T_I = \frac{4 Z_1 Z_2}{(Z_1 + Z_2)^2}$$

- Intensity reflection R_I at an interface between medium 1 and medium 2:

$$R_I = \left(\frac{Z_2 - Z_1}{Z_1 + Z_2} \right)^2$$

- Doppler effect:
  - for moving receiver and stationary source:

$$f_{receiver} = f_0 \left(1 \pm \frac{v_{receiver}}{c} \right)$$

  - for moving source and stationary receiver:

$$f_{source} = f_0 \left(\frac{1}{1 \pm v_{source}/c} \right)$$

  - for receiver and source moving relative to medium:

$$f_{combined} = f_0 \left(\frac{1 \pm v_{reciever}/c}{1 \pm v_{source}/c} \right)$$

MULTIPLE-CHOICE QUESTIONS

MC–16.1. When sound is absorbed in a medium, its intensity level IL decreases with distance travelled through the medium x as (*note:* β is a constant)
(a) IL $\propto e^{-\beta x}$.
(b) IL $\propto -x$.
(c) IL $\propto \beta$.
(d) IL $\propto \ln(-x)$.

MC–16.2. A sound travels from medium I into medium II. Consider the following four conditions: (I) For the speed of sound $c_I = c_{II}$ applies. (II) For the density of the medium $\rho_I = \rho_{II}$ applies. (III) For the wavelengths in the two media $\lambda_I = \lambda_{II}$ applies. (IV) For the frequencies in the two media $f_I = f_{II}$ applies. No reflection of sound intensity at the interface between media I and II occurs if the following condition(s) is/are fulfilled.
(a) only (I)
(b) only (II)
(c) both (I) and (II)
(d) both (I) and (III)
(e) both (II) and (IV)

MC–16.3. Ultrasound cannot be heard by humans because
(a) its intensity is too low.
(b) its frequency is too low.
(c) its amplitude is too high.
(d) its pressure variations are too high.
(e) its frequency is too high.

MC–16.4. Fig. 16.14 shows a bat using echolocation to detect its prey. The animal uses the reflected frequency to analyze the state of motion of the insect. This is possible because of

Figure 16.14 A bat catching prey in the dark.

(a) the formation of beats.
(b) the formation of standing waves.
(c) the Doppler effect.
(d) the adiabatic processes during sound propagation.
(e) the formation of a second harmonic.

MC–16.5. A moth flies along a path perpendicular to the flight path of a bat. While the moth is within a narrow range of angles in front of the bat, the bat detects a reflected frequency that is
(a) less than its emitted frequency.
(b) the same as its emitted frequency.
(c) more than its emitted frequency.
(d) no longer in the range it can hear.
(e) in a range that attracts dogs, like a dog whistle.

MC–16.6. Doppler ultrasound is used in medicine to detect the following physiological feature:
(a) bone fractures.
(b) blood flow velocity.
(c) blood pressure.
(d) nervous breakdowns.
(e) respiration rate under stress.

CONCEPTUAL QUESTIONS

Q–16.1. As a wave and its reflected wave move through each other in a tube that is aligned with the x-axis, there is an instant when the gas in the tube shows no displacement from equilibrium: $D = 0$ for all positions x. At that instant, where is the energy carried by the wave?

Q–16.2. Why is it not possible for two divers to communicate by talking underwater?

Q–16.3. You are moving toward a stationary wall while emitting a sound. Is there a Doppler shift in the echo you hear? If so, is it the case of a moving source or the case of a moving receiver?

Q–16.4. Explain how bats use "sonar" to detect objects and map their environment.

Q–16.5. How is the "echo" of a sound wave generated?

Q–16.6. What happens to a sound wave when it encounters an object and travels from one medium to another?

ANALYTICAL PROBLEMS

P–16.1. A technician tries to image with ultrasound but has not applied the jelly-like fluid for a proper contact of the transducer to the skin. To see the implications for imaging, we study an air-to-muscle tissue interface and calculate (a) the intensity reflection R_I and (b) the intensity transmission T_I. (c) Express the result in part (b) in unit dB. (d) Why does this attempt to image not lead to useful images? Why does the jelly-like fluid improve the image?

P–16.2. (a) We operate with an incident ultrasound pulse of 50 mW/cm^2 and receive an echo pulse with an intensity of 4 μW/cm^2. What is the intensity loss of the pulse in dB? (b) An ultrasound pulse is attenuated by 8.7 dB when passing through a depth d of tissue. If the detected (attenuated) pulse intensity is 15 mW/cm^2, what is the incident pulse intensity?

P–16.3. We consider a 2.6-MHz ultrasound pulse sent into an object of alternating tissue layers of different acoustic impedance. Each subsequent layer is 1 cm thick. Both types of tissue have a sound attenuation coefficient $\alpha =$ 0.5 dB/(cm A MHz). (a) We assume that the speed of sound in tissue 1 is 1450 m/s, while for tissue 2 it is 1678 m/s. The density of tissue 1 is 1.1 g/cm^3, while the density of tissue 2 is 1210 kg/m^3. Calculate the acoustic impedance of each type of tissue. (b) What are the reflected and transmitted intensities at any of the interfaces, expressed as a fraction of the intensity incident on the interface? Express the results also in unit dB. (c) Assume that the object is composed of 20 layers, 10 layers of each type of tissue. Further assume that the last layer ends at an interface to air. What is the intensity loss of the sound wave in dB just before it reaches the final air interface? How many echoes are received by the transducer when we ignore secondary effects such as reflected reflections? *Hint:* The transducer cut-off attenuation is 60 dB; i.e., signals attenuated by 60 dB and more are no longer detected.

P–16.4. (a) For an unknown tissue type we observe the ultrasound echo 38 μs after the signal of $f = 4.0$ MHz is emitted. What is the depth of the interface at which the ultrasound was reflected? *Hint:* For unknown tissue, a speed of sound of 1540 m/s is used as a typical value. (b) By what distance does the value in part (a) underestimate the actual value if the tissue through which the signal travels is muscle tissue?

P–16.5. Ultrasound echolocation is used by bats to enable them to fly and hunt in the dark. The ultrasound used by bats has frequencies in the range from 60 kHz to 100 kHz. We consider a bat that uses an ultrasound frequency of 90 kHz and flies with a speed of 10 m/s. What is the frequency of the echo the bat hears which is reflected off an insect that moves towards the bat with a speed of 3 m/s?

P–16.6. We consider an ultrasound imaging application with an 8 MHz transducer that has a cut-off attenuation of 60 dB. The transducer emits pulses that consist of three complete wavelength cycles. The imaged tissue has an average sound attenuation coefficient of $\alpha = 0.5$ dB/(cm A MHz) and carries sound at a speed of 1540 m/s. The image is formed with 200 pulses (lines). (a) What is the maximum depth from which an echo can be measured if the overlaying tissue layer is uniform? (b) What is the maximum time delay of an echo? (c) What is the maximum pulse repetition frequency with which the transducer may operate? (d) What is the best spatial resolution in the direction of sound propagation? (e) What is the maximum image frame rate?

P–16.7. The sound waves generated by a locomotive engine and a jet engine of an airplane are 100 and 900 microwatts/cm^2 respectively. How many decibels is the louder sound above the other?

P–16.8. A source emits sound energy uniformly in all directions. A sound-level meter records 97 dB when situated 3.0 m from the source. Given that I_0, the threshold intensity of hearing is 2.0×10^{-12} Js/m^2, calculate the total sound power emitted by the source.

P–16.9. The "rule of thumb" has it that in order to estimate the distance to the source of a lightning flash, one has to divide the time in seconds between the moment the light is seen and the moment the sound of thunder is heard by the factor three (3). The result is the distance to the lightning flash source in kilometres (km). Explain how this estimate of the distance works?

P–16.10. Describe, using diagrams when needed, what is heard when (a) a vibrating tuning fork is placed above a long vertical tube containing water, which slowly runs out at the lower end; (b) two sound waves of slightly different frequencies reach the ear simultaneously; and (c) a car, sounding its horn continuously, passes close by a stationary observer.

P–16.11. The speed of sound c_s in an ideal gas satisfies the following relation:

$$c_s = \alpha\sqrt{T},$$

where T is the temperature and α is a constant with the proper units. A parallel beam of sound passing through an ideal gas at 15°C makes an angle of 40° on a plain thin membrane separating the gas from another sample of the same ideal gas at 90°C. (a) What will be the angle of refraction of the beam? *Note:* You can assume that the membrane itself does not induce any deviation of the beam. (b) If the incident beam is to be totally internally reflected at the boundary, calculate the minimum temperature to which the gas on the other side of the membrane must be raised to.

P–16.12. (a) Define what frequency is, and explain why an observer moving toward or away from a stationary source of sound perceives an apparent frequency, which is not the true frequency of the source. (b) Derive an expression for this apparent frequency f in terms of the true frequency f_0, the speed of the observer v_0, and the speed of sound c_s.

P–16.13. A model aircraft with an engine producing vibrations of constant frequency of 500 Hz flies at constant speed in a horizontal circle of radius 15 m and completes one revolution in 5.0 s. An observer situated in the plane of the circle and 50 m from its center monitors the frequency of the sound from the engine. (a) Explain why the observed frequency shows periodic variations. (b) Derive the relation for the minimum observed frequency in terms of f, the true frequency of the engine; v, the speed of the aircraft; and c_s, the speed of the sound in air. Write down the corresponding relation for the maximum observed frequency. (c) Taking c_s to be 340 m s^{-1}, calculate the maximum and minimum observed frequencies and determine the time interval between the occurrence of a maximum frequency and the next minimum frequency.

P–16.14. (a) Sound waves are longitudinal waves. What is meant by longitudinal? (b) Two sources of sound S_1 and S_2 are placed at a distance of 6.0 m apart at either end of a narrow pipe. Both sources are emitting waves of wavelength 1.5 m and of similar amplitude, which travel along the pipe. (i) Draw on a diagram how the amplitude of the resultant wave varies along the line joining the two sources. (ii) In what two ways would the resultant wave be different if the sources were replaced by microwave sources?

P–16.15. The intensity of a sound as heard by an individual depends on the frequency and amplitude of the sound. (a) Give definitions for frequency and intensity. (b) Draw a graph to show how loudness of a sound, as heard by a person with normal hearing, depends on frequency when the intensity of the sound is constant. Point out any special features of the graph. (c) Describe the response of the ear to different intensities of sound with the frequency maintained constant. Define the intensity level.

Part Four

ELECTRIC PHENOMENA

In this part we study a new fundamental force. So far, the only non-contact force we discussed was gravity, which is called within the biosphere more specifically the weight. It is interesting to put this step a bit into historical context. The fundamental concepts of mechanics and the gas laws go back to the late 1600s with names like Newton, Pascal, and Boyle. When we moved on to thermodynamics we jumped relatively fast to the time of the French Revolution, in which individuals like Lavoisier, Charles, and Gay-Lussac became embroiled. The conservation of energy then brings us to the early and mid-1800s, with names such as Carnot, Helmholtz, Joule, and Clausius attached. The empiric description of the transport phenomena developed in parallel, driven by Fourier and Fick in the early 1800s.

With electricity, we observe focus on the late 1700s to mid-1800s again, and the individuals we see at work are Coulomb, Faraday, Ampère, and Ohm. They have in common a more practical thinking, with questions such as What can I use my discoveries for?, which is a different approach to that of the earlier scientists who focussed more on a fundamental understanding of nature. For this reason, Newton and Pascal are also cited as philosophers, while people such as Faraday appear like modern entrepreneurs to us. Read these chapters also a bit with this in mind, and enjoy the different dynamics of thought, illustrating one strength of us humans: we have very flexible minds and adapt rapidly to new angles of view of the world around us.

CHAPTER 17

Electric Force and Field

Biological systems are electrically active at the molecular level. Many observations can be explained only when electric effects are included, from the unique role of water in aqueous solutions to the conduction of a nerve signal through an axon.

Electric charge and mass are two different properties of matter. Different from mass, charge is quantized and comes in two types: two equal electric charges repel each other, while two dissimilar charges attract each other. The electric force is a long-range contact-free force that varies in magnitude with the inverse square of the distance between the interacting charges (Coulomb's law).

The electric field is a new concept that is introduced to allow us to handle the many possible variations of charge arrangements. For stationary charge distributions it can be derived from Coulomb's law only, and represents at each point in space the net electric force per unit charge due to the studied stationary charges in the system.

The magnitude of the electric field due to a stationary point charge is inversely proportional to the square of the distance from the point charge; the magnitude of the electric field due to a dipole is inversely proportional to the cube of the distance to the dipole. Therefore, dipoles interact strongly at close proximity, whereas they show a significantly weaker electric interaction at longer distances. This is evident in the tight hydration shells that form around ions in aqueous solutions.

The electric field has its simplest possible form for a parallel plate arrangement with equal but opposite surface charge densities on the two plates: the field is constant in magnitude and direction for all positions between the plates. This simplicity makes the charged parallel plate arrangement a preferred model for biological membrane systems.

Water defines the conditions for life on Earth. It is the one chemical compound scientists seek on other planets to determine whether they may bear life. Water receives this extraordinary attention in the life sciences because it has a long list of unique physical properties.

The water molecule is shown in Fig. 17.1. It consists of one oxygen atom and two hydrogen atoms, which are arranged at an angle of 104.5°. The bonds that hold the atoms together in the molecule are polar covalent bonds; i.e., strong chemical bonds exist between hydrogen and oxygen, but the electrons that form these bonds are shifted toward the oxygen atom. As a result, the oxygen end of the molecule carries a partial negative charge δ_- and the hydrogen end carries a partial positive charge δ_+. Opposite charges separated by a fixed distance define a dipole.

- **Hydrogen-bond formation.** Dipoles interact electrically with ions and other dipoles. The interaction is strongest at short range and more effective along the dipole axis than in other directions (directional anisotropy). In liquid water, where molecules are sufficiently mobile, the positive end of one dipole and the

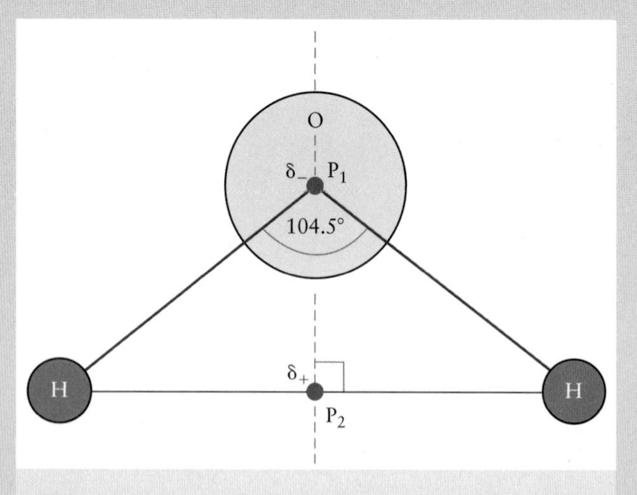

Figure 17.1 The water molecule consists of two hydrogen atoms and one oxygen atom. The hydrogen atoms are connected to the oxygen atom with covalent bonds. The electrons in these bonds are drawn closer to the oxygen atom, leading to a net negative charge δ_- near the oxygen end of the molecule at point P_1, and a net positive charge δ_+ near its hydrogen end at point P_2.

negative end of another dipole approach each other. In close proximity they then form a hydrogen bond, which is an *intermolecular* sharing of some of the excess electric charge carried by the oxygen atom.

- **Hydrogen bonds lead to cohesion.** Water molecules stick to each other as a result of hydrogen bonding. We observe this when water droplets form, such as on the branch shown in Fig. 17.2. Hydrogen bonds are more fragile than covalent bonds: they are only about 5% to 10% as strong. In liquid water, with its enhanced mobility of molecules relative to the solid state (ice), each hydrogen bond lasts only for about 1×10^{-12} s, but so many of these bonds exist in a droplet of water at any given time that the net effect is profound. Thus, we have found a collective property of matter: hydrogen bonds hold water together at a macroscopic scale. This effect is called **cohesion.**

© Photos.com

Figure 17.2 Water cohesion is seen in the formation of water droplets on leaves and branches.

Related to cohesion is the effect of surface tension, which is a measure of how much effort is needed to increase the surface area of a liquid. Water has a larger surface tension than most other liquids due to the large cohesion between the water molecules. We note the large surface tension of water when we observe animals such as water striders standing on water without breaking the surface.

Large values of cohesion and surface tension allow water to be liquid at room temperature. Most other small molecules are in the gaseous state at room temperature; the closest to being liquid is ammonia, which boils at −33°C. In an ammonia solution, all chemical processes are significantly slower because chemical reaction rates roughly double for every increase of 10 K in temperature.

- **The high latent heat of vaporization of water enables evaporative cooling.** The latent heat of vaporization is the amount of energy required for the phase transition of one mol of a material from the liquid to the vapour state at its boiling point. Liquids with hydrogen bonds have a large latent heat of vaporization because the hydrogen bonds have to be broken during evaporation. Evaporation takes place at all temperatures when the vapour phase is undersaturated. Thus, a large amount of energy is required when water evaporates from a surface; this energy is taken from the internal energy of the substrate.

On a global scale, evaporative cooling helps moderate the climate: a significant fraction of the solar heat absorbed by equatorial oceans is stored in water vapour that forms during evaporation of surface water. This moisture then circulates toward the poles, where it releases the stored thermal energy as it condenses to form rain or snow. On the level of individual organisms, evaporative cooling allows perspiration to be an effective mechanism to prevent terrestrial animals from overheating. Water evaporation from leaves also keeps plant tissues from overheating in direct sunlight.

- **Ice floats on the surface of liquid water.** Water is one of the few materials that are less dense in solid form than in the liquid state. While most materials contract when they solidify, water expands to accommodate the hydrogen-bond structure. At temperatures above 4°C, water behaves like other liquids: it expands when heated and it contracts when cooled. Below 4°C, more and more hydrogen bonds remain stable, which requires a greater intermolecular distance. At 0°C water then freezes as the thermal energy of the molecules is no longer sufficient to break hydrogen bonds. Ice is fully ordered with a wide-open molecular arrangement, leading to a 10% reduction in **ice density** compared to liquid water at 4°C. This effect is ecologically important because if ice were to sink, ponds, lakes, and oceans would eventually freeze solid, making life as we know it impossible. To the contrary, floating ice thermally insulates the liquid water below, allowing life to exist under the frozen surface.
- **Water is an effective solvent in chemistry.** Water forms in an exothermal reaction from the elements (which is the energy-supplying process in the production of ATP in the mitochondria) and is the by-product of many chemical reactions, including the main metabolic processes in our body. However, its chemical importance is based on its role as a solvent. A sodium chloride crystal would not dissolve if the water molecules could not form a hydration shell that stabilizes the ions in solution. A hydration shell consists of a large number of water molecules that form a layer around a charged particle. This shell is energetically favoured because of the electric interaction between the charged particle and the water molecule.

Aqueous solutions are widespread: from seawater to the cytoplasm in cells, a great variety of dissolved ions are found. Water is a very versatile solvent. A compound does not need to be ionic to dissolve in water; sugar, for example, dissolves because it is a polar molecule (dipole structure). Even molecules as large as proteins dissolve in water as they often have ionic and/or polar regions. The impressive amount of salt dissolved in seawater is illustrated in Fig. 17.3, which depicts salt farming in Southeast Asia.

All these phenomena point to the electric dipole structure of the water molecule in Fig. 17.1 as a common cause. In this chapter, we want to establish the fundamental laws that characterize electric systems. Choosing the interactions of a molecule as a focal point for this chapter also suggests that the discussion of electric phenomena will take a different path than the discussion of forces in mechanics. In mechanics, the fundamental force of gravity was simplified to a single application called weight because we are limiting our discussion to processes in the biosphere; weight is always directed downward and its magnitude is proportional to the mass of the object. The discussion of the electric force won't be that short because there is no single dominant charge that plays a role equivalent to the mass of Earth.

It is therefore important that we avoid becoming overwhelmed by the multitude of electric phenomena. An introduction-level text must take a well-organized and selective approach to achieve this goal: we will start by defining the concept of charge and the law that describes the interaction of just two charges as the simplest possible case. But then we do not expand in what would be a futile effort to develop a complete overview of electricity. Rather, we pause after the introduction of the electric force and reflect on what we want to accomplish with the basic concept established. That discussion will lay out the plan for the remainder of this and the next two chapters: we find that the water dipole we referred to above and the charged membrane system we discuss in detail in the next two chapters for the nervous system are two versatile key models. Once we have studied these two models, most other electric systems can be discussed analogously.

Sigit Pamungkas/Reuters/Landov

Figure 17.3 Salt farming in Southeast Asia.

17.1: Electric Charge and Force

17.1.1: Beyond Mass: The Particle Property Electric Charge

A wide range of phenomena exists in which particles are not sufficiently characterized by their mass. Such phenomena include acidity regulation of the blood, salt dissolution in water, and salt counter-current filtration in the kidneys. In order to describe these phenomena, a second, mass-independent property of matter has to be introduced: the **electric charge**. The physical laws that govern the behaviour of objects carrying electric charge are distinct from the laws of mechanics because a new fundamental force is associated with electric charges: the electric force. This force is the second fundamental force we discuss in this textbook, following our earlier discussion of gravity and weight.

In the same manner that mass is an intrinsic property of particles, electric charge is an intrinsic property of the same particles that comprise matter. Because particles carrying single charges are usually very small, the concept of a **point charge** is introduced.

KEY POINT

A point charge is a charged particle of negligible size.

The mass and the charge of a particle are independent from each other. In our discussion of mechanics we established that there is only one type of mass; objects may have more or less of it, but none can have a different type of mass, e.g., a negative mass. The idea of a negative mass had indeed been discussed for a while in the scientific community. It was postulated in 1697 as part of Georg Ernst Stahl's *phlogiston theory* to explain combustion. He claimed that combustion is the loss of particles with negative mass; he called these particles phlogiston. This theory was finally discredited in 1777 by Antoine Laurent de Lavoisier when he properly described combustion as a chemical reaction with oxygen.

For charges, on the other hand, two different types exist: two charges of the same type repel each other and two opposite charges attract each other. To distinguish these two kinds of charges we call one type of charge a **positive charge**, q_+, and the other type a **negative charge**, q_-. One could have called them blue and red charges instead, but invoking the notation of mathematical signs turned out to be convenient as opposite charges indeed offset each other in their physical effects: a given amount of positive charge is shielded by the same amount of negative charge in close proximity, which explains the apparent electric neutrality of matter.

To see point charges of either type display their physical properties at a macroscopic level, the electric charges in a system must be separated. This can be achieved in physical experiments but also happens inadvertently in our daily life, e.g., when your shoe soles rub against plastic surfaces on a dry day. When you later touch another person or a conducting surface such as a doorknob, you feel a tingle as electric charges leave your skin. Hospital personnel therefore must wear special conducting shoes to avoid sparking when working with oxygen.

17.1.2: The Magnitude of the Electric Force

We introduce the property charge quantitatively in terms of the force that occurs between separated point charges. The relation between electric force and charge was discovered by Charles Augustin de Coulomb in 1784 and is called Coulomb's law. He experimented with metal spheres carrying electric charges. He observed that:

KEY POINT

The magnitude of the electric force between two charged spheres is proportional to the absolute amount of charge on each sphere, and is proportional to $1/r^2$, where r is the distance between the spheres.

His observations are summarized in a formula for the electric force. We consider initially the magnitude of this force, F_{el}, and postpone the discussion of directional features to the next section:

$$F_{el} \propto \frac{|q_1||q_2|}{r^2}.$$

Note that F_{el} is proportional to the absolute values of the two interacting charges q_1 and q_2. We will include the signs of the charges only in the next section, when introducing the electric force as a vector. We also note that the force is reduced to 1/4 when the distance r between the spheres is doubled. The standard unit of charge is C (coulomb), named in honour of Coulomb's work.

To write the electric force above as an equation, a proportionality factor is required. This factor is $1/(4\pi\varepsilon_0)$, in which ε_0 is a fundamental physical constant called **permittivity of vacuum**: $\varepsilon_0 = 8.85 \times 10^{-12}\ C^2/(N\ m^2)$. How ε_0 is related to the idea of *reaching through vacuum* becomes clear later in this chapter when we discuss the concept of electric field. The **electric force** is then written in the form:

$$F_{el} = \frac{1}{4\pi\varepsilon_0}\frac{|q_1||q_1|}{r^2}. \qquad [17.1]$$

This is called **Coulomb's law**. Quantitatively, two equal point charges of 1 C each, placed at a distance of 1 m, attract each other with a force of 9×10^9 N. Compared to forces we saw in earlier chapters, this force is very large; we will develop an intuitive sense for the strength of the electric force in the many examples we discuss below.

An important difference between mass and charge is the fact that the latter is **quantized**. What does this mean? The concept of quantization can be illustrated with our use of money. Goods you buy at a store cannot cost any amount of money; they can cost only a multiple of the smallest unit of currency, which in North America is 1 cent. Even when sales taxes are added to a bill, you will never be asked to pay 7.5 cents. Any payment will always be an amount that is an integer multiple of 1 cent. In nature, no such limitation exists for mass; any amount of mass can occur. However, with respect to charge, all processes are based on the transfer of an integer multiple of the smallest amount of charge, which we call the **elementary charge** $e = 1.6 \times 10^{-19}$ C.

Table 17.1 illustrates this for the mass and charge of the fundamental particles in atoms: the electron (e^-), the proton (p^+), and the neutron (n^0). While the masses of elementary particles vary, their charges have the same value regardless of whether it is positive or negative. Charge quantization and the value of the elementary charge were first determined by Robert Andrews Millikan in 1909 with the experiment sketched in Fig. 17.4. His set-up consisted of a chamber into which oil is sprayed from the metallic nozzle of an oil vaporizer. The oil mist consists of microscopic droplets that often carry an electric charge due to friction of the oil with the inner surface of the metallic nozzle during spraying. All oil droplets in the chamber sink slowly due to their weight. Some fall through a hole in a plate that separates the upper and lower parts of the chamber. In the lower part the oil droplets can be observed with a microscope. The separating plate with the hole and a second plate at the bottom of the chamber are electrically isolated and form a parallel plate arrangement when charged with a battery. As we see later in

TABLE 17.1

Elementary particles in the atom

Particle	Mass (kg)	Charge (C)
Electron	9.11×10^{-31}	-1.6×10^{-19}
Proton	1.673×10^{-27}	$+1.6 \times 10^{-19}$
Neutron	1.675×10^{-27}	0

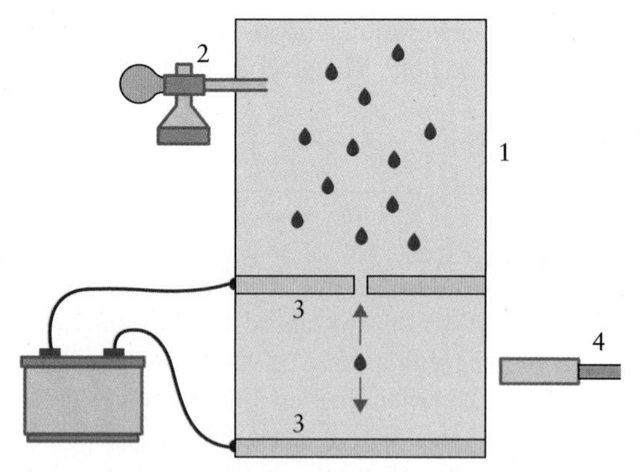

Figure 17.4 Robert Millikan's experiment. Oil mist, consisting of microscopic oil droplets, is sprayed into a chamber (1). Many of the oil droplets carry charges due to the friction of the oil when leaving the metallic nozzle of the vaporizer (2). The droplets sink due to their weight. It is possible to levitate individual oil droplets between two charged metallic plates (3) that form a horizontal, parallel plate arrangement. A mechanical equilibrium can be observed with a microscope (4) and is due to the balance between the weight and the electric force acting upward, as indicated on one lower oil droplet by two arrows. From this experiment the electric charge on the oil droplet is quantified and is always found to be a multiple of the elementary charge.

this chapter, parallel charged plates allow us to exert an electric force on a point charge that is located between them. In the case of Fig. 17.4, we choose the electric force such that it is directed upward, i.e., counteracting the weight. The electric force due to the two plates is adjusted such that a particular oil droplet, chosen with the microscope, levitates at a fixed level in the chamber. This means that the magnitude of the electric force F_{el} (acting upward) and the magnitude of the weight of the droplet W (acting downward) are balanced as required by Newton's first law. The charge of the droplet is determined from the amount of charges needed on the two plates in Fig. 17.4. This experiment has been done numerous times, and an oil drop with a charge other than an integer multiple of the elementary charge has never been found. We quantify Millikan's experiment later in this chapter in Example 17.9.

CONCEPT QUESTION 17.1

Two point charges attract each other with an electric force of magnitude F_{el}. If we triple one of the charges and increase the distance between the points by a factor of 3, what does the magnitude of the force between the particles become?

(A) $F_{el}/9$

(B) $F_{el}/3$

(C) 0 (zero)

(D) $3\ F_{el}$

(E) $9\ F_{el}$

EXAMPLE 17.1

(a) Find the magnitude of the electric force F_{el} exerted on a point charge +3 e by a point charge −5 e that is located 7 nm away. (b) Compare the electric and gravitational forces between the electron and the nucleus (proton) in a hydrogen atom.

Solution

Solution to part (a): We use Coulomb's law. To find the magnitude of the force only the absolute values of both charges are included:

$$F_{el} = \frac{|q_1||q_2|}{4\ \pi\varepsilon_0 r^2}$$

$$= \frac{\left(9\times10^9 \dfrac{\text{N m}^2}{\text{C}^2}\right)(4.8\times10^{-19}\text{C})(8\times10^{-19}\text{ C})}{(7\times10^{-9}\text{ m})^2}$$

$$= 7.1\times10^{-11}\text{ N.}$$

The result in Example 17.1 seems to represent a small force, particularly when you take the short distance between the two point charges into account. However, the elementary charge is a small amount of charge.

Solution to part (b): It becomes evident that the electric force is very strong when we compare it to the gravitational force between the electron and the nucleus in a hydrogen atom. We have all the necessary data for this comparison in Table 17.1, except for the separation distance between the two particles. This distance is called the *Bohr radius* (Chapter 20), with $r_{Bohr} = 5.3 \times 10^{-11}$ m = 0.053 nm. Eq. [17.1] yields:

$$F_{electric} = \frac{1}{4\ \pi\varepsilon_0}\frac{e^2}{r_{Bohr}^2} = 8\times10^{-8}\text{ N,}$$

while Eq. [3.1] yields:

$$F_{gravity} = G * \frac{m_e m_p}{r_{Bohr}^2} = 4\times10^{-47}\text{ N;}$$

i.e., the two forces are different by almost 40 orders of magnitude!

17.1.3: The Direction of the Electric Force

As the discussion of Millikan's experiment suggests, the electric force, like any other force, is characterized by magnitude and direction. Including the direction based on vector notation, Coulomb's law expresses the force a point charge q_1 exerts on a point charge q_2:

$$\vec{F}_{12} = \frac{1}{4\ \pi\varepsilon_0}\frac{q_1 q_2}{r^2}\vec{r}_{12}^{\,0}, \qquad [17.2]$$

in which $\vec{r}_{12}^{\,0}$ is a vector of unit length (denoted by the superscript 0) that points from point charge q_1 to point charge q_2. Note that Eq. [17.2] no longer is limited to the absolute amount of the two charges. The direction of the electric force is therefore determined by two factors:

- the relative positions of the two point charges, and
- the signs of the charges of the two interacting particles.

This is illustrated in Fig. 17.5(a) for two positive point charges, and in Fig. 17.5(b) when the two point charges carry opposite signs. Also included is the force $\vec{F}_{21}$, which is the reaction force to $\vec{F}_{12}$ and acts on point charge q_1.

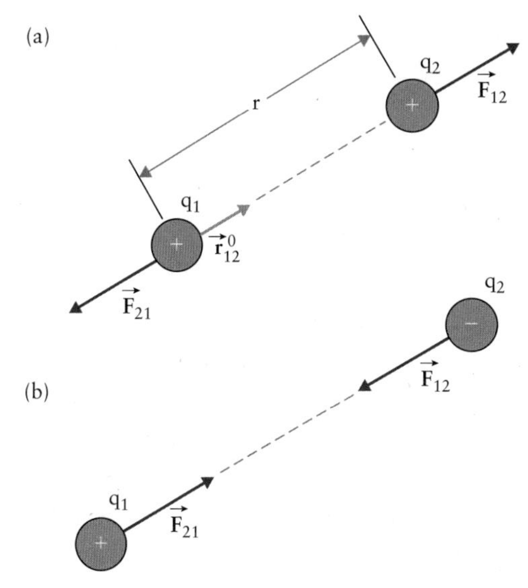

Figure 17.5 The direction of the electric force as a function of the relative positions of two point charges and the signs of their charges. (a) Two positive point charges. (b) Point charges that carry opposite signs.

CONCEPT QUESTION 17.2

Fig. 17.6 shows eight possible cases for the horizontal arrangement of two point charges. The figure further shows the direction of the electric force vector acting in each case on the point charge represented by the larger circle. To describe these arrangements with Coulomb's law the unit vector $\vec{r}^{\,0}$ in Eq. [17.2] has to be identified. Do all eight unit vectors in Fig. 17.6 point in the same direction?

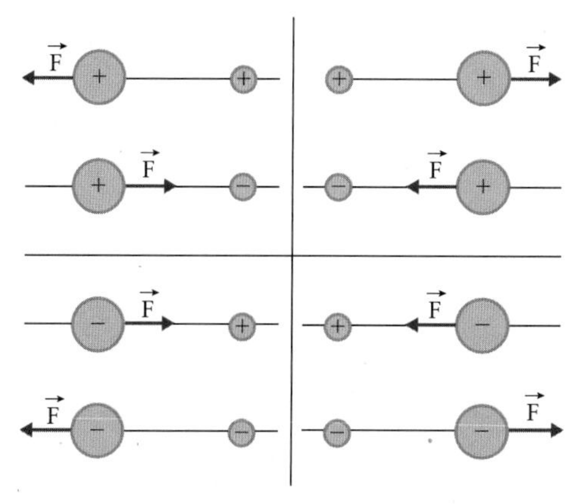

Figure 17.6 Eight cases showing the direction of the electric force on a point charge of interest (large circle) due to a second point charge (smaller circle) in its vicinity.

17.2: Newton's Laws and Charged Objects

In the Newtonian approach the electric force is treated like any other force we have discussed in the previous chapters of this textbook. In particular, if point charges are present in a system, free body diagrams must also include the relevant electric forces. The net force acting on a point charge is calculated as the sum of all forces, including electric forces, that act on the object of interest. When N electric forces act on an object of charge Q, the net electric force is written in vector notation:

$$\vec{F}_{\text{el,net}} = \sum_{i=1}^{N} \vec{F}_{\text{el},i} = \frac{1}{4\pi\varepsilon_0} \sum_{i=1}^{N} \frac{Qq_i}{r_i^2} \vec{r}_i^{\,0}, \qquad [17.3]$$

in which r_i is the distance between the i-th point charge q_i and point charge Q, and $\vec{r}_i^{\,0}$ is the unit vector that is directed from the i-th point charge to point charge Q.

EXAMPLE 17.2

We consider three point charges that are positioned along an axis as illustrated in Fig. 17.7(a). Two of these point charges are positive—$q_1 = +20\ \mu C$, $q_2 = +5\ \mu C$—and are separated by a distance $L = 1.5$ m. At what distance x_0 from q_2 must a negative point charge q_3 be positioned such that the resulting force on it is zero?

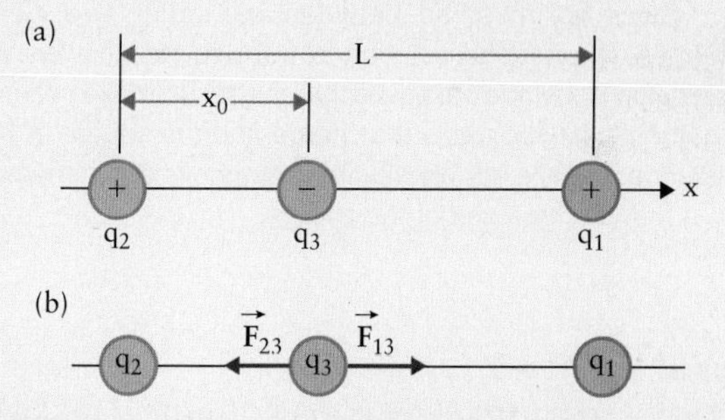

Figure 17.7 (a) Three point charges are positioned along the x-axis. (b) Two forces act on point charge q_3, which is positioned between the other two point charges.

Solution

The sketch in Fig. 17.7(b) shows the same point charge arrangement as Fig. 17.7(a) but highlights the two forces acting on q_3. In that sketch q_3 is attracted toward each of the other two point charges since q_3 is negative while q_1 and q_2 each are positive. As a consequence, the two forces shown in Fig. 17.7(b), $\vec{F}_{13}$ and $\vec{F}_{23}$, point in opposite directions. The circle for charge q_3 and the two forces acting it constitute the free body diagram for this problem since the vertical components of both forces are zero. The absolute values of the horizontal components of the forces are equal to their magnitudes. Mechanical

continued

equilibrium is established when the horizontal component of the net force on q_3 is zero:

$$F_{\text{el,net}} = +\frac{1}{4\pi\varepsilon_0}\frac{|q_1||q_3|}{r_{13}^2} - \frac{1}{4\pi\varepsilon_0}\frac{|q_2||q_3|}{r_{23}^2} = 0,$$

i.e., specifically when $F_{23} = F_{13}$ due to Newton's first law. We substitute the given values for the charges and distances and cancel q_3 on both sides of the equation:

$$\frac{20\times10^{-6}\,\text{C}}{(L-x_0)^2} = \frac{5\times10^{-6}\,\text{C}}{x_0^2}.$$

Note that we need not define the origin of the axis along which the three point charges are located because only their distances enter Coulomb's law. The last equation leads to a quadratic equation that is solved for the distance x_0 between q_2 and the central point charge:

$$5(1.5-x_0)^2 = 20\,x_0^2.$$

This yields:

$$0 = 15x_0^2 + 15x_0 - 11.25.$$

This equation has two solutions:

$$x_0 = \frac{-15 \pm \sqrt{15^2 + 4\cdot15\cdot11.25}}{2\cdot15};$$

i.e.,

$$x_{0,1} = -1.5\text{ m}$$

$$x_{0,2} = +0.5\text{ m}.$$

Since q_3 must lie between the other two point charges, the second solution is the desired answer. We can easily convince ourselves that this result makes sense. Coulomb's law contains the charge and the square of the distance. Thus, a doubling of the distance is compensated by a fourfold increase of the charge.

EXAMPLE 17.3

Three point charges q_1, q_2, and q_3 are located at the corners of an equilateral triangle with side length d, as shown in Fig. 17.8. Assume $q_1 = q_2 = q_3 = +20\ \mu$C and $d = 1.5$ m. What is the magnitude of the net force on point charge q_1?

Solution

We determine first the magnitude of force between any two of the point charges using Coulomb's law (with $r = d$):

$$F_{\text{el}} = \frac{1}{4\pi\varepsilon_0}\frac{|q_1||q_2|}{r^2}$$

$$= \frac{\left(9\times10^9\,\frac{\text{N m}^2}{\text{C}^2}\right)(2\times10^{-5}\,\text{C})^2}{(1.5\,\text{m})^2}$$

$$= 1.6\text{ N}.$$

continued

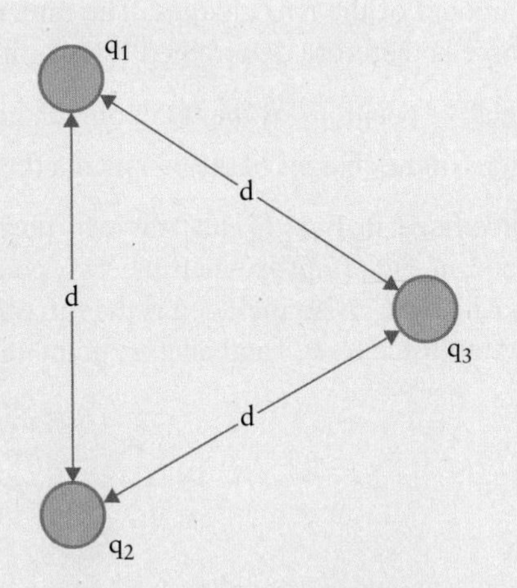

Figure 17.8 Three point charges q_1, q_2, and q_3 are located at the corners of an equilateral triangle with side length d.

The net force on point charge q_1 consists of two forces, one due to point charge q_2 and one due to point charge q_3. These two forces are not parallel to each other, as illustrated in Fig. 17.9. We solve the problem using Eq. [17.3]. We first note that the magnitudes of the two forces $\vec{F}_{13}$ and $\vec{F}_{23}$ are equal because the charges q_2 and q_3 are equal and each is at a distance d from charge q_1. We also note in Fig. 17.9 that $\theta = 60°$, because the triangle formed by the three point charges is an equilateral triangle.

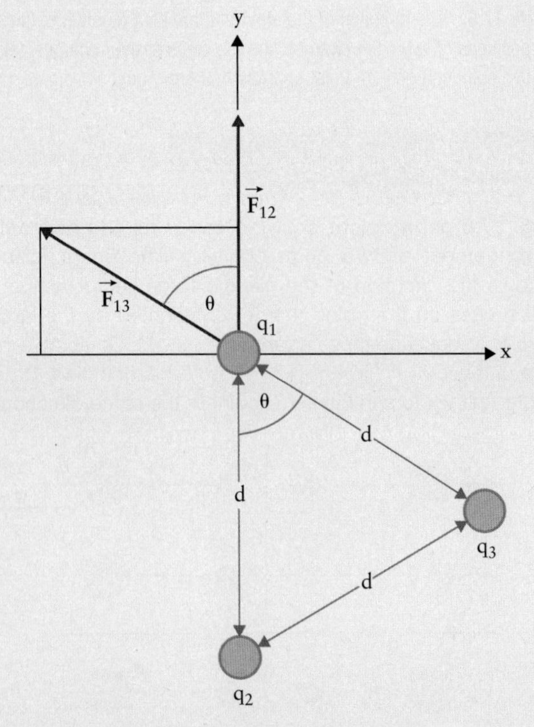

Figure 17.9 The same three point charges as shown in Fig. 17.8, forming an equilateral triangle. Two electric forces act on the point charge q_1 at the top, labelled $\vec{F}_{12}$ and $\vec{F}_{13}$ to indicate which point charge exerts the respective force. These two forces form an angle θ; vector algebra is used to calculate the net force acting on point charge q_1.

continued

Next, we choose a coordinate system to write the two forces in component form. Our choice is shown in Fig. 17.9: the *x*-axis is directed toward the right and the *y*-axis is directed up. With this coordinate system we express the components of the net force based on the components of the two forces $\vec{F}_{13}$ and $\vec{F}_{23}$. For the *x*-component, we find:

$$F_{\text{net},x} = -F_{13}\sin\theta = -(1.6\text{ N})\sin(60°)$$
$$= -1.4\text{ N},$$

and for the *y*-component we get:

$$F_{\text{net},y} = F_{12} + F_{13}\cos\theta$$
$$= 1.6\text{ N} + (1.6\text{ N})\cos(60°)$$
$$= +2.4\text{ N}.$$

The Pythagorean theorem is used to combine the two components:

$$F_{\text{el,net}} = \sqrt{F_{\text{net},x}^2 + F_{\text{net},y}^2} = 2.78\text{ N}.$$

17.3: How Do We Approach Electric Phenomena in Life Science Applications?

Faced with a huge number of possible charge arrangements, we want to organize our approach in two ways before we proceed:

- We divide the force calculations based on Eq. [17.3] into three steps. This is done by distinguishing the different roles the various point charges play in an electric arrangement.
- We select a few representative model systems that allow us to establish the important physical concepts without getting lost in numerous applications.

To address the first point, we follow a method that Michael Faraday originally developed and that has proven very effective in all practical cases. The net force, as introduced in Eq. [17.3], requires a simultaneous inclusion of all point charges in a system because the electric force is a contact-free force and non-negligible interactions take place across some distance. Faraday reasoned that it would be easier to study electric phenomena if we no longer had to consider all point charge interactions at once but could confine the discussion to a point charge of interest that interacts with what he called an electric field. The electric field then represents the effect of all point charges in the system at the position of the point charge of interest.

The success of Faraday's approach lies in the fact that it works for all electric systems. However, we choose to restrict the cases we study to diminish the risk of confusion during a first discussion of the electric field concept: we study only those systems in which the point charge sources of the electric field are stationary (i.e., do not move relative to each other). This then leads to a **stationary electric field**, and this approach is defined in the physical literature as **electrostatics**. This is a restrictive choice because, in real life, charges other than the charge of interest may move. We consider such cases later: in Chapters 18 and 19 we talk about changing electric fields at nerve membranes or in the heart, and in Chapter 21 we explore how moving charges and changing electric fields are related to magnetic fields and electromagnetic radiation. Our choice to confine the discussion in the current chapter to stationary electric fields has two important justifications: (i) it includes the most important systems in the life sciences, and (ii) it is conceptually easier to grasp and allows us therefore to lay a strong foundation for the discussion of more general cases later. Thus, our approach for the current chapter is as follows:

Step (i): We divide the system into two groups of point charges: First, a single point charge that we consider the point charge of interest: conceptually this point charge should be *mobile* since otherwise no interesting electric processes could occur. We follow common practice in the literature and call the mobile charge of interest the **probe charge**. Second, an arrangement of point charges that constitute the **source point charges** for an electric field: we require (in this chapter) the source point charges to be stationary, and therefore usually refer to them as the **stationary point charges**.

Step (ii): For the stationary point charge arrangement, we determine its **stationary electric field**. The term *field* means mathematically that we assign a numerical value to each position in space within a system. In the case of an electric field, these values represent the strength of electric interaction that the probe charge would encounter if placed at the respective position. The electric field is a vector field since its values are calculated with Coulomb's law, in which the force enters as a vector.

Step (iii): Dynamic properties of the probe charge are studied by inserting it into the system. Its motion is fully determined by the stationary electric field and Newton's second law, which then determines the magnitude and direction of the acceleration of the mobile probe charge at every position.

Both steps (ii) and (iii) may require extensive mathematical calculations; however, once the stationary electric field is calculated for a particular arrangement of charges, this field can be used for a large number of dynamic calculations in step (iii) without a need to repeat step (ii). Were we not to introduce this procedure adopted from Faraday's reasoning, all such calculations would be a combination of steps (ii) and (iii); i.e., the calculations for every new position of the probe charge would include the entire field calculation.

For every new arrangement of stationary charges a new calculation of the stationary electric field is required. Thus, any means of limiting the number of arrangements we need to study reduces the overall effort. Specifically, we want to pick a small number of charge arrangements that are representative of as many practical cases as possible. These we consider our **model systems**. To arrive at their selection we can ask ourselves any of the following questions:

- Which models describe most electric systems that occur in nature?
- Which models describe most electric devices engineered for applications in life science laboratories and health care facilities?
- Which models allow us to study basic electricity with the least mathematical effort?

Interestingly, all three questions lead to the same three basic systems. This is not accidental—engineering is an effort to mimic and exploit the approach taken by nature, and nature in turn usually finds the simplest solutions that also tend to be mathematically easiest to describe. Our three model systems are summarized in Table 17.2.

TABLE 17.2

Stationary point charge systems of interest in this textbook

Application in:	Stationary point charge	Electric dipole	Charged parallel plates
Physics	electrons in a metal	antenna	capacitors
Biology	ions in solution	water molecule	nerve membrane

17.3.1: Model System I: Stationary Point Charge

This is the most basic system. We already used it when we introduced Coulomb's law in Eq. [17.2]. Once we identify one of the two point charges in that equation as a probe charge, a single stationary point charge results as the system. Stationary point charges are frequently observed; e.g., each ion in an ionic salt can be treated as a point charge. We will use this approach in P–17.5 to study the electric properties of a void defect in a CsCl crystal, and in P–17.3 for interactions between elementary particles within an atomic nucleus. For the point charge model, the force concept has been used in Examples 17.1 to 17.3; electric fields will be calculated in the next section.

The simplest dynamic system for an arrangement of stationary point charges is an electron in a solid **metal**, because this system allows us to apply concepts we already developed in Chapter 8 for the ideal gas. Metal atoms [i.e., all elements left of a line from boron (B) to iodine (I) in the periodic table] consist of a tightly bound shell of inner electrons and one or a few loosely bound outer electrons, called **valence electrons**. When a large number of metal atoms are brought together, the atoms form an ordered lattice by releasing these outermost electrons into a cloud of **quasi-free electrons**. These electrons are mobile within the stationary lattice of metal ions. When no external force exists, the electron moves randomly like gas particles in a container, as discussed for the kinetic gas theory. Instead of collisions with other gas particles, collisions with the metal ions occur as illustrated in Fig. 17.10. We use the electron in a metal system in Chapter 19 to introduce electric conduction.

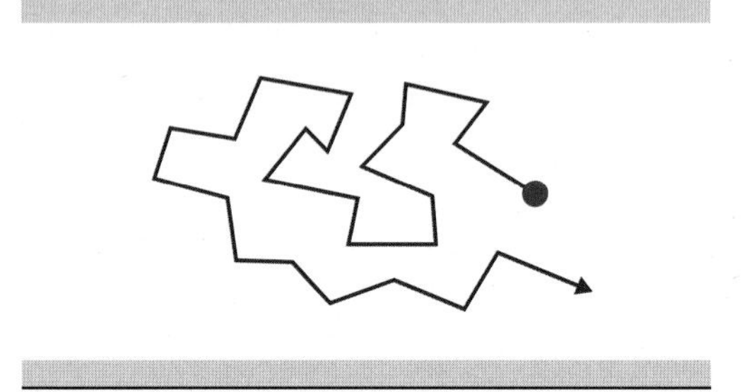

Figure 17.10 Model of an electron in a piece of metal. The electron moves freely, similar to a particle in a box filled with gas. However, the electrons scatter frequently off the densely packed immobile metal ions.

17.3.2: Model System II: Electric Dipole

Even though one may argue that the electric dipole is just a special application of model system I with two stationary point charges, it is identified here as a separate model system due to its widespread applications.

For this model, the electric field will again be calculated in the next section and will be used to introduce the dipole moment as a parameter that identifies the strength of dipoles. Hydrogen bonds and the chemical concepts of electronegativity and ionic character of a chemical bond are linked to this discussion in Case Study 17.1, Concept Question 17.5 and Example 17.6.

There are two important cases of mobile particles in the vicinity of stationary dipoles: an approaching point charge and an approaching dipole. The latter leads to dipole–dipole interactions that we refer to only qualitatively in this textbook due to the extensive nature of force calculations. In turn, the dipole–point charge interaction is addressed as a model for the interaction of ions in aqueous (water-based) solutions. Positively charged ions are called **cations** and negatively charged ions are **anions**. When a salt is dissolved in water, the water dipoles attack the salt crystal and attach themselves to the released cations and anions. The water molecule is of comparable size to most inorganic ions, which allows this interaction to be very effective. Choosing rock salt as an example, this process is written as:

$$NaCl + n\,H_2O \rightarrow Na_{aq}^{\;+} + Cl_{aq}^{\;-},$$

where the index "aq" stands for aqueous. The index indicates that each ion is embedded in a **hydration shell**. The hydration shell has a specific morphology for each of the salt ions: as illustrated in Fig. 17.11, the positive sides of water molecules point toward the chlorine anion, and their negative sides point toward the sodium cation.

When the ion moves, the hydration shell is dragged along, as indicated for Na^+ and Cl^- ions in the figure. Dragging the hydration shell slows the ion diffusion because the moving entity is much bigger. The hydration shell also causes a screening of the charge of the ion by smearing the charge over a much larger surface. As a result, the electric interactions of cations and anions in aqueous solutions are significantly reduced.

17.3.3: Model System III: Charged Parallel Plate Arrangement

All systems with more than a few stationary charges require extensive mathematical analysis to determine the electric field. In the case of parallel plates with equal but opposite charges, however, a simple form of the electric field is found. This is one reason why the parallel plate arrangement is an often used model. The other reason is its widespread applicability. It describes all biological membrane systems where charges are separated onto their opposite faces at a fixed distance determined by the lipid bilayer thickness. This model allows us to discuss in particular charged membranes, such as nerve membranes and pacemaker cell membranes in the heart, both in Chapter 19. The electric field for this model is discussed below.

Semipermeable membranes allow ions to pass, which establishes a major application of the interaction of a mobile point charge with the charged parallel plate systems, as illustrated in Fig. 17.12. The figure shows a membrane with stationary charges—cations on the extracellular side (labelled EXT) and anions on the internal side (labelled INT)—with a potassium cation transported across the membrane. This process is quantified in Chapter 19. Point charges also move between charged parallel plates in other systems, such as in Millikan's experiment (see Example 17.9 and P–17.12), or in the atmosphere between a differently charged ground surface and cloud cover.

17.4: Electric Field

We now develop the concept of a static electric field. Keep in mind that Michael Faraday did not intend to find a simplifying mathematical trick when he developed this concept. This means that we should not focus on the mathematical aspect of writing a field. We have to focus, rather, on its conceptual purpose: disconnecting ourselves from Coulomb's law, which requires we study all individual interactions simultaneously by describing a field that summarizes the effect of an arrangement of charges at any position, in particular the position we are interested in. Note for this discussion that a position is a point in space. There is no requirement for a charged particle to actually be located at that position!

The best way to capture Faraday's method is based on rewriting Coulomb's law in the form:

$$\vec{F}_{el} = q_{probe}\,\vec{E}. \qquad [17.4]$$

This equation defines the electric field numerically, with the unit for $\vec{E}$ as N/C.

KEY POINT

At every position within an arrangement of stationary charges the stationary electric field is the force vector that acts on a positive probe charge of unit charge if placed at that position.

What is great about the electric field is that it can be determined without the probe charge actually being present in the system. The electric field measures how

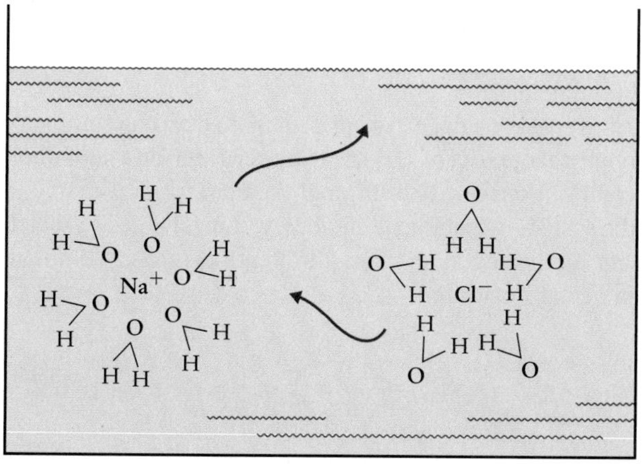

Figure 17.11 Sketch of a positive sodium ion and a negative chlorine ion in an aqueous solution. Water dipoles form a hydration shell around each ion while it moves through the solution, screening the ion's charge and thus reducing the interaction between ions.

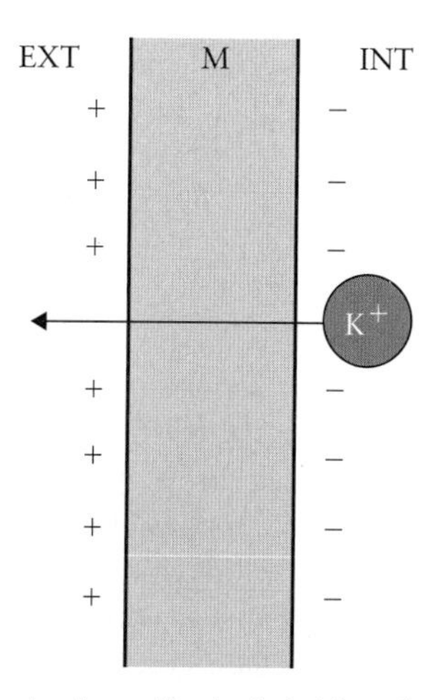

Figure 17.12 A potassium cation is ejected from the interior of a cell through the cell membrane (M), which carries surface charges.

capable the stationary charge arrangement is of interacting electrically; i.e., we predict an interaction that will occur if a point charge reaches that position in the future.

17.4.1: The Electric Field of a Stationary Point Charge

We combine Eq. [17.4] with Eq. [17.2]:

$$\vec{F}_{el} = \frac{1}{4\pi\varepsilon_0}\left(\frac{q_{\text{stationary}}\,\vec{r}^{\,0}_{\text{stationary to probe}}}{r^2_{\text{stationary to probe}}}\right) q_{\text{probe}},$$

in which $r_{\text{stationary to probe}}$ is the distance between the two point charges in the system. In the next step we remove the probe charge. The distance, which we labelled $r_{\text{stationary to probe}}$, then refers to the distance between the stationary point charge and the position at which the probe charge was previously. Dropping the index on the distance parameter r means that it now represents the distance r from the stationary point charge to any position in space:

$$\vec{E} = \frac{1}{4\pi\varepsilon_0}\frac{q_{\text{stationary}}}{r^2}\vec{r}^{\,0}. \qquad [17.5]$$

Eq. [17.5] represents a **vector field**. This means that it assigns to each position relative to the stationary point charge a vector that represents both the magnitude and direction of a force acting on a probe charge at that position if it were brought there.

The magnitude of the electric field vector, $|\vec{E}|$, results from Eq. [17.5]:

$$|\vec{E}| = \frac{1}{4\pi\varepsilon_0}\frac{|q_{\text{stationary}}|}{r^2}. \qquad [17.6]$$

Eq. [17.6] represents the magnitude of the electric field due to a single stationary point charge: the field varies with distance from the stationary point charge as $1/r^2$. The magnitude of the electric field is sufficient for applications in which we study only one stationary point charge or when we study several stationary point charges located along a common axis. For more general cases, we need to use Eq. [17.5].

Eq. [17.5] shows that the electric field of a stationary point charge is a radial field; i.e., the electric field vector $\vec{E}$ points at every position in space along the connection line of that point and the stationary point charge. The electric field for a positive stationary point charge is illustrated in Fig. 17.13: each arrow in the figure represents the electric field vector at the point where the foot of the arrow is placed. Note the directional variation of the field and its variation in magnitude, which is indicated by the length of the arrows in the figure. In applications it will not be necessary to derive the direction of the electric field from Coulomb's law. We can use instead Fig. 17.14.

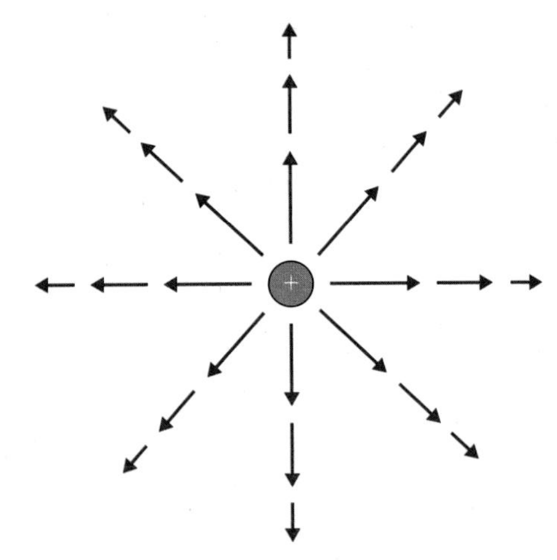

Figure 17.13 The electric field of a positive stationary point charge. The arrows represent the vectors of the electric field. The length of each vector indicates the magnitude of $\vec{E}$ at the respective position. The electric field of a point charge has a radial (in three dimensions therefore spherical) symmetry.

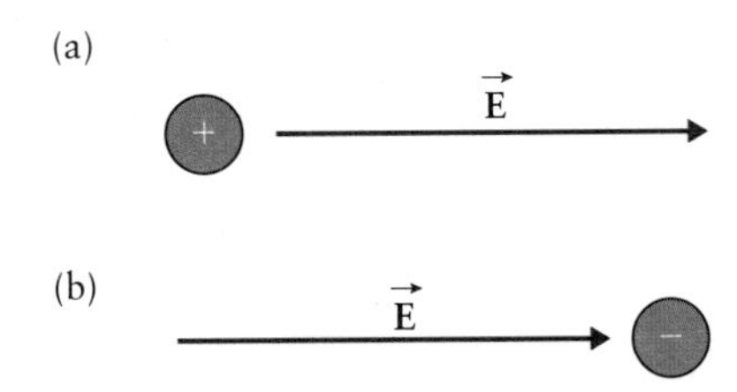

Figure 17.14 The electric field vector is always pointing (a) away from a positive charge or (b) toward a negative charge. If a positive mobile point charge is brought into the system, a force in the direction of the electric field will act on it.

KEY POINT

The electric field vector for a single point charge is directed away from a positive point charge or toward a negative point charge.

Thus, a positive probe charge brought into an electric field will feel a force in the direction of the field; a negative probe charge will feel a force in the direction opposite to the field.

If we consider a system with more than one stationary point charge, Eq. [17.5] has to be rewritten as a sum of all contributions due to each single stationary point charge in the system. Assuming N stationary point charges q_i, which can be located at as many different positions, we obtain in vector notation:

$$\vec{E}_{\text{net}} = \frac{1}{4\pi\varepsilon_0}\sum_i q_i \frac{\vec{r}_i^{\,0}}{r_i^2}, \qquad [17.7]$$

in which we consider all particles with index from $i = 1$ to N to form the stationary configuration of charges. To apply Eq. [17.7], we use the same approach we discussed previously for vector sums: the equation consists of three component equations for the three Cartesian coordinates, each representing the respective component of the net electric field:

$$E_{x,\text{net}} = \sum_i E_x(q_i)$$
$$E_{y,\text{net}} = \sum_i E_y(q_i) \qquad [17.8]$$
$$E_{z,\text{net}} = \sum_i E_z(q_i).$$

To apply this equation, the electric field due to each stationary point charge has to be written as three Cartesian components. In Example 17.4, we see how the components of the electric field are first determined for a single point charge. This is followed by a special example with $N = 2$ for the electric field of a dipole.

CONCEPT QUESTION 17.3

What is the direction of the electric field at point P in Fig. 17.15?

(A) to the upper right

(B) to the lower right

(C) to the upper left

(D) to the lower left

(E) none of the above

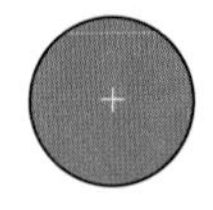

P

Figure 17.15 A stationary point charge and a point P in its vicinity.

CONCEPT QUESTION 17.4

What is the direction of the electric field in the water molecule in Fig. 17.1?

EXAMPLE 17.4

Fig. 17.16 shows a positive stationary point charge at the origin. Calculate the electric field at point $P = (x_0, y_0)$. Note that the problem is set up in two dimensions to reduce the required mathematical effort.

continued

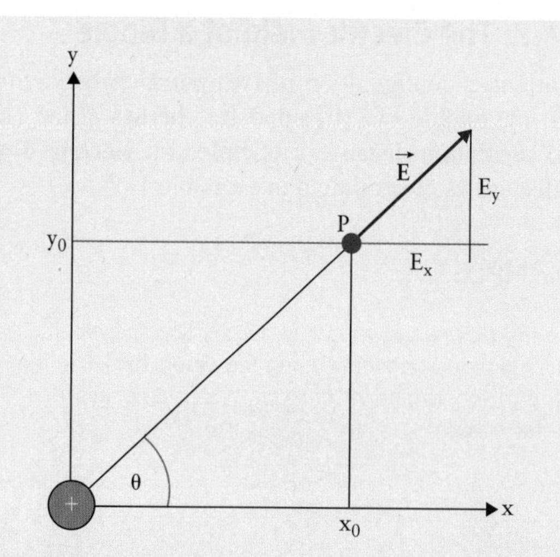

Figure 17.16 The electric field and electric field components in a Cartesian coordinate system at point $P = (x_0, y_0)$, with a positive point charge at the origin. Fig. 17.14 is used to determine the direction of the electric field.

Solution

We determine the two components of the electric field in two steps. First, the magnitude of the electric field $|\vec{E}|$ is calculated at point P; next, the geometric properties of the system in Fig. 17.16 are used to write its x- and y-components.

The magnitude of the electric field at point P is calculated from Eq. [17.6]. If we label the charge of the stationary point charge q, we obtain:

$$|\vec{E}| = \frac{1}{4\pi\varepsilon_0} \frac{|q|}{x_0^2 + y_0^2},$$

in which, according to the Pythagorean theorem, the denominator in the second term on the right-hand side equals the square of the distance between the charge q and the point P.

The components E_x and E_y are derived from the previous equation using trigonometric considerations based on Fig. 17.16. We find for the x-component:

$$E_x = \frac{1}{4\pi\varepsilon_0} \frac{|q|}{x_0^2 + y_0^2} \cos\theta,$$

and for the y-component:

$$E_y = \frac{1}{4\pi\varepsilon_0} \frac{|q|}{x_0^2 + y_0^2} \sin\theta,$$

with:

$$\sin\theta = \frac{y_0}{\sqrt{x_0^2 + y_0^2}}$$

$$\cos\theta = \frac{x_0}{\sqrt{x_0^2 + y_0^2}}.$$

17.4.2: The Electric Field of a Dipole

A stationary configuration of two point charges with (i) equal magnitude but (ii) opposite charges q and (iii) a fixed separation distance d is called an **electric dipole**. We discuss its electric field in Example 17.5.

EXAMPLE 17.5

For the dipole shown in Fig. 17.17, find the electric field as a function of position along the dipole axis, i.e., the line through the two point charges. We define this direction as the x-axis.

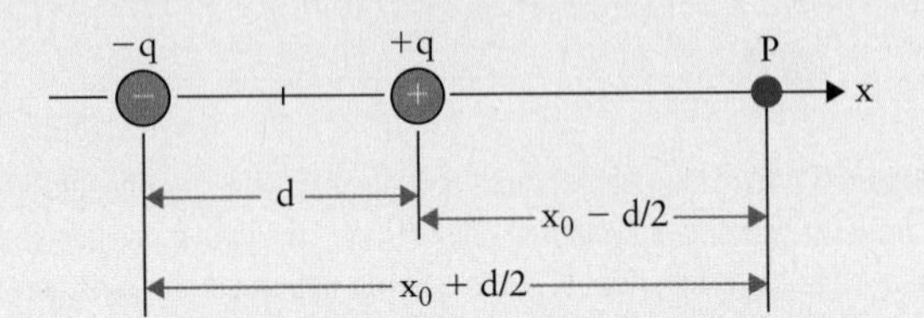

Figure 17.17 Dipole configuration: a dipole consists of two point charges separated by a fixed distance d. The point charges carry opposite charges of equal magnitude. The figure defines the axis of the dipole (x-axis) and identifies a point P at a distance x_0 from the centre of the dipole. The electric field is calculated at point P. In the calculation, the distance x_0 is variable.

Supplementary chemical information: The restriction to study only the electric field along the axial direction of the dipole does not affect our ability to discuss the physical consequences of the role of the water molecule in aqueous solutions, particularly the formation and structure of the hydration shell. This is evident from Fig. 17.1. A net negative charge is present near the oxygen atom (at point P_1) and a net positive charge near the hydrogen atoms (at point P_2). Due to Coulomb's law, this dipole always approaches an anion with its positive end and a cation with its negative end. Thus, it is sufficient to know the electric field along the axial direction of the dipole because water molecules can freely rotate in liquid water.

Solution

Eq. [17.6] is sufficient to determine the components of the electric field at point P because both point charges and the point P in Fig. 17.17 are located along a common axis. The directional information is straightforward in this case as all vector components are aligned with this axis: the net electric field is therefore also directed along the x-axis. The remaining directional issue is then resolved with Fig. 17.17. Choosing the origin of the axis at the centre of the two dipole charges, we find for the two separate field components at P, which is located at position x:

$$E_{+q} = \frac{+q}{4\pi\varepsilon_0 x_{+q}^2} = \frac{q}{4\pi\varepsilon_0\left(x - \frac{d}{2}\right)^2}$$

$$E_{-q} = \frac{-q}{4\pi\varepsilon_0 x_{-q}^2} = \frac{-q}{4\pi\varepsilon_0\left(x + \frac{d}{2}\right)^2},$$

continued

in which the first formula is the x-component of the electric field due to the positive charge, $+q$, and the second formula is its x-component due to the negative charge, $-q$. The net electric field is the sum of these two components:

$$E_{net} = \frac{q}{4\pi\varepsilon_0}\left[\frac{1}{\left(x - \frac{d}{2}\right)^2} - \frac{1}{\left(x + \frac{d}{2}\right)^2}\right].$$

Several algebraic operations are applied to simplify this equation. First, the two terms in the brackets are combined with a single denominator:

$$E_{net} = \frac{q}{4\pi\varepsilon_0}\left[\frac{\left(x + \frac{d}{2}\right)^2 - \left(x - \frac{d}{2}\right)^2}{\left(x - \frac{d}{2}\right)^2\left(x + \frac{d}{2}\right)^2}\right].$$

Then the numerator and denominator in the brackets are further analyzed. For the numerator, we write:

$$\left(x + \frac{d}{2}\right)^2 - \left(x - \frac{d}{2}\right)^2$$
$$= x^2 + xd + \frac{d^2}{4} - x^2 + xd - \frac{d^2}{4}$$
$$= 2xd$$

and for the denominator:

$$\left(x + \frac{d}{2}\right)^2\left(x - \frac{d}{2}\right)^2$$
$$= \left[\left(x + \frac{d}{2}\right)\left(x - \frac{d}{2}\right)\right]^2 = \left[x^2 - \left(\frac{d}{2}\right)^2\right]^2$$
$$= x^4 - \frac{x^2 d^2}{2} + \frac{d^4}{16}.$$

Thus, the x-component of the electric field of a dipole at point P is:

$$E_{net} = \frac{qxd}{2\pi\varepsilon_0}\frac{1}{x^4 - \frac{x^2d^2}{2} + \frac{d^4}{16}}. \qquad [17.9]$$

Eq. [17.9] is the proper answer to the question asked in Example 17.5. For most applications of dipoles, Eq. [17.9] can be simplified because we are interested only in the electric field at large distances from the dipole. In this case, we rewrite Eq. [17.9] for $x >> d$, i.e., for the case in which the point P is much farther from the dipole than length d. Mathematically this means that the leading x^4 term dominates the sum in the denominator on the right-hand side of Eq. [17.9]. We can therefore neglect the other two terms; i.e., $x^4 >> x^2 d^2/2$ and $x^4 >> d^4/16$. This yields a formula for the x-component of the electric field that is applicable far from the dipole:

$$\lim_{x \gg d} E_{net} = \frac{q}{2\pi\varepsilon_0}\frac{d}{x^3}. \qquad [17.10]$$

KEY POINT

The electric field of a dipole drops proportionally to $1/r^3$, i.e., more rapidly than the electric field of a single charge, for which the field is proportional to $1/r^2$.

The x-dependence in Eq. [17.10] is not surprising, since a dipole is the same as two very close point charges that compensate each other when looking from a position far away. Had the text in Example 17.5 not limited the position of point P to the axis of the dipole, electric field components would have had to be calculated at each point and would have had to be combined into a net effect, as outlined in Eq. [17.8]. This leads to an extensive mathematical effort, which we leave to P–17.21.

Fig. 17.18 shows electric field lines for three cases to illustrate the variety of possible directional patterns of electric fields for just two charges. A plot of electric field lines illustrates the direction of the electric field: the electric field at any given point is directed tangentially to the field line at that position. The plot shows the magnitude of the electric field in only an indirect manner; it is represented by the local density of field lines.

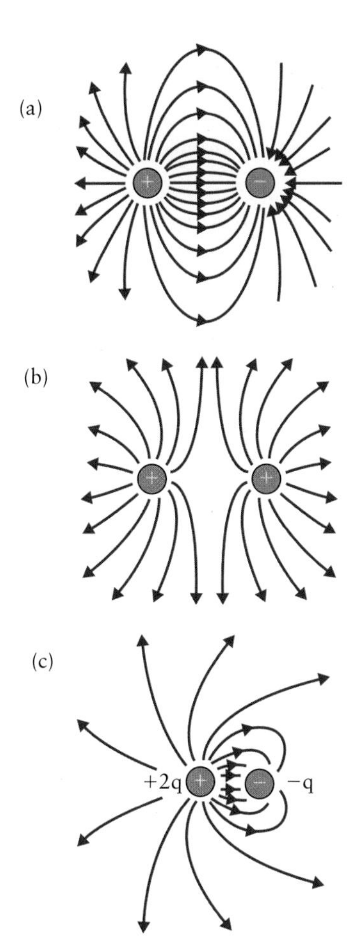

Figure 17.18 Plots of the electric field lines for three charged systems. An electric field line illustrates the direction of the electric field. At any given point the field points in the direction tangential to the field line. Note the variability of the directional patterns of electric fields. (a) The electric dipole. The electric field lines point away from the positive charge and toward the negative charge, as required by Fig. 17.14. (b) The electric field lines for two equal positive point charges. (c) The electric field lines for two unequal point charges. This plot indicates that electric field lines can easily form complex patterns.

Fig. 17.18(a) illustrates the case of a dipole. The electric field lines point away from the positive charge and/or toward the negative charge, as required by Fig. 17.14. The electric field of a dipole shows a significant directional variation; it approaches the radial symmetry of the electric field of a point charge only very close to either of the two point charges. The electric field lines along the dipole axis are straight lines, as discussed in Example 17.5.

Fig. 17.18(b) shows the electric field lines for two equal positive point charges. The electric field lines between these charges do not form closed curves as in Fig. 17.18(a). Fig. 17.18(c) indicates that electric fields easily become complex patterns. In the figure, the electric field lines are shown for a positive and a negative point charge, where the positive charge is twice the magnitude of the negative charge. Far from the pair of point charges the positive charge dominates the pattern, as is evident from following the electric field lines leaving the positive point charge.

17.4.3: The Electric Dipole Moment

The product of charge and distance defines the **electric dipole moment** μ in Eq. [17.11]:

$$\mu = qd. \qquad [17.11]$$

Electric dipole moments characterize the chemical and physical properties of molecules. Table 17.3 lists the electric dipole moment, the melting temperature, and the boiling temperature for several hydrogen-containing molecules. Fig. 17.19 is a plot of the dipole moment of these molecules versus the range between melting temperature and boiling temperature at atmospheric pressure. The data show that higher melting and boiling temperatures

TABLE 17.3

The correlation of the dipole moment and the temperature interval between melting point and boiling point at 1 atm pressure for several small molecules

Molecule	Dipole moment (C m)	Melting temperature (°C)	Boiling temperature (°C)
HF	6.37×10^{-30}	−84	+20
HCl	3.57×10^{-30}	−117	−85
H_2O	6.17×10^{-30}	±0	+100
H_2S	3.67×10^{-30}	−86	−61
NH_3	4.80×10^{-30}	−78	−33
PH_3	1.83×10^{-30}	−134	−88
AsH_3	7.33×10^{-31}	−117	−62
CH_4	0	−182	−161
SiH_4	0	−185	−112

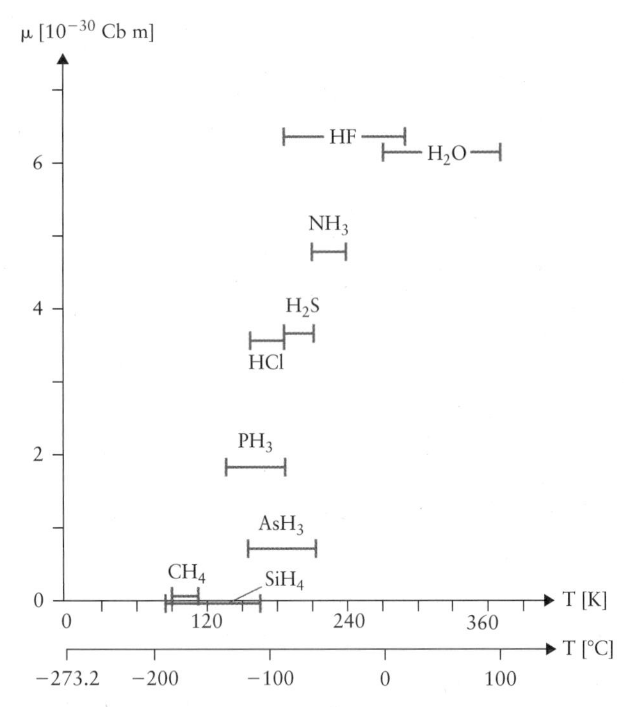

Figure 17.19 Graphic representation of the data of Table 17.3. The temperature axis is shown in units K and °C. For each molecule, the horizontal bar shown stretches from the melting point to the boiling point. Both phase transition temperatures increase with increasing dipole moment of the molecule.

are associated with larger dipole moments. In particular, water is distinct from the other molecules because its very large electric dipole moment yields melting and boiling points that are unusually high for such small molecules.

The electric dipole moment in Fig. 17.19 is one way to demonstrate quantitatively the direct relation between the electric properties of the water molecule and its applications as a key ingredient in life processes. Physically, this relation is established by the direct relation of the electric dipole moment to the strength of **hydrogen bonds**. A hydrogen bond between water molecules forms when a hydrogen atom of one water molecule approaches the oxygen atom in another water molecule. The hydrogen bond is therefore a dipole–dipole interaction that leads to the molecular arrangement illustrated in Fig. 17.20. In the figure, the large spheres represent oxygen atoms and the small spheres hydrogen atoms. Covalent bonds in each molecule are shown as dashed lines; hydrogen bonds are present where the spheres of hydrogen and oxygen atoms from neighbouring molecules touch. The covalent bonds of each water molecule correspond to the H–O lines in Fig. 17.1. Note that every oxygen atom in Fig. 17.20 is surrounded symmetrically by four hydrogen atoms; i.e., it forms two hydrogen bonds. Hydrogen bonds require only 5% to 10% of the energy needed to break a covalent bond; however, this energy still exceeds the thermal energy available at sufficiently low temperatures. The structure shown in Fig. 17.20 therefore illustrates the local organization of water molecules in liquid water: it becomes the long-range order in solid ice.

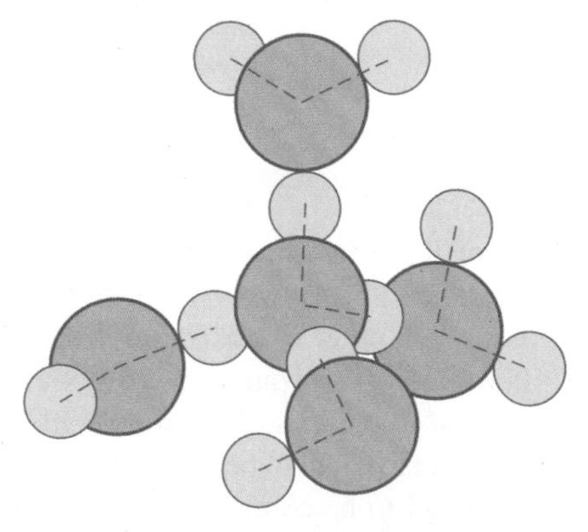

Figure 17.20 Typical arrangement of five neighbouring water molecules (each identified by dashed lines) in solid water (ice). The larger spheres represent oxygen atoms and the smaller spheres hydrogen atoms. Four hydrogen atoms are arranged tetrahedrally around each central oxygen atom. In liquid water, hydrogen bonds break frequently and the local order of molecules is less regular. The hydrogen bonds are about 5% to 10% as strong as chemical O–H bonds (dashed lines).

CASE STUDY 17.1

Do hydrogen bonds also occur elsewhere, e.g., in biomolecules? Try a literature or Internet search.

Answer: *Yes. The DNA molecule consists of nucleotides, which in turn contain a nitrogenous base, a pentose sugar, and a phosphate group. The nitrogenous groups are cytosine, thymine, adenine, or guanine. Pairwise they connect with hydrogen bonds, as shown in Fig. 17.21. These bonds stabilize the DNA double-helix structure.*

Adenine Thymine

Guanine Cytosine

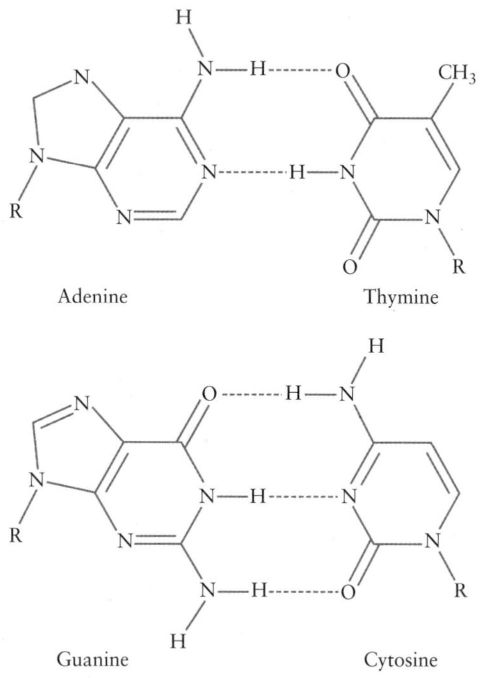

Figure 17.21 Hydrogen bonds also play a crucial role in the formation of DNA molecules. The nitrogenous bases of the nucleotides include either cytosine or thymine (pyrimidine compounds), or adenine or guanine (purine compounds). These groups bond in pairs to form the DNA double-helix structure.

When comparing the various molecules in Table 17.3 we note that the ability to form dipoles varies. In water, oxygen succeeds in drawing electrons from hydrogen to form a strong dipole; carbon in methane is much less efficient in this way. We quantify this observation with two new variables:

- **Electronegativity**, which is calculated from the energy of formation of the molecule. It is primarily used for systems with dominant covalent bonds.
- **Ionic character**, which is calculated from the dipole moment. It is used when the bond is partially ionic.

Electronegativity was introduced by Linus Pauling to quantify the propensity of an atom within a molecule to draw electric charges along a chemical bond toward itself and away from its partners (a property called electron affinity). To quantify the concept, Pauling used a comparison of the actual energy of a chemical bond to the energies of the respective bonds between the two atoms independently. The energy discrepancy is explained by the transfer of electric charge to the more electronegative atom. Table 17.4 shows electronegativity values for several physiologically important elements.

In methane, the electronegativity difference between carbon and hydrogen is 0.3. Such small differences favour non-polar covalent bonds. In water, the respective difference is 1.3, which points to a polar covalent bond. Rock salt (NaCl) has an electronegativity difference of 2.1, which represents an ionic bond.

CONCEPT QUESTION 17.5

We study the acetylene molecule. This is a linear molecule with the chemical formula HC≡CH. Using the data in Table 17.4, which statement about the C_2H_2 molecule is correct?

(A) Acetylene is not a dipole because the electronegativity values of carbon and hydrogen are close to each other.

(B) Acetylene is not a dipole because the angle between the C–H single bond and the C≡C triple bond is 180°.

(C) Acetylene is not a dipole because the charge distribution within the molecule is symmetric.

(D) Whether a given molecule is a dipole or not can be determined only in an experiment.

TABLE 17.4

Selected electronegativity values for elements important in human physiology

Element	Electronegativity	Element	Electronegativity
H	2.2	Li	1.0
B	2.0	C	2.5
N	3.0	O	3.5
F	4.0	Na	0.9
Mg	1.2	P	2.1
S	2.5	Cl	3.0
K	0.8	Ca	1.0
As	2.0	I	2.5
Ba	0.9		

EXAMPLE 17.6

Calculate the ionic character for the water molecule. Determine the maximum dipole moment for the molecule by allowing each of the two hydrogen atoms to transfer one electron to the oxygen atom. The bond length O–H is 0.096 nm and the bond angle H–O–H is 104.5°.

Supplementary chemical information: Using the dipole moment from Table 17.3 and the distance between the positive and negative centres of a molecule allows us to define the **ionic character** Ic for the molecule:

$$\mathrm{Ic} = 100\frac{\mu_{\mathrm{actual}}}{\mu_{\mathrm{max}}}, \quad [17.12]$$

in which Ic is given as a percentage value. For example, for the HCl molecule the maximum dipole moment is based on the transfer of a single electron (elementary charge) from hydrogen to chlorine across the separation distance of the atoms in the molecule. The bond length is 0.127 nm. This leads to:

$$\mu_{\mathrm{max}}(\mathrm{HCl}) = (1.6\times10^{-19}\,\mathrm{C})(1.27\times10^{-10}\,\mathrm{m})$$
$$= 2.05\times10^{-29}\,\mathrm{C\,m}.$$

Thus, with the measured dipole moment of HCl from Table 17.3, the ionic character of the HCl bond is:

$$\mathrm{Ic(HCl)} = 100\frac{3.57\times10^{-30}\,\mathrm{C\,m}}{2.05\times10^{-29}\,\mathrm{C\,m}}$$
$$= 17.4\%;$$

i.e., the HCl molecule is polar, but with a predominantly covalent bond character.

continued

Solution

Fig. 17.1 illustrates the geometry of the water molecule and allows us to determine the maximum separation of the positive and the negative charges in the polarized molecule. This maximum separation would be reached when the single electron of each hydrogen atom is fully shifted to the oxygen atom. This corresponds to a double negative charge (2 e) at point P_1 and a double positive charge at point P_2 due to the symmetry of the molecule. The distance between points P_1 and P_2 is obtained geometrically from the figure:

$$P_1P_2 = r\cos\left(\frac{104.5^\circ}{2}\right) = 0.059 \text{ nm}.$$

Next, we determine the maximum dipole moment of the water molecule using this distance:

$$\begin{aligned}\mu_{max}(H_2O) &= (2\,e)P_1P_2\\ &= (3.2\times10^{-19}\text{C})(0.059\times10^{-9}\text{ m})\\ &= 1.89\times10^{-29}\text{C m}.\end{aligned}$$

The ionic character of the water molecule then follows from the definition in Eq. [17.12], i.e., the ratio of the value given in Table 17.3 to the value determined above:

$$\begin{aligned}\text{Ic}(H_2O) &= 100\frac{6.17\times10^{-30}\text{ C m}}{1.89\times10^{-29}\text{ C m}}\\ &= 32.6\%.\end{aligned}$$

This value significantly exceeds the value for HCl.

17.4.4: The Electric Field of Charged Parallel Plates

It is no longer practical to calculate the electric field from Eq. [17.7] when a significant number of stationary point charges make up the system. The sum in Eq. [17.7] must be rewritten using more advanced mathematical tools, leading to intermediate results that then require numerical methods to solve. These calculations are greatly simplified when two conditions are met: the stationary charges have a uniform density in the system, and the arrangement of these charges is symmetric in space. We are interested in only one such arrangement: a system with two parallel charged plates. This system is important for applications and leads to a mathematically simple electric field.

Our approach to systems with many charges is analogous to the approach taken when we studied extended objects or fluids: we no longer identified the system as a large number of particles but expressed the mass of the system as its volume and density. In a similar manner, one can introduce two charge density terms:

- The volume charge density, or **charge density**, with $\rho = Q/V$ where Q is the sum of all charges, $Q = \Sigma_i\, q_i$, V is the volume of the system in which the charges reside, and ρ is the charge density. This parameter is needed for systems in which free point charges cannot move (insulator systems) and remain distributed in three dimensions. The unit of ρ is C/m^3.
- The **surface charge density** is σ, with:

$$\sigma = \frac{Q}{A} = \frac{\sum_{i=1}^{N} q_i}{A}, \qquad [17.13]$$

in which A is the surface area of the system on which the charges are located; σ has unit C/m^2. This parameter is frequently used because mobile point charges of equal sign drift apart as far as possible in a conducting system such as aqueous solutions. An example is the nerve membrane, where the charges that cause the signal transport are located on the inner and outer surfaces, leading to a uniform surface charge density of 700 $\mu C/m^2$.

The resulting electric field between two infinitely large parallel plates separated by a vacuum, and with each plate carrying a uniform surface charge density σ, has the magnitude:

$$\left|\vec{E}_{||}\right| = \frac{\sigma}{\varepsilon_0}. \qquad [17.14]$$

The direction of this field is uniform, with the vector $\vec{E}_{||}$ pointing toward the negative plate along a line perpendicular to the plates.

KEY POINT

The electric field between charged parallel plates is independent of the position and is proportional to the surface charge density of the plates.

EXAMPLE 17.7

What is the electric field between two parallel flat plates where one plate is charged positively and one negatively, with $\sigma = 700\ \mu C/m^2$? This value is a typical surface charge density for nerve membranes.

Solution

Using Eq. [17.14], we find:

$$\left|\vec{E}_{||}\right| = \frac{7\times10^{-4}\,\frac{\text{C}}{\text{m}^2}}{8.85\times10^{-12}\,\frac{\text{C}^2}{\text{N m}^2}} = 7.9\times10^{7}\,\frac{\text{N}}{\text{C}}.$$

Note that this value is not the proper value for the electric field across a nerve membrane. The reason is that Eq. [17.14] is derived with the assumption that no other charges interfere with the electric interaction between the plates, which is an incorrect assumption for a lipid bilayer. To calculate the correct value the membrane material has to be taken into account, as discussed in section 18.5.

EXAMPLE 17.8

A flat plate has a surface charge density value of $\sigma = +5\ \mu C/m^2$. What is the electric field in close proximity to the surface of the plate?

Solution

The electric field of a single, flat plate is half the value of the parallel plate arrangement discussed in Eq. [17.14]—i.e., $|\vec{E}| = e_0/(2\sigma)$—and is directed perpendicular away from a positive plate. Thus:

$$|\vec{E}| = \frac{\sigma}{2\varepsilon_0} = \frac{5.0\times 10^{-6}\ \frac{C}{m^2}}{2\cdot 8.85\times 10^{-12}\ \frac{C^2}{N\ m^2}}$$

$$= 2.8\times 10^5\ \frac{N}{C}.$$

This is the correct value for the electric field at any point near the surface of the flat plate except close to the edges, where fringe effects require corrections.

Air breaks down electrically when an electric field of 3×10^6 N/C is reached, with air molecules ionizing and the gas becoming a conductor. These are the conditions that occur in the atmosphere during a lightning storm.

EXAMPLE 17.9

In Robert Millikan's experiment as illustrated in Fig. 17.4, a constant electric field along the vertical axis is obtained with charged parallel plates, one located above and one below the observation chamber. The electric field is directed downward. An oil droplet of radius 1.4 μm and density 0.85 g/cm^3 levitates in the chamber when an electric field of 1.5×10^5 N/C is applied. Find the charge on the droplet as a multiple of the elementary charge *e*.

Solution

A levitating droplet is in mechanical equilibrium. Two forces that act on the droplet balance each other based on Newton's first law: the electric force upward and the weight downward:

$$m_{\text{droplet}}\, g = q_{\text{droplet}}\, |\vec{E}|.$$

We isolate the charge of the droplet:

$$q_{\text{droplet}} = \frac{m_{\text{droplet}}\, g}{|\vec{E}|} = \frac{4\pi r_{\text{droplet}}^3\, \rho g}{3|\vec{E}|},$$

in which the mass of the droplet has been replaced by its volume and density. We substitute the given values:

$$q_{\text{droplet}} = \frac{4\pi (1.4\times 10^{-6}\ \text{m})^3 \left(850\ \frac{kg}{m^3}\right)\left(9.8\ \frac{m}{s^2}\right)}{3(1.5\times 10^5)\ \frac{N}{C}}$$

$$= 6.4\times 10^{-19}\ \text{C}.$$

continued

The oil droplet must carry a net negative charge because the electric field $\vec{E}$ is directed down and the electric force $\vec{F}_{el}$ is directed up. In terms of the elementary charge, we find:

$$\frac{q_{\text{droplet}}}{e} = \frac{6.4\times 10^{-19}\text{C}}{1.6\times 10^{-19}\text{C}} = 4;$$

i.e., the droplet carries four elementary charges. The benefit of using a constant, position-independent electric field in this experiment is self-evident.

SUMMARY

DEFINITIONS

- Dipole moment: $\mu = q\, d$, in which d is distance between charges $+q$ and $-q$ in a dipole.

UNITS

- Charge q: C
- Electric force $\vec{F}_{el}$: N
- Electric field $\vec{E}$: N/C
- Dipole moment μ: C m

LAWS

- Electric force between two point charges (Coulomb's law):

$$\vec{F} = \frac{1}{4\pi\varepsilon_0}\frac{q_1 q_2}{r^2}\vec{r}^{\,0},$$

in which $\vec{r}^{\,0}$ is the unit vector that is directed to the charge on which the force acts, and r is the distance between charges q_1 and q_2

- Electric field:
  (i) For a stationary point charge:

$$\vec{E} = \frac{q_{\text{stationary}}}{4\pi\varepsilon_0}\frac{\vec{r}^{\,0}}{r^2}.$$

The electric field vector is directed toward a negative charge or away from a positive charge.
  (ii) For a small number of stationary point charges:

first calculate the field components for each charge. Then the components are added for all charges:

$$E_{x,\text{net}} = \sum_i E_x(q_i)$$

$$E_{y,\text{net}} = \sum_i E_y(q_i)$$

$$E_{z,\text{net}} = \sum_i E_z(q_i)$$

  (iii) Magnitude far from a dipole in axial direction (axial direction chosen to be along the x-axis):

$$\lim_{x \gg d}\left|\vec{E}\text{ at }P\right| = \frac{q}{2\pi\varepsilon_0}\frac{d}{x^3}$$

(iv) Magnitude for charged parallel plates:

$$|\vec{E}| = \frac{\sigma}{\varepsilon_0},$$

σ is surface charge density $\sigma = q/A$.

MULTIPLE-CHOICE QUESTIONS

MC–17.1. Water consists of
(a) positively charged atoms.
(b) negatively charged atoms.
(c) positively charged molecules.
(d) negatively charge molecules.
(e) neutral molecules.

MC–17.2. Which feature of water cannot be explained with its dipole character and the related formation of hydrogen bonds?
(a) high melting point
(b) high freezing point
(c) open crystal structure of ice (low density)
(d) formation of hydration shells when mixed with salts
(e) all four choices can be explained with the dipole character

MC–17.3. Two sodium ions in an aqueous solution
(a) repel each other.
(b) do not interact at any separation distance.
(c) attract each other.
(d) seek each other to form metallic sodium precipitates.

MC–17.4. The following is *not* a property of a hydrogen bond.
(a) It connects a hydrogen atom to an oxygen atom in a different molecule.
(b) It connects two hydrogen atoms between two different molecules.
(c) It is a bond that is weaker than a covalent OH bond.
(d) Its formation allows water to remain liquid at temperatures where most other small molecules are already in the gaseous state.

MC–17.5. The water molecule is an electric dipole. The following statement about the water molecule is therefore wrong.
(a) The distance between the centres of positive and negative charges in the molecule is a stationary distance.
(b) The amount of charge we assign to the positive end and the amount of charge we assign to the negative end of the molecule are equal but opposite in sign.
(c) The electric field points from the hydrogen atoms toward the oxygen atom.
(d) If we choose the electric potential at infinite distance from the water molecule to be zero, $V = 0$, then the electric potential of the water dipole does not vanish anywhere within a distance of $5d$ from the water molecule, where d is the distance between the charged centres of the dipole.

MC–17.6. The force between a water dipole with electric dipole moment μ and charge separation distance d, and a negative ion of charge q located along the dipole axis of the water molecule is calculated using the following formula (with r the distance from the centre of the dipole to the negative charge):

(a) $F = -\frac{1}{2\pi\varepsilon_0}\frac{\mu q}{r^2}$

(b) $F = -\frac{1}{2\pi\varepsilon_0}\frac{\mu q}{r^3}$

(c) $F = -\frac{1}{2\pi\varepsilon_0}\frac{\mu q}{d^2}$

(d) $F = -\frac{1}{2\pi\varepsilon_0}\frac{\mu q}{d^3}$

(e) $F = -\frac{1}{2\pi\varepsilon_0}\frac{\mu q}{rd^2}$

MC–17.7. Two point charges repel each other with an electric force of magnitude f_0. If we double both charges, the magnitude of the force between the point charges becomes
(a) $f_0/4$.
(b) $f_0/2$.
(c) f_0 (i.e., it remains unchanged).
(d) $2 f_0$.
(e) $4 f_0$.

MC–17.8. Which statement about the electric dipole moment μ is correct?
(a) A molecule is a strong dipole if the net charge separation occurs over a short distance within the molecule.
(b) The unit of μ is C/m.
(c) A dipole exerts a larger force on a point charge than on another dipole at the same distance.
(d) For strong dipole molecules, μ is typically a value on the order of 1×10^{-30} C m, while the elementary charge is 1.6×10^{-19} C. Thus, the charge separated in strong dipole molecules is many orders of magnitude less than an elementary charge.
(e) The dipole moment can be zero even if neither q nor d is zero.

MC–17.9. Two dipoles at close proximity
(a) always repel each other.
(b) always attract each other.
(c) never interact electrically.
(d) attract or repel each other based on their relative orientation.

MC–17.10. A dipole and a point charge at close proximity
(a) always repel each other.
(b) always attract each other.
(c) never interact electrically.
(d) attract/repel each other based on the location of the point charge relative to the dipole axis.

MC–17.11. What do Newton's law of gravity and Coulomb's law have in common?
(a) force dependence on mass
(b) force dependence on electric charge
(c) force dependence on distance

(d) the magnitude of the proportionality constant
(e) nothing

MC–17.12. A mobile positive point charge is located between two parallel charged plates, with the upper plate carrying a positive surface charge density $+\sigma$ and the lower plate carrying a negative surface charge density $-\sigma$. How does the magnitude of the electric field at the position of the mobile point charge change if it moves to twice the distance from the negative plate (i.e., toward the positive plate)?
(a) It remains unchanged.
(b) It increases by a factor of 4.
(c) It doubles.
(d) It becomes half the initial value.
(e) It becomes one-quarter the initial value.

MC–17.13. In electrophoresis,
(a) ions are separated by diffusion.
(b) atoms and molecules are separated in an electric field.
(c) an electric field is applied to an ionic solution.
(d) charged oil droplets are levitated in an observation chamber.
(e) none of the above

MC–17.14. The electric field of a dipole
(a) is the same as for a point charge.
(b) cannot be described with a formula.
(c) decreases faster than for a point charge as we move away from it.
(d) decreases slower than for a point charge as we move away from it.

MC–17.15. The HF molecule is a dipole with an electronegativity of 4.0 assigned to the fluorine atom and 2.2 to the hydrogen atom. In which direction does the electric field point between these two atoms in the molecule?
(a) perpendicular to the line between the two atoms
(b) along the line between the two atoms, in the direction of the hydrogen atom
(c) along the line between the two atoms, in the direction of the fluorine atom
(d) at an angle of 104.5° relative to the line between the two atoms

MC–17.16. In Millikan's experiment,
(a) charged oil droplets are separated by diffusion.
(b) neutral oil mist droplets are separated in an electric field.
(c) an electric field is applied to an ionic solution.
(d) charged oil droplets are levitated in an observation chamber.

MC–17.17. The electric field of a point charge
(a) increases proportional to r^2 with distance.
(b) decreases proportional to r^2 with distance.
(c) increases linearly with distance.
(d) decreases linearly with distance.
(e) none of the above

CONCEPTUAL QUESTIONS

Q–17.1. In which direction does the electric field point at point P for the two charges in Fig. 17.22? Assume that $q_1 = +1\ \mu$C and $q_- = -1\ \mu$C. *Hint:* Draw the field due to each of the charges separately at P, then see for the resulting direction.

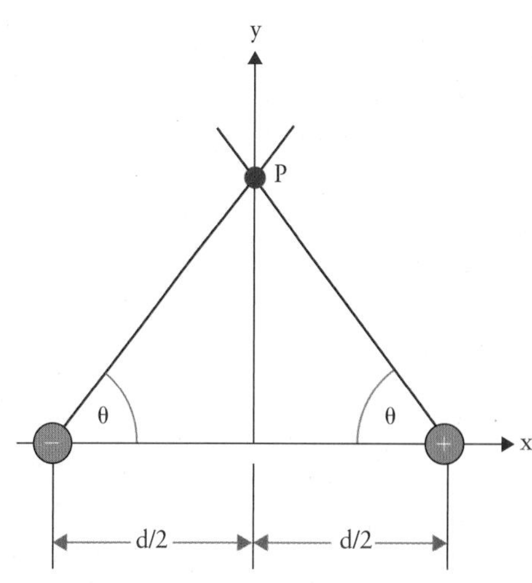

Figure 17.22 A system of two point charges for determining the electric field at point *P*.

Q–17.2. Can electric field lines ever cross? Explain your answer.

Q–17.3 Can the electric field amplitude vanish at a point in space where the electric potential from which it is derived is non-zero? Justify your answer.

Q–17.4. Two identical conducting spheres, each carrying a charge $+Q$, are placed in a vacuum with their centres a distance d apart. Explain why the net force between them is not given by:

$$F = \frac{1}{4\pi\varepsilon_0}\frac{Q^2}{d^2}.$$

Q–17.5. Estimate the ratio of the electrostatic force to the gravitational force between the protons in a helium nucleus.

ANALYTICAL PROBLEMS

P–17.1. We study three point charges at the corners of a triangle, as shown in Fig. 17.23. Their charges are $q_1 = +5.0 \times 10^{-9}$ C, $q_2 = -4.0 \times 10^{-9}$ C, and $q_3 = +2.5 \times 10^{-9}$ C. Two distances of separation are also given, $l_{12} = 4$ m and $l_{13} = 6$ m. Find the net electric force on q_3.

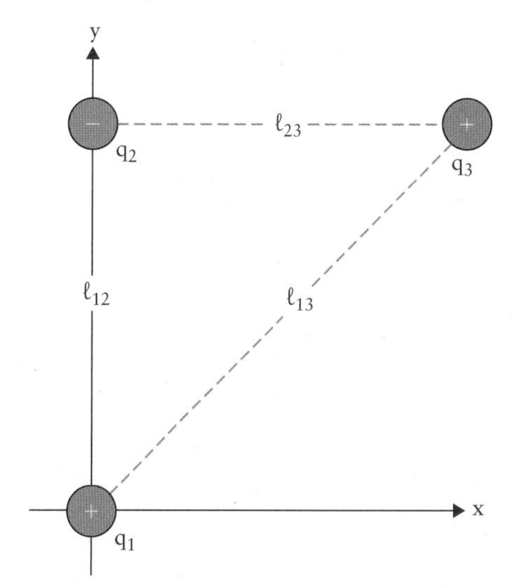

Figure 17.23

P–17.2. In Fig. 17.24(a) we study two particles, A and B. They are separated by distance a. Particle A initially has a positive charge of $+Q$ and particle B is electrically neutral ($q = 0$). Thus, no electric force acts between them. The particles are then connected, as shown in Fig. 17.24(b). This allows for an equal distribution of charges between the two particles. What is the electric force between the particles after the connection is removed?

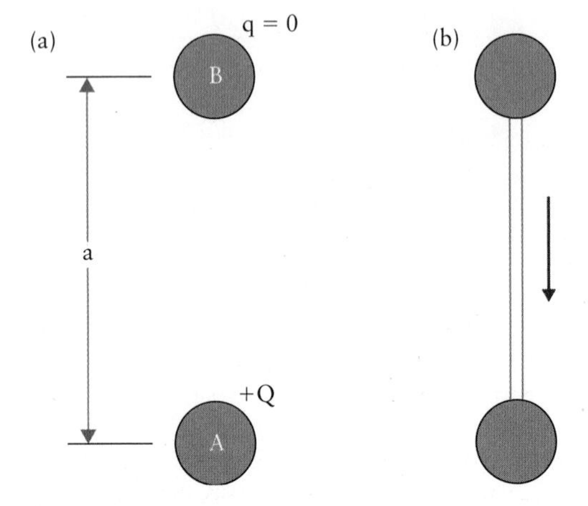

Figure 17.24

P–17.3. The radii of atomic nuclei closely follow the formula:

$$r = 1.2 \times 10^{-15}\ A^{1/3},$$

in which r has unit m and A is the atomic mass in unit g/mol. (a) Confirm that the density of nuclear matter is independent of the type of atom studied. This density is 2×10^{17} kg/m^3! (b) Using the given equation for the radius and $A(\text{Bi}) = 209.0$ g/mol, find the magnitude of the repulsive electrostatic force between two of the protons in a bismuth nucleus when they are separated by the diameter of the nucleus.

P–17.4. How much negative charge is in 1.0 mol of neutral helium gas? Each He atom has two electrons in its atomic shell.

P–17.5. A CsCl (cesium chloride) salt crystal is built from the unit cells shown in Fig. 17.25. Cl^- ions form the corners of a cube and a Cs^+ ion is at the centre of the cube. The edge length of the cube, which is called the lattice constant, is 0.4 nm. (a) What is the magnitude of the net force exerted on the cesium ion by its eight nearest Cl^- neighbours? (b) If the Cl^- in the lower left corner is removed, what is the magnitude of the net force exerted on the cesium ion at the centre by the seven remaining nearest chlorine ions? In what direction does this force act on the cesium ion?

P–17.6. A piece of aluminium has a mass of 27 g and is initially electrically neutral. It is then charged with $q = +0.5\ \mu$C. (a) How many electrons were removed from the piece of aluminium in the charging process? (b) What fraction of the total number of electrons in the piece of aluminium does the number in part (a) represent?

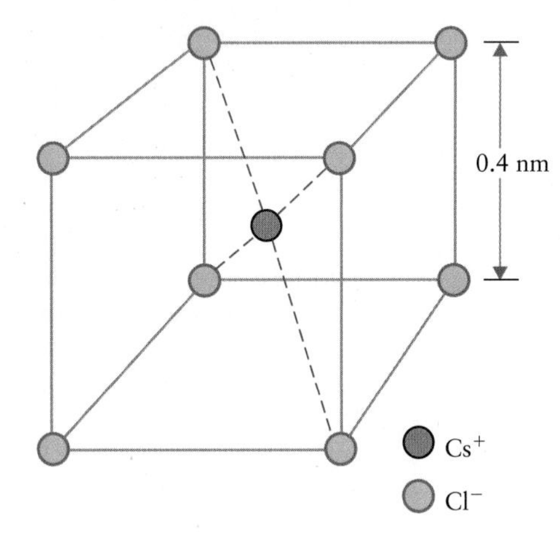

Figure 17.25

P–17.7. An alpha particle (carrying two elementary charges) approaches at high speed a gold nucleus with a charge of 79 e. What is the electric force acting on the alpha particle when it is 2.0×10^{-17} m from the gold nucleus?

P–17.8. Two metal spheres, each of mass 0.25 g, are suspended by massless strings from a common pivot point at the ceiling, as shown in Fig. 17.26. When both spheres carry the same electric charge we find that they come to an equilibrium when each string is at an angle of $\theta = 6.0°$ with the vertical. If each string is 25 cm long, what is the amount of the charge on each sphere?

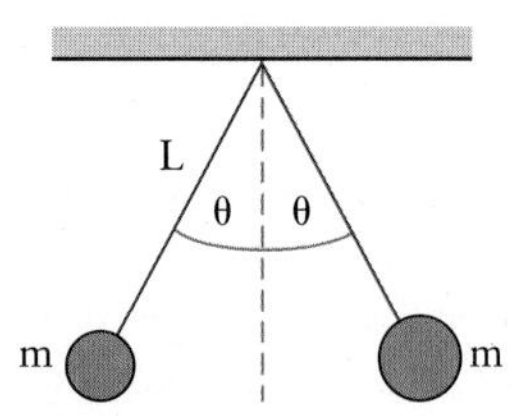

Figure 17.26

P–17.9. Determine the magnitude of the force between an electric dipole with a dipole moment of 3×10^{-29} C m and an electron. The electron is positioned $r = 20$ nm from the centre of the dipole, along the dipole axis. *Hint:* Assume that $r >> d$, with d the charge separation distance in the dipole.

P–17.10. Fig. 17.27 shows three positive point charges, with two charges of magnitude q at a distance d along the negative x- and the positive y-axes and one charge of magnitude $2q$ at the origin. Calculate the electric field at point P for $q = 1.0$ nC and a distance $d = 1.0$ m.

P–17.11. Calculate the electric field halfway between two point charges, one carrying $+10.0 \times 10^{-9}$ C and the other (a) -5.0×10^{-9} C at a distance of 20 cm and (b) -5.0×10^{-9} C at a distance of 20 cm.

P–17.12. In Millikan's experiment in Fig. 17.4, a droplet of radius $r = 1.9$ mm has an excess charge of two electrons. What are the magnitude and direction of the electric field that is required to levitate the droplet? Use for the density of oil $\rho = 0.925$ g/cm^3.

P–17.13. An electron is released into a uniform electric field of magnitude 1.5×10^3 N/C. Calculate the acceleration of the electron, neglecting gravity.

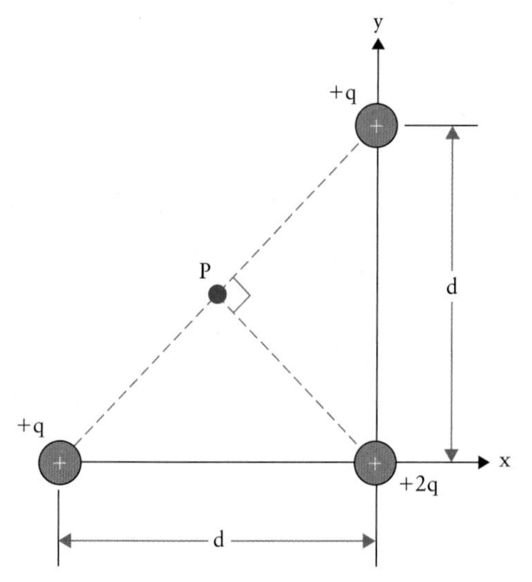

Figure 17.27

P–17.14. Humid air breaks down electrically when its molecules become ionized. This happens in an electric field of magnitude $E = 3.0 \times 10^6$ N/C. In that field, calculate the magnitude of the electric force on an ion with a single positive charge.

P–17.15. A constant electric field is obtained experimentally with the set-up shown in Fig. 17.28: a 12-V battery is connected to two parallel metal plates separated by a distance of $d = 0.25$ cm. Calculate the magnitude of the electric field between the plates.

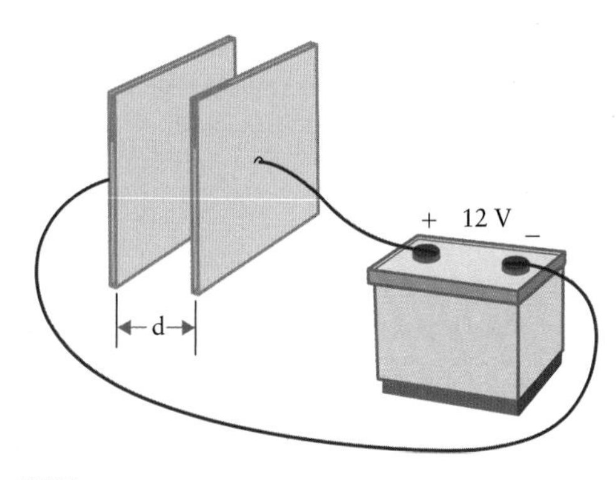

Figure 17.28

P–17.16. Three positive point charges are located at the corners of a rectangle, as illustrated in Fig. 17.29. Find the electric field at the fourth corner if $q_1 = 3$ nC, $q_2 =$ 6 nC, and $q_3 = 5$ nC. The distances are $d_1 = 0.6$ m and $d_2 = 0.2$ m.

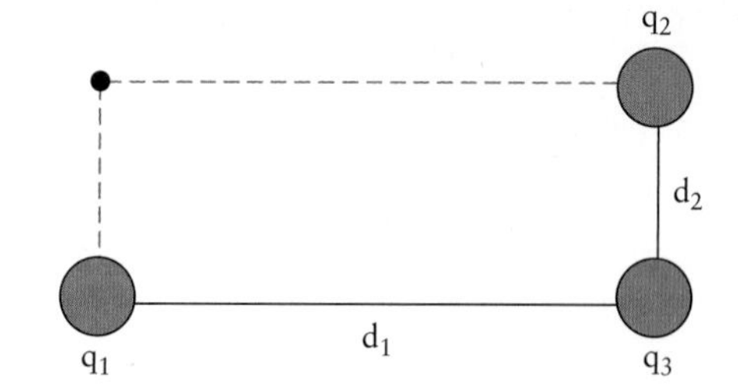

Figure 17.29

P–17.17. In a hydrogen atom, what are the magnitude and direction of the electric field due to the proton at the location of the electron, which we assume is 5.0×10^{-11} m away from the proton?

P–17.18. An electron is accelerated by a constant electric field of magnitude 250 N/C. (a) Calculate the magnitude of the acceleration of the electron. (b) Calculate the speed of the electron after 10 ns (nanoseconds), assuming it starts from rest.

P–17.19. A piece of aluminium foil has a mass of 50 g. It is suspended by a massless string in an electric field directed vertically upward. If the charge carried on the foil is 2.5 μC, calculate the strength of the field that will reduce the tension in the string to zero.

P–17.20. A proton accelerates from rest in a uniform electric field of 640 N/C. At some later time, its speed is found to be 1.2×10^6 m/s. (a) Calculate the magnitude of the acceleration of the proton. (b) How long does it take the proton to reach the given speed?

P–17.21. (a) Determine the formula for the electric field at an arbitrary position in the x,y-plane for the dipole shown in Fig. 17.17. (b) Simplify your result in part (a) with the assumption that $x,y >> d$.

ANSWERS TO CONCEPT QUESTIONS

Concept Question 17.1: (B). The electric force is linear in the absolute amount of each charge but non-linear in their distance.

Concept Question 17.2: The unit vector in Eq. [17.2] is directed from the point charge that is the source of the force to the point charge on which the force acts. Thus, the unit vector points to the left for the four cases in the left column, and it points to the right for the four cases in the right column.

Concept Question 17.3: (B) based on Fig. 17.14.

Concept Question 17.4: An electric field can be assigned to each of the two partial charges at points P_1 and P_2; they both point in the same direction and form a net electric field based on Fig. 17.14 with its direction from P_2 to P_1.

Concept Question 17.5: (A) and (C). (C) is also correct because even a molecule with large electronegativity differences can contain opposing dipoles that cancel each other, e.g., the linear CO_2 molecule.

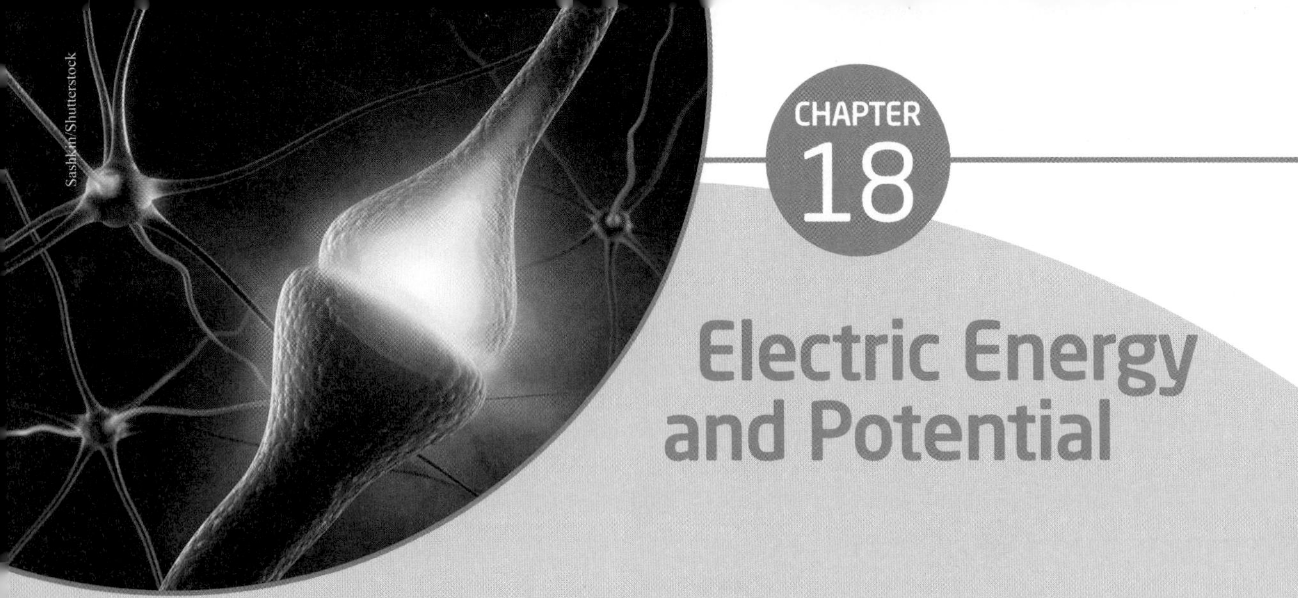

CHAPTER 18

Electric Energy and Potential

The electric potential energy is introduced in analogy to the gravitational potential energy. It resumes its simplest form for the parallel plate arrangement, for which the electric energy is linearly proportional to the distance of the charged particle to the oppositely charged plate. For two point charges the electric potential energy varies with the inverse distance. It drops off more significantly, with $1/r^2$, for a dipole as the two charges in a dipole compensate each other at great distance.

In analogy to the electric field, the potential is developed from the electric potential energy as a scalar field, at each position representing the electric potential energy a positive unit charge (probe charge) would encounter.

A capacitor is an arrangement of parallel plates that are charged with equal but opposite surface charge densities. The capacitance of this arrangement is defined as the proportionality factor between the charge on the parallel plates, q, and the potential difference between the plates, ΔV. In addition, non-conducting materials called dielectrics are allowed between the plates. This alters the capacitance by a materials-specific factor called the dielectric constant.

The resting nerve is modelled as a capacitor, with the phospholipid bilayer as the dielectric material between an outer positively charged surface and a parallel inner surface charged negatively. The Na/K pump generates a concentration gradient for K^+ and Na^+ with a membrane potential difference of −70 mV.

Nerves were identified surprisingly early in the scientific and medical history as important systems in our body. Galen, the physician to Emperor Marcus Aurelius in ancient Rome, noted the existence of the spinal cord and its connection to muscle action. Luigi Galvani demonstrated in the late eighteenth century that electricity could trigger the muscles of frogs.

18.1: Nerves as a Physical and Physiochemical System

18.1.1: Microscopic Nerve Anatomy

Nerves serve two purposes: (i) to communicate stimuli to the control centres of the nervous system when they are registered by organs sensitive to the environment, and (ii) to communicate commands from these control centres to the organs in turn, either in response to a stimulus or without any external stimulus, e.g., when controlling the heart beat.

Both types of information transfer must occur within very short times. If you are driving your car and an obstacle suddenly appears on the road, your brain's command, "Hit the brakes," reaches the muscles in your leg in less than 0.1 second. In this case, the command has to travel through two of the longest nerves in your body: the nerve from your brain to the lower back synapse and the nerve from there to the foot. These two nerves have a total length of up to two metres. Thus, an impulse speed in nerves in excess of 20 m/s is vital in our everyday lives. The evolutionary solution to the problem of communicating information at such high speeds is based on electric and electrochemical effects.

A nerve is a strand of up to 1000 nerve cells *(neurons)*. These strands contain two types of nerves, *myelinated nerve cells* (about 30%) and *unmyelinated nerve cells* (about 70%). We will discuss the physical and physiological differences between these two types in detail later in this chapter.

Fig. 18.1 shows a microscopic sketch of a single myelinated nerve cell. Nerves have tentacle-like receptors called *dendrites*. These respond to a wide range of stimuli, ranging from temperature change to motion parameters. The stimulus causes an electric response that travels along the nerve to the synapse. At the synapse, vesicles are triggered to release neurotransmitters, which pass through the synaptic cleft to a secondary receptor. This cleft is typically 10 nm to 20 nm wide if the secondary receptor is a dendrite (neural junction) and 50 nm to 100 nm wide if the secondary receptor is a muscle (myoneural junction).

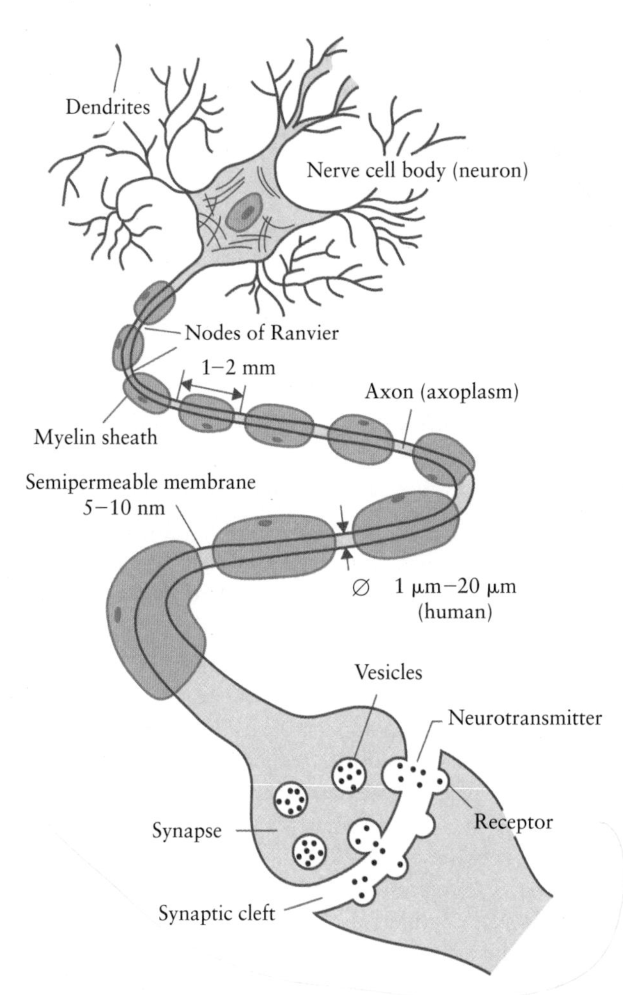

Figure 18.1 Sketch of a single myelinated nerve cell. The axon, containing the axoplasm, carries the nerve impulse from the dendrites to the synapse. The myelin sheath plays an important role in the mechanism of nerve impulse propagation. Note that myelin sheaths are interrupted about every 1 mm to 2 mm by nodes of Ranvier.

The neuron and the synapse are connected by the axon. Axons of both myelinated and unmyelinated nerves are cylindrical tubes that contain a solution called axoplasm. The axoplasm is separated from the extracellular fluid (also called interstitium) by a 5-nm to 10-nm-thick semipermeable membrane. As illustrated in Fig. 18.1, myelinated nerves are further encapsulated by a myelin sheath (also called Schwann cells for their discoverer Theodor Schwann). Schwann cells are 1 mm to 2 mm long and fully surround the axon. The myelin sheath is interrupted by short gaps, called nodes of Ranvier (named for Louis Ranvier).

Table 18.1 lists some of the properties of human nerve cells for myelinated and unmyelinated nerves. The table indicates that the presence of the myelin sheath makes a significant difference in the function of the nerves. Myelinated nerves are vital when high signal speeds are required, e.g., when attempting an emergency stop of a car. On the other hand, temperature information is processed by slower, unmyelinated nerves. You can establish this in a simple self-experiment. Consciously reach for a mug of coffee. You become aware that you touch the mug clearly before you can tell how warm it is.

TABLE 18.1

Comparison of physiologically relevant data for myelinated and unmyelinated human nerve cells

Human nerve type	Unmyelinated	Myelinated
Fraction of nerves	70%	30%
Cross-section of nerve cell	≈1.5 μm	≤ 20 μm*
Axon walls	semipermeable membrane (5–10 nm thick)	myelin sheath (2000 nm thick)
Impulse speed	0.6–10 m/s	10–100 m/s
Purpose	slow information (e.g., temperature stimulation)	motor information

*Giant axon of squid with a diameter 500 μm = 0.5 mm.

18.1.2: Electrochemical Processes in Nerves

To describe the electric mechanism of a nerve, a few additional microscopic observations are needed. We will find that two types of processes are essential for the understanding of nerves: non-equilibrium chemical processes and electric phenomena. On the other hand, neither fluid dynamics nor mechanics are required since signal conduction in nerves occurs without the nerve itself changing shape.

Fig. 18.2 shows the concentrations of key chemical components in both the axoplasm and the extracellular fluid. The data are given in unit mmol/L. Most important in both solutions are cations and anions, with NaCl (rock salt) the dominant compound outside the nerve, and potassium salts (potassium bound by protein and phosphate ions, shown as miscellaneous anions in the figure) the major component inside. Cations and anions are balanced on either side of the membrane, but the axoplasm carries the larger total salt concentration of 165 mmol/L.

We have to describe two states of nerves to fully understand their function: the resting nerve and the nerve

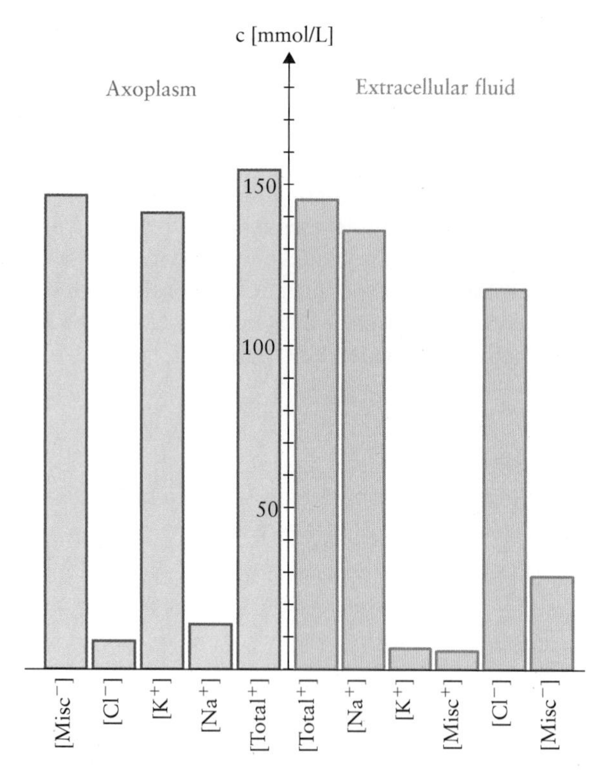

Figure 18.2 The concentration of various electrolyte components in the axoplasm (left of centre) and in the extracellular fluid (right of centre). The given values are based on mammalian spinal cord motor neurons.

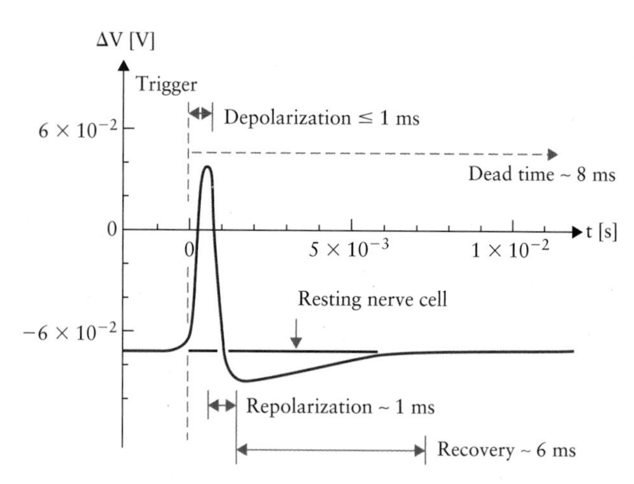

Figure 18.3 The changes in the electric potential across the membrane of a nerve cell when an impulse passes. Most of the changes occur within a millisecond, followed by a period of about 6 ms recovery time before the nerve can be stimulated again (i.e., a total of about 8 ms of dead time).

carrying an impulse. The resting nerve is not simply a nerve that is shut off; rather, this describes a dead nerve cell. Diffusion and ion drift in an electric field are continuous dynamic processes that establish a unique state for the resting nerve that we have to quantify before we can address nerve impulses. The adjective "resting" is still justified for two reasons: the physical processes occurring in a resting nerve are minimal, and no time dependence of these processes can be observed macroscopically. Thus, the **resting nerve** is a nerve in a steady state.

The potential change during the passage of an impulse through a nerve is illustrated in Fig. 18.3. This is the main data set we want to explain in the last part of the chapter. The electric potential is a measure of the strength of electric effects, and is often the first information available for a system because potential differences are relatively easy to measure. The existence of a non-zero electric potential difference across the nerve membrane indicates that cations and anions are separated across the nerve membrane. In Fig. 18.3 we follow the electric potential difference across the nerve membrane while a **nerve impulse** passes a particular point along the nerve as a function of time. The data in the figure can be described as a sequence of four steps:

(i) **Trigger:** An initial change in the potential difference across the membrane of 10 mV causes an impulse.

(ii) **Depolarization:** The nerve is polarized to a positive axoplasm potential within a millisecond.

(iii) **Repolarization:** The depolarization is followed immediately by a repolarization to a potential difference that is more negative than the resting potential difference.

(iv) **Recovery:** The nerve returns during a period of about 6 ms to its resting potential difference. It can now carry another impulse.

Each impulse leads to the same time profile of the cross-membrane potential difference, i.e., the same shape and peak height of the impulse. Thus, information is communicated by the repetition frequency of impulses.

To understand these observations we need to develop a model of the nerve. We start with the unmyelinated nerve. A very simple model of such a nerve is shown in Fig. 18.4. The figure shows a cross-section, with the axoplasm inside a cylindrical tube and the extracellular fluid outside. The nerve membrane separates ions that we treat as electric point charges. The membrane is modelled as charged parallel plates, for which we studied some of the electrostatic properties in the previous chapter. There is an excess number of positive ions outside and an excess number of negative ions inside the nerve. This model is a good starting point to describe a nerve when no impulse is passing through, i.e., a resting nerve.

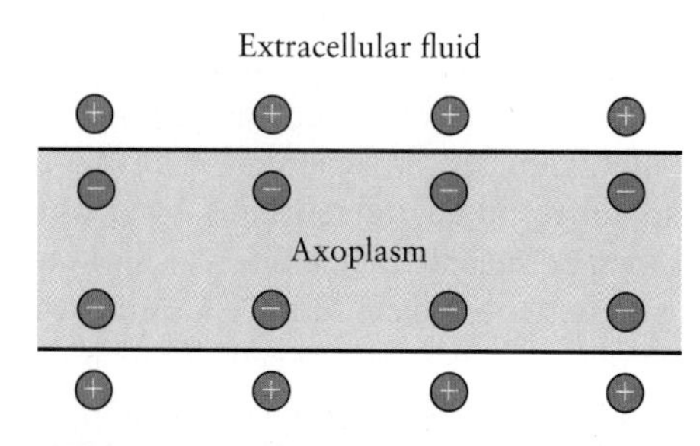

Figure 18.4 Simplified model of a nerve. The inner surface of the membrane (axoplasm side) carries negative charges; the outer surface of the membrane (extracellular fluid or interstitium side) carries positive charges.

To fully quantify this model we have to develop additional physics concepts: first, the concept of energy has to be introduced for electric systems. This will lead to the related concept of potential, which is the quantity plotted in Fig. 18.3. We then have to discuss in greater detail the properties of parallel plates, which we do by introducing the concepts of capacitance and a dielectric placed between the plates of the capacitor. We are then in a position toward the end of this chapter to model the nerve as a capacitor. This model proves to be insufficient because the actual nerve membrane allows for electric leakage currents across the membrane. Electric currents are discussed in the next chapter, when we combine all physics concepts from Chapters 17 and 18 to quantify signal transport in nerves.

18.2: The Electric Energy

Earlier in this textbook we found it very useful to introduce the concept of energy. It simplified many calculations, particularly due to the conservation law, but also allowed us to move beyond the boundaries of mechanics, which are set by the force concept. Again, the force concept has dominated the discussion of electricity to this point. Thus, we may ask the question whether energy concepts can be as useful in electricity as they were in mechanics and thermal physics. After you read through the rest of this textbook you will agree that the answer is "yes," for the same reasons we discussed in Chapter 7 and due to two additional arguments:

- Note that we have been quiet so far on how we actually measure electric forces or fields. This is because we cannot measure them directly. What we measure directly is an electric potential difference, a concept we derive below from energy concepts; electric fields or forces are then calculated only from the electric potential.
- Most electric phenomena of interest in the life sciences occur at the microscopic and atomic scale. We discuss the structure of atoms and their physical properties in Chapter 20. We will note then that energy concepts are much more suitable to describe these systems, while the application of the force concept becomes elusive.

Describing electricity with energy concepts is straightforward when we follow the approach we took in Chapter 7 in deriving the gravitational potential energy. At that time we relied on the fact that a gravitational force is assigned to each point in space, which allowed us to develop the potential energy since it is defined as the form of energy stored in the relative position of objects. Replacing objects of a given mass with point charges that carry a particular amount of charge and replacing gravity with the electric force, allows us to develop formulas for the electric potential energy in the same fashion.

The electric potential energy can be combined with kinetic and other forms of energy to determine the total or internal energy of a system. The possibility of converting electric energy into other forms of energy, the possibility of adding or removing it from a system in the form of work, and the conservation of the total energy of a system allow us to apply the many tools we developed earlier in thermal physics and mechanics. Thus, we have a greatly enhanced arsenal of physical tools available once electric potential energy is defined.

Electric charges can be arranged in a system in many ways. We selected in Chapter 17 for the discussion of electric phenomena three model systems that met the following three conditions: (i) they are suitable to describe most natural systems that are electrically active, (ii) they build the foundation of most engineered electric systems, and (iii) they are mathematically simplest to quantify. These systems are listed in Table 17.2. We discuss energy concepts in the current chapter for the same three systems but start with the case of parallel plates, as it leads to the simplest mathematical formalism.

18.2.1: The Potential Energy for Charged Parallel Plates

In this section, we follow the approach taken in Chapter 7 for the gravitational potential energy to introduce the electric potential energy. We start with the definition of work W. Since work is determined by a force $\vec{F}$ and a displacement $\Delta\vec{r}$, we can write $W = f(\vec{F}, \Delta\vec{r})$, in which the notation $f(\ldots)$ means "a function of."

In the electric case we consider a probe charge, which we move from an initial to a final equilibrium position within a system of stationary charges. An external force $\vec{F}_{ext}$ is needed to ensure mechanical equilibrium. Once the work is determined, the electric potential energy E_{el} is calculated from:

$$W = E_{el,final} - E_{el,initial}.$$

Work is written specifically as the product $W = \pm F_{ext}\Delta r$ if

- the external force and the displacement are co-linear, and
- the external force is constant at every position along the displacement.

If one or both of these conditions do not apply, more complicated formulas have to be applied: if the external force is constant but not co-linear with the displacement vector, the dot product of the two vectors is used, $W = \vec{F}_{ext} \bullet \Delta\vec{r}$. If the force varies but is co-linear with the displacement vector, the work is based on graphic analysis of the area under the $F_{ext}(r)$ curve, as discussed in section 9.3. Since the product $F_{ext}\Delta r$ is easiest to evaluate, we start with the parallel plate arrangement, as we saw in Chapter 17 that parallel plates yield an electric force that is constant at any position between the two plates.

We use Fig. 18.5 for the quantitative calculation. The plate at the top is charged positively and the plate at the bottom is charged negatively. This leads to a downward-directed electric field. We further assume that a positive probe charge moves from close to the positive plate to a position close to the negative plate, i.e., from position y_i to position y_f. Moving a probe charge from one equilibrium position to another requires an external force F_{ext} to prevent it from accelerating toward the negative plate. Consequently, in Fig. 18.5 the external force is positive and the displacement is negative. A negative work follows from anti-parallel external force and displacement; i.e., the probe charge in Fig. 18.5 releases work to the source of the external force:

$$W = F_{ext}\,\Delta y = q_{mobile}\left|\vec{E}\right|(y_f - y_i) < 0. \qquad [18.1]$$

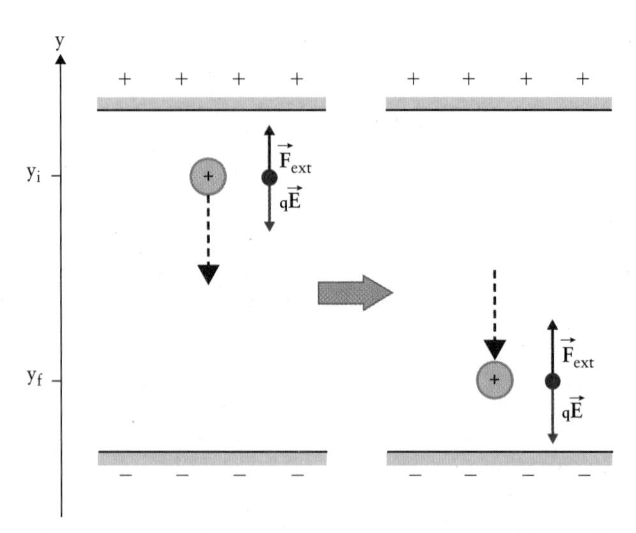

Figure 18.5 A parallel plate arrangement with an electric field $\vec{E}$ pointing downward. The forces acting on a positive probe charge are shown in a free body diagram at its right side. The sketch at the left side shows the probe charge at its initial position y_i and the sketch at the right at its final position y_f. Note that the probe charge is in both cases in mechanical equilibrium.

Using the work–energy relation reiterated above, we determine the electric potential energy:

$$E_{el} = q_{mobile}\left|\vec{E}\right| y,$$

in which we use Eq. [17.14] to specify the magnitude of the electric field:

$$E_{el} = q_{mobile}\frac{\sigma}{\varepsilon_0}y. \qquad [18.2]$$

KEY POINT

The electric energy of a mobile point charge in a parallel plate arrangement is a linear function of distance from the plate that carries a charge with sign opposite to that of the mobile charge.

Note that Eq. [18.2] is mathematically similar to the formula derived for gravitational potential energy, which is a linear function of height above ground ($E_{pot,grav} = m\,g\,h$). This simple form of Eq. [18.2] is the reason why parallel plate arrangements are an often used model for studying electric phenomena. The analogy to the gravitational case goes even further. If we release an object held at a given height, it accelerates toward the ground due to the attractive gravitational pull. In the same manner, if you remove the external force that holds the mobile point charge in Fig. 18.5, it will accelerate toward the plate with the opposite charge due to the attractive electric force.

CONCEPT QUESTION 18.1

In the six processes shown in Fig. 18.6, a probe charge (green circle with positive or negative charge indicated in the circle) is moved from an initial to a final position (see arrow). In which cases is the work positive (energy is received by the probe charge during the displacement)?

(A) iii and vi

(B) i and iii

(C) ii and iv

(D) i and iv

(E) ii and v

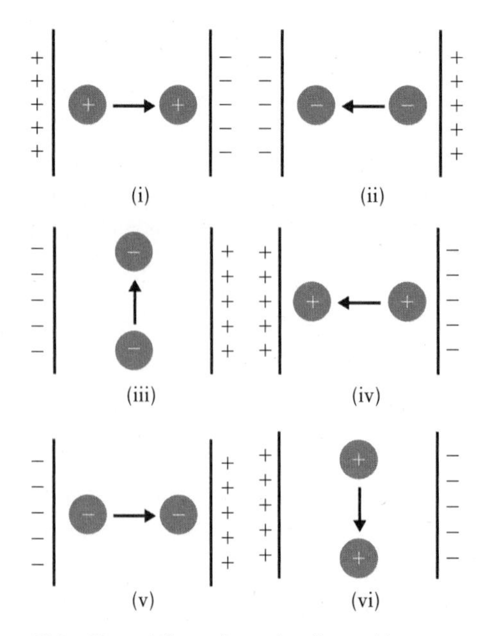

Figure 18.6 Six variations of a probe charge (shown as a small open circle with positive or negative charge indicated in the circle) that moves from an initial to a final position (see arrow) in a charged parallel plate arrangement.

EXAMPLE 18.1

In Fig. 18.7 we are given a single, infinite, non-conducting sheet with a positive surface charge density σ. How much work is done by the external force as a positive probe charge q_{probe} is moved perpendicularly from the surface of the sheet at z_i to a final position at z_f?

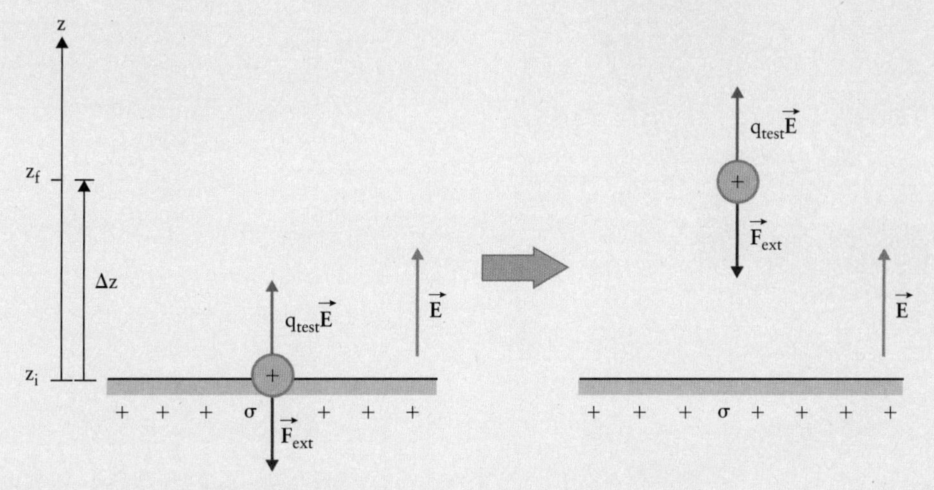

Figure 18.7 A positive probe charge is moving from the surface of a single plate (position z_i) with a positive surface charge density σ (at left) to a final position z_f (at right). The electric field of the plate points upward. The total displacement of the probe charge is Δz. It is in mechanical equilibrium in both positions.

Solution

The electric field of a single flat plate is given in the form $|\vec{E}| = \sigma/(2\varepsilon_0)$. The external force needed to create an equilibrium is directed opposite the Coulomb force and is therefore anti-parallel to the displacement. Thus, the work is negative; i.e., work is done by the mobile point charge on the source of the external force. With the magnitude of the external force given by $F_{ext} = q\,|\vec{E}|$, we find:

$$W = -\frac{q_{probe}\,\sigma}{2\,\varepsilon_0} z_f < 0.$$

18.2.2: Electrophoresis

EXAMPLE 18.2

A blood sample from a patient with cirrhosis of the liver is analyzed by electrophoresis. During the protein separation, a γ-globulin molecule from this blood sample moves a distance of 7 cm toward the negative plate of the electrophoresis set-up. The electric field between the parallel plates is $|\vec{E}| = 2000$ N/C. How much work has been done on the γ-globulin molecule?

Supplementary physiological information: Electrophoresis is widely used to separate and identify charged components in solutions such as blood. Two types of experimental set-ups exist. In forensic science, DNA fragments are routinely separated across a gel after the original DNA molecule has been split by restriction enzymes. This set-up is shown in Fig. 18.8. Electrophoresis separates ionic molecules based on their mass.

Fig. 18.9 shows a second set-up that is more suitable for recovering the separated components. This approach is used for blood sample analysis of protein composition. The test sample travels downward through a cellulose matrix in a vertical arrangement, driven by the steady flow of a buffer solution from a reservoir. Concurrently, these components drift toward the left or the right, driven by the electric field applied horizontally across the matrix. Thus, the test sample separates as it is washed downward, and the various components can be collected in tubes at the bottom of the cellulose sheet.

Fig. 18.10 shows scans of electrophoretically separated blood samples for (a) normal and (b–f) pathological cases. The components of blood identified are albumin (labelled A) and five different globulin proteins, labelled α_1, α_2, β_1, β_2, and γ. The relative concentration of these proteins in the blood sample allows the physician to detect various diseases, including the ones highlighted.

Solution

The electrophoresis set-up is based on a charged parallel plate arrangement. Eq. [18.1] gives us the work done on a point charge between the two plates when it moves closer to one of the plates. The electric field between the plates in Fig. 18.10 points toward the negative plate at the right due to the convention of Fig. 17.14. We know the γ-globulin ion is positively charged because it drifts toward the negative plate. Therefore, the electric force on the protein ion is directed toward the right and the external force to hold it in mechanical equilibrium, $\vec{F}_{ext}$, must be directed toward the

continued

Figure 18.8 Electrophoresis set-up used in forensic science. DNA fragments are separated across a gel after the original DNA molecule has been split by restriction enzymes. The sample is brought onto the gel close to the negative plate at right. Electrophoresis separates the charged molecule fragments on the basis of their rate of movement through the gel in a given electric field. Typically, larger fragments travel slower than smaller ones. The DNA sample is brought close to the negative plate because the fragments carry a negative charge due to their phosphate groups.

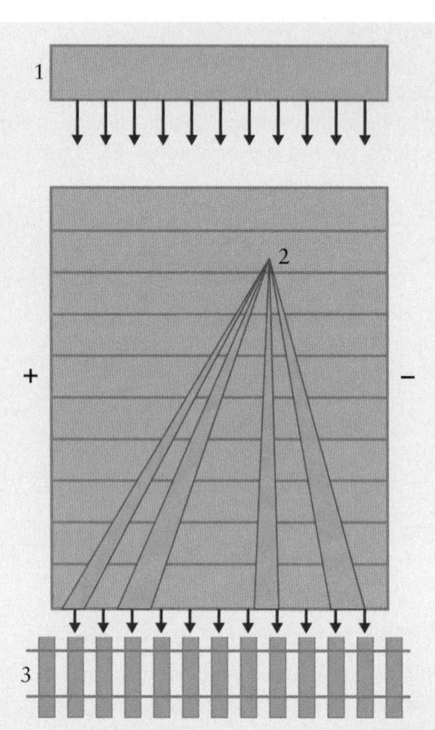

Figure 18.9 The proteins in the sample, added at point (2), travel vertically downward in a cellulose matrix, driven by the steady flow of a buffer solution supplied from a reservoir at the top (1). The components of the blood sample drift toward the left or the right, driven by the electric field applied horizontally across the matrix. The negative plate, shown at the right, causes positive ions to drift to the right. Separation of equally charged ions occurs as the result of their molecular sizes. The various components of the sample can be collected in test tubes at the bottom (3) for further analysis.

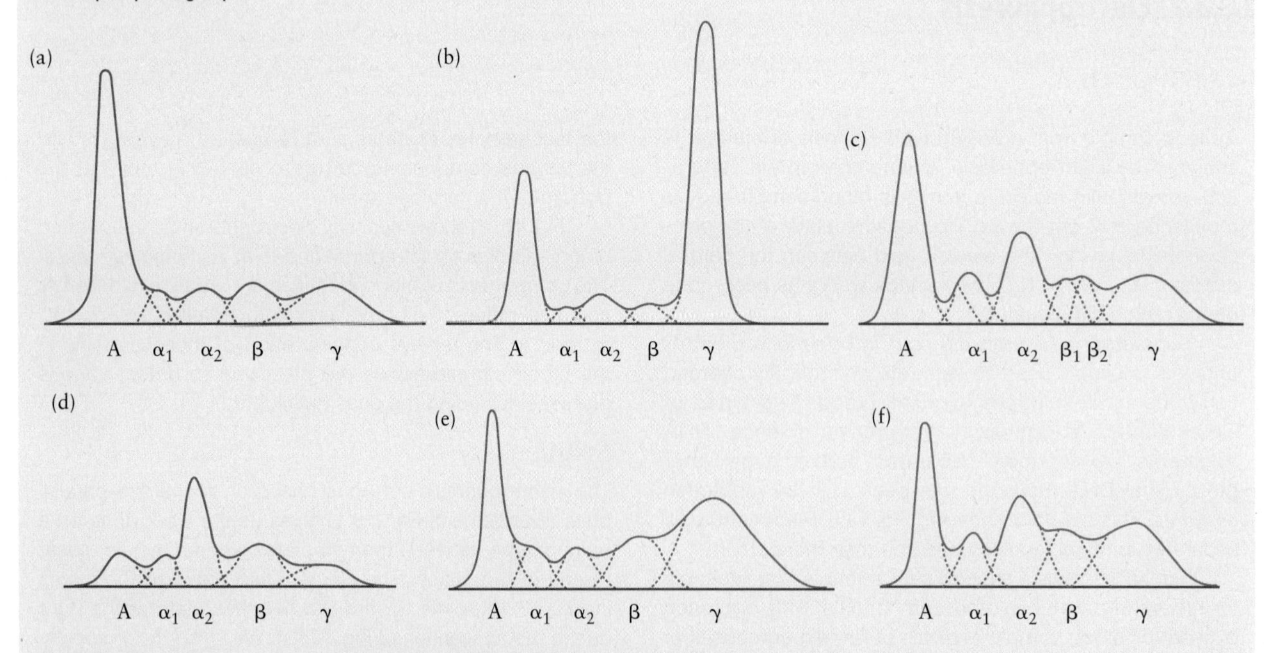

Figure 18.10 Typical scan results for blood samples that have been separated by electrophoresis using the set-up in Fig. 18.9. The components of blood identified are albumin (labelled A) and five different globulin proteins, labelled α_1, α_2, β_1, β_2, and γ. (a) A comparison sample for a healthy patient. (b–f) Pathological cases: (b) plasma cell tumour, (c) acute inflammation, (d) severe nephrosis, (e) cirrhosis of the liver, (f) liver parenchyma damage (parenchyma refers to the essential and distinctive tissue of an organ).

continued

left. The displacement of the protein ion occurs toward the right. If we define the positive y-axis toward the right, then $F_{ext,y} = -F_{ext} < 0$ and $\Delta y > 0$. Thus, the work calculated from $W = \pm F_{ext}\,\Delta y$ is negative, $W < 0$.

For the numerical value of the work, we substitute the given values into Eq. [18.1]. An elementary charge is carried by the γ-globulin: $q_{mobile} = e = 1.6 \times 10^{-19}$ C. The magnitude of the field is $|\vec{E}| = 2.0 \times 10^3$ N/C and the displacement is $\Delta y = 7.0 \times 10^{-2}$ m:

$$W = -(1.6\times10^{-19}\,\mathrm{C})\left(2\times10^{3}\,\frac{\mathrm{N}}{\mathrm{C}}\right)(0.07\ \mathrm{m})$$

$$= -2.2\times10^{-17}\ \mathrm{J}.$$

We want to discuss the sign in this equation a little further. Why is the work negative? Let's start with a positive ion between two charged plates in vacuum; i.e., we remove the cellulose matrix of the electrophoresis set-up. In that case the ion would accelerate toward the negative plate. The ion and the charged plate represent an isolated system for which the conservation of energy applies. Thus, the potential energy of the ion decreases during the acceleration, but its kinetic energy increases by the same amount so that the internal energy (total energy) remains unchanged.

In the case of the electrophoresis experiment, the positive ion does not accelerate. It moves with a small constant speed toward the negative plate. This is due to the cellulose matrix, which acts as the origin of an external force. Thus, in the electrophoresis experiment, the system ion plus charged plates is not an isolated system and the conservation of energy is written in the form $\Delta U = Q + W$; i.e., internal energy changes are due to exchange of work and heat with the environment. In electrophoresis, no heat exchange is observed and the point charge changes only its potential energy, not its kinetic energy. Thus, $\Delta E_{el} = W$: the negative work is due to the decrease in potential energy of the ion.

18.2.3: The Potential Energy for a Stationary Point Charge

We are interested in two other electric arrangements in the current chapter: a single point charge and a dipole. We saw earlier that the force varies in both cases with distance from the stationary charges; i.e., calculating work is slightly more complicated than in the case of charged parallel plates. We use the case of a single stationary point charge in Case Study 18.1 to illustrate a graphical approach.

CASE STUDY 18.1

Fig. 18.11 shows a positive probe charge (small circle) that is brought closer to a positive stationary point charge Q (large circle) at position $r = 0$. Is work required in this process?

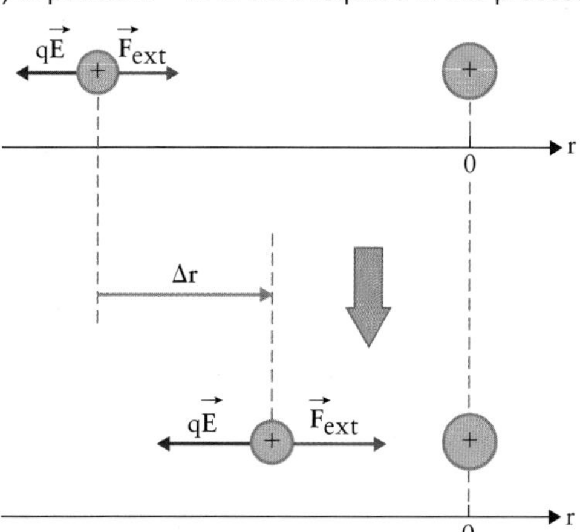

Figure 18.11 Sketch to illustrate the work associated with moving a positive probe charge (small circle) by a distance Δx closer to a positive stationary point charge (large circle, positioned at $r = 0$). Both the initial position (top) and the final position (bottom) of the probe charge are at negative positions along the axis. Note that the electric force $q\,\vec{E}$ and the external force $\vec{F}_{ext}$, which is required to establish a mechanical equilibrium, increase in magnitude as the probe charge moves closer to the stationary point charge.

Answer: *The electric force acting on the probe charge is written as q $\vec{E}$ at both the initial and the final positions. To bring the probe charge closer to the stationary point charge, an external force is needed to transfer it through the displacement Δr. Bringing the probe charge closer requires work; i.e., the work is positive.*

Note that we set up Fig. 18.11 such that the probe charge moves directly toward the stationary point charge. This simplifies the discussion because the force and the displacement are parallel and we do not need to include vector notation to calculate the work. We draw a graph of the magnitude of the external force as a function of position in Fig. 18.12, and then determine the work as the area under the curve from r_i to r_f:

$$W = \frac{q_{mobile}Q}{4\pi\varepsilon_0}\left(\frac{1}{|r_f|} - \frac{1}{|r_i|}\right). \qquad [18.3]$$

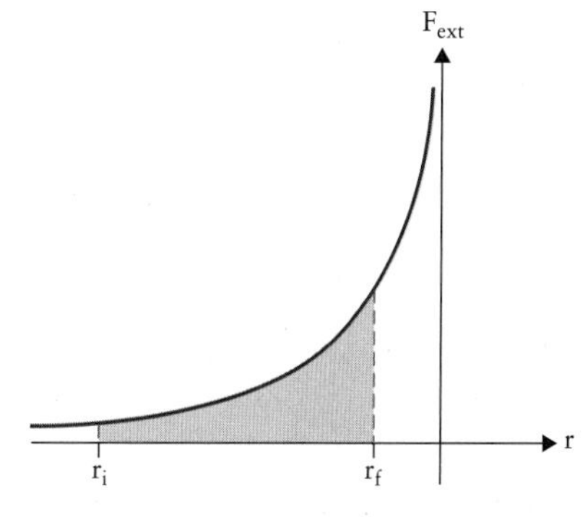

Figure 18.12 The electric potential energy for the system shown in Fig. 18.11 is obtained from the area under the curve for the electric force as a function of position.

From Eq. [18.3] we derive the electric potential energy:

$$E_{el} = \frac{1}{4\pi\varepsilon_0}\frac{q_{mobile}Q}{|r|}. \qquad [18.4]$$

Note that the result is independent of the sign of the position of the probe charge. This is indicated by the absolute bars of the variable r in the denominators of Eqs. [18.3] and [18.4].

The existence of positive and negative charges distinguishes the electric potential energy from the gravitational potential energy. The gravitational force is always attractive; therefore, the external force vector establishing mechanical equilibrium always points away from Earth. However, whether the electric force is attractive or repulsive depends on whether the two interacting charges are respectively positive or negative. In Fig. 18.11 both charges are positive and the electric force is repulsive. The external force points toward $r = 0$, as shown in the figure. If the stationary point charge were in turn negative, then the electric force would be attractive and the external force needed to establish a mechanical equilibrium would point away from the origin. Eq. [18.4] illustrates that the electric interaction between two particles with opposite charges always leads to a negative electric potential energy, which is an energy we have to overcome to separate the two point charges. This energy is particularly large if the initial separation of the two point charges is small.

CASE STUDY 18.2

Separating two opposite point charges requires work. Why is it, then, that sodium and chlorine ions of rock salt (NaCl) almost always occur separated in our body? Where does the energy to break up a digested rock salt crystal come from?

Answer: *Water, with its large electric dipole moment, provides the means of separating Na^+ and Cl^- ions. If we break apart sodium and chlorine in a water-free environment, the separating Na^+ and Cl^- ions turn neutral (via an electron exchange) at a distance of about one nanometre, as illustrated in Fig. 18.13. The figure compares, as a function of separation distance, the total energy of a pair of neutral Na and Cl atoms (dashed line) with the total energy of a Na^+ ion and a Cl^- ion (red line). At large distances, the energy difference results from the following thermodynamic relations:*

$$\begin{array}{lll} Na & \rightarrow Na^+ + e^- & \Delta U = +498\ \text{kJ/mol} \\ Cl + e^- & \rightarrow Cl^- & \Delta U = -351\ \text{kJ/mol} \\ \hline Na + Cl & \rightarrow Na^+ + Cl^- & \Delta U = +147\ \text{kJ/mol.} \end{array}$$

Remember, a positive change in the internal energy means that the system requires energy from its environment. Thus, the transfer of 1 mol electrons from 1 mol sodium to 1 mol chlorine requires +147 kJ. However, Fig. 18.13 shows that when the atoms are at a distance of less than one nanometre from each other the ions in turn are energetically favoured due to the electric potential energy.

As a consequence, rock salt ions can be separated by distances greater than 1 nm only if this neutralizing electron exchange is prevented. A separation of ions across a typical 6-nm nerve membrane would be impossible. In order to do this, the energy of the separating ions must be lowered. In an aqueous solution, this is achieved by the formation of the hydration shell. The hydration shell screens an ion, an effect that reduces the interaction with other ions. It also lowers the energy by redistributing the charge over a larger volume, which stabilizes the ion.

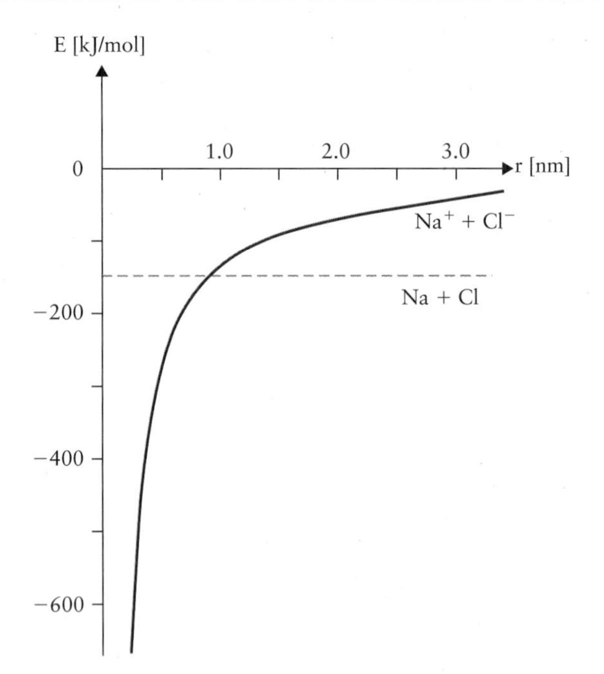

Figure 18.13 Comparison of the potential energy of a neutral sodium atom and a neutral chlorine atom (dashed line) with the potential energy of a positive sodium ion and a negative chlorine ion (red line) as a function of distance. Neutral atoms are favoured when the separation exceeds 1 nm.

EXAMPLE 18.3

Calculate the distance between the two ions in Fig. 18.13 at the point at which the two curves cross, i.e., the distance at which the potential energy of the neutral Na/Cl pair equals the electric potential energy of the Na^+/Cl^- pair.

Solution

The electric energy of two point charges at infinite distance is set to the reference value of 0 J, as follows from Eq. [18.4]. At any closer distance, the energy between a positive and a negative point charge is then negative. We want to find in particular the distance at which the electric energy is equal to the change of the internal energy in the neutralization process of a Na/Cl pair, $\Delta U = -147$ kJ/mol. This energy is converted into an energy for a single pair of neutral atoms, $\varepsilon_{Na/Cl}$, by dividing ΔU by the Avogadro number. This yields a value of $\varepsilon_{Na/Cl} = -2.45 \times 10^{-19}$ J. We rewrite Eq. [18.4] to find the distance at which this energy equals the electric potential energy of the system:

$$r = \frac{q_{Na^+} q_{Cl^-}}{4\pi\varepsilon_0} \frac{1}{E_{el}}$$

$$= \frac{(1.6\times10^{-19}\text{C})(-1.6\times10^{-19}\text{C})}{4\pi\left(8.85\times10^{-12}\frac{\text{C}^2}{\text{Nm}^2}\right)(-2.45\times10^{-19}\text{J})}$$

$$= 9.4\times10^{-10}\text{ m},$$

which is $r = 0.94$ nm. This calculation confirms a value close to 1 nm, which we read from Fig. 18.13 earlier.

For systems with N stationary point charges, the electric energies of each pair interaction with the probe charge are added to obtain its total energy:

$$E_{el} = \sum_{i=1}^{N} E_{el,i}. \qquad [18.5]$$

18.3: The Electric Potential

When studying electric force in the previous chapter, we noted that the great variability of electric systems caused mathematical complexity. We dealt with this problem by using Faraday's method to split the electric force, defining the electric field. The same issues apply for electric energy. Thus, an equivalent approach is taken to address energy-related issues: we find a field, called electric potential, that represents at every position in space the electric potential energy that a positive probe charge of unit charge would have if brought to that point.

This means that we have two routes to the electric potential: we can proceed from the electric force to electric potential energy and then eliminate the probe charge, or we can proceed from the electric force to the electric field by eliminating the probe charge and then make the step from electric field to potential. Below we choose the first approach, since we just completed the discussion of the various forms of electric potential energy and want to use those results to establish the potential with the least mathematical effort.

Note that we continue to limit the discussion to systems in which all charges are stationary except for a single probe charge. Thus, the resulting potential will be a stationary field (time-independent field).

18.3.1: Calculating the Electric Potential

We separate the electric energy into

- a probe charge which carries a positive unit charge unless specified otherwise, and
- a field due to all stationary point charges in the system. This field is called the **electric potential**, which should not be confused with the electric potential energy.

Different from the electric field, the potential is a **scalar field**. It is labelled V and is defined by dividing the electric potential energy by the probe charge:

$$E_{el} = q_{mobile} V. \qquad [18.6]$$

KEY POINT

The electric potential energy is the product of the electric potential and the charge of the probe charge. Both energy and potential are scalars; the potential is therefore a scalar field.

A new unit is introduced for the potential, called **volt** V = J/C, named for Alessandro Count Volta. Two specific cases follow immediately from Eq. [18.6] and the energies we established in the previous section for the parallel plate arrangement and a stationary point charge. For the parallel plate arrangement, we obtain:

$$V_{\parallel} = \frac{E_{el}}{q_{test}} = \left|\vec{E}\right| y = \frac{\sigma}{\varepsilon_0} y, \qquad [18.7]$$

in which σ is the surface charge density on the plates. For a stationary point charge, we get:

$$V = \frac{E_{el}}{q_{test}} = \frac{1}{4\pi\varepsilon_0} \frac{q}{|r|}. \qquad [18.8]$$

Note that both equations seem to allow us to calculate absolute values for the potential. While this is correct, we have to remind ourselves that absolute potential values, like absolute values for a potential energy, are physically meaningless. It makes no difference whether you consider yourself to be at 0 V potential right now or whether you prefer to define your current potential as 1000 V. What matters are potential differences in the same way we discussed that only potential energy differences matter. There is no danger in touching an object that has the same potential as yourself; but do not touch objects that have significantly different potentials from the ground under your feet! In short, we always seek potential differences, ΔV, when discussing real physical phenomena. Example 18.4 highlights this aspect.

EXAMPLE 18.4

A potential difference of 80 mV exists between the inner and outer surfaces of a membrane of a cell. The inner surface is negative relative to the outer surface. How much work is required to eject a positive potassium ion (K^+) from the interior of the cell?

Solution

Fig. 17.12 shows the environment of the diffusing potassium ion. Initially the ion is inside (as shown), where it is in close proximity to the negative surface charge on the membrane. Work is required to transfer the potassium ion through the membrane as the ion leaves a negative environment and approaches a positive environment. The work is calculated using Eq. [18.7], noting that a potassium ion carries a single positive elementary charge $q = e$:

$$W = q\Delta V = (1.6\times10^{-19}\ \text{C})(80\times10^{-3}\ \text{V})$$
$$= +1.28\times10^{-20}\ \text{J}.$$

The work is positive for a process into which work has to be invested.

CASE STUDY 18.3

We consider the potential curve $V(x)$ in Fig. 18.14(b). Which of the four charged parallel plate arrangements in Fig. 18.14(a) is consistent with this potential curve?

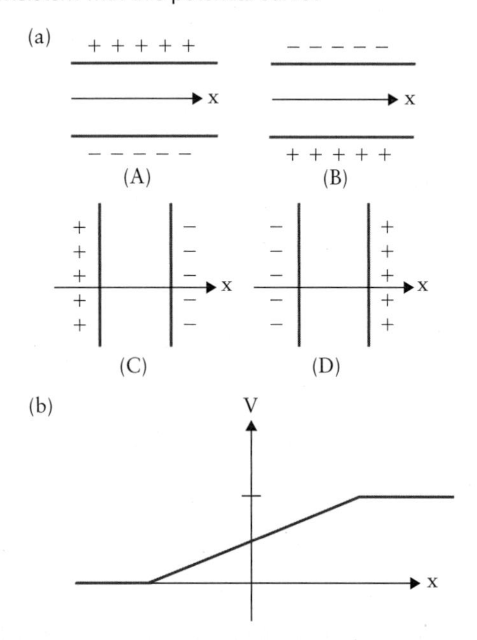

Figure 18.14 (a) Four charged parallel plate arrangements. (b) A potential curve $V(x)$, i.e., a curve that shows potential as a function of position along the x-axis.

Answer: *Choice (D). Options (A) and (B) can be ruled out because the x-axis runs parallel to the electrically charged plates. In that direction the electric energy does not change, nor does the potential. Options (C) and (D) differ in that the x-axis points in one case toward the plate with positive charges, in the other toward the plate with negative charges. We can use the result from Example 18.4 to decide which case applies: the potassium ion is a positive probe charge. We found a positive work in the example, which means that the electric energy for the potassium ion increases as it moves in Fig. 17.12 toward the left. Dividing an increase in energy by the positive charge in Eq. [18.6] yields an increase in the potential. We can combine this result with Fig. 17.14 to state below the relation of the direction of the electric field to the direction of increasing or decreasing potential.*

KEY POINT

The potential increases when we follow a line that moves from an area with negative charges to an area with positive charges. The electric field points in the direction of decreasing potential.

CONCEPT QUESTION 18.2

If the electric field has a constant magnitude and points in the positive x-direction, which of the following formulas is correct to describe the potential if a is constant and b and c are positive non-zero constants (i.e., $b > 0$ and $c > 0$)?

(A) $V = a$

(B) $V = a + bx$

(C) $V = a - bx$

(D) $V = a + bx + cx^2$

(E) $V = a - bx - cx^2$

Next we address the energy and the potential of our third model system, a dipole. It is introduced as an example based on two opposite point charges.

EXAMPLE 18.5

Find the potential at point P, which is located at an arbitrary position relative to a dipole, as shown in Fig. 18.15(a). Assume that the point P is far from the dipole.

Solution

For more than one stationary point charge, electric potentials have to be added in the same manner as we discussed for the electric potential energy in Eq. [18.5]. In particular, the electric potential at point P, V_P, in Fig. 18.15(a) is:

$$V_P = V_{P,+q} + V_{P,-q}$$
$$= \frac{1}{4\pi\varepsilon_0}\left(\frac{q}{r_+} + \frac{-q}{r_-}\right),$$

continued

with P a distance r_+ from the positive point charge and a distance r_- from the negative point charge of the dipole. This equation is rewritten with a common denominator for the term in the parentheses:

$$V_P = \frac{q}{4\pi\varepsilon_0} \frac{r_- - r_+}{r_- r_+}.$$

If P is close to the dipole, this result cannot be simplified further and is therefore final. However, if P lies far from the dipole with $r_+, r_- >> d$, then both the numerator and the denominator of the last term are rewritten using geometrical relations obtained from Fig. 18.15(b). In that figure, the two lines that extend from the dipole charges to point P are treated as parallel. The difference in distance from the two charges to point P is a side in the right-angle triangle shown in the figure. This triangle allows us to express the term $r_- - r_+$ as a function of the length d and the angle θ between the dipole axis and the position vector to point P:

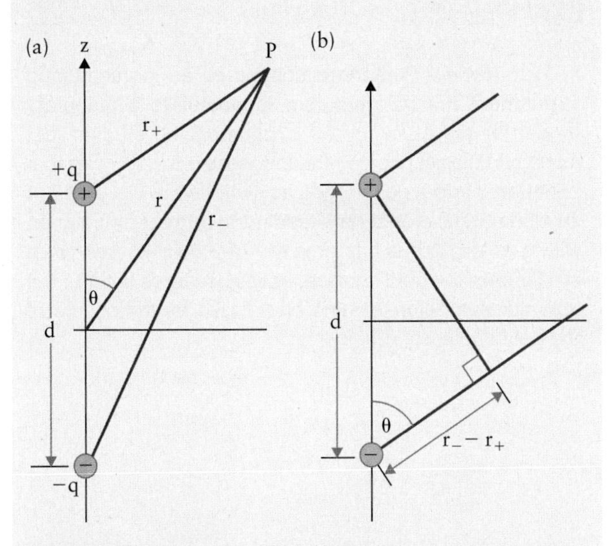

Figure 18.15 (a) A dipole oriented along the z-axis. The point P is located at an arbitrary position relative to the dipole. (b) Geometric plot to determine the difference in the distances from each of the two point charges of the dipole in part (a) to a distant point P.

$$r_- - r_+ = d\cos\theta.$$

We further use $r_- \cong r_+ \cong r$, with r the distance of the point P from the centre of the dipole. This simplifies the denominator to:

$$r_- r_+ = r^2.$$

Inserting the last two equations in the formula for V_P we find:

$$\lim_{r \gg d} V_P = \frac{q}{4\pi\varepsilon_0} \frac{d\cos\theta}{r^2} = \frac{\mu}{4\pi\varepsilon_0} \frac{\cos\theta}{r^2}, \qquad [18.9]$$

in which μ is the electric dipole moment of the dipole in Fig. 18.15. Thus, the electric potential of a dipole diminishes more rapidly with distance than the electric potential of a point charge.

18.3.2: Why Is the Electric Potential an Important Concept?

In your future work, should you expect to use the force or the energy concept? In the earlier chapters, the energy concept allowed us to move beyond a mechanical view of nature. Thus, in the context of the life sciences you expect arguments based on energy to occur more often, as this approach links to the thermodynamic description of systems. But is this also true when electric charges are involved?

The answer is that, again, the energy approach is more useful and therefore more often applied. From a life scientist's point of view, the first reason is the same as we developed in earlier chapters: in order to describe nature at the molecular level, force is a more elusive concept while energy connects to the thermal physics properties of the system. The link between electric and thermodynamic properties is established in the next chapter; its key physical law will be Nernst's equation. In the case of electricity, a second reason for the priority of energy over force has to be added: potential differences can be measured, while electric field values are experimentally inaccessible. The instrument allowing us to measure the potential is called a **voltmeter**. All we have to do is position two metallic electrodes (wires) at two chosen points in a system and the voltmeter registers the **potential difference** between these two points. We do not discuss the actual instruments here as this requires the concept of magnetism, which we discuss in Chapter 21.

18.3.3: Equipotential Lines

The electric potential for more complex arrangements of charges is often illustrated using equipotential lines in a graphical sketch. Equipotential lines are lines that connect points of equal value of the electric potential. These lines are always perpendicular to electric field lines. Fig. 18.16 illustrates the equipotential lines for the three types of systems we discuss in the current chapter: Fig. 18.16(a) shows horizontal electric field lines (dashed arrows) and vertical equipotential lines (solid lines) for a uniformly charged parallel plate arrangement. Neither the electric field lines nor the equipotential lines indicate a more complicated position dependence in this case.

Fig. 18.16(b) shows radial electric field lines and concentric equipotential lines for a positive point charge. To illustrate the non-linear r dependence, the equipotential lines are chosen with a constant potential difference ΔV between neighbouring lines.

Fig. 18.16(c) shows the more complex structure of electric field lines and equipotential lines for a dipole. Note that the electric field lines are straight only along the dipole axis, while a straight equipotential line exists perpendicular to the dipole axis at the centre of the dipole.

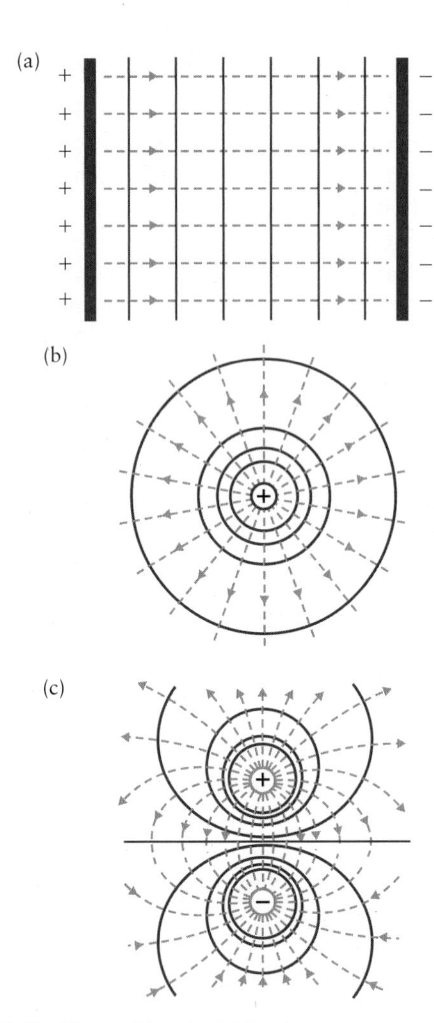

Figure 18.16 The electric potential for charge arrangements is often visualized by using equipotential lines in a sketch. Equipotential lines are lines that connect points of equal value of the electric potential. They are shown as solid lines in the figure. These lines are always perpendicular to electric field lines, shown as dashed lines. Illustrated are (a) a uniformly charged parallel plate arrangement, (b) a positive stationary point charge, and (c) a dipole.

18.4: Conservation of Energy

Electric energy is a form of energy that was not considered in Chapter 7. Therefore, we want to determine whether it has the same properties as the previously defined forms of energy. We are interested in seeing how key concepts of thermodynamics are applied to electric systems, particularly

- that the internal energy of a system is governed by the first law of thermodynamics, and
- that the Gibbs free energy is equal for two systems in chemical equilibrium.

We consider energy conservation here; the electrochemical equilibrium across a semipermeable membrane is discussed in Chapter 19. The conservation of energy is valid in all experiments that include electric effects; i.e., electricity does not contradict the previously introduced laws of thermodynamics. In a closed system of point charges that interact with their environment, the total work exchanged with the environment includes an electric work contribution in the form:

$$W = \Delta E_{el} = E_{el,f} - E_{el,i}. \qquad [18.10]$$

The inclusion of a new contribution to the total energy does not complicate the calculations, as the number of energy terms that are relevant for a given system is usually small. For example, in most physiological systems gravitational potential energy is negligible. Kinetic energy plays a role only when parts of the system may accelerate. For example, kinetic energy need not be included in the discussion of signal transport in nerves because essentially no acceleration of point charges along the nerve is involved.

EXAMPLE 18.6

In 1911, Ernest Rutherford conducted an ion scattering experiment that allowed him to postulate a planetary model for the atom. In the experiment he studied fast alpha particles of charge +2*e* and mass 6.6×10^{-27} kg that penetrated into a gold target, as sketched in Fig. 18.17. If an alpha particle on a path leading to a head-on collision with a gold nucleus has a speed of 2.0×10^7 m/s while still far from the gold nucleus, how close does it get to the gold nucleus before turning back? Use for the charge of the gold nucleus +79*e*.

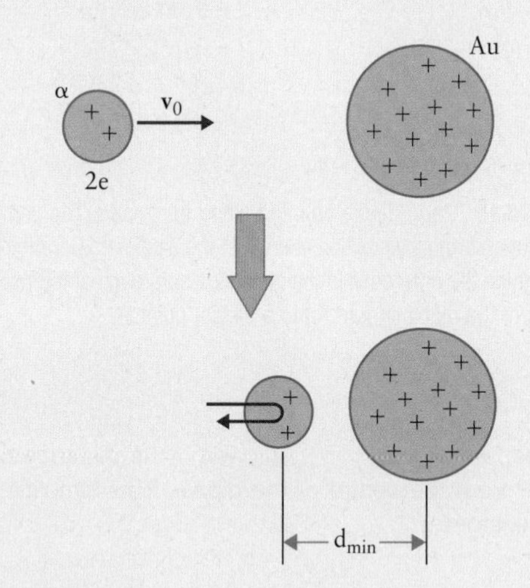

Figure 18.17 Sketch of Ernest Rutherford's experiment: a helium nucleus (alpha particle with two positive charges) approaches at high speed a gold nucleus at rest in the target.

Solution

We use the conservation of energy. In this particular case, the kinetic energy and the electric potential energy are the only two energy forms that vary during the process. Thus, we write the conservation of energy in the form:

$$E_{kin,i} + E_{el,i} = E_{kin,f} + E_{el,f},$$

continued

in which the kinetic energy is related to the speed of the moving alpha particle as $E_{kin} = ½\ mv^2$ and the electric potential energy is given by Eq. [18.4] since the interaction occurs between two point charges.

We choose the initial state at a point where the alpha particle is still far away from the gold nucleus. In this context "far" essentially means farther than the radius of the gold atom. This very short distance is sufficient, as the negative electrons in the atomic shell of a neutral gold atom screen the positive charge centred in the nucleus. In the initial state the electric potential energy of the alpha particle is zero, as obtained from Eq. [18.4] for $d = \infty$. The speed of the particle is determined from the acceleration process available in the laboratory. (Usually, the speed of the alpha particle would not be given as such but would be calculated from the energy transferred to the particle during acceleration.)

The final state is given where the alpha particle reaches the closest proximity to the gold nucleus (distance d_{min}). The particle is now well within the innermost shell of atomic electrons and thus facing the unscreened charge of the Au nucleus. At that point, the kinetic energy of the particle becomes zero since the particle comes momentarily to rest before accelerating away from the nucleus. Its entire energy has been shifted into its electric potential energy. Substituting these two states into the energy conservation formula yields:

$$\frac{1}{2}mv_i^2 = \frac{1}{4\pi\varepsilon_0}\frac{q_\alpha q_{Au}}{d_{min}},$$

which is equivalent to:

$$d_{min} = \frac{1}{4\pi\varepsilon_0}\frac{2q_\alpha q_{Au}}{m_\alpha v_i^2}.$$

With the data given in the example text, this yields:

$$d_{min} = \frac{2\left(9\times10^9\frac{\mathrm{Nm}^2}{\mathrm{C}^2}\right)2e79e}{(6.6\times10^{-27}\mathrm{kg})\left(2\times10^7\frac{\mathrm{m}}{\mathrm{s}}\right)^2}$$

$$= 2.8\times10^{-14}\mathrm{m}.$$

We compare this value to the radius of the gold nucleus, which is $r = 7 \times 10^{-15}$ m. Thus, the alpha particle approaches the gold nucleus to within a distance of twice the diameter of the nucleus! Still, the interaction is entirely electric; the nuclear force does not reach that far beyond the nucleus.

18.5: Capacitors

It is useful to note that we have already laid a good foundation for the discussion of nerve membranes at this point. For example, Fig. 18.18 gives an overview of the potential difference (top panel) and the electric field across an axon (bottom panel). In the interior of the axon the potential is −70 mV relative to the potential outside; i.e., the potential changes by 70 mV between the inside and outside surfaces of the membrane, which are 6 nm apart. Inside the axoplasm and in the interstitium no uncompensated point charges occur. This allows us to determine the magnitude of the electric field in the membrane in Example 18.7.

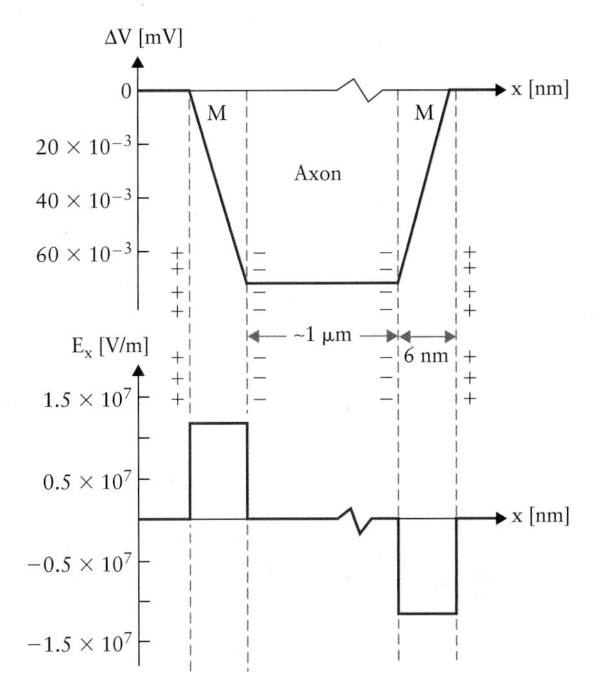

Figure 18.18 The potential (top panel) and the vector component of the electric field (bottom panel) across a human nerve cell in the x-direction. The electric fields everywhere in the membrane point toward the inside of the axon. The potential decreases in the direction in which the electric field vector points.

EXAMPLE 18.7

Confirm the value of the electric field across a human nerve membrane, as shown in Fig. 18.18.

Solution

The magnitude of the electric field across the membrane is calculated from the measured potential difference. We use the formula $\Delta V = |\vec{E}|\,y$ for the variation of the potential across the gap of a parallel plate arrangement and set $y =$ 6 nm for the full membrane width:

$$|\vec{E}| = \frac{\Delta V}{y} = \frac{7\times10^{-2}\ \mathrm{V}}{6\times10^{-9}\ \mathrm{m}} = 1.17\times10^7\ \frac{\mathrm{V}}{\mathrm{m}}.$$

As a vector, the electric field points toward negative charge, i.e., $E_x = +|\vec{E}|$ for the membrane on the left side in Fig. 18.18, and $E_x = -|\vec{E}|$ for the membrane on the right side. The calculated value is used in the lower panel of Fig. 18.18 to sketch the electric field quantitatively. Note that the notion of positive or negative electric fields in Fig. 18.18 is of no practical meaning as the choice of the direction of the x-axis is arbitrary. The relevant conclusion from the figure is that the electric field points everywhere in the membrane toward the axoplasm, i.e., into the nerve.

18.5.1: Capacitance

The potential and electric field curves of Fig. 18.18 are the direct consequence of the arrangement of the point charges at the interior and exterior membrane surfaces, with the surface charge density on the outside equal but opposite to the inside. This arrangement corresponds to two parallel conducting plates that are separated by an insulator of thickness b, as illustrated in Fig. 18.19(a). This arrangement is called a **capacitor** when the two conducting plates are charged with an equal amount of charge of opposite signs, as sketched in Fig. 18.19(b). When two initially neutral plates are connected to the two terminals of a battery, positive charge builds up on the plate connected to the positive terminal and an equal amount of negative charge builds up on the plate connected to the negative terminal. Note that we have so far assumed that the insulator is a vacuum, but other insulators, such as the membrane in Fig. 18.18, are possible and have to be considered later in this section.

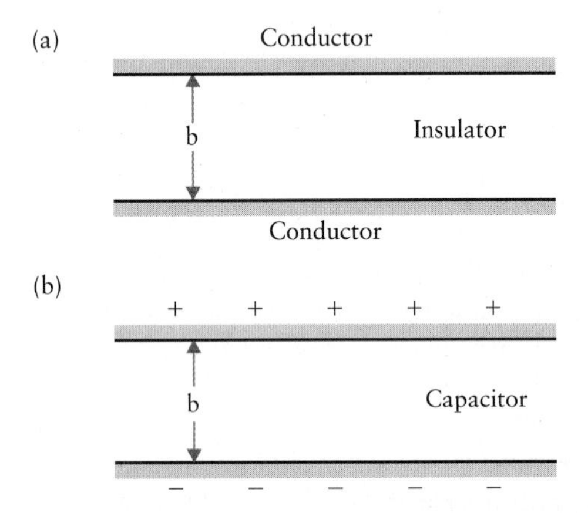

Figure 18.19 (a) The geometric configuration of a parallel plate capacitor with gap width b. Both plates must be electrically conducting, while the gap contains either a vacuum or an insulating material. (b) The electric configuration of a charged parallel plate capacitor.

Three quantities characterize a capacitor: the surface charge density σ, the capacitance C, and the dielectric constant κ. Of these, only the surface charge density has been introduced so far. Here we define the other two quantities.

To define capacitance we start with a parallel plate capacitor, as shown in Fig. 18.20. In contrast to previous arrangements, a specific common area A is defined for the plates. The upper plate is given a charge $+q$ and the lower plate a corresponding charge $-q$. The gap between the plates has a width b and contains a vacuum.

For such a parallel plate capacitor we know that the potential difference between the plates is linear in the surface charge density and linear in the plate separation: $\Delta V = \sigma\, b/\varepsilon_0$. Using the definition of the surface charge density, $\sigma = q/A$, we conclude that the potential difference is directly proportional to the charge q: $\Delta V \propto q$. This

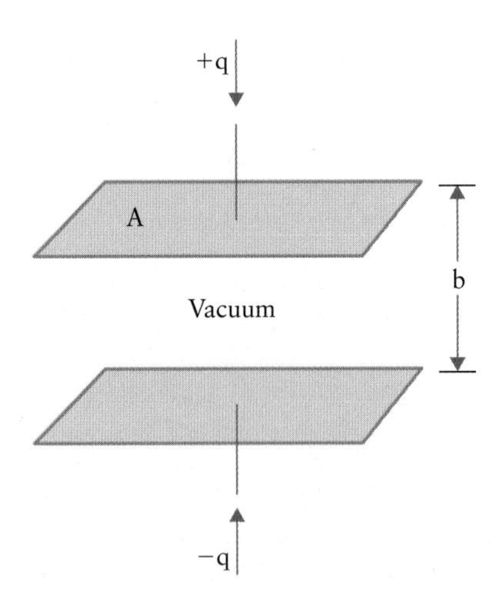

Figure 18.20 Three-dimensional sketch of a parallel plate capacitor illustrating its finite area A.

proportionality is written as an equation by introducing the capacitance C as the proportionality constant:

$$q = C\,\Delta V. \qquad [18.11]$$

The unit of the capacitance is C/V = F (F stands for **farad**, named in honour of Michael Faraday).

KEY POINT

The potential difference across a parallel plate capacitor is proportional to the charge on the plates. The proportionality factor is the capacitance.

Therefore, the capacitance of a capacitor is the ratio of the charge on each plate to the potential difference across the capacitor. The term **capacitance** is a fitting choice for C: a large capacitance means that a large amount of charge causes only a small potential difference; i.e., a system with a large capacitance can accommodate a lot of charge before a significant potential difference develops between its plates.

Starting with the potential as a function of position between charged parallel plates, we derive an explicit expression for the capacitance. Above we found $V_{\|} = (\sigma/\varepsilon_0)\, y$, with σ the surface charge density and y the position along the axis perpendicular to the parallel plates. The potential difference between the plates results when we set $\Delta V_{\|} = V_{\|}(y = b) - V_{\|}(y = 0)$:

$$\Delta V_{\|} = \frac{\sigma}{\varepsilon_0} b = \frac{q}{\varepsilon_0 A} b = \frac{q}{C} \Rightarrow C = \frac{\varepsilon_0 A}{b}. \qquad [18.12]$$

KEY POINT

The capacitance of a parallel plate capacitor depends only on geometric properties: it is proportional to the area of the plates and inversely proportional to the width of the gap between the plates.

Eqs. [18.11] and [18.12] also explain the name of the constant ε_0, which is called the **permittivity of vacuum**. For a capacitor of fixed unit area ($A = 1\ m^2$) and fixed unit plate gap width ($b = 1$ m), ε_0 is the proportionality factor of the charge brought onto the plates and the resulting potential difference between the plates. Thus, ε_0 is a measure of the ability of charge to reach through a vacuum to affect charge on the other side.

18.5.2: Dielectric Materials

We can change the permittivity of the capacitor by replacing the vacuum with a material that we place between the plates of a capacitor. This material cannot cause a transfer of charges between the plates as this would short-circuit the device. Michael Faraday determined in 1837 that all electrically insulating materials qualify. We call an insulating material placed in a capacitor a **dielectric**. Faraday showed that inserting a dielectric into a capacitor can be accounted for with a correction factor to the formulas we have introduced so far: we change the permittivity of vacuum ε_0 to $\kappa\,\varepsilon_0$ in all formulas in which ε_0 appears. The dimensionless correction factor κ is called the **dielectric constant**. It is a materials constant because it has a well-defined value for each material. Table 18.2 lists several values at room temperature. The product $\kappa\,\varepsilon_0$ represents, then, the **permittivity of the dielectric**.

TABLE 18.2

Dielectric constants for various materials at 25°C

Material	Dielectric constant κ
Vacuum	1.0
Air at 1.0 atm	1.0005
Polystyrene	2.6
Paper	3.5
Pyrex glass	4.7
Porcelain	6.5
Nerve membrane	7.0
Silicon	12.0
Ethanol	25.0
Water	78.5

We study a layer of water that is inserted in a parallel plate capacitor to illustrate Faraday's approach. Assuming that the capacitor has been charged and that the charging battery has then been disconnected, we know that the amount of charge on the plates is fixed. How the capacitance of the capacitor changes as the layer of water is inserted is derived from Eq. [18.12]:

$$C_{\text{vacuum}} = \frac{\varepsilon_0 A}{b}$$

$$\Rightarrow C_{\text{water}} = \frac{k\varepsilon_0 A}{b} = \kappa C_{\text{vacuum}};$$

i.e., the capacitance increases by a factor of 78.5 based on Table 18.2. The consequence of this change is derived from Eq. [18.11]:

$$q = C_{\text{vacuum}}\ \Delta V_{\text{i}} \Rightarrow q = C_{\text{water}}\ \Delta V_{\text{f}};$$

i.e., the potential difference across the capacitor drops by a factor of 78.5. Inserting a dielectric with a large dielectric constant means that the potential difference of a given parallel plate capacitor with a fixed charge is significantly decreased. Thus, materials with large dielectric constants effectively screen the effect of the charge on the plates.

Note that there is a second way in which the experiment to insert a dielectric can be performed. If the capacitor is not disconnected from its charging device, e.g., a battery with a given potential difference between its terminals, the potential difference remains fixed while the amount of charge on the plates must vary. In this case, the amount of charge on the parallel plates has to increase significantly when a layer of water is inserted. Note that this case also applies to the membranes of live nerves, because Na/K pumps maintain a −70-mV potential difference. The role of the Na/K pump (sodium–potassium pump) is discussed in the next chapter.

18.5.3: Energy Stored in a Capacitor

Charged capacitors store electric potential energy. This is illustrated in Fig. 18.21. To determine this potential energy quantitatively, we start with an uncharged capacitor. In the first step a small amount of charge, Δq, is transferred from one plate to the other. In this step no work is done during the charge transfer because the capacitor still has a zero potential difference. However, as we continue to transfer the same amount of charge step by step, work is required. The formula for the work done when moving an amount of charge Δq against an electric force is

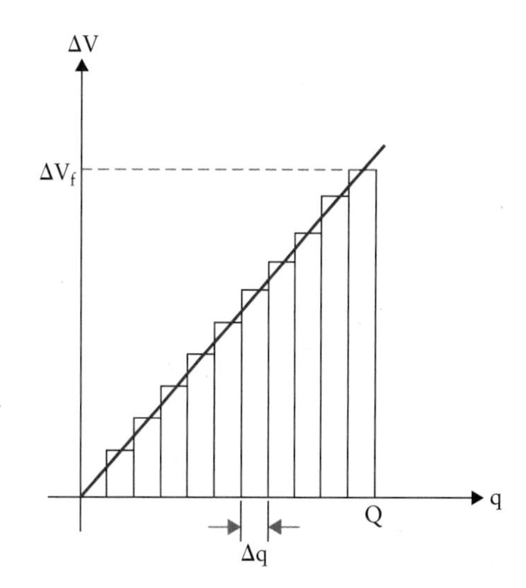

Figure 18.21 A sketch illustrating how the successive transfer of small amounts of charge from one plate to the other yields a charged parallel plate capacitor. The work required for this charging process is determined from the figure by determining the area under the curve.

$W = \Delta q\, \Delta V$. The potential difference ΔV is due to the previously transferred charge, leading to the step function shown in Fig. 18.21. The total work is then the area under the curve in the figure. This area is described by a triangle for which we find:

$$W = \frac{1}{2}\, Q\Delta V_{\mathrm{f}}, \qquad [18.13]$$

where ΔV_{f} is the final potential difference and Q is the total amount of charge transferred. The result in Eq. [18.13] represents also the work stored in the capacitor, i.e., the electric potential energy of the device.

CONCEPT QUESTION 18.3

Fig. 18.22 shows four undissociated HCl molecules in an acid solution. The solution is placed in a parallel plate capacitor that is indicated by the charge on its opposite plates. The small HCl molecule is able to rotate freely in the solution. Which of the four orientations shown is the preferred orientation of the molecule?

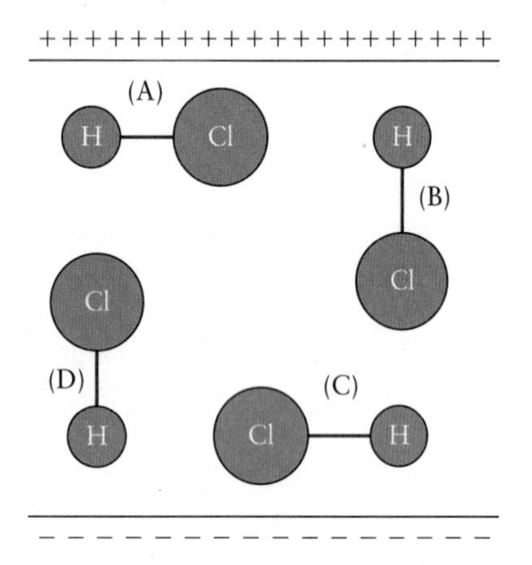

Figure 18.22 An HCl solution is inserted into a capacitor gap. We consider undissociated HCl molecules that can rotate freely in the solution. Four suggested orientations of the molecule are illustrated.

CASE STUDY 18.4

Why is the dielectric constant of water in Table 18.2 so large?

Answer: *The large dielectric constant of water is due to its dipole moment, which we discussed in Chapter 17. Water molecules orient themselves in a parallel plate capacitor in the same fashion as discussed for HCl in Concept Question 18.3: the positive end of the molecule is directed toward the negatively charged plate, and vice versa. The notable screening of charge in this case happens because the polarized molecules in the dielectric interact at close proximity more effectively with the electric charge on the capacitor plates than with the electric charge at the faraway opposite plate. The large dielectric constant of ethanol in Table 18.2 is a result of the same polarization effect due to the dipole moment of the OH group in the alcohol molecule.*

EXAMPLE 18.8

A parallel plate capacitor has an initial capacitance C = 10 pF (the unit pF stands for picofarad). It has been charged with a 9.0-V battery. With the battery disconnected, a piece of Pyrex glass is placed between the plates. (a) What is the potential energy of the device before the glass piece is inserted? (b) What is its potential energy afterward?

Solution

Solution to part (a): The electric potential energy is determined from Eq. [18.13]. The work term in the equation is equal to the change of the electric potential energy, $W = \Delta E_{\mathrm{el}}$, and thus we find for the charging process:

$$\Delta E_{\mathrm{el}} = \frac{1}{2}\, Q\Delta V_{\mathrm{f}},$$

To quantify the energy in this equation, we use Eq. [18.11] to replace the unknown charge by the fixed capacitance and potential difference:

$$\begin{aligned}\Delta E_{\mathrm{el}} &= \frac{1}{2}\, Q\Delta V_{\mathrm{f}} = \frac{1}{2}(C\Delta V_{\mathrm{f}})\Delta V_{\mathrm{f}} \\ &= \frac{1}{2}\, C(\Delta V_{\mathrm{f}})^2.\end{aligned} \qquad [18.14]$$

Next, we substitute the given values:

$$\begin{aligned}\Delta E_{\mathrm{el}} &= \frac{1}{2}(1.0\times10^{-11}\,\mathrm{F})(9.0\ \mathrm{V})^2 \\ &= 4.1\times10^{-10}\ \mathrm{J}.\end{aligned}$$

Solution to part (b): The next step is the disconnecting of the battery. This means that now the charge on the plates of the capacitor is fixed while its potential difference may vary. Eq. [18.14] is rewritten for this case:

$$\Delta E_{\mathrm{el}} = \frac{1}{2}\, Q\Delta V = \frac{1}{2}\, Q\left(\frac{Q}{C}\right) = \frac{Q^2}{2C}. \qquad [18.15]$$

Using Faraday's approach, placing a Pyrex glass dielectric between the plates of the capacitor requires that we rewrite ε_0 as $\kappa\,\varepsilon_0$, with κ the dielectric constant. In Eq. [18.15], the capacitance in the denominator is affected. Identifying the initial state of the capacitor when it is air-filled and the final state when the glass dielectric is placed between the plates, we find:

$$\begin{aligned}C_{\mathrm{i}} &= \varepsilon_0 \frac{A}{b} \\ \Rightarrow C_{\mathrm{f}} &= \kappa\varepsilon_0 \frac{A}{b} = \kappa C_{\mathrm{i}}.\end{aligned} \qquad [18.16]$$

Substituting Eq. [18.16] into Eq. [18.15] yields:

$$\begin{aligned}\Delta E_{\mathrm{el,f}} &= \frac{1}{\kappa}\Delta E_{\mathrm{el,i}} \\ &= \frac{4.1\times10^{-10}\,\mathrm{J}}{4.7} = 8.7\times10^{-11}\ \mathrm{J},\end{aligned}$$

in which the dielectric constant for Pyrex glass is taken from Table 18.2. Note that the energy with the glass dielectric inserted is less than the energy without it. The energy difference was dissipated into the piece of glass.

18.5.4: The Nerve Membrane as a Capacitor with Dielectric

We discuss the nerve membrane in this section as a parallel plate capacitor with dielectric. The dielectric is the lipid bilayer, and the plates are the internal and external surfaces of this bilayer. Quantitative conclusions are drawn in the following example.

EXAMPLE 18.9

For a resting nerve, determine (a) the capacitance per unit area of membrane and (b) the surface charge density on the nerve membrane in unit C/m^2.

Solution

Solution to part (a): We use the dielectric constant of a nerve membrane from Table 18.2. For the capacitance per unit area we find from Eq. [18.16]:

$$\frac{C}{A} = \frac{\kappa\varepsilon_0}{b} = \frac{7.0\left(8.854\times10^{-12}\,\frac{C^2}{Nm^2}\right)}{6\times10^{-9}\ \text{m}}$$
$$= 0.01\frac{\text{F}}{\text{m}^2},$$

in which b is the membrane thickness of 6 nm. The unit C^2/(N m^3) is equivalent to F/m^2 because J/C = V, N m = J, and F = C/V.

What does this result mean? We know from Eq. [18.11] that a smaller capacitance allows a smaller amount of charge to sustain a fixed potential difference. We combine this observation with Eq. [18.12], which states that the capacitance is inversely proportional to b: $C \propto 1/b$. This has an important consequence for the comparison of myelinated and unmyelinated nerves in the next chapter: the membrane of myelinated nerves is 300 times thicker (axon membrane and myelin sheath together); myelinated nerves therefore have a significantly smaller capacitance than unmyelinated nerves. This means that myelinated nerves maintain the same –70-mV potential difference with significantly smaller surface charge densities.

Solution to part (b): Next, we determine the surface charge density on the resting nerve membrane. We start with the magnitude of the electric field for charged parallel plates, $|\vec{E}| = \sigma/\varepsilon_0$. Taking the dielectric into account by including a dielectric constant and using the electric field across the membrane from Example 18.7 as $|\vec{E}| = 1.17 \times 10^7$ V/m, we find:

$$\sigma = \kappa\varepsilon_0|\vec{E}| = 7.0\left(8.85\times10^{-12}\frac{\text{C}^2}{\text{N}\cdot\text{m}^2}\right)\times\left(1.17\times10^7\frac{\text{V}}{\text{m}}\right)$$
$$= 7\times10^{-4}\frac{\text{C}}{\text{m}^2}.$$

Compare this calculation with Example 17.7; in particular, you should now understand why we got a wrong electric field value in that example even though we used the correct surface charge density for nerves.

continued

We want to discuss this value a little bit further. Na$^+$, K$^+$, and Cl$^-$ ions at the nerve membrane are all singly charged; i.e., they carry a single elementary charge of $e = \pm1.6 \times 10^{-19}$ C. The number of ionized particles per unit area on the nerve membrane, c_{ions}, can therefore be determined from the surface charge density as:

$$c_{\text{ions}} = \frac{\sigma}{e} = \frac{7\times10^{-4}\,\frac{\text{C}}{\text{m}^2}}{1.6\times10^{-9}\,\frac{\text{C}}{\text{ion}}}$$
$$= 4.4\times10^{15}\frac{\text{ions}}{\text{m}^2}.$$

This surface concentration is small compared with the density of atoms in the surface of a nerve membrane, which is about 10^{20} atoms/m^2; only 1 in every 20 000 sites on the membrane is occupied by an ion.

CASE STUDY 18.5

In a hospital or at an emergency scene, patients are often revived with a defibrillator (see Fig. 18.23). How does this machine work? Use a literature or Internet search.

Answer: *While the amount of charge stored in a large capacitor is sufficient to kill a person, the flow of this charge through the heart can, under appropriate conditions, save the life of a heart attack victim.* ***Fibrillation*** *is a process in which the heart produces a rapid, out-of-control pattern of heartbeats in response to which the heart muscles contract uncoordinatedly. A fast discharge of electric energy through such a heart is the only medical approach known to return the heart to its normal rhythm and thereby save the patient.*

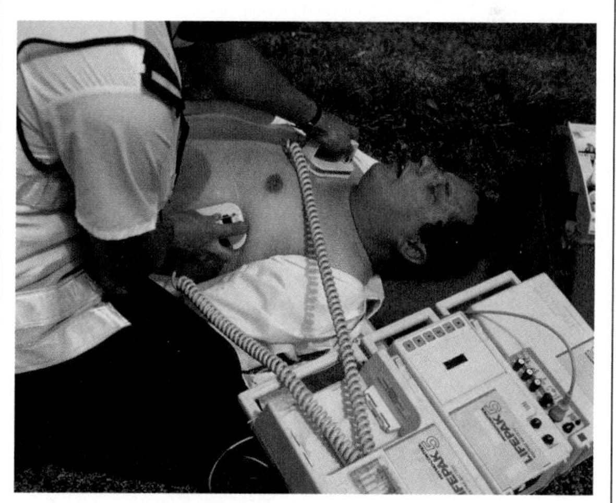

Adam Hart-Davis/Science Photo Library/Custom Medical Stock Photo

Figure 18.23 A portable defibrillator in use.

The device designed to allow a controlled discharge from a capacitor through a patient's body is called a defibrillator. It exists in stationary form in hospitals and is carried in portable form by emergency medical units. The portable defibrillator consists of a large capacitor and a series

continued

of batteries capable of charging that capacitor to a large potential difference. When the defibrillator is discharged the charge flows through two electrodes (paddles) that are placed on both sides of the patient's chest.

The emergency paramedics have to wait between successive discharges due to the time delay for charging the capacitor (less than a minute). Typical defibrillators have a capacitance of about 70 μF and a potential difference of about 5 kV, and thus release an electric potential energy of ΔE_{el} = ½ C $(\Delta V)^2$ = 875 J. In an emergency discharge, about 200 J of this energy flows through the heart of the patient, carried by an impulse of about 2 ms duration.

The time that elapses between the occurrence of fibrillation and the application of the defibrillator is critical: if the response time is between 6 and 10 minutes, typically only 8% of patients survive. With response times of about 3 minutes, the survival rate can be improved to about 20%. To shorten the response time, semi-automatic defibrillators have been developed that release an electroshock only after positively identifying fibrillation. Such instruments can in principle be operated by untrained personnel and have been installed in casinos (Las Vegas), department stores, and subway stations (Germany). As such instruments become more common, some basic training—e.g., with respect to the placement of the electrodes—is beneficial: training of personnel in Las Vegas achieved a survival rate of 80% of cases occurring in the casino!

SUMMARY

DEFINITIONS

- Capacitance: $C = q/V$
- Dielectric constant κ: $\kappa = 1$ vacuum; multiplied by permittivity of vacuum to obtain the permittivity of a material $\varepsilon_0 \Rightarrow \kappa\,\varepsilon_0$

UNITS

- Electric potential energy E_{el}: J
- Potential V: V
- Capacitance C: F = C/V (farad)
- Dielectric constant κ: dimensionless materials constant

LAWS

- Electric energy for two point charges at distance r:

$$E_{el} = \frac{1}{4\pi\varepsilon_0}\frac{q_1 q_2}{|r|}$$

- Electric potential:

  (i) for a stationary point charge:

$$V = \frac{1}{4\pi\varepsilon_0}\frac{q_{stationary}}{|r|}$$

  (ii) for a dipole (at a position far from the dipole, $r >> d$):

$$V = \frac{\mu}{4\pi\varepsilon_0}\frac{\cos\theta}{r^2}$$

  μ is the dipole moment and θ is the angle between the dipole axis and the line from the centre of the dipole to point P.

  (iii) for charged parallel plates:

$$V = |\vec{E}|\,y = \frac{\sigma}{\varepsilon_0}y$$

  the y-axis is perpendicular to the parallel plates.

- Capacitance of charged parallel plates: $C = \varepsilon_0\, A/b$, with ε_0 the permittivity of vacuum, b the plate separation, and A the plate area.
- Work stored in a parallel plate capacitor:

$$W = \frac{1}{2}Q\Delta V$$

  with Q the capacitor charge and ΔV the final potential difference.

MULTIPLE-CHOICE QUESTIONS

MC–18.1. In the four processes shown in Fig. 18.24 a probe charge (small green circle with positive or negative charge indicated) is moving in the direction of the red arrow shown in the vicinity of an arrangement of stationary charges (either a stationary point charge, depicted as a large green circle with its charge indicated, or a pair of charged plates). In which case is the work negative (energy is released by the probe charge during the displacement)?

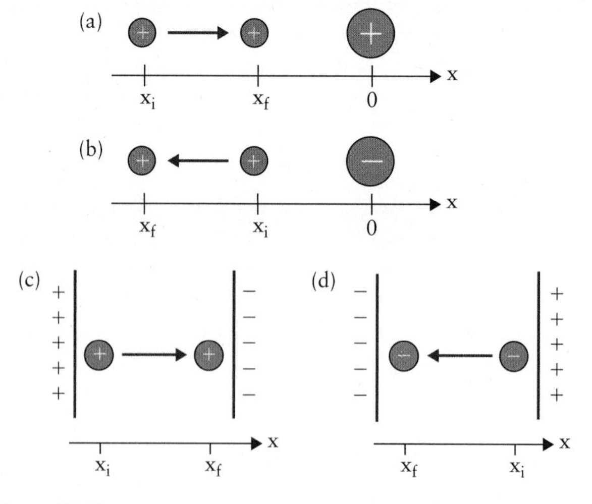

Figure 18.24

MC–18.2. Fig. 18.25 shows the potential energy for a mobile electron in a hydrogen atom (the nucleus is a positive stationary point charge at $x = 0$). Three total energies are considered and are shown in the figure by horizontal dashed lines labelled 1, 2, and 3. Which statement is correct?

(a) The electron with total energy at level 3 oscillates back and forth.
(b) The electron with total energy at level 1 is the only electron that will never travel away from the nucleus.
(c) The electron with total energy at level 2 is bound but can travel farther from the nucleus than the electron with total energy at level 1.

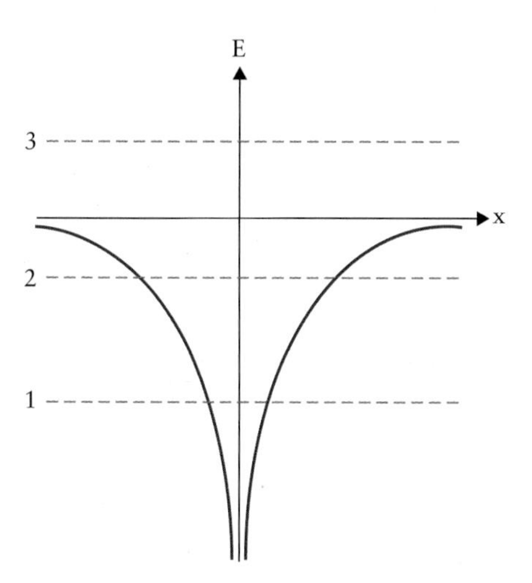

Figure 18.25

MC–18.3. Fig. 18.26 shows the electric potential energy of a proton (positive point charge) as a function of distance from a positively charged atomic nucleus, which is located at $x = 0$. Let's assume that the atomic nucleus is very heavy, e.g., the nucleus of a lead atom. We consider two values for the total energy of the proton: proton (1) with total energy $E_{total}(1)$ and proton (2) with total energy $E_{total}(2)$. Which statement about this system is *wrong*?
(a) Neither proton (1) nor proton (2) will travel straight through the nucleus at $x = 0$.

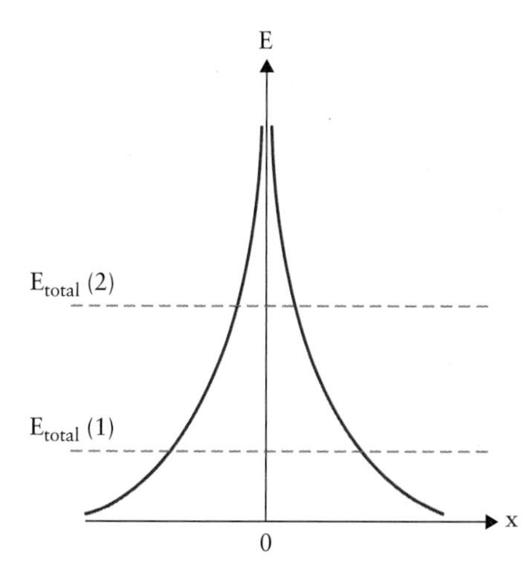

Figure 18.26

(b) Proton (2) will approach the nucleus to closer proximity than proton (1).
(c) At the same distance from the nucleus, protons (1) and (2) have the same total energy.
(d) At any given position along the x-axis, the electric potential energies for proton (1) and proton (2) are the same.
(e) Proton (1) reaches the same electric potential energy as proton (2) when at infinite distance from the nucleus.

MC–18.4. The absolute value of the potential difference between the interior and exterior surfaces of a particular membrane is 65 mV. What additional information will allow you to determine which surface of the membrane carries an excess of positive ions?
(a) I need the thickness of the membrane.
(b) I need to know which positive ions are involved (e.g., K^+, or Na^+, or Ca^{2+}).
(c) I need to establish a Cartesian coordinate system.
(d) I need to know the magnitude of the electric field across the membrane.
(e) None of the above would allow us to determine which is the positive plate.

MC–18.5. The water molecule shown in Fig. 17.1 is an electric dipole. Given the orientation of the molecule in the figure, in which direction does the electric potential point within the molecule?
(a) up
(b) down
(c) left
(d) right
(e) There is no direction because the electric potential does not contain directional information.

MC–18.6. The electric potential cannot be given in the following unit:
(a) J/C (joule per coulomb)
(b) V (volt)
(c) N/C (newton per coulomb)
(d) (N m)/C (newton metre per coulomb)
(e) More than one of these cannot be used.

MC–18.7. If the potential is constant in a certain volume around a given point, what does this mean for the electric field in that volume?
(a) The electric field is constant and has a negative value.
(b) The electric field is inversely proportional to the distance to the nearest charges.
(c) The electric field is zero.
(d) The electric field depends linearly on the distance to the nearest charges.
(e) The electric field is constant and has a positive value.

MC–18.8. Four point charges are positioned on a circular line. Their respective charges are $+1.5\ \mu C$, $+0.5\ \mu C$, $-0.5\ \mu C$, and $-1.0\ \mu C$. Assume that the potential at the centre of the circle due to the $+0.5$-μC point charge is 45 kV. What is the potential at the centre of the circle due to all four point charges combined?
(a) 180 kV
(b) 45 kV
(c) 0 V
(d) −45 kV

MC–18.9. How does the capacitance of a parallel plate capacitor change when its plates are moved to twice their initial distance and a slab of material with dielectric constant $\kappa = 2$ is placed between the plates to replace air?

(a) It is increased to four times the original value.
(b) It doubles.
(c) It remains unchanged.
(d) It is halved.
(e) It is reduced to one-quarter the original value.

MC–18.10. We treat an unknown biological membrane as a parallel plate capacitor with a dielectric material between its plates. If you know the capacitance per unit area in unit F/m^2, what additional parameter do you have to measure to determine the dielectric constant of the membrane material?

(a) We need no other information.
(b) We need to measure the area of the capacitor.
(c) We need to measure the thickness of the membrane.
(d) We need to measure the permittivity of a vacuum (ε_0).
(e) We need to measure the resistance of the membrane.

MC–18.11. A capacitor provides an electric field that points from left to right. How will the water molecule shown in Fig. 17.1 rotate if it can rotate freely as in liquid water? (Assume its orientation as shown; i.e., left in the context of this question is left on the page.)

(a) It will not rotate.
(b) It will rotate such that the oxygen atom is left and the two hydrogen atoms are right.
(c) It will rotate such that the oxygen atom is right and the two hydrogen atoms are left.
(d) It will rotate such that the oxygen atom is up and the two hydrogen atoms are down.
(e) It will rotate such that the oxygen atom is down and the two hydrogen atoms are up.
(f) It will spin continuously, like a top.

MC–18.12. We consider the capacitor shown in Fig. 18.27. At which of the five positions (A–E) does the potential have its smallest value? *Hint:* Note that $+2 < +5$ and $-3 < -1$.

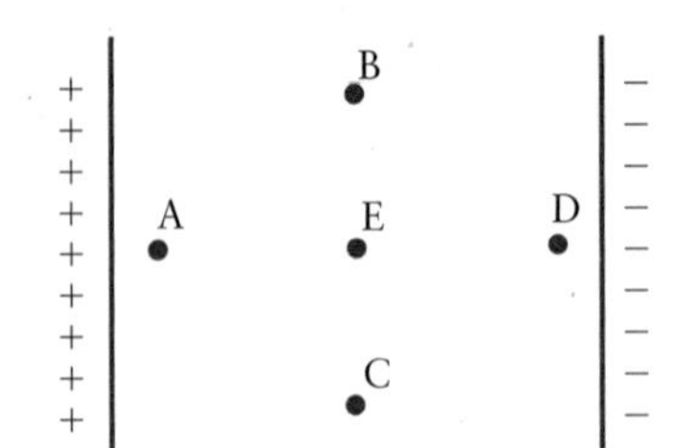

Figure 18.27

MC–18.13. Fig. 18.28 shows two capacitors that are rotated 90° relative to each other. In what direction does the electric field point at point *P* at the centre of the arrangement?

(a) up
(b) down
(c) left
(d) right
(e) in another direction

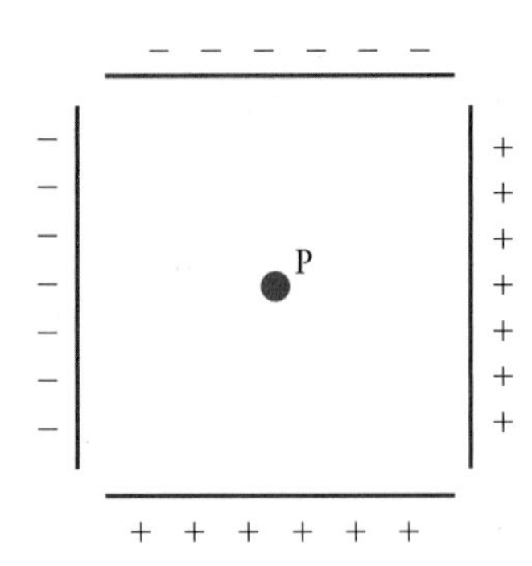

Figure 18.28

MC–18.14. How does the capacitance of an air-filled parallel plate capacitor change when we place a material with dielectric constant $\kappa = 1.0$ between its plates? (Choose the closest value.)

(a) It triples.
(b) It doubles.
(c) It remains unchanged.
(d) It is halved.
(e) It is reduced to one-third the original value.

MC–18.15. We use a solid material with a large dielectric constant. We predict the following observation when a dielectric made of this material is removed from between the plates of a parallel plate capacitor:

(a) If we keep the potential difference across the capacitor constant, the amount of charge on the plates increases.
(b) If we keep the amount of charge on the capacitor plates constant, the potential difference between the plates decreases.
(c) If we keep the amount of charge on the plates of the capacitor constant, the potential difference between the plates increases.
(d) We cannot keep either the potential difference across the capacitor or the amount of charge on its plates constant. Thus, the first three answers are not valid.

MC–18.16. We compare myelinated and unmyelinated nerves. Note that the latter have a much smaller membrane thickness. What consequence does this have for human nerves, which all have a resting potential difference of −70 mV?

(a) The capacitance per unit membrane area of the myelinated nerve is larger than the capacitance per unit membrane area of the unmyelinated nerve.
(b) The dielectric constant is much smaller for the myelinated nerves.
(c) The potential difference across the myelinated membrane must be larger.
(d) The amount of charge separated across the membrane per unit area is larger for the unmyelinated nerve.
(e) The amount of charge separated across the membrane per unit area is larger for the myelinated nerve.

MC–18.17. You construct a parallel plate capacitor with a 1.0-mm-thick rutile dielectric layer ($\kappa_{rutile} = 100$). If the area of the capacitor plates is 1.0 cm^2, what is its capacitance? (Choose the closest value.)
(a) 90 pF
(b) 180 pF
(c) 9 mF
(d) 100 mF

CONCEPTUAL QUESTIONS

Q–18.1. We find an electric field of magnitude $E = 1.0 \times 10^7$ V/m across a particular nerve membrane of thickness 5 nm. What is the potential difference between the interior and exterior surfaces of the membrane?

Q–18.2. If the potential at a particular point is 0 V, can you correctly state that no point charges exist in the vicinity of that point?

Q–18.3. If a proton is released from rest in a uniform electric field, does the electric potential at its position increase, stay the same, or decrease?

Q–18.4. Why is it dangerous to touch the terminals of a capacitor with a high potential difference even after the charging source has been disconnected? What would you do to make the capacitor safer to handle?

Q–18.5. What design options do you have if you need a small-sized capacitor with a large capacitance?

Q–18.6. If the potential difference across a capacitor is doubled, by what factor does the stored electric energy change?

Q–18.7. What happens to the charge on the plates of a capacitor if the potential difference between the plates is doubled?

ANALYTICAL PROBLEMS

P–18.1. A large number of energetic cosmic-ray particles reach Earth's atmosphere continuously and knock electrons out of the molecules in the air. Once an electron is released, it responds to an electrostatic force that is due to an electric field $\vec{E}$ produced in the atmosphere by other point charges. Near the surface of Earth this electric field has a magnitude of $|\vec{E}| = 150$ N/C and is directed downward, as shown in Fig. 18.29. Calculate the change in electric potential energy of a released electron when it moves vertically upward through a distance $d = 650$ m.

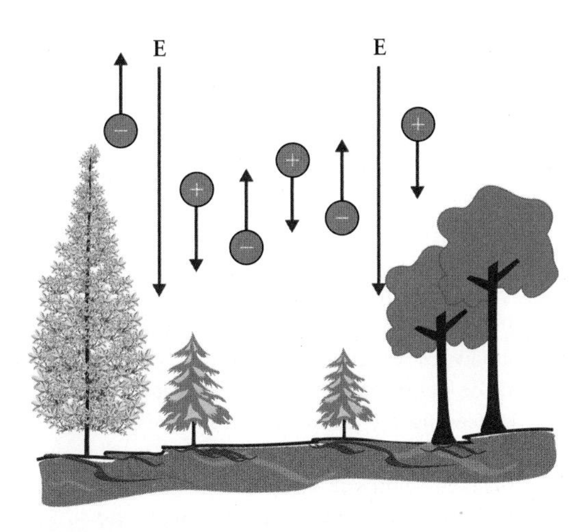

Figure 18.29

P–18.2. (a) What is the electric potential V at a distance $r = 2.1 \times 10^{-8}$ cm from a proton? (b) What is the electric potential energy in units J and eV of an electron at the given distance from the proton? (c) If the electron moves closer to the proton, does the electric potential energy increase or decrease?

P–18.3. (a) For the arrangement of charges in Fig. 18.30, calculate the electric potential at point P. Use $q = 1.0$ nC and $d = 1.0$ m, and assume that $V = 0$ V at infinite distance. (b) If a charge $-2\,q$ is brought to point P, what is the electric energy of this charge? Assume again that the electric potential energy is zero at infinite distance.

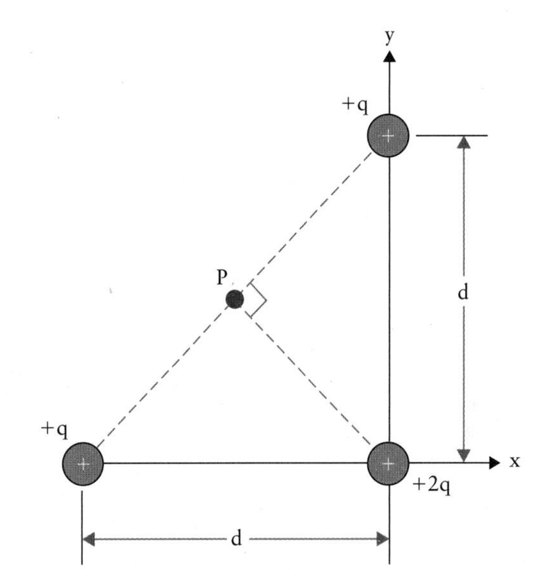

Figure 18.30

P–18.4. We study the three point charges shown in Fig. 18.31. They are held at the corners of an equilateral triangle with $l = 0.2$ m. What is the electric potential energy of the system of three point charges? Use for the three charges $q_1 = +2Q$, $q_2 = -3Q$, and $q_3 = +Q$, where $Q = 100$ nC. *Hint:* The solution is done in steps. Assume that you first bring one of the point charges from a very large (infinite) distance to its position, then repeat this procedure for the second and third point charges.

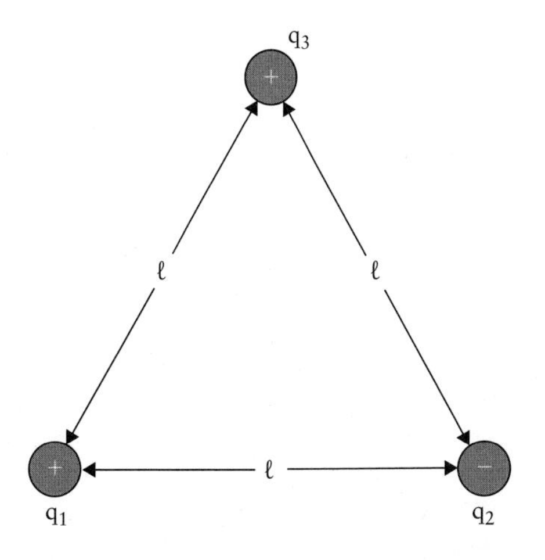

Figure 18.31

P–18.5. An ion is accelerated through a potential difference of 60 V, causing a decrease in its electric potential energy of 1.92×10^{-17} J. Calculate the charge the ion carries.

P–18.6. Fig. 18.32 shows three positive point charges at the corners of a rectangle. Find the electric potential at the upper right corner if $q_1 = +8\ \mu$C, $q_2 = +2\ \mu$C, and $q_3 = +4\ \mu$C. The distances are $d_1 = 6.0$ cm and $d_2 = 3.0$ cm. The potential is defined such that it is 0 V at infinite distance from the point charge arrangement shown.

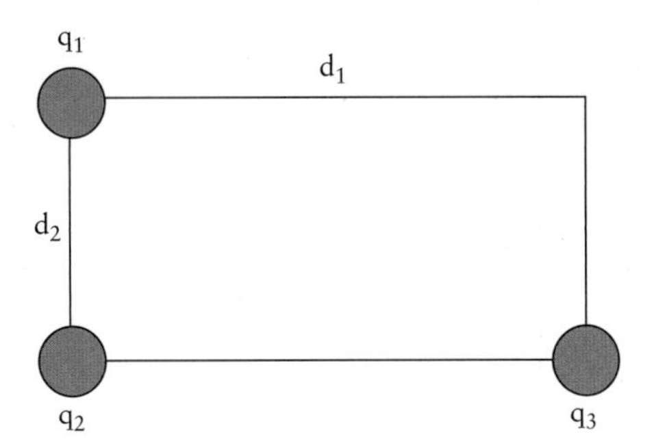

Figure 18.32

P–18.7. A uniform electric field of magnitude 300 N/m is aligned with the positive x-axis. A point charge of $+10\ \mu$C moves from the origin to a position of $(x, y) = (30$ cm, 40 cm$)$. (a) What is the change in the potential energy of the point charge? (b) Through what potential difference did the point charge move?

P–18.8. An ion is accelerated through a potential difference of 50; its potential energy decreases by 2.4×10^{-17} J. How many elementary charges does the ion carry?

P–18.9. To recharge a 9-V battery, a charging device must move 2.7×10^5 C of charge from the negative terminal to the positive terminal. How much work is done by this device?

P–18.10. (a) An air-filled parallel plate capacitor has a plate separation of $b = 1.5$ mm and an area $A = 4.0$ cm^2. Find its capacitance. (b) A capacitor with capacitance of $C = 4.5\ \mu$F is connected to a 9-V battery. What is the amount of charge on each plate of the capacitor?

P–18.11. The plates of a parallel plate capacitor are 3 cm wide and 4 cm long. The plates are separated by a 1.5-mm-thick layer of paper. (a) Calculate the capacitance of the device using the dielectric constant of paper from Table 18.2. (b) Any dielectric material other than a vacuum has a maximum electric field that can be generated in the dielectric material before it physically or chemically breaks down and begins to conduct. This maximum electric field is called dielectric strength. The dielectric strength for paper is reached at a value of 15×10^6 V/m. Calculate the maximum charge that can be placed on the capacitor at this dielectric strength.

P–18.12. An air-filled parallel plate capacitor has a capacitance of 60 pF. (a) What is the separation of the plates if each plate has an area of 0.5 m^2? (b) If the region between the plates is filled with a material with $\kappa = 4.5$, what is the final capacitance?

P–18.13. An air-filled parallel plate capacitor has a plate area of 2.0 cm^2 and plate separation of 2.0 mm. How much charge does this device store when charged with a 6.0-V battery?

P–18.14. An air-filled parallel plate capacitor has a plate separation of 0.1 mm. What plate area is required to provide a capacitance of 2.0 pF?

P–18.15. An air-filled parallel plate capacitor has a plate area of 5.0 cm^2 and plate separation of 1.0 mm. It stores a charge of 0.4 nC. (a) What is the potential difference across its plates? (b) What is the magnitude of the electric field between its plates?

P–18.16. An air-filled parallel plate capacitor has a plate area of 2.0 cm^2 and plate separation of 5.0 mm. If a 12.0-V battery is connected to its plates, how much energy does the device store?

P–18.17. A water-filled parallel plate capacitor has a plate area of 2.0 cm^2 and plate separation of 2.0 mm. The potential difference between its plates is held at 6.0 V. Calculate (a) the magnitude of the electric field between its plates, (b) the charge stored on each plate, and (c) the charge stored on each plate after water is replaced by air.

P–18.18. A parallel plate capacitor carries a charge Q on plates of area A. A dielectric material with dielectric constant κ is located between its plates. We can show that the force each plate exerts on the other is given by:

$$F = \frac{Q^2}{2\kappa\varepsilon_0 A}.$$

When a potential difference of 0.1 kV exists between the plates of an air-filled parallel plate capacitor of $C = 20\ \mu$F capacitance, what force do the two plates exert on each other if they are separated by 2.0 mm?

P–18.19. Fig. 18.33 shows an electron at the origin that is released with initial speed $v_0 = 5.6 \times 10^6$ m/s at an angle $\theta_0 = 45°$ between the plates of a parallel plate capacitor of plate separation $D = 2.0$ mm. If the potential difference between the plates is $\Delta V = 100$ V, calculate the closest proximity, d, of the electron to the bottom plate.

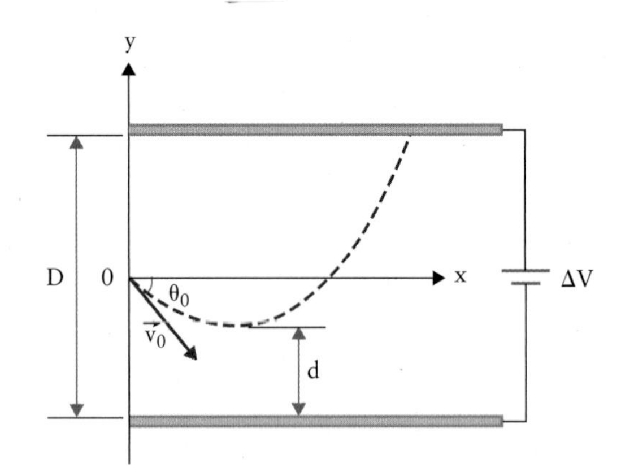

Figure 18.33

P–18.20. We model the surface of Earth and a cloud layer at 900 m height as a parallel plate capacitor. (a) If the clouds cover an area of 1.5 km^2, what is the capacitance of our model? (b) If an electric field strength exceeding 3.0×10^6 N/C causes lightning (the air becomes conducting), what is the maximum charge that can be present on the cloud layer? (c) Assume the capacitor discharges in a single lightning strike. How much energy is released in the strike?

ANSWERS TO CONCEPT QUESTIONS

Concept Question 18.1: (C). ii and iv are the cases in which the mobile point charge is moved toward the plate that carries the same charges.

Concept Question 18.2: (C). The potential decreases linearly along the positive *x*-axis.

Concept Question 18.3: (D). HCl is a dipole with a negative charge centre near the chlorine atom. The molecule rotates such that its negative end points toward the positive plate of the capacitor. This effect is called polarization of the dielectric and is generally observed for mobile dipole molecules.

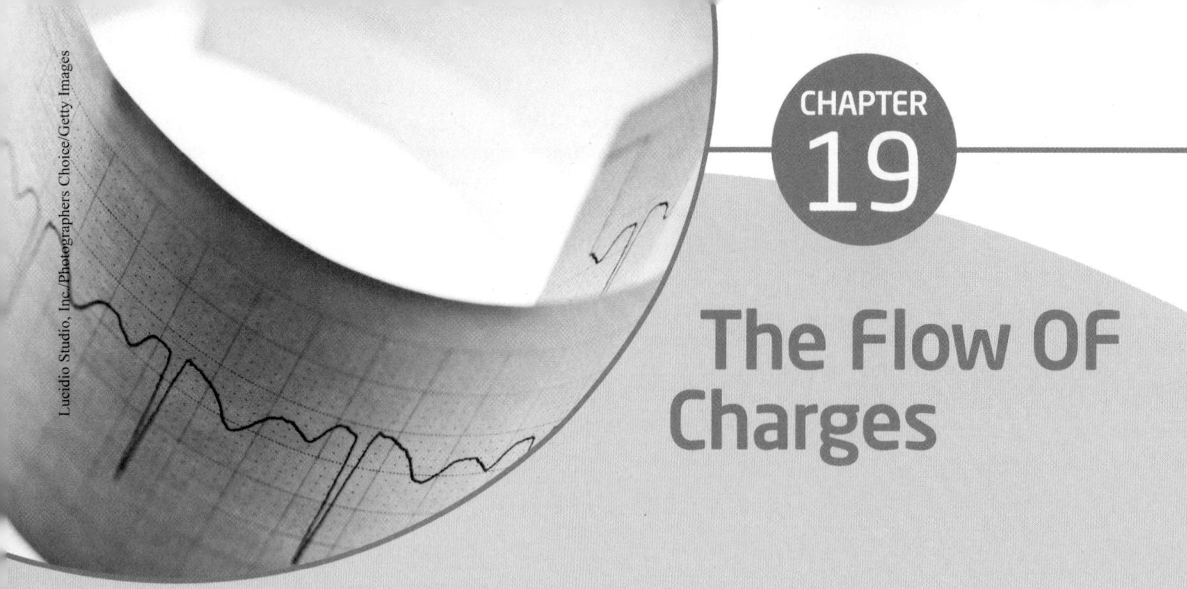

CHAPTER 19

The Flow OF Charges

Electric currents flow when an electric potential difference is established along a conducting path. Point charges move such that they compensate the potential difference. If the moving particles are positive they travel in the direction of the electric field along the conductor. The current is proportional to the density of mobile point charges in the material, the number of elementary charges carried by each moving particle, the average velocity (drift velocity) of these particles, and the cross-sectional area of the conductor. The current per unit area is a materials-specific property and is called the current density.

Flowing point charges usually encounter resistance against their motion in a conductor. The resistance depends on the material, the length and cross-section of the conductor, and the proportionality factor between the electric current and the applied potential difference. Near the electrochemical equilibrium the relation between current and potential is linear and is called Ohm's law. The resistivity is the corresponding materials-specific property. Resistivities at room temperature vary from good conductors (e.g., copper) to good insulators (e.g., quartz glass) by more than 20 orders of magnitude.

The resting nerve is modelled as an electrically active non-equilibrium system. The nerve membrane is a capacitor with the phospholipid bilayer as the dielectric material between an outer positively charged surface and a parallel inner surface charged negatively. The Na/K pump generates a concentration gradient for K^+ and Na^+ with a membrane potential difference of −70 mV. Diffusion and electric drift across the membrane establish a leakage current.

We started considering nerve membranes already in the past two chapters. While Chapter 14 primarily focused on the dipole character of the water molecule, charged parallel plates were discussed in the context of nerve membranes. We then solidified this model further in the previous chapter, where the nerve membrane was modelled as a capacitor. We have to expand on this model in the current chapter for two reasons: even resting nerves are actually non-equilibrium systems with a leakage current, and propagating nerve signals have to be explained.

Which physical concepts we are still missing to characterize human nerves is outlined in this introduction. Since we can identify these concepts for the simpler case of a resting nerve, the basic model for the resting human nerve is developed in Fig. 19.1. We have to go through this complex figure step by step to understand all the details of the model.

The first of the five panels represents a passive (dead) nerve. The **potassium** and **sodium** concentrations inside and outside the nerve cell are respectively equal; the extracellular fluid has a slightly higher **chlorine** concentration because negative chlorine ions can permeate the nerve membrane, but negative phosphate and protein ions inside cannot. There is no potential difference across the membrane, as indicated at the top of the panel; nor are any ion transport processes active in the nerve membrane. The term "dead" applies in two ways: (i) this nerve cannot carry a nerve impulse, and (ii) the system nerve/extracellular fluid is in thermal and chemical equilibrium as the only remaining concentration gradient cannot be compensated because the membrane is impermeable to large protein and phosphate ions.

Starting from the dead cell, we build up a live cell in four steps. First, as shown in the second panel, a non-equilibrium process separates the positive potassium and sodium ions. The resting nerve needs an excess potassium concentration inside and an excess sodium concentration outside. To establish these concentrations, a transport of Na^+ and K^+ against their concentration gradients is required; therefore, a simple diffusion process cannot achieve the ion separation across the membrane. Instead, an active transport process is needed. This process is called the **Na/K pump (sodium–potassium pump)**. The energy for the ion pump is provided by ATP molecules. A resting nerve cell consumes 30% of its ATP to maintain the Na/K ion imbalance across the membrane.

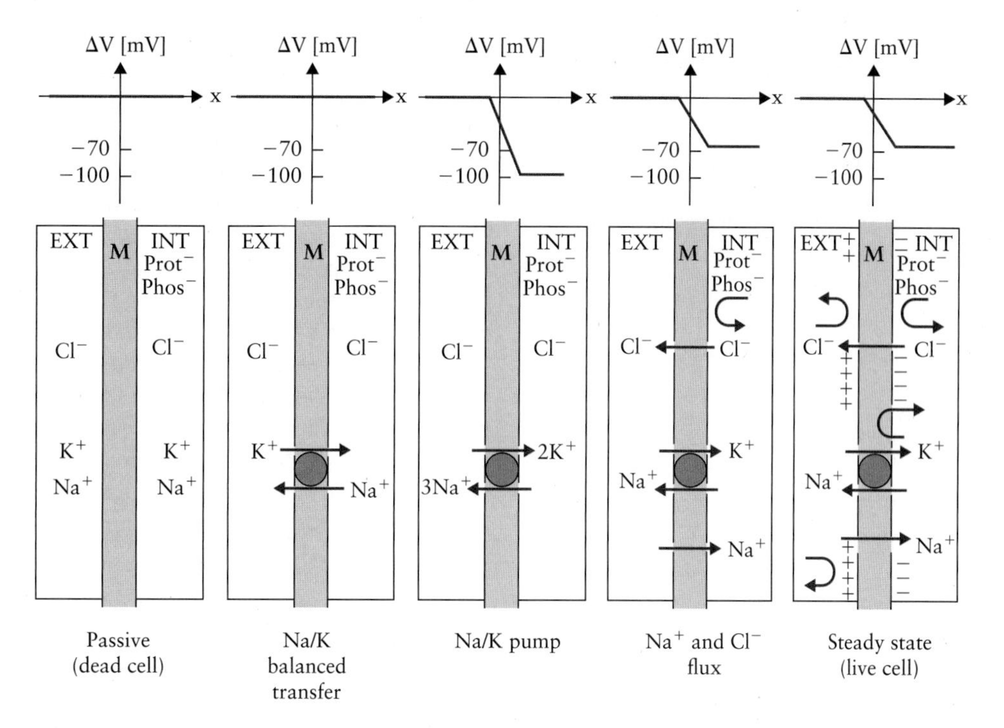

Figure 19.1 Non-equilibrium model of a nerve cell: a live cell is built up from a dead cell in equilibrium in several steps. We use the following abbreviations: EXT for extracellular fluid (interstitium); INT for interior fluid (axoplasm); M for membrane; *Prot* for protein; and *Phos* for phosphate. Note that the concentration of a chemical species is higher on the side on which its chemical symbol is boldfaced. The Na/K pump separates Na^+ and K^+ across the membrane. If the pump would transfer sodium and potassium at an equal ratio, no electric potential would emerge (second panel). The actual biological Na/K pump transfers ions with a ratio $Na^+:K^+ = 3:2$, which yields an axoplasm potential of −100 mV (third panel) relative to the extracellular potential. As a result of the electric field Na^+ and Cl^- drift across the membrane, lowering the axoplasm potential difference to the resting nerve value of −70 mV (fourth panel). The various transport processes lead to a steady state as further cross-membrane drift of ions is hindered by the surface charge densities on the membrane surfaces.

If the same number of ions of each type pass through the membrane per time unit, as shown in the second panel of Fig. 19.1, the system remains electrically neutral. However, this is not the net result of the actual biological Na/K pump. The net effect of the active ion transport is illustrated in the third panel: for every three sodium ions transferred to the extracellular fluid, only two potassium ions are transferred to the axoplasm. This imbalance of charge transfer leads to a potential difference between the axoplasm and the extracellular fluid. Defining the potential of the extracellular fluid as 0 V, we observe the development of a −100-mV potential on the axoplasm side. This potential is negative because the number of positive charges is reduced as a result of the imbalanced pumping.

The −100-mV potential difference of the axoplasm subsequently has an effect on the chlorine ions that is similar to the effect of the emerging potential in a battery. Unlike the larger negative ions in the axoplasm (phosphate and protein ions), the chlorine ions are driven by the electric field to permeate the membrane, drifting against their concentration gradient toward the more positive potential. Some amount of chlorine passes through the membrane, effectively lowering the potential difference to −70 mV, as shown in the fourth panel of Fig. 19.1. In the same fashion a minor electric drift occurs for the sodium ions toward the inside of the nerve.

At this stage a steady state of the system is reached. It is maintained by the Na/K pump and leads to a −70-mV potential difference between the axoplasm and the extracellular fluid. An imbalance of the potassium, sodium, and chlorine ion concentrations across the membrane is also established. The accumulated charge on the membrane surfaces (negative on the side of the axoplasm and positive on the side of the extracellular fluid) blocks further net diffusive flow or net electric drift of all ion types across the membrane (indicated by curved arrows in the last panel of Fig. 19.1). However, there is a continuous flow of Na^+, K^+, and Cl^- across the membrane, operating against the effect of the Na/K pump.

19.1: Moving Point Charges in a Resting Nerve

We take one more step toward a model for the resting nerve. The capacitor model developed in the previous section is not quite sufficient because it does not address a continuous flow of ions across the membrane. This leakage is added in Fig. 19.2(a). The various processes are modelled by their respective electric counterparts in Fig. 19.2(b). Shown is the parallel plate capacitor with capacitance C and surface charge Q, establishing

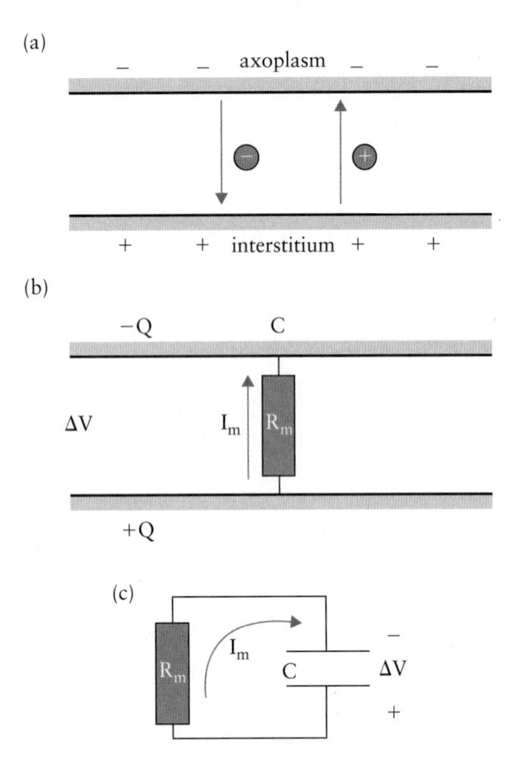

Figure 19.2 Model for the passive processes in a nerve membrane. (a) A real nerve is not a perfect capacitor, as ions continuously flow across the membrane, leading to a leakage current. (b) This phenomenon is modelled electrically by introducing a conducting connection across the membrane. This connection has a resistance R_m leading to a leakage current of I_m, with the subscript m for membrane. (c) The electric model, as sketched in panel (b), is equivalent to an electric circuit, which we are able to analyze with physics methods developed in this chapter.

an electric potential difference ΔV. In addition, a flow of charge across the membrane occurs, defined as a **leakage current** I_m (the index m stands for membrane). To allow charge to flow, an electrically conducting connection between the plates of the capacitor must exist. This cannot be a simple conductor, however, as this would short-circuit the capacitor—which means that the entire charge would flow across the capacitor gap and neutralize (destroy) the nerve. Thus, along the electric connection between the capacitor plates must be a resistor R_m, as shown. Finally, in Fig. 19.2(c) these electric elements are arranged to form a circuit.

To understand the dynamic properties of point charges moving across the membrane in a resting nerve, we have to establish two new concepts in this section, as suggested in Fig. 19.2: (i) the electric current and (ii) the electric resistance.

19.1.1: Electric Current

The flow of charge is called electric current. To introduce this concept, we return to our model of an electron in a metal that we introduced in Fig. 17.10. The metal in that figure consists of a lattice of stationary positive ions and very loosely bound electrons. These electrons are moving continuously on a microscopic-length scale similar to molecules in a stationary gas.

An electric current is established when these electrons move collectively in a common direction. This requires an electric force acting on the electrons. Such a force causes an acceleration of each electron according to the equation of motion (Newton's second law), $F = m\ a$. Surprisingly, we will find that this leads only to a very small electron speed in metallic conductors; e.g., in household wires electrons typically move at 0.1 mm/s!

This small velocity will also allow us to use the model we develop for the electric current to describe the motion of ions in a solution when an electrochemical force is applied, e.g., in a battery where anions and cations drift toward opposite electrodes, or for ions moving across the membrane of a resting nerve, as illustrated in Fig. 19.2.

KEY POINT

The electric current is defined as the amount of charge transferred through a cross-sectional area of a conductor per time unit.

We write the electric current as:

$$I = \frac{\Delta Q}{\Delta t}, \qquad [19.1]$$

in which ΔQ is the amount of charge transferred and Δt is a time interval. The unit **ampere** (A) is introduced for the current, named in honour of André-Marie Ampère. Based on Eq. [19.1], it is related to the unit of charge as 1 A = 1 C/s. Note that A is the only electric unit identified as one of the fundamental standard (SI) units because it can easily be measured using the same instrument we referred to as a voltmeter, now rearranged as an ampere-meter. Since charge cannot be measured directly, it is preferable to define the current unit as fundamental rather than the unit of charge.

The actual speed of point charges in a conductor with an applied electric field is determined from Fig. 19.3. The figure shows a cylindrical conductor with a cross-sectional area A along the x-axis. For simplicity we assume that positive point charges are mobile within the conductor, i.e., they move toward the right if an electric field is applied as shown. Note that this is a choice of convention. A consequence of the convention is that the electrons in

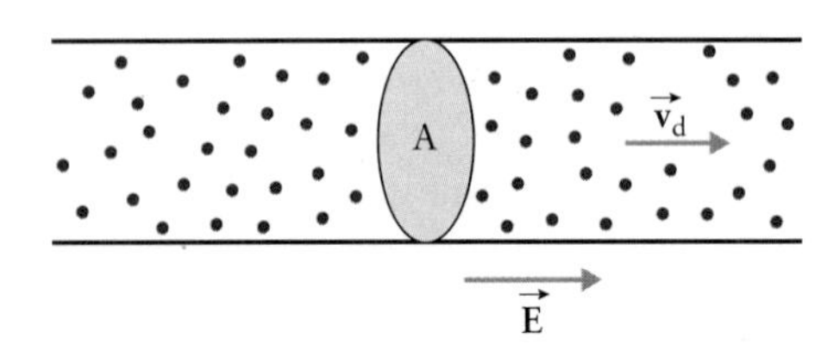

Figure 19.3 Simple, atomic scale model for a metallic conductor of cross-sectional area A with an external electric field. In response to the electric field, positively charged particles move with the drift velocity $\vec{v}_d$.

metallic conductors actually move in the direction opposite to the current I.

The motion of point charges is indicated in Fig. 19.3 by an arrow labelled $\vec{v}_d = \Delta x/\Delta t$, where $\vec{v}_d$ is the **drift velocity**. By introducing a single drift velocity we assume that point charges move through the conductor with an average speed, although they are constantly accelerated by the electric field. This assumption will be justified after the magnitude of the drift velocity is found.

We now develop a formula for the current through area A for Fig. 19.3. During the time interval Δt all mobile point charges that are closer than a distance $|\vec{v}_d|$ Δt pass through A because point charges can neither be created nor eliminated in the conductor. This defines the volume V, from which point charges pass through the area A:

$$V = A\Delta x = Av_d\Delta t.$$

We allow each point charge to carry an elementary charge e. The density of mobile point charges is labelled n with unit $1/m^3$. The total amount of charge ΔQ passing through the area A during Δt is given by the product of the number of point charges and their individual charge:

$$\Delta Q = enV = enAv_d\Delta t.$$

In order to apply Eq. [19.1], we divide this equation by t to obtain the current:

$$I = \frac{\Delta Q}{\Delta t} = nev_dA. \qquad [19.2]$$

We find that the current is proportional to the density of the mobile point charges, the drift velocity, and the geometric cross-sectional area of the wire. The parameter n is a materials constant and is large for good conductors. Note that the current I is positive in Eq. [19.2], which corresponds to the case shown in Fig. 19.3.

Due to its dependence on the cross-sectional area A, the current I is a parameter for a particular conductor; i.e., it is a practical parameter but does not characterize a conductor in a fundamental fashion. If we want to classify conductors according to their ability to carry electric charge, we define the **current density** J with unit A/m^2. J follows from Eq. [19.2] after division by the area A:

$$J = \frac{I}{A} = nev_d \qquad [19.3]$$

KEY POINT

The current density is defined as the current per unit cross-sectional area of the conductor.

Although J does not depend on any geometrical factors, and thus is a fundamental property of a given conductor, many calculations and studies use current instead because the current is measured directly with an ampere-meter while the current density must be calculated using the definition in Eq. [19.3]. J is not a materials constant since the drift velocity depends on the applied electric field.

EXAMPLE 19.1

A copper wire (cross-sectional area 1.0 mm^2) carries a current of I = 2.0 A. Calculate (a) the density of mobile electrons in copper n_{Cu}, (b) the current density J, and (c) the drift velocity v_d.

Solution

Solution to part (a): The density of mobile electrons is estimated from basic chemical information about copper, in particular its molar mass M_{Cu} = 63.5 g/mol and its density ρ_{Cu} = 8.95 g/cm^3. The density of mobile point charges is:

$$n_{Cu} = \frac{ZN_A}{V_{mol}} = \frac{ZN_A}{\frac{M_{Cu}}{\rho_{Cu}}} = \frac{ZN_A\rho_{Cu}}{M_{Cu}},$$

in which V_{mol} is the volume of 1 mol of copper, N_A is the Avogadro number, and Z is the number of electrons released per atom. With $Z = 1$ for copper, we find:

$$n_{Cu} = \frac{(6.0\times10^{23}\,\mathrm{mol}^{-1})\left(8.95\times10^3\,\frac{\mathrm{kg}}{\mathrm{m}^3}\right)}{0.0635\,\frac{\mathrm{kg}}{\mathrm{mol}}} = 8.5\times10^{28}\,\mathrm{m}^{-3}.$$

Solution to part (b): Eq. [19.3] is used to determine the current density:

$$J = \frac{I}{A} = \frac{2.0A}{1.0\times10^{-6}\,\mathrm{m}^2} = 2\times10^6\,\frac{\mathrm{A}}{\mathrm{m}^2}.$$

Solution to part (c): The drift velocity follows from the last term in Eq. [19.3]:

$$v_d = \frac{J}{ne} = \frac{2\times10^6\,\frac{\mathrm{A}}{\mathrm{m}^2}}{\left(8.5\times10^{28}\,\frac{1}{\mathrm{m}^3}\right)(1.6\times10^{-19}\,\mathrm{C})},$$

which leads to:

$$v_d = 1.5\times10^{-4}\,\frac{\mathrm{m}}{\mathrm{s}} = 0.15\,\frac{\mathrm{mm}}{\mathrm{s}}.$$

CASE STUDY 19.1

A drift velocity of 0.15 mm/s for electrons in a metal wire is a surprisingly small value when we think, for example, of turning on a light switch and the light fixture operating immediately despite a distance of several metres between the switch and the light fixture. Explain the small value of the drift velocity by referring to Figs. 17.10 and 19.3.

Answer: *The same density of mobile electrons is present throughout the metallic wire. Closing a light switch creates a continuous line of conducting electrons from the power plant, through the switch and the light fixture, and back to the power plant. The power plant causes an electric field in this loop, and all electrons simultaneously start to drift along the loop. The light fixture brightens up regardless of the speed of an individual electron in the wire: it operates due to the local motion of electrons in the fixture and not due to the flow of electrons from the switch to the light fixture.*

The electric field generated by the power plant causes a significant acceleration of each electron. However, before the electron can pick up any significant speed it is scattered by one of the positive core ions in the metal, as illustrated in Fig. 17.10. Since the electron is much lighter than the core ion, the impact causes the electron to bounce backward. It has to slow down and then accelerate again in the direction of the field, only to be scattered again almost immediately by the same core ion or by one close by. In a way, the progress of the electron resembles that of an overly aggressive sports-car driver during rush hour in a big city!

The electric potential and current are not sufficient to characterize the flow of point charges. To see why, we compare it with the flow of fluids. For fluids we identified the pressure difference Δp along a tube as the cause of flow in Chapter 13. The volume flow rate $\Delta V/\Delta t$ is then related to Δp by the linear equation $\Delta p = R(\Delta V/\Delta t)$ in which R is the flow resistance.

A resistance against the flow of point charges through a conductor is needed to relate current and potential difference in the same fashion. This can be confirmed experimentally. Eq. [19.2] does not contain the length of the conductor, yet when connecting a light bulb to a battery (which provides a potential difference) with variable lengths of wires, the bulb is brighter for the shorter wires. Comparing this situation to that for the flow of fluids, we note that the **flow resistance** in Eq. [13.10] introduces a length dependence. We expect the length dependence in the electric experiment to be caused by an electric flow resistance in the same fashion.

In this section we define three ways to express the flow resistance in electric systems: the resistivity ρ, the resistance R, and the conductivity γ. Resistivity and resistance are introduced as separate parameters to allow us to distinguish experimentally measurable and fundamental materials parameters.

19.1.2: Resistivity

We observe a current density when we apply an electric field across a piece of conducting material. If the electric field is not too large the current density increases linearly with the electric field, as illustrated in Fig. 19.4. The slope of this curve is constant and defines the **resistivity** ρ. Georg Simon Ohm wrote for the data in Fig. 19.4 the linear relation we call **Ohm's law**:

$$\left|\vec{E}\right| = \rho\, J. \qquad [19.4]$$

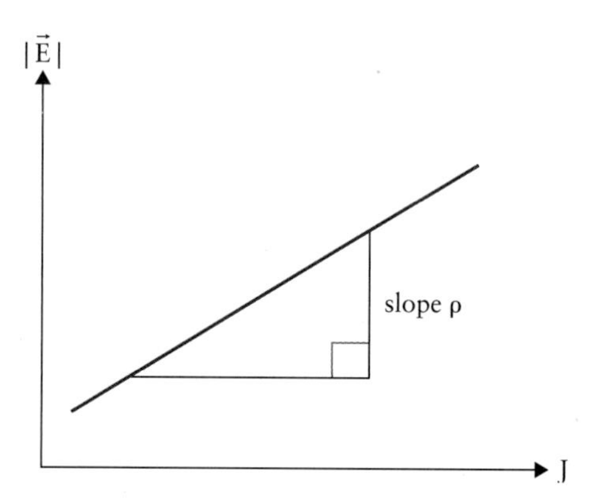

Figure 19.4 Illustration of Ohm's law: the magnitude of the electric field is proportional to the current density with the slope defining the resistivity ρ.

KEY POINT

Resistivity is the proportionality factor between the magnitude of the electric field, which causes point charges to move, and the current density, which represents the charge flow rate.

Both the magnitude of the electric field $\left|\vec{E}\right|$ and the current density J are not directly measurable, and thus the resistivity ρ is also not a quantity accessible to direct measurement.

Technically, Eq. [19.4] should be written as a vector equation, with the current density a vector in the direction of the flow of the charges. In the current chapter, we need the current density only as a magnitude, and restrict Eq. [19.4] therefore to its scalar form. The current density will be used as a vector in Chapter 21 when we discuss magnetism.

If we extend the experiment that led to Eq. [19.4] to much larger electric fields or to other classes of materials, such as semiconductors, we find that Ohm's law does not apply universally: the current density in those cases is not linearly proportional to the applied electric field. When the relation between electric field and current density is linear, i.e., follows Ohm's law, we call the material to behave **ohmically**. This is the case we discuss first, and it is the only case we discuss quantitatively in this textbook. However, the electric properties of nerve membranes require us later in the chapter to discuss also non-ohmic behaviour.

The unit of the resistivity follows from Eq. [19.4] as $(V/m)/(A/m^2) = V\ m/A$. For the ratio volt to ampere a new unit is introduced, called **ohm**, and is abbreviated by the Greek letter Ω. Thus, the unit of the resistivity is Ω m. The resistivity is a materials constant with a very wide range of values, as illustrated in Table 19.1. At very low temperatures some materials become superconductors, i.e., perfect conductors in which the flow of point charges encounters no resistivity at all, $\rho = 0\ \Omega$ m. On the other side, good insulators have extremely large resistivities against charge conduction. At room temperature, resistivities cover a range of 22 orders of magnitude.

Of particular interest is the value of the resistivity for the axoplasm and nerve membranes. The axoplasm is a fair conductor, with a resistivity of 1.1 Ω m. For the nerve membrane, a value measured for thin, artificial lipid membranes was initially proposed: $\rho = 1 \times 10^{13}\ \Omega$ m. This is a very high value that would make the mechanism of impulse transport in nerves as we describe it later impossible. Luckily, the actual nerve membrane contains additional proteins that reduce its resistivity significantly to $\rho = 1.6 \times 10^7\ \Omega$ m. The nerve membrane is still an insulator, allowing us to use the capacitor model developed in Fig. 19.2.

19.1.3: Resistance

Resistance is used instead of resistivity when an actual conductor is described rather than an intrinsic property of a material. To define resistance, we revisit Eq. [19.4]. With measurability in mind, the electric field is replaced by the respective potential difference, and the current density is replaced by the current:

$$|\vec{E}| = \rho J \Rightarrow \frac{\Delta V}{l} = \rho \frac{I}{A},$$

in which l is the length of the conductor and in which we use the definition of the current density from Eq. [19.3]. Next, the various parameters are grouped such that a new constant, the **resistance** R, is introduced as the proportionality factor of electric potential difference and electric current:

$$\Delta V = \frac{\rho l}{A} I = RI, \qquad [19.5]$$

with:

$$R = \frac{\rho l}{A}. \qquad [19.6]$$

The unit of resistance is V/A = Ω. Eq. [19.5] is a second way to write **Ohm's law**.

KEY POINT

Resistance is the proportionality factor of the potential difference along a given conductor and the current it carries. It increases with increasing length and decreases with increasing cross-sectional area of the conductor.

TABLE 19.1

Resistivity values for various materials at $T = 20$°C

Material	Resistivity (Ω m)
Insulators and semiconductors	
Yellow sulphur	2.0×10^{15}
Artificial lipid membrane	1.0×10^{13}
Quartz	1.0×10^{13}
Nerve membrane	1.6×10^{7}
Silicon	2.5×10^{3}
Axoplasm	1.1×10^{0}
Germanium	5.0×10^{-1}
Metals	
Mercury	1.0×10^{-6}
Iron	1.0×10^{-7}
Gold	2.4×10^{-8}
Copper	1.7×10^{-8}

CONCEPT QUESTION 19.1

How does the resistivity of a piece of metal change if both its length and its diameter are reduced to one-half their original value?

(A) It increases to four times the original value.

(B) It doubles.

(C) It remains unchanged.

(D) It is halved.

(E) It reduces to one-quarter of the original value.

19.1.4: Conductivity

The last quantity we introduce in this context is the conductivity γ. The **conductivity** is defined as the inverse value of the resistivity:

$$\gamma = \frac{1}{\rho}, \qquad [19.7]$$

where γ has the unit $1/(\Omega\ m)$. It is used when we prefer to think in terms of a current passing through a resistor rather than in terms of hindrance to the motion of point charges.

EXAMPLE 19.2

Considering again the copper wire we studied in Example 19.1, find (a) the electric field, (b) the potential difference along 10 m of wire, and (c) the resistance of the same 10 m of wire. Use (from Example 19.1) $A = 1.0\ \text{mm}^2$, $J = 2 \times 10^6\ \text{A/m}^2$, and $I = 2.0$ A. The resistivity of copper is given in Table 19.1.

Solution

Solution to part (a): The electric field is calculated from Eq. [19.4]:

$$\begin{aligned}\left|\vec{E}\right| &= \rho J \\ &= (1.7\times10^{-8}\ \Omega\text{m})\left(2\times10^{6}\frac{\text{A}}{\text{m}^2}\right) \\ &= 0.034\frac{\text{V}}{\text{m}}.\end{aligned}$$

Note that this value does not depend on the length or the diameter of the wire.

Solution to part (b): The potential difference ΔV along 10 m of wire is:

$$\Delta V = \left|\vec{E}\right| l = \left(0.034\frac{\text{V}}{\text{m}}\right)(10\ \text{m}) = 0.34\ \text{V},$$

i.e., a small drop in potential for an appreciable length of wire.

Solution to part (c): The resistance can be obtained in two different ways. One possibility is to use Ohm's law in the form given in Eq. [19.5]:

$$R = \frac{\Delta V}{I} = \frac{0.34\ \text{V}}{2.0\ \text{A}} = 0.17\ \Omega.$$

A second way is to use the definition of the resistance in Eq. [19.6]:

$$\begin{aligned}R &= \frac{\rho l}{A} = \frac{(1.7\times10^{-8}\ \Omega\text{m})(10\text{m})}{1.0\times10^{-6}\ \text{m}^2} \\ &= 0.17\ \Omega.\end{aligned}$$

EXAMPLE 19.3

(a) Calculate the resistance along an axon 1 cm long with an inner radius of 0.5 μm (a value typical for an unmyelinated nerve). (b) Repeat the same calculation for an inner axon radius of 10 μm (a value near the upper limit of human myelinated nerves). Use the resistivity value for the axoplasm from Table 19.1.

Solution

Solution to part (a): For axoplasm we obtain from Table 19.3 a resistivity $\rho = 1.1\ \Omega$ m. The cylindrical axon has a cross-sectional area $A = \pi r^2$ with the radius given

continued

as $r = 5 \times 10^{-7}$ m. For a length of $l = 1.0 \times 10^{-2}$ m, we find for the resistance:

$$\begin{aligned}R &= \frac{\rho l}{A} = \frac{(1.1\ \Omega\text{m})(1\times10^{-2}\ \text{m})}{\pi(5\times10^{-7}\ \text{m})^2} \\ &= 1.4\times10^{10}\ \Omega = 14\ \text{G}\Omega,\end{aligned}$$

in which GΩ is the unit giga-ohm. This value is much larger than the result in Example 19.2 for two reasons:

- the resistivity of axoplasm is several orders of magnitude higher than that of typical metals, and
- the cross-sectional area A of a nerve is orders of magnitude smaller than that of typical metal wires.

Solution to part (b): The only change in the second part of the problem is the modified radius, with a value of $r = 1 \times 10^{-5}$ m leading to a resistance of:

$$\begin{aligned}R &= \frac{\rho l}{A} = \frac{(1.1\ \Omega\text{m})(1\times10^{-2}\ \text{m})}{\pi(5\times10^{-5}\ \text{m})^2} \\ &= 3.5\times10^{7}\ \Omega = 35\ \text{M}\Omega,\end{aligned}$$

in which MΩ is the unit mega-ohm. A bigger nerve leads to a smaller resistance of the axoplasm.

19.2: Electrochemistry of Resting Nerves

At this point all basic physics concepts required to describe resting nerves have been introduced. Additional information we require from electrochemistry is presented in the current section; then, we will be prepared to discuss how both physics and electrochemistry concepts are combined to describe resting nerves and nerves carrying impulses.

19.2.1: Nernst's Equation

We established in section 10.6.3 that a closed or open system is in thermodynamic equilibrium when the Gibbs free energy (G) is equilibrated between system parts that are separated by membranes. This equilibrium condition remains true when the electric energy contributes to the enthalpy. Applying the concept of equilibrium across a semipermeable membrane for the specific system of a nerve—i.e., a thin biological membrane with an extracellular fluid outside and an axoplasm inside—we write:

$$(G^* + E_{\text{el}})_{\text{interstitium}} = (G^* + E_{\text{el}})_{\text{axoplasm}}. \quad [19.8]$$

In this equation, G^* is the part of the Gibbs free energy that is due to electrically neutral system parts (chemical energy) and E_{el} is the part that is due to electric charge (electric energy). The reason we keep G^* and E_{el} separate is that this approach allows us to specifically track the role of the electric energy. Eq. [19.8] allows us to derive Nernst's equation, which governs all electrochemical phenomena at a membrane.

Walther Nernst used Eq. [19.8] for a system that contains electrolytes (i.e., ionic solutions) of different chemical species on the two sides of a semipermeable membrane. The interaction between electrically neutral solutions across a membrane has already been discussed in Chapter 11. We make the assumption that both solutions are ideal solutions, i.e., their components do not interact.

We eliminate the logarithm function to write the concentration ratio as the independent variable. From this model **Nernst's equation** is found:

$$\frac{c_{i,\text{interstitium}}}{c_{i,\text{axoplasm}}} = \exp\left\{-\frac{e}{kT}(V_{i,\text{interstitium}} - V_{i,\text{axoplasm}})\right\}. \quad [19.9]$$

KEY POINT

Nernst's equation defines the electrochemical equilibrium of a system separated by a semipermeable membrane. It states that the concentration ratio of the i-th component in solution on both sides of the membrane is proportional to an exponential term containing the difference of the electric potential between the two sides of the membrane.

Of course, this formula applies only if an equilibrium across the membrane has been established for the i-th component of the solution. A necessary condition for this equilibrium is that the membrane is permeable for the respective component.

CONCEPT QUESTION 19.2

How can Eq. [19.9] be rewritten using Faraday's constant?

19.2.2: Test of an Equilibrium Model for the Resting Nerve: Batteries

We use Nernst's equation to determine whether nerves are in electrochemical equilibrium. As noted before, a resting nerve is not transporting an impulse; thus, its state is not changing with time. However, the nerve is an open system in which it is possible for energy and chemical components to be exchanged with the surrounding extracellular fluid. Radioactive tracer experiments have established that sodium and potassium ions penetrate the nerve membrane continuously. Therefore, the nerve is either in equilibrium or in a time-independent non-equilibrium state, i.e., a steady state. Our first task is to determine which of these two possibilities is actually the case. In the current section we test the possibility of describing resting nerves with an equilibrium model; we find that this model is not suitable. In the next section a non-equilibrium model will then be tested.

An equilibrium is expected if only passive processes occur in a system. Concentration differences exist across the nerve membrane for most ion species. If we focus for example on potassium, which has a much higher concentration in the axoplasm than in the extracellular fluid, then we expect the following processes to occur passively: potassium ions diffuse through the membrane into the extracellular fluid along the concentration gradient, following Fick's law of diffusion. Since potassium is present in the system as a cation, each potassium ion diffusing outward causes the axoplasm to become more negatively charged and the extracellular fluid to become more positively charged. Due to the competing electric and diffusive phenomena, we expect that the motion of potassium ions across the membrane is a self-terminating process: each potassium ion that diffuses in the direction of the concentration gradient increases the electric potential difference in that direction, which makes it harder for the next potassium ion to diffuse as it must move from a negatively charged environment toward a positively charged environment.

These two competing passive processes occur in this form at the surface of **electrodes** in a **battery**. A simple type of battery arrangement is shown in Fig. 19.5. Shown are two **electrochemical half-cells**: at the left side a zinc metal strip immersed in an aqueous $ZnSO_4$ solution and at the right side, separated by a porous barrier (1), a copper strip immersed in an aqueous $CuSO_4$ solution. Thus, we follow the standard discussion in the electrochemical literature to evaluate our equilibrium model.

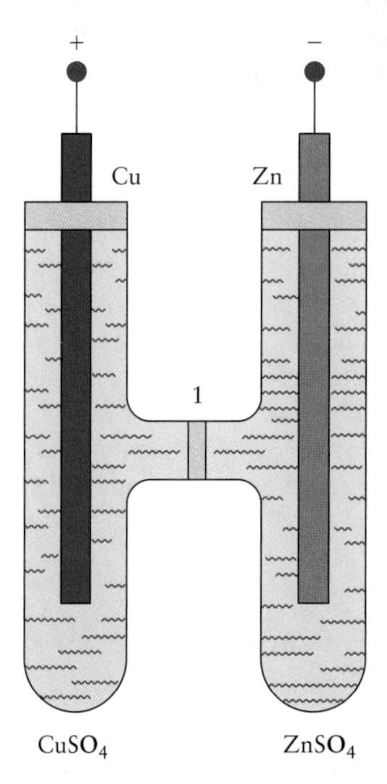

Figure 19.5 An electrochemical cell that consists of a Zn/Zn^{2+} and a Cu/Cu^{2+} half-cell. A porous barrier (1) separates the two half-cells.

To study the phenomena occurring at the metal/solution interfaces, we focus first on the surface of the immersed zinc metal. The fundamental step that takes place at this surface is illustrated in Fig. 19.6; neutral zinc atoms dissolve into the solution as cations, leaving behind two electrons each:

$$Zn \rightleftharpoons Zn^{2+} + 2e^{-}. \qquad [19.10]$$

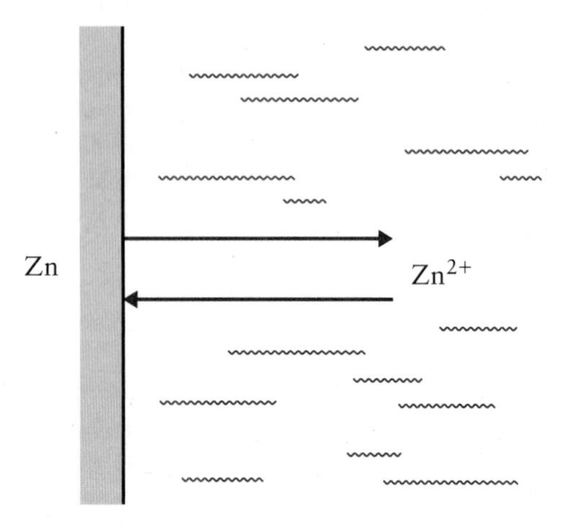

Figure 19.6 Conceptual sketch of the microscopic dynamic equilibrium at the Zn metal/solution interface.

This process is **self-terminating**, as every zinc atom that becomes an ion in the solution increases the negative charge of the metal, causing an increasing electric potential barrier for subsequent zinc atoms leaving the metal. It is important to note that the term *self-terminating* is somewhat misleading because microscopically Zn atoms transfer as ions into the solution all the time. However, once the potential difference reaches a certain value, the electric potential difference causes an electric drift of ions toward the metal strip, as indicated by the double-arrow in Fig. 19.6 and in Eq. [19.10]. Zn ions, which reach the metal surface, pick up two electrons and condense onto the Zn strip as neutral atoms. The final state of the processes at the metal/solution interface is therefore called a **dynamic equilibrium**. When the dynamic equilibrium is reached, we can use Nernst's equation to quantify the concentration difference and the potential difference across the interface:

$$\frac{c(Zn)}{c(Zn^{2+})} = \exp\left\{-\frac{2e(V_{Zn} - V_{Zn^{2+}})}{kT}\right\}, \qquad [19.11]$$

in which the factor 2 in the exponent on the right-hand side is due to the double charge of the Zn cations, $Z_{Zn} = 2$. While the concentration in a pure metal does not vary, the concentration of the ions in solution, the temperature, and the potential difference across the metal/solution interface are variable in Eq. [19.11].

However, Eq. [19.11] does not describe an experimental set-up since Eq. [19.10] does not represent a proper chemical reaction. Studying Fig. 19.5, we note that an electrochemical cell always consists of two half-cells, each based on a reaction similar to the one in Eq. [19.10]. While the potential difference in Eq. [19.11] is experimentally not measurable, the combination of two half-cells leads to an actual potential difference that we call the **electromotive force (emf)**, $\mathscr{E}$. Since we can choose which half-cells we combine, listed in the literature you find the standard electrode potentials, in which the term *standard* refers to $T = 25°C$ and a concentration of the solution of $c = 1.0$ mol/L. The reference potential, $V = 0$ V, is assigned to the **hydrogen gas half-cell**, which is shown in Fig. 19.7. This half-cell consists of an inert platinum foil immersed in a one-molar acid solution and is flooded with hydrogen gas at atmospheric pressure. A few examples of half-cell electrode potentials are listed in Table 19.2.

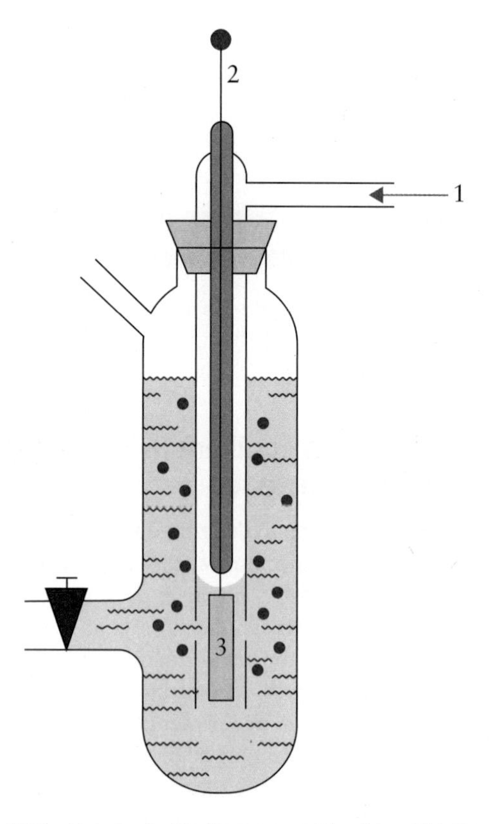

Figure 19.7 The standard hydrogen gas half-cell to which the potential of 0 V is assigned. An inert Pt metal strip (3) extends an isolated metal electrode (2) and is flooded with hydrogen gas at 1 atm (1) and surrounded by an acid solution. Blue dots indicate H_2 gas bubbles. Combining this half-cell with any other half-cell allows us to measure the electromotive force (emf) $\mathscr{E}$ of the second half-cell.

TABLE 19.2

Selected standard electrode potentials for electrochemical half-cells

Electrode	Electrode process	V (V)
Li^+/Li	$Li^+ + e \rightleftharpoons Li$	−3.045
K^+/K	$K^+ + e \rightleftharpoons K$	−2.925
Ca^{2+}/Ca	$Ca^{2+} + 2e \rightleftharpoons Ca$	−2.865
Na^+/Na	$Na^+ + e \rightleftharpoons Na$	−2.715
Zn^{2+}/Zn	$Zn^{2+} + 2e \rightleftharpoons Zn$	−0.763
Pb^{2+}/Pb	$Pb^{2+} + 2e \rightleftharpoons Pb$	−0.125
$H^+/H_2/Pt$	$2H^+ + 2e \rightleftharpoons H_2$	0.0
Cu^{2+}/Cu	$Cu^{2+} + 2e \rightleftharpoons Cu$	+ 0.337
Ag^+/Ag	$Ag^+ + e \rightleftharpoons Ag$	+ 0.800

Acid solutions at 25°C, calibrated against a standard hydrogen cell $H^+/H_2/Pt$ (as shown in Fig. 19.7)

CONCEPT QUESTION 19.3

Figure 19.8 shows two connected electrochemical half-cells. Let's assume that one electrode is made of potassium metal and is immersed in a 1.0 mol/L K^+ solution and the other electrode is made of silver and is immersed in a 1.0 mol/L Ag^+ solution. Considering the ions indicated in the figure (which we treat as positive point charges) and the electrons (which we treat as negative point charges), (a) which electrode has the higher electric potential and (b) which electrode is the silver electrode?

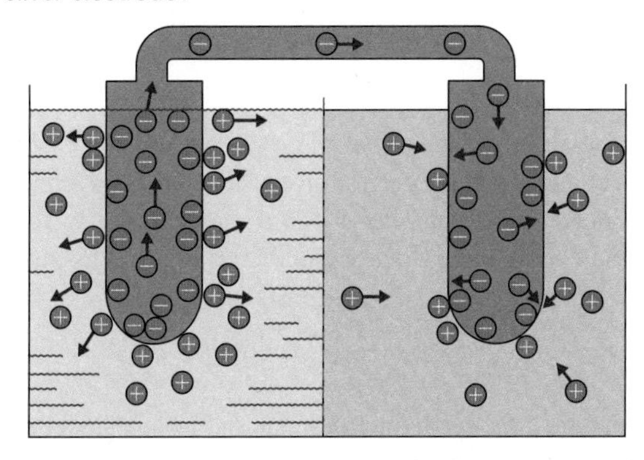

Figure 19.8 Two electrochemical half-cells are connected. Electrons are shown in the electrodes and their electric connection as small circles with – signs; metal ions are shown in the solutions surrounding the electrodes as circles with + signs.

EXAMPLE 19.4

Calculate the electromotive force $\mathscr{E}$ of the electrochemical cell at T = 25°C, as in Fig. 19.5. The concentration of the $ZnSO_4$ solution is 1.0 mol/L and is 0.1 mol/L for the $CuSO_4$ solution.

continued

Solution

The underlying chemical processes of the two half-cells are:

$$\begin{array}{ll} \text{(i) Zn} & \rightleftharpoons Zn^{2+} + 2e^- \\ \text{(ii) } Cu^{2+} + 2e^- & \rightleftharpoons Cu \\ \hline Zn + CuSO_4 \rightleftharpoons Cu + ZnSO_4, & \end{array} \quad [19.12]$$

which together constitute a valid chemical reaction. Assuming first that both half-cells are standard cells, we calculate from Table 19.2 the electromotive force. Note that the reaction for the Zn cell is included backward in Eq. [19.12]:

$$\mathscr{E} = V_{Cu^{2+}/Cu} - V_{Zn^{2+}/Zn}$$
$$= +0.337\text{ V} - (-0.763\text{ V}) = +1.1\text{ V}.$$

Next we correct for the non-standard concentration of the Cu solution with Nernst's equation applied along the electron connection at the top:

$$\frac{c_{Cu^{2+},\text{standard}}}{c_{Cu^{2+},\text{actual}}} = \exp\left\{\frac{2e(V_{\text{standard}} - V_{\text{actual}})}{kT}\right\},$$

in which V_{standard} stands for the standard potential from Table 19.2 and V_{actual} is the actual potential for the given system. This yields:

$$V_{\text{actual}} - V_{\text{standard}} = \frac{kT}{2e}\ln\left(\frac{c_{Cu^{2+},\text{standard}}}{c_{Cu^{2+},\text{actual}}}\right)$$
$$= \frac{\left(1.38\times10^{-23}\,\frac{\text{J}}{\text{K}}\right)(298\text{ K})}{2(-1.6\times10^{-19}\text{ C})}\ln 10$$
$$= -0.03\text{ V}.$$

This leads to a correction to the 1.1 V-value for the electromotive force, and thus the actual electromotive force is $\mathscr{E}$ = 1.1 V – 0.03 V = 1.07 V. The electromotive force of an electrochemical cell is primarily governed by the choice of half-cells. Variations in temperature or concentration of the solutions cause only minor corrections.

19.2.3: Electrochemistry of the Resting Nerve

Which of the two models describes the live nerve cell better, the equilibrium model developed in analogy to Fig. 19.5, or the non-equilibrium model of Fig. 19.1? The method of presentation of both models obviously favours Fig. 19.1. But can we prove that this model is a better description than Fig. 19.5?

The proof is provided by Nernst's equation. The measured potential difference across the nerve membrane is −70 mV. At a core human body temperature of 37°C = 310 K, a nerve in equilibrium would have the following concentration ratios based on Eq. [19.9]:

- for cations:

$$\frac{c_{\text{cation,interstitium}}}{c_{\text{cation,axoplasm}}} = \exp\left\{-\frac{(1.6\times10^{-19}\,\text{C})(7\times10^{-2}\,\text{V})}{\left(1.38\times10^{-23}\,\frac{\text{J}}{\text{K}}\right)(310\,\text{K})}\right\},$$

which yields:

$$\frac{c_{\text{cation,axoplasm}}}{c_{\text{cation,interstitium}}} = 13.7$$

- for anions:

$$\frac{c_{\text{anion,interstitium}}}{c_{\text{anion,axoplasm}}} = \exp\left\{-\frac{(-1.6\times10^{-19}\,\text{C})(7\times10^{-2}\,\text{V})}{\left(1.38\times10^{-23}\,\frac{\text{J}}{\text{K}}\right)(310\,\text{K})}\right\},$$

which yields:

$$\frac{c_{\text{anion,axoplasm}}}{c_{\text{anion,interstitium}}} = 0.073.$$

Now we compare these electrochemical equilibrium values with the actual concentration ratios. Using Fig. 18.2, we obtain:

$$\frac{c_{\text{Na}^+\text{,axoplasm}}}{c_{\text{Na}^+\text{,interstitium}}} = \frac{15\ \text{mmol/L}}{145\ \text{mmol/L}} = 0.103$$

$$\frac{c_{\text{K}^+\text{,axoplasm}}}{c_{\text{K}^+\text{,interstitium}}} = \frac{150\ \text{mmol/L}}{5\ \text{mmol/L}} = 30.0$$

$$\frac{c_{\text{Cl}^-\text{,axoplasm}}}{c_{\text{Cl}^-\text{,interstitium}}} = \frac{9\ \text{mmol/L}}{125\ \text{mmol/L}} = 0.072.$$

This result applies to most mammalian nerve cells. The ratios for potassium and sodium deviate significantly from the respective equilibrium values, while the chlorine value is close to the equilibrium ratio. The deviations are due to the Na/K pump, which affects the two positive ion species, but not chlorine. Thus, the battery model cannot be used for the nerve and the more complex model shown in Fig. 19.1 is used for further study of a resting nerve.

19.3: The Signal Decay Time of a Resting Nerve

We are now in a position to analyze the electric model of a resting nerve in Fig. 19.2. Qualitatively, we expect that the membrane, acting as a resistor, allows the separated charges to recombine, thus neutralizing the capacitor. However, since the resistivity of the nerve membrane is quite high, the recombination might be a slow process that could then be neglected. The aim of this section is therefore to establish the time it takes to neutralize the nerve membrane electrically, which is called the **decay time** of the nerve. Deriving the decay time requires calculus based operations. The reader not familiar with this approach from mathematics can skip to Eq. [19.17]. Basic concepts in calculus, as used in the later chapters in this textbook, are reviewed in the section entitled Calculus as part of the Math Review located after Chapter 27 in this textbook.

Note that Fig. 19.2 does not include the active effect of the Na/K pump, as we concentrate only on passive processes. We obtain the decay time by combining three equations that govern the motion of point charges through the membrane in Fig. 19.2:

- The amount of charge on the plates varies when an electric current flows according to Eq. [19.1]:

$$\frac{dQ}{dt} = -I_m, \qquad [19.13]$$

in which I_m is the current across the membrane. Note the minus sign on the right-hand side. A positive current I_m means that the quantity of uncompensated charge Q on both sides of the membrane in Fig. 19.2 diminishes; i.e., the change of charge with time dQ/dt is negative.

- A change in the number of point charges also leads to a change in the capacitor properties of the nerve membrane. We use the definition of capacitance to relate the charge term to the capacitance and the potential:

$$\frac{dQ}{dt} = \frac{d(C\Delta V)}{dt} = C\frac{d}{dt}\Delta V, \qquad [19.14]$$

in which the second term on the right-hand side means that we study the time change of a potential difference. The capacitance has been separated from this term because we know that it is constant (time independent).

CASE STUDY 19.2

When you observe fast electric processes at a nerve membrane, which ions do you suspect are involved?

Answer: *Chemical systems react to perturbations faster when they are already outside the thermodynamic equilibrium. We showed that an electrochemical equilibrium exists across a nerve membrane neither for potassium nor for sodium, but that the chlorine concentration is close to equilibrium. Thus, we suspect potassium and sodium to play a role in fast processes. We will confirm this later when discussing nerve signal propagation in the Hodgkin–Huxley model.*

- We assume that the actual flow of point charges across the membrane due to its resistance is governed by Ohm's law. We use it in the form introduced in Eq. [19.5]:

$$I_m = \frac{\Delta V}{R_m}, \qquad [19.15]$$

in which R_m is the resistance of the nerve membrane.

These three equations are combined to eliminate the charge and the current and provide us with an equation for the time change of the potential difference across the membrane. This is mathematically achieved in two steps. We first combine Eqs. [19.13] and [19.14] to eliminate the charge term *dQ/dt:*

$$-I_m = C\frac{d}{dt}\Delta V.$$

Next, we use Eq. [19.15] to eliminate the current:

$$\frac{d}{dt}\Delta V = -\frac{1}{R_m C}\Delta V. \qquad [19.16]$$

This equation means that the time change of the potential difference is linearly proportional to the potential difference itself. Eq. [19.16] is called a **differential equation**. We find this particular differential equation, in which the derivative is directly proportional to the function itself, frequently in the sciences: the time change of a variable is proportional to its absolute value. Life sciences examples in which you will see this relation again are population growth studies (epidemiology) and applications of radioactive decay (radiology). Due to this widespread importance, we want to study Eq. [19.16] in detail as we continue our discussion of the time dependence of the nerve potential.

What does Eq. [19.16] imply? The rate of change of the potential difference is proportional to the potential difference at each instant of time. This linearity leads to an exponential function:

$$\Delta V(t) = \Delta V_i \exp\left\{-\frac{t}{R_m C}\right\} = \Delta V_i e^{-t/\tau}, \qquad [19.17]$$

where ΔV_i is the potential difference at the time $t = 0$, which is $\Delta V_i = -70$ mV for a live nerve cell. The negative sign in the exponent means that ΔV decays with time. We introduce τ as a new parameter in Eq. [19.17]. The unit of τ is s, thus τ is a time constant. We find out what happens at time τ by substituting $t = \tau$ in Eq. [19.17]: $\Delta V(\tau) = \Delta V_i/e$ with e the Euler number. This means that the potential difference drops to $1/e \cong 37\%$ of its initial value after this time. After 2τ the potential drops to about 14%, and after 3τ to about 5%. Thus, τ is a measure of the time to observe a significant but not yet total decay; it is a **decay time constant**. The decay time constant follows from Eq. [19.17]:

$$\tau = R_m C.$$

This result for the decay time constant is not satisfactory. The membrane resistance is a quantity that depends on an arbitrary reference area (e.g., 1 m^2). It is preferable to replace the resistance with the resistivity as this quantity contains only fundamental properties of the nerve membrane material:

$$\tau = R_m C = \frac{\rho_m b}{A}\frac{\kappa\varepsilon_0 A}{b} = \kappa\varepsilon_0\rho_m. \qquad [19.18]$$

Eq. [19.18] confirms that the decay time constant is a materials constant. We emphasize that this time constant in particular does not depend on the thickness of the membrane, *b*. From this we conclude that nature did not develop myelinated nerves—with the membrane thickness the main difference from unmyelinated nerves—to change the decay time constant. The reason why myelinated nerves work faster must lie elsewhere! Using Eq. [19.18], we quantify the time constant τ for a nerve:

$$\tau = 7\left(8.85\times10^{-12}\frac{C^2}{Nm^2}\right)(1.6\times10^7\,\Omega m)$$
$$= 1.0\times10^{-3}s = 1.0\ ms.$$

Thus, the potential difference across a nerve membrane may decay in the very short time of one millisecond! The Na/K pump prevents this, consuming energy (provided by ATP molecules) to maintain a steady state. We note again how thin a line nature has drawn between life and death: we saw in Chapter 11 that we are within a micrometre of suffocating with the given width of the membrane between alveoli and pulmonary blood capillaries; now we have found that we are always just a millisecond away from a catastrophic failure of our nervous system!

19.4: Stimulated Nerve Impulses

We have fully characterized the resting nerve with the preceding discussion. We now move on to the response of a nerve to a stimulus. An impulse is triggered when an external stimulus causes the electric potential difference of a nerve to collapse momentarily at the stimulated point (usually a dendritic receptor). We will study two basic models for this response:

(I) In **electrotonus spread** or **passive spread** the nerve membrane behaves ohmically; i.e., the resistivity of the membrane remains constant. This model is valid for smaller potential changes. We will find that electrotonus spread alone is not sufficient to explain impulse transport in unmyelinated nerves. However, the mechanism still plays a contributing role in the propagation of nerve impulses.

(II) The **Hodgkin–Huxley model** is based on experimental evidence, originally obtained by Alan Lloyd Hodgkin and Andrew Fielding Huxley. Their observations of the nerve impulse transport in an unmyelinated nerve showed that a significant variation in the permeability of the membrane occurs during a potential change. This causes a non-ohmic, non-linear dependence of the membrane current I_m on the potential difference ΔV. We will see that the Hodgkin–Huxley model is sufficient to explain the nerve impulse transport in unmyelinated nerves.

19.4.1: Electrotonus Spread

Let us assume a localized drop in the potential difference across an unmyelinated nerve membrane as shown in Fig. 19.9. To quantify the response to such a drop, we assume that the membrane resistivity does not change, i.e., that the membrane acts as a resistor obeying Ohm's law. The response of the nerve to a perturbation of the potential difference in this case is quantified in two ways: as a function of time and as a function of distance from the perturbation along the nerve.

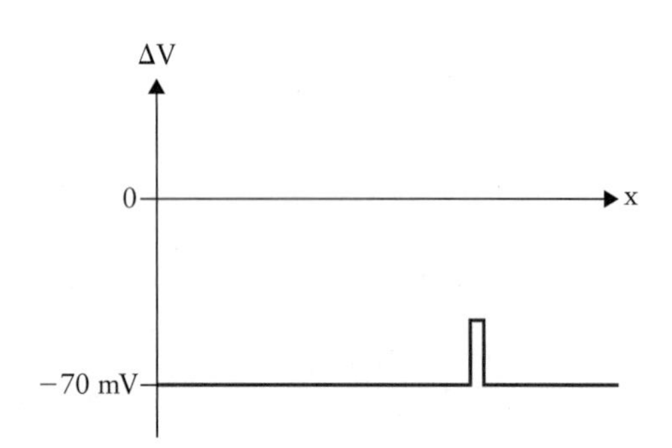

Figure 19.9 Sketch of a localized, minor perturbation in the electric potential difference across the nerve membrane, as a function of position along a nerve cell.

19.4.1.1: Response as a Function of Time

We established in the previous sections that a resting nerve has a potential difference of $\Delta V_{rest} = -70$ mV and is thermodynamically in a steady state due to the continuous operation of the Na/K pump. The electrotonus response to a perturbation in the potential difference is quantitatively described in the same manner as we derived Eqs. [19.17] and [19.18] from Eqs. [19.13] to [19.15], except that the potential difference does not drop to zero this time but returns to the steady-state value of −70 mV. Therefore, a perturbation diminishes as a function of time as:

$$\Delta V = \Delta V_i e^{-t/\tau} \text{ with } \tau = \kappa \varepsilon_0 \rho_m = 1.0 \text{ ms.}$$

This change in the potential difference as a function of time is shown in Fig. 19.10(a). The perturbation reduces to $1/e \cong 37\%$ of its initial value after time τ.

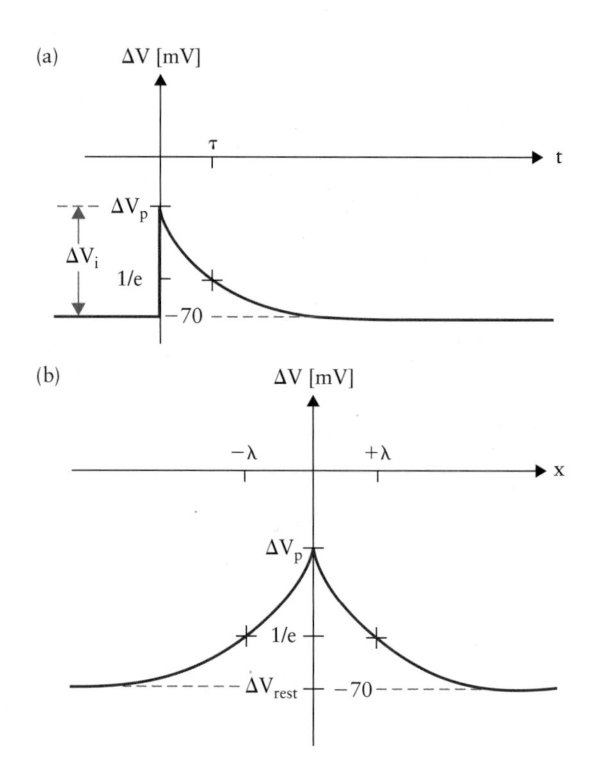

Figure 19.10 Electrotonus spread for the perturbation in Fig. 19.9 (a) as a function of time, and (b) as a function of distance from the perturbation along the nerve cell. $\Delta V_{perturbation}$ is the potential difference of the perturbation, $\Delta V_{rest} = -70$ mV is the potential across the membrane of a resting nerve, and $\Delta V_i = \Delta V_{perturbation} - \Delta V_{rest}$.

19.4.1.2: Response as a Function of Distance to the Perturbation

To evaluate the spatial spread of a localized perturbation, we assume that the perturbation itself is kept fixed, i.e., that the deviation ΔV_i from the value of −70 mV at the place of the perturbation does not change with time. Consider, for example, the case qualitatively shown in Fig. 19.11. There is a potential of −65 mV on the inside (axoplasm) of the membrane relative to a potential of −5 mV on the

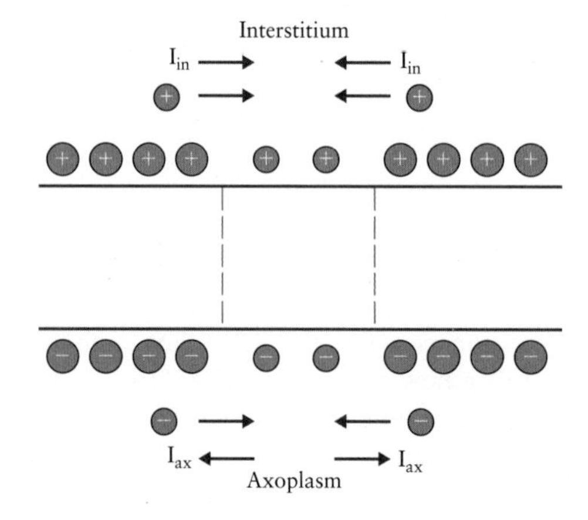

Figure 19.11 The motion of charged particles and the corresponding electric currents along a nerve cell in response to a potential perturbation between the dashed lines. I_{in} is the current in the interstitium and I_{ax} is the current in the axoplasm.

outside of the membrane (extracellular fluid). This means that $\Delta V_i = 10$ mV. Because this potential difference is maintained, mobile point charges in the vicinity of the perturbation respond. As indicated in Fig. 19.11, positive point charges move toward the more negative potential at the perturbation on the interstitium side, causing a current $I_{interstitium}$ toward the perturbation. At the same time, negative point charges move toward the more positive potential at the perturbation on the axoplasm side, causing a current $I_{axoplasm}$ along the nerve away from the perturbation. The second current is in the direction away from the perturbation since we defined a current to be in the direction of moving positive point charges. The currents along the axoplasm or the extracellular fluid flow much more easily than a current across the membrane, as the resistivity of these fluids is about seven orders of magnitude lower than that of the membrane (see Table 19.1). As point charges move along the nerve membrane, the perturbation of the potential difference broadens spatially.

It is of interest to see which final profile of the potential difference results from this broadening. In particular, does the perturbation broaden as much as several tens of centimetres to a metre? If so, electrotonus spread would be sufficient to transport a signal (i.e., the perturbation of the potential difference) from the dendrite to the synapse of the nerve, and thus to the brain.

Instead of fully quantifying the case of electrotonus spread we study a semi-quantitative model, which is sufficient for understanding the important aspects. The model is based on Fig. 19.12, which illustrates in the top sketch the directions and the amount of point charges flowing near the membrane surface in the interstitium when a final, time-independent profile of the potential difference is reached, as shown in the lower part of the figure. An equivalent model with currents in the opposite directions can be sketched on the axoplasm side of the nerve membrane.

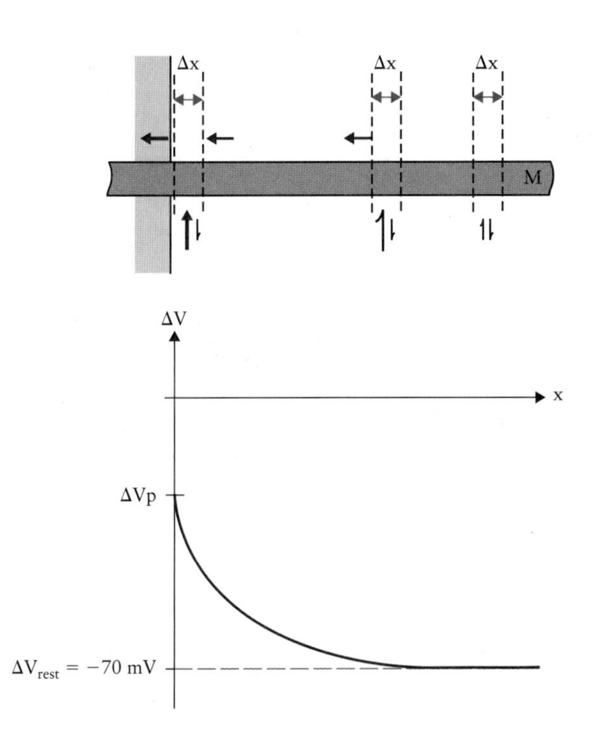

Figure 19.12 Illustration of how a time-independent potential profile due to electrotonus spread along a nerve cell can be maintained by a fixed perturbation (grey). Both charge flow across and along the nerve membrane are required.

We focus in Fig. 19.12 on three arbitrarily chosen zones, each with a width of Δx. For the zone directly adjacent to the perturbation area, a large deviation of the potential difference from the steady-state value of −70 mV causes the Na/K pump to produce a strong net flow of point charges across the membrane. This is indicated by a pair of vertical arrows, with the arrow directed upward much more prominent than the downward arrow. To maintain a potential difference of less than −70 mV across the membrane in this zone, a larger amount of charge must flow out of the Δx-zone along the nerve membrane (in Fig. 19.12 horizontally toward the left into the perturbation) compared to the amount of charge entering the Δx-zone along the nerve membrane from the right.

Δx-zones farther away from the perturbation, such as the one shown at the centre of Fig. 19.12, maintain a potential difference closer to −70 mV because fewer point charges can flow toward the perturbation zone. In a Δx-zone far away from the perturbation no net flow occurs through the membrane; i.e., charge exchange across the membrane is balanced, as indicated by a pair of equally strong vertical arrows.

The diminishing potential difference with distance from the perturbation is shown in Fig. 19.10(b) and is quantified by the formula:

$$\Delta V(x) = \Delta V_{x=0}\, e^{-x/\lambda}, \qquad [19.19]$$

in which the x-axis is defined along the nerve and the perturbation is located at $x = 0$. The constant λ has unit m and is called a **decay length**; it is given as:

$$\lambda = \sqrt{\frac{ab\,\rho_{membrane}}{2\,\rho_{axoplasm}}};$$

λ depends on the axon radius, a, the membrane thickness, b, the resistivity of the membrane, $\rho_{membrane}$, and the resistivity of the axoplasm, $\rho_{axoplasm}$. λ is called a decay length since a perturbation ΔV has dropped to $1/e \cong 37\%$ of its original value at that distance.

EXAMPLE 19.5

Calculate the decay length for an unmyelinated nerve.

Solution

We use the formula for the decay length. The resistivities in the equation are obtained from Table 19.1; the membrane thickness and axon diameter are given in

continued

Table 18.1. We use $b = 6$ nm and $a = 0.6$ μm, i.e., values for a large human unmyelinated nerve:

$$\lambda = \sqrt{\frac{(6\times10^{-7}\text{m})(6\times10^{-9}\text{m})(1\times10^{7}\Omega\text{m})}{2(1.1\Omega\text{m})}}$$

$$= 1.28\times10^{-4}\text{m} = 0.13 \text{ mm}.$$

This mechanism cannot carry a nerve impulse along an entire nerve of about 10 cm to 1 m length because the electrotonus spread has a decay length of less than a millimetre.

19.4.2: Hodgkin-Huxley Model

The discussion in the previous section demonstrated that the electrotonus spread is not sufficient to transport a nerve impulse. Since the electrotonus spread is based on Ohm's law, it is a model with a (charge) transport that has a linear relation between the driving force (the electric field) and the resulting flow (the current density). In the current section, we discuss a more successful model for unmyelinated nerves, which is called the Hodgkin–Huxley model, in which the charge transport does not obey Ohm's law.

The Hodgkin–Huxley model was introduced only about 50 years ago because it is based on the results of extensive and difficult experiments.

The experiments use the set-up sketched in Fig. 19.13: two electrodes are inserted into the axoplasm of a nerve and a third electrode runs along the nerve in the extracellular fluid. One of the electrodes inside and the electrode outside are used to measure the potential difference across the nerve membrane. The way the experiment is then conducted is called a **voltage-clamp measurement**: the potential difference between the two electrodes is kept constant by the controller via a feedback connection to a current generator. Whenever a change in the potential difference occurs, it is compensated by a current from the current generator, which is sent through the third electrode. This way, we can measure the membrane current directly by recording the current output of the current generator.

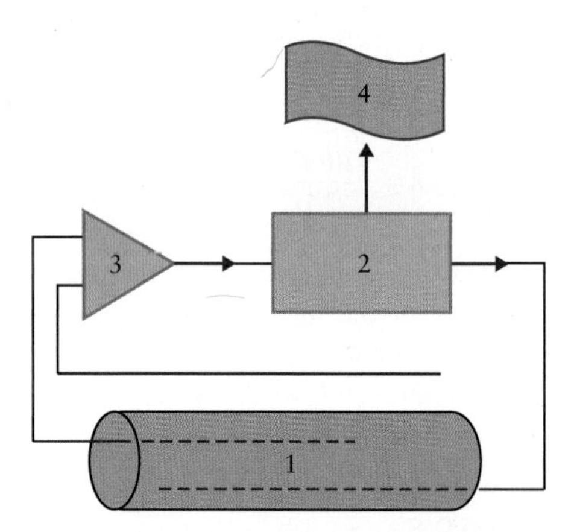

Figure 19.13 Experimental set-up for the voltage-clamp experiment conducted by Hodgkin and Huxley. (1) Axon with two inserted electrodes. One electrode runs parallel to a second electrode outside the nerve cell. Both are connected to a potentiometer (3), an instrument that measures the potential difference between the electrodes. The potentiometer also serves as a controller because it is connected to a current generator (2), which allows the potential to be clamped (fixed) by supplying a current through the second intracellular electrode. The current output of the generator is recorded (4).

It is not easy to insert two non-touching electrodes into an axon of a cross-section of 1.5 μm or less. It was only after the discovery of the giant axon of the squid (with diameters of up to 1 mm) in 1936 that experiments with single nerve cells became feasible. However, it is not possible to conduct these experiments on a live squid specimen. Therefore, the first issue was whether the nerve could be kept functional after the animal had been destroyed. After dissection, the axoplasm and the extracellular fluid had to be replaced by laboratory electrolytes. It turned out that the nerve continued to operate for a time period that was sufficient to conduct the experiments. The most important result of the early electrolyte experiments was that the crucial processes in nerve signal propagation occur across the membrane and not within the axoplasm. Several electrolytes were tested, identifying the ion species that are important for the nerve impulse transport process: sodium, potassium, and chlorine.

EXAMPLE 19.6

(a) As a reference for the processes associated with sodium and potassium flow across a nerve membrane, calculate their respective electrochemical equilibrium potential difference. (b) What do these values mean physically?

Solution

Solution to part (a): We define electrochemical equilibrium potential differences for each ion species by using the concentration values from Fig. 18.2 and inserting these in Nernst's equation (Eq. [19.9]) at a temperature of 310 K.

- We find an equilibrium potential difference for sodium ions:

$$\Delta V_{\text{Na}^+,\text{equilibrium}} = \frac{kT}{e}\ln\left(\frac{c_{\text{Na}^+,\text{interstitium}}}{c_{\text{Na}^+,\text{axoplasm}}}\right)$$

$$= \frac{kT}{e}\ln\left(\frac{145\frac{\text{mmol}}{\text{L}}}{15\frac{\text{mmol}}{\text{L}}}\right),$$

which yields:

$$\Delta V_{\text{Na}^+,\text{equilibrium}} = +60 \text{ mV}.$$

continued

- The equilibrium potential difference for potassium ions is:

$$\Delta V_{K^+,\text{equilibrium}} = \frac{kT}{e}\ln\left(\frac{c_{K^+,\text{interstitium}}}{c_{K^+,\text{axoplasm}}}\right)$$

$$= \frac{kT}{e}\ln\left(\frac{5\frac{\text{mmol}}{\text{L}}}{150\frac{\text{mmol}}{\text{L}}}\right),$$

which yields:

$$\Delta V_{K^+,\text{equilibrium}} = -90 \text{ mV}.$$

Solution to part (b): Nernst's equation expresses the electrochemical equilibrium. For a given concentration step across a membrane, this means that an electric current must occur unless the equilibrium potential difference is established. The results above also enable us to predict the direction of a current when it occurs. This is illustrated for the resting nerve potential difference at −70 mV. Since neither of the equilibrium potential differences for sodium and potassium are equal to −70 mV, a net current of both ion species across the membrane is observed in a resting nerve:

$$\Delta V_{\text{actual}} < \Delta V_{Na^+,\text{equilibrium}}$$

$$\Rightarrow Na^+ \text{flow to axoplasm}$$

$$\Delta V_{\text{actual}} > \Delta V_{K^+,\text{equilibrium}}$$

$$\Rightarrow K^+ \text{flow to axoplasm.}$$

From the five last equations we conclude that the actual potential difference during an impulse should not exceed a potential difference of +60 mV and should not drop below a potential difference of −90 mV since beyond these values both ion species flow in the same direction across the nerve membrane. We note further that the flow of ions as found in the last equation is exactly compensated by the Na/K pump for ΔV_{actual} = −70 mV, i.e., for resting nerve conditions. For any other potential difference, different sodium and potassium currents follow; these are then not compensated by the Na/K pump because the pump operates independently of the potential difference.

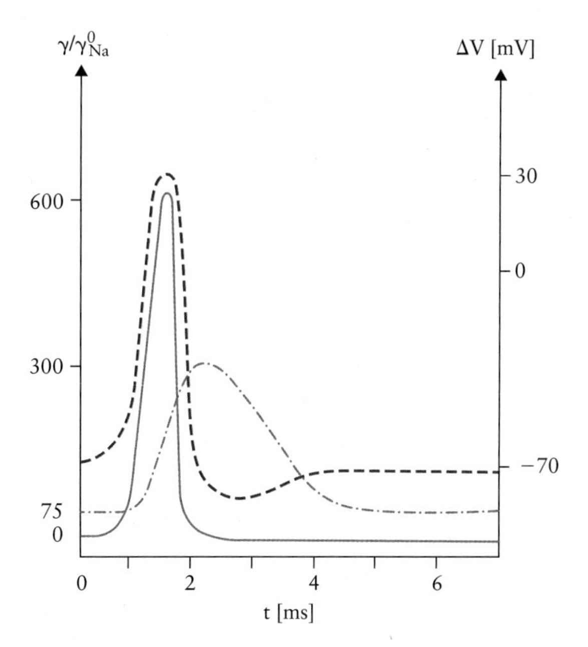

Figure 19.14 Response of a squid axon to an externally triggered impulse at 20°C (room temperature). The dashed line (with the corresponding potential axis at the right) is consistent with the same curve in Fig. 18.3, confirming that a nerve impulse is triggered by an initial 10-mV perturbation of the potential difference. The solid curve (for sodium) and dash-dotted curve (for potassium) show their respective conductivity within the nerve membrane as a function of time. The conductivity is given as a value relative to the nerve membrane sodium conductivity for a resting nerve, γ^0_{Na}; the corresponding axis is shown at the left.

The specific result we focus on to illustrate the non-linear character of the Hodgkin–Huxley mechanism is shown in Fig. 19.14. The figure shows the response of the giant axon of a squid to an artificial nerve impulse at 20°C (room temperature). The dashed line with the corresponding axis at the right is the potential difference across the nerve membrane and is consistent with the curve in Fig. 18.3. Note that a minor initial change from −65 mV to a value of about −55 mV is sufficient to trigger an impulse. During the impulse the potential difference swings all the way to a positive value of +30 mV, and returns in less than 0.5 ms to negative values, falling first to a negative potential difference of about −75 mV before slowly recovering to the resting nerve value. The solid and dash-dotted lines with the corresponding axis at the left show the conductivity of sodium and potassium relative to the sodium conductivity for the resting nerve, γ^0_{Na}. Recall that the conductivity and the resistivity are inversely proportional: $\gamma \propto 1/\rho$. Thus, an increase of the conductivity means a decrease in the resistivity, which yields an increased current and vice versa. The data show that the immediate response to the onset of an impulse is a strong but short-lived peak in the conductivity of sodium, followed by a slower but longer peak in the conductivity of potassium. Note that an ohmic response would be a fixed conductivity value for both sodium and potassium at all times (horizontal lines).

To see how this non-ohmic response leads to a propagating nerve impulse, we use Figs. 19.15 and 19.16 to focus on the profile of an impulse as a function of position along the nerve. Fig. 19.15 shows the change in the potential difference, and Fig. 19.16 shows the corresponding change in the current along the axon surface on the axoplasm side.

We start with Fig. 19.15. This is again a detailed figure and we have to go through it step by step to understand what happens. The top graph in the figure shows the potential difference as a function of position for a nerve impulse travelling with a speed of about 1 m/s toward the

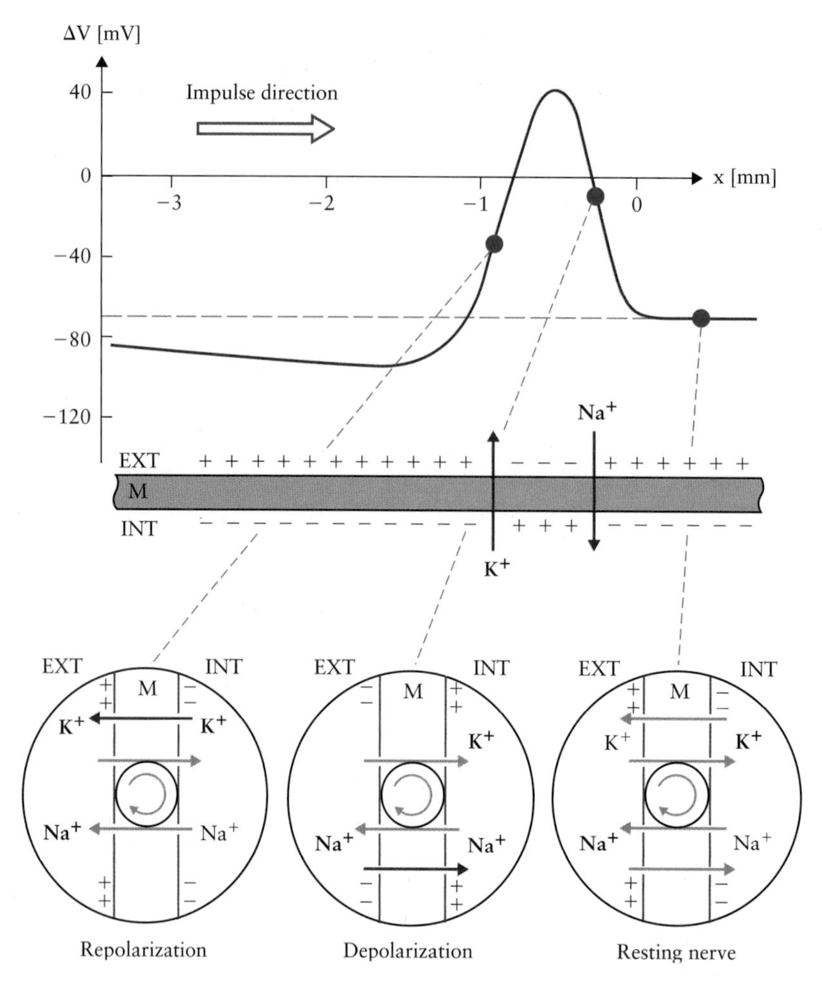

Figure 19.15 The potential difference across the nerve membrane as a function of the position along the nerve cell while a nerve impulse passes. The sketch below the curve shows the polarity of the corresponding charge on either side of the membrane. The three circular panels at the bottom illustrate the microscopic processes that dominate across the membrane at the respectively highlighted positions.

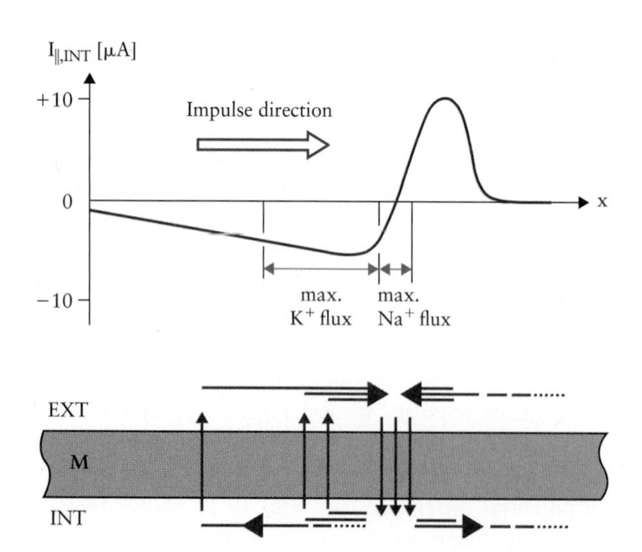

Figure 19.16 The electric current along the inner surface of a nerve membrane as a function of position along the nerve while an impulse passes. The sketch below the curve illustrates the formation of a current loop behind the crest of the impulse. A reduction in the amount of charge ahead of the impulse lowers the potential there, causing a Hodgkin–Huxley type of response. This leads to a forward motion of the impulse.

right (positive x-direction). As noted in Table 18.1, this is an average speed for an unmyelinated human nerve. The spatial profile of the potential difference shown can be determined from the potential difference profile as a function of time in Fig. 18.3, and by taking into account that nerve impulses travel in only one direction along a given nerve cell. The origin of the x-axis is chosen such that the leading edge of the impulse has just reached it. The nerve impulse has a steep leading slope, with the potential difference rising to a maximum 0.5 mm behind the onset. At 1 mm behind the leading slope the potential difference drops sharply back below the initial resting nerve potential difference. For several more millimetres behind the leading slope, the potential difference is more negative than the resting value. In Fig. 19.15 we also highlight three points along this curve:

- **Resting nerve before the impulse arrives (circle at right).** A potential difference of −70 mV exists between the axoplasm and the extracellular fluid because the operation of the Na/K pump and the passive diffusion of sodium and potassium are balanced, as discussed in Fig. 19.1. This potential difference is achieved by excess negative charges on the extracellular side of the membrane and excess positive charges on the axoplasm side.

- **The impulse front arrives, depolarization occurs (centre circle).** An initial, small drop in the potential difference triggers a sudden increase in the conductivity of sodium in the membrane (see solid line in Fig. 19.14), leading to a net sodium ion current inward. The sodium concentration in the axoplasm is sufficiently raised to reverse the charge distribution across the membrane, which temporarily establishes a negative surface charge density on the extracellular side of the membrane and a corresponding positive surface charge density on the axoplasm side.
- **The tail end of the impulse passes, repolarization occurs.** When the membrane is fully depolarized the sodium conductivity falls off sharply and the potassium conductivity increases. As a result, the inward current of sodium ceases, but an outward current of potassium emerges. This quickly returns the charge distribution across the membrane to its original polarity, with excess positive charges on the extracellular side. The potential difference drops accordingly. After these fast processes are completed, a slower recovery of the resting nerve conditions follows as the Na/K pump transports potassium and sodium back to their original side. This is associated with the slow approach of the potential difference from negative values as low as −90 mV to the resting nerve value of −70 mV. While the potential difference has not recovered, the ability to trigger a new impulse is diminished, which reduces the risk of unintentional secondary impulses.

It becomes apparent how the impulse travels forward along the nerve when the current along the axon membrane is studied as a function of the position of the impulse at a given instant. This is shown in Fig. 19.16. Note carefully that the curve shows a different property of the system compared to Fig. 19.15 even though the two profiles appear superficially similar. The sketch of the membrane below the current density diagram in Fig. 16.16 once again illustrates the currents occurring across the membrane, which are also shown in Fig. 19.15: when the impulse front arrives, a strong sodium ion current occurs toward the axoplasm accompanied by a weaker potassium ion current across a broader zone toward the extracellular fluid farther upstream.

We focus first on the point at which the sodium current occurs across the membrane. The associated reduction of positive charge on the extracellular fluid side of the membrane causes other positive charge in the vicinity to move toward this point. In the same fashion, the sudden increase in positive charge on the axoplasm side of the membrane leads to an outflow of positive charge along the membrane in both directions. We studied a similar case in Fig. 19.11. The current parallel to the membrane in the axoplasm is shown in the top diagram of Fig. 19.16. The current is positive in the direction downstream and is negative in the direction upstream. These currents are very small, e.g., typically below 10 μA.

The current components in the downstream and upstream directions have different consequences. The current toward the left in Fig. 19.16—i.e., toward the side which the impulse front has already passed—is short-circuited by the countercurrent of potassium, leading to current loops trailing the impulse front. On the side where the parallel current flows ahead of the impulse front, it serves as the trigger for the impulse front to move forward by lowering the potential difference. As we noted before, a small drop in the potential difference is sufficient to cause the sodium conductivity to shoot up, pushing the nerve impulse front forward. Note that this current is the same we used in Fig. 19.12 to explain electrotonus spread. Thus, the signal propagates forward by electrotonus spread, triggering the adjacent segment downstream along the nerve.

19.4.3: Impulse Transport Mechanism in a Myelinated Nerve

The Hodgkin–Huxley model explains most aspects of the impulse transport in unmyelinated nerves, which comprise about 70% of the nerve cells in the human body. As Table 18.1 shows, the remaining 30% of myelinated nerves play an important part as they are responsible for transmission of motor information and response. They are designated for motor response because impulse transport in these nerve cells is much faster than in unmyelinated nerves.

Fig. 19.17, a double-logarithmic plot of the same data as in the previous two figures, addresses the question of

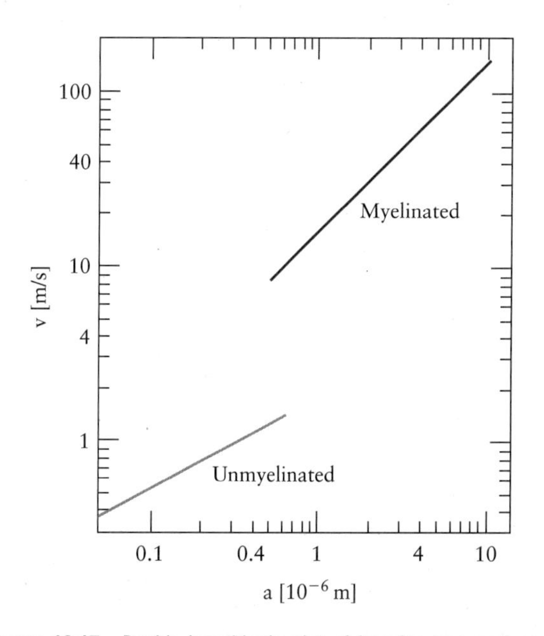

Figure 19.17 Double-logarithmic plot of impulse transport speed versus axon radius for myelinated and unmyelinated nerves in the human body. Different slopes for the two nerve types suggest that different mechanisms have to be found for the impulse transport in myelinated and unmyelinated nerves.

whether the two types of nerve share the same mechanism. The answer is that, for both the myelinated and the unmyelinated nerve, the impulse transport speed and the axon radius obey a power law; however, the power-law exponents are different:

$$\text{unmyelinated nerve: } v \propto a^{0.5} = \sqrt{a}$$

$$\text{myelinated nerve: } v \propto a^{1.0} = a.$$

Thus, three main aspects of the impulse transport speed of a myelinated nerve must be explained:

- the dependence of the signal speed on the axon radius, i.e., the fact that the signal travels faster for larger nerves;
- the increase of speed by about an order of magnitude at the same axon radius as for an unmyelinated nerve due to the myelin sheath; and
- the different mechanism of signal transport as indicated by the different functional dependence on the axon radius in comparison to the unmyelinated nerve. This aspect implies that a model must be found for the myelinated nerve that is distinct from the Hodgkin–Huxley model described in the previous section.

The best way to develop a model for the impulse transport in a myelinated nerve is to return to Fig. 18.1: the myelin sheath is too thick to allow ion diffusion or ion currents to pass through it. Therefore, the 1-mm to 2-mm-long myelinated sections are passive elements in which, electrically, only electrotonus spread is possible. However, the myelinated sections are interrupted by the nodes of Ranvier. At these nodes a membrane only 5 nm to 10 nm thick separates the axoplasm from the extracellular fluid, and the electric behaviour is the same as that of an unmyelinated nerve.

Based on these observations a mechanism has been established that combines electrotonus spread in the myelinated sections with Hodgkin–Huxley-type impulses in the nodes of Ranvier. This mechanism is called **saltatory conduction** because this mechanism is based on impulse stimulation jumping from one node of Ranvier to the next. To confirm this model, we establish first that electrotonus spread is sufficient to cause the next node to be triggered in response to an impulse at the previous node for a myelinated nerve. We cannot assume this to be the case without calculation: remember that in Example 19.5 we found for the unmyelinated nerve a decay length of only about 0.1 mm. If this value applied to myelinated nerves as well, the myelin sheath length of 1 mm to 2 mm might be too long for a Hodgkin–Huxley impulse in one node to trigger such an event in the next one.

The calculation starts from Eq. [19.19], which provides the decay length of a perturbation of the potential difference. For a typical myelinated nerve we use $a = 5\ \mu\text{m}$ and $b = 2\ \mu\text{m}$. Combined with the resistivity data from Table 19.1, we find:

$$\lambda = \sqrt{\frac{ab\rho_{\text{membrane}}}{2\rho_{\text{axoplasm}}}}$$

$$= \sqrt{\frac{(5\times10^{-6}\text{m})(2\times10^{-6}\text{m})(1\times10^{7}\Omega\text{m})}{2(11\,\Omega\text{m})}}$$

$$= 7\times10^{-3}\text{m} = 7\ \text{mm}.$$

This decay length is much longer than the corresponding value in the unmyelinated nerve. We compare the value for λ more carefully with the distance D between subsequent nodes of Ranvier. For a myelinated nerve with $a = 5\ \mu\text{m}$, this distance is about $D = 1.4$ mm. Thus, the distance between two nodes of Ranvier is 20% of the decay length; not only will the next node of Ranvier be triggered, but several nodes farther downstream also should be stimulated in response to a Hodgkin–Huxley impulse at a given node. Exactly how many is calculated in Case Study 19.3.

CASE STUDY 19.3

How many nodes of Ranvier are triggered as a result of a Hodgkin–Huxley impulse at a given node upstream?

Answer: *We note that the threshold for a node to be triggered is about 10% of the total change of the potential difference during a Hodgkin–Huxley impulse: the maximum potential difference variation is $\Delta V_{HH} = 100$ mV, as shown in Fig. 19.14. We saw above that a node of Ranvier is triggered when its potential difference changes to a value of $V_{actual} \geq -60$ mV, which is a variation of $\Delta V = 10$ mV. Based on Eq. [19.19], we find the following values for the fraction of the change of the potential difference due to electrotonus spread for the next 20 nodes of Ranvier:*

$$\Delta V/\Delta V_{\text{HH}} = e^{-D/\lambda} = 0.80$$

$$\Delta V/\Delta V_{\text{HH}} = e^{-2D/\lambda} = 0.67$$

$$\Delta V/\Delta V_{\text{HH}} = e^{-5D/\lambda} = 0.37$$

$$\Delta V/\Delta V_{\text{HH}} = e^{-10D/\lambda} = 0.13$$

$$\Delta V/\Delta V_{\text{HH}} = e^{-20D/\lambda} = 0.02.$$

Thus, at least the next 10 nodes of Ranvier are triggered as a result of an impulse at a given node. This is an effective design; with all the little things that can go wrong, it is better that we do not depend on each of the several hundred nodes of Ranvier along a nerve firing in sequence. Procaine, a common local anaesthetic, blocks the nodes of Ranvier in the vicinity of its application; clearly, this drug must act over more than a few centimetres to be effective.

19.5: Extended Case Study in Medicine: Electrocardiography

The electrocardiogram (ECG) is a medical diagnosis and observation tool for time-resolved measurements of the electric activity of the cardiac muscle. The heart muscle of higher vertebrates is stimulated by the rhythmic action of pacemaker cells in the sino-atrial node (SA-node) of the autonomous nervous system of the heart.

The complex human heart is a relatively recent evolutionary development we share only with mammals, birds, and some reptiles. Its physiological role has been understood since William Harvey discovered the circulatory flow of blood in the cardiovascular system in 1628. Almost 300 years later, in 1906, Willem Einthoven developed a method to observe the heart while it beats in the human chest. This method is called electrocardiography (ECG) and is based on the electric properties of the heart.

In this Extended Case Study, we first explore the nerve signals and muscle cell responses in the heart. Then we relate these to the corresponding electric potential variations on a patient's skin. The time dependence of electric potential differences on the skin is illustrated for the most commonly used type of ECG as developed by Einthoven.

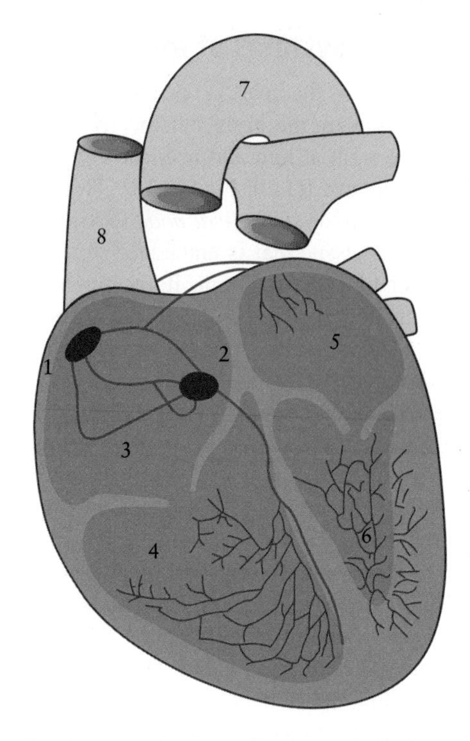

Figure 19.18 Anatomic sketch of the human heart. Shown are (1) the sino-atrial node (SA-node), (2) the atrioventricular node (AV-node), (3) the right atrium, (4) the right ventricle, (5) the left atrium, (6) the left ventricle, (7) the aorta with aortic bend, and (8) the two venae cavae.

19.5.1: Anatomic Overview of the Heart

The human heart serves as the pump for two blood circulation systems: the systemic circulation and the pulmonary circulation. Blood flow in these circulations has been discussed in Chapter 13, with Fig. 13.1 providing an overview. To operate as an effective pump, the heart is divided into two halves, each consisting of a pair of chambers that are in turn separated by valves to ensure blood flow is one-directional. Its main anatomical components are shown in Fig. 19.18. Fig. 19.19 illustrates how these components cooperate. On the left side, the pulmonary veins feed oxygenated blood, which arrives from the lungs, into the left *atrium.* The blood then passes through the mitral valve into the left *ventricle.* From there it leaves through the aortic valve into the aorta. In turn, the two venae cavae supply deoxygenated blood back to the right atrium. From there, it passes the tricuspid valve into the right ventricle. The right ventricle ejects blood into the pulmonary artery toward the lungs.

An interesting feature of the anatomy of the healthy human heart is the complete separation of both sides of the heart by the *septum.* This allows the heart to serve as a single pump for two circulatory systems simultaneously. Both halves of the heart have a similar set of heart muscles with a common electric trigger system to keep their operation synchronous. This indicates the uniqueness of the nervous system of the heart: the nerves are the only anatomical component passing through the septum, with nerve bundles extending from the right side, where the nervous centres are located, to the left side.

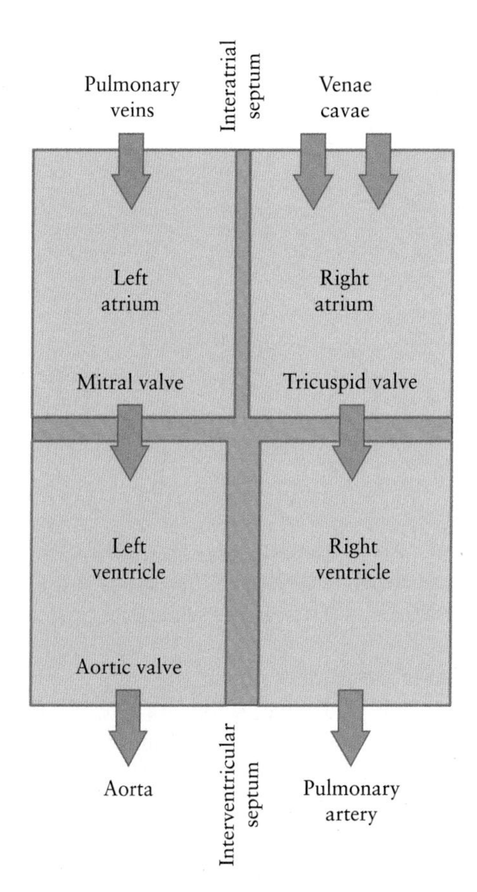

Figure 19.19 Sketch of blood flow patterns through the various anatomic components of the heart.

19.5.2: Electric Systems of the Heart

Since the operation of the heart is unconditional, it is not controlled remotely by the brain but locally by a nerve centre above the right atrium and near the entry point of the superior vena cava [(1) in Fig. 19.18]. This centre is called the *sinus node* or *sino-atrial node* (SA-node). It is a small amount of tissue that is embedded in the muscle with a diameter of 1 cm to 2 cm. Its rhythmic electric action is controlled by *pacemaker cells*. The electric properties of pacemaker cells are one main topic of this Extended Case Study.

The electric action caused by pacemaker cells is carried to the muscle cells of the heart through nerves at speeds of 0.5 m/s to 1.0 m/s in the same way we discussed earlier in the chapter. The sketch in Fig. 19.20 highlights the hierarchy of the signal processing. The impulse from the SA-node travels along the internodal tracts to the *atrioventricular node* (AV-node), which is a secondary nerve centre located at the bottom of the right atrium near the tricuspid valve. The internodal tract is illustrated in Fig. 19.18 by several nerve lines leading from the SA-node (1) to the AV-node (2).

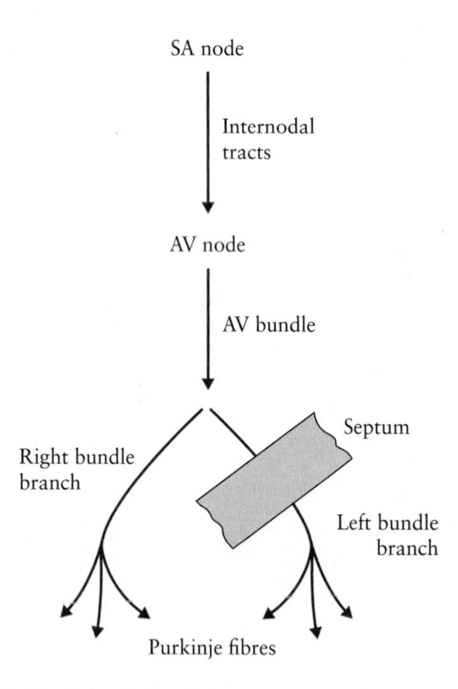

Figure 19.20 Paths of electric impulses from the SA-node. Note that the impulse is delayed in the AV-node. The impulse then passes the septum, spreading on both sides to the Purkinje fibres (heart muscles).

The AV-node operates similarly to the SA-node, except at a slower rate of depolarization. As a result, a healthy SA-node triggers the AV-node every time before it triggers an impulse on its own. However, the AV-node still serves an important purpose in that it delays the nerve impulse by about 0.1 to 0.2 seconds in reaching the ventricular muscle cells, thus allowing the atria to empty into the ventricles before the blood is pushed out of the ventricles into the blood vessels.

Once the AV-node is triggered, the electric impulse leaves through the AV-bundle and splits into two major branches, one spreading along the right side of the heart (right-bundle branch) and the other, after passing through the interventricular septum, spreading along the left side of the heart (left-bundle branch). The impulses reach their final destination in the *Purkinje fibres* (named for the Czech physiologist Jan Purkyně, with the name usually shown in German spelling), which are the muscle cells of the heart that contract as a result of the stimulation. The mechanism of contraction is the same as for skeletal muscles. The impulse speed from the AV-node to the Purkinje fibres, 1.0 m/s to 4.0 m/s, is slightly faster than from the SA-node to the AV-node.

We noted above that the heart serves two circulatory systems. With two atria and two ventricles, one could argue that the heart is actually a combination of two adjacent pumps. The fact that only one nervous system operates the heart dismisses the two-pump model. Indeed, the existence of only one nervous control system is vital as the synchronization of the contractions provides additional mechanical support for the pumping action in the larger systemic circulation, which can obviously benefit from a stronger pushing of the blood. This additional effect is achieved when the right atrium contracts slightly ahead of the left atrium (due to the location of the SA-node on the right side); but the left ventricle contracts between 10 ms and 30 ms ahead of the right ventricle, where the unit ms stands for milliseconds.

19.5.3: Pacemaker Cells and Muscle Cell Response

To understand the properties of the heart, we must first establish the mechanism of the pacemaker cells. We know that these cells cannot be usual nerve cells: if they were, they could cause an impulse only if triggered externally. Fig. 19.21 illustrates how the potential difference of the pacemaker cells in the SA-node develops as a function of time. In contrast to the electric profile of normal nerve

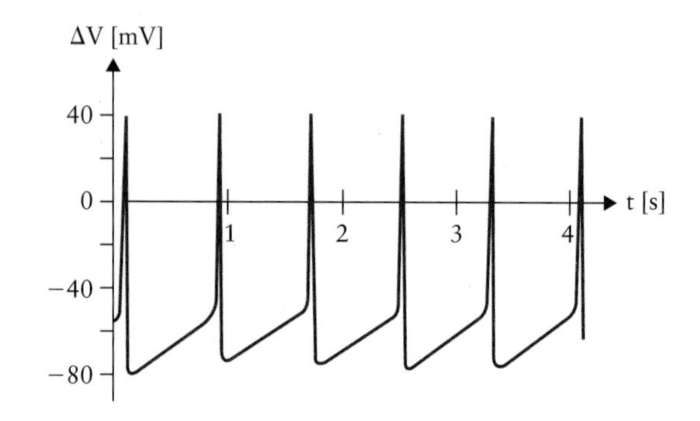

Figure 19.21 Rhythmic action of the SA-node. Pacemaker cells do not have a steady-state potential difference to which the nerve cell returns after stimulation. We associate this behaviour with an additional small calcium ion concentration in the interstitium.

cells in Fig. 19.20, no resting potential difference exists for the pacemaker cells. Instead, the potential difference increases steadily from −80 mV to about −55 mV after the pacemaker cell has been triggered, i.e., undergoes a depolarization. This steady increase takes just under one second. When a potential difference of about −55 mV is reached, a new depolarization cycle is triggered. This different behaviour is achieved by adding a new ion species to those that are involved in cross-membrane exchange: *calcium ions* (Ca^{2+}) prevent the nerve cells from reaching their resting nerve state, while Na^+ and K^+ play the same role in the pacemaker cells as they do in regular nerve cells.

The *rhythmic action* of the pacemaker cells in the SA-node triggers the response of the heart throughout its nervous system. Fig. 19.22 shows the profile of the potential difference at various points throughout the system in comparison to the profile of the potential difference of the pacemaker cells in the SA-node, which is shown in the top panel. Every time the threshold near −55 mV is reached in the pacemaker cells, a depolarization is sent toward both the AV-node and the muscle cells of the two atrial chambers of the heart. The atrial muscle cells depolarize within 50 ms of the depolarization of the SA-node [Fig. 19.22(b)]. This is indicated by a slight shift of the profile of the potential difference to the right. The shape of the depolarization profile of the atrial muscle cells adds another delay of about 50 ms following the initial overshoot to +30 mV. This shoulder-type feature of the profile of the potential difference is typical for heart muscle cells and is due to Ca^{2+} ions, as we will describe for the ventricular muscle cells below.

The impulse from the SA-node also reaches the AV-node and causes a depolarization, which is further delayed relative to the SA-node, as shown in Fig. 19.22(c). Fig. 19.22(d) illustrates the electric response of the Purkinje fibres (red curve) and their contraction (blue curve with a separate scale of relative contraction in percent at the right side of the figure). The delay varies from 100 ms to 150 ms for the cells close to the septum, which respond first, and those that are located on the outside wall of the heart. Note the characteristic shape of the depolarization and repolarization profile of the ventricular muscle cells as it prominently features a delay of up to 300 ms occurring near the 0-mV potential difference level, compared with a 1-ms period between depolarization and repolarization for regular nerve cells.

This unusual profile of the potential difference is explained in more detail in Fig. 19.23. The top part of the figure shows again a single cycle of the potential

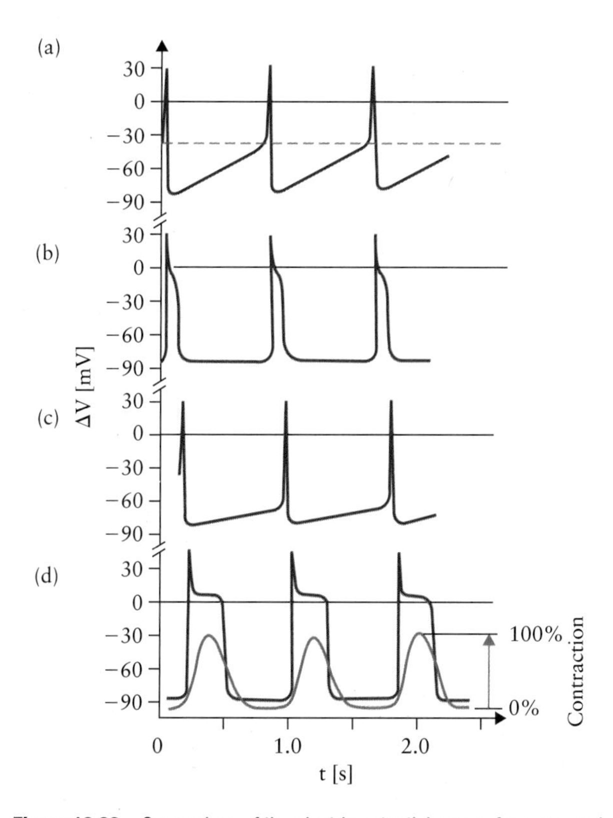

Figure 19.22 Comparison of the electric potential curves for nerve and muscle cells in the heart. (a) Pacemaker cells in the SA-node (see also Fig. 19.21), (b) atrial muscle cells, (c) AV-node cells, and (d) Purkinje fibre cells (red curve). Panel (d) shows also the muscle contraction profile of the fibre cells (blue curve with scale at the right).

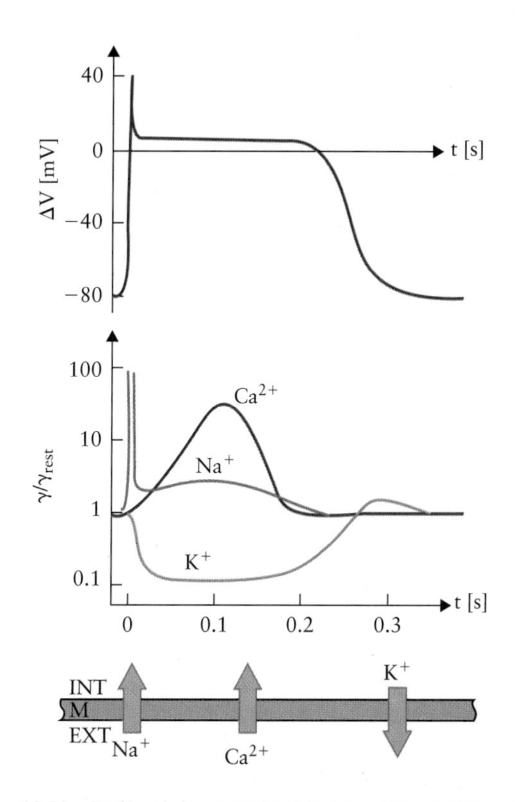

Figure 19.23 Profile of the potential difference (top), relative conductivity of the cell membrane (middle), and sketch of the dominant transmembrane ion flux (bottom) for the Purkinje fibre cells. The figure illustrates how a small calcium ion concentration causes a significant delay in the repolarization of the cells.

difference of a heart muscle cell as a function of time, taken from Fig. 19.22(d). The middle part shows the time profile of the relative conductivity (relative to a value of 1 for the resting muscle cell) for sodium, calcium, and potassium ions. Note that the relative conductivity is shown on a logarithmic scale. The sketch below the two plots identifies the dominant ion transport across the muscle cell membrane for each stage of the impulse.

The fast depolarization, initially taking place when the nerve impulse arrives, is due to a steep increase in the conductivity of sodium. This leads to the diffusion of sodium along its concentration gradient from the extracellular fluid into the cell (with the concentrations given in Table 19.3). This step is equivalent to the depolarization of a regular nerve cell, as discussed earlier in the chapter.

TABLE 19.3

Ion concentrations on the extracellular and intracellular sides of the membrane of a heart muscle cell

Ion species	Intracellular concentration (mmol/L)	Extracellular concentration (mmol/L)
Na^+	12	145
Ca^{2+}	0.0015	1.25
K^+	150	4

The intracellular calcium value represents its maximum value during heart action; the corresponding value at rest is 0.1 μmol/L.

The next step is unique to the muscle cells of the heart. The changed potential difference triggers an increased calcium transport across the membrane, leading to the plateau phase of the action potential difference near 0 mV for about 150 ms, i.e., more than a tenth of a second. The sodium transport across the membrane is significantly reduced at this potential difference, in the same fashion as we already saw for the nerve cells. However, the calcium diffusion suppresses the membrane conductivity of potassium, while transporting little charge across the membrane due to much smaller concentration levels of calcium inside and outside the cell (see Table 19.3). Note that the very low intracellular calcium concentration increases by up to 50% during this period, to a maximum of 1.5 μmol/L. The calcium diffusion into the cell is notably slower than the diffusion of either sodium or potassium, thus leading to the long delay at this stage.

The continuous change in the calcium concentration eventually causes a reduction of the calcium diffusion rate, favouring a steep increase in the potassium conductivity. Once potassium flows across the membrane into the interstitium, the potential difference quickly drops in the repolarization stage to −80 mV. This last stage is analogous to the repolarization step for regular nerve cells. The electric effect due to the potassium flow is much larger than the effects during calcium diffusion because of the larger potassium concentrations involved, as shown in Table 19.3.

19.6: The Electrocardiogram

The ECG is widely used in medicine to identify disturbances of the heart rhythm, the extent and location of myocardial damage, and the effects of drugs. It is routinely used to monitor patients during surgery.

19.6.1: The ECG Signal

The long period of the depolarized state of the muscle cells in the heart that we found in Fig. 19.23 is important to the external electric detection of the heart action in electrocardiography. Before we can measure and interpret an electrocardiogram, we have to understand how a measurable potential difference outside the cell membrane is formed, such as on a patient's skin.

Earlier in the chapter, our focus was on the potential difference across a nerve membrane, i.e., an experiment where one electrode is placed inside the axoplasm and the second one outside in the extracellular fluid. This corresponds to the experimental set-up of Hodgkin and Huxley in Fig. 19.13. To measure this potential difference requires an intrusive technique—e.g., dissection, in the case of a squid. If such an experiment is done, as illustrated in Fig. 19.24(b), a potential difference of −70 mV is measured for a resting nerve cell. However, if the measurement is done non-intrusively, i.e., with both electrodes placed outside the nerve cell, no potential difference is observed, as indicated for a second voltmeter in Fig. 19.24(a).

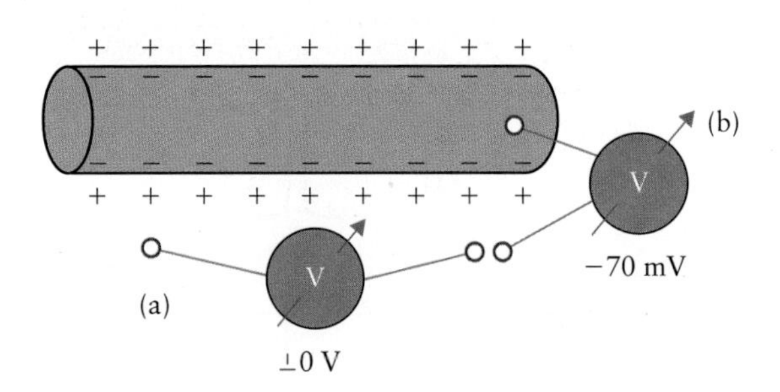

Figure 19.24 Comparison of a potential measurement (a) along a nerve membrane and (b) across the nerve membrane for a resting nerve cell.

The non-intrusive measurement in Fig. 19.24(a) can still lead to the observation of a non-zero potential difference when a nerve impulse passes through the section along which the two electrodes are placed. This is illustrated in Fig. 19.25, which shows in five steps the electric signal that is observed from a passing nerve impulse with two external electrodes placed alongside the nerve cell. In the first panel, the two electrodes are both located ahead of the nerve section currently carrying the nerve impulse (simplified as a grey area). Since the nerve is

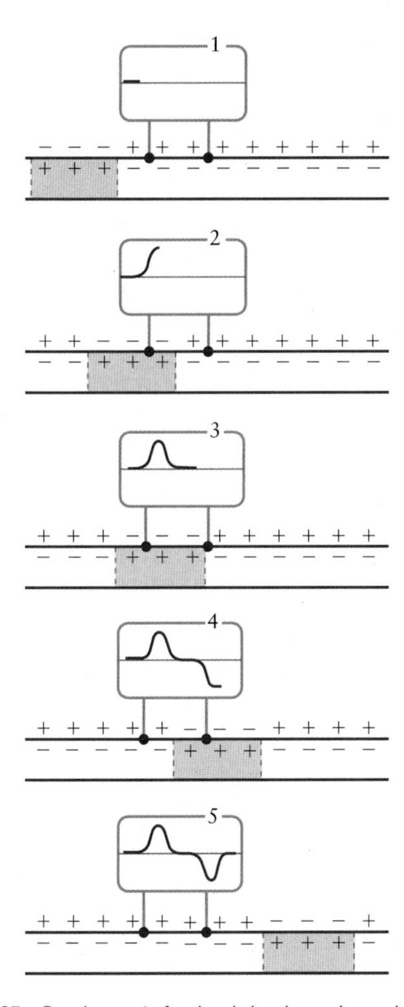

Figure 19.25 Development of a signal showing a change in the potential difference for a passing nerve impulse with two external electrodes. The location of the nerve impulse is shown as the grey section. The signal on the oscilloscope (above the nerve) is taken between the two electrode positions indicated by green wires. It is positive above and negative below the thin baseline (zero potential). No potential difference is detected in panels 1, 3, or 5 as the nerve depolarization (panel 2) and the nerve repolarization (panel 4) have respectively passed both electrodes.

anatomically and electrically in the same stage at both electrode positions—i.e., the same potential difference applies—a zero potential difference is measured, shown as a red line on the oscilloscope sketched above the nerve.

In the second panel, the travelling nerve impulse has passed the first electrode but has not reached the second electrode farther downstream. Thus, the electric effect at the two electrodes differs, with positive ions near the electrode where the nerve cell is still at rest and negative ions near the electrode where the nerve impulse has already arrived. A potential difference is measured between the two electrodes and is seen on the oscilloscope. Whether a positive or negative deviation from the reference line occurs depends on the polarity of the signal. The polarity of the potential difference for a passing nerve impulse in Fig. 19.25 is based on the definition of the electric field direction as a function of the electric charges, as shown in Fig. 17.14, and the connection between the direction of the field vector and the polarity of the potential difference, as illustrated in Fig. 18.18. From Fig. 18.18 we find that the potential difference is lower at the point where negative charges are present. The oscilloscope in the second panel of Fig. 19.25 therefore shows a positive value as it represents the potential at the leading electrode minus the potential at the trailing electrode.

In the third panel of Fig. 19.25 both electrodes are located along the depolarized section of the nerve, causing a zero potential difference. Looking at the fourth and the fifth panels illustrates that the measured potential difference has a peak with opposite polarity as the repolarization passes the electrodes in sequence.

Fig. 19.25 illustrates that the polarity of the signal matters, and that the depolarization and repolarization of a cell lead to signals of opposite polarity. Although a potential difference is a scalar quantity, the polarity and the orientation of the studied nerve tissue allow us to interpret the quantity we observe on the oscilloscope in Fig. 19.25 as the projection of a vector onto the line between the two electrodes. To simplify the later discussion of the ECG we define this vector as the **depolarization vector.**

KEY POINT

The depolarization vector is a vector pointing from depolarized tissue toward polarized tissue.

It is important to distinguish the depolarization vector from electric field vectors. This is highlighted in Fig. 19.26, which shows a nerve (at the top) and the interstitium (bottom) in which the electric potential measurement takes place. For this nerve, the left part is depolarized (negative excess charges in the interstitium) and the right part is polarized or in the resting state, with positive excess charges in the interstitium. Based on Fig. 17.14, the electric field vector in the interstitium

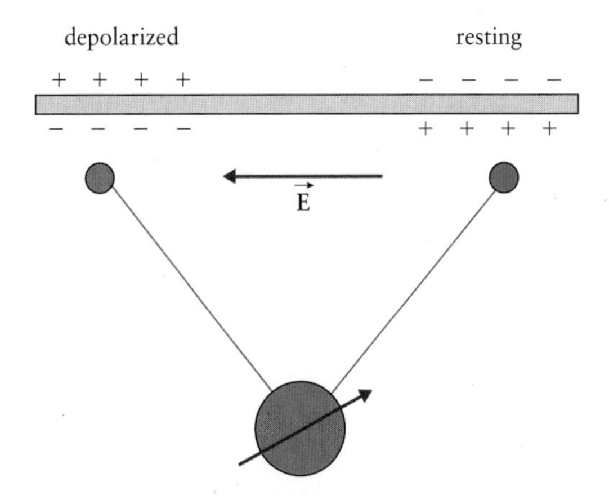

Figure 19.26 Illustration of the direction of the electric field between two electrodes when a depolarization has passed the first but not the second electrode.

points from the resting/polarized section toward the depolarized section. We see that the depolarization vector and the electric field vector in the interstitium are anti-parallel.

The discussion of Fig. 19.25 illustrates that, in general, non-zero potential differences occur during nerve depolarization if measured with electrodes outside the nerve tissue itself. This establishes the possibility for non-intrusive measurements of nervous processes in the heart. We also note that the long delay of the repolarization in the muscle cells of the heart improves the measurability, as it guarantees that significant fractions of the tissue are depolarized for extended periods during the heart's cycle. However, Fig. 19.25 indicates that the potential difference measurement depends not only on time but also on the location of the two electrodes relative to each other and relative to the depolarized tissue.

What potential differences do we expect to measure on the patient's skin? Fig. 19.27 illustrates the equipotential lines for the human heart at a given instant in time, specifically during the R-peak of the ECG as defined below. The figure quantifies the absolute potential values in unit mV for each equipotential line shown when determined on the skin of the patient. If we now assume that two electrodes are attached to the skin of the patient, the potential difference measured for the R-peak corresponds to the difference of the potential values shown for the locations of the two electrodes. Comparing the equipotential lines of Fig. 19.27 with those of an electric dipole (solid lines in Fig. 19.28) shows that the dipole is a useful model for approximating the skin surface potential for the heart action.

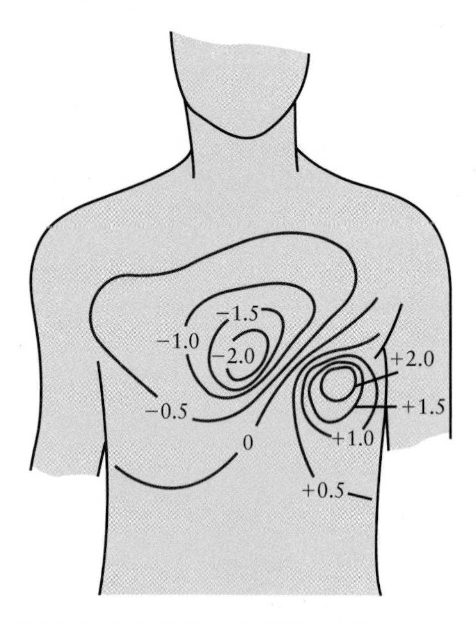

Figure 19.27 Equipotential lines for the human heart as measured on the skin surface during the R-peak of an ECG. The values given in the figure are in unit mV.

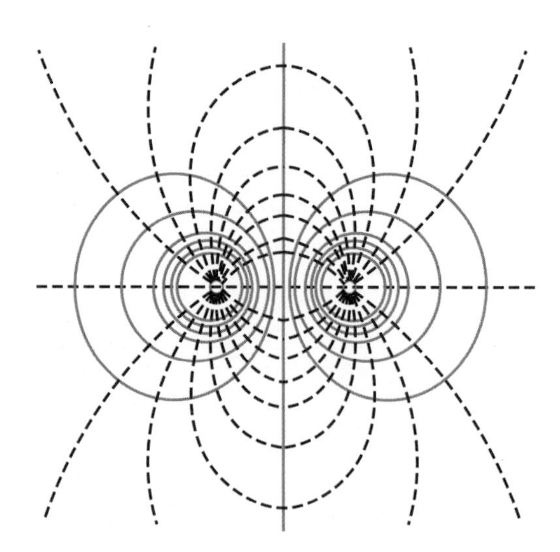

Figure 19.28 Electric field lines (dashed curves) and equipotential lines (green lines) for an electric dipole with the two charges indicated by open circles. Comparison with Fig. 19.27 allows us to approximate the electric field of the heart as a changing dipole field.

19.6.2: Einthoven's ECG Method

Fig. 19.27 shows that the measured potential difference depends on the location of the two electrodes attached to the patient's skin. Since equipotential line patterns at other stages of the heart's beating cycle are similar to those of Fig. 19.27 but are rotated due to the change of the direction of the depolarization vector as discussed below, it is important to establish a convention for the placing of the electrodes to ensure a consistent reading on the oscilloscope. The universally used convention is based on the work of Willem Einthoven, who developed the ECG in 1906. Einthoven noted that the potential of the heart action could be detected as far away as the limbs. These are therefore chosen as the locations for the electrodes since their positions relative to the heart are fixed.

Einthoven further decided to connect three electrodes to the patient, one each to the right arm, left arm, and left leg. This allowed him to measure three different signals between any pair of electrodes, which are called *leads*. Fig. 19.29(a) illustrates these leads in a conceptual sketch; Fig. 19.29(b) is a historical photograph illustrating Einthoven's original experimental arrangements. The triangle shown overlays a person in front view; i.e., the upper left corner is the lead on the right arm, the lower corner is the lead on the left leg.

The potential difference between electrodes 1 and 2 can be defined in two ways:

$$(\mathrm{I})\ \Delta V = V_1 - V_2$$

$$(\mathrm{II})\ \Delta V = V_2 - V_1.$$

To distinguish between the two cases, we define the term **polarity** for an electrode such that the potential at positive polarity is added and the potential at negative polarity is subtracted. The electrodes cannot be chosen to have

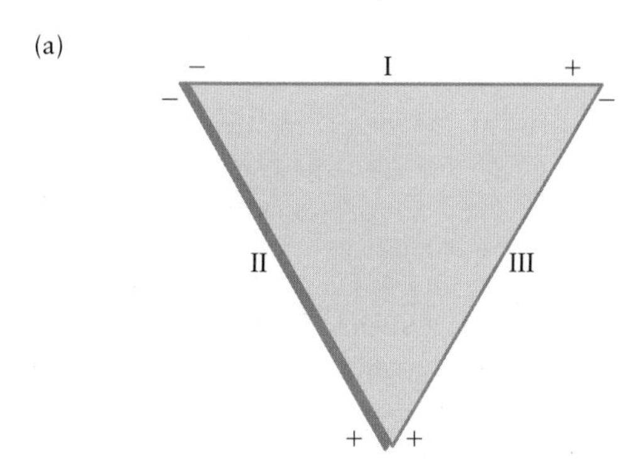

(b)

Figure 19.29 (a) Three leads as defined by Einthoven for the ECG measurement. Lead II is selected for further discussion in this chapter. (b) A person connected to an early electrocardiogram machine.

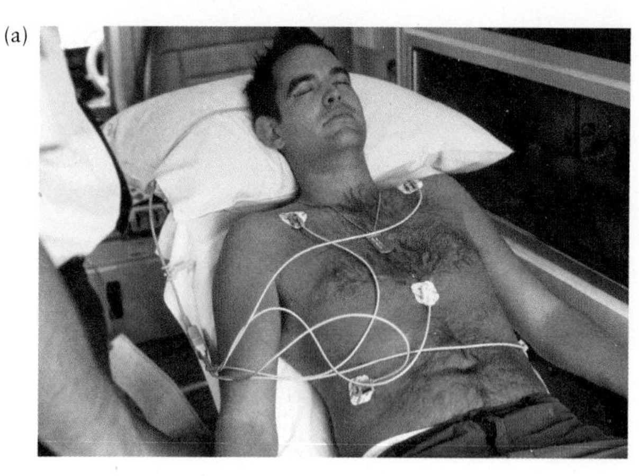

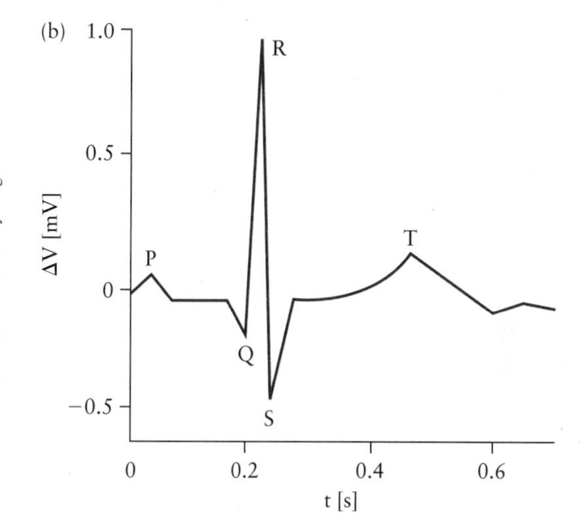

Figure 19.30 (a) A contemporary application of electrocardiography. (b) The repeating unit of an ECG signal using lead II from Fig. 19.29. Note the P-, Q-, R-, S-, and T-peaks, which are explained in detail in Fig. 19.31.

a fixed polarity for all three leads in Fig. 19.29 because each lead requires one electrode with positive polarity and one with negative polarity. The standard polarities for the three ECG leads are indicated by the respective + and − signs in Fig. 19.29. Highlighted as a thicker line in the figure is lead II, which is used in the modern application in Fig. 19.30(a) and for which the electrocardiogram for a full cycle of a healthy human heart is shown in Fig. 19.30(b).

We want to understand each of the features of the electrocardiogram in Fig. 19.30(b) and thus follow the electric processes of the heart step by step, as provided in Fig. 19.31. This figure consists of five panels, each illustrating the present direction of depolarization as an arrow overlaying a sketched heart (depolarization vector). Also depicted is a dashed line representing the reference line for lead II, onto which the depolarization vector is projected in this measurement. To the left of each panel is an oscilloscope, shown with the ECG signal as it develops at each stage. The figure starts at a time when the heart is fully polarized.

The first panel shows the stage when the SA-node has triggered an impulse and the resulting depolarization progresses to the two atrial chambers. The SA-node is the upper dot on the sketch of the heart. The depolarization vector points in this stage toward the apex of the heart. This direction is not perpendicular to the direction of lead II and, therefore, the projection of the depolarization vector onto the direction of lead II is needed to quantify the measured potential difference. Due to the polarity of lead II, as shown in Fig. 19.31, a positive potential difference is associated with this stage. The resulting positive peak in the ECG is called the *P-peak* and represents the atrial depolarization. This peak lasts for about 0.1 seconds.

The second panel contains two features. First, following the P-peak the two atrial chambers are fully depolarized. The depolarization of the AV-node (lower dot in the sketch of the heart) occurs concurrently. Since the AV-node represents a small amount of tissue, no apparent progress of the depolarization occurs during this stage. This leads to a depolarization vector of zero length, and thus no projected length is present along the direction of lead II. As a consequence, the ECG signal is back to the zero potential line, which is called the PQ-interval. The *PQ-interval* coincides with the depolarization of the AV-node.

The same panel also shows the start of depolarization of the ventricles. The initial depolarization begins on the left side of the septum and progresses initially toward

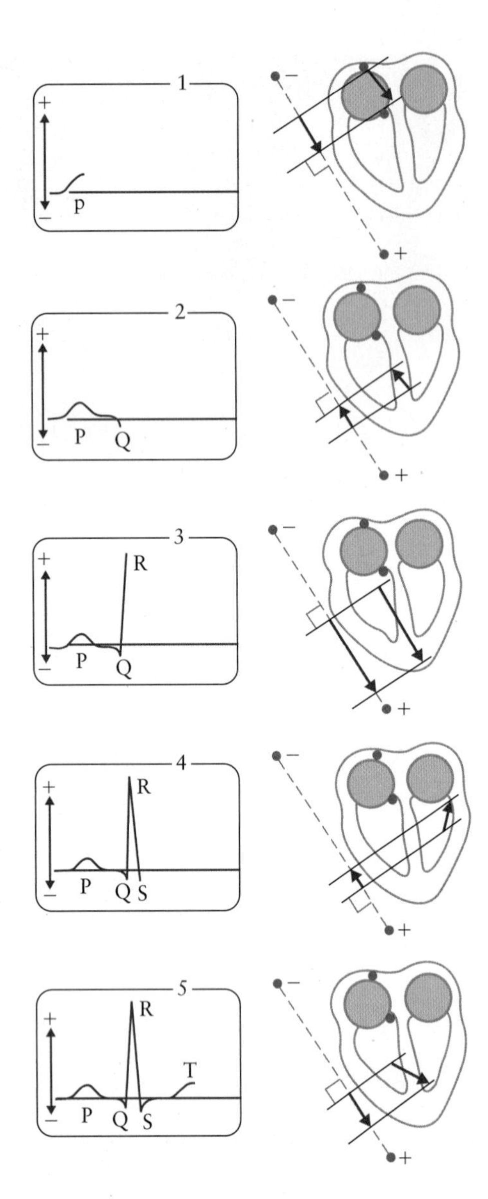

Figure 19.31 Development of an ECG profile in five steps. Oscilloscope images are shown at the left; the depolarization vector in the heart is shown in the sketches at the right. The dashed line represents lead II as defined by Einthoven. The two solid dots on each heart represent the SA- and AV-nodes respectively.

the upper end of the heart. Projecting this direction onto lead II causes a negative potential difference (backward projection). This stage is called the *Q-peak*, which corresponds to the depolarization of the upper end of the ventricles.

Once the depolarization of the upper end of the ventricles is complete, the primary direction of the progressing depolarization is toward the apex of the heart. Projecting this depolarization vector in the third panel onto lead II shows that the sign of the measured potential difference switches to a positive value. The resulting peak is particularly strong as a large amount of tissue is present in this part of the ventricles. This peak is called the *R-peak* and represents the major depolarization of the ventricles. Note that the equipotential lines during the R-peak are shown in Fig. 19.29.

Not all parts of the heart are depolarized when the depolarization of the lower part of the ventricles is complete. A small part of the back left side of the heart (posterobasal region) depolarizes after the R-peak is complete, as shown in the fourth panel. Since the respective depolarization vector is associated with a projection onto lead II, which has again turned in direction, a second peak with negative potential difference is obtained. This peak is called the *S-peak* and represents the completion of the ventricle depolarization. The Q- and S-peaks are smaller than the R-peak because a smaller amount of tissue is affected by their respective depolarization.

The last panel shows the repolarization of the ventricles (note the Ca^{2+} transport-related time delay discussed above). The repolarization starts at the apex of the heart and progresses upward. This constitutes the opposite direction to the R-peak. However, we also have a reversed process: a repolarization instead of a depolarization. Since we define the direction of the vector we project onto lead II as a depolarization vector, the inversion of its direction must be taken into account. Thus, the repolarization leads again to the measurement of a positive potential difference. The corresponding feature in the ECG is the *T-peak*, which represents the repolarization of the ventricles.

Now the repolarization of the heart is complete and a new cycle can begin. Note that we did not identify a feature in the ECG that corresponds to the repolarization of the atrial chambers. This is due to the fact that the respective signal is overshadowed by the much stronger *QRS structure*, indicating that the repolarization of the atrial chambers occurs less than 0.2 seconds after the onset of the P-peak.

KEY POINT

P-peak: Atrial depolarization

PQ-interval: Depolarization of the AV-node

Q-peak: Depolarization of upper end of ventricles

R-peak: Major depolarization of the ventricles

S-peak: Completion of ventricle depolarization

T-peak: Repolarization of the ventricles

SUMMARY

DEFINITIONS

- Current $I = \Delta q/\Delta t$
- Current density $J = I/A$, with A the cross-sectional area of the conductor

UNITS

- Current I: A = C/s (ampere)
- Resistance R: Ω = V/A (ohm)
- Resistivity r: Ω m = V m/A

LAWS

- Nernst's equation for a semipermeable membrane (separating systems 1 and 2) and assuming only singly charged ions:

$$\frac{c_{i,1}}{c_{i,2}} = \exp\left\{\frac{e}{kT}(V_{i,1} - V_{i,2})\right\},$$

with e the elementary charge, k the Boltzmann constant, c_i the concentration of the i-th component, and ΔV the potential difference across the membrane.
- Ohm's law:
- relation between magnitude of electric field $|\vec{E}|$ and magnitude of current density J:

$$|\vec{E}| = \rho J$$

with ρ the resistivity of the material carrying the current density.
  - relation between potential difference ΔV and current I:

$$\Delta V = RI$$

with $R = \rho\, l/A$ in which l is the length, A is the cross-sectional area of the conductor, and R is the resistance.

MULTIPLE-CHOICE QUESTIONS

MC–19.1. We study a current passing through a metallic wire. Leaving everything else unchanged, how does this current change when we reduce the magnitude of the electric field along the wire to 50% of its initial value?
(a) The current remains unchanged.
(b) The current becomes half of its initial value.
(c) The current becomes one-quarter of its initial value.
(d) The current becomes twice its initial value.
(e) The current becomes four times its initial value.

MC–19.2. Which property causes point charges to move along a conductor?
(a) the conductor's resistance
(b) the conductor's resistivity
(c) the electric current
(d) the electric potential
(e) the electric charge of the electron

MC–19.3. We assume that a negative mobile point charge is placed at point E in Fig. 19.32. What happens to it when it is released?
(a) It accelerates to the left.
(b) It moves with constant speed to the left.
(c) It accelerates to the right.
(d) It moves with constant speed to the right.
(e) Nothing; it remains at the same position.

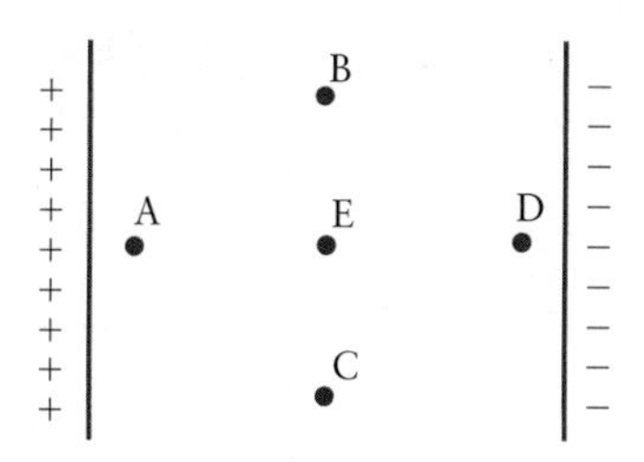

Figure 19.32

MC–19.4. Which of these units is the standard unit for resistance?
(a) Ω
(b) Ω/m^2
(c) Ω m^2
(d) Ω/m
(e) Ω m

MC–19.5. How does the resistance change if we exchange a conductor for a piece of metal with double the conductivity and double the cross-sectional area but leave its length unchanged?
(a) It is increased to four times the original value.
(b) It doubles.
(c) It remains unchanged.
(d) It is halved.
(e) It is reduced to one-quarter the original value.

MC–19.6. Which of these is the standard unit for resistivity?
(a) V/A
(b) A/V
(c) none; it is a unitless materials constant
(d) same as for resistance
(e) none of the above

MC–19.7. What does the term drift velocity refer to?
(a) the current that passes the cross-sectional area of a conductor per second
(b) the velocity with which point charges buoy to the surface of a conductor in an external electric field
(c) the volume flow rate of charged particles in a conductor when it carries a current of 1.0 A
(d) the mass flow rate of charged particles in a conductor when it carries a current of 1.0 A
(e) none of the above

MC–19.8. Which of these is the standard unit of the drift velocity?
(a) V/Ω^2
(b) V/A
(c) Ω/m
(d) m/s
(e) V^2/C

MC–19.9. Which of the following parameters—(i) the cross-sectional area of the wire, (ii) the metal used for the wire, and (iii) the length of the wire—do/does NOT contribute to the conductivity of a metal wire?
(a) only (i)
(b) only (ii)
(c) only (iii)
(d) all three
(e) only (i) and (iii)

MC–19.10. The electric system in Fig. 19.33 is often used to model a nerve membrane. It consists of a series of identical resistors aligned in two parallel rows with identical capacitors bridging the two rows following each resistor. Which of the following assumptions in order (a) to (d) has to be revised for Fig. 19.33 to be a reasonable model for a nerve (rendering the subsequent assumptions invalid as well)?

(a) The membrane can be divided into segments.
(b) The segments can be chosen such that they have the same capacitance.
(c) The resistances in each row are the same.
(d) For each single segment, the resistances on both sides of the membrane are the same.
(e) None of these assumptions needs to be revised.

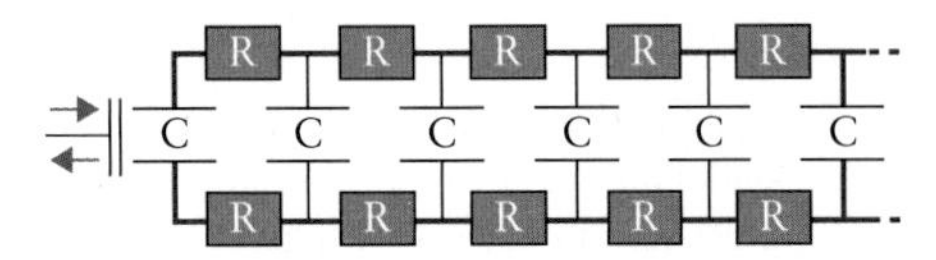

Figure 19.33 Electric axon model based on serial segments of capacitors and resistors.

CONCEPTUAL QUESTIONS

Q–19.1. Explain why all points in a conductor must have the same potential under stationary conditions.

Q–19.2. A potential difference of 1.0 V is maintained along a conductor with resistance of 10 V for a period of 20 seconds. What total charge passes through the conductor during that time interval?

Q–19.3. (a) Why don't free electrons in a piece of metal fall to the bottom due to gravity? (b) Why don't free electrons in a piece of metal all drift to the surface as a result of mutual electric repulsion based on Coulomb's law?

ANALYTICAL PROBLEMS

P–19.1. All commercial electric devices have identifying plates that specify their electrical characteristics. For example, a typical household device may be specified for a current of 6.0 A when connected to a 120-V source. What is the resistance of this device?

P–19.2. If a current of 60 mA flows through a metal wire, how many electrons move past a given cross-section of the wire in 5 minutes?

P–19.3. A total charge of 8.0 mC flows through the cross-section of a metallic wire in 4 s. What is the current in the wire?

P–19.4. In the Bohr model of the hydrogen atom the electron in the lowest energy state (ground state) moves with a speed of 2.2×10^6 m/s along a circular orbit of radius 5.3×10^{-11} m. What is the equivalent current associated with the orbiting of the electron?

P–19.5. When used at 120 V, a resistor carries a current of 0.6 A. What current is carried if the potential difference is lowered to 70 V?

P–19.6. Calculate the diameter of a 2.0-cm long tungsten filament that is used in a small light bulb if its resistance is 0.05 Ω.

P–19.7. A potential difference of 12 V causes a current of 0.4 A in a 3.2-m long metallic wire with uniform radius of 4.0 mm. What are (a) the resistance of the wire and (b) the resistivity of the wire?

P–19.8. A minor shock is perceived by a person when an electric current through the thumb and index finger of the same hand exceeds 80 μA. Compare the maximum allowable potential difference for the hand when the person has (a) dry skin with a resistance of $R = 4.0 \times 10^5$ V, and (b) wet skin with a resistance of $R = 2000$ V.

P–19.9. A rectangular piece of copper is 2 cm long, 2 cm wide, and 10 cm deep. (a) What is the resistance of the copper piece as measured between the two square ends? (Use the resistivity of copper from Table 19.1.) (b) What is the resistance between two opposite rectangular faces?

P–19.10. A current of 6.0 A flows through a 20-Ω resistor for t = 3 minutes. What total amount of charge passes through any cross-section of the resistor in this time? (a) Express your result in unit C. (b) Express your result as the number of electrons passing through the cross-sectional area.

P–19.11. A conducting, cylindrical wire has a diameter of 1.0 mm, a length of 1.67 m, and a resistance of 50 mΩ. What is the resistivity of the material? Using Table 19.1, identify the material of which this conductor is made.

P–19.12. A mass of 3.25 g of gold is deposited on the negative electrode of an electrolytic cell in a period of 197 minutes. What current was flowing through the cell in that time period? Assume that the solution contained Au^+ ions.

P–19.13. You often see birds resting on power lines that carry currents of 50 A. The copper wire on which the bird stands has a radius of 1.1 cm. Assuming that the bird's feet are 4.0 cm apart, calculate the potential difference across its body.

P–19.14. We develop a simplified model for Fig. 19.15, which shows at a given instant the potential along an axon while a nerve impulse travels through. This model is shown in Fig. 19.34. Assume that the axon radius is

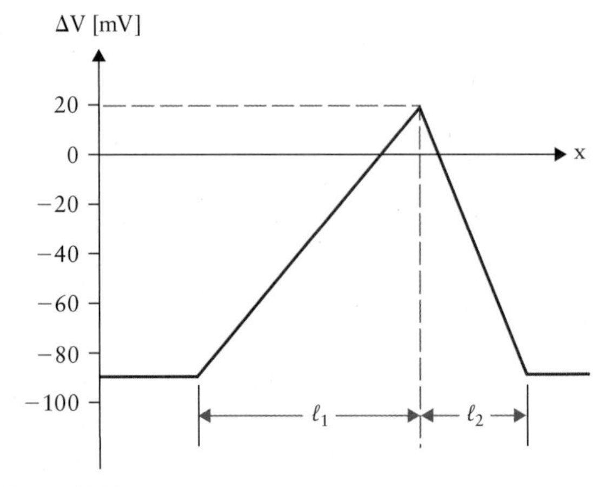

Figure 19.34

8 μm. What is the current along the axon as a function of position? Use for the lengths the values $l_1 = 0.8$ mm and $l_2 = 0.4$ mm.

P–19.15. A current density of 0.8×10^{-4} A/cm^2 stimulates a 6-nm-thick nerve membrane for 150 μs. How does the potential across the membrane change as a result of this current density?

P–19.16. For the myelinated nerve the axon radius is 10 μm, the resistivity of the membrane is $r_m = 1.0 \times 10^7\ \Omega$ m, the axoplasm resistivity is $r_a = 0.5\ \Omega$ m. Assume that the myelin sheath thickness b is related to the axon radius a as $a + b = 1.4\ a$. (a) What is the electrotonus spread decay length l for this nerve? (b) Using $D = 280\ a$ to quantify the distance D between nodes of Ranvier for this nerve, how many nodes of Ranvier are triggered along the nerve as a result of a certain node being stimulated? Use a potential difference of $\Delta V = 100$ mV for the maximum potential change in the node that is initially stimulated, and allow other nodes of Ranvier to be triggered if electrotonus spread at their site causes at least a change from –70 mV to –60 mV.

P–19.17. Using Fig. 19.17, graphically confirm the following two relations:

$$\text{unmyelinated nerve: } v \propto a^{0.5} = \sqrt{a}$$

$$\text{myelinated nerve: } v \propto a^{1.0} = a.$$

P–19.18. Table 19.4 provides approximate values for the intracellular and extracellular concentrations in unit mmol/L for sodium, potassium, and chlorine ions in a frog muscle with a resting potential difference of –98 mV and for the squid axon with a resting potential difference of –70 mV. Calculate the equilibrium potential difference for each ion species in both cases at 20°C.

P–19.19. A simplified model for an erythrocyte is a spherical capacitor with a positively charged liquid interior of surface area A. The interior fluid is separated by a membrane of thickness b from the surrounding, negatively charged, plasma fluid. The potential difference across the membrane is 100 mV and the thickness of the membrane is about 100 nm with a dielectric constant of $\kappa = 5.0$. (a) Calculate the volume of the blood cell assuming that an average erythrocyte has a mass 1×10^{-12} kg. From the volume determine the surface area of the erythrocyte. (b) Calculate the capacitance of the blood cell. For this calculation, model the membrane as a parallel plate capacitor with the area found in part (a). (c) Calculate the charge on the surface of the membrane. How many elementary charges does this represent? Use 1.06 g/cm^3 as the density of blood.

TABLE 19.4

Intracellular and extracellular ion concentrations for muscle cells of frogs and nerve cells of squids, each in unit mmol/L

Ion species	Intracellular	Extracellular
Frog muscle		
Na^+	9–13	120
K^+	140	2.5
Cl^-	3.5	120
Squid axon		
Na^+	50	440
K^+	400	20
Cl^-	40–100	560

P–19.20. An X-ray tube used in cancer treatment operates at a potential difference of 4 MV, with a beam current of 25 mA striking the metal electrode. The energy deposited by this beam in the electrode has to be transferred to the cooling water flowing through the system. What mass flow rate in unit kg/s is needed if the temperature rise in the water cannot exceed $\Delta T = 50$ K?

P–19.21. In the literature, you find axoplasm resistivities reported from $\rho_{axoplasm} = 0.5\ \Omega$ m to $1.1\ \Omega$ m. We want to test what consequence this variability of the value has. (a) By how much does the time constant for electrotonus spread vary? (b) Assume a perturbation changing the potential difference across a nerve membrane from −70 mV to −60 mV. What is the potential difference across a nerve with axoplasm resistivity of 0.5 Ω m at the same distance from the perturbation at which the potential difference has fallen to 20% for a nerve with axoplasm resistivity of 1.1 Ω m?

ANSWERS TO CONCEPT QUESTIONS

Concept Question 19.1: (C). Recall that the resistivity is a materials constant. It does not change unless we switch materials.

Concept Question 19.2: In the exponent, the Boltzmann constant is replaced by the gas constant, with $k = R/N_A$. This yields a factor $e\,N_A = F$ in the numerator:

$$\frac{c_{i,\text{interstitium}}}{c_{i,\text{axoplasm}}} = \exp\left\{-\frac{F}{RT}(V_{i,\text{interstitium}} - V_{i,\text{axoplasm}})\right\}.$$

This formula is equivalent to Eq. [19.9] but contains only macroscopic parameters.

Concept Question 19.3(a): The electrically conducting bridge at the top between the two electrodes illustrates that electrons move toward the right. Electrons move toward positive potentials; thus, the electrode at the right is at a higher potential.

Concept Question 19.3(b): Table 19.2 shows that the potential of a potassium cell is about −2.9 V and the potential of a silver cell is about +0.8 V. Thus, the higher potential is associated with the half-cell containing the silver electrode. Based on part (a) we conclude that the right electrode is the silver electrode.

Part Five

ATOMIC, ELECTROMAGNETIC, AND OPTICAL PHENOMENA

Friedrich Schiller discussed in the late 1700 in his *Letters upon the Aesthetic Education of Man* that humans are the only irrational creature on Earth because animals live cautiously to satisfy their preservation drive, while we are willing to risk it all as we give in to our curiosity drive. The golden age of physics, the century from 1850 to 1950, with its extraordinary discoveries and theories, is probably the most impressive example in history for the triumph of human curiosity. It started with zeroing in on the questions that seemed to elude answers during the previous two centuries of inquiry that started during the Renaissance and the period of Enlightenment. What is light, a particle or a wave? Where does it come from? Is matter made of atoms as Avogadro's and Dalton's laws suggest? What are they? Do they have a structure?

We start in the 1910s with Niels Bohr's radical break with traditional physical thought, as represented by Rutherford's hydrogen model of a light negative particle (called electron) orbiting a light positive particle (called proton). Postulating a new axiom governing nature, the quantum condition, Bohr was able to describe not only the hydrogen atom, but also found the elusive source of light: transitions of the electron between the discrete energy levels allowed in the atom.

Even though Bohr's model has to be seen as outdated following the development of quantum mechanics in the 1920s, it allows us to discuss light as an electromagnetic wave phenomenon, which leads to the discussion of optics and motivates us to ask about the atomic nucleus, the realm of Nuclear Physics. Nuclear Physics is a separate field because it is only in the nucleus that we encounter the third fundamental force discussed in this textbook, which is the nuclear force.

At the same time though, Physics was already a mature area of inquiry, and new discoveries were almost immediately applied in the life sciences. X-rays, which are a form of electromagnetic radiation from atoms, were discovered in the late 1890s by Röntgen. By the time he got the first Nobel prize for Physics for this discovery, medical applications had already begun.

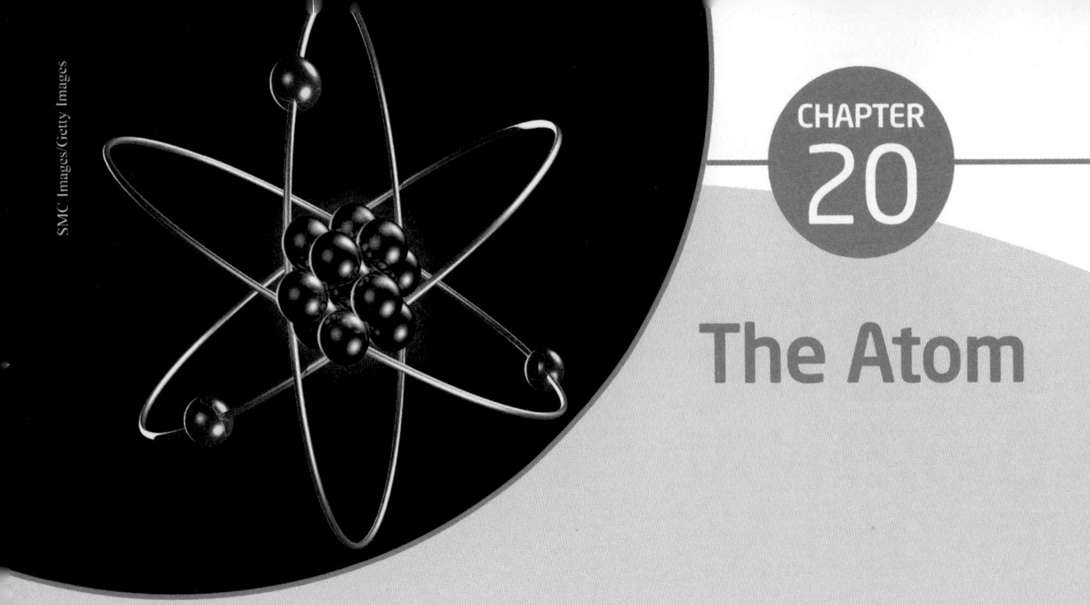

CHAPTER 20

The Atom

Matter consists of discrete units called atoms. Each electrically neutral atom has a positively charged nucleus and a shell of negatively charged electrons. Rutherford's model of the atom describes the motion of the electrons as circular motion in the central electrostatic potential of the more massive nucleus. Such atoms would radiate constantly. Bohr's model of the hydrogen atom is based on two quantum conditions: that electrons in certain orbits do not lose energy through radiation, and that radiation absorbed by or emitted from an atom is due to intra-atomic electron transfer between two allowed orbits. The model predicts energy levels for single-electron systems correctly, but does not allow us to quantify multi-electron systems or determine actual orbitals.

The quantum mechanical model is based on the particle–wave dualism of atomic and subatomic particles. Its axioms include Heisenberg's uncertainty relation that eliminates deterministic reasoning in the realm of quantum phenomena. Schrödinger developed an equation that predicts the probability to find an electron at any position in the vicinity of an atomic nucleus. Correct energy terms, a justification for three independently defined quantum numbers, and the structure of the atomic orbitals are obtained.

In this chapter we focus on the processes of light generation and absorption in matter. These processes play an important role in medical imaging, for example CT scans (computed tomography) with X-rays, and in laboratory-based diagnostic procedures, as illustrated with the analysis of amniotic fluid. One of the original applications of *amniocentesis* (see the opening photo for this chapter) was to diagnose *fetal erythroblastosis* based on a sample of amniotic fluid. Erythroblastosis is the term used to describe all those illnesses that lead to a presence of immature red blood cells in the blood. In the case of a fetus, this usually occurs as an immune system reaction due to the transfer of antibodies across the placenta, triggered by an incompatibility of blood types between mother and fetus.

To diagnose fetal erythroblastosis, a ray of white light (including all wavelengths from 300 nm to 700 nm) is sent through the amniotic fluid and the intensity of the transmitted light is analyzed as a function of wavelength. This method is called spectroscopic analysis. Fig. 20.1 shows the fraction of light absorbed from an incident light intensity plotted versus wavelength. The four spectra correspond, from left to right, to a normal pregnancy and minor, medium, and severe cases of fetal erythroblastosis.

The relative light absorption in Fig. 20.1 is shown with a logarithmic scale. This means that the amniotic fluid is transparent for red light around 700 nm, where almost 90% of the incident light passes through the sample, while it is almost opaque for violet light around 350 nm, where only 20% to 25% of the incident light passes through the sample. In addition to the steady decrease in transparency

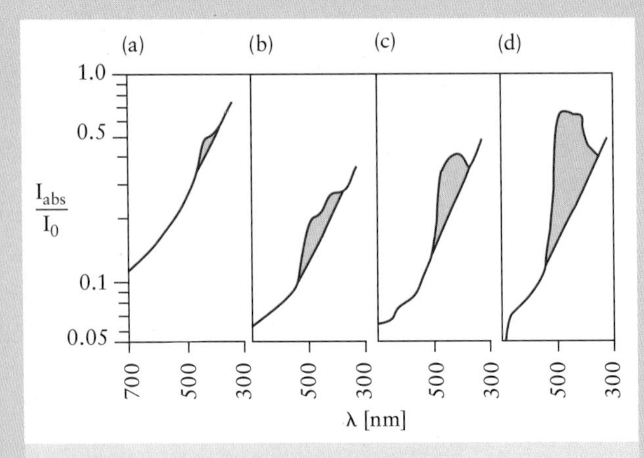

Figure 20.1 Spectroscopic analysis of the amniotic fluid obtained in an amniocentesis for (a) a healthy baby, (b) a light case, (c) a medium case, and (d) a severe case of fetal erythroblastosis. The shaded peak at wavelength $\lambda = 450$ nm is a measure of the severity of the illness. The ordinate is a logarithmic axis and shows the relative absorbed intensity, I_{abs}/I_0, with I_0 the incident intensity.

with decreasing wavelength, a characteristic absorption of light occurs in the range between 410 nm and 460 nm if the fetus suffers from fetal erythroblastosis. This additional absorption is quantified by the difference between the peak height and the corresponding level along the extrapolated background at 450 nm (Liley's method). For example, for the severe case in Fig. 20.1 the peak value is 64.5% and the corresponding extrapolated base value is 19.8%, leading to an additional absorption of 64.5% − 19.8% = 44.7%. This value is then used by a physician to determine from Fig. 20.2 the severity of the case: 44.7% corresponds as a fraction to 0.447 in this plot, which is indeed a severe case at any stage of the pregnancy.

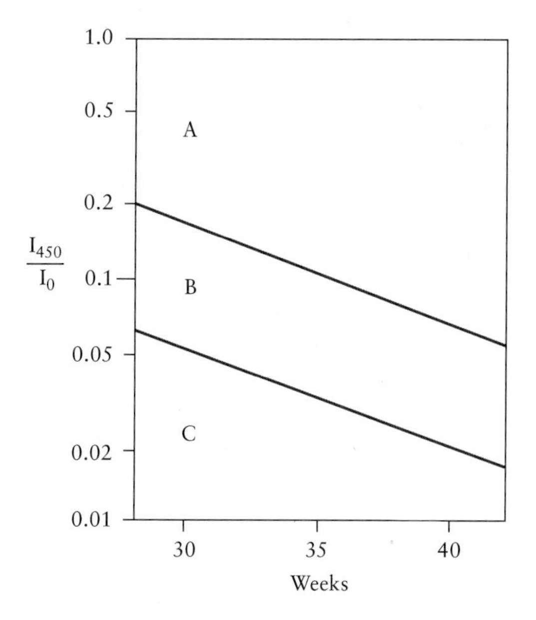

Figure 20.2 Liley's method to allow a physician to determine the severity of fetal erythroblastosis as a function of the week of gestation in a pregnancy. The ordinate shows the difference between the peak height at $\lambda = 450$ nm from Fig. 20.1 and the corresponding level along the extrapolated background. The chart allows us to distinguish light cases (C), medium cases (B), and severe cases (A). During the early years of amniocentesis, a severe case would have led to induction of labour or an intrauterine blood transfusion.

But how is the light absorption of the amniotic fluid linked to fetal erythroblastosis; i.e., why does the above analysis yield a diagnosis? In an affected fetus, heme, which is a component of the hemoglobin molecule of the red blood cells, is chemically degraded to bilirubin, which the fetus's body in turn disposes of into the amniotic fluid. The higher the bilirubin concentration in this fluid, the higher its absorption of light in the wavelength range around 450 nm. This means that the more severe the fetal erythroblastosis, the higher the absorption in the interval from 410 nm to 460 nm of the electromagnetic spectrum.

Implicit in the discussion above is the assumption that the bilirubin molecule is able to pick light of a certain wavelength out of a ray of white light passing through the solution. How does the molecule do this? Can the bilirubin molecule in turn generate light of the same wavelength? Are there other molecules that can do the same in other parts of the electromagnetic spectrum? In order to answer these questions, we must develop a model for a molecule. We can no longer treat molecules as structureless particles, since these effects hint at an internal structure that must be characteristic for each type of molecule.

20.1: The Atom in Classical Physics: Rutherford's Model

In 1897, Sir Joseph Thomson discovered the **electron** with an apparatus in which electric and magnetic fields deflected a beam emerging from a hot metallic filament in a vacuum chamber. He was able to demonstrate that the particles in the beam carried a single negative elementary charge and had a very small but finite mass.

Thomson's experiment was the first direct evidence of an elementary particle. Since the electron is negatively charged and the matter from which it came is electrically neutral, he concluded that atoms must consist of something positively charged and electrons. Calculating the typical mass and size of an atom using Avogadro's hypothesis, he further concluded that the electron is much lighter and smaller than the atom it emerged from. Thus he proposed the model shown in Fig. 20.3: electrons oscillate back and forth within an area defined by a diffusely spread positive charge. The positive charge distribution defines the boundaries of the atom, which has a diameter of the order of 0.1 nm. While this model describes a stable state for the electron in Sir Isaac Newton's classical framework, it is inconsistent with the experimental evidence that had been observed in the late 1800s, in particular the extensive spectroscopic data from Gustav Kirchhoff and Robert Bunsen. Thus, it had to be abandoned.

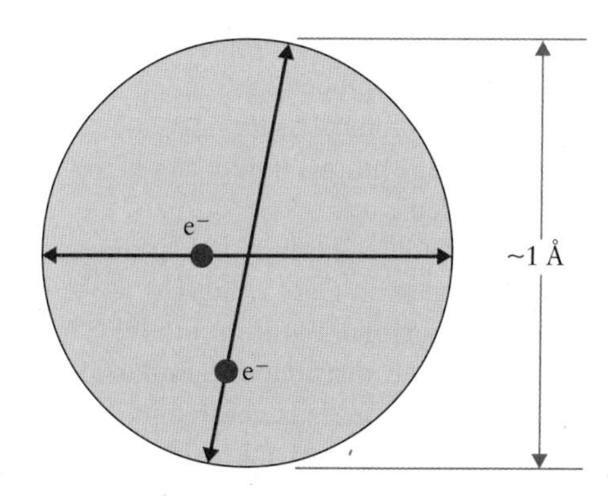

Figure 20.3 Thomson's atomic model: electrons oscillate in a uniform positive charge distribution.

20.1.1: The Size of the Atom and Its Inner Structure

The first model based on experimental interactions between atomic-sized particles was proposed by Ernest Rutherford in 1911. Bombarding a gold foil with α-particles, which are energetic helium nuclei emitted from natural sources such as thorium or uranium ores, he found that the mass and the positive charge of an atom are concentrated at its centre. Note that we can quantify Rutherford's experiment with Coulomb's law because the α-particle and the gold nucleus in the foil interact electrically.

CASE STUDY 20.1

Quantify the difference in size between the atom and its nucleus, using for the nuclear radius:

$$r_{\text{nucleus}} = (1.2\times10^{-15}\frac{\text{m}}{(\text{g/mol})^{1/3}})A^{1/3},$$

in which A is the atomic mass of the atom in unit g/mol.

Answer: *The huge difference between the size of the nucleus and the size of the atom is illustrated for a carbon atom with atomic mass 12.01 g/mol. Substituting values in the formula for the nuclear radius we get* $r_{nucleus} = 2.75 \times 10^{-15}$ *m. The atomic radius of carbon atoms in turn is* $r_{atom} = 7.7 \times 10^{-11}$ *m, i.e., 28 000 times the nuclear radius. This is a tremendous difference when compared for example with the solar system. Using the Sun, with its radius of* 6.96×10^5 *km, as the nucleus and the orbit of Pluto for the radius of the solar system (Sun–Pluto distance is* 5.91×10^9 *km), we find a ratio of 8500; i.e., the atomic nucleus is smaller in the atom than is the Sun in the solar system.*

The huge size difference in the atom is also the reason that atomic concepts were developed and understood by the 1920s, while our understanding of the internal structure of the nucleus developed only after World War II. We follow this chronological order as we develop nuclear concepts only after the atomic model and its ramifications have been introduced.

Rutherford tried to explain his findings with the same analogy we used in Case Study 20.1. We now call this approach **Rutherford's atomic model**. The mass of the solar system has a distribution similar to that of an atom: almost the entire mass is centred in the Sun, while the planets define the size of the solar system by their orbits. Thus, he proposed that the atomic nucleus carries most of the mass and all the positive charge of the atom; the electrons, which carry the negative charge, orbit the nucleus like planets. This model is sketched in Fig. 20.4. To understand this classical model quantitatively, a detailed discussion of circular motion is required. This includes the kinematic motion in section 20.1.2, followed by the interpretation of circular motion in the context of Newton's laws in section 20.1.3.

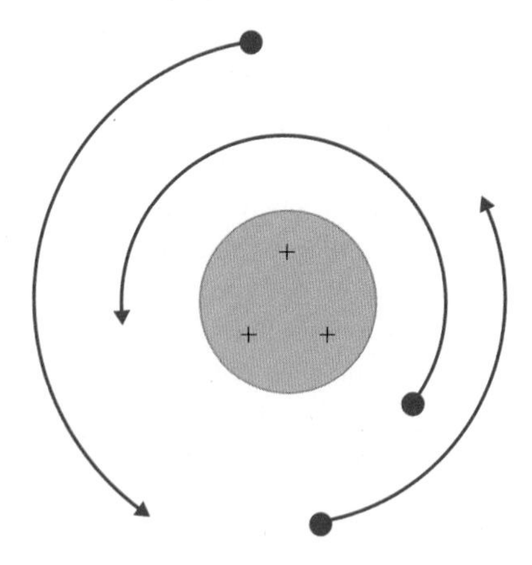

Figure 20.4 Rutherford's planetary atomic model. The electrons circle around a massive nucleus like planets move around the Sun.

20.1.2: Electrons in Uniform Circular Motion

In Rutherford's model, electrons circle the atomic nucleus. The comparison to planetary motion seems reasonable since the underlying electric force has the same radial dependence as gravity. The electron and the nucleus are both much smaller than the atom itself and can therefore be treated as point-sized particles. Coulomb's law states for the magnitude of the force between the electron and the stationary nucleus:

$$F_{\text{Coulomb}} = \frac{1}{4\pi\varepsilon_0}\frac{(-e)\cdot(+e)}{\text{r}^2}.$$

This is the only force between the electron and the nucleus because gravity can be neglected. At every position the electron is attracted straight toward the nucleus. For this reason, Coulomb's force due to a stationary point charge is called a **central force**, in the same way as Newton's gravity.

The best way to apply Newton's second law, $\vec{F}_{\text{net}} = \text{m}\,\vec{a}$, to describe the motion of the electron starts with rewriting the right-hand side for an anticipated uniform circular motion based on the discussion in section 2.6 in the form:

$$a_\perp = \frac{\text{dv}}{\text{dt}} = \frac{v}{r}\frac{\text{dr}}{\text{dt}} = \frac{v^2}{r} \qquad [20.1]$$

in which $dr/dt = v_{\parallel} = |\vec{v}|$ is the speed of the object. The resulting acceleration $a_\perp$ is called the **centripetal acceleration** because it is always directed toward the centre of the circular path. It is constant in magnitude because both v and r are constant for uniform circular motion. It is important to note that the acceleration in Eq. [20.1] is not an x- or a y-component in the xy-plane of the circular motion; in respect to fixed x- and y-axes the components of centripetal acceleration continuously change, while $a_\perp$ is constant.

With the definition of the period T and the circumference of a circle as 2 π r, speed is rewritten as $2\pi r/T$. Substituting this in Eq. [20.1] leads to a second formula for the acceleration perpendicular to the direction of motion (compare Eq. [2.16]):

$$a_{\perp} = \frac{4\pi^2 r}{T^2}. \quad [20.2]$$

We define further the **angular frequency** ω as the angle (in unit radians) through which an object on a cyclic path passes per second:

$$\omega = \frac{2\pi}{T}. \quad [20.3]$$

This allows us to rewrite the acceleration perpendicular to the path in Eq. [20.2] in a third way:

$$a_{\perp} = \omega^2 r. \quad [20.4]$$

CONCEPT QUESTION 20.1

Which of the following statements is wrong? Centripetal acceleration increases with

(A) the period.

(B) the distance from the centre of the path.

(C) the angular frequency of the circular motion.

(D) the frequency of the circular motion.

20.1.3: Centripetal Acceleration and Newton's Second Law

In Newton's mechanics, accelerations are caused by unbalanced forces, represented by a non-zero net force:

$$\vec{F}_{net} = \sum_{i=1}^{n} \vec{F}_i = m\,\vec{a}.$$

This equation is written for n forces acting on the object. Using again the coordinate system with perpendicular and parallel components we rewrite $\vec{a} = (a_{\parallel}, a_{\perp})$ and obtain:

$$\begin{aligned} F_{net,\perp} &= \sum_{i=1} F_{i,\perp} = m\frac{v^2}{r} \\ F_{net,\parallel} &= \sum_{i=1} F_{i,\parallel} = 0 \end{aligned} \quad [20.5]$$

in which Eq. [20.1] was used to substitute for $a_{\perp}$. Without the simplifying assumptions regarding $a_{\parallel}$ the right-hand side of the second formula in Eq. [20.5] would be written as $m\,a_{\parallel}$. If further the third spatial dimension is of interest (always then defined as the z-axis, assuming that the circular motion occurs in the xy-plane), a third formula must be added: $F_{net,z} = m\,a_z$.

KEY POINT

The net force causing a uniform circular motion acts perpendicular to the path within the plane in which the circular motion occurs.

It is important to note that Eq. [20.5] does not contain a separate force that we could call the centripetal force; i.e., there is no additional force in a system to separately account for the circular motion. Still, the sum of forces on the left-hand side of the first formula in Eq. [20.5] is sometimes misleadingly referred to as centripetal force.

EXAMPLE 20.1

How fast has blood to be ejected from the human heart into the aorta for it to exert a force directed upward while passing the top of the aortic arch? Use 3 cm for the radius of curvature of the aortic arch. *Note:* this process leads to the so-called Windkessel effect.

Solution

The forces causing the Windkessel effect are the result of the circular motion of blood through the 180° bend of the aorta. Three cases are possible, as illustrated in Fig. 20.5: in Fig. 20.5(a) the shaded blood element at the top of the aortic arch moves slowly. In this case, the blood can be compared to a person in a seat at the top of a slow-moving Ferris wheel: the weight of the blood element is balanced by an upward-directed normal force exerted by the aortic wall from below. This is not the case we expect to apply as the main stress on the blood vessel wall: the reaction force to the normal force would act downward.

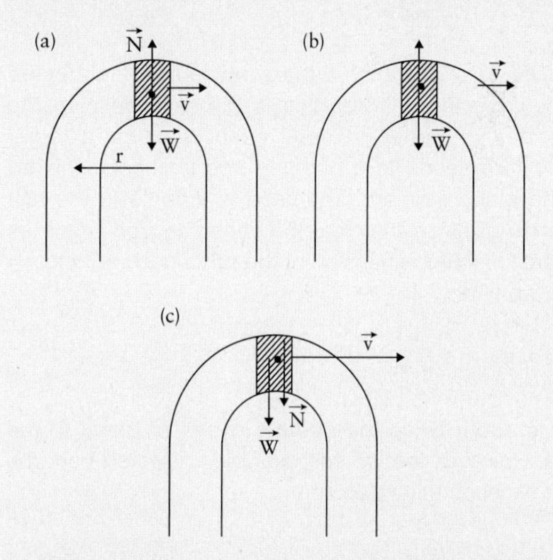

Figure 20.5 The forces causing the Windkessel effect are the result of the circular blood motion in the aortic bend. Three possible cases are distinguished: (a) The shaded blood element moves slowly; its weight dominates the free body diagram. (b) The blood element moves at an intermediate speed; its weight provides the needed centripetal acceleration for the circular motion with no normal force acting on the blood element. (c) The blood element moves much faster; it is kept on its circular path by a downward-oriented normal force exerted by the aortic wall from above.

continued

In Fig. 20.5(c) the blood element moves much faster. This case compares to a person seated in a roller coaster moving at high speed through a vertical loop. The blood element moves through the bend despite a strong outward acceleration; it is kept on its circular path by a downward-oriented normal force exerted by the aortic wall from above.

The transition case is shown in Fig. 20.5(b): the weight of the blood element provides the needed centripetal acceleration for the circular motion with no normal force acting on the blood element. This case compares to airplane flights along a parabolic path to simulate weightlessness in the cabin. This type of blood flow would be ideal as it causes the least stress on the blood vessel; however, weight and centripetal acceleration are collinear only at the top of the arch; at any other angle, normal forces would still have to act on the blood element to keep it on a circular path.

All three cases can be quantified with the same equation when using the direction of the forces as shown in Fig. 20.5(c). Choosing further the positive *x*-axis downward we write Eq. [20.5] for the shaded blood element:

$$F_{net,\perp} = \sum_i F_{i,\perp} = N + W = m\frac{v^2}{r}$$

The three cases in Fig. 20.5 are then distinguished by the following conditions:

$$\text{Fig. 20.5(a): } W > m\frac{v^2}{r} \Rightarrow N < 0$$

$$\text{Fig. 20.5(b): } W = m\frac{v^2}{r} \Rightarrow N = 0$$

$$\text{Fig. 20.5(c): } W < m\frac{v^2}{r} \Rightarrow N > 0.$$

The case in which we are numerically interested is shown in Fig. 20.5(b); this is the case defining the threshold for the Windkessel effect because a larger $m\,v^2/r$ term leads to a normal force acting downward on the blood element. The second formula in the last equation gives us therefore the speed beyond which an upward-directed expansion of the aorta occurs. With r = 3 cm we find:

$$v = \sqrt{mg\frac{r}{m}} = \sqrt{rg} = \sqrt{(3.0\times10^{-2}\,\text{m})(9.8\frac{m}{s^2})} = 0.54\,\frac{\text{m}}{\text{s}}.$$

This is about twice the *average speed* of blood in the aorta. However, the *top speed* of blood ejected from the heart exceeds this value.

20.1.4: The Basic Equation of Rutherford's Atomic Model

Rutherford's model is based on his experimental observation that the atomic nucleus carries most of the mass and all the positive electric charge of the atom. The electrons move in uniform circular fashion around the atomic nucleus. Coulomb's electrostatic force provides the centripetal acceleration. Quantitatively, Rutherford used the first formula in Eq. [20.5]. Limiting the discussion for simplicity to the hydrogen atom with one electron (with elementary charge e) and a single positive elementary charge in the nucleus, we write:

$$F_{net,\perp} = \frac{1}{4\pi\varepsilon_0}\frac{|-e||+e|}{r^2} = m\frac{v^2}{r} \qquad [20.6]$$

in which r is the radial distance between the electron and the nucleus and m is the mass of the electron.

KEY POINT

Rutherford's atomic model is quantified by Newton's second law for uniform circular motion, with the net force specified as Coulomb's force.

We need not solve Eq. [20.6] since we already know its solutions: the radial dependence on the left-hand side in Coulomb's force is the same as in Newton's force of gravity, $F \propto r^{-2}$. Eq. [20.6] has previously been solved to describe planetary motion.

Different from the mechanical applications in astronomy, Rutherford's model is fatally flawed when we attempt to draw conclusions about the atom's properties. The problem stems from the centripetal acceleration of the electron. Accelerated electrons radiate energy, as you probably know from the use of antennas in radios and as we discuss quantitatively in the next chapter when introducing basic electromagnetic phenomena. An electron that continuously emits radiation loses energy; i.e., the total energy of the electron steadily diminishes. As a result, the electron would spiral into the nucleus and the atom would self-destruct. While this is obviously not the fate of the atom, it has been the fate of classical physics as we pass into the realm of atoms: we can twist them any way we like, the classical concepts of mechanics and electromagnetism simply do not allow us to develop a working model for the atom.

20.2: Semi-Classical Model: Niels Bohr's Hydrogen Atom

In 1913, Niels Bohr became convinced that no classical explanation of the hydrogen atom would work. To fix this problem, he formulated two postulates that stipulate the observations of actual atoms onto classical physics. These postulates lie outside the classical approach, but will be combined with concepts from classical physics to yield a quantitative description of the hydrogen atom. We introduce these postulates and the predictions Bohr derived from them in this section. Note that Bohr's postulates have no immediate physical justification. It only follows from quantum mechanical principles that we discuss later in the chapter.

KEY POINT

Bohr's two postulates are

- *Electrons do not lose energy via radiation while they are in certain orbits around the atomic nucleus. These orbits are called* ***allowed orbits*** *because electrons can be found only in these.*

- *An electron loses or gains energy when it transfers between two allowed orbits. The energy lost or gained is equal to the energy difference of the initial and final orbit of the transition.*

Bohr quantified his first condition by stating that the angular momentum of the electron can resume only values that are integer multiples of $h/(2\ \pi)$ with h the Planck constant:

$$|\vec{L}| = m\ v\ r = n\frac{h}{2\pi}, \quad [20.7]$$

in which $\vec{L}$ is the angular momentum, m the mass, v the speed of the electron, and r the radius of its circular path. Bohr's second condition is quantified in the form:

$$\Delta E = E_{\mathrm{f}} - E_{\mathrm{i}} = h\,f, \quad [20.8]$$

in which f is the frequency of the electromagnetic radiation absorbed or emitted by the atom during the transition of the electron from one orbit to another. First we discuss Eqs. [20.7] and [20.8] in detail and then use them to describe the hydrogen atom quantitatively.

CASE STUDY 20.2

How can you recover the results of classical physics from Bohr's postulates?

Answer: *The classical limit is obtained when we set Planck's constant h = 0. The classical results follow in all cases when terms including h are small enough to be neglected.*

20.2.1: Motivation for Bohr's Second Postulate: The Photoelectric Effect

Eq. [20.8] was first written by Max Planck in 1900 to explain the frequency spectrum of the thermal radiation of a black body, which is an ideal system that absorbs all radiation that it receives from its environment. It is a system that can be realized experimentally reasonably easy and reasonably close to the ideal state with an isolated cavity that has a small hole to analyze its emissions. However, the formalism is complicated and it is better to use **Millikan's experiment of the photoelectric effect** to motivate Bohr's thinking.

Fig. 20.6(a) shows the experimental set-up used by Millikan. Monochromatic light, that is, light of only a single wavelength as transmitted by a monochromator, is incident from the left on an alkali metal cathode [positively charged electrode; the particular data shown in Fig. 20.6(b) apply to sodium]. Two reasons required Millikan to place the electrode in a vacuum chamber: (i) he wanted to observe electrons that are released from the metal surface, and these would not travel far enough in air to be detected; and (ii) alkali metals are not stable in air because they oxidize exothermally. The vacuum system was sealed with a quartz window that permitted the incident light to pass into the vacuum chamber. The anode was designed such that it collected all electrons that escaped from the cathode surface using a variable potential difference ΔV. Millikan then varied the frequency of the incident light and plotted the kinetic energy of the released electrons. Their kinetic energy was measured by varying the potential difference ΔV between anode and cathode until a value ΔV_0 was reached at which the released electrons no longer were able to reach the anode. At that point we know:

$$E_{\mathrm{kin}} = e\,\Delta V_0.$$

(a)

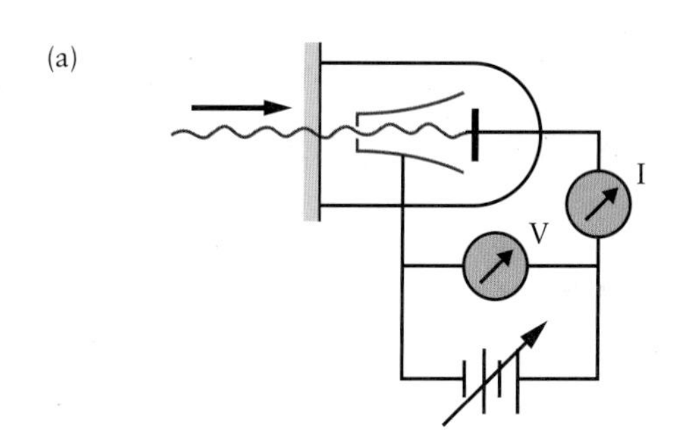

(b)

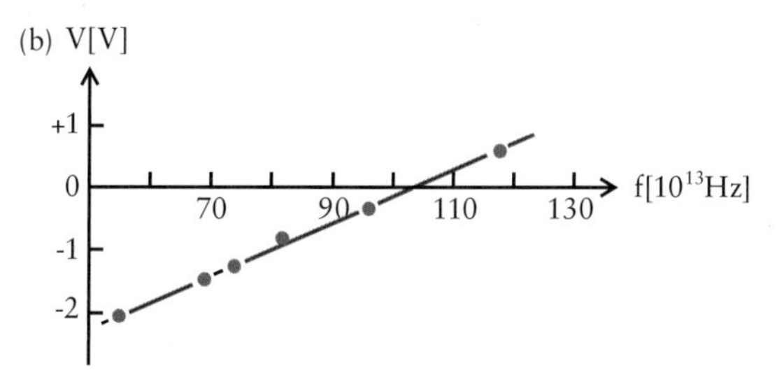

Figure 20.6 (a) Millikan's experimental set-up. Monochromatic light is incident from the left on a sodium cathode. The anode collects all electrons that escape from the cathode surface using a variable potential difference ΔV. (b) Millikan's original result: the energy of the released electrons depends only on the frequency of the incident light; it does not depend on the intensity of that light.

Fig. 20.6(b) shows Millikan's original result: the energy of the released electrons depends only on the frequency of the incident light; it does not depend on the intensity of that light, which is the energy light deposits on the cathode per second. Analyzing the linear data in Fig. 20.6(b) we write:

$$E_{\text{kin}} = h f - E_{\text{A}}, \qquad [20.9]$$

in which the constant h is determined from the slope of the curve and E_{A} is the **work function**, which is the binding energy of the electron that it has to overcome to leave the free electron gas in the metal. This measurement was done by Millikan in 1916, testing theoretical predictions made by Albert Einstein in 1905. Thus, the most important feature of Fig. 20.6(b) was indeed to confirm that the slope factor in Eq. [20.9] is identical to Planck's constant. This result confirmed that $h\,f$ is the total energy the electron picked up from the radiation.

The result in Eq. [20.9] is inconsistent with classical physics: In electromagnetism we will find in Chapter 21 that the intensity of irradiance is proportional to the square of the electric field vector, $I \cdot |\vec{E}|^2$, and the force acting on the electron in the metal is $\vec{F}_{\text{el}} = e\,\vec{E}$. Thus, the kinetic energy of the electron would have to depend on the light intensity. Einstein interpreted Eq. [20.9] such that energy is not transferred to the electron in a classical process, but as a quantum of energy, $h\,f$. In this non-classical interpretation, the light intensity affects the number of electrons released but not their individual energy. Bohr picked up on Einstein's interpretation of the photoelectric effect when using Planck's constant in his quantum condition. Bohr then argued that $h = 0$ would be a classical situation in which the electron in the atom can have any energy value. However, $h > 0$ means that energy can only be transferred in finite quanta. Once electrons can have only certain energy values, only orbits with discrete radii should be possible, and this leads to a quantum condition for the angular momentum.

20.2.2: Motivation for Bohr's First Postulate: Angular Momentum and Its Conservation

We introduced the concept of circular motion of objects in Chapter 2 and further earlier in this chapter. Here we expand on this type of motion. The angular momentum for circular motion is related to the (linear) momentum discussed in Chapter 5 in the same fashion as the torque $\vec{\tau}$ is related to the force $\vec{F}$. In vector notation, torque is written as a vector product:

$$\vec{\tau} = [\vec{r} \times \vec{F}],$$

in which $\vec{r}$ is the position vector from the rotation axis to the object.

The **angular momentum** is defined as:

$$\vec{L} = [\vec{r} \times \vec{p}]. \qquad [20.10]$$

CONCEPT QUESTION 20.2

How does Fig. 20.7 capture both the directional information and the magnitude of the torque?

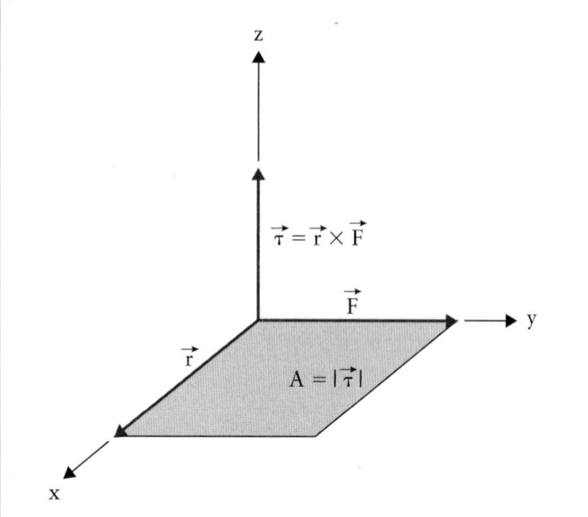

Figure 20.7 The torque $\vec{\tau}$ is perpendicular to the plane defined by the force $\vec{F}$ and the displacement $\vec{r}$ that points from the fulcrum to the position on the extended object at which the force acts. The area defined by the vectors $\vec{F}$ and $\vec{r}$ equals the magnitude of the torque.

The change of momentum with time is equal to the force, $\vec{F} = d\vec{p}/dt$; therefore the change of the angular momentum with time is equal to the torque:

$$\vec{\tau} = \frac{d\vec{L}}{dt}. \qquad [20.11]$$

The magnitude of the angular momentum follows from Eq. [20.10]:

$$|L| = p\,r \sin\varphi = m\,v\,r \sin\varphi, \qquad [20.12]$$

in which p is the magnitude of the linear momentum of the object, r is the distance from the axis of the circular motion and φ is the angle between the momentum vector and the radius vector, $\varphi = \sphericalangle\,(\vec{p},\vec{r})$.

Eq. [20.11] states that the angular momentum of a system changes only when a torque is exerted on the system; for a system free of external forces the angular momentum is therefore conserved.

20.2.3: The Radius of the Hydrogen Atom

Bohr combined Rutherford's classical equation for the electric interaction in the atom, Eq. [20.6], with his first postulate in the form of Eq. [20.7]. This allows us to derive a semi-classical model for the hydrogen atom. In this section we show how these two equations lead to discrete allowed energy levels for the electron in an atom, as Bohr's first postulate requires. We start with Eq. [20.6]. Note that this equation is written with absolute values to consider only the magnitude. Multiplying both sides by $m\,r^2$ leads to:

$$\frac{m\,e^2}{4\pi\varepsilon_0}=\frac{m^2v^2r^2}{r}.$$

The numerator on the right-hand side is then replaced by Niels Bohr's first postulate, which reads in quadratic form $(m\,v\,r)^2 = (n\,h/2\,\pi)^2$:

$$\frac{m\,e^2}{4\pi\varepsilon_0}=\frac{n^2\,\mathrm{h}^2}{4\,\pi^2 r}\Rightarrow r=n^2\frac{\varepsilon_0\,h^2}{m\,e^2\pi}. \qquad [20.13]$$

Since the right-hand side of Eq. [20.13] contains only the integer number n as a variable, it describes discrete radii. We call n therefore a **quantum number**. The smallest radius, that for which the quantum number n is $n = 1$, is called the **Bohr radius**:

$$r_{\mathrm{Bohr}}=\frac{\varepsilon_0\,h^2}{m\,e^2\pi}=5.29\times10^{-11}\ \mathrm{m}.$$

This radius defines the size of the hydrogen atom. It is usually referred to as 0.05 nm or 0.5 Å, in which 1 Å = 1.0×10^{-10} m is the non-standard length unit Ångström, named for the Swedish physicist Anders Ångström. The found value is consistent with experimentally measured sizes of hydrogen atoms.

CONCEPT QUESTION 20.3

Which of the following changes leads to a change in the radius of a system with a stationary positive point charge and a mobile negative point charge in Bohr's lowest energy state? *Note:* More than one answer might be correct.

(A) a variation of the mass of the stationary point charge

(B) a variation of the mass of the mobile point charge

(C) a variation of the charge of the stationary point charge

(D) a variation of the charge of the mobile point charge

(E) switching positive and negative charges

20.2.4: The Energy of the Hydrogen Atom

With the radius of the atom calculated we determine next the energy of the electron. We start with the electric potential energy between electron and nucleus:

$$E_{\mathrm{el}}=-\frac{1}{4\pi\varepsilon_0}\frac{e^2}{r}.$$

We use Eq. [20.13] to substitute for the radius r:

$$E_{\mathrm{el}}=-\frac{me^4}{4\pi\varepsilon_0^2\,h^2}\frac{1}{n^2}.$$

Since the electron is in circular motion at a distance r, its total energy is larger than this electric potential energy. The kinetic energy is easiest obtained by multiplying both sides of Eq. [20.6] by r:

$$\frac{e^2}{4\pi\varepsilon_0}\frac{1}{r}=-E_{\mathrm{el}}=m\,v^2=2\,E_{\mathrm{kin}}.$$

This leads to the total energy of the electron in the hydrogen atom, written as a function of the quantum number n of the orbit:

$$E_{\mathrm{total}}=E_{\mathrm{el}}+E_{\mathrm{kin}}=\frac{1}{2}E_{\mathrm{el}}=-\frac{me^4}{8\varepsilon_0^2\,h^2}\frac{1}{n^2}. \qquad [20.14]$$

Note the inverse quadratic dependence on the quantum number n.

CONCEPT QUESTION 20.4

The dashed horizontal lines (labelled 1 to 3) in Fig. 20.8 represent the total energy of an electron in a hydrogen atom. Which electron is capable of escaping the atom?

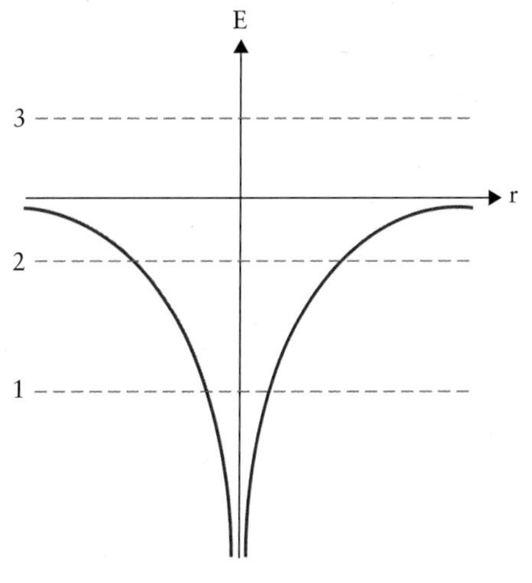

Figure 20.8 Coulomb potential for a stationary positive point charge (nucleus) and a mobile negative point charge (electron) in the hydrogen atom. The dashed horizontal lines (labelled 1 to 3) represent three possible total energy values of the electron.

20.2.5: The Hydrogen Atom's Spectrum

Eq. [20.14] is particularly important as it provides us with a direct experimental test for Bohr's model: the radiation frequency in Bohr's second postulate is calculated from Eq. [20.14] in the form:

$$hf = \Delta E_{\text{total}} = \frac{me^4}{8\varepsilon_0^2 h^2}\left(\frac{1}{n_{\text{i}}^2} - \frac{1}{n_{\text{f}}^2}\right). \qquad [20.15]$$

This equation is rewritten for the frequency of the emitted or absorbed light by dividing by Planck's constant:

$$f = R_{\text{H}}\left|\left(\frac{1}{n_{\text{i}}^2} - \frac{1}{n_{\text{f}}^2}\right)\right|, \qquad [20.16]$$

which defines **Rydberg's constant** R_{H} as:

$$R_{\text{H}} = \frac{me^4}{8\varepsilon^2{}_0 h^2} = 3.29 \times 10^{15}\ \text{Hz}.$$

Frequencies of emitted or absorbed light have been measured since the 1880s. Gustav Kirchhoff and Robert Bunsen introduced this method, called **spectral analysis**. It is still used widely in inorganic chemical analysis.

In **absorption spectroscopy**, light is sent through a transparent hydrogen-filled gas cylinder. The frequency of the incident light is scanned across a range that includes the frequencies that are solutions to Eq. [20.16]. The intensity of the transmitted light beam is detected. When the frequency of the incident light is consistent with Eq. [20.16], hydrogen atoms absorb the radiation as energy, $E = hf$. Electrons are transferred from an initial to a final orbit, where $n_{\text{i}} < n_{\text{f}}$. The energy difference in Eq. [20.15] is positive in this case because the total energy of the system is increased by the energy absorbed from the incident light ray. The energy lost by the light ray to the system corresponds to a loss of light intensity as recorded by the detector.

The second spectroscopic method is called **emission spectroscopy**. In this case, light with a wide range of frequencies passes through a hydrogen-filled gas cell. A detector that measures the intensity of light as a function of frequency is placed at an angle off the straight direction of the incident beam. Thus, the detector records no light when the gas cell is empty. However, when hydrogen atoms absorb energy from the incident light beam, the electrons that have moved into orbits with higher energy (leading to an **excited atom**) will quickly drop back to the lowest energy state $n = 1$. When the electrons return to the lower orbit, Eq. [20.15] leads to a negative energy, which is the energy lost by the system. This energy leaves the atom in the form of a massless particle called a photon. The emitted photons travel in random directions, including the direction to the detector.

Absorption spectra show a constant intensity across the observed frequency range with intensity gaps at certain frequencies. Sunlight reaching Earth has such a spectrum. Light from deeper, hotter zones of the Sun's surface is radiated in the direction of Earth. It passes through the outer, cooler zones of the Sun where atoms absorb some of the light at frequencies determined by Eq. [20.16]. Note that the Sun's outer zones do not consist of atomic hydrogen alone. Therefore, the dark lines in the Sun's spectrum reveal the chemical composition of that zone. The element helium was discovered this way. Its name indicates that it was initially believed to be a metallic element present exclusively in the Sun.

Emission spectra, in turn, consist of complementary information, i.e., single intensity lines at characteristic frequencies. These correspond to the transitions of electrons from higher energy orbits (excited states) to lower energy orbits, shown in Fig. 20.9. These transitions are grouped in so-called spectroscopic series: (a) transitions to the lowest orbit, called the **Lyman series**; (b) transitions to the orbit with $n = 2$, called the **Balmer series**; (c) transitions to the orbit with $n = 3$, called the **Paschen series**; and (d) transitions to the orbit with $n = 4$, called the **Brackett series**. Several of these series of emission lines had been observed before Bohr's model was proposed. The excellent agreement between spectroscopic measurements and the model helped Bohr's non-classical postulates to be accepted quickly.

In principle, the quantum number n can take any value between 1 and ∞. From a practical point of view, however, only small values are of interest as the energy difference between higher n values becomes smaller and smaller, as seen in Eq. [20.15]. This is indicated in Fig. 20.9 by the dashed line labelled ∞, which we interpret as the ionization energy level, i.e., the energy level beyond which the electron leaves the atom. When an electron leaves, the atom becomes a positive ion. The energy

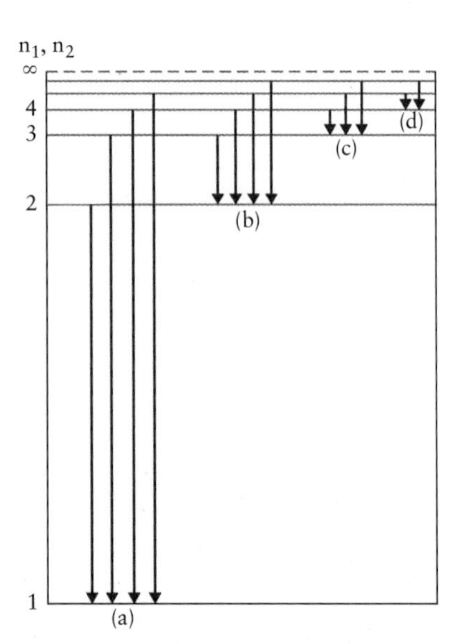

Figure 20.9 The allowed electronic states in Bohr's atomic model. The ground state has quantum number $n = 1$; the excited states have quantum numbers $n > 1$. The frequency of emitted electromagnetic radiation (photon energy) is related to the energy difference of various levels in this term scheme. (a) Lyman series, (b) Balmer series, (c) Paschen series, (d) Brackett series.

values calculated in Eq. [20.15] are, therefore, energies relative to the ionization energy of the atom (with E = 0.0 eV at $n = \infty$). For the hydrogen atom we find:

$$E_{total}\,(n{=}1) = -13.6 \text{ eV}$$

$$E_{total}\,(n{=}2) = -3.4 \text{ eV}$$

$$E_{total}\,(n{=}3) = -1.51 \text{ eV}$$

$$E_{total}\,(n{=}4) = -0.85 \text{ eV}$$

$$E_{total}\,(n{=}5) = -0.54 \text{ eV}.$$

EXAMPLE 20.2

What are the wavelength and the frequency of the light emitted by a hydrogen atom in which the electron makes a transition from the orbit n = 2 to the orbit n = 1 (called the **ground state**)?

Solution

We use Eq. [20.16] for the frequency emitted as a result of this transition:

$$f = R_H \left|\left(\frac{1}{2^2} - \frac{1}{1^2}\right)\right| = \frac{3}{4} R_H = 2.47 \times 10^{15} \text{ Hz}.$$

The corresponding wavelength is calculated from $c = \lambda f$ since the speed of light is known:

$$\lambda = \frac{c}{f} = \frac{3\times10^8 \frac{\text{m}}{\text{s}}}{2.47\times10^{15}\text{Hz}} = 1.21 \times 10^{-7} = 121 \text{ nm}.$$

Thus, the lowest Lyman series transition in hydrogen is not visible; the emitted light is in the ultraviolet (UV) part of the spectrum.

CONCEPT QUESTION 20.5

Does the hydrogen atom have transitions in the visible range of the electromagnetic spectrum?

EXAMPLE 20.3

How much energy is required to ionize hydrogen atoms when they are (a) in the ground state and (b) when they are in an excited state with quantum number n = 3?

Solution

Ionizing an atom means to remove an electron. This is achieved when the atom absorbs at least an amount of energy such that the electron is lifted into the state with $n_f = \infty$.

continued

Solution part (a): We insert n_i = 1 and n_f = ∞ in Eq. [20.15]:

$$\Delta E_{n=1} = h\,R_H \left(\frac{1}{n_i^2} - \frac{1}{n_f^2}\right) = h\,R_H = 2.17 \times 10^{-18} \text{ J} = 13.6 \text{ eV}.$$

Solution part (b): We insert n_i = 3 and n_f = ∞ in Eq. [20.15]:

$$\Delta E_{n=1} = h\,R_H \left(\frac{1}{n_i^2} - \frac{1}{n_f^2}\right) = \frac{h}{9} R_H = 2.4 \times 10^{-19} \text{ J} = 1.5 \text{ eV}.$$

20.3: Quantum Mechanical Model of the Atom

Quantum mechanics and classical (Newtonian) mechanics are two fundamentally different views of nature. Each is based on a **self-consistent set of axioms**, which are *ab initio* statements that allows us to make predictions for systems that operate within the given axiomatic framework. Which of these or several other sets of axioms is most suitable in a particular case is decided by comparing the theoretical predictions with experimental observations. For phenomena observed at the atomic and subatomic levels, the predictions of quantum mechanics are more precise than those of classical mechanics, while Newtonian mechanics makes sufficiently precise predictions for mechanical systems at the mesoscopic and macroscopic levels.

20.3.1: Wave–Particle Dualism

Quantum mechanics addresses the key problem with Bohr's model: the assumption that the electron is a particle that acts similar to a planet circling the nucleus. We pinpoint the problem with this idea by studying one of the most famous experiments in modern physics: the **double-slit experiment** shown in Fig. 20.10. In part (a) of the figure we see the intensity pattern at the right side behind the double slit if parallel light waves approach. If we open one slit at a time, the parallel wave becomes a cylinder-symmetric wave behind the slit due to Huygens' wave model, which stipulates that a wave can be represented by a radial wave from each point that it reaches during propagation. On the screen a bit farther to the right, the intensity of the wave has a maximum for a straight ray and tails off to both sides as shown. You can observe this with a small hole in a cardboard through which you allow the (essentially parallel) light from the Sun to fall onto a surface.

When we open both slits, an interference pattern results since the cylindrical waves from both slits overlap. Depending on the distances to both slits, these waves superimpose constructively or destructively. To best illustrate this effect, create two waves on a water surface by dropping two small pebbles at the same time at some distance.

Imagine now that the pattern you see is frozen along a vertical plane you submerge in the water. On that screen you observe the interference pattern shown in Fig. 20.10(a). The formation of this interference pattern is explained classically. The physical properties that allow waves to propagate (e.g., the electric field vector in a light wave or small air pockets that vibrate about their equilibrium position in acoustic waves) can be described by sinusoidal functions with given amplitudes. The square of the amplitude provides us with the intensity. We discussed this for example for sound waves when we studied ultrasound earlier in this textbook, but the intensity–amplitude relation shouldn't be surprising since the elastic energy in a vibration of amplitude A is $E = \frac{1}{2}\, k\, A^2$ (k is the spring constant), and intensity is a measure of energy per area and per time. In summary, a wave passing a barrier with slits produces on the screen behind the barrier a pattern such that:

$$\begin{aligned} &\textit{with only slit 1 open: } I \propto |A_1|^2 \\ &\textit{with only slit 2 open: } I \propto |A_2|^2 \\ &\textit{with both slits open: } I \propto |A_1 + A_2|^2. \end{aligned} \qquad [20.17]$$

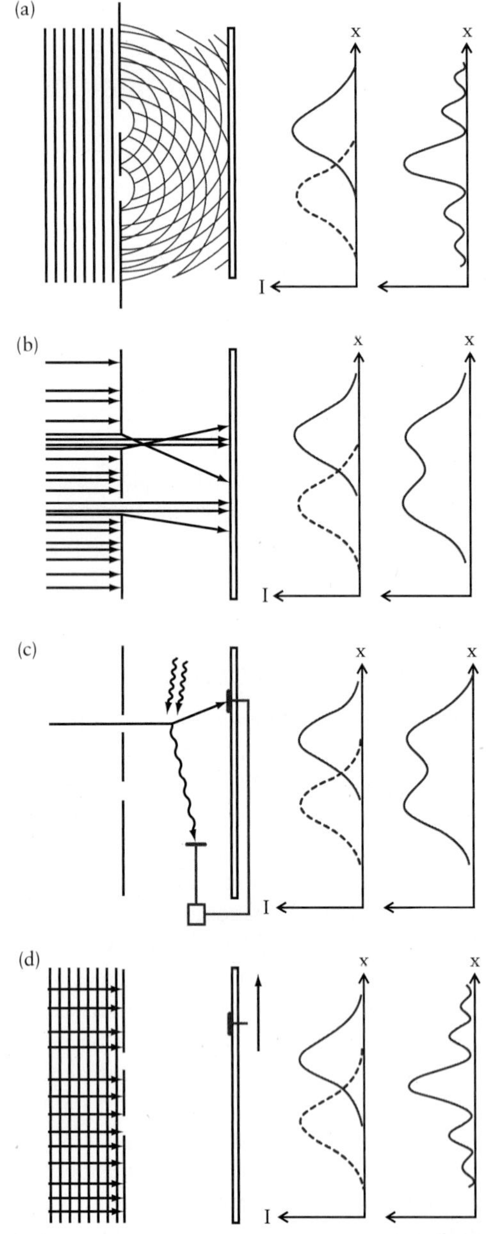

Figure 20.10 Double-slit experiment. (a) The intensity interference pattern at the right side behind the double slit for parallel light waves. The left pattern results with one slit open; the right pattern results when both slits are open. (b) The experiment done with classical particles, e.g., billiard balls on a table with one or two gaps in a barrier. In (c) and (d) electrons are used. In (c) a light source is installed behind the barrier to locate each electron as it passes through the slits. In (d) the electron is not observed before reaching the screen.

If we repeat the experiment with classical particles (e.g., billiard balls on a table with one or two gaps in a barrier) we observe the result in Fig. 20.10(b): for a single hole in the barrier, billiard balls reach with highest frequency the screen along a straight line, but there are also balls arriving toward both sides since some scatter at the edges of the gap. Opening both gaps, we observe a different result than in part (a) of the figure: the two intensity curves are just added as one after the next billiard ball can pass through either of the two gaps. Thus, the intensity of particle registration on the screen is given as:

$$\begin{aligned} &\textit{with only slit 1 open: } I = I_1 \\ &\textit{with only slit 2 open: } I = I_2 \\ &\textit{with both slits open: } I = I_1 + I_2. \end{aligned} \qquad [20.18]$$

Comparing Eqs. [20.17] and [20.18], or Figs. 20.12(a) and (b), we find that the double-slit experiment is suitable to distinguish particles and waves. In Figs. 20.12(c) and (d) we now test electrons as to whether they are particles or waves. In part (c) we install a light source behind the barrier to locate each electron as it passes through the slits. As a result, we observe the same intensity curves as in Eq. [20.18], i.e., the electron acts like a particle, which is consistent with Bohr's assumption.

If, in turn, we do not install a detection system between barrier and screen, then we have no way of telling for any particular electron which slit it actually passed. In this case, we find the result shown in Fig. 20.10(d): the electron acts like a wave with the same result as given in Eq. [20.17].

Does this mean that the electron is both a wave and a particle? No! Note that the two experiments in Figs. 20.12(c) and (d) are not the same.

KEY POINT

In any experiment we can do, an electron will act as either a wave or a particle, never both at the same time. This is expressed with the term ***particle–wave dualism****.*

What is unique about the electron and other atomic size particles is that this dualism applies. This is different from macroscopic experiments; e.g., the billiard balls above won't ever act as if they are waves, even if you turn off the light, eliminating the possibility to observe them between the barrier and the screen.

Louis de Broglie used this dualism in 1923 to quantitatively relate particle and wave properties. He studied the photon as the corpuscle of light. We know the most basic properties of a wave from studying light in optics. Its propagation speed and energy are given as:

$$v_{\text{photon}} = c = \lambda f$$

$$E_{\text{photon}} = h f,$$

in which c is the speed of light, f is the frequency, and λ is the wavelength. For the photon as a particle, the relation of its momentum and energy requires relativistic considerations, as outlined in the appendix of this chapter:

$$E_{\text{photon}} = c\, p,$$

with c the speed of light in a vacuum.

If both sets of equations are suitable to describe a photon, then de Broglie stated that we can further write:

$$p = \frac{E_{\text{photon}}}{c} = \frac{hf}{\lambda f} = \frac{h}{\lambda}; \qquad [20.19]$$

i.e., the momentum property of a particle and the wavelength property of a wave are related inversely.

EXAMPLE 20.4

Determine de Broglie's wavelength for an electron with energy 1.5 eV.

Solution

We use the equation for kinetic energy, rewritten for momentum with $p = m\, v$:

$$E_{\text{kin}} = \frac{1}{2} m v^2 = \frac{p^2}{2m} \Rightarrow p = \sqrt{2mE_{\text{kin}}}\,.$$

Using de Broglie's equation in Eq. [20.19] we find:

$$\lambda = \frac{h}{\sqrt{2\, m\, E_{\text{kin}}}} = 1.0 \text{ nm}.$$

A correction for possible relativistic properties, i.e., properties due to velocities close to the velocity of light in a vacuum, are needed for electrons of kinetic energies larger than 10 keV. For example, the wavelength calculated for an electron at 50 keV is off by 2.5% if calculated as shown in this example. See the appendix to this chapter for a more detailed discussion of relativistic corrections.

The same considerations apply to electrons, whether they travel through vacuum or whether they are bound to an atom. They may act like particles (e.g., when we remove them from the atom using the photoelectric effect described earlier in Fig. 20.6) or they may act like waves, which we will find a useful assumption when trying to identify atomic orbitals.

What is different between an electron in the double-slit experiment in Fig. 20.10 and an electron in an atom is that the electron in the atom is spatially confined. Thus, a planar wave in open space is not suitable to describe the electrons in an atom.

We need to apply de Broglie's particle–wave dualism in a confined space. We need not start this discussion anew because we studied waves in a confined space already in Chapter 15. In particular, we observed that waves can be sustained indefinitely in a confined space when they form standing waves.

20.3.2: The Electron as a Standing Wave in the Atom

A standing wave is a time-independent state of the system, i.e., a state in which the system parameters do not continuously change. This time-independence brought de Broglie to the idea of using standing waves to describe electrons because the atom does not change with time. Assuming the electron is located on a circular path around the nucleus, he argued that a multiple of the length of the wave describing the electron must fit onto the length of the circumference of the electron's path. Allowing n wavelengths to be fitted on the circumference in the manner illustrated for two complete wavelengths in Fig. 20.11, we find the condition for a standing wave:

$$2\, \pi\, r = n\, \lambda.$$

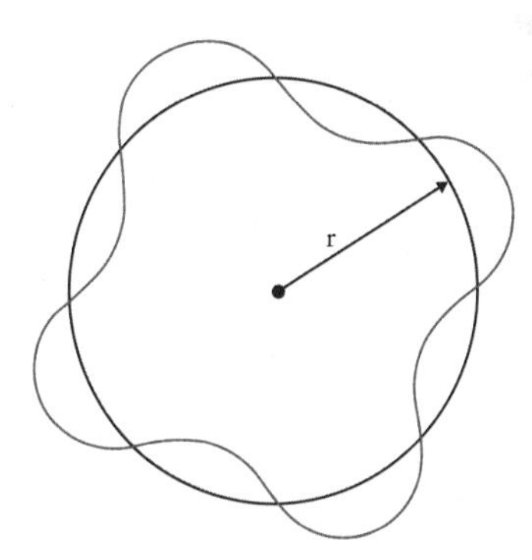

Figure 20.11 An electron forms a standing wave in an orbit of radius r around the nucleus.

in which n is a positive integer: n = 1, 2, 3,.... We combine this condition with de Broglie's wavelength equation due to the particle–wave dualism of the electron from Eq. [20.19] and rewrite the linear momentum as $p = m\, v$, with m the mass and v the speed of the electron (when the electron is considered to be a particle). This yields:

$$2\, \pi\, r = n \frac{h}{mv} \Rightarrow m\, v\, r = \frac{nh}{2\pi} = n\, \hbar,$$

in which h is Planck's constant and $\hbar = h/(2\pi)$. This result is identical to the quantum condition Bohr used to quantify his first postulate in Eq. [20.7]. While Bohr wrote Eq. [20.7] without justification, de Broglie's corresponding result is rooted in the concept of a particle–wave dualism.

20.3.3: The Axioms of Quantum Mechanics: Uncertainty Relation

Particle–wave dualism itself is a consequence of the axiomatic set of quantum mechanics. These axioms are, however, mathematically abstract. To develop the important concepts in a transparent manner, we omit this formalism and start instead with illustrating the most important of these axioms based on an experiment.

Werner Heisenberg motivated the **uncertainty relation** by studying an attempt to measure the position and momentum of a particle as precisely as possible under a microscope. When looking through the microscope we try to pin down the position of the particle to within an uncertainty of a small distance Δx. To do this, we must interfere with the particle. To be sure that it is within a distance interval Δx we need to use light of wavelength $\lambda \leq \Delta x$ for the observation. If the wavelength of the light is greater, then we cannot observe the particle. For example, a typical virus has a size of 100 nm or less. We cannot see them in a compound light microscope because visible light has wavelengths between 400 nm and 700 nm.

Based on de Broglie's arguments, light of wavelength λ carries a momentum p since $\lambda = h/p$, where h is Planck's constant. Thus, the momentum light carries into the system during observation is at least $p = h/\lambda$. Since we cannot know what fraction of this momentum is transferred to the particle we observe, we find that the product of uncertainty of the position of the particle after the observation, Δx, and the uncertainty of the momentum after the observation, Δp, is given by:

$$\Delta p\, \Delta x \geq \frac{h}{\lambda}\lambda = h. \qquad [20.20]$$

Uncertainty of this kind does not exist in the realm of classical physics because in classical physics it is implicitly assumed that $h = 0$. Instead, we have to accept that we cannot know precisely both the momentum (from which we can calculate the energy when the mass of the particle is known) and the position of a particle; the better we know the position, the greater is the uncertainty in momentum and energy, and vice versa.

KEY POINT

The uncertainty principle states that position and momentum of a particle cannot be measured precisely at the same time.

EXAMPLE 20.5

We apply Heisenberg's uncertainty principle to the hydrogen atom to determine both its radius and its ground state energy from quantum mechanical principles rather than Bohr's atomic model. Assume that two contributions play a role in the potential energy of the electron as illustrated in Fig. 20.12:

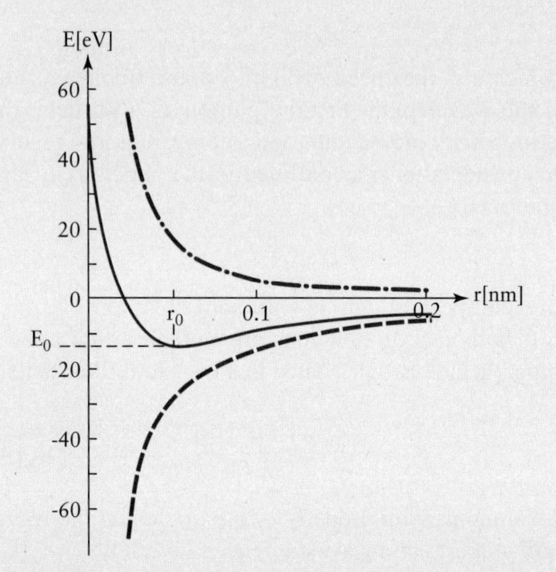

Figure 20.12 We apply Heisenberg's uncertainty principle to the hydrogen atom. Two contributions play a role in the potential energy of the electron: (i) the electrostatic attraction toward the nucleus (dashed curve) and (ii) the uncertainty of energy and position within the spherical space accessible to the electron, which leads to a repulsive contribution (dash-dotted line).

- the electrostatic attraction toward the nucleus:

$$E_{el} = -\frac{e^2}{4\pi\varepsilon_0 r}$$

- the uncertainty of energy and position within the spherical space accessible to the electron.

Solution

We start with Heisenberg's uncertainty for momentum and position. We set the position uncertainty Δr equal to the radius r of the accessible sphere, and the momentum uncertainty Δp equal to the total momentum p of the electron. This yields $\Delta p\, \Delta r = p\, r = h/(2\pi)$, and for the kinetic energy of the electron:

$$E_{kin} = \frac{p^2}{2m} = \frac{h^2}{8m\pi^2 r^2}.$$

The total energy of the electron is the combination of kinetic and electric potential energy:

$$E_{total} = E_{kin} + E_{el} = \frac{h^2}{8m\pi^2 r^2} - \frac{e^2}{4\pi\varepsilon_0 r}.$$

continued

Applying calculus, we can find the minimum of the total energy as a function of position by setting $dE_{\text{total}}/dr = 0$:

$$\frac{dE_{total}}{dr} = -\frac{h^2}{4m\pi^2 r^3} + \frac{e^2}{4\pi\varepsilon_0 r} = 0,$$

which yields:

$$r = \frac{h^2\varepsilon_0}{m\pi e^2}.$$

This radius is consistent with Eq. [20.13] for the ground state, $n = 1$, from Bohr's model.

20.3.4: Particles in Boxes and Atomic Orbitals

The remaining axioms of quantum mechanics are abstract: (I) A physical system is described by a function of state ψ. (II) A measurable physical quantity corresponds to a particular type of operator that can be applied to the function of state. (III) To obtain a precise value for a physical quantity in a system, the function of state must be a particular function, called eigenfunction of the operator, with the corresponding eigenvalue representing the measurable value of the quantity of interest. An example is the **Schrödinger equation**, first written by Erwin Schrödinger in 1926, which evaluates the energy operator (Hamilton operator H) for a system to identify the measurable **energy eigenvalues** E of a system:

$$H\,\psi = E\,\psi. \qquad [20.21]$$

For an atom, the Hamilton operator H contains terms due to the kinetic and electric potential energies. Solutions to this equation yield certain values of E, which are the energy values that yield stationary wave functions, i.e., functions ψ that are not changing with time. We will not follow this approach since it requires advanced mathematical concepts that are better presented in a dedicated course on quantum mechanics. Instead, we use the results of Eq. [20.21] for three key systems to illustrate its usefulness to conceptualize atomic physics.

20.3.4.1: Particle in a Heaviside Potential

A particle in a one-dimensional box is described as an electric system by a **Heaviside potential function**, as shown in Fig. 20.13:

$$V = V_0 \; for{:}\; -x_0 \le x \le x_0$$

$$V = \infty \; for{:}\; x < -x_0 \text{ and } x > x_0.$$

This is the simplest possible potential form and is therefore often used to test calculations with

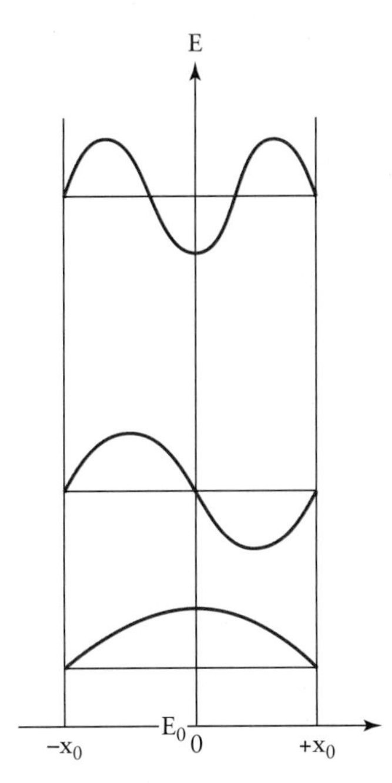

Figure 20.13 A particle in a box is described by a Heaviside potential function. This is the simplest possible potential form. Further shown are three wave functions for the stationary solutions of Schrödinger's equation with the lowest allowed energy values. Compare with the standing waves in the three lowest harmonics in a closed tube in Chapter 15.

potentials. The figure shows further three wave functions for the stationary solutions of Schrödinger's equation with the lowest allowed energy values. We include this case because it shows nicely how the quantum mechanics formalism relates to the concept of standing waves in a confined space. Note that the same standing waves result for the three lowest sound harmonics in a closed tube (see Chapter 15).

The figure also allows us to discuss an additional feature of quantum mechanical wave functions. Look at the second harmonic. It includes positive wave function values for the left half of the potential well and negative values for the right half. This is easy to interpret if the wave function is an acoustic standing wave: a negative value means a displacement of an air pocket to the left and a positive value respectively to the right. However, if the wave function represents an electron, it is not obvious what a negative versus a positive value is supposed to mean. It turns out they don't have a physical meaning. The wave function in quantum mechanics is only a mathematical construct. Physical meaning has the probability function that is obtained from the wave function by an operation that we can compare to squaring of a number: the probability function represents the probability for the particle to be at a given position in space. Like for square numbers, only positive values and zero occur for the probability function. We can motivate the step from wave function to probability function by studying once more Fig. 20.10: when comparing waves and particles in that case, we used the intensity, not the amplitude of waves, with the intensity proportional to the square of the amplitude. We use probability functions to define orbitals in the atom below.

CASE STUDY 20.3

The two upper wave functions in Fig. 20.13 yield points where their probability functions are zero (nodes). How can these probability functions represent a particle?

Answer: *After obtaining the probability function, we indeed still find a zero probability (node) at the centre of the potential well for the second lowest wave function in Fig. 20.13. If the particle is therefore never at the centre, how does it ever get from the left to the right side, or vice versa? This is a typical case in which the classical approach to the electron as a particle fails. Energy in waves travels through nodes all the time, as seen for example for higher standing sound waves in closed tubes.*

The respective energy levels in Fig. 20.13 are drawn to scale, indicating that a Heaviside potential leads to increasing energy gaps for higher levels. Fig. 20.14 illustrates that this characterizes the applied potential. In the figure, the Heaviside potential (V = const) is compared to the elastic potential ($V \propto r^2$), which has equidistant allowed energy levels, and to the Coulomb potential ($V \propto 1/r$), which has energy level steps narrowing at higher energies. All three potential functions play an important role in nature at the atomic and subatomic levels: the Heaviside potential is a good approximation for the **nuclear potential** we discuss in Chapter 23, the elastic potential describes molecular vibrations (chemical bonds) in their ground state, and the Coulomb potential applies to the electron in the hydrogen atom.

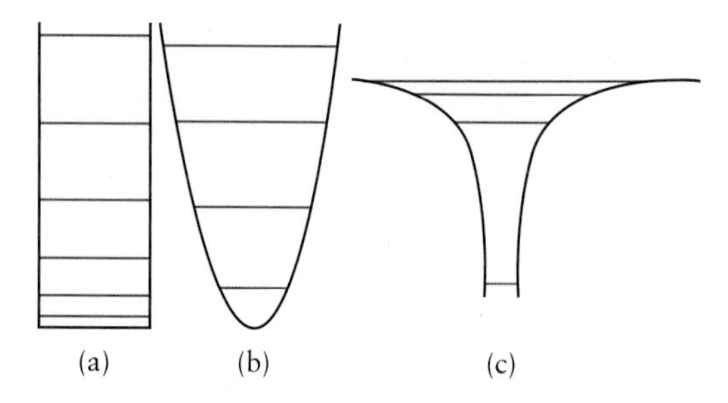

Figure 20.14 The energy distance between neighbouring orbitals in the Heaviside potential (V = const), the elastic potential ($V \propto r^2$), and the Coulomb potential ($V \propto 1/r$).

20.3.4.2: Particle in a Harmonic Oscillator Potential

Fig. 20.15 compares the classical and quantum mechanical probability to find an **object attached to a spring** along the x-axis in a one-dimensional vibrating system. The quantum mechanical values (solid lines) result from calculating the stationary wave function solutions for the system. The classical values (dashed lines) can be derived for a point-like object attached to a horizontal spring with frictionless motion. Shown are in particular the five lowest stationary solutions with their respective classical case, with the lowest allowed total energy of the system shown at bottom (ground state). We note that the classical and quantum mechanical solutions asymptotically approach each other for increasing quantum numbers. All but the lowest harmonics in the quantum mechanical case display the nodes we discussed for the Heaviside potential.

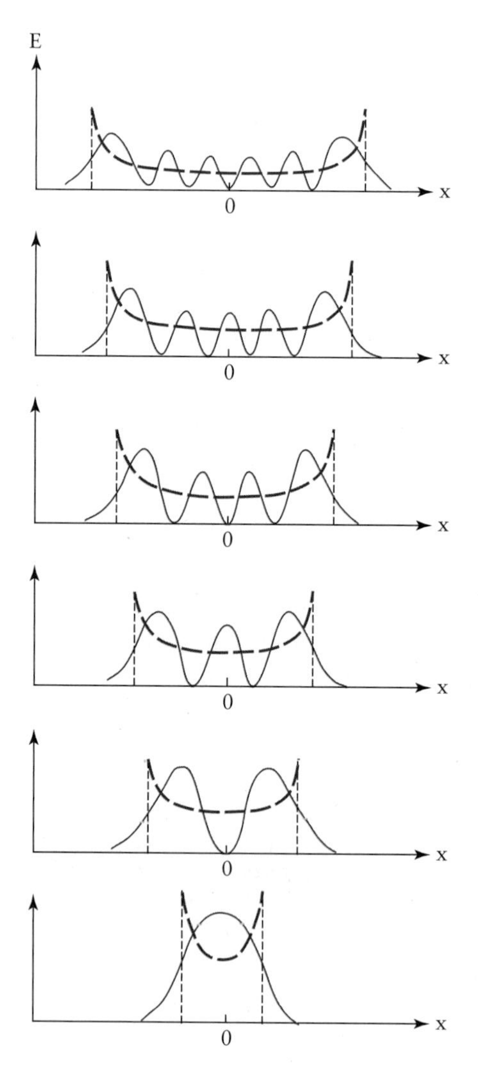

Figure 20.15 Comparison of the classical and quantum mechanical probability to find a particle attached to a spring along the x-axis in a one-dimensional vibrating system. The quantum mechanical values (solid lines) result from calculating the stationary wave function solutions for the system. The classical values (dashed lines) can be derived for a point-like mass attached to a horizontal spring with frictionless motion.

The most intriguing result in quantum mechanics is the non-zero probability for the particle to move beyond the position where its potential energy exceeds its total energy. Classically this would require a negative kinetic energy and is impossible. The uncertainty relation allows this penetration to a point that lies within the uncertainty of the measurable values; i.e., if Δx is small the energy constraint can be violated to a notable degree. The phenomenon can be observed experimentally when the forbidden energy barrier is thin enough for the tail of the probability function to reach into an adjacent allowed zone. This is schematically illustrated in Fig. 20.16: the wave function decays exponentially where the particle cannot be found classically. In the figure there is a finite probability for the particle to escape through the wall, a phenomenon called **tunnelling**. We need this process later when explaining observations in nuclear physics.

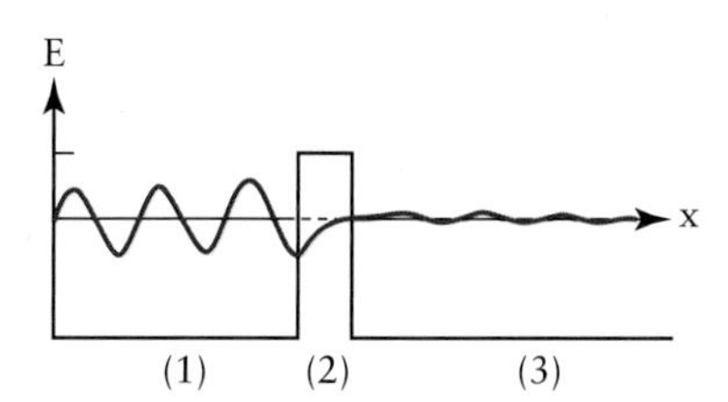

Figure 20.16 A wave function decays exponentially where the particle cannot be found classically; i.e., its kinetic energy would have to be negative because its potential energy exceeds its total energy. Quantum mechanically, there is a finite probability for the particle to escape through such an energy barrier (wall). This phenomenon is called tunnelling.

20.3.4.3: Electron in a Three-Dimensional Coulomb Potential

The stationary solutions to Schrödinger's equation require not one but three quantum numbers in this case, as Bohr and Sommerfeld had already predicted based on the dimensionality of the system. Thus, Schrödinger's solutions for the orbitals of a hydrogen atom are not all of spherical symmetry. The three independent quantum numbers needed to characterize all solutions are

(I) The **principal quantum number** n, which is due to a quantization of the orbitals with certain total energy values for the electron. This quantum number has already been discussed above in the context of Bohr's original model.

(II) The **orbital quantum number** l, which is due to the quantization of the angular momentum of the electron. This means that different electrons can move in orbitals with discrete values of eccentricity.

(III) The **magnetic quantum number** m_l, which is due to a quantization of the orientation of the orbitals of the electron in an external magnetic field. This quantum number is called the magnetic quantum number because a non-uniform magnetic field allows us to split an atomic beam into as many components as there are allowed values for the magnetic quantum number.

Later a fourth quantum number was added to make the model consistent with the periodic system of elements. This quantum number is called **spin** and allows for two values (up or down). The classical concept behind the spin is a rotation of the electron about its own axis. This is another possible motion of the electron, similar to the rotation of Earth about its own axis while it revolves around the Sun. Two possible values of the spin correspond in this classical picture to the possible East–West or West–East rotation of a planet. Note that such classical descriptions for the different quantum numbers have to be used cautiously.

The orbitals differ in shape when the values of the quantum numbers are different. To distinguish the many possibilities, the orbitals have been given names, which were taken from earlier spectroscopic observations of electron transitions between different orbitals. Table 20.1 summarizes the possible combinations of quantum numbers, which follow two selection rules:

TABLE 20.1

Quantum numbers and orbitals

n	*l*	*ml*	Name
1	0	0	1s
2	0	0	2s
2	1	−1, 0, +1	2p
3	0	0	3s
3	1	−1, 0, +1	3p
3	2	−2, −1, 0, +1, +2	3d
4	0	0	4s
4	1	−1, 0, +1	4p
4	2	−2, −1, 0, +1, +2	4d
4	3	−3, −2, −1, 0, +1, +2, +3	4f

The quantum numbers are n, the principal quantum number; l, the orbital quantum number; and m_l, the magnetic quantum number. The name of an orbital combines the number of the principal quantum number and a letter representing the orbital quantum number: s = sharp, p = principal, d = diffuse, f = fundamental. The magnetic quantum number is identified as an index:

$$\begin{aligned} &\text{(I)} \quad 0 \le l \le n-1 \\ &\text{(II)} \quad -l \le m_l \le +l. \end{aligned} \qquad [20.22]$$

These orbitals can be illustrated in various ways:

(I) Fig. 20.17 illustrates two-dimensional graphs that show the probability of finding the electron as a function of distance r from the nucleus. The figure includes plots for six of the orbitals in Table 20.1.

(II) Fig. 20.18 is an artist's sketch of the 99% surfaces in three-dimensional display. This form of representation is useful when trying to envisage processes in stereo-chemistry.

Once the orbitals of the atom are defined, quantum theory allows us to predict the order in which these orbitals are filled with electrons. For this, the orbitals are

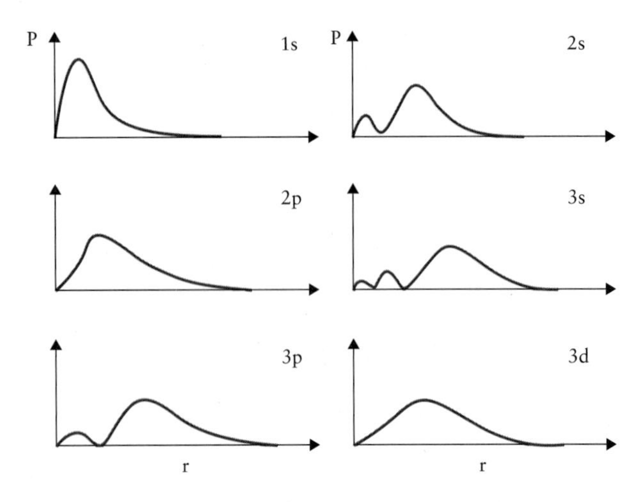

Figure 20.17 The probability to find an electron as a function of distance *r* from the nucleus in various orbitals of an atom. Note that only the s-orbitals are spherically symmetric.

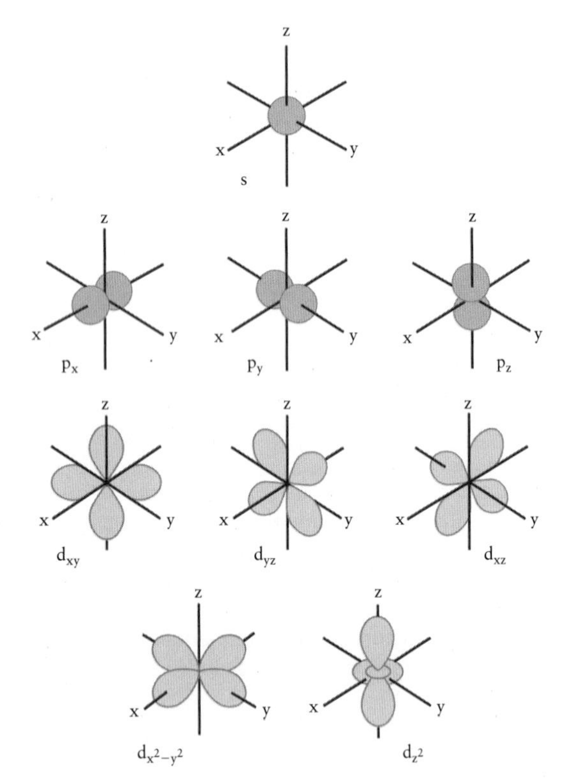

Figure 20.18 Artist's sketch of the three-dimensional orbitals of an atom, representing the surfaces within which the electron is found 99% of the time.

first ordered along the energy axis based on their energy eigenvalues. This order is then called a **term scheme**. The filling process is governed by two principles:

(I) The **minimum energy principle** states that any additional electron must occupy the free orbital with the least total energy.

(II) **Pauli's principle** (named for Wolfgang Pauli) states that no two electrons in the same atom can have the same four quantum numbers (including spin).

Based on Pauli's principle, each atomic orbital in Table 20.1 can accommodate two electrons due to the existence of two possible values for the spin quantum number. Thus, up to 2 electrons can be placed in each s-orbital, up to 6 electrons in p-orbitals of a given quantum number *n*, up to 10 electrons in d-orbitals, and up to 14 electrons in f-orbitals.

20.4: Appendix: Relativistic Speed, Energy, and Momentum of an Electron

In 1905, Albert Einstein addressed a discrepancy between Newton's classical mechanics and Maxwell's equations of electromagnetism. Maxwell's equations identify the speed of light in vacuum as a unique parameter that governs the speed of propagation of any electromagnetic wave. On the other side, Newton's laws do not contain any reference to this speed; if an object is accelerated its speed should eventually exceed the speed of light.

Einstein noted that the discrepancy is associated with the fact that Newton's laws apply only at speeds well below the speed of light and that corrections are required at higher speeds. He built his theory of relativity on this and similar observations by stating two principles:

(I) **Relativity principle:** Absolute values of velocity cannot be measured. The laws of nature apply equally in all inertial systems, i.e., systems that move relative to each other with constant speed along straight lines.

(II) The **speed of light** in vacuum is independent of the speed of the light source. It is $c = 3.0 \times 10^8$ m/s, regardless of direction.

These two principles have extensive ramifications as they affect our view of space and time: length scales shorten when moving relative to the observer (Lorentz contraction), and time scales slow down when observed in a system moving relative to the observer (time dilatation).

Einstein's relativity principle has further consequences on physical parameters describing a moving system. For example, the mass of an object that moves with speed *v* has increased in its own inertial system relative to an observer at rest. Thus, a fast electron beam (e.g., in an X-ray tube) is stiffer, i.e., harder to collimate or deflect, than a beam at lower speed because the electrons appear to be heavier.

Due to this effect, Einstein also reinterpreted the conservation of energy law. In order to remain correct, mass and energy must be two aspects of the same quantity. The equivalence of mass and energy is expressed in the form $E = m c^2$, in which *m* is the actual mass of the system in its own inertial system. If we call m_0 the mass at rest, then the kinetic energy of the system is the increase of the energy due to the energy–mass equivalence, i.e., the energy due to motion on top of the energy at rest:

$$E_{kin} = E - m_0 c^2 = m_0 c^2 \left[\frac{1}{\sqrt{1 - \frac{v^2}{c^2}}} - 1 \right], \quad [20.23]$$

in which v is the speed with which the system moves relative to the observer (reference system at rest). This formula is used to calculate the speed of an object with a given kinetic energy. This affects Example 20.6 for electrons at higher energy.

EXAMPLE 20.6

What is the speed of an electron with kinetic energy E = 100 keV?

Solution

First, Eq. [20.23] is rearranged for v. We multiply both sides of the equation with the energy at rest, then add 1 on both sides:

$$\frac{E_{kin}}{m_0 c^2} + 1 = \frac{1}{\sqrt{1 - \frac{v^2}{c^2}}}.$$

Next, the equation is inverted and squared:

$$\left[\frac{E_{kin}}{m_0 c^2} + 1 \right]^{-2} = 1 - \frac{v^2}{c^2}.$$

This yields:

$$\frac{v}{c} = \sqrt{1 - \left(\frac{m_0 c^2}{E_{kin} + m_0 c^2} \right)^2}.$$

We substitute the kinetic energy with a value of 100 keV = 1.6×10^{-14} J. With the mass at rest for an electron and the speed of light in vacuum we find $v/c = 0.548$; i.e., the electron moves with a speed just above 50% of the speed of light.

Table 20.2 compares for electrons with kinetic energies between 1 keV and 1 MeV the ratio of the speed of an electron to the speed of light, v/c, as calculated with the classical formula, $E_{kin} = ½\ m\ v^2$, and as calculated with the relativistic v/c formula. Discrepancies are notable for energies above 10 keV; i.e., for these kinetic energies of the electron a relativistic correction is needed.

TABLE 20.2

Classical and relativistic speeds of electrons with kinetic energies between 1 keV and 1 MeV

Kinetic energy	$(v/c)_{classical}$	$(v/c)_{relativistic}$
1 keV = 1.6×10^{-16} J	0.063	0.062
5 keV = 8.0×10^{-16} J	0.140	0.138
10 keV = 1.6×10^{-15} J	0.198	0.195
50 keV = 8.0×10^{-15} J	0.442	0.412
100 keV = 1.6×10^{-14} J	0.625	0.548
500 keV = 8.0×10^{-14} J	1.398	0.863
1 MeV = 1.6×10^{-13} J	1.977	0.941

In classical physics, momentum was introduced as the mass times the velocity of an object. This formula has to be corrected for the changing mass as the object approaches the speed of light:

$$p = mv = \frac{m_0 v}{\sqrt{1 - \frac{v^2}{c^2}}},$$

in which v is again the speed with which the system moves relative to the observer (reference system at rest). This equation can be combined with Eq. [20.23] to express the energy of a fast moving object as a function of its momentum (and its mass at rest):

$$E = \sqrt{m_0^2 c^4 + c^2 p^2}.$$

For an object with rest mass $m_0 = 0$, such as a photon, this equation simplifies to $E = c\ p$, which we used in section 20.3.1 to quantify de Broglie's idea of the wave–particle dualism.

SUMMARY

DEFINITIONS

- Uniform circular motion:
  - tangential velocity:

$$v_{\parallel} = \frac{2\pi r}{T}$$

  - perpendicular velocity: $v_{\perp} = 0$
  - tangential acceleration: $a_{\parallel} = 0$
  - centripetal acceleration:

$$a_{\perp} = \frac{v^2}{r} = \omega^2 r = \frac{4\pi^2 r}{T^2}$$

- Bohr radius of the hydrogen atom:

$$r_{Bohr} = \frac{\varepsilon_0 h^2}{m e^2 \pi} = 5.29 \times 10^{-11}\ \text{m}$$

- Rydberg constant:

$$R_H = \frac{m e^4}{8 \varepsilon_0^2 h^3} = 3.29 \times 10^{15}\ \text{Hz}$$

e is elementary charge, m mass of electron.

LAWS

- Newton's law for a system with uniform circular motion:
  - perpendicular to path:

$$\sum_i F_{i,\perp} = m \frac{v^2}{r}$$

- parallel to path:

$$\sum_i F_{i,\parallel} = 0$$

- perpendicular to plane of motion:

$$\sum_i F_{i,z} = 0$$

- Bohr's postulates
  - Electrons do not lose energy while they are in certain orbits.
  - Electrons lose or gain energy when they transfers between two allowed orbits:

$$\Delta \mathrm{E} = \mathrm{E_f} - \mathrm{E_i} = \mathrm{h}\, f$$

h is Planck's constant, f is frequency of photon absorbed or emitted.

- Bohr's quantum condition for the atom:

$$m\ v\ r = \frac{nh}{2\pi} = n\,\hbar$$

with $\hbar = h/(2\ \pi)$ and n the quantum number (integer).

- De Broglie's wavelength of a particle:

$$\lambda = \frac{h}{p} = \frac{h}{mv}$$

p is momentum, m mass, v speed of the particle.

- Energy levels of the hydrogen atom (ionization energy level is at $E = 0$):

$$E_{\text{total}} = -\frac{me^4}{8\ \varepsilon_0^2\ h^2}\frac{1}{n^2}$$

- Electronic transitions in the hydrogen atom:
  - written as an energy:

$$h\,f = \Delta E_{\text{total}} = \frac{me^4}{8\ \varepsilon_0^2\ h^2}\left[\frac{1}{n_i^2} - \frac{1}{n_f^2}\right]$$

  - written as a frequency:

$$f = R_H \left|\left[\frac{1}{n_i^2} - \frac{1}{n_f^2}\right]\right|$$

with R_H the Rydberg constant.

- Selection rules for atomic orbitals:

$$\text{(I)}\quad 0 \le l \le n - 1$$

$$\text{(II)}\quad -l \le m_l \le +1$$

n is principal quantum number, l is orbital quantum number, m_l is magnetic quantum number. The spin quantum number m_s can take two values: +½, −½.

- Order of occupying atomic orbitals:
  - minimum energy principle: an electron added to an atom occupies a free orbital with the least total energy.
  - Pauli principle: no two electrons in the same atom may have the same four quantum numbers.
- Uncertainty relation:

$$\Delta p\ \Delta x \ge \frac{h}{\lambda}\lambda = h$$

with h is Planck's constant, Δx is the uncertainty of the position of the particle after observation, and Δp is the uncertainty of the momentum after the observation.

- Relativistic energy of a photon:

$$E_{\text{photon}} = c\,p$$

with p the momentum of the photon and c the speed of light in a vacuum.

MULTIPLE-CHOICE QUESTIONS

MC–20.1. What is wrong with Rutherford's atomic model (see Fig. 20.4), which is based on an analogy to the planetary system?

(a) The planetary system has eight planets, but the atom may have more or fewer electrons.
(b) The planetary system formed after a supernova ejected large amounts of gas/dust into space. No such event precedes the formation of an atom.
(c) The planets are not charged electrically like the electron.
(d) An electron orbiting a positive charge must lose energy via electromagnetic radiation.
(e) The International Astronomic Society can decide how many objects in the solar system qualify as planets, but it has no jurisdiction to decide how many electrons are present in the shell of an atom.

MC–20.2. Why did Thomson postulate a diffuse positive charge as the background for oscillating electrons in his atomic model?

(a) because he needed to define the size of the atom independently
(b) because such diffuse positive charges had been observed experimentally
(c) because he needed the atom to be an ion
(d) because the electron wouldn't oscillate without it
(e) because he needed this so that his model would match the spectroscopic data available at the time

MC–20.3. Bohr postulated that an electron in a particular orbit does not radiate. What consequence does this assumption have?

(a) The linear momentum of the electron is conserved.
(b) The total energy of the electron is conserved.
(c) The kinetic energy of the electron is conserved.
(d) The angular momentum of the electron varies only with time.

MC–20.4. De Broglie's wavelength of a subatomic particle depends on its kinetic energy E as
(a) $\lambda \propto E$.
(b) $\lambda \propto E^{1/2}$.
(c) $\lambda \propto E^{-1/2}$.
(d) $\lambda \propto E^{2}$.
(e) $\lambda \propto E^{-2}$.
(f) $\lambda \propto E^{-1}$.

MC–20.5. Niels Bohr calculated the radius of the smallest stable electron orbit in the hydrogen atom. It is called the Bohr radius and depends on the following variable.
(a) the kinetic energy of the electron
(b) the total energy of the electron
(c) the wavelength of the electron
(d) the speed of the electron
(e) none of the above

MC–20.6. Bohr used Newton's second law in the form $F_{net} = m\,v^2/r$, not in the form $F_{net} = m\,a$ (a is acceleration, m is mass, v is speed, and r is radius). When is Newton's second law used in this form?
(a) when objects move with constant velocity
(b) when objects move along circular paths
(c) when the total energy is conserved
(d) in cases of uniform circular motion
(e) when the system consists of two interacting objects

MC–20.7. When an electron makes a transition from $n = 1$ to $n = 2$ in a hydrogen atom, which of the following is a correct statement about this transition?
(a) The electron emits UV radiation.
(b) The electron releases thermal energy.
(c) The electron transfers to an orbit that lies closer to the nucleus.
(d) The nucleus must pick up the difference in angular momentum and spin faster.
(e) none of the above

MC–20.8. On which variable does the frequency of radiation emitted from a hydrogen atom depend linearly?
(a) the quantum number of the initial orbit
(b) the quantum number of the final orbit
(c) the Rydberg constant
(d) the wavelength of the emitted radiation in a vacuum
(e) none of the above

MC–20.9. What does a transition to $n = \infty$ (infinity) imply?
(a) that the atom explodes (atomic bomb)
(b) that the electron becomes electrically neutral
(c) that the atom loses an (almost) infinite amount of energy (big bang)
(d) that the electron moves to an orbit that is larger than the radius of the universe
(e) that the atom becomes an ion

MC–20.10. When a hydrogen atom absorbs a photon of energy $h\,f$, the kinetic energy of the electron that transfers to an excited state changes by what factor?
(a) zero
(b) $\frac{1}{2}\,h\,f$
(c) $h\,f$
(d) $-2\,h\,f$
(e) $-h\,f$

MC–20.11. When a hydrogen atom absorbs a photon of energy $h \cdot f$, the potential energy of the electron that transfers to an excited state changes by what factor?
(a) zero
(b) $-\frac{1}{2}\,h\,f$
(c) $-h\,f$
(d) $-2\,h\,f$
(e) $+2\,h\,f$

MC–20.12. Krypton has atomic number 36. How many electrons does this noble gas hold in its next-to-outer shell, i.e., in the orbit with $n = 3$?
(a) 2
(b) 4
(c) 8
(d) 18
(e) none of the above

MC–20.13. Lithium, sodium, and potassium display similar chemical properties. Which of the following atoms should vary chemically the most from the others? (Consult the periodic table.)
(a) chlorine
(b) oxygen
(c) bromine
(d) fluorine

CONCEPTUAL QUESTIONS

Q–20.1. Once a helicopter takes off from its landing platform, one would expect that the non-zero friction between the rotor axis and its bearings would cause the helicopter cabin to spin out of control. Look at a photograph of a helicopter and determine how this effect is prevented.

Q–20.2. Cats are very agile animals; a cat will usually land on its feet following a fall from any position. If you have ever seen a cat fall out of a tree, you may have noticed that the animal's upper body twists one way while its lower part rotates the other way. Why does this counter-rotation occur?

Q–20.3. Why do figure skaters pull their arms close to their chests when they want to spin fast, e.g., during a fast spin or a triple jump?

Q–20.4. A figure skater finishes a performance with a fast spin, then suddenly stops the rotation and bows to the audience. In the process, the angular momentum of the athlete changed from a finite value to zero. Did the athlete violate the conservation of angular momentum?

Q–20.5. The quantum number n can increase to infinity in Bohr's hydrogen atom. Does this mean that the possible frequencies of its spectral lines also increase without limit?

Q–20.6. Can a hydrogen atom in the ground state absorb a photon of energy less than 13.6 eV? Can it absorb a photon of energy greater than 13.6 eV?

Q–20.7. Must an atom first be ionized to emit light?

Q–20.8. The ionization energies for Na, K, and Rb are 5.14 eV, 4.34 eV, and 4.18 eV, respectively. Why are the values decreasing in this order?

ANALYTICAL PROBLEMS

P–20.1. Calculate the angular momentum of Earth as it moves around the Sun.

P–20.2. Halley's Comet moves around the Sun along an elliptical path. Its closest point to the Sun is 0.59 A.U., and its farthest point is 35 A.U. (A.U. is the astronomical unit: 1 A.U. = distance Earth to Sun). What is the speed of Halley's Comet at its farthest point from the Sun if it moves at 54 km/s through the closest point? Neglect any changes in the comet's mass while it orbits.

P–20.3. Fig. 20.19 shows four point-like objects that are connected by massless strings and rotate about the origin with an angular speed of two revolutions per second. Use $L = 1.0$ m. If we shorten their distances to the origin to 0.5 m by pulling the strings shorter, what is their new angular speed? Can you use the result to draw conclusions about the rotation of a collapsing star?

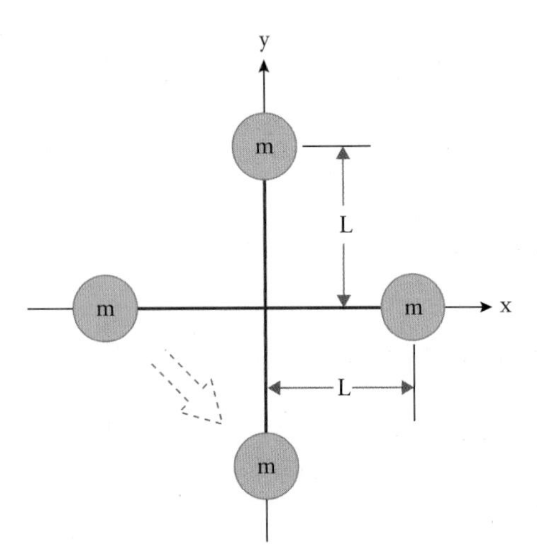

Figure 20.19

P–20.4. In the spectrum of helium ions, a series of absorption lines exists (Pickering series) for which every other line coincides with a Balmer series line of the hydrogen atom (the remaining lines fall between the Balmer series lines). Which transitions in the helium ion are responsible for the Pickering series?

P–20.5. A hydrogen atom is in its first excited state ($n = 2$). Using Bohr's atomic model, calculate (a) the radius of the electron's orbit, (b) the potential energy of the electron, and (c) the total energy of the electron.

P–20.6. The size of Rutherford's atom is about 0.1 nm. (a) Calculate the attractive electrostatic force between an electron and a proton at that distance. (b) Calculate the electrostatic potential energy of that atom. Express the result in unit eV. (c) The size of Rutherford's atomic nucleus is about 1 fm. Calculate the repulsive electrostatic force between two protons at that distance. (d) Calculate the electrostatic potential energy of a pair of protons in such a nucleus. Express the result in unit MeV.

P–20.7. Calculate the electric force on the electron in the ground state of the hydrogen atom.

P–20.8. What is the wavelength of light that can cause a transition of an electron in the hydrogen atom from the orbit with $n = 3$ to $n = 5$?

P–20.9. A hydrogen atom emits a photon of wavelength $\lambda =$ 656 nm. Which transition did the hydrogen atom undergo to emit this photon?

P–20.10. Calculate the wavelength of an electron in a hydrogen atom that is in the orbit with $n = 3$.

P–20.11. The size of Rutherford's atom is about 0.1 nm. (a) Calculate the speed of an electron that moves around a proton, based on their electrostatic attraction, if they are separated by 0.1 nm. (b) Calculate the corresponding de Broglie wavelength for Rutherford's electron.

P–20.12. Show that the speed of the electron in the n-th orbit of a hydrogen atom in Bohr's model is

$$v = \frac{1}{4\pi\varepsilon_0} \frac{e^2}{nh}$$

P–20.13. How much energy is required to ionize hydrogen when it is in the state with $n = 3$?

P–20.14. Calculate for Bohr's atomic model the speed of the orbiting electron in the ground state.

ANSWERS TO CONCEPT QUESTIONS

Concept Question 20.1: (A). The centripetal acceleration is larger for an object moving with a shorter period, and it is larger for an object at a greater distance from the centre of the path, as shown in Eq. [20.2]. It also increases with the angular frequency in Eq. [20.4]. The frequency is defined as $f = 1/T = \omega/(2\pi)$ with unit 1/s = Hz (Hertz), thus the same dependence on the frequency applies.

Concept Question 20.2: The vectors are related by the right-hand rule. If we represent their lengths proportional to the magnitudes F and r, then does the area enclosed by vectors $\vec{F}$ and $\vec{r}$ represent the magnitude of the torque, τ.

Concept Question 20.3: (B) to (D). Only the mass of the mobile object in uniform circular motion enters Newton's second law. Both charges are present in Bohr's formula in the form of the term e^2. Coulomb's law contains both charges symmetrically; i.e., a switching of the charge between the two particles makes no difference. In the particular case, an anti-hydrogen antimatter atom would result.

Concept Question 20.4: For electrons 1 and 2, the horizontal line for the total energy and the solid curves for the potential energy cross. At distances greater than this crossover point the electron would require a negative kinetic energy because the potential energy exceeds the total energy. This is not possible, and the electron remains confined to positions between the two crossover points. In turn, electron 3 has a positive kinetic energy at any distance from the nucleus.

Concept Question 20.5: The visible range of the electromagnetic spectrum lies between 380 nm and 750 nm. Several lines of the Balmer series, which are transitions to the second lowest state in the hydrogen atom with $n = 2$, fall within this interval. These Balmer series lines are listed in Table 20.3.

TABLE 20.3

Wavelengths of the lowest four transitions in the Balmer series of the hydrogen atom

$n_i \rightarrow n_f$	Wavelength
$3 \rightarrow 2$	$\lambda = 655$ nm
$4 \rightarrow 2$	$\lambda = 485$ nm
$5 \rightarrow 2$	$\lambda = 433$ nm
$6 \rightarrow 2$	$\lambda = 409$ nm

CHAPTER 21

Magnetism and Electromagnetic Waves

Magnetism is closely related to electricity: the magnetic force is caused by and acts on electric currents in the same manner as Coulomb's law characterizes the interactions between stationary electric charges. We first exploit this analogy to develop the basic properties of magnetism, starting with the magnetic field of a current-carrying conductor. Like electric force and field, their magnetic equivalents are vector properties; the situation is slightly more complicated though because the charge in Coulomb's law is a scalar, but is now replaced by the vector current density (or the current combined with a unit vector that points in the direction of the motion of positive charges).

In a second view, the conductor is removed and charged particles are allowed to travel individually in space; i.e., the current density is replaced by the charge and the velocity vector of the moving particle. The vector character of the velocity, an externally acting magnetic field and the magnetic force on the moving point charge, leads to a uniform circular motion of such particles; the cyclotron frequency of this motion is governed by the mass and the speed of the particle and the magnitude of the magnetic field. Mass spectrometry is an analytical technique in organic chemistry and biochemistry that exploits this motion.

With magnetism introduced, Faraday's and Maxwell's reasoning is discussed that the close relation with electricity allows us to combine both phenomena in a single formalism. Faraday demonstrated that Ampère and Oersted's observations of an electric current causing a magnetic field can be supplemented with an opposing effect, i.e., that a changing magnetic field causes an electric current. Maxwell extended this symmetry by showing that Ampère's law, like Faraday's law of induction, is not limited to a conductor, but that a changing electric field can cause a magnetic field in empty space. This allows us then to write four equations that correctly predict all phenomena due to electric and magnetic fields, including Maxwell's new conclusion that these equations of electromagnetism apply to visible light (and that optics is therefore just an application of electromagnetism).

Maxwell's equations in empty space lead to a wave equation, with the wave propagation speed the speed of light. Thus, the speed of light in vacuum is a universal constant, and its presence in Maxwell's equations sets the electromagnetic theory apart from Newton's laws of mechanics in which no distinct speed occurs. This was immediately realized when Maxwell published his results in 1865; it took until 1905 to resolve this discrepancy, when Einstein's theory of relativity introduced corrections to Newton's laws at speeds close to the speed of light.

Medical diagnostics evolved rapidly during the last third of the twentieth century toward non-invasive techniques. This development is driven by four objectives: (i) to improve the confidence level of the diagnosis, (ii) to minimize the pain and suffering for the patient, (iii) to reduce the risks due to complications associated with the diagnostic procedure, and (iv) to contain the costs to the health care system. The traditional diagnostic methods, for example surgical removal of tissue in a biopsy, define a reference standard against which progress is measured. For example, traditional diagnostics may include anaesthetics and wound healing (due to surgery), which prevent the patient from returning to everyday life for some period; a biopsy-related cut represents an infection risk (portal of entry for pathogens), which is further enhanced when the patient has to be hospitalized; and professional fees, hospital fees and/or cost of out-patient care, and time lost at work define the economic impact. Biopsy was still a major method in classical diagnostics and remains important because it allows for effective laboratory tests that cannot be performed on the patient, thus significantly increasing the confidence in a final diagnosis.

Radiology and nuclear medicine-based imaging methods, which we discuss later in the textbook, eliminate the need of hospitalization and minimize (in angiography) or eliminate the need to cut through the patient's skin. These imaging modalities provide a high level of confidence in diagnosing a wide range of pathological changes, and are cost-effective, despite the high price tag of the imaging facility, because of a reduction in associated physician hours and an enhanced patient throughput. However, there are new risks caused by the required radiation energy and dose that have to be weighted against the risks of traditional methods.

Combining the benefits of radiological methods with an elimination of radiation risks is the main driving force behind the development of modalities that operate with low energy radiation. Both ultrasound methods and magnetic resonance imaging (MRI) represent no radiation risk to the patient, as seen by the routine use of ultrasound in obstetrics to image the unborn child.

There are several fundamental issues we need to address before we can interpret MRI images, such as Fig. 21.1. We start with the first letter in the abbreviation MRI: magnetism. When magnetism and its connection to electric phenomena are established, two more steps are needed. In Chapter 27, we introduce **nuclear magnetic resonance** (NMR), an analytical method extensively used in chemistry. The principles of NMR are used in MRI for imaging applications. As the technical term implies, NMR is based on (i) the properties of atomic nuclei, (ii) the use of magnetic fields, and (iii) a resonance phenomenon. Thus, the basic concepts required to understand NMR include resonances as a wave phenomenon from Chapter 15, magnetism from this chapter, and nuclear properties from Chapter 24.

The step from NMR to MRI requires us to identify the medically relevant information in an NMR signal. For example, when MRI is used to locate tumours in the body of a patient, the NMR signal from malignant tissue must differ from the signal from healthy tissue. This discussion takes place in Chapter 27.

Fig. 21.1 shows a magnetic resonance image (MRI) of the human brain. While MRI is used to detect malignant tumours and sports-related injuries, it is its ability to obtain contrast between the various tissues in the brain that illustrates the unique potential of this imaging technique. MRI is sensitive to small changes in local blood flow in the brain; increases in blood flow occur when localized areas of the brain become active and require an increased amount of oxygen. Thus, MRI allows us to perform non-invasive studies of mental activity.

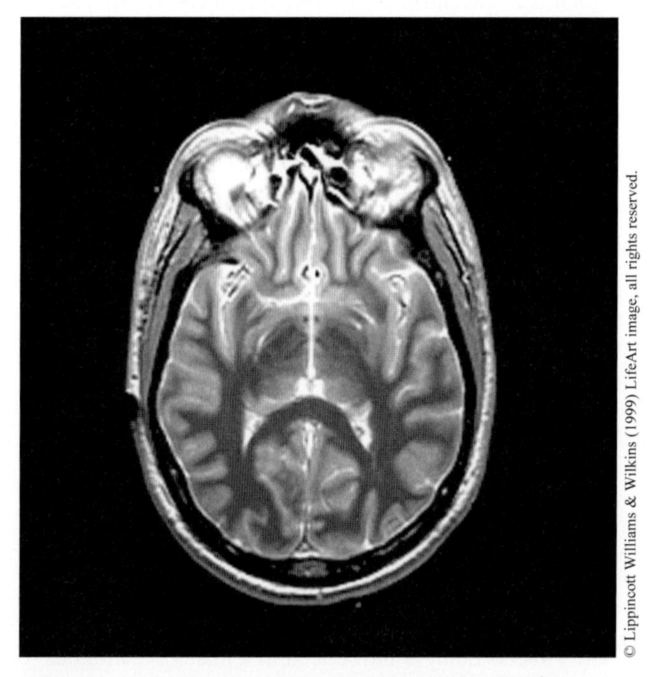

Figure 21.1 Spin–density weighted MRI scan of the human head.

MRI is the method of choice when we seek scientific answers to the most fundamental questions about ourselves: how do I see what I see? how do I hear what I hear? and, ultimately how do I think? It was René Descartes' statement "cogito ergo sum" (I think, therefore I am) that marks the beginning of the modern era of scientific thought, implying that thinking is the most basic process associated with our identity. With MRI we are now closer than ever to understanding what constitutes a thought at the physical and chemical level.

21.1: Magnetic Force and Field Due to an Electric Current

The introduction of **magnetism** is divided into two parts in this chapter: First, we develop its fundamental properties, following the same approach we took when we introduced electricity to highlight the close relation of both phenomena: magnetism is caused by electric currents in the same way electricity is caused by electric charges. This discussion is based on a current in a conductor. We then expand on the applicability of magnetism by moving in two directions beyond the concept of electric current as confined to a conductor. First, we model the motion of individual charged particles as a current outside a conductor and identify their interaction with an external magnetic field. This allows us to generalize magnetism to physiological systems such as nerve and muscle tissues, but also allows us to discuss mass spectrometry as an analytical method in organic chemistry.

In a second step we then no longer think of moving charges as the origin of a current but as the origin of a variable electric field (like in an antenna). The interaction of variable electric and magnetic fields leads to Maxwell's equations, which in turn lead to electromagnetic waves. This step is important in explaining the nature of electromagnetic waves (such as light in optics), but is also needed for our later discussion of the coupling of electromagnetic waves with nuclear spins in MRI.

What may appear unusual about our approach to magnetism is that we do not start with well-known magnetic materials such as permanent iron magnets. This has a very fundamental reason: in analogy to the electric system of an ionic salt, one would expect a magnetic material to contain magnetic charges (which we would call **magnetic monopoles**). However, magnetic monopoles do not exist; all magnetic phenomena stem from electric currents. This is not trivial to see for magnetic materials since they employ electric current loops at the atomic length scale.

21.1.1: Ampère's Law

The standard discussion of electricity starts from Coulomb's experiment with charged spheres, which led to the definition of the electric force. Since there are no magnetic monopoles to define the magnetic force, this force must instead be the result of materials properties we already introduced.

In 1819, Hans Christian Oersted and André Marie Ampère established the magnetic force as the interaction between two electric currents. Fig. 21.2 shows two conductors of length l at a distance d. Varying the distance, the lengths, and the electric currents I_1 and I_2 in both conductors, Ampère found that the magnitude of the **magnetic force** is proportional to the two currents and the length, and inversely proportional to the distance between the conductors:

$$\left|\vec{F}_{\text{mag}}\right| \propto \frac{I_1 I_2}{d} l \Rightarrow \frac{\left|\vec{F}_{\text{mag}}\right|}{l} \propto \frac{I_1 I_2}{d} \qquad [21.1]$$

The second formulation represents the force per unit length of the conductor. Note that the two forces shown in Fig. 21.2, $\vec{F}_{1\text{ on }2}$ and $\vec{F}_{2\text{ on }1}$, are an action–reaction pair as defined by Newton's third law.

KEY POINT

The magnetic force between two current-carrying conductors is proportional to each current and inversely proportional to the distance between the conductors.

We note from Fig. 21.2 that the magnetic force points perpendicular to the direction in which the current flows through the conductors. Since the force is a vector, we have to be able to write the right-hand side of Eq. [21.1] also as a vector. If the right-hand side contains only one vector, the force would have to point in the same direction as that vector. However, none of the variables on

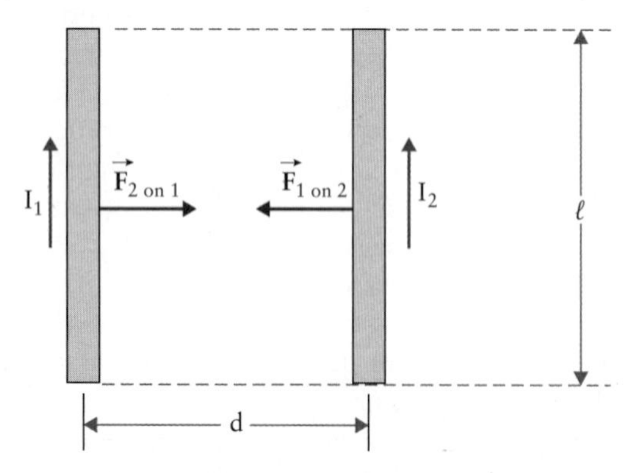

Figure 21.2 Two parallel conductors of length l and distance d carry currents I_1 and I_2. As a result, magnetic force is observed: $\vec{F}_{2\text{on}1}$ acting on conductor 1, and $\vec{F}_{1\text{on}2}$ acting on conductor 2. The two forces shown are an action–reaction pair as defined by Newton's third law.

the right-hand side is a vector pointing in the direction observed for the force. Expecting therefore a more complicated relation between the directions of the variables involved, we postpone the discussion of direction until after we introduce the magnetic field.

The **magnetic field** is defined by following the same approach Faraday introduced when defining the electric field. The electric field was obtained by eliminating from Coulomb's law one point charge that is seen as the probe charge. The probe charge is sampling the electric field of the remaining arrangement of charges. In this textbook we limited this arrangement of charges to systems with stationary charges. This approach was therefore restrictive as it can be used only for stationary electric fields. The restriction is not inherent in Faraday's original approach and we need to lift this restriction later when we want to study time-dependent electric fields. However, we use it at this stage because it allows us to simplify the present discussion.

Analogous to the electric case, we assign a stationary magnetic field to stationary currents. We do this by eliminating one of the single, long, straight conductors in Fig. 21.2 by identifying it as a probe current that serves only to sample the magnetic field. This is a reasonable line of reasoning because it is consistent with the experiment in which Oersted discovered the magnetic force due to a current: he observed that a compass needle reacts when you send an electric current through a conductor nearby.

The definition of the magnitude of the magnetic field B then follows from Eq. [21.1], with I_2 identified and eliminated as the probe current. If the remaining stationary straight conductor carries a current I, we find:

$$\left|\vec{B}\right| = \frac{\mu_0}{2\pi}\frac{I}{d} \qquad [21.2]$$

which is a special form of **Ampère's law**. We discuss its more general form when we study Maxwell's equations later in this chapter.

KEY POINT

The magnitude of the magnetic field of a straight conductor at a nearby point is proportional to the current and inversely proportional to the distance from the conductor to the point.

We note the following for the magnetic field definition in Eq. [21.2] and its relation to the magnetic force in Eq. [21.1]:

- In Fig. 21.3, the magnitude of the magnetic field of a long straight conductor is shown as a function of distance r from the conductor. It has a cylindrical symmetry; i.e., the magnitude of the field varies in the same manner in all directions from the conductor.
- We wrote Eq. [21.2] as an equation, which required the introduction of a proportionality factor, $\mu_0/(2\pi)$. This constant is introduced in a manner analogous to the permittivity of vacuum ε_0 in electricity: the factor

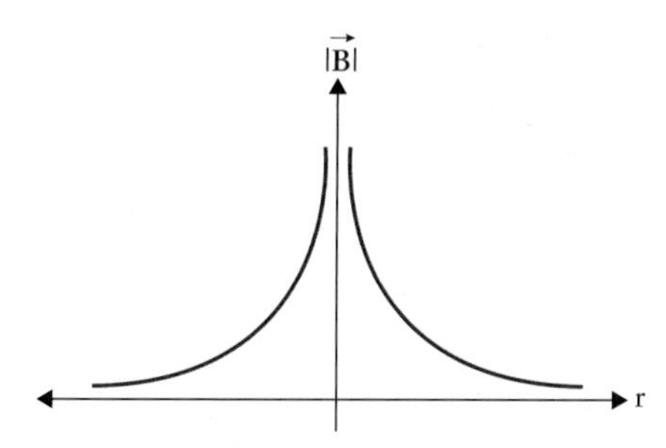

Figure 21.3 Sketch of the magnitude of the magnetic field $|\vec{B}|$ as a function of distance r from a straight conductor carrying an electric current.

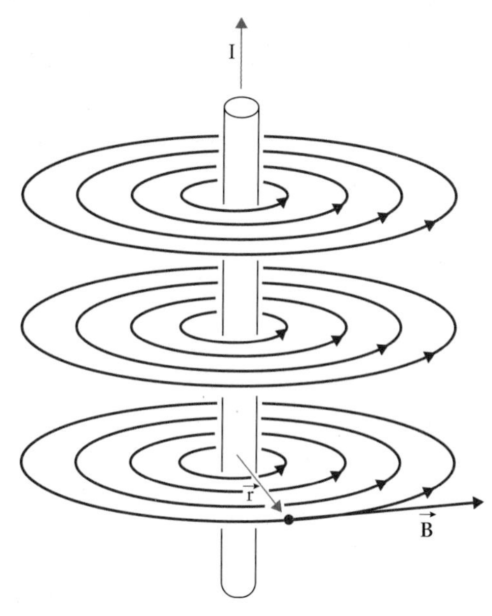

Figure 21.4 The magnetic field is perpendicular to the direction of the current at any point in the vicinity of a current-carrying conductor. The field is also perpendicular to the radius vector from the conductor to the point at which the magnetic field is determined. Instead of drawing each magnetic field vector separately, the figure illustrates the magnetic field with thin lines connecting all points in a plane perpendicular to the conductor that have magnitude equal to the magnetic field.

2π takes the cylindrical symmetry of the magnetic field into account. μ_0 is called the **permeability of vacuum** and has a value of $\mu_0 = 1.26 \times 10^{-6}$ N/A^2. It applies strictly only in a vacuum, otherwise a materials-dependent factor (analogous to the dielectric constant in the electric case) has to be included.

- The unit of the permeability of vacuum follows from Eq. [21.1] as N/A^2. The unit of the magnetic field $\vec{B}$ then follows from Eq. [21.2]:

$$\frac{N}{A^2}\frac{A}{m} = \frac{N}{Am} = \frac{N}{C\frac{m}{s}} = T$$

The new unit T is called **tesla** in honour of Nikola Tesla. You find several non-standard units still in use in the literature, including the unit gauss (G, named after Carl Friedrich Gauss) with the conversion 1 G = 1×10^{-4} T, and the unit oersted (Oe) for $|\vec{B}|/\mu_0$ for which 1.0 Oe = 79.59 A/m. Typical values include Earth's magnetic field at the surface, with $B = 5 \times 10^{-5}$ T; a standard bar magnet with $B \approx 1 \times 10^{-2}$ T; and the largest superconducting magnets with fields of up to 20 T.

- The direction of the magnetic field has to be discussed carefully, as we already noted in the context of Eq. [21.1]. We illustrate in Fig. 21.4 the direction of the magnetic field for a long straight conductor. The magnetic field is perpendicular to the direction in which the current flows and to the radial direction pointing from the conductor to the point in space at which we determine the magnetic field. The circular lines in the figure connect positions of equal magnitude of the magnetic field. The direction of the magnetic field vector is at every point tangential to the field lines, as illustrated at one point in the figure. The direction of the magnetic field can be determined with a (modified) **right-hand rule**: when the thumb points in the direction of the current (direction of flow of positive charges in the conductor) the remaining fingers of the right-hand curl and point like the magnetic field.

Before we include the directional information in a mathematical form, it is worthwhile to note how Fig. 21.4 is related to Fig. 21.2. For this we consider in Fig. 21.2 the left conductor with current I_1 as the stationary current causing a magnetic field, and we treat the conductor with current I_2 as the probe current. Projecting Fig. 21.4 onto the left conductor in Fig. 21.2 we note that the magnetic field at the position of the probe current (at the right) is pointing into the plane of the textbook page. For the force $\vec{F}_{1\text{ on }2}$ to result from this magnetic field and the current I_2, the three vectors $\vec{B}$, $\vec{F}_{1\text{ on }2}$, and $\vec{j}_2$ must be perpendicular to each other. $\vec{j}_2$ is the current density vector, which represents the current and the direction of the probe current. As long as we work with current-carrying conductors, we do not use the current density, but write the current I_2 combined with a unit vector 1^0, which points in the direction in which the current flows; thus, 1^0 is directed upward for the probe current in Fig. 21.2. This leads to Fig. 21.5: we note that the directions of the three perpendicular vectors $\vec{B}$, $\vec{F}_{1\text{ on }2}$, and $\vec{j}_2$, adhere to the right-hand rule. Thus, we need the vector product notation to fully describe their relation:

$$\vec{F}_{\text{on wire}} = I\,l[\vec{l}^{\,0} \times \vec{B}]. \qquad [21.3]$$

We are not performing vector algebra operations in this chapter because we can work with the magnitude of the magnetic force near a long straight conductor carrying a current. This magnitude follows from Eq. [21.3] as:

$$|\vec{F}_{\text{on wire}}| = |\vec{B}|\,I\,l\sin\varphi, \qquad [21.4]$$

in which φ is the angle between the direction of the magnetic field and the direction of the current-carrying conductor affected by the magnetic field. The directional information contained in the vector $\vec{B}$ is then established with Figs. 21.4 and 21.5.

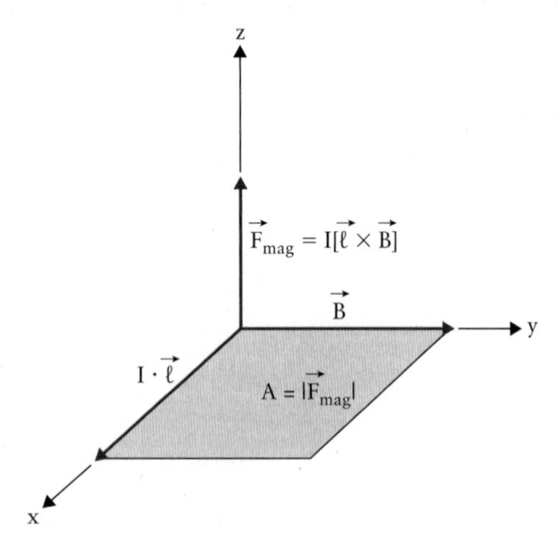

Figure 21.5 Right-hand rule relating the magnetic force, the orientation of the current-carrying conductor, and the magnetic field.

EXAMPLE 21.1

A straight conducting wire is placed between the poles of a permanent horseshoe magnet as shown in Fig. 21.6. The magnet produces a uniform magnetic field of 2.0 T. The wire runs through the gap between the poles of the magnet, perpendicular to the direction of the magnetic field. The length of the wire is 0.3 m in the gap of the magnet. When the wire is connected to a battery, it carries a current. What current must flow to obtain a force of 0.98 N?

Solution

The horseshoe magnet provides the magnetic field acting on the wire. We use Eq. [21.4] with $\sin \varphi = 1$. This leads to:

$$I = \frac{F}{IB} = \frac{0.98\ \text{N}}{(0.3\ \text{m})(2.0\ \text{T})} = 1.63\ \text{A}.$$

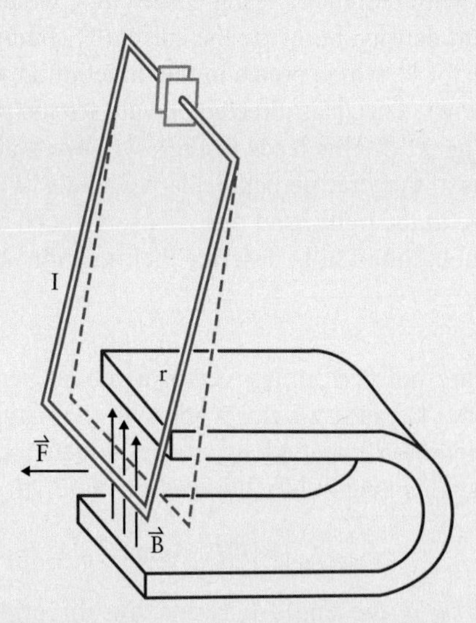

Figure 21.6 A current runs through a conductor that is mounted such that a force acting on the conductor can cause it to move in the direction of the force. The conductor responds to a uniform magnetic field generated between the poles of a horseshoe magnet.

EXAMPLE 21.2

Two long stationary conducting wires cross at the origin, as shown in Fig. 21.7. One wire runs along the *x*-axis and carries a current of $I_1 = 2$ A. The other wire runs along the *y*-axis and carries a current of $I_2 = 3$ A. What are magnitude and the direction of the magnetic field at point P, which is located 4 cm from the *x*-axis and 6 cm from the *y*-axis?

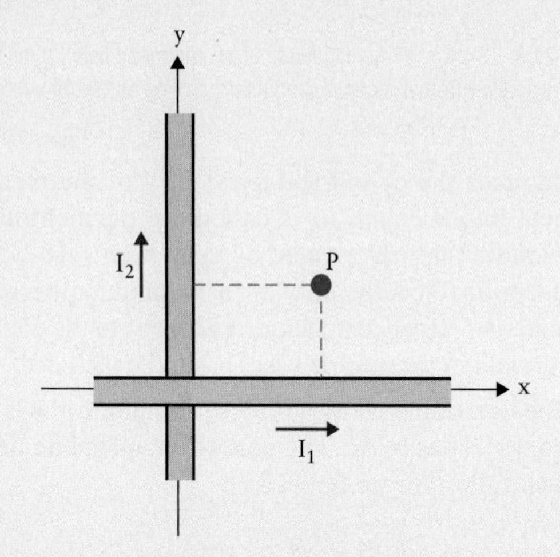

Figure 21.7 Two straight conducting wires are positioned perpendicular to each other, each carrying a current as indicated. The magnetic field at point P is calculated in the text.

Solution

If several currents are present in a system their magnetic fields are added at each point in space. Thus, we use Eq. [21.2] to calculate the magnitude of the magnetic field at point P for each of the two wires separately. For wire 1:

$$|\vec{B}_1| = \frac{\mu_0 I_1}{2\pi d_1} = \frac{\left(1.26\times10^{-6}\ \frac{\text{N}}{\text{A}^2}\right)(2\ \text{A})}{2\pi(0.04\ \text{m})} = 1.0\times10^{-5}\ \text{T};$$

and for wire 2:

$$|\vec{B}_2| = \frac{\mu_0 I_2}{2\pi d_2} = \frac{\left(1.26\times10^{-6}\ \frac{\text{N}}{\text{A}^2}\right)(3\ \text{A})}{2\pi(0.06\ \text{m})} = 1.0\times10^{-5}\ \text{T}.$$

The respective directions of the two contributions to the magnetic field at point P are determined with the right-hand rule: for the current along the *x*-axis the magnetic field at P is directed out of the plane of the paper, and for the current running along the *y*-axis the magnetic field at point P is directed into the plane of the paper. Thus, the two contributions have to be subtracted from each other. The net magnetic field at point P is therefore $B_{net} = 0$ T.

21.1.2: Biot-Savart's Law

This section requires calculus; if you are not familiar with calculus-based calculations, you may skip to section 21.1.3.

Eq. [21.2] applies specifically to a long straight conductor as chosen in Fig. 21.4. Different formulas describe the magnetic field for other arrangements of electric currents. In the electric case we derived the field for an arbitrary arrangement of charges as the sum of the field contributions of each single charge. When this approach was no longer feasible due to the large number of charges, for example in the case of a parallel plate capacitor, the electric field was calculated from the contributions of small segments into which the system had been divided.

The same approach is used to determine the magnetic field of an arbitrary arrangement of currents. This can be either a large number of currents (like in Example 21.2) or small segments of a single current that flows along an arbitrarily curved conductor. The general formula used to calculate the magnetic field in such cases is called **Biot–Savart's law**. Jean-Baptiste Biot and Félix Savart based their reasoning on the arrangement shown in Fig. 21.8. A short conductor segment $d\vec{l}$ is highlighted along an arbitrary conductor in the *xy*-plane. The segment contributes to the magnetic field at point P, which we have chosen in the figure at the origin. If we label this contribution $d\vec{B}$, then we find for its magnitude:

$$|d\vec{B}| = \frac{\mu_0}{4\pi}\frac{I|d\vec{l}|}{r^2}\sin\varphi \qquad [21.5]$$

in which φ is the angle between the direction of $d\vec{l}$ and $\vec{r}$, with $\vec{r}$ the vector pointing from the current element to point P. The magnetic field at point P is obtained by integrating along the entire conductor. For conductors that wind through space arbitrarily, Eq. [21.5] has to be rewritten to include directional information. This yields Biot–Savart's law in the form:

$$d\vec{B} = \frac{\mu_0}{4\pi}\frac{I}{r^3}[d\vec{l} \times \vec{r}] \qquad [21.6]$$

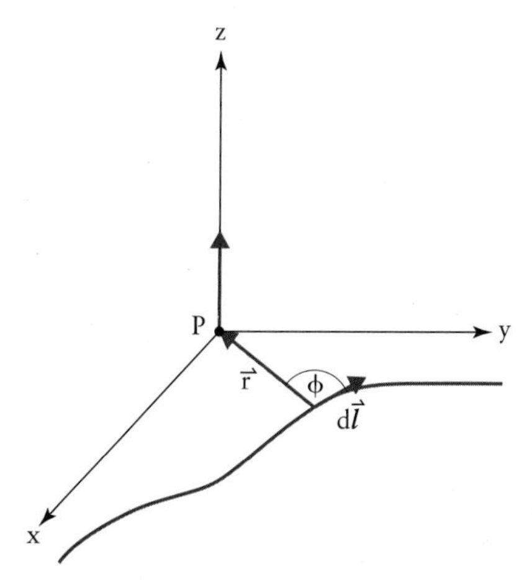

Figure 21.8 Conceptual sketch used to define the parameters in Biot–Savart's law. An arbitrary conductor, chosen in the *xy*-plane, is subdivided into short straight elements of length $d\vec{l}$. The magnetic field at point P (chosen at the origin) is calculated as the sum of all $d\vec{B}$ contributions of the conductor segments. The calculation takes the angle between the conductor segment $d\vec{l}$ and the vector to the point, $\vec{r}$, into account.

CASE STUDY 21.1

How is Eq. [21.2] derived from Eq. [21.5]; i.e., how do we obtain Ampère's law for a long straight conductor from Biot–Savart's law?

Answer: *We use Fig. 21.9, which shows a long straight conductor carrying a current I at the left and the point P at which we want to calculate the magnetic field at the right. The direction of the arbitrarily chosen conductor segment $d\vec{l}$ indicates the direction in which a positive current flows. The closest distance between point P and the conductor is d. We discuss in this case the relation between Eqs. [21.2] and [21.5] since the directional aspect is straightforward: For all segments along the conductor the contributions to the magnetic field point in the same direction out of the plane of the paper (as indicated by the symbol ⊙ at the position P).*

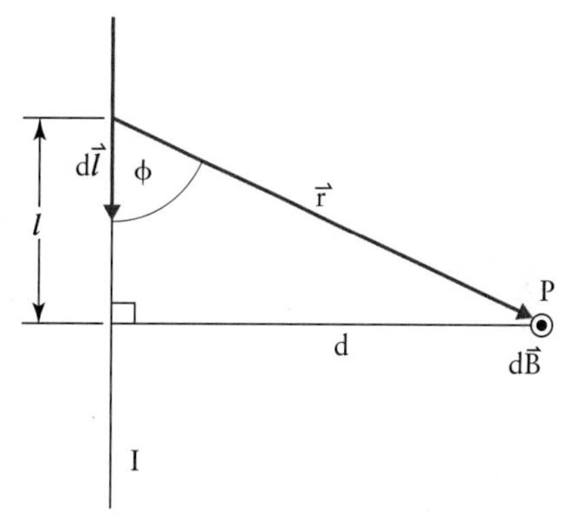

Figure 21.9 A long straight conductor carries a current *I* (at left). P is a point in space at which the magnetic field is sought. An arbitrary conductor segment $d\vec{l}$ is used for the calculation; its direction shows the direction of a positive current. The closest distance between point P and the conductor is labelled distance *d*. The resulting magnetic field points out of the plane of the paper as indicated by the symbol ⊙ at P.

We first determine the contribution to the magnetic field due to the highlighted conductor segment in Fig. 21.9. Using the triangle seen in the figure, we replace the distance r and sin φ with:

$$r = \sqrt{l^2 + d^2}; \ \sin\varphi = \frac{d}{r} = \frac{d}{\sqrt{l^2 + d^2}},$$

which yields:

$$dB = \frac{\mu_0}{4\pi}\frac{I\,d\,dl}{(d^2 + l^2)^{3/2}}$$

Next we integrate from $l = -\infty$ to $l = +\infty$, which covers all contributions of an infinitely long conductor:

$$B(\mathrm{P}) = \frac{\mu_0 Id}{4\pi}\int_{l=-\infty}^{l=+\infty}\frac{dl}{(d^2 + l^2)^{3/2}},$$

which yields Eq. [21.2]. You may use tabulations of integrals to confirm this. Note that such calculations are extensive even though the final result is a simple formula.

continued

This example illustrates that choosing arbitrary conductor arrangements, either for engineering applications or as a model for a physiological system, is not advisable. Instead, very few, highly symmetric geometries are chosen for which we already know the resulting field; this in particular includes the arrangements discussed in the next section.

21.1.3: Solenoid and Current Loop

When we developed the concept of the electric field we noted that one benefit of the approach is to allow us to calculate it for systems with a large number of charges prior to studying the properties of a probe charge in the field. While the calculation of the field may be a difficult mathematical task, once a formula for the field has been obtained it can be used for many cases. This has proven particularly valuable for the parallel plate arrangement: it allows us to describe most experimental capacitor arrangements (including the nerve membrane) with a very simple formula for the electric field.

We proceed with the magnetic field concept again in analogy to the electric case. Instead of a straight conductor, many other arrangements of one or several conductors of practical interest have been studied with Eqs. [21.5] and [21.6]. One particular arrangement was found that has a magnetic field with a formula simpler than Eq. [21.2]. We study this arrangement in more detail as it provides us with a convenient model system for applications of magnetism.

The arrangement is called a **solenoid**, which is a single conductor that is curled as shown in Fig. 21.10. The radii of the conductor loops of the solenoid are constant, as is the number of windings N per length l of the solenoid. When a current is sent through the conductor, a magnetic field develops in its vicinity in the same manner we discussed before for the straight conductor (Fig. 21.4). Both the magnitude and the direction of the magnetic field within a solenoid take particularly simple forms. The magnitude is given by:

$$\left|\vec{B}_{\text{solenoid}}\right| = \mu_0 \frac{N}{l} I; \qquad [21.7]$$

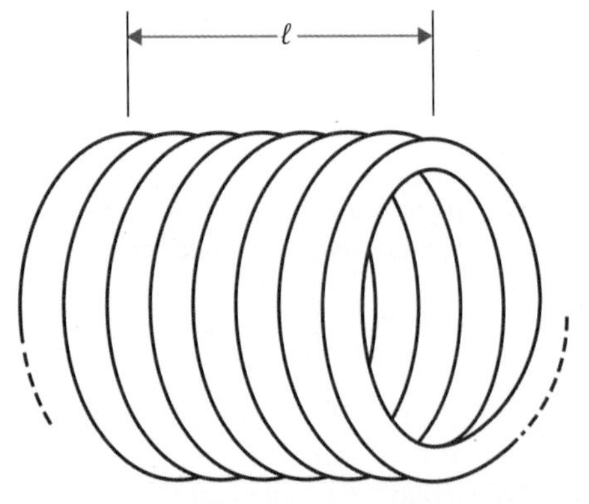

Figure 21.10 A solenoid is a coiled conductor with a fixed radius of the coils and a constant number N of coils per length l.

i.e., the magnitude of the magnetic field in a solenoid has a constant value for a given device. The direction of the magnetic field is shown by the thin lines in Fig. 21.11: the direction of the field in the solenoid does not vary and runs parallel to its axis.

Eq. [21.7] describes a much simpler case than Eq. [21.2] since the magnitude of the magnetic field varies with distance from a long straight conductor; i.e., the magnetic field in Eq. [21.2] is not position independent. Also, the magnetic field of a straight conductor changes its direction from point to point, as illustrated in Fig. 21.4, while the magnetic field in a solenoid is directed everywhere along the solenoid axis.

Due to these simple properties, solenoids are widely used as electromagnets (i.e., devices that act as magnets when they carry an electric current). The lenses used in state-of-the-art electron microscopes are solenoids, so are the magnets in MRI facilities.

A second arrangement with practical applications in medicine, in particular for transcranial magnetic stimulations as discussed in more detail later in this chapter, is a **single current loop** as illustrated in Fig. 21.12. If the current loop has a radius R we find for the magnitude of the magnetic field along the loop's axis (labelled z-axis with $z = 0$ at the centre of the loop):

$$\left|\vec{B}_{\text{loop}}\right| = \frac{\mu_0 I}{2} \frac{R^2}{(z^2 + R^2)^{3/2}}, \qquad [21.8]$$

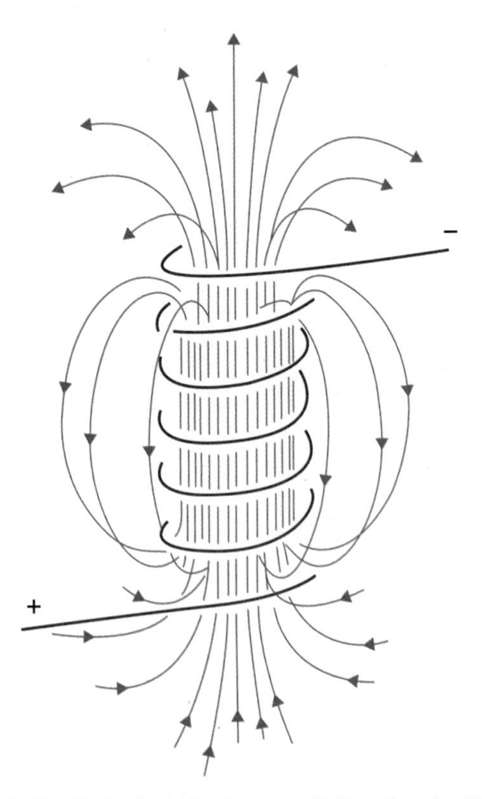

Figure 21.11 Sketch illustrating the magnetic field of a solenoid. Inside the coil exists a magnetic field that is position independent in both direction and magnitude. The direction of the magnetic field inside the solenoid is parallel to the axis of the solenoid.

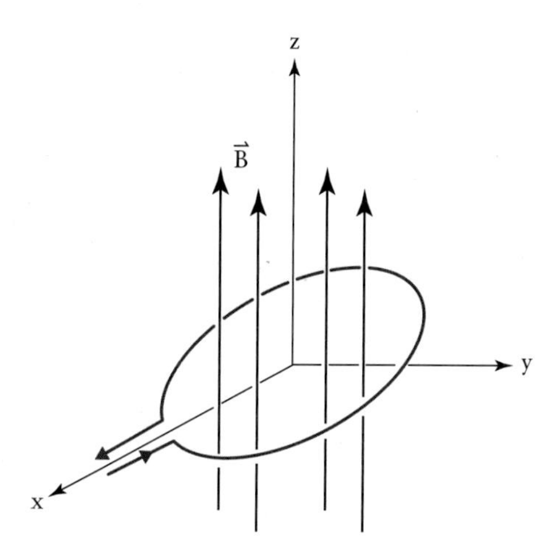

Figure 21.12 Magnetic field of a single current loop in the *xy*-plane.

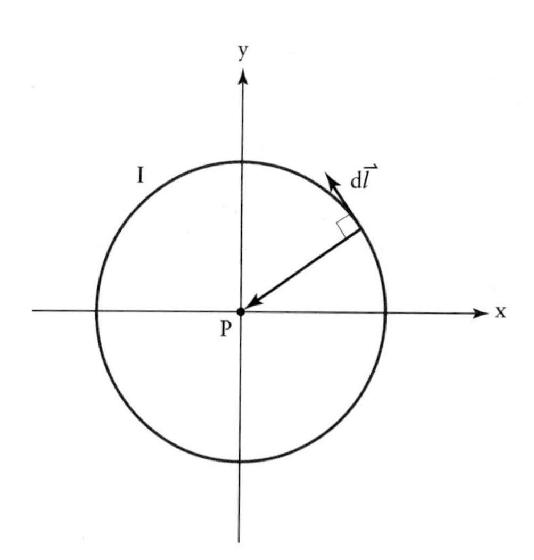

Figure 21.13 A conductor loop of radius *R* in the *xy*-plane at position $z = 0$. A conductor segment $d\vec{l}$ along the loop is identified such that it points in the direction of a positive current *I*. Point P lies at the centre of the loop.

Both I and R^2 have been pulled out of the integral since they are constant. The remaining integral is a geometrical term that is equal to the length of the loop, $2\pi R$, which is the circumference of the circle. Thus:

$$B(P) = \frac{\mu_0 I}{4\pi R^2} 2\pi R = \frac{\mu_0 I}{2R},$$

which is Eq. [21.9].

which simplifies at the centre of the loop, i.e. at $z = 0$, to:

$$\left|\vec{B}_{\text{center of loop}}\right| = \frac{\mu_0 I}{2R}. \qquad [21.9]$$

Note that Eqs. [21.7] and [21.9] are the two asymptotic cases of the general formula for the magnetic field along the axis of N parallel current loops of total length l:

$$\left|\vec{B}_{\text{axial}}\right| = \frac{\mu_0 N I}{\sqrt{4R^2 + l^2}}, \qquad [21.10]$$

with Eq. [21.9] following for length $l = 0$ and $N = 1$, and Eq. [21.7] following for $l >> R$.

CASE STUDY 21.2

Derive Eq. [21.9] from Biot–Savart's law. *Note:* This Case Study requires calculus. If you are not familiar with calculus-based operations, skip to section 21.2.

Answer: *Fig. 21.13 shows the conductor loop of radius R in the xy-plane (i.e., at position z = 0). A conductor segment $d\vec{l}$ is identified that points in the direction in which the positive current I flows. The point P, for which we want to calculate the magnitude of the magnetic field, is at the centre of the loop. Since the conductor segment and the radius vector are perpendicular to each other, we use in Eq. [21.5] $\varphi = 90°$; i.e., $\sin\varphi = 1$. Since this applies to any segment along the loop, Eq. [21.5] is sufficient as all contributions to the magnetic field at P point in the same direction along the z-axis:*

$$dB = \frac{\mu_0}{4\pi} \frac{I dl}{R^2}$$

The net magnetic field then results for an integration along the loop:

$$B(P) = \frac{\mu_0 I}{4\pi R^2} \int_{\text{loop}} dl.$$

continued

21.2: Magnetism Due to Charged Particles in Motion

21.2.1: The Magnitude of the Magnetic Force

We start with the two parallel conductors in Fig. 21.2, but use a microscopic model for the current, in which it is identified as the flow of individual charged particles (we derived this formula in Chapter 19):

$$I = n_e e v_d A,$$

where n_e is the density of mobile charges in the conductor, e is the elementary charge, v_d is the drift velocity, and A the cross-sectional area of the conductor. We substitute this current into Eq. [21.4], which applies to Fig. 21.2 with $\sin\varphi = 1$. This yields for the magnetic force on either of the parallel conductors:

$$\frac{F_{\text{mag}}}{l} = BI = Bn_e e v_d A.$$

To abandon the need for a current-carrying conductor, this equation is generalized by replacing the elementary charge e of the electron with the generic charge variable q, which also applies to ions. The product of length l and cross-sectional area A is the volume of interest, $V = l\,A$. It can be combined with the density of the moving charges, n_e, to yield the total number of charged particles, N_q:

$$lAn_e = Vn_e = N_q.$$

Thus, the magnetic force per length is rewritten for a single charged particle as:

$$\frac{F_{mag}}{N_q} = Bqv, \qquad [21.11]$$

in which the index d for "drift" has been dropped as the velocity of any moving charged particle obeys this formula, not only the electrons drifting in a metallic wire.

We introduce a new notation for the force on a single particle in the form $\vec{f}_{mag} = \vec{F}_{mag}/N_q$. Note that Eq. [21.11] is written in scalar form, i.e., all vector quantities are represented by their respective magnitude. When the direction of the magnetic field vector, $\vec{B}$, and the direction of the velocity of the particle, $\vec{v}$, are not perpendicular to each other, Eq. [21.11] is modified according to Eq. [21.4] to take into account the angle φ between these vectors:

$$\frac{F_{mag}}{N_q} = f_{mag} = qvB\sin\varphi. \qquad [21.12]$$

The standard units of the quantities in Eq. [21.12] are C for the charge, m/s for the velocity, T for the magnetic field, and N for the force. The following example illustrates that Eq. [21.12] is sufficient for a wide range of magnetic phenomena even though it does not express the directional information.

KEY POINT

The magnetic force on a single, moving particle is proportional to its charge, its speed, and the magnetic field. It also depends on the angle between the magnetic field and the path of the particle.

EXAMPLE 21.3

A 10.0-keV proton beam (i.e., protons extracted from a vacuum ion source with a potential difference of 10 kV) enters a magnetic field of magnitude 2 T perpendicular to the field (2 T is the magnitude of the magnetic field of a strong conventional laboratory magnet and is currently the upper limit of magnets used for MRI). (a) What is the speed of a proton? (b) What force acts on each proton in the beam?

Solution

Solution to part (a): 10 keV is the kinetic energy of each proton. We use for the energy conversion and for the mass of the proton:

$$10\ \text{keV} = 1.6 \times 10^{-15}\ \text{J}; \qquad m_p = 1.67 \times 10^{-27}\ \text{kg}.$$

continued

With these values we calculate the velocity of a proton in the beam:

$$v = \sqrt{\frac{2E}{m}} = \sqrt{\frac{2(1.6\times10^{-15}\text{J})}{1.67\times10^{-15}\text{kg}}} = 1.38\times10^{6}\,\frac{\text{m}}{\text{s}}.$$

We applied the classical formula for the kinetic energy, $E_{kin} = ½\, mv^2$. This is justified because the speed of a proton turns out to be only of the order of 0.5% of the speed of light.

Solution to part (b): We substitute the given values, including the result in part (a), in Eq. [21.12]. The angle between the velocity direction and the magnetic field direction is $\varphi = 90°$. Thus $\sin\varphi = 1$ and:

$$f_{mag} = (1.6\times10^{-19}\ \text{C})\left(1.38\times10^{6}\,\frac{\text{m}}{\text{s}}\right)(2.0\ \text{T}) = 4.4\times10^{-13}\text{N}.$$

It is interesting to compare this force with typical mechanical and electric forces. While the numerical value of the force is small, we have to keep in mind that this is a force acting on an subatomic sized particle. It is certainly smaller than the electric force between a proton and an electron in a hydrogen atom; however, it is much stronger than the gravitational force represented by the weight of a hydrogen atom.

21.2.2: The Direction of the Magnetic Force

The directional information is included in Eq. [21.12] when it is rewritten as a vector product:

$$\vec{f}_{mag} = q\left[\vec{v}\times\vec{B}\right]. \qquad [21.13]$$

In this form the magnetic force is called the **Lorentz force**, named to honour Hendrik Lorentz. Three vectors are included in Eq. [21.13]: $\vec{f}_{mag}$, $\vec{v}$, and $\vec{B}$. These vectors must be perpendicular to each other. Based on the right-hand rule, the direction of the magnetic force for a positive charge results as the direction of the middle finger of your right-hand if you point the thumb in the direction of the velocity and the index finger in the direction of the magnetic field.

The directions related in magnetic interactions are illustrated in Fig. 21.14. Note that the sketch is two-dimensional; i.e., it shows the plane defined by the force and the velocity. The directional information of the magnetic field in the third dimension is given in a standard notation for vectors pointing into or out of a plane: a circle with a cross (⊗) stands for a vector pointing into the plane and a circle with a dot (⊙) stands for a vector pointing out of the plane. This notation is easiest to remember by using the inset in the lower right corner of Fig. 21.14, which shows an arrow as used with a bow in the Middle Ages. When you shoot the arrow, you see the cross of feathers at the end of the arrow moving away (i.e., into the plane of the paper); if you are shot at, you see the sharp tip of the arrow approaching (i.e., coming out of the plane of the paper).

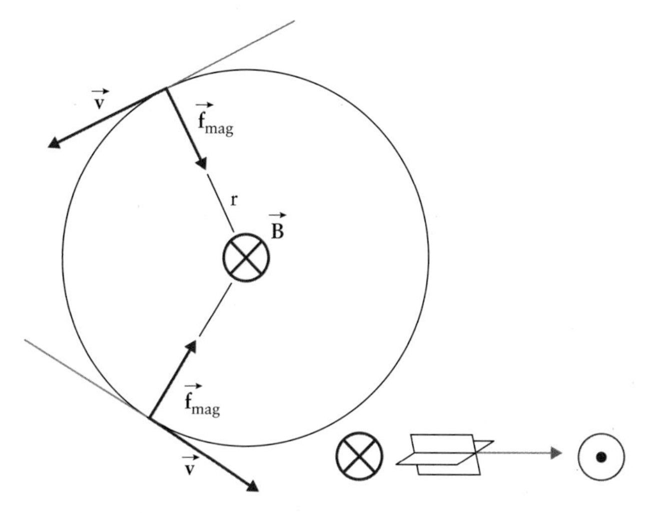

Figure 21.14 The magnetic force, $\vec{f}_{mag}$, acting on a particle with a positive charge moving on a circle of radius *r* in a magnetic field that is directed into the plane. The arrow at the lower right indicates the standard notation of the vector direction into the plane (⊗) and the direction out off the plane (⊙).

Thus, the magnetic field in Fig. 21.14 is directed into the plane of the figure. On a positively charged particle moving with a given velocity $\vec{v}$ toward the lower left, a force is exerted toward the lower right. When the particle moves toward the lower right, a force is exerted toward the upper right. Confirm these directions using the right-hand rule.

21.2.3: Motion of Charged Particles in a Uniform Magnetic Field

Fig. 21.14 shows that the magnetic force and the velocity vector of a charged particle in a magnetic field B are perpendicular to each other. A force with constant magnitude that points perpendicular to the direction of motion of the object leads to uniform circular motion. We use Newton's second law to quantify the circular motion. The net force acts in the direction perpendicular to the path of the object and is due to the magnetic force:

$$\sum_i F_{i,\perp} = qvB = m\frac{v^2}{r}. \qquad [21.14]$$

This equation relates the magnitude of the magnetic field, the mass and charge of the particle, and the radius and speed of the motion of the particle. The radius of the circular motion is given by:

$$r = \frac{mv}{qB}. \qquad [21.15]$$

From this equation an angular frequency of the point charge can be calculated. For this we start with the speed of the object in circular motion and divide by the radius of the circular path:

$$v = \frac{2\pi r}{T} \Rightarrow \frac{v}{r} = \frac{2\pi}{T} = \omega.$$

Thus, isolating the term v/r in Eq. [21.15] yields:

$$\omega = \frac{qB}{m}. \qquad [21.16]$$

This angular frequency is called the **cyclotron frequency**. The frequency, $f = \omega/(2\pi)$, and the period, $T = (2\pi)/\omega$, of the circular motion can then be derived.

KEY POINT

A charged particle moves in a constant magnetic field with uniform circular motion. The path of the particle curls around the magnetic field lines. Its radius is proportional to the mass and speed of the particle. The radius is further inversely proportional to the charge of the particle and the magnetic field.

EXAMPLE 21.4

Food in a microwave oven gets heated because the water molecules in the food absorb energy from radiation of frequency at $f = 2450$ MHz. What strength of the magnetic field is needed for a microwave oven in which the radiation is obtained from electrons circling with frequency f?

Supplementary information: Electromagnetic radiation of a given frequency f is associated with an energy $E = h\,f$ in which h is Planck's constant. This energy is carried by a photon. Photons with frequency $f = 2450$ MHz lie in the microwave region of the electromagnetic spectrum, i.e., between radio waves and infrared radiation. Microwave energies are sufficient to cause molecules to rotate. At a frequency of 2450 MHz, water has a resonance; i.e., it absorbs energy efficiently from the radiation field. Thus, during microwave cooking the water molecules in the food are set into fast rotation. As a result of inter-molecular collisions the surrounding molecules slow that rotation down, turning the rotation energy into heat, and thereby raising the temperature of the food.

Solution

To find the magnetic field, we first convert the given frequency into an angular frequency:

$$\omega = 2\pi f = 2\pi(2.45 \times 10^9\ \text{s}^{-1}) = 1.54 \times 10^{10}\ \text{Hz}.$$

Using Eq. [21.16] with the electron mass $m = 9.11 \times 10^{-31}$ kg and the elementary charge $q = 1.6 \times 10^{-19}$ C yields:

$$B = \frac{m\omega}{q} = \frac{(9.11\times10^{-31}\ \text{kg})(1.54\times10^{10}\ \text{Hz})}{1.6\times10^{-19}\ \text{c}} = 0.09\ \text{T}.$$

This is a moderate magnetic field, but the magnet still represents a major fraction of the weight of the microwave oven. Note that electrons move in the opposite direction of positive charges in a magnetic field. However, Eq. [21.16] remains valid as it relates only the *magnitudes* of vector quantities.

EXAMPLE 21.5

A proton moves along a circular path with radius 20 cm in a uniform magnetic field. The magnitude of the magnetic field is 0.5 T and the field is directed perpendicularly to the velocity vector of the proton. (a) What is the speed of the proton along its path? (b) If the proton is replaced with an electron moving with the same speed, what is the radius of the circular path of the electron?

Solution

Solution to part (a): We solve Eq. [21.15] for the speed of the proton:

$$v = \frac{qBr}{m} = \frac{(1.6\times10^{-19}\,\text{C})(0.5\ \text{T})(0.2\ \text{m})}{1.67\times10^{-27}\,\text{kg}} = 9.6\times10^{6}\,\frac{\text{m}}{\text{s}},$$

in which we used the elementary charge for the charge of the proton and the mass of the proton.

Solution to part (b): We could use Eq. [21.15] directly or, as done here, rewrite it relative for the two cases:

$$\frac{r_{e-}}{r_{p+}} = \frac{\dfrac{m_{e-}\,v}{qB}}{\dfrac{m_{p+}\,v}{qB}} = \frac{m_{e-}}{m_{p+}},$$

which leads to:

$$r_{e-} = \frac{m_{e-}}{m_{p+}} r_{p+} = 1.1\times10^{-4}\ \text{m},$$

in which the mass of the electron is $m = 9.11 \times 10^{-31}$ kg.

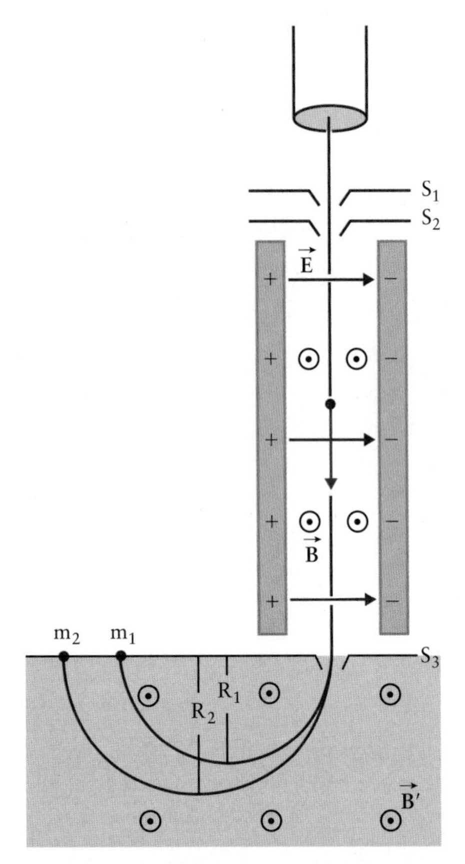

Figure 21.15 Aston's mass spectrometer. The ions, collimated with slits S_1 and S_2, first pass through a velocity selector (Wien filter) with perpendicular electric and magnetic fields and then enter a mass selector with magnetic field $\vec{B}'$.

21.3: Aston's Mass Spectrometer

Francis William Aston developed the first mass spectrometer in 1919 by combining an ion source with two devices:

- a velocity selector based on a magnetic and an electric field, and
- a mass selector based on a magnetic field.

In this section we discuss how this instrument measures the mass of molecules and their fragments. Fig. 21.15 is a conceptual sketch of the mass spectrometer developed by Aston. The instrument consists of an ion source, which is a hot filament that ionizes molecules as they leave a gas chromatography set-up. Gas chromatography is used to separate the components of a vapour phase sample. The ion source is shown at the top of Fig. 21.15. The generated ions travel through a set of slits. These slits serve two purposes:

- they confine the beam of particles, and
- they are electrically charged to operate as a parallel plate capacitor to accelerate the ions.

Two devices then separate the components of the ion beam on the basis of their masses: a velocity selector (called Wien filter) and a magnetic mass analyzer.

In Aston's set-up the final mass separation was obtained when the charged ions hit a photographic plate that was developed afterward. In modern instruments the charged ions are measured as a current with a device called a Faraday cage. The main components in Fig. 21.15 are discussed in more detail first, followed by an application of mass spectrometry in organic chemistry.

21.3.1: The Wien Filter

The Wien filter (named for Wilhelm Wien) serves as a velocity selector. The ionization process of a typical mass spectrometer produces almost exclusively ions with a single positive charge. After these ions pass through the acceleration section, the beam contains a range of ionized molecular fragments, each with a range of different velocities. All of these ions enter the Wien filter parallel to each other. The Wien filter contains a magnetic and an electric field that are arranged perpendicular to each other and perpendicular to the direction of the motion of the charged particles. This is illustrated in Fig. 21.16(a). The magnetic field points into the plane and the electric field is directed downward.

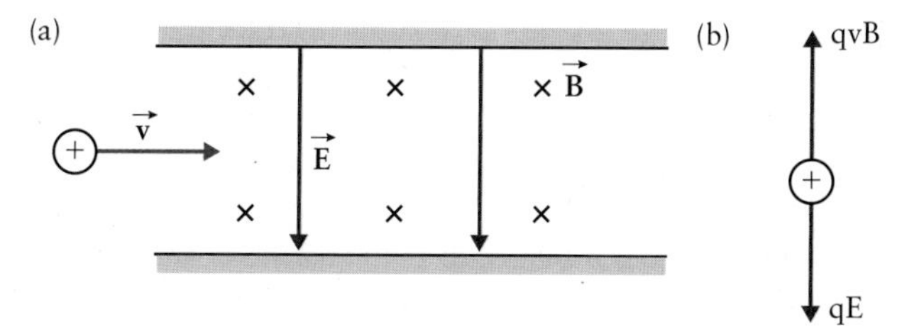

Figure 21.16 (a) Sketch of a Wien filter. The positively charged ion enters horizontally, with the electric field due to a parallel plate capacitor directed vertically and the magnetic field (usually provided in the gap between two coaxial solenoids) directed perpendicular to the plane of the paper. (b) Free body diagram for the particle shown in part (a) after it has entered the Wien filter. Forces are labelled by their magnitude.

When a charged particle enters a region where an electric and a magnetic field are present, two contributions, the electric and the magnetic force, are added. The resulting net force is given as:

$$\vec{F}_{\text{net}} = q\vec{E} + q[\vec{v} \times \vec{B}] \qquad [21.17]$$

Fig. 21.16(b) shows the free body diagram for a positively charged ion: the electric force acts downward and the magnetic force acts upward. The Wien filter, which includes a narrow slit at its exit (S_3 in Fig. 21.15), allows us to select particles of a single velocity using Newton's first law based on the free body diagram in Fig. 21.16(b). Newton's first law defines a mechanical equilibrium, i.e., the condition under which the particle is not accelerated and moves straight through the Wien filter. Particles that are not in mechanical equilibrium accelerate, drifting away from the straight line and get blocked at the exit slit. The condition of mechanical equilibrium for the charged particle in Fig. 21.16(a), with the direction upward labelled the $+y$-direction, is derived from the previous vector equation:

$$F_{\text{net},y} = -qE + qvB = 0,$$

which yields:

$$v = \frac{E}{B}. \qquad [21.18]$$

Thus, to select a certain velocity with a Wien filter, the ratio of the magnitude of the electric field to the magnitude of the magnetic field must be chosen according to Eq. [21.18].

21.3.2: The 180° Sector Magnet

Once a single velocity is selected with the Wien filter, the beam of particles enters a **180° sector magnet** in Aston's set-up. The charged particles move in a circular manner through the magnetic field, with the radius of the path given in Eq. [21.15]. Because all particles enter the magnet with the same speed, the only variable term on the right-hand side of Eq. [21.15] is the mass of the ion. Thus, the heavier the ion, the farther to the left it hits the photographic plate in Fig. 21.15. The analysis of the obtained data is very simple as the distances along the photographic plate are directly proportional to the ion's mass, $R \propto m$.

21.4: Interacting Electric and Magnetic Fields

The successful use of the analogy between electric and magnetic phenomena establishes a first hint that both are closely related. In this section, we study how electric and magnetic fields relate to each other. The observations and results ultimately lead to the formulation of Maxwell's equations, which combine all electric and magnetic phenomena in a single formalism called **electromagnetic theory**.

21.4.1: Faraday's Law

We have already established one definitive link between electric and magnetic phenomena: Ampère's law states that an electric current causes a stationary magnetic field. This observation can be extended easily to a variable current that will cause a variable magnetic field. We now focus on the reverse causality: can a magnetic field cause an electric phenomenon, such as a current? Michael Faraday answered this question when he conducted his famous series of experiments in 1831 that led to the formulation of Faraday's law. Like Faraday, we establish this law by analyzing his experimental evidence.

Faraday's basic set-up is shown in Fig. 21.17: A single conductor loop (e.g., bent from a copper wire) is connected to a galvanometer that can measure a steady current, but will also indicate a short current peak of the form $I(t)$. If the width of the current peak is shorter than the response time of the instrument (inertia of a needle

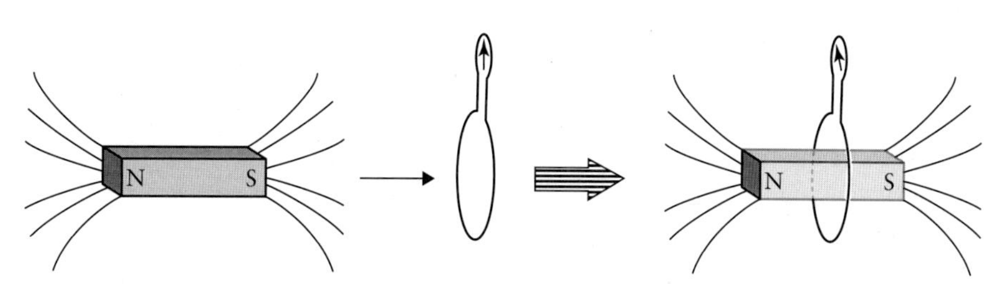

Figure 21.17 Michael Faraday's experiment. A bar magnet is moved into a single loop of a conductor that is connected to a galvanometer (measuring the electric current that flows through the loop). In this experiment the magnetic field in the conductor changes as the magnet is moved. While the magnetic field changes a current is detected.

attached to a spring) then the time profile of the observed current peak does not represent the actual time profile in the conductor; however, the area under the recorded peak is still a measure of the total current that was flowing in the experiment. Since there is no battery or other current-generating device included in the conductor loop, the galvanometer shows a zero current before we conduct an experiment with this set-up.

In Faraday's original experiment the North Pole of a permanent bar magnet (a magnetized piece of iron) is moved through the conductor loop. As a result, the galvanometer displays a current peak. This phenomenon is called **induction** and is interpreted as a current that is induced by a changing magnetic field in the plane of the conductor loop. If further experiments confirm this interpretation, we have established that magnetic fields can cause an electric phenomenon. Faraday's modifications of the experiment in Fig. 21.18 not only provide this proof but allow us to write a formula to describe induction quantitatively. We summarize these additional experiments in seven groups, indicating in each case what additional insight the particular experiments provide:

- **Switch to the South Pole in the original experiment.** The galvanometer records the same current peak, regardless of the speed of motion of the magnet, but the needle swings in the opposite direction. This shows that our experiment is limited by the instrument response time, but that the direction of the current in the loop is opposite for the two poles of a bar magnet. Thus, induction is a vector-based phenomenon.
- **Pull the magnet out from the current loop.** This reverses the experiment as shown in Fig. 21.18. Again, the observed galvanometer response is in magnitude equal to that of the original experiment, but the needle of the instrument swings in the opposite direction. This confirms further the directional interpretation of the two previous experiments: induction is sensitive to the direction in which the magnetic field is changing.
- **Double the number of bar magnets and move them together.** The area under the current peak of the galvanometer in all previous experiments doubles as well. This shows that induction is proportional to the magnitude of the changing magnetic field. If we further replace permanent bar magnets with current-carrying solenoids, we find the same results but can vary the changing magnetic field easier and more quantitatively.
- **Replace the conductor loop by one with *N* windings.** An *N*-fold increase in the galvanometer response is observed. This means that induction is also proportional to the magnitude of the total area of the loop through which the magnetic field change occurs.
- **Place a solenoid in the loop and switch it on and off.** The same effect is observed as for the moving bar magnet, indicating that induction is not the result of mechanical motion of a magnet but indeed due to the changing magnetic field.
- **Place a large magnet or solenoid in the vicinity of the loop, then spin the conductor loop in the stationary magnetic field.** An equivalent response of the galvanometer to the one seen in the previous experiments shows that induction is linked to a combination of the direction of the magnetic field and the normal vector representing the area of the loop. Even if both area and magnetic field are constant, rotating the normal vector in a uniform magnetic field leads to a changing field within the plane of the loop. The response of the galvanometer for a slow enough rotation indicates that the two vectors combine as a dot product in the formula for induction. The combination of the vectors representing the area of the loop and the magnetic field leads to the definition of a magnetic flux.
- **Replace the galvanometer with a voltmeter in the original experiment.** The voltmeter responds in the same manner as the galvanometer did before. This shows that the initial effect of induction is a change in the potential along the conductor loop, which in turn causes a current to flow.

Faraday's experiment leads to the formulation of **Faraday's law of induction**. We formulate this law initially not for the general case, but for the simple case of the arrangement shown in Fig. 21.18. The required geometric parameters are specified in Fig. 21.18. In that figure the conductor loop is circular with radius *R*. The figure contains several simplifying assumptions for the magnetic field:

- The magnetic field lines are parallel to the axis of the conductor loop at all times. If we choose the axis of the conductor loop along the *z*-axis, this means that the magnetic field in the loop simplifies from its generic form $\vec{B} = (B_x, B_y, B_z)$ to $\vec{B} = (B_x = 0, B_y = 0, B_z)$; i.e., the *x*- and *y*-components are zero.
- The B_z component is uniform across the conductor loop; i.e., it has the same value everywhere in the area defined by the loop.
- The *z*-component of the magnetic field can vary with time, $B_z(t)$, but will do so again uniformly across the area enclosed by the conductor loop; i.e., also dB_z/dt is uniform at any time *t*.

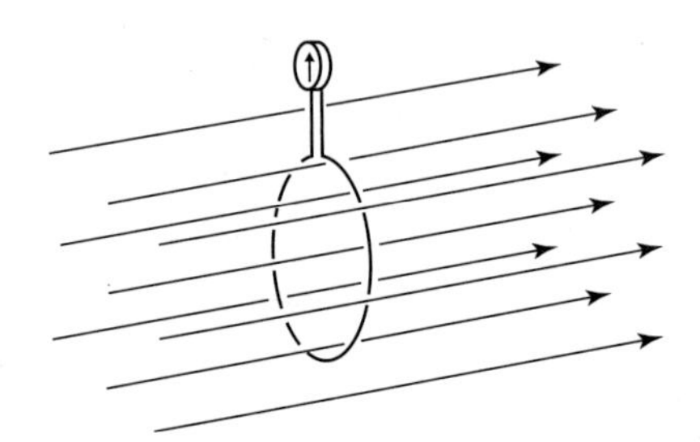

Figure 21.18 A circular conductor loop with radius *R* in the *xy*-plane. The magnetic field lines are parallel to the axis of the conductor loop in the *z*-direction. This B_z component is uniform across the conductor loop, but can vary with time in the form dB_z/dt.

Faraday's experiments showed that induction is quantified by the dot product of the changing magnetic field and the area defined by the conductor loop. The specific case in Fig. 21.18 allows us to neglect vector notation. We define the **magnetic flux Φ** through the current loop of area A as:

$$\Phi = B_z A = B_z \pi R^2. \quad [21.19]$$

KEY POINT

The magnetic flux is the magnetic field component that reaches perpendicular through the area of a current loop.

Faraday's observations mean that the change of the magnetic flux with time, $d\Phi/dt$, is equal to the total induced electric field along the loop, which is the field at the position of the loop multiplied by the length of the loop:

$$\left|\vec{E}\right|(2\pi R) = \frac{d\Phi}{dt} = \pi R^2 \frac{dB_z}{dt},$$

in which $2\pi R$ is the circumference of a circular current loop of area πR^2. This equation rearranges to:

$$E(t) = \frac{R}{2}\frac{dB(t)}{dt},$$

in which we dropped the subscript z to allow for any orientation of the loop with the magnetic field perpendicular to the loop area. Note that the magnetic and electric fields are always perpendicular to each other because the electric field occurs along the loop.

KEY POINT

A changing magnetic flux through a conductor loop causes an electric field along the conductor. This electric field is proportional to the radius of the circular loop. It causes a measurable current in the conductor.

EXAMPLE 21.6

A current of 1000 A is sent through a circular coil with 100 windings of radius of 5 cm. The magnet is 50 cm long. (a) What is the magnitude of the axial magnetic field in the coil? (b) The current is switched on in 100 μs. What is the rate of change of the magnetic field during the switching of the current? (c) The changing magnetic field in part (b) causes an induced electric field. How strong is this electric field?

Solution

Solution to part (a): We use Eq. [21.10] with N = 100, I = 1000 A, l = 0.5 m, and R = 0.05 m:

$$B = \frac{\mu_0 N I}{\sqrt{4R^2 + l^2}} = \frac{\left(1.26\times10^{-6}\ \frac{N}{A^2}\right)100(1000\ \text{A})}{\sqrt{4(0.05\ \text{m})^2 + (0.5\ \text{m})^2}},$$

continued

which yields:

$$B = 0.21\frac{\text{N}}{\text{A m}} = 0.21\ \text{T}$$

Solution to part (b): Electromagnetic fields propagate with the speed of light; therefore, the switching time for the current is essentially equal to the switching time for the magnetic field. Thus:

$$\frac{\Delta B}{\Delta t} = \frac{0.21\ \text{T}}{1.0\times10^{-4} s} = 2100\frac{T}{s}.$$

Solution to part (c): We use Eq. [21.20]

$$E(t) = \frac{0.05\ \text{m}}{2}\left(2100\frac{T}{s}\right) = 52.5\frac{\text{V}}{\text{m}}.$$

21.4.2: Transcranial Magnetic Stimulation

Fig. 21.19 shows the experimental set-up for transcranial magnetic stimulation (TMS). The equipment creates rapidly changing magnetic fields near the scalp of the patient. These pass safely through skin and bone because each short pulse contains only a very small amount of energy. The magnitude of the magnetic field $\left|\vec{B}\right|$ tails off rapidly with distance z from the coil as shown in Eq. [21.8]: $B \propto z^{-3}$ when $z >> R$, with R the radius of the coil. Thus, the magnetic field penetrates only a few centimetres into the outer cortex of the brain. The shape and radius of the coil allow us to localize the magnetic field; it induces in nearby neurons an electric field that activates the targeted region.

We use the discussion in the previous section to establish the relation between signal localization and signal depth in TMS. The localization is inversely proportional to the radius of the coil: the smaller R the better localized the changing magnetic field and, in turn, the induced

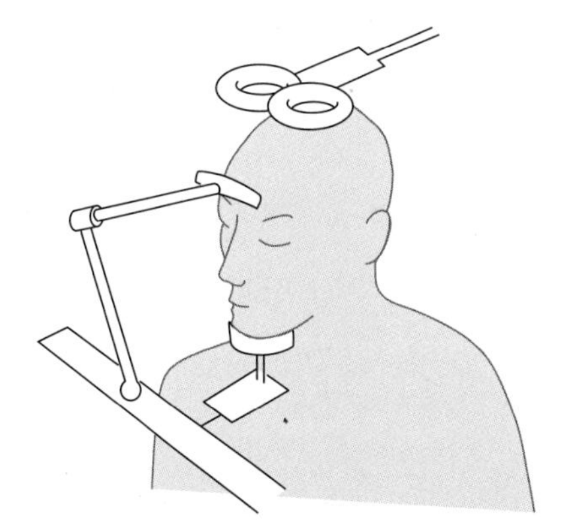

Figure 21.19 Set-up for transcranial magnetic stimulation. The head of the patient rests on a supporting frame. The stimulation device consists of two conductor coils acting as current loops. They are placed on the head such that the magnetic field is perpendicular to the skull. The magnetic field does not penetrate deep into the brain, stimulating mostly the outer sections. The conductor coil design localizes the magnetic field laterally.

electric field. Note that this result seems at odds when we insert Eq. [21.9], which applies to a single current loop, in Eq. [21.20]:

$$E(t) = \frac{R}{2}\frac{dB(t)}{dt} = \frac{R}{2}\frac{d}{dt}\left(\frac{\mu_0 I}{2R}\right) = \frac{\mu_0}{4}\frac{dI(t)}{dt}.$$

Thus, the induced electric field is independent of the radius R of the coil. Note, however, that Eq. [21.9] applies only at the centre of the loop ($z = 0$), which is above the patient's skull. At points z further along the axis of the loop, Eq. [21.8] has to be used instead:

$$E(t) = \frac{\mu_0}{4}\frac{R^3}{(z^2+R^2)^{3/2}}\frac{dI(t)}{dt}.$$

The easiest way to read the implications of this result is to study the case $z = R$. For a fixed current change in the loop, we find that the electric field at $z = R$ is the same at any R:

$$[E(t)]_{z=R} = \frac{\mu_0}{2^{7/2}}\frac{dI(t)}{dt}.$$

Thus, the larger radius R allows the same induced electric field deeper in the brain than a smaller radius R. At a smaller radius R we obtain better localization of the TMS signal, but cannot penetrate deep into the brain.

21.4.3: A First Attempt at Maxwell's Equations

We have established several connections between electric and magnetic phenomena by now, and should be able to combine these into one formalism. Coulomb's, Ampère's, and Faraday's laws have to be the three cornerstones and seem to be a complete set when combined with the observation that there are no magnetic monopoles. Even though it will turn out that we are still missing an important aspect, which will force us back to the discussion of Maxwell's equations later, it is useful to introduce a first attempt at this formalism as it contains several features that we need to familiarize ourselves with.

What we definitively need to do first is to find the generalized form of the three laws mentioned above. Coulomb's law assumes the interaction between two point charges, but we may rather want to have a formulation that allows for any possible charge distribution in space. Also, Ampère's law was only written for a current in a long straight conductor, and Faraday's law was written for a circular conductor loop. These have to be written for more general geometries as well.

To prevent this chapter from becoming a discussion of advanced integral or differential calculus, we reduce the discussion by stating Maxwell's (incomplete) equations in integral form first, then derive the three laws we introduced earlier by applying the specific, simplified geometries we used before. This approach omits the differential form of Maxwell's equations, but allows us to get a good idea what purpose these equations serve, and what physical concepts they contain. We also exclude cases in which materials are present in the system that are either dielectric or magnetic; i.e., we assume that everywhere the permittivity of vacuum (ε_0) and the permeability of vacuum (μ_0) are sufficient to characterize the system. In such a system *Maxwell's (incomplete) equations* are written in the form:

(I) *Gauss' law:* $\oint_{\text{surface}} \vec{E} \bullet d\vec{A} = \frac{q}{\varepsilon_0}$

(II) $\oint_{\text{surface}} \vec{B} \bullet d\vec{A} = 0$ [21.21]

(III) *Faraday's law:* $= \oint_{\text{loop}} \vec{E} \bullet d\vec{s} = 0 = -\oint_{\text{surface}} \frac{d\vec{B}}{dt} \bullet \overrightarrow{dA}$

(IV) *Ampere's law:* $= \oint_{\text{loop}} \vec{B} \bullet d\vec{s} = \mu_0 I_{\text{enclosed}}.$

We have to first give a physical meaning to the various new terms these equations contain. Note that all integral symbols contain a circle halfway down. We call such integrals **line integrals** because they are along arbitrary lines or surfaces, not along basic geometric axes. Note carefully which line or surface we integrate over.

The first two equations contain on the left-hand side an integral with the integration interval called *surface* (also indicated by the differential vector term $d\vec{A}$). When solving such integrals, a specific surface has to be chosen based on the problem at hand. This area must be closed because it is the surface that encloses a given volume. Both equations then state that the integral over the field across the enclosing surface is equal to the total amount of charge within the enclosed volume. Specifically

KEY POINT

Gauss' law states that the integral over the electric field reaching through a closed surface is equal to the total amount of electric charge within the enclosed volume. This is due to the fact that electric charges are the source of the electric field. Maxwell's second equation states that magnetic monopoles do not exist, i.e., that the integral over the magnetic field reaching through any closed surface is zero. Thus, no sources of magnetic field lines exist in any chosen volume.

Maxwell's third and fourth equations connect the integral of a field along a closed line (which we call a loop) to a changing second field that reaches through the surface defined by the loop (which we call a changing flux). In particular

KEY POINT

Faraday's law states that the time change of the magnetic flux through a given surface is negative equal to the integral of the electric field vector along the loop around the surface, and Ampère's law states that the electric current is equal to the integral of the magnetic field vector along the loop around the surface through which the current flows.

Note that we comment on the minus sign on the right-hand side of Maxwell's third equation when we discuss the complete set of equations in vacuum below. We illustrate that Maxwell's four equations are consistent with Coulomb's law and the special forms of Ampère's and Faraday's laws we introduced earlier in this chapter. We do this in the form of three Case Studies.

CASE STUDY 21.3

Show that Maxwell's first equation (Gauss' law) is consistent with Coulomb's law.

Answer: *We know from electrostatics how Coulomb's law applies to point charges. In the simplest case just a single point charge with charge q is considered. Note that Maxwell's first two equations do not require the choice of a particular shape of the volume we study. Thus, we choose in Fig. 21.20 the simplest possible surface to minimize the mathematical effort involved. This is a sphere of radius R surrounding the point charge at its centre, as illustrated by the dashed line in Fig. 21.20. Note that the arrows in that figure are drawn in the direction of the electric field at each point, and that the length of the arrows is proportional to the magnitude of the electric field at the respective position. We recall from electrostatics that the electric field of a positive point charge is directed radially outward (or inward if q is a negative charge), and its magnitude depends only on the distance from the point charge.*

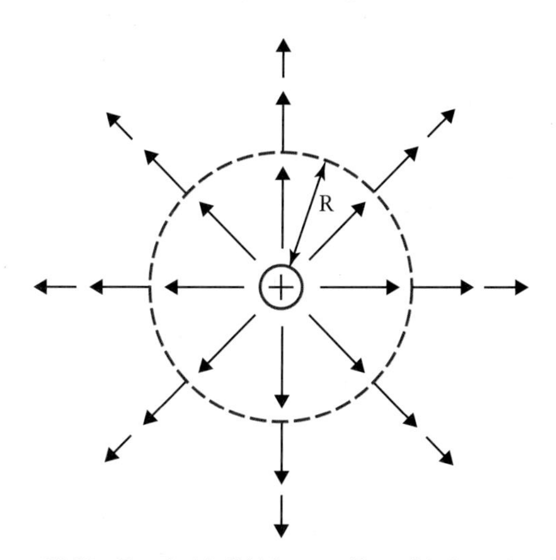

Figure 21.20 The electric field for a positive point charge is radial, with the magnitude of the electric field decreasing with the square of the distance from the point charge. The dashed line indicates the spherical surface at radius R for which we evaluate Gauss' law. Note that the direction of the electric field is everywhere perpendicular to the dashed surface, with a constant magnitude $|\vec{E}(R)|$.

The spherical choice of the dashed surface in Fig. 21.20 allows us therefore to pull the electric field out of the integral as it is directed everywhere perpendicular to the dashed surface, with a constant magnitude $|\vec{E}(R)|$ The remaining argument of the integral is reduced to the geometric term dA and yields the surface of the sphere $4\pi R^2$.

continued

Thus, the left side of Maxwell's first equation reduces to $4\pi R^2\,|\vec{E}(R)|$. With the right-hand side of Maxwell's first equation already given in explicit form, we can substitute this result to find:

$$\oint_{\text{surface}} \vec{E}\bullet d\vec{A} = 4\pi\, R^2 |\vec{E}(R)| = \frac{q}{\varepsilon_0},$$

which is rearranged in the form in which the electric field for a point charge is usually written (Coulomb's law):

$$|\vec{E}(R)| = \frac{q}{4\pi\varepsilon_0}\frac{1}{R^2}.$$

Thus, Gauss' law is indeed a generalization of Coulomb's law: the electric field through a closed surface represents the enclosed charges. Since there are no magnetic monopoles, the corresponding law for the magnetic field yields zero (Maxwell's second equation). This law is referred to as Gauss' law because Karl Friedrich Gauss developed the mathematical formalism that is required to analyze an arbitrary charge distribution.

In the next Case Study we want to address in a similar manner Maxwell's third equation. For this, we generalize first the concept of a magnetic flux and connect it with the terms in Maxwell's third equation. Eq. [21.19] defines the magnetic flux for the particular case of Fig. 21.18. We generalize this result by re-establishing the vector character of the magnetic field $\vec{B}$ and the area $\vec{A}$ on the right-hand side. We also allow the magnetic field to vary across the area, thus requiring an integration:

$$\varphi = \oint_{\text{surface}} \vec{B}\bullet d\vec{A}. \quad [21.22]$$

This is the general definition of the **magnetic flux** as it appears in Maxwell's second equation explicitly in this form. It is also present as a time derivative in Maxwell's third equation. Eq. [21.22] allows us therefore to rewrite Maxwell's third equation with the magnetic flux:

$$\oint_{\text{loop}} \vec{E}\bullet d\vec{s} = -\frac{d\Phi}{dt} \quad [21.23]$$

CASE STUDY 21.4

Show that Maxwell's third equation is consistent with the specific form of Faraday's law we introduced in Eq. [21.20].

Answer: *In Maxwell's third equation the two integrations are related in that the loop on the left-hand side encloses the surface on the right-hand side. Thus, choosing a suitable surface to minimize the required mathematical formalism is usually the first step in applying Faraday's law. In the particular case, the choice of surface is determined by the set-up of Fig. 21.18 that led to the specific formulation in Eq. [21.20].*

continued

The loop in Fig. 21.18 has radius R and circumference $2\pi R$, and encloses an area πR^2. Faraday showed that the changing magnetic field penetrates this area in the axial direction, which yields a force on point charges in the conductor that is everywhere collinear with the conductor loop. The corresponding electric field points in the same direction as this force. Thus, the two dot products in Faraday's law reduce for this geometry to products of the respective magnitudes:

$$\oint_{\text{conductor loop}} E\,ds = \oint_{\text{circle}} \frac{dB}{dt}\,dA.$$

We further used in Fig. 21.18 that the rate of change of the magnetic field, dB/dt, is uniform across the entire loop area. Therefore, the magnitude of the caused electric field is the same at every point along the conductor loop. Thus, the two field terms can be taken out of their integrals respectively. The remaining arguments of the integrals are reduced to the geometric terms ds and dA, which yield the circumference at the left side and the circular area on the right side:

$$2\pi R E = \frac{dB}{dt}\pi R^2,$$

which is Eq. [21.20].

CASE STUDY 21.5

Show that Maxwell's fourth equation predicts the magnetic field inside a long solenoid.

Answer: Fig. 21.21 shows a cross-section though a horizontal solenoid that carries current I. The dots near the top and at the bottom of the figure indicate the windings passing through the plane of the paper. We know from experiments that the magnetic field lines within the solenoid are coaxial with the device and that the magnetic field inside is uniform. Fig. 21.21 also indicates that the magnetic field directly outside a solenoid is much smaller than within the device (much reduced density of field lines). Thus, we neglect the external field in the present calculation (as indicated with the notation $\vec{B} = 0$ in Fig. 21.22).

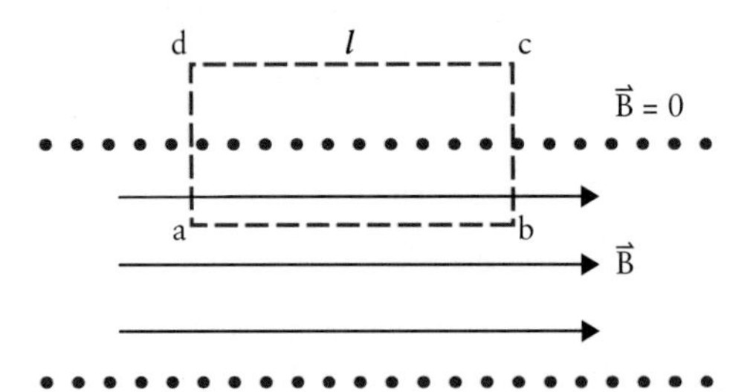

Figure 21.21 Cross-section though a horizontal solenoid with current I. The dots near the top and at the bottom are the windings passing through the plane of the paper. The magnetic field lines within the solenoid are coaxial with the device; outside the magnetic field is much smaller and is neglected as indicated with $\vec{B} = 0$. A dashed rectangle of length l along the solenoid is used to evaluate Maxwell's fourth equation.

continued

Fig. 21.21 also shows a dashed rectangle of length l along the solenoid, with its four corners labelled by lower case letters. This rectangle defines the loop we use in Maxwell's fourth equation:

$$\oint_{\text{rectangle}} \vec{B} \bullet d\vec{s} = \int_a^b \vec{B} \bullet d\vec{s} + \int_b^c \vec{B} \bullet d\vec{s} + \int_c^d \vec{B} \bullet d\vec{s} + \int_d^a \vec{B} \bullet d\vec{s} = \mu_0 N I.$$

This equation illustrates an advantage of choosing a rectangular path: the line integral breaks down into four regular integrals. The right-hand side of the equation results because we assume that the solenoid has N windings per length l.

The four integrals are easy to evaluate individually. The integrals from b to c and from d to a are both zero because the magnetic field $\vec{B}$ and the path element $d\vec{s}$ are perpendicular to each other. The integral from c to d is zero because the magnetic field outside the solenoid is negligible. This leaves only the integral from a to b with a finite value; along this path the magnetic field is parallel to the path and uniform in value. This allows us to drop the dot product in favour of a product of the magnitudes, and allows us further to pull B out of the integral:

$$B\int_a^b ds = \mu_0 N I.$$

The remaining argument of the integral is ds and results in the length l. Thus, we find:

$$Bl = \mu_0 N I,$$

which is Eq. [21.7].

21.4.4: Maxwell's Displacement Current

We called our first attempt at writing Maxwell's equations *incomplete* because there was one term missing; indeed, the incomplete set of equations summarizes the work of several scientists who had made their contributions before James Clerk Maxwell's work. Maxwell himself added this missing term and thereby was first to complete the electromagnetic theory. In this section we motivate this missing term. This allows us to review Maxwell's equations in complete form in the next section, and then complete this chapter with the most important conclusion Maxwell drew from his equations: the prediction of electromagnetic waves.

Maxwell revisited Ampère's law because he wondered about the need to have a conductor present. His reasoning was that a changing magnetic field should be felt at the position of the conductor, whether or not the conductor is present. He tested this idea with the experiment we show in Fig. 21.22. Shown is a current loop for which we know from Fig. 21.4 that a magnetic field

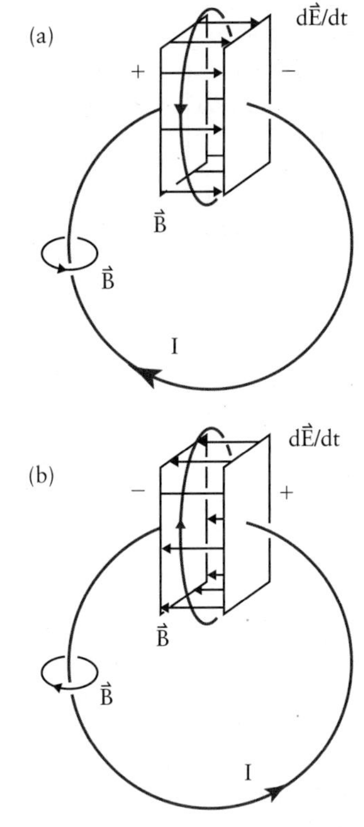

Figure 21.22 A current loop that connects the two plates of a parallel plate capacitor. When a current flows through the loop, it charges and discharges the capacitor, representing a changing electric field between the plates. Frames (a) and (b) assume that the charges on the plates flow back and forth, causing all vector properties to switch directions as the current flows in opposite directions. Maxwell postulated the existence of a magnetic field associated with the changing electric field between the capacitor plates.

is present when a current flows. We assume the current flows back and forth between two parallel plates that form a capacitor; the two directions of flow are shown as parts (a) and (b) of the figure.

Maxwell then focused on the space between the parallel plates. We know that a changing electric field must exist in this gap because the alternating charging and discharging of the capacitor causes such a field. Maxwell suggested that the changing electric field would be associated with a magnetic field as drawn in the figure. He used an indirect argument based on the effect we observe when a dielectric is placed between the parallel plates. For example, if there were liquid water, we know that the water molecule dipoles align themselves with the external electric field such that the positive end of the molecule (where the two hydrogen atoms are located) points toward the negative capacitor plate. Thus, for every discharge and charge of the capacitor, each water molecule rotates by 180°. Since the water molecule has a finite size, this corresponds to a minute shift of charges in the direction perpendicular to the plates. A shift of charges can be interpreted as an electric current; Maxwell called this particular current a **displacement current**. The displacement current is then associated with a magnetic field in the same manner as a normal electric current, e.g. as discussed in Fig. 21.4.

Finally, Maxwell reasoned that, even if there is no dielectric between the capacitor plates (e.g., if there is a vacuum), the changing electric field should nevertheless be accompanied by a magnetic field. Thus, Ampère's basic idea that a current causes a magnetic field can be generalized by complementing the current, in its ability to generate a magnetic field, with a changing electric field, such as present between the parallel plates in Fig. 21.22. This magnetic field can be measured to confirm Maxwell's idea.

21.4.5: A Second Attempt at Maxwell's Equations

Maxwell's equations in their complete integral form read as follow:

$$\begin{aligned} &\oint_{\text{surface}} \vec{E} \bullet d\vec{A} = \frac{q}{\varepsilon_0} \\ &\oint_{\text{surface}} \vec{B} \bullet d\vec{A} = 0 \\ &\oint_{\text{loop}} \vec{E} \bullet d\vec{s} = -\oint_{\text{surface}} \frac{d\vec{B}}{dt} \bullet d\vec{A} \\ &\oint_{\text{loop}} \vec{B} \bullet d\vec{s} = \mu_0 \varepsilon_0 \oint_{\text{surface}} \frac{d\vec{E}}{dt} \bullet d\vec{A} + \mu_0 I \text{ enclosed.} \end{aligned} \qquad [21.24]$$

Recall that we excluded the presence of magnetic or dielectric materials in the system of interest. In cases when such materials are present, Maxwell's equations are slightly more complex than Eq. [21.24]. The new term Maxwell added is the first term on the right-hand side of the fourth equation. This equation therefore states that two phenomena cause a magnetic field: a current, as previously stated by Ampère, and a changing electric field, as observed by Maxwell himself.

Note that Maxwell's equations are now intriguingly symmetric in the magnetic and electric fields except for the fact that electric charges exist, but not magnetic monopoles. Thus, electric field lines always originate in a charged object and end in another charged object, while all magnetic field lines must be closed loops without a source or a sink. This apparent symmetry becomes even more evident when we consider Maxwell's equations in empty space, i.e., when no electric charges or electric currents are in the vicinity. In this case we write for **Maxwell's equations in vacuum**:

$$\begin{aligned} &\oint_{\text{surface}} \vec{E} \bullet d\vec{A} = 0 \\ &\oint_{\text{surface}} \vec{B} \bullet d\vec{A} = 0 \\ &\oint_{\text{loop}} \vec{E} \bullet d\vec{s} = -\oint_{\text{surface}} \frac{d\vec{B}}{dt} \bullet d\vec{A} \\ &\oint_{\text{loop}} \vec{B} \bullet d\vec{s} = \mu_0 \varepsilon_0 \oint_{\text{surface}} \frac{d\vec{E}}{dt} \bullet d\vec{A} \end{aligned} \qquad [21.25]$$

KEY POINT

Maxwell's equations in vacuum state that a changing magnetic field induces an electric field, and a changing electric field induces a magnetic field.

There are only two minor differences between magnetic and electric field terms:

- The product of two natural constants, $\mu_0\,\varepsilon_0$, occurs in the last equation. This product occurs due to the choice of the SI unit system for all parameters in Eq. [21.25]. Note that one can find a different unit system in which this term is one; however, we will see in the next section that this term contains important information about a system that is governed by these equations.
- The third equation carries a minus sign that the fourth equation does not contain. The single minus sign is important when we discuss the interplay of electric and magnetic fields in the next section. When you allow the third equation to generate a changing electric field due to a changing magnetic field, and substitute this electric field in the fourth equation, then a secondary changing magnetic field results. Thanks to the minus sign, this secondary field is directed in the opposite direction to the initial field.

21.4.6: Electromagnetic Waves

Maxwell's equations predict all phenomena in electromagnetic theory. Discussion of many of these phenomena is usually the subject of specialized lectures in electromagnetic theory, but that discussion also plays a role in philosophical treatise (when we discuss the theory of science based on its complete axiomatic systems, of which Maxwell's equations are one) and science history courses, where discrepancies between Maxwell's and Sir Isaac Newton's axiomatic systems are used to introduce Albert Einstein's special theory of relativity.

We limit our discussion here to the one point that immediately intrigued Maxwell himself when he looked at Eq. [21.25]. He felt that this set of equations and their physical implications looked very similar to the equations that led to the wave equation for sound. Like the interplay of pressure variations causing air to accelerate, and the resulting speed variations in turn causing changes in pressure, Maxwell's third and fourth equations suggest the same kind of interplay between electric and magnetic fields. We illustrate this with Fig. 21.23: A changing magnetic field, $d\vec{B}_1/dt$, causes a perpendicular electric field $\vec{E}$ to curl around the magnetic field lines; the variations of this field in turn cause a magnetic field $\vec{B}_2$, which is opposed in direction to the original magnetic field. As in the case of sound, a harmonic wave results if there is a harmonic excitation. Such an excitation can be achieved with the device we illustrated in Fig. 21.22, except that we can further simplify it to a single metallic rod with vibrating charges, called an **antenna**. The harmonic vibration of charges in an antenna is illustrated in Fig. 21.24.

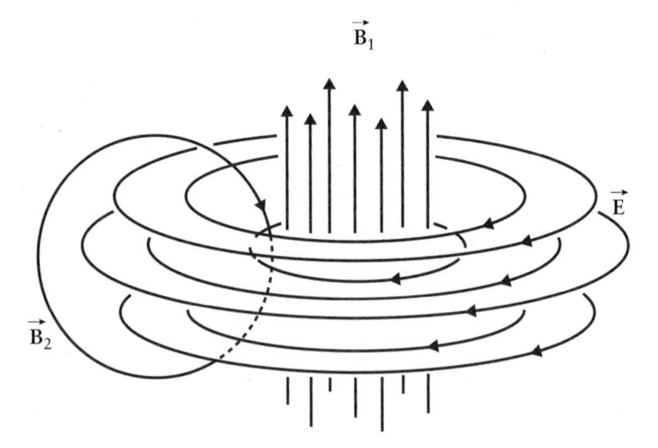

Figure 21.23 Maxwell discovered that a changing magnetic field, $d\vec{B}_1/dt$, can not only cause a current in a conductor, but can also cause a changing electric field, $d\vec{E}/dt$, outside a conductor. The changing magnetic and electric fields stipulate each other as illustrated with the secondary magnetic field $\vec{B}_2$.

At time $t = 0$ the antenna is charged like a dipole, with the positive charges at the upper end and the same number of negative charges at the lower end. Associated with the separated charges is an electric field. At time $t = T/4$, the charges recombine (the antenna is a conductor), which requires the motion of charges along the conductor,

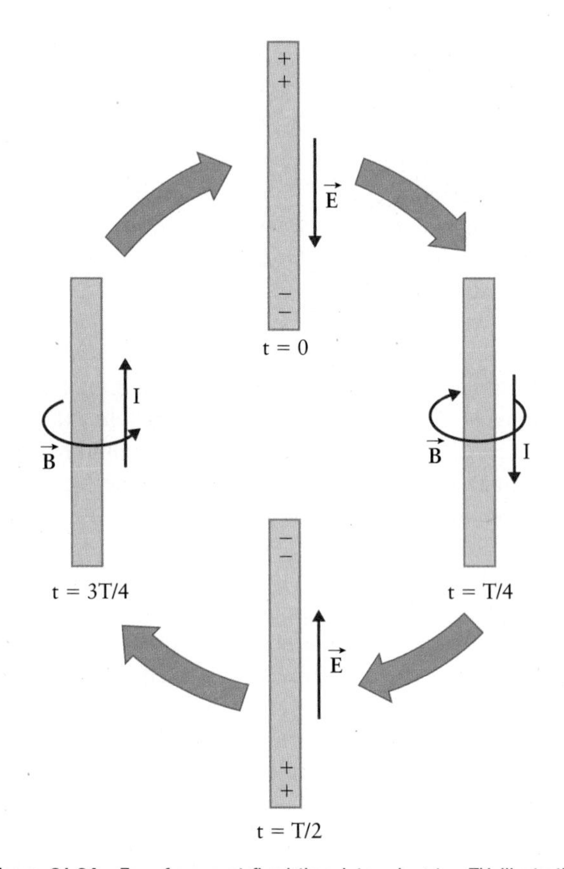

Figure 21.24 Four frames at fixed time intervals $\Delta t = T/4$ illustrating the concept of an antenna. Positive and negative charges are separated along the antenna in a periodic manner. When the charges move along the conductor, a magnetic field forms.

yielding a current. Associated with the current is a magnetic field. The current continues to flow until, at $t = T/2$, the charges are again separated; however, this time the negative charges are at the top and the positive charges are at the bottom. Therefore, the electric field points now in the opposite direction. At the next time frame (at $t = 3T/4$) the charges are once again recombining. The current flows now up, causing a magnetic field opposite to the one observed at time $t = T/4$. A harmonic oscillation of charges in the antenna follows if we force the charges to move in a periodic manner with period T as shown in Fig. 21.24.

The changing magnetic field shown in the figure causes a changing electric field, which in turn causes a changing magnetic field, as illustrated in Fig. 21.24. These interacting electric and magnetic fields travel outward from the antenna like the sound waves caused by a vibrating piston. For electromagnetic waves, no medium is needed because the electric and magnetic fields maintain each other. This is illustrated in Fig. 21.25, where the perpendicular magnetic and electric fields are indicated by arrows. The perturbation travels in the direction perpendicular to both the electric and the magnetic fields (**transverse wave**) with a speed Maxwell calculated from Eq. [21.25] as $c = (\varepsilon_0 \mu_0)^{-1/2}$, in which ε_0 is the permittivity of vacuum from Coulomb's law and μ_0 is the permeability of vacuum from Eq. [21.2]. We quantify c by substituting the given values for the two natural constants:

$$c = \frac{1}{\sqrt{\mu_0 \varepsilon_0}} = \frac{1}{\sqrt{\left(1.2566 \times 10^{-6} \frac{N}{A^2}\right)\left(8.854 \times 10^{-12} \frac{As}{Vm}\right)}}$$

$$= 2.998 \times 10^8 \ \frac{\text{m}}{\text{s}}.$$

Maxwell recognized this as the **speed of light**.

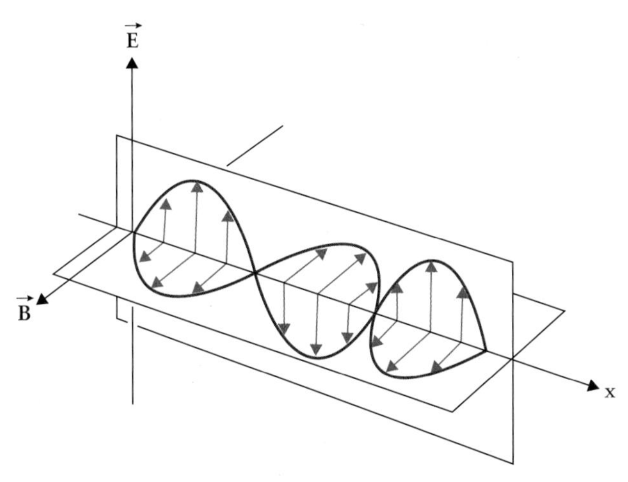

Figure 21.25 Maxwell's theory allows changing magnetic and electric fields to sustain each other as they travel outward from the vibrating charges of an antenna. The resulting electromagnetic waves are characterized by magnetic and electric field vectors that are perpendicular to each other and to the direction of wave propagation. The speed of the wave is the speed of light.

KEY POINT

Visible light is an electromagnetic wave that propagates in vacuum with the speed of light. Thus, all phenomena discussed in optics can be explained by Maxwell's four equations.

21.5: Physics and Physiology of Colour

21.5.1: Wavelength Dependence on Medium

Are there further effects on a propagating electromagnetic wave due to the medium? Since atoms and molecules have internal electric fields, as we discussed in Chapter 20, and the travelling electromagnetic wave is carried by its own electric field, light will always be slowed down when passing through matter. Therefore we know that at least one of the quantities wavelength λ and frequency f must vary as light passes through an interface because:

$$v_{\text{light}} = \lambda f.$$

Fig. 21.26 determines which of the parameters on the right-hand side varies from medium to medium. Shown in the figure are two media with different speeds for light. Light of wavelength λ_1 and frequency f_1 approaches the interface from the top. We assume that light travels slower in medium 2, $v_{\text{light},2} < v_{\text{light},1}$.

Let's assume that two observers, at positions A and B, count the number of wave maxima passing through their respective dashed observation planes. If observer A were to count a smaller number of wave maxima than observer B, then eventually there would

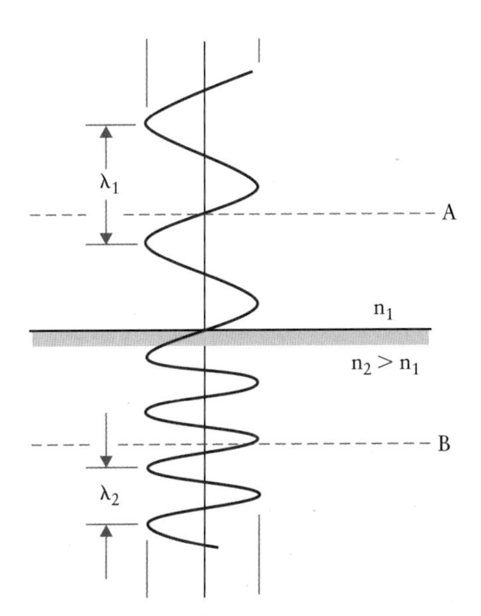

Figure 21.26 Light wave passing through an interface. The wavelength varies; the frequency remains unaltered.

be no wave maxima left in the range between the two observers and the whole experiment would somehow collapse. If, in turn, observer A were to count a larger number of wave maxima than observer B, then wave maxima would pile up between the two dashed lines. Since neither case makes sense, we conclude that the same number of wave maxima must pass observers at A and B during any time interval; i.e.,

$$f_1 = f_2 \Rightarrow \lambda_2 < \lambda_1.$$

Thus, the speed of light and the wavelength of light change when light passes through an interface (as illustrated for the wavelength in Fig. 21.26), but the frequency of the light remains fixed. From this we conclude that the frequency of light is the most fundamental of the three quantities frequency, wavelength, and speed of light in a medium, as nature conserves it for light travelling through different media.

Finding the frequency more fundamental than the wavelength or the speed of light in a particular medium is consistent with the corpuscle theory of light, which we referred to in Chapter 20 when discussing the atom. Light corpuscles are called **photons**. If a corpuscle travels as indicated in Fig. 21.26, we expect its energy to remain unchanged. Thus, the corpuscle energy should not be linked to the wavelength or its speed. Indeed, we note that the energy of a photon is given by:

$$E_{\text{photon}} = h f \text{ with } h = 6.6 \times 10^{-34} \text{ J s}, \qquad [21.26]$$

where h is the **Planck constant**, named in honour of Max Planck.

CASE STUDY 21.6

The angle of refraction depends on the wavelength of the incident light. Shouldn't the angle of reflection also depend on the wavelength?

Answer: *No. Dispersion is the result of electric interaction between the light ray and the medium through which it travels. The reflected beam interacts to only a very limited extent with the material surface.*

Dispersion affects any transparent material, including the lenses in our eyes. This leads to several colour-related illusions. An example is the perception that the red and blue bars of the French flag differ in width. For a red and a blue field at the same distance, the lens must be bent more to focus the red light as it is refracted less by a given lens. The focusing of the eye is associated with adjustments of the lens caused by action of the ciliary muscle. The brain notices the degree of work the ciliary muscle is doing and interprets this information as a measure of distance to the observed object. Thus, while you focus on the red area of the flag your brain thinks *near*, and while you focus on the blue area of the flag your brain thinks *farther away*. A lot of motion occurs in your eyes while looking at an object such as the French flag in order to get the picture right. The apparent difference in width of the two bars is an illusion that is caused by the brain as it tries to correct for the obviously inconsistent depth information; you know that the three bars of the flag are at the same distance since they are woven together.

Why then do we not have the same problem with the Italian flag? The dispersion effect is greatest for colours at opposite ends of the visible spectrum, such as red and blue. The Italian flag has a red and a green bar; thus, the same effect applies but is less noticeable. Note that the flag of Haiti has a blue and a red bar, like the French flag. Why does the problem discussed for the French flag not occur with this flag? For a while, French legislators had a law that required flag manufacturers to compensate for the dispersion effect by using uneven widths for the three bars; no such law ever existed in Haiti. Historically, the Haitian flag is the result of the removal of the white bar in the French flag when this former colony separated from France. Physically removing the white bar eliminated a neutral visual separation between the two colour bars. Our eyes are in constant motion and can adjust for optical illusions more easily when the two affected areas are adjacent to each other. This is why another good example of the dispersion effect is stained-glass church windows, in which coloured panels are separated by lead strips.

21.5.2: Colour as a Physical Concept

The fact that electromagnetic waves have different wavelengths and frequencies is not sufficient to explain why we see colours. To understand colour vision, we must discuss two additional phenomena. The first is the range of visible wavelengths of the electromagnetic spectrum. This also explains why we see white light emerging from the Sun and why white light is a mixture of all colours of the rainbow. These issues can be addressed with physical concepts and are discussed in this section.

The other necessary ingredient for understanding colour vision is the mechanism by which our eye and our brain convert wavelengths into colour impressions. This second issue requires us to study the interplay of physics and physiology, which we reserve for the last section of this chapter.

The complete **spectrum** of electromagnetic waves is shown in Fig. 21.27. It reaches from radio waves with wavelengths in the centimetre and metre range, to microwave and infrared radiation in the micrometre-to-centimetre wavelength range, to visible and ultraviolet light with wavelengths in the nanometre range, and finally to X-rays at the shortest wavelengths. The fraction of this spectrum that is visible is very small, reaching from about 370 nm (violet) to about 760 nm (red). This visible part of the electromagnetic spectrum is highlighted in Fig. 21.28, which correlates the names of various colours to the respective wavelengths, frequencies, and energies of the light. Note that the reason

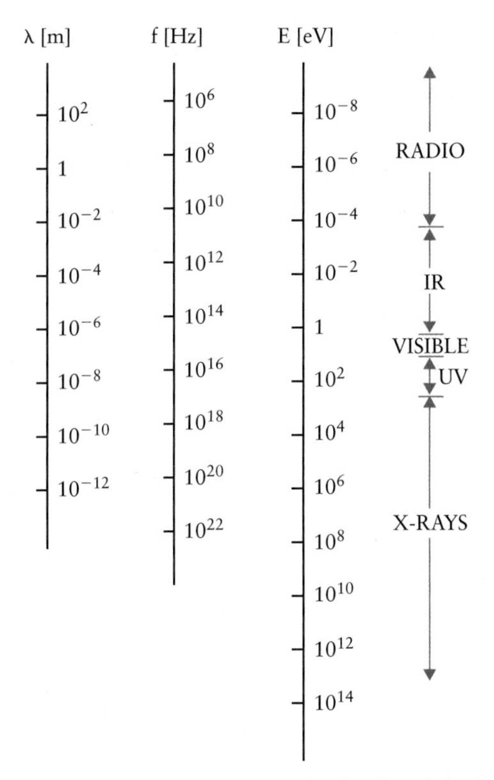

Figure 21.27 Electromagnetic spectrum showing the relation between frequency f, wavelength λ, and photon energy $E = hf$. IR is infrared and UV is ultraviolet.

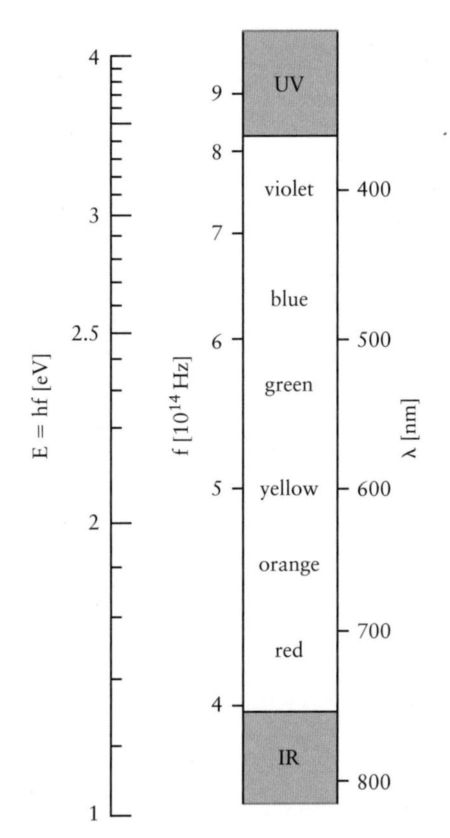

Figure 21.28 Visible part of the electromagnetic spectrum. The colours indicated correspond to the respective wavelengths.

we can give an energy scale in Figs. 21.27 and 21.28 is due to the corpuscle theory of light, which defines the energy of light photons of frequency f as $E = h f$, with h the Planck constant. That E is indeed an energy is evident when you let sunlight shine on your skin. The warmth you feel is the energy of the photons deposited in your skin, retained as the result of light absorption.

Do we have to conclude from Figs. 21.27 and 21.28 that our vision is ill adapted to the real world around us because it cannot detect most of the electromagnetic spectrum? No, because only a small fraction of the entire electromagnetic spectrum reaches our eye. To understand why, we need to introduce the concept of **blackbody radiation**.

A blackbody is defined as an object that perfectly absorbs all the light that reaches it; i.e., it does not reflect any of that light. As a blackbody absorbs the incoming light, the light's energy is converted into thermal energy, raising the temperature of the blackbody. If light shines on the blackbody continuously, it is not in thermal equilibrium but becomes hotter and hotter. When a body becomes hotter, however, we know that it starts to emit light. Examples include the metal filament in an incandescent light bulb. By balancing light absorption, this light emission establishes a thermal equilibrium for a blackbody immersed in a fixed light radiation field.

Fig. 21.29 illustrates the intensity of light of various wavelengths emitted from a blackbody at three different temperatures. Both the overall intensity (area under the curve) and the wavelength of the maximum intensity vary with temperature. The peak in Fig. 21.29 shifts to shorter

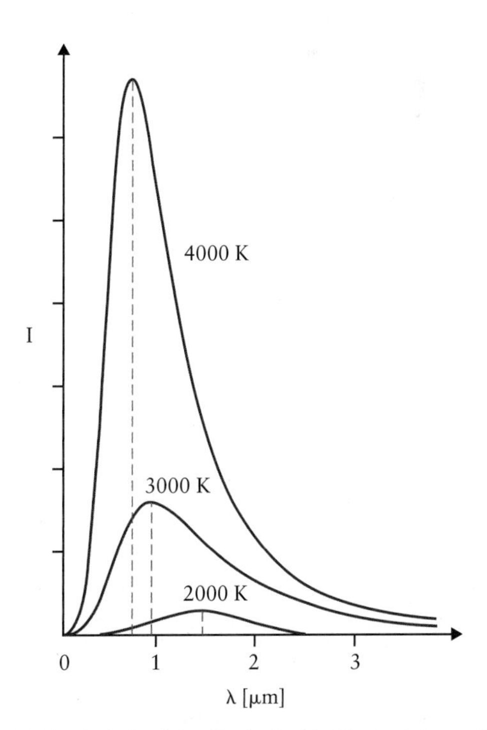

Figure 21.29 Emission intensity, I, of a blackbody at three different temperatures. The vertical dashed lines indicate the wavelengths of the respective maximum intensities.

wavelengths with increasing temperature. This observation is called **Wien's displacement law** and is quantified in the form:

$$\lambda_{max} T = 0.29 \times 10^{-2} \text{ mK}. \quad [21.27]$$

Let us assume the Sun is a blackbody. Due to its high temperature it radiates light with a spectrum like that shown in Fig. 21.29. The actual spectrum (total intensity and wavelength of radiation peak) depends on the Sun's surface temperature. While the interior of the Sun reaches temperatures of 1×10^7 K, the electromagnetic spectrum is determined by the surface temperature of the Sun, which is 5800 K. Light from greater depths does not leave the Sun because the Sun's surface layer is not transparent to that light. Using Fig. 21.29, we predict a maximum intensity of the radiation reaching Earth from the Sun in the visible wavelength range. As a consequence, many processes on Earth are tuned in to this maximum, including photosynthesis and our vision.

Fig. 21.29 also shows that the total intensity of light emission from a blackbody depends strongly on the temperature of the blackbody. The total intensity is the area underneath the curve and can be calculated analytically using the correct formula for the curves in the figure. The intensity curves in Fig. 21.29 follow from a quantum-mechanical calculation done initially by Max Planck. In 1879, Josef Stefan wrote **Stefan's law for the blackbody**:

$$\frac{\Delta Q/\Delta t}{A} = -\frac{2\pi^5 k^4}{15 c^2 h^3} T^4 = -\sigma T^4,$$

in which $\Delta Q/\Delta t$ is total rate of emission of heat in unit J/s and A is the surface area of the blackbody. The term $(\Delta Q/\Delta t)/A$ is the rate of energy loss per unit area, i.e., a term we calculated for Earth based on heat conduction in Chapter 11. The right-hand side contains the Boltzmann constant k because it is part of the thermal energy of a system in the form $k\,T$; the speed of light c because it converts energy density to intensity, as discussed in Eq. [15.14]; and the Planck constant h because of the discrete nature of transitions in the blackbody that cause the radiation. The constants on the right-hand side of the last equation are combined in a single constant, called s, the **Stefan–Boltzmann constant**, with $\sigma = 5.67 \times 10^{-8}$ J/(m^{2}s K^4). Note that the equation contains a negative sign on the right-hand side following our convention that energy lost by the object of interest is negative.

Insofar as Stefan's law applies only to a true blackbody, it is a model equation. However, it can be applied to actual objects if the right-hand side is corrected with an object-specific constant, called ε, the **emissivity**. In most cases it is sufficient to know the material of the object to find a tabulated value for ε, with $\varepsilon = 1$ for a true blackbody and $0 < \varepsilon \leq 1$ for any other object. Once we include the emissivity, **Stefan's law for arbitrary objects** is written as:

$$\frac{\Delta Q/\Delta t}{A} = -\varepsilon\sigma T^4. \quad [21.28]$$

Energy loss by radiation is also significant for objects near room temperature, as shown in Table 11.2 for the energy loss from the human body.

CASE STUDY 21.7

On a clear cold night in Northern Europe, the Northern United States, or Canada, why does frost tend to form on the tops of objects but not on their sides?

Answer: *Objects lose some fraction of their energy through radiation. During a clear night little radiation comes from above (outer space), while objects exchange radiative heat sideways with other objects in their environment. This effect is not observed farther south because the air temperature at night remains warm enough to compensate the energy loss by radiation with energy gain due to heat conduction.*

EXAMPLE 21.7

A person with a skin temperature of 37°C is in a room at 20°C. How much heat does the person's body lose per hour if the human body has a surface area of 1.5 m^2 and is modelled as a blackbody with emissivity $\varepsilon = 0.9$?

Solution

Modelling the human body as a blackbody means that the human body is in a radiative equilibrium when placed in an environment of the same temperature. As a consequence, the emissivity ε for radiative loss and radiative gain must be the same; otherwise, the body's temperature would eventually differ from the 37°C temperature of the environment.

Placing the human body in an environment with a different temperature means no radiative equilibrium is established and the warmer object loses energy to the colder object. In the given case, we determine the radiative energy loss and radiative energy gain from the surroundings using Eq. [21.28]:

$$\left(\frac{\Delta Q}{\Delta t}\right)_{loss} = -\sigma\varepsilon A T_{body}^4$$

$$\left(\frac{\Delta Q}{\Delta t}\right)_{gain} = +\sigma\varepsilon A T_{environment}^4,$$

which leads to a net heat balance of:

$$\Delta Q_{net} = \sigma\varepsilon\Delta t A\left(-T_{body}^4 + T_{environment}^4\right).$$

Inserting the numerical values in this equation leads to:

$$\Delta Q = \left(5.67\times10^{-8}\frac{\text{J}}{\text{m}^2\text{sK}^4}\right)0.9(3600 \text{ s})$$
$$\times(1.5 \text{ m}^2)(293^4\text{K}^4 - 310^4\text{K}^4)$$
$$= -5.1\times10^5 \text{ J}.$$

Note that the heat loss discussed in this example is due to radiation at electromagnetic wavelengths, and thus is not dependent on a medium to carry the heat. However, heat cannot pass through opaque interfaces radiatively. This is the reason for the **greenhouse effect**. In a greenhouse, incoming energy from the Sun passes through the glass surfaces in the visible range. The radiation of the plants in the greenhouse occurs at much longer wavelengths (infrared) since the temperature of the plants is much lower than the surface temperature of the Sun. At infrared wavelengths glass is opaque; i.e., the radiation cannot escape from the greenhouse, thus increasing the temperature in the greenhouse beyond the temperature outside.

21.5.3: Colour as a Physiological Concept

As complex and subjective as the process of interpretation of wavelengths in our brain may be, interesting physics are included in the physiological aspects of colour vision. We want to establish these in this section. Once the light from an object reaches the retina, the colour you see depends on the relative excitation of three **colour-sensitive receptor** types embedded at each point in your retina. We already mentioned that Young postulated three different receptors are needed. Their respective sensitivity as a function of the wavelength of the incoming light is shown in Fig. 21.30. Corresponding to the range of greatest sensitivity, the three receptors are labelled R, G, and B for **red receptor**, **green receptor**, and **blue receptor**. The significant overlap of the peaks in Fig. 21.30 rules out a simple colour composition concept. Instead, each colour we perceive in our brain is associated with light that has excited at least two different cones, and most colour impressions result from an excitation of all three types of cones.

How we use Fig. 21.30 to predict a colour impression is illustrated next. We start with the definition of the **colour triangle**. For an object with a given colour, let the absolute excitation intensity of the red receptor be R, the absolute excitation intensity of the green receptor be G, and the absolute excitation intensity of the blue receptor be B. The total excitation intensity is then $R + G + B$. The total excitation intensity determines whether the object appears bright or dim and duplicates the information the brain receives from the retinal rods. Using the total excitation intensity, we define a relative excitation intensity for each of the three receptor types:

$$r = \frac{R}{R+G+B} \quad \text{with } 0 \le r \le 1$$

$$g = \frac{G}{R+G+B} \quad \text{with } 0 \le g \le 1$$

$$b = \frac{B}{R+G+B} \quad \text{with } 0 \le b \le 1,$$

with

$$r + g + b = 1$$

The red and the green relative excitation intensities, r and g, are used to form the two perpendicular sides of the colour triangle in Fig. 21.31. With r and g given in the figure, the value of b is derived from $b = 1 - g - r$.

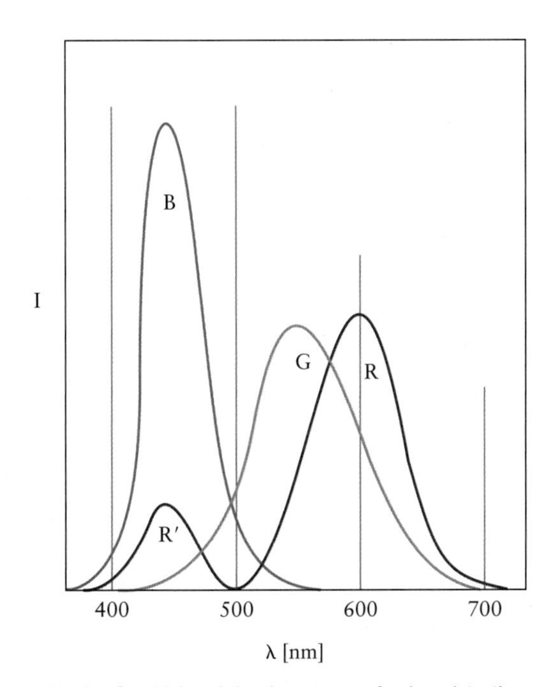

Figure 21.30 Sensitivity of the three types of colour-detecting cones as a function of the wavelength in the visible range of the electromagnetic spectrum. The cones are labelled R (red), G (green), and B (blue), according to the colour at the wavelength of the maximum of each curve.

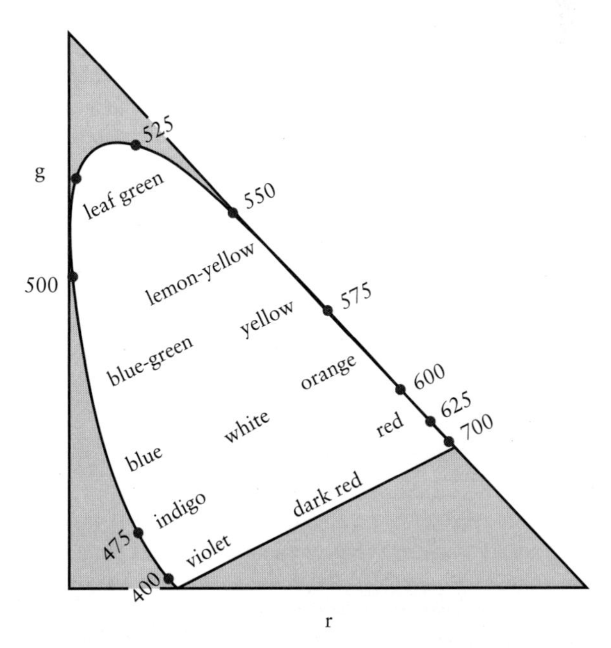

Figure 21.31 The visible part of the colour triangle (white area) and the respective colour impressions. No colour impressions are possible in the grey areas due to the overlap of the sensitivity peaks in Fig. 21.30.

Not all relative intensity combinations represented in the colour triangle are accessible. For example, the point $g = 1$ and $r = b = 0$ is inaccessible due to the overlap of the three receptors in Fig. 21.30, which shows that no wavelength of light excites only the green receptor. In Fig. 21.31 the accessible part of the colour triangle is shown as a white area, with the corresponding colour impressions included. Points near the centre of the triangle are perceived as white (additive colour sensitivity). Fig. 21.32 illustrates the use of the colour triangle for sunflower yellow and the blue colour of the sky on a beautiful day.

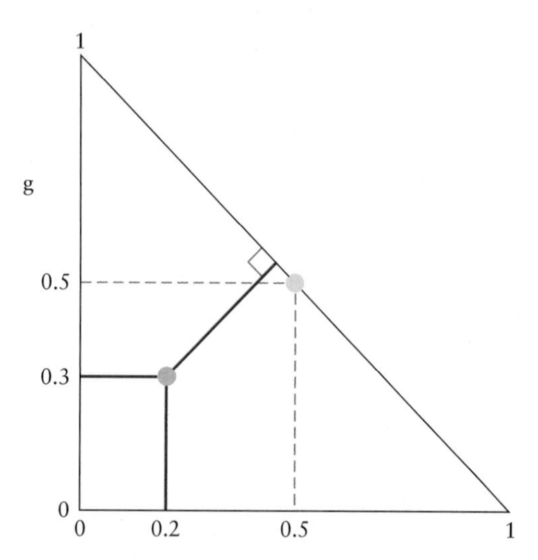

Figure 21.32 The position of sunflower yellow ($r = g = 0.5$, $b = 0$) and sky blue ($r = 0.2$, $g = 0.3$, $b = 0.5$) in the colour triangle.

CASE STUDY 21.8

Do all humans see colours the same way? More specifically, did Homer, who lived around 750 B.C. to 700 B.C., see colours as we do? Homer wrote works including the *Iliad* and the *Odyssey*. In these he repeatedly referred to the Aegean Sea (part of the Mediterranean Sea) as "a sea coloured like violets" or a "wine-coloured sea," descriptions most of us would disagree with. So, did he see colours differently?

Answer: *Homer most likely saw colours the same way you see them! For him to have seen colours differently at the blue end of the spectrum, the sensitivity of the eyes of the ancient Greek people must have stretched slightly farther toward shorter wavelengths into ultraviolet. To allow for this, Figs. 21.30 and 21.31 must be modified for Homer's eyes. This could occur in two ways:*

- *A first possibility for Homer's vision is to shift the sensitivity peak of the blue receptor toward the left in Fig. 21.30, and shift the minor peak of the red receptor accordingly. Using Fig. 21.31, we find that Homer would have seen in the UVA band as we see blue today. The only difference would be that he might*

continued

have perceived the colour of the Aegean Sea as more intense due to the high reflection of UVA light from a water surface.

- *A second possibility for Homer's eye would be to again shift the sensitivity peak of the blue receptor toward the left in Fig. 21.30, but leave the minor peak of the red receptor where it is in our eye. In this case we simply cannot say what Homer would have seen. While looking at the Aegean Sea, his brain would have received signals corresponding to a point in the grey area near b = 1 in Fig. 21.31. We don't know what the brain would do with such a signal, as none of our brains has ever had to deal with it. The only thing we know is that the* ***physiological rule of specific sensory perception*** *applies; i.e., the retina responds even to inadequate stimuli (e.g., electrical or mechanical stimulation) with a visual perception.*

Of course, Homer, as a poet, might just have been exercising his artistic freedom when describing the sea beloved by the Greeks.

SUMMARY

DEFINITIONS

- Permeability of vacuum: $\mu_0 = 1.26 \times 10^{-6}$ N/A^2.
- Magnetic flux Φ:

$$\Phi = B_z A = B_z \pi R^2$$

A is area of the current loop of radius R, B_z is the magnetic field component perpendicular to the loop surface.

UNITS

- Magnetic field $|\vec{B}|$: T = N/(A m)

LAWS

- Magnetic force
  - between two parallel currents in conductors of length l and distance d:

$$\left|\vec{F}_{\text{mag}}\right| \propto \frac{I_1 I_2}{d} l$$

  - force magnitude on a straight conductor due to an external magnetic field:

$$\left|\vec{F}_{\text{on wire}}\right| = \left|\vec{B}\right| I l \sin\varphi$$

φ is the angle between the direction of the magnetic field and the direction of the current I carrying conductor of length l.

  - force on a straight conductor due to an external magnetic field (vector notation):

$$\vec{F}_{\text{on wire}} = I\,l[\vec{I}^{\,0} \times \vec{B}]$$

- Magnetic field
  - for a current in a straight conductor:

$$\left|\vec{B}\right| = \frac{\mu_0}{2\pi}\frac{1}{d}$$

  - in a solenoid:

$$\left|\vec{B}\right| = \mu_0 \frac{N}{l} I$$

  N is number of windings.
  - at the centre of a single current loop of radius R:

$$\left|\vec{B}_{\text{centre of loop}}\right| = \frac{\mu_0 I}{2R}$$

  I is the current.
  - along the axis (z-axis) of a single current loop of radius R:

$$\left|\vec{B}_{\text{loop}}\right| = \frac{\mu_0 I}{2}\frac{R^2}{(z^2+R^2)^{3/2}}$$

  - general form for magnitude of magnetic field (Biot–Savart's law):

$$\left|d\vec{B}\right| = \frac{\mu_0}{4\pi}\frac{I\left|d\vec{l}\right|}{r^2}\sin\varphi$$

  φ is the angle between the direction of $d\vec{l}$ and $\vec{r}$, with $\vec{r}$ the vector pointing from the current element to point P at which the field is determined.
  - general form for magnetic field vector (Biot–Savart's law):

$$d\vec{B} = \frac{\mu_0}{4\pi}\frac{I}{r^3}[d\vec{l} \times \vec{r}]$$

  $d\vec{l}$ is an infinitesimally short conductor segment.
- Magnetic force per particle:
  - magnitude:

$$\frac{F_{\text{mag}}}{N_q} = f_{\text{mag}} = qvB\sin\varphi$$

  F_{mag} is the magnitude of the total magnetic force, N_q is the number of charged particles with charge q, φ is the angle between the magnetic field $\vec{B}$ and the velocity vector of the particle $\vec{v}$.
  - in vector notation, $\vec{f}_{\text{mag}}$:

$$\vec{f}_{\text{mag}} = q\left[\vec{v} \times \vec{B}\right]$$

- Motion of a charged particle in a uniform magnetic field:
  - radius of circular trajectory:

$$r = \frac{mv}{qB}$$

  - cyclotron frequency:

$$\omega = \frac{qB}{m}$$

- Velocity selected in a Wien filter: $v = E/B$, with $\vec{E}$ the electric field and $\vec{B}$ the magnetic field.
- Maxwell's equations in vacuum:

$$\oint_{\text{surface}} \vec{E} \bullet d\vec{A} = 0$$

$$\oint_{\text{surface}} \vec{B} \bullet d\vec{A} = 0$$

$$\oint_{\text{loop}} \vec{E} \bullet d\vec{s} = -\oint_{\text{surface}} \frac{d\vec{B}}{dt} \bullet d\vec{A}$$

$$\oint_{\text{loop}} \vec{B} \bullet d\vec{s} = \mu_0\varepsilon_0 \oint_{\text{surface}} \frac{d\vec{E}}{dt} \bullet d\vec{A}$$

- Stefan's law for blackbody radiation:

$$\frac{\Delta Q}{\Delta t} = -\sigma A T^4$$

where $\Delta Q/\Delta t$ is the total energy emitted, A is the surface area, and σ is the Stefan–Boltzmann constant.
  - for arbitrary objects:

$$\frac{\Delta Q}{\Delta t} = -\sigma\varepsilon A T^4$$

  with ε the emissivity, $0 < \varepsilon \leq 1$.

MULTIPLE-CHOICE QUESTIONS

MC–21.1. In a presentation, you use Fig. 21.2 to describe the magnetic force between two current-carrying wires. Someone in the audience challenges your statement that $\vec{F}_{2\text{ on }1}$ and $\vec{F}_{1\text{ on }2}$ are an action–reaction pair of forces. Which of the following statements would you NOT make in response?
(a) The two forces act on different objects.
(b) The forces are equal in magnitude.
(c) The forces are opposite to each other in direction.
(d) The electric currents flow in parallel directions.

MC–21.2. Which of the following is NOT a suitable unit for the magnitude of a magnetic field?
(a) G (gauss)
(b) T (tesla)
(c) Oe (oersted)
(d) N/(A m)
(e) (N m)/(C s)

MC–21.3. Fig. 21.33 shows a long wire that carries a current I. In what direction does the magnetic field point at point P, which lies in a common yz-plane with the wire? Answer the question based on the coordinate system shown in the figure.
(a) along the positive x-axis
(b) along the negative x-axis
(c) along the positive y-axis
(d) along the negative y-axis
(e) along the positive or negative z-axis

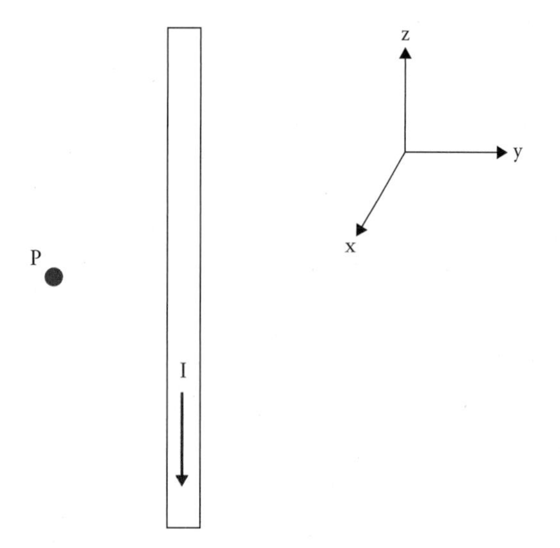

Figure 21.33 A long wire carries current *I*. A point P is located in the common *yz*-plane with the wire. The coordinate system is shown at the right.

MC–21.4. In which of the following locations is the magnetic field uniform in direction and constant in magnitude?
(a) inside a solenoid through which a current flows
(b) between the plates of a parallel plate capacitor
(c) outside a solenoid through which a current flows
(d) between two parallel current-carrying wires
(e) far from an antenna operated as a high-frequency sender

MC–21.5. Consider Fig. 21.34. In which quadrants does one conductor contribute to the magnetic field in the positive *z*-direction and the other in the negative *z*-direction?
(a) in quadrants I and II
(b) in quadrants I and III
(c) in quadrants I and IV
(d) in quadrants II and III
(e) in quadrants II and IV
(f) in quadrants III and IV

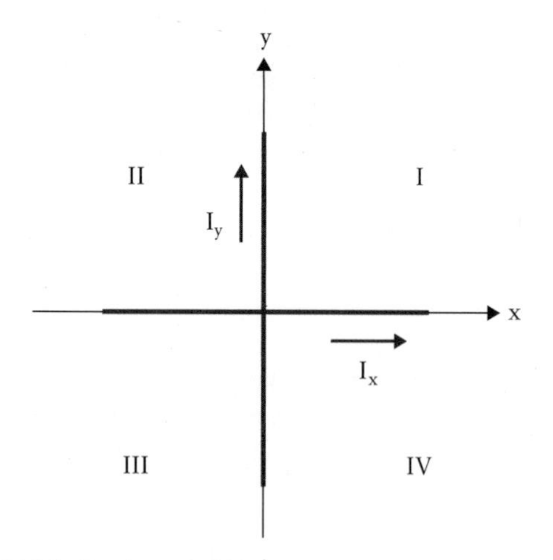

Figure 21.34 Two long, straight conductors (red), one running along the *y*-axis and the other along the *x*-axis.

MC–21.6. A thin cylinder of copper has mass $m = 50$ g and is one metre long. What is the minimum current that has to flow through the cylinder for it to levitate in a magnetic field of 0.1 T? Choose the closest value.
(a) 1.2 A
(b) 2.5 A
(c) 4.9 A
(d) 9.8 A

MC–21.7. In Michael Faraday's experiment a permanent magnet is moved through a wire loop. If you measure the electric current in the wire loop, you find a non-zero current when
(a) the South Pole of the magnet is held in the plane of the loop.
(b) the North Pole of the magnet is held in the plane of the loop.
(c) the bar magnet is held such that its North and South Poles are equally far from the plane of the loop.
(d) the bar magnet moves with constant speed through the wire loop.
(e) the bar magnet is moved toward the wire loop from below with its acceleration vector in the plane defined by the loop.

MC–21.8. Light travels in the direction that is
(a) parallel to the electric field vector.
(b) parallel to the magnetic field vector.
(c) at an angle of 45° with the electric field vector.
(d) at an angle of 45° with the magnetic field vector.
(e) perpendicular to the electric field vector.

MC–21.9. Light waves and sound waves have the following in common:
(a) Both travel in a vacuum.
(b) Both are transverse waves.
(c) Both travel at the same speed.
(d) When they reach the water surface from air, both refract and reflect.

MC–21.10. When white light passes through a prism it is split into the colours of the rainbow. The following feature does not contribute to this observation:
(a) White light contains light of all visible frequencies (between 360 nm and 760 nm).
(b) The index of refraction of the prism material depends on the wavelength of the light.
(c) The prism material displays a non-constant dispersion relation.
(d) The speed of light for the various wavelengths varies in the glass body of the prism.
(e) The frequencies of the different light components change unequally as the light enters the prism.

MC–21.11. The peak of the electromagnetic radiation intensity from the Sun lies
(a) in the microwave range.
(b) in the range of radio frequencies.
(c) in the far infrared range.
(d) in the visible/near-ultra-violet range.
(e) in the range of X-rays, because the Sun is a cosmic object.

CONCEPTUAL QUESTIONS

Q–21.1. In which arrangement is a conductor with a current not subjected to a magnetic force?

Q–21.2. Which way does a compass point when you are at the North Magnetic Pole?

Q–21.3. We move with the drift speed parallel to the electrons in a current-carrying conductor. Do we measure a zero magnetic field?

Q–21.4. (a) Two parallel conductors are located one above the other, with both carrying currents in opposite directions. As a result, they repel each other. Is the upper conductor in a stable state of levitation? (b) Now we reverse one current so that the two conductors attract each other. Is the lower conductor in a stable state of levitation?

Q–21.5. Parallel conductors exert magnetic forces on each other. What about two current-carrying conductors that are oriented perpendicular to each other?

Q–21.6. Describe the path of a charged particle that travels with velocity straight toward the North Pole along the north–south axis of a bar magnet.

Q–21.7. What happens to a mobile but initially stationary point charge if a magnet travels past its location?

Q–21.8. What changes take place in Fig. 21.14 if the direction of the magnetic field is reversed, i.e., if $\vec{B}$ points out of the plane of the page?

Q–21.9. Why does the picture on a computer screen become distorted when a magnet is brought close to it? *Note:* Don't try this, because it may cause permanent damage.

Q–21.10. Two charged particles travel into a region in which a magnetic field acts perpendicular to the particles' velocity vectors. What do you conclude if they are deflected in opposite directions?

Q–21.11. What work does the magnetic force do on a charged particle that moves through a uniform magnetic field?

Q–21.12. Why do cosmic-ray particles strike Earth more often near the poles than near the equator?

Q–21.13. Can a constant magnetic field set into motion a charged particle from rest?

Q–21.14. What happens in Fig. 21.15 if the three slits S_1 to S_3 are opened to twice their width?

Q–21.15. Fig. 21.35 shows the mass spectrum of benzoic acid methylester (the molecule is shown as an inset of the figure). What functional groups have been split off the molecule to generate the peaks at 105, 77, and 51 mass units, respectively?

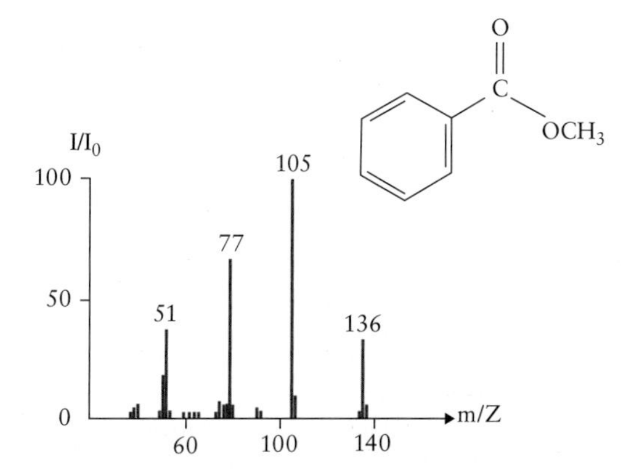

Figure 21.35 The mass spectrum of benzoic acid methylester. The chemical formula of the molecule is shown as an inset in the figure.

Q–21.16. We use a conducting line as a receiving antenna. What should the orientation of this antenna be relative to the antenna that emits electromagnetic waves?

Q–21.17. A vibrating object is the source of sound. What is the physical source of an electromagnetic wave?

Q–21.18. What is it that actually moves when a light wave travels through outer space?

Q–21.19. Assume your eyes were sensitive in the infrared wavelength range. Look around the room you are in. What would you see?

Q–21.20. Why does an infrared photograph taken of a person look different from a photograph taken with visible light?

Q–21.21. Fig. 21.36 shows that our eye is much more sensitive to absolute intensities of green light in comparison to absolute intensities of red light. Why then do the green and red lights of a traffic light still look roughly equally bright?

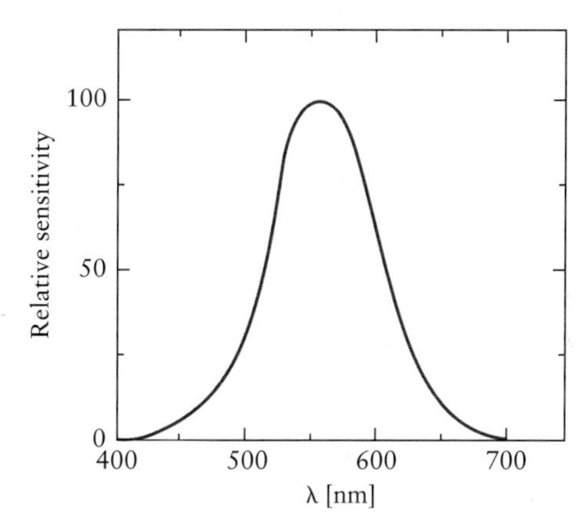

Figure 21.36 Relative sensitivity of the human eye as a function of wavelength.

Q–21.22. Describe the colour vision of an alien if the sensitivity of the three colour receptors in the alien's eye is as shown in Fig. 21.37.

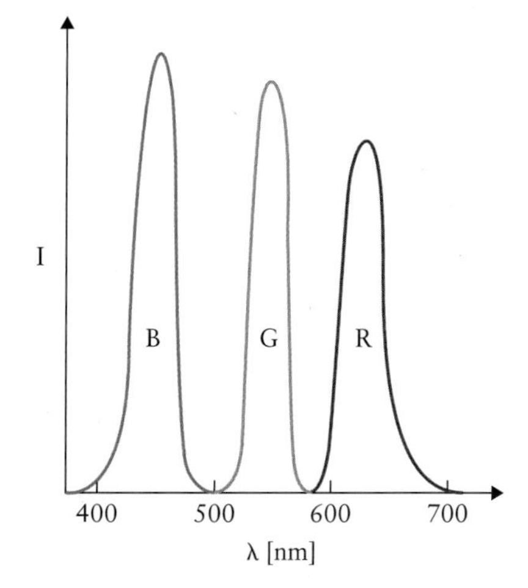

Figure 21.37

Q–21.23. Why is a rainbow red at the top and blue at the bottom?

Q–21.24. A light ray of given wavelength travels from air into glass ($n > 1$). Does the wavelength of the light change? Does its frequency change? Does its speed change? Does its colour change?

Q–21.25. A mixed light beam of two colours, X and Y, is sent through a prism. In the prism, the X component is bent more than the Y component. Which component travelled more slowly in the prism?

ANALYTICAL PROBLEMS

P–21.1. Two long, parallel wires are separated by a distance of $l_2 = 5$ cm, as shown in Fig. 21.38. The wires carry currents $I_1 = 4$ A and $I_2 = 3$ A in opposite directions. Find the direction and magnitude of the net magnetic field (a) at point P_1, which is a distance $l_1 = 6$ cm to the left of the wire carrying current I_1; and (b) at point P_2, which is a distance $l_3 = 5$ cm to the right of the wire carrying current I_2. (c) At what point is the magnitude of the magnetic field zero; i.e., $|\vec{B}| = 0$?

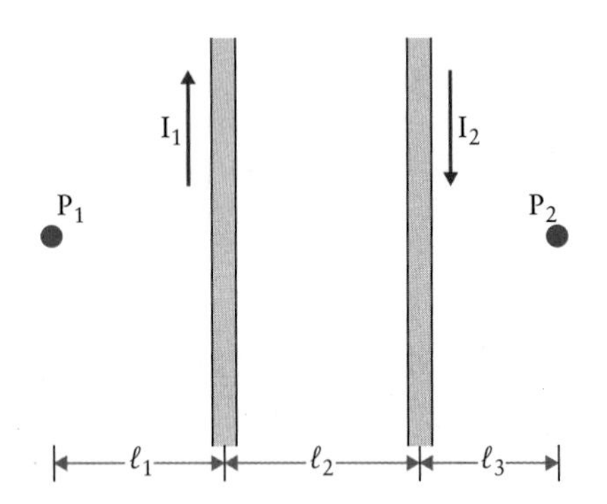

Figure 21.38

P–21.2. At what distance from a long, straight conductor that carries a current of 1 A is the magnitude of the magnetic field due to the wire equal to the magnitude of Earth's magnetic field at the surface of Earth; i.e., $|\vec{B}| = 50\ \mu$T?

P–21.3. A conducting wire has a mass of 10 g per metre of length. The wire carries a current of 20 A and is suspended directly above a second wire of the same type that carries a current of 35 A. How far do you have to close the separation distance between the wires so that the upper wire is balanced at rest by magnetic repulsion?

P–21.4. Two parallel conductors each carry a current of 2 A and are 6 cm apart. (a) If the currents flow in opposite directions, find the force per unit length exerted on either of the two conductors. Is the force attractive or repulsive? (b) How do the results in part (a) change if the currents flow parallel to each other?

P–21.5. Fig. 21.39 shows two parallel wires that carry currents $I_1 = 100$ A and I_2. The top wire is held in position; the bottom wire is prevented from moving sideways but can slide up and down without friction. If the wires have a mass of 10 g per metre of length, calculate current I_2 such that the lower wire levitates at a position 4 cm below the top wire.

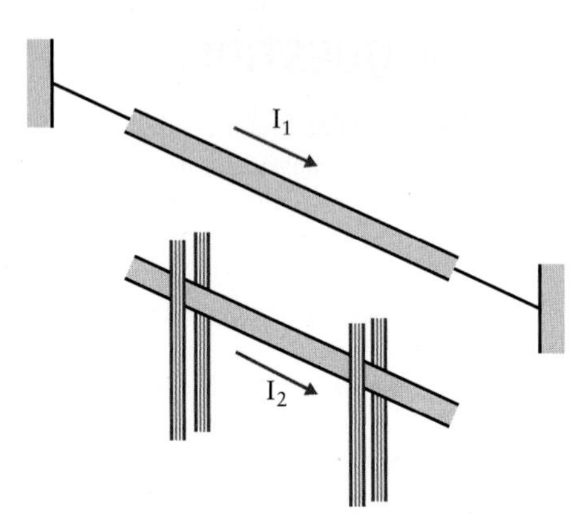

Figure 21.39

P–21.6. A current of $I = 15$ A flows through a conductor in the positive x-direction. Perpendicular to the current is a magnetic field, causing a magnetic force on the conductor per unit length of 0.12 N/m in the negative y-direction. Calculate the magnitude and determine the direction of the magnetic field in the region through which the current flows.

P–21.7. A conductor carries a current of 10 A in a direction that makes a 30° angle with the magnetic field of strength $B = 0.3$ T. What is the magnitude of the magnetic force on a 5-m segment of the conductor?

P–21.8. Fig. 21.40 shows two conductors that carry currents in opposite directions. The right conductor carries a current $I_1 = 10$ A. Point A is the midpoint between the conductors, which are separated by $L = 10$ cm, and point B is located 5.0 cm to the right of I_1. The current I_2 is adjusted such that the magnetic field at point B is zero. (a) Find I_2. (b) Find the magnitude of the magnetic field at point A.

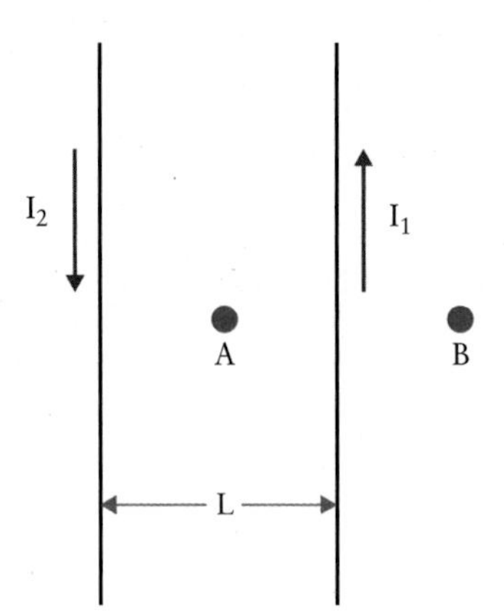

Figure 21.40

P–21.9. Fig. 21.41 shows the cross-sections of four long, parallel, current-carrying conductors. Each current is 4.0 A, and the distance between neighbouring conductors is $L = 0.2$ m. A dot on the conductor means the current is flowing out of the plane of the paper and a cross means it flows into the plane of the paper. Calculate the magnitude and determine the direction of the magnetic field at point P at the centre of the square shown.

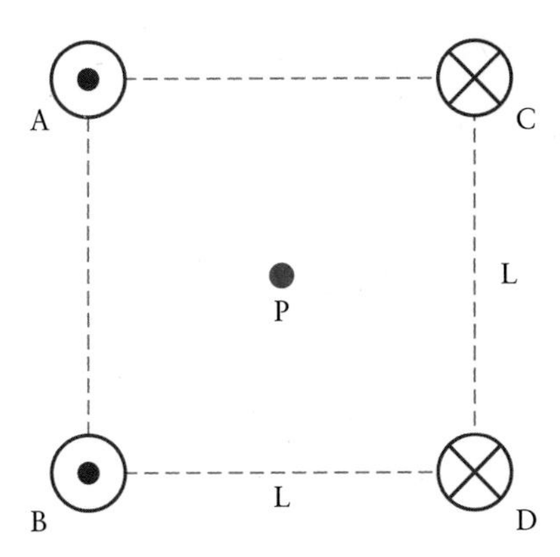

Figure 21.41

P–21.10. A proton moves with a speed of 100 km/s through the magnetic field of Earth, which has at a particular location a magnitude of 50 μT. What is the ratio of the gravitational force to the magnetic force on the proton when the proton travels perpendicular to the magnetic field?

P–21.11. A long, straight wire in a vacuum system carries a current of 1.5 A. A low density, 20-eV electron beam is directed parallel to the wire at a distance of 0.5 cm. The electron beam travels against the direction of the current in the wire. Find (a) the magnitude of the magnetic force acting on the electrons in the electron beam, and (b) the direction in which the electrons are deflected from their initial direction.

P–21.12. A proton travels with $v = 3.0 \times 10^6$ m/s at an angle of 37° with the direction of the magnetic field of a magnet of strength $B = 0.3$ T. The magnetic field is oriented along the y-axis. (a) Calculate the magnitude of the magnetic force acting on the proton. (b) Calculate the magnitude of the acceleration of the proton.

P–21.13. Sodium ions Na^+ move at 0.5 m/s through a blood vessel. The blood vessel is in a magnetic field of $B = 1.0$ T. The blood flow direction subtends an angle of 45° with the magnetic field. What is the magnetic force on the blood vessel due to sodium ions if the blood vessel contains 0.1 L blood with a sodium concentration of $c = 70$ mmol/L?

P–21.14. A positive charged particle carries 0.2 μC and moves with a kinetic energy of 0.09 J. It travels through a uniform magnetic field of $B = 0.1$ T. What is the mass of the particle if it moves in the magnetic field in circular manner with a radius $r = 3.0$ m?

P–21.15. A proton moves in uniform circular motion perpendicular to a uniform magnetic field with $B = 0.8$ T. What is the period of its motion?

P–21.16. We consider Aston's mass spectrometer, as illustrated in Fig. 21.15. The magnitude of the electric field is $E = 1.0$ kV/m and the magnitude of the magnetic fields in both the Wien filter and the mass selector are 1.0 T. Calculate the radius of the path in the mass selector for an ion with a single positive charge and with a mass of $m = 2.0 \times 10^{-26}$ kg.

P–21.17. A mass spectrometer is used to separate isotopes. If the beam emerges with a speed of 250 km/s and the magnetic field in the mass selector is 2 T, what is the distance between the collectors for (a) ^{235}U and ^{238}U, and (b) ^{12}C and ^{14}C?

P–21.18. An ion carrying a single positive elementary charge has a mass of 2.5×10^{-23} g. It is accelerated through an electric potential difference of 0.25 kV and then enters a uniform magnetic field of $B = 0.5$ T along a direction perpendicular to the field. What is the radius of the circular path of the ion in the magnetic field?

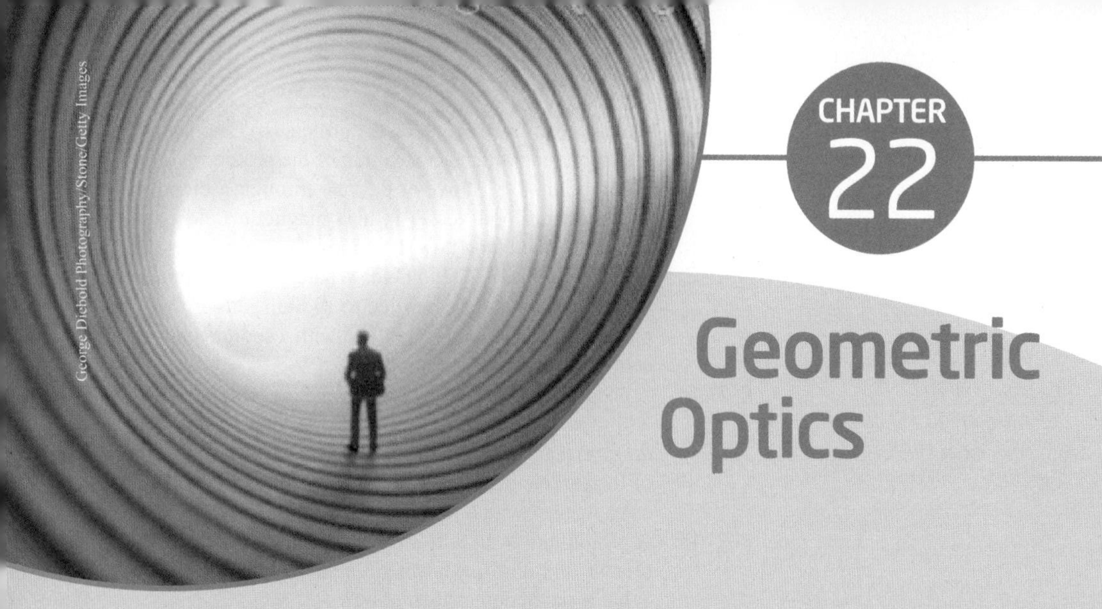

CHAPTER 22

Geometric Optics

The physics of light is described in the field of optics. It is divided into three major branches: geometric optics is concerned with phenomena that can be modelled with light rays, wave optics characterizes light as an electromagnetic wave, and photon optics describes light as corpuscles. In the ray model that is the subject of this chapter, light is assumed to travel along straight lines until it reaches an interface between two different media. At such an interface it will be reflected and/or pass the interface in a process called refraction.

Reflection is studied with mirrors. The angles of light rays that reach a mirror are defined with the normal direction of the mirror surface. The angles of an incoming light ray and the corresponding reflected light ray are equal. For a spherical mirror, sharp images of objects in front of the mirror form when the mirror has a focal point. The mirror equation, then, relates the inverse focal length to the sum of the inverse object and image distances. The image is magnified if the object is closer to the mirror than its radius of curvature.

When a light ray passes through an interface it travels closer to the normal direction of the interface in the medium with the higher index of refraction, which is usually the denser medium. If the interface is spherical, an object in front of the interface forms a sharp image if the interface has a focal point. Lenses are combinations of two spherical interfaces. The thin-lens formula and the equation for the magnification of a lens are identical to the respective equations for the spherical mirror. The inverse focal length of a lens is defined as the refractive power of the lens, measured in diopters.

The path of light through the human eye and the formation of images on the retina can be described with the ray model. The cornea is modelled as a transparent, single spherical interface. Cornea and lens contribute to the refractive power of the eye. Eye defects, such as myopia (nearsightedness) and hyperopia (farsightedness), are corrected with prescription lenses that are customized based on the optical properties of the eye, and that are manufactured by using the lens maker's equation.

The geometric optics concepts introduced in the previous chapter are applied to the magnifying glass and the compound microscope. To properly quantify the magnification of a microscope, the observer's eye has to be included as part of the optical setup. This leads to the definition of the angular magnification. Variable combinations of eyepieces and objective lenses are studied for their total angular magnification, which is the product of the angular magnification of the eyepiece and the magnification of the objective lens.

One of the three necessary conditions for life is the recognition of external stimuli and the ability to respond. Arguably one of the most astonishing achievements of the evolutionary process in satisfying this condition is vision. The human eye's complexity in design and versatility in function is unmatched by engineered imitations. The anatomy of the human eye, shown as a cross-sectional side view in Fig. 22.1, identifies at least eight individual components required for us to see: light reaches the eye at the cornea (3), then passes through the anterior chamber (2). Its intensity is adjusted by the iris (4) and the light rays are focused by the lens (1). To accomplish the focusing, the lens must be adjusted, which is achieved by the ciliary muscles (6). Before forming an image on the retina (7),

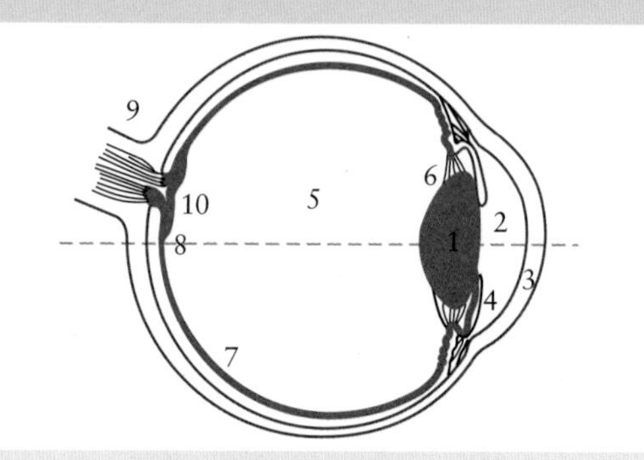

Figure 22.1 Cross-sectional sketch of a human eye: (1) lens, (2) anterior chamber, (3) cornea, (4) iris, (5) vitreous body, (6) ciliary muscle, (7) retina, (8) fovea centralis, and (9) optic nerve. (10) Locates the blind spot, which is due to the optic nerve passing through the retina to the brain.

the light passes through the vitreous body (5). The retina then converts the image into electrical signals, which are sent to the brain through the optic nerve (9). The interplay of these components allows us to clearly see structures on an object as far away as the Moon, or to read small letters in a book at just 20 centimetres in front of the eye.

But does the complex human eye indeed work as precisely as an instrument? What does it measure, and how does it measure? Even though we return with these questions to the psychophysical boundary between the exact physical sciences and the subjective psychological perception we discussed in the context of hearing, it is necessary to investigate the physics of our vision to understand its limitations.

An interesting hint concerning the imperfections of human vision is provided by the many optical illusions we easily fall victim to. Optical illusions are never just imaginings, but are most often linked to an attempt to correct an adverse physical effect. A well-known example is shown in Fig. 22.2. Close your left eye and look at the cross with your right eye. Now bring the textbook slowly toward you. You will notice that the black spot disappears when the distance from your eye to the book is about 30 cm! Continue to bring the textbook closer to your eye. When you come closer than 20 cm the spot reappears. This self-experiment clearly illustrates that part of the image we see is the result of intelligent image extrapolation rather than fact.

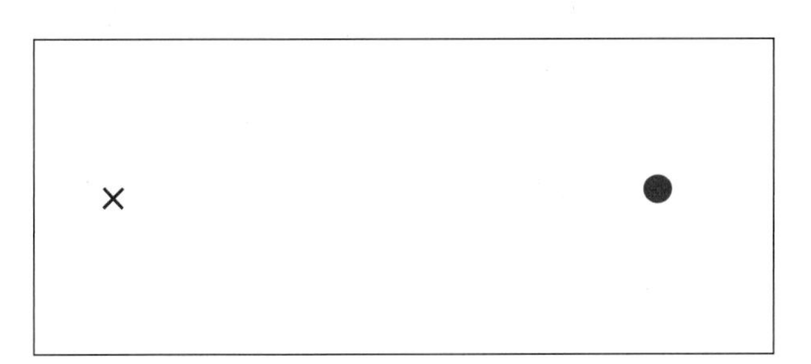

Figure 22.2 Self-test to illustrate the existence of the blind spot in our field of vision. With the left eye closed, look at the cross with your right eye. Move the textbook toward you from about a metre (arm's length). When the page is about 20 cm to 30 cm from your eye the blue dot disappears.

Let's briefly explain the effect of Fig. 22.2. The position of the blind spot in our field of vision is shown in Fig. 22.3. The plot applies to the left eye. Each concentric circle represents an angle increment of 10°, with the outermost circle corresponding to 90°. Plots like Fig. 22.3 are called **polar plots**. The black line encloses the field of vision for black/white vision; the other lines enclose the field of vision for various colours. The solid dot at about 15° on the temporal side is the **blind spot** that we noticed in the experiment from Fig. 22.2.

You can easily verify Fig. 22.3: close your right eye, and with your left eye look straight ahead. Stretch your left arm, bending it slightly behind your shoulder line. Rapidly flutter the fingers on your left hand as you move the arm slowly forward in an arc. You should see your finger movement at the periphery of your field of vision when your arm reaches an

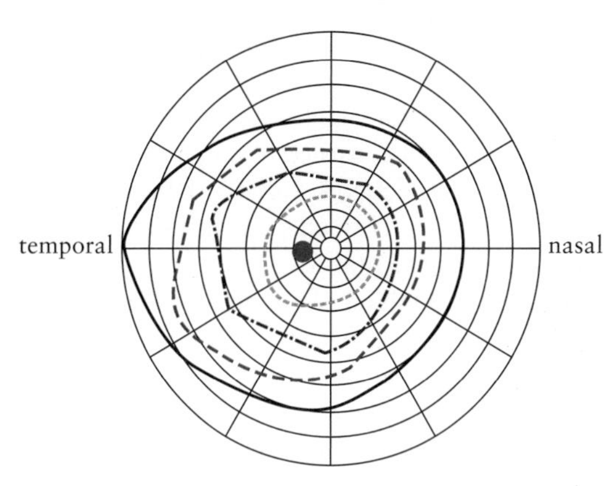

Figure 22.3 A polar plot illustrating the field of vision for the left eye. The field of vision depends on the viewed colour: the largest field applies to black/white vision (black line), the blue line is for blue colour vision, the red line is for red vision, and the green line is for green vision. The concentric circles correspond to angle increments of 10°, with the centre as the direction in which the person looks and the outermost circle at 90° to that direction.

angle of about 80° to 90° with the direction of vision. Now repeat the same experiment with the right arm, but still using the left eye. This time you have to bring it much farther forward before you notice the finger movement, to about 60° with the direction of vision. The asymmetry of the field of vision is because of your nose. The combined field of vision of both eyes compensates for this asymmetry.

With the issue of colours discussed already in Chapter 21, here we want to understand the physiological origin of the blind spot, particularly because it is inconveniently located near the middle of our field of vision. For this, we look at the anatomy of the human eye in Fig. 22.1. The spot on the retina that light reaches when travelling straight through the centre is called the fovea centralis (8) and is the most light-sensitive area in the eye. This point is located at the centre in Fig. 22.3. The blind spot (10 in Fig. 22.1) is the point at which the optic nerve is bundled and leaves the eye. This leads to an area with no vision because the optic nerve has to interrupt the retina to pass through it to the brain.

As in other cases of optical illusions, our brain corrects for otherwise confusing signals from the eye. In this case, the brain does not allow us to have a missing spot near the middle of our field of view. Instead, it modifies the received image by filling in the blind spot based on a best guess before allowing the image to reach our consciousness. Usually this works without problems as we move our head and eyes constantly. However, in the case of Fig. 22.3, the uniform white area around the spot is too tempting not to correct for the blind spot by adding a uniform white!

22.1: What Is Optics?

Three different models have been developed to describe the physics of light: the ray model, the wave model, and the corpuscle model. The initial development of the field came in the 17th century when René Descartes, Christiaan Huygens, and Sir Isaac Newton tried to interpret their experiments with visible light. Already the ancient Greeks had thought that light consisted of corpuscles, an idea further developed by Newton. Newton believed that light consists of a stream of small particles that interact with matter like mechanical objects. At the same time, Huygens promoted a **wave model** for light, treating light as a propagating wave similar to surface waves on water. He encountered problems with that model when comparing it to sound waves, mainly because he couldn't identify a medium that carried the light waves. On balance, however, the wave model appeared to his contemporaries as more consistent with an increasing body of experimental observations, and Huygens's theory was widely accepted from about 1800 to 1905—in particular after James Clerk Maxwell derived the properties of visible light from electromagnetic wave equations in 1865, as shown in Chapter 21.

The **corpuscle theory** was revived in the early 1900s when Albert Einstein used a corpuscle model (in which light particles are called **photons**) to explain the photoelectric effect, which describes the ability of light to knock electrons out of solid matter. Our modern view of light is that it has a dual, wave-and-corpuscle character. This is discussed with the double-slit experiment in Chapter 20.

In this chapter, we start our discussion of optics with a simpler model, called the **ray model**. This greatly simplified model is applicable as long as the objects involved in an optical study are not smaller than the wavelength of light, which lies in the vicinity of 500 nm. The physical laws that we derive with the ray model are summarized by the term **geometric optics**, because we can construct its features with geometric methods.

KEY POINT

In the ray model, the assumption is made that light moves along straight lines while travelling within a homogeneous medium. It may change its direction when reflected by and/or passing through an interface into another medium.

The concepts of geometric optics are established in the current chapter. These include reflection off a mirror and refraction when light passes through a transparent interface.

22.2: Reflection

22.2.1: Flat Mirror

Fig. 22.4 shows a light ray that is reflected off a planar mirror. The direction of the travelling light ray is shown by arrows. We will frequently use the following two features:

- The incoming and reflected rays are in the same plane as the vector directed perpendicular to the mirror surface. This vector is called the **normal vector**, in which the word "normal" is synonymous with "perpendicular" (in the same manner as we used the term when we defined a normal force in mechanics).

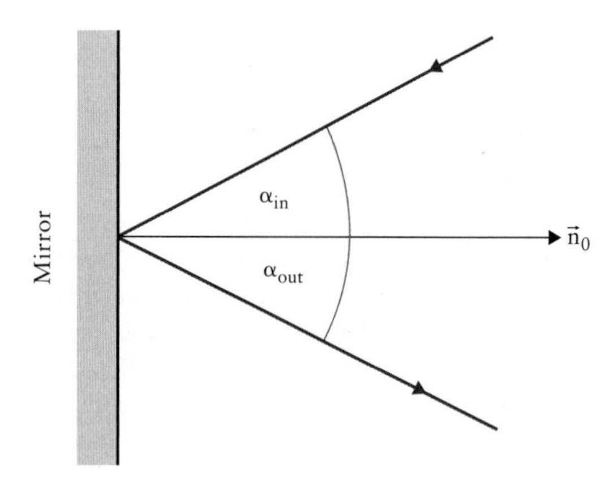

Figure 22.4 A light ray reflects off a flat mirror: $\vec{n}^0$ indicates the direction perpendicular to the mirror surface (i.e., the normal vector); α_{in} is the angle of the incoming light ray with the normal; and α_{out} is the angle of the outgoing, reflected light ray with the normal. The two angles are related by the law of reflection.

- The angle between the incoming ray and the normal vector, α_{in}, is equal to the angle between the reflected ray and the normal vector, α_{out}.

These two conditions constitute the **law of reflection**. Note that we already derived this law in Chapter 16 for sound waves using Huygens's principle. It is quantitatively written in the form:

$$\alpha_{in} = \alpha_{out}. \quad [22.1]$$

The angle α_{out} is called the **specular angle**.

KEY POINT

The law of reflection states that the angle of an incoming light ray with the normal direction of the mirror is equal to the angle of the reflected light ray with the normal direction of the mirror.

We use Fig. 22.5 to illustrate how the law of reflection leads to the formation of an image, as we are used to seeing it in a flat mirror. We start with a point-like light source (solid dot at top right). The light source emits light rays that travel along straight lines in all directions. When such a straight line reaches the mirror surface (shown for three rays in the figure) the law of reflection is applied, as illustrated for the centre ray in Fig. 22.5. After reflection, the rays continue to travel along straight lines until they reach the eye of the observer. Observers can interpret the light rays reaching the eye in two ways: Either they are aware of the presence of the mirror and draw the rays as shown in front of the mirror, or they are not aware of the mirror and extrapolate the rays straight to the dashed point behind the mirror at the left in the figure. At the point where these lines cross, the image of the point-like light source forms. The position of the light source (object) and its image are both at a distance d from the mirror. This is a direct consequence of the law of reflection.

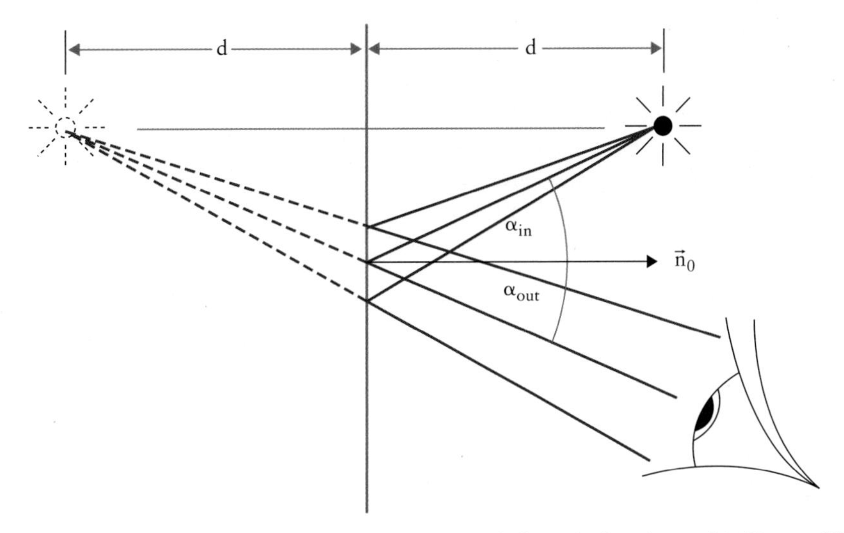

Figure 22.5 When a person observes a point-like light source located a distance *d* in front of a flat mirror, a virtual image of the light source is seen at the same distance *d* behind the mirror surface. The image of the light source is constructed using the law of reflection, as indicated for the light ray at the centre. Solid lines indicate the actual paths of light rays; dashed lines represent extrapolations of the light rays behind the mirror (vertical blue line).

We call an image a **real image** when light rays actually reach the image. In Fig. 22.6, light cannot physically reach the image since a mirror contains a metallic layer that prevents light from passing through. In this case, the image is called a **virtual image**.

In Fig. 22.6, we generalize our choice of object, replacing the point-like light source of Fig. 22.6 with an extended object of height h (reaching from point P_2 to point P_1). The object is still at distance d from the mirror. Using the law of reflection for light rays coming separately from points P_2 and P_1, i.e., $\alpha_1 = \alpha'_1$ and $\alpha_1 = \alpha'_2$, we find that the image in Fig. 22.7 is a virtual image, forming at distance d behind the mirror. For comparison with later cases, note also that the image is upright; i.e., the corresponding image points P'_2 and P'_1 are at positions $y = 0$ and $y = h$ with a vertical y-axis, respectively.

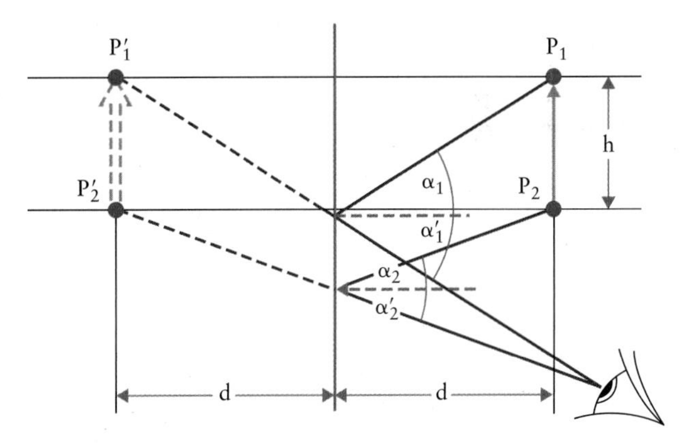

Figure 22.6 Formation of the image of an extended object in a flat mirror. Light rays from both the top and the bottom ends of the object are used to construct a virtual image (green dashed arrow).

CASE STUDY 22.1

Flat mirrors allow us only to see the object the same way it looks when viewing it directly. Why, then, do we use flat mirrors at all?

Answer: *One object exists that we cannot view directly: ourselves, in particular our own face. People have been obsessed with their reflections since the dawn of humankind. The first human-made mirrors came from Turkey and were polished volcanic glass. The ancient Egyptians switched to bronze metal surfaces, and the ancient Greeks had schools in which sand polishing of reflective surfaces was taught. The modern design of a thin metal film (mercury–tin alloy) on the back of a glass plate emerged in the 14th century in Venice, Italy.*

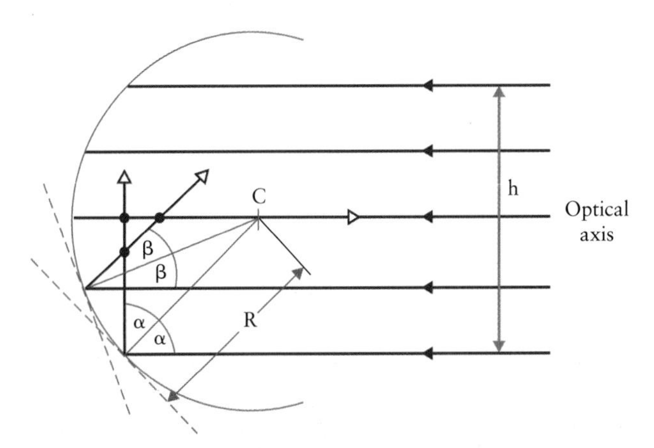

Figure 22.7 Concave spherical mirror. Replacing a flat mirror with a spherical mirror does not yield a useful image if light rays reach the mirror far from the optical axis (*h*/2) compared to the radius of curvature of the mirror *R*. The figure illustrates that the light from an object at infinite distance (parallel incoming light rays) does not form a focal point, in contrast to the setup in Fig. 22.8.

22.2.2: Spherical Mirror

The applications of mirrors expand greatly if we do not require them to be flat. Of the many possible mirror shapes, you will find only two in actual instruments: the parabolic mirror in astronomical facilities, and the spherical mirror in all other applications. We limit our discussion to spherical mirrors, i.e., mirrors for which the reflecting surface is shaped as a partial sphere. Spherical mirrors can be arranged in two ways:

- Light rays approach from the side of the centre of curvature (point C) of the mirror, as shown in Fig. 22.9. In this case the mirror is called a **concave mirror**.
- Light rays approach from the opposite side of the mirror. In this case the mirror is called a **convex mirror**.

For convenience, we mostly use concave mirrors for the discussions in this chapter. This is done with no loss of generality, because every relation we introduce applies to convex mirrors in an analogous manner.

The centre of curvature in Fig. 22.7 further allows us to introduce the **optical axis**: an incoming light ray defines the optical axis if it passes through the centre of curvature (point C). To establish the key physical properties of a spherical mirror, we follow several parallel light rays in Fig. 22.7 as they reflect off its surface. We notice that the reflected rays do not intersect at a common point. We will find such an intersection point very useful in developing the concepts of geometric optics and therefore dismiss the setup of Fig. 22.7 as not sufficient. To correct for this problem we need to introduce a further restriction on the mirrors we use. This additional restriction follows from comparing Fig. 22.7 and Fig. 22.8(a). In both cases we assume that the light source is at a very large distance from the mirror (an infinite distance) and, therefore, that the light rays from the light source (object) approach the mirror parallel to each other. The two figures vary only in the spread of the light rays that are allowed to reach the mirror: in Fig. 22.8(a) the separation of the incoming light rays from the optical axis is small compared to the **radius of curvature** R of the mirror. As a result, the reflected light rays in Fig. 22.8(a) intersect at a common point F, as illustrated by the experimental setup in Fig. 22.8(b). The point at which the reflected rays cross is called the **focal point**; the **focal length** f is then the distance from the mirror surface to the focal point.

If in practice we observe the outcome shown in Fig. 22.7—i.e., a case with no focal point—we refer to it as **spherical aberration**. To avoid such cases we use **apertures**, which confine the spread of the incoming light. Another way to state this restriction is to say that a focal point must exist, i.e., that incoming parallel light rays must intersect at a single point F after reflection.

Where is the focal point located for a given mirror? First, the focal point lies on the optical axis. Second, the focal point of a spherical mirror is at half the distance

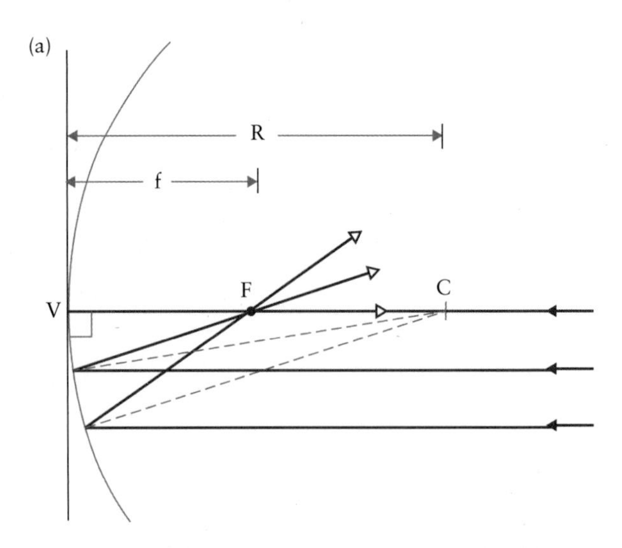

Figure 22.8 (a) Spherical mirrors can be used to form images if the incoming light rays travel close to the optical axis, i.e., at a distance that is small compared to the radius of curvature of the mirror. Light from an object at infinite distance (parallel light) allows us to define the focal point at the position on the optical axis where all reflected light rays cross. V is the point at which the optical axis intersects the mirror surface. F is the focal point with f the focal length. C is the centre of curvature, with R the radius of the spherical surface. (b) The crossing of the reflected light rays at the focal point is illustrated with an experimental setup. The mirror is the circular piece at the bottom.

between the centre of curvature and the point V at which the optical axis intersects with the mirror:

$$f = \frac{R}{2} \quad [22.2]$$

Eq. [22.2] is derived geometrically using Fig. 22.9. The two horizontal lines in the figure are the optical axis and an incoming light ray that is reflected at point P, then passes through point F. The law of reflection states that $\alpha_1 = \alpha_2$. Further, we know that $\alpha_3 = \alpha_1$ because the line PC intersects two parallel lines to form these two angles. Thus, the triangle PFC is isosceles; i.e., the lines PF and FC are equally long. If we now bring the incoming light ray closer and closer to the optical axis, all three angles α approach 0° and $VF = FC = R/2$. Since VF is the focal length f, we have confirmed Eq. [22.2]. The need to look at very small angles α reinforces once more the point made for Figs. 22.7 and 22.8 that a focal point exists only when we limit our considerations to light rays that move very close to the optical axis.

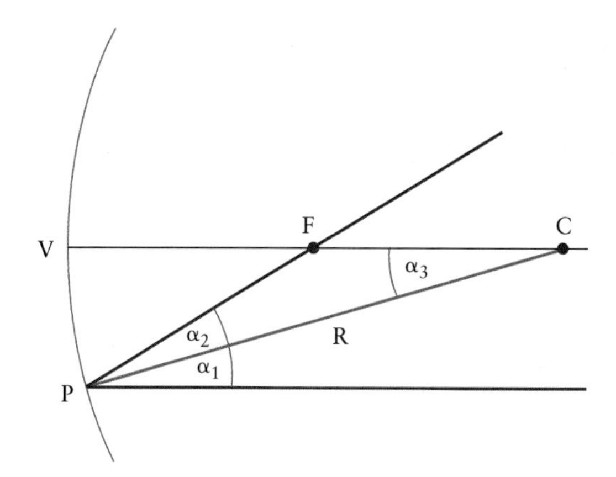

Figure 22.9 Geometric construction of the relation between the focal length (length *VF*) and the radius of curvature (length *VC*) for a spherical mirror with *C* its centre of curvature.

Comparing Figs. 22.4 and 22.8, we note that concave mirrors can form images at distances that differ from the object distance. In the particular case of Fig. 22.8, a point-size light source at infinite distance has generated an image at the focal point. To develop the properties of a concave mirror further, we consider in Fig. 22.10 an object at finite distance p from the mirror. To find the image in this and other cases geometrically, three different light rays must be followed, one along the optical axis from the lower end of the object (which is always placed on the optical axis) and two from the upper end of the object. The optical axis is used for convenience to reduce the amount of graphic construction needed. We need two rays from the upper end of the object as the position of the corresponding upper end of the image does not lie on the optical axis. The upper end of the image is then defined as the point at which the two light rays from the upper end of the object intersect after reflection.

We choose two light rays such that one reaches the mirror parallel to the optical axis and the other reaches

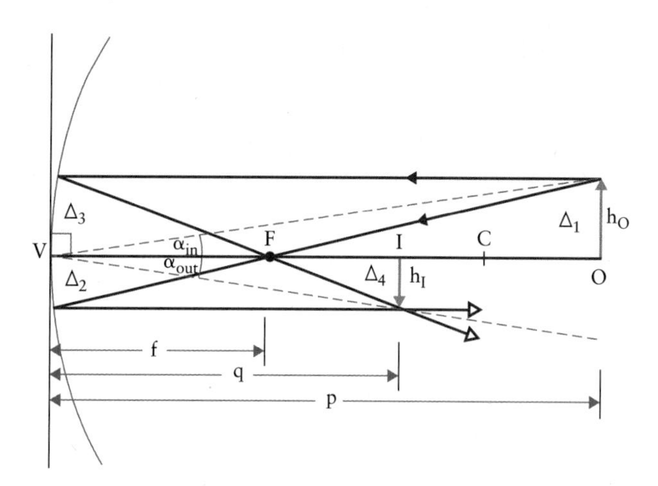

Figure 22.10 Formation of an image I for an object O at finite distance p from a concave mirror. Three light rays are used to construct the image, one along the optical axis (line *CFV*) since the bottom end of the object is placed on the optical axis. The second and third rays come from the top of the object of size h_O, with one ray travelling parallel to the optical axis and one ray passing through the focal point F. The image has a size h_I, is inverted, and is located at the image distance q. The various triangles labelled in the figure are used for geometric constructions because of their similarity, e.g., Δ_1 and Δ_2.

it after passing through the focal point F. Using these two light rays, we can construct their path after reflection using the observations we made in Fig. 22.8: the light ray moving parallel to the optical axis passes through the focal point after reflection and vice versa—the light ray passing through the focal point becomes a ray that moves parallel to the optical axis.

Using these principles, we find that the image in Fig. 22.10 is inverted (upside down) and real. Defining the distance between the image and the mirror as **image distance** q, we are now able to develop a general relation between object distance p, image distance q, and focal length f. This formula is called the mirror equation and originally was derived by Newton.

We call h_O the height of the object and h_I the height of the image. The derivation is based on the geometrical similarity of two pairs of triangles in Fig. 22.10, Δ_1, Δ_2 and Δ_3, Δ_4, respectively, leading to:

$$\Delta_1 \text{ and } \Delta_2 \quad f:(p-f) = h_I : h_O$$

$$\Delta_3 \text{ and } \Delta_4 \quad f:(q-f) = h_O : h_I$$

From these equations we extract two relations for h_I/h_O:

$$\frac{h_I}{h_O} = \frac{f}{p-f} = \frac{q-f}{f}.$$

The last formula in this equation leads to:

$$f^2 = (p-f)(q-f) = f^2 - (p+q)f + pq,$$

in which the f^2 term is dropped on both sides:

$$pq = (p+q)f.$$

This equation is further rewritten with $1/f$ as the dependent variable:

$$\frac{1}{f} = \frac{p+q}{pq},$$

which leads to the **mirror equation**:

$$\frac{1}{f} = \frac{1}{p} + \frac{1}{q} \qquad [22.3]$$

KEY POINT

The mirror equation relates the focal length of a mirror with the object and image distances.

An interesting feature of the spherical mirror that is not a feature of the flat mirror is the possibility of obtaining a magnified image. This is not the case in Fig. 22.10, since $h_I < h_O$. However, if the object in Fig. 22.10 is removed and instead placed where the figure shows the image, then object and image switch places as all light rays can travel along the same paths in the opposite direction. In this case a magnified image has formed.

The ratio of the image and object height in Fig. 22.10 allows us also to quantify the magnification M because it is directly defined as the ratio of the size of the image to the size of the object:

$$M = \frac{h_I}{h_O} = -\frac{f}{p-f}. \qquad [22.4]$$

The negative sign is due to the fact that the image in Fig. 22.10 is inverted. The magnification can alternatively be expressed in terms of the object distance p and the image distance q, again using Fig. 22.10. Due to the law of reflection, the two angles α_{in} and α_{out}, which are formed by the dashed line in the figure, are equal. From geometry we find:

$$\tan\alpha_{in} = \frac{h_O}{p}$$

$$\tan\alpha_{out} = -\frac{h_I}{q}$$

which leads to:

$$M = \frac{h_I}{h_O} = -\frac{q}{p} \qquad [22.5]$$

Eq. [22.4] allows us to establish the cases for which the magnification is larger than 1; i.e., when the image is larger than the object: $M > 1$ follows for $p - f < f$, which is equivalent to $p < 2 \cdot f = R$. Thus, the image is larger than the object if the object is placed closer to the mirror than the centre of curvature.

KEY POINT

A spherical mirror produces a magnified image when the object distance is smaller than the image distance. This requires the object distance to be smaller than twice the focal length.

If the object is closer to the mirror than the focal point, a magnified but virtual image is formed that is no longer inverted. This case is illustrated in Fig. 22.11. Again, to construct the image, two rays are followed from the upper end of the object, one that travels parallel to the optical axis and then reflects through the focal point, and one that travels in the direction away from the focal point and becomes parallel to the optical axis after reflection. These two light rays do not intersect anywhere on the right (real) side of the mirror, but they do intersect on the left (virtual) side of the mirror. This is therefore the point where the image forms.

To use the mirror equation consistently in all such cases, Table 22.1 summarizes the sign conventions for mirrors. In the table, as throughout the chapter, p is the

TABLE 22.1

Sign conventions for mirrors

p is positive	Object is in front of the mirror (real object)
p is negative	Object is behind the mirror (virtual object)
q is positive	Image is in front of the mirror (real image)
q is negative	Image is behind the mirror (virtual image)
f and R are positive	Centre of curvature is in front of the mirror (concave mirror)
f and R are negative	Centre of curvature is behind the mirror (convex mirror)
M is positive	Image is upright
M is negative	Image is inverted

These conventions are used when the mirror equation and mirror magnification formulas are applied.

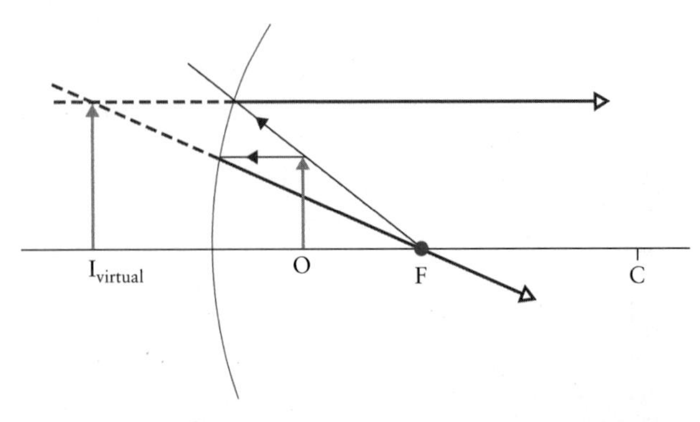

Figure 22.11 Formation of an upright virtual image for an object that is closer to the mirror than to the focal point. C is the centre of curvature of the mirror.

object distance, q is the image distance, f is the focal length, R is the radius of curvature of the mirror, and M is the magnification as defined in Eq. [22.5].

You can qualitatively verify several of the possible combinations of parameters in Table 22.1 by doing the following experiment at home: Take a well-polished tablespoon and a small object such as the tip of a pencil. Hold the spoon at arm's length and move the pencil closer and closer to the spoon while observing the changes to its image: the magnification, whether the image is inverted or upright, and whether it is real or virtual. By using either the back or front side of the spoon as a mirror you can switch between concave and convex mirrors. The spoon's use as a concave mirror is illustrated in Fig. 22.12(a).

CONCEPT QUESTION 22.1

Fig. 22.12(b) shows the centre of curvature C, the focal point F, and an object O for a concave mirror. Which of the four proposed image constructions is correct?

EXAMPLE 22.1

A dentist uses a mirror to examine a tooth. The tooth is 1.0 cm in front of the mirror, and the image is formed 10.0 cm *behind* the mirror. Determine (a) the radius of curvature of the mirror and (b) the magnification of the image.

Solution

Solution to part (a): The focal length of the mirror is found from the mirror equation with p = 11.0 cm and q = –10.0 cm. The image length is negative due to the sign conventions of Table 22.1. Thus:

$$\frac{1}{f}=\frac{1}{p}+\frac{1}{q}=\frac{1}{0.01\ \mathrm{m}}-\frac{1}{0.1\ \mathrm{m}},$$

which yields:

$$f=+1.11\ \mathrm{cm}.$$

Thus, the radius of curvature of the mirror is $R = 2f = 2.22$ cm.

Solution to part (b): We obtain the magnification from the mirror equation in part (a):

$$M=-\frac{q}{p}=-\frac{-10.0\ \mathrm{cm}}{1.0\ \mathrm{cm}}=+10.0.$$

The image is upright because the magnification is positive. This is convenient, since this is the way the dentist prefers to look at the image.

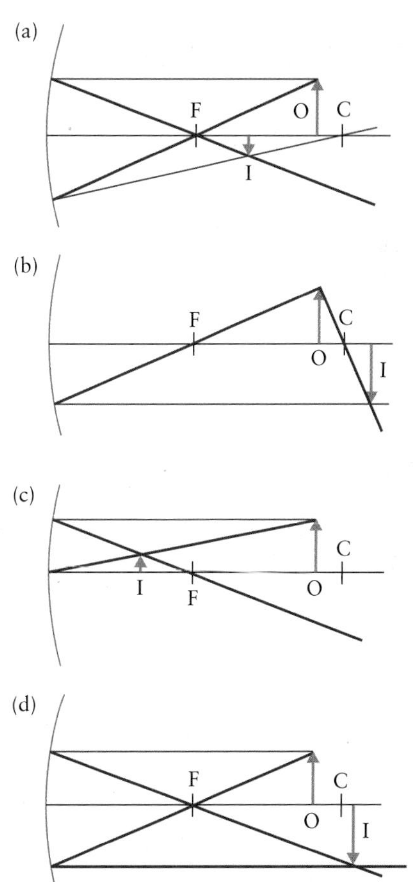

Figure 22.12 (a) A spoon is used as a concave mirror. (b) The centre of curvature C, focal point F, and object O are shown for a concave mirror. Four attempts to construct the image are included.

22.3: Refraction

22.3.1: Flat Interface

Light can pass through transparent media; e.g., visible light passes through window glass. Except in a vacuum, the intensity of light attenuates as it passes through any medium, leading to the definition of the **optical depth** of a medium. As an example, you have no problem seeing an object at the bottom of a beaker filled with water, but you cannot see the bottom surface of a deep lake. Also, you can see the ground through Earth's atmosphere from outer space (e.g., from the International Space Station), but you cannot see the surface of Jupiter during a fly-by mission. Still, for short distances, the gases in Jupiter's atmosphere (helium and hydrogen) are transparent to visible light. In the remainder of this chapter we refer to transparent materials with the assumption that their thickness is chosen such that light travels through the material without a noticeable loss in intensity.

When light is incident upon an interface between two transparent media under a not too steep angle, we observe that a fraction of the light is reflected and a fraction of the light passes through the interface into the second medium. This is illustrated in Fig. 22.13, with a sketch of a light ray approaching a glass surface from a vacuum in part (a) and with an experimental setup in part (b). The reflected ray obeys the law of reflection. The angle between the direction of the normal vector of the glass surface (along the thin vertical line) and the light ray that has passed through the interface, labelled β in Fig. 22.13(a), depends on the material forming the interface with the vacuum. The relation of this angle to the incoming ray's angle is **Snell's law** (named for Willebrord Snell):

$$\frac{\sin\alpha_{in}}{\sin\beta} = n, \qquad [22.6]$$

with n the **index of refraction**, which is a dimensionless materials constant. Table 22.2 lists indices of refraction for a range of materials. Note that gases have values very close to the value of the vacuum. This is convenient from a practical point of view: we need not distinguish whether we do the experiment in air or in a vacuum. The wavelength restriction noted below Table 22.2 is due to the effect of dispersion.

As Fig. 22.13 indicates, refraction causes light to travel closer to the normal direction in denser materials. Since the path of a light ray is reversible, light can of course also be sent across an interface approaching from the denser medium. Fig. 22.13(b) shows that in this case the light is refracted away from the normal direction. This allows us to choose an angle β^* such that the angle on the vacuum side becomes 90°. This is called the threshold angle for **total reflection**, since light approaching the interface from the denser side at angles larger than β^* cannot leave into the less dense medium. The threshold angle for total reflection is calculated from Eq. [22.6].

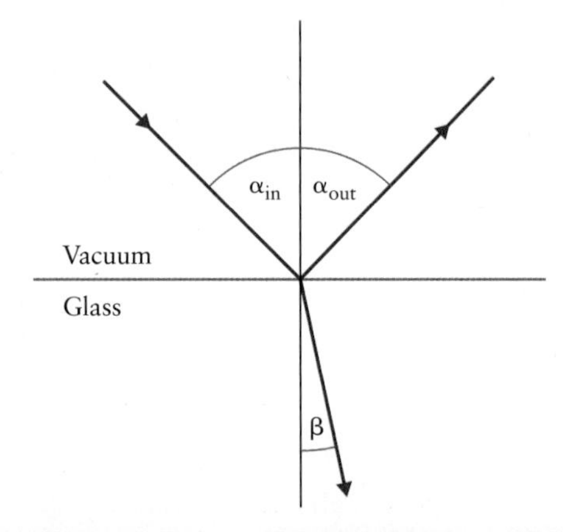

Figure 22.13 (a) A light ray arriving through a vacuum reflects off a transparent glass surface. Part of the light passes through the interface and forms a refracted light ray in the glass. The angles α_{in} and β are related by Snell's law. (b) Photograph of an experimental setup illustrating Snell's law.

Note that Snell's law is not in the simplest possible mathematical form, which would be $n = \alpha_{in}/\beta$. The underlying cause becomes evident when comparing to the case of sound waves in Chapter 15: the speed of light varies between different materials. When light passes through a medium its electric field interacts with that of the atoms and molecules in the medium. This slows the light down to speeds less than the speed of light in a vacuum at $c = 3 \times 10^8$ m/s, $v_{light} < c$. From Huygens' principle in Chapter 16 we find:

$$\frac{\sin\alpha}{\sin\beta} = \frac{c}{v_{light}}. \qquad [22.7]$$

The subscript *incoming* of the angle a has been dropped, because the reversibility of the light ray means that we do not need to specify whether α is associated with an incoming or outgoing beam. Eq. [22.7] shows that the ratio of the two sine terms is constant, since c and v_{light} are both constant across a given interface. Due to Eq. [22.6], the index of refraction represents the factor by which the speed of light is lowered in a medium in comparison to

TABLE 22.2

Index of refraction for various materials

Material	Index of refraction
Solids at 20°C	
Diamond (C)	2.42
Sapphire (Al_2O_3)	1.77
Fluorite (CaF_2)	1.43
Fused quartz (SiO_2)	1.46
Crown glass	1.52
Flint glass	1.61
Ice (H_2O, at 0°C)	1.31
Sodium chloride (NaCl)	1.54
Liquids at 20°C	
Benzene (C_6H_6)	1.50
Carbon tetrachloride (CCl_4)	1.46
Ethanol (C_2H_5OH)	1.36
Glycerine	1.47
Water (H_2O)	1.33
Sugar solution (30%)	1.38
Sugar solution (80%)	1.49
Gases at 20°C and 1 atm	
Air	1.00027
Carbon dioxide (CO_2, at 0°C)	1.00045
Vacuum	1.0

The data are measured with light rays of vacuum wavelength 589 nm.

a vacuum. This index is never smaller than 1, as light cannot be faster in any medium than in a vacuum. It can be slowed down considerably—e.g. in diamond, by more than a factor of 2!

If we replace the vacuum in Fig. 22.13(a) with a second medium with another index of refraction $n_2 > 1$, Snell's law is generalized as the **law of refraction**:

$$n_1 \sin\alpha_1 = n_2 \sin\alpha_2. \qquad [22.8]$$

KEY POINT

A light ray passing through an interface is refracted. The law of refraction relates the angles of the light ray with the normal on both sides of the interface to the two indices of refraction. Snell's law is a special case where one medium is a vacuum (or air).

EXAMPLE 22.2

A light ray travelling through air is incident on a flat slab of transparent solid material. The incident beam makes an angle of 40° with the normal, and the refracted beam makes an angle of 26° with the normal. Find the index of refraction of the transparent material.

continued

Solution

We use the law of refraction to solve for the unknown index of refraction of the transparent material with the given data and the index of refraction of air as $n = 1$ (see Table 22.2):

$$n_{slab} = n_{air}\frac{\sin\alpha_{air}}{\sin\alpha_{slab}} = 1.00\frac{\sin(40°)}{\sin(26°)} = 1.47.$$

If we want to identify the material based on this result, we might suggest fused quartz with $n = 1.46$ from Table 22.2, since glycerine is a liquid.

22.3.2: Single Spherical Interface

As in the case of reflection, the more interesting applications of refraction result for spherical interfaces. Spherical interfaces of transparent materials include all types of lenses, e.g., in optical instruments and corrective eyeglasses.

Before studying lenses, which have two spherical interfaces, we first establish the basic relations for a single, spherical interface between two different transparent media. We start with Fig. 22.14, showing a spherical slab of material with radius R. The indices of refraction are n_2 at the right side and n_1 at the left side. We choose $n_1 < n_2$. When introducing a point light source O at a distance p from the interface, we automatically define the optical axis as the line passing through O and C, the centre of curvature of the spherical slab.

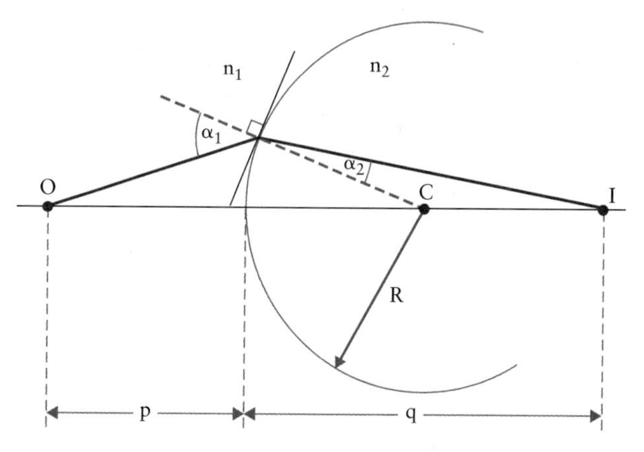

Figure 22.14 Refraction of the light from a point source O at distance p from a spherical interface separating two media of indices of refraction n_1 and n_2. C is the centre of curvature of the interface; its radius is R. An image I is formed at distance q from the interface.

Following the path of light rays from O that are incident on the interface, we observe that they pass through a common point on the optical axis as long as the incident angle α_1 is not too large. This point defines the image I. We obtain the distance q between this point and the interface using the law of refraction for the spherical interface:

$$\frac{n_1}{p} + \frac{n_2}{q} = \frac{n_2 - n_1}{R}. \qquad [22.9]$$

This equation relates the image and object distances with the two indices of refraction and the radius of curvature of the interface.

Based on Eq. [22.9], the refractive effect of the interface is determined by the term $\Delta n/R$ on the right-hand side. This term is called the **refractive power** of a single interface and has unit **diopter** (dpt), which corresponds in SI units to dpt = m^{-1}. We will use this term when light rays pass single spherical interfaces in an optical setup, e.g., when light passes through the cornea into the human eye.

KEY POINT

The refractive power of a single interface is equal to the difference in indices of refraction across the interface divided by its radius of curvature.

When using Eq. [22.9], a new set of sign conventions for p, q, and R is required, analogous to those for mirrors in Table 22.1. The sign conventions for refracting surfaces are given in Table 22.3.

TABLE 22.3

Sign conventions for single refracting surfaces

p is positive	Object is in front of the surface (real object)
p is negative	Object is behind the surface (virtual object)
q is positive	Image is behind the surface (real image)
q is negative	Image is in front of the surface (virtual image)
R is positive	Centre of curvature is behind the surface
R is negative	Centre of curvature is in front of the surface

Eq. [22.9] can also be applied to a flat refracting surface, which is important for judging the depth of an object in another medium. We extrapolate Eq. [22.9] for $R \to \infty$. We find for the relation between object and image distance:

$$\frac{n_1}{p} + \frac{n_2}{q} = 0,$$

which leads to:

$$q = -\frac{n_2}{n_1}p.$$

Following the sign conventions in Table 22.3, the image and the object are on the same side of the refracting surface as shown in Fig. 22.15. Their respective distance to the interface depends on the difference of the two indices of refraction, with the case $n_1 < n_2$ illustrated in (a) and the case $n_1 > n_2$ illustrated in (b).

In addition to Eq. [22.9], we introduce a formula that allows us to calculate the magnification for the image size h_I of an object of height h_O:

$$M = \frac{h_I}{h_O} = -\frac{n_1 q}{n_2 p}. \qquad [22.10]$$

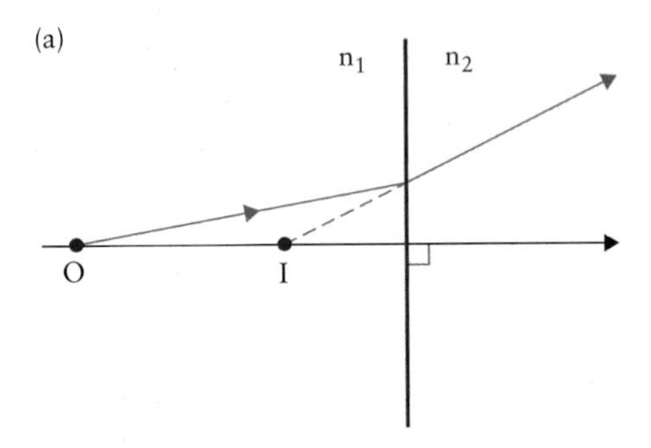

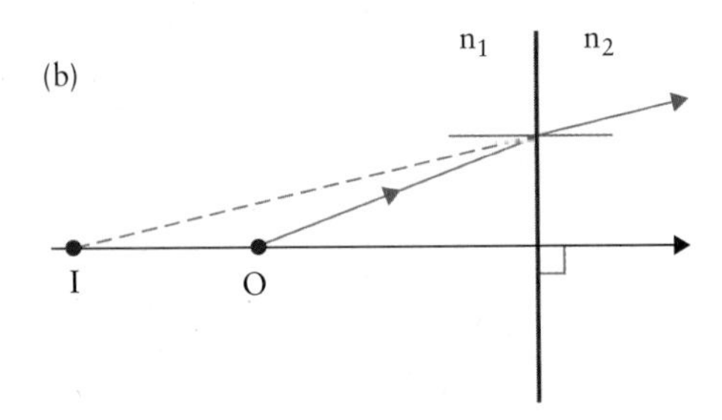

Figure 22.15 Formation of image I for a point source O for a flat refracting surface. (a) $n_1 > n_2$; (b) $n_1 < n_2$. Actual light rays are shown as solid lines. The dashed lines allow us to construct the image.

We do not derive this formula in this textbook. However, you can do a simple experiment at home to illustrate that Eq. [22.10] indeed allows for a magnification of objects seen across a single spherical interface. Cut off the bottom and top lids of a food can. Then cover one end of the can with transparent plastic wrap. Fill your kitchen sink with water and drop a penny in the water. Push the can below the water surface with the plastic wrap down. The water pressure forces the plastic wrap upward, which creates a spherically shaped interface. If you now observe the penny through the can, you will see it magnified.

EXAMPLE 22.3

A small fish is swimming at a depth d below the surface of a fishbowl, as shown in Fig. 22.16. What is the apparent depth of the fish as viewed directly from above?

Solution

In this problem, the refracting interface is flat and the object is in the denser medium, which is the medium with the higher index of refraction. This case corresponds to Fig. 22.15(a). We apply Eq. [22.10] with $n_2 = 1$ (for air) and $n_1 = 1.33$ (for water) to find the image distance:

$$q = -\frac{n_2}{n_1}p = -\frac{1}{1.33}d = -0.75d.$$

Thus, the apparent depth of the fish is three-quarters of its actual depth.

continued

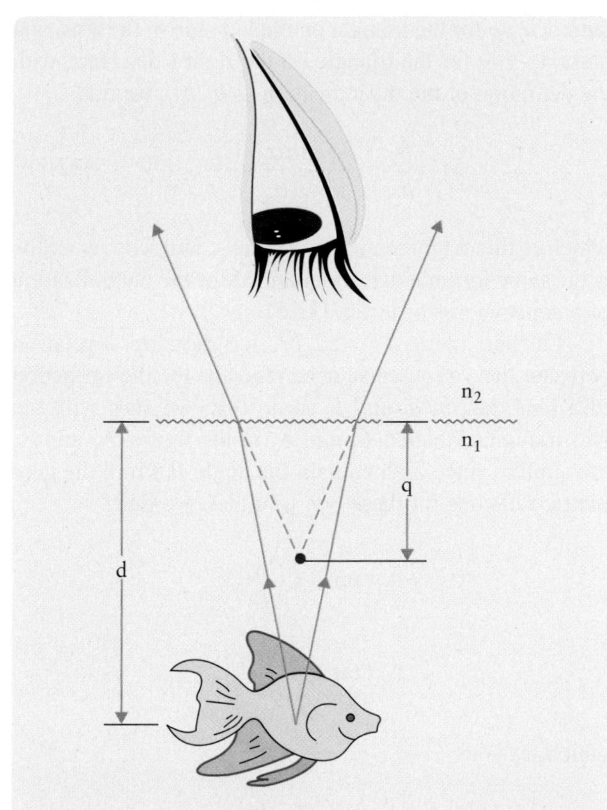

Figure 22.16 A fish is observed from a position vertically overhead

22.3.3: Lenses: Thin-Lens Formula and Magnification

Lenses are transparent objects with either two partially spherical refracting surfaces or a combination of a flat and a spherical refracting surface. Light passes through both surfaces before an image is formed. Lenses come in many shapes, as illustrated in Fig. 22.17. They are characterized by two physical parameters:

- Lenses are grouped as either **converging lenses** [Fig. 22.17(a)] or **diverging lenses** [Fig. 22.17(b)], based on their effect on incoming light rays; and
- Lenses are distinguished as **thick lenses** or **thin lenses**, based on their thickness in relation to other lengths, such as the object and image distances. A thin lens is modelled by a single refracting plane that combines the contributions of both surfaces.

In this textbook, we limit our discussion to thin lenses and more often choose converging lenses for examples. The discussion of diverging lenses is entirely analogous, though, and therefore duplication of the presentation is omitted without loss of generality. We do need to consider diverging lenses when discussing eye defects later in the chapter. Discussion of thick lenses would lead to mathematically complicated formulas but does not contribute additional insight. Also, applications of lenses in the life sciences are based almost exclusively on thin lenses.

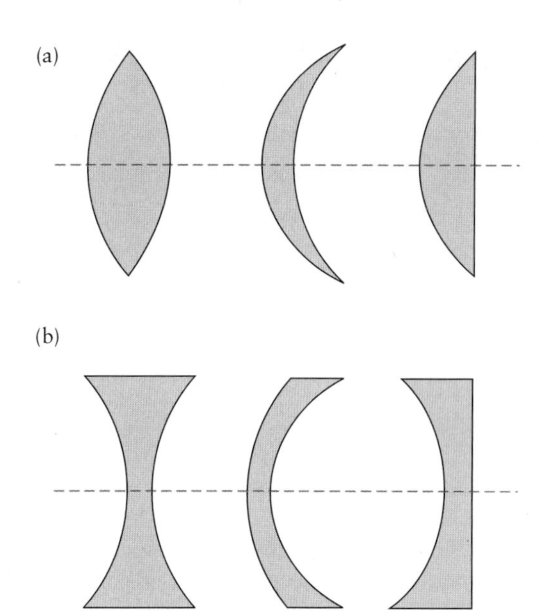

Figure 22.17 Various types of commonly used lenses: (a) converging lenses, which are thicker at the optical axis than toward the edges; and (b) diverging lenses, which are thinner at the optical axis. The dashed lines are the respective optical axes.

We first reintroduce the concept of a focal point for a thin lens. Fig. 22.18 shows several parallel incoming light rays as in Fig. 22.8(a), where we defined the focal point for a mirror. Incoming parallel light rays correspond to a light source (object) at infinite distance. The focal point is the point (if it exists) at which all these light rays intersect, i.e., where they form a point image of the light source. This approach is sufficient to define a focal point *F* for a thin lens if the light rays travel not too far from the optical axis, as chosen in Fig. 22.18.

Fig. 22.18 illustrates the thin lens simplification: a dashed line is drawn perpendicular to the optical axis at the centre of the lens, allowing us to limit the ray refraction to a single step in geometric image constructions.

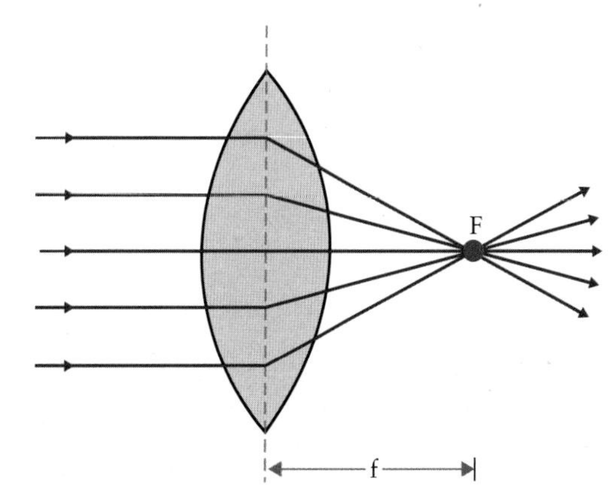

Figure 22.18 Definition of the focal point for a converging lens. As in the case of the mirror, parallel incoming light rays are used, such as light rays from an object at infinite distance. *F* is the focal point and *f* the focal length. The lens is a thin lens, which is indicated by the dashed vertical line at which the refraction is drawn in a single step.

We use three light rays to determine the position and size of images, as we did in the case of mirrors:

- a light ray travelling along the optical axis,
- a ray incident on the lens along a path parallel to the optical axis, and
- a ray incident on the lens after passing through its focal point.

We show how this approach allows us to construct an image for an object at finite distance from the lens in Fig. 22.19. The object O of height h_O is placed at distance p to the left of a thin converging lens. The image is constructed by placing one end of the object on the optical axis. Two light rays are followed from the upper end of the object. The ray parallel to the optical axis is refracted through the focal point on the right side of the lens (i.e., the side of the lens opposite the object) and passes through the optical axis at that point with an angle θ. A second ray, emerging from the upper end of the object, passes through the focal point at an angle ϕ on the left side, which is the same side of the lens on which the object is located. This ray becomes a light ray travelling parallel to the optical axis beyond the lens. The light rays that emerge from the upper end of the object intersect at a distance q from the lens, defining the upper point of the image I. This point determines the height of the image, h_I.

We first derive the magnification of a thin lens from Fig. 22.19. To allow for a simple geometric derivation, a light ray is drawn from the top of the object to the top of the image (dashed line), crossing the optical axis at the centre of the lens and defining an angle α on both sides. We find $\tan\alpha = h_O/p$ for the triangle on the left side of the lens, and $\tan\alpha = -h_I/q$ for the triangle on the right side. Thus, with the definition of the magnification as h_I/h_O, we find:

$$M = \frac{h_I}{h_O} = \frac{-q\tan\alpha}{p\tan\alpha} = -\frac{q}{p}, \qquad [22.11]$$

which is the magnification of the lens, and, conveniently, is the same formula that we obtained for the magnification of a concave mirror in Eq. [22.5].

Further, using Fig. 22.19, we develop a relation between the various distances relevant for the refraction of a lens, i.e., p, q, and f. To do this, we start with the two triangles labelled Δ_1 and Δ_2 in the figure. Δ_1 and Δ_2 are similar, since both contain the angle θ. From the geometric relations for these two triangles, we find:

$$\Delta_1 : \tan\theta = \frac{h_O}{f}$$

$$\Delta_2 : \tan\theta = \frac{-h_I}{q-f},$$

which leads to:

$$\frac{h_I}{h_O} = \frac{q-f}{f}.$$

Combining this equation with Eq. [22.11] we find:

$$-\frac{q}{p} = -\frac{q-f}{f},$$

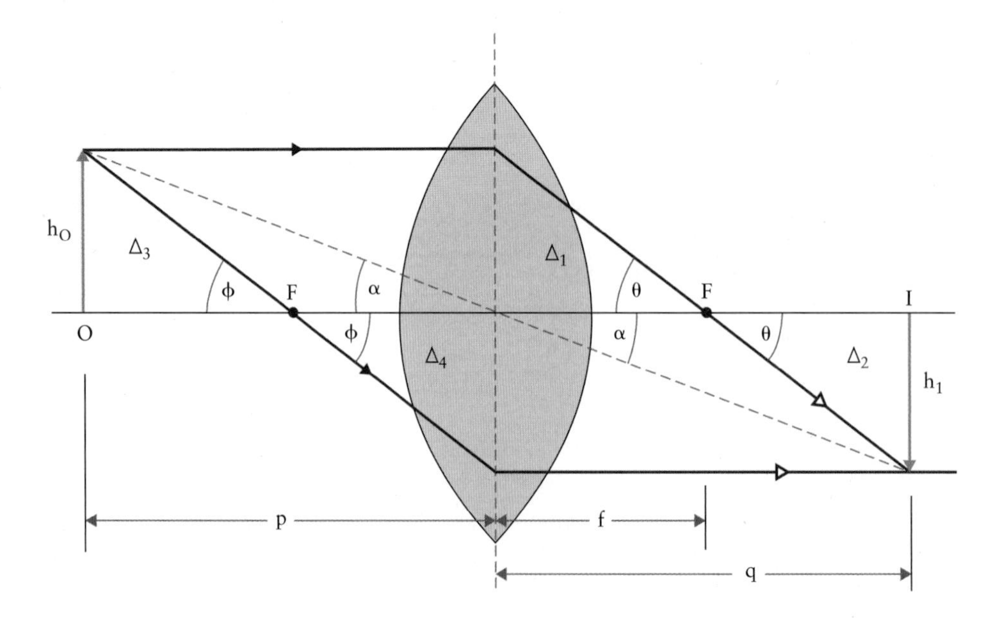

Figure 22.19 Construction of the image I at distance q for an object O at distance p from a converging lens. The lens is a thin lens with the refraction drawn at the dashed vertical line. F indicates the focal point. Note that only three rays are needed to find the image size h_I: (1) a light ray along the optical axis from the bottom end of the object; (2) a light ray travelling from the top of the object parallel to the optical axis, then refracting through the focal point; and (3) a light ray travelling through the focal point and proceeding parallel to the optical axis after refraction.

which is rewritten as:

$$\frac{q}{p} = \frac{q}{f} - 1$$

and finally leads to:

$$\frac{1}{p} + \frac{1}{q} = \frac{1}{f} \qquad [22.12]$$

Eq. [22.12] is called the **thin-lens formula**. Note that it is the same formula we found for the spherical mirror in Eq. [22.3].

KEY POINT

The formulas for the magnification and for the relation between focal, object, and image lengths are the same for spherical mirrors and thin lenses.

All formulas derived in this section for converging lenses apply to diverging lenses as well. A set of sign conventions must be followed, as summarized in Table 22.4.

TABLE 22.4

Sign conventions for thin lenses

p is positive	Object is in front of the lens
p is negative	Object is behind the lens
q is positive	Image is behind the lens
q is negative	Image is in front of the lens
R_1 and R_2 are positive	Centre of curvature for each surface is behind the lens
R_1 and R_2 are negative	Centre of curvature for each surface is in front of the lens
f is positive	Converging lens
f is negative	Diverging lens

These conventions are used when the thin-lens formula and the lens magnification formula are applied. R_1 is the radius of curvature of the front surface of the lens and R_2 is the radius of curvature of its back surface. These are used when the lens maker's equation is applied.

22.3.4: Lenses: The Lens Maker's Equation

We can approach the quantitative treatment of a lens in a second manner, starting with the results above for a single spherical interface. Fig. 22.20 is introduced to quantify the optical properties of two consecutive refracting surfaces. The first interface separates material 1, with index of refraction n_1, and material 2, with index of refraction n_2. The two materials are chosen such that this interface alone does not allow the formation of an image on the right side. This means specifically that the change in the index of refraction at the first interface is not sufficient to cause diverging light rays from the object *O* at distance *p* in front of the interface to converge after passing the interface. A second interface, from material 2 to material 3, with index of refraction n_3, is needed so that an image *I* at distance *q* is formed. Note that the distance *q* is defined

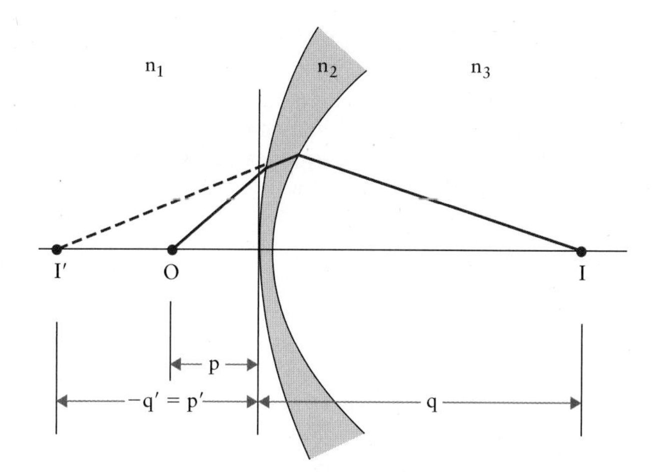

Figure 22.20 Formation of an image *I* for a point-like light source *O* located on the optical axis. Light passes two refracting surfaces travelling through three media, with indices of refraction n_1, n_2, and n_3. Note that the two interfaces are located very close to each other to allow us to develop relations applicable to thin lenses.

with reference to the same point along the optical axis as the distances p, p', and $-q'$. This is a good approximation for the case of a thin lens where the thickness of material 2 is negligible when compared with p, q, R_1, and R_2.

A formula describing the relations among the relevant parameters in Fig. 22.20 is developed. For this, we apply Eq. [22.9] at the first interface:

$$\text{(I)} \quad \frac{n_1}{p} + \frac{n_2}{q'} = \frac{n_2 - n_1}{R_1},$$

where q' is the image distance, which is negative as shown in Fig. 22.20. R_1 is the radius of curvature of the first interface, i.e., the interface that light from the object hits first. Eq. [22.9] further applies at the second interface:

$$\text{(II)} \quad \frac{n_2}{p'} + \frac{n_3}{q} = \frac{n_3 - n_2}{R_2}.$$

Note that the image distance of the first interface has become the object distance for the second interface, $p' = -q'$:

$$\text{(II}'\text{)} \quad -\frac{n_2}{q'} + \frac{n_3}{q} = \frac{n_3 - n_2}{R_2}.$$

The intermediate term n_2/q' is eliminated by combining the equations at the first and second interfaces:

$$\frac{n_1}{p} + \frac{n_3}{q} = \frac{n_2 - n_1}{R_1} + \frac{n_3 - n_2}{R_2}. \qquad [22.13]$$

The left-hand sides of Eqs. [22.13] and [22.9] are the same, containing the parameters of the medium left and right of the studied interfaces. Thus, several interfaces do not change the left-hand side of the equation. Additional interfaces have an effect on the right-hand side, as we see when comparing Eqs. [22.9] and [22.13]: a term of the form $\Delta n/R$ is added for each interface. We defined the term $\Delta n/R$ earlier as the refractive power of an interface.

EXAMPLE 22.4

Can you see clearly underwater without diving goggles? For the discussion, combine the cornea and the lens of the human eye to be a single symmetric lens with $n = 1.5$. Take $n = 1.3$ for the vitreous body.

Solution

Eq. [22.13] connects the object distance, the image distance, and the radii of curvature for two consecutive refractive interfaces. The setup is shown in Fig. 22.20, illustrating that the light travels from a medium with refractive index n_1, through a medium of refractive index n_2, and finally into a medium of refractive index n_3.

In the current example we want to compare a tourist first looking at other people at the beach and then looking at an approaching shark underwater. When looking at the beach, medium 1 is air, medium 2 is the lens of the eye, and medium 3 is the vitreous body behind the lens, with $n_1 = 1.0$, $n_2 = 1.5$, and $n_3 = 1.3$. With these values, Eq. [22.13] yields for a symmetric lens (i.e., when $-R_1 = R_2 = R$):

$$\frac{1}{p}+\frac{1.3}{q}=\frac{2(1.5-1.0-1.3)}{R}=\frac{0.7}{R}.$$

If people at the beach are practically at infinite distance, $p = \infty$, they generate a focused image on the tourist's retina (using $q = 2.8$ cm for a typical distance between lens and retina) if the effective radius of curvature of the lens is $R = 1.5$ cm. This is a value well within the range of accommodations of the human eye, which can reach a maximum effective accommodation of $R = 1.0$ cm.

Once below the water's surface, the index of refraction of the medium containing the object—now the shark—changes to $n_1 = 1.33$. This changes the last equation to:

$$\frac{1.33}{p}+\frac{1.3}{q}=\frac{2(1.5-1.33-1.3)}{R}$$

$$=\frac{0.37}{R}.$$

A shark at infinite distance ($p = \infty$) leads to an image distance of $q = 3.5$ cm for the maximum effective accommodation of the eye with $R = 1.0$ cm, Thus, the shark's image is significantly blurred on the retina, which lies only 2.8 cm behind the lens.

The situation does not improve when the object comes closer. If we choose for example $p = 25$ cm, we find $q = 4.1$ cm at the maximum effective accommodation of the eye. Thus, the image of a near object underwater is even more blurred. All these optical effects, of course, will escape the tourist's attention.

Eq. [22.13] simplifies when media 1 and 3 are identical. This is also the most common case as it applies specifically to artificial lenses, where media 1 and 3 are usually air ($n_1 = n_3 = 1$) and medium 2 is a transparent material with $n_2 = n$. In this case, Eq. [22.13] becomes the **lens maker's equation** for thin lenses:

$$\frac{1}{p}+\frac{1}{q}=(n-1)\left(\frac{1}{R_1}-\frac{1}{R_2}\right). \qquad [22.14]$$

Again, note that p and q are measured to the same point along the optical axis; i.e., the transparent material 2 is of negligible thickness.

22.3.5: Lenses: Refractive Power

Note that the left-hand side of Eq. [22.12] connects the object and image distances of a lens in the same manner as the left-hand side of Eq. [22.9]; the right-hand side is therefore a measure of the ability of the lens to refract the light. Using this observation, the **refractive power of a lens** $\mathfrak{R}$ is defined:

$$\mathfrak{R}=\frac{1}{f}. \qquad [22.15]$$

Eq. [22.15] combines the effect of two refractive interfaces; for single interfaces the refractive power was defined earlier as $\Delta n/R$, with R the radius of curvature of a spherical interface. Eq. [22.15] is consistent with that definition; i.e., the combined refractive power of two spherical interfaces, which is the sum of two $\Delta n/R$ terms, is equal to the inverse focal length of the lens formed by the two interfaces.

KEY POINT

The refractive power of a thin lens, $\mathfrak{R}$ in unit diopters (dpt), is equal to its inverse focal length.

EXAMPLE 22.5

A converging lens with $f = 10$ cm (i.e., a lens of refractive power $\mathfrak{R} = +10$ dpt) forms images of objects at (a) 30 cm, (b) 10 cm, and (c) 5 cm from the lens. In each case, find the image distance and describe the image.

Solution

Solution to part (a): Substituting the given values in the thin-lens formula, we find:

$$\frac{1}{10\ \text{cm}}=\frac{1}{30\ \text{cm}}+\frac{1}{q},$$

which corresponds to $q = +15$ cm. Further substitution in the magnification formula yields:

$$M=-\frac{q}{p}=-\frac{15\ \text{cm}}{30\ \text{cm}}=-0.5.$$

A positive q value means that a real image is formed on the side of the lens opposite the object. The image is half the height of the object and is inverted due to the negative sign of M. This case is shown in Fig. 22.21(a).

continued

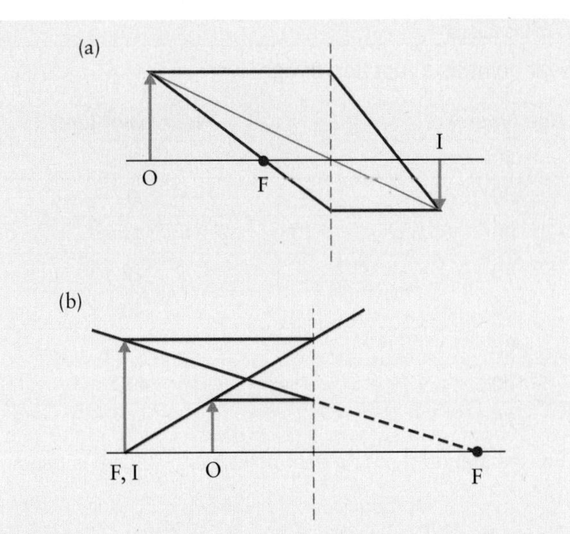

Figure 22.21 (a) Sketch of an image *I* formed for an object *O* located farther away than the focal length. For convenience, the thin lens is not drawn but is represented by a dashed vertical line. The three rays shown allow us to construct the image. (b) Sketch of an image *I* formed for an object *O* located closer than the focal length. Again, three rays are used to construct the image.

Solution to part (b): Replacing $p = 30$ cm with $p = 10$ cm in part (a) leads to $q = \infty$. Note that this case is equivalent but with reverse light rays to Fig. 22.18, where an object at the focal length has an image at infinite distance.

Solution to part (c): In the third case, the object lies inside the focal length. Replacing $p = 30$ cm with $p = 5$ cm in part (a) leads to:

$$\frac{1}{10\text{ cm}} = \frac{1}{5\text{ cm}} + \frac{1}{q},$$

which yields $q = -10$ cm. Further substituting the new values in Eq. [22.11] leads to:

$$M = -\frac{q}{p} = \frac{10\text{ cm}}{5\text{ cm}} = +2.0.$$

This result is illustrated in Fig. 22.21(b). The negative image distance represents a virtual image, i.e., an image on the same side of the lens as the object. A positive magnification $M > 1$ means that the image is enlarged and upright.

CONCEPT QUESTION 22.2

In Fig. 22.21(a) you see an object *O* and its image *I* that formed for a converging lens we represented by a vertical dashed line. Which of the following quantities is a negative number in this case?

(A) object distance *p*

(B) image distance *q*

(C) focal length *f*

(D) refractive power of the lens $\mathfrak{R}$

(E) magnification *M*

22.4: Applications in Optometry and Ophthalmology

Two main applications of the ray model of optics exist in the life sciences: vision and microscopy. We discuss both in this textbook, beginning in this section with the healthy eye. We then proceed to the most common eye defects and their corrections. The discussion of light microscopes, which allow us to see objects too small to observe with the naked eye, follows in the next section.

22.4.1: The Eye

Fig. 22.1 is a sketch of the side view cross-section of a human eyeball. The optically active parts of the eye are the cornea and the lens. The **cornea** contains a convex external interface (*facies externa*) and a concave internal interface (*facies interna*). It has a small radius of curvature of about 8 mm and bridges the biggest difference in indices of refraction, from $n = 1.0$ for air to $n = 1.33$, which is close to the value for water. Thus, the cornea provides the biggest fraction of the refractive power of the eye, with:

$$\mathfrak{R}_{\text{cornea}} = \frac{\Delta n}{R} = \frac{1.33 - 1.0}{0.008\text{ m}} = 41\text{ dpt}.$$

Due to the variability of the curvature of the cornea, a value of $\mathfrak{R}_{\text{cornea}} = 40$ dpt is generally adopted in the literature.

The **lens** is suspended by fibres (suspensory ligament of the lens, or *zonula ciliaris*) that are stretched or loosened by **ciliary muscles**. The **iris** defines the opening of the lens, allowing light to pass through only the visible area of the lens, which is called the **pupil**. The iris can vary the diameter of the pupil to adjust the total light intensity reaching the **retina**. We do not emphasize light-intensity-related issues in the current context since the related topic of sound intensity is extensively discussed in section 15.3.5.

The lens is a transparent, pliable, biconvex body with an index of refraction of $n = 1.41$. The elastic variation of the lens is illustrated in Fig. 22.22, which allows for a change of refractive power between $\mathfrak{R} = 18$ dpt and $\mathfrak{R} = 32$ dpt. When the ciliary muscle is relaxed (lower part of Fig. 22.22) the suspension fibres of the lens are stretched and the lens is elongated. This leads to a flatter surface with increased radius of curvature and reduced refractive power $\mathfrak{R}$. When the relaxed eye looks at an object at infinite distance (approximated by $p \geq 60$ m), its focal length is calculated from the two contributions to the refractive power of the eye:

$$\mathfrak{R}_{\text{relaxed eye}} = \mathfrak{R}_{\text{cornea}} + \mathfrak{R}_{\text{lens}} = (40\text{ dpt}) + (18\text{ dpt})$$
$$= 58\text{ dpt},$$

which results in $f = 1.7$ cm.

While focusing up close, as illustrated in the upper part of Fig. 22.22, the ciliary muscle is contracted,

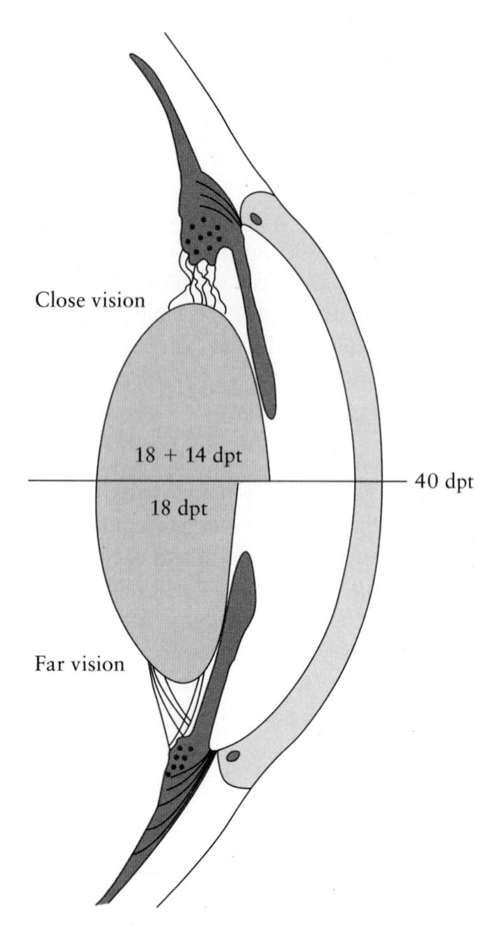

Figure 22.22 Anatomy of the lens, ciliary muscles, and ligaments of the human eye. Note the changes of these three components between close vision (top half of figure) and far vision (bottom half of figure). The cornea has a fixed refractive power of 40 dpt, to which the lens adds a refractive power between 18 dpt and 32 dpt.

relaxing the suspension fibres and allowing the lens to contract in response to its elasticity. The surfaces of the contracted lens have smaller radii of curvature, and thus a larger refractive power:

$$\Re_{\text{focused eye}} = \Re_{\text{cornea}} + \Re_{\text{lens}} = (40\ \text{dpt}) + (32\ \text{dpt})$$

$$= 72\ \text{dpt},$$

which results in a focal length of less than 1.4 cm. The process of changing the refractive power of the lens due to ciliary muscle action is called **accommodation**.

The ability to view objects close up deteriorates with age due to the sclerosing effects of the lens. Physiologically, this is quantified by defining the **near point**. The near point is the shortest object distance p for which the human eye produces a sharp image on the retina. Table 22.5 illustrates the change of the near point distance with age.

22.4.2: Eye Defects and Diseases

Two commonly occurring eye defects are discussed in greater detail as examples of how prescription eyeglasses are used to correct vision deficiencies.

TABLE 22.5

Near point as a function of age

Age (years)	Near point (cm)
10	7
20	9
30	12
40	22
50	40
60	100
70	400
>75	∞

For the standard man a near point of $s_0 = 25$ cm is used.

22.4.2.1: Hyperopia (Farsightedness)

Hyperopia is an eye defect associated with an insufficient elasticity of the lens, leading to an incomplete reshaping when the suspension fibres of the lens are relaxed. As a result, the maximum refractive power of the lens of $\Re = 32$ dpt is not reached and the eye cannot form an image of nearby objects on the retina.

Hyperopia is illustrated in Fig. 22.23. Part (a) shows the optical properties of a hyperopic eye when observing an object at great distance. The ciliary muscle is relaxed,

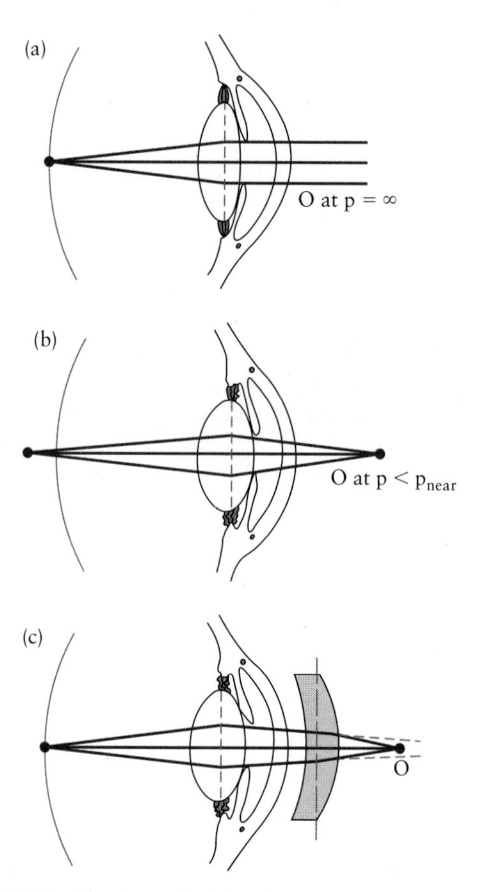

Figure 22.23 Hyperopia or farsightedness: (a) far vision, (b) close vision, (c) correction with prescription glasses.

the suspension fibres and the lens are stretched, and an image is formed properly on the retina. Thus, the person is called farsighted. Part (b) illustrates the problem of the patient when focusing on a nearby object, i.e., when the ciliary muscle contracts and the lens should relax toward its most spherical shape. If the object is closer than the near point, the image is formed behind the retina. If this near point is too far from the eye the patient has a problem, for example, when reading. This is a typical effect of old age, but may also occur when the eyeball is too short. The correction is done with prescription glasses, as illustrated in Fig. 22.23(c). The corrective lens is a convex lens with a positive refractive power $\mathfrak{R}$ to add to the too small refractive power of the eye. As indicated in the figure, the glasses cause an apparent shift of the object to greater distance, a distance at which the defective eye is able to see it clearly.

EXAMPLE 22.6

The near point of a particular person is at 50 cm. What focal length must a corrective lens have to enable the eye to clearly see an object 25 cm away?

Solution

We use the thin-lens formula with an object distance of p = 25 cm. The lens we want to prescribe must form an image on the same side of the lens as the object but at a distance of 50 cm. The eye of the person looking at that image then sees it clearly. Due to the sign in Table 22.4, we write q = –50 cm:

$$\frac{1}{f}=\frac{1}{p}+\frac{1}{q}=\frac{1}{0.25\ \text{m}}+\frac{1}{-0.5\ \text{m}}.$$

This leads to a focal length of f = +0.5 m. $\mathfrak{R} = 1/f$ = +2.0 dpt is the refractive power of the prescribed glasses.

22.4.2.2: Myopia (Nearsightedness)

Myopia is an eye defect that is due to an insufficient stretching of the lens when a person tries to obtain a lower refractive power of $\mathfrak{R}$ = 18 dpt. Myopia is illustrated in Fig. 22.24. Part (a) shows the eye trying to observe an object at great distance. The lens is not sufficiently elongated and thus the image is formed in front of the retina. The same person can see an object at the near point of the standard man (at 25 cm distance) without any problem as the elasticity of the lens is sufficient to reshape the lens to form an image on the retina [illustrated in 22.24(b)]. Fig. 22.24(c) shows how myopia is corrected with prescription glasses: parallel light rays reaching the eye from an object at great distance are refracted away from the optical axis such that they form

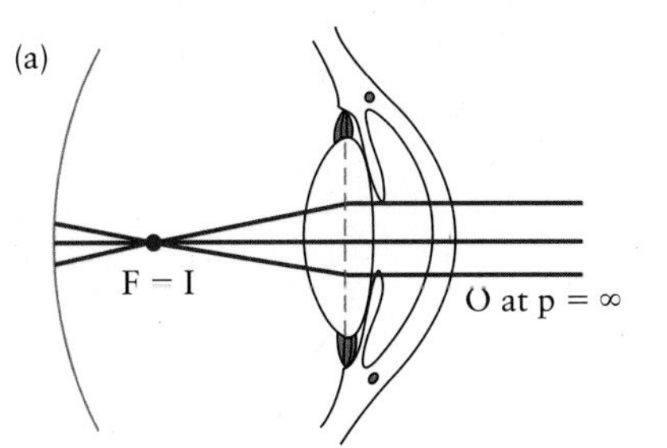

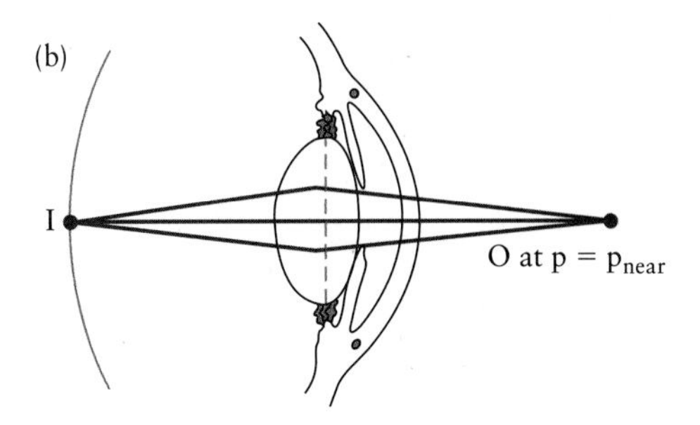

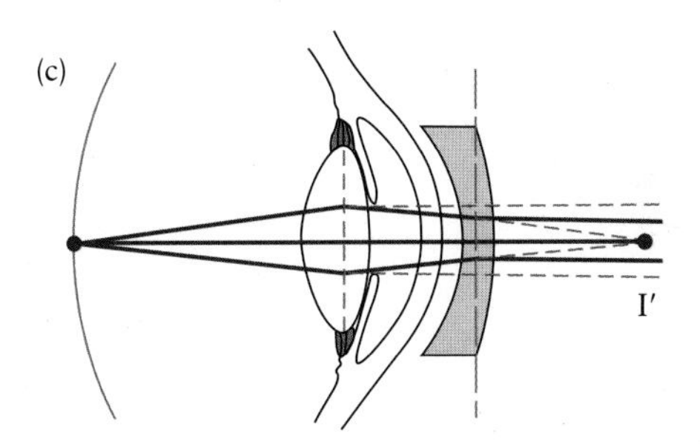

Figure 22.24 Myopia or nearsightedness: (a) far vision, (b) close vision, (c) correction with prescription glasses.

an image I' at a point closer to the eye. This image is observed with the myopic eye, forming the final image on the retina.

Typical causes of myopia are elongated eyeballs or weakened ligaments and muscles, e.g., due to diabetes mellitus. The prescription glasses are concave to lower the too high refractive power $\mathfrak{R}$ of the eye.

EXAMPLE 22.7

A certain person cannot see objects clearly when they are beyond a distance of 50 cm. What focal length should the prescribed lens have to correct this problem?

Solution

We choose the object distance as infinite, $p = \infty$, since we want to enable the eye to see anything beyond 50 cm, including objects very far away. The image of the prescription lens must be on the same side of the lens as the object and cannot be further than 50 cm; that means q = –50 cm (the negative sign results from the sign conventions in

continued

Table 22.4). If the lens accomplishes this, then the eye can look at the intermediate image and see it properly. The thin-lens formula reads:

$$\frac{1}{f}=\frac{1}{p}+\frac{1}{q}=\frac{1}{\infty}+\frac{1}{-0.5\text{ m}},$$

which yields for the focal length $f = -0.5$ m. A negative focal length means a concave lens must be prescribed (based on the sign conventions in Table 22.4). The lens has a refractive power $\Re = -2.0$ dpt.

EXAMPLE 22.8

An artificial lens is implanted in a patient's eye to replace a diseased lens. The distance between the artificial lens and the retina is 2.8 cm. In the absence of the lens, the image of a very distant object (formed by the refraction of the cornea) is formed 2.53 cm behind the retina. The lens is designed to put the image of the distant object on the retina. What is the refractive power $\Re$ of the implanted lens? *Hint:* Consider the image formed by the cornea as a virtual object.

Solution

Following the hint, we consider the image formed by the eye without the implanted lens as a virtual object for the implanted lens. To later use the thin-lens formula to determine the focal length of the implanted lens, we need to determine the object distance for the implanted lens from the image distance of the cornea without a lens. The image distance is q = (2.53 cm + 2.8 cm) = 5.33 cm, as shown in Fig. 22.25(a). The two lengths are added since the original image is formed behind the retina, i.e., farther away from the location of the missing lens.

When the new lens is implanted, the image distance calculated above becomes the object distance for the implanted lens. The object distance is p = −5.33 cm, where the negative sign indicates that this is a virtual

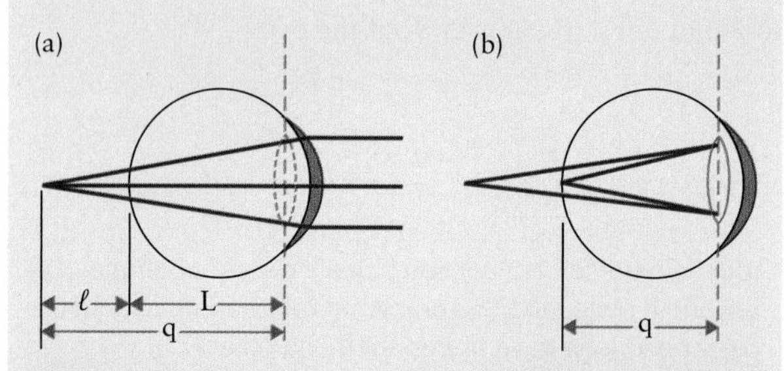

Figure 22.25 (a) Eye with a surgically removed lens (dashed lines indicate the missing lens). The cornea (light red crescent) is the only focusing component of this eye, but it does not have a sufficient refractive power to focus light from a source at infinite distance. Thus, the image forms behind the retina and the person would see only a very blurred image. (b) The artificial lens implanted into the eye must correct the position of the image such that the image is formed on the retina.

continued

object, i.e., an object that appears behind the lens. The implanted lens must now form a final image on the retina as indicated in Fig. 22.25(b), i.e., at a distance q = +2.8 cm behind the lens. The thin-lens formula yields:

$$\frac{1}{f}=\frac{1}{p}+\frac{1}{q}=\frac{1}{-5.33\text{ cm}}+\frac{1}{2.8\text{ m}},$$

which allows us to calculate the focal length as f = +5.9 cm. Using Eq. [22.15], we find the refractive power of the implanted lens to be $\Re$ = +17.0 dpt. Note that this is a much larger value than a typical refractive power for a prescription lens, but close to the value of the natural lens in Fig. 22.22.

22.5: The Light Microscope

22.5.1: From Lenses to Microscopes

No other technological development influenced the early history of biology and medicine as much as the development of the microscope. The first light microscopes were introduced in the late 1600s. Modern instruments, such as the one shown in Fig. 22.26, allow us to see small cells such as human erythrocytes and bacteria. For biologists and medical researchers, microscopes like the one shown in Fig. 22.26 are essential tools used on a daily basis. It is likely you will develop a great level of familiarity with this instrument. It is dangerous, though, to confuse familiarity with a good understanding of technical specifications and limitations: microscopes can easily deceive you, as many examples of faulty discoveries, particularly during the 19th century, illustrate. While modern instruments are designed to minimize the occurrence of artifacts that caused such mistakes, it is the user's knowledge of the physical properties of the instrument that can prevent unwarranted embarrassment. We build a solid foundation of the properties of the light microscope in this section, using the optics concepts developed in the previous section. Some issues, such as diffraction effects, spectral resolution, and Abbe's theory of the resolving power of a microscope, are not included as their discussion requires more advanced concepts from wave optics. For these, the reader should consult advanced texts on optics.

In the previous chapter, we saw that a single lens allows us to obtain magnified images. The formula describing the magnification of a single lens was introduced in the form:

$$M=-\frac{q}{p}=\frac{f}{f-p}=\frac{f-p}{f},$$

with p the object distance, q the image distance, and f the focal length. To obtain a magnified image ($M > 1$), the object must be closer to the lens than the image, which is possible when $p - f < f$ or $p < 2 \cdot f$: an object placed closer

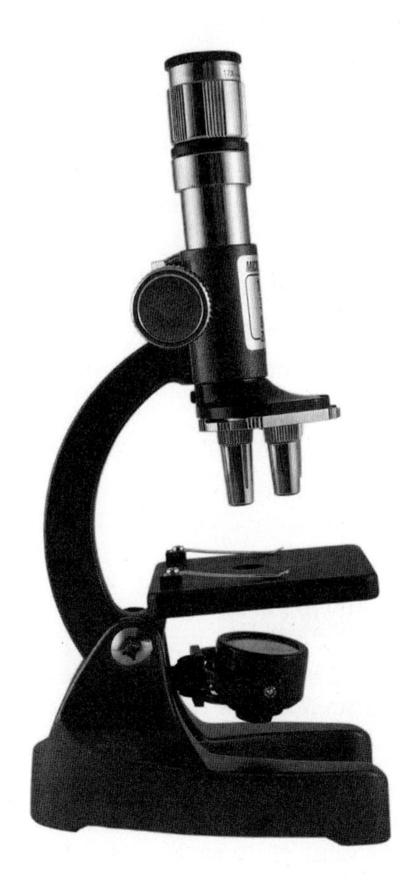

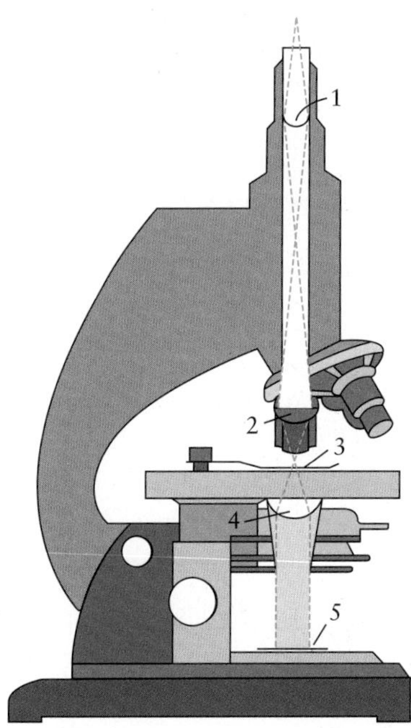

Figure 22.26 Sketch of a light microscope. In this instrument, visible light from a light source (5) is focused on a sample (3) by a condenser lens (4). This light is used to form an image that is magnified by an objective lens (2) and an eyepiece (1).

than twice the focal length generates a magnified image. Note that the magnification is only a function of *f*, *p*, and *q*—it does not depend on the position of the observer! Thus, the magnification is a property of the physical lens but does not tell us what an observer actually sees.

A microscope, in turn, is not just a device, but also a process that allows the observer to obtain a particular outcome: to see an object larger than it is. Thus, the step from lens to microscope requires us to include the observer to quantify the apparent size of an object. This is illustrated in Fig. 22.27. The position of the observer defines the angle θ between two light rays reaching the eye from opposite ends of the object. The figure compares the size of the image on the retina for two identical objects, one at object distance p_1 and the other one at object distance p_2. The figure defines the **angular magnification** m:

$$m = \frac{\theta}{\theta_0}, \qquad [22.16]$$

in which θ_0 is the angle subtended by the object when it is placed at $p = 25$ cm, which we define as the **standard near point** s_0 of a healthy adult eye.

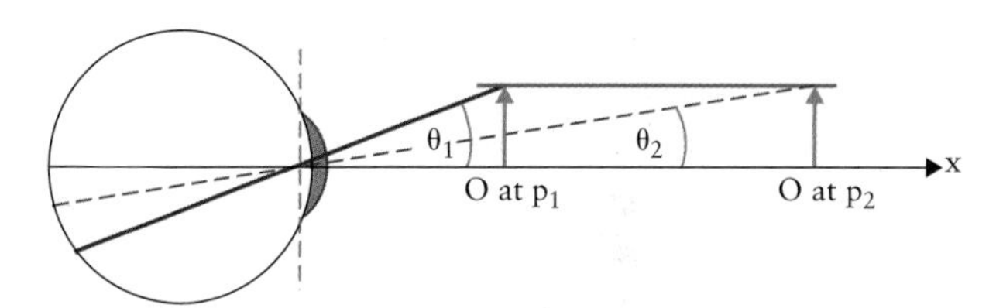

Figure 22.27 When the observer's eye becomes part of the optical system, we define an angular magnification based on the angle under which an object appears to the observer. The figure shows a geometrical sketch to illustrate that the apparent size of an object 0 varies with its distance from the observer's eye.

KEY POINT

The angular magnification is the ratio of the angle subtended by a given object and the angle subtended by the same object when placed at the standard near point s_0.

Based on this definition, an object you hold at a distance of 25 cm from your eye has an angular magnification of $m = 1$. Any object farther away appears smaller ($m < 1$), while any object closer appears bigger ($m > 1$).

The justification for defining the angular magnification as a new parameter is that this is the quantity in which we are ultimately interested. However, it also illustrates the limitations of our ability to see things larger than they are with the naked eye. Even the juvenile eye cannot focus on an object closer than about 7 cm before the eye. We can easily illustrate this with a self-test: touch your nose with the ball of your right thumb and try to see the palm of your hand in focus.

Using a small object of size h_O in Fig. 22.28, e.g., a human hair, we determine the maximum angular magnification that a juvenile eye can achieve relative to the eye of the standard man. We start with the definition in Eq. [22.16]. For a small object, the two angles are small, allowing us to use the approximation $\tan\theta = \theta$. The trigonometric terms are analyzed from Fig. 22.28 as $\tan\theta = h_O/p$

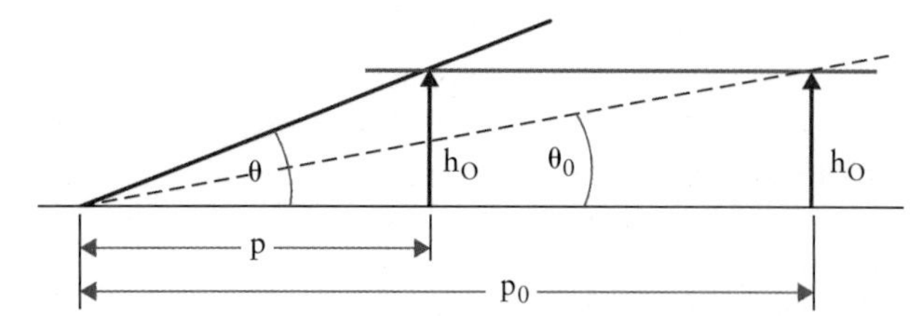

Figure 22.28 Sketch illustrating the angular magnification of the human eye.

and $\tan\theta_0 = h_O/p_0$. We use $p_0 = s_0$; i.e., the reference object distance is the standard near point:

$$m = \frac{\theta}{\theta_0} = \frac{\tan\theta}{\tan\theta_0} = \frac{p_0}{p} = \frac{s_0}{p}. \qquad [22.17]$$

This yields m = 25 cm/7 cm = 3.6.

We are, of course, not satisfied with this result, particularly not for research in the life sciences. An entire world exists at microscopic-length scales that we cannot see, as illustrated in Fig. 22.29. The lower limit of objects we can see without optical instruments is about 100 μm (e.g., a human oocyte). Below we will discuss the simple magnifying glass (with angular magnification of $m = 5 - 10$) and then the light microscope. These allow us to see objects as small as 200 nm, a value determined by the wavelengths of visible light.

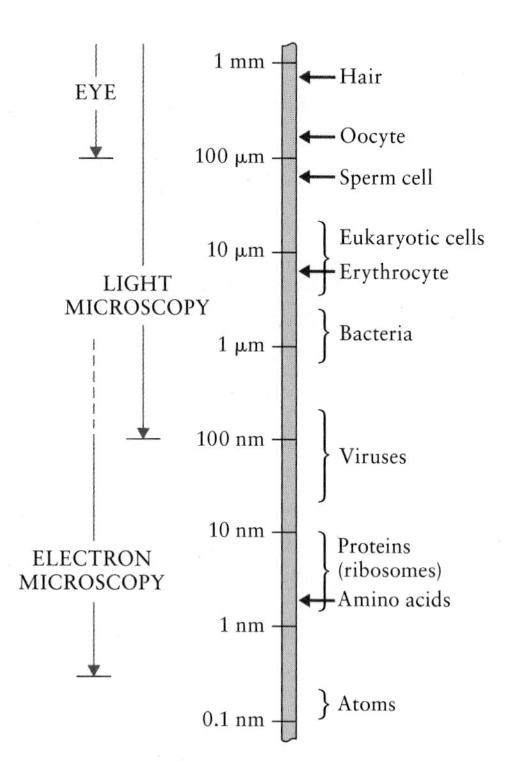

Figure 22.29 Size range of typical objects of biological interest. The vertical range indicators at the left-hand side illustrate what is visible with the eye, a light microscope, and an electron microscope.

CONCEPT QUESTION 22.3

Fig. 22.27 shows an observer's eye representing the optical system. The figure is used to define the angular magnification. What fact about the figure did we use for that definition?

(A) that the object at p_1 is larger than the object at p_2

(B) that angles θ_1 and θ_2 are the same

(C) that the object moves back and forth between positions p_1 and p_2

(D) that we use the same object at p_1 and p_2

(E) none of the above

CASE STUDY 22.2

Fig. 22.29 shows that a standard man can see objects as small as 100 μm. When the same person uses a microscope with an angular magnification of m = 1000, the person can observe objects as small as (choose closest value) (A) 100 μm (e.g., an oocyte), (B) 50 μm (e.g., a sperm cell), (C) 1 μm (e.g., an average bacterium), (D) 100 nm (e.g., a large virus), (E) 10 nm (e.g., large biomolecules, such as DNA).

Answer: *Choice (D). We extend Fig. 22.28 as shown in Fig. 22.30, which includes additional geometric details. Using Fig. 22.30 means that we interpret the use of the microscope as if it allows us to move the object much closer to the eye. How close the object would have to be is calculated from Eq. [22.17]:*

$$p = \frac{p_0}{m} = \frac{25\text{ cm}}{1000} = 2.5\times10^{-4}\text{m}.$$

This distance is obviously not practical and indicates why a microscope is needed. However, based on Fig. 22.30 we can further rewrite the ratio p_0/p based on the similarity of the two triangles shown:

$$\frac{h_O}{p} = \frac{h}{p_0}.$$

We substitute p in this equation and find:

$$h_O = \frac{h}{m} = \frac{100\ \mu\text{m}}{1000} = 100\text{ nm}$$

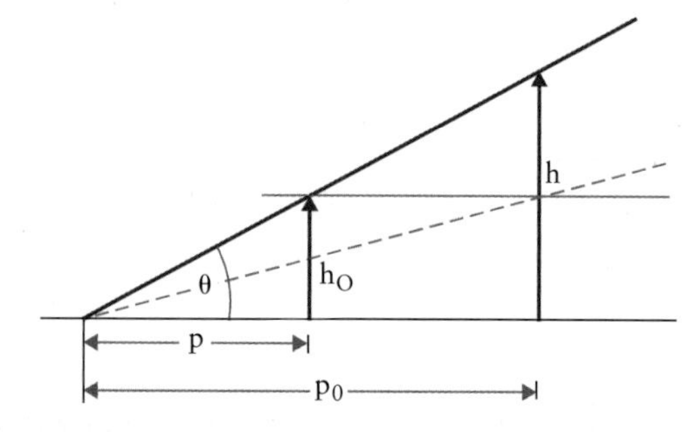

Figure 22.30 Sketch extending Figure 22.28 with additional geometric details.

22.5.2: The Magnifying Glass

As shown earlier, a lens produces an image larger than the object size if the object is placed closer than twice the focal length. Using the magnification formula, we see that the object has to be placed very close to the focal point to obtain a large magnification M. Thus, we arrange the lens such that $p \approx f$. Next we include the observer; i.e., we determine the angular magnification m for this arrangement. We illustrate this approach assuming that the observer looks at the object as if it is at infinite distance (with a relaxed eye).

In Fig. 22.31 an object O of size h_O is placed at the focal point F of a lens because this causes light rays emerging from the object to travel parallel behind the lens. This simulates the case where the object is at infinite distance from the observer's eye. More specifically, the lens forms an intermediate image at infinite distance, which in turn is the object for the observer's eye. Thus, the observer looks at the intermediate image with relaxed eyes. The angular magnification is determined from Fig. 22.31: θ is the angle under which the object appears for the observer. We further need the distance of the lens from the eye. This distance is the focal length of the magnifying glass, f, because a light ray from the top of the object that travels parallel to the optical axis crosses the optical axis at a distance f behind the lens. We use the small-angle approximation $\theta = \tan\theta$ and find $\theta = h_O/f$. We substitute this angle in the definition of the angular magnification. For the angle θ_0 we use the standard man looking at the object with the naked eye:

$$m = \frac{\theta}{\theta_0} = \frac{h_O/f}{h_O/s_0} = \frac{s_0}{f}. \quad [22.18]$$

The shorter the focal length f of the magnifying lens the larger the angular magnification.

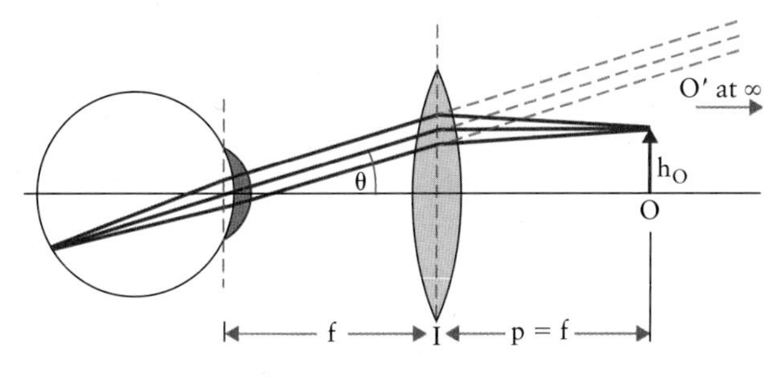

Figure 22.31 A magnifying glass is used to observe an object O with relaxed eyes. The lens simulates an object O' at infinite distance.

An observer with eyes focused can do better than Eq. [22.18]. In this case, the object is placed closer to the lens, $p < f$, and the intermediate image is at a finite distance. The most the observer can do, though, is bring the intermediate image to his/her near point at s_0. A full calculation yields $m = 1 + s_0/f$ for this case, which is called the **maximum angular magnification**. Example 22.8 illustrates that using focused eyes with a magnifying lens does not provide a notable improvement, and is therefore neglected in the further discussions.

EXAMPLE 22.9

(a) What is the angular magnification when an observer with relaxed eyes uses a lens with focal length $f = 5$ cm to observe an object? (b) By what factor does an observer with a focused eye increase the angular magnification at most?

Solution

Solution to part (a): We use Eq. [22.18] for the angular magnification:

$$m = \frac{25 \text{ cm}}{5 \text{ cm}} = 5.0.$$

Solution to part (b): As expected, the magnifying glass works better for the relaxed eye:

$$m_{max} = 1 + \frac{25 \text{ cm}}{5 \text{ cm}} = 6.0.$$

However, this is only a factor of 6/5 = 1.2, or 20%, better than the observation with relaxed eyes. Instead of focusing the eye all day with a magnifying glass, biologists are better served using more powerful microscope arrangements, as discussed in the next two sections.

22.5.3: The Compound Microscope

To achieve angular magnifications larger than a value of about $m = 10$, a single magnifying lens is not sufficient; compound microscopes are used that have two or more lenses. An instrument with two lenses is sketched in Fig. 22.32. It combines an **objective lens**, which has a very short focal length of $f_O < 1$ cm, and an **ocular lens (eyepiece)**, which has a focal length f_E of a few centimetres. The two lenses are separated by a distance L with

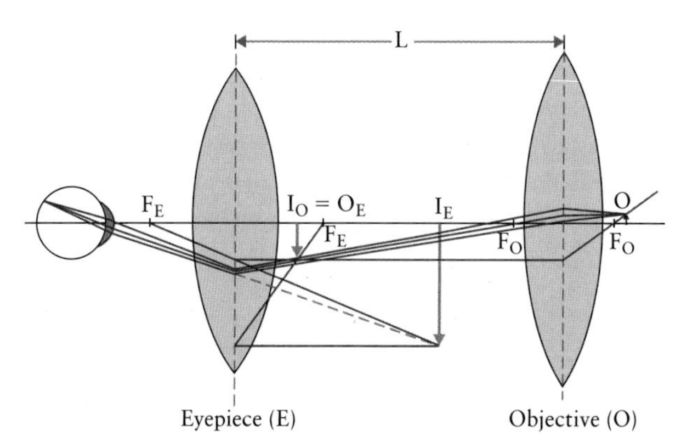

Figure 22.32 A typical compound microscope with objective lens and eyepiece at a distance L that coincides with the length of the tube of the microscope. The final image I_E is constructed with the same three light rays that we used previously for single lenses and mirrors.

$L \gg f_O, f_E$. The instrument allows the observer to look with the eyepiece at the image of the objective lens; i.e., $I_O = O_E$. The small object that is to be viewed is positioned just outside the focal length of the objective lens. This generates a real, enlarged image I_O far from the lens. This image lies within the focal distance of the eyepiece. Therefore, treating the image of the objective lens as the object for the eyepiece leads to a virtual image I_E. The eye then looks at the image I_E. Note that several lines in Fig. 22.32 are drawn to illustrate the construction of the two images, I_O and I_E. As for single lenses in the previous chapter, construction is based on three light rays. One is defined by the optical axis, and the other two emerge from the top of the object: one ray is incident on the lens parallel to the optical axis, and one ray is incident through its focal point.

We determine the **total angular magnification** for this arrangement from two contributions: the magnification of the objective lens, M_O, and the angular magnification of the eyepiece, m_E:

$$m_{total} = M_O\, m_E. \qquad [22.19]$$

Note that we use the magnification for the objective lens because the size of the image I_O does not depend on the position of the observer. In turn, the angular magnification of the eyepiece is used since we want to know the total angular magnification seen by the observer. For m_E, the value for the relaxed eye is used in the literature. We find from Eqs. [22.11] and [22.18]:

$$\text{(I)} \quad M_O = -\frac{q_O}{p_O} \cong -\frac{L}{f_O}$$

$$\text{(II)} \quad m_E = \frac{s_0}{f_E}$$

with $s_0 = 25$ cm for the near point of the standard man. The length of the microscope tube in formula (I) is introduced as a simplification, because neither the image nor the object distance of the objective lens is easily quantified for Fig. 22.32. We find the first formula by inserting the approximation $q_O - f_O \cong L$ in Eq. [22.11]. M_O is negative, since the image of the objective lens is inverted. The total angular magnification of the two-lens arrangement is then:

$$m_{total} = M_O\, m_E \cong -\frac{L}{f_O}\frac{s_0}{f_E}. \qquad [22.20]$$

Both relations in Eq. [22.20] can be used to calculate the total angular magnification of a compound microscope with two lenses. As illustrated in Examples 22.12 and 22.13, the second, approximate formula is much faster to apply.

CONCEPT QUESTION 22.4

The total angular magnification quoted for a compound microscope does not depend on the following parameter:

(A) the distance from the objective lens to the eyepiece

(B) the focal length of the objective lens

(C) the focal length of the eyepiece

(D) the near point of the observer

EXAMPLE 22.10

The length of a microscope tube is 15.0 cm. The focal length of the objective lens is 1.0 cm and the focal length of the eyepiece is 2.5 cm. What is the total angular magnification of the microscope if the observer's eye is relaxed? (a) Calculate the exact result using Eq. [22.20], and (b) calculate the approximate result using Eq. [22.20].

Solution

Solution to part (a): To apply the exact formula in Eq. [22.20], we need to determine the angular magnification of the eyepiece, m_E, and the magnification of the objective lens, M_O. The parameter m_E is obtained from Eq. [22.18] using the focal length of the eyepiece, f_E:

$$m_E = \frac{s_0}{f_E} = \frac{25.0 \text{ cm}}{2.5 \text{ cm}} = 10.$$

The magnification of the objective lens is defined by the object and image distances of the lens, $M_O = -q_O/p_O$. Both of these terms have to be calculated separately. We begin with the image distance of the objective lens. For the compound microscope, q_O is related to the object distance of the eyepiece via $q_O = L - p_E$, in which L is the distance between the two lenses in the microscope. p_E is determined from the thin-lens formula for the eyepiece (note that $q_E = \infty$ because the observer looks at the image with a relaxed eye):

$$\frac{1}{p_E} = \frac{1}{f_E} - \frac{1}{q_E} = \frac{1}{2.5 \text{ cm}} - \frac{1}{\infty},$$

which yields:

$$p_E = 2.5 \text{ cm}$$

and:

$$q_O = L - p_E = 12.5 \text{ cm}.$$

The object distance p_O is then found from the thin-lens formula applied to the objective lens:

$$\frac{1}{p_O} = \frac{1}{f_O} - \frac{1}{q_O} = \frac{1}{1.0 \text{ cm}} - \frac{1}{12.5 \text{ cm}},$$

continued

which yields:

$$p_O = 1.09 \text{ cm}.$$

We know now all the data needed to determine the magnification of the objective lens. Note that we do not calculate the angular magnification of the objective lens, as the observer is not involved in the process of its image formation. We find with Eq. [22.11]:

$$M_O = -\frac{q_O}{p_O} = -\frac{12.5 \text{ cm}}{1.09 \text{ cm}} = -11.5.$$

Thus, the total angular magnification of the microscope is:

$$m_{total} = M_O \, m_E = -11.5 \cdot 10.0 = -115.$$

Solution to part (b): The problem becomes a simple substitution problem when applying Eq. [22.20] in its approximate form:

$$m_{total} = -\frac{L}{f_O}\frac{s_0}{f_E} = -\frac{(0.15 \text{ m})(0.25 \text{ m})}{(0.01 \text{ m})(0.025 \text{ m})} = -150.$$

The difference between the results parts (a) and (b) illustrates the extent to which the second approach yields an approximate result. For most applications, the result in part (b) is sufficient.

SUMMARY

DEFINITIONS

- Focal length of a spherical mirror: $f = R/2$; R is the radius of curvature of the mirror.
- Magnification of a spherical mirror or thin lens:

$$M = \frac{h_I}{h_O},$$

with h_I the height of the image and h_O the height of the object.

- Index of refraction n (Snell's law: the incoming ray must travel through a vacuum or air):

$$\frac{\sin\alpha_{in}}{\sin\beta} = n$$

with α_{in} and β the respective angles of the light ray with the normal.

- Refractive power
  - for a single interface: $\mathfrak{R} = \Delta n/R$
  - for a thin lens: $\mathfrak{R} = 1/f$
  - Standard near point: $s_0 = 25$ cm.
- Angular magnification: $m = \theta/\theta_0$, with θ the angle subtended by the object and θ_0 the angle subtended by the same object placed at the standard near point.

UNITS

- Refractive power $\mathfrak{R}$: dpt (diopters) = m^{-1}

LAWS

- Reflection, flat mirror:

$$\alpha_{in} = \alpha_{out},$$

where angles are measured to the normal of the mirror surface.

- Law of refraction at the interface between media 1 and 2:

$$n_1 \sin\alpha_1 = n_2 \sin\alpha_2$$

where angles are measured to the normal of the interface.

- Mirror equation and thin-lens formula:

$$\frac{1}{f} = \frac{1}{p} + \frac{1}{q},$$

where p is the object distance and q is the image distance.

- Magnification of spherical mirror or thin lens:

$$M = -\frac{q}{p}$$

- Law of refraction for a spherical interface between media 1 and 2:

$$\frac{n_1}{p} + \frac{n_2}{q} = \frac{n_2 - n_1}{R}$$

- Magnification of a spherical interface between media 1 and 2:

$$M = \frac{h_1}{h_O} = \frac{n_1 q}{n_2 p}$$

- Angular magnification m of a lens
  - for a relaxed eye:

$$m = \frac{s_0}{f}$$

  - for an eye focused at the near point:

$$m = 1 + \frac{s_0}{f}$$

- Total angular magnification of a compound microscope:

$$m_{total} = M_O \, m_E \cong -\frac{L}{f_O}\frac{s_0}{f_E}$$

in which index O stands for objective lens and index E stands for eyepiece. L is the distance between the two lenses.

MULTIPLE-CHOICE QUESTIONS

MC–22.1. A candle produces an image of 0.2 m in height with the wick pointing downward when observed in a flat mirror. What statement is true about the candle in the observer's hand?

(a) The candle is 2 cm high and is held upside down.
(b) The candle is 2 cm high and is held upright.
(c) The candle is 20 cm high and is held upside down.
(d) The candle is 20 cm high and is held upright.

MC–22.2. An object is placed 35 cm in front of a flat mirror. Where does the image form? Choose the closest value.
(a) 1 m in front of the mirror
(b) 0.35 m in front of the mirror
(c) at the focal length behind the mirror
(d) 0.35 m behind the mirror

MC–22.3. A spherical mirror has a radius of 25 cm. What is the focal length f of the mirror?
(a) $f = 0.125$ m
(b) $f = 0.25$ m
(c) $f = 0.5$ m
(d) $f = 1$ m
(e) $f = 5$ m

MC–22.4. A spherical mirror has a focal length of 20 cm. What is the radius of curvature r of the mirror?
(a) $r = 0.04$ m
(b) $r = 0.1$ m
(c) $r = 0.4$ m
(d) $r = 1$ m
(e) $r = 4$ m

MC–22.5. An object at distance $p = +0.3$ m in front of a spherical mirror forms a virtual image at a distance of 15 cm from the mirror. What is the focal length of the mirror? Choose the closest value.
(a) $f = -30$ cm
(b) $f = -10$ cm
(c) $f = +10$ cm
(d) $f = +30$ cm

MC–22.6. An object is placed at $p = +20$ cm in front of a spherical mirror. The image is magnified by a factor of 2 and is inverted. What is the image distance? Choose the closest value.
(a) $q = +0.4$ m
(b) $q = +0.1$ m
(c) $q = -0.1$ m
(d) $q = -0.4$ m

MC–22.7. Why does the fish in Fig. 22.16 appear closer to the observer than it actually is?
(a) because we look at the fish in the direction perpendicular to the water surface
(b) because the fish is in water and the observer is in air
(c) because the index of refraction of the water is smaller than the index of refraction of the air
(d) because the fish floats toward the surface
(e) because the water surface is flat, not spherically shaped

MC–22.8. Light is incident on a flat horizontal interface from a vacuum to an unknown type of glass. The light travels at 45° with the normal of the glass surface in the vacuum and at 35° in the glass. What is the refractive index n of the glass? Choose the closest value.
(a) $n = 0.8$
(b) $n = 1.0$
(c) $n = 1.2$
(d) $n = 1.5$
(e) $n = 2.0$

MC–22.9. Light is incident at an angle of 45° with the vertical on a flat horizontal interface from a vacuum to an unknown type of glass. What is the speed of light in that glass if it travels in the glass with an angle of 27° with the vertical? Choose the closest answer.
(a) $v = 4 \times 10^8$ m/s
(b) $v = 3 \times 10^8$ m/s
(c) $v = 2 \times 10^6$ km/s
(d) $v = 2 \times 10^5$ km/s
(e) $v = 2 \times 10^4$ km/s

MC–22.10. A light ray travelling through air is incident on a flat slab of crown glass ($n = 1.52$) at an angle of 30° to the normal. What is the angle of refraction? Choose the closest value.
(a) $\alpha_{glass} = 50°$
(b) $\alpha_{glass} = 40°$
(c) $\alpha_{glass} = 30°$
(d) $\alpha_{glass} = 20°$
(e) $\alpha_{glass} = 10°$

MC–22.11. Which of the four bodies in Fig. 22.33, each showing a thin lens made of flint glass ($n = 1.61$), has the smallest refractive power? *Note:* Choose (e) if two bodies tie for the smallest value.

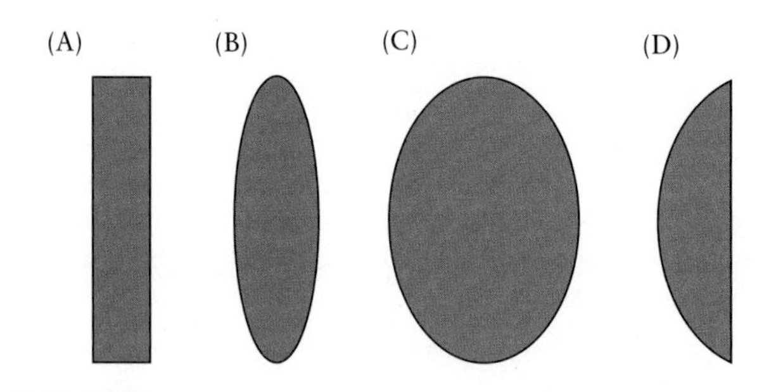

Figure 22.33

MC–22.12. In Fig. 22.21(b), you see an object (O, at left) and its image (I, farther at the left), the latter forming as the result of a converging lens represented by the vertical dashed line. Which of the following statements is false?
(a) The magnification is larger than 1.
(b) The image forms in front of the lens.
(c) The image is upright.
(d) The image distance q is a positive length, $q > 0$.
(e) The object is closer to the lens than the focal length.

MC–22.13. A converging lens with focal length $f = 20$ cm is used to view an object 50 cm from the lens. How far from the lens does the object appear? Choose the closest value.
(a) 120 cm
(b) 90 cm
(c) 30 cm
(d) 20 cm
(e) 5 cm

MC–22.14. For a thin lens, we find the image distance $q = +10$ cm for an object placed at $p = +20$ cm. What is the refractive power of the lens in diopters? Choose the closest value.
(a) $\Re = 2$ dpt
(b) $\Re = 5$ dpt
(c) $\Re = 10$ dpt
(d) $\Re = 15$ dpt

MC–22.15. A lens forms an observable magnified image (an image that is larger than the object) if
(a) the image is on the same side of the lens as the object.
(b) the object is placed at the focal point of the lens.
(c) the object is placed closer than twice the focal length in front of the lens.
(d) the image distance is smaller than the object distance.

MC–22.16. A diverging lens with focal length $f = 30$ cm is used to view an object 90 cm from the lens. How far from the lens does the object appear? Choose the closest value.
(a) 120 cm
(b) 90 cm
(c) 30 cm
(d) 22.5 cm
(e) 18.5 cm

MC–22.17. Two coaxial converging lenses, with focal lengths f_1 and f_2, are positioned a distance $f_1 + f_2$ apart, as shown in Fig. 22.34. This arrangement is called a **beam expander** because it is often used for widening laser beams. If h_1 is the size of the incident beam, the size of the emerging beam is
(a) $h_2 = \frac{f_2}{f_1} h_1$
(b) $h_2 = \frac{f_1}{f_2} h_1$
(c) $h_2 = (f_2 + f_1) h_1$
(d) $h_2 = (f_2 - f_1) h_1$
(e) $h_2 = (f_2 f_1) h_1$

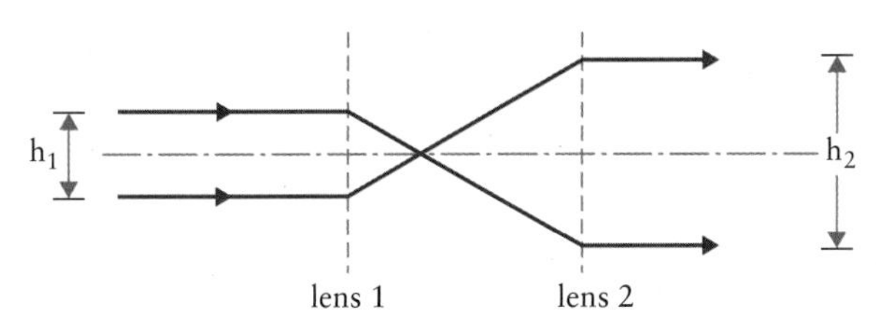

Figure 22.34 Two coaxial converging lenses with focal lengths f_1 and f_2 are positioned a distance $f_1 + f_2$ apart. h_1 is the size of the incident light beam and h_2 is the size of the emerging beam.

MC–22.18. What is the closest distance between an object and a screen such that an image is formed on the screen using a diverging lens with focal length $-f$?
(a) $f/2$
(b) f
(c) $2f$
(d) $4f$
(e) No image will form.

MC–22.19. A person's face is 30 cm in front of a concave mirror. What is the focal length of the mirror if it creates an upright image that is 1.5 times as large as the actual face? Choose the closest value.
(a) 12 cm
(b) 20 cm
(c) 70 cm
(d) 90 cm

MC–22.20. Many nocturnal or crepuscular mammals have eyeshine, like the lesser bush baby shown in Fig. 22.35: when a bright light is shone into their eyes, a reflection comes back. The reflection is due to a crystalline layer behind the retina, called **tapetum lucidum**. This layer increases the amount of light that passes across the retina and thereby assists in night vision. What optical system is a good model for the *tapetum lucidum* based on Fig. 22.30?
(a) a single, thin lens
(b) a flat mirror
(c) a flat refractive interface
(d) a spherical mirror
(e) a double-lens system as in a microscope

Figure 22.35 A lesser bush baby looking at a photographer's camera at night.

MC–22.21. Stereoscopic vision is very important to us. A person may lose vision in one eye due to an accident. How can that person judge with just one eye the distance to a nearby object? Choose the statement that best describes how it is done.
(a) The lens has a fixed refractive power; the size of the eyeball is variable. Thus, sensing the pressure in the brain behind the eye when a focused image is obtained on the retina provides a measure of the object distance.
(b) The size of the eyeball is fixed, but the lens is pliable to vary the refractive power. The extent of contraction of the ciliary muscles is used to judge the distance.
(c) The eyes must produce a thick water coat on the cornea (as when the person cries). The eyelid then senses the thickness of the water layer when a focused image is obtained on the retina. That value is a measure of the object distance.
(d) If we open and close the eye very fast, the rods in the retina measure the time it takes for the light to reach the retina. The longer the light travels through the eye, the farther away the object is.

MC–22.22. The near point of a particular person is at 50 cm. This is due to a defect of the person's eye; an image forms behind the retina for objects closer than 50 cm, as indicated in Fig. 22.23(b). To correct this problem, what refractive power must a corrective lens have [shown in Fig. 22.23(c)] to enable the eye to clearly see an object at 25 cm?

(a) $\mathfrak{R} = -2.0$ dpt
(b) $\mathfrak{R} = -1.0$ dpt
(c) $\mathfrak{R} = 0.0$ dpt
(d) $\mathfrak{R} = +1.0$ dpt
(e) $\mathfrak{R} = +2.0$ dpt

MC–22.23. A person with a near point of 35 cm tries to see a text with small print better by bringing the page closer to the eye. The person will achieve what angular magnification?

(a) none
(b) $m = 1$ (no gain)
(c) $m < 1$ (the person does worse than the standard man)
(d) $m > 1$ but $m < 2$—a moderate angular magnification is achieved
(e) $m \gg 1$ (for this person, this is the way to go to see small objects)

MC–22.24. Assume you use a converging lens as a magnifying glass. Initially, you hold the lens far from a page with small print. Then you move the lens closer and closer to the text until the lens lies on the page, as shown in Fig. 22.36. What do you observe?

(a) The text is always upright, no matter how far the lens is held.
(b) The text is initially inverted, then blurs and becomes upright.
(c) The text is initially upright, then blurs and becomes inverted.
(d) When the magnifying glass is held far enough from the page the text will run from right to left.

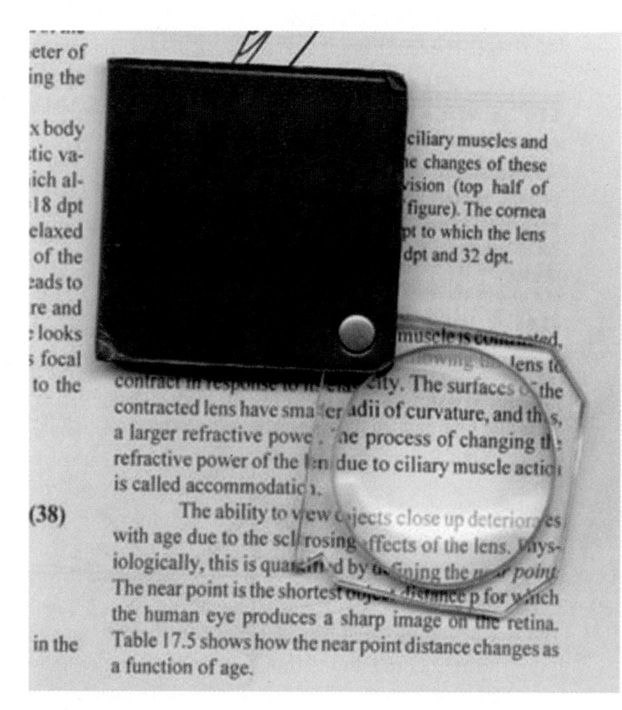

Courtesy of Martin Zinke-Allmang

Figure 22.36

MC–22.25. Which of the following statements about the angular magnification is correct?

(a) The angular magnification can be calculated for an optical device without taking the observer's eye into account.
(b) A larger angular magnification can be obtained when the observer uses the optical device with a relaxed eye.
(c) The eye of a juvenile usually has the same angular magnification as the eye of a standard man.
(d) The angular magnification of a compound microscope depends on the near point of the particular person using the microscope.
(e) Two different values for the angular magnification result when we distinguish between an observer with a relaxed eye and one with a focused eye.

MC–22.26. Which of the following total angular magnifications m_{total} can be selected with a compound microscope with an eyepiece of angular magnification $m_E = 20$ and three switchable objective lenses with magnifications M_O of 10, 30, and 50?

(a) $m_{\text{total}} = 100$
(b) $m_{\text{total}} = 300$
(c) $m_{\text{total}} = 500$
(d) $m_{\text{total}} = 700$
(e) None of the four total angular magnifications above can be selected with this microscope.

MC–22.27. What happens if the object in Fig. 22.32 shrinks to one-twentieth of its current size?

(a) The image the observer sees becomes larger.
(b) The image the observer sees becomes smaller.
(c) The observer can see only a part of the object magnified.
(d) The microscope cannot produce an image of the object.
(e) No light-ray construction is possible, as shown for the larger object in Fig. 22.32.

MC–22.28. Most commercial microscopes have an additional lens, called the condenser lens, which is located between the light source and the object [see (4) in Fig. 22.26(b)]. What does this lens do?

(a) enhances the overall angular magnification
(b) inverts the image so that we don't see everything upside down
(c) substitutes for the eyepiece when the eyepiece becomes defective due to poor upkeep of the instrument
(d) focuses light from a light source on the object
(e) illuminates the image for faster photographic exposure

MC–22.29. We study again the condenser lens discussed in MC–22.28. What focal length would you choose for this lens?

(a) It doesn't matter as long as the lens is transparent.
(b) For a microscope of length 20 cm, we would choose a focal length of about 20 cm to focus the light from the light source into the eye.
(c) A focal length would be best that focuses the light of the light source at the intermediate image, i.e., the image that we observe through the eyepiece.
(d) We would use a short focal length to focus the light from the light source on the object.

MC–22.30. You are unhappy with the overall magnification you achieve with your homemade microscope. Which alteration will improve the results?

(a) Shorten the distance between the objective lens and the eyepiece.
(b) Exchange the objective lens for a lens with a larger focal length.
(c) Exchange the eyepiece for a lens with a larger focal length.
(d) Loosen up and look through the microscope with a relaxed eye.
(e) None of the above.

MC–22.31. Two thin lenses, one with focal length f and the other with focal length $2f$, are placed very close to each other along the optical axis. Their combined effect is the same as if a thin lens of the following focal length replaces them:

(a) $f_{combined} = 0$.
(b) $f_{combined} = f/2$.
(c) $f_{combined} = 2f$.
(d) $f_{combined} = -2f$.
(e) $f_{combined}$ is infinite.

MC–22.32. We study two thin lenses, where lens 1 with $f_1 = 15$ cm is placed a distance of $L = 35$ cm to the left of lens 2 with $f_2 = 10$ cm. An object is then placed 50 cm to the left of lens 1. What is the magnification of the final image taken with respect to the object? *Note:* The magnification you determine is not an angular magnification.

(a) $M = 0.6$
(b) $M = 1.0$ (no magnification)
(c) $M = 1.2$
(d) $M = 2.4$
(e) $M = 3.6$.

MC–22.33. An optical compound microscope has an objective lens with $f_O = 0.8$ cm and an eyepiece with $f_E = 4.0$ cm. If the microscope is 15 cm long, what is the total angular magnification? Choose the closest answer.

(a) 3.5
(b) 6.5
(c) 50
(d) 120
(e) 500

CONCEPTUAL QUESTIONS

Q–22.1. When you look at yourself in a flat mirror, you see yourself with left and right sides switched, but not upside down. Why? *Hint:* Remember that you are a three-dimensional body. Study the image of the following three vectors: (I) head to foot, (II) left to right hand, and (III) nose to back of head. The remainder of the puzzle is perception of the brain!

Q–22.2. Can a virtual image be photographed?

Q–22.3. Tape a picture of a person on a flat mirror. Approach the mirror to within 20 cm to 25 cm. Can you focus on the picture and your image at the same time?

Q–22.4. In this line, the word DECEITFUL is capitalized and printed in a *sans serif font*. Take a transparent rod (e.g., a water-filled test tube in your chemistry or biology lab) and read this word through the rod. Why are the first five letters unchanged, while the last four letters are upside down when using the plastic rod?

Q–22.5. Sunlight refracts while passing through the atmosphere due to a small difference between the indices of refraction for air and vacuum. We define dawn optically as the instant when the top of the Sun just appears above the horizon, and we define dawn geometrically when a straight line drawn from the observer to the top of the Sun just clears the horizon. Which definition of dawn occurs earlier in the morning?

Q–22.6. Put a straw in a glass of water. Why does the straw look bent, as in Fig. 22.37?

Figure 22.37

Q–22.7. (a) Some gardeners advise against watering flowers in full sunshine to avoid burns to leaves due to the focusing effect of water droplets. Is this advice reasonable? *Hint:* Treat the water droplet as a sphere placed on the leaf, as shown in Fig. 22.38, and use the thin-lens formula. (b) *If you are interested:* Do you know why it is still not a good idea to water flowers in full sunlight?

Figure 22.38

Q–22.8. Optometrists use the **Snellen test** to evaluate their patients' vision. The Snellen test consists of letters of different sizes that a person with healthy eyes can read at particular distances. The patient is placed 6.1 m (20 feet) from the chart and asked to read the letters. If the patient's eyes are healthy, he/she will read the same line without errors that the healthy reference group was able to read at that distance. We therefore call this 20/20 vision. A juvenile may have 20/10 vision, which means that he/she can read a line that a healthy adult can read only at a distance of 3.05 m (10 feet). Vision-impaired patients may score as low as 20/200, which corresponds to the single, largest letter at the top of the Snellen test. A person with healthy eyes can read that letter as far away as 61 m (200 feet), which coincides with the distance at which the eye is accommodated for vision of objects at infinite distance. Many optometrists have offices in a mall with high rent. To keep the cost down, the examination room may have a length of only 4 m, with the patient sitting at the examination instruments near the centre of the room. Suggest an appropriate setup for the Snellen test in this room.

Q–22.9. The optic nerve and the brain invert the image formed on the retina. Why then do we not see everything upside down?

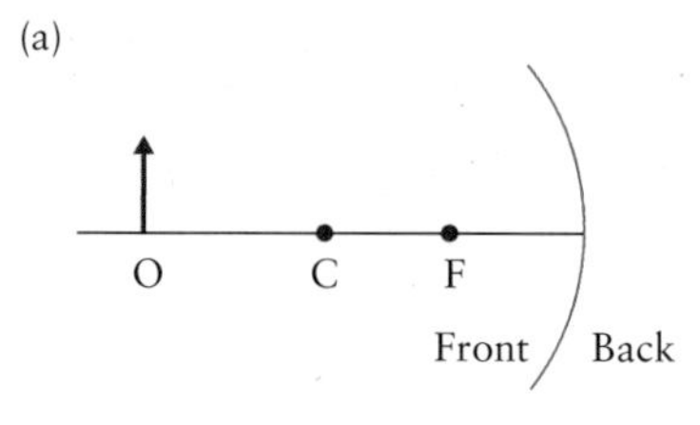

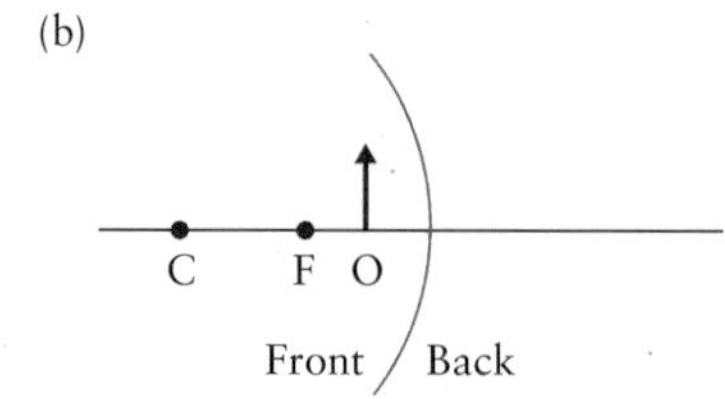

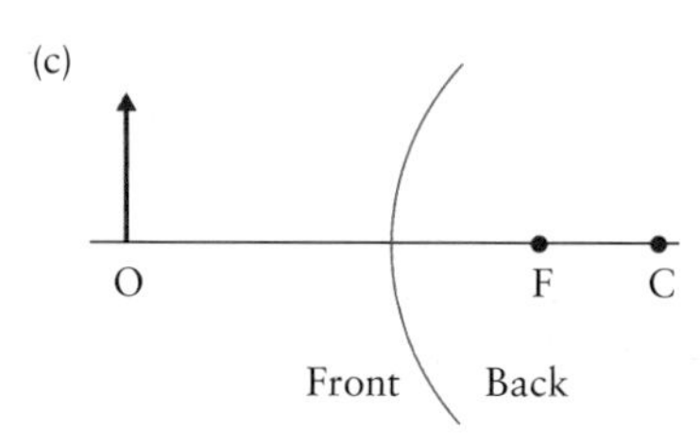

Figure 22.39

ANALYTICAL PROBLEMS

P–22.1. When you look at your face in a small bathroom mirror from a distance of 40 cm, the upright image is twice as tall as your face. What is the focal length of the mirror?

P–22.2. A concave spherical mirror has a radius of curvature of 20 cm. Locate the images for object distances as given below. In each case, state whether the image is real or virtual and upright or inverted, and find the magnification. (a) $p = 10$ cm; (b) $p = 20$ cm; (c) $p = 40$ cm.

P–22.3. Construct the images for the three objects shown in Fig. 22.39.

P–22.4. A light ray enters a layer of water at an angle of 36° with the vertical. What is the angle between the refracted light ray and the vertical?

P–22.5. A light ray strikes a flat, $L = 2.0$-cm-thick block of glass ($n = 1.5$) in Fig. 22.40 at an angle of $\theta = 30°$ with the normal. (a) Find the angles of incidence and refraction at each surface. (b) Calculate the lateral shift of the light ray d.

P–22.6. In Fig. 22.41, an ultrasonic beam enters an organ (grey) at $\theta = 50°$, then reflects off a tumour (green) in the surrounding organ and leaves the organ with a lateral shift $L = 12$ cm. If the speed of the wave is 10% less in the organ than in the medium above, determine the depth of the tumour below the organ's surface.

P–22.7. A light ray travels through air and then strikes the surface of mineral oil at an angle of 23.1° with the normal to the surface. What is the angle of refraction if the light ray travels at 2.17×10^8 m/s through the oil?

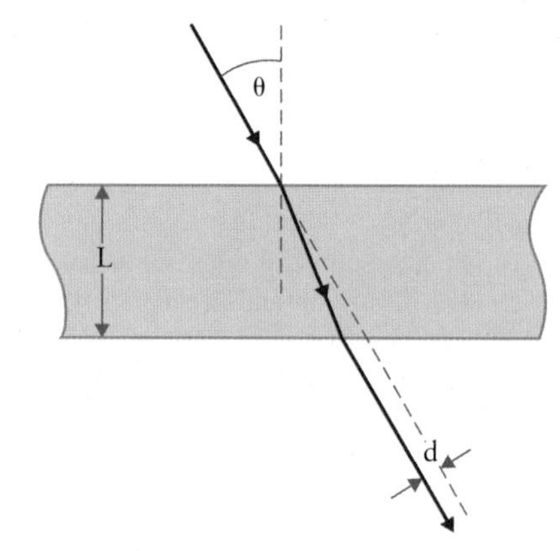

Figure 22.40

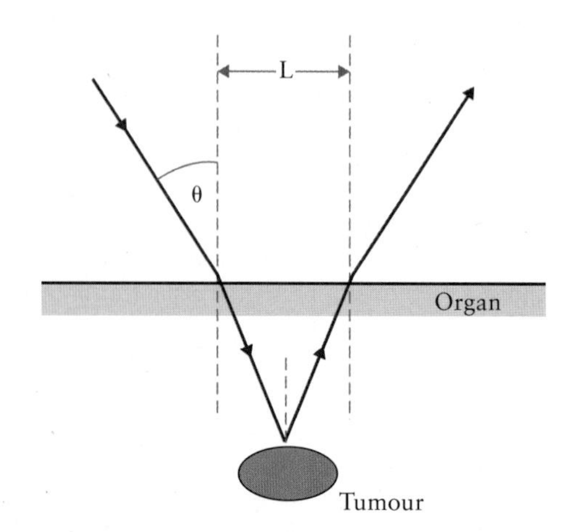

Figure 22.41

P–22.8. A light source at the bottom of a 4.0-m-deep water pool sends a light ray up at an angle so that the ray strikes the surface 2.0 m from the point straight above the light source. What is the emerging ray's angle with the normal in air?

P–22.9. The laws for refraction and reflection are the same for light and sound. If a sound wave in air approaches a water surface at an angle of 12° with the normal of the water surface, what is the angle with the normal of the refracted wave in water? Use for the speed of sound in air 340 m/s and 1510 m/s in water.

P–22.10. A slab of ice with parallel surfaces floats on water. What is the angle of refraction of a light ray in water if the light ray is incident on the upper ice surface with an angle of 30° to the normal?

P–22.11. A light ray is incident from air onto a glass surface with index of refraction n = 1.56. Find the angle of incidence for which the corresponding angle of refraction is one-half the angle of incidence. Both angles are defined with the normal to the surface.

P–22.12. Construct the images for the three lenses shown in Fig. 22.42. Note that the third case is a diverging lens.

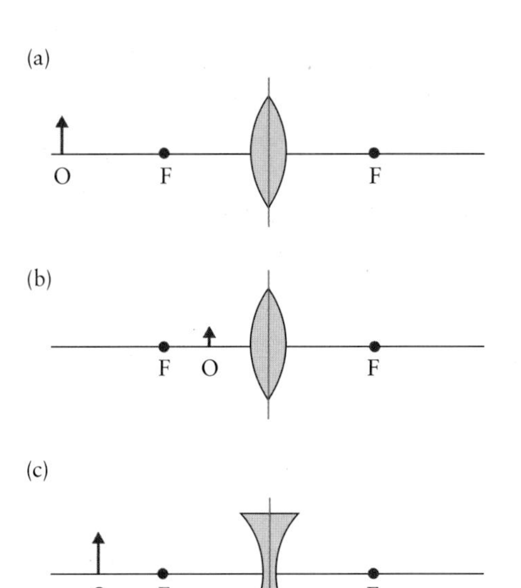

Figure 22.42

P–22.13. A converging lens has a focal length f = 20.0 cm. Locate the images for the object distances given below. For each case state whether the image is real or virtual and upright or inverted, and find the magnification. (a) 40 cm; (b) 20 cm; (c) 10 cm.

P–22.14. Where must an object be placed to have no magnification ($|M|$ = 1.0) for a converging lens of focal length f = 12.0 cm?

P–22.15. Fig. 22.43 shows an object at the left, a lens at the centre (vertical dashed line), and a concave mirror at the right. The respective focal lengths and the distance between lens and mirror are indicated at the bottom of the figure. Construct the image that forms after light from the object has passed through the lens and has reflected off the mirror.

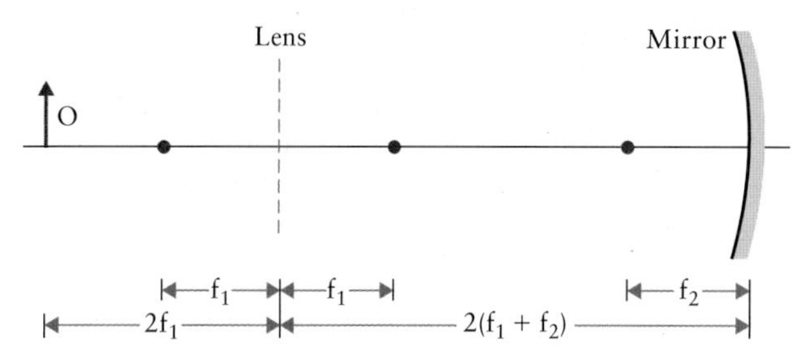

Figure 22.43

P–22.16. An object is placed in front of a converging lens with f = 2.44 cm. The lens forms an image of the object 12.9 cm from the object. How far is the lens from the object if the image is (a) real or (b) virtual?

P–22.17. A contact lens is made of plastic with an index of refraction of n = 1.58. The lens has a focal length of f = +25.0 cm and its inner surface has a radius of curvature of +22.0 mm. What is the radius of curvature of the outer surface?

P–22.18. A person can see an object in focus only if the object is no farther than 30 cm from the right eye and 50 cm from the left eye. Write a prescription for the refractive powers $\Re$ (in diopters) for the person's corrective lenses.

P–22.19. The near point of an eye is 100 cm. A corrective lens is to be used to allow this eye to focus clearly on objects 25 cm in front of it. (a) What should be the focal length of the lens? (b) What is the refractive power $\Re$ of the lens?

P–22.20. A person who can see clearly when objects are between 30 cm and 1.5 m from the eye is to be fitted with bifocals. (a) The upper portion of the corrective lenses is designed such that the person can see distant objects clearly. What refractive power $\Re$ does that part of the lenses have? (b) The lower portion of the lenses has to enable the person to see objects comfortably at 25 cm. What refractive power $\Re$ does that part of the lenses have?

P–22.21. The near point of a patient's eye is 75.0 cm. (a) What should be the refractive power $\Re$ of a corrective lens prescribed to enable the patient to clearly see an object at 25.0 cm? (b) When using the new corrective glasses, the patient can see an object clearly at 26.0 cm but not at 25.0 cm. By how many diopters did the lens grinder miss the prescription?

P–22.22. A magnifying glass is used to examine the structural details of a human hair. The hair is held 3.5 cm in front of the magnifying glass and the image is 25.0 cm from the lens. (a) What is the focal length of the magnifying glass? (b) What angular magnification is achieved?

P–22.23. Two converging lenses that have focal lengths of f_1 = 10.0 cm and f_2 = 20.0 cm are placed L = 50 cm apart. The final image is shown in Fig. 22.44. (a) How far to the left of the first lens is the object placed if l = 31 cm? (b) What is the combined magnification (not the total angular magnification in this case!) of the two lenses using the same data as in part (a)?

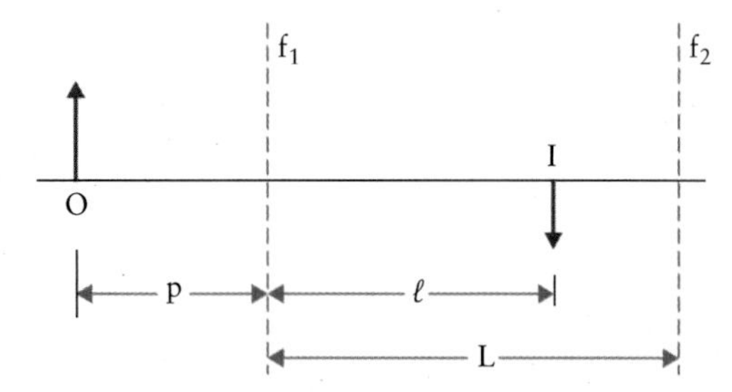

Figure 22.44

P–22.24. A microscope has an objective lens with $f = 16.22$ mm and an eyepiece with $f = 9.5$ mm. With the length of the microscope's barrel set at 29.0 cm, the diameter of an erythrocyte's image subtends an angle of 1.43 mrad with the eye. If the final image distance is 29.0 cm from the eyepiece, what is the actual diameter of the erythrocyte? *Hint:* Start with the size of the final image then use the thin-lens formula for each lens to find their combined magnification. Use this magnification to calculate the object size in the final step.

P–22.25. Fig. 22.45 shows two converging lenses placed $L_1 = 20$ cm apart. Their focal lengths are $f_1 = 10.0$ cm and $f_2 = 20.0$ cm. (a) Where is the final image located for an object that is $L_2 = 30$ cm in front of the first lens? (b) What is the total magnification of the lens system? *Note:* Do not calculate a total angular magnification in this case since we are not dealing with a microscope.

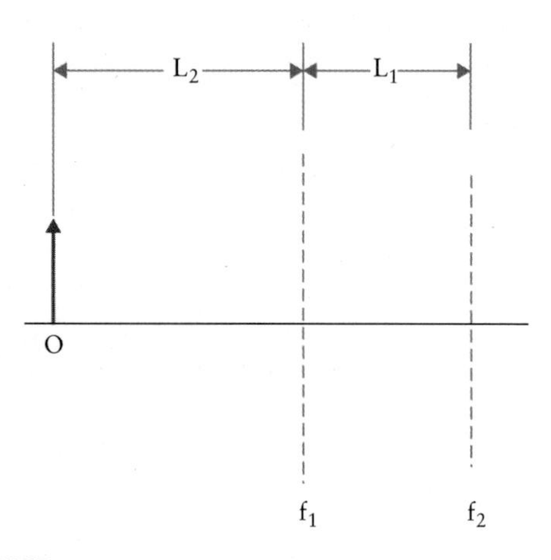

Figure 22.45

P–22.26. An object is located 20 cm to the left of a converging lens of focal length 25 cm. A diverging lens with focal length 10 cm is located 25 cm to the right of the converging lens. Find the position of the final image.

P–22.27. If light is incident at an angle θ from a medium of index of refraction n_1 to a medium with n_2 such that the angle between the reflected and refracted beams is β, show that:

$$\tan\theta = \frac{n_2 \sin\beta}{n_1 - n_2 \cos\beta}.$$

Hint: Use the formula for $\sin(\alpha_1 + \alpha_2)$ from the Math Review on Trigonometry found after Chapter 27.

P–22.28. Light of wavelength λ_0 in vacuum has a wavelength of $\lambda_w = 438$ nm in water and a wavelength of $\lambda_b = 390$ nm in benzene. (a) What is the wavelength λ_0 in a vacuum? (b) Using only the given information, determine the ratio of the index of refraction of benzene to that of water.

P–22.29. A 400-nm-wavelength light ray is incident at an angle of 45° on acrylic glass and is refracted. What wavelength of light, also incident at an angle of 45° but on fused quartz, refracts at the same angle? *Hint:* Use Fig. 22.46.

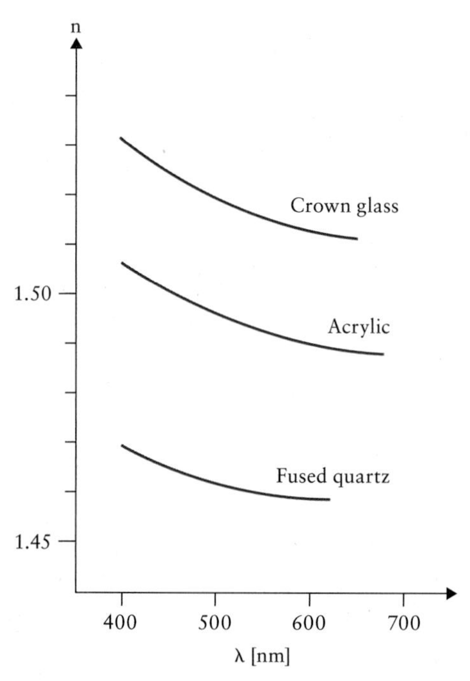

Figure 22.46 Dispersion relation, $n(\lambda)$, in the visible part of the electromagnetic spectrum for three types of glass: crown glass, acrylic, and fused quartz.

P–22.30. The index of refraction of red light in water is $n = 1.331$, and for blue light it is $n = 1.340$. If a ray of white light enters the water at an angle of incidence of 83°, what are the underwater angles of refraction for the two light components?

P–22.31. A glass has an index of refraction for blue light at $\lambda = 430$ nm of $n = 1.650$, and for red light at $\lambda = 680$ nm it is $n = 1.615$. If a light ray containing these two colours is incident at an angle of 30° on the glass, what is the angle between the two light components inside the glass?

ANSWERS TO CONCEPT QUESTIONS

Concept Question 22.1: (D). In (A) the light ray from *O* through *F* to the mirror violates the law of reflection; in (B) the light ray from *O* through *C* does not reflect off the mirror; and in (C) the light ray from *O* to *I* contributes to the image before it has been reflected off the mirror.

Concept Question 22.2: (E). You confirm this result with Table 22.4 and Eq. [22.11].

Concept Question 22.3: (D). The same object is required to make a relative statement at two distances.

Concept Question 22.4: (D). Note that s_0 is the near point of the standard man, not of any particular person using the microscope.

CHAPTER 23

The Atomic Nucleus

The atom contains a positively charged nucleus that carries the predominant fraction of its mass. Rutherford's scattering experiments showed that this nucleus is many orders of magnitude smaller than Bohr's radius of the hydrogen atom.

The nucleus consists of protons and neutrons; the large electrostatic repulsion between the protons in the nucleus is overcompensated by the attractive nuclear force in which both protons and neutrons participate. Yukawa's nuclear potential shows that the nuclear force has only a very short range, limiting the maximum size of nuclei.

Two models have been developed to quantify nuclear energy for the various elements of the periodic system. The liquid drop model treats protons and neutrons as a condensed fluid. It predicts well those nuclear properties that vary continuously with the atomic number. Alternatively, a shell model approaches the nucleus in the same manner as we discussed for the atomic shell, except that neutrons and protons have to be treated separately, and slightly different potential curves apply.

Radioactive decay occurs when a nucleus becomes metastable or unstable. Several decay paths exist; which one applies for a particular nuclide depends primarily on its location in an N-Z plot (neutron number versus atomic number, called nuclide chart). Decay for all but the heaviest nuclides results in emission of γ-ray photons, electrons, positrons, α-particles, neutrinos, and/or antineutrinos.

The radioactive decay law is a first-order differential equation that results from the proportionality between the number of radioactive isotopes in the sample and their rate of decay (activity). The decay constant in that equation—or the closely related half-life—are decay process-specific parameters.

The angular momentum of a rotating extended object combines the linear momentum of its parts with their respective distance to the rotation axis. It is a vector that points along the rotation axis. If a net external force causes a torque on a system, it responds with an angular acceleration and/or a change of the alignment of its rotation axis. A system with a zero net torque conserves the angular momentum.

The same angular momentum formalism is employed in atomic systems; we used it to develop Bohr's atomic model. We also referred to rotational concepts when we introduced the spin of the electron as a fourth quantum number. This discussion is now pursued further with a focus on the nuclear spin.

The nuclear spin is quantized; for example, a proton has a spin of magnitude $\hbar/2$. The nuclear spin interacts with external magnetic fields based on its magnetic dipole moment. The field causes a torque on the magnetic dipole, allowing it to occupy only two possible states, one with the spin parallel to the magnetic field or anti-parallel.

In thermal equilibrium, i.e., when the ensemble of spins in the sample is in thermal equilibrium with its environment, the lower energy state is more populated. However, the population difference is small because the energy difference between the two states in typical magnetic fields is much smaller than the thermal energy of the system.

Once the atomic hypothesis had been accepted in the late nineteenth century, questions arose about the composition and structure of the atom. This question was split into two separate parts early on: Heinrich Hertz (1891) and Philipp Lenard (1900) showed that atoms consist of a tiny nucleus, which contains essentially all the mass of the atom, and a large space around the nucleus, which they said is "as empty as the outer space in the universe."

In Chapter 20 we established the structure and composition of the atom excluding the nucleus: the electron is the elementary particle present in the atomic shell and, by the 1930s, quantum mechanics properly had described the structure of the orbitals. The nucleus, in turn, remained at that time elusive, primarily due to its small size in comparison to the atom as a whole. The first indication of its composition came from studying cosmic rays and unstable, i.e., radioactive, elements. Since the primary cosmic rays travel vast distances in outer space before reaching Earth, they consist of stable protons and α-particles. The interaction of the primary cosmic rays with molecules of the atmosphere at altitudes of 20 to 30 km causes the secondary cosmic rays that reach Earth's surface. Due to the short period of time that it takes for the secondary cosmic rays to penetrate Earth's atmosphere, they include unstable particles, such as neutrons, and antimatter particles, such as positrons.

23.1: Stable Atomic Nuclei

In the 1930s the first conclusive discoveries and theoretical models for the atomic nucleus emerged. The field was defined as nuclear physics, a discipline that investigates the stable nuclei in the atoms of regular matter. It built on intensive studies of the limits of stability evident in radioactivity, which was first observed by Henri Becquerel in 1896, and by Marie and Pierre Curie in 1898.

Nuclear physics has therefore been a major branch of the physical sciences for the past 100 years. As with other major advances in the fundamental sciences, it should not be surprising to find that nuclear physics concepts are widely applied in the life sciences. While we ultimately focus on short-lived nuclei due to their applications in radiology and nuclear medicine, medium-, long-lived, and stable nuclei play important roles in forensics, human evolution, and palaeontology to establish the age of fossils, and in paleoclimatology to determine environmental conditions during the development of life on Earth.

23.1.1: Which Particles Form Atomic Nuclei?

From the onset of atomic research it was clear that the nucleus must contain as many positive charges as there are electrons in its shell because atoms are electrically neutral. Let us assume that we want to build a nucleus from fundamental particles, which we call nucleons. We first consider the case of only one type of nucleon that carries a positive charge. This model would be sufficient to explain the hydrogen atom, with this single nucleon then called proton. We run immediately into a problem with this approach, though, because the nucleus of a helium atom has two positive charges but four times the mass of a hydrogen nucleus.

Since a single, positive nucleon is, therefore, not sufficient to explain the mass and charge properties of all atomic nuclei, the question was whether two different nucleons would do. There were two possible models considered in the 1920s: either a combination of positive and negative particles, or positive and electrically neutral particles.

The first model was based on combining protons and electrons to make up the nucleus. Both particles were known to exist: the electron (e^-) had been discovered in 1897 by Thomson and the proton (p^+) had been observed in primary cosmic rays. In turn, no neutral particles had at that time been found. To disprove the feasibility of this model we need to return to a key quantum mechanical principle: Werner Heisenberg's uncertainty relation. It states that it is impossible to know precisely both the position and the momentum of a particle at any given time instant. Several key concepts of the atomic nucleus are based on this principle. Uncertainty is a fundamental property of particles because they also have wave properties, as postulated by de Broglie. Momentum and position is not the only parameter pair affected in this manner. A second, equally often applied uncertainty relation applies to the energy of the particle:

$$\Delta E \Delta t \geq h, \qquad [23.1]$$

in which Δt is the duration of observation of a particle and ΔE is the precision with which we can measure the energy of the particle during Δt. In nuclear physics, Δt is often the lifetime of an unstable particle. Thus, the faster a particle decays, the higher is the upper limit, which we have to accept for its energy. Eq. [23.1] follows again from the wave model of the particle. To measure the energy of a particle, $E = hf$, we need to determine its frequency. The frequency is determined by counting the number of 2π-wavelengths passing the observer. During a period of Δt, the number of such wavelengths is $f\Delta t$. Since we are counting only complete wavelengths, the minimum error we have to allow in counting is 1. Therefore, $\Delta f \Delta t \geq 1$, or $\Delta f \geq 1/\Delta t$. The error in the energy measurement is then $\Delta E = h \Delta f \geq h/\Delta t$. This leads to Eq. [23.1].

KEY POINT

The energy of a particle cannot be measured more precisely than $h/\Delta t$ when Δt is the duration of the measurement.

EXAMPLE 23.1

(a) Calculate the uncertainty in momentum and energy for an electron in a hypothetical helium nucleus that consists of four protons and two electrons. You obtain the radius of the helium nucleus from:

$$r_{\text{nucleus}} = \left(1.2 \times 10^{-15} \frac{\text{m}}{(\text{g/mol})^{1/3}}\right) A^{1/3},$$

in which r is the radius and A is the atomic mass. (b) The binding energy per nucleon in atomic nuclei is about 7 MeV/nucleon (see Example 23.3). What conclusion do you draw when comparing the result of your calculation in part (a) with this value?

Solution

Solution to part (a): We calculate a radius of 1.9×10^{-15} m for the helium nucleus. Substituting this in Heisenberg's uncertainty equation for the momentum, we find:

$$\Delta p \geq \frac{\hbar}{\Delta r} = \frac{1.05 \times 10^{-34}\,\text{Js}}{1.9 \times 10^{-15}\,\text{m}} = 5.5 \times 10^{-20}\,\frac{\text{kg m}}{s}.$$

continued

This equation contains $\hbar = h/(2\pi)$ and not Planck's constant h because we assume spherical symmetry for the nucleus. Using the classical formula $p = mv$ for the momentum, the result for Δp corresponds to a speed of $v = 6.1 \times 10^{10}$ m/s when using $m = 9.1 \times 10^{-31}$ kg for the electron. Since this value by far exceeds the speed of light we know that the electron in our hypothetical nucleus has to be treated as a relativistic particle. For relativistic particles, the momentum and the energy are related in Chapter 20 as $E_{photon} = cp$. Assuming that the energy at rest of the electron, m_0c^2, is only a small fraction of its total energy, $m_0c^2 << cp$, we can write:

$$\Delta E = c\Delta p \geq 1.7 \times 10^{-11} \text{ J} = 1.0 \times 10^{8} \text{ eV}.$$

Our assumption $m_0c^2 << cp$ is therefore justified because $m_0c^2 = 5.11 \times 10^5$ eV. The electron in the nucleus would have an energy as high as 100 MeV.

Solution to part (b): With the binding energy of a nucleon in the atomic nucleus of about 7 MeV and a total binding energy of the helium nucleus of 28.4 MeV (see Example 23.3), the result exceeds the binding energy per nucleon by more than a factor of 10. An electron confined to the nucleus would have to have an energy much less than the value we calculated in part (a), and thus would violate Heisenberg's uncertainty relation, or it cannot be confined to the nucleus as it has more than enough energy to escape.

The same problem does not apply to the protons in the nucleus due to their much larger mass. Indeed, any other particle confined to within the atomic nucleus should have a mass at least similar to that of a proton.

As a result of the calculation in Example 23.1, it was theoretically postulated that the atomic nuclei must contain protons and an electrically neutral nucleon, called the neutron. While the neutron does not interact electrically with matter, it does interact with protons via the nuclear force. The neutron (n^0), however, remained elusive until 1932, when James Chadwick discovered it indirectly by observing protons set free from hydrogen atoms in collisions with neutrons. There are two reasons why the neutron is hard to observe experimentally:

- A neutral particle interacts with matter much less than a charged particle does. Most interactions we observe are based on the electrostatic force because it is the only strong and far-reaching fundamental force; gravity is too weak, and the nuclear and weak force act only across distances of the size of the atomic nucleus.
- Free neutrons are not stable, they are only stable as part of a nucleus. We quantify stability later with the concept of the half-life. This is the time by which 50% of an initial amount of unstable particles have decayed. Protons and electrons are stable since their half-life at least exceeds by many orders of magnitude the age of the universe. This also applies to a neutron in a nucleus; however, when the neutron is separated from a nucleus, its half-life is only 12.5 min, i.e., after a few hours essentially no isolated neutrons are left. What happens when a neutron decays is written in the following form:

$$n^0 \rightarrow p^+ + e^- + \bar{v}. \quad [23.2]$$

The neutron decays into a proton, an electron, and an antineutrino. There is also energy released in the neutron decay: $\Delta E = 770$ keV.

The fact that the neutron decays quickly when isolated but is stable when it is part of a nucleus indicates that nucleons act differently as part of nuclear matter. This is due to the nuclear force and the short distance across which it acts.

23.1.2: Notation of Nuclides

With variations in both the number of protons and number of neutrons in the nucleus, a wide range of different nuclei can be formed. We know more than 1700 different nuclei, of which 271 are stable. Thus, we cannot simply distinguish the nuclei by the chemical element to which they belong. Instead, we must report two of three numbers to unequivocally identify a nucleus:

(i) the mass number, A, which corresponds to the number of nucleons in the nucleus,

(ii) the atomic number, Z, which corresponds to the number of protons in the nucleus, and

(iii) the number of neutrons, N.

These numbers are related in a simple manner:

$$A = Z + N. \quad [23.3]$$

When we refer to a nucleus for which A, Z, and N are defined, we call it a nuclide. The term nuclide may also refer to the complete atom, i.e., the nucleus and the electronic shell.

It is common practice to identify the atomic number Z and the mass number A, and to connect them with the familiar chemical symbol X. Two examples illustrate this notation for the most abundant forms of carbon and oxygen:

$${}^{A}_{Z}\text{X e.g.: } {}^{12}_{6}\text{C}, {}^{16}_{8}\text{O}.$$

Various terms have been introduced to identify relations between different nuclides. The most frequently used terms are isobaric nuclides and isotopic nuclides (isotopes):

- Isobaric nuclides refer to nuclei with the same mass number, A = const. An example are the three nuclei ${}^{40}_{18}\text{Ar}$, ${}^{40}_{19}\text{K}$, and ${}^{40}_{20}\text{Ca}$, in which the argon and calcium isotopes are, respectively, the most abundant nuclides of the element. The potassium nuclide is radioactive and is connected through a radioactive decay mechanism with argon, called the potassium/argon clock in palaeontology.

- Isotopic nuclides (isotopes) refer to nuclides with the same atomic number, $Z = \text{const}$. Two examples are ${}^{1}_{1}\text{H}$, ${}^{2}_{1}\text{H}$, and ${}^{3}_{1}\text{H}$, or ${}^{11}_{6}\text{C}$, ${}^{12}_{6}\text{C}$, ${}^{13}_{6}\text{C}$, or ${}^{14}_{6}\text{C}$, where the names given to the isotopes in the first set are, from left to right, hydrogen, deuterium, and tritium.

Eighty-one elements in the periodic table have stable nuclides. When referring to nuclides, the atomic number is often omitted, since the information is already contained in the chemical symbol. For example, we discuss later a carbon-dating method for determining the age of biological samples based on the carbon isotope with $A = 14$. Instead of referring to this isotope in the form shown in the equation above, we often find the abbreviated notation ${}^{14}\text{C}$.

Note that the terms nuclide and isotope are sometimes used interchangeably in the literature. In this textbook, we maintain the definition of the current section: the term isotope is used when we refer to nuclides of the same element.

- **Isomer** refers to identical nucleus composition (i.e., $A = \text{const}$ and $Z = \text{const}$) but different energy states of the nuclide. Nuclei, which result from radioactive decay or are the result of a nuclear reaction, are often not in the ground state (labelled g) but in an excited state (labelled m for metastable). Several such nuclides are important in nuclear medicine, in particular the metastable technetium isomer ${}^{99m}_{43}\text{Tc}$ versus ${}^{99g}_{43}\text{Tc}$.

CONCEPT QUESTION 23.1

How are the following two nuclides related to each other: ${}^{14}_{7}\text{N}$ and ${}^{14}_{6}\text{C}$?

23.2: Nuclear Force and Energy

When we expanded our discussion of physical phenomena from mechanics to electricity, we had to introduce a new force to explain the observed phenomena. From the electric force concept we then derived the electric energy in analogy to the introduction of the gravitational potential energy for Newton's force of gravity. The same approach is followed in the current chapter: the atomic nucleus is obviously held together by a new force because gravity is much too weak and the electric force would disintegrate a nucleus that contains many positively charged nucleons. We first address the forces that hold the nuclei together. Using the nuclear force, the nuclear potential energy is then derived.

23.2.1: The Nuclear Force

Once we recognize how small the atomic nucleus is in comparison to the entire atom, we can immediately conclude that large amounts of energy must be involved in holding the nucleus together. We establish this quantitatively in analogy to our knowledge of chemical processes. Atoms form molecules with radii of about $r = 1 \times 10^{-10}$ m. Typical chemical processes (e.g., explosions of dynamite) release binding energies of the order of 1 eV or less. The size of the atomic nuclei is a factor of 10^5 smaller than the atomic size. Coulomb's law states that the repulsive interaction energy of charged particles is proportional to $1/r$. Thus, an exploding nucleus (e.g., energy released in an atomic bomb or in a nuclear reactor) has to be at least 10^5 times higher, i.e., in the range of 1 MeV per decaying nucleus. The nuclear force, which prevents the protons in a nucleus from repelling each other and cause the nucleus to explode, must be of this order of magnitude as it must exceed the Coulomb repulsion.

KEY POINT

The nuclear force is a fundamental contact-free force like gravity and the electric force. Different from those two forces, it allows particles to interact only at distances of the order of the size of the atomic nucleus. At that distance, it is stronger than the electric force. It is distinct from the electric force as it acts also between neutral particles: both neutrons and protons are affected by the nuclear force

The last statement must apply since, otherwise, neutrons would steadily evaporate from nuclei, which in turn would render all but the hydrogen nucleus unstable in a very short time.

CASE STUDY 23.1

What evidence have we already discussed that confirms the fact that the nuclear force acts over only a very short distance?

Answer: *In electrostatics, we discussed Rutherford's scattering experiments (alpha particles penetrating a thin gold foil). We can quantitatively describe the intensity of scattered α-particles as a function of scattering angle in that experiment when using Coulomb's law. No other force needs to be considered. We also showed that the α-particles in Rutherford's experiment approach the nucleus to within 3 to 4 times the nuclear radius; i.e., the two particles approach each other to within 1×10^{-14} m. Thus, the nuclear force does not affect two nucleons when they are separated by 1×10^{-14} m or more.*

It was this last observation that caused a significant revision of our view of natural phenomena. The Japanese physicist Hideki Yukawa proposed that all fundamental contact-free forces (called field forces) are based on the exchange of interaction particles. In particular, the electrostatic force is due to the exchange of photons between particles that carry electric charges, and the nuclear force is due to the exchange of pions between nucleons (electrons are not nucleons, only protons and neutrons qualify). The range of these forces is then related to the mass of the exchanged particle because Heisenberg's uncertainty relation applies. The photon has a zero mass at rest and, therefore, the electrostatic force is not limited. The pion in turn has a mass of 139.6 MeV, which is 273 times the mass of an electron, or almost 15% of the mass of a proton. Such a particle can escape a nucleus only for a very short time before violating Heisenberg's uncertainty principle. We calculate the distance to which it is allowed to travel from the uncertainty principle:

$$r_Y = \frac{\hbar}{m_0 c}, \qquad [23.4]$$

in which the index Y indicates that this is the interaction radius based on Yukawa's argument, m_0c is the momentum of the particle. Eq. [23.4] contains the constant $\hbar = h/(2\pi)$ because we again assume spherical symmetry of the nucleus.

EXAMPLE 23.2

Estimate the mass of a pion from the size of a deuteron, which is stable and contains one proton and one neutron.

Solution

We use the radius formula from Example 23.1 to calculate the radius of the deuteron, using A = 2 g/mol: $r = 1.5 \times 10^{-15}$ m. Assuming that this distance is about the range of the nuclear force, i.e., $r = r_Y$, we rewrite Yukawa's condition in Eq. [23.4] in the form:

$$m_0 c = \frac{\hbar}{r_Y} = \frac{1.05 \times 10^{-34}\,\mathrm{J\,s}}{1.5 \times 10^{-15}\,\mathrm{m}} = 7.0 \times 10^{-20}\,\frac{\mathrm{kg\,m}}{\mathrm{s}}$$

Dividing by the speed of light in vacuum, this leads to $m_0 = 2.3 \times 10^{-28}$ kg.

23.2.2: Nuclear Binding Energy

The next step in establishing the basic concepts in nuclear physics is to connect the newly introduced force to the concept of energy through the definition of a potential energy. In the current case, this is the energy we attribute to each nucleon due to the nuclear force caused by the other nucleons in the vicinity. We expect this to be a large energy because the nuclear force is a strong force.

In introducing the nuclear energy, we specifically follow the approach we used when defining the potential energy of electrons in the atomic shell: we define the potential energy of a free nucleon as E_{pot} = 0 J and measure the potential energy of a bound nucleon relative to this value. This is the reason why the energy we introduce is called the binding energy. Nuclear binding energies can be estimated from the type of observations discussed in Example 23.3.

EXAMPLE 23.3

(a) Compare the mass of a helium (^{4}He) nucleus with the sum of the masses of its nucleons. (b) Calculate the binding energy per nucleon in a ^{4}He nucleus.

Solution

Solution to part (a): A ^{4}He consists of two protons and two neutrons. The comparison is based on the calculation of the difference between the mass of the ^{4}He nucleus and its four nucleons. The individual masses can be measured very precisely with a mass spectrometer. They are, $m_{p^+} = 1.6736 \times 10^{-27}$ kg, $m_{n^0} = 1.6750 \times 10^{-27}$ kg, and $m_{4\mathrm{He}} = 6.6466 \times 10^{-27}$ kg, This yields:

$$\Delta m = 2m_{p^+} + 2m_{n^0} - m_{^4\mathrm{He}}$$
$$= (3.3472 + 3.350 - 6.6466) \times 10^{-27}\ \mathrm{kg}$$
$$= 5.05 \times 10^{-29}\ \mathrm{kg}$$

Thus, the helium nucleus is slightly lighter than the four nucleons of which it is made.

Solution to part (b): We interpret the result in part (a) in a manner similar to the formation of an atom. When an ion catches an electron and becomes neutral, the electron's energy is lowered. The energy difference is released as a photon and is no longer present in the atom. In the same manner, when four nucleons fuse to form a stable ^{4}He, each releases some energy. Using Einstein's mass-to-energy conversion formula, $E = mc^2$, we calculate the energy difference corresponding to the result in part (a):

$$E(^4\mathrm{He}) = (5.05 \times 10^{-29}\,\mathrm{kg})\left(3 \times 10^8\,\frac{\mathrm{m}}{\mathrm{s}}\right)^2$$
$$= 28.4\ \mathrm{MeV}.$$

Interpreting this value as the total binding energy of the nucleons, we can attribute a fraction of ¼ to each nucleon; i.e., $E_{nucleon}$ = 7.1 MeV.

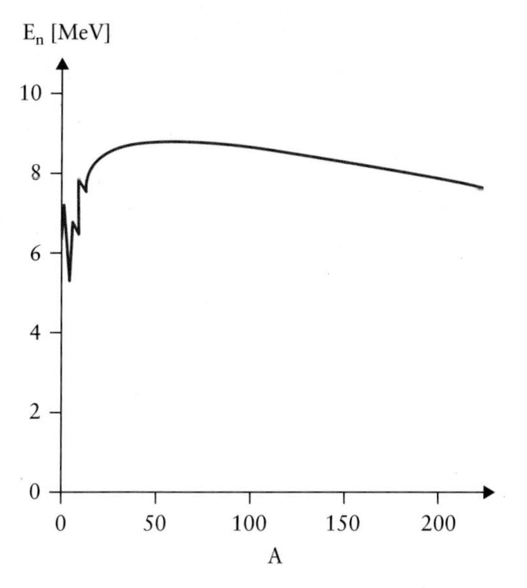

Figure 23.1 Binding energy per nucleon in nuclear matter as a function of the mass of the nucleus. The binding energy per nucleon varies only slightly with an average value of about 8 MeV.

Calculating binding energies for all stable elements allows us to compare the contribution of each nucleon to the total binding energy. Fig. 23.1 shows that this value holds roughly constant across the periodic table; the binding energy per nucleon $E_{nucleon}$ varies between 5.5 MeV and 8.5 MeV from deuterium to uranium.

With these values, we draw the Yukawa potential energy as a function of distance from the centre of the nucleus. Fig. 23.2 shows the potential energy for a proton

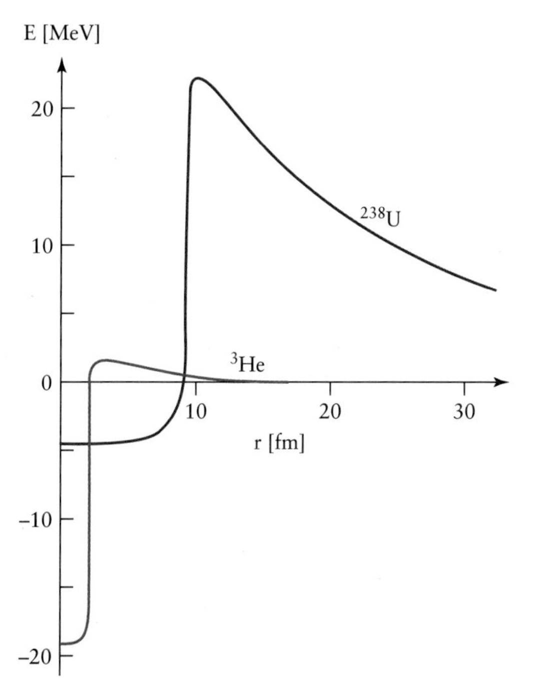

Figure 23.2 Potential energy profile for an approaching proton as a function of distance from the centre of a ^{3}He and a ^{238}U nucleus. The high-energy barrier is due to electrostatic repulsion; the low energy within the nucleus is due to the attractive nuclear force. The length scale of the abscissa is given in femtometres.

approaching one of the lightest nuclei, a ^{3}He nucleus (which is a light helium isotope), or one of the heaviest nuclei, a uranium (^{238}U) nucleus. Note the unit of the distance axis: a femtometre is 10^{-15} m. An approaching proton at first encounters the repulsive Coulomb force, as we observed in Rutherford's experiments. However, when the proton penetrates the nuclear matter, it suddenly feels an attractive force, which is essentially constant across the nucleus.

23.3: Radioactive Decay

Of the 1700 nuclides that have been studied, more than 1400 are unstable. These are called radioactive nuclides. Many of these occur naturally: some decay so slowly that they are still around from the time the solar system formed, such as uranium; others are formed in nuclear reactions of stable nuclides with particles of cosmic rays, such as ^{14}C. During a radioactive decay, the nuclide emits energy and/or particles; these decay processes are called radiation.

Radiation was first observed in 1896 by Henri Becquerel when he noticed that a uranium salt emits invisible particles that darken a photographic plate. Systematic studies by Marie and Pierre Curie led to the discovery of several radioactive nuclides by the beginning of the 20th century.

23.3.1: Decay Mechanisms

A nuclide is radioactive if there is a non-zero probability that it decays or transforms after a finite time. We distinguish six basic processes by which nuclides transform.

KEY POINT

A nuclide is stable with respect to a particular radioactive decay process if its mass is smaller than the sum of the masses of all the particles that result from the decay.

The most general way to describe a nuclear decay is in the form

X	→ Y	+ z	+ ΔE
mother	→ daughter	+ emitted	+ energy or
nuclide	nuclide	particle(s)	γ-ray.

In this schematic representation, the parent nuclide is labelled X (this is the nuclide that decays) and the daughter nuclide is labelled Y (this is a nuclide that resulted in the decay process). We further identify other particles that are emitted, such as α, β^+, and β^- particles. Due to energy and momentum conservation, decay processes also include high-energy photons (γ-rays) and neutral particles. For example, β^-- and β^+-decays are always accompanied by the emission of neutrinos (indicated by a Greek letter ν) or antineutrinos, for which we use the same symbol but add an overstrike. Neutrinos and antineutrinos are neutral particles that usually pass through the entire planet Earth without interaction. An interesting side note on neutrinos: they are now believed to play a major role during supernovas to provide the internal pressure that explodes the dying star. Since our solar system formed from the cloud of a supernova 4.6 billion years ago, one could claim that we owe our existence to these most elusive particles. Energy is carried from a decaying nuclide in the form of kinetic energy of emitted particles; excess energy is emitted in the form of X- and/or γ-rays, leaving the transformation event either directly or in a secondary process when the daughter nuclide decays from a metastable state to the ground state.

KEY POINT

We define as the Q-value of the nuclear decay the total energy released in the process; this is calculated from the mass difference of the particles:

$$E_Q = \Delta mc^2 = (m_X - \sum_i m_{Y_i} - \sum_i m_{Z_i})c^2.$$

Note, that a calculation of the Q-value from the energy of the emitted particles z_i and the released energy must also include the kinetic energy of the daughter nucleus (which cannot be neglected due to momentum conservation).

Next we introduce the six basic decay mechanisms: In the first process, called γ-decay, the nuclide emits only energy in the form of energetic photons (the nucleus is thermodynamically a closed system in this case). In the next three processes, the radioactive nuclide emits well-defined particles: emission of a helium nucleus is called α-decay, emission of an electron is called β^--decay, and emission of a positron is called β^+-decay. The fifth process, electron or K-shell capture, is a process in which the nucleus absorbs an electron from its own atomic shell, leading to the same nuclear transition as a β^+-decay. The sixth process is called spontaneous fission and occurs only for the heaviest nuclei, such as thorium and beyond.

23.3.1.1: Gamma Decay

Gamma decay is written in the form ${}^{Am}_{Z}X \rightarrow {}^{A}_{Z}Y + \gamma$; i.e., a γ-decay occurs between two isomers and is therefore called an isomeric transition. Radiopharmaceutical labelling with subsequent γ-ray imaging is most frequently done with ^{99m}Tc because this technetium isotope is versatile and the emitted gamma energy at 140.5 keV occurs in an ideal energy window for imaging. Its decay formula is:

$${}^{99m}_{43}Tc \rightarrow {}^{99}_{43}Tc + \gamma_{141keV}.$$

Note that 88% of ^{99m}Tc decays emitting a 140.5 keV gamma photon, while the energy in 12% of decays is taken up by an electron in the technetium shell, leading to the emission of a conversion electron.

It is important to note that the formation of a conversion electron is not a two-step process of gamma-emission followed by an internal photoelectric effect, but is a single-step process due to a Coulomb interaction between the nucleus and the electrons in the shell. Thus, the two possible pathways for the decay of ^{99m}Tc, gamma emission or conversion electron emission, occur independently from each other. The existence of these two pathways leads to an increase of the rate of decays for the metastable technetium state. This is confirmed by a dependence of the technetium decay rate on its chemical environment, which in turn alters the interaction of the electrons in the technetium shell with the nuclear charge (electron density at the location of the nucleus).

23.3.1.2: Beta Decay, Release of an Electron (β^-)

Beta decay is written in the form ${}^{A}_{Z}X \rightarrow {}^{A}_{Z+1}Y + e^- + \bar{\nu}$. We cannot use the emitted electron in β^--decay for medical imaging because it does not carry enough energy to leave the tissue and reach a detector. The range limitations of particles emitted in nuclear decay processes are discussed in the next chapter. However, this limitation does not preclude the use of β^--emitters: they can be applied in biological experiments done in Petri dishes, and they can be applied in medical imaging if the decaying nuclide also emits γ-rays in a detectable energy window.

An example of a β^--emitter with no concurrent γ-ray emission is ^{32}P, which is often used in phosphate molecules to study metabolic processes in cell cultures. Its decay is written as:

$${}^{32}_{15}P \rightarrow {}^{32}_{16}S + e^- + \bar{\nu}.$$

The combined kinetic energy of the emitted electron and antineutrino is 1.7 MeV. The kinetic energy of the electron varies because the division of energy between the antineutrino and the electron changes from one decay event to the next. Fig. 23.3 shows the energy

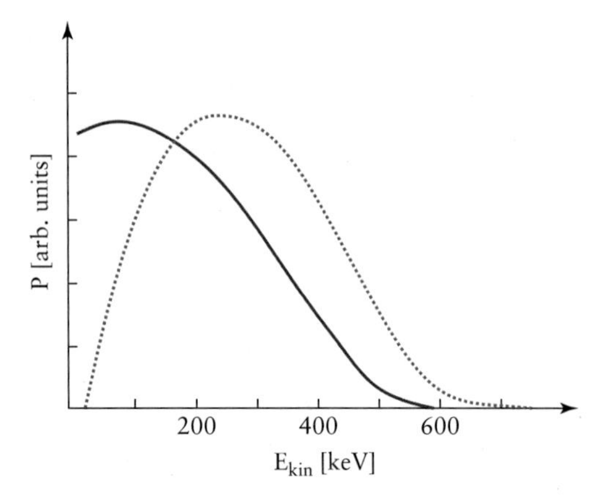

Figure 23.3 Probability distribution of the kinetic energy of the emitted electron and positron in β-decay of ^{64}Cu. The balance of energy is carried away by a neutrino or an antineutrino respectively. The distribution of the kinetic energy of the electron in the β^--decay to ^{64}Zn is shown as the solid line, and the kinetic energy of the positron emerging from the β^+-decay to ^{64}Ni is shown as the dashed line. Positron decay leads to a slightly higher maximum energy due to an additional Coulomb repulsion contribution between e^+ and the positive nucleus.

distribution of a large number of electrons emitted from the decay of ^{64}Cu. This nuclide is interesting in this context because it decays through both a β^-- and a β^+-channel, allowing us to compare the kinetic energy spectra of the positrons and electrons. The spectra are very similar except for a shift to higher energies in the case of the positron decay. This is due to an additional energy term caused by the electrostatic repulsion between positron and nucleus. In the case of the copper isotope decay, the maximum energies differ by 86 keV (571 keV for the electron emission versus 657 keV for the positron emission). The figure shows that the most probable value of the kinetic energy of the electron in a β^--decay is about 1/3 the combined kinetic energy carried by electron and antineutrino. In the case of ^{32}P, this means a typical electron has a kinetic energy in the range of 0.5 MeV to 1.0 MeV, which is sufficient to leave the cell culture and reach a detector.

In nuclear medicine, ^{131}I was used for many years in thyroid imaging as its decay is accompanied by γ-photons at 284 keV, 364 keV, and 637 keV. Of these, the 364-keV emission occurs in 81% of all decays:

$$^{131}_{53}\mathrm{I} \rightarrow {}^{131}_{54}Xe + e^- + \bar{\nu} + \gamma.$$

Due to a high dose to the patient, alternate iodine isotopes are now used, in particular iodine-123.

CONCEPT QUESTION 23.2

^{131}I is still used in radiotherapy, as are the β^--emitters ^{153}Sm and ^{89}Sr. Which daughter nuclides are formed by these samarium and strontium isotopes?

23.3.1.3: Beta Decay, Release of a Positron (β^+)

We write ${}^{A}_{Z}X \rightarrow {}_{Z-1}^{\ A}Y + e^+ + \nu$. β^+-decay processes are more important than β^--processes for medical imaging. The reason is that the emitted positron slows down in the tissue and then annihilates with its antimatter partner, the electron. This results in the emission of two characteristic γ-rays of 511 keV, which are used in positron emission tomography (PET). The antimatter properties of the positron are discussed later in this chapter. The most frequently applied isotope in positron emission tomography is a fluor isotope:

$$^{18}_{9}\mathrm{F} \rightarrow {}^{18}_{8}\mathrm{O} + e^+ + \nu.$$

The additional particle emitted this time is a neutrino; i.e., in both β-decays one antimatter particle is emitted, in one case the positron, in the other the antineutrino. The patient receives the radioactive fluor isotope in the form of a biochemical compound called **fluorodeoxyglucose** (FDG).

The β^+-decay represents 97% of the decay loss of ^{18}F. Another 3% of the nuclei transform instead through electron capture, as illustrated in the term scheme in Fig. 23.4. The branching ratio strongly favouring positron emission is important for medical imaging applications where the electron capture route is useless for imaging but contributes to the dose to the patient due to X-ray emission.

CONCEPT QUESTION 23.3

Positron-emitting nuclides recently investigated for PET applications are ^{11}C, ^{13}N, ^{68}Ga (gallium), and ^{82}Rb (rubidium). Which daughter nuclides are formed by these parent nuclides?

23.3.1.4: Electron Capture

Nuclei can transform further by capture of an electron from an inner atomic shell. This is not a decay process as the nucleus does not lose any particles (except for a neutrino). For consistency in term scheme labelling, it is often denoted with a Greek letter epsilon (ε). However, the process leads to a nuclear transformation equivalent to a β^+-decay; it occurs for neutron-deficient nuclides and the same daughter nuclide is formed ${}^{A}_{Z}X + e^- \rightarrow {}_{Z-1}^{\ A}Y + \nu$. A neutrino is released for momentum conservation; energy may be released in the form of γ-rays, but the capture event is, in every case, accompanied by an X-ray cascade due to the creation of a K-shell vacancy. Electron capture occurs exclusively when the energy difference between parent and daughter nuclide is less than 1.02 MeV. Above this energy, which is the minimum energy needed to create an electron/positron pair, positron emission becomes possible.

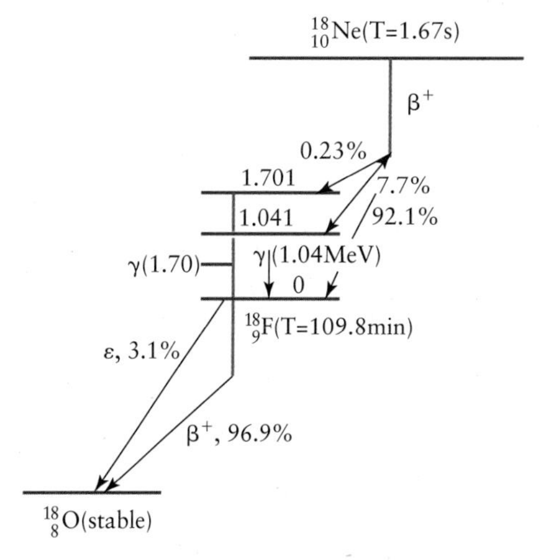

Figure 23.4 Decay schematics of ^{18}F, showing that it also is itself the result of a β^+-decay of ^{18}Ne.

The electron capture process often does not lead to γ-ray emission useful in medical imaging. In particular, no γ-rays at 511 keV are emitted. However, X-ray cascades from the daughter nuclide filling its K-shell vacancy may allow for imaging if the emitted X-rays lie in an accessible energy window. An example is ^{201}Tl (thallium), which is used in cardiac perfusion studies:

$$^{201}_{81}\text{Tl} + e^- \rightarrow \; ^{201}_{80}\text{Hg} + \nu.$$

The X-rays emitted from the mercury nuclide lie in the range of 70 keV to 80 keV.

Note that using ^{201}Tl for medical imaging is superior to X-ray imaging for two reasons: (i) the X-rays are generated only in metabolically active zones; i.e., the imaging with thallium-201 is a functional imaging method as defined in the introduction; and (ii) the emitted X-rays are monoenergetic instead of spread across a wide range of energies. Still, ^{201}Tl is only cautiously used as the low X-ray energies cause significant attenuation in the patient, which in turn leads to imaging artifacts, particularly in the female breast located near the diagnosed heart.

CONCEPT QUESTION 23.4

Medical applications of nuclides that transform by electron capture include imaging with ^{67}Ga, ^{111}In (indium), and ^{123}I, and non-imaging in vivo analysis with ^{51}Cr (chromium) and ^{125}I. The indium-111 nuclide has a long lifetime (half-life 2.83 days) for nuclides suitable for imaging; it is therefore often used when imaging is possible only more than 24 hours after injection. Which daughter nuclides are formed by the five parent nuclides listed above?

23.3.1.5: Alpha Decay

Alpha decay is written in the form $^{A}_{Z}X \rightarrow \; ^{A-4}_{Z-2}\text{Y} + \; ^{4}_{2}\text{He}$. Nuclides emitting α-particles are not used in nuclear medicine, due partially to the short range of the emitted alpha particle and partially because of the strong absorbed dose effect. We are still affected by the α-decay in our daily life because the alpha emitter radon-222 contributes more than 50% of our annual radiation dose due to natural and artificial sources. The radon isotope is created during the radioactive decay of uranium-238. It is a heavier than air noble gas and collects in basements due to a long lifetime (half-life 3.8 days). Typical indoor concentrations in North America are about 50 decays per second per cubic metre, i.e., 10 times above average outdoor levels; the indoor level may increase by a factor 50 in poorly ventilated rooms. The radiation we receive is from the radon decay directly, but also from its short-lived daughter nuclides. The radon-222 isotope decays with the emission of a 5.5-MeV α-particle:

$$^{222}_{86}\text{Rn} \rightarrow \; ^{218}_{84}\text{Po} + \alpha + \gamma.$$

CONCEPT QUESTION 23.5

The following are daughter nuclides that result from α-decays. Which were the respective mother nuclides?

(a) ^{182}W

(b) ^{209}Bi

(c) ^{213}At

23.3.2: Activity and the Radioactive Decay Law

When an individual radioactive nucleus decays cannot be predicted. Radioactive decay is a stochastic process, which means that each nucleus decays randomly and independently of the others. As in other stochastic processes that require statistical methods, we can make only quantitative statements about a large number of radioactive nuclei.

We assume that N radioactive nuclei (with $N >> 1$) is a sufficiently large number. For these we define a rate of radioactive decays (which is the same as decays per time interval), $-dN/dt$, where the negative sign indicates that the number of nuclei is decreasing with time. The term:

$$A = -\frac{dN}{dt} \qquad [23.5]$$

represents the number of nuclear transformations per second in the sample. The quantity A is called the **activity**. The standard unit for the activity is Bq (**becquerel**, named for Henri Becquerel), with the definition 1 Bq = 1 decay/s. Typical radioactive sources have activities in the range of several MBq. The negative sign in the definition of the activity guarantees that the activity itself is a positive number.

KEY POINT

The activity of a radioactive sample is defined as the negative value of the decay rate (number of decays per time unit).

For any type of radioactive decay, the decay rate is observed to be proportional to the number of radioactive nuclei present in the sample at time t:

$$-\frac{dN}{dt} = \lambda N, \qquad [23.6]$$

in which we introduced the proportionality factor λ, which is called the **decay constant** and has unit 1/s. Using the activity A, we can alternatively write Eq. [23.6] in the form:

$$A = \lambda N. \qquad [23.7]$$

The linear proportionality between decay rate dN/dt and the number of remaining nuclides N in Eq. [23.6] leads, after integration, to an exponential law. To illustrate this, we first separate the variables in Eq. [23.6]:

$$\frac{dN}{N} = -\lambda\, dt.$$

The left side leads to a logarithmic term in N, and the right side to a linear term in time:

$$\ln N \Big|_{No}^{N(t)} = -\lambda t \Big|_{t=0}^{t},$$

with the boundary condition that the sample contains N_0 radioactive nuclei at time $t = 0$. We keep the upper boundary variable to obtain a function $N(t)$. Substituting the boundary conditions yields:

$$\ln\left(\frac{N(t)}{N_0}\right) = -\lambda t$$

which is equivalent to:

$$N(t) = N_0 e^{-\lambda t} \qquad [23.8]$$

after both sides of the equation are raised to the power of e (Euler number). Eq. [23.8] describes the time dependence of the number of radioactive nuclei remaining in a sample; it is called the **law of radioactive decay** and is illustrated in Fig. 23.5, highlighting that the law is not a linear but an exponential function. To linearize Eq. [23.8], we return to the logarithmic equation from which it was derived. This equation shows that we obtain a linear plot

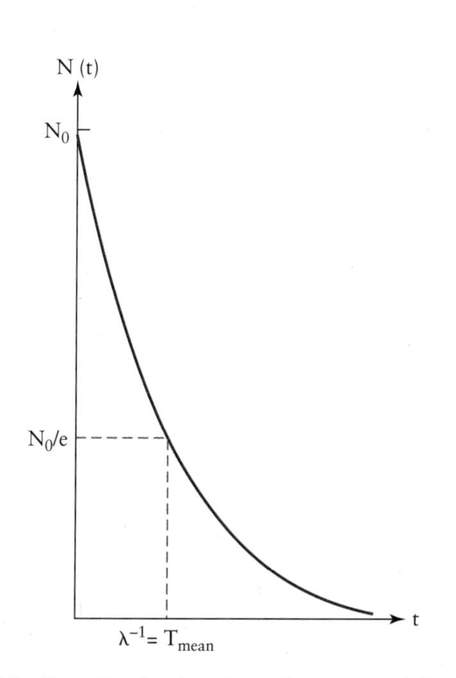

Figure 23.5 The radioactive decay law in linear representation.

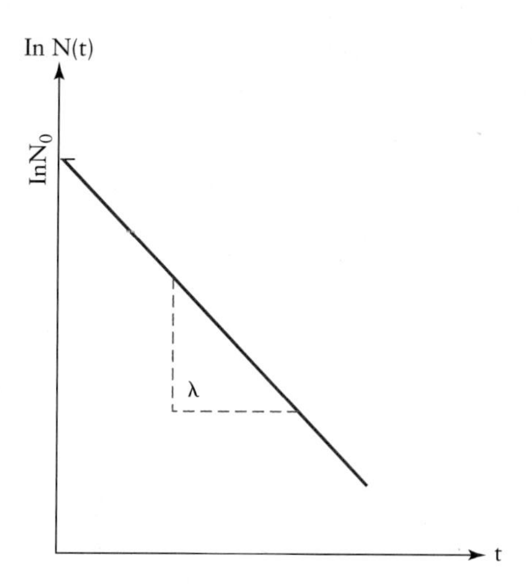

Figure 23.6 The radioactive decay law in logarithmic representation.

if we use a logarithmic ordinate for $N(t)/N_0$, which is the fraction of particles remaining after time t. Alternatively, we can separate the logarithm term in the form:

$$\ln N(t) = \ln N_0 - \lambda t,$$

which is illustrated in Fig. 23.6. In this plot, $\ln N_0$ is the intercept with the ordinate, and λ is the slope.

KEY POINT

The radioactive decay law quantifies the fraction of an initial radioactive sample that is still present after a time period t has elapsed. The time dependence is exponential with the decay constant the proportionality factor between the decay rate and the number of remaining radioactive nuclides.

We can rewrite Eq. [23.8] also for the activity, noting that $A(t) \propto N(t)$ in Eq. [23.7] for all times t. This leads to:

$$A(t) = A_0 e^{-\lambda t} \qquad [23.9]$$

or, in linearized form:

$$\ln = \left(\frac{A(t)}{A_0}\right) = -\lambda t.$$

23.3.3: Half-Life

Instead of reporting the decay constant λ, or the **mean lifetime** $T_{mean} = \lambda^{-1}$, the **half-life** $T_{1/2}$ is more often introduced in the scientific literature.

KEY POINT

The half-life is a nuclear decay-specific time constant that represents the time after which 50% of the initial number of radioactive nuclei in a sample have decayed.

We relate the half-life to the decay constant λ by rewriting Eq. [23.8] for $t = T_{1/2}$, with $N(T_{1/2}) = N_0/2$:

$$\frac{N_0}{2} = N_0 e^{-\lambda T_{1/2}}.$$

After dividing by N_0 and taking the natural logarithm on both sides, we find:

$$\ln\frac{1}{2} = -\lambda T_{1/2},$$

which is equivalent to:

$$\ln 2 = \lambda T_{1/2}.$$

Thus, half-life is written as:

$$T_{1/2} = \frac{\ln 2}{\lambda} \qquad [23.10]$$

This equation shows that the half-life and decay constant are inversely proportional to each other. Note in particular that the decay of 3/4 of the initial amount of radioactive nuclei requires a time of $2T_{1/2}$. To represent the decay law with half-life, we substitute the result from Eq. [23.10] in Eq. [23.8]:

$$N(t) = N_0 e^{-\ln 2 \frac{t}{T_{1/2}}}.$$

This relation is shown in Fig. 23.7, which highlights the half-life and twice the half-life of the given decay. This equation further reads in linearized form:

$$\ln\left(\frac{N(t)}{N_0}\right) = -\ln 2\left(\frac{t}{T_{1/2}}\right),$$

which, for the activity, is equivalent to:

$$\ln\left(\frac{A(t)}{N_0}\right) = -\ln 2\left(\frac{t}{T_{1/2}}\right).$$

This is illustrated in Fig. 23.8. When we introduce a $t/T_{1/2}$-abscissa, we find a *generic* plot for radioactive decay processes, i.e., a plot that applies in the shown form to every decay process. We can further relate the mean lifetime and the half-life using Eq. [23.10]:

$$T_{\text{mean}} = \frac{1}{\lambda} = 1.44\,T_{1/2}. \qquad [23.11]$$

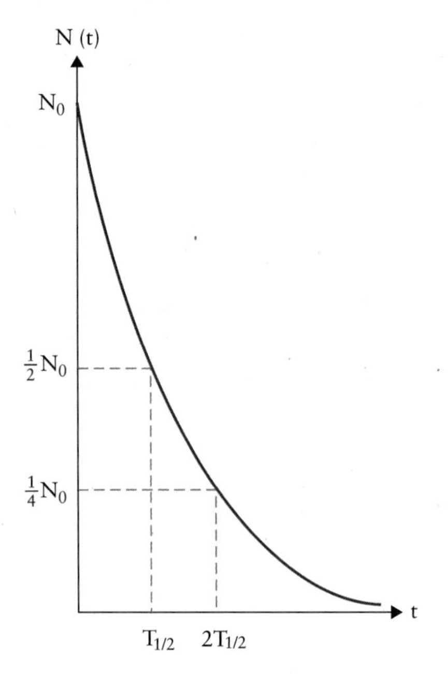

Figure 23.7 Sketch of the number of nuclei of a given radioactive isotope in a sample as a function of time. The dashed lines indicate the reduction after one half-life at $t = T_{1/2}$ and two half-lives at $2T_{1/2}$.

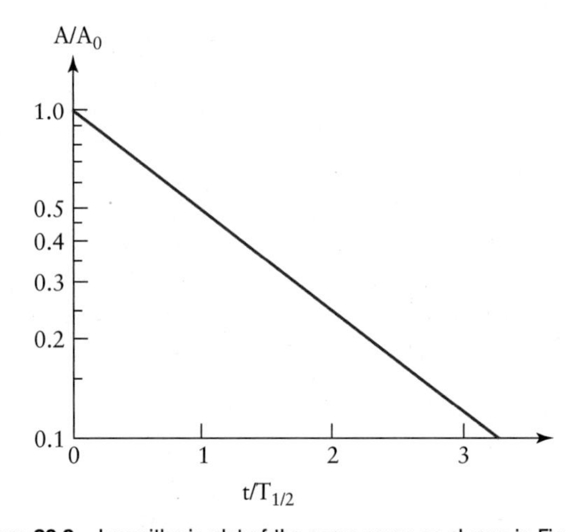

Figure 23.8 Logarithmic plot of the same curve as shown in Fig. 23.7. The ordinate shows the relative activity, A/A_0; the abscissa is a relative time scale, $t/T_{1/2}$, with $T_{1/2}$ the half-life.

EXAMPLE 23.4

We consider a sample that contains 3×10^{16} nuclei of radium-226 (^{226}Ra), which has a half-life of 1600 years. Note that 226-radium is an α-emitter and is the parent nuclide to 222-radon, which is the major natural indoor radiation source in North America. What is the activity of the sample at $t = T_{1/2}$?

Solution

This is not a laboratory experiment since the two reference times are 1600 years apart. In the first step we

continued

determine the decay constant λ. For this, we express the half-life in the standard unit s: $T_{1/2} = 5 \times 10^{10}$ s. Thus:

$$\lambda = \frac{\ln 2}{T_{1/2}} = 1.4 \times 10^{-11}\ \mathrm{s}^{-1}.$$

We substitute this value into Eq. [23.7] to determine the initial activity A_0, i.e., the activity at $t = 0$:

$$A_0 = \lambda N_0 = (1.4 \times 10^{-11}\ \mathrm{s}^{-1})(3 \times 10^{16})$$
$$= 4.1 \times 10^5\ \mathrm{Bq} = 0.41\ \mathrm{MBq}.$$

We use the same formula to calculate the activity at the half-life of the sample:

$$A = \lambda \frac{N_0}{2} = 2.05 \times 10^5\ \mathrm{Bq} = 0.21\ \mathrm{MBq}.$$

EXAMPLE 23.5

^{81m}Kr has a half-life of $T_{1/2}$ = 13 seconds. An initially generated gas sample can be used for a lung ventilation measurement as long as its activity has not dropped below 1% of the initial activity. What is the maximum delay before the experiment has to start?

Solution

We use Eq. [23.9] for the time dependence of the activity and substitute λ by $T_{1/2}$ using Eq. [23.10]:

$$\frac{A(t_{max})}{A_0} = 0.01 = \exp\left\{-\ln 2 \frac{t_{max}}{T_{1/2}}\right\}$$

and we solve for t_{max}:

$$\ln(0.01) = -\ln 2 \frac{t_{max}}{T_{1/2}},$$

which yields:

$$t_{max} = -\frac{\ln(0.01)}{\ln 2} T_{1/2} = 6.64 \cdot 13.0\ \mathrm{s} = 86.3\ \mathrm{s}.$$

We find t_{max} for ^{81m}Kr samples to be 1.5 minutes. This illustrates that the major obstacle in nuclear medicine with short-lived nuclides is the need to do diagnostic measurements very fast.

23.4: Angular Momentum

23.4.1: Angular Momentum in Classical Physics

The physical quantity used to study both the orbiting point-like object and the spinning extended object is the angular momentum. Since we referred to the angular momentum before, we briefly review the properties we continue to use. But the discussion is also taken in two directions beyond a review:

- we allow for a non-zero external torque, which leads to a change of the angular momentum with time; and
- we study a rotating extended object, leading to the definition of the moment of inertia.

To characterize objects that undergo uniform circular motion, we introduced the **angular momentum** with magnitude L:

$$L = p r \sin\varphi = m v r \sin\varphi, \qquad [23.12]$$

in which $p = m\,v$ is the linear momentum of the object, r is the distance from the axis of the circular motion and φ is the angle between the momentum vector and the radius vector, $\varphi = \sphericalangle(\vec{p}, \vec{r})$. The unit of the angular momentum is kg m²/s = J s. We know from our earlier studies that both the linear momentum and the position are vectors. We expect the angular momentum to be a vector as well because rotational phenomena are sensitive to the orientation of the rotation axis in space. Combined with the sine term on the right-hand side of Eq. [23.12], we conclude the three vectors are connected through a vector product:

$$\vec{L} = [\vec{r} \times \vec{p}] \qquad [23.13]$$

in which $\vec{L}$ is parallel to the axis of rotation. The relative orientation of the vectors in Eq. [23.13] is again determined with the right-hand rule.

The angular momentum does not change when the term on the right-hand side of Eq. [23.13] remains unchanged. Cases where this occurs lead to a conservation of angular momentum. We have to evaluate the right-hand side of Eq. [23.13] carefully in this respect because of its vector product form. In particular, not every change to the linear momentum will result in a change of the angular momentum. For example, we can add a component to $\vec{p}$ along the direction of vector $\vec{r}$ that will not yield a change in $\vec{L}$. On the other side, if $\vec{p}$ changes in a direction other than along the vector $\vec{r}$, the angular momentum must change as well. Changes to the linear momentum are always caused by a force: $\vec{F} = d\vec{p}/dt$. To distinguish forces that act along the vector $\vec{r}$ from those that act in a different direction, we combine the external force acting on the system with vector $\vec{r}$ in vector product form; i.e., we study the term $[\vec{r} \times \vec{F}]$. This term is zero when $\vec{F}$ is parallel to $\vec{r}$; in all other cases a non-zero value results. Thus, the vector product $[\vec{r} \times \vec{F}]$ is suitable to selectively discuss the forces that lead to a change in the angular momentum.

We recognize the term $[\vec{r} \times \vec{F}]$, at least when it is written as a magnitude of the resulting vector as $Fr\sin\varphi$, as the torque $\vec{\tau}$ acting on the system due to the external force $\vec{F}$:

$$\vec{\tau} = [\vec{r} \times \vec{F}], \qquad [23.14]$$

which is illustrated in Fig. 23.9 to emphasize the right-hand rule that governs the directional relation between the three vectors. Thus, a system with a net torque is a system in which the angular momentum is expected to change:

$$\vec{\tau} = \frac{d\vec{L}}{dt}. \qquad [23.15]$$

We state the opposite observation as the **conservation law of angular momentum**:

KEY POINT

The angular momentum of a closed system is conserved if no net torque acts on the system. The system needs not to be isolated, as is required for the conservation of energy, because, for example, a force acting toward the rotation axis does not contribute a torque term.

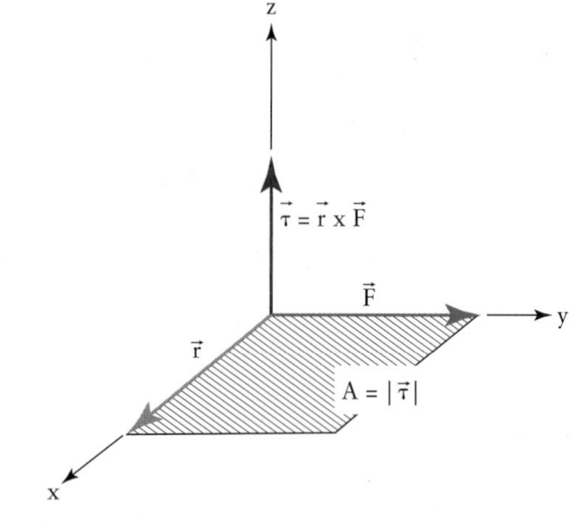

Figure 23.9 Relative directions of the force, the position vector, and the torque. The torque results from the vector product of the other two vectors. Their directions are determined with the right-hand rule.

Eq. [23.15] also indicates the direction in which the change of the angular momentum occurs when an external torque acts on the system: the change occurs in the direction of the torque, i.e., *not* in the direction of the external force that causes the torque. We note from Fig. 23.9 that these two vectors are perpendicular to each other. A nice way to illustrate the effect predicted by Eqs. [23.13] to [23.15] is a top on a pole. In Fig. 23.10(a) the top consists of a disk and a rod, and the pole is shown as a dotted triangle. If the top spins perfectly upright, there is no torque acting on it because such a torque has to be caused by the top's own weight, and for the perfectly upright top, the top's weight acts along its axis. However, once we tilt the spinning top, for example by 90° as shown in Fig. 23.10(a), the weight $\vec{W}$ acts no longer along the axis of the top. If the top were not spinning, the gravitational force would cause it to fall. However, because the top is spinning, we need to study the torque

CASE STUDY 23.2

We focussed on the linear momentum $\vec{p}$ and not the position vector $\vec{r}$ in the discussion of the conservation of angular momentum. Does this mean that we captured only part of the argument?

Answer: *No. The reason lies in the vector product form of Eqs. [23.13] and [23.14], and in the fact that the linear momentum is defined as $\vec{p} = m\,\vec{v}$, with m the mass and $\vec{v}$ the velocity of the object. To see this, we start with Eq. [23.14] and rewrite the force as the time change of the linear momentum:*

$$[\vec{r}\times\vec{F}] = \left[\vec{r}\times\frac{d}{dt}\vec{p}\right] = \left[\vec{r}\times\frac{d}{dt}(m\vec{v})\right] = m\left[\vec{r}\times\frac{d}{dt}\vec{v}\right],$$

in which we assume mass is time independent. If we replace torque on the other side and angular momentum in Eq. [23.15] with Eqs. [23.14] and [23.13] respectively, we find:

$$[\vec{r}\times\vec{F}] = \frac{d}{dt}[\vec{r}\times\vec{p}].$$

The right-hand sides of the last two equations have to be the same; otherwise we would have overlooked some mechanical features when stating the conservation law above. To check this identity we apply the product rule of calculus to the right-hand side of the last equation:

$$\frac{d}{dt}[\vec{r}\times\vec{p}] = m\left\{\left[\frac{d}{dt}\vec{r}\times\vec{v}\right] + \left[\vec{r}\times\frac{d}{dt}\vec{v}\right]\right\},$$

in which we pulled the constant mass again out of the two terms in the bracket. With the definition of velocity as the change of position in time, the first term in the bracket leads to $\vec{v}\times\vec{v} = 0$ because the velocity vector is obviously parallel to itself.

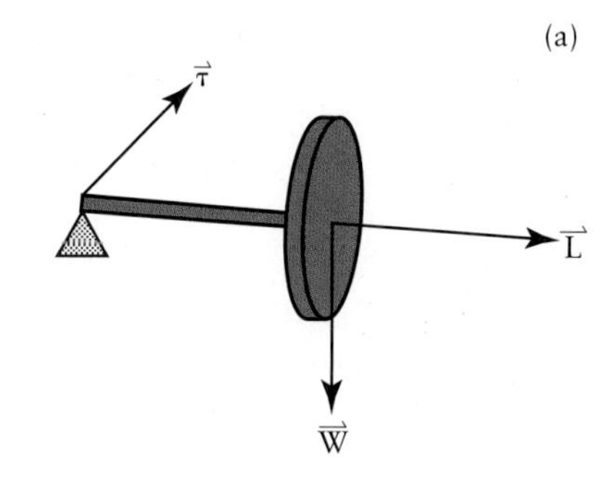

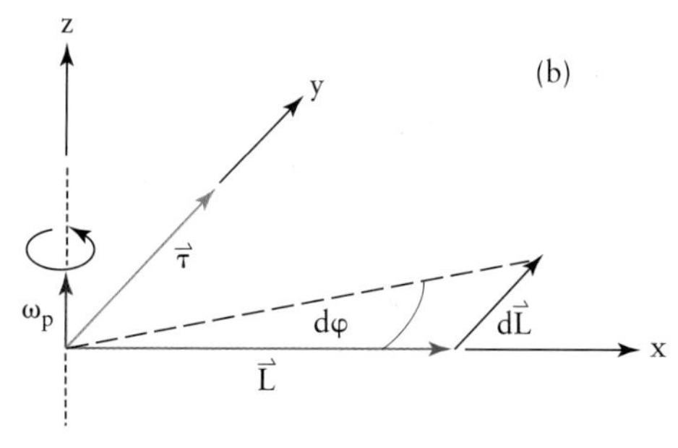

Figure 23.10 (a) A tilted top with angular momentum $\vec{L}$ does not drop due to gravity, but precesses about the pivot point (triangle). (b) This motion is due to the direction of the torque $\vec{\tau}$, which is parallel to the direction of the changing angular momentum, $d\vec{L}/dt$. The choice of a top tilted by 90° simplifies the calculations because the weight $\vec{W}$ and the angular momentum $\vec{L}$ are perpendicular to each other.

and not the force. Fig. 23.10(b) illustrates the direction of the torque $\vec{\tau}$. Choosing the top's axis along the x-axis, and the vertical axis as the z-axis, the torque points in the y-direction. Eq. [23.15] requires that the change dL of the angular momentum occurs in the same direction and leads therefore to a sideways change of the top's axis. The top does not fall off the pole, but its axis moves in a circular manner in the xy-plane. This motion of the top is called **precession**.

With precession defined as a circular motion of the axis of the angular momentum $\vec{L}$ in Fig. 23.10, we can rewrite the magnitude of terms in Eq. [23.15] in the form:

$$\tau = L\frac{d\phi}{dt}, \qquad [23.16]$$

where the angle φ is taken from Fig. 23.10. Eq. [23.16] recognizes that the angular momentum does not change in magnitude, but only in direction. Limiting the change of the angle φ to a small value, we used in Eq. [23.16] the approximation $\sin\varphi \cong \varphi$. Since precession is a uniform circular motion, the term $d\varphi/dt$ is constant and equal to the angular frequency of the circular motion.

We want to generalize the result in Eq. [23.16] for tops that are tilted by angles other than 90°. In Fig. 23.11 we consider a top that is tilted by an angle α. The torque causes a component $d\vec{L}$ along the direction of $\vec{\tau}$ to be added to the angular momentum $\vec{L}$. From geometry for the shaded

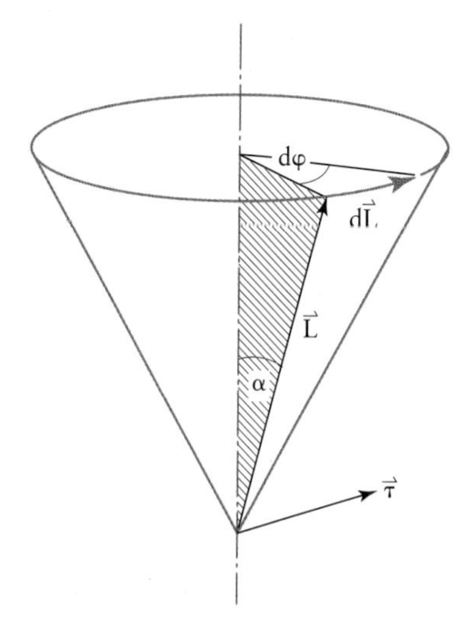

Figure 23.11 The geometric relations between the various variables for a precessing spin with $\vec{\tau}$ the torque, $d\vec{L}$ the change of the angular momentum, α the angle between the external force and the axis of the top, and $d\varphi/dt$ the angular velocity of the precession.

triangle and the triangle in the top circle of the figure, we know that $dL = L\,(\sin\alpha)\,d\varphi$ because only the component of the angular momentum perpendicular to the top's axis is affected by Eq. [23.15]. Thus Eq. [23.16] becomes:

$$\tau = L\sin\alpha\frac{d\phi}{dt}.$$

This causes again a uniform circular motion and we write $d\varphi/dt = \omega_p$ for the **angular frequency** of the precession. Thus:

$$\tau = L\sin\alpha\,\omega_p. \qquad [23.17]$$

Rearranging for the angular frequency we note:

$$\omega_p = \frac{\tau}{L\sin\alpha}.$$

KEY POINT

Precession occurs the faster and larger the torque, the smaller the tilt angle of the top's axis, and the smaller the angular momentum. We will see below that the angular momentum is small when the mass distribution of the top is close to its axis and when it spins slowly.

We have to discuss one more aspect for systems that consist of more than one object. For such systems, the question arises whether internal forces have an impact on the angular momentum of the system. To answer this question, we study the simplest possible system, one that consists of two objects that interact, let us say through gravity. In this case, four forces have to be considered when we calculate torque: Each object is subject to an external force, which we call $\vec{F}_1$ and $\vec{F}_2$ respectively. Further, the internal force $\vec{F}_{1\,on\,2}$ acts on object 2 and the internal force $\vec{F}_{2\,\text{on}\,1}$ acts on object 1 due to the other object. Thus, net torque has four contributions:

$$\vec{\tau}_{\text{net}} = [\vec{r}_1 \times (\vec{F}_1 + \vec{F}_{2\,\text{on}\,1})] + [\vec{r}_2 \times (\vec{F}_2 + \vec{F}_{1\,\text{on}\,2})].$$

The terms on the right hand side can be rearranged to separate the sums of forces in each vector product term:

$$\vec{\tau}_{\text{net}} = [\vec{r}_1 \times \vec{F}_1] + [\vec{r}_2 \times \vec{F}_2] + [(\vec{r}_1 - \vec{r}_2) \times \vec{F}_{2\,\text{on}\,1}],$$

in which we used Newton's third law, which requires $\vec{F}_{1\,\text{on}\,2} = -\vec{F}_{2\,\text{on}\,1}$. Fig. 23.12 shows that $\vec{F}_{2\,\text{on}\,1}$ and the separation vector $\vec{r}_1 - \vec{r}_2$ are parallel, and thus their vector product is zero.

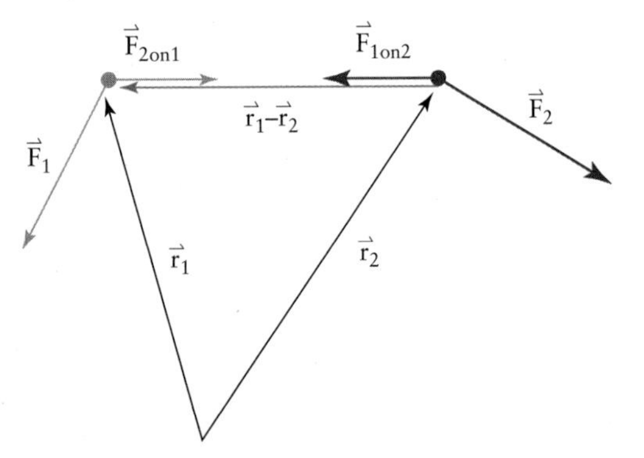

Figure 23.12 A system of two particles with an internal interaction. Net torque is calculated from four forces: $\vec{F}_1$ and $\vec{F}_2$ are external forces acting on objects 1 and 2; the internal force $\vec{F}_{1\,\text{on}\,2}$ acts on object 2, and the internal force $\vec{F}_{2\,\text{on}\,1}$ acts on object 1, with $\vec{F}_{1\,\text{on}\,2} = -\vec{F}_{2\,\text{on}\,1}$ due to Newton's third law. We note that the difference in the two position vectors $\vec{r}_1 - \vec{r}_2$ is parallel to $\vec{F}_{2\,\text{on}\,1}$.

KEY POINT

Internal forces do not contribute to the torque or the time change of the total angular momentum of a system.

23.4.2: The Nuclear Spin

With the concept of spin defined for a macroscopic system, we turn our attention to spin on the atomic scale. In Bohr's model of the atom, electrons are pictured to orbit the nucleus. An angular momentum can be attributed to this motion as used by Bohr to treat his model quantitatively.

At the centre of the atom, the nucleus can be considered to rotate about its own axis. The rotation of the nucleus leads to a nuclear spin contribution to the total angular momentum of the atom.

We note for the atomic system that the orbiting electrons each carry a spin of their own. We already used this fact in Chapter 20 when we allowed two electrons (differing only in their spin quantum number) to fill each of the atomic orbitals to describe all elements in the periodic table. For this reason, Pauli's principle is written such that any two electrons in the same atom must differ in at least one of four quantum numbers, including the spin quantum number m_s. Helium, for example, has two electrons that have the same three quantum numbers, $n = 1$, $l = 0$, $m_s = 0$, but they do not violate Pauli's principle since they differ in their spin. This can be pictured in a classical model: one electron spins in the direction east to west, the other west to east. Since we want to develop the basic physical principles of MRI, which are based on nuclear spin, the spin of the electrons is not discussed further. Note, however, that there are interesting experimental methods using the electron's spin. In particular, electron spin resonance (ESR) is used to identify unpaired electrons, e.g., in electrically neutral molecules called **radicals**.

In Chapter 20 we found we must use quantum mechanical models to describe properly physical properties at the atomic level. This remains true in the present discussion: nuclear spins have features that classical spin systems do not have. Of these, we have to consider the quantization of the spin: the spin of a nucleus can be only zero or an integer multiple of the value $h/(4\pi) = \hbar/2$, in which h is Planck's constant with unit J s. Values for various important nuclei are shown in Table 23.1. Nuclei with a zero spin cannot be used for nuclear magnetic resonance techniques such as MRI. Therefore, of the most common elements in bioorganic molecules, only hydrogen responds to the processes discussed in Chapter 20.

TABLE 23.1

Nuclear spin values for selected elementary particles and atomic nuclei

Nucleus	Spin	Nucleus	Spin
neutron	$\hbar/2$	proton (^{1}H)	$\hbar/2$
deuteron (^{2}H)	$\hbar$	α (^{4}He)	0
^{12}C	0	^{13}C	$\hbar/2$
^{14}N	$\hbar$	^{16}O	0
^{19}F	$\hbar/2$	^{31}P	$\hbar/2$

23.5: Nuclear Spins in a Magnetic Field

23.5.1: Magnetic Dipole Moment

Nuclear spin is not observed directly, but indirectly due to the magnetic dipole moment of the nucleus, which is a consequence of the presence of the spin. A spin leads to a magnetic dipole because the nucleus is electrically charged. We saw earlier in the chapter that nuclei contain neutrons and protons. When a charged extended object spins about its axis, individual charges within the object describe a circular motion; i.e., they form an electric current loop. How this current loop causes a magnetic dipole moment is shown in analogy to the electric case.

Based on Faraday's interpretation of electricity, an electric force results when a point charge is placed in an electric field. In turn, this allows us to define an electric field as the electric force per unit charge. This approach cannot be used to define a magnetic field because there are no separate magnetic monopoles. Thus, to develop an analogy to electricity, we need to look at the next simplest charge arrangement, which is the electric dipole with two equal but opposite charges at a separation distance d. The combination of charge q and distance d characterizes the electric dipole in the form:

$$\vec{\mu}_{el} = q\vec{d},$$

in which the vector $\vec{\mu}_{el}$ is called the **electric dipole moment** with unit C m. The position vector $\vec{d}$ points from the positive to the negative charge. For the electric dipole, the electric field is determined as illustrated in Fig. 23.13: the field vector points at each position toward the negative charge. The electric field lines form the characteristic structure shown in the figure's right-hand sketch. The electric field of a dipole is **anisotropic**; i.e., it is not varying in the same manner in all directions away from the dipole: perpendicular to the dipole the electric field increases faster than in any other direction.

Magnetic fields displaying the same structure of field lines occur, for example in Fig. 23.14 for the magnetic field of planet Earth. The similarity between the fields in Figs. 23.13 and 23.14 leads to the (oversimplifying) model of Earth as a bar magnet. The magnetic field of a bar magnet is shown in Fig. 23.15(a). The sketch illustrates why the magnetic field in this case is equivalent to the electric field of a dipole: a bar magnet consists of a north pole and a south pole at a fixed distance d from each other, i.e., a structure very similar to the two separated charges of an electric dipole.

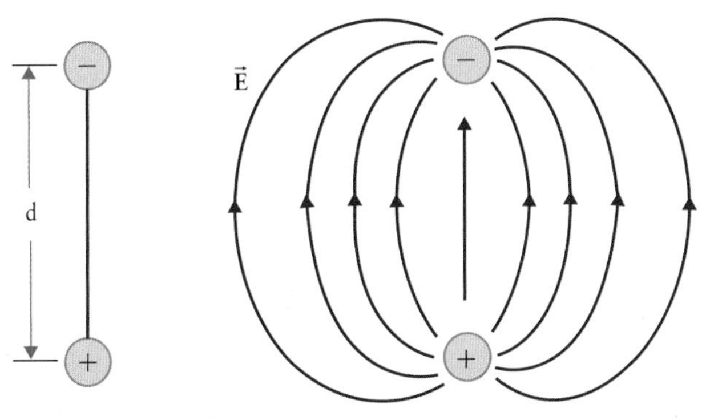

Figure 23.13 Electric field lines for an electric dipole.

A magnetic field of the same structure for a solenoid is found in Chapter 21. Reducing a solenoid to a single current loop leads to a similar field, as illustrated in Fig. 23.15(b). The similarities of the magnetic field with the electric field of an electric dipole allow us to identify bar magnets and current loops as **magnetic dipoles**.

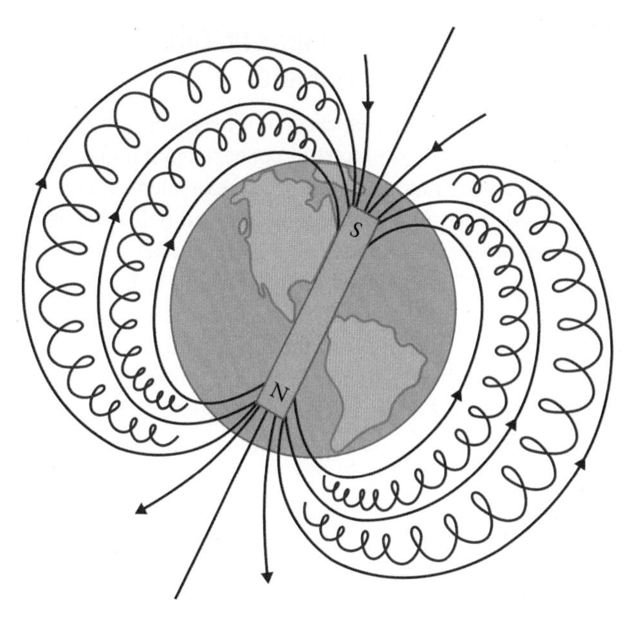

Figure 23.14 The magnetic field of Earth, for simplicity indicated as if it were generated by a bar magnet located in the planet. The real origin of the magnetic field must differ from the sketch, as no permanent magnet can exist at the temperatures found below Earth's crust.

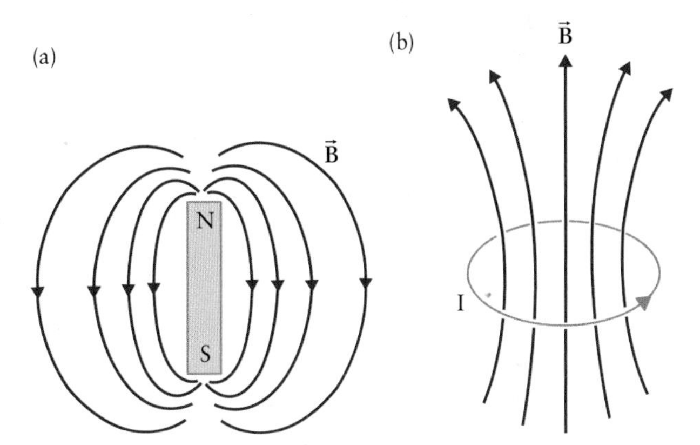

Figure 23.15 Magnetic field lines (a) of a bar magnet and (b) of a circular current loop indicated by current I.

In analogy to the electric case, a **magnetic dipole moment** is identified. For the current loop, the current replaces the charges in the electric case, and the area (expressed as a magnitude A and a normal vector $\vec{n}^0$) replaces the distance vector along the axis:

$$\vec{\mu}_{mag} = IA\vec{n}^0 \qquad [23.18]$$

The area A is defined as the area enclosed by the current loop. In the case of an atomic nucleus, a current loop is a good model because the protons describe circular paths around the axis of the spin.

KEY POINT

The magnetic dipole concept is developed in analogy to the electric dipole.

23.5.2: Torque on a Magnetic Dipole in an External Magnetic Field

Next we quantify the interaction of a magnetic dipole with an external field. We develop this concept again for the electric case first. A dipole represents an extended object. An external field causes a force acting on the dipole. We know from our earlier discussions that a force can act in two different ways on an extended object, causing either a linear acceleration or a rotation. The latter case occurs when the external force does not act in the direction toward the axis of the extended object, therefore causing a torque.

A torque acts on an electric or magnetic dipole when the external field and the axis of the dipole are not parallel. This is illustrated in Fig. 23.16. In part (a) of the figure, an electric dipole is shown with an external electric field $\vec{E}_{ext}$ acting on the dipole such that vector $\vec{\mu}_{el}$ and the electric field vector form an angle φ. With the torque, the dipole moment, and the electric field all representing vectors, a proper formulation of the relation of these three quantities is again based on a vector product. We write the magnitude of the torque first:

$$\tau = \mu_{el} E_{ext} \sin\varphi = qdE_{ext} \sin\varphi.$$

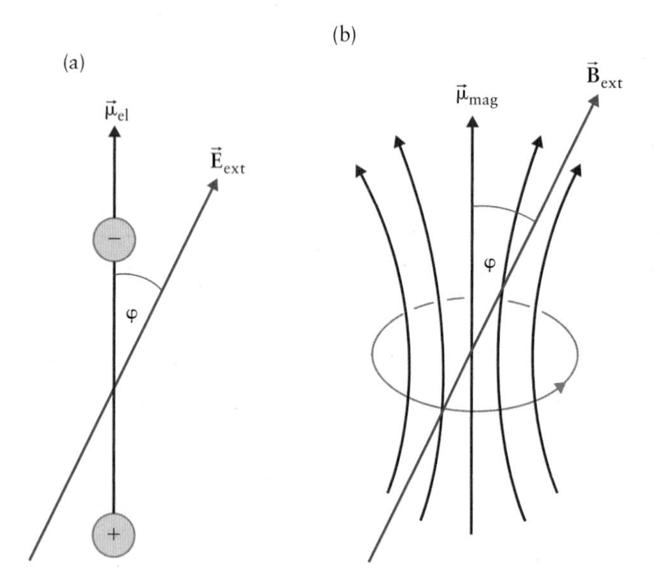

Figure 23.16 (a) Electric dipole moment $\vec{\mu}_{el}$, oriented with an angle φ relative to an external electric field $\vec{E}_{ext}$. (b) Magnetic dipole moment $\vec{\mu}_{mag}$ of a current loop, which is oriented with an angle φ relative to an external magnetic field $\vec{B}_{ext}$.

Thus, the magnitude of the torque $\vec{\tau}$ is zero when $\varphi = 0°$, i.e., when the dipole is fully aligned with the external electric field. For any other orientation of the dipole, the net torque should cause a rotation of the dipole until it is aligned with the external field. The torque is written in vector notation as:

$$\vec{\tau} = [\vec{\mu}_{el} \times \vec{E}_{ext}].$$

The vector product connects the three vectors via the right-hand rule.

The magnetic case is analogously discussed in Fig. 23.16(b). Shown is a current loop that causes a magnetic dipole moment as quantified in Eq. [23.18]. An external magnetic field $\vec{B}_{ext}$ (for example, caused by a horseshoe magnet) acts under an angle φ on the magnetic dipole $\vec{\mu}_{mag}$. The torque $\vec{\tau}$ is again a vector with magnitude:

$$\tau = \mu_{mag} B_{ext} \sin\varphi = IAB_{ext} \sin\varphi. \qquad [23.19]$$

The torque disappears when the external magnetic field and the normal vector, representing the area enclosed by the loop, are parallel; i.e., $\varphi = 0°$. Thus, as we saw for water molecule dipoles that align themselves with an external electric field, we expect magnetic dipoles (be they small iron pieces as in a compass, or small circular current loops as in atomic nuclei) to align with the external magnetic field. Eq. [23.19] is written in vector notation as:

$$\vec{\tau} = [\vec{\mu}_{mag} \times \vec{B}_{ext}]. \qquad [23.20]$$

The magnetic dipole moments of atomic nuclei with a spin also align in external magnetic fields. This is illustrated in Fig. 23.17 for hydrogen. Note that part (b) of the figure emphasizes that two orientations occur: one with parallel and one with anti-parallel alignment. The sketch of Fig. 23.17 oversimplifies the situation. For an external magnetic field along the z-axis, the actual quantum mechanical orientations of the magnetic dipole moment are shown in Fig. 23.18 for a nucleus with spin $\hbar/2$: one orientation has a z-component parallel to the magnetic field and the other orientation has a z-component anti-parallel to the magnetic field. The quantum mechanical presence of just two states for a spin-$\hbar/2$ system has been illustrated experimentally by Stern and Gerlach for the atomic spin.

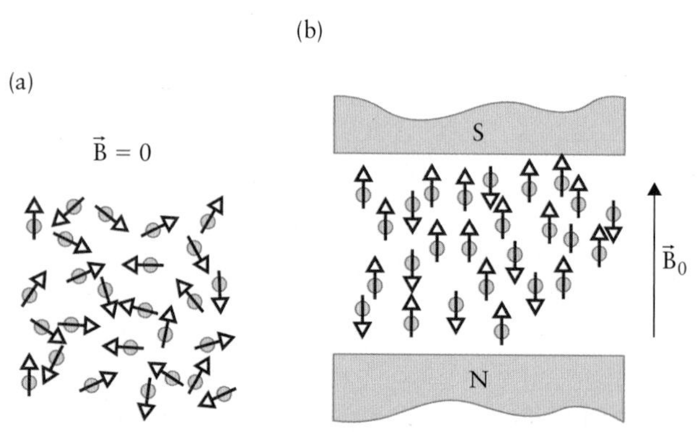

Figure 23.17 Sketch of the alignment of the magnetic dipole moments of a macroscopic sample (a) in space free of a magnetic field, and (b) in an external magnetic field $\vec{B}_0$.

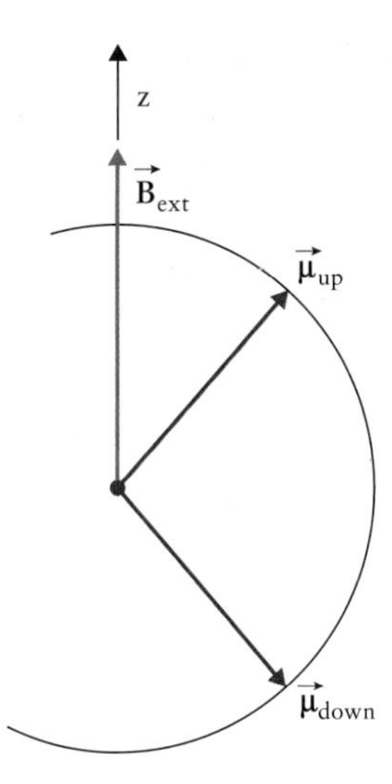

Figure 23.18 Quantum mechanically allowed orientations of a spin-$\hbar/2$ nucleus with the external magnetic field oriented along the $+z$-direction.

23.5.3: Larmor Precession and Gyromagnetic Ratio

We apply the mechanical properties of the top from the beginning of this chapter to those of a magnetic dipole. We start with a spinning object responding to an external torque by changing its angular momentum in the direction of the torque. We quantified the angular frequency of this motion, called precession, in Eq. [23.17]. Substituting Eq. [23.19] for the torque, and noting that the angle φ in Fig. 23.16(b) is the same as angle α in Fig. 23.11, we find:

$$\omega_{\mathrm{p}} = \frac{\tau}{L\sin\alpha} = \frac{\mu_{\mathrm{mag}} B_{\mathrm{ext}}}{L} = \gamma B_{\mathrm{ext}}, \qquad [23.21]$$

in which we substituted $\gamma = \mu_{\mathrm{mag}}/L$, where γ is called the **gyromagnetic ratio**. In the nuclear case, the angular frequency ω_{p} is called **Larmor frequency** (named for Sir Joseph Larmor). The gyromagnetic ratio is a constant and depends only on the specific nucleus considered. For example, for protons we find:

$$\frac{\gamma}{2\pi} = 42.58\frac{\mathrm{MHz}}{\mathrm{T}}.$$

EXAMPLE 23.6

Determine the Larmor frequency of protons in a magnetic field of (a) 1 T, and (b) 2.349 T [typical nuclear magnetic resonance (NMR) magnet].

Solution

Solution to part (a): Substituting the magnitude of the magnetic field in Eq. [23.21] yields ω_{p} = 267.5 rad/s. We

continued

can read from the gyromagnetic ratio for protons that the magnetic dipoles precess with a frequency of 42.58 MHz about the direction of the external magnetic field.

Solution to part (b): We multiply 42.58 MHz/T by B = 2.349 T. This yields 100 MHz. This even value is the reason this particular magnitude of the magnetic field is chosen in NMR.

KEY POINT

The Larmor frequency describes the precession of a nuclear spin in an external magnetic field. It is independent of the angle between the magnetic moment and the external field, but depends for a specific nucleus on the magnitude of the external field.

23.5.4: Energy of the Nuclear Spin in a Magnetic Field

We establish the energy for a nucleus with a magnetic dipole moment in a magnetic field in analogy to an electric dipole that is brought into an electric field. For simplicity, we confine both discussions to uniform fields. The energy of an electric dipole in a uniform electric field is established with Fig. 23.19. The electric field lines are horizontal, with the electric field directed toward the right. The direction of the electric dipole forms an angle α with the direction of the field lines. Using the definition of the electric energy of a point charge for each of the two point charges forming the dipole we find:

$$E_{\mathrm{el}} = q_+V_+ + q_-V_- = q(V_+ - V_-),$$

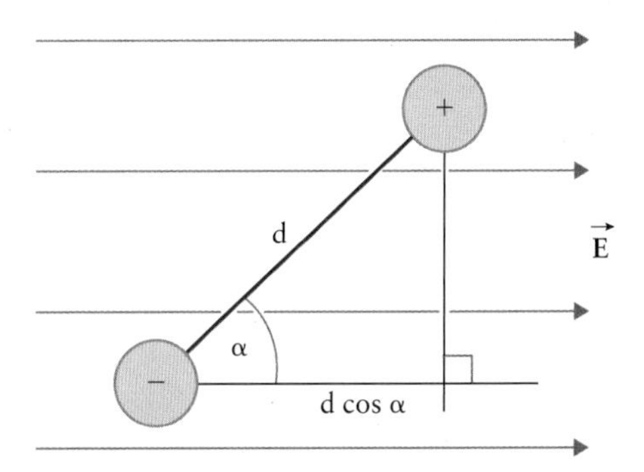

Figure 23.19 An electric dipole in a uniform external electric field. This sketch allows us to calculate the electric energy of the dipole as a function of its orientation relative to the field.

in which V_+ and V_- are the electric potential at the positions of the positive and negative charge respectively. The electric energy, and therefore the electric potential, varies linearly in a uniform electric field. Thus, we need to find the difference in the position along the direction of the external field for the two charges forming the dipole in Fig. 23.19. As the figure illustrates, the difference in position is $d\cos\alpha$. With this distance we can rewrite the electric energy in the form:

$$E_{el} = q(|\vec{E}|\,d\cos\alpha) = \vec{\mu}_{el} \bullet \vec{E},$$

in which the charge, q, and the distance between the charges, d, are combined to the electric dipole moment. The $\cos\alpha$ term is an indication of a dot product between two vectors, in the present case the electric field and the electric dipole moment. The **energy of a magnetic dipole in a uniform magnetic field** is written in analogy:

$$E_{mag} = \vec{\mu}_{mag} \bullet \vec{B} = IAB\cos\alpha. \qquad [23.22]$$

In this case, α is the angle between the direction of the magnetic field $\vec{B}$ and the unit vector $\overline{n^0}$ that is normal to the area of the current loop.

Using Eq. [23.22], we calculate the difference in energy for two nuclei that are respectively in the two allowed orientations of the spin in a magnetic field. We limit the discussion again to nuclei with spin $\hbar/2$, for which Fig. 23.18 applies. The two orientations of a spin-$\hbar/2$ nucleus are directed such that the difference in the two terms from Eq. [23.22] is:

$$\Delta E_{mag} = \mu_{mag} B_{ext}, \qquad [23.23]$$

in which B_{ext} is the magnitude of the applied magnetic field.

Eq. [23.23] is illustrated in Fig. 23.20. The figure shows the magnetic energy of a nucleus with spin $\hbar/2$ as a function of the external magnetic field for both the parallel and the anti-parallel orientations. The two orientations are labelled with the respective spin quantum numbers m_s. The energy difference for a particular external field of magnitude B_0 is highlighted by a vertical double arrow.

KEY POINT

The energy difference between the parallel and the anti-parallel states of a spin-$\hbar/2$ nucleus increases linearly with the magnitude of the external magnetic field.

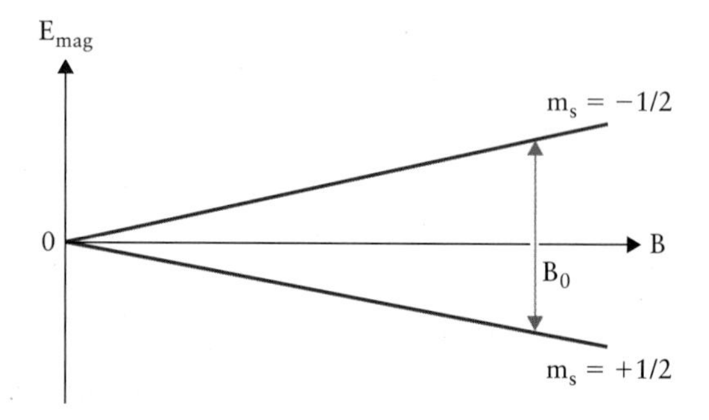

Figure 23.20 The magnetic energy of a magnetic dipole splits into two energy levels for a nucleus with spin $\hbar/2$ when brought into a uniform external magnetic field of magnitude *B*. The energy difference is proportional to the magnitude of the magnetic field.

23.6: Two-Level Systems

In Chapter 20 we introduced Bohr's model of the atom. The major difference to classical systems, such as the Maxwell–Boltzmann distribution of velocities in an ideal gas, is the discrete nature of the allowed energy levels in atoms or molecules. We expect a system that has only two allowed energy levels to be mathematically easier to treat than the atomic model. An extensive discussion of the properties of two-level systems exists in the literature; they serve as model systems in a wide range of applications, including for example glasses and polymers at low temperatures.

Fig. 23.20 defines a microscopic system that can occupy one of two energy levels at any given external magnetic field. Choosing the particular value $B_{ext} = B_0$, we illustrate this system in Fig. 23.21: the lower energy level E_{low} represents a magnetic dipole aligned parallel to the external magnetic field, e.g., μ_{up} in Fig. 23.18. This energy level is called the **ground state**. The higher energy level in Fig. 23.21 represents a magnetic dipole aligned anti-parallel to the external magnetic field, e.g. μ_{down} in Fig. 23.18. This is called the **excited state**.

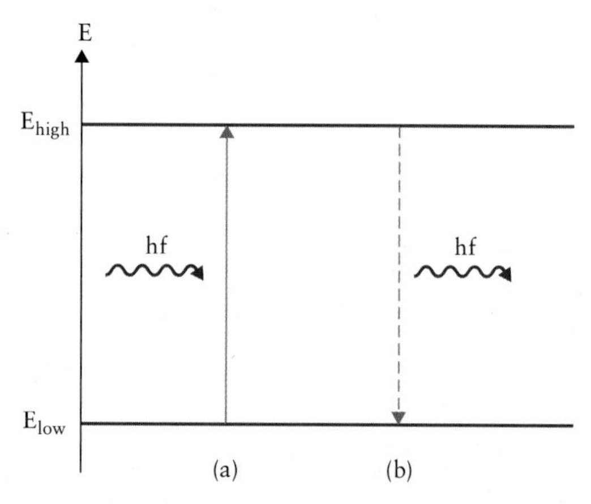

Figure 23.21 Illustration of the transition processes for a nuclear spin in an external magnetic field: absorption of radiation allows us to excite the nucleus (spin flips into anti-parallel orientation), and emission of radiation allows the nucleus to return to the ground state (spin flips into parallel orientation).

In the same manner as we discussed for electrons in atomic shells in Chapter 20, a system that occupies the excited state in Fig. 23.21 must eventually emit a photon to transfer to the ground state. Equally, external photons can be absorbed to excite a system from the ground state to the excited state. If a photon with the right amount of energy is used, a nucleus in the parallel state can be brought into the anti-parallel state; this corresponds to the flipping of the spin. The energy of the photon emitted or absorbed is hf; the actual value is determined by the energy difference between the two states as given in Eq. [23.23]:

$$hf = \hbar\omega = \mu_{mag} B_{ext}, \qquad [23.24]$$

with $\hbar = h/(2\pi)$. The angular frequency in this formula is the Larmor frequency. We use the definition of the gyromagnetic ratio for protons to rewrite Eq. [23.24] in the form:

$$\Delta E_{mag} = \hbar \gamma B_{ext}. \qquad [23.25]$$

EXAMPLE 23.7

What is the magnitude of the magnetic dipole moment for a proton?

Solution

From Eqs. [23.24] and [23.25] we know $\mu_{mag} = \hbar\gamma$. We substitute the gyromagnetic ratio for protons in this formula:

$$\mu_{mag} = \hbar\gamma = \left(42.58 \tfrac{\mathrm{MHz}}{\mathrm{T}}\right)(6.63 \times 10^{-34}\,\mathrm{J\,s})$$
$$= 2.82 \times 10^{-26}\,\mathrm{A\,m^2}.$$

The analysis of large ensembles of two-level systems leads to predictions of their macroscopically observable properties. Two steps are required to establish these predictions: In the first step, the occupation ratio of the two states in thermodynamic equilibrium is determined. This allows us to determine the excess occupation of the ground state. In the second step, macroscopic properties of the system are calculated by multiplying the excess ground state occupation with the difference in the microscopic property of interest. For a magnetic spin system, a macroscopic property of interest is the magnetization of a particular sample. The magnetization of a system in thermodynamic equilibrium is determined in the remainder of this chapter. Then, in Chapter 27, we consider the dynamics of the two-level system outside this equilibrium.

23.6.1: Occupation of the Two Energy Levels

A macroscopic system with two energy levels in thermodynamic equilibrium is fully determined when we know the relative probability to find microscopic systems in either of the two possible energy states. Ludwig Boltzmann showed that the **occupation ratio** of the two states depends only on the **transition energy** of the states, $\Delta E_{transition}$, and the temperature T:

$$\frac{N_{excited}}{N_{ground}} = \exp\left\{-\frac{\Delta E_{transition}}{kT}\right\}, \qquad [23.26]$$

in which kT is the product of the Boltzmann constant and the temperature. The energy difference of the two states, ΔE, is a system-specific constant in most cases but depends on the external magnetic field in the case of nuclear spin states, as shown in Fig. 23.20.

In conceptual discussions in physics and chemistry, the transition energy is compared to kT, which is a measure of the thermal energy available per particle. We distinguish two limiting cases: in case 1, $\Delta E_{transition} \ll kT$; and in case 2, $\Delta E_{transition} \gg kT$. In case 1, enough thermal energy is present that every particle can easily undergo a transition from the ground to the excited state without the need of externally provided energy. Thus, the macroscopic system is fully excited with an (almost) equal number of particles in the excited and the ground states in Fig. 23.21:

$$\frac{N_{excited}}{N_{ground}} = e^0 = 1 \Rightarrow N_{excited} = N_{ground}.$$

In case 2, the thermal energy is insufficient to allow particles to obtain the energy to undergo an excitation. This means they are essentially all in the ground state:

$$\frac{N_{excited}}{N_{ground}} = e^{-\infty} = 0 \Rightarrow N_{excited} = 0.$$

We find examples for both cases in physics and chemistry: rotational excitation states for gas molecules are separated from their ground state by very small energies, falling into the microwave range. Thus, most molecules occupy excited rotation states at room temperature. On the other hand, electronic excitation states we discussed in Chapter 20 are separated by large energy gaps from the ground state (in the visible to UV range), and thus atoms and molecules at room temperature occupy the ground state exclusively.

The transition energy ΔE between the two states of a spin-$\hbar/2$ nucleus in an external magnetic field is small in comparison to transitions between electronic states in atoms. Even in strong magnets with $B = 10$ T, the frequency of the photon causing a transition in Fig. 23.21 is only 500 MHz, and for typical magnetic fields of $B = 1$ T the transition occurs at 50 MHz.

EXAMPLE 23.8

Confirm quantitatively that the nuclear spins in a typical external magnetic field are almost fully excited. Assume a large magnetic field, e.g. 10 T, with a corresponding excitation frequency of 500 MHz. Further assume that the sample is at human body temperature, 37°C, during the experiment. This assumption is justified in MRI since cooling of human patients is not possible without causing damage to live tissue.

continued

Solution

We evaluate the two terms we need to compare in Boltzmann's exponential factor separately:

$$hf = (500\ \text{MHz})(6.63\times10^{-34}\ \text{Js}) = 3.3\times10^{-25}\ \text{J}$$

for the energy of the transition, and:

$$kT = \left(1.38\times10^{-23}\ \tfrac{\text{J}}{\text{K}}\right)(310\ \text{K}) = 4.3\times10^{-21}\ \text{J}$$

for the thermal energy per microscopic system. Thus, the thermal energy exceeds the energy needed for the transition between the two states of a spin in an external magnetic field by more than a factor of 10^4. As a consequence, both states in Fig. 23.21 are almost equally occupied in a tissue sample, with a difference in favour of the ground state of less than 10 ppm (parts per million), as derived from Boltzmann's equation in the next example.

23.6.2: Equilibrium Tissue Magnetization

With the excess number of nuclear spins that populate the ground state established in the previous section, we now determine the macroscopic **magnetization** of a tissue sample. The nuclear spins in the ground state are those that align parallel to the magnetic field in Fig. 23.18; the nuclear spins in the excited state align anti-parallel. Even though the two populations are very close, a system without external excitation has a small excess number of spins in the ground state; these constitute an overall magnetic effect in a macroscopic sample. Magnetization can be understood in analogy to the electric case of the polarization of dielectric materials.

In Chapter 18 on electricity, we discussed the effect of a static electric field on materials placed between the plates of a capacitor. We found that all materials respond to the external electric field because they can be polarized; in particular, we noted a very strong effect in the case of water because the water molecule is polar and aligns itself in an external field, such that the partially positive end near the two hydrogen atoms points toward the negative capacitor plate. As proposed by Michael Faraday, this effect is quantified by introducing a dielectric constant κ, which is a materials constant. Faraday observed that the electric properties of a system that includes dielectric materials can be described with the same formalism derived for electric systems in vacuum, except that the permittivity of vacuum, ε_0, has to be replaced by the permittivity of the used material, $\kappa\varepsilon_0$. All dielectric constants except for a vacuum are greater than 1, $\kappa > 1$.

The analogous magnetic case is slightly more complex. Again, materials brought into a magnetic field respond to that magnetic field, often in the same manner as the electric case by aligning their magnetic dipoles along the external magnetic field. Materials that behave in this way are called **paramagnetic**: they display a magnetization in the external field that is parallel to that field. However, there are also two other types of materials: diamagnetic and ferromagnetic materials. **Diamagnetic materials** display a magnetization that is anti-parallel to the external magnetic field; ferromagnetic materials behave like paramagnetic materials but show a much larger magnetization that no longer is proportional to the magnitude of the external magnetic field.

Materials magnetization is quantified in a manner similar to Faraday's approach to the dielectric effect: the permeability of vacuum μ_0 is supplemented by the **susceptibility** χ in the form:

$$\mu_0 \Rightarrow (1+\chi)\mu_0.$$

Table 23.2 shows a few values. Diamagnetic materials have values of $\chi < 0$, indicating their opposite behaviour compared to the paramagnetic materials: to insert a diamagnetic material such as bismuth in a magnetic field, work has to be done against a force that tries to push the metal out. In turn, a paramagnetic material such as platinum releases energy when it is drawn into a magnetic field.

TABLE 23.2

Magnetic susceptibility values χ for diamagnetic and paramagnetic materials. Note that χ is a dimensionless quantity.

Material	χ
Diamagnetic materials	
Bismuth	-1.4×10^{-5}
Water	-7.2×10^{-7}
Nitrogen gas	-3.0×10^{-10}
Paramagnetic materials	
Platinum	$+1.93 \times 10^{-5}$
Liquid oxygen	$+3.6 \times 10^{-4}$
Oxygen gas	$+1.4 \times 10^{-7}$

Diamagnetic materials have typically small χ values that do not depend on the temperature. For paramagnetic materials, the susceptibility is proportional to $1/T$; this is called **Curie's law** in honour of Pierre Curie. Ferromagnetic materials depend on the temperature in a more complex fashion, showing a phase transition to paramagnetic behaviour above a transition temperature (called Curie temperature).

The magnetization of a macroscopic sample is defined as its total magnetic moment divided by its volume:

$$M = \frac{1}{V}\Sigma_n \mu_{\text{mag},n} = \frac{1}{V}|\Delta N|\,\hbar\,\gamma.$$

The last term estimates the magnetization by multiplying the number of excess magnetic moments in a volume by the value of each single magnetic moment, $\mu_{\text{mag}} = \hbar\gamma$.

Substituting $|\Delta N|$ on the right-hand side allows us to study the temperature dependence of the magnetization:

$$M = \frac{\hbar^2 \gamma^2 B_{ext}}{kT} \frac{N_{ground}}{V},$$

in which N_{ground}/V is very close to ½ the number density of protons in the sample. This equation is only an estimate of the magnetization; a more exact, quantum mechanical treatment of the magnetization for a spin-$\hbar/2$ system (e.g. protons) leads to a proper formulation of Curie's law:

$$M = \frac{\hbar^2 \gamma^2 B_{ext}}{4kT} \frac{N_{ground}}{V} \quad [23.27]$$

KEY POINT

The magnetization of a sample is proportional to the magnitude of the external magnetic field, depends on the spin density in the sample, and is inversely proportional to the temperature.

SUMMARY

DEFINITIONS

- Notation of nuclei: ${}^A_Z X$

  A is the mass **number** (number of nucleons in nucleus), Z is the atomic number (number of protons in nucleus), N is the number of neutrons, with $A = Z + N$

- Decay processes
  - α-decay:

    $${}^A_Z X \rightarrow {}^{A-4}_{Z-2}\text{Y} + {}^4_2\text{He}$$

  - β-decay mechanism (leading to an electron or a positron):

    $$\beta^+ \text{ decay: } {}^A_Z X \rightarrow {}_{Z-1}^{A} X + e^+$$

    $$\beta^- \text{ decay: } {}^A_Z X \rightarrow {}^A_{Z+1} X + e^-$$

  - γ-decay:

    $${}^{Am}_Z X \rightarrow {}^A_Z X + \gamma$$

  - Electron capture (labelled ε-process):

    $${}^A_Z X + e^- \rightarrow {}_{Z-1}^{A} X + v$$

- Angular momentum
  - Magnitude:

    $$L = pr \sin\varphi = mvr \sin\varphi$$

    $p = mv$ is the linear momentum of the object, r is the distance from the axis of the circular motion, and φ is the angle between the momentum vector and the radius vector, $\varphi = \sphericalangle(\vec{p}, \vec{r})$.
  - In vector notation:

    $$\vec{L} = (\vec{r} \times \vec{p})$$

- Magnetic dipole moment:

  $$\vec{\mu}_{mag} = IA\vec{n}^0$$

- The area is given by its magnitude A (area enclosed by loop) and a normal vector $\vec{n}^0$, I is the current.
- Gyromagnetic ratio γ (for protons):

  $$\frac{\gamma}{2\pi} = 42.58 \frac{\text{MHz}}{\text{T}}$$

UNITS

- Activity: Bq (becquerel) with 1 Bq = 1 decay/s.
- Angular momentum L: kg m^2/s = J s
- Torque τ: N m

LAWS

- Yukawa radius:

  $$r_Y = \frac{\hbar}{m_0 c}$$

  $m_0 c$ is the momentum of the particle, $\hbar = h/(2\pi)$
- Radioactive decay law:
  - for activity A as a rate equation (differential form):

    $$A = -\frac{dN}{dt} = \lambda N$$

    λ (in unit 1/s) is decay constant.
  - for activity A in integrated form:

    $$A(t) = A_0 e^{-\lambda t}$$

  - for number of radioactive nuclei in sample, N:

    $$N(t) = N_0 e^{-\lambda t}$$

    N_0 is number of radioactive nuclei at time $t = 0$.
- Half-life $T_{1/2}$: time after which 50% of initial radioactive nuclei have decayed:

  $$T_{1/2} = \frac{\ln 2}{\lambda}$$

- Mean lifetime: $T_{mean} = \lambda^{-1}$

  $$T_{mean} = \frac{1}{\lambda} = 1.44 T_{1/2}$$

- Torque:
  - Condition for the conservation of the angular momentum:

$$\vec{\tau} = \frac{d\vec{L}}{dt}$$

  Angular momentum is conserved if no external torque acts on the system.
  - Angular frequency of precession (mechanical system):

$$\omega_p = \frac{\tau}{L \sin\alpha}$$

  α is angle of tilt of the precessing system.
  - Larmor frequency (magnetic system):

$$\omega_p = \frac{\tau}{L\sin\alpha} = \frac{\mu_{mag} B_{ext}}{L} = \gamma B_{ext}$$

  with $\gamma = \mu_{mag}/L$ the gyromagnetic ratio.
  - On a magnetic dipole in a uniform magnetic field B_{ext} (magnitude):

$$\tau = \mu_{mag} B_{ext} \sin\phi = I\,A\,B_{ext} \sin\phi$$

  - On a magnetic dipole in a uniform magnetic field (in vector notation):

$$\vec{\tau} = [\vec{\mu}_{mag} \times \vec{B}_{ext}]$$

- Magnetic energy of a magnetic dipole in a uniform magnetic field $\vec{B}$:

$$E_{mag} = \vec{\mu}_{mag} \bullet \vec{B} = I\,A\,B\cos\alpha$$

  α is the angle between the direction of the magnetic field $\vec{B}$ and the unit vector $\overrightarrow{n^0}$, which is normal to the area A of the current loop. I is the current.
- Energy split of a two-level spin system (spin $\hbar/2$) in a uniform magnetic field:

$$\Delta E_{mag} = \mu_{mag} B_{ext} = \hbar\gamma B_{ext}$$

  B_{ext} is the magnitude of the applied magnetic field, $\hbar = h/(2\pi)$.
- Boltzmann's occupation ratio of a two-level system:

$$\frac{N_{excited}}{N_{ground}} = \exp\left\{-\frac{\Delta E_{transition}}{kT}\right\}$$

  kT is the product of the Boltzmann constant and the temperature, ΔE is the energy difference of the two states of the system.
- Magnetization of a given volume of a spin-$\hbar/2$ system (Curie's law):

$$M = \frac{\hbar^2\,\gamma^2 B_{ext}}{4kT}\frac{N_{ground}}{V}$$

  with $\mu_{mag} = \hbar\gamma$, N_{ground}/V is very close to ½ the number density of protons in the sample.

CONCEPTUAL QUESTIONS

Q–23.1. (a) What does one mean when one says that two nuclides are isotopes? (b) What are the numbers of protons and neutrons in the single nucleus of the nuclide $^{14}_{6}C$. (c) Write the symbol for a different carbon isotope.

Q–23.2. (a) A figure skater begins to spin at an angular velocity of 9 rad/s with his arms extended. If his moment of inertia with arms folded is 60% of that with arms extended, what is his angular velocity when he folds his arms? (b) By how much does his kinetic energy change during the process of folding his arms?

Q–23.3. How is the stability of a radioactive nucleus toward spontaneous decay measured?

Q–23.4. Does the quantum spin of a particle have a classical analog?

Q–23.5. What is (are) the physical difference(s) between a paramagnetic material and a diamagnetic material?

ANALYTICAL PROBLEMS

P–23.1. Two uranium isotopes, ^{235}U and ^{238}U, and ^{234}Th are the only three primordial nuclides heavier than bismuth. The decay sequences in all four cases lead ultimately to stable lead isotopes. Including only decay events that occur with 10% or higher branching ratio, the decay sequences are

(i) uranium-238 ($T_{1/2} = 4.47 \times 10^9$ years)
(ii) uranium-235 ($T_{1/2} = 7.04 \times 10^8$ years)
(iii) thorium-232 ($T_{1/2} = 1.405 \times 10^{10}$ years).

(a) Which sequence contains the most α-decays? (b) How much of the uranium-238 formed during the supernova event immediately preceding the formation of our solar system 4.6 billion years ago and then accumulated in planet Earth is still there? What about uranium-235?

P–23.2. The nucleus of the deuterium atom consists of one proton and one neutron. What is the binding energy of this nucleus if the mass of the deuterium nucleus is given as 2.014102 u? Note that the atomic unit u has a value u = $1.6605677 \times 10^{-27}$ kg.

P–23.3. Nuclear waste from power plants may contain ^{239}Pu, a plutonium isotope with a half-life of 24 000 years. How long does it take for the stored waste to decay to 10% of its current activity level?

P–23.4. We assume a 500 MBq ^{99m}Tc source with $T_{1/2} = 6.02$ h and the emitted γ-energy at 140.5 keV. (a) Derive a general expression for the total amount of energy a radioactive source emits over the first three half-lives if it decays via isomeric transition. (b) Using the expression in part (a), calculate the specific value for the given source.

P-23.5. A five-day therapeutic dose is prepared using a radon seed. For circumstances beyond the control of the radiology department, the administration of the treatment is delayed by a full day (24 hours). Given that the decay constant of radon is $2.1\times10^{-6}\ \text{s}^{-1}$, how long should the radon seed be left in position to deliver the necessary dose?

P-23.6. The first nuclear reaction performed by Rutherford in 1919 can be written in the following form: ${}^{14}_{7}\text{N} + {}^{4}_{2}\text{He} \rightarrow {}^{17}_{8}\text{O} + {}^{1}_{1}\text{H}$. (a) Describe the nucleus ${}^{17}_{8}\text{O}$. (b) The total rest masses of both the right and the left hand sides of the equation are given by ${}^{14}_{7}\text{N} + {}^{4}_{2}\text{He} = 18.00568\ \text{amu}$ and ${}^{17}_{8}\text{O} + {}^{1}_{1}\text{H} = 18.00696\ \text{amu}$. Describe the reaction, and explain under what conditions this reaction can occur. (*Note:* amu stands for atomic mass unit.)

P-23.7. Calculate the mass of the neutron produced during the reaction ${}^{11}_{5}\text{B} + {}^{4}_{2}\text{He} \rightarrow {}^{14}_{7}\text{N} + {}^{1}_{0}n$. In order to find the mass of the neutrons you are given the masses of boron, nitrogen, and alpha particle, which are 11.01280 amu, 14.00752 amu, and 4.00387 amu, respectively. Moreover, you are also given the kinetic energy of the incident α-particle, 5.250 MeV, and the energies of the resultant neutron and nitrogen atom are to be 3.260 MeV and 2.139 MeV.

P-23.8. A magnetic dipole moment vector with amplitude $\mu = 6.28 \times 10^{-5}\ \text{Am}^2$ is placed perpendicular ($\theta_i = 90°$) to a magnetic field $\vec{B} = 24$ T. The dipole is allowed to rotate until it reaches a final angle θ_f with the magnetic field $\vec{B}$. If the interaction energy is $E = -5.16\times10^{-4}$ J, calculate the magnitude of the torque experienced by the dipole moment.

P-23.9. The magnetic moment of a small current-carrying wire is $4.2 \times 10^{-4}\ \text{Am}^2$. The plane containing the wire makes an angle θ with an applied magnetic field $\vec{B}$ of amplitude 0.65 T. The magnitude of the torque exerted by the magnetic field on the loop is 9.0×10^{-5} Nm. Calculate the potential energy of the system.

P-23.10. Rutherford designed an experiment through which 7.69 MeV collimated alpha particles from a ${}^{214}\text{Po}$ source were scattered by a thin gold foil. Determine the distance of closest approach of an alpha particle to a gold nucleus.

P-23.11. The nuclear alpha decay of ${}^{239}_{94}\text{Pu}$ produces ${}^{235}_{92}\text{U}$ plus an alpha particle and some energy. (a) How much energy Q is released in this nuclear reaction? (b) What are the energies of the alpha particle and the daughter nucleus? (c) What is the speed of the alpha particle?

P-23.12. ${}^{41}\text{Ar}$ decays by β^--emission to an energy state of ${}^{41}\text{K}$ 1.293 MeV above the ground state. Determine the maximum kinetic energy of the β^--particle emitted.

P-23.13. The activity of a radioisotope is found to decrease by 40% of its original value in 20 days. (a) Calculate the decay constant. (b) What is the half-life? (c) What is the mean life?

P-23.14. (a) How energetic must a photon be in order to excite the innermost hydrogen electron into the first excited state in the Bohr model? (b) What is the classical orbital velocity of the excited electron?

P-23.15. A nuclear physicist would like to conduct several experiments on the stable element ${}^{56}_{26}\text{Fe}$. (a) The first experiment conducted attempts to remove a single neutron. What is the energy required to strip a single neutron? (b) How about removing a single proton: how much energy is required? (c) How much energy is required to break down the iron into its most basic elements? (d) How much energy would be necessary to fission the iron atom into to identical lighter nuclides ${}^{28}_{13}\text{Al}$? (*Note:* ${}^{56}_{26}\text{Fe}$: 55.934942 amu; ${}^{28}_{13}\text{Al}$: 27.981910 amu.)

ANSWERS TO CONCEPT QUESTIONS

Concept Question 23.1: The first nuclide is the stable and most abundant nitrogen isotope and the second nuclide is well known due for its role in carbon dating. The two are related to each other as isobaric nuclides; isobaric nuclides for two neighbouring elements in the periodic system can often transform into each other through a β-decay process.

Concept Question 23.2: ${}^{153}\text{Sm} \rightarrow {}^{153}\text{Eu}$ ($Z = 63$), and ${}^{89}\text{Sr} \rightarrow {}^{89}\text{Y}$ ($Z = 39$).

Concept Question 23.3: ${}^{11}\text{C} \rightarrow {}^{11}\text{B}$ ($Z = 5$), ${}^{13}\text{N} \rightarrow {}^{13}\text{C}$ ($Z = 6$), ${}^{68}\text{Ga} \rightarrow {}^{68}\text{Zn}$ ($Z = 30$), and ${}^{82}\text{Rb} \rightarrow {}^{82}\text{Kr}$ ($Z = 36$).

Concept Question 23.4: ${}^{67}\text{Ga} \rightarrow {}^{67}\text{Zn}$, ${}^{111}\text{In} \rightarrow {}^{111}\text{Cd}$ ($Z = 48$), ${}^{123}\text{I} \rightarrow {}^{123}\text{Te}$ ($Z = 52$), ${}^{125}\text{I} \rightarrow {}^{125}\text{Te}$, and ${}^{51}\text{Cr} \rightarrow {}^{51}\text{V}$ ($Z = 23$).

Concept Question 23.5(a): ${}^{182}_{74}\text{W}$ from: ${}^{186}_{76}\text{X} = {}^{186}_{76}\text{Os}$.
(b) and (c): ${}^{209}_{83}\text{Bi}$ from: ${}^{213}_{85}\text{X} = {}^{213}_{85}\text{At}$ from: ${}^{217}_{87}\text{X} = {}^{217}_{87}\text{Fr}$.

Part Six

APPLIED CLINICAL PHYSICS

In this last Part of the text, we focus on the clinical applications of some of the concepts that were introduced earlier. We do this in order to underscore the substantial role physics plays in modern medical practice, with the caveat that time and space allow only a very limited sampling of such applications. We begin with the study of X-rays in Chapter 24 and start by looking at the two physical mechanisms responsible for X-ray production. We then examine the two most relevant physical mechanisms by which X-rays interact with biological tissue, which results in the exponential decay of the X-ray beam. From there, we focus on the special role that one of these interactions—the photoelectric effect—plays in the formation of X-ray images. We close the chapter by briefly addressing the concepts of image contrast and the dose associated with clinical X-ray procedures. In Chapter 25 we introduce nuclear medicine imaging, a fundamentally physiological diagnostic tool. Aside from the basic physics involved in producing a nuclear image of the human anatomy, the single most important idea we emphasize is that we are dealing with **functional** or physiological information, rather than with anatomical details. Examples of different methods of obtaining nuclear images are reviewed. We end the chapter with an example of a nuclear image from an actual patient with advanced multiple bone cancer. In Chapter 26, we introduce the use of ionizing radiation in the treatment of malignant disease, or cancer. This technique, called **radiotherapy**, along with surgery and chemotherapy constitute the standard treatment methods for cancer today. We begin by examining elemental concepts in radiobiology, i.e., the dynamic response of cellular systems to escalating doses of ionizing radiation. We then move on to some details of the two most commonly used devices in radiotherapy today: the Co-60 unit and the linear accelerator. We also introduce some underlying physical concepts of the radiation beams used in clinical practice. We close the chapter with an actual clinical example of a patient who was treated for an aggressive brain cancer. In Chapter 27 we explore the laboratory and medical diagnostic applications of nuclear magnetic resonance (NMR). We begin by describing how single energy levels of an atomic nucleus can be split into an excited state and a ground state by imposing a strong external magnetic field. The precise magnitude of the energy splitting is a function of the electronic environment of the nucleus. In NMR, this energy difference—and hence the chemical structure of a sample—can be probed by pumping in precise bundles of radio waves such that they exactly match this difference. We say that the radio waves are in **resonance** with the nucleus, hence the technique's name. Furthermore, the time it takes for nuclei in the excited state to fall back to the ground state is unique to each different tissue type. This difference in **relaxation** times between tissues is exploited in magnetic resonance imaging (or MRI) to produce tomographic images of the anatomy. The chapter closes with examples of anatomical MRI images.

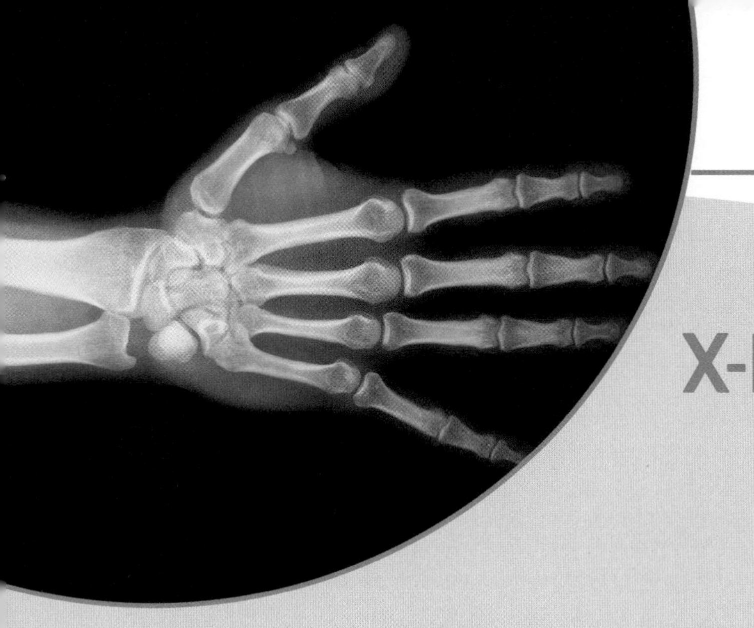

CHAPTER 24

X-Rays

X-rays are electromagnetic radiation with energies exceeding 1 keV. They are generated when fast electrons hit a metal surface, such as the anode in an X-ray tube. Both bremsstrahlung and characteristic X-rays contribute to a broad X-ray emission that is accompanied with a significant loss to thermal energy. Bremsstrahlung results when the incident electron interacts strongly with a single atom, causing it to lose a significant fraction of its kinetic energy. Characteristic X-rays occur when the incident electron collides with an electron in an inner shell such that it is removed from the atom.

X-rays are high-energy photons that in turn interact with matter through Rayleigh scattering, Compton scattering, the photoelectric effect, or pair production. The relative contribution of each of these processes depends on the photon energy and the atomic number of the target matter, yielding case-specific linear attenuation coefficients in Beer's law. The mass attenuation coefficient is introduced when we want to separate the density of the medium from the attenuation coefficient. This allows us also to define the half-value layer, a thickness at which half the incident X-rays have been removed from the beam.

Contrast and image resolution are the two parameters defined to quantify the relative yield of radiation for different parts of a non-uniform sample. Both parameters depend on the physical properties of the incident X-ray beam and can therefore be optimized for specific applications.

The interaction of X-rays with the penetrated tissue may cause tissue damage. Thus, the amount of radiation required in a clinical study, called the dose, and the rate at which this dose is administered, called the dose rate, are additional parameters that determine the selection and set-up of the imaging modality used.

The atomic model we established in Chapter 20 allows us now to approach radiology, which is a major field in medical imaging. We will look deep into the atom to establish the origin of X-rays, which are the tool of radiology due to their capability to penetrate living tissue and yet interact with the typical non-uniformity of human tissues to an extent that allows for the formation of images with sufficient contrast and spatial resolution.

At the same time, our understanding of the atomic model and the ability of atoms to form biomolecules allows us to answer basic questions about safety that underlie the practice of radiology. What effect do X-rays have on the human body? Do we have to fear the high-energy parts of the electromagnetic spectrum? If so, does this radiation reach dangerous intensity levels anywhere, e.g. near an X-ray machine in the hospital?

Based on the simple grounds of experience, the answer is both yes and no. The widespread use of X-rays for radiography, mammography, and CT scans (computed axial tomography) implies that safety in handling X-rays is possible for both the radiologist and the patient. At the same time, we know high doses of X-rays can be lethal.

X-rays cause alterations to living organisms also at lower non-lethal doses, as shown in Hermann Joseph Muller's 1927 experiment that proved genes are (artificially) mutable. Once Gregor Mendel's experiments had been rediscovered in the early years of the 20th century, an apparent discrepancy between Mendel and Darwin was noted, in that Darwin's theory needs a steady change for evolutionary progress while Mendel's theory is based on the stability of the genetic code. To test the apparent immutability of genes, Muller bombarded fruit flies with X-rays and linked new deformities in the offspring to the radiation, proving that he had artificially altered genes in the insects. In 1940, George Beadle and Edward Tatum used the same X-ray technique on a species of bread mould (*Neurospora*) to prove the correlation between genes and enzymes: the mutation of certain genes led to a lack of certain enzymes in the offspring.

We know from the medical use of X-rays, in which energies in the range of 25 keV to 150 keV are used, that this type of electromagnetic radiation penetrates biological matter more readily than light. Thus, whatever adverse effect radiation has on living organisms, the effect of light is limited to the surface (e.g., skin cancer due to

UV radiation), while the damage done by X-rays reaches much deeper. To control this effect, radiation dose would be introduced as a quantitative parameter that allows us to specify the risks involved with the various radiological methods.

24.1: Origin of X-Rays

X-rays are a type of **electromagnetic radiation** with high frequencies and small wavelengths (when we consider them to be waves) or photons with high energies (when we treat them as particles). Fig. 24.1 allows us to be more specific: the figure provides corresponding values of photon energies, wavelengths, and frequencies for all types of electromagnetic radiation. Adjacent to the visible range, the short wavelength range is called ultraviolet radiation (UV). We define radiation as γ- or X-rays when the energy of the photon exceeds 1 keV. This includes X-rays in medical applications, which lie between 25 keV (mammography) and 150 keV (high kilovoltage radiography).

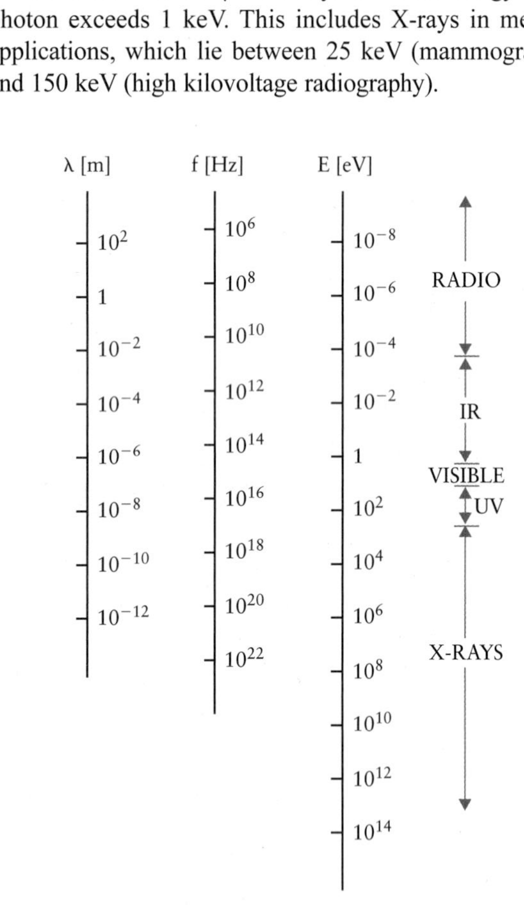

Figure 24.1 Electromagnetic spectrum showing the relation between the frequency f, the wavelength λ, and the photon energy $E = h\,f$. Note that photons with energy above 1 keV are called X-rays when they originate in the atomic shell, and γ-rays if they are the result of an event in a nucleus.

KEY POINT

X-rays are photons of energy $E > 1$ keV that originate from atomic transitions of electrons in inner shells; γ-rays are photons in the same energy range that originate from nuclear processes.

CONCEPT QUESTION 24.1

(a) Photon energies in Fig. 24.1 are given in unit electron volt (eV). Why do we use this non-standard unit? (b) What is the energy in standard units of an X-ray photon of 25 keV in mammography?

EXAMPLE 24.1

We call photons X-ray photons when they have an energy of 1.0 keV or more. (a) What is the lowest frequency of an X-ray? (b) What is the longest wavelength of an X-ray?

Solution

Solution to part (a): We use the energy–frequency relation $E = h\,f$ to calculate the frequency for a photon of 1.0 keV energy. Based on the energy conversion between units eV and joule (J), the energy of this photon is 1.6×10^{-16} J, which yields:

$$f_{\text{photon}} = \frac{E_{\text{photon}}}{h} = \frac{1.6\times10^{-16}\,\text{J}}{6.6\times10^{-34}\,\text{Js}} = 23.4 \times 10^{17}\ \text{Hz}.$$

Higher energy X-ray photons have larger frequencies because $E \cong f$.

Solution to part (b): We use the result in part (a) and the speed of light in a vacuum to determine the wavelength of the photon:

$$c = \lambda_{\text{photon}}\, f_{\text{photon}};\ \lambda_{\text{photon}} = \frac{c}{f_{\text{photon}}} = \frac{3.0\times10^{8}\ \text{m/s}}{2.4\times10^{17}\ \text{Hz}} = 1.25\,\text{nm}.$$

Other X-rays will have shorter wavelengths because they have larger frequencies, and $\lambda \cong 1/f$. Recall from Chapter 20 that Bohr's radius of the hydrogen atom is 0.05 nm. Thus, X-ray wavelengths are of magnitude comparable to the sizes of atoms. For this reason X-rays are used when analyzing crystal structures in condensed-matter physics or geophysics. Watson and Crick's discovery of the double-helix structure of the DNA molecule was based on X-ray crystallographic data.

24.1.1: Why Can Single-Electron Atoms or Ions Not Emit X-Rays?

Roentgen's original discovery showed that X-rays are generated when high-energy electrons interact with matter, or, more specifically, when these electrons interact with individual atoms. Thus, we assume X-rays should be associated with Bohr's second postulate: the X-ray photon is the result of a transition of an electron within the atom.

The first approach then is to use the calculations from Examples 24.2 and 24.3 to identify a transition that generates X-rays. The highest-energy photon we obtain from a hydrogen atom has an energy of just over 10 eV, e.g. when a positive hydrogen ion catches an electron that falls to the lowest energy state (inverse process to ionization). This energy lies in the near-UV and is a factor of more than 1000 smaller than the photon energies used in mammography.

Next we could propose using a heavier atom for which we allow the electron capture. We developed the equation:

$$E_{\text{total}}(Z) = Z^2 E_{\text{total}}(Z = 1)$$

in Example 24.4 for this case to estimate the energy released by an electron. If we boldly take the largest feasible value of Z (i.e., $Z = 83$ for bismuth), we predict a photon of energy $Z^2 E(Z = 1) = 93$ keV. An X-ray released during electron capture of this type would be sufficient to do mammography, but several other radiographic applications would not yet be possible. However, the real problem with this calculation is that we assumed the bismuth ion has no electrons before the electron capture; otherwise, the caught electron could not end up in the lowest energy state with $n = 1$. A Bi^{83+} ion is impossible to generate, let alone an amount that would provide a sufficiently intensive X-ray beam. From inorganic chemistry you know we can expect something like Bi^{3+} at best in a salt compound. When a Bi^{3+} catches an electron, photons of very small energy are emitted because the electron drops into an empty orbital in the outmost shell of the ion (the other 80 electrons of the ion fill the lowest energy states due to the minimum energy principle) and it will not be attracted by 83 positive elementary charges in the nucleus, but only by about 3 of these charges as all others are screened by the 80 electrons already present in the ion.

Thus, X-rays for medical applications must be generated in a different manner; simple electron capture by a positive heavy ion is either not possible or would not release enough energy.

24.1.2: Generating X-Rays: The X-Ray Tube in Radiography

For medical applications, X-rays are generated in **X-ray tubes**. Such a set-up is sketched in Fig. 24.2. The X-ray tube consists of an evacuated glass tube and two electrodes. The vacuum is needed because the technique is based on accelerating free electrons toward a solid target electrode. In air, an electron would not be able to travel more than one millimetre. To generate the free electrons, a current I_T is sent through a thin metallic electrode (called a **cathode**), causing thermal evaporation of electrons (Richardson effect). The electrons are then extracted and accelerated through a large potential difference, ΔV, toward a metallic **anode**. When the electrons strike the anode, X-rays are generated. The glass tube is transparent for these rays (in the same manner as it is transparent to visible light). The X-rays then travel through a collimator (and filters) toward the patient. After penetrating the tissue, they expose a detector such as photographic film behind the patient. The film is developed and used by physicians for their diagnoses.

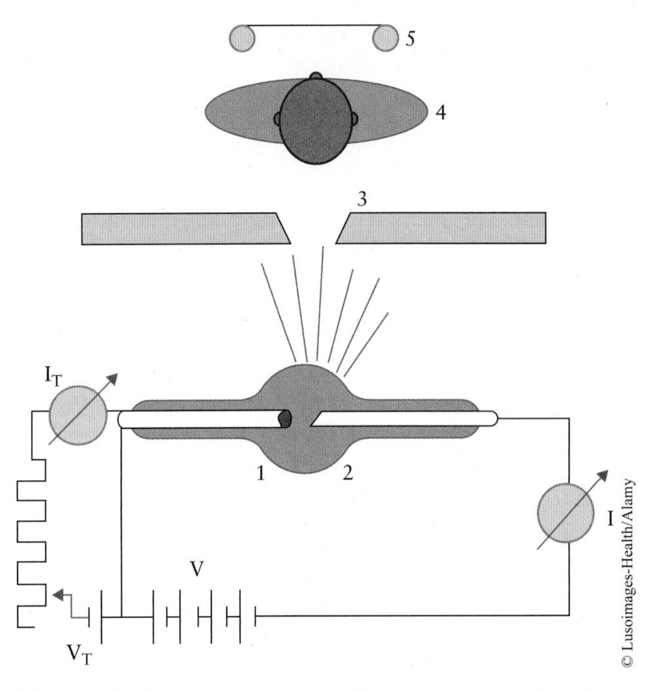

Figure 24.2 X-ray tube and sketch of its standard use in medicine. Electrons are evaporated from a hot negative electrode in a vacuum tube. The free electrons are then accelerated by a high potential difference to strike the positive electrode. The generated X-rays leave the tube, are confined by a collimator, pass through the body of the patient, and reach a photographic film.

This process generates X-rays with a wide range of energies, as illustrated in Fig. 24.3. The figure shows the result of using a tungsten anode and an X-ray tube potential difference of $\Delta V = 100$ kV between cathode and anode. This potential leads to X-rays of energy up to 100 keV

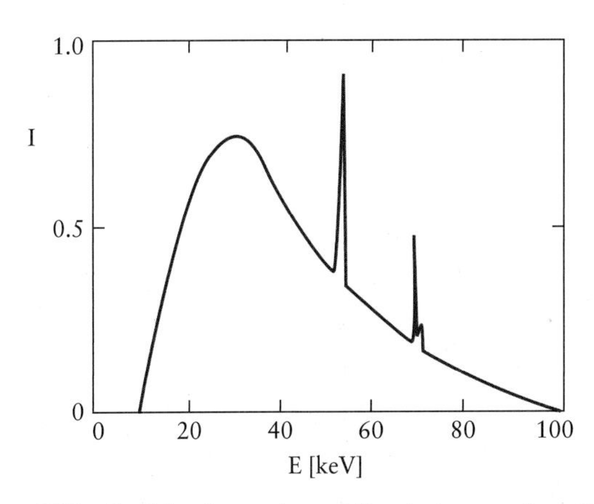

Figure 24.3 The intensity spectrum of the electromagnetic radiation emitted from a tungsten electrode bombarded by 100 keV electrons. Note the cut-off at the X-ray energy of 100 keV. Two main features characterize the spectrum: a broad peak called bremsstrahlung, and several sharp peaks labelled characteristic X-rays.

with sufficient intensity. The intensity spectrum contains several distinguishable features:

- a continuous distribution of X-ray energies from about 10 keV to a maximum energy (**cut-off energy**) at $E = 100$ keV;
- a broad peak that tails off toward larger energies; this radiation is called **bremsstrahlung**; and
- several sharp peaks around 55 keV and 70 keV. These peaks are called **characteristic X-rays**.

Both processes are based on physical phenomena we have not yet introduced. In the next two sections we use the atomic model to quantify these processes.

Figure 24.3 indicates that we do not obtain monoenergetic X-ray beams for medical imaging, but rather a wide range of energies with a non-uniform intensity distribution (**polyenergetic beam**). This means certain types of applications cannot be performed or require extensive additions or alterations to the equipment used. For example, synchrotrons provide intensive monoenergetic X-ray beams for surface chemistry studies and radioactive γ-sources allow for radiographic calibrations with a monoenergetic beam, e.g., 70 keV from ^{201}Tl or 140 keV from ^{99m}Tc.

As discussed later in this chapter, the X-ray spectrum of Fig. 24.3 is sufficient requiring only minor modifications with filters for standard radiographic applications. We have to keep the polyenergetic nature of radiographically used X-rays in mind; referring to the radiation in Fig. 24.3 as a 100 keV X-ray beam is obviously misleading. One way to use proper nomenclature in this context is to refer to the X-rays in Fig. 24.3 as a 100 kVp beam. This indicates that a **peak voltage** of 100 kV was used to accelerate electrons in the X-ray tube.

EXAMPLE 24.2

What is the speed of the electron when it strikes the tungsten anode to generate the X-ray spectrum shown in Fig. 24.3?

Solution

We discuss first how we *cannot* solve this problem. In mechanics, we discussed the conservation of energy. Applied to the electron in the X-ray tube of Fig. 24.2, the decreasing electric potential energy of the electron travelling from cathode to anode must be turned into kinetic energy. This yields:

$$\Delta E_{\mathrm{el}} = 100\ \mathrm{keV} = \tfrac{1}{2} m v^2 \Rightarrow v = \sqrt{\frac{2\Delta E_{\mathrm{el}}}{m}}$$

$$= \sqrt{\frac{2(1.6\times10^{-14}\ \mathrm{J})}{9.1\times10^{-31}\ \mathrm{kg}}} = 1.9\times10^{8}\ \mathrm{m/s}.$$

What is wrong with this result? The result, 1.9×10^8 m/s, is roughly 65% of the speed of light in a vacuum. When

continued

an object moves at a considerable fraction of the speed of light, relativistic corrections have to be applied in the calculation. These are quantified with Albert Einstein's theory of relativity, as outlined in the Appendix of Chapter 24. In the current case, the actual velocity of the electron is calculated from the equation derived for *v/c* in Example 24.8:

$$v/c = \sqrt{1 - \left[\frac{m_o c^2}{E_{\mathrm{kin}} + m_o \times c^2}\right]^2} = 0.548;$$

i.e., only about 55% of the speed of light with the remaining energy taken up by the mass of the electron. Note that the classical formula for the kinetic energy, $E_{\mathrm{kin}} = \frac{1}{2} m v^2$, should not be used for electrons with kinetic energies exceeding 10 keV, when the discrepancy of classical and relativistic speeds exceeds 1.5%.

24.1.3: Bremsstrahlung

The observation of a sharp cut-off energy at 100 keV allows us to develop a simple model for the origin of the broad peak of X-rays in Fig. 24.3. Electrons slow down as they penetrate the metallic anode. Their kinetic energy must be converted into other forms of energy in this process due to the conservation of energy. If we compare to classical systems, such as a bullet fired into a sandbag, or a billiard ball shot across a billiard table with other billiard balls, we expect at least two types of energy loss for the electrons: energy transfer to particles in the sample through collisions, and conversion of kinetic energy to thermal energy.

In the case of a billiard ball, we know that the loss of kinetic energy can be sudden, e.g. in a head-on collision with an object of same mass. If that object absorbs the energy and moves forward (elastic collision), then the total kinetic energy of the incoming particle is released in one step. For the electrons causing the spectrum in Fig. 24.3 this means we are looking for events in which 100 keV of energy is released. These exist at the upper end of the spectrum: the energy is released in the form of a photon of 100 keV, which is an X-ray photon. The conservation of energy therefore explains why this is the largest energy of X-rays observed for a 100 kVp beam; none of the electrons has a larger amount of energy to release.

Extending the corpuscle collision reasoning, the origin of the broad peak of X-ray photons with energies between about 10 keV and 100 keV (**bremsstrahlung spectrum**) can be explained. Not all electrons will be stopped suddenly in a single step. Most, like the speeding bullet in a sandbag, are slowed down more slowly, losing their kinetic energy in steps. Fig. 24.4 illustrates a typical process in which an electron loses only some fraction of its energy. When the electron comes in close proximity to a nucleus within the anode, the direction of its motion changes due to the electrostatic attraction to the positive nucleus. Deviation from a straight path constitutes an acceleration. As we saw in the context of Rutherford's model, such an accelerating electron emits electromagnetic radiation. In the case of strong acceleration near a nucleus, that radiation has short wavelengths, lying in the X-ray part of the electromagnetic spectrum. The term *bremsstrahlung* was adopted as only electrons with a significant negative acceleration contribute to X-ray intensity. Most other interaction processes, such as the interaction with valence electrons in the anode atoms, lead to loss of energy in small amounts, which is converted into heat rather than radiation.

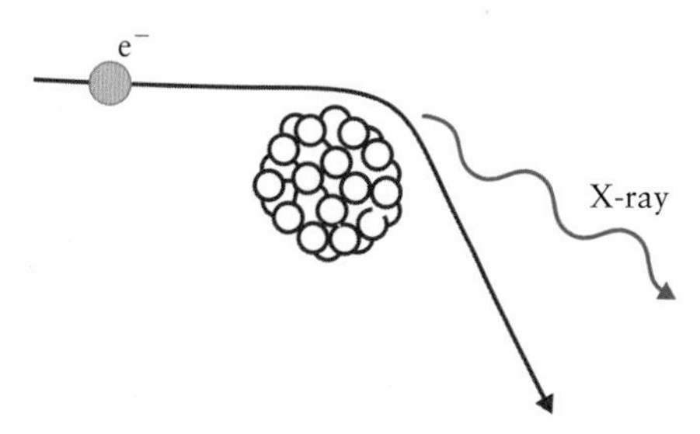

Figure 24.4 Sketch of the process causing bremsstrahlung. An electron passes by a nucleus at close proximity and is attracted by the positive nuclear charge. The deflection occurs with a large angle. The electron emits X-rays while accelerating along a curved trajectory.

KEY POINT

Bremsstrahlung represents X-rays that are generated in a sudden interaction of a fast electron and the electrostatic potential of an atomic nucleus.

The intensity of bremsstrahlung emission is inversely proportional to the energy of the emitted X-ray photon, shown as a dashed line in Fig. 24.5 for an incident electron beam of 90-keV energy. The actual distribution of X-rays (solid line in Fig. 24.5) decreases to below 30 keV and is completely cut off at about 10 keV due to absorption of the emitted X-rays in the anode and filter.

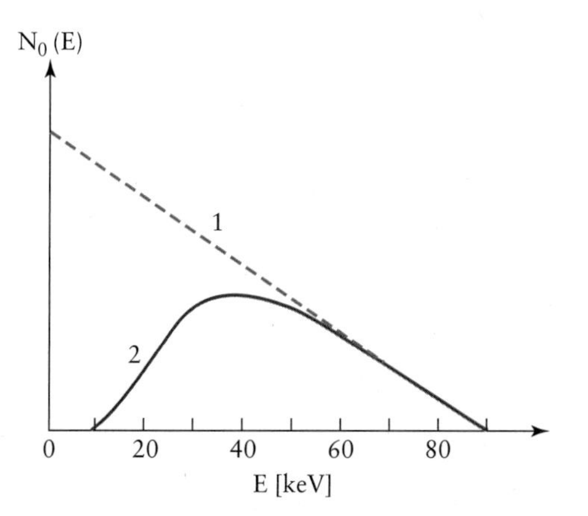

Figure 24.5 The bremsstrahlung energy distribution for 90-keV electrons. Line 1 shows the unfiltered spectrum with a major contribution of small energy X-rays. Spectrum 2 illustrates that filtering reduces preferentially low energy X-rays.

Thus, an X-ray tube generates X-ray photons with a broad peak of energies, but primarily causes a significant thermal heating of the anode. This is an important design factor for X-ray tubes in radiological instruments as the cooling rate of the anode metal often limits the intensity of the X-ray beam that can be generated with a particular type of X-ray tube. Rotating anodes allow for the metal to be exposed to the electron beam for short periods on an intermittent basis. Fig. 24.6 shows a modern set-up of the X-ray source: the cathode filament is located off-axis and the tungsten anode disk is mounted on a rotating molybdenum stem, which in turn is driven by a rotor (3000 – 10 000 revolutions per minute).

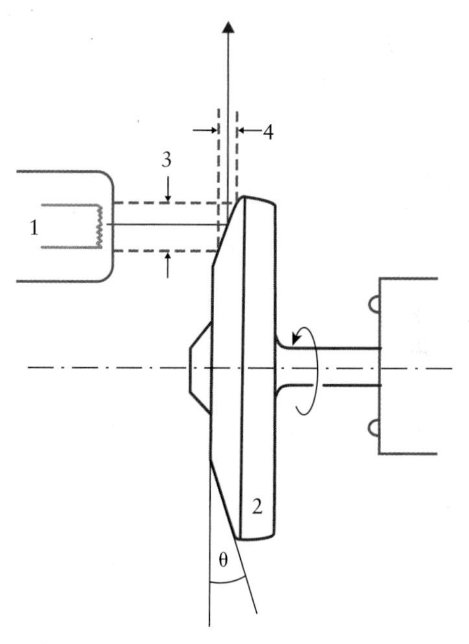

Figure 24.6 Technical set-up of an X-ray source. The cathode is indicated at the left, with a filament from which electrons are evaporated. The anode at the right rotates. The cathode is arranged off-axis with the anode such that the accelerated electrons hit the edge, which is machined at an angle θ with the vertical. θ is called the target angle. The dashed lines indicate the focal spot size; a small angle θ reduces the true spot size of the electron beam from the cathode to an effective spot size of the X-rays leaving the source.

Tungsten is used due to its high melting point and high atomic number ($Z = 74$). Using an anode with high atomic number is important as the ratio of energy loss due to desired X-ray radiation (radiative energy loss) to undesired thermal energy loss through anode heating, Q, is proportional to the product $Z\,E_{kin}$:

$$\frac{E_e \Rightarrow \text{X-ray}}{E_e \Rightarrow Q} = \left[1.22 \times 10^{-7}\ \%/\text{eV}\right] Z\,E_{kin}, \quad [24.1]$$

in which Z is the atomic number of the anode material, and the kinetic energy of the electrons is given in unit electron volts (eV). The energy fraction on the left-hand side is then given in percent (%).

CONCEPT QUESTION 24.2

What fraction of the energy of the cathodic electrons is radiated as X-rays in an X-ray tube used for (a) typical chest X-ray radiograms, for which we use a tungsten anode with an operating potential of 100 kV; (b) typical mammograms, for which we use a molybdenum anode ($Z = 42$) and an operating potential of 30 kV; and (c) typical radiation therapy, for which we use a tungsten anode with an operating potential of 6 MV?

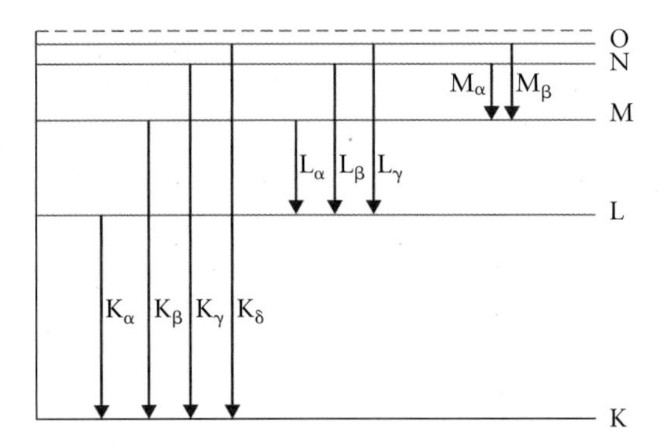

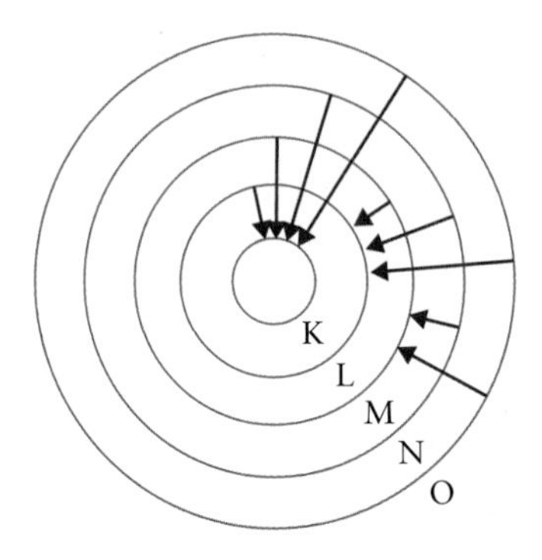

Figure 24.7 Electron transitions in an atom in which an electron has been removed from an energetically lower orbital. Electrons from energetically higher orbitals make transitions to the vacancy, in either one or several steps. This corresponds to a range of possible transitions in heavier atoms, with several of the transitions in the X-ray part of the electromagnetic spectrum (characteristic X-rays). The figure indicates the standard notation used for these transitions, with the capital letter indicating the principal quantum number of the orbital with the vacancy (K-shell corresponds to $n = 1$). The index indicates the difference in the principal quantum numbers of the initial and final orbital of the electron that makes the transition (α corresponds to $\Delta n = 1$).

24.1.4: Characteristic X-Rays

Note that the spectrum in Fig. 24.3 displays several sharp X-ray peaks between 50 keV and 70 keV. This radiation is characteristic for the metal used in the anode, as we confirm experimentally by changing the material in the anode: while the bremsstrahlung part of the X-ray spectrum varies little, the sharp peaks shift to new energies. As a result, we refer to these spikes as characteristic X-rays.

We saw already in the discussion of Bohr's model for elements heavier than hydrogen that the same transitions require more energy for heavier elements; the transition from $n = 2$ to $n = 1$ releases 10.2 eV in hydrogen, but 40.8 eV in He^+. However, these values apply only to atoms and ions with a single electron. To remove a valence electron from heavier atoms requires less energy. The situation during the bombardment with energetic electrons is different. The energetic electrons act like particles and can kick electrons out from one of the inner orbitals of the anode atoms. Once an inner electron is removed, electrons in higher orbitals fill the vacancy. The energy difference for the electrons involved in such transitions is large due to the large positive charge in the nucleus and due to the large energy difference between the lower lying orbitals.

KEY POINT

Characteristic X-rays originate from atoms in which a collision of a fast incident electron removes an electron from an inner shell. The photon is released with an energy that represents the energy change of an outer shell electron as it drops into the vacancy below.

There are several possible transitions of electrons into a lower orbital vacancy, as illustrated in Fig. 24.7. For an electron removed from the innermost orbital (the orbitals with quantum number $n = 1$ are called the **K-shell**), four possible transitions are shown, labelled with Greek letter indexes. Other transitions leading to X-ray emission involve the initial removal of an electron from the **L-shell** (i.e., orbitals with $n = 2$) or from the **M-shell** (i.e., orbitals with $n = 3$).

EXAMPLE 24.3

(a) Estimate the photon energy of the characteristic X-rays emitted from a tungsten anode when an electron from the M-shell ($n = 3$) drops into a vacancy created in the K-shell ($n = 1$). (b) What is the corresponding wavelength of the X-rays?

Solution

Solution to part (a): The best way to calculate this energy is to start from:

$$E_{\text{total}}(Z) = Z^2 E_{\text{total}}(Z = 1).$$

We cannot simply substitute the value $Z = 74$, which is the number of positive charges in the tungsten nucleus, since the electrostatic force between an electron and the positive charge in the nucleus is partially screened by other electrons in the atom. In 1914, Henry Moseley showed that the effective value for Z is given by the number of

continued

protons in the nucleus, diminished by the number of electrons in shells at lower energy than the electron that makes the transition. In the K-shell we use the above equation modified in the form $E_K = (Z - 1)^2 E_1$, where E_1 is the ionization energy for the hydrogen atom. In the M-shell, we use $E_M = (Z - 9)^2 E_3$ since there are 8 electrons in the L-shell and 1 electron in the K-shell (the other one has been kicked out), shielding the M-shell electron before the transition. Further, we know from Chapter 20 that $E_3 = E_1/3^2$. Using $E_1 = -13.6$ eV, we find for the transition:

$$\Delta E = E_K - E_M = (-72.5 \text{ keV}) - (-6.4 \text{ keV}) = -66.1 \text{ keV}.$$

This transition contributes the second peak in Fig. 24.7.

Solution to part (b): The wavelength is calculated from the relation $\Delta E = h\,f = h\,c/\lambda$. We find $\lambda = 0.019$ nm.

24.2: Photon Interaction Processes with Matter

To judge what consequences X-rays have on the penetrated tissue, the physical effects of the interaction of the X-ray photon with matter have to be identified. We introduce in this section four processes by which X-rays interact with matter: (1) Rayleigh scattering, (2) Compton scattering, (3) the photoelectric effect, and (4) pair production. Combining these four effects allows us to quantify the loss of X-ray intensity as the radiation passes through biological tissues.

24.2.1: Rayleigh Scattering

In Rayleigh scattering (named for J. W. Strutt, Lord Rayleigh), the **incident photon** interacts with the entire atom, causing all electrons to oscillate together. The photon is instantaneously re-emitted, causing a slight change in direction of the photon but no change in its energy. The process is quantified in a classical electromagnetic model, in which the photon is described as a wave and the electrons in the atom respond to the electric field vector of the wave, which forces the oscillations.

For X-rays, Rayleigh scattering occurs mainly for very low-energy diagnostic X-rays, as used in mammography. At 30 keV photon energy, Rayleigh scattering contributes about 12% of the interactions with tissue. In medical imaging, the Rayleigh-scattered X-rays have a negative effect on image quality because X-rays of the same energies as the incident beam reach the detector but not along a straight line from the X-ray source. However, this type of interaction has a low probability of occurring at typical diagnostic X-ray energies: it contributes less than 5% of the X-ray interactions with tissue for photon energies above 70 keV. Even at low X-ray energies, Rayleigh scattering contributes less effectively to photon–tissue interactions than the photoelectric effect, and is therefore neglected in the further discussion.

CONCEPT QUESTION 24.3

In the visible part of the electromagnetic spectrum, the Rayleigh scattering probability is proportional to the fourth power of the frequency of the light, f^4. Can this type of scattering explain the scattering of visible light in the atmosphere, causing the sky to be bright in all directions, not only in the direction of the Sun?

24.2.2: Compton Scattering

Compton scattering (named for Arthur Compton, who observed it for the first time in 1922) is the main interaction of X-rays in the energy range from about 25 keV to 30 MeV, which includes the diagnostic energy range with soft tissue. Compton scattering is a process that occurs when an X-ray photon interacts with a **valence electron**, as illustrated in Fig. 24.8. The electron is ejected from the atom and the photon is scattered, with reduced energy. The Compton scattering event results in ionization of the atom, with a division of the incident photon energy between the scattered photon and the ejected electron. The ejected electron then loses its kinetic energy in the surrounding material producing further ionizations. The scattered photon can interact further with the tissue sample or it can reach the detector, which is detrimental for image quality because it did not travel along a straight line from source to detector.

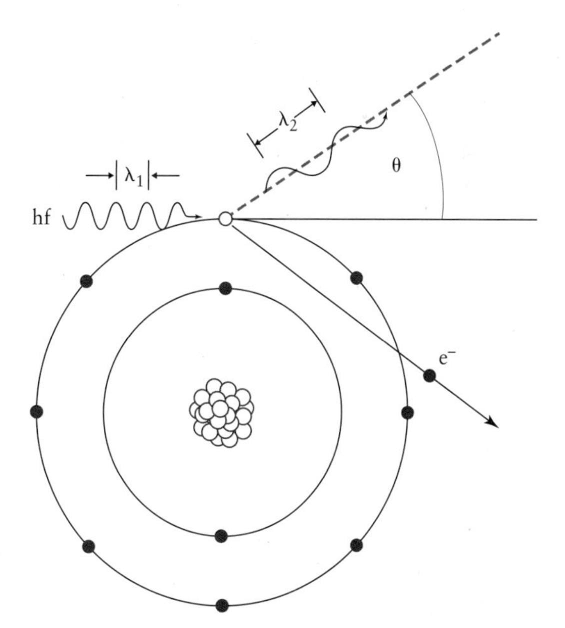

Figure 24.8 Principle of Compton scattering: the incident photon of energy $h\,f$ interacts with a valence electron of an atom, removing the valence electron. Simultaneously, a Compton scattered photon with an energy less than the incident energy ($\lambda_2 > \lambda_1$) leaves the atom with a scattering angle θ. The scattering angle is defined as the angle between the trajectory of the incoming photon and the trajectory of the scattered photon.

Due to energy conservation, the energy of the incident photon $E_{\text{photon, initial}}$ is equal to the sum of the energy of the scattered photon $E_{\text{photon, final}}$ and the kinetic energy of the ejected electron E_{electron}:

$$E_{\text{photon, initial}} = E_{\text{photon, final}} + E_{\text{electron}}.$$

In this formula, the binding energy of the ejected electron is neglected because it is small compared to the three energy terms shown.

The scattering event is also subject to the conservation of momentum. The two conservation laws allow us to calculate the energy of the scattered photon as a function of the energy of the incident photon and the angle of the scattered photon with respect to the incident trajectory (for the geometry see Fig. 24.8):

$$E_{\text{photon, final}} = \frac{E_{\text{photon, initial}}}{1 + \frac{E_{\text{photon, initial}}}{m_e c^2}(1 - \cos\theta)}. \quad [24.2]$$

This formula applies when the energy of the initial and the scattered photon are similar, i.e., the energy of the electron after the collision is small. Eq. [24.2] is derived in the Appendix.

As the incident photon energy increases, both scattered photon and electron are more likely scattered in the forward direction. Thus, in X-ray transmission imaging, the scattered photons are much more likely to be detected in the image, reducing image contrast, as discussed later in this chapter.

As the energy of the incident photons increases (high kilovoltage radiography), an increasing fraction of the incident photon energy is transferred to the scattered electron. This is evident from Eq. [24.2] for a given angle θ: as the term $E_{\text{photon, initial}}$ increases in the denominator on the right-hand side, the second term in the denominator increases compared to 1, and thus decreases $E_{\text{photon, final}}$.

KEY POINT

A good agreement between theory and experimental measurements in Compton scattering confirmed the applicability of the corpuscle model of the X-ray photon. Both conservation of energy and momentum are obeyed in the interaction of radiation with matter.

CASE STUDY 24.1

Is Compton scattering an elastic collision?

Answer: *We defined elastic collision in classical mechanics as a collision in which both the kinetic energy and the momentum are conserved. Momentum is always conserved; however, conservation of kinetic energy does not always apply (e.g., in inelastic collisions).*

continued

When studying subatomic particles that require quantum mechanical considerations, such as the various equations we derived for the particle–wave dualism, the term elastic collision needs to be extended. In particular, the classical case of an incident particle of equal mass hitting head-on a particle at rest cannot be extended to a photon hitting an electron. The relativistic mass of the photon may match that of the electron, but the photon cannot be at rest after the collision.

If we assume the definition includes a quantum mechanically behaving X-ray photon (with zero mass at rest), then Compton scattering is quasi-elastic, as only a negligible energy contribution due to the binding energy of the electron is not accounted for in the kinetic energy formula. One may still be cautious using the term elastic, with its classical mechanics implications.

EXAMPLE 24.4

What fraction of the energy of the incident photon is taken up by the released electron for a $\theta = 60°$ scattering angle when using (a) 100-keV X-rays in chest radiography, and (b) 5-MeV X-rays in radiation therapy?

Solution

Solution to part (a): We use Eq. [24.2] with $mc^2 = 511$ keV, which represents the energy of an electron at rest. For 100-keV X-rays we find that the constant 1 dominates the denominator on the right-hand side of the equation, and that the scattered photon picks up 91 keV (i.e., 91% of the incident energy). This leaves only 9% of the energy for the electron, or 9 keV.

Compton scattering at the lower X-ray energies used in diagnostic imaging (18–150 keV) causes the majority of the incident X-ray energy to be transferred to the scattered photon. These photons may still reach the detector and contribute to image degradation. For example, an 80-keV X-ray photon leads to a minimum energy of about 60 keV for the scattered photon in a Compton event, which is still high enough to penetrate tissue.

Solution to part (b): We use again Eq. [24.2]. A 5-MeV X-ray photon loses most of its energy, with the scattered photon carrying only 850 keV. The difference, about 4.2 MeV, is the energy picked up by the electron (representing more than 80% of the incident energy). Note that Eq. [24.2], however, cannot be applied in this case as the initial and scattered photon energies differ greatly; thus, the numerical results in part (b) can be used as an estimate at best. This problem is therefore a good example to remind ourselves that formulas such as Eq. [24.2], which are derived with certain assumptions, should not be applied without checking the validity of the original assumptions.

The incident photon energy must be substantially greater than the binding energy of the electron before Compton scattering becomes likely. Thus, the probability of Compton scattering increases as the incident photon energy increases up to energies of about 50 keV. This is illustrated in Fig. 24.9. Were the electrons in the sample free, a much lesser dependence of the probability for Compton scattering below 50 keV would be observed. The probability for Compton events then decreases for X-ray energies above 100 keV. The decrease is a reflection of the cross-section for the photon–electron interaction: the higher the energy the more directly the X-ray must hit the electron to interact.

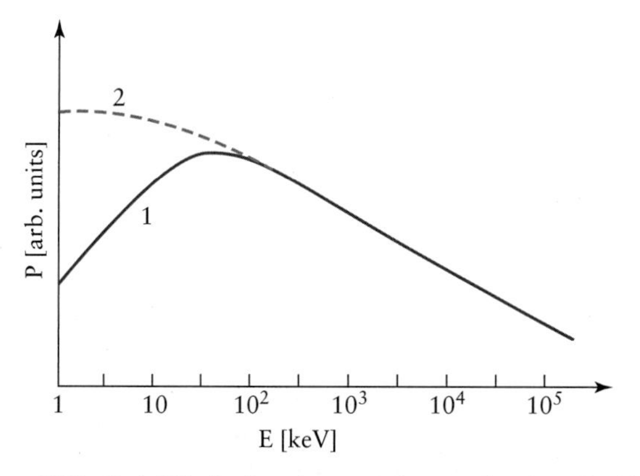

Figure 24.9 Probability for Compton scattering as a function of incident photon energy. The solid curve 1 assumes interaction with a bound electron. The incident photon energy must be substantially greater than the binding energy of the electron before Compton scattering becomes likely, leading to a peak near 50 keV. For free electrons (dashed curve 2), Compton scattering depends much less on the incident energy below 50 keV. The probability for Compton events decreases for X-ray energies above 100 keV.

At a given X-ray energy, the probability of Compton scattering also depends on the number density of electrons in the tissue. The number density of electrons contains two factors: the number of electrons per kilogram of tissue, times the density in unit kg/m^3. With the exception of hydrogen, the number of electrons per kilogram is fairly constant in tissue. As a result, the probability of Compton scattering *per unit mass* is almost independent of the atomic number of the elements in the tissue, *Z*, and the probability of Compton scattering *per unit volume* is approximately proportional to the density of the material. Compared to other elements, the absence of neutrons in the hydrogen atom results in an approximate doubling of electron density. Thus, hydrogenous material has a higher probability of Compton scattering than hydrogen-free material of equal mass.

KEY POINT

Rayleigh scattering is the interaction between a classical wave with an entire atom, and Compton scattering is the interaction of a light corpuscle (photon) with an individual valence electron within the atom.

24.2.3: Photoelectric Effect

In the photoelectric effect, the entire energy of the incident X-ray photon is transferred to an electron, which in turn is ejected from the atom, as illustrated in Fig. 24.10.

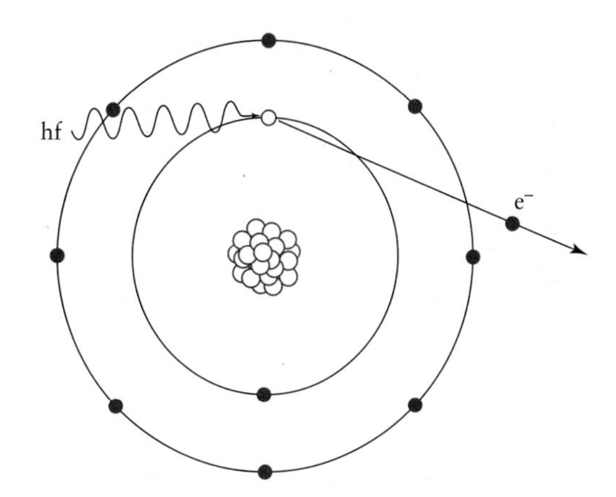

Figure 24.10 Principle of photoelectric absorption: an incident photon of energy *h f* interacts with an electron of an inner shell. The released electron has a kinetic energy equal to the energy of the incident photon minus the binding energy of the released atomic electron. In the case shown in the figure, the binding energy is the K-shell binding energy.

KEY POINT

Rayleigh scattering is exclusively a scattering effect, Compton scattering is a scattering effect with a fraction of the energy absorbed in the tissue sample, and the photoelectric effect is exclusively an ***absorption process****.*

A second difference to Compton scattering is the greater probability in the photoelectric effect that electrons deeper in the atomic shell are involved. Thus, the conservation of energy in the photoelectric effect is written as:

$$E_{\text{kin,electron}} = E_{\text{photon}} - E_{\text{binding}}; \quad [24.3]$$

i.e., the kinetic energy of the ejected electron is equal to the incident photon energy minus the **binding energy** of the electron in its orbital. The binding energy is now a major contribution because the photoelectric effect represents an inelastic collision between photon and electron. Obviously, photoelectric absorption of an X-ray photon can occur only when the incident photon carries a greater energy than the binding energy of the electron that is to be ejected.

The reason that electrons in deeper shells are involved is that X-ray energies far exceed the binding energy of outer valence electrons, but are more closely matched with the binding energy of electrons in lower shells. The probability that an electron will be ejected through a photoelectric event increases the closer the X-ray photon energy is to the binding energy of the electron. For example, if the photon energy exceeds the K-shell binding energy in the atom, then the photoelectric interaction with a K-shell electron is most probable.

The photoelectric effect itself does not produce a secondary X-ray photon and therefore allows only primary photons to reach the detector in medical imaging techniques. This is an advantage in comparison to Compton scattering, which results in loss of image contrast due to scattered photons reaching the detector.

The **probability for photoelectric absorption** to occur in a sample of unit mass is approximately proportional to Z^3/E^3, where Z is the average atomic number of the analyzed material and E is the energy of the incident photon.

CASE STUDY 24.2

We compare the photoelectric absorption of soft tissue in mammography at 25 keV with the photoelectric absorption of bone in chest radiography at 150 keV. What is the ratio of photoelectric absorption in these two cases?

Supplementary radiological information: the average atomic number for a soft tissue is $Z_{tissue} = 7$ (containing mostly elements such as C, N, O, and H). Bone consists primarily of calcium ($Z = 20$) and phosphorus ($Z = 15$), leading to an atomic number of $Z_{bone} = 13$.

Answer: *We use the proportionality to Z^3/E^3 for the probability for the photoelectric effect, P, and write:*

$$P_{mammography} = \text{const}\frac{Z^3_{mammography}}{E^3_{mammography}}$$

$$P_{radiography} = \text{const}\frac{Z^3_{radiography}}{E^3_{radiography}},$$

with the constant independent of energy and atomic mass. To obtain the ratio, we divide the two formulas:

$$\frac{P_{mammography}}{P_{radiography}} = \frac{Z^3_{mammography}}{Z^3_{radiography}}\frac{E^3_{radiography}}{E^3_{mammography}} = \left(\frac{Z_{mammography}}{Z_{radiography}}\frac{E_{radiography}}{E_{mammography}}\right)^3.$$

We substitute the given values:

$$\frac{P_{mammography}}{P_{radiography}} = \left(\frac{7 \cdot 150\ \text{keV}}{13 \cdot 25\ \text{keV}}\right)^3 = 35.$$

Thus, the fraction of incident X-ray photons absorbed in a photoelectric event is much greater in mammography than in chest radiography. Photoelectric absorption has the benefit in X-ray transmission imaging that it yields no photons of the same energy as the incident beam; i.e., no scattered photons reach the detector to degrade the image. The strong inverse energy dependence of the probability for photoelectric absorption explains to some extent why image contrast decreases at higher X-ray energies.

Can the photoelectric effect lead to X-ray emissions as well? The answer is yes, although such emission is at significantly lower energy. The production of the X-ray is a secondary effect and occurs for both the electron ejected in the primary photoelectric event and the ionized atom with an electron missing in a core shell.

Using Eq. [24.3] we determine the kinetic energy of the ejected electron for two examples. A 100-keV incident X-ray photon can eject an electron from all shells in a barium atom ($BaCO_3$ is frequently used as a radiographic contrast material since $Z(\text{Ba}) = 56$): the binding energy of the K-shell electrons is 37.5 keV. An electron ejected from the K-shell has an energy of 100 keV − 37.5 keV = 62.5 keV. It will lose its energy, most likely in the form of thermal energy, but may alternatively generate some bremsstrahlung, with a peak at an energy well below 62.5 keV (compare with Fig. 24.5).

Next, we consider a lighter element present in tissue. The binding energy for an electron in the K-shell of carbon is given as 0.29 keV. The electron ejected in a photoelectric event with an incident photon energy of 100 keV generates, therefore, electrons with a kinetic energy exceeding 99 keV. A bremsstrahlung spectrum with photon energies closer to the incident energy should be generated; however, the initial photoelectric event is much less likely due to the large energy difference between the incident photon and the binding energy of the K-shell electron. Thus, the electron ejected in the photoelectric effect does not contribute significantly to the X-ray emission from the sample.

Next we consider the radiation from the ionized atom with an inner shell electron vacancy. This ion undergoes one of two processes, which are illustrated in Fig. 24.11 for a carbon atom.

- In Fig. 24.11(a) the vacancy is filled by an electron from the L-shell (which is the outmost shell carrying electrons in carbon). In heavier atoms, the newly created vacancy in the L-shell is then filled by an electron with an even lower binding energy, leading to an electron cascade from outer to inner shells. In carbon, the difference in binding energy of the K- and L-shells is released as a photon of energy 280 eV; i.e., the emitted photon does not lie in the X-ray range. For barium, the highest energy photon emission is generated by the direct transition of an outer shell electron into the K-shell vacancy, yielding a photon with an energy around 37 keV. The more likely transition of an electron from the L-shell yields a photon of energy 31.8 keV. These photons have much lower energy than the incident beam and can be prevented from reaching the detector using filters.
- In Fig. 24.11(b) the vacancy in the core shell is filled by an outer electron, but the released energy is transmitted in a **radiationless process** to another electron, which leaves the atom with a characteristic kinetic energy. This process is called the **Auger process** and the ejected electron is called an Auger electron (named for Pierre Victor Auger, who discovered this process in 1925). The Auger process does not yield X-ray emission and therefore does not cause a reduction in contrast in medical imaging techniques. Note that Auger electron emission is favoured for samples with small atomic numbers, such as soft tissues, while characteristic X-ray emission occurs for heavier atoms.

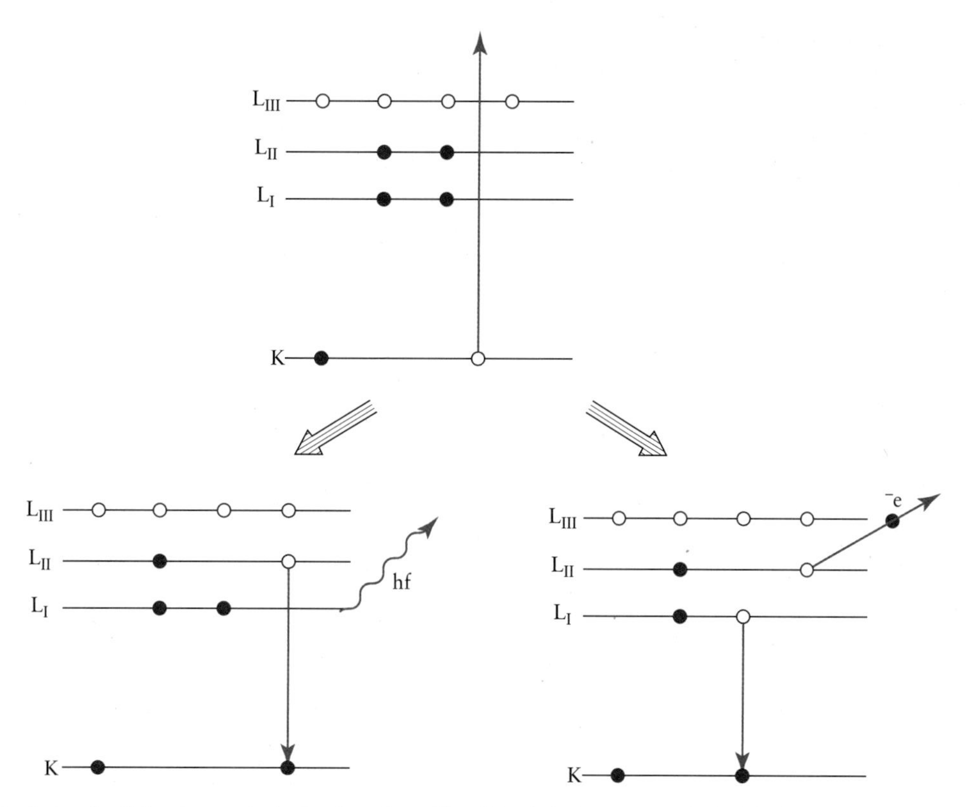

Figure 24.11 Two types of radiation from an ionized carbon atom with a K-shell electron vacancy. At the left, the vacancy is filled by an electron from the L-shell. The difference in binding energy of the K- and L-shells is released as a photon of energy 280 eV. At the right (Auger process), the vacancy in the core shell is filled by an outer electron, but the released energy is transmitted in a radiationless process to another electron, which leaves the atom with a characteristic kinetic energy.

CONCEPT QUESTION 24.4

Image contrast is a major concern for all X-ray imaging techniques. Let us consider the contrast between bone and soft tissue in chest radiography using 100-keV X-rays. What is the ratio of X-rays stopped by bone and soft tissue based on the photoelectric effect?

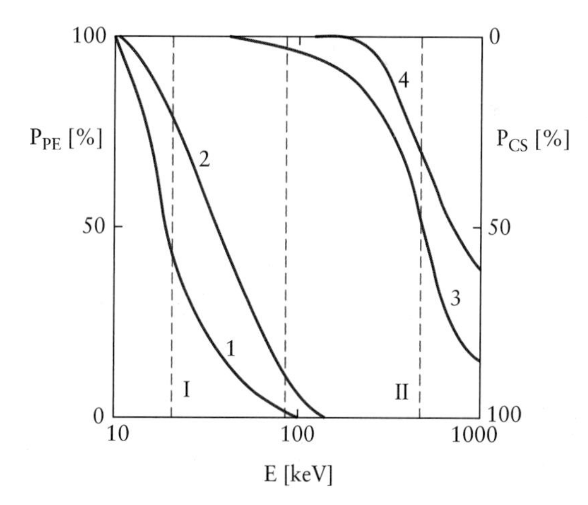

Figure 24.12 The relative contributions of the photoelectric effect (left vertical axis) and the Compton effect (right vertical axis) to the X-ray attenuation process in various materials as a function of X-ray energy. The contributions are expressed as probabilities P in unit percent (%), the index CS refers to Compton scattering and PE to photoelectric effect. The curves shown correspond to tissue (1), bone (2), NaI ((3); most frequently used detector material), and lead ((4); Pb, used for shielding). Note that the energy range I (between the left dashed vertical lines) is applied in diagnostic radiology, and the energy range II (between the right dashed vertical lines) is used in nuclear medicine for radiation treatment.

At photon energies below 50 keV, the photoelectric effect plays an important role in imaging soft tissue. The photoelectric absorption process is often exploited to amplify differences in attenuation between tissues with slightly different atomic numbers to improve image contrast. The photoelectric process is most important when lower-energy X-ray photons interact with high Z materials, as illustrated in Fig. 24.12. The figure shows that, at 50 keV, photoelectric absorption is the dominant interaction of diagnostic X-rays with screen phosphors, radiographic contrast materials, and bone. In turn, Compton scattering dominates in tissues.

Although Case Study 24.2 illustrated how the probability of the photoelectric effect decreases with increasing

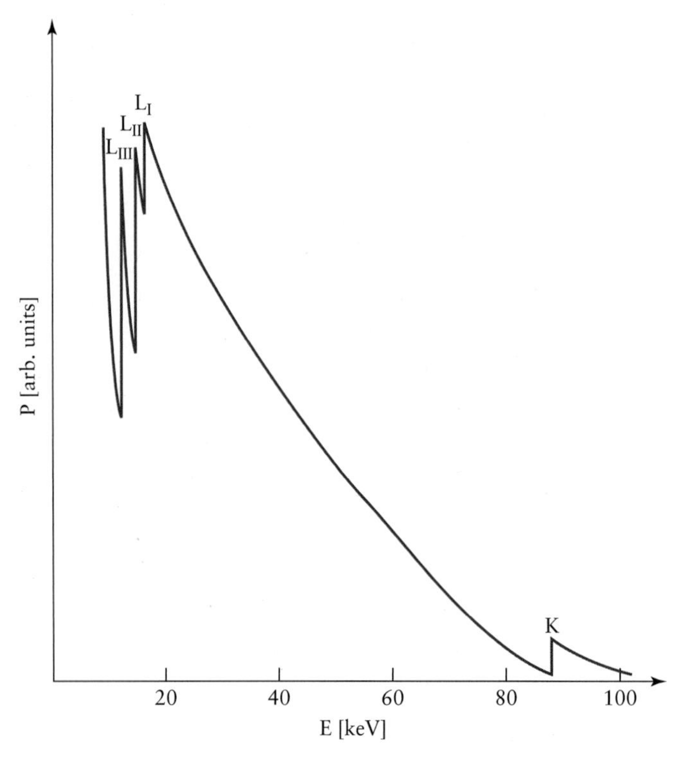

Figure 24.13 X-ray absorption in lead. The vertical axis shows the absorption coefficient in arbitrary units. Four absorption edges are visible, with the K-edge at 88 keV and the three L-edges below 20 keV.

X-ray photon energy, there is an additional opposing effect we need to take into account. This effect is illustrated for a lead (Pb, $Z = 82$) sample in Fig. 24.13. The figure shows absorption of X-rays of increasing energy. The sharp **absorption edges** in the otherwise decreasing probability of the photoelectric effect occur when the photon energy crosses a value at which electrons in the next deeper shell can be ejected; i.e., the X-ray energy exceeds the binding energy for these electrons. The change in the probability may be as large as a factor of 10 across an absorption edge.

The absorption edges are labelled according to the atomic shell of the affected electrons, e.g., in Fig. 24.13 as K, L_1, L_2, and L_3. The subscripts distinguish various orbitals of slightly varying energy. From the equation in Section 24.1.1 we conclude that the X-ray photon energy corresponding to a particular absorption edge increases with the atomic number Z of the sampled element. The primary elements comprising soft tissue (H, C, N, and O) have their absorption edges all below 1 keV, i.e., at energies below the range used in medical imaging. However, elements such as iodine and barium, with Z values near 50, are commonly used in radiographic contrast agents to provide enhanced X-ray attenuation. These elements have K-absorption edges between 30 keV and 40 keV.

24.2.4: Pair Production

With increasing photon energy, the probability for the photoelectric effect decreases rapidly and proportionally to E^{-3}. The probability of Compton scattering decreases as well. Still, X-ray absorption of matter increases in the MeV photon energy range. This absorption is due to a new, energy-requiring process: the generation of an **electron–positron pair**.

A positron is the antimatter particle of an electron; it is identical to the electron but carries a positive charge. Positrons are stable in isolation but they are annihilated when they encounter electrons. Thus, a newly generated positron slows down in the material in which it is formed. When caught electrically by an electron, the annihilation occurs in less than 1.5×10^{-7} s. The annihilation results in two X-ray photons of 511 keV energy each.

An electron–positron pair is generated when a photon of sufficient energy interacts with matter. Conservation of energy requires that the photon carries enough energy to create the two particles, i.e., $E_{\text{photon}} \geq 2\ m_e\ c^2$, where m_e is the mass of the electron. Thus, E_{photon} must exceed 1.02 MeV. If the initial photon energy is higher, the remaining energy is distributed as kinetic energy to electron and positron:

$$E_{\text{kin}}(e^+) + E_{\text{kin}}(e^-) = E_{\text{incident photon}} - 1.02\ \text{MeV}. \quad [24.4]$$

Conservation of momentum further requires that the pair formation occurs in the vicinity of another particle. This other particle is usually an atom, causing the probability for pair production to depend on the matter in which it occurs.

Pair production does not play a role in medical diagnostics, as X-ray energies are well below 1 MeV. We neglect this effect, therefore, in the discussion of medical imaging. However, the effect contributes significantly in radiotherapy.

24.3: Photon Attenuation

24.3.1: The Linear Attenuation Coefficient

The ability to penetrate various materials depends on the actual energy of the X-rays and the consistency of the penetrated matter. When a beam of X-rays of energy of less than 1 MeV enters a sample, it loses photons primarily by two of the processes we discussed above: (i) through absorption of photons in the photoelectric effect, and (ii) through a combination of energy loss and scattering in the Compton effect.

These two effects combine to yield the **total attenuation** of the incident beam. The total attenuation depends on the material and the energy of the incident beam. At a given photon energy, the number of photons in the transmitted beam, N, depends on the depth in the sample in the following way:

$$N(x) = N_0 e^{-\mu x}, \quad [24.5]$$

where N_0 is the number of incident X-ray photons, $N(x)$ is the number of photons transmitted to a depth x, and μ is the **linear attenuation coefficient** in units of m^{-1}. Eq. [24.5] states that the number of X-ray photons decreases with an exponential law (**Beer's law**). The same law applies to sound and light absorption in matter.

KEY POINT

Beer's law states that the probability scattering or absorbing a photon at each depth in a sample is proportional to the number of incident photons reaching that depth.

If we label with dN the number of photons scattered or absorbed (i.e., removed) at depth x and N the number of photons reaching that depth, then the basic principle expressed by Beer's law is written in the form:

$$dN(x) \propto N(x),$$

which is rewritten in equation form as:

$$-dN = \mu N \, dx,$$

where μ is a constant and dx is the depth interval in which the scattering/absorption takes place. We rewrite this equation by dividing by dx:

$$-\frac{dN}{dx} = \mu N.$$

This can be integrated to find the depth dependence of N, i.e., the function $N(x)$. For the integration, we need to find a function for which both itself and its change with depth have the same mathematical form. The only mathematical function of this type is the exponential function. Integration leads then to Eq. [24.5].

The linear attenuation coefficient is the physically important factor in Eq. [24.5]. While it is a constant with respect to the number of photons and the depth, it does vary with other parameters, particularly with incident X-ray photon energy. This is illustrated in Fig. 24.14, which shows μ in lead for energies between 100 keV and 100 MeV. At lower energies, the photoelectric effect and Compton scattering dominate; above 5–10 MeV, pair production is dominant. μ also depends on the composition of the irradiated sample. This effect is discussed below when we focus on the use of X-rays in medical imaging.

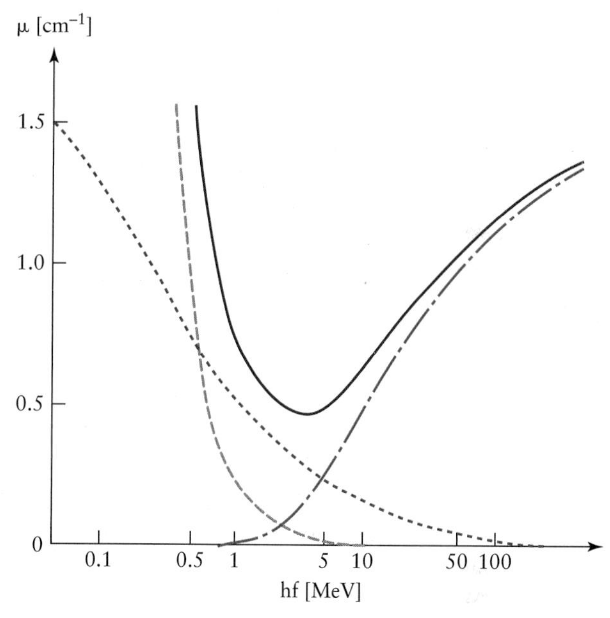

Figure 24.14 Linear absorption coefficient μ (solid line) as a function of incident photon energy in lead. Three contributing processes are shown separately, absorption due to the photoelectric effect (long-dash curve), due to Compton scattering (short-dash curve), and due to pair formation (dash-dotted curve).

EXAMPLE 24.5

X-rays at 50 keV have a linear attenuation coefficient of 0.193 cm^{-1} in fat. What fraction of an incident X-ray beam at this energy is absorbed or inelastically scattered in a 3.0-cm thick fat sample?

Solution

We substitute the given data in Eq. [24.5]:

$$\frac{N(3\text{ cm})}{N_0} = \exp[(-0.193\text{ cm}^{-1})(3\text{ cm})] = 0.56;$$

i.e., 56% of the beam passes through a 3 cm thick sample, with the remaining 44% absorbed or scattered in the fat tissue.

The fraction of X-rays passing through the sample is used in medical imaging techniques. Detecting these photons allows us to measure $N(L)/N_0$ in Eq. [24.5], with $x = L$ the total thickness of the sample. The fraction $N(L)/N_0$ is called the **yield** Y and depends on the sample material:

$$Y(L) = \frac{N(L)}{N_0}. \quad [24.6]$$

KEY POINT

The yield is the fraction of primary photons transmitted through a sample of given thickness. It depends on the linear attenuation coefficient.

24.3.2: The Mass Attenuation Coefficient: Definition and Motivation

Note that the parameter μ in Eq. [24.5] is called the linear attenuation coefficient. This name is sensible because the coefficient describes the loss of X-ray photons per unit path length in a sample. In medical applications, however, μ is not exclusively used because it depends on both the chemical composition (atomic number Z) and the density ρ of the sample. The dependence on these two parameters is separated by introducing the **mass attenuation coefficient**: we rewrite Eq. [24.5] by multiplying the argument of the exponential function with ρ/ρ, where ρ is the density in units of kg/m^3:

$$N(x) = N_0 e^{-(\mu/\rho)\rho x}. \quad [24.7]$$

The unit of the mass attenuation coefficient is $m^{-1}/(kg/m^3) = m^2/kg$ or the cm^2/gm. Table 24.1 lists mass attenuation coefficients as a function of X-ray energy for five tissues found in the human body.

KEY POINT

The mass attenuation coefficient does not depend on density when isolated in Beer's law. Thus, it is convenient to use μ/ρ for the chemical composition of the sample, ρ for the density, and x for the depth in the sample as three independent variables in Beer's law.

To confirm that the mass attenuation coefficient is independent of density, we study the function $N(x)/N_0$ for two biological systems in Fig. 24.15 to test the validity of Eq. [24.7]. The figure shows in logarithmic representation the fraction of 100-keV (solid line) and 50-keV (dashed line) X-rays reaching various depth d in bone (1) and muscle tissue (2). Note that the abscissa is given as the product of depth x and density ρ; i.e., $d = x\rho$, with the units of d as g/cm^2. For muscle tissue, the transmitted

TABLE 24.1

Mass attenuation coefficient as a function of X-ray photon energy in water, air, muscle, bone, and fat. Average densities, atomic numbers and electron number densities are included, with values for air at $p = 1.0$ atm.

	Water	Air	Muscle	Fat	Bone
	μ/ρ	μ/ρ	μ/ρ	μ/ρ	μ/ρ
Energy (keV)	(cm^2/g)	(cm^2/g)	(cm^2/g)	(cm^2/g)	(cm^2/g)
5	42.1	40.3	42.0	25.6	138
10	5.07	4.91	5.15	3.08	19.8
15	1.57	1.52	1.60	1.01	6.19
20	0.761	0.733	0.778	0.533	2.75
30	0.361	0.340	0.365	0.296	0.953
40	0.263	0.243	0.264	0.235	0.509
50	0.225	0.205	0.224	0.210	0.347
80	0.183	0.166	0.182	0.179	0.208
100	0.171	0.154	0.169	0.168	0.180
150	0.151	0.136	0.149	0.150	0.149
200	0.137	0.123	0.136	0.137	0.133
1000	0.071	0.064	0.070	0.071	0.068
5000	0.030	0.028	0.030	0.030	0.030
Density (kg/m^3)	1000	1.20	1040	915	1650
Electron density (e/kg)	3.34×10^{26}	3.01×10^{26}	3.31×10^{26}	3.34×10^{26}	3.19×10^{26}
Atomic number $\langle Z \rangle$	7.5	7.8	7.6	6.5	12.3

beam is reduced to 10% at a tissue thickness $d = \rho/\mu \cong 8 - 10$ g/cm^2, which corresponds to an actual depth of about 9 cm. The same beam reduction is reached for bone at a thickness of $d = \rho/\mu \cong 4 - 5$ g/cm^2, which corresponds to a depth of about 2–3 cm.

CASE STUDY 24.3

Are the data in Fig. 24.15 consistent with the assumption that the mass attenuation coefficient is independent of the sample density?

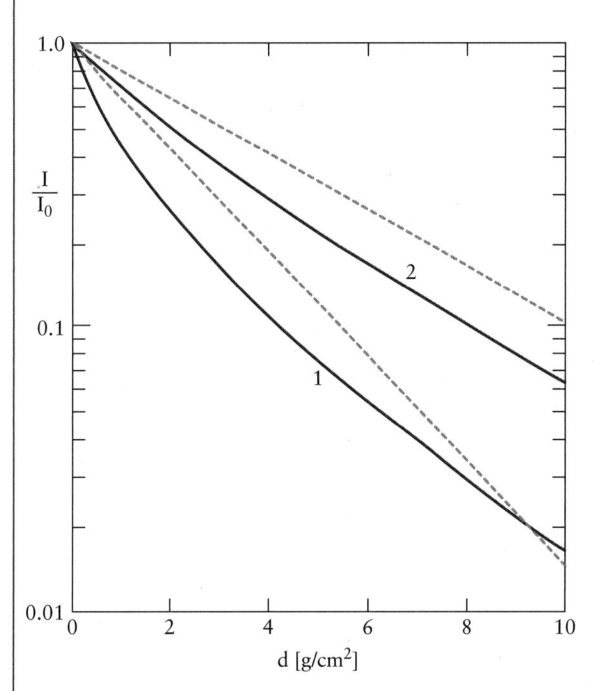

Figure 24.15 Intensity attenuation of X-ray beams of 100 keV (solid lines) and 50 keV (dashed lines) in human bones (lower curves, 1) and muscle tissue (upper curves, 2). The depth d is given in unit g/cm^2, which represents the density multiplied by the path length in the tissue. The vertical axis is the ratio of the number of photons transmitted to depth d to the number of incident photons; this ratio is called the yield.

Answer: *The four curves in Fig. 24.15 have to be straight lines if the mass attenuation coefficient, μ/ρ, were independent of the sample density. This is evident when we rewrite Eq. [24.7] in logarithmic form:*

$$\ln\left(\frac{N(x)}{N_0}\right) = -(\mu/\rho)\rho\, x,$$

which yields $\ln(N(x)/N_0) \propto -\rho\, x = -d$ when $\mu/\rho = \text{const}$. Inspecting the figure, we note that Eq. [24.7] with a density-independent mass attenuation coefficient applies reasonably well for biological systems.

24.3.3: Energy Dependence of the Mass Attenuation Coefficient

The mass attenuation coefficient is the sum of the individual attenuation contributions due to the four X-ray interaction processes we discussed in the previous section. We saw that we can neglect the Rayleigh effect as it is weaker than the photoelectric effect and Compton scattering throughout the X-ray energy range. Pair production contributes only above 1.02 MeV, i.e., at energies exceeding the range of diagnostic applications. Thus, for all practical purposes, the mass attenuation coefficient is determined by:

$$(\mu/\rho)_{\text{total}} = (\mu/\rho)_{\text{CS}} + (\mu/\rho)_{\text{PE}}, \qquad [24.8]$$

in which CS denotes Compton scattering and PE denotes the photoelectric effect. We know from the discussion in the previous section how each of the two terms on the right-hand side of Eq. [24.8] depends on the incident X-ray energy of the sample. For the photoelectric effect we found:

$$(\mu/\rho)_{\text{PE}} \propto (1/E_{\text{photon}})^3, \qquad [24.9]$$

and for the Compton effect we use to first order:

$$\begin{aligned}(\mu/\rho)_{\text{CE}} &\propto \text{const for } E_{\text{photon}} < 100 \text{ keV} \\ (\mu/\rho)_{\text{CE}} &\propto (1/E_{\text{photon}}) \text{ for } E_{\text{photon}} > 300 \text{ keV.}\end{aligned} \qquad [24.10]$$

Eqs. [24.8] to [24.10] are illustrated in Fig. 24.16, showing the energy dependence of the mass attenuation coefficient for soft tissue with an effective Z value of 7. The figure covers the energy range from 10 keV to 50 MeV. The graph also highlights the relative contribution of each of the four individual interaction processes we discussed above. We note that Compton scattering dominates the X-ray loss in soft tissue from 30 keV to 30 MeV, i.e., essentially across the entire range of medical diagnosis. The mass attenuation coefficient decreases fast at the left side of the figure due to Eq. [24.9], then decreases at a smaller rate due to Eq. [24.10] in the range dominated by the Compton effect.

24.3.4: Atomic Number Dependence of the Mass Attenuation Coefficient

Table 24.2 illustrates that the mass attenuation coefficient varies with the atomic number of the sample. The table reports values of μ/ρ for aluminum and copper at four different X-ray energies. The mass attenuation coefficients for both metals are not the same, indicating the role of the atomic number Z, with $Z = 29$ for copper and $Z = 13$ for aluminum.

This dependence is again the result of the combination of the photoelectric effect and Compton scattering, as indicated in Eq. [24.8]. In the case of the Z-dependence though, only the photoelectric effect term is sensitive to variations:

$$(\mu/\rho)_{\text{PE}} \propto Z^3. \qquad [24.11]$$

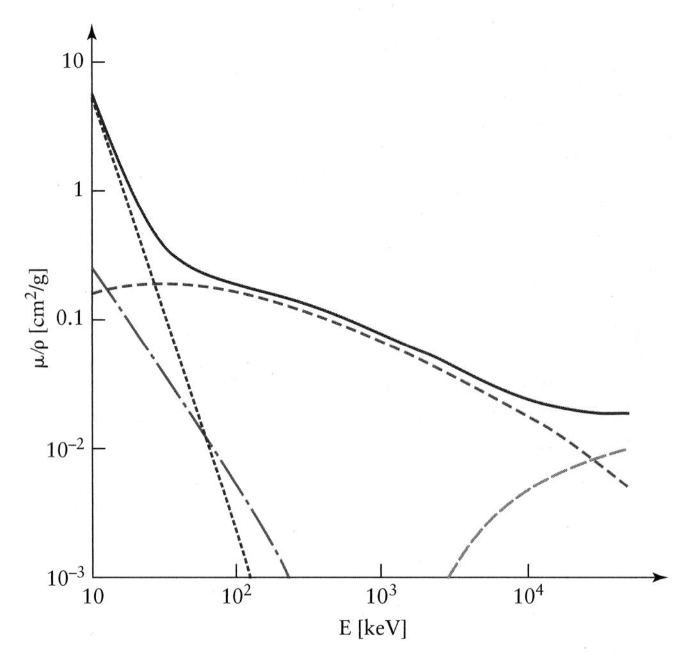

Figure 24.16 The mass attenuation coefficient of soft tissue as a function of incident photon energy (solid curve). Shown are also the individual contributions of four processes: Rayleigh scattering (dash-dotted curve), the photoelectric effect (short-dashed curve), Compton scattering (medium-dashed curve), and pair production (long-dashed curve). Compton scattering dominates attenuation between 50 keV and 10 MeV photon energies.

TABLE 24.2

Mass attenuation coefficient μ/ρ (in unit m^2/kg) as a function of X-ray energy for Al and Cu

E (keV)	μ/ρ (Al)	μ/ρ (Cu)
6.2	10.2	18.8
12.4	1.5	13.1
24.8	0.2	1.9
123.8	0.016	0.033

The contribution due to the Compton effect does not vary with *Z*:

$$(\mu/\rho)_{CE} \propto \text{const.} \qquad [24.12]$$

Exceptions to Eq. [24.12] occur for samples with significant hydrogen content.

EXAMPLE 24.6

Confirm that the observations we made for aluminum and copper in Table 24.2 are consistent with the data for soft tissue in Fig. 24.16.

Solution

The lowest energy in Table 24.2 is below the range of energies shown in Fig. 24.16. We could try to extrapolate the data in Fig. 24.16 to energies below 10 keV

continued

because we know that the photoelectric effect dominates in this range and that it is proportional to E^{-3}. However, for simplicity, we evaluate this problem only at 12.4 keV (an energy dominated by the photoelectric effect in soft tissue), at 123.8 keV (an energy dominated by Compton scattering in soft tissue), and at the intermediate energy of 24.8 keV.

In the first step, the mass attenuation coefficients for soft tissue at these energies are read off Fig. 24.16. We add these values to Table 24.2 after conversion to the standard unit of m^2/kg. This yields Table 24.3. The data in Table 24.3 confirm that the mass attenuation coefficient for soft tissue follows the same trend as the data for the two metals. For better comparison, we draw the data of Table 24.3 in the form $(\mu/\rho) = f(Z)$. This graph is shown in Fig. 24.17. It shows that no simple relation can be given for the *Z*-dependence. We expect this result when studying Fig. 24.16: for different sample materials the crossover from attenuation dominated by the photoelectric effect to attenuation dominated by the Compton effect occurs at different energies.

TABLE 24.3

Mass attenuation coefficient (in unit m^2/kg) as a function of X-ray energy, compared for Al, Cu, and soft tissue

E (keV)	Al	Cu	Soft tissue
12.4	1.5	13.1	0.32
24.8	0.2	1.9	0.05
123.8	0.016	0.033	0.02

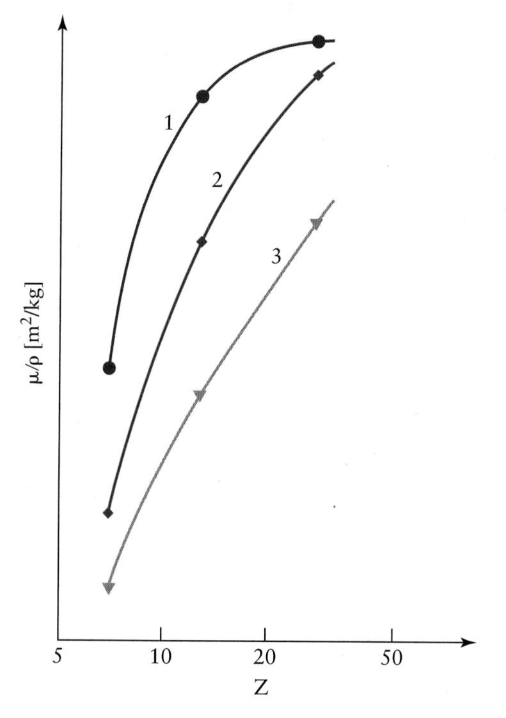

Figure 24.17 Mass attenuation coefficients as a function of atomic number Z for soft tissue (data points at the left), aluminum (data point at the centre), and copper (data points at the right). The data are taken from Table 24.3 at various incident photon energies (dots at 12.4 keV, diamonds at 24.8 keV, and triangles at 123.8 keV). The graph illustrates that no simple Z-dependence of the mass attenuation coefficient applies because different sample materials display a crossover from an attenuation dominated by the photoelectric effect to an attenuation dominated by the Compton effect at different energies.

The energy dependence of (μ/ρ) is exploited in dual energy methods. These methods allow us, for example, to emphasize either the soft tissue contribution or the bone contribution in an image by weighted subtraction of two images obtained with 50-kVp and 120-kVp X-ray beams. **Dual-energy X-ray absorptiometry** (DEXA) is such a method recently developed for bone mineral density measurements. This method is used in the diagnosis of **osteoporosis**. Osteoporosis is a disorder in which bones lose their density and become weak and brittle. The probability for osteoporosis increases with age and is observed most frequently in post-menopausal women. Osteoporosis-related bone loss is most prominent in bones containing a large fraction of spongy bone, e.g., in the vertebral column, the hips, and the wrist. In this condition, bones lose calcium, phosphate, and the matrix of the bone consisting of connective tissue.

24.3.5: Half-Value Layer

Both the linear attenuation coefficient and the mass attenuation coefficient are intuitively less accessible than a quantity that has units of length, describing how far the X-rays get. However, such a quantity is physically meaningless as the exponential function in Eq. [24.5] implies that some fraction of X-rays gets to any chosen depth in a sample. As an alternative, the **half-value layer** is defined as a thickness related to the penetration depth of X-rays. Specifically, the half-value layer (HVL) is defined such that 50% of the incident X-ray photons are stopped at that depth. This definition of a depth is useful (e.g., when designing X-ray filters) and easy to relate to the linear attenuation coefficient. To illustrate this point, we substitute HVL into Eq. [24.5], requiring that $N(HVL) = N_0/2$:

$$N_0/2 = N_0\, e^{-\mu \cdot HVL}.$$

Dividing both sides of the equation by N_0 leads to:

$$\ln(1/2) = -0.693 = -\mu \cdot HVL,$$

which yields an inverse relation between the linear attenuation coefficient and the half-value layer (as is consistent with the unit cm^{-1} for the linear attenuation coefficient and unit cm for the half-value layer):

$$HVL = (\ln 2)/\mu = 0.693/\mu. \qquad [24.13]$$

KEY POINT

The half-value layer is inversely proportional to the mass attenuation coefficient.

EXAMPLE 24.7

Calculate the thickness of a layer of soft tissue that stops 75% of an incident X-ray beam (a) of 30-keV X-rays with $\mu = 0.35\ \mathrm{cm}^{-1}$, and (b) of 100-keV X-rays with $\mu = 0.16\ \mathrm{cm}^{-1}$.

Solution

We use Eq. [24.13]. At HVL only 50% of the incident beam is stopped. However, at 2 HVL, 75% is stopped as another fraction ½ is removed for the depth range from HVL to 2 HVL. Thus, we multiply the result of Eq. [24.13] by the factor 2.

Solution to part (a): Substituting μ for 30-keV photons, we find:

$$HVL = 0.693/(0.35\ \mathrm{cm}^{-1}) = 1.98\ \mathrm{cm}.$$

Thus, 75% of 30-keV photons get stopped at about 4 cm depth in soft tissue.

Solution to part (b): Next, the case of photons at 100 keV is evaluated:

$$HVL = 0.693/(0.16\ \mathrm{cm}^{-1}) = 4.33\ \mathrm{cm}.$$

To stop 75% of 100-keV photons, a soft tissue thickness of more than 8.5 cm is required. Results such as the ones in this example are the reason why the HVL value is seen as a measure of the penetrability of photons in matter.

24.4: Contrast in X-Ray Images

Visible light is not useful in medical imaging (except for studying superficial features in dermatology) because light rays cannot penetrate tissue. In turn, neutrinos, which are generated in huge amounts in the Sun, penetrate the entire planet Earth with only a rare case of interaction. Therefore, neutrino radiation is useless in medical imaging for the opposite reason than visible light: the image taken of a person in transmission with visible light is black (no radiation reaching the film), and the image taken with neutrinos would be uniformly bright as the neutrinos do not interact with the patient's tissue. Thus, the availability of penetrating radiation as such is not sufficient for medical imaging.

X-ray exposure of a non-uniform sample should yield spatial differences in the transmitted X-rays because the mass attenuation coefficient varies with sample composition. If these differences are sufficiently large and sufficiently spaced to cause varying exposure across the detector, an image will form. In the current section we study the parameters that influence the contrast of such images.

KEY POINT

Contrast is a quantitative measure of the difference between the lightest and the darkest parts of an image. Physically, it is governed by variations in the yield, $Y(L)$, which in turn is based on the number of photons transmitted through the patient's body: $Y(L) = N(L)/N_0$.

Contrast between two points, A and B, of the image is quantified in the form:

$$C = \frac{N_A - N_B}{N_A} = \frac{Y_A - Y_B}{Y_A}. \quad [24.14]$$

Points A and B are chosen such that $N_A \geq N_B$. The contrast in Eq. [24.14] depends on the same parameters on which N or Y depends: the sample thickness L, the sample density ρ, and its mass attenuation coefficient (μ/ρ).

24.5: Radiation Dose

Note that the lost X-ray intensity in Example 24.5 is not reflected, but is either scattered and absorbed (Compton effect) or directly absorbed (photoelectric effect) by the bone and the tissue. This absorption can cause adverse biological effects in two ways: (i) due to the thermal energy deposited in the tissue and (ii) due to the ionization of chemical bonds as a result of the interaction with an X-ray photon. This dual impact is reflected in the way X-ray loss is measured and reported:

- as the amount of ionization occurring in the material due to the radiation (**exposure dose**), and
- as the energy deposited by the radiation in the material (**absorbed dose**).

Before we can further quantify the biological impact of X-rays, we must define the unit systems used to measure the respective dose or dose rate.

KEY POINT

*The **dose** is the total amount of ionization or energy deposited in a given amount of material, and the **dose rate** is the amount of ionization or energy deposited in a given amount of material per time unit (often reported per hour (h)).*

The first unit of dose we defined was also the first system used historically: the exposure dose is defined as the total charge generated by the ionizing radiation per kilogram of air (unit C/kg). The reference to air at standard conditions (sea level pressure) stems from the original observations of cosmic radiation. Cosmic radiation at sea level generates about 1 ion/(cm^3 · s). As a result, there is a steady state concentration of ions present; this concentration is about 1×10^3 ions/cm^3. A non-standard unit still used for the exposure dose is the unit roentgen (R). It is defined in the form:

$$1\ \text{R} = 23.08 \times 10^9\ \text{ions/cm}^3$$

$$1\ \text{R} = 23.58 \times 10^{-4}\ \text{C/kg}.$$

Therefore, cosmic rays cause an ionization dose rate of $1.0\ \text{ion/(cm}^3 \cdot \text{s)} = 1.2 \times 10^{-13}\ \text{C/(kg} \cdot \text{s)} = 1.7 \times 10^{-6}\ \text{R/h}$, which is about 10% of the total ionization dose rate on Earth's surface, with the rest due to natural radioactive elements in the environment. The ionization dose rate in the van Allen belt is about 23.9×10^7 ions/(cm^3 · s), which equals 3.6×10^{-6} C/(kg · s) = 50 R/h; strong samples in therapeutic nuclear medicine (cobalt source) have ionization dose rates as high as 6×10^9 ions/(cm^3 · s), which equals 7×10^{-4} C/(kg · s) = 10^4 R/h at a distance of 1.0 m from the source.

The unit now more commonly used is the energy dose. The energy dose is the amount of energy deposited per kilogram of air, in unit J/kg. The unit **gray** is introduced as a derived standard unit for the energy dose, 1 Gy = 1 J/kg. This unit has replaced an older, non-standard unit that you may still find in the literature (and which should not be mixed up with the unit radians), the unit rad with a conversion factor 1 Gy = 100 rad. Note that for most biological materials, both dose definitions are roughly equivalent with 1 Gy = 0.026 C/kg.

EXAMPLE 24.8

We study once more the case introduced in Example 24.5. As additional information, we assume a photon fluence of $\phi = 1.0 \times 10^8$ photons/cm^2, in which we define **photon fluence** as the number of photons entering the tissue per unit area, $\phi = N_0/A$. Calculate (a) the energy fluence Ψ at the skin's surface, when the **energy fluence**

continued

is defined as the photon fluence multiplied by the energy of the incoming photons; i.e., $\Psi = \phi\, E_{photon}$; and (b) the energy dose per cm^2 skin for the 3.0 cm thick fat sample.

Solution

Solution to part (a): With a photon energy of 50 keV, we find:

$$\Psi = \phi\, E_{photon} = (1.0\times 10^{8}\ \text{cm}^{-2})(5.0\times 10^{4}\ \text{eV}) = 8.0\times 10^{-7}\ \text{J/cm}^2,$$

in which we used 1.0 eV = 1.6×10^{-19} J. The given photon fluence is typical for a chest X-ray radiogram; with a typical area of A = 1000 cm^2 exposed, we find an energy fluence of almost 1 mJ for the entire chest. Note that the maximum entrance exposure of X-rays is legally regulated.

Solution to part (b): We found in Example 24.5 that only 44% of the photon fluence is stopped in the fat sample. Thus, to obtain the energy dose, we multiply the result in part (a) by 0.44. We must also determine the mass of the fat sample that is affected because the energy dose is defined per kilogram of tissue. The result in part (a) applies to an area of 1 cm^2. The volume of the fat tissue affected is therefore V = (1 cm^2)(3 cm) = 3 cm^3. Using a density of 0.915 g/cm^3 for fat, we find:

$$D_E = \frac{\Psi}{\rho V} = \frac{0.44(8.0\times 10^{-7}\ \text{J/cm}^2)}{(3.0\times 10^{-6}\ \text{m}^3)(915\ \text{kg/m}^3)} = 1.28\times 10^{-4}\ \text{J/kg}$$

$$= 0.128\ \text{mGy}.$$

This is, of course, a small dose compared to whole body doses that affect the health of the patient. Nausea and vomiting would be expected after 3 hours to a day if the dose were increased by a factor of 4000 to 0.5 Gy.

X-rays can damage biological tissue through ionizations or through energy deposition. An example is illustrated in Fig. 24.18, showing the surviving fraction of exposed cells (with a logarithmic scale) as a function of the X-ray energy dose. Curve 1 represents thyroid cells, curve 2 mammary cells, and curve 3 bone marrow. While there is an interval of reduced sensitivity for cells at low doses, all three curves have the same steep slope at larger energy doses. The initial non-exponential behaviour is due to DNA repair mechanisms operating in living cells. UV radiation does not destroy tissue with the same efficiency, due primarily to the limited penetration depth of UV radiation, which does not allow it to reach many body cells, such as thyroid cells.

As dangerous as X-rays are for biological tissue, we usually do not have to worry about them, as far as the environment is concerned. In particular, the X-ray portion of the solar spectrum is not too intensive. This is due partially to the fact that the Sun consists mainly of very light elements and partially to the outer gas layers of the Sun, which absorb most of the more intense radiation generated in the nuclear fusion processes that fuel

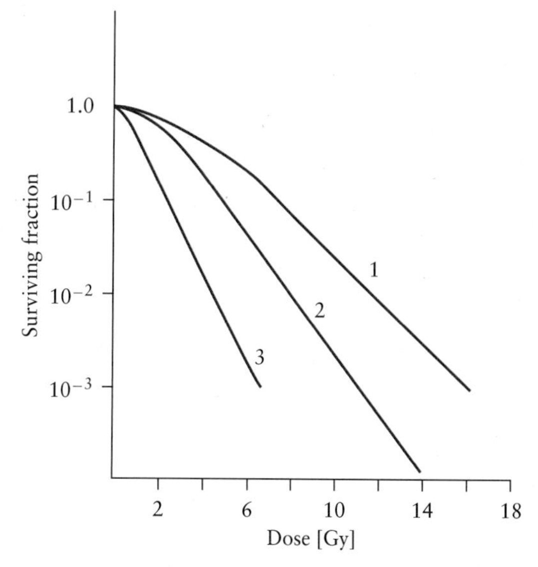

Figure 24.18 The surviving fraction of three types of human cells as a function of energy dose in unit Gy. The energy dose is the energy deposited by the radiation per kilogram of tissue. Note the lower steepness at doses below 1 Gy, which is due to self-repair mechanisms in living cells. Various cells respond with different sensitivity to radiation: (1) thyroid cells, (2) mammary cells, and (3) bone marrow.

the Sun's core. The remaining radiation is absorbed by the upper layers of Earth's atmosphere. For an astronaut, the danger of exposure to X-rays is significantly higher. This applies particularly to the van Allen radiation belt between 1000 and 25 000 km above ground, where intensive streams of energetic particles, mostly electrons, are coupled with an extremely low concentration of gas molecules.

24.6: Appendix: Energy of Scattered Photon for the Compton Effect

A Compton scattering event is the elastic collision of a photon with a free electron at rest. Elastic collisions are governed by the laws of conservation of energy and momentum. We consider the geometry shown in Fig. 24.8, with θ the scattering angle of the photon in the collision. The energy conservation is given as:

$$E_{photon,\ initial} = E_{photon,\ final} + E_{electron}.$$

The three momentum vectors for the electron, $\vec{p}_{electron}$, the initial photon, $\vec{E}_{photon,i}/c$, and the final photon, $\vec{E}_{photon,f}/c$ are shown in Fig. 24.19. Note that the momentum of the photon is expressed as its energy divided by the speed of light. Thus, the momentum conservation in vector notation reads:

$$\vec{E}_{photon,i}/c = \vec{E}_{photon,f}/c + \vec{p}_{electron},$$

where $\vec{p}_{electron} = m\vec{v}_{electron}$ is the momentum of the electron.

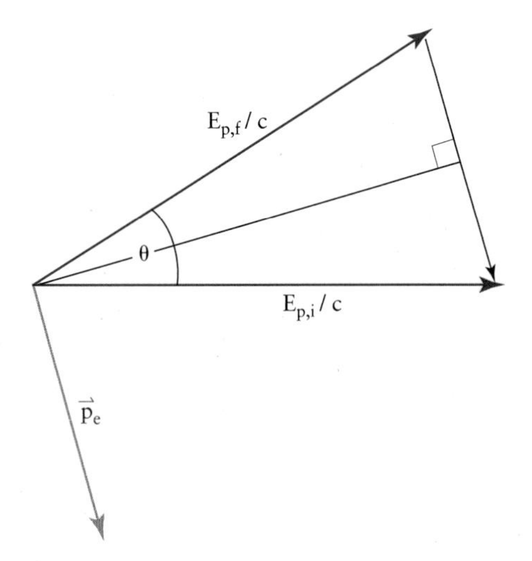

Figure 24.19 Sketch illustrating the conservation of momentum during Compton scattering.

Combining these two equations is possible, but leads to a cumbersome result. Instead, we focus on a limiting case that is often observed: the energy of the photon changes only by a small amount compared to the total energy of the photon; i.e., the electron picks up only a small amount of kinetic energy:

$$\vec{E}_{\text{photon,i}} \cong \vec{E}_{\text{photon,f}} \quad \text{and} \quad \vec{E}_{\text{electron}} = 0.$$

Thus the angle θ in Fig. 24.19 is small and we read from the geometry of the figure that:

$$\frac{1}{2}mv = \frac{E_{\text{photon}}}{c}\sin\frac{\theta}{2}.$$

Multiplying the left side by the speed of the electron, v, allows us to rewrite this equation for the kinetic energy of the electron, E_{electron}:

$$\frac{1}{2}mv^2 = \frac{1}{2}\frac{(mv)^2}{m} = \frac{4E^2_{\text{photon}}}{2mc^2}\sin^2\frac{\theta}{2}.$$

Combining this result with the energy conservation formula yields:

$$E_{\text{photon,i}} - E_{\text{photon,f}} = \frac{2E^2_{\text{photon}}}{mc^2}\sin^2\frac{\theta}{2}.$$

Next, we divide both sides of the equation by $(E_{\text{photon}})^2$ and use the condition $E_{\text{photon,i}} = E_{\text{photon,f}}$:

$$\frac{E_{\text{photon,i}} - E_{\text{photon,f}}}{E^2_{\text{photon}}} \cong \frac{1}{E_{\text{photon,f}}} - \frac{1}{E_{\text{photon,i}}} = \frac{2}{mc^2}\sin^2\frac{\theta}{2}.$$

Last, we solve for $E_{\text{photon,f}}$. This yields Eq. [24.2] when we take the following trigonometric identity into account:

$$2\sin^2\frac{\theta}{2} = 1 - \cos\theta.$$

SUMMARY

DEFINITIONS

- (Photon) Yield Y:

$$Y(L) = \frac{N(L)}{N_0}$$

  $N(L)$ is the number of photons passing a sample of thickness $x = L$, N_0 is the number of incident photons.

- Half-value layer (HVL) is the depth at which 50% of the number of incident X-ray photons are stopped. It relates to the linear attenuation coefficient μ:

$$HVL = \frac{\ln 2}{\mu}$$

- Contrast C between any two points A and B of an image is:

$$C = \frac{N_A - N_B}{N_A} = \frac{Y_A - Y_B}{Y_A}$$

  Points A and B are chosen such that $N_A \geq N_B$. Y is the yield and N is the number of photons passing the sample of given thickness.

UNITS

- Linear attenuation coefficient μ: m^{-1}
- Mass attenuation coefficient μ/ρ: m^2/kg
- Exposure dose D_E: R (roentgen) with:

$$1\ \text{R} = 2.08 \times 10^9\ \frac{\text{ion pairs}}{\text{cm}^3}$$

$$1\ \text{R} = 2.58 \times 10^{-4}\ \frac{\text{c}}{\text{kg}}$$

LAWS

- Ratio of energy loss due to X-ray radiation to thermal energy in the form of heat Q:

$$\frac{E_e \Rightarrow \text{X-ray}}{E_e \Rightarrow Q} = \left(1.22 \times 10^{-7}\ \frac{\%}{\text{eV}}\right) ZE_{\text{kin}}$$

- Z is the atomic number, E_{kin} is the kinetic energy of the incident electron in unit eV. The pre-factor in the bracket leads to the energy fraction in percent (%).
- Compton scattering
  - Energy balance:

$$E_{\text{photon,i}} = E_{\text{photon,f}} + E_{\text{electron}}$$

    $E_{\text{photon,i}}$ is the energy of the incident photon, $E_{\text{photon,f}}$ is the energy of the scattered photon, E_{electron} is the energy of the ejected electron.

- Scattered photon energy:

$$E_{\text{photon, f}} = \frac{E_{\text{photon, i}}}{1 + \frac{E_{\text{photon, i}}}{mc^2}(1 - \cos\theta)}.$$

$E_{\text{photon,i}}$ is the energy of the incident photon, m is the electron mass, c is the vacuum speed of light, and θ is the angle of the scattered photon with respect to the incident trajectory. This formula is an approximation for the case when the energy of the initial and the scattered photons are similar.

- Photon energy and atomic number Z-dependence of the mass attenuation coefficient μ/ρ

$$(\mu/\rho)_{\text{CE}} \propto \text{const for } E_{\text{photon}} < 100 \text{ keV}$$

$$(\mu/\rho)_{\text{CE}} \propto (1/E_{\text{photon}}) \text{ for } E_{\text{photon}} > 300 \text{ keV}$$

CE stands for Compton effect. The coefficient is independent of Z.

- Photoelectric effect
  - Energy balance:

$$E_{\text{kin,electron}} = E_{\text{photon}} - E_{\text{binding}}$$

$E_{\text{kin,electron}}$ is the kinetic energy of the electron, E_{photon} is the incident photon energy, E_{binding} is the binding energy of the electron in its orbital.

  - Photon energy and atomic number Z dependence of the mass attenuation coefficient:

$$(\mu/\rho)_{\text{PE}} \propto \left(\frac{Z}{E_{\text{photon}}}\right)^3$$

- Pair production
  - Energy balance: we require $E_{\text{photon}} \geq 2\ m_e c^2$, i.e., $E_{\text{photon}} \geq 1.02$ MeV. m_e is the mass of the electron

$$E_{\text{kin}}(e^+) + E_{\text{kin}}(e^-) = E_{\text{incident photon}} - 1.02 \text{ MeV}$$

- Attenuation
  - Total attenuation with the linear attenuation coefficient (Beer's law):

$$N(x) = N_0 e^{-\mu \cdot x}$$

N_0 is the number of incident X-ray photons, $N(x)$ is the number of photons transmitted to a depth x, μ is the linear attenuation coefficient.

  - Total attenuation with the mass attenuation coefficient:

$$N(x) = N_0 e^{-(\mu/\rho)\rho \cdot x}$$

μ/ρ is the mass attenuation coefficient, ρ is the density of the tissue.

CONCEPTUAL QUESTIONS

Q–24.1. Explain the difference between Rayleigh scattering and Compton scattering.

Q–24.2. Describe with the help of a diagram the generation of X-rays by the bremsstrahlung mechanism.

Q–24.3. Describe briefly the difference between the photoelectric effect and Compton scattering.

Q–24.4. How are electron–positron pairs produced?

Q–24.5. Explain what is meant by contrast of an X-ray image.

ANALYTICAL PROBLEMS

P–24.1. Using the approach taken in Example 24.3, estimate the energy of K_α X-rays emitted from a gold anode.

P–24.2. The K-shell ionization energy of Cu is 8980 eV and the L-shell ionization energy is 950 eV. Determine the wavelength of the K_α X-rays emission of Cu.

P–24.3. (a) For an imaging application using monoenergetic X-rays, the energy of the incident X-rays has to be changed to increase the probability for the photoelectric effect by a factor of 10. By what factor has the energy to be lowered? (b) If the initial X-ray energy was 50 keV, how would the Compton scattering probability change for the X-ray energy change calculated in part (a)?

P–24.4. Assuming that Eq. [24.2] can be approximately applied at all scattering angles θ, (a) what is the minimum energy of a scattered X-ray photon if the incident X-ray energy is 80 keV? (b) What is the largest possible energy of a scattered X-ray emitted at a scattering angle of 90°?

P–24.5. Soft tissue has a linear attenuation coefficient in the range of $\mu = 0.35$ cm^{-1} at 30 keV and $\mu = 0.16$ cm^{-1} at 100 keV. For this problem we use $\mu = 0.21$ cm^{-1}, which applies at around 50 keV incident X-ray energy. (a) What fraction of X-ray photons at 50 keV are passing through a person's body? *Hint:* the person's body thickness is about 20 cm. Use a soft tissue approximation, i.e., neglecting bones. (b) We compare bone and soft tissue of 2 cm thickness each. Using $\mu_{\text{bone}} = 0.57$ cm^{-1} at 50 keV, what fraction of a 50 keV incident X-ray beam is stopped in bone and soft tissue respectively?

P–24.6. A 180-keV X-ray photon coherently scatters off one of the valence electrons of a nitrogen atom. Assume that the scattering angle of the photon is $\theta = 14°$. (a) Calculate the energy of the scattered photon. (b) Calculate the velocity of the ejected electron. (c) What is the final velocity of the scattered photon? (d) By what angle would the photon have to be scattered in order for its energy to be reduced by 10%? Derive the result as a general formula first, then calculate the particular angle for an incident photon of 180 keV.

P–24.7. (a) For a monoenergetic incident photon beam, the linear attenuation coefficient in a given material is $\mu = 0.35$ cm^{-1}. What is the half-layer value (*HVL*) in that case? (b) In a second experiment, a half-layer value of 24.5 mm is found for the same beam. What is the linear attenuation coefficient in the material studied in this case?

P–24.8. The *HVL* concept is most frequently used when we design a filter to modify the energy distribution of an X-ray beam through beam hardening. Alternatively, the concept can be applied to calculate the thickness of a material required to protect an X-ray technician from X-ray exposure. (a) Find a general expression for the **1% value layer**, i.e., the thickness required to stop 99% of the X-ray photons incident on the material, as a function of the linear attenuation coefficient of that material. (b) Calculate the *1% value-layer* for an 80-keV X-ray photon energy for (i) air, (ii) lucite (which is a plastic used as mammography paddles with $\mu/\rho = 0.1748$ cm^2/g and $\rho = 1.18$ g/cm^3), and (iii) lead (which is used in radiology aprons with $\mu/\rho = 2.316$ cm^2/g and a density $\rho = 11.35$ g/cm^3).

P–24.9. (a) Living cells can be damaged by radiation mechanisms. Describe these mechanisms. (b) Explain the probable impact of the damage on the living cells exposed to radiation. (c) Explain how the impact of the damage depends on (i) the type of radiation the cells are exposed to, (ii) the total dose of radiation, and (iii) the dose rate of the radiation.

P–24.10. The accelerating voltage V and the filament current of an X-ray tube, along with the exposure time t, can be tuned to properly expose an X-ray film for imaging purposes. The expression for X-ray beam exposure at a given point in space is given by $E = kV^2 It$, where k is a constant. (a) Explain, using physical principles, why the X-ray exposure depends on filament current. (b) In order to obtain an X-ray image of the chest of an average-sized individual, the voltage V is adjusted to 75 kV and the tube current to 460 mA. What changes would a radiologist need to make to obtain a correctly exposed X-ray film for the chest of a very large person?

P–24.11. An X-ray tube with a chromium target is being operated by applying an acceleration potential difference of 70 kV. (a) Draw the graph representing the X-ray spectrum emitted by this tube. (b) Label three frequencies on the spectrum and give their numerical values.

P–24.12. What is the element that gives rise to the K_α emission with a wavelength of $\lambda = 2.51 \times 10^{-10}$ m?

P–24.13. Empirical results have revealed that for every unit of radiation dose (1 rad) there are 5×10^{11} inactivating events per cubic centimeter treated. (a) If one were to administer a dose of 10 millirads, how many inactivating events would this dose produce? (b) The approximate volume of a cell is estimated to be 10^{-11} cm^3. Calculate the fraction of the cells of the tissue treated by X-rays that will be hit?

P–24.14. The frequency of the X-rays generated by a laboratory source is 5×10^{19} Hz. (a) Determine the energy of these X-rays in Joules. (b) Determine the energy of these X-rays in electron volts. (c) To ionize a hydrogen atom it takes 13.6 eV. How many hydrogen atoms could be ionized by one of these X-rays?

ANSWERS TO CONCEPT QUESTIONS

Concept Question 24.1(a): The conversion 1 eV = 1.6×10^{-19} J indicates which cases the non-standard unit eV is mostly used: when we deal with particles at the atomic scale and do not want to express energies with large negative exponents.

Concept Question 24.1(b): The energy in standard units for a photon of 25 keV is:

$$E_{\text{photon}} = 25 \text{ keV} = (2.5 \times 10^4 \text{ eV})(1.6 \times 10^{-19} \text{ J/eV})$$
$$= 4.0 \times 10^{-15} \text{ J}.$$

It is more convenient to refer to 25 keV than to 4×10^{-15} J.

Concept Question 24.2 (a): Substituting the given values into Eq. [24.1] yields a value of 0.9% for the ratio. Thus, more than 99% of the energy deposited in the anode has to be dissipated thermally.

Concept Question 24.2 (b): The electrons accelerate to a kinetic energy of 30 keV in this case. Substituting the given values into Eq. [24.1] yields a value of 0.15% for the ratio. Thus, the source in mammography is less suitable for high X-ray doses than a standard chest X-ray set-up.

Concept Question 24.2 (c): Substituting the given values into Eq. [24.1] yields a ratio of 54.2%. Anode heating is a lesser concern for sources used in radiation therapy.

Concept Question 24.3: Yes. Since blue light has a higher frequency than red light, the preferential scattering of blue light causes the sky to be blue.

Concept Question 24.4: Assume that a radiogram is obtained with a particular X-ray photon energy. Only the difference in atomic number Z between soft tissue and bone contributes to the contrast because X-ray photons are stopped proportional to Z^3. With the probability of photons stopped as P we find:

$$\frac{P_{\text{bone}}}{P_{\text{soft tissue}}} = (13/7)^3 = 6.5.$$

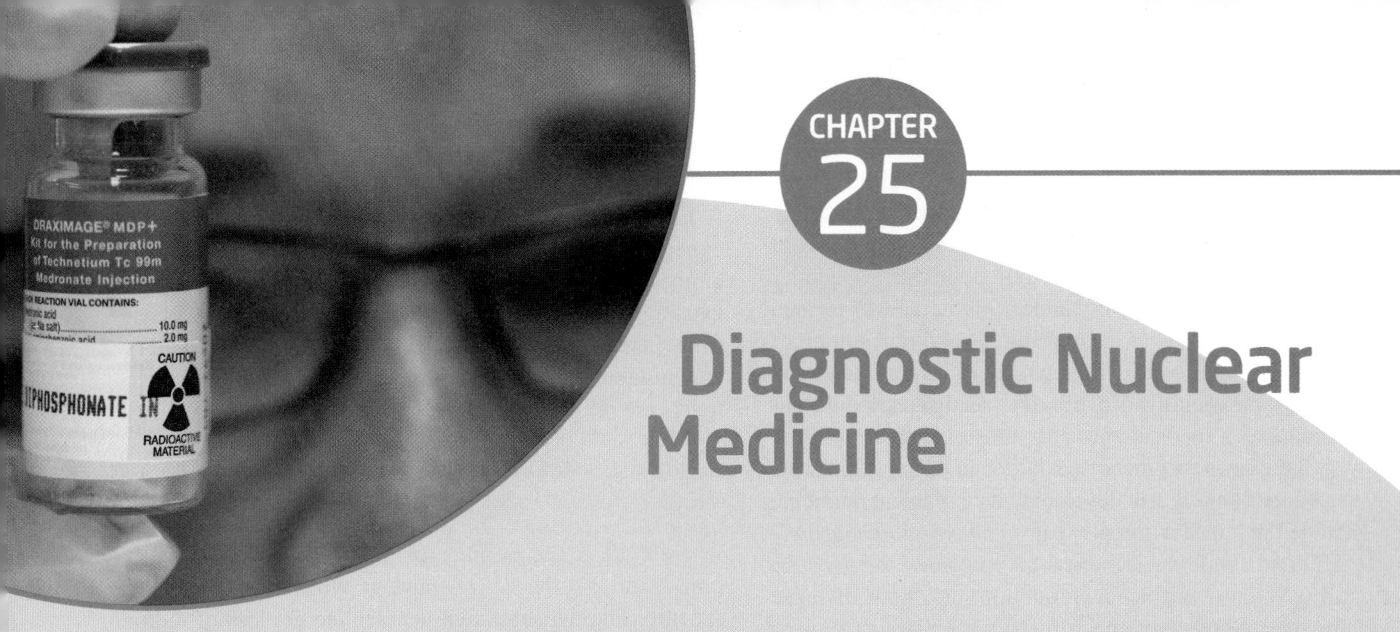

CHAPTER 25

Diagnostic Nuclear Medicine

Suppose you wish to understand why a sophisticated piece of equipment such as a modern automobile experiences a mechanical malfunction. In general, there are two approaches: you could visually inspect the components you suspect are the cause of the problem and then perhaps repair or change them, or you could interrogate the automobile's electronic computer to search for clues to the cause of the malfunction. Both approaches are valid as they complement one another, and technicians will often employ both methods to properly diagnose a malfunction. The first method relies on visually detecting *structural* abnormalities such as a broken belt, a detached gear, etc. The second method relies on the analysis of *functional* information, such as whether the engine is operating at the proper temperature, or whether the fuel meets certain requirements of purity or octane content. Why do we bring this example up? Because the practice of modern radiology, the branch of medicine that concerns itself with the diagnosis of disease through images, uses the same two approaches. On the one hand we have such techniques as conventional X-rays, computed tomography, ultrasound imaging, and mammography, which look at structural or anatomical abnormalities in the body. On the other hand, nuclear medicine focuses specifically on functional or physiological abnormalities. Magnetic resonance imaging is a special in-between technique as it can be used for either anatomical or functional imaging, depending on the specific application. Of course the optimal clinical strategy is to combine the strengths of both anatomical and physiological imaging methods to increase the probability of making a correct diagnosis. This is precisely what the modern practice of radiology is all about. In this chapter we focus on some of the mathematical, physical, and clinical principles of functional or physiological imaging, i.e., nuclear medicine.

25.1: Historical Introduction

Nuclear medicine is a branch of medicine that employs *radioactive isotopes*, also referred to as *radioisotopes* or *radionuclides*, for the diagnosis and treatment of disease. In this chapter, we will concentrate exclusively on the diagnostic aspects of nuclear medicine; we leave the consideration of therapeutic applications aside. In nuclear medicine, radionuclides are chemically bound to physiologically important molecules to form radiopharmaceuticals. These radiopharmaceuticals are then administered to the patient via intravenous injection or oral ingestion, after which they will localize in specific organs or anatomical structures. External detectors can then capture the radiation emitted by the radiopharmaceuticals, and this information can be used to mathematically reconstruct an image. The physiological nature of radiopharmaceuticals allows nuclear medicine to diagnose disease based on the status cellular function or physiology, rather than relying on anatomical changes as do other well-established imaging modalities such as X-rays and ultrasound imaging, which were discussed in previous chapters. Typically in disease, physiological changes precede anatomical changes by weeks or months. Hence, in many instances an earlier diagnosis is possible with the use of nuclear medicine, allowing treatment at an earlier stage, which can translate into a higher cure probability. This advantage is especially significant in the management of malignant tumours, or *cancers*.

The discovery of artificially produced radioisotopes in February 1934 by Frédéric and Irène Joliot–Curie can be considered a most significant milestone in the history of nuclear medicine. An equally important turning point was reached in December 1946 when an article describing the successful treatment of a patient with thyroid cancer using a radioisotope of iodine (I-131) was published in the *Journal of the American Medical Association*. Discovered in 1937 by Carlo Perrier and Emilio Segrè, the most important radionuclide used in nuclear medicine today is technetium-99m (Tc-99m). These were years of remarkable growth for the fledgling medical specialty and, in 1954 the Society of Nuclear Medicine was formed with the objective of promoting the advancement of the discipline in North America. In 1971, the American Medical Association officially recognized nuclear medicine as a medical specialty, thus placing it on a par with the other recognized specialties in terms of prestige and status. More recent developments in nuclear medicine include the introduction of *positron emission tomography* (PET) and *single photon emission computed tomography* (SPECT). In a further evolutionary step, both of these imaging methods have been fused with *computed tomography* (CT) for the purpose of properly dealing with tissue attenuation of the radiation, as well as maximizing diagnostic information by combining physiological images with the superb anatomical detail of CT. Technical aspects of PET and SPECT imaging will be dealt with in some detail in this chapter.

One method of producing radioisotopes for use in nuclear medicine is through the recovery of partially spent fuel in a nuclear reactor. Inside the core of a reactor, nuclei of a so-called fissile material such as U-235 are continuously undergoing fragmentation, or "fission." The process results in the release of thermal energy as well as the production of lighter and highly radioactive nuclei called fission products. One such fission product is molybdenum-99 (Mo-99), which decays to Tc-99m. The Tc-99m then further decays inside the patient, releasing a photon that can be detected with suitable equipment such as a *gamma camera*. Most of the Mo-99 used in North America is produced at the National Research Universal (NRU) reactor at the Chalk River Laboratories in Ontario, Canada. An unanticipated shut down of this reactor led to a serious shortage of the global Tc-99m supply in December of 2007. Another medically useful isotope, fluorine-18 (F-18), is produced with a circular particle accelerator called a *cyclotron*. F-18 decays by the emission of a positive electron, or positron, and the use of this radioisotope provides the foundation for PET. Typically, most hospitals that perform PET procedures will either have their own cyclotrons or be located relatively close to one to ensure availability of this radioisotope.

A patient undergoing a nuclear medicine procedure will receive a small radiation dose comparable to what is received during an X-ray examination. The dose received during a nuclear medicine procedure is measured in SI units of sieverts, or millisieverts (mSv). The effective dose depends on many factors, including the amount of radioactivity administered, the physical properties of the radiopharmaceutical, its distribution in the body, and its rate of clearance. Typical doses for common nuclear medicine procedures range between 1 mSv and 50 mSv.

25.2: Radioactive Decay

It has been verified in countless carefully conducted experiments over the past 100 years that the rate at which atoms in a radioisotope decay is proportional to the existing number of atoms of the radioisotope at any given time. This rate is independent of such external factors as temperature, pressure, humidity, magnetic or electric fields, etc. Expressed in mathematical terms:

$$\frac{\Delta N}{\Delta t} = -\lambda N(t), \qquad [25.1]$$

where $N(t)$ is the number of atoms of radioisotope at time t, and λ is the *decay constant*, which is radioisotope specific. The negative sign denotes the net loss of radioactive atoms per unit time. This decay rate is referred to as the *activity* of the radioisotope, and it is given in SI units of $\sec^{-1}$ or *becquerels (Bq)*. It is useful to solve this equation for $N(t)$ to understand the time behaviour of the radioisotope as a whole. The equation may be solved by taking the limit as Δt approaches zero and then applying standard methods in calculus. The solution to this equation is:

$$N(t) = N_0 e^{-\lambda t}, \qquad [25.2]$$

where N_0 is the initial (i.e., at $t = 0$) number of atoms in the radioisotope and e is the base of the natural logarithm. This type of mathematical relation is called a decaying exponential function. A more useful form of Eq. [25.2] can be obtained by multiplying both sides by λ, giving:

$$A(t) = A_0 e^{-\lambda t}, \qquad [25.3]$$

where A_0 is the initial activity and $A(t)$ is the activity at time t. The amount of time necessary to reduce the activity of the isotope by 50% is known as the *half-life* and is denoted by the symbol $T_{1/2}$. A little algebra shows that $\lambda \cdot T_{1/2} = \ln 2$. An interesting case with practical applications arises when isotope A (called parent) decays to isotope B (daughter), which itself decays according to Eq. [25.3]. In this case, the rate of change of atoms in B is equal to the rate at which they are being produced minus the rate at which they are decaying. But note that atoms B are being produced at the rate of decay of atoms A, or $\lambda_A N_A(t)$, and decay at rate $\lambda_B N_B(t)$, where λ_A and λ_B are the decay constants for isotopes A and B respectively. We can then write the following equation:

$$\frac{\Delta N_B}{\Delta t} = \lambda_A N_A(t) - \lambda_B N_B(t). \qquad [25.4]$$

If we assume the initial activity of B is zero and take the limit as Δt approaches zero, the methods of calculus yield the following expression for the activity of B as a function of time:

$$A_B(t) = A_A(t)\left[\frac{\lambda_B}{\lambda_B - \lambda_A}\right]\left[1 - e^{-(\lambda_B - \lambda_A)t}\right]. \qquad [25.5]$$

Using standard methods in calculus, it can be shown that $A_B(t)$ reaches a maximum at time:

$$t_m = \frac{\ln(\lambda_B/\lambda_A)}{(\lambda_B - \lambda_A)}. \quad [25.6]$$

After a sufficiently large interval of time, note from Eq. [25.5] that the ratio:

$$\frac{A_B(t)}{A_A(t)} \approx \frac{\lambda_B}{\lambda_B - \lambda_A}, \quad [25.7]$$

which is a constant. Here we assumed that $\lambda_B > \lambda_A$. This condition, referred to as *secular equilibrium*, implies that the daughter isotope decays at the same apparent rate as the parent isotope, a phenomenon with important practical applications in nuclear medicine. The case that occupies our attention is that of the isotope Mo-99, with $T_{1/2}$ = 66.7 h decaying to Tc-99m ,which itself decays with a $T_{1/2}$ = 6.02 h, emitting a clinically useful 140-keV photon. The Tc-99m can be chemically separated from the Mo-99 in a portable, self-contained apparatus called a *generator.* This process is called "eluting" or "milking" the generator, and it is carried out repeatedly in a hospital environment according to clinical requirements. The largest activity of Tc-99m is obtained at time t_m, and waiting longer than this is counterproductive as the activity begins to drop. Once eluted, the Tc-99m activity in the generator goes to zero and the process of build-up starts again, until a new maximum is reached after another interval of time t_m. Typically, the process can be repeated some five to ten times before the generator becomes unsuitable for clinical use. Fig. 25.1 shows the relative activities of Mo-99 and Tc-99m as a function of multiple milkings over a period of several days.

EXAMPLE 25.1

Compute the ratio of Tc-99m to Mo-99 activities at secular equilibrium.

Solution

From Eq. [25.7]:

$$\frac{A_B(t)}{A_A} \approx \frac{\lambda_B}{(\lambda_B - \lambda_A)} = \frac{\left[\frac{ln2}{6.02\,h}\right]}{\left[\frac{ln2}{6.02\,h} - \frac{ln2}{66.7\,h}\right]} = 1.999$$

In reality, the correct ratio turns out to be 0.945 due to the fact that Mo-99 decays to Tc-99m only 86% of the time, a subtle phenomenon we will not address further.

25.3: The Detection of Radiation

The detection of radiation in nuclear medicine is a two-step process in which high-energy gamma photons (or rays) are first converted into low energy visible photons, and then the visible photons are converted into electrical pulses. The first step is accomplished in a *scintillator crystal*, or simply crystal; we address this process first. The high-energy gamma photons that originate in the nuclei of radioisotopes make their way easily through the human anatomy and interact with the crystal by exciting electrons, which quickly relax to their ground state by emitting monoenergetic visible light photons in a process called *scintillation*. The energy of these visible photons is a characteristic of the crystal used. A single incoming

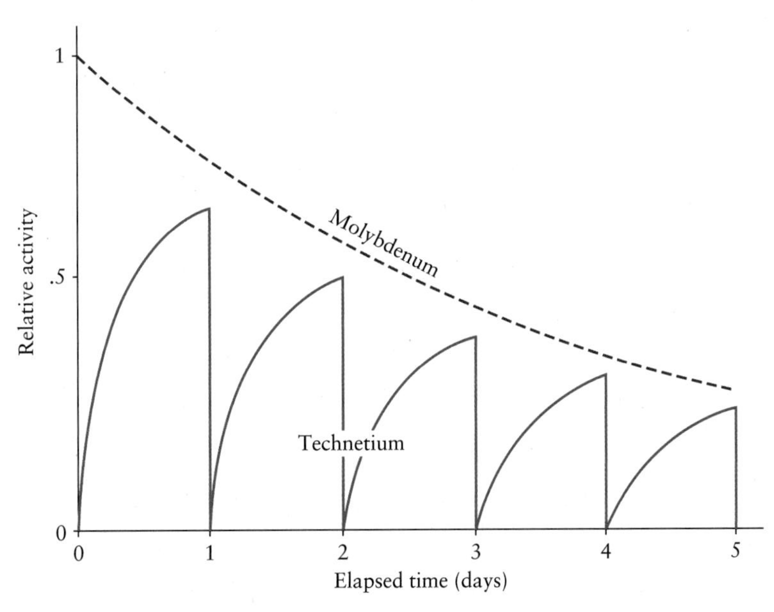

Figure 25.1 Relative activities of Mo-99 and Tc-99m as a function of multiple elutions over several days.

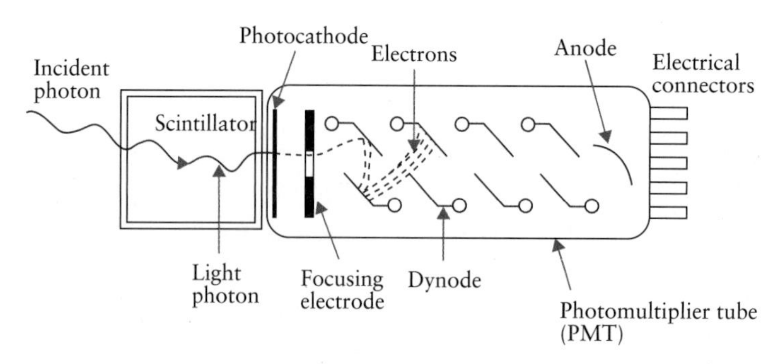

Figure 25.2 Schematic diagram of a typical crystal–PMT assembly employed in nuclear medicine.

gamma ray is capable of producing thousands of these visible photons, but the important concept is that the number of photons produced is always proportional to the energy of the original incoming gamma ray. This is a direct consequence of energy conservation. Crystals are typically constructed of NaI, to which minute amounts of thallium are added for improved scintillation performance. There are several reasons why NaI is chosen for this application, among them its relatively high density (3.67 g/cm^3) and the high atomic number of iodine ($Z = 73$). Both factors increase the probability of capturing the incoming gamma rays, which gives NaI a high gamma-to-visible scintillation conversion efficiency. These crystals do however have the disadvantages of being very sensitive to mechanical and thermal loads (i.e., rapid temperature changes) that can result in fracture. Their performance is also degraded by humidity, thus they need to be sealed hermetically.

The second step in the process, the conversion of visible photons into electrical pulses, takes place further downstream in the *photomultiplier tube* (PMT), which is a glass-encased structure under vacuum. The photons strike the front end, or *photocathode*, of the PMT and, through the photoelectric effect, electrons are ejected. The ratio of electrons ejected per incoming photon is referred to as the *quantum efficiency* of the PMT, and it is always less than unity. Some distance away is a metallic surface called a *dynode* kept at a positive voltage such that the electrons are attracted to it. A focusing electrode may be placed between the photocathode and the dynode to limit electron divergence due to electrostatic repulsion. When each incoming electron strikes this dynode, several secondary electrons are ejected. The ratio of secondary to incoming electrons is called the *multiplication factor*, which is constant for all dynodes and it is always greater than unity. The secondary electrons are attracted to a second dynode, and the process is repeated through some 9 to 12 dynodes such that the overall electron multiplication factor is exponential. The final electron bundle ejected from the last dynode is collected at the PMT's anode and sent downstream for further processing. The total charge collected at the anode is proportional to the number of photons striking the photocathode, which is proportional to the energy of the original gamma ray striking the crystal. Hence, the crystal–PMT assembly can be used not only to detect incident gamma rays, but also to determine their energy and therefore reject pulses from clinically irrelevant rays such as those originating in outer space. Fig. 25.2 shows a schematic diagram of a typical crystal–PMT assembly employed in nuclear medicine.

25.4: The Gamma (Anger) Camera

In the late 1950s, the gamma or Anger camera was developed by Hal Anger, an electrical engineer working at the University of California. It is the most widely used instrument in nuclear medicine today, and hence we will describe its basic operating principles here. The gamma camera can simultaneously record and measure the radioisotope concentration in many organs in its field of view as a function of time. The physiological relevance of this is the ability to examine the rates at which a radiopharmaceutical is accumulated or eliminated from a particular organ. The first component in a gamma camera is a large circular NaI scintillation crystal of the type described previously, with a diameter of around 40 cm and a thickness of 10 mm. Directly behind the crystal is an array of 19, 37, or 61 PMTs, depending on the detector size and other considerations, that detect scintillation flashes induced in the crystal by incident gamma rays. On the outer surface of the camera and directly between the gamma source organ and the crystal, an external removable *collimator* is mounted. This collimator is made of a heavy metal, such as lead, and it contains many thousands of tiny parallel holes whose function it is to allow only the transmission to the crystal of gamma rays entering parallel to the holes, while absorbing rays entering at slant angles. This results in a reduction of image blur produced by scattered radiation associated with gamma rays incident at slanted angles, as illustrated in Fig. 25.3.

When a gamma ray is incident in a vertical or near vertical direction, it will pass through a hole in the collimator and produce a scintillation in the crystal. The positioning circuits then determine where in the crystal the scintillation event took place. When a scintillation takes place, light will be emitted isotropically such that, to some extent, all PMTs will be excited. The intensity of light reaching a given PMT will depend on the distance between it and the scintillation event. Since the signal collected at the anode of each PMT

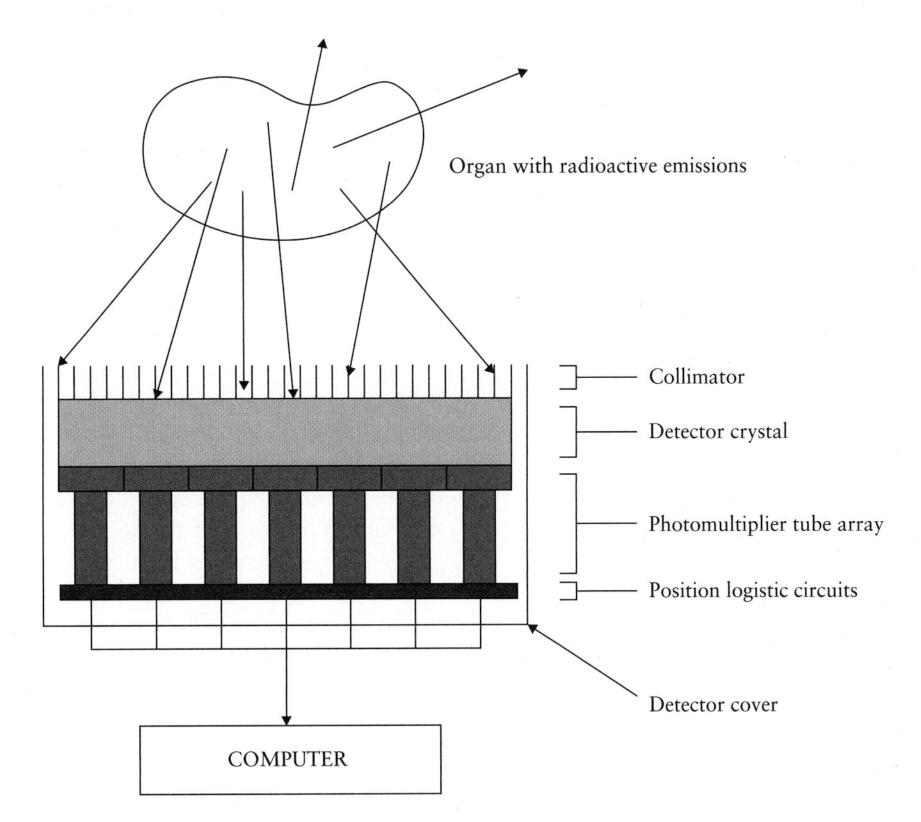

Figure 25.3 Major components of a gamma camera.

is proportional to the intensity of the light reaching it, by the clever use of electronic circuits the distance between each PMT and the scintillation event can be determined. With this information, the exact location of the scintillation event can always be computed; in the early gamma cameras this process was accomplished with analog electronics, in the more modern units it is done digitally. By adding the intensities of all the signals of all the PMTs, the energy of the original gamma ray can be computed. A special energy filter called a *pulse height analyzer* is then used to selectively allow only gamma rays of a specific energy to be recorded by the camera, while rejecting all other gamma energies. This is particularly important because if an organ is being imaged with Tc-99m with 140-keV photons for example, then all other incoming photons of different energies such as cosmic rays and other terrestrial background sources, must be eliminated as they create blur and degrade image quality. Finally, once the proper position of an incoming gamma ray has been determined and the pulse height analyzer determines it is of the proper energy, the gamma ray can be recorded as a valid count on a photographic film, or a computer monitor, or it can be stored in a suitable electronic format for later analysis. By properly placing counts on an *xy*-plane to correspond with the corresponding position where the scintillation took place in the crystal, a two dimensional map of the radioisotope activity distribution may be obtained. This map is referred to clinically as a nuclear medicine *scan*. An example of such a scan is presented later in the chapter.

25.5: Single-Photon Emission Computed Tomography (SPECT) Imaging

A fundamental limitation of the gamma camera is that it produces flat, or *planar*, images in which different organs are superimposed, or "fused," on each other. It may therefore be difficult at times to establish which anatomical structures are actually contributing counts to an image. One method of overcoming this limitation is by combining the use of the gamma camera with the principles of computed tomography. In this technique, called single-photon emission computed tomography (SPECT) imaging, the camera is rotated around the patient so that many images, or *projections*, are obtained, each at a different angle. A special mathematical procedure called *back-projection* is then used to reconstruct two-dimensional images or slices of selected planes inside the anatomy, completely eliminating complicating contributions from all anterior and posterior planes. The process can be loosely compared to studying the interior of a loaf of bread by slicing it into many thin slices and studying each slice carefully while disregarding all other slices momentarily. We present here a very qualitative approach to SPECT imaging, as the mathematical details of the image acquisition and reconstruction processes are much beyond the aims of this text, and are typically encountered in a graduate course in medical physics.

We begin our approach by considering a gamma camera with a collimator that allows only perpendicular photons to pass through, as seen in Fig. 25.4. Note that the patient has three distinct regions of activity, with black representing the highest and white the lowest. The camera then records a radiation intensity, or *projection profile*, which is simply the sum of all the radiation passing through the collimator at each given location. Note that where there is more radioactivity (black), the projection profile is highest, it takes a dip in the region with low radioactivity (white), and vanishes at the edges of the patient.

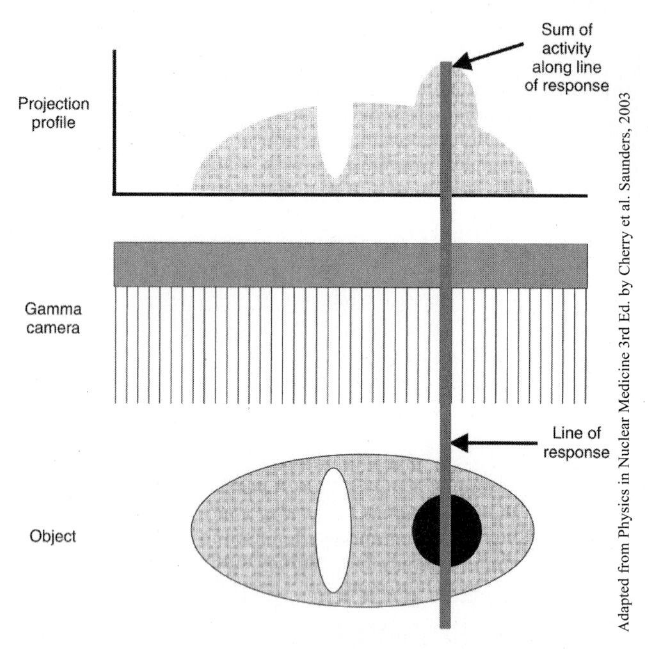

Figure 25.4 Generation of a projection profile in a SPECT system.

Now, if the camera is allowed to rotate around the patient, as seen in Fig. 25.5, the projection profile will change as a function of the angle of rotation. Typically, the camera acquires a projection profile at angular intervals of between 5° and 10°, such that between 36 and 72 projection profiles will be collected for a complete 360° orbit. The data from all profiles are then processed by an onboard computer using the back-projection algorithm to produce a reconstructed image of a single plane, or slice, through the patient.

In order to appreciate at least intuitively how the back-projection algorithm works, suppose we begin with a simple point source of radiation, as illustrated in Fig. 25.6(A). Note that we follow the camera over a 180° degree arc and acquire eight projection profiles at 22.5° intervals. With respect to the fixed camera coordinate system, the profile shifts from the right edge on the upper acquisition, to the left edge on the lower acquisition. If an image matrix is defined such that it corresponds roughly to the dimension of the patient slice itself, as depicted in Fig. 25.6(B), then the profile for each angle can be "back-projected" so to speak into this matrix. Typical matrix sizes used in clinical practice are 64×64 or 128×128 pixels. It is important to remember this back-projection happens in a virtual or mathematical space rather than in any physical dimension. We see the first such back-projection corresponding to the upper acquisition angle in the top illustration of Fig. 25.6(B). Also, in the same figure but below, we see back-projections of 2, 8, and 256 angles respectively. Note that as the number of angles—or profiles—increases, the quality of the reconstruction is improved. A fundamental limitation of the back-projection algorithm is that there will always be some image blur present due

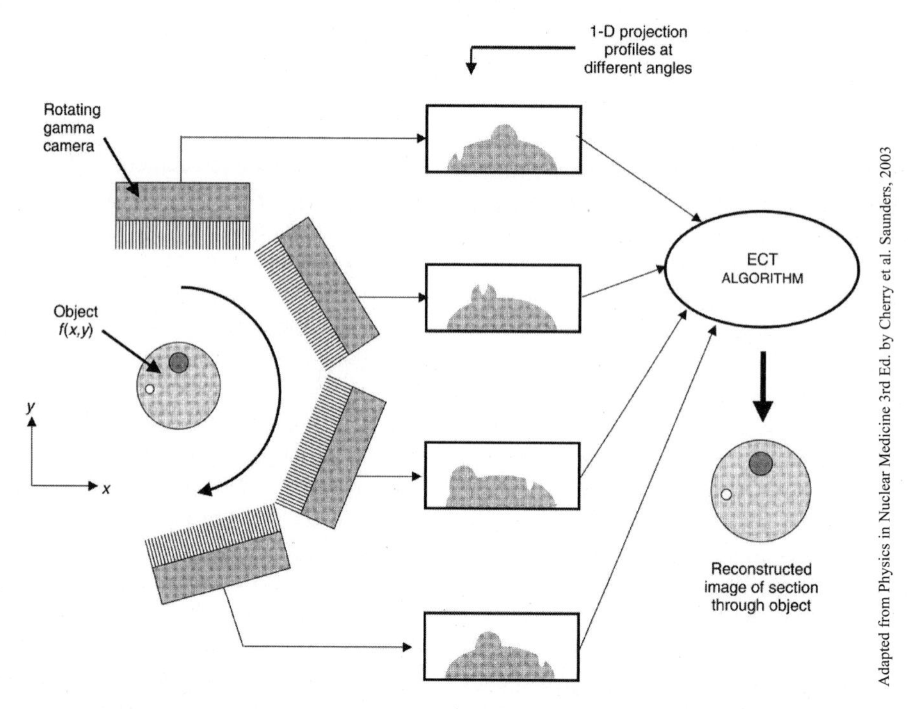

Figure 25.5 Generation of multiple projection profiles from different angles that are then used to reconstruct the image of the slice through the patient.

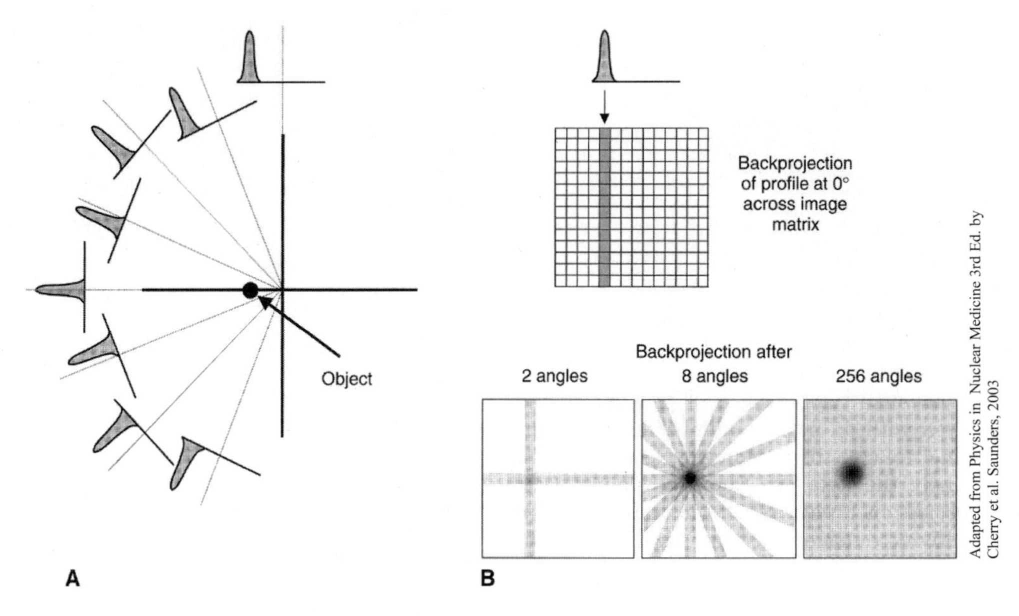

Figure 25.6 The steps in back-projection. A. Generation of projection profiles at different angles. B. Back-projection of a single projection profile (top), and (bottom) generation of the image using multiple back-projections.

to some counts being incorrectly assigned to locations outside the true location of the object. In practice, very sophisticated filtering algorithms are used to eliminate these undesirable blurring effects, but we will not address such advanced topics here. In order to save time in the acquisition of images, modern SPECT systems can have two or three cameras attached to the same rotating gantry, such that the data are gathered two or three times faster than in a single camera system.

25.6: Positron Emission Tomography (PET) Imaging

Positron emission tomography (PET) is another form of nuclear medicine imaging that has similarities, as well as differences, with SPECT imaging. Like SPECT, it relies on back-projection for the mathematical reconstruction of tomographic slices of the anatomy, but both the physical and the biochemical principles employed in generating the data differ significantly from SPECT. In PET, radioisotopes with half-lives of the order of minutes, which decay by emitting positive electrons, or *positrons*, are used. This process in which a proton in the nucleus ejects a positron and turns into a neutron is referred to as *positron emission* and gives the technique its name. Once the positron is ejected within an organ, it travels a fraction of a centimetre before it interacts with a nearby electron by a process called *positron annihilation*. In this process, both the positron and the electron disappear and, in order to conserve charge, energy, and momentum, two 511 keV photons are emitted at nearly 180° to each other, i.e. along a nearly straight line. The energy of each photon is exactly equal to the rest mass energy equivalent of the electron or positron (both are identical), as required by conservation of energy and relativity theory. Because the photons are emitted opposite each other, in order to detect the photon pairs a PET camera consists of a ring of stationary crystal–PMT assemblies surrounding the patient. Based on which two crystal–PMT assemblies are activated, the diameter (or secant) line along which a given annihilation event took place can always be determined. As opposed to SPECT, no collimation is necessary since only events that simultaneously activate diametrically opposite pairs of crystal–PMT assemblies are taken into consideration. The lack of collimators makes PET much more sensitive than SPECT; in other words, less radioactivity is needed to record counts. Such events are referred to as *coincidences*, and all other signals are electronically rejected. PET also has a higher spatial resolution than SPECT, meaning smaller anatomical structures can be resolved. Finally, the data set consisting of all coincidences is mathematically reconstructed using back-projection, much like it is in SPECT. The data acquisition process is illustrated in Fig. 25.7.

One significant clinical advantage of PET with respect to SPECT and conventional nuclear medicine (i.e., the Anger camera) is that radioisotopes of biologically important elements such as carbon, nitrogen, and oxygen are used. These radioisotopes are used to label biologically occurring molecules that then serve as exquisite indicators of physiology, since the body cannot differentiate between a radioactive and a non-radioactive carbon, nitrogen, or oxygen atom. In SPECT and conventional nuclear medicine, non-biologically occurring radioisotopes such as Tc-99m are employed, which may not be fully representative of natural physiology, as the body considers them foreign substances. The single most often used radiopharmaceutical in PET is fluorodeoxyglucose

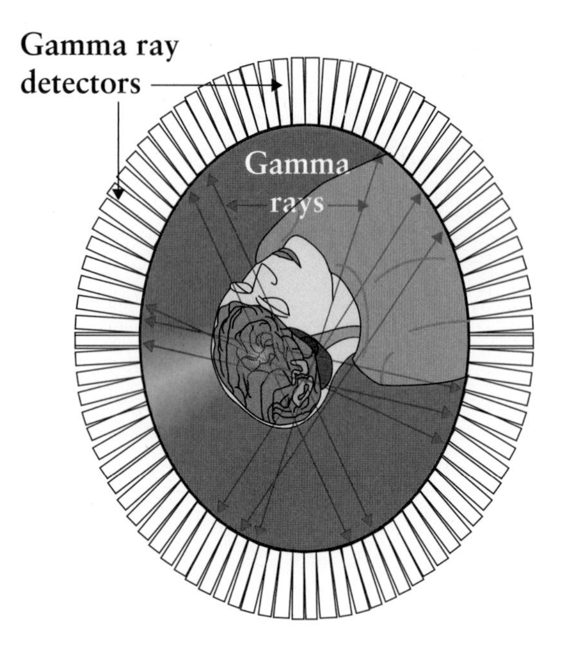

Figure 25.7 The principle behind PET imaging. Note the multiple opposite pairs of photons generated that are detected in coincidence by diametrically opposing pairs of detectors.

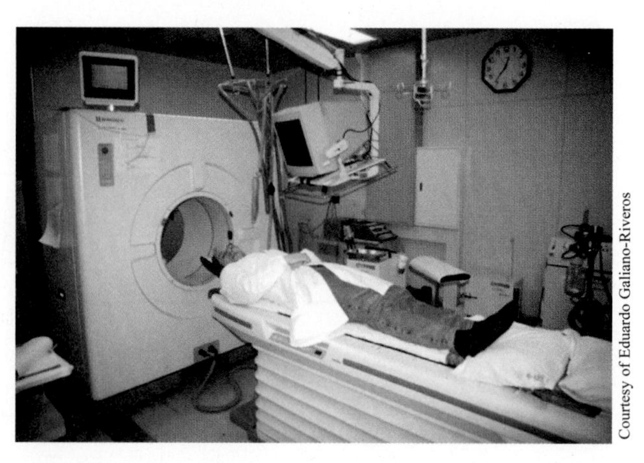

Figure 25.8 The author about to undergo a PET scan for research purposes.

(F-18 FDG), in which one oxygen atom in the glucose molecule is exchanged for a positron-emitting F-18 atom. The resulting glucose is structurally very similar to natural glucose and has a physiological behaviour that is essentially indistinguishable from it. Since glucose is the primary fuel used by cells to power biological processes, its rate of consumption is an excellent indicator of the state of health of a given tissue or organ. A practical example of this is the use of F-18 FDG in the diagnosis of cancer. In cancer, the rate at which malignant cells divide is increased, thus increasing the power requirements since cell division requires energy and thus glucose. By quantifying the rate at which F-18 FDG is consumed in tissues, and comparing these numbers with baseline rates for healthy tissues—along with other clinical considerations—a probable diagnosis of cancer can be made. Other clinical applications of PET include cardiology and neurology. A distinct disadvantage of PET with respect to SPECT and conventional nuclear medicine is the high cost of the camera and cyclotron combination, typically running into the several millions of dollars, thus limiting this technology to larger institutions with appropriate budgets. A picture of the author about to undergo a PET scan for research purposes is shown in Fig. 25.8.

Contrary to Tc-99m and other radioisotopes used in SPECT and conventional nuclear medicine that are produced in a nuclear reactor as described previously, the positron-emitting isotopes used in PET are produced in a special circular accelerator called a cyclotron. Due to the extremely short half-lives of these isotopes, cyclotrons are typically located either on-site or within a few kilometres from the PET imaging centre. Because the method of production of PET isotopes differs fundamentally from that of isotopes used in conventional nuclear medicine and SPECT, we now turn our attention to some specific physical and mathematical aspects of their production. The typical production scheme involves high-energy protons accelerated in a cyclotron to tens of mega-electron volts of kinetic energy and colliding them against neutrons in a suitable target element. The neutrons are knocked off the nuclei while the protons are captured, thus turning into "proton rich" positron emitters. We define the number of protons/sec striking the target as the proton current I, the number of neutrons/cm^2 of area in the target element as the radiological thickness n, and the area of each neutron in cm^2 as the cross section σ. Then the number of radioactive nuclei or atoms N_{rad} created in a time interval Δt will be given by:

$$N_{rad} = In\sigma\Delta t, \qquad [25.8]$$

but they are lost due to radioactive decay at the rate of λN_{rad}. So in a time interval Δt, the net change in the number of radioactive atoms ΔN_{rad} is given by the number produced minus the number lost, or:

$$\Delta N_{rad} = In\sigma\Delta t - \lambda N_{rad}\Delta t. \qquad [25.9]$$

This is a differential equation and it can be solved by taking the limit as Δt approaches zero and using calculus. The solution is:

$$N(t)_{rad} = \frac{In\sigma(1-e^{-\lambda t})}{\lambda} \qquad [25.10]$$

or, since what we are really interested in is the activity $A(t)$ as a function of time:

$$A(t) = \lambda N(t)_{rad} = In\sigma(1 - e^{-\lambda t}). \qquad [25.11]$$

The quantity $In\sigma$ is referred to as the saturation activity A_{sat} since, after a sufficiently long period of time, this will be the maximum value the activity will approach. Other physical phenomena that are described by the same equation—and hence exhibit the same saturation characteristics—are the saturation or final velocity of an object in free fall through the atmosphere, and the charge of a charging capacitor in an RC circuit. Fig. 25.9 shows a plot of $A(t)$ for the production of a typical PET isotope.

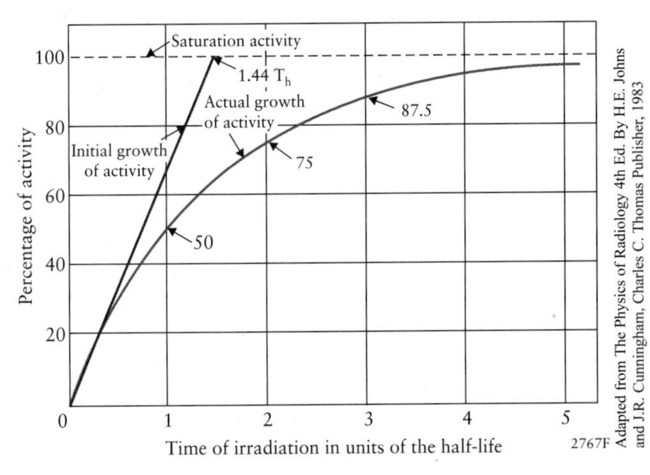

Adapted from The Physics of Radiology 4th Ed. By H.E. Johns and J.R. Cunningham, Charles C. Thomas Publisher, 1983

Figure 25.9 Plot of activity $A(t)$ as a function of time during cyclotron irradiation for the production of a typical PET isotope. Time is expressed in units of half-lives, and activity as a percent of A_{sat}.

EXAMPLE 25.2

Suppose a cyclotron with $I = 1.25 \times 10^{14}$ prot/sec is used to irradiate a suitable target with $n = 10^{20}$ neutrons/cm^2 for the production of F-18 ($T_{1/2} = 2$ h). Furthermore, suppose $\sigma = 3 \times 10^{-26}$ cm^2 for the reaction. Compute (a) A_{sat} for this reaction, (b) the activity after 6 hours of irradiation, and (c) the number of F-18 nuclei (or atoms) after 6 hours of irradiation.

Solution

a) $A_{sat} = In\sigma = (1.25 \times 10^{14}$ prot/sec$)(10^{20}$ neutrons/cm$^2)(3 \times 10^{-26}$ cm$^2) = 3.7 \times 10^8$ Bq.

b) From Eq. [25.11], $A(t) = In\sigma(1 - e^{-\lambda t}) = 3.7 \times 10^8$ Bq$(1 - e^{-(.693)(3)}) = 3.23 \times 10^8$ Bq.

c) Also from Eq. [25.11], $N(t)_{rad} = A(t)/\lambda = (3.23 \times 10^8$ Bq$)(7200$ s$/0.693) = 3.35 \times 10^{12}$ atoms.

25.7: Clinical Applications

Nuclear medicine in any of its three modalities is fundamentally a functional or physiological imaging method, rather than anatomic imaging method. One of the most frequently performed procedures in nuclear medicine is the *bone scan*, with the intent of assessing the state of health of bony structures. Because of its clinical importance, we will examine some basic aspects of this procedure. The procedure starts with a molecule called methylene diphosphonate (MDP), which has a biochemical behaviour similar to the Ca^{+2} ion. The Ca^{+2} ion is collected and metabolized by the internal structure of healthy bone—playing an important role in bone physiology—particularly in bone growth. Therefore, regions with active bone growth will have increased uptake of Ca^{+2}, and hence MDP. The MDP molecule is labelled with Tc-99m and administered to the patient intravenously. It takes about 3 hours after administration for the Tc-99m–MDP molecule to be properly deposited in the skeleton, and imaging is performed typically at this time. The usual injected activity is around 700 MBq. Certain types of cancers, such as those of the breast and prostate, have an unfortunate tendency to spread quickly to distant organs, predominantly the bones. Such spread is called *metastasis* and is typically associated with terminal disease, which can be treated, but not cured. Metastatic deposits in bone are associated with rapid but abnormal bone growth that can be detected easily with a bone scan weeks or months before they can be detected with conventional radiological methods such as radiographs or computed tomography. This early detection capability occurs because, in disease, physiological changes precede structural anatomic changes by weeks or months. When an early diagnosis of metastasis is made, appropriate treatment can be initiated earlier, which may be associated with longer and better quality survival. It also avoids the unnecessary burden of surgical procedures when it is clear the disease is terminal. Because of its inexpensive, safe, effective, and non-invasive nature, combined with its early diagnostic ability, the bone scan has played a historic role in establishing the clinical importance of nuclear medicine. Fig. 25.10 shows a bone scan obtained at the author's nuclear medicine laboratory of a patient with primary breast cancer that has metastasized to the skeleton. It is a posterior view of the thoracic region. The multiple yellow areas in the ribs and vertebral bodies correspond to metastatic sites, indicative of advanced disease.

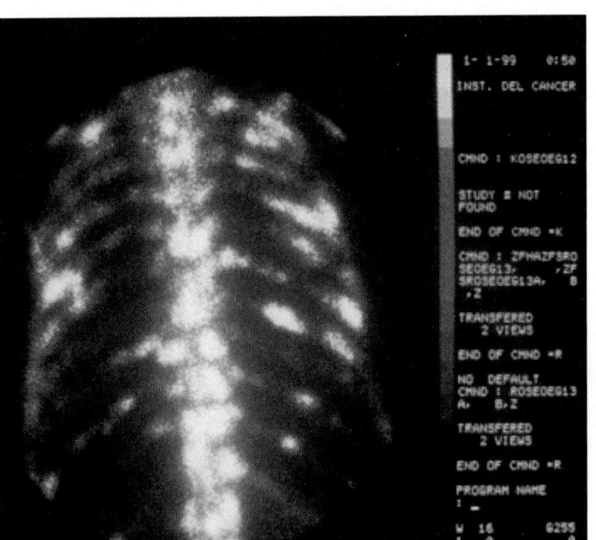

Courtesy of Eduardo Galiano-Riveros

Figure 25.10 Bone scan of patient with primary breast cancer that has metastasized to the skeleton. The multiple yellow areas in the ribs and vertebral bodies correspond to metastatic sites. Technical scan data appear on the right side of the image.

SUMMARY

DEFINITIONS

- Nuclear medicine is a branch of medicine that employs radioactive isotopes—also referred to as radioisotopes or radionuclides—for the diagnosis and treatment of disease.
- The physiological nature of nuclear medicine allows the diagnosis of disease based on the status of cellular function, rather than relying on anatomical changes, as other imaging modalities such as X-rays, CT, and ultrasound imaging.
- The gamma camera is an instrument that can simultaneously record and measure the radioisotope concentration in many organs as a function of time, which allows study of the rates at which a radiopharmaceutical is accumulated or eliminated from a particular organ.
- SPECT imaging is an evolution of the gamma camera and involves collecting data over a 360° angle and reconstructing a tomographic image using a mathematical technique called back-projection.
- PET imaging is a different modality that involves detecting annihilation photons from a positron emitting isotope. Its main advantage is that isotopes of naturally occurring elements such as carbon, oxygen, and nitrogen can be imaged. It is the only method currently available to image glucose metabolism.

UNITS

- The decay rate of a radioisotope is referred to as the activity, and it is given in SI units of sec^{-1} or becquerels (Bq).
- The decay constant λ of each isotope is unique and is given in units of sec^{-1}.

LAWS

- The activity of a radioisotope as a function of time, $A(t)$, is given by the equation $A(t) = A_0 e^{-\lambda t}$, where A_0 is the initial activity.
- If parent radioisotope A decays to daughter radioisotope B, then the activity of B as a function of time is given by

$$A_B(t) = A_A(t)\left[\frac{\lambda_B}{\lambda_B - \lambda_A}\right]\left[1 - e^{-(\lambda_B - \lambda_A)t}\right],$$

where $A_A(t)$ is the activity of the parent at time t.

- After a sufficiently long time, the ratio of parent-to-daughter activities is given by the approximate expression

$$\frac{A_B(t)}{A_A(t)} \approx \frac{\lambda_B}{\lambda_B - \lambda_A}.$$

- The activity of a cyclotron-produced isotope as a function of irradiation time is given by the expression $A(t) = In\sigma(1 - e^{-\lambda t})$, where $In\sigma$ is the saturation activity.

CONCEPTUAL QUESTIONS

Q–25.1. Explain how nuclear medicine differs from other medical imaging techniques, discussing relative advantages and disadvantages.

Q–25.2. Differentiate by structure and function, a radioisotope from a radiopharmaceutical.

Q–25.3. Explain the advantage of employing a Mo-99/Tc-99m generator for clinical use in nuclear medicine.

Q–25.4. Invoking energy conservation, explain why the number of monoenergetic visible photons produced in a scintillation crystal must be proportional to the energy of the incoming gamma ray. Can the constant of proportionality be greater than unity? Explain.

Q–25.5. With the aid of a diagram, explain how a collimator works to reduce the scattered radiation reaching the detector, hence improving image quality.

Q–25.6. Suggest a method by which the distance between a scintillation event and a particular PMT can be determined in an Anger-type camera.

Q–25.7. Suggest an explanation for how the use of multiple detectors in a SPECT system can save time in the image-acquisition process.

Q–25.8. Determine the minimum total angle that a SPECT camera has to rotate through in order to acquire sufficient data to reconstruct a full image of a slice through a patient. In working out this problem, a series of sketches may prove useful.

Q–25.9. In PET, the use of collimators is not required, which results in a greater sensitivity compared to SPECT. Explain with a diagram why collimators are not required in PET.

Q–25.10. In PET, is it possible to have a coincidence from an annihilation event that does not occur along the straight line joining the two activated crystal–PMT assemblies? If so, explain with a diagram.

ANALYTICAL PROBLEMS

P–25.1. Show that $\lambda T_{1/2} = \ln 2$.

P–25.2. Show that, in a generator at $t = t_m$, $A_B(t) = A_A(t)$.

P–25.3. Show that, in a generator, the maximum activity of Tc-99m is reached at $t \approx 23$ hours.

P–25.4. For a 50-mg Ra-226 sample ($T_{1/2} = 1600$ years) with an activity of 1.83×10^9 Bq, compute (a) λ; (b) the number of atoms in the sample, based on mass considerations; and (c) the number of atoms in the sample based on its activity.

P–25.5. In a special generator, Te-131 ($T_{1/2} = 30$ hours) decays to I-131 ($T_{1/2} = 8$ days). Initially, the activities of Te-131 and I-131 are 1.85×10^8 Bq and zero respectively. (a) Compute t_m for the I-131 activity. (b) What are the Te-131 and I-131 activities at t_m? (c) How many atoms of each exist at t_m?

P–25.6. Suppose a given crystal–PMT assembly has a quantum efficiency of 0.3 and a multiplication factor of 3. Furthermore, the crystal is capable of generating a visible photon for each 30 eV of energy deposited on it. The PMT has 12 dynodes in it. Compute the charge collected at the anode when a single Tc-99m 140-keV photon strikes the crystal.

P–25.7. A cyclotron with $I = 1.00 \times 10^{14}$ prot/sec is used to irradiate a 1 g O-18 target with $n = 5 \times 10^{20}$ neutrons/cm^2 for the production of F-18 ($T_{1/2} = 2$ h). Furthermore, suppose $\sigma = 3 \times 10^{-26}$ cm^2 for the reaction. Compute (a) A_{sat} for this reaction, (b) the activity after 6 hours of irradiation, and (c) the ratio of F-18 to O-18 atoms after 6 hours of irradiation.

P–25.8. For the situation described in problem 7, compute (a) the irradiation time required for the F-18 activity to equal 10^9 Bq and (b) the additional activity generated by irradiating for another 6 hours, i.e. for a total of 12 hours. (c) Comment on whether this constitutes a reasonable method of generating more F-18 activity.

P–25.9. The human body normally has some amount of radioactivity due to a variety of factors, most notably the uptake of fallout from atomic tests in the atmosphere. Assume 32 million atoms undergo radioactive decay in your body every day; compute your activity.

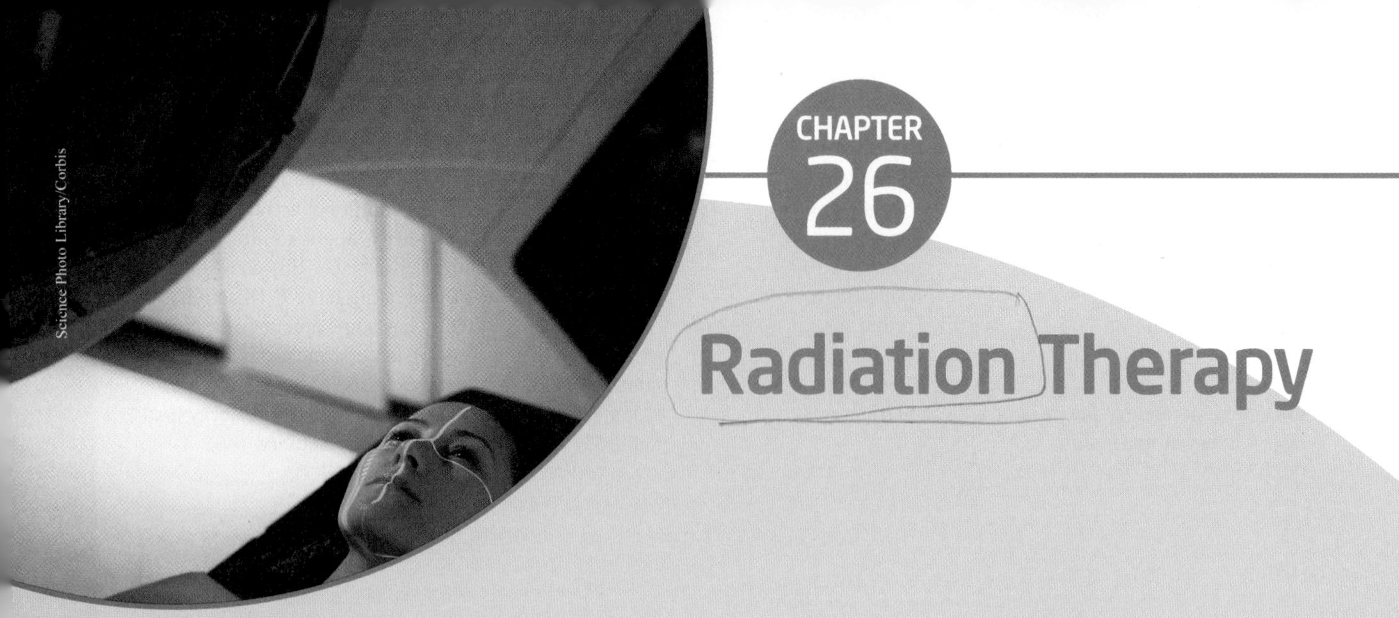

CHAPTER 26

Radiation Therapy

In the automobile malfunction analogy discussed in the introductory paragraph of the previous chapter, we looked at the problem of properly diagnosing the malfunction. In general, this is only part of the picture, if we assume that the overall intent is to correct the fault. In other words, proper malfunction diagnosis is a necessary but not sufficient step in the overall fault correction process. Once proper information on the nature of the fault is obtained, a decision must be made on how to correct the fault. The same logical process is typically followed in the diagnosis and treatment of human disease. Frequently, imaging modalities such as nuclear medicine are used to diagnose malignant tumours, or cancer, which are abnormal growths that can compress or invade adjacent structures and, more ominously, have the capacity to extend to distant organs. In deciding how to treat cancer, physicians, physicists, and other healthcare professionals, have three tools generally available to them: surgery, drugs (chemotherapy), and radiation. The application of radiation, or radiotherapy, to treat cancer consists of the precise and deliberate use of ionizing radiation on malignant tumours to arrest their growth, with the intent of curing the patient or at least relieving the symptoms of the disease to improve quality of life. The central problem in the practice of radiotherapy can be articulated in the following way: how can a prescribed dose of radiation be precisely delivered to a malignant tumour to sterilize it, while delivering only a minimal tolerable dose to the surrounding healthy structures? In this chapter we will introduce some of the physical concepts necessary to understand and manage this problem, along with some tools and techniques in widespread clinical use throughout the world.

26.1: Introduction

The purpose of radiotherapy is the clinical control of *malignant tumours*, or *cancers*. It is therefore important, before we start our study of radiotherapy, to give a precise definition of what a cancer is. Cancer is defined as an abnormal growth anywhere in the body that (a) has the capacity to grow uncontrollably and therefore compress or invade adjacent structures, and/or (b) has the capacity to extend to distant organs in a process called *metastasis*. Cancer presents a significant medical problem because it has the ability to either kill a patient or substantially degrade his quality of life. Presently, three different treatment modalities are employed in the treatment of cancer: surgery, chemotherapy (the use of anticancer drugs), and radiation. Typically, these treatments are used in combination, and about half of all cancer patients require radiation as part of their treatment. To put things in perspective, any given individual has a one in three probability of being diagnosed with cancer at some point in their lifetime, and then a roughly 50% chance of requiring radiation as part of their treatment. Thus on average, one-sixth of the general population will undergo radiation treatment during their lifetime—not an insignificant number!

We can now define radiotherapy as the careful and deliberate use of ionizing radiation on malignant tumours to arrest their growth to either cure the patient or, when cure is out of reach, at least relieve the symptoms of his disease. Aside from the radiation oncologist, who is a medical doctor specializing in radiotherapy, a key individual on the radiotherapy team is the medical physicist, who applies his/her knowledge of radiological physics to ensure the patient receives the correct dose in the proper anatomical locations as specified by the radiation oncologist. Hence, physicists are very interested—and play an important role—in the delivery of radiation treatments. With very few exceptions, radiation treatments in North America are delivered with photons, ranging in energies between 6 MeV and 20 MeV produced by linear accelerators, which are also capable of producing electron beams. Thus, in this chapter we will focus our attention on the operation of linear accelerators and for historical purposes, its predecessor the Co-60 unit. In very few instances, patients may be treated with either protons or neutrons produced by cyclotrons, but such treatments are typically available only at large research institutions and we will not address these further.

The clinical foundations of radiotherapy are better understood by considering the quantitative effects of a given dose, D, of radiation on a specific cell population representing a malignant tumour. The surviving fraction of cells, S, is given by the expression:

$$S = e^{-(\alpha D+\beta D^2)} \qquad [26.1]$$

where α and β are tissue-specific constants expressed in units of Gy^{-1} and Gy^{-2} respectively. It follows that the ratio α/β is also tissue specific and is associated with the tissue's sensitivity to radiation, ranging from 5 Gy to 25 Gy for malignant tissues and from 2 Gy to 5 Gy for normal tissues. This fundamental radiobiological difference between malignant and normal cells is the crucial foundation of modern radiotherapy; it typically results in a greater survival (or lower killing) of normal cells with respect to those that are malignant, for any given dose. The ratio of normal to malignant cell survival for a given dose is a measure of the *therapeutic ratio*, and it is a function of, among other factors, the specific tumour type and the type of radiation employed. It is imperative that this ratio always be greater than unity for the treatment to be effective. More loosely, we can refer to malignant cells as being more *radiosensitive* than normal cells or, alternatively, we can say normal cells are more *radioresistant* than malignant cells. We need to carefully consider the effects of radiation on both normal and malignant tissue because, in any treatment, some amount of dose will inevitably be absorbed by normal tissues surrounding the tumour, and this additional dose may cause clinical complications. Eq. [26.1] is referred to as the *linear-quadratic* (LQ) model of radiobiological cell survival, since the exponential term contains both a linear and a quadratic term in D.

Another equally important phenomenon exploited in modern radiotherapy is the general ability of normal cells to repair radiation damage more quickly than malignant cells. In order to maximize the clinical benefits of this preferential repair capability of normal cells, modern treatments are delivered in daily doses, or *fractions*, such that about a 24-hour repair interval is allowed between fractions. Knowing that after a given fraction more normal than malignant cells survive, and that after a repair interval more normal than malignant cells repair damage, we can combine those two effects to actually enhance the therapeutic ratio. Now consider what can happen to a given cell, malignant or normal, during the repair interval after receiving a fraction of radiation. In this case, exactly one of three outcomes will follow: the cell will die, it will survive without dividing, or it will survive and divide (more specifically it will undergo *mitosis*). Which outcome a given cell will undergo is a completely stochastic matter, and therefore cannot be predicted with absolute certainty. However, statistical methods can be applied to this problem to compute mean fractional survival during the repair interval.

In an attempt to better understand cell population dynamics during radiotherapy treatments, a research group of which this author is a member has coupled the LQ model, which predicts survival after a fraction, with a statistical technique called a *Markov Chain* to analyze survival during the repair interval. To see how the model works, consider the following realistic case: We take a typical human cancer with α and β set at 0.4 Gy^{-1} and 0.003 Gy^{-2} respectively, and a typical fractionated schedule of 2 Gy per fraction, 5 days per week, with breaks on weekends. The initial tumour cell population was set at 10^9 cells, which is representative of a tumour at the time of diagnosis. Fig. 26.1 shows the number of surviving cells as a function of treatment time. Twenty-seven fractions (denoted by solid points) delivered on every weekday for a total of 38 days were required for the population to reach zero, i.e., to "cure" the patient. In Fig. 26.2, the data are displayed in semilogarithmic manner; they clearly exhibit an attenuation in cell death during weekends, as expected since there is no cell killing and only cell proliferation during those intervals. Also note from Fig. 26.1 that, in absolute terms, most cell killing occurred during the first week of treatment, a fact well known to clinicians and patients alike. In other words, tumours tend to shrink rapidly during the first week, but there is a reduction in shrink rate in the later stages of treatment. The model correctly predicts the clinical outcome of actual patient treatments.

26.2: The Co-60 Unit

X-rays were discovered in 1895 by Wilhelm Roentgen; their ability to damage biological tissue was recognized almost immediately. Within one year of their discovery, X-rays were being routinely used in Europe and North America for the treatment of a variety of diseases, including cancer. The ability of these early units to effectively treat deep-seated tumours however was seriously compromised due to the low energy of the photons they produced, which could reach depths of the order of only 1 cm. It wasn't until the 1950s with the advent of high energy, or megavoltage, machines that the effective treatment of tumours located deep within the anatomy was made possible. We will focus our attention on the two most important such devices: the cobalt-60 (Co-60) unit and the medical linear accelerator.

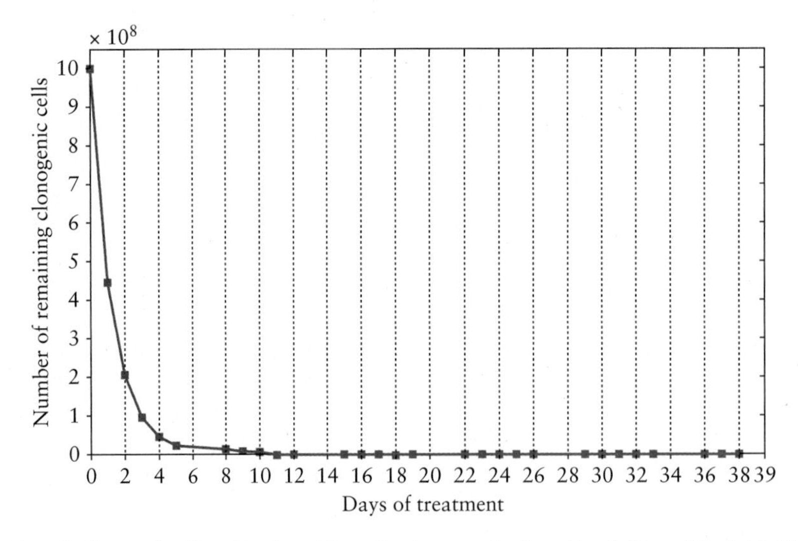

Figure 26.1 Surviving number of cells as a function of treatment time. Twenty-seven fractions (denoted by solid points) delivered on every weekday for a total of 38 days are required for the tumour cell population to reach zero, i.e., to "cure" the patient. Note that most gains in cell killing occur during the first week of treatment.

The Co-60 unit is being phased out in the industrialized world due primarily to the risks associated with housing a strong radioactive source in a clinical setting; however, the reliability, simplicity, and low cost of these machines guarantee they will continue to see widespread use for many years to come in the developing world.

As the name implies, a Co-60 unit has as its most important component a Co-60 source. It is produced by neutron bombardment of stable ^{59}Co in a reactor. The isotope has a mean photon energy of 1.25 MeV and a 5.25 year half-life, attractive physical characteristics for the treatment of deeper tumours. The first such source was produced in 1951 at Chalk River's NRX reactor in Ontario and shipped to Prof. H.E. John's laboratory at the University of Saskatchewan, in Saskatoon. The first patient was treated in November of that year, thus initiating the modern era of megavoltage, external-beam radiotherapy. The Co-60 in the form of pellets is encased in a 1.0 to 2.0 cm diameter stainless steel cylinder to prevent any leakage of radioactivity. The source, consisting of the pellets and the steel container, is housed in the *treatment head*, which consists of lead for shielding purposes, surrounded by a steel outer case for structural integrity. Within the treatment head, a sliding or rotating mechanism is provided to move the source from the "off" or stored position, to the "on" or treatment position, in which the radiation is allowed to escape through an orifice referred to as the *primary collimator*. All Co-60 units incorporate by design a failsafe mechanism that automatically retracts the source to the "off" position in case of a power failure. Further downstream, the *secondary* or *adjustable collimators* are located. Typically, they are sliding rectangular jaws made of a very dense metal, such

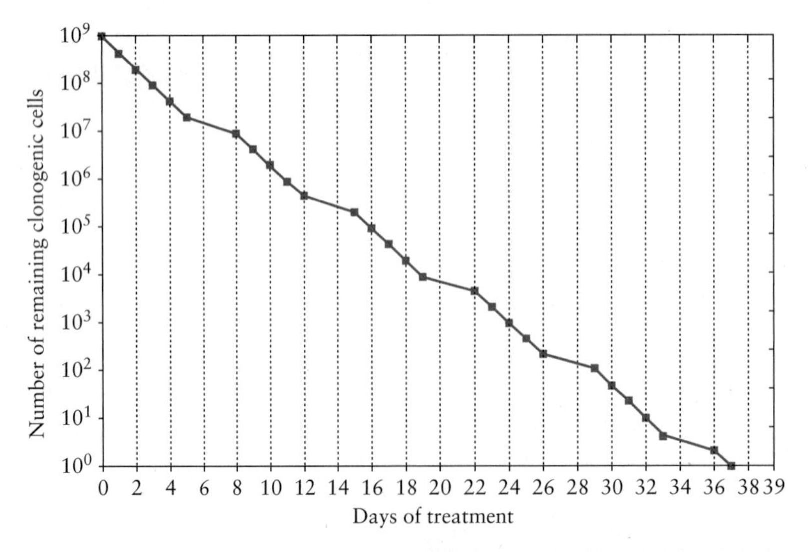

Figure 26.2 Data displayed in semilogarithmic manner clearly exhibit an attenuation in cell death during weekends, since there is no cell killing and only cell proliferation during those intervals.

as depleted uranium, that allow the final shaping of the beam into squares (or rectangles) ranging in size from about 4 cm × 4 cm to about 40 cm × 40 cm at the proper treatment distance referred to as the *isocentre*. One distinct physical characteristic of all Co-60 units is the existence of the *penumbra* effect. Because of the physical dimensions of the actual source, at the edges of the radiation field the collimators only partially block the source and thus the dose changes rapidly from a maximum value to nearly zero as a function of distance from the beam axis. This edge region with high dose gradients is referred to as the penumbra, and is physically analogous to the penumbra effect—the partial darkening of Earth during a solar eclipse. The penumbra, which effectively results in the radiation field edges being somewhat "fuzzy," can have desirable or undesirable clinical consequences. For example, in a large tumour with poorly defined borders such as those in the abdominal cavity, such fuzzy field edges can actually be an advantage in sterilizing undetected malignant cells surrounding the known tumour. On the other hand, if one is dealing with a small, sharply defined tumour such as those in certain structures of the brain, fuzzy field edges cannot be tolerated, as some dose will inevitably be absorbed by healthy critical surrounding structures.

Virtually all Co-60 units manufactured in the past 50 years are of the *isocentric* variety. In this design, the whole treatment head is mounted on a rotating gantry that allows the treatment head to rotate 360° around the patient, who is supported on a special couch. The centre of rotation of the treatment head is an imaginary point in space called the *isocentre*, and the source to isocentre distance is referred to as the treatment distance. Co-60 machines have a treatment distance of either 80 cm or 100 cm. The clinical advantage of the isocentric design is that the tumour is placed precisely at the isocentre, and then the treatment head can be rotated to deliver beams from multiple angles without ever moving the patient. This strategy then results in multiple beams converging at the tumour, thus depositing a very high dose while keeping the dose to surrounding tissues to tolerable levels. Fig. 26.3 shows a modern isocentric 100 cm machine built by Atomic Energy of Canada Limited (AECL) installed at the National Cancer Hospital in Paraguay, where the author maintains a clinical appointment. This machine incorporates all the technical features described above.

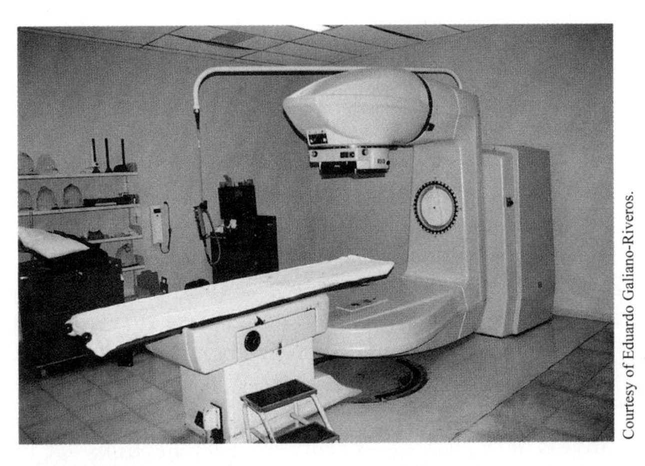

Courtesy of Eduardo Galiano-Riveros.

Figure 26.3 A modern isocentric Co-60 treatment unit.

26.3: The Medical Linear Accelerator

The medical linear accelerator, or *linac*, is a device that employs high frequency electromagnetic waves to accelerate electrons to speeds comparable to the speed of light, and kinetic energies of many millions of electron volts (eV) or mega electronvolt (MeV). The acceleration phase takes place in a straight or linear cavity called the *waveguide*, hence the name "linear" accelerator. The high energy electron beam is then dumped on a high density metallic target constructed typically of tungsten, which results in the production of highly forward-scattered bremsstrahlung X-rays. After careful shaping of the X-ray beam—a process referred to as *collimation*—the beam can be directed toward regions of the anatomy where malignant tumours are seated. A less frequently employed technique involves the removal of the metallic target, allowing the electron beam to treat the tumour directly.

The electrons are accelerated using the electric field component of the high frequency, high power, electromagnetic (EM) wave or field that exists in the waveguide. The EM field, which is in the microwave region of the EM spectrum, is generated by a special power electronic component called a *magnetron* for low energy accelerators of up to about 6 MeV, or a *klystron* for high energy machines of up to 25 MeV. Both of these devices produce microwaves by the cyclic acceleration of electrons under the influence of electric fields. The microwaves are then dumped into the waveguide, or accelerator proper, where electrons are introduced with an initial kinetic energy of about 50 keV. The electrons then progressively gain kinetic energy as they glide down the waveguide structure, much like a surfer gains kinetic energy by climbing aboard an incoming wave on the sea. When the electrons emerge from the distal end of the waveguide, they are focused onto the X-ray target, which is cooled by running chilled water through it to prevent it from melting down. The resulting bremsstrahlung X-rays have energies ranging from effectively zero all the way up to the kinetic energy of the electrons striking the target; their mean energy however is approximately one third of the electron beam's kinetic energy. The accelerating waveguide structure and the target assembly are maintained at a high vacuum to minimize the probability of interaction of the electrons with air molecules. Because the bremsstrahlung X-rays produced at the target are forward directed, with a pronounced intensity peak in the beam centreline, a *flattening filter* is typically positioned immediately downstream of the target. This filter, which consists of a conically shaped metallic piece, absorbs photons in the beam centreline such that the resulting beam has a more uniform intensity profile. The photon beam is then transmitted through an ion chamber, which monitors beam diagnostics such as dose rate, cumulative dose, and uniformity. If any clinically significant deviation from pre-established parameters is detected, the beam is immediately shut off! Finally, before exiting the accelerator, the beam is shaped to the required clinical specifications by a set of metallic "jaws" or collimators. In older systems, the beam could be collimated only in rectangular shapes ranging in lateral dimensions from 4 cm to 40 cm, but in newer systems the beam can be collimated in arbitrary shapes to exactly match the size and shape of the actual tumours. The beam size and shape must be such as to cover the whole extension of the tumour, but not much larger in order to minimize the dose to the surrounding healthy tissues. Finally, as is the case with modern Co-60 units, all modern linear accelerators are isocentrically mounted on a rotating gantry capable of a 360° rotation. Fig. 26.4 shows a patient being positioned for treatment in a 6 MeV accelerator that incorporates all the technical features described above.

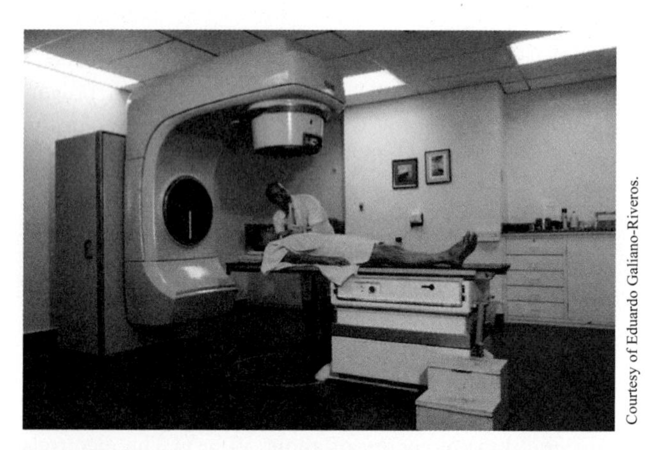

Figure 26.4 Patient being positioned for treatment in a 6 MeV linac.

EXAMPLE 26.1

Focusing back on the basic physics of electron acceleration and invoking the theory of relativity, the ratio Y of the total relativistic energy E_{rel} to the electron rest mass energy equivalent E_0 is defined as:

$$\gamma = E_{rel}/E_0 = (E_{kin} + E_0)/E_0 = [1 - (v/c)^2]^{-1/2}, \qquad [26.2]$$

where E_{kin} is the electron's kinetic energy, v is the electron's velocity, and c is the speed of light in a vacuum. Compute the velocity of an electron that strikes the target of an accelerator with an E_{kin} of 15 MeV.

continued

Solution

First note that the electron rest mass energy equivalent in MeV is given by:

$$E_0 = m_0c^2 = (9 \times 10^{-31})(3 \times 10^8)^2(6.25 \times 10^{-14}) \text{ MeV/J} = 0.51 \text{ MeV}.$$

Now, from Eq. [26.2]:

$$[(E_{kin} + E_0)/E_0]^{-2} = 1 - (v/c)^2$$
$$1 - [(E_{kin} + E_0)/E_0]^{-2} = (v/c)^2$$
$$\{1 - [(E_{kin} + E_0)/E_0]^{-2}\}^{0.5} = v/c$$
$$c\{1 - [(E_{kin} + E_0)/E_0]^{-2}\}^{0.5} = v.$$

Substituting values:

$$3 \times 10^8\{1 - [(15 + 0.51)/0.51]^{-2}\}^{0.5} = 2.99 \times 10^8 \text{ m/s}.$$

This electron is travelling at more than 99% of the speed of light!

26.4: The Percent Depth Dose Function

Since from a molecular point of view human tissue is composed primarily of water (80% or so), it has been established experimentally that radiation transport in water closely mimics radiation transport in tissue. Therefore typically all physical measurements on clinical radiation beams are performed in specially designed water containers called *phantoms*, as it is impractical to obtain these measurements on actual patients. In fact, the most physically relevant parameters of any clinical radiation beam are defined based on measurements in water phantoms. Perhaps the single most important of these is the *percent depth dose* (PDD) function. By definition, it is given by the expression:

$$\text{PDD} = [D(d)/D_{max}] \times 100\%, \qquad [26.3]$$

where $D(d)$ is the dose at depth d in the phantom, and D_{max} is the maximum dose, both along the central beam axis as seen in Fig. 26.5. Note that the maximum possible value of the PDD function is 100% (see Conceptual Question 4).

In general, as a function of depth d in the phantom, the PDD function starts at an intermediate value and increases monotonically until D_{max} is reached. After this point, the function decreases more or less exponentially, asymptotically reaching zero at infinite depth. The initial region of monotonic increase is referred to as the *build-up region*; it is clinically important because it results in a lower dose to the skin, which is a very radiosensitive structure. This effect—referred to clinically as *skin sparing*—occurs because electrons set in motion by Compton interactions must travel a certain

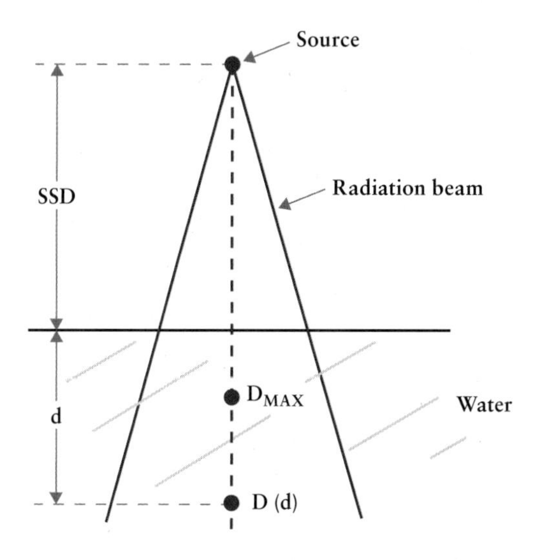

Figure 26.5 The water phantom geometry used in the definition of the PDD function.

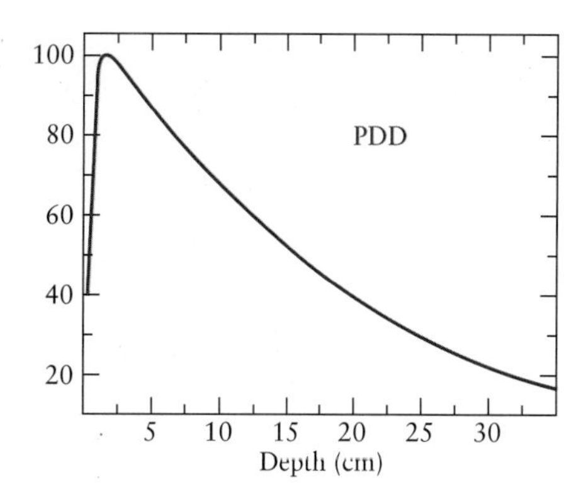

Figure 26.6 PDD curve for a 6 MeV, 10 × 10 cm beam with a 100 cm SSD. D_{max} is reached at a depth of 1.5 cm.

distance before they begin depositing dose by ionizations. This results in a low entrance dose that increases gradually until D_{max} is reached. For a typical Co-60 unit, D_{max} is reached at a depth of 0.5 cm, and for a 6 MeV accelerator it is about 1.5 cm. Beyond this depth, the exponential attenuation of the photons results in the creation of fewer Compton electrons and, in general, the dose falls more or less exponentially. Aside from the depth in phantom, the PDD function depends on three other factors: beam size, beam energy, and source to surface distance (SSD). The SSD is defined as the distance between the radiation source—such as the Co-60 source in a cobalt unit or the target in a linac—and the surface of the water, as shown in Fig. 26.5. It is generally the case that for a given depth, the PDD will increase with increasing field size due to more dose being deposited at the beam axis from Compton photons set in motion in the distal regions of the beam. However, after a certain field size of about 25 × 25 cm is reached, there is no further increase in PDD with field size increase because the distal electrons cannot reach the central axis. In terms of beam energy, it is generally the case that for a given depth, the PDD will increase with increasing energy. This can be understood based on the fact that a more energetic beam will undergo less attenuation for a given depth, and therefore will deposit a greater dose. Finally, an increase in the SSD typically results in an increase in PDD for a given depth. General tables of PDD as a function of *d*, SSD, beam energy, and beam size have been published, but these tables should be used as general guidelines only since each treatment machine is unique. It is therefore important that the physicist in charge of a given treatment machine generate specific PDD tables for that machine before it is used clinically to treat patients. A representative PDD curve for a 6 MeV, 10 × 10 cm beam with a 100 cm SSD is shown in Fig. 26.6. Note that D_{max} is reached at a depth of about 1.5 cm.

The PDD function is a very useful tool in understanding the physical aspects of a radiation beam, but it does have one important limitation: it provides no information on the dose deposited off the central beam axis. As an off-axis extension of the PDD function, we introduce the concept of the isodose curves. This corresponds to a graphical representation of the relative dose at any point within the beam normalized in percentage form to D_{max}. The clinical advantage of isodose curves is that they provide an instant and very accurate picture of the dose anywhere within the beam, which is fundamental in estimating the dose to healthy tissues typically off-axis. Fig. 26.7 shows the isodose curves for a 6 MeV beam for a 10 × 10 cm field with an SSD of 100 cm.

EXAMPLE 26.2

A dose of 2.0 Gy is to be delivered to a tumour located at a depth of 8 cm. A 6 × 6 cm, 6 MeV beam is to be used with a 100 cm SSD. The PDD for this beam at the given depth is 72.7%. Compute the maximum dose received at any location within the patient's anatomy.

Solution

By the definition of PDD function:

$$D_{max} = [D(d)/\text{PDD}] \times 100\%$$
$$= [2 \text{ Gy}/72.7\%] \times 100\%$$
$$= 2.75 \text{ Gy}.$$

Note that this dose is delivered to tissue at a point about 1.5 cm below the skin; the radiosensitivity of anatomical structures at this depth must be taken into consideration before delivering such a treatment.

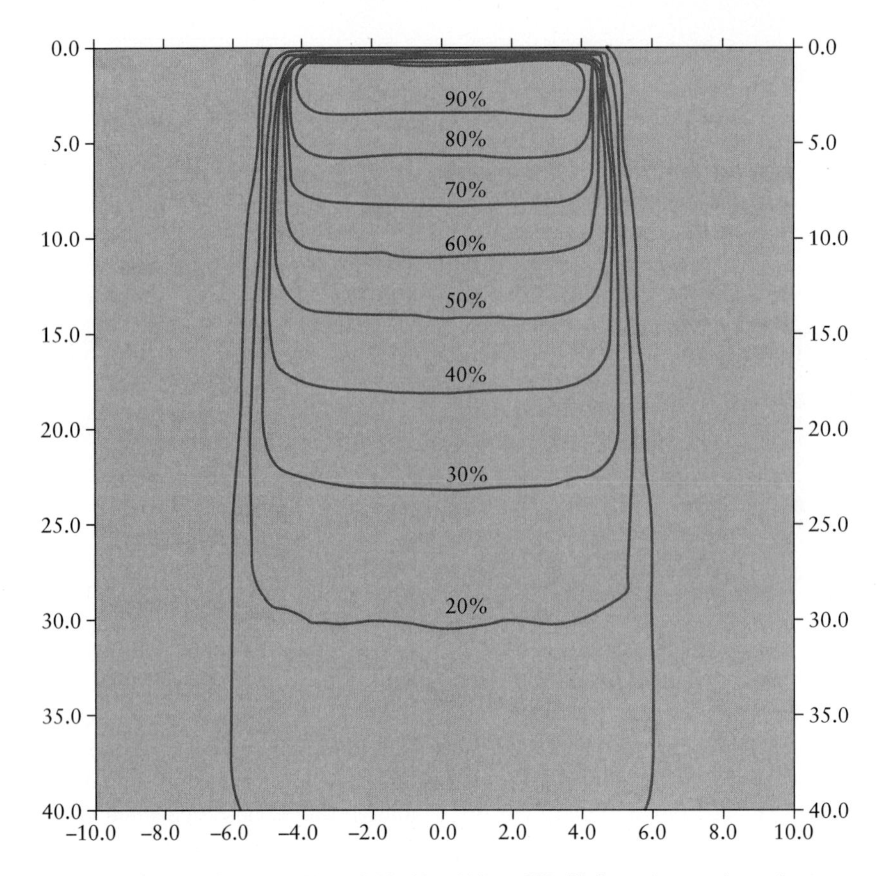

Figure 26.7 Isodose curves for a 6 MeV linac for a 10 × 10 cm field with a 100 cm SSD. All dimensions are in centimetres.

26.5: The Tissue Air Ratio Function

Of equal importance in radiotherapy calculations is the *tissue to air ratio*, or TAR function. It is defined as the ratio of the dose at depth d in a phantom, to the dose under identical conditions in air, with the proviso that both measurements are made at the isocentric radius of rotation of the beam. The geometry used in this definition is illustrated in Fig. 26.8.

In mathematical form:

$$\text{TAR} = D(d)/D_{\text{air}}. \qquad [27.4]$$

The definition was introduced in 1953 by H. E. Johns at the University of Saskatchewan, in Saskatoon, specifically

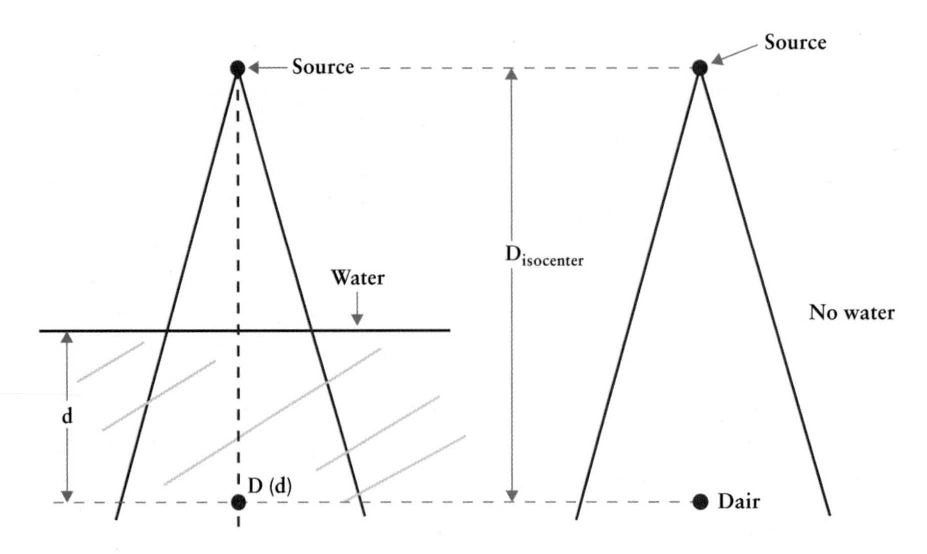

Figure 26.8 The water phantom geometry used in the definition of the TAR function.

to facilitate treatments with the newly introduced isocentric Co-60 units. A physical explanation of isocentric treatments was given in the section on the Co-60 unit, but note that using the PDD function would be cumbersome because as the beam rotates around the patient, both the SSD and tumour depth *d* change. Since the PDD function depends on both SSD and *d*, it would have to be recalculated for every beam angle used in the treatment. In introducing the TAR function—which he originally referred to as the tumour air ratio—Prof. Johns realized that the SSD dependence could be eliminated if the tumour was located at the isocentre. In other words, the TAR function is independent of the SSD. It does, however, depend on the tumour depth *d*, field size, and beam energy. We now examine how TAR depends on these three factors.

Let's look first at the effect of tumour depth *d* on TAR. Because the photon beam is exponentially attenuated between the phantom surface and the axis of rotation but D_{air} remains constant, in general for a given beam, TAR decreases more or less exponentially with tumour depth *d*. Much like the case for the PDD function, TAR will increase with increasing field size due to more dose being deposited at the beam axis from Compton photons set in motion in the distal regions of the beam in the phantom. D_{air} increases very slowly with field size increases, thus TAR follows the increase in $D(d)$ with field size. Again, much like the case for PDD, after a certain field size of about 25 × 25 cm is reached, there is no further increase in TAR with field size increases because the distal electrons cannot reach the central axis. Finally, for a given field size and depth *d*, it is generally the case that TAR increases with beam energy. This can be understood on the basis of a reduced in-phantom attenuation for a more energetic beam resulting in a higher value of $D(d)$, with a slower increase in D_{air}. The combination of these factors results in an increase in TAR. Even though general TAR tables have been published, just as in the case for PDD, the physicist should always generate TAR tables for each specific treatment machine before it is used clinically.

EXAMPLE 26.3

A tumour is to be treated isocentrically with a Co-60 unit to a dose of 1.8 Gy. The tumour is situated at a depth of 8 cm and the field size to be used is 6 × 6 cm. The physicist measured the TAR for this beam under the given conditions to be 0.74. The machine's in-air dose rate at isocentre is measured to be 1.5 Gy/min. Compute the treatment time needed.

Solution

Since the treatment is isocentric, from Eq. [26.3]:

$$\begin{aligned} D_{air} &= D(d)/\text{TAR} \\ &= 1.8\ \text{Gy}/0.74 \\ &= 2.43\ \text{Gy}, \end{aligned}$$

continued

but note that:

$$\begin{aligned} t_{treat} &= D_{air}/\text{dose rate @ isocentre} \\ &= 2.43\ \text{Gy}/1.5\ \text{Gy/min} \\ &= 1.62\ \text{min} \\ &= 97\ \text{s}. \end{aligned}$$

Due to subtle but important physical considerations, the TAR function is not used to characterize beams of energies above those of Co-60. For such higher energy beams, a modified version of the TAR function called the *tissue maximum ratio* (TMR) function is used. The differences between the TAR and TMR functions are beyond the focus of this book.

26.6: Clinical Applications

We choose as a clinical example to end this chapter with, a case of a brain tumour. In North America about 50 000 new cases of primary (i.e., non-metastatic) malignant brain tumours are diagnosed on a yearly basis. This accounts for only about 1.5% of all cancers but for about 3% of annual cancer deaths, making these tumours very deadly. In an especially cruel statistical twist, these tumours account for about 25% of all pediatric cancer cases, thus children carry an inordinate mortality burden from this disease. Brain tumours are invariably serious because of their invasive character, and their detection typically occurs in advanced stages when the presence of the tumour has side effects that cause unexplained symptoms. Among such symptoms are tissue compression that may cause vomiting or headaches, neurological deficits that may result in cognitive and behavioural changes, or physical impairment such as partial paralyses. The most common type of brain tumour is called *glioblastoma multiforme* (GBM), a tumour that affects the *glial* cells of the brain. This is also the deadliest form of brain cancer; mean survival after diagnosis is measured typically in a handful months. The treatment, which is palliative in nature, consists of a multi-modality approach, i.e., surgery, radiation, and chemotherapy used in combination. A typical strategy if the tumour at the time of diagnosis exceeds a certain size (for example 7 cm in diameter) is to administer preoperative radiotherapy in order to reduce the tumour size such that surgical resection is safer and more effective. After surgical removal of the tumour, postoperative radiotherapy may be administered to sterilize any malignant tissue left behind in the surgical bed. Chemotherapy is usually administered to sterilize any additional malignant cells flowing in the circulatory system.

In Fig. 26.9 we see an MRI image of an adult patient with a GBM, with an approximate diameter of 6 cm, in the left temporal lobe. This is a voluminous tumour that, among other symptoms, was causing a compressive paralysis in the patient. The surgical resection of such a

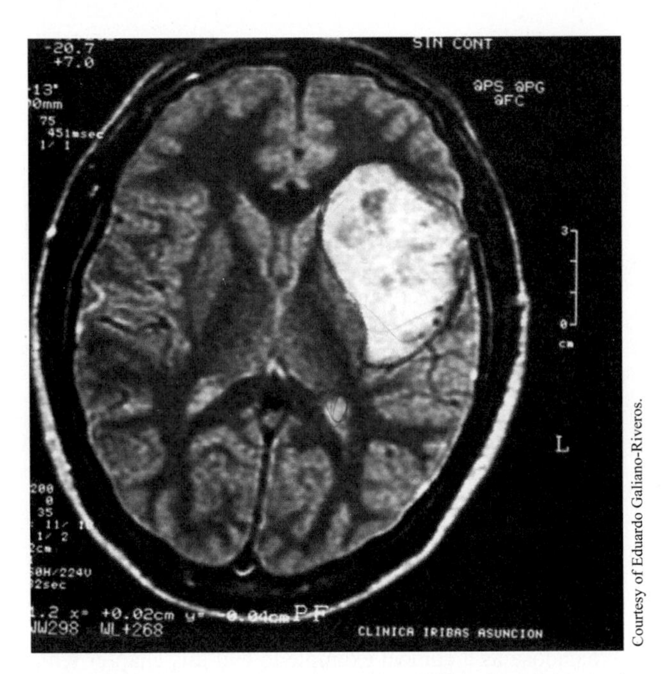

Courtesy of Eduardo Galiano-Riveros.

Figure 26.9 MRI image of a patient with a GBM in the left temporal lobe (outlined manually with red marker) with an approximate diameter of 6 cm. This massive tumour was causing a compressive paralysis in the patient.

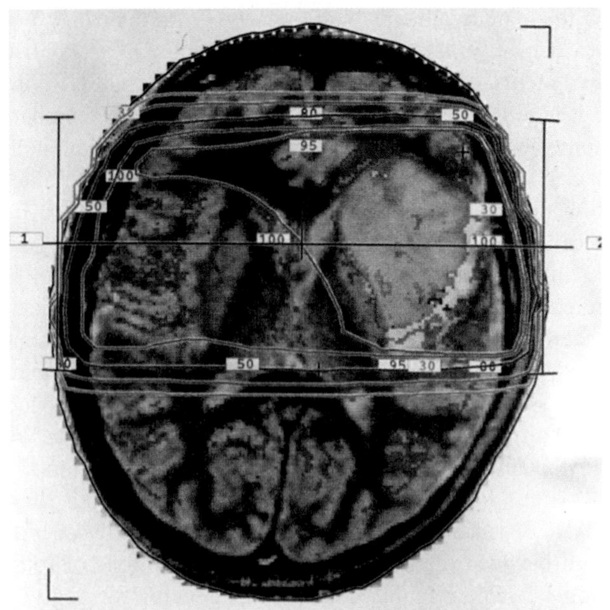

Figure 26.10 Mathematical simulation of the dose distribution. The two, parallel opposed, lateral, collinear beams are labelled as 1 and 2. Note the weighing of the dose distribution toward the left hemisphere, to better match the location of the tumour.

large mass was considered very risky for the patient and instead a preoperative course of radiotherapy was considered the best option. When this patient was seen at the author's institution, the radiation oncologist in consultation with the medical physicist, decided to treat the patient to a dose of 60 Gy in 30 daily fractions, with two lateral opposing fields using a 6 MeV photon beam. The dose distribution in the brain was simulated through computational techniques for different beam angles and weighing factors until a reasonable compromise was reached, *before* the patient was ever irradiated! By "reasonable compromise" we mean a high enough dose to the tumour, a surrounding margin, and its likely routes of spread, with a tolerable dose to the other healthy structures of the brain. Fig. 26.10 shows the results of the mathematical simulation called a *treatment plan*, of the dose distribution superimposed on an image of the brain. Labelled as 1 is the right beam (the patient's feet are toward the reader) and as 2 the left beam, both delineated by bars. Note that both beams are opposed but collinear, this is a standard geometry in radiotherapy referred to as a *parallel opposed* geometry. The dose distribution is normalized to 100 (in orange), such that 100 corresponds to 60 Gy or maximum dose, etc. Note that the region of maximum dose is heavily biased toward the left hemisphere to account for the location of the tumour mass, which brings us to a more subtle point. In this case, in order to better match the dose distribution to the anatomical location of the tumour, more weight is given to beam 2. In other words, for every Gy deposited at the centre of the tumour from beam 1, beam 2 will deposit 2 Gy, thus shifting the dose distribution curves to the left. This is a simple 2 to 1 weighing ratio, and in practice it is achieved by leaving beam 2 on longer than beam 1 (although it is not as simple as leaving it on twice as long, as other non-linear factors come into play). Also note that the region of maximum dose has a lengthy extension to the right hemisphere; this is a deliberate attempt to treat probable areas of microscopic disease and/or likely routes of spread of the disease. When looking at these pictures, it is important to keep in mind that the human brain and the tumour are three-dimensional structures that we are representing in two-dimensional figures. State-of-the-art imaging technologies now available are capable of simulating the three-dimensional nature of a tumour by rotating it in virtual space.

SUMMARY

DEFINITIONS

- Radiotherapy is defined as the precise use of ionizing radiation on malignant tumours to arrest their growth to either cure the patient or improve his/her quality of life.
- The fundamental problem in radiotherapy is how to selectively deliver a sterilizing dose of radiation to a malignant tumour while avoiding radiation damage to the surrounding healthy tissues.
- The Co-60 unit is a treatment machine that produces high energy photon beams suitable for cancer treatment through the nuclear decay of a Co-60 radioisotope source.
- The linear accelerator, or linac, produces high energy photon beams by accelerating electrons in a radio frequency field and bringing them to a sudden stop against a heavy metallic target, a process called bremsstrahlung radiation.

- An isocentric machine is one that is designed to rotate around an imaginary point called the isocentre, which is typically placed to coincide with the centre of the tumour being treated. Virtually all modern treatment machines are of the isocentric variety.

UNITS

- The unit of absorbed dose used in clinical radiotherapy is the gray (Gy), or its subunit, the centigray (cGy).
- The energy of high energy clinical photon beams is expressed in units of mega electronvolt (MeV).

LAWS

- When subjected to a dose of radiation D, the linear quadratic model predicts a surviving fraction S of cells given by the expression:

$$S = e^{-(\alpha D + \beta D^2)},$$

where α and β are tissue-specific constants and are expressed in units of Gy^{-1} and Gy^{-2} respectively.

- The percent depth dose (PDD) function is defined as:

$$\text{PDD} = [D(d)/D_{max}] \times 100\%,$$

where $D(d)$ is the dose at depth d in the phantom and D_{max} is the maximum dose, both along the central beam axis.

- The tissue air ratio (TAR) function is defined as:

$$\text{TAR} = D(d)/D_{air},$$

where $D(d)$ is the dose in the phantom corresponding to a dose in air D_{air} at an equivalent point in free space, with $D(d)$ and D_{air} measured at the isocentre.

CONCEPTUAL QUESTIONS

Q–26.1. In the LQ model, explain which term in the exponential dominates cell survival at low doses, and which term is likely to control survival at higher doses.

Q–26.2. Explain why an isocentric strategy in which multiple beams are targeted onto a tumour would be clinically desirable.

Q–26.3. Explain why the maximum possible value of PDD is 100%.

Q–26.4. Can the value of TAR be greater than unity? Explain.

Q–26.5. The *back scatter factor* (BSF) is defined as the TAR at the depth of maximum dose on the central axis. Can the BSF be greater than unity? Explain.

Q–26.6. In determining the effective TAR to be used in rotational therapy, is an average depth first determined and then the effective TAR taken simply as the TAR corresponding to that average depth, or is the effective TAR simply the average TAR for the different beam angles? Explain your reasoning.

Q–26.7. In the isocentric treatment technique, it is of utmost importance that the tumour be precisely centred on the machine's isocentre. Explain why this is so. In your explanation, you may wish to consider what would happen if this condition were not met.

Q–26.8. A patient is to be treated with a single beam of radiation. Is it possible that two anatomical sites at different depths below the skin receive the same dose? Explain.

Q–26.9. Discuss qualitatively the clinical advantage of asymmetrically weighing multiple beams in a treatment.

ANALYTICAL PROBLEMS

P–26.1. Compute the instantaneous fractional loss of cells with respect to dose. What are the SI units of this quantity?

P–26.2. A tumour with an α/β ratio of 5, where the α/β ratio for the surrounding normal tissue is 1, is to be treated. Assume also that the β's for both tissues are equivalent. Show that for any arbitrary dose, the therapeutic ratio is always greater than 1, and hence that the treatment will be effective.

P–26.3. In a Co-60 unit, suppose that the source has a diameter of 1 cm, the source to collimator distance is 30 cm, and the source to tumour distance is 100 cm. Using geometric arguments, determine the size of the penumbra at the tumour.

P–26.4. Typically, the limit of applicability of classical physics in the engineering design of linacs occurs when $\gamma = 1.01$. If this value is exceeded, relativistic corrections need to be incorporated into the design. How large can the velocity of an electron be such that no relativistic corrections need to be accounted for?

P–26.5. A dose of 10 Gy is to be delivered with a 6 MeV beam to a tumour located at a depth of 12 cm. An 11 × 11 cm field is to be used with a 100 cm SSD. For this beam in this geometry, you are given the following additional data: PDD for 10 × 10 cm field at depth of 10 cm = 67%, PDD for 12 × 12 cm field at depth of 10 cm = 67.9%, PDD for 10 × 10 cm field at depth of 15 cm = 51.6%, and PDD for 12 × 12 cm field at depth of 15 cm = 52.5%. Compute the maximum dose received at any location within the patient's anatomy.

P–26.6. A dose of 2 Gy is to be isocentrically delivered to a tumour with an 8 × 8 cm, Co-60 beam using a technique called rotational therapy. In this technique, the treatment head rotates continuously around the patient while delivering the treatment. The dose rate in air at the isocentre is measured at 0.61 Gy/min. Under these conditions, the average or effective TAR is 0.765. For clinical reasons, three complete revolutions of the machine are required to deliver this treatment. Compute the rotational speed of the treatment head in rev/min.

P–26.7. A superficial tumour at a depth of 0.5 cm is to be treated isocentrically with a 4 × 4 cm field with a Co-60 unit, to a dose of 3 Gy. The BSF (see Conceptual Question 26.5) for this beam is 1.015. If the machine output in air at isocentre is measured to be 1.5 Gy/min, compute the time needed to deliver this treatment.

P–26.8. Assuming a linear attenuation coefficient $\mu = 0.1$/cm (see Chapter 23) for a Co-60 beam in tissue, compute the approximate value of the PDD at a depth of 10 cm by making the simplifying assumption that the beam is exponentially attenuated. Compare your value with the published value of 55.6% and suggest possible ways to account for the discrepancy.

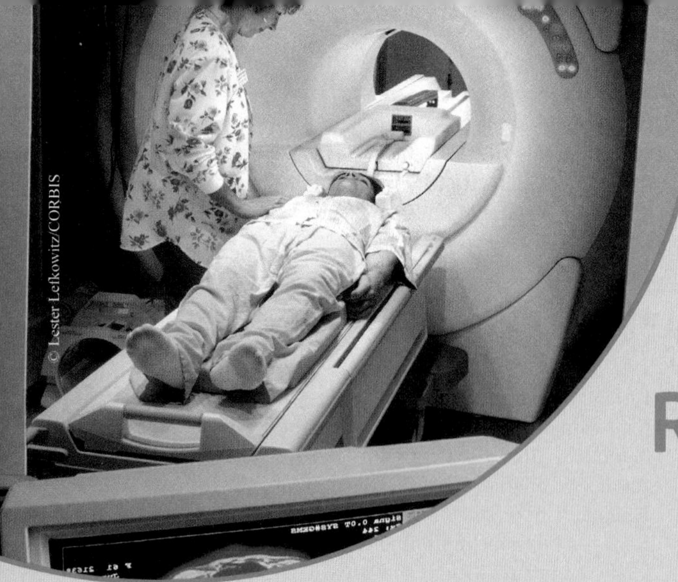

CHAPTER 27

Nuclear Magnetic Resonance

The magnetic dipole moments of spin-$\hbar/2$ nuclei constitute a two-level system in an external magnetic field. High frequency radiation that matches the energy difference of the two states causes resonant absorption, an effect called nuclear magnetic resonance (NMR).

NMR is a preferred method in structural analysis of a pure sample of a single organic compound. This is possible because individual hydrogen nuclei absorb high-frequency radiation at slightly different frequencies as they interact not only with the external magnetic field, but also with a field contribution internal to the molecule. This internal contribution is characteristic for hydrogen atoms in each functional group because it is caused by the local electronic density and the presence of identical spins in close proximity.

The same type of analysis is not applicable in vivo due to the wide range of complex biomolecules present in the sampled volume. NMR is still of interest in medicine due to two tissue-specific effects: the signal intensity is a measure of the spin density (hydrogen atom density) and the time profile of the signal is a measure of the decay time constants which are tissue-specific.

In magnetic resonance imaging (MRI), these tissue-specific effects are exploited to obtain contrast. Image resolution is achieved by applying tomographic methods to localize the origin of a signal within a stationary three-dimensional frame of reference that is defined by modifications of the external magnetic field.

Two-level spin systems in thermodynamic equilibrium were introduced in Chapter 23. We study these systems now under thermodynamic *non*-equilibrium conditions. The thermodynamic non-equilibrium is established through external high-frequency radiation, altering the occupation ratio of the ground and excited states. Fig. 27.1 indicates, in a quantum mechanical view, how this interaction takes place: photons of energy hf couple resonantly with the nuclear spins in a sample located in a static magnetic field. The photon energy is absorbed by the nuclear spin in the ground state, causing it to transfer to the excited state.

The result in Example 23.8 means that a resonance measurement using the transition between orientations of spins in an external magnetic field differs significantly from other well-established resonance techniques, such as infrared (IR) and ultraviolet (UV) spectroscopy. In the latter techniques, externally provided photons of the resonance energy led to strong absorption, as practically all molecules are in the ground state, and thus up to half of them can be excited at any time. In the nuclear spin resonance case, we get much weaker signals, as almost all spins that can be excited are excited already.

Figure 27.1 illustrates the experimental set-up with which we can excite resonant transitions in the orientation of nuclear spins in an external magnetic field. A constant and uniform external magnetic field is needed [indicated by the poles of a horseshoe magnet (shaded areas) and the vector labelled $\vec{B}_0$]. This static field causes the two orientations of the spin to split into two separate energy

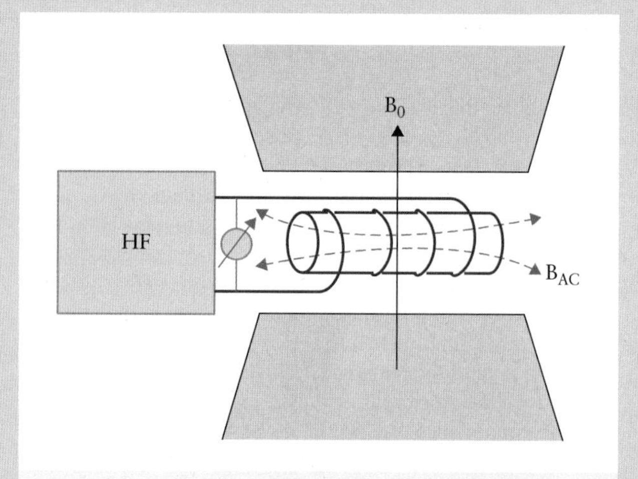

Figure 27.1 Experimental set-up of a nuclear magnetic resonance (NMR) experiment in chemistry.

levels. For resonance experiments with hydrogen nuclei in chemical molecules, a magnetic field of 2.349 T is used. We saw in Example 23.6(b) that a resonance frequency of 100 MHz corresponds to this magnetic field. The 100-MHz resonance radiation is provided by an external, high-frequency generator (box labelled HF in Fig. 27.1).

If all hydrogen nuclei in the sample were identical and independent from each other, the system would absorb an amount of energy from high-frequency radiation that is proportional to the hydrogen density in the sample. Both the resonance frequency and the time constant for the sample to return to thermodynamic equilibrium would be fixed values. This though would not warrant the use of **nuclear magnetic resonance** (NMR) as an analytical or medical diagnostic tool because exclusively determining the average density of hydrogen in a sample can be achieved more easily using other methods. However, nuclear spins are not isolated. They interact with each other at close proximity (spin–spin interaction) and they are sensitive to variations in the local magnetic field (spin–lattice interaction).

There are two fundamentally different approaches to exploiting these interactions: In organo-chemical analysis, the spin–lattice interactions lead to slightly different resonance frequencies, and the spin–spin interactions lead to splitting of the resonance signal into multiplets. NMR analysis requires a generator with variable high-frequency radiation to sample the different resonances concurrently. In medical imaging, spin–lattice and spin–spin interactions lead to different relaxation times of non-equilibrium states of the magnetized tissue. In **magnetic resonance imaging** (MRI), these are measured by providing the high frequency in the form of variable energy pulses.

27.1: Nuclear Magnetic Resonance in Organic Chemistry

In the NMR set-up of Fig. 27.1, the high frequency can be scanned across a range of ±500 Hz, which is a $1 \times 10^{-3}\%$ variation of the primary high frequency of 100 MHz. In a chemically pure sample, this small range is sufficient to observe protons' various resonances, which are caused by nuclear spin interactions. Resonances for particular protons are shifted in frequency because local magnetic fields overlap with the external magnetic field. The magnetic field contribution at the position of a given hydrogen nucleus is the result of the motion of the molecular electrons in the proton's vicinity. Electrons in molecular orbitals forming standing wave solutions to Schrödinger's equation represent current loops that cause magnetic fields characteristic for the arrangement of the local chemical structure. Thus, NMR is suitable to distinguish the chemical environment of protons in a molecule.

The recording technique of NMR spectra changed significantly as more powerful computers became available. Initially, Edward Purcell and Felix Bloch developed the cw-mode (continuous-wave mode), in which the external high frequency (HF) is swept through a ±500-Hz range around the 100 MHz signal. It is easy to imagine how a spectrum results in this case: whenever the external frequency sweeps across an actual resonance frequency, nuclear spins absorb energy from the high-frequency field and change their orientation to the excited state. The energy absorption in the sample leads to the need of a higher power input of the high-frequency generator. A separate instrument, indicated as a circle with an arrow in Fig. 27.1, records the power input of the HF unit as a function of frequency.

Purcell and Bloch's technique has significant drawbacks. With individual resonances as narrow as 1 Hz, most of the sweep passes through frequency intervals without resonances, not leading to useful data. Even worse, the small difference in occupation between the lower and higher energy levels requires many sweeps to be overlapped before a statistically significant signal is obtained.

To address these issues and accelerate the recording technique, Fourier transform NMR (FT-NMR) was developed, which is based on a short pulse in real time sent through the solenoid in Fig. 27.1. Such a pulse represents a wide range of frequencies sent through the system at once, in a manner similar to that in our discussion in the context of sound waves in Chapter 16. The resulting signal then requires a **Fourier transformation** to reveal the individual frequencies. Recall that a Fourier transformation is a mathematical operation that links a function of time, $F(t)$, with its corresponding function of frequency [i.e., its spectrum $F^*(f)$]. Fourier transformations require extensive computing capacity. Typical operational conditions for a pulsed FT-NMR are pulse lengths of about 50 μs, followed by a data acquisition period of one second.

CASE STUDY 27.1

A ^{1}H-NMR spectrum for a pure sample of 1-thiophenyl-propanone-(2) is shown in Fig. 27.2. The chemical

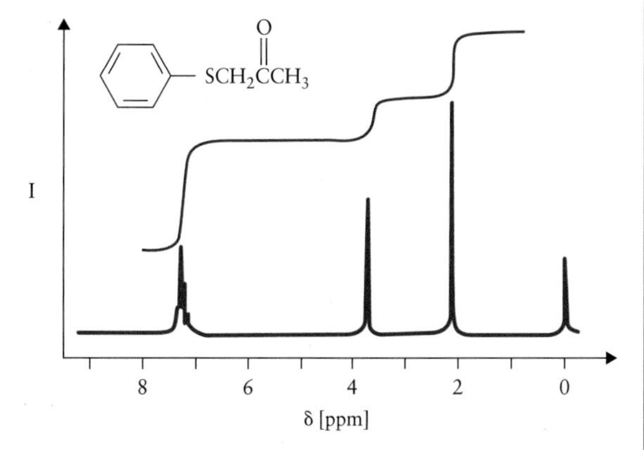

Figure 27.2 A typical NMR spectrum, showing peaks that correspond to various resonances (thick line) and the corresponding integrated signal (thin line) used to quantify the number of equivalent hydrogen atoms involved in the respective peak.

continued

compound in the inset of the figure is dissolved in an NMR-inert solvent, i.e., a solvent that does not contain hydrogen. Typically, CCl_4 is used as a non-polar solvent (as in the present case) or D_2O (deuterated water) is used as a polar solvent. The sample is then brought into the magnet of the NMR apparatus and a resonance measurement is obtained. How do you interpret the various features of the spectrum shown?

Answer: *the spectrum contains a range of features that we need to include to obtain a proper analysis:*

- *The abscissa is not shown as a frequency axis but provides the* ***chemical shift*** *δ. This value is a shift because it is recorded relative to a calibration compound arbitrarily assigned the chemical shift $\delta = 0$ (see Fig. 27.2). The calibration compound is tetramethylsilane (TMS), with the stoichiometric formula $(CH_3)_4Si$. TMS has 25 equivalent hydrogen atoms that lead in NMR to a single peak at frequency f_{TMS}. The chemical shift δ of other peaks in the spectrum is then calculated:*

$$\delta = 10^6 \frac{f_{\mathrm{sample}} - f_{TMS}}{f_0}.$$

The primary frequency, f_0, for the spectrum shown in Fig. 27.2 is 100 MHz. The factor 10^6 is used so that chemical shifts are given in parts per million (ppm).

- *Two curves are shown in Fig. 27.2, a step function at the top and a spectrum of discrete peaks at various resonance frequencies. The step function is the result of integrating the area under each of the peaks. This measurement is provided since the step height is proportional to the number of equivalent hydrogen atoms in the molecule.*

- *The hydrogen atoms in the thiophenyl-propanone molecule give rise to three peaks at different chemical shifts. This is the most important aspect of the application of the NMR technique in structural analysis studies in chemistry. As discussed in general terms above, the external magnetic field $\vec{B}_0$ determines only to first order the magnetic field at the positions of hydrogen nuclei within a molecule. The actual magnetic field may differ from this value due to magnetic field contributions at the location of the hydrogen nucleus caused by its electronic environment. For example, in the thiophenyl-propanone molecule, three types of electronic environments can be identified: We obtain peaks for (i) five hydrogen atoms attached to the benzene ring at $\delta = 7.25$ ppm, (ii) two hydrogen in the double-substituted methyl group ($-CH_2-$) at the centre of the molecule at $\delta = 3.73$ ppm, and (iii) three hydrogen atoms in the end-standing methyl group ($-CH_3$) with $\delta = 2.16$ ppm. The step height of the integral measurement confirms the ratio 5:2:3, in this order, from lower to higher resonance frequency in the spectrum.*

Chemical shifts, and signal splitting, as can be seen for the five hydrogen atoms of the benzene group in Fig. 27.2, are important tools to identify unknown chemical compounds based on their NMR spectra. At the same time, these fingerprint-type details are useful only when a single chemical compound has been isolated for the NMR analysis. We do not discuss the chemical shift or signal splitting further since these features cannot be used in MRI where a separation of chemical compounds in the human body is not possible.

27.2: Interactions of Nuclear Spins in Condensed Matter

27.2.1: The Spin-Lattice Interaction: *T1* Recovery of the *z*-Direction Magnetization

At first glance one may expect little useful information from a nuclear magnetic resonance signal obtained from a random sample of human tissue with an apparatus like the one shown in Fig. 27.1. The range of complex molecules and the multitude of hydrogen atoms in widely varying local magnetic environments should lead to a confusing multitude of peaks. Indeed, no chemical shift information is collected. However, it has been experimentally established that the time constants associated with the relaxation process in Fig. 23.21 vary with tissue composition.

- **Describing the two-level system with a macroscopic thermodynamic approach.** The time dependence of the relaxation process following a high-frequency pulse is illustrated in Fig. 27.3. The figure shows the difference in occupation of the excited and ground states for a macroscopic ensemble of nuclear spins in a tissue sample as a function of time after the pulse. In the previous section, we found that the occupation difference in thermal equilibrium is due to a slightly higher occupation of the ground state: $N_{\mathrm{excited}} - N_{\mathrm{ground}} = \Delta N_{\mathrm{eq}} < 0$. The high-frequency pulse then causes an increase in the number of nuclear spins occupying the excited state. The figure shows the

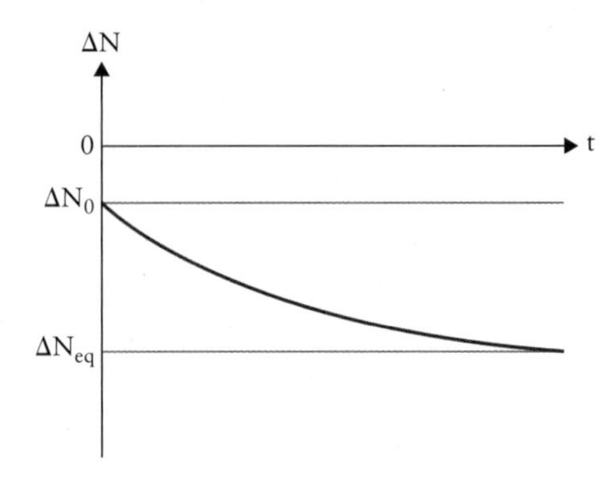

Figure 27.3 Decay of the population change of a two-level system of spins in a magnetic field after a high-frequency pulse has been sent through the sample. An exponential decay law determines the curve. The characteristic decay time constant is calculated from the curve.

case in which, at time $t = 0$ (instant of the pulse), the occupation difference is $N_{upper} - N_{lower} = \Delta N_0$. The decay of the occupation difference toward the equilibrium value ΔN_{eq} is labelled $\Delta N(t)$ and can be written in the form:

$$\Delta N(t) = \Delta N_{eq} + (\Delta N_0 - \Delta N_{eq})\, e^{\left\{\frac{-t}{t_{relax}}\right\}}. \qquad [27.1]$$

This formula is consistent with Fig. 27.3 as it is easily confirmed by substituting $t = 0$ and $t = \infty$. For $t = 0$ we note that $e^0 = 1$; i.e., we obtain $\Delta N(t = 0) = \Delta N_0$. For $t = \infty$ we note that $e^{-\infty} = 0$; i.e., we obtain $\Delta N(t = \infty) = \Delta N_{eq}$.

The experimental set-up for FT-NMR operates with short HF pulses. Data analysis is easier than in NMR measurements because the time-dependent relaxation is directly recorded; i.e., no Fourier transformation is needed. The relaxation times, T_{relax}, differ for various tissues in their healthy and malignant states, as illustrated in Table 27.1.

TABLE 27.1

Relaxation time of hydrogen nuclear spins in an external magnetic field of 2 T as a function of tissue with and without tumours

	Relaxation time T_{relax} (s)	
Tissue	**Healthy**	**malignant**
Breast	0.37	1.08
Skin	0.62	1.05
Muscle	1.02	1.41
Liver	0.57	0.83
Stomach	0.77	1.24
Lung	0.79	1.10
Bone	0.55	1.03
Water	3.6	—

- **Describing the two-level system with a microscopic mechanical model.** While Fig. 27.3 explains the basic physical principle of MRI, it falls short of a satisfactory description: the thermodynamic approach does not explain what actually happens in the sample when a high-frequency pulse is applied, and it therefore does not provide any insight into the actual method of imaging. For this we need a microscopic model that focuses on individual nuclear spins. A suitable model is a mechanical top on which an external torque is exerted. Even though a nuclear spin is not a classical system, we will see that the top model is convenient to illustrate the technique when we allow for magnetic properties to substitute for the equivalent mechanical properties.

We study an ensemble of tops precessing in an external magnetic field, with the external magnetic field chosen parallel to the z-axis. The magnitude of the initial equilibrium magnetization of the tissue sample is given in Eq. [27.2] with the magnetization in the z-direction labelled as component M_z. At time $t = 0$ a high-frequency pulse is sent into the sample. Fig. 27.1 indicates what effect such a high-frequency pulse has in a magneto-mechanical picture: the high-frequency signal is provided as an alternating current through a solenoid. This causes an alternating magnetic field as indicated by the dashed lines labelled B_{AC}. In this picture, the resonance effect is due to a magnetic torque applied to the spin, which flips the nuclear spin away from the z-axis. The external push could flip the tops all the way into the anti-parallel direction. However, for MRI applications, the high-frequency pulse is chosen such that it tilts the tops into the xy-plane. This 90° pulse is illustrated as the top curve, labelled P in Fig. 27.4. The spins in the xy-plane are shown at the bottom, below the graph.

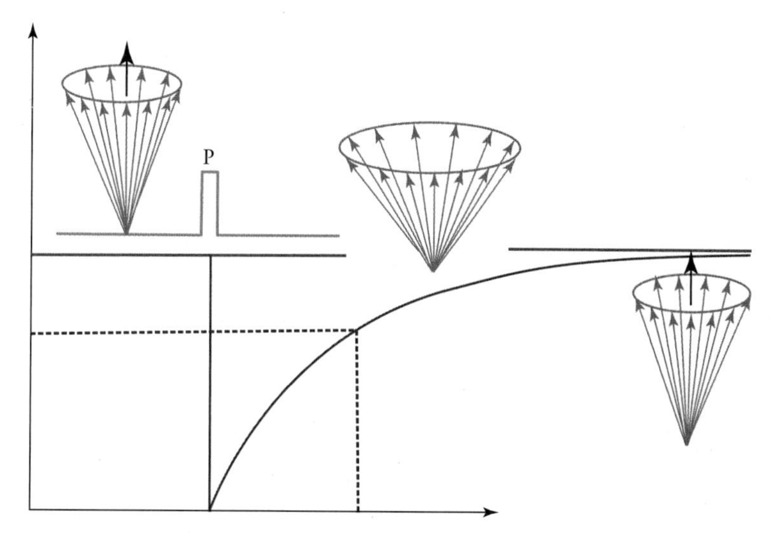

Figure 27.4 A 90° pulse tilts the equilibrium magnetization of a sample into the *xy*-plane. The equilibrium magnetization is then recovered exponentially with a time constant *T1* determined by spin–lattice interactions in the sample because energy can only be dissipated by the spin system into the surrounding molecular lattice.

For such a 90° pulse, the sample magnetization in the z-direction becomes zero at $t = 0$; this represents a non-equilibrium state and the individual tops will return to their equilibrium orientation after a short time period. We write for the sample magnetization in the z-direction in analogy to Eq. [27.1]:

$$M_z(t) = M_z(eq)(1 - e^{-t/T1}), \qquad [27.2]$$

in which M_z(eq) is given in Eq. [23.27] and $T1$ is a time constant. $T1$ depends on the interaction of the spin with the surrounding material (referred to as lattice) because the spin has taken up energy from the high-frequency pulse and can lower its energy only by dissipating it to the environment. Therefore, $T1$ is called the **spin–lattice relaxation constant**. Fig. 27.4 illustrates the classical picture associated with $T1$: the spins turn in a continuous motion back into their equilibrium position, with 63% returning to equilibrium after time $T1$. $T1$ is equivalent to the relaxation time we introduced in Table 27.1 and it contains medically useful information since it depends on the type of tissue analyzed. Before we discuss the use of the spin–lattice relaxation time for contrast in medical imaging, a second, concurrent, relaxation process has to be introduced.

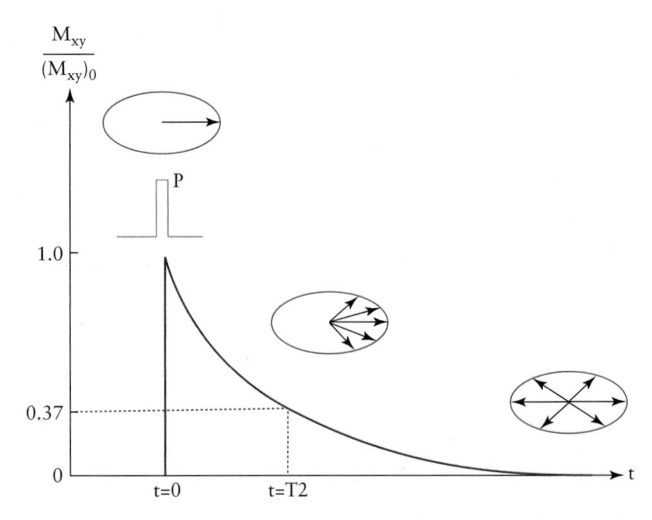

Figure 27.6 Magnetization in the *xy*-plane is illustrated in a rotating frame of reference. The spin rotation due to the Larmor frequency is eliminated in this representation. All the spins tilt into the same orientation as a result of the 90° pulse. They drift apart (dephasing) due to spin–spin interactions, causing the *xy*-plane magnetization to decrease exponentially with time constant *T2*.

27.2.2: The Spin-Spin Interaction: *T2* Decay of the Phase Coherence

The 90° pulse in Fig. 27.4 tips the individual spins into the xy-plane, perpendicular to the direction of the external magnetic field $\vec{B}_0$. Figs. 27.5 and 27.6 illustrate the time-dependent behaviour of the spins after the 90° pulse has been applied: Fig. 27.5 shows the spins in a laboratory frame, which is a frame of reference with fixed x- and y-axes in the laboratory room; and Fig. 27.6 shows the spins in a rotating frame that follows a spin rotating with the Larmor frequency such that it points always to the right. The magnetization of interest in the laboratory system depends on the location of the receiver antenna. Assuming the receiver is located along the y-axis, the magnetization component M_y is recorded. In the rotating frame, the magnetization is studied in the plane of the tilted spins. This magnetization is labelled M_{xy} because its position rotates in the xy-plane with the Larmor frequency.

Beginning with Fig. 27.5, we note that the 90° pulse rotates the spins not randomly into the xy-plane, but tilts all the spins initially into the same orientation. We refer to the angle between such spins in the xy-plane as their relative phase, using this term synonymously to the phase difference between harmonic waves. When all spins point in the same direction, we refer to them as *in phase*, or say their phases are **coherent**. In Fig. 27.5(a) we assume the common direction of the coherent spins is the x-direction. This means that all spins precess about the magnetic field direction (z-axis) in phase at $t = 0$. The precession occurs with the Larmor frequency [part (b)], causing a sinusoidal magnetization measurement when recorded with a stationary receiver [part (c)]. The plot shows magnetization in the y-direction as a function of time. We note that the maximum magnetization, which corresponds to the envelope of the oscillation, decreases exponentially with time. The physical process causing this signal decrease is illustrated in Fig. 27.6.

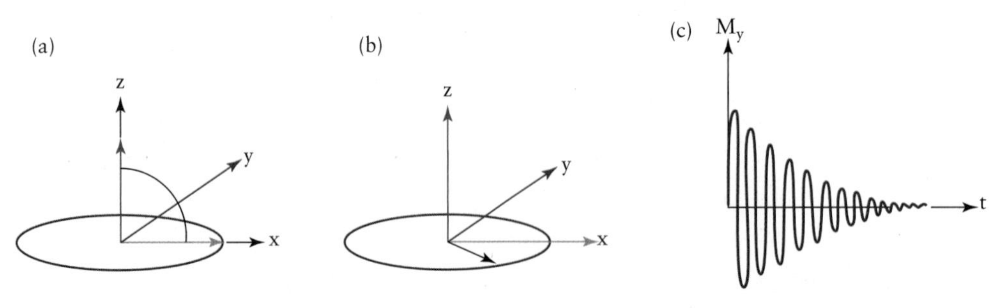

Figure 27.5 Response of an ensemble of spins tilted into the *xy*-plane by a 90° pulse. The spins continue to precess about the *z*-axis with the Larmor frequency. An antenna stationary in the laboratory measures a sinusoidal signal. The envelope signal strength diminishes because the magnetization in the *xy*-plane is reduced by dephasing of the spins in the ensemble.

Due to interactions between neighbouring spins in the sample, individual spins precess with frequencies that differ slightly from the Larmor frequency. As time progresses, the spins drift increasingly out of phase in the *xy*-plane (dephasing). The time constant of this process is defined as *T2*, which is the **spin–spin dephasing constant**. The dephasing process leads to an exponential reduction in the magnetization in the *xy*-plane:

$$M_{xy}(t) = M_{xy}(t = 0)(e^{-t/T2}), \qquad [27.3]$$

in which $M_{xy}(t = 0)$ is the magnetization at time $t = 0$. The time dependence in Eq. [27.3] is illustrated in Fig. 27.6; the exponential decay corresponds to the envelope of the decreasing sinusoidal signal in Fig. 27.5.

The time constant *T2* is a tissue-specific constant like *T1*, with $T2 < T1$, as illustrated in Table 27.2. Thus, both *T1* and *T2* are suitable to obtain contrast in imaging.

KEY POINT

Following a 90° pulse the z-direction, magnetization M_z recovers its equilibrium value slowly. The xy-component of the magnetization precesses with the Larmor frequency but diminishes fast due to dephasing of the individual spins.

Figures 27.5 and 27.6 are idealizations of the actual experiment because they assume perfect instrumentation. In practice, it is impossible to provide a large magnetic field with a homogeneity that would not affect spins. Using a real magnet, magnetic field inhomogeneities cause an accelerated dephasing, which is described by a time constant *T2**. Since $T2^* < T2$, the decrease of the magnetization in the *xy*-plane in Fig. 27.6 is actually governed by *T2**, and therefore is not providing sample-specific information. We discuss in the next section how the spin echo technique circumvents the *T2** effect, allowing *T2*-sensitive measurements.

27.3: Pulse Repetition Time and Time of Echo

The processes following a 90° pulse lead to two time constants, *T1* and *T2*, with characteristic values for various tissues. However, direct measurements of *T1* and *T2* are not used for magnetic resonance imaging for three reasons:

- Eqs. [27.2] and [27.3] depend on the time *t*; thus, *T1* and *T2* measurements would require a time-dependent data analysis,
- *T1* and *T2* require separate measurements of the magnetization in the *z*-direction and in the *xy*-plane, and
- *T2* cannot be measured directly because of magnetic field inhomogeneities.

We omit the last issue in the current section and assume that Eq. [27.3] represents the measurable magnetization in the *xy*-plane. We return to the complication due to *T2** in the following section.

To circumvent the need to measure *T1*- and *T2*-dependent effects separately, repetition of the 90° pulse is introduced with a **pulse repetition period** *TR*. The first pulse tilts all excess spins into the *xy*-plane, subsequent pulses tilt only a fraction of these spins into the *xy*-plane if *TR* is short enough that the thermodynamic equilibrium has not been recovered. *TR* is usually chosen such that it is shorter or of the order of *T1*. This means *z*-direction magnetization, M_z, does not fully recover its equilibrium value between pulses; the tissue magnetization saturates at a lower value, which depends on *T1* and *TR*:

$$M_z(TR) = M_z(eq)(1 - e^{-TR/T1}). \qquad [27.4]$$

This expression is derived from Eq. [27.2] and quantifies the sample magnetization that tilts into the *xy*-plane at each subsequent 90° pulse. These spins then dephase from an initial magnetization of $M_{xy}(0) = M_z(TR)$ to zero according to Eq. [27.3]. If we measure the magnetization in the *xy*-plane at time $t = TE$ (*TE* stands for **time of echo**), we find from Eqs. [27.3] and [27.4]:

$$M_{xy}(TR, TE) \propto \frac{N_{ground}}{V}(1 - e^{-TR/T1})e^{-TE/T2,} \qquad [27.5]$$

in which the equilibrium magnetization M_z(eq) is substituted by Eq. [23.21], but all material-independent constants are omitted due to the proportional symbol (∝). We conclude from Eq. [27.5] that a measurement of the *xy*-plane magnetization with a repetitive 90° pulse and data recording at a fixed time after each pulse leads to a signal that depends on the proton density in the sample (N_{ground}/V), *T1*, and *T2*.

If the magnetization measurement can be done for a particular voxel in a patient's body, then these three parameters should allow us to identify the composition of the voxel using Eq. [27.5]. In MRI, we are more interested in contrast between different tissues than in the composition of a particular tissue. For imaging it is therefore more interesting that Eq. [27.5] yields different values for different tissues. With the multitude of tissues involved, a single choice of *TR* and *TE* may often not provide the desired contrast. Imaging is therefore done in three modes: one that emphasizes contrast due to proton density, one that emphasizes contrast due to relaxation time *T1*, and one that emphasizes contrast due to relaxation time *T2*. Examples are discussed below, after the spin echo technique is introduced and is shown to eliminate effects due to *T2**.

27.4: Spin Echo Technique

The term *time of echo* indicates that we are actually *not* measuring magnetization in the *xy*-plane at a given time, but an echo sent from the system. This echo is the result of a second pulse we send into the system. This

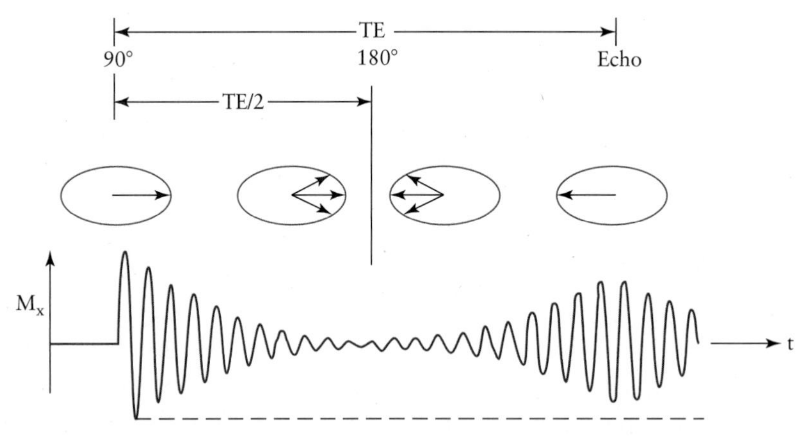

Figure 27.7 Illustration of the spin echo technique. To obtain a signal independent of the short time constant *T2**, which masks the time constant *T2* we are interested in a 180° pulse is sent into the sample at time $t = TE/2$ following the 90° pulse. This causes the spins to flip 180° and to drift back together. The envelope of the Larmor frequency signal in the *xy*-plane shows that the signal becomes a maximum at time *TE* (time of echo). All magnet inhomogeneity effects are self-compensating in this approach, but the signal from maximum to maximum decreases with time constant *T2* because spin–spin interactions occur continuously and are not affected by the 180° pulse.

pulse is a 180° pulse, inverting each spin by 180° in the *xy*-plane. Its physical consequences are illustrated in Fig. 27.7. At the left, the figure shows the 90° pulse and the resulting single direction of spins in the *xy*-plane (coherent phase). Using the rotating frame we introduced with Fig. 27.6, the second sketch indicates the dephasing of the spins. At time $t = TE/2$, the 180° pulse is sent into the system. This inverts all spins and they now converge toward the direction at 180° relative to the initial *xy*-plane magnetization. After time *TE*, the spins return to point in a single direction with a maximum magnetization. This causes a detectable signal called the **echo**.

Dephasing effects due to inhomogeneities of the external magnet are self-compensating in the spin echo technique: a spin that drifted away faster from the initial magnetization direction will drift back faster to the echo magnetization direction because the inhomogeneity acts on the spin in an inverse manner during dephasing and re-phasing. Thus, the *T2** effect is eliminated in the spin echo technique. True spin–spin interaction in turn is not affected by the 180° pulse. Thus, the spin echo after time $t = TE$ is a signal with reduced magnetization; the reduction in magnetization is due to the last exponential term in Eq. [27.5]. This term represents a slowly decreasing envelope that overlays the fast increasing and decreasing signal in the graph at the bottom of Fig. 27.7. This graph, therefore, shows how the *xy*-plane magnetization in the laboratory system varies due to the 90° and 180° pulses. Thus, the spin echo contains only the signal decrease due to the true *T2* effect.

Assuming we can distinguish the magnetization signal from neighbouring voxels in the spin echo technique, the following imaging applications are made possible by Eq. [27.5]. We focus on an example of imaging of the human brain, with a quantitative discussion, and supplement the data with a qualitative example of the human chest.

27.4.1: Contrast in *T1*-Weighted Imaging

Table 27.2 provides *T1* data for fat, grey, and white matter of the brain, and cerebrospinal fluid. We note that fat has the shortest *T1* value, grey and white matter have similar values, and the value is the largest for cerebrospinal fluid. Fig. 27.8 shows the relaxation curves of the *z*-direction magnetization after a 90° pulse. The various tissues differ most at short times. Choosing therefore a short time *TR* allows for the greatest variation in *z*-direction magnetization. Eq. [27.5] indicates that a short *TE* time affects the resulting *xy*-plane magnetization the least; for $TE \ll T2$ we find approximately $e^{-TE/T2} = 1$ and:

$$M_{xy}(TR,TE) \propto \frac{N_{\text{ground}}}{V}(1 - e^{-TR/T1}).$$

KEY POINT

A combination of a short TR and a short TE leads to a signal that is T1 weighted.

TABLE 27.2

Typical time constants, *T1* and *T2*, and proton densities for healthy tissues

Tissue	*T1* (ms) at $B_0 = 0.5$ T	*T1* (ms) at $B_0 = 1.5$ T	*T2* (ms)	Proton density (g/cm³)
Fat	250	260	80	
Muscle	550	860	45	
Brain, white matter	500	750	90	0.82
Brain, grey matter	650	900	100	0.90
Brain, cerebrospinal fluid	1800	2400	160	

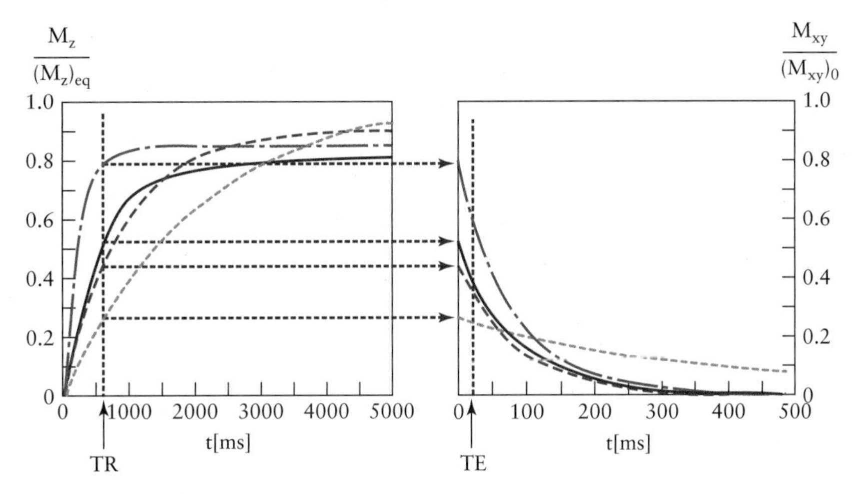

Figure 27.8 Equilibration of z-direction magnetization after a 90° pulse for four tissues of the head (left) and corresponding dephasing of the magnetization in the xy-plane (right). The tissues are fat (dash-dotted curve), white brain matter (solid curve), grey brain matter (long-dashed curve), and cerebrospinal fluid (short-dashed curve). The repetition time TR defines the maximum magnetization in the xy-plane for each tissue. The time of echo TE defines the brightness of each tissue, with white at $M_{xy} = 100\%$ and black at $M_{xy} = 0\%$. The shown choice of TR and TE leads to a $T1$-weighted image.

Figure 27.9 is a scan of the human brain in the $T1$-weighted mode, with $TR = 500$ ms and $TE \leq 15$ ms. In the grey-scale image, fat is brightest, followed by white and grey matter. The cerebrospinal fluid is darkest. Fig. 27.10 is a $T1$-weighted scan of the human thorax and abdominal cavity. Note that we know the plane of the image lies near the back because the vertebrea are visible.

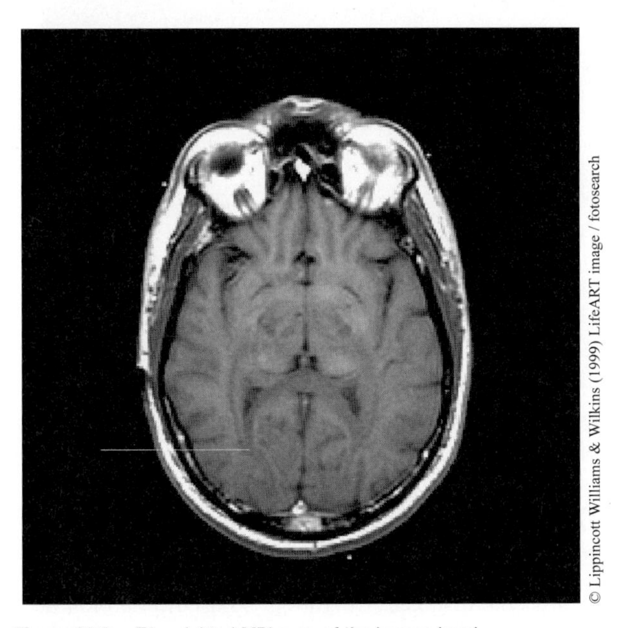

Figure 27.9 $T1$-weighted MRI scan of the human head.

27.4.2: Contrast in *T2*-Weighted Imaging

Table 27.2 also provides $T2$ data for the same tissues. We note that the various values relate with different ratios to each other compared to $T1$ times. Fig. 27.11 illustrates

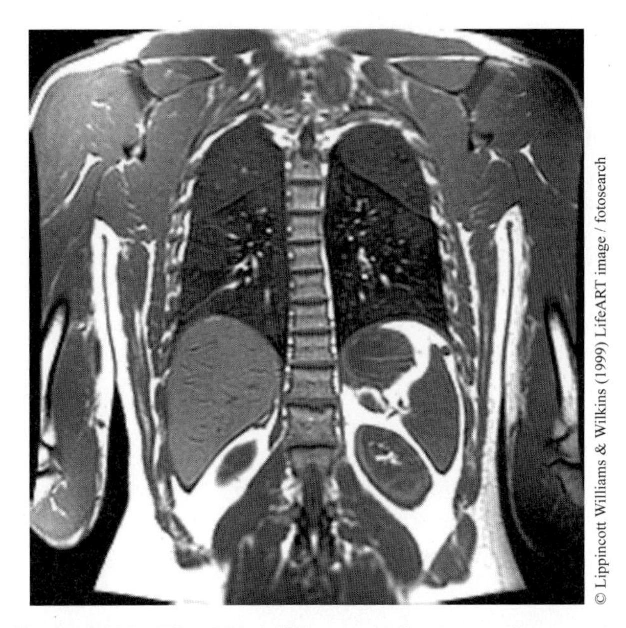

Figure 27.10 $T1$-weighted MRI scan of the human thorax and abdominal cavity.

how $T2$-weighted imaging is achieved for the tissues in the brain. We start with the same relaxation curves of the z-direction magnetization after a 90° pulse at the left. Choosing a long TR time diminishes the differences due to $T1$ differences. Choosing a longer TE time then emphasizes the differences in the plot at the right, which shows the exponential decay of the magnetization in the xy-plane after the 90° pulse. We use again Eq. [27.5] to establish the $T2$-weighted signal: for $TR >> T1$ we find that $e^{-TR/T1} = 0$ approximately, and:

$$M_{xy}(TR,TE) \propto \frac{N_{\text{ground}}}{V} e^{-TE/T2}.$$

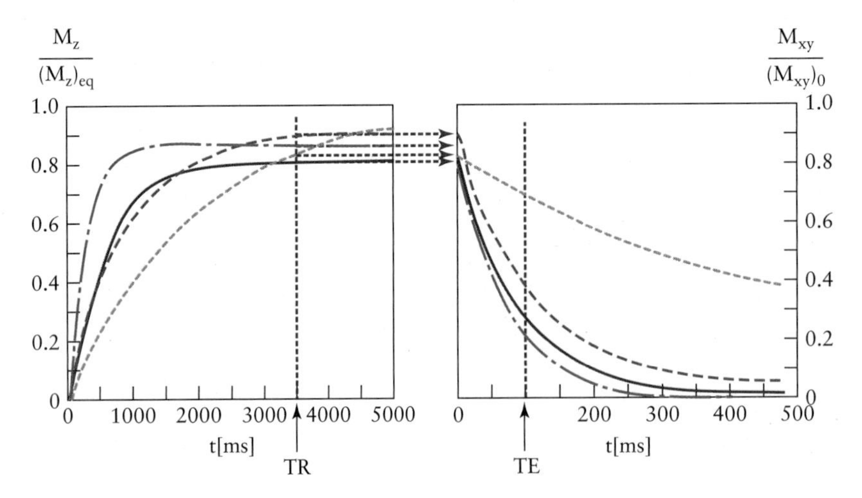

Figure 27.11 Equilibration of *z*-direction magnetization after a 90° pulse for four tissues of the head (left), and corresponding dephasing of the magnetization in the *xy*-plane (right). The tissues are fat (dashed-dotted curve), white brain matter (solid curve), grey brain matter (long-dashed curve), and cerebrospinal fluid (short-dashed curve). The repetition time (*TR*) defines the maximum magnetization in the *xy*-plane for each tissue. The time of echo (*TE*) defines the brightness of each tissue, with white at $M_{xy} = 100\%$ and black at $M_{xy} = 0\%$. The choices of *TR* and *TE* shown led to a *T2*-weighted image.

KEY POINT

A combination of a long TR and a long TE leads to a signal that is T2 weighted.

Figure 27.12 is a scan of the human brain in the *T2*-weighted mode, with $TR > 2500$ ms and $TE > 80$ ms. In this scan, fat is darkest and cerebrospinal fluid is brightest. Fig. 27.13 is a *T2*-weighted scan of the human chest. The scan shows a coronal cross-section to be compared to Fig. 27.10.

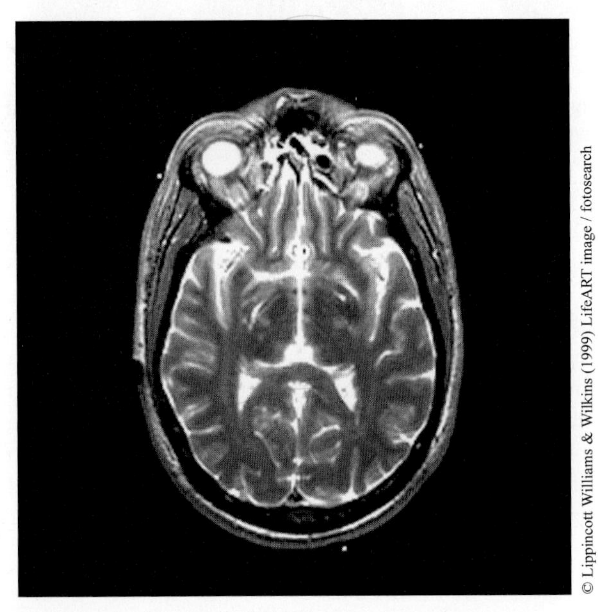

Figure 27.12 *T2*-weighted MRI scan of the human head.

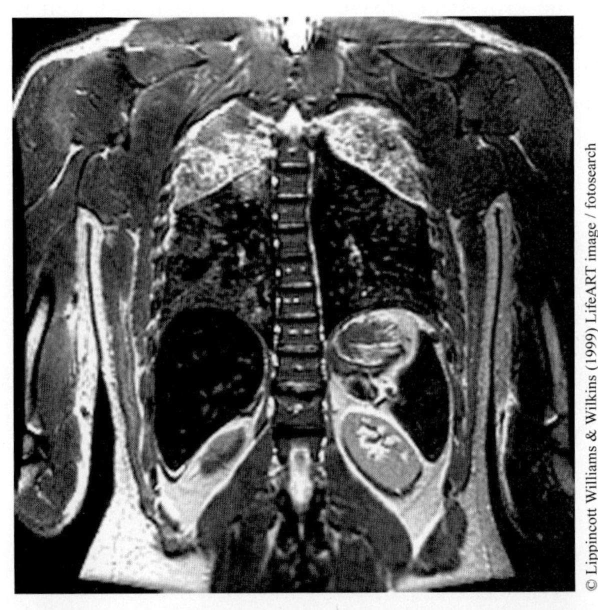

Figure 27.13 *T2*-weighted MRI scan of the human thorax and abdominal cavity.

27.4.3: Contrast in Spin Density-Weighted Imaging

A third type of image weighting is based on proton density. For this, both the *T1* and *T2* effects have to be eliminated: the *T1* effect is diminished with a long *TR* time, and the *T2* effect is diminished with a short *TE* time, as illustrated in Fig. 27.14. As in Figs. 27.8 and 27.11, the left side shows *z*-direction magnetization after a 90° pulse and the right side shows the dephasing of the spins, which captures the *T2* effect when measured with the spin echo technique.

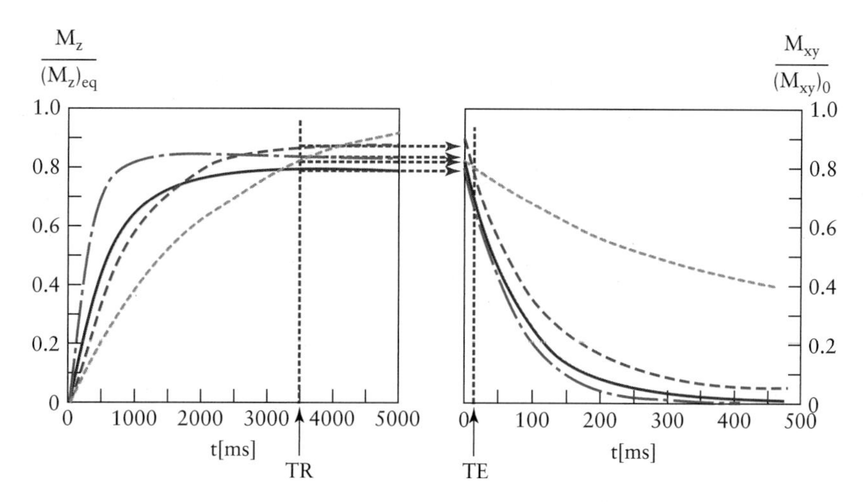

Figure 27.14 Equilibration of *z*-direction magnetization after a 90° pulse for four tissues of the head (left), and corresponding dephasing of the magnetization in the *xy*-plane (right). The tissues are fat (dash-dotted curve), white brain matter (solid curve), grey brain matter (long-dashed curve), and cerebrospinal fluid (short-dashed curve). The repetition time (*TR*) defines the maximum magnetization in the *xy*-plane for each tissue. The time of echo (*TE*) defines the brightness of each tissue, with white at $M_{xy} = 100\%$ and black at $M_{xy} = 0\%$. The choices of *TR* and *TE* shown led to a spin density-weighted image.

Quantitatively, we use Eq. [27.5] with $TE << T2$, yielding approximately $e^{-TE/T2} = 1$, and $TR >> T1$, yielding approximately $e^{-TR/T1} = 0$. Thus:

$$M_{xy}(TR, TE) \propto \frac{N_{\text{ground}}}{V}.$$

KEY POINT

A combination of a long TR and a short TE leads to a spin density-weighted signal.

This imaging mode distinguishes white and grey matter differently from the *T1*- and *T2*-weighted modes. Fig. 27.15 is a scan of the human brain in the proton density-weighted mode, with $TR > 2500$ ms and $TE \leq 35$ ms. Conceptually, proton density-limited MRI is closest to CT scans, except that the contrast in CT is based on total density, and in MRI it is based only on proton density. Note that *T1*- and *T2*-weighted measurements are also sensitive to proton density. Fig. 27.16 is a spin density-weighted scan of the human thorax and abdomen. Compare this scan with Figs. 27.10 and 27.13.

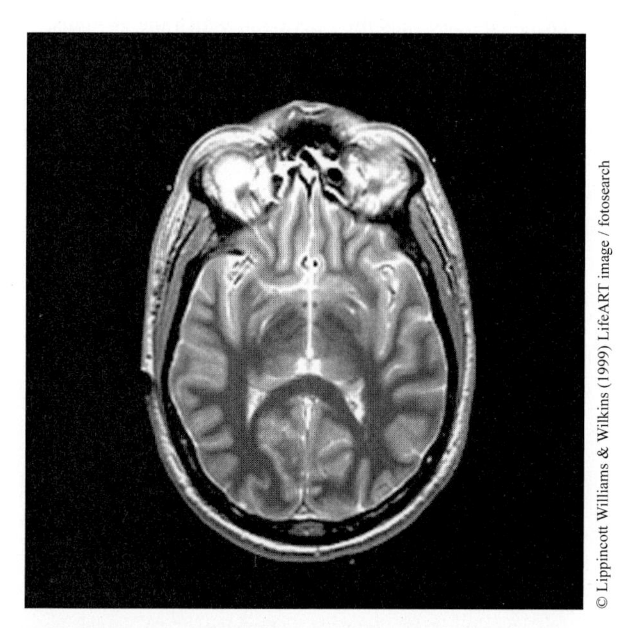

Figure 27.15 Spin density-weighted MRI scan of the human head.

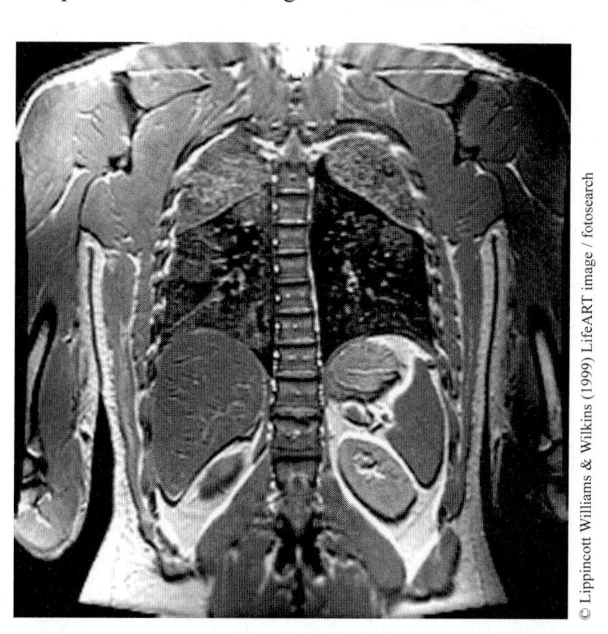

Figure 27.16 Spin density-weighted MRI scan of the human thorax and abdominal cavity.

27.5: Imaging Utilizing the Gradient Field Method

The MRI images in this chapter illustrate good spatial resolution of anatomical features of the order of 1 mm.

The use of larger magnetic fields can improve image resolution even further. However, we still need to establish how images are actually recorded, i.e., how neighbouring voxels in the human body are analyzed independently. The equipment components we have discussed so far do not alone allow imaging because they interact with the entire sample throughout the measurement.

The technical modification that enables imaging is a linear variation of the static magnetic field across the sample along each of the three Cartesian axes. For example, the magnetic field in the z-direction varies as a function of the x-position:

$$B_Z(x) = B_0 + G_x(x - x_0),$$

with

$$G_x = \frac{dB_z}{dx},$$

in which G_x is a **gradient**. A constant gradient leads to a linear variation in the magnetic field. The varying magnetic field causes, at each location, a different Larmor frequency; this frequency variation can be used to trace the signal back to its x-position. Typical gradients are of the order of a few milli-Tesla per metre (mT/m), and the corresponding variations of the Larmor frequency lie in the range of a few tens of kHz.

In practice, the frequency variations are used to determine the **field-of-view** (*FOV*) by presetting lower and upper frequency thresholds. Combining Eq. [23.21] with the gradient formula above we find for the threshold frequencies:

$$f_{\min} = \frac{\gamma}{2\pi} B_{\min} = \frac{\gamma}{2\pi}[B_0 + G_x(x_{\min} - x_0)]$$

$$f_{\max} = \frac{\gamma}{2\pi} B_{\max} = \frac{\gamma}{2\pi}[B_0 + G_x(x_{\max} - x_0)].$$

The frequency difference between the two lines is:

$$\Delta f = f_{\max} - f_{\min} = \frac{\gamma}{2\pi} G_x x_{\max} - \frac{\gamma}{2 \cdot \pi} G_x x_{\min},$$

which yields for the field-of-view as $FOV_x = x_{\max} - x_{\min}$:

$$FOV_x = \frac{\Delta f}{\frac{\gamma}{2\pi} G_x} = \frac{\Delta\omega_p}{\gamma G_x}.$$

Thus, the field of view is proportional to the range of Larmor frequencies we analyze. This frequency range is called the **signal bandwidth**. A particular combination of field-of-view and signal bandwidth can be chosen if we adjust the gradient G_x accordingly.

When we use multiple gradients and analyze the data with a computer, crossed planes produce stripes, and crossed stripes produce single voxels of signals. This information is processed such that relaxation times are usually shown in false-colour codes or grey-scales, such as in the MRI images shown in this chapter.

A second method that can be employed as an alternative to the gradient field method is based on applying a gradient field only for a short time period after the 90° pulse. While the gradient field is being applied, the spins in the sample precess with various Larmor frequencies based on the local magnetic field. When the gradient field is switched off, all spins continue to precess with the same frequency, but the effect of the gradient field is encoded in the phase of the spins. For example, spins at minimum or maximum gradient field precess with frequencies:

$$f_{\min} \propto \cos\left(2\pi\left[f_0 - \frac{\gamma}{2\pi} G_x \frac{FOV_x}{2}\right] t\right)$$

$$f_{\max} \propto \cos\left(2\pi\left[f_0 + \frac{\gamma}{2\pi} G_x \frac{FOV_x}{2}\right] t\right),$$

while the gradient field is turned on. The signal frequency at the centre of the field-of-view is represented as f_0. If the gradient field is applied for a total time period $t = \tau$, the phase difference $\Delta\varphi$ between spins at both ends of the field-of-view is:

$$\Delta\varphi = \gamma G_x FOV_x \tau.$$

SUMMARY

DEFINITIONS

- Spin–lattice relaxation constant *T1*: *T1* is the time required by spins to return to their equilibrium orientation after a 90° pulse (a fraction of 63% return after time $t = T1$).
- Spin–spin dephasing constant *T2*: *T2* is the time spins require to lose their phase information after the 90° pulse.
- *TR* is the pulse repetition period at which the 90° pulse is repeated periodically.
- *TE* is time of echo when the sample magnetization is recorded after a 180° pulse was applied at time *TE*/2 following each 90° pulse.

LAWS

- Magnetization after 90° high-frequency pulse:

$$M_z(t) = M_z(eq)(1 - e^{-t/T1})$$

  M_z is the magnetization in the z-direction with the external magnetic field oriented along the z-axis. At $t = 0$, the high-frequency 90° pulse is sent, transferring the spins into the xy-plane. M_z(eq) is the equilibrium magnetization of the sample and *T1* is the spin–lattice relaxation constant.

- Spin–spin dephasing in the xy-plane after 90° pulse:

$$M_{xy}(t) = M_{xy}(t = 0)(e^{-t/T2})$$

$T2$ is the spin–spin dephasing constant of magnetization in the xy-plane, $M_{xy}(t = 0)$ is the magnetization at time $t = 0$. Note that $T2 < T1$.

- z-direction magnetization M_z as a function of $T1$ and TR:

$$M_z(TR) = M_z(eq)(1 - e^{-TR/T1})$$

TR is the pulse repetition period.

- Magnetization in the xy-plane at time $t = TE$ after each 90° pulse (twice the time at which the 180° pulse was applied):

$$M_{xy}(TR, TE) \propto \frac{N_{\text{ground}}}{V}(1 - e^{-TR/T1})e^{-TE/T2}$$

with initial magnetization $M_{xy}(t = 0) = M_z(TR)$. (N_{ground}/V) is the spin density in the sample, $T1$ is the spin–lattice relaxation constant, $T2$ is the spin–spin dephasing constant, TR is the pulse repetition time, and TE is the time of echo.

CONCEPTUAL QUESTIONS

Q–27.1. (a) Why is spin important in NMR? (b) How does one choose active nuclei for NMR?

Q–27.2. Define the Larmor frequency, and explain the need for a strong magnetic field in NMR.

Q–27.3. An undergraduate student working over the summer at the MRI laboratory is asked to figure out what RF pulse should be used to excite spins located in an xy-plane at $z = -6.5$ cm. The resonance frequency at the isocentre is known to be 63.85 MHz and the slice-selection gradient is set at 3 Gauss/cm. Describe the RF pulse used by the student.

Q–27.4. What is the physical meaning of the $T1$ relaxation time?

Q–27.5. What is the physical difference between the $T2$ and $T2^*$ relaxation times?

Q–27.6. What is the physical mechanism responsible for the generation of spin echoes?

Q–27.7. Why are magnetic field gradients used in MRI experiments?

ANALYTICAL PROBLEMS

P–27.1. We consider spin echo technique-based MRI imaging of the human brain with a magnet of magnetic field $B_0 = 1.5$ T. We are interested in the contrast between white and grey matter. (a) Which of the two tissues is brighter on a $T1$-weighted and a $T2$-weighted image respectively? Which of the two tissues is brighter on a spin density-weighted image? Use data from Table 27.2. (b) If $TR = 600$ ms and $TE = 5$ ms, what type of weighting is expected? Calculate the contrast between the two tissue types. (c) Repeat part (b) with $TR = 3000$ ms and $TE = 90$ ms. Which tissue feature ($T1$, $T2$, or spin density) is mostly responsible for the resulting contrast in this case?

P–27.2. (a) We work with a signal bandwidth of 100 kHz. If we want a field-of-view of 24 cm (typical field-of-view for the human head), what gradient field strength in unit mT/m must be applied? (b) If the maximum available gradient field strength is 40 mT/m (representative value for the best clinical systems at present), what is the minimum field-of-view that can be achieved? Use a maximum signal bandwidth of 500 kHz. (c) A gradient field of what strength is necessary to reduce the field-of-view to 2 cm? *Note:* This *FOV* allows high-resolution anatomical scanning of the carotid artery in the neck. Use a 250-kHz signal bandwidth.

P–27.3. In an MRI experiment, a graduate student is asked to estimate the time it will take a magnetic sample, with a spin–lattice relaxation time $T_1 = 2.0$ seconds and an initial magnetization set to zero, to recover 95% of its equilibrium value.

P–27.4. A graduate student at the university MRI laboratory has prepared a magnetic sample with a spin–spin relaxation time $T_2 = 200$ ms. The student wants to find out how long it would take the transverse magnetization of the sample to decay to 40% of its starting value.

P–27.5. A spin system with an equilibrium magnetization $M_z = M_0$ is exposed to an RF pulse B_1. Immediately after the application of the pulse, the ratio $M_{xy}(0^+)/M_z(0^+) = 0.8$. If the duration of the RF pulse $t_p = 100/\gamma$, where γ is the gyromagnetic ratio of the spin system, calculate the amplitude of the RF pulse B_1.

P–27.6. Assuming a 10-cm field of view, a 10-ms phase encoding gradient, and 8 phase encoding steps, what are the phase angles for each phase encoding step?

P–27.7. The MRI laboratory technician has prepared a hydrogen sample that he/she wants to test. The sample is immersed in a magnet providing a 2.5 Tesla magnetic field. A constant RF magnetic field B_1 of amplitude 1.25×10^{-4} Tesla is applied along the $+x$-axis for 75 μs. Determine the direction of the net magnetization vector after turning off the RF pulse.

P–27.8. The states of a two-state system are populated such that the difference in numbers of occupation of the upper and lower states is expressed as $\Delta N = N_u - N_d$. The time rate of change of ΔN is given by

$$\frac{d\Delta N}{dt} = -\frac{1}{T_1}(\Delta N - \Delta N_0)$$

where T_1 is the spin–lattice relaxation time and ΔN_0 is the population difference at equilibrium. (a) Determine the expression for the population difference at time t after placing the system in the magnet. (b) How long does it take for the distribution to get to 90% of the Boltzmann distribution if T_1 is 800 ms?

P–27.9. A MRI sequence that consists of two RF pulses, with a time delay TR in between, and with tip angles θ_1 and θ_2 is applied to a homogeneous system with single valued T_1 and T_2, and with an initial magnetization M_0. We also know that $TR >> T_2$, and that $T_1 = 5\ TR$. (a) Calculate the relative signal amplitudes immediately after the RF excitations. (b) Determine θ_1 such that the signal amplitude from the second excitation is always zero. (c) What value should θ_2 take in order to produce equal signal amplitudes, if $\theta_1 = 25°$?

(d) Determine the tip angle combination (θ_1, θ_2) that leads to the maximum possible signal amplitudes.

P–27.10. Consider the following sequence: (1) the sequence starts with a 180° pulse, (2) followed by a delay time *TI*, (3) followed by a 90° pulse and a delay time *TR* – *TI*, such that the total repetition time for this sequence is *TR*. (a) Using the longitudinal magnetization present before the 180° pulse, and assuming a homogeneous T_1 and T_2 for the system, determine the longitudinal and the transversal components of the magnetization after the 180° pulse and before and after the 90° pulse. (b) Where, after the second pulse, will the signal maximize? (c) If one repeats the pulse sequence, what will be the steady state signal amplitude?

P–27.11. (a) A new imaging system has been installed at the hospital and it now operates a 7.0 T superconducting magnet. If the nuclei of interest are protons and if the background magnetic field is B_0 = 7.0 T, calculate the ratio of parallel to anti-parallel spins at room temperature. (b) How does the ratio compare to those of imaging systems with 1.5 T and 3.0 T magnets? (c) What would be the values of these ratios if the nuclei of interest were carbon 13?

P–27.12. (a) If one were to perform an equivalent of the basic NMR experiment on a free electron (ESR), what would be the resonance frequency at 1.7 T? (b) Is this a practical experiment? (c) What limitations might there be for doing this in humans? (d) What field strength would you need to produce the same electron resonance frequency as you get from a proton NMR experiment? (e) What B_1 field strength would you need to apply to obtain a 90° RF pulse in 10 μs?

P–27.13. A graduate student at the MRI laboratory wants to test the gradient coils. He prepares a sample containing water at two very specific locations. He then runs the frequency-encoding gradient of 3 Gauss/cm along the *y*-axis and collects the NMR spectrum. The spectrum shows to peaks (lines) at the frequencies +1500 Hz and −750 Hz relative to the isocentre frequency. What locations can the graduate student deduce from this experiment?

P–27.14. (a) For a 10-turn solenoid of radius 10 cm, what current would need to be applied to generate a magnetic field $B_1 = 10^{-5}$ T? (b) Estimate the power needed to apply this current.

Index

Page numbers with italicized f's and t's refer to figures and tables.

G

N

O

P

Math Review

Powers and Logarithms

Definitions: $a^1 = a$, $a^0 = 1$, and $a^{-n} = 1/a^n$.
Calculation rules with powers:

$$a^x a^y = a^{x+y}$$
$$a^x b^x = (ab)^x$$
$$\frac{a^x}{a^y} = a^{x-y}$$
$$\frac{a^x}{b^x} = \left(\frac{a}{b}\right)^x$$
$$(a^x)^y = a^{xy}$$

Logarithms were introduced by John Napier in 1614 to simplify mathematical operations. As you see below, when using logarithms we can use multiplications instead of powers, and additions instead of multiplications. The use of logarithms also proves useful when we analyze data graphically, as discussed in the section on "Graph Analysis Methods."

Definitions: The term $c = \log_b a$ represents the number c, which satisfies the equation $b^c = a$; i.e., the logarithm is the inverse operation to raising to a power. In the textbook, only two types of logarithm functions are used: logarithms with base $b = 10$ (usually labelled log without a subscript), and natural logarithms with base $b = e = 2.71828\ldots$, in which e is Euler's number. This logarithm function is labelled ln. Thus:

$$\log_{10} a = \log a = c \Leftrightarrow a = 10^c$$
$$\log_e a = \ln a = c \Leftrightarrow a = e^c = \exp(c)$$

The following rules apply to logarithms with any base; here, we write them specifically for the case of base $b = 10$:

$$\log(xy) = \log x + \log y$$
$$\log\left(\frac{x}{y}\right) = \log x - \log y$$
$$\log x^y = y \log x$$

and for the case of base $b = e$:

$$\ln(xy) = \ln x + \ln y$$
$$\ln\left(\frac{x}{y}\right) = \ln x - \ln y$$
$$\ln x^y = y \cdot \ln x$$

For the same variable, the two logarithm functions are related by a factor of ln 10:

$$\ln x = \ln 10 \log x$$
$$\Rightarrow \ln x = 2.3026 \log x$$

Graph Analysis Methods

Scientific progress is based on experimental data. These data often are plotted and presented without a (yet) conclusive model that would provide a mathematical formula for the relation between the shown parameters. The three most frequently used methods of representing such data are the linear plot, the logarithmic plot, and the double-logarithmic plot.

Linear Plots

In these most frequently used graphs, the variation of a dependent variable (y) is illustrated as a function of an independent variable (x) by plotting y versus x, as shown in Fig. 1. We say that "y is linear in x" in the x-interval $[x_1, x_2]$ if the data in that interval can be fitted with a straight line, such as the one in the figure. Mathematically, this linear behaviour is described by the equation:

$$y = ax + b,$$

in which a and b do not depend on the variables x and y. a is called the slope of the curve and is determined from:

$$a = \frac{y_4 - y_3}{x_4 - x_3}.$$

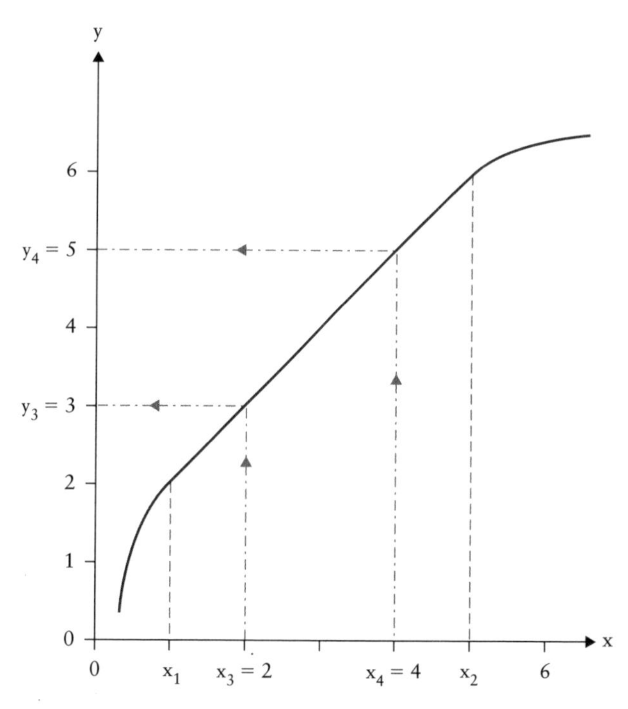

Figure 1 Linear plot of a function $y = f(x)$ with a linear dependence of y on x in the interval $[x_1, x_2]$.

Where the values x_3 and x_4 must lie within the interval $[x_1, x_2]$, b is called the intercept of the y-axis and is the value of y at $x = 0$. The value of b is read from the plot directly if the point $x = 0$ lies within the interval $[x_1, x_2]$. If this is not the case, like in Fig. 1, then we choose a particular point on the curve and substitute it together with the result for a in the general linear equation.

Example 1

Determine the slope and the intercept with the y-axis for the curve shown in interval $[x_1, x_2]$ of Fig. 1.

Solution: We choose $x_3 = 2$ and $x_4 = 4$. The corresponding y-values are read from the graph (using the dash-dotted lines) as $y_3 = 3$ and $y_4 = 5$. Thus, we find $a = (5 - 3)/(4 - 2) = +1$. For b, we choose the point with $x_3 = 2$ and $y_3 = 3$. These data are then substituted in the general linear equation $y = a\,x + b$, leading to $3 = 1\;2 + b$. Thus, $b = +1$.

Logarithmic Plots

Graphs other than linear plots are usually used for two purposes. Either a large variation in the data leads to an undesirable appearance of the plot, or the data are tested for an exponential dependence. In an exponential dependence the independent variable x and the dependent variable y are connected in the form:

$$y = ae^{bx} = a\exp(bx).$$

How can you convince yourself that this equation describes a given experimental set of data? Often, data sets show an increasing upward trend if plotted in a coordinate system with a y- and an x-axis. However, such a trend is not sufficient. You have two options: If you have logarithmic paper available or your graphics software allows you to use logarithmic axes, you show the data in the form y versus x using a logarithmic scale for the y-axis (ordinate). Alternatively, we can plot $\ln y$ versus x where the natural logarithm of the y-data has been calculated first.

If the data points can be represented by a straight-line segment in a logarithmic plot, such as in the interval $[x_1, x_2]$ in Fig. 2, then we know that the exponential function applies to the data in that interval. From the logarithmic plot we determine the coefficients a and b, since the logarithmic plot corresponds mathematically to $\ln y = \ln a + b\,x$; i.e., the slope of the straight-line segment in Fig. 2 gives the pre-factor in the exponent, b, and the intercept with the y-axis gives the logarithm of the pre-factor of the exponential term, $\ln a$.

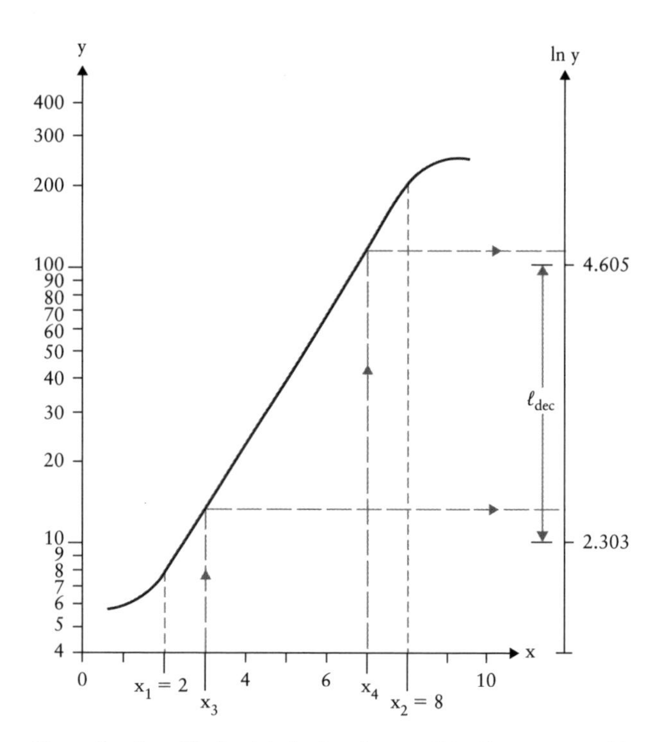

Figure 2 Logarithmic plot of a function $y = f(x)$ with an exponential dependence of y on x in $[x_1, x_2]$.

Example 2

For Fig. 2, determine the parameters a and b in the interval where the exponential function applies.

Solution: We need two data points from within the interval $[x_1, x_2]$, since two unknown parameters a and b exist. For these we choose $x_3 = 3$ and $x_4 = 7$. Notice that the values given on the y-axis are y-values, not $\ln y$-values. You can tell this first because the axis is labelled y and not $\ln y$, and also because the scale is not linear; i.e., the distance between the $y = 0$ and $y = 100$ tick-marks is not equal to the distance between the $y = 100$ and $y = 200$ tick-marks. In other plots it might be the case that the y-data are given as $\ln y$ with a linear scale. In that case we would read the corresponding $\ln y$ values directly off the y-axis of the graph.

However, in the present case an intermediate step in the data analysis is needed. In this step, we add a new ordinate—done in Fig. 1 at the right side—where the y-data are converted into $\ln y$-data. Notice that the new axis is linear (the distance between the $y = 0$ and the $y = 2.303$ tick-marks is the same as that between the $y = 2.303$ and $y = 4.605$ tick-marks). With the new support scale established, we read the $\ln y$-values from this new axis and list them in Table 1.

We illustrate how we arrive at the value $\ln(y_3) = 2.600$ as an example. Each decade on a ln-grid corresponds to an increment of $\ln 10 = 2.303$. This increment corresponds to the length l_{dec} (expressed in unit cm as measured from Fig. 2). The value of $\ln(y_3)$ is a distance of $0.129\;l_{dec}$ above the next lower full decade (in this case, $y = 10$). Thus $\ln(y_3) = 0.129\;2.303 + 2.303 = 2.600$.

TABLE 1

Data sets from Fig. 2

i	x_i	$\ln(y_i)$
3	3.0	2.600
4	7.0	4.740

With the data in Table 1 we determine in the next step the coefficient b from $\ln y = \ln a + b\,x$:

$$b = \frac{\ln(y_4) - \ln(y_3)}{x_4 - x_3}.$$

Substituting the data from Table 1 for the coefficient b leads to $b = (4.74 - 2.6)/(7.0 - 3.0) = 0.54$. In the last step we substitute the data pair x_3 and $\ln(y_3)$: $2.6 = \ln a + 0.54\;3.0$, which yields $a = 2.66$.

Double-Logarithmic Plots

If logarithmic plots do not lead to straight-line segments, a double-logarithmic plot can be used to reveal whether the y-data depend on the x-variable in the form of a power law. Even if the real dependence is more complicated, a double-logarithmic plot often leads to straight-line segments as shown for the interval $[x_1, x_2]$ in Fig. 3, as power laws are often good approximations to the actual physical or biological law.

Again, a double-logarithmic plot can take one of two equivalent forms: either the x and y data are plotted directly on paper, with logarithmic grids for the abscissa and ordinate, as done in Fig. 3, or $\ln x$ and $\ln y$ can be plotted on paper with a linear grid.

For the straight line-segment in Fig. 3, the mathematical dependence of the dependent variable y on the independent variable x is then given in the form of a power law with a and b constant:

$$y = ax^b.$$

When we rewrite both sides of this equation as the respective logarithmic values, we find:

$$\ln y = \ln a + b \ln x;$$

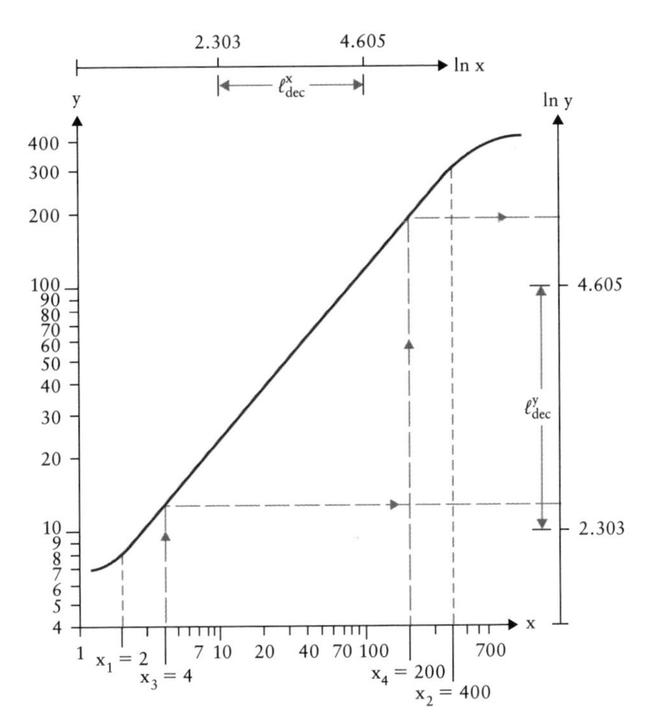

Figure 3 Double-logarithmic plot of a function $y = f(x)$ with a power law dependence of y on x in $[x_1, x_2]$.

i.e., the slope of the straight-line segment corresponds to b and the intercept of the ordinate equals $\ln a$. For practice, analyze Fig. 3 yourself. You follow the same approach we took for the logarithmic axis in Fig. 2, except the procedure must be applied to both axes in Fig. 3, as shown with the additional axes at the right side and at the top of the figure. For the two coefficients, you should find values close to $b = 0.68$ and $a = 5.07$.

Example 3

The molecular masses (M) and the radii (R) of some molecules are given in Table 2. Plot the data in double-logarithmic representation and develop an empirical relationship between the two quantities in the form $R = f(M)$, in which the notation $f(\cdots)$ means "function of."

Solution: The double-logarithmic plot of the data in Table 2 is shown in Fig. 4, where the abscissa values are taken from the first column of the table and the ordinate values from the second column in unit metre (m). Note that the single data points deviate slightly from a straight line. It is possible, however, to draw a reasonably straight line through the points (mathematically, this is called the *best fit*). Using this straight line for the analysis of the constants a and b, instead of any particular data points, reduces the statistical error in the result.

TABLE 2

Molecular mass and radius of various molecules

Substance	M (g/mol)	R (10^{-10} m)
Water	18	1.5
Oxygen	32	2.0
Glucose	180	3.9
Mannitol	180	3.6
Sucrose	390	4.8
Raffinose	580	5.6
Inulin	5000	12.5
Ribonuclease	13 500	18
β-lactoglobulin	35 000	27
Hemoglobin	68 000	31
Albumin	68 000	37
Catalase	250 000	52

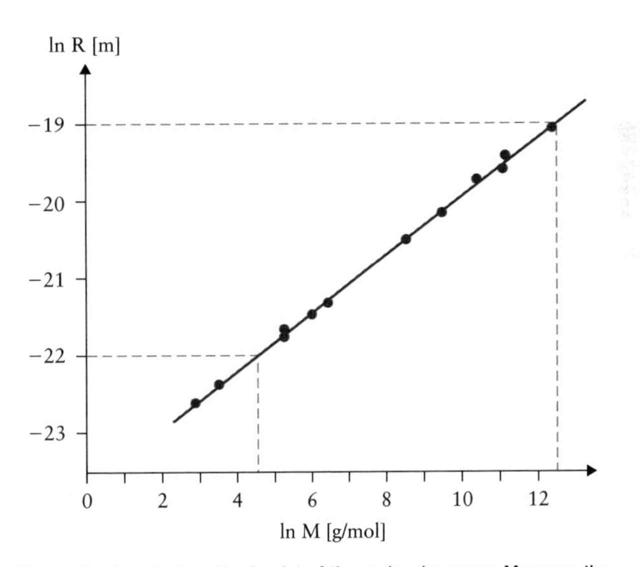

Figure 4 Double-logarithmic plot of the molecular mass M versus the molecular radius R for a wide range of molecules, based on Table 2.

To determine the constants a and b in the power law $R = a\,M^b$, we again analyze the problem in the form $\ln R = \ln a + b \cdot \ln M$ as described above. From Fig. 4 we read two data pairs $\ln R$ and $\ln M$ (dashed lines), as listed in Table 4. This table allows us to write two linear formulas in the form $\ln y = \ln a + b \ln x$:

(I) $\quad -22.0 = b4.5 + \ln a$

(II) $\quad -19.0 = b12.5 = \ln a$

(II) $-$(I) $+3.0 = b(12.5 - 4.5)$

TABLE 3

Data sets from Fig. 4

Data set	ln[M (g/mol)]	ln[R (m)]
#1	4.5	−22.0
#2	12.5	−19.0

Thus, $b = 0.375$ and $\ln a = -23.69$. The value for $\ln a$ is obtained by substituting the result for b in one of the two formulas. From $\ln a$ we calculate the parameter a as $a = 5.2 \times 10^{-11}$ m.

Problem-Solving Strategy

Problems in the many sub-disciplines of physics can be phrased in an almost infinite number of ways. Therefore, no simple problem-solving procedure exists that we can follow and expect to succeed in each case. Still, providing a structured approach to problem solving will often save time. Thus, we devise three general steps that are useful in most contexts.

Schematic Approach

In the first step you compile the known facts about the problem and note what you do not know. If a sketch is given with the problem, familiarize yourself with it during this step. If one is not given you may want to make your own sketch.

Physical Model

In the second step you address the physical aspects of the problem, including the physical parameters that play a role and the physical laws you need to solve it. A physical model also includes simplifying assumptions. Make sure you are aware of the assumptions you make and test whether they are valid. In this step you may have to draw additional sketches.

Quantitative Treatment

In the last step the physical model is transformed into mathematical equations and the known parameters are substituted such that an explicit solution is provided.

Geometry

Often you will evaluate angles in sketches accompanying a problem. To do this, it is necessary to know which angles in a given situation are equal, and which angles add up to 90° or 180°. Inspecting Figs. 5 and 6, confirm that $\alpha = \gamma$, $\alpha + \beta = 180°$, $\varepsilon = \phi$, $\varepsilon = \theta$, and $\delta + \theta = 180°$.

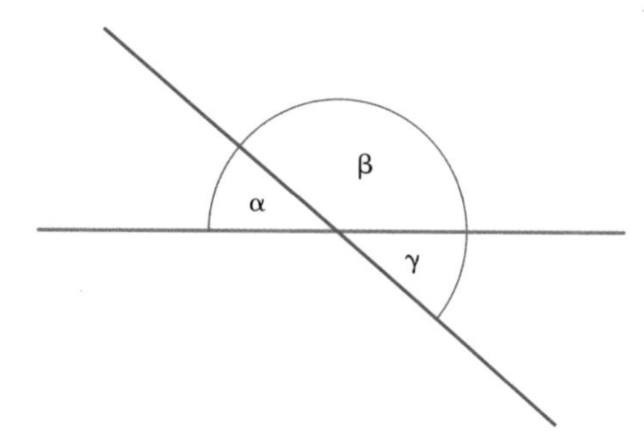

Figure 5 The three angles adjacent to an intersection of two straight lines are related in the form $\alpha = \gamma$, $\alpha + \beta = 180°$, and $\beta + \gamma = 180°$.

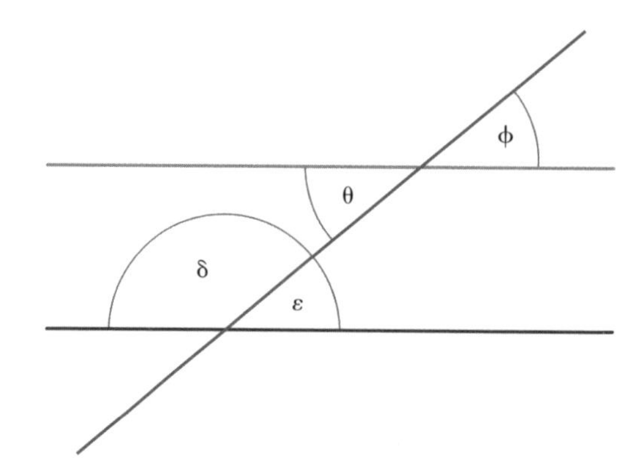

Figure 6 The four angles δ, ε, θ, and ϕ, which are adjacent to a straight line that intersects with two parallel straight lines, are related in the form $\varepsilon = \phi$, $\varepsilon = \theta$, and $\delta + \theta = 180°$.

We use three trigonometric functions in the text, the sine function ($\sin\theta$), the cosine function ($\cos\theta$), and the tangent function ($\tan\theta$). It is important to know how they are connected for a right triangle. Defining one angle other than the 90° angle in the triangle as θ, the trigonometric functions are given as follows:

$$\sin\theta = \frac{\text{length of side opposite to } \theta}{\text{length of hypotenuse}}$$

$$\cos\theta = \frac{\text{length of side adjacent to } \theta}{\text{length of hypotenuse}}$$

$$\tan\theta = \frac{\text{length of side opposite to } \theta}{\text{length of side adjacent to } \theta}$$

The following relations apply for negative angles:

$$\sin(-\theta) = -\sin\theta$$

$$\cos(-\theta) = -\cos\theta$$

$$\tan(-\theta) = -\tan\theta$$

Relations between the sine and cosine functions follow from the basic definitions:

$$\sin(90° - \theta) = \cos\theta$$

$$\cos(90° - \theta) = \sin\theta$$

Example 4

In the triangle given in Fig. 7 identify the trigonometric functions for angles ψ and ϕ.

Solution: The following relations hold:

$$\sin\psi = \frac{a}{c} \quad \cos\psi = \frac{b}{c} \quad \tan\psi = \frac{a}{b}$$

$$\sin\phi = \frac{b}{c} \quad \cos\phi = \frac{a}{c} \quad \tan\phi = \frac{b}{a}$$

The Pythagorean theorem states that, for a right triangle in which c is the length of the hypotenuse and a and b are the lengths of the sides opposite and adjacent to the angle θ:

$$c^2 = a^2 + b^2.$$

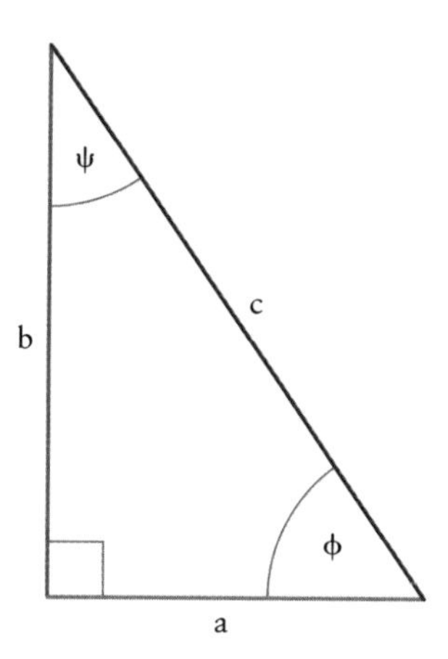

Figure 7 A right triangle with all three sides and angles labelled.

This is illustrated in the left sketch of Fig. 8. The two sketches on the right side of the figure demonstrate how simply the Pythagorean theorem is proven: The four triangles in the left box are arranged to leave open the area c^2. They are then rearranged in the same box at right to leave open two areas, a^2 and b^2.

Example 5

(a) Confirm that the corners of the area labelled c^2 in the middle panel of Fig. 8 are right angles. (b) Use the Pythagorean theorem to prove the following formula:

$$\sin^2\phi+\cos^2\phi=1.$$

(c) Use Fig. 7 to prove the following formula, which applies as long as $\cos\phi \neq 0$:

$$\tan\phi=\frac{\sin\phi}{\cos\phi}.$$

Solution to part (a): The sum of the three angles in a triangle is 180°, and therefore the sum of the angles excluding the right angle in a right triangle is equal to 90°. Choosing the lower corner of the tilted area within the larger box in the middle panel of Fig. 8, we see that the two angles between the side of the larger box and the sides of the inner box are equal to the sum of the two angles (excluding the right angle) in any one of the four identical triangles. Thus, the angle of the corner is 90° since this sum equals 90°, and the larger box describes an angle of 180° at the corner point.

Solution to part (b): We calculate $\sin^2\phi$ and $\cos^2\phi$ from Fig. 7. Then we use the Pythagorean theorem to simplify the result:

$$\sin^2\phi+\cos^2\phi=\frac{b^2}{c^2}+\frac{a^2}{c^2}=\frac{a^2+b^2}{c^2},$$

which yields:

$$\sin^2\phi+\cos^2\phi=\frac{c^2}{c^2}=1.$$

Solution to part (c): The proof is based on Fig. 7:

$$\frac{\sin\phi}{\cos\phi}=\frac{b/c}{a/c}=\frac{b}{a}=\tan\phi.$$

Vectors and Basic Vector Algebra

The Cartesian coordinate system (named after René Descartes, *La Géométrie,* 1637) is suitable for describing a three-dimensional mathematical space (Fig. 9). It is based on three orthogonal axes, which are labelled, in order, x-axis, y-axis, and z-axis. The **right-hand rule** was developed to confirm that a coordinate system is labelled properly: take your right hand and stretch the thumb and the index finger. They are automatically forming a right angle between them. Use your middle finger to point in a direction perpendicular to the thumb and the index finger. You can do this in only one direction. Now point the thumb in the direction of the x-axis. Then turn your hand such that the index finger points in the y-direction. At this point your middle finger points automatically in the z-direction.

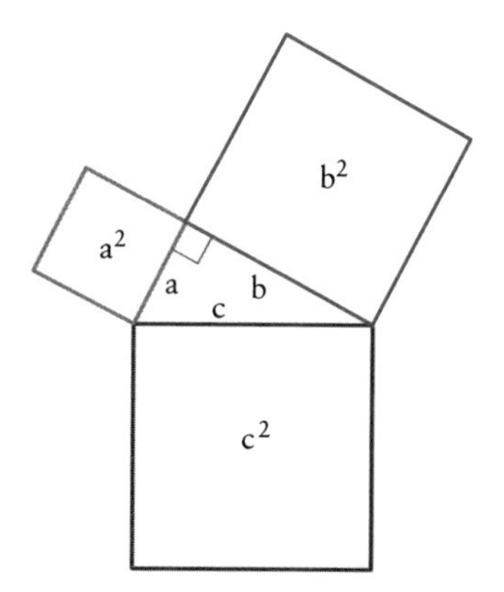

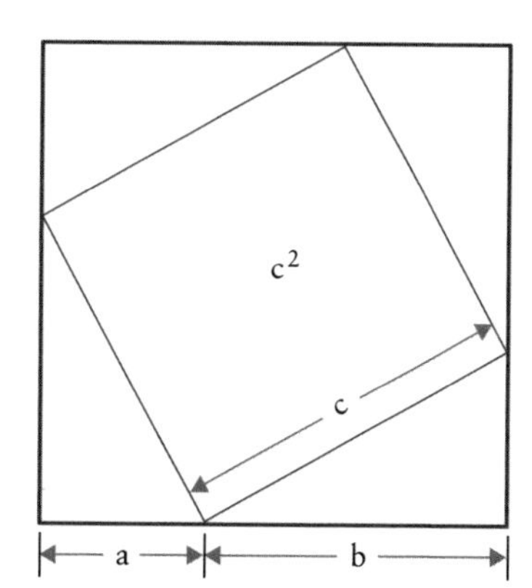

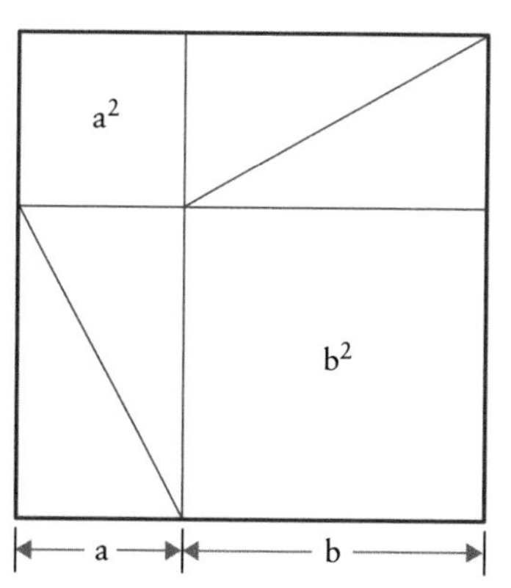

Figure 8 Left panel: Illustration of the Pythagorean theorem for a right triangle with hypotenuse c. Middle and right panel: Simple geometric proof of the Pythagorean theorem. Each box has the total area $(a+b)^2$. Inserting four right triangles, with sides a and b meeting at the right angle, an area c^2 is left open in the middle panel and, after rearranging the four triangles in the right panel, the areas a^2 and b^2 are left open.

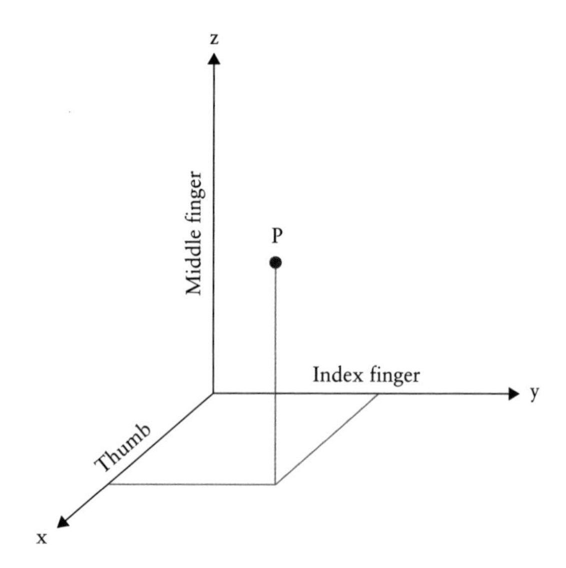

Figure 9 Three dimensional, perpendicular coordinates (Cartesian coordinate system). The order of the axes is determined by the right-hand rule described in the text.

The three axes intersect at the **origin**. Three numbers are assigned to any point *P* that are proportional to the distances from the origin along each axis. These numbers are called the *coordinates*, and are labelled using indices to identify the axis: p_x, p_y, and p_z. This is illustrated in Fig. 10, where for clarity only a two-dimensional space is shown.

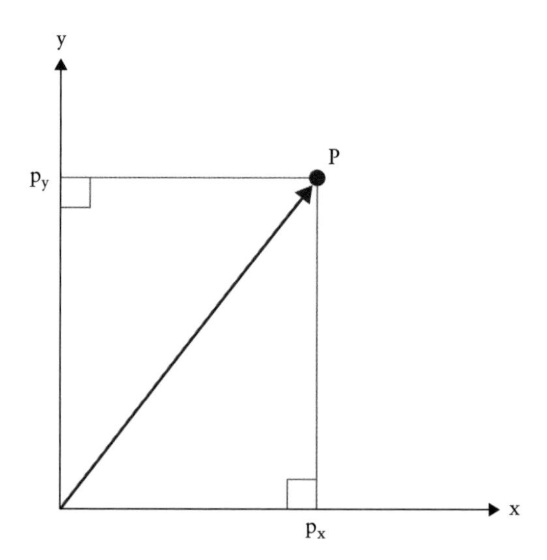

Figure 10 Coordinates of point *P* in a two dimensional, Cartesian coordinate system. The coordinates are proportional to the lengths of the axes from the origin to the points of perpendicular projection of point *P* onto the axes.

Associated with point *P* is an arrow reaching from the origin to point *P*, called vector $\vec{p}$. A vector is represented by two or three numbers (with the number of coordinates depending on the dimensionality of the considered space), $\vec{p} = (p_x, p_y, p_z)$ or, as in Fig. 10, $\vec{p} = (p_x, p_y)$.

Vectors must be distinguished from *scalars,* which are just simple numbers. In contrast, vectors have both magnitude (i.e., a simple number) and direction. The physical quantities discussed in this textbook are described by either a scalar (e.g., temperature) or a vector (e.g., force). Even if a problem is one dimensional, vector quantities retain their directional information, then carrying a + or a − sign.

Example 6

We use the methane molecule as an example of the vector algebra concepts introduced in this section. The methane molecule, CH_4, is placed in a cube of side length *l* in Fig. 11. Express the positions of the four hydrogen atoms and the carbon atom in the methane molecule in Cartesian coordinates.

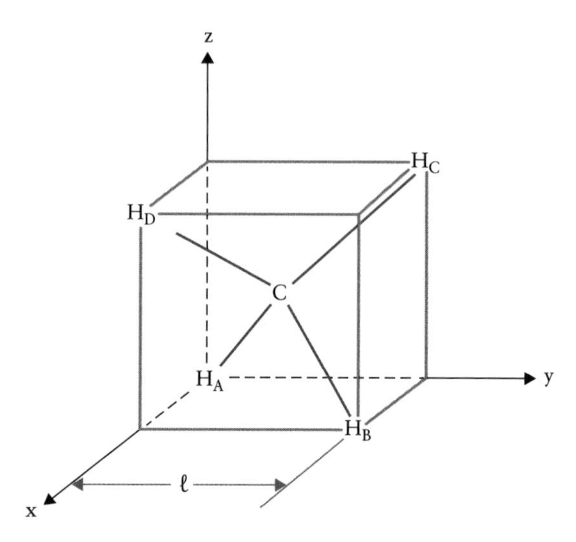

Figure 11 The geometry of the tetrahedral methane molecule CH_4 is best described by placing the molecule in a cube of side length *l* in a Cartesian coordinate system. The four hydrogen atoms form four corners of the cube as shown. They are indistinguishable in a real molecule but have been labelled in the sketch with different indices for calculation purposes.

Solution: The positions of the five atoms are: C (0.5*l*, 0.5*l*, 0.5*l*); H_A (0, 0, 0); H_B (*l*, *l*, 0); H_C (0, *l*, *l*); and H_D (*l*, 0, *l*).

Polar Coordinates

An alternative way to describe vectors is to use polar coordinates. We apply these in this textbook only for two-dimensional systems. The length $|\vec{p}|$ and the angle θ between the vector and the positive *x*-axis replace the two Cartesian coordinates p_x and p_y, as shown in the top sketch of Fig. 12. The bottom sketch of Fig. 12 illustrates how the polar coordinates and the Cartesian coordinates are related to each other: $\sin\theta = p_y/|\vec{p}|$ and $\cos\theta = p_x/|\vec{p}|$. With the basic vector definitions established, the fundamental vector operations can be introduced.

Magnitude or Length of a Vector

With the Pythagorean theorem, we find for the triangle in the bottom sketch of Fig. 12:

$$|\mathrm{p}| = \sqrt{p_x^2 + p_y^2}.$$

For example, the length of the two-dimensional vector $\vec{p} = (3, 4)$ is $|\vec{p}| = 5$.

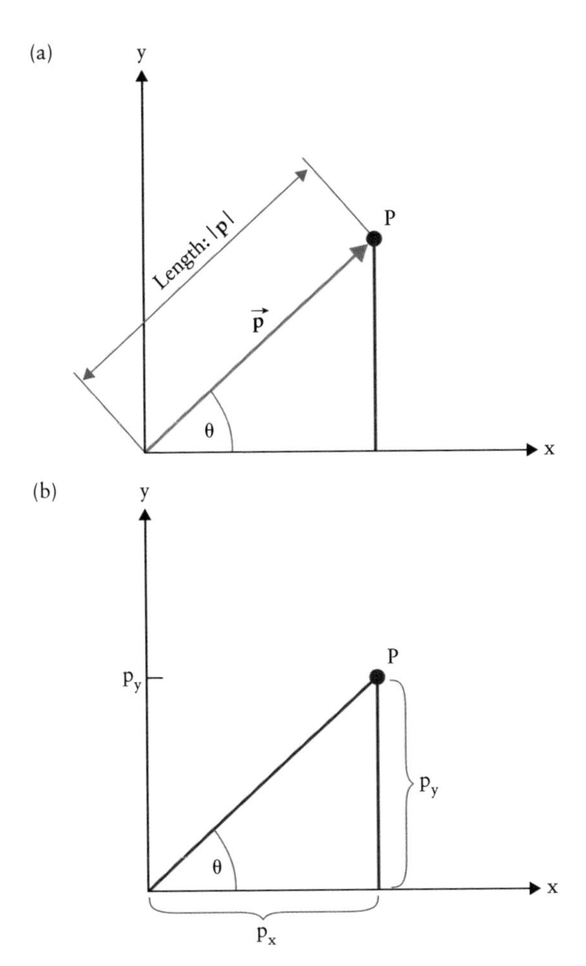

Figure 12 Representation of the position of a point P using polar coordinates. (a) Definition of angle θ and length of vector, $|\vec{p}|$. (b) Sketch highlighting the trigonometric relations between the polar coordinates and the Cartesian coordinates.

Vector Addition: $\vec{a} + \vec{b} = \vec{r}$

The vector addition is sketched in Fig. 13. For practical applications, each component of a Cartesian coordinate system is added separately:

$$x\text{-component}: \quad a_x + b_x = r_x$$

$$y\text{-component}: \quad a_y + b_y = r_y$$

$$z\text{-component}: \quad a_z + b_z = r_z$$

Thus, all algebra rules apply; e.g., the commutative law in the form $\vec{a} + \vec{b} = \vec{b} + \vec{a}$.

Multiplication of a Vector with a Scalar: $n\vec{a} = \vec{r}$

Again, the operation is done for each component separately:

$$x\text{-component}: \quad na_x = r_x$$

$$y\text{-component}: \quad na_y = r_y$$

$$z\text{-component}: \quad na_z = r_z$$

Combining the multiplication and addition of vectors, we introduce the subtraction of vectors in the form $\vec{a} - \vec{b} = \vec{a} + (-1)\vec{b}$.

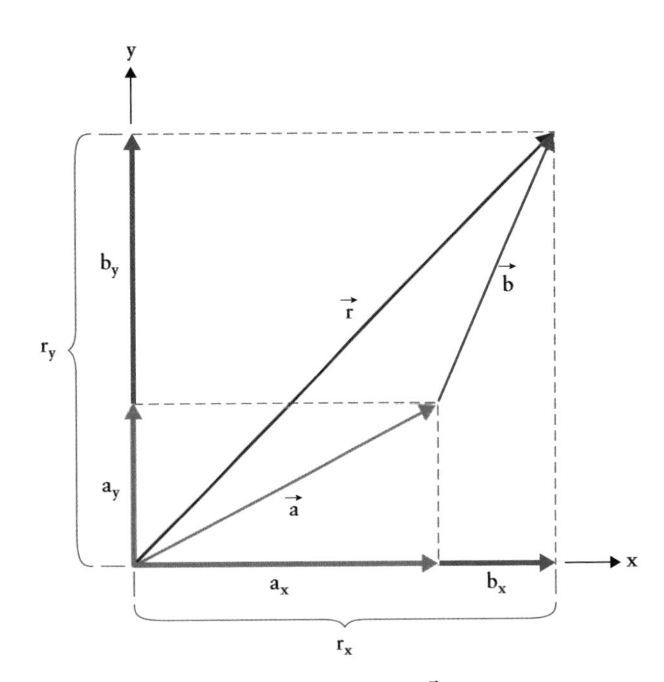

Figure 13 Vector addition of vectors $\vec{a}$ and $\vec{b}$ and the relation of the components of the resulting vector $\vec{r}$ to the components of $\vec{a}$ and $\vec{b}$; i.e., $a_x + b_x = r_x$ and $a_y + b_y = r_y$.

Vector Multiplication: Scalar Product

Two ways exist to multiply two vectors with each other. The scalar product or dot product leads to a scalar ($\vec{a} \bullet \vec{b} = r$), and the vector product or cross-product leads to a vector ($\vec{a} \times \vec{b} = \vec{r}$). Both products are discussed in this Math Review. Note that we use a large dot (•) to indicate a dot product.

Fig. 14 illustrates that the scalar product of two vectors is related to the product of the lengths of the two vectors after one vector has been projected onto the direction of the other:

$$\vec{a} \bullet \vec{b} = |\vec{a}||\vec{b}|\cos\varphi.$$

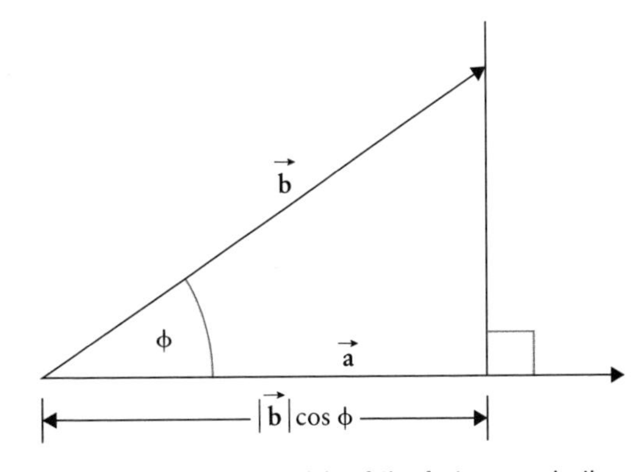

Figure 14 Illustration of the origin of the factor $\cos\phi$ in the scalar product of vectors $\vec{a}$ and $\vec{b}$.

Special cases are $= 90°$, with $\vec{a} \bullet \vec{b} = 0$; $\phi = 0°$, with $\vec{a} \bullet \vec{b} = |\vec{a}||\vec{b}|$ and $\vec{a} \bullet \vec{a} = |\vec{a}|^2$.

The scalar product is calculated using the components of the vectors:

$$\vec{a} \bullet \vec{b} = a_x\, b_x + a_y\, b_y (+a_z b_z)$$

The benefit of using the Cartesian coordinate system with its orthogonal axes lies in the fact that the different components along the x-, y-, and z-axes do not mix up in the last equation. Thus, vector algebra (with the exclusion of the vector product below) is simple, as any operation is equivalent to the same algebraic operation for numbers except that the operation is repeated for each component separately.

Example 7

(a) Calculate the angle between any two CH bonds in the methane molecule, shown in Figs. 11 and 5.4. This angle is called the tetrahedral angle for the sp^3-hybridization of carbon atoms in organic molecules. (b) Using the length of 0.11 nm for a CH bond, determine the side length of the cube, l, and (c) the distance between any two hydrogen atoms.

Solution to part (a): First we determine the vectors connecting the C atom with the hydrogen atoms $\vec{H}_A$ and $\vec{H}_B$ in Fig. 5.4. These vectors are:

$$\overline{\mathrm{CH}}_A = -\vec{\mathrm{C}} + \vec{\mathrm{H}}_A;$$

written in component form:

$$x\text{-direction:} \quad -\frac{l}{2} + 0 = -\frac{l}{2}$$

$$y\text{-direction:} \quad -\frac{l}{2} + 0 = -\frac{l}{2}$$

$$z\text{-direction:} \quad -\frac{l}{2} + 0 = -\frac{l}{2}$$

and:

$$\overline{\mathrm{CH}}_B = -\vec{\mathrm{C}} + \vec{\mathrm{H}}_B$$

written in component form:

$$x\text{-direction:} \quad -\frac{l}{2} + l = +\frac{l}{2}$$

$$y\text{-direction:} \quad -\frac{l}{2} + l = +\frac{l}{2}$$

$$z\text{-direction:} \quad -\frac{l}{2} + 0 = -\frac{l}{2}$$

i.e., to get from the carbon atom to the hydrogen atom H_A, you have first to travel the vector $\vec{C}$ backward to the origin, and then from the origin to the H atom forward along the vector $\vec{H}_A$. The angle θ between the vectors follows from the dot product:

$$\overline{\mathrm{CH}}_A\,\overline{\mathrm{CH}}_B = \left|\overline{\mathrm{CH}}_A\right|\left|\overline{\mathrm{CH}}_B\right|\cos\theta.$$

The vector magnitudes on the right-hand side are:

$$\left|\overline{\mathrm{CH}}_A\right| = \left|\overline{\mathrm{CH}}_B\right| = \sqrt{\left(\frac{l}{2}\right)^2 + \left(\frac{l}{2}\right)^2 + \left(\frac{l}{2}\right)^2}$$

$$= \frac{\sqrt{3}}{2}l$$

and the dot product on the left-hand side is calculated from the basic definition of the dot product in component form:

$$\overline{\mathrm{CH}}_A\,\overline{\mathrm{CH}}_B = \left(-\frac{l}{2}\right)\frac{l}{2} + \left(-\frac{l}{2}\right)\frac{l}{2} + \left(-\frac{l}{2}\right)\left(-\frac{l}{2}\right)$$

$$= -2\frac{l^2}{4} + \frac{l^2}{4} = -\frac{l^2}{4}$$

Thus, $\cos\theta = (-l^2/4)/(3l^2/4) = -\frac{1}{3}$ and $\theta = 109.47°$.

Solution to part (b): We calculated above that $\left|\overline{CH}_B\right| = ½\, l\sqrt{3}$. With this length given as 0.11 nm, we find that $l = 2\left|\overline{\mathrm{CH}}_B\right|/\sqrt{3} = 0.127\ \mathrm{nm}$.

Solution to part (c): Since the distance between any two hydrogen atoms is equal to the magnitude of the vector connecting the two H atoms, we find for the example of the hydrogen atoms labelled A and B that the distance:

$$\left|\vec{\mathrm{H}}_A\vec{\mathrm{H}}_B\right| = \left|-\vec{\mathrm{H}}_A + \vec{\mathrm{H}}_B\right| = (l^2 + l^2)^{1/2} = \sqrt{2}l = 0.180\ \mathrm{nm}.$$

Vector Product

The vector product is written in the form

$$\vec{r} = [\vec{a} \times \vec{b}]$$

with $\vec{a}$ and $\vec{b}$ two non-parallel vectors and $\vec{r}$ the resulting vector. You apply it in its component form:

$$x\text{-component:} \quad r_x = a_y b_z - a_z b_y$$

$$y\text{-component:} \quad r_y = a_z b_x - a_x b_z$$

$$z\text{-component:} \quad r_z = a_x b_y - a_y b_x$$

Note that the components of vectors $\vec{a}$ and $\vec{b}$ are no longer separated in the three component formulas. These equations determine the direction of the resulting vector: it is directed perpendicular to the plane defined by vectors $\vec{a}$ and $\vec{b}$. This is illustrated in Fig. 15, establishing the **right-hand rule**: stretch

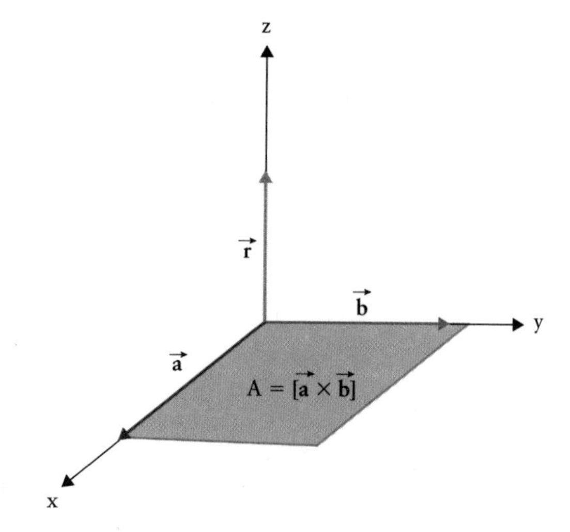

Figure 15 Relative directions of the vectors forming a vector product.

your thumb, index finger, and middle finger of the right hand such that they form pair-wise right angles with each other. The thumb represents the first vector of the vector product $\vec{a}$; the index finger points in the direction of the second vector $\vec{b}$. The middle finger then represents the direction of the resulting vector $\vec{r}$.

Summations

When adding more than two terms, it is convenient to use a condensed notation for the summation (sigma notation):

$$\sum_{i=1}^{N} F_i = F_1 + F_2 + F_3 + \cdots + F_N,$$

in which i is an index that runs from 1 to N so that N terms of the quantity F are added. As an example, let us add the squares of numbers 1 to 5. This can be written in two ways:

$$1^2 + 2^2 + 3^2 + 4^2 + 5^2 = 55$$

$$\text{or} \sum_{i=1}^{5} i^2 = 55$$

Symmetric Objects

(I) *Circle:* Defining C as the circumference and A as the area for a two-dimensional circle of radius r, we get $C = 2\,\pi\,r$ and $A = \pi\,r^2$.

(II) *Triangle:* For a triangle with base a and height h the area A is given as $A = a\,h/2$.

(III) *Sphere:* Defining A as the surface area and V as the volume for a three-dimensional sphere of radius r we get $A = 4\,\pi\,r^2$ and $V = 4\,\pi\,r^3/3$.

(IV) *Cylinder:* Defining A as the mantle surface area and V as the volume for a right circular cylinder of radius r and height h we get $A = 2\,\pi\,r\,h$ and $V = \pi\,r^2\,h$.

Binomials and Quadratic Equations

The following algebraic relations are used frequently:

$$(a+b)^2 = a^2 + 2ab + b^2$$

$$(a-b)^2 = a^2 - 2ab + b^2$$

$$a^2 - b^2 = (a+b)(a-b)$$

$$a^3 - b^3 = (a-b)(a^2 + ab + b^2)$$

For a quadratic equation written in the form $ax^2 + bx + c = 0$, with a, b, and c constant, there are a maximum of two real solutions, labelled x_1 and x_2:

$$x_{1,2} = \frac{-b \pm \sqrt{b^2 - 4ac}}{2a}$$

Degrees and Radians

In the physical sciences, we often use the approximation $\sin\theta = \theta$. This approximation holds generally for small angles θ. However, to be more quantitative we need to first review the definition of an angle. An angle is measured as the ratio of the length of the arc it subtends in a circle to the radius of that circle. For example, in Fig. 16 the angle θ is given by $\theta = s/L$. This angle carries the unit **radians** (rad). In physical applications angles must always be used in radians; other units of angles, such as degree (°) or revolutions (rev) have to be converted to radians to obtain proper quantitative results. It is worthwhile to study your pocket calculator to figure out how it deals with the unit radians. Most calculators allow you to switch between a degree mode and a radians mode. In the degree mode (DEG), it provides you with the following results:

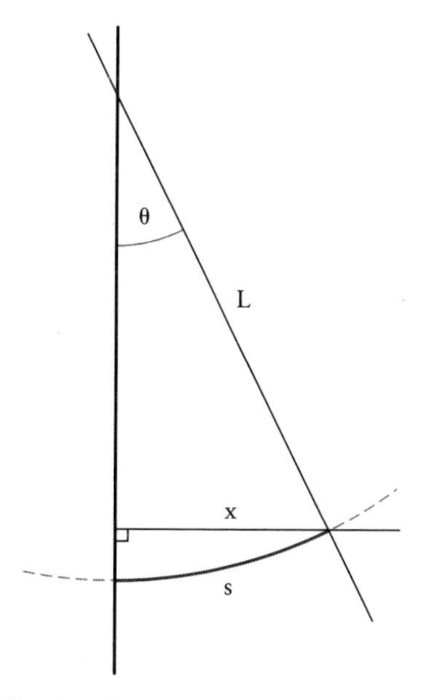

Figure 16 Sketch defining the path segment *s* and the horizontal displacement *x* from the equilibrium position for an object on a massless string of length *L* at an angle *θ* with the vertical.

$$\sin(0°) = 0.0 \qquad \cos(0°) = 1.0$$

$$\sin(90°) = 1.0 \qquad \cos(90°) = -0.0$$

$$\sin(180°) = 0.0 \qquad \cos(180°) = 1.0$$

$$\sin(270°) = -1.0 \qquad \cos(270°) = 0.0$$

In the radians mode (RAD) it instead provides you with the following results:

$$\sin(0) = 0.0 \qquad \cos(0) = 1.0$$

$$\sin(\pi/2) = 1.0 \qquad \cos(\pi/2) = 0.0$$

$$\sin(\pi) = 0.0 \qquad \cos(\pi) = -1.0$$

$$\sin(3\pi/2) = -1.0 \qquad \cos(3\pi/2) = 0.0$$

in which the conversion is given as π rad = 180°, and 1 rad = 57.3°.

Now we can test the approximation $\sin\theta = \theta$ with a pocket calculator. Let's start with an angle of 10°. We can confirm that 10° qualifies as a small angle because 10° = 0.1745 rad and sin(10°) = 0.1736; i.e., both values vary by only 0.5%. Thus, the range of small angles we referred to in the text isn't too restrictive; angles well beyond 10° can be included.

Trigonometry

The following theorems for the sine function apply:

$$\sin(\alpha+\beta) = \sin\alpha\cos\beta + \cos\alpha\sin\beta$$

$$\sin(\alpha-\beta) = \sin\alpha\cos\beta - \cos\alpha\sin\beta.$$

For the cosine function, we find:

$$\cos(\alpha+\beta) = \cos\alpha\cos\beta - \sin\alpha\sin\beta$$

$$\cos(\alpha-\beta) = \cos\alpha\cos\beta + \sin\alpha\sin\beta.$$

And for the tangent function we use:

$$\tan(\alpha+\beta) = \frac{\tan\alpha + \tan\beta}{1 - \tan\alpha\cdot\tan\beta}$$

$$\tan(\alpha-\beta) = \frac{\tan\alpha - \tan\beta}{1 + \tan\alpha\cdot\tan\beta}$$

Further, we note for the sum or difference of two sine functions:

$$\sin\alpha + \sin\beta = 2\sin\left(\frac{\alpha+\beta}{2}\right)\cos\left(\frac{\alpha-\beta}{2}\right)$$

$$\sin\alpha - \sin\beta = 2\cos\left(\frac{\alpha+\beta}{2}\right)\sin\left(\frac{\alpha-\beta}{2}\right)$$

and for the sum or difference of two cosine functions:

$$\cos\alpha + \cos\beta = 2\cos\left(\frac{\alpha+\beta}{2}\right)\cos\left(\frac{\alpha-\beta}{2}\right)$$

$$\cos\alpha - \cos\beta = 2\sin\left(\frac{\alpha+\beta}{2}\right)\sin\left(\frac{\alpha-\beta}{2}\right)$$

Differential Calculus

For a function $y = f(x)$, which is defined in an interval of width ξ around $x = x_0$, we define two differentials for a variable h with $0 < h < \xi$:

$$\frac{f(x_0+h) - f(x_0)}{h} \quad \text{right-sided differential}$$

$$\frac{f(x_0-h) - f(x_0)}{-h} \quad \text{left-sided differential.}$$

If each differential converges toward the same value for $\lim h \to 0$, then the function $f(x)$ is called differentiable at $x = x_0$. The value of the differential for $\lim h \to 0$ is called the derivative at $x = x_0$:

$$\lim_{h\to 0}\left(\frac{f(x_0+h) - f(x_0)}{h}\right) =$$

$$\lim_{h\to 0}\left(\frac{f(x_0-h) - f(x_0)}{-h}\right) \equiv \frac{df(x)}{dx}\Big|_{x_0}.$$

Frequently used notations include:

$$\frac{df(x)}{dx} \equiv f'(x) \equiv \frac{d}{dx}f(x).$$

The derivative is usually again a function of x. Geometrically, it represents the *slope* of $y = f(x)$ at each value of x. If the derivative function is differentiable, a second derivative is calculated:

$$f''(x) \equiv \frac{d}{dx}f'(x) \equiv \frac{d^2}{dx^2}f(x).$$

The second derivative represents the *curvature* of the original function. Table 4 presents the most frequently used functions and their derivatives. Using this table and the following rules, any differentiable function can be differentiated:

(I) $y = c \cdot f(x)$, with c = const $\Rightarrow y' = c f'(x)$

(II) $y = f(x) + g(x)$, where $f(x)$ and $g(x)$ are two separate functions $\Rightarrow y' = f'(x) + g'(x)$

(III) *Product rule:* $y = f(x)\, g(x)$, where $f(x)$ and $g(x)$ are separate functions $\Rightarrow y' = f(x)\, g'(x) + f'(x)\, g(x)$; e.g., $y = x\sin(x)$ yields $y' = x\cos(x) + \sin(x)$.

(IV) *Quotient rule:* assuming $g(x) \neq 0$, then:

$$y = \frac{f(x)}{g(x)} \Rightarrow$$

$$y' = \frac{f'(x)g(x) - f(x)g'(x)}{(g(x))^2};$$

TABLE 4

Some fundamental mathematical functions and their derivatives

Function	Derivative
y = const	$y' = 0$
$y = x^n$	$y' = n\,x^{n-1}$
$y = \sin(x)$	$y' = \cos(x)$
$y = \cos(x)$	$y' = -\sin(x)$
$y = \tan(x)$	$y' = l/\cos^2(x)$
$y = e^x = \exp(x)$	$y' = c^x$
$y = \ln x$	$y' = 1/x$

e.g.,

$$y = \frac{\sin x}{\cos x} \Rightarrow y' = \frac{\cos^2 x + \sin^2 x}{\cos^2 x} = \frac{1}{\cos^2 x}.$$

This is indeed the derivative of tan(x).
(V) *Chain rule:* Given $v = g(x)$ and $y = f(v)$, then:

$$y = f(g(x)) \Rightarrow$$

$$y' = f'(v)g'(x) = \frac{dy}{dv}\frac{dv}{dx};$$

e.g., $y = \sin x^3 \Rightarrow y' = 3x^2 \cos x^3$ with $v = x^3$.

Diagnostic Nuclear Medicine

The activity of a radioisotope as a function of time $A(t)$ is given by the equation:

$$A(t) = A_0 e^{-\lambda t},$$

where A_0 is the initial activity. This is a special type of mathematical function called a decaying exponential.

If parent radioisotope A decays to daughter radioisotope B, then the activity of B as a function of time is given by:

$$A_B(t) = A_A(t)\left[\frac{\lambda_B}{\lambda_B - \lambda_A}\right]\left[1 - e^{-(\lambda_B - \lambda_A)t}\right],$$

where $A_A(t)$ is the activity of the parent at time t.

After a sufficiently long time, it can be shown by letting $t \rightarrow \infty$ that the ratio of parent to daughter activities is given by the approximate expression:

$$\frac{A_B(t)}{A_A(t)} \approx \frac{\lambda_B}{\lambda_B - \lambda_A}.$$

The activity of a cyclotron-produced isotope as a function of irradiation time is given by the expression:

$$A(t) = In\sigma(1 - e^{-\lambda t}),$$

Where I is the proton current, n is the radiological thickness of the target material, and σ is the reaction cross section. It can be shown by letting $t \rightarrow \infty$ that, after a sufficiently long irradiation time, the maximum activity is approximately equal to In σ. This activity is therefore referred to as the "saturation" activity.

Radiotherapy

When subjected to a dose of radiation D, the linear quadratic model predicts a surviving fraction S of cells given by the expression:

$$S = e^{-(\alpha D + \beta D^2)},$$

where α and β are tissue specific constants and are expressed in units of Gy^{-1} and Gy^{-2} respectively. This is a two-exponent model with a linear and quadratic term in the variable D.

The percent depth dose (PDD) function for a radiation beam is defined as:

$$\text{PDD} = [D(d)/D_{max}] \times 100\%,$$

where $D(d)$ is the dose at depth d in the phantom and D_{max} is the maximum dose also in the phantom, both along the central beam axis. It is a unitless quantity.

The tissue air ratio (TAR) function for a radiation beam is defined as:

$$\text{TAR} = D(d)/D_{air},$$

where $D(d)$ is the dose in the phantom corresponding to a dose in air D_{air} at an equivalent point in free space. $D(d)$ and D_{air} are measured at the beam isocentre and, like the PDD, it is a unitless quantity.

Units Review

CHAPTER 2

- Displacement $\vec{x}$: m
- Velocity $\vec{v}$: m/s
- Acceleration $\vec{a}$: m/s^2

CHAPTER 3

- Force: N (newton), 1N = 1 kg m/s^2
- Charge q: C (coulomb)
- Pressure p: Pa (pascal), 1 Pa = 1 N/m^2

CHAPTER 4

Force and force components: N (newton), with 1 N = 1 kg m/s^2

Density: ρ: kg/m^3

CHAPTER 5

- Linear momentum $\vec{p}$: kg m/s

CHAPTER 6

Torque τ with N · m = kg · m^2/s^2

CHAPTER 7

- Work W, energy E: J = N m = kg m^2/s^2

CHAPTER 8

- Volume V: L (litre with 1.0 L = 1.0×10^{-3} m^3); m^3 (standard unit)
- Temperature T: K (standard unit), °C (frequently used non-standard unit)
- Pressure or gauge pressure p: Pa = N/m^2 (standard unit); mmHg, torr, atm, bar (frequently used non-standard units)

CHAPTER 9

- Work W, heat Q, energy E: J = N m = kg m^2/s^2
- Specific heat capacity c: J/(kg K)
- Molar heat capacity C: J/(mol K)

CHAPTER 10

- Entropy S: J/K
- Gibbs free energy G: J

CHAPTER 11

- Molar mass M: kg/mol
- Amount of matter n: mol
- Concentration c: mol/m^3
- Density ρ: kg/m^3

CHAPTER 12

- Surface tension σ: J/m^2 = N/m

CHAPTER 13

- Volume flow rate $\Delta V/\Delta t$: m^3/s
- Mass flow rate $\Delta m/\Delta t$: kg/s
- Viscosity coefficient η: N s/m^2
- Flow resistance R: Pa s/m^3

CHAPTER 14

- Stress σ: N/m^2 = Pa
- Strain ε: no units
- Spring constant k: N/m
- Amplitude A: m
- Period T: s
- Frequency f: 1/s = Hz
- Angular frequency ω: rad/s

CHAPTER 15

- Wave number κ: m^{-1}
- Wavelength λ: m
- Frequency f: Hz
- Energy density ε: J/m^3
- Intensity of a sound wave I: J/(m^2 s)
- Sound pressure level (SPL) and intensity level (IL): dB

CHAPTER 16

- Acoustic impedance Z: kg m^{-2} s^{-1} = rayl

CHAPTER 17

- Charge q: C
- Electric force $\vec{F}_{el}$: N
- Electric field $\vec{E}$: N/C
- Dipole moment μ: C m

CHAPTER 18

- Electric potential energy E_{el}: J
- Potential V: V
- Capacitance C: F = C/V (farad)
- Dielectric constant κ: dimensionless materials constant

CHAPTER 19

- Current I: A = C/s (ampere)
- Resistance R: Ω = V/A (ohm)
- Resistivity r: Ω m = V m/A

CHAPTER 21

- Magnetic field $|\vec{B}|$: T = N/(A m)

CHAPTER 22

- Refractive power $\Re$: dpt (diopters) = m^{-1}

CHAPTER 23

- Activity: Bq (becquerel) with 1 Bq = 1 decay/s.
- Angular momentum L: kg m^2/s = J s
- Torque τ: N m

CHAPTER 24

- Linear attenuation coefficient μ: m^{-1}
- Mass attenuation coefficient μ/ρ: m^2/kg
- Exposure dose D_E: R (roentgen) with:

$$1\ \text{R} = 2.08 \times 10^9\ \frac{\text{ion pairs}}{\text{cm}^3}$$

$$1\ \text{R} = 2.58 \times 10^{-4}\ \frac{c}{\text{kg}}$$

CHAPTER 25

- The decay rate of a radioisotope is referred to as the activity, and it is given in SI units of sec^{-1} or becquerels (Bq).
- The decay constant λ of each isotope is unique and is given in units of sec^{-1}.

CHAPTER 26

- The unit of absorbed dose used in clinical radiotherapy is the gray (Gy), or its subunit, the centigray (cGy).
- The energy of high energy clinical photon beams is expressed in units of mega electronvolt (MeV).

Standard Units

These basic seven units were internationally adopted in 1969 under the title *Système International* (SI units).

Length	m	metre
Time	s	second
Mass	kg	kilogram
Temperature	K	kelvin
Amount of material	mol	mole
Electric current	A	ampere
Luminous intensity	Cd	candela

The Greek Alphabet

Greek capital and lower case letters

Alpha	A, α	Beta	B, β	Gamma	Γ, γ
Delta	Δ, δ	Epsilon	E, ε	Zeta	Z, ζ
Eta	H, η	Theta	Θ, θ	Iota	I, ι
Kappa	K, κ	Lambda	Λ, λ	Mu	M, μ
Nu	N, ν	Xi	Ξ, ξ	Omicron	O, o
Pi	Π, π	Rho	P, ρ	Sigma	Σ, σ
Tau	T, τ	Upsilon	Υ, υ	Phi	Φ, ϕ
Chi	X, χ	Psi	Ψ, ψ	Omega	Ω, ω

Standard Prefixes

Standard prefixes for terms of the form 10^n

$n = 9$: G for "giga-"	$n = 6$: M for "mega-"
$n = 3$: k for "kilo-"	$n = -1$: d for "deci-"
$n = -2$: c for "centi-"	$n = -3$: m for "milli-"
$n = -6$: μ for "micro-"	$n = -9$: n for "nano-"
$n = -12$: p for "pico-"	$n = -15$: f for "femto-"

Examples: 1.0 μm = 1.0×10^{-6} m; 6.5 cm^2 = 6.5 $(10^{-2}\ m)^2$ = $6.5 \times 10^{-4}\ m^2$; 1.0 L (litre) = 1.0 dm^3 = 1 $(10^{-1}\ m)^3$ = $1 \times 10^{-3}\ m^3$. The radius of Earth is written as 6370 km and the diameter of a chlorine ion is 0.181 nm.

Physical Constants

Fundamental physical constants. The columns are the name and symbol for each constant, the generally accepted value, the value to use for your calculations, and its unit. Based on NIST SP 959 (2005).

Name	Symbol	Value	Value Used	Unit
Gravitational constant	G*	6.6742×10^{-11}	—	$N \cdot m^2/kg^2$
Gravitational acceleration on Earth	g	—	9.8	m/s^2
Avogadro's constant	N_A	6.0221415×10^{23}	6×10^{23}	mol^{-1}
Universal gas constant	R	8.314472	8.314	$J/(mol \cdot K)$
Boltzmann's constant	k	$1.3806505 \times 10^{-23}$	1.38×10^{-23}	J/K
Molar volume of ideal gas		—	22.4	L
Electric force constant	k	—	9×10^9	$N \cdot m^2/C^2$
Permeability of a vacuum	μ_0	$4\pi \times 10^{-7}$	$4\pi \times 10^{-7}$	$T \cdot m/A$
Permittivity of a vacuum	ε_0	$8.854187817 \times 10^{-12}$	8.85×10^{-12}	$C^2/(N \cdot m^2)$
Elementary charge	e	$1.60217653 \times 10^{-19}$	1.6×10^{-19}	C
Mass of electron	m_e	$9.1093826 \times 10^{-31}$	9.1×10^{-31}	kg
Mass of proton	m_p	$1.67262171 \times 10^{-27}$	1.67×10^{-27}	kg
Vacuum speed of light	c	2.99792458×10^8	3×10^8	m/s
Planck's constant	h	$6.6260693 \times 10^{-34}$	6.6×10^{-34}	$J \cdot s$
Stefan-Boltzmann constant	σ	5.670400×10^{-8}	5.67×10^{-8}	$J/(m^2 \cdot s \cdot K^4)$
(Unified) atomic mass unit	u	$1.66053886 \times 10^{-27}$	1.66×10^{-27}	kg

Frequently Used Conversion Factors

Frequently used conversion factors

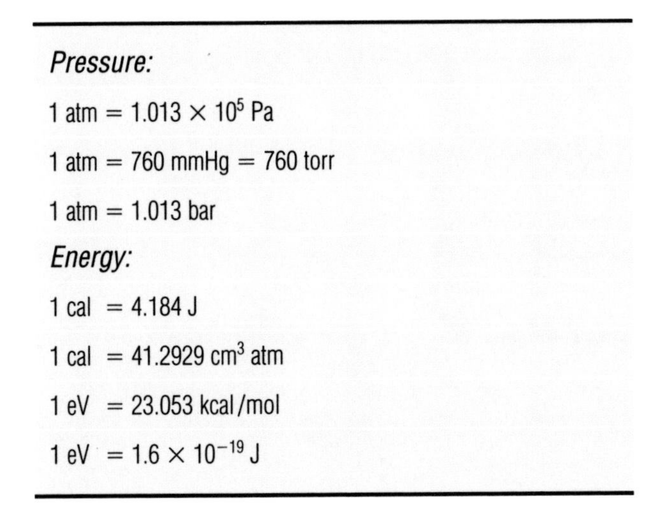

Pressure:

1 atm = 1.013×10^5 Pa

1 atm = 760 mmHg = 760 torr

1 atm = 1.013 bar

Energy:

1 cal = 4.184 J

1 cal = 41.2929 cm^3 atm

1 eV = 23.053 kcal/mol

1 eV = 1.6×10^{-19} J

Periodic Table of the Elements

Metals

Nonmetals

Metalloids

	1A (1)	2A (2)	3B (3)	4B (4)	5B (5)	6B (6)	7B (7)	8B (8)	8B (9)	8B (10)	1B (11)	2B (12)	3A (13)	4A (14)	5A (15)	6A (16)	7A (17)	8A (18)
1	1 **H** 1.0079																1 **H** 1.0079	2 **He** 4.0026
2	3 **Li** 6.941	4 **Be** 9.0122											5 **B** 10.811	6 **C** 12.0107	7 **N** 14.0067	8 **O** 15.9994	9 **F** 18.9984	10 **Ne** 20.1797
3	11 **Na** 22.9898	12 **Mg** 24.3050											13 **Al** 26.9815	14 **Si** 28.0855	15 **P** 30.9738	16 **S** 32.065	17 **Cl** 35.453	18 **Ar** 39.948
4	19 **K** 39.0983	20 **Ca** 40.078	21 **Sc** 44.9559	22 **Ti** 47.867	23 **V** 50.9415	24 **Cr** 51.9961	25 **Mn** 54.9380	26 **Fe** 55.845	27 **Co** 58.9332	28 **Ni** 58.6934	29 **Cu** 63.546	30 **Zn** 65.409	31 **Ga** 69.723	32 **Ge** 72.64	33 **As** 74.9216	34 **Se** 78.96	35 **Br** 79.904	36 **Kr** 83.798
5	37 **Rb** 85.4678	38 **Sr** 87.62	39 **Y** 88.9059	40 **Zr** 91.224	41 **Nb** 92.9064	42 **Mo** 95.94	43 **Tc** (98)	44 **Ru** 101.07	45 **Rh** 102.9055	46 **Pd** 106.42	47 **Ag** 107.8682	48 **Cd** 112.411	49 **In** 114.818	50 **Sn** 118.710	51 **Sb** 121.760	52 **Te** 127.60	53 **I** 126.9045	54 **Xe** 131.293
6	55 **Cs** 132.9055	56 **Ba** 137.327	57 **La** 138.9055 *	72 **Hf** 178.49	73 **Ta** 180.9479	74 **W** 183.84	75 **Re** 186.207	76 **Os** 190.23	77 **Ir** 192.217	78 **Pt** 195.078	79 **Au** 196.9665	80 **Hg** 200.59	81 **Tl** 204.3833	82 **Pb** 207.2	83 **Bi** 208.9804	84 **Po** (209)	85 **At** (210)	86 **Rn** (222)
7	87 **Fr** (223)	88 **Ra** (226)	89 **Ac** (227) **	104 **Rf** (261)	105 **Db** (262)	106 **Sg** (266)	107 **Bh** (264)	108 **Hs** (277)	109 **Mt** (268)	110 **Ds** (271)	111 **Rg** (272)	112 **Uub** (285)	113 **Uut** (284)	114 **Uuq** (289)	115 **Uup** (288)	116 **Uuh** (292)		

*Lanthanide Series

58 **Ce** 140.116	59 **Pr** 140.9076	60 **Nd** 144.24	61 **Pm** (145)	62 **Sm** 150.36	63 **Eu** 151.964	64 **Gd** 157.25	65 **Tb** 158.9253	66 **Dy** 162.50	67 **Ho** 164.9303	68 **Er** 167.259	69 **Tm** 168.9342	70 **Yb** 173.04	71 **Lu** 174.967

** Actinide Series

90 **Th** 232.0381	91 **Pa** 231.0359	92 **U** 238.0289	93 **Np** (237)	94 **Pu** (244)	95 **Am** (243)	96 **Cm** (247)	97 **Bk** (247)	98 **Cf** (251)	99 **Es** (252)	100 **Fm** (257)	101 **Md** (258)	102 **No** (259)	103 **Lr** (262)

Note: Atomic masses are IUPAC values (up to four decimal places).